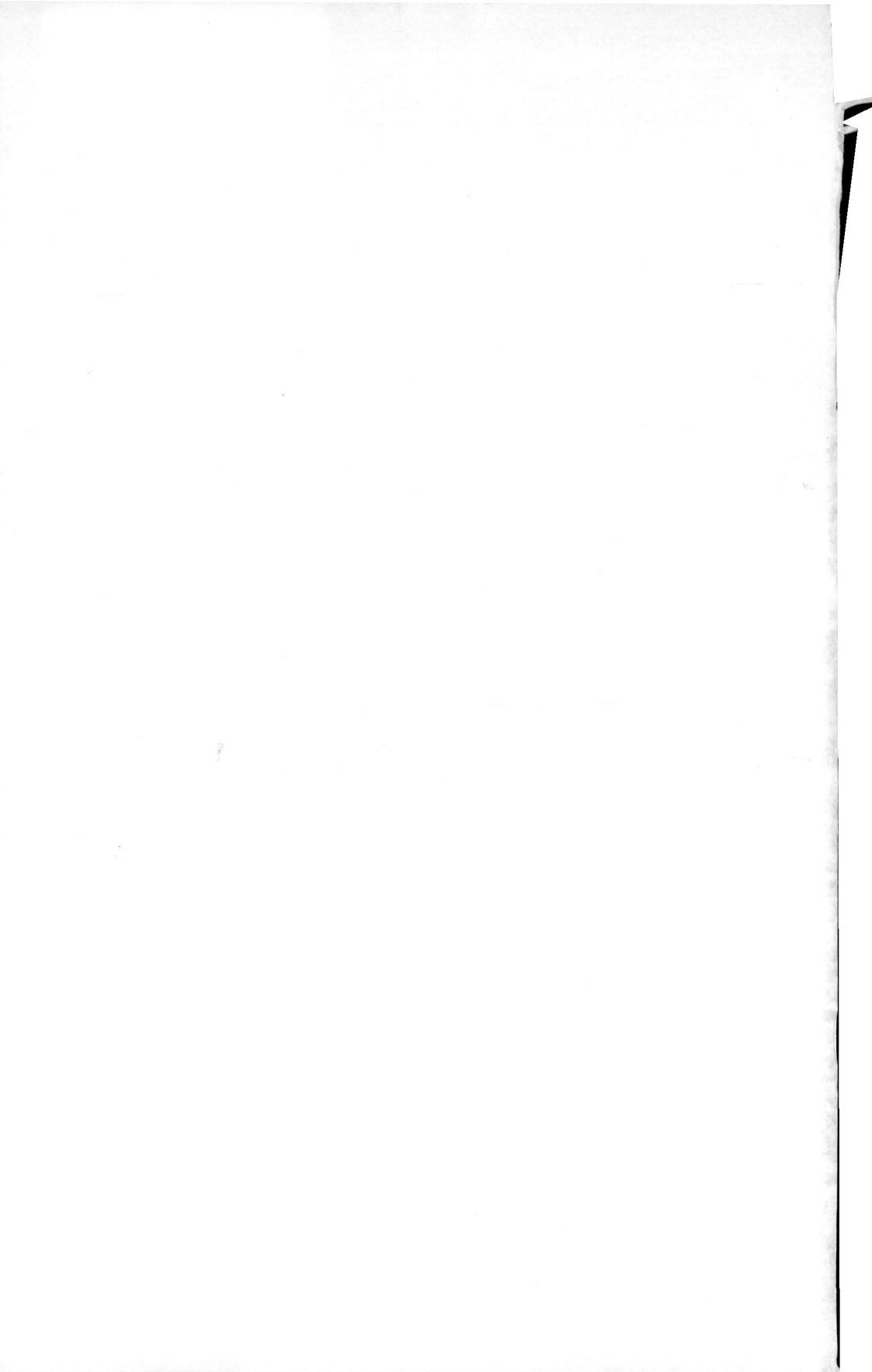

INTELLIGENT SYSTEMS

THEORY AND DECISION LIBRARY

General Editors: W. Leinfellner (*Vienna*) and G. Eberlein (*Munich*)

Series A: Philosophy and Methodology of the Social Sciences

Series B: Mathematical and Statistical Methods

Series C: Game Theory, Mathematical Programming and Operations Research

Series D: System Theory, Knowledge Engineering and Problem Solving

SERIES D: SYSTEM THEORY, KNOWLEDGE ENGINEERING AND PROBLEM SOLVING

VOLUME 15

Scope: Design, study and development of structures, organizations and systems aimed at formal applications mainly in the social and human sciences but also relevant to the information sciences. Within these bounds three types of study are of particular interest. First, formal definition and development of fundamental theory and/or methodology, second, computational and/or algorithmic implementations and third, comprehensive empirical studies, observation or case studies. Although submissions of edited collections will appear occasionally, primarily monographs will be considered for publication in the series. To emphasize the changing nature of the fields of interest we refrain from giving a clear delineation and exhaustive list of topics. However, certainly included are: artificial intelligence (including machine learning, expert and knowledge based systems approaches), information systems (particularly decision support systems), approximate reasoning (including fuzzy approaches and reasoning under uncertainty), knowledge acquisition and representation, modeling, diagnosis, and control.

The titles published in this series are listed at the end of volume II.

INTELLIGENT SYSTEMS

THIRD GOLDEN WEST INTERNATIONAL CONFERENCE

Edited and Selected Papers

VOLUME I

edited by

E. A. YFANTIS
University of Nevada, Las Vegas, U.S.A.

Springer-Science+Business Media, B.V.

A C.I.P. Catalogue record for this book is available from the Library of Congress

ISBN 978-94-011-7110-6 ISBN 978-94-011-7108-3 (eBook)
DOI 10.1007/978-94-011-7108-3

Printed on acid-free paper

TABLE OF CONTENTS

VOLUME 1

Preface xiii

List of Participants xv

ARTIFICIAL INTELLIGENCE 1

The Kolmogorov Metric and Classifying Linear Cellular Automata
L. D'Alotto and C. Giardina 3

A Hybrid Learning Rule for a Feedforward Network
W.-M. Lippe, Th. Feuring and A. Tenhagen 13

Logic Programming and Default Logic
G. Antoniou 19

Abductive Behaviour of Concurrent Logic Programs
G.A. Papadopoulos 27

Explanation of Designs Through Heuristic Reasoning
B.D. Britt and T. Glagowski 39

Search Reduction in Linguistic Geometry
B. Stilman 45

Providing Intelligent Support for Predicting Costs of Waste Water Treatment Systems
M. Wood, Y. Wan, M. Kaye, Bruce Bowhill and T. Pe 55

Planning Text for Interactive Plan Explanations
S. Haller 61

Knowledge-Based Uncertainty Management: An IDSS Approach
N. Ayre and T.J. Anderson 69

Collective Learning as an Efficient Learning and Function Optimization Procedure
R. Salomon 77

Scientific Analysis Language for Retrieval and Analysis of Knowledge
P. Lingras 85

Towards a Broader Theory for Abstract Interpretation
K. Musumbbu 95

A Tale of Three Tutoring Protocols: The Implications for Intelligent Tutoring Systems
R.A. Khuwaja, A.A. Rovick, J.A. Michael and M.W. Evens 109

Quantum Mechanical Wave Function Collapse and the Problem of Free Will in Physics
F. Winterberg 119

Evaluating Language Acquisition Models - A Common Framework
B.S.N. Cheung and R.C. Uzgalis 135

Annotations, Signs, and Generally Paraconsistent Logics
J.J. Lu and E. Rosenthal 143

A Type-Free System of Higher Order Logic for PROLOG
R. Butrick 159

Search Reduction Techniques in Jumping Games
I. Pevac and L. Pevac 167

An Intelligent Terminal for Local and Remote Access to the Office Environment
V. Cappellini, A. Ferri, L. Lastrucci, A. Mecocci and A. Raggioli 175

An Implementation of the Faceted Classification System for Software Reuse
R. Mareddy and M. Samadzadeh 181

Investor Counseling and Financial Market Screening: An Experimental Study to Extend Traditional Means in Portofolio Management
J. Geiger and H.F. Hofmann 195

A First-Order Analogue of Higher-Order Unification
W. Fang 209

Using String Logic Programming for Morphological Analysis: A Case Study on Indonesian Language
H.R. Yusuf 221

A Knowledge Organization Framework for Knowledge-Based Pattern Recognition
Y.F. Chen and N.A. Warsi 231

An Interactive Front-End for Fuzzy Databases
C. Liu, Q. Yang, J. Wu and C. Yu 239

Extending Occam's Razor
K.S. Van Horn and T.R. Martinez 249

Modelling Occam2 Programs with Meta Logic Programming
K. Zhang 261

Performance Comparison of Different Learning Methods for Weather Forecasting Operations
V.R. Kumar, P. Guignard and C.Y.C. Chung 275

Quantitative Evaluation of Classification Performance for Machine Learning ROC Analysis and Sensitivity/Specificity Plots
P.A. Guignard 283

Semantic Trees for Disjunctive Logic Programs
J.R. Fisher 291

The OK BDI Architecture
D. Kumar and S.C. Shapiro 307

Evolving the Size of Rule-Based Fuzzy Systems
M.G. Cooper and J.J. Vidal 319

ANALOG: A Knowledge Representation System for Natural Language Processing
S.S. Ali 327

Conversion of Complete Logic Programs to Double Defined Logic Programs
F. Liu and D.H. Moore 333

An Efficient Metric for Heterogeneous Inductive Learning Applications in the Attribute-Value Language
C. Giraud-Carrier and T. Martinez 341

Improving Decision Support Through Hypermedia
V.R. Kumar and C.A. Lindley 351

The Design of Object-Oriented Meta-Architectures for Programming Languages
G. Banavar and G. Lindstrom 363

Non-Well-Founded Set Theory and the Circular Semantics of Semantic Networks
R.K. Hill 375

A New System Approach to Study Program Properties
P.A. Venkatachalam and P.V. Raja 387

Evolutionary Processing
P.A.I. Wijkman 393

Model-Based Diagnosis of the Human Body
J.E. Larsson 405

Intelligent Decision Making in Two-Level Active Systems
R.A. Wasniowski 413

Automatic Generation of C Program Code from Data Flow Diagrams
Y.K. Nam, J. Kim, K. Lee and L.J. Henschen 421

Approach C-C_d, A New Contradiction Handling Strategy and Extended Logic Programs
S. Ghosh 437

NEURAL NETWORKS AND GENETIC ALGORITHMS 443

Stabilizing Techniques in Training Feedforward Neural Networks
C.G. Looney 445

Two Genetic Algorithm Enhancements
J. Paxton and J. Evans 451

Chaos, Neural Networks and Gaming
S. Walczak and J. Krause 457

Application of Genetic Algorithms to 2D Velocity Inversion of Seismic Refraction Data
L. Li, S.J. Louis and J.N. Brune 467

Reverse Engineering: A Case Study on Neural Network Softwork
Y.B. Reddy 475

A Genetic Algorithm Approach to Solving the Battlefield Communication Network Configuration Problem
T.F. Chang and W.D. Potter 483

Evolving Cascade-Correlation Networks for Time-Series Forecasting
J.R. McDonnell and D.E. Waagen 497

Adaptive User Models for Intelligent Information Filtering
K.J. Mock and V. Rao Vemuri 507

Using Genetically Evolving Multi-Layer Cellular Automata for Image Processing
P. Sahota, M.F. Daemi and D.G. Elliman 517

Speech Recognition Using Neural Networks(Hearnet): Mapping from a set of Phoneme Strings to Character/Word Recognition
B.J. Lee 525

An Adaptive Training Method of Back-Propagation Algorithm
J.-H. Yoo, J.-W. Kim and J.-U. Choi 531

Multiversion Information Retrieval: Performance Evaluation of Neural Networks vs. Dempster-Shafer Model
G.V. Meghabghab and D.B. Meghabghab 537

A Study of Genetic Algorithms to Find Approximate Solutions to Hard 3-SAT Problems
J. Frank 547

Fingerprint Image Coding by a Clustering Learning Network
W. Chang, H.S. Soliman and A.H. Sung 555

Parallel Genetic Processes
K.P. Kratzer and R.A. Scholze 565

An Intelligent Neural Network Based System for 3-D Motion Analysis
T. Chen 573

Neural Networks for Rainfall Forecasting
L. Lastrucci and M. Maggini 579

Receiver Function Inversion Using Genetic Algorithms
S. Ozalaybey, M.K. Savage and S.J. Louis 583

An Automatic Transcript Evaluation System
R. Burke, B. Cong and Y. Li 589

Extrapolation of Vibration Data Using Neural Networks
M. Karam and A.M. Trzynadlowski 599

Design Strategies for Evolutionary Robotics
A. Murray and S.J. Louis 609

Efficient Construction of Networks for Learned Representations with General to Specific Relationships
C. Barker and T. Martinez 617

Air Pollution Source Apportionment Using Genetic Algorithms
L.C. Pritchett, G. Mekala and D.N. Wittorff 627

A Transformation for Implementing Neural Networks With Localist Properties
G.L. Rudolph and T.R. Martinez 637

The Application of Neural Network to Fault Diagnosis
J.-L. Chen, H.-F. Sun and R. Tsai 647

VOLUME 2

EXPERT SYSTEMS 655

RSCL3: A Learning Expert System for Intelligent Tutoring
S.H. Rubin 657

No Causality in Function: Building a Function-Centered Knowledge Base
D.J. Russomanno and R.D. Bonnell 665

Methodology for Expandable Expert System Development
W. Dai 681

Intelligent Telecommunication Services: Adaptive and Demonstrational User Interfaces
M. Yvon, F. Lefèvre and P. Piernot 691

Multi-Layered Hybrid Architecture to Solve Complex Tasks of an Autonomous Mobile Robot
F. Tièche, C. Facchinetti and H. Hügli 699

ALINSPEC Project: An Intelligent Vision System for Automatic Inspection of Alimentary Products
A. Barducci, M. Barni, V. Cappellini, S. Livi and A. Mecocci 705

An Intelligent System for GIS Information Mapping
G.M. Gallitano, E.A. Yfantis and A. Pitchford 713

PITS: An Intelligent Tutoring System Loosely Coupled to External Database Systems
Y.D. Yoo 719

DATA BASE MANAGEMENT 729

A Finely-Interleaved Consistency Checking Method for Knowledge-Bases
J.V. Harrison 731

Heuristic Query Analysis Within a Distributed Deductive Database
K. Neumann 741

Methodology for Implementing the Access Control in Behaviorally Object-Oriented Database Systems
C. Radu, M. Vandenwauver, R. Govaerts and J. Vandewalle 749

Prospective View on Intelligent Databases
S.C. Suh, R.D. Crieder and V. Kandula 757

Adaptive Query Reformulation in Attribute Based Image Retrieval
G.S. Jung and V.N. Gudivada 763

Implicit Representation for Extensional Answers in Object-Oriented Systems
S.-C. Yoon and I.-Y. Song 775

A General Framework for Building Intelligent Database Applications
Z. Cui and J. Fox 789

Retrieval in Image Databases
N. Dimitrova and F. Golshani 801

Design Issues of Object Manager for the TOS
T. Al-Sayegh, A. Shah, I. Faraj, F. Fotouhi and W. Grosky 813

Issues in Management of Class-History in Object-Oriented Databases
A. Al-Khudair, A. Shah and H. Mathkour 823

Methodology to Convert a Traditional Design to an Object-Oriented Design
S.H. Al-Harbi, A. Shah, H. Mathkour and J.C. Agrawal 833

COMPUTER GRAPHICS AND IMAGE PROCESSING 843

Deformation of Volumes Using Scattered Landmark Points
T.A. Foley 845

A Hybrid Mountain Generation Algorithm Using Subdivision and B-Splines
B.L. Hagstrom and E.A. Yfantis 853

View Variations in Angles
R. Malik and T. Whangbo 863

A Simple and Efficient Thinning Method
C.-E. Wang 873

Using Proper Object-Oriented Design to Enhance Portability of Graphical Applications
J. Brown and D. Eisenberger 877

A Fast Fourier Method for Mountain Generation
S.P.V. Pallati and E.A. Yfantis 885

A New Differential Pulse Code Modulation Compression Algorithm
T. Pike, E.A. Yfantis and Z. Psyllakis 897

ALGORITHMS 903

The Problem of Partitioning With Duplications and its Applications
J. Haralambides and S. Tragoudas 905

Pr/t Net Method for Robot Planning
J. Yim and T. Murata 913

Recognition of Hand Printed Digits Using Multiple Parallel Methods
J.R. Parker 923

Designing Fail-Soft Systems for Distributed Computing
C. Stivaros 933

Distributed Programming Using Objects - A Case Study
D. Molaro and J.R. Parker 945

A Distributed Deadlock Detection Algorithm
B.M. Johnston and A.K. Datta 953

Index 965

PREFACE

The 1994 GWIC was held June 6th , 7th and 8th, 1994, on the Campus of the UNLV. It was sponsored by UNLV, UNR, and ACM-SIGART. The keynote speakers were Prof. Bonnie Weber of the University of Pennsylvania, Prof Stuart Shapiro, Director of the Center for Cognitive Science at SUNY at Buffalo, and Prof. Nicolas Bourbakis of SUNY at Binghamton.

Dr. Bonnie Webber, the first keynote speaker, presented the first talk of the conference Monday morning June 6th, entitled "Instructing Animated Agents: Natural Language and Human Figure Animation". Her one hour lecture and the computer graphics video in which figures emulating realistically humans were able to successfully perform a number of human motions and functions, were very well received by the participants.

Dr. Stuart Shapiro, presented his keynote speech, entitled "Formalizing English", Tuesday morning, June 7th. His objective was to construct a natural language using an intelligent agent. His talk was of great interest and drew a great deal of discussion and questions by the participants.

"The Role of AI in Multimedia Information Systems", was the topic presented by the third keynote speaker, Dr. N. Bourbakis, Wednesday morning June 8th. He addressed the changes in computing with the introduction of Multimedia and the usage of AI to store and retrieve intelligently massive visual, audio, and other data.

The parallel track sessions included the new trends in AI and computing, along with the more traditional subjects. The sessions were very professionally and successfully chaired by Profs. Carl G. Looney, Tony Martinez, John Paxton, Charles Giardina, Steven Walczak, Louis D'Alotto, James Haralambides, Yuh Jeng Lee, Stuart Rubin, Sushil J. Luis, Mansur Samadzadeh, W. M. Lippe, John T. Minor, David Russomano, Kia Makki, Ajoy K. Datta, John V. Harrison, Thomas A. Foley, Michael Wood, George A. Papadopoulos, Boris Stilman, and Gail M. Gallitano.

From the list of participants it is evident that they represented all the corners of the earth. Approximately 50% of the extended abstracts and papers received were accepted for publication. We would like to extend our sincere thanks to the organizing committee of the conference, UNLV officials, our gracious keynote speakers, and especially to the participants and speakers that helped make this conference a success. We would like to express our appreciation to Dr. C. Looney for his encouragement, and immense contribution to the Conference. Thanks are also due to our reviewers and especially to Drs. J. Harrison, G. M. Gallitano, C. Giardina, and G. Antoniou for their valuable contributions.

E. A. Yfantis
Computer Science Department
University of Nevada, Las Vegas
Las Vegas, Nevada, 89154

LIST OF PARTICIPANTS

S. S. Ali, Department of Computer Science, Southwest Missouri State University, 901 South National Avenue, Springfield, MO 65804

S. H. Al-Harbi, Directorate of Computing, Royal Saudi Air Force, P. O. Box 59787, Riyadh-11535, Kingdom of Saudi Arabia

A. Al-khudair, Department of Information Systems and Exchanges, Telecom College, P.O. Box 2067, Riyadh 11451, Saudi Arabia

T. Al-Sayegh, National Information Center, Ministry of Interior, P.O. Box 69910

G. Antoniou, Department of Management, The University of Newcastle, Callaghan, NSW 2308, Australia

N. Ayre, Department of Information Systems, University of Ulster at Jordanstown, Newtownabbey, Co Antrim, BT37 0QB, N. Ireland

G. Banavar, Department of Computer Science, University of Utah, Salt Lake City, Utah, USA

C. Barker, Computer Science Department, Bringham Young University, Prove, Utah, 84602, USA

B. D. Britt, Computer Science Department, Eastern Washington University, Cheney, WA 99004-2495

J. Brown, Center for Communications and Information Technology, Duquesne University, Pittsburgh, PA 15282, USA

N. Bourbakis, Department of Electrical Engineering, Binghamton University, P.O. Box 600, Binghamton, N.Y. 13902-6000, USA

R. Burke, South Dakota State University, Brookings, SD, 57007, USA

R. Butrick, Computer Science Department, Ohio University, Athens, Ohio, 45701, USA

J.-L. Chen, Advanced Technology Center Computer and Communication Research Laboratories, Industrial Technology Research Institute, Hsinchu, Taiwan, R.O.C.

T. Chen, Argonne National Laboratory, 9700 S. Cass Ave., Argonne, IL 60439, USA

Y. F. Chen, Army Center of Excellence in Information Sciences, Clark Atlanta University, Atlanta, GA, 30314, USA

T. F. Chang, VerTec Solutions Inc., 5601 Roanne Way, Suite 606, Greensboro, NC, 27409, USA

W. Chang, Department of Computer Science, New Mexico Institute of Mining and Technology, Socorro, New Mexico, 87801-4682, USA

B. S. N. Cheung, Department of Computer Science, The University of Hong Kong

M. G. Cooper, University of California, Los Angeles, 4531 Boelter Hall, Los Angeles, California, 90024, USA

Z. Cui, Advanced Computational Laboratory, Imperial Cancer Research Fund, 61 Lincoln's Inn Fields, London, WC2A 3PX, UK

W. Dai, Artificial Intelligence Systems, Telecom Australia Research Laboratories, 770 Blackburn Road, Clayton, Victoria 3168, Australia

L. D'Alotto, Department of Mathematics and Computer Science, St. John's University, 300 Howard Evenue, Staten Island, NY 10301, USA

N. Dimitrova, Department of Computer Science and Engineering, Arizona State University, Tempe, AZ 85287-5406

D. Eisenberger, Duquesne University, 600 Forbes Ave., Pittsburgh, PA. 15282, Rockwell Hall, CCIT, USA

Th. Feuring, Institut fuer Numerische und Instrumentelle Mathematik- Informatik, Westfaelische Wilhelms-Universitaet Muenster Einsteinstrasse 62 - 48149 Muenster - Germany

J. R. Fisher, Computer Science Department, California State Polytechnic University, Pomona, CA, 91768, USA

T. A. Foley, Department of Computer Science and Engineering, Arizona State University, Tempe, AZ 85287-5406, USA

J. Frank, Division of Computer Science, University of California at Davis, Davis, CA 95616, USA

G. M. Gallitano, Department of Mathematics and Computer Science West Chester University, 13-15 University Ave, West Chester Pennsylvania, 19383

J. Geiger, Department of Computer Science, University of Zurich, Winterthurerstr, 190, CH-8057 Zurich, Switzerland

C. Giardina, Department of Mathematics and Computer Science, CUNY/College of Staten Island, Staten Island, NY 10314 USA

C. Giraud-Carrier, Department of Computer Science, Brigham Young University, Provo, UT 8462

D. Glasser, Computer Science Department, University of Nevada, Las Vegas, 89154

V. N. Gudivada, Department of Computer Science, Ohio University, Athens, OH 45701

P. A. Guignard, Knowledge Based Systems Laboratory, CSIRO Division of Information Technology, Locked Bag 17, North Ryde, NSW 2113, Australia

B. L. Hagstrom, Computer Science Department, University of Nevada, Las Vegs, 89154, USA

S. Haller, Computer Science and Engineering Dept., University of Wisconsin-Parkside, Kenosha, WI 53141

J. Haralambides, Department of Mathematics and Computer Science, Barry University, Miami Shores, FL 33161

J. V. Harrison, Department of Computer Science, University of Queensland, Brisbane, QLD, 4072, Australia

R. K. Hill, Computer Science Department, State University of New York at Buffalo, 226 Bell Hall, Buffalo, NY, 14260, USA

B. M. Johnston, System Computing Services, University and Community College System of Nevada, Las Vegas, NV, 89154, USA

M. Karam, Department of Electrical Engineering, University of Nevada, Reno, Reno, NV, 89557-0153, USA

R. A. Khuwaja, Computer Science Department, Illinois Institute of Technology, Chicago, Illinois, 60616, USA

K. P. Kratzer, Fachhochschule Ulm, Prittwitzstr 10, 89075 Ulm, FRG

J. Krause, University of Tampa, 401 W. Kennedy Blvd., Tampa, FL 33606-1490, USA

D. Kumar, Department of Mathematics, Bryn Mawr College, Bryn Mawr, PA, 19010, USA

V. R. Kumar, Knowledge-Based Systems Laboratory, CSIRO, Division of Information Technology, Locked Bag 17, North Ryde, NSW 2113, Australia

J. E. Larson, Knowledge Systems Laboratory, Stanford University, 701 Welch Road, Building C, Palo Alto, CA 94304

L. Lastrucci, Dipartimento di Electronica, Universita di Firenze, Via di Santa Marta 3-50139 Firenze-Italy

B. J. Lee, Department of Computer Science, University of Iowa, Iowa City, IA 52240

P.-Y. Li, School of Engineering and Technology, Computer Science old Westbury Campus, P. O. Box 800, Old Westbury, NY 11568-8000

L. Li, Seismo. Lab. University of Nevada, Reno, Reno, NV 89557, USA

P. Lingras, Department of Computer Science, Algoma University College, Sault Ste. Marie, Ontario, P6A 2G4, Canada

W. M. Lippe, Institut fuer Numerische und Instrumentelle Mathematik- Informatik, Westfaelische Wilhelms-Universitaet Muenster Einsteinstrasse 62 - 48149 Muenster - Germany

C. Liu, Department of Computer Science and Information Systems, DePaul University, Chicago, Illinois, USA

F. Liu, Department of Computer Science and Computer Engineering, La Trobe University, Bundoora Vic 3083, Australia

C. G. Looney, Computer Science Department, University of Nevada, Reno, NV 89557, USA

S. J. Louis, Department of Computer Science, University of Nevada, Reno, NV, 89557, USA

R. Malik, Department of Electrical Engineering and Computer Science Stevens Institute of Technology, Hoboken, N.J. 07030, USA

R. Mareddy, Mirage Resorts, Inc., Las Vegas, NV 89109, USA

T. Martinez, Department of Computer Science, Brigham Young University, Provo, Utah, 84602, USA

M. Maskarinec, Department of Computer Science, California State University, Fullerton, CA, 92634, USA

J. R. McDonnel, NCCOSC, RDT and E Div., San Diego, CA 92152

J. Minor, Department of Computer Science, University of Nevada, Las Vegas, Las Vegas, Nevada, 89154, USA

K. J. Mock, Department of Computer Science, University of California at Davis, California, 95616

D. Molaro, Laboratory for Computer Vision, Department of Computer Science, University of Calgary, Calgary, Alberta, Canada

H. Mueller, Computer Science Department, University of Calgary, Alberta, Canada

A. Murray, Department of Computer Science, University of Nevada, Reno, Reno, NV, 89557, USA

K. Musumbu, LaBri, CNRS/Universite Bordeaux I, 351, cours de la Liberation, 33405 Talence Cedex, France

K. Neumann, Department of Mathematics / PIC, UCLA, Los Angeles, CA 90024-1555, USA

S. Ozalaybey, Seismological Laboratory, University of Nevada, Reno, 89557, USA

S. P. V. Pallati, Department of Computer Science, University of Nevada, Las Vegas, Las Vegas, Nevada, 89154, USA

G. A. Papadopoulos, Department of Computer Science, University of Cyprus, 75 Kallipoleos Str., Nicosia, T.T. 134, P.O. Box 537, Cyprus

J. R. Parker, Laboratory for Computer Vision, Department of Computer Science, University of Calgary, Calgary, Alberta, Canada

J. Paxton, Computer Science Department, Montana State University, Bozeman, MT 59717

I. Pevac, Computer Science Department, Central Conn. State University, New Britain, CT 06050, USA

T. Pike, Computer Science Department, University of Nevada, Las Vegas, 89154, USA

L. C. Pritchett, Department of Computer Science, University of Nevada, Reno, Reno, NV, 89557, USA

M. Quafafou, Universite de Nantes, IRIN 2, rue de la houssiniere, 44072 Nantes Cedex 03, France

Y. B. Reddy, Department of Mathematics and Computer Science, Grambling State University, Grambling LA, 71245,

E. Rosenthal, Department of Mathematics, University of New Haven, West Haven, CT, 06516, USA

S. H. Rubin, Department of Computer Science, Central Michigan University, Mt. Pleasant, MI 48859

G. L. Rudolph, Computer Science Department, Brigham Young University, Provo, Utah, 84602, USA

D. J. Russomanno, Department of Electrical Engineering, The University of Memphis, Memphis, TN 38152

P. Sahota, Department of Computer Science, University of Nottingham, Nottingham, NG7 2RD, UK.

R. Salomon, International Computer Science Institute, 1947 Center Sr., Suite 600, Berkeley, CA 94704-1198

M. Samadzadeh, Oklahoma State University, Computer Science Department, Stillwater, OK, 74078, USA

E. Schwalb, Department of Information and Computer Science University of California at Irvine, CA, 92717. USA

S. C. Shapiro, Department of Computer Science, State University of New York at Buffalo, Buffalo, NY, 14260

B. Stilman, Department of Computer Science and Engineering, University of Colorado at Denver, Campus Box 109, Denver, CO 80217-3364, USA

C. Stivaros, Computer Science Department, Fairleigh Dickinson University, Madison, NJ, 07940

S. C. Suh, Department of Computer Science, East Texas State University, Commerce, Texas 75429

F. Tièche, University of Neuchâtel, Rue de Tivoli 28, CH-2003, Neuchâtel, Switzerland

K. S. Van Horn, Computer Science Department, Bringham Young University, Prove, UT, 84602, USA

V. R. Vemuri, Department of Applied Science, University of California at Davis, California, 95616

M. Vandenwauver, Katholicke Universiteit Leuven Laboratorium ESAT Kardinaal Mercierlaan 94, B-3001 Heverlee, Belgium

P. A. Venkatachalam, School of Electrical and Electronic Engineering, Universiti Sains Malaysia, 31750 Tronoh, Malaysia

R. Venkatachalam, Knowledge-Based Systems Laboratory, Knowledge-Based Systems Laboratory, CSIRO Division of Information Technology, Locked Bag 17, North Ryde, NSW 2113, AUSTRALIA

S. Walczak, University of South Florida, 4202 East Foweler Ave., CIS 1040, Tampa, FL 33620-7800, USA

C.-E Wang, Department of Computer Science, California State University, Sacramento, Sacramento, CA 95819-6021

R. A. Wasniowski, New Mexico Highlands University, 405 New Mexico Ave. Las Vegas, NM 87701, USA

B. Webber, Computer Science Department, University of Pennsylvania, USA

P. A. I. Wijkman, Royal Institute of Technology and University of Stockholm, Department of Computer and System Sciences, Electrum 230, 164 40 Kista, Sweden

F. Winterberg, Desert Research Institute, University and Community College System of Nevada, Reno, Nevada, 89506, USA

M. Wood, University of Portsmouth, Locksvay Road, Milton, Southsea, Hants, PO4 8jF, UK

E. A. Yfantis, Department of Computer Science, University of Nevada, Las Vegas, NV, 89154, USA

J. Yim, Department of Computer Science, Dong-Kook University at Kyungju, Kyungju City, Kyung-Book, Korea, 780-714

J.-H. Yoo, Artificial Intelligence Division, Systems Engineering Research Institute, P.O. Box 1, Yoosung, Taejeon, 305-600, Korea

Y. D. Yoo, Department of Information Systems, Dongguk University, Kyungju, Korea

S-C. Yoon, Department of Computer Science, Widener University, Chester University, Chester, PA 19013, USA

K. Nam Young, Systems Engineering Research Institute, KIST Taejon Korea

H. R. Yusuf, Department of Computer Science, New Mexico State University Las Cruces, NM 88003, USA

M. Yvon, Laboratoire d'intelligence Artificielle de Paris 5, UFR de Mathematique et d'Informatique, Universite Rene Descartes, 45, Rue des Sains-Peres, 75006 Paris-France

K. Zhang, Department of Computing, Macquarie University, Sydney, NSW 2109, Australia

ARTIFICIAL INTELLIGENCE

The Kolmogorov Metric and Classifying Linear Cellular Automata

Louis D'Alotto*
Department of Mathematics/Computer Science
St. John's University
300 Howard Avenue, Staten Island, NY 10301 USA

Charles Giardina
Departments of Mathematics and Computer Science
CUNY/College of Staten Island
Staten Island, NY 10314 USA

Abstract

By introducing a new and useful metric on the domain space of bi-infinite strings, this paper suggests a classification approach for linear (one-dimensional) cellular automata according to their dynamical behavior.

Keywords. Cellular Automata, Classification, Metric, Lattice, Probability, Dynamical Systems, Periodicity, Equicontinuity, Digital Signal Processing, Strings.

1 Introduction

The idea of classifying linear cellular automata by means of their dynamic behavior was initiated by Wolfram [13], [14], [15]. Gilman [7] later constructed a probabilistic classification of linear automata. The classification results presented here, utilizing a new and applicable metric, follow those obtained in [7]. Linear automata are thus partitioned into three classes, where each class seems to correspond to a different type of dynamical behavior. For brevity we omit the proofs of the theorems and lemmas. These proofs and additional examples, can be found in [3].

2 Definition of Cellular Automata and The Metric

Let S be an alphabet of size s such that $2 \leq s < \infty$ and let $X = (S \cup \{*\})^{\mathbf{Z}}$. X is the set of all maps from the integers to $S \cup \{*\}$. That is, for $f \in X$, $f : \mathbf{Z} \rightarrow S \cup \{*\}$. The set $S \cup \{*\}$ is compact and hence the product space X is compact. The restriction of a map $f \in X$ to a non-empty interval $[i, j]$ of $\mathbf{Z}$, where $i \leq j$, is called a *word*. Words are written as $f[i, j]$

*\footnote{This work was, in part, accomplished and funded while the author was at the Department of Computer Science, Lafayette College, Easton, PA USA. }

E. A. Yfantis (ed.), Intelligent Systems, 3–12.

for $-\infty < i \leq j < \infty$ or $f[k,\infty)$, $f(-\infty,k]$ and $f(-\infty,\infty)$ for right infinite, left-infinite and bi-infinite intervals of $\mathbf{Z}$, respectively. $f(-\infty,\infty)$ is simply denoted by f. $\aleph$ will be used to represent the natural numbers and $\aleph_0 = \aleph \cup \{0\}$. It will be convenient to employ bound vector notation of digital signal processing to represent elements of X. In general, write: $f = (a_0\ a_1\ a_2 .... \ a_n)_p^q$ whenever

$$f(i) = \begin{cases} q & if \quad i < p \\ a_0 & if \quad i = p \\ a_1 & if \quad i = p+1 \\ a_2 & if \quad i = p+2 \\ . & \\ . & \\ a_n & if \quad i = p+n \\ q & if \quad i \geq p+n+1 \end{cases}$$

Hence, $f = (1\ 2\ 3)_{-1}^{*}$ represents $(....*\ *\ *\ 1\ 2\ 3\ *\ *\ * ....)$ where $f(-1) = 1$, $f(0) = 2$, $f(1) = 3$ and all other entries are $*$ valued.

Definition 1 *For any f, $y \in X$ define the binary infimum operator $\wedge$ on X to be:*

$$f \wedge y = \begin{cases} f & if \quad f = y \quad and \quad f(i) \neq * \quad \forall i \in \mathbf{Z} \\ * & if \quad f(0) \neq y(0) \quad or \quad f(0) = * \\ (f(-m)....f(m))_{-m}^{*} & if \quad f(i) = y(i) \quad and \quad f(i) \neq * \ \forall i \in [-m,m] \\ & \quad and \quad f(-m-1) \neq y(-m-1) \quad or \\ & \quad f(m+1) \neq y(m+1) \quad or \\ & \quad f(-m-1) = * \quad or \quad f(m+1) = * \end{cases}$$

Thus $f \wedge y$ consists of the largest center stretch of values from S where f and y agree (with no $*$) and is $*$ valued outside this center. It can be easily verified that the binary relation $\leq$, on X, defined by $f \leq y$ if and only if $f \leq f \wedge y$, forms a partial order on X and that the pair $< X, \wedge >$ forms a lower (infimum) semilattice with the lower unit $*$.

Let $Y = S^{\mathbf{Z}}$ be the subspace of X defined as the set of all maps from the integers to the alphabet S. As the space X is compact so is Y. The space Y can also be considered as the space of all bi-infinite strings with elements taken from S. Define a metric on the space Y as follows:

$$d(f,y) = \begin{cases} 1 & if \quad f(0) \neq y(0) \\ 0 & if \quad f = y \\ \prod_{i=-m}^{m} \lambda_i & if \quad (f \wedge y) = (....*\ *f(-m)...f(0)...f(m)\ *\ *....) \end{cases}$$

where λ is any real-valued function defined on S and taking values in the open interval $(0,1)$, i.e. $\lambda : S \rightarrow (0,1)$ where $\lambda_i = \lambda(f(i))$, so $0 < \lambda_i < 1$. (Y,d) forms a metric space where the metric d is non-archimedean, i.e. $d(f,y) \leq \max\{d(f,z), d(z,y)\}$. The metric just defined will be referred to as *The Kolmogorov Metric.*

Linear cellular automata are induced by arbitrary (local) maps $G : S^{2r+1} \rightarrow S$ (these maps are sometimes called *local rules* or *block maps* in the literature) where $r \in \aleph_0$. The value r is called the *range* of the cellular automaton. The automaton map g induced by G is defined by $g(f) = z$ with $z(n) = G(f(n-r), ..., f(n), ..., f(n+r))$.

Example 1 *The right 1-shift (or just shift) map σ is the automaton of range 1 induced by $G(\alpha, \beta, \gamma) = \alpha$.*

Definition 2 $C_\varepsilon(f) = \{z \mid d(f,z) \leq \varepsilon, \text{ for } \varepsilon > 0\}$, *$C_\varepsilon(f)$ is the open ball of radius ε around f.*

Since the metric is non-archimedean, given any two disks $C_\varepsilon(f)$, $C_\alpha(y)$, either $C_\varepsilon(f) \cap C_\alpha(y) = \emptyset$ or one contains the other. In this topology, the C_ε sets are also closed. for fixed $\varepsilon > 0$, the relation $f \sim y$ if $d(f,y) \leq \varepsilon$ is an equivalence relation with equivalence classes $\{C_\varepsilon(f)\}$.

Example 2 *Consider the alphabet $S = \{1,2,3\}$ and let $\lambda(1) = 0.5$, $\lambda(2) = 0.2$, $\lambda(3) = 0.1$. If $f = (2\ 2\ 3)^1_{-1}$ and $z = (1\ 2\ 2\ 3\ 1)^3_{-2}$ then since $f \wedge z = (1\ 2\ 2\ 3\ 1)^3_{-2}$ it follows that $d(f,z) = (.5)(.2)(.2)(.1)(.5) = 0.001$. Moreover, $z \in C_\varepsilon(f)$ for any $\varepsilon \geq 0.001$. Notice that $y = (1)^1_0$ is not in $C_\varepsilon(f)$ for any $\varepsilon < 1$. Additionally, observe that if $x = (1\ 1\ 2\ 2\ 3\ 1\ 1)^2_{-3}$ then x is also in $C_\varepsilon(f)$ since $d(x,f) = (.5)(.5)(.2)(.2)(.1)(.5)(.5) = 0.00025$.*

Another useful equivalence class is obtained in the same manner using the following definition:

Definition 3 $B_\varepsilon(f) = \{y \mid d(g^i(f), g^i(y)) \leq \varepsilon,\ \forall i \in \aleph_0 \text{ and } \varepsilon > 0\}$.

The $\{B_\varepsilon(f)\}$ are equivalence classes with the equivalence relation $x \approx f$ if and only if $d(g^i(f), g^i(y)) \leq \varepsilon$, $\forall i \in \aleph_0$. The notation $g^i(x)$ will always be used to represent the compositions of g with itself, i.e. $g^i(x) = g \circ g \circ .... \circ g(x)$. Note that $g^0(x) = x$. It is also important to note that $B_\varepsilon(f) \subset C_\varepsilon(f)$.

The behavior of f, that is the forward iterations (compositions) of f under a linear automaton map g, can be visualized as an array $(a_{i,j})$ with rows called i and columns j. Here each entry $a_{i,j} = (g^i(f))(j)$ $\forall i \in \aleph_0$ and $j \in \mathbf{Z}$. Then $B_\varepsilon(f)$ equals the set of all y whose behavior agrees with the behavior of f on the infinite *vertical jagged edge strip* defined by the intervals $[-m_i, m_i]$ such that $d(g^i(f), g^i(y)) \leq \varepsilon$ $\forall i \in \aleph_0$. That is, $B_\varepsilon(f)$ is the set of y where $(g^i(f))[-m_i, m_i] = (g^i(y))[-m_i, m_i]$ for each $i \in \aleph_0$. Given $\lambda : S \to (0,1)$, the minimum length of the $[-m_i, m_i]$ intervals is determined by ε and the intervals themselves are determined by the places where f and y agree. It will be said that y has behavior $B_\varepsilon(f)$ on the intervals $[-m_i + k, m_i + k]$ whenever $\sigma^{-k}(y) \in B_\varepsilon(f)$ (σ is the right-shift map) for $k \in \aleph_0$ and all $i \in \aleph_0$. For a fixed automaton g and fixed ε, the shape of the jagged edge strips will depend on the elements y in $B_\varepsilon(f)$. Many computer experiments of linear automata, as well as applications, investigate a certain finite portion of the array $(a_{i,j})$. Therefore the sets $B_\varepsilon(f)$ become natural structures to study. The classification scheme used here will be based on the analysis of the sets $B_\varepsilon(f)$.

3 Periodicity and Equicontinuity

Definition 4 *Let (Y,d) be the metric space previously defined. Let $\Im$ be a subset of the collection of elements of the space Y^Y which are automata. If $y_0 \in Y$, the family of functions $\Im$ is equicontinuous at y_0 if given $\varepsilon > 0$, there exists a δ-neighborhood C_δ of y_0 such that for all $y \in C_\delta$ and all $g \in \Im$, $d(g(y), g(y_0)) \leq \varepsilon$.*

The subset $\Im$, of functions from Y^Y, that is of study here is the set of all forward iterates of an automaton g, i.e. the set $\{g^i \mid i \in \aleph_0\}$. Instead of saying the set of iterates $g^i \; \forall i \in \aleph_0$ is equicontinuous at f, it is simply said that g is equicontinuous at f.

Definition 5 *A non-empty word $y[-q,q]$ is ultimately periodic if there exists $k \geq 0$ and $q \geq 0$ such that $(g^{n+i}(y))[-q,q] = (g^n(y))[-q,q]$ for some $n \geq 1$ and all $i \geq k$. The smallest n such that $(g^{n+i}(y))[-q,q] = (g^n(y))[-q,q]$ is called the period of the word, i.e. $(g^i(y))[-q,q]$ has period n for $i \geq k$.*

Definition 6 *For $\varepsilon \leq (\min_{j \in S}\{\lambda(j)\})^{2q+1}$, $B_\varepsilon(f)$ is ultimately periodic if the word $f[-q,q]$ is ultimately periodic.*

The forward orbit of f under the automaton map g, that is the set of points

$$f, g(f), g^2(f), ..., g^i(f), ...$$

will be denoted by $O^+(f)$. Similarly, the backward orbit of f, that is

$$f, g^{-1}(f), g^{-2}(f), ..., g^{-i}(f), ...$$

will be denoted by $O^-(f)$. $\overline{O^+(f)}$ is the closure of $O^+(f)$ and for any $A \subseteq X$, A^o is the interior of A. $\omega(f)$ is used to denote the set of contact (adherence) points of $O^+(f)$.

lemma 1 *Let Y be a compact metric space, $g : Y \to Y$ a continuous function and $f \in Y$. Then $g(\omega(f)) \subset \omega(f)$.*

lemma 2 *Consider any $f \in Y$ and $\varepsilon > 0$, then the following statements hold:*

1. *$B_\varepsilon(f)$ is closed.*
2. *g is equicontinuous at f iff $f \in B_\varepsilon(f)^o$ for all $\varepsilon > 0$.*
3. *the restriction of g to $\overline{O^+(f)}$ is equicontinuous iff $B_\varepsilon(f)$ is ultimately periodic for all $\varepsilon > 0$.*

As a result g is equicontinuous at f iff for all $\varepsilon > 0 \; \exists \; \delta > 0$ such that $C_\delta(f)$, the open ball of radius δ around f, is contained in $B_\varepsilon(f)$.

An example of the previous lemma demonstrates the fact that given $B_\varepsilon(f)$ is ultimately periodic then g, restricted to $\overline{O^+(f)}$, is equicontinuous.

Example 3 *Use the following automaton rule defined on the alphabet $S = \{0,1\}$ induced by the local rule (Wolfram totalistic rule 24 [13],[14]): $G : S^5 \to S$ where*

$$G(a,b,c,d,e) = \begin{cases} 1 & if \quad a+b+c+d+e = 3 \text{ or } 4 \\ 0 & otherwise \end{cases}$$

Choose $f = (1\,1\,1\,0\,0\,0\,1\,1\,0\,1\,0\,0\,0\,1\,1\,1)^{0}_{-6}$ *Hence,*

$$\begin{array}{ll} f & = \ldots.0\,0\,0\,1\,1\,1\,0\,0\,0\,1\,1\,0\,1\,0\,0\,0\,1\,1\,1\,0\,0\,0\ldots. \\ g(f) & = \ldots.0\,0\,0\,1\,1\,1\,0\,0\,0\,0\,1\,1\,0\,0\,0\,0\,1\,1\,1\,0\,0\,0\ldots. \\ g^2(f) & = \ldots.0\,0\,0\,1\,1\,1\,0\,0\,0\,0\,0\,0\,0\,0\,0\,0\,1\,1\,1\,0\,0\,0\ldots. \\ \cdot & \cdot \\ \cdot & \cdot \\ g^n(f) & = \ldots.0\,0\,0\,1\,1\,1\,0\,0\,0\,0\,0\,0\,0\,0\,0\,0\,1\,1\,1\,0\,0\,0\ldots. \\ \cdot & \cdot \\ \cdot & \cdot \end{array}$$

The sequence is fixed for all iterates $n \geq 2$, *and therefore ultimately fixed. Without loss of generality,* λ *can be chosen equal to* $\frac{1}{2}$ *for both* 0 *and* 1. *Thus for* $\varepsilon \leq \left(\frac{1}{2}\right)^{2q+1}$, *where in this case* $q \leq 11$, *and* $\alpha = \left(\frac{1}{2}\right)^{23}$ *the open ball*

$$C_\alpha((0\,0\,0\,0\,0\,1\,1\,1\,0\,0\,0\,1\,1\,0\,1\,0\,0\,0\,1\,1\,1\,0\,0\,0))^{*}_{-11} \subset B_\varepsilon(f)$$

(note that here $* = 0$ *or* 1*). Hence,* g *is equicontinuous at* f. *For any* $\varepsilon \leq \left(\frac{1}{2}\right)^{2q+1}$, *an open ball* $C_\alpha(f)$ *can be found that is contained in* $B_\varepsilon(f)$. *It is easily seen that any point* x *belonging to the open ball*

$$C_\alpha((0\,0\,0\,0\,0\,0\,1\,1\,1\,0\,0\,0\,1\,1\,0\,1\,0\,0\,0\,1\,1\,1\,0\,0\,0))^{*}_{-12}$$

also belongs to $B_\varepsilon(f)$.

4 Classes

The set $B_\varepsilon(f)$ consists of elements from $Y = S^{\mathbf{Z}}$, hence the shift map σ can be defined on $B_\varepsilon(f)$. Referring to the previously used terminology, y has behavior $B_\varepsilon(f)$ on $[-m_i + k, m_i + k]$ if, for some $k \in \mathbf{Z}$, $\sigma^{-k}(y) \in B_\varepsilon(f)$ for all $i \in \aleph_0$. Since σ is a homeomorphism, y has behavior $B_\varepsilon(f)$ on $[-m_i + k, m_i + k]$ can be restated as $y \in \sigma^k(B_\varepsilon(f))$ for all $i \in \aleph_0$. That is the behavior of y is known on the vertical jagged edge strip determined by the intervals $[-m_i + k, m_i + k]$ for all $i \in \aleph_0$. Using the array visualization for the behavior of f, if g has range r and if $(a_{i,j})$ is known on two vertical jagged edge strips, each of width at least r, and the top row of $(a_{i,j})$ between the two strips is known, then that part of $(a_{i,j})$ between the two strips is uniquely determined. This yields as a consequence.

lemma 3 *Given a finite alphabet* S, $\sigma : Y \to Y$ *is the right shift map and* $f \in Y$. *Let* $\lambda_\wedge = \min_{j \in S}\{\lambda(j)\}$ *and* $\lambda_\vee = \max_{j \in S}\{\lambda(j)\}$. *If* g *is an automaton map of range* $r \leq 2n + 1$ *for some* $n \in \aleph_0$ *and*

1. $f \in \sigma^{-k}(B_\varepsilon(y)) \cap \sigma^u(B_\varepsilon(z))$ *for* $0 < \varepsilon \leq (\lambda_\wedge)^{2n+1}$ *and* $k, u \in \aleph_0$
2. *let* $c, q \in \aleph_0$ *where* $0 \leq c \leq k, u \leq q$, $0 < \alpha \leq (\lambda_\wedge)^{2q+1}$ *and* $0 < (\lambda_\vee)^{2c+1} \leq \delta$

Then $\sigma^{-k}(B_\varepsilon(y)) \cap \sigma^u(B_\varepsilon(z)) \cap C_\alpha(f) \subset B_\delta(f)$.

Remarks:

- The upper bound of $(\lambda_\wedge)^{2n+1}$, for ε, places a minimum on the size of the intervals of the jagged edge strips.
- The upper bound of $(\lambda_\wedge)^{2q+1}$, for α, in condition 2, places a minimum on the size of the intervals for elements of $C_\alpha(f)$. That is, if $x \in C_\alpha(f)$ then at least $x[-q,q] = f[-q,q]$. This insures that the top row between the two jagged edge strips is known.
- The lower bound of $(\lambda_\vee)^{2c+1}$, for δ, insures that the widths of the intervals $[-c_i, c_i]$, for $B_\delta(f)$ do not necessarily extend outside the two jagged edge strips.

To develop a measure theoretic classification we first need to choose a probability distribution on the alphabet S such that each element $a \in S$ has positive probability. That is, let $n \geq 2$ be a fixed integer and let $(p_0, p_1, p_2, \ldots, p_{n-1})$ be a probability vector whose entries $p_i > 0$ for each i. μ is the corresponding product measure on Y.

The previous lemma is crucial in the proofs of the classification theorems; the illustration is given below in figure 1:

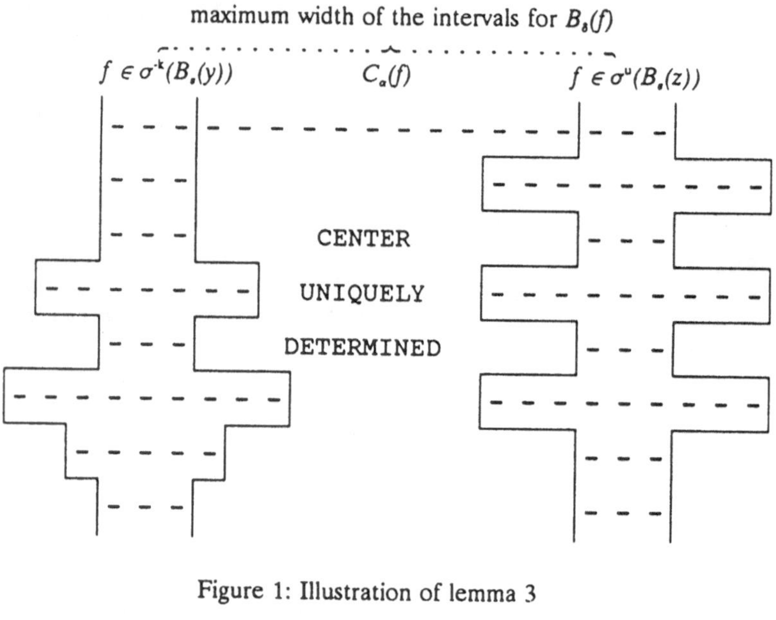

Figure 1: Illustration of lemma 3

Definition 7 *T_σ is the set of $f \in Y$ with dense forward and backward orbits under the shift σ. Using more formal notation:*

$$T_\sigma = \{y \in Y \mid \overline{O^+(y)} = Y\} \cap \{y \in Y \mid \overline{O^-(y)} = Y\}$$

Note that $f \in T_\sigma$ if and only if every finite word can be written as $f[i,j]$ for $i,j \in \mathbb{Z}$ with $0 \leq i \leq j$ and also every finite word can be written as $f[i,j]$ for $i,j \in \mathbb{Z}$ with $i \leq j \leq 0$.

lemma 4 *$\mu(T_\sigma) = 1$, in particular σ is topologically transitive (see also [12],theorems 5.15 and 5.16).*

Having defined equicontinuous linear cellular automata, other possible types of linear automata will now be defined. The following definition uses the product measure to define a stochastic analogue of equicontinuity.

Definition 8 *g is almost equicontinuous at f if for all $\varepsilon > 0$*

$$\lim_{\alpha \to 0} \frac{\mu(C_\alpha(f) \cap B_\varepsilon(f))}{\mu(C_\alpha(f))} = 1$$

Figure 2 displays the implication of the definition of almost equicontinuous.

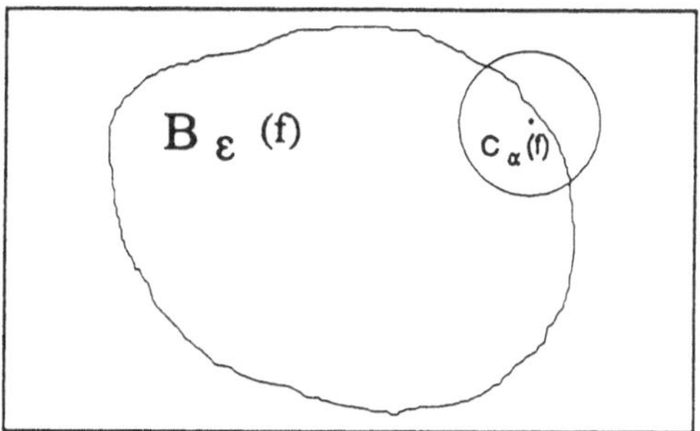

Figure 2: Pictorial representation of the definition of almost equicontinuous at f

Definition 9 *g is almost expansive if there is $\varepsilon > 0$ such that for all $f \in Y$, $\mu(B_\varepsilon(f)) = 0$.*

Definition 10 *g is expansive if there exists $\varepsilon > 0$ such that for all $f \in Y$, $B_\varepsilon(f) = \{f\}$.*

Definition 11 *The three classes of linear automata are defined:*

1. *$g \in \mathcal{A}$ if g is equicontinuous at some $f \in Y$.*

2. *$g \in \mathcal{B}$ if g is almost equicontinuous at some $f \in Y$ but $g \notin \mathcal{A}$.*

3. *$g \in \mathcal{C}$ if g is almost expansive.*

Clearly, all expansive linear automata belong to class $\mathcal{C}$. The following theorems give a description about the classes $\mathcal{A}$, $\mathcal{B}$ and $\mathcal{C}$. In particular, they imply that these classes form a partition of linear automata.

Theorem 5 *The following are equivalent:*

1. *$g \in \mathcal{A}$*

2. *g is equicontinuous at some $f \in Y$.*

3. g is equicontinuous on a set of measure 1

4. g is equicontinuous on T_σ

5. for some $n \geq (r-1)/2$ *and* $0 < \varepsilon \leq (\min_{j \in S}\{\lambda(j)\})^{2n+1}$ *there exists a class* $B_\varepsilon(f)$ *with* $B_\varepsilon(f)^o \neq \emptyset$

6. for all $\varepsilon > 0$ *there exists a class* $B_\varepsilon(f)$ *with* $B_\varepsilon(f)^o \neq \emptyset$.

The following example has interesting implications:

Example 4 *For the alphabet* $S = \{1,2,3\}$ *let g be the automaton induced by* $G : S^3 \to S$ *where* $G(2,2,2) = 2$, $G(3,3,3) = 3$ *and* $G(a,b,c) = 1$ *otherwise. Using bound vector notation, choose* $f = (0)_0^0$ *and let* $\lambda(j) = \frac{1}{2}\ \forall j \in S$. *Choose* $\alpha \leq \frac{1}{8}$ *and* $\varepsilon = \frac{1}{2}$. *Now* $C_\alpha(f) \subset B_\varepsilon(f)$, *hence* $g \in \mathcal{A}$. *Note that* $(2)_0^2$ *and* $(3)_0^3 \notin T_\sigma$ *and it is easily seen that* $\lim_{n\to\infty} g^n(y) = (1)_0^1$ *except for* $y = (2)_0^2$ *or* $(3)_0^3$. *Hence, it is shown that class* $\mathcal{A}$ *can contain members which are not equicontinuous at all* $f \in Y$.

It is seen that if $g \in \mathcal{A}$ and $f \in T_\sigma$, then $f \in B_\varepsilon(f)^o$ for all $\varepsilon > 0$. Thus $C_\alpha(f) \subset B_\varepsilon(f)$ for some $\alpha > 0$. This demonstrates an important characteristic of class $\mathcal{A}$ cellular automata. The approximate behavior of f, on the infinite vertical jagged edge strip defined by the intervals $[-m_i, m_i]$ for $i \in \aleph_0$, is decided by a finite amount of information about f, namely $f(n)$ for $|\, n \,| \leq q$ for a finite positive integer q.

Theorem 6 *The following are equivalent:*

1. $g \in \mathcal{A} \cup \mathcal{B}$

2. g is almost equicontinuous at some $f \in Y$

3. g is almost equicontinuous on a set of measure 1

4. for some $n \geq (r-1)/2$ *and* $0 < \varepsilon \leq (\min_{j \in S}\{\lambda(j)\})^{2n+1}$ *there exists a class* $B_\varepsilon(f)$ *with* $\mu(B_\varepsilon(f)) > 0$

5. for all $\varepsilon > 0$ *there exists a class* $B_\varepsilon(f)$ *with* $\mu(B_\varepsilon(f)) > 0$.

Theorem 7 *The following are equivalent:*

1. $g \in \mathcal{C}$

2. g is almost expansive

3. there exists $\varepsilon > 0$ *such that* $\mu(B_\varepsilon(f)) = 0$ *for all* $f \in Y$

4. for all $0 < \varepsilon \leq (\min_{j \in S}\{\lambda(j)\})^{2n+1}$ *where* $n \in \aleph_0$, $n \geq (r-1)/2$, *and all* $f \in Y$, $\mu(B_\varepsilon(f)) = 0$.

The following is a generalization of an example which can be found in [7].

Example 5 *Let $S = \{0,1,2\}$. Define $G : S^3 \to S$ in the following manner (Here $* = 0, 1,$ or 2):*

$*00$	$*01$	$*02$	$*10$	$*11$	$*12$	$*20$	$*21$	$*22$
0	1	0	0	1	0	2	0	2

If the array visualization $(a_{i,j})$ is used 0, is considered a background element and in passing from one row of $(a_{i,j})$ to the next row it is seen that the 1's move left, the 2's move straight down and a 1 and 2 which collide annihilate each other (that is they produce the background element 0). The behavior of g can be connected with a random walk on the integers (see [7] or [9] for a similar analysis). Let (p_0, p_1, p_2) be the probability vector that assigns p_0, p_1, p_2 to 0, 1, 2, respectively, in defining the measure μ. That is $p(0) = p_0$, $p(1) = p_1$ and $p(2) = p_2$. Consider the random walk with $p(1)$, $p(2)$, and $p(0)$ being the probability of moving left, moving right, and staying stationary, respectively. Suppose $f(0) = 2$, then the probability that this 2 is never annihilated equals the probability of no return to 0 starting at 1 in the random walk. In a random walk, the probability of returning to 0, starting at 1 is:

$$Q_1 = \begin{cases} 1 & if \quad p(2) \leq p(1) \\ \frac{p(1)}{p(2)} & if \quad p(2) > p(1) \end{cases}$$

Hence the probability of no return to 0, starting at 1, in the random walk is:

$$P_1 = 1 - Q_1 = \begin{cases} 0 & if \quad p(2) \leq p(1) \\ 1 - \frac{p(1)}{p(2)} & if \quad p(2) > p(1) \end{cases}$$

As seen, the probability of no return to 0, starting at 1, in the random walk is positive if and only if $p(2) > p(1)$. Let $f = (2)_0^2$ and let $\lambda(j) = \frac{1}{2}\ \forall j \in S$. Thus $p(2) > p(1)$ implies that $\mu(B_{\frac{1}{2}}(f)) > 0$. Hence $g \in \mathcal{A} \cup \mathcal{B}$. To show that g actually belongs in $\mathcal{B}$ take any $n \in \aleph_0$, if f is chosen such that $f(k) = 0$ for all $k > n$ then $(g^i(f))[0]$ is eventually 0 or 2. If $f(k) = 1$ for all $k > n$ then $(g^i(f))[0]$ is eventually 1. Therefore no finite amount of information about f determines $B_{\frac{1}{2}}(f)$ hence $g \notin \mathcal{A}$.

Example 6 *The shift map σ, induced by the local rule $G(a,b,c) = c$, is not expansive since for $\varepsilon > 0$ $B_\varepsilon(f) \neq f$ for some f. However σ does belongs to class $\mathcal{C}$. To see this consider $S = \{0,1\}$ and suppose $\lambda(j) = \frac{1}{4}\ \forall j \in S$. Let $n = 0$ and $\varepsilon = \frac{1}{4}$. Choose a probability distribution p on S in the following manner: $p(1) = p$ where $0 < p < 1$, $p(0) = 1 - p = q$ and, without loss of generality, assume $p \geq q$. Using the definition of the product measure, it can easily be shown that there exists $\varepsilon > 0$ such that $\mu(B_\varepsilon(f)) = 0$ for all $f \in Y$ where $\varepsilon = (\min_{j \in S}\{\lambda(j)\})^{2n+1} = (\frac{1}{4})^{2n+1}$.*

References

[1] S. Burris and H. P. Sankappanavar, (1981) *A Course in Universal Algebra*, Springer-Verlag, NY.

[2] K. Culik II, L. P. Hurd and S. Yu (1990), "Computation Theoretic Aspects of Cellular Automata." *Physica D.* 45 357-378.

[3] L. D'Alotto (1993), *The Kolmogorov Metric and a Classification of Linear Cellular Automata*, Ph.D. Dissertation, CUNY/Graduate Center, NY.

[4] W. Feller, (1968) *An Introduction to Probability Theory and its Applications*, John Wiley & Sons, NY.

[5] M. Gardner (1971), "On Cellular Automata, Self-Reproduction, the Garden of Eden and the Game of 'Life'," *Scientific American* 224(2), 112-117.

[6] C. Giardina, (1991) *Parallel Digital Signal Processing*, Regency Publishing Co., NJ.

[7] R. Gilman (1987), "Classes of Linear Automata". *Ergodic Theory and Dynamical Systems*, 7, 105-118.

[8] Kolmogorov and Fomin (1957), *Functional Analysis Vol. 1 Metric and Normed Spaces*, Graylock Press, NY.

[9] P. Grassberger (1984), "Chaos and Diffusion in Deterministic Cellular Automata", *Physica 10D* 52-58.

[10] Narici, Beckenstein and Bachman (1971), *Functional Analysis and Valuation Theory*, Marcel Dekker, Inc., NY.

[11] I. P. Natanson (1961) *Theory of Functions of a Real Variable*, Frederick Unger Pub. Co., NY.

[12] P. Walters (1982), *An Introduction to Ergodic Theory*, Springer-Verlag, NY.

[13] S. Wolfram (1983), "Statistical Mechanics of Cellular Automata." *Reviews of Modern Physics*. Vol. 55, No. 3, 601-644.

[14] S. Wolfram (1984), "Universality and Complexity in Cellular Automata." *Physica 10D* 1-35.

[15] S. Wolfram (1984), "Computation Theory of Cellular Automata". *Communications in Mathematical Physics*, 96, 15-57.

A HYBRID LEARNING RULE FOR A FEEDFORWARD NETWORK

W.-M. Lippe
Th. Feuring
A. Tenhagen
Institut fuer Numerische und instrumentelle Mathematik/Informatik
Westfaelische Wilhelms-Universitaet Muenster
Einsteinstrasse 62 – 48149 Muenster – Germany

August 23, 1994

Abstract. In feedforward networks trained by the classic backpropagation algorithm, the weights are modified according to the method of steepest descent. We point out drawbacks of steepest descent and suggest improvements on it. These yield a feedforward network, which adjusts its weights according to a (globally convergent) parallel coordinate descent method. A hybrid learning rule from $\bar{\delta}$-δ-rule and momentum version is introduced next. For adjusting the parameters of this rule a Sugeno/Tagaki fuzzy controller is used. We conclude that this algorithm is very suitable for fast training.

Keywords. Backpropagation, Coordinate Descent, Delta-bar-delta Rule, Fuzzy Controlled Learning Law, Fuzzy-Neural-Net, Momentum-Version, Steepest Descent, Sugeno-Controller.

1 Classic Backpropagation and Steepest Descent

The backpropagation net is a multilayer neural network consisting of one input-layer, one output-layer and at least one hidden-layer. The weights $\vec{w} = (w_1, \ldots, w_n)$ of this net are modified by means of the *backpropagation learning rule*, which is supposed to perform steepest descent with the *mean-squared-error*-function $F(\vec{w})$.
To accomplish this, [Rumelhart et al., 1986] introduced the *generalized delta rule*:

$$\vec{w}^{\text{new}} = \vec{w}^{\text{old}} - \eta \nabla_{\vec{w}} F(\vec{w}) \tag{1}$$

$\nabla_{\vec{w}} F(\vec{w})$ is the gradient of F with respect to $\vec{w}$ and gives the direction of maximum increase relative to $\vec{w}$. $\eta > 0$ is a constant, which is referred to as the *learnrate*.

`Image`(F) can be interpreted as a surface over the space of weight vectors. The goal of all weight adjustments is to find a $\vec{w}^*$ for which F takes on a global minimum.

The method of steepest descent has some weak points, that take effect on typically shaped error surfaces. Jacobs [Jacobs, 1988] examined the behaviour of steepest descent on backpropagation error surfaces and reached these conclusions:

- If the error surface is flat in the dimension of one of the weights, the corresponding derivative is (absolutely speaking) small. Because of the gradient-component being that small, the corresponding weight may only be slightly adjusted.

E. A. Yfantis (ed.), Intelligent Systems, 13–18.

- If — on the other hand — the curvature of the error surface is high for some weight-dimensions, a related problem with (absolutely speaking) large gradient-components arises. The weight vector may be moved too far – thus overshooting the minimum.

2 Parallel Coordinate Descent

The learnrate η determines decisively by what amount each weight is adjusted. Because one learnrate for all weights cannot allow for the different curvature of the error surface in each dimension, each weight should be equipped with an individual learnrate.

By using individual learnrates, the learning law no longer moves a point on the error surface in the direction of the negative gradient and therefore no longer performs steepest descent. A parallel coordinate descent, which adjusts all components of $\vec{w}$ simultaneously is performed instead.

It is not sure that this parallel coordinate algorithm is still a descent method because the weights are not changed in the direction of the negative gradient. So we need the following lemma [Ortega et al., 1970]:

Lemma: *Supose that $g : D \subset \mathbb{R}^n \to \mathbb{R}^1$ is differentiable at $x \in \int(D)$ and that, for some $p \in \mathbb{R}^n$, $g'(x)p > 0$ (p stands for the direction of descent). Then is a $\delta > 0$ so that*

$$g(x - \alpha p) < g(x), \qquad \forall \alpha \in (0, \delta).$$

With this lemma we can conclude the global convergence of the parallel coordinate descent. The following theorem holds:

Theorem: *If the preliminaries for the convergence of the classic backpropagation algorithm are given, the parallel coordinate descent method will also converge to a minimum of the error surface.*

It is easy to prove, that the method of parallel coordinate descent satisfies the preliminaries of the Lemma.

3 Delta-bar-delta Rule

Introducing individual learnrates does only half of the job: Because the error surface curvature (in each dimension) does not stay the same as we move $\vec{w}$ around, every learnrate η_i should be allowed to vary.

This variation can be controlled by the following heuristics:

When the sign of the derivative of a weight is the same on several consecutive steps, the corresponding learnrate should be increased (we suppose the error surface to be flat in this situation). When the sign alternates on consecutive steps, the learnrate should be decreased.

Note: In situations, where these heuristics are wrong, the control of the learnrates will not work properly.

The delta-bar-delta rule was developed by Jacobs [Jacobs, 1988]. It is a variation of the generalized delta rule and implements the heuristics mentioned above. In fact it consists of

two rules: one for weight adjustment and the other for control of the learnrates.
The weights are modified analogously to the parallel coordinate descent:

$$\vec{w}(t+1) = \vec{w}(t) - \sum_{i=1}^{n} \eta_i(t) E_{i,i} \nabla_{\vec{w}} F(\vec{w(t)}), \tag{2}$$

where n is the dimension of $\vec{w}$ (= number of weights in the net); $\vec{w}(t)$ is the weight vector's value at time step t; $\eta_i(t) > 0$ is the value of the learnrate corresponding to w_i at time step t; $E_{i,i}$ is a $n \times n$ Matrix with every component = 0, except one component with row = column = i, which is = 1, $(1 \leq i \leq n)$.

Every learnrate is adjusted according to $\eta_i(t+1) = \eta_i(t) + \Delta\eta_i(t)$, with:

$$\Delta\eta_i(t) = \begin{cases} \kappa & \text{, if } \bar{\delta}(t-1)\delta(t) > 0 \\ -\phi\, \eta_i(t) & \text{, if } \bar{\delta}(t-1)\delta(t) < 0 \\ 0 & \text{else} \end{cases} \tag{3}$$

where $\delta(t) = \dfrac{\partial F(t)}{\partial w_i(t)}$ and

$$\begin{aligned} \bar{\delta}(t) &= (1-\theta)\delta(t) + \theta\bar{\delta}(t-1) \\ &= (1-\theta)\sum_{i=0}^{t} \theta^i\, \delta(t-i) \end{aligned}$$

$\kappa > 0$ and $\phi \in [0,1]$ are constants; θ is $\in [0,1]$; κ, ϕ, θ are the same for every learnrate.

When the sign of the current derivative and that of the exponential average of the past derivatives are the same the learnrate η_i is increased by a constant $\kappa > 0$. When the signs are different the learnrate is decreased by a portion of itself.

The η_i are decreased exponentially by the $\bar{\delta}$-δ rule, thereby guaranteeing fast decrease and $\eta_i > 0$. The increase of the learnrates is done linearly to prevent them from increasing too quickly.

The effectiveness of the net depends decisively on κ: Set to an inadequately small value, the increase of the learnrates will take place too slow. If κ is set too large, the algorithm will become very inaccurate. Taking into consideration the existence of extensive flat areas on error surfaces, we see the importance of a good choice for κ. Ideally, κ should be set to a different (appropriate) value for each weight and each step. Therefore we introduce a fuzzy control of κ in **(6)**.

4 Momentum Version

The momentum version of the generalized delta rule leaves the (single!) learnrate unchanged. At step t each weight $w_i(t)$ is adjusted according to:

$$w_i(t+1) = w_i(t) + \Delta w_i(t) \tag{4}$$

$$\Delta w_i(t) = -(1-\alpha)\,\eta \frac{\partial F(t)}{\partial w_i(t)} + \alpha\, \Delta w_i(t-1) \tag{5}$$

where $\alpha \in [0,1]$ (referred to as the *momentum term*) and η is the (single!) learnrate; $\Delta w_i(t-1)$ gives the amount by which the weight w_i was changed during the previous step. Typically, α is set ≈ 0.9. This is an arbitrary choice and may have to be revised after a couple of experiments (if $\alpha = 0$, we have steepest descent).

5 Hybrid Rule

The prime advantage of the momentum version is its ability to speed up "learning" on flat areas of the error surface. This is exactly the situation in which the $\bar{\delta}$-δ rule may be too slow, (if κ is chosen small).

So the idea of combining the two rules to one hybrid rule seems quite obvious: The hybrid rule uses individual, variable learnrates and one momentum term (i.e. there is only one momentum term for all weights). The learnrates are adjusted according to the learnrate updating rule of the $\bar{\delta}$-δ rule. The momentum version (with individual learnrates) is used as weight modification rule.

Without further changes both methods do not cooperate ideally (which is what Jacobs observed in his comparison of pure $\bar{\delta}$-δ rule with the hybrid rule).

6 Fuzzy Control

The hybrid rule can be improved strongly by allowing the parameters κ (of the $\bar{\delta}$-δ rule) and α (of the momentum version) to vary. The goal of these variations is to make use of the advantage of the momentum version, still allowing the $\bar{\delta}$-δ rule to have a lot of control over the progress.

To control α and κ we use a fuzzy controller which is based on the same heuristics as the $\bar{\delta}$-δ rule: The longer the weight vector is in a flat area of the error surface, the larger κ and α may be set. κ should be allowed to become large enough, so it can take effect despite of a large α. In areas of high curvature κ and α should be small.

Using a fuzzy controller is an easy way to implement these rules. Also, fuzzy control yields flexible outputs and we do not have to think about exponential and/or linear de- or increases.

We employ a Sugeno-type fuzzy controller [Sugeno et al., 1985], which is based upon the 'familiar' `IF...THEN` rules but uses *crisp* functions in the `THEN`-instructions.

6.1 Details

We now describe the controller in detail:

For each weight w_i of the net, we introduce a new variable $c[i]$, in which we record how often each case in (3) is selected. This new variable is then used to represent the curvature of the error surface in its i-th dimension (this is where the $\bar{\delta}$-δ rule-heuristics are used).

$$c[i] := \begin{cases} c[i]+1 & \text{,if } \eta_i \text{ was increased by } \kappa \text{ ,} \\ c[i]-5 & \text{,if } \eta_i \text{ was decreased by } \phi\,\eta_i(t) \text{ .} \end{cases}$$

κ and ϕ are the parameters of the $\bar{\delta}$-δ rule. Additionally, we make certain, that $c[i] \in [-1, 100]$

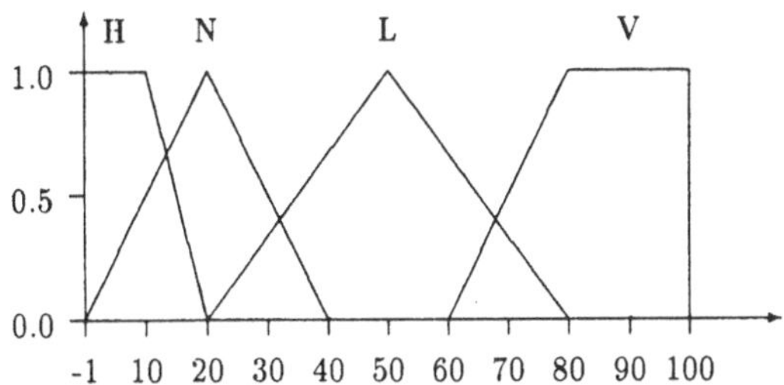

Figure 1: Membership-functions of the fuzzy sets VERYLOW, LOW, NOTSURE and HIGH.

IF (c[i] is)	THEN $\alpha :=$	$\kappa :=$
(V)ERYLOW	0.9	$100\kappa_0$
(L)OW	0.7	$10\kappa_0$
(N)OTSURE	0.3	κ_0
(H)IGH	0.01	$\kappa_0/10$

Table 1: Rules of the fuzzy controller.

We use a *singleton fuzzifier*, which performs a mapping from the observed crisp data $c[i]$ to fuzzy sets $A_{c[i]}$ on $[-1, 100]$. Four fuzzy sets which describe the linguistic term 'curvature of the error surface' are defined next. The sets are called VERYLOW, LOW, NOTSURE and HIGH. The corresponding membership functions can be seen in **Fig.1**.

The rule-base of the controller contains the four rules, by which we implement the heuristics described above. The rules are illustrated in **Tab.1**.
The output of the controller is determined by computing all four rules:

First, the IF-clause 'IF (c[i] is V)' is computed by calculating $\mu_V(c[i])$ – a crisp value! This corresponds to the fuzzy-minimum-intersection of the singleton fuzzy set $A_{c[i]}$ with V. The other IF-clauses are treated in an analogous way. The average of all values of the (crisp) functions for κ (or α) yield the output of the controller for κ (or α), for example:

$$\kappa = \frac{(\mu_V * 100 + \mu_L * 10 + \mu_N + \mu_H * 0.1)\,\kappa_0}{\mu_V + \mu_L + \mu_N + \mu_H}$$

where $\mu_X = \mu_X(c[i])$ (for convenience) and κ_0 is a starting value that has to be set by the user.

κ and α are computed newly for each weight at each step, so their values do not have to be stored.

As a result of the controller producing crisp output, no defuzzification has to be performed – it works faster than the 'familiar' controller.

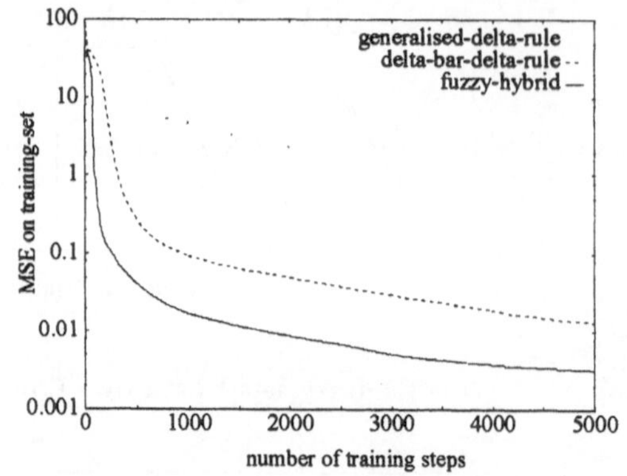

<u>learning law parameters:</u>

gen.-δ-rule	$\eta = 0.1$			
$\bar{\delta}$-δ rule	$\eta_0 = 0$	$\kappa = 0.0005$	$\phi = 0.85$	$\theta = 0.5$
fuzzy-hyb.	$\eta_0 = 0.1$	$\kappa_0 = 0.001$	$\phi = 0.8$	$\theta = 0.5$

Figure 2: The net was trained to learn $f : \mathbb{R}^3 \rightarrow \mathbb{R}$, $(x, y, z) \mapsto xy + z$. The training set was $\{1, \ldots, 5\}^3$. The average of 10 training runs is shown

7 Test and Conclusion

We present a test of our new learning law (referred to as *fuzzy hybrid*) in **Fig.2**. We compare the results of fuzzy hybrid with the performance of the generalized delta rule and the $\bar{\delta}$-δ rule. The parameters of each rule were set to values that lead to fastest convergence. η_0 and κ_0 are starting values that have to be set by the user.

We modified the classic backpropagation algorithm to perform a parallel coordinate descent, based upon Jacobs' ideas. Heuristics about the properties of the error surface have also been implemented in a fuzzy control of the parameters κ and α. The modified net needs far less training steps to 'learn' a training set, without too much extra computations.

References

[Ortega et al., 1970] Ortega, J. M., Reinboldt W. C., (1970) *Iterative Solution of Nonlinear Equations in Serveral Variables*, Computer Sciences and Applied Mathematics, Academic Press.

[Sugeno et al., 1985] Sugeno, M., Takagi, T.,(1985) *Fuzzy identification of systems and its applications to modeling and control*, IEEE Trans. Systems Man Cyb., 15, 116-132.

[Rumelhart et al., 1986] Rumelhart, D. E., Hinton, G. E., Williams, R. J.,(1986) *Parallel Distributed Processing: Explorations in the Microstructure of Cognition*, 1, 318-362.

[Jacobs, 1988] Jacobs, R. A.,(1988) *Increased Rates of Convergence Through Learning Rate Adaption*, Neural Networks, 1, 295-307.

LOGIC PROGRAMMING AND DEFAULT LOGIC

Grigoris Antoniou
University of Osnabrueck, FB 6,
49069 Osnabrueck, Germany,
ga@informatik.uni-osnabrueck.de

Abstract. We present several ideas of increasing complexity how to translate default theories to normal logic programs that make direct use of the deductive capacity of logic programming. We show the limitations of simple, ad hoc approaches, and arrive at a more general construction; its main property is that the answer substitutions computed by the logic program via its standard operational semantics correspond exactly to the extensions of the default theory.

1 Introduction

This paper belongs to a continuing effort to establish relationships between logic programming and default logic [13]. In recent years, much work has been done to establish relationships between nonmonotonic reasoning and logic programming [4, 6, 7, 9, 10, 11]. This is usually done by defining new (nonmonotonic) semantics for logic programming (in particular of negation) and deriving a relationship to some nonmonotonic logic.

In this paper we go the other way around: We maintain the classical standard operational semantics of logic programming (SLDNF-resolution) and try to translate a nonmonotonic logic into appropriate logic programs.

What are the benefits of such an approach? Standard semantics of logic programming is well analyzed and understood and has in Prolog a powerful implementation. Therefore, we provide an implementational paradigm for nonmonotonic reasoning that might prove valuable in practice. For example, we might use parallel logic programming systems to achieve efficient reasoning systems.

We describe some simple ideas of translating default theories into logic programs as well as their limitations. This way we show why the more complicated approach first presented in [2] is necessary for most cases. Throughout the paper we assume familiarity with notation and basic notions of predicate logic and logic programming [8] and of default logic [13].

2 Operational interpretation of default logic

In [1] we provided the following operational characterization of extensions. Let $T = (W, D)$ be a default theory and $\Pi = (\delta_0, \delta_1, \delta_2, \ldots)$ a finite or infinite sequence of defaults from D not containing any repetitions (modeling an application order of defaults from D). We

E. A. Yfantis (ed.), Intelligent Systems, 19–25.

denote by $\Pi[k]$ the initial segment of Π of length k, provided the length of Π is at least k. Then we define the following concepts:

- $In(\Pi)$ is $Th(M)$ (the first-order deductive closure of M), where M is the set of formulas of W united with all consequents of defaults occurring in Π.
- $Out(\Pi)$ is the set of the negations of justifications of defaults in Π.
- Π is called a *process of* T iff δ_k is applicable to $In(\Pi[k])$ w.r.t. belief set $In(\Pi[k])$, for every k such that δ_k occurs in Π.
- Π is called *successful* iff $In(\Pi) \cap Out(\Pi) = \emptyset$, otherwise it is called *failed.*
- Π is *closed* iff every $\delta \in D$ which is applicable to $In(\Pi)$ with respect to belief set $In(\Pi)$ already occurs in Π.

$In(\Pi)$ collects all formulas in which we believe after application of the defaults in Π, while $Out(\Pi)$ consists of all those formulas which we should avoid to believe for the sake of consistency. The following theorem states the relationship between the extensions of a default theory T and the closed successful processes of T.

Theorem Let $T = (W, D)$ be a default theory. If Π is a closed successful process of T, then $In(\Pi)$ is an extension of T. Conversely, for every extension E of T there exists a closed, successful process Π of T with $E = In(\Pi)$. —

3 Direct implementations: simple approaches

The process approach to default logic can be implemented in Prolog in a straightforward way [1]. Every such implementation, though, requires a theorem prover of classical logic. But why use an external prover and not make use of the deductive power of logic programming (Prolog) instead? This is the main theme of the rest of the paper. We call these approaches direct implementations of default logic. Note that merely implementing default logic in Prolog is trivial as the latter is a universal programming language.

From now on we shall restrict attention to default theories whose set of truths is a Horn logic program. This is a direct implication of the fact that we want to use SLDNF-resolution as the underlying deductive calculus.

3.1 Taxonomic default theories

The first idea of 'encoding' default theories as logic programs is to make use of Prolog's negation operator 'not'. Already this simple idea is sufficient to handle some default theories like taxonomic default theories that are motivated from isa-hierarchies including negative information [5]. The defaults have the form $p{:}\neg q/r$ (with positive ground literals p, q, r), and the truths are rules of the form $p \leftarrow q$. The defaults are translated to logic rules

$$r\ :-\ p,\ not\ q.$$

A goal $?-r$ succeeds iff r belongs to all extensions of the default theory (note that taxonomic default theories have only one extension). For more details see [5]. We would like to point out that in this very simple class there was no need to represent classical negation in the logical language.

3.2 Dealing with negation

If we consider more general default theories, we must cope with negation. A simple (and common) idea is to represent negated predicates $\neg p$ by new predicates $\overline{p}$. Unfortunately, a straightforward, naive approach as in 3.1 does not work. Consider, for example, the Nixon diamond; we would be tempted to ex- press the theory in Prolog as follows:

pac :- qua, not $\overline{pac}$. ('Quakers are usually pacifists')
$\overline{pac}$:- rep, not pac. ('Republicans are usually non-pacifists')
rep.
qua.

This representation obviously leads to an infinite loop. We have the problem of maintaining the current knowledge base which changes after application of a default. A simple solution to this problem consists of enumerating the defaults of the theory and maintaining a list L to avoid circular arguments as above. The Nixon example can be treated as follows:

$pac(L)$:- not $member(1, L)$, $qua(L)$, not $\overline{pac}([1|L])$.
$\overline{pac}(L)$:- not $member(2, L)$, $rep(L)$, not $pac([2|L])$.
$rep(\_)$.
$qua(\_)$.

Unfortunately, this naive approach is insufficient even for normal theories: Consider, for example, the theory with no truths and the defaults $\{true{:}r/r,\ true{:}\neg r/\neg r,\ r{:}p/p,\ \neg r{:}p/p\}$. It has two extensions: $Th(r, p)$ and $Th(\neg r, p)$; p belongs to both of them. The corresponding program according to the approach above is

$r(L)$:- not $member(1, L)$, not $\overline{r}([1|L])$.
$\overline{r}(L)$:- not $member(2, L)$, not $r([2|L])$.
$p(L)$:- not $member(3, L), r(L)$, not $\overline{p}([4|L])$.
$p(L)$:- not $member(4, L), \overline{r}(L)$, not $\overline{p}([4|L])$.

Unfortunately, a call of the goal $? - p([\ ])$ fails: It leads to a call of $? - r([\ ])$ and then of $? - \textit{not}\ \overline{r}([1])$. A call of $? - \overline{r}([1])$ leads to $? - \textit{not}\ r([1,2])$. Now, $? - r([1,2])$ fails, so $? - \textit{not}\ \overline{r}([1])$ succeeds, therefore $? - r([\ ])$ fails. The same result is obtained when using the fourth rule as alternative.

This indicates that the approach works for the Nixon diamond rather accidentally. Indeed, soundness and completeness of the approach can only be shown for finite, supernormal, propositional default theories with atomic prerequisites, justifications and consequents, which is a very restricted class.

4 Adding control structure: motivation and examples

4.1 Informal description

We saw in the last section that ad hoc methods of implementing default logic in Prolog fail for many interesting theories. This shows that we have to compute extensions rather than try to answer queries (anyway, the latter is, in general, only possible for normal default theories).

The approach taken in [2] is strongly motivated by the process model of default logic presented in section 2. It also uses the ideas developed in section 3, as there are negation as failure, representation of negative literals $\neg p$ by new predicates $\overline{p}$, and enumeration of defaults. [2] describes an automated translation of default theories (P, D) to normal logic programs $LogProg(P, D)$ according to these ideas, such that the computed answer substitutions of $LogProg(P, D)$ correspond exactly to the extensions of (P, D). To be more precise, the logic program computes the set of generating defaults of each extension.

In the following subsection we give an example illustrating the approach. For its understanding, we have to briefly discuss how negative literals are treated. When developing processes by applying defaults, the following situation may arise: Check whether a literal L follows from $P \cup \{\neg G_1, \ldots, \neg G_k\}$, where P is a Horn logic program, and $G_1, \ldots, G_k$ are positive literals ($\neg G_i$ may be consequents of defaults that have already been applied). The following result can be shown (for a proof, see [3]).

Definition and Theorem Let P be a definite logic program. We define $\overline{P}$ as the definite logic program

$$\{\overline{B_i} \leftarrow \overline{A_0}, B_1, \ldots, B_{i-1}, B_{i+1}, \ldots, B_n \mid A_0 \leftarrow B_1, \ldots, B_n \in P,\ n \geq 1\}$$

Let P be a definite logic program, and let $A, G_1, \ldots, G_k$ be positive literals. Further, let $P \cup \{\leftarrow G_1, \ldots, \leftarrow G_k\}$ be consistent. Then, assertions (1) and (2) are equivalent.

1. $P \cup \{\leftarrow G_1, \ldots, \leftarrow G_k\} \models \exists(\neg A)$
2. $P \cup \overline{P} \cup \{\overline{G_1} \leftarrow, \ldots, \overline{G_k} \leftarrow\} \cup \{A \leftarrow\} \models \exists(\overline{A})$.

Also, assertions (3) and (4) are equivalent.

3. $P \cup \{\leftarrow G_1, \ldots, \leftarrow G_k\} \not\models \exists(A)$
4. $P \cup \overline{P} \cup \{\overline{G_1} \leftarrow, \ldots, \overline{G_k} \leftarrow\} \models \exists(A)$. —

This means that to check derivability of a negative literal $\underline{A}$ using SLD-resolution, we need contraposition of rules in the logic program P, and the positive literal A. The latter leads to some seemingly strange rules in the $Supplement(D)$ part of the logic programs $LogProg(P, D)$ (see the example below; further examples can be found in a longer version of the paper).

4.2 The Nixon example reconsidered

In section 3.2 we showed a simple logic program that adequately solves the Nixon example. Here we shall show which program is obtained by the general approach to enable a comparison. The default theory for the Nixon diamond is (P, D) with $D = \{qua{:}pac/pac, rep{:}\neg pac/\neg pac\}$ and $P = \{rep \leftarrow, qua \leftarrow\}$. The corresponding logic program $LogProg(P, D)$ is constructed as follows:

P_L: $rep(L) \leftarrow$
(truths are always included in the current knowledge)
$qua(L) \leftarrow$

Supplement(D):
$pac(L) \leftarrow member(0, L)$
(default No. 0 applied, so *rep* is in the current knowledge base)
$\overline{pac}(L) \leftarrow member(1, L)$
$pac(L) \leftarrow member([conscheck, 0], L)$
(implication of the discussion in 4.1)

Process(D): (expand a process by applying a default)

$process(Lold, L) \leftarrow$
$consistent(Lold),$
$not\ member(0, Lold),$
$qua([[pre, 0]|Lold]),$
$not\ \overline{pac}([[conscheck, 0]|Lold]),$
$process([0|Lold], L)$

$process(Lold, L) \leftarrow$
$consistent(Lold),$
$not\ member(1, Lold),$
$rep([[pre, 1]|Lold]),$
$not\ pac([[conscheck, 1]|Lold]),$
$process([1|Lold], L)$

$process(L, L) \leftarrow consistent(L), closed(L), successful(L)$
... (plus program for *consistent*, *closed*, *successful*)

We now show some intermediate steps in the SLDNF-refutation of

$$LogProg(P, D) \cup \{\leftarrow process([], L)\}.$$

$\leftarrow process([\,], L)$

$\leftarrow consistent([\,]), notmember(0, [\,]), qua([[pre, 0]]),$
$not\ \overline{pac}([[conscheck, 0]]), process([0], L)$

$\leftarrow qua([[pre,0]])$ succeeds due to the presence of $qua(L) \leftarrow$
$\leftarrow \overline{pac}([[conscheck,0]])$ fails; the only rule for $\overline{pac}$ is
$\overline{pac}(L) \leftarrow member(1,L)$, but 1 is not member of $[[conscheck,0]]$.

$\leftarrow process([0],L)$

$\leftarrow consistent([0]), notmember(1,[0]), rep([[pre,1],0]),$
$not\ pac([[conscheck,1],0]), process([1,0],L)$

$\leftarrow not\ pac([[conscheck,1],0]), process([1,0],L)$

$pac([[conscheck,1],0])$ succeeds because of the rule
$pac(L) \leftarrow member(0,L)$. So, this branch of the SLDNF-tree fails.
Backtracking leads to

$\leftarrow process([0],[0])$
which succeeds using the last rule for process

$\{L/[0]\}$ is thus a computed answer substitution. It corresponds to the extension $Th(rep,qua,pac)$. In a similar way we obtain another answer substitution $\{L/[1]\}$ representing the other extension $Th(rep,qua,\neg pac)$.

5 Conclusion

In this paper, we presented several ideas of increasing complexity on how to directly encode default theories in normal logic programs that make direct use of the deductive capacity of logic programming. We showed the limitations of simple, ad hoc approaches, and arrived at a more general construction with the main property that the answer substitutions computed by the logic program via its standard semantics correspond exactly to the extensions of the default theory. Soundness and completeness of the approach is proven in [2].

We are working on weakening some restrictions to the default theories. We are thinking especially of admitting conjunction(this can indeed be done by a generalization of the Theorem in subsection 4.1) and investigating cases where infinite theories and extensions could be treated (using unification).

We are also planning to make an experimental comparison with other methods of implementing default logic like truth maintenance systems, graphs, the system Theorist [12] etc. One of the most promising ideas is to exploit the parallelism in the logic programs given in this paper and thus in providing an efficient, parallel implementation of (parts of) default logic.

References

[1] Antoniou, G. and Sperschneider, V. (1993) 'Computing Extensions of Nonmonotonic Logics', Proc. 4th Scandinavian Conference on AI, IOS Press.

[2] Antoniou, G. and Langetepe, E. (1994) 'Soundness and Completeness of a Logic Programming Approach to Default Logic', Proc. AAAI-94, MIT Press.

[3] Antoniou, G., Langetepe, E. (1994) 'Applying SLD-resolution to a class of non-Horn logic programs', IGPL Bulletin (submitted).

[4] Bidoit, N. and Froidevaux, C. (1991) 'Negation by Default and Unstratifiable Logic programs', Theoretical Computer Science 78, 85-112.

[5] Froidevaux, C. (1986) 'Taxonomic Default Theory', Proc. ECAI-86.

[6] Gelfond, M. and Lifschitz, V. (1989) 'The Stable Model Semantics for Logic Programming', Proc. 5th Int. Conference/Symposium on Logic Programming.

[7] Gelfond, M. and Lifschitz, V. (1991) 'Classical Negation in Logic Programs and Disjunctive Databases', New Generation Computing 9, 365-385.

[8] Lloyd, J.W. (1987) Foundations of Logic Programming 2. edition. Springer.

[9] Marek, W. and Truszczynski, M. (1989) 'Stable Semantics for Logic Programs and Default Theories', Proc. North American Conference on Logic Programming, 243-256.

[10] Marek, W. and Subrahmanian, V.S. (1992) 'The Relationship between Stable, Supported, Default and Autoepistemic Semantics for General Logic Programs', Theoretical Computer Science 103, 365-386.

[11] Pereira, L.M. and Nerode, A. (1993) 'Logic Programming and Non-monotonic Reasoning', Proc. of the 2nd International Workshop, MIT Press.

[12] Poole, D. (1988) 'A Logical Framework for Default Reasoning', Artificial Intelligence 36.

[13] Reiter, R. (1980) 'A Logic for Default Reasoning', Artificial Intelligence 13.

ABDUCTIVE BEHAVIOUR OF CONCURRENT LOGIC PROGRAMS

GEORGE A. PAPADOPOULOS
Department of Computer Science, University of Cyprus, 75 Kallipoleos Str., Nicosia, T.T. 134, P.O. Box 537, Cyprus. EMAIL: george@turing.cs.ucy.ac.cy

Abstract

We present a possible way to model abduction within the framework of concurrent logic programming. In particular, we describe an extension to the concurrent logic programming language PARLOG and an associated computational model which allows a user to specify an abductive behaviour for a concurrent logic program. The proposed model therefore exploits the inherent parallelism of concurrent logic programs. We discuss some of the problems we faced and the solutions we adopted. A prototype implementation of the model has been built and is also discussed.

Keywords. Abductive Logic Programming, Concurrent Logic Programming.

1 Abduction

Abduction was introduced by the philosopher Pierce as one of the three main forms of reasoning (the other two being deduction and induction). Recently, the importance of abductive reasoning has been demonstrated in many areas of Artificial Intelligence such as diagnosis, temporal reasoning, planning and semantic networks but also elsewhere such as in the field of databases and linguistics. In logic programming, in particular, abduction is achieved by means of finding conditional answers to queries. As a result, it is useful to study ways for making the computation of abduction more effective. It has been argued ([5]) that abductive inference and its parallel realisation should be one of the future research themes in parallel logic programming.

The development of an abductive framework in logic programming has been proposed in [4] and further developed, among others, in [7,8]. An abductive logic program is a triple `<P,A,I>` where `P` is a general logic program, `A` is a set of abducible atoms and `I` a set of constraints. For simplicity a number of restrictions are usually imposed: there are no rules for abducible atoms, integrity constraints are compiled into denials with at least one abducible and the hypotheses generated are variable free. In the abductive proof procedure for logic programming (see [7] for a review of the main ideas), the computation interleaves between *abductive* phases that generate and collect abductive hypotheses with *consistency* phases that incrementally check these hypotheses for consistency with respect to the integrity constraints. The operational semantics for sequential execution of abduction is well defined within this framework and has been used in building meta interpreters on top of Prolog systems. However, as in usual deductive logic programming, abductive inference mechanisms have several sources of parallelism of many forms (OR-parallelism, independent and dependent AND-parallelism). In [9] we examine the introduction of these forms of parallelism into an abductive logic program, we study the operational behaviour of such a program enhanced with parallelism and we highlight its effect on the efficiency of execution compared with the corresponding sequential version.

An alternative idea however has been proposed in [2] within the general context of concurrent constraint programming which encompasses concurrent logic and constraint programming. The foundamental move is that in the case of a concurrent constraint program deadlocking, this is interpreted as a need to generate some hypothetical values for the benefit of the suspended agents, some of which should be able to resume execution. This generation

E. A. Yfantis (ed.), Intelligent Systems, 27–37.

of hypothetical values can then correspond to an abductive phase whereas the execution of agents in the ordinary way can correspond to the deductive phase.

In this paper we apply this idea to the class of the so called concurrent logic languages ([15]) and in particular the language PARLOG ([6]). More to the point, we extend PARLOG with suitable annotations to indicate abductive behaviour and we show how a concurrent logic program can exhibit such an abductive behaviour in a way that can be supported by the underlying "ordinary" deductive implementation. The rest of the paper is organised as follows: the next section introduces *abductive PARLOG* and shows how abductive PARLOG programs can be coded up in this extended language whereas the third section discusses a prototype implementation of the extended language on top of an ordinary PARLOG system; the final section comprises the conclusions and suggestions for further work.

2 Abductive behaviour of a concurrent logic program

Concurrent logic languages ([15]) have enjoyed a widespread use and have undergone major development over the past decade, not least due to their paramount importance and role they played in the Japanese FGCS project. A concurrent logic program is a set of guarded Horn clauses of the form

```
H :- G1,…,Gm | B1,…,Bn              m,n ≥ 0
```

where 'H' is the head, '|' is the commit operator, 'G1,…,Gm' is the guard part and 'B1,…,Bn' is the body part. Declaratively, the meaning of the above clause is that H is true if both G1,…,Gm and B1,…,Bn are true. Operationally, the guard calls G1 to Gm are evaluated first in parallel and upon successful termination the computation commits to the body of the clause. The head H is of the form p(t1,…,tn) where p/n is a predicate name of arity n and t1,…,tn are its arguments. There may be more than one rule with the same name p and arity n, in which case they form a group definition of the process p.

Computation starts with a set of cooperating processes (goals) executing in parallel and communicating by means of shared variables. The clauses of a program specify the behaviour and the various transitions possible for each process. If for a certain goal to be reduced there are more than one candidate clauses to select from, the first one to perform head unification and solve its guard successfully will be chosen and the computations in the head or guard of the other candidate clauses will be abandoned. Thus concurrent logic languages incorporate the concept of committed choice "don't care" non determinism from CSP.

In this paper we show how the notion of abduction can be modelled in the framework of concurrent logic programming by applying the ideas proposed in [2] to the case of the concurrent logic programming language PARLOG ([6]). In PARLOG the clauses that form a procedure are associated with a mode declaration that states which arguments are input (i.e. the values specified in the head of one of the procedure's clauses have to be present for that clause to be candidate for selection) and which are output (i.e. the corresponding values will be produced upon commitment to that clause). We now extend the language with a third mode annotation, '@', which states that the corresponding argument can be abduced. In particular, a PARLOG clause for a procedure p/n now takes the form

```
p(i1?,…,ik?,ab1@,…,abm@,o1^,…,on^)  :-  Guard | Body
```

where the arguments i1 to ik are input arguments, o1 to on are output arguments and ab1 to abm are abducible arguments whose value *can be assumed* if the need arises (i.e. if they are not present and the clause cannot be reduced). As an example the arguments of the following procedure

```
equipment(ok@,Signals^)  :-  produce_signals(Signals).
equipment(not_ok@,Signals^)  :-  error_condition(Signals).
```

comprise an abducible parameter and an output one.

As for the case of the traditional abductive logic programming framework, computation interleaves between an ordinary deductive phase and an abductive one. For the case of a PARLOG program, the deductive phase corresponds to the performing of the input unifications, guard evaluations and reductions to the bodies of the selected clauses. The need for an abductive phase emerges when the whole computation has suspended with no remaining process able to reduce. In this case an "abductive" phase can be initiated whose purpose is to abduce one or more of the abducible arguments of some suspended process so that the computation can resume. The abductive phase effectively assumes that the selected abducibles have indeed been instantiated to their indicated values, thus activating all the processes that are suspending on them. Considering again the above example, assume the existence of the clause

```
monitor_equipment(ok?)  :- ...
monitor_equipment(not_ok?)  :- ...
```

and the following suspended AND-conjunction

```
..., monitor_equipment(Status), equipment(Status,Signals), ...
```

After the suspension of the computation, the abductive phase should select the suspended literal `equipment` and assume either of the two values `ok` or `not_ok` by instantiating the variable `Status` to either of them. This will now activate the process `monitor_equipment` and ordinary deductive computation will resume. If it is discovered in retrospect that the assumption was wrong, the second of the two values should be tried. Computation will either succeed if an assumed value succeeds in causing the successful termination of all processes involved in the computation or fail if all assumptions lead to failed derivations.

The following example, taken from [2] and adapted to the syntax of abductive PARLOG, illustrates a configuration comprising three light bulbs connected in sequence to a battery. Note that the first argument of the procedures `battery`, `wire` and `bulb` that denote the state of the corresponding component are abducible arguments whose value may be assumed if necessary. The operational meaning of the procedure `wire`, for instance, is that the call `wire(ok,plus,Out)` is reduced with the variable `Out` being instantiated to the value of the second parameter (`plus`) whereas the call `wire(broken,plus,Out)` is reduced with the variable `Out` being instantiated to `0`. On the other hand the call `wire(State,In,Out)` will initially suspend; if however the computation eventually enters an abductive phase, it is possible to abduce the first argument, i.e. instantiate `State` to either `ok` or `broken` and continue from that point onwards. The procedure `circuit` is, in fact, the one forming the actual circuit configuration where the last three arguments correspond to the observations that were made regarding the state of the three bulbs and the rest denote the explanations that must be generated.

```
mode battery(charging_level@,left_wire^,right_wire^).
battery(empty,0,0).
battery(ok,plus,plus).

mode wire(state@,left_connection?,right_connection^).
wire(ok,Connect,Connect).
wire(broken,_,0).

mode bulb(condition@,light^,left_wire?,right_wire?).
bulb(ok,on,plus,plus).
bulb(ok,off,0,_).
bulb(ok,off,_,0).
bulb(damaged,off,_,_).
```

```
mode circuit(@,@,@,@,@,@,@,@,@,@,?,?,?).
circuit(S,B1,B2,B3,W1,W2,W3,W4,W5,W6,L1,L2,L3) <-
   battery(S,Sl,Sr),
   wire(W1,Sl,B1l), wire(W2,Sr,B1r), wire(W3,B1l,B2l),
   wire(W4,B1r,B2r), wire(W5,B2l,B3l), wire(W6,B2r,B3r),
   bulb(B1,L1,B1l,B1r), bulb(B2,L2,B2l,B2r),
   bulb(B3,L3,B3l,B3r).
```

The way a query is formulated in abductive PARLOG is illustrated below.

```
<- abductive_parlog(Goal,Constraints),
   Goal=[circuit(S,B1,B2,B3,W1,W2,W3,W4,W5,W6,on,on,on),
         print([S,B1,B2,B3])],
   Constraints=[...].
```

where the role of `Constraints` will be explained later on. Incidentally, note that the above query produces the single explanation `[ok,ok,ok,ok]` whereas the following one

```
<- abductive_parlog(Goal,Constraints),
   Goal=[circuit(S,B1,B2,B3,W1,W2,W3,W4,W5,W6,off,off,off),
         print([S,B1,B2,B3])],
   Constraints=[...].
```

has multiple explanations such as `[empty,ok,ok,ok]`, `[empty,damaged,broken,ok]`, `[ok,damaged,broken,broken]`, etc.

Note that by extending PARLOG with the abductive mode annotation '`@`', we have allowed a programmer to specify in a procedure which arguments, if any, can be abduced, thus controlling the extent to which abduction will be performed and consequently reducing the search space. Later on we discuss the possibility of enhancing an abductive mode annotation with extra information that will help in selecting during the abductive phase the best candidate for abduction from the suspended processes. Note also that a procedure mode declaration without any abductive annotations cannot possibly take part in the abductive phase.

3 Implementation of abductive PARLOG

In implementing the abductive extensions to PARLOG, the following problems must be solved:

— Detection of a globally suspended state of computation that indicates the end of the current deductive phase. Note that in general the processes involved in a computation are scattered around in a parallel system, or even a distributed one over a number of machines. Note also that for an ordinary concurrent logic program a globally suspended state is, in fact, an erroneous situation because it indicates deadlock.

— Upon detection of a suspended state, a decision must be taken as to which abducibles will be abduced. The decision is critical in computing efficiently the solutions since "bad guesses" will lead to unnecessary computations.

— If the abduction of some abducible(s) proves to be unsuccessful in deriving a solution (or more solutions are sought) the computation should backtrack to the point before the abduction and try a different path. Note here that an ordinary concurrent logic language implementation supporting only committed choice "don't care" parallelism has no machinery to backtrack.

In the sequel we discuss the solutions we adopted for each one of the above points, showing in the process how the original program is transformed into one having the additional functionality required for supporting an abductive behaviour. We describe also the most important parts of the implementation itself resorting to the use of '...' to hide details which could obscure the presentation of the model.

3.1 DETECTION OF DEADLOCK

In solving the first problem we are interested in a solution which requires no extra machinery in the underlying PARLOG implementation (namely a suitably modified WAM) and which can work even in the case where an abductive PARLOG program may be running in a distributed environment. The solution we have adopted is based on an *all-declarative* approach using the power of short circuits, a programming technique first introduced by Takeuchi and used, among other applications, for detecting deadlock and termination or getting snapshots of systems of processes running concurrently ([14,15]).

In particular, all messages produced by a concurrently executing system of processes are connected by means of a short circuit, the ends of which are held by a monitoring process. Upon receiving and consuming a signal, a process unifies the left and right switches of the circuit for the particular message. If a process produces more messages that what it consumes, it extends the circuit accordingly. Deadlock is detected when the monitoring process observes that the two ends of the circuit that it holds have been unified, i.e. no more messages are in transit ([15]). We illustrate how an abductive PARLOG procedure can support this functionality by showing the transformed version of `wire`.

```
mode  wire(state@,left_connection?,right_connection^).
wire(m(L,R,ok),ConnectIn,ConnectOut)  <-
   L=R,  ConnectOut=ConnectIn.
wire(m(L,R,broken),_,ConnectOut)  <-  ConnectOut=m(L,R,0).
```

Note that while in the first clause we close the switch because we absorb a message (`ok`), in the second we simply propagate the switch from the consumed message (`broken`) to the produced one (`0`). In general, for every abductive or input argument of a procedure there exists a clause enhanced with the functionality just described. The monitoring process is of the form

```
mode  monitor(messSCL?,messSCR?,...).
monitor(MessSC,MessSC,...)  <-  resolve deadlock
```

where its exact functionality will be described later on.

3.2 ABDUCTIVE PHASE

The second phase comprises two parts: first, the deadlocked global state of the computation must be examined and second, a decision must be taken as to which abducible argument(s) must be abduced.

3.2.1 Examining the Deadlocked Global State of the Computation. The global state of a set of concurrently executing processes can be examined by using again techniques based on short circuits, in particular the one for collecting snapshots of the executing processes ([14]). More to the point, all processes involved in a computation are connected by means of left and right switches which, collectively, form a short circuit. When the monitoring process detects deadlock it sends the message `collect_states([])` to all the deadlocked processes using one of the ends of the short circuit and then it waits for the message to reappear at the other end. Each deadlocked process that receives the message via one of its local switches propagates it to the rest of the processes using the other switch after enhancing it with the current values of its arguments (i.e. its state). Note that because this mechanism is initiated after the detection of deadlock it is not possible for a process to change its state after it has reported it to the monitoring process by means of the `collect_states` message. The exact way a process's state is represented within a `collect_states` message is shown below for the case of the procedure wire which, when transformed, must now be extended accordingly to handle this new functionality.

```
mode  wire(leftS?,rightS?,state@,l_connect?,r_connect^).
wire(LSC,RSC,m(L,R,ok),ConnectIn,ConnectOut)  <-
  LSC=RSC,  L=R,  ConnectOut=ConnectIn.
wire(LSC,RSC,m(L,R,broken),_,ConnectOut)  <-
  LSC=RSC,  ConnectOut=m(L,R,0).
wire([collect_states(S)|LSC1],RSC,State,ConIn,ConOut)  <-
  RSC=[collect_states(wire(LSC1,RSC1,abd(State),
                               ConIn,ConOut)|S])|RSC1],
  wire(LSC1,RSC1,State,ConIn,ConOut).
```

Each abducible argument is indicated by means of enclosing it into the structure abd whereas the rest of the arguments (input and output) are included as they are. Note that upon terminating, a process closes both circuits (the one for messages and the one for processes); had it been reduced to a number of processes it would have splitted the corresponding circuit accordingly. In general, a transformed procedure comprises two additional arguments (the left and right local switches of the processes' short circuit and an extra clause for reporting its state to the monitoring process. We now describe the monitoring process in more detail noting its enhancement with the second short circuit.

```
mode  monitor(procSCL?,procSCR?,procmessSCL?,messSCR?,...).
monitor(PSC,PSC,_,_,...)  <-  end computation
monitor(PSCL,PSCR,MessSC,MessSC,...)  <-
  PSCL=[[collect_states([])|PSCL1],
  monitor_wait(PSCL1,PSCR,...).

mode  monitor_wait(procSCL?,procSCR?,...).
monitor_wait(PSCL,[collect_states(States)|PSCR],...)  <-
  resolve_deadlock(States,NStateAbd,NStateRest),
  next_deductive_phase(PSCL,PSCR,NStateAbd,NStateRest,...).
```

The first clause detects the end of the computation by checking whether the left and right ends of the short circuit for processes have been unified. Similarly, the second detects deadlock, i.e. the end of the current deductive phase, and invokes the mechanism for examining the current state of the computation by sending the triggering mechanism collect_states([]) to all suspended processes and then invoking the process monitor_wait which waits for the message collect_states to reappear with a snapshot of the system of deadlocked processes. The snapshot is then passed to the process resolve_deadlock which examines the state of the computation and decides which abducibles to abduce. This will lead to the instantiation of some abducible variables and the formation of a new state of computation which is passed to next_deductive_phase for the third phase of the computation.

3.2.2 Deciding What to Abduce. The top level definition of resolve_deadlock is shown below.

```
mode  resolve_deadlock(state?,selected_abd^,n_state_rest^).
resolve_deadlock(State,SelectAbds,NStateRest)  <-
  filter_states(State,CandidateAbds,RestState),
  abduce(CandidateAbds,RestState,Select_Abds,NStateRest).

mode  filter_states(state?,selected_abds^,rest_of_state^).
filter_states(...)  <-  ....
```

```
mode  abduce(list_abds?,rest_goals?,sel_abd^,n_state_rest^).
abduce(CandAbds,RestState,Sel_Abds,NStateRest)  <-  ....
```

Finding the best combination of abducibles to abduce among the set of candidate ones is a popular area of current research in the field of abductive reasoning and its solution usually involves the use of heuristics. In our model we try to minimise the guesses taken and we choose the following simple heuristic: the process chosen to have its abdicible parameters abduced is the one with the smallest number of abducible parameters that can still be abduced. Furthermore, if a number of processes satisfy the above heuristics, those for which reduction is deterministic (i.e. only one of a process's defining clauses can be selected to reduce that process) are given preference. The ability to use more sophisticated heuristics is discussed later on.

More to the point, the process `filter_states` is responsible for filtering from the AND-conjunction of suspended goals those that cannot be abduced. A goal that cannot be abduced is one comprising non-abducible parameters or ground abducible parameters (namely parameters that have already been abduced). Here the structure `abd` mentioned earlier on assists in the filtering process. That part of the suspended AND-conjunction comprising the literals that can be abduced is then passed to the process `abduce` which forms the new state. This latter process is responsible for selecting those abducibles which will actually be abduced using the heuristics discussed already. As an example, if the snapshot of the deadlocked state is

```
States=[p(abd(X),3),q(abd(Y),Z),r(X,W),s(abd(2),Z,W)]
```

then `filter_states` will instantiate its two output arguments as follows.

```
CandidateAbs=[p(abd(X),3),q(abd(Y),Z)],
RestState=[r(X,W),s(abd(2),Z,W)]
```

Assuming that the reduction of `p` is deterministic, `abduce` will instantiate its two output arguments as follows.

```
SelectedAbs=[p(abd(X),3)],
NewStateRest=[q(abd(Y),Z),r(X,W),s(abd(2),Z,W)]
```

3.3 GENERATING NEW DEDUCTIVE PHASES

The process `next_deductive_phase` is responsible for implementing the third phase, i.e. the initiation of new deductive phases corresponding to the abduction of the selected abducibles. There may be more than one such deductive phase attributed to the possibility of producing multiple explanations. In order to be able to produce multiple explanations every new deductive phase must be executed as a different OR-branch. The top level machinery to do that is shown below and it is an adaptation of well known techniques for implementing OR-parallel interpreters in concurrent logic languages ([15]) using operations such as *copy*, *freeze* and *melt*. In particular, for each different set of selected abducibles a new deductive phase commences with a copy of the generated new state of computation whereas the old suspended state is saved. Upon failure (or need for further explanations), the system backtracks to the saved state and another copy is constructed and executed, this time with a different set of selected abducibles.

```
mode  next_deductive_phase(prSCL?,prSCR?,abds?,rest?,...).
next_deductive_phase(SCL,SCR,NewStateAbd,NStateRest,...)  <-
   get_clauses(NewStateAbd,Clauses),
   generate_explanations(SCL,SCR,NewStateAbd,
                              Clauses,NStateRest,...).
```

```
mode generate_explanations(pSCL?,pSCR?,sel_abs?,
                              cls?,rest_state?,…).
generate_explanations(SCL,SCR,Abd,[Cl|Cls],RestState,…) <-
  generate_explanation(SCL,SCR,Abd,Cl,RestState,…),
  generate_explanations(SCL,SCR,Abd,Cls,RestState,…).

mode generate_explanation(pSCL?,pSCR?,sel_abs?,
                             clause?,rest_state?,…) <-
generate_explanation(SCL,SCR,Abds,Clause,RestState,…) <-
  form_query(SCL,SCR,Abds,Clause,RestState,…,Query),
  run_query(Query,…).

mode form_query(pSCL?,pSCR?,sel_abs?,cls?,
                    rest_state?,query^…).
form_query(SCL,SCR,Abds,Clause,RestState,…,Query) <- ….

mode run_query(query?,…).
run_query([ProcSCL,ProcSCR,MessSCL,MessSCR,Query],…) <-
  Query, monitor(ProcSCL,ProcSCR,MessSCL,MessSCR,…).
```

Note that `form_query` is responsible for setting up the mechanisms for detecting deadlock and examining the global state of the computation (i.e. forming new short-circuits for messages and processes).

3.4 TOP-LEVEL QUERY AND USE OF CONSTRAINTS

The top-level predicate `abductive_parlog(Goal,Constraints)` commences the computation and it is defined as follows.

```
mode abductive_parlog(query?,constraints?).
abductive_parlog(Query,Constraints) <-
  run_query([PSCL,PSCR,MessSCL,MessSCR,Query,Constraints]
          ,…).
```

The parameter `Constraints` which is effectively added to the first AND-conjunction is used to restrict even more the extent to which the system resorts to abduction in breaking the deadlock. This parameter states the conditions under which abductive hypotheses can be generated. The simplest form of such *integrity constraints* is negation combined with the restriction that hypotheses should be variable free ([7,8]). As an example, consider the following query.

```
<- abductive_parlog(Goal,Constraints),
     Goal=[circuit(S,B1,B2,B3,W1,W2,W3,W4,W5,W6,off,on,off),
          print([S,B1,B2,B3])],
     Constraints=[not((S=empty,B1=on)),not((S=empty,B2=on)),
                  not((S=empty,B3=on))].
```

The user has imposed the constraint that if at least one bulb is observed to be working then it is not possible for the battery to be empty. The inclusion of these constraints into the computation reduces considerably the search space since those abducible values that do not satisfy the constraints will not be tried.

3.5 OTHER HEURISTIC MECHANISMS

It is important to note that, if desired, the model can be enhanced with other heuristics. It is possible, for instance, to associate with each head argument in an abducible position an attribute indicating an assumption cost ([16]) or a probability value ([12]) as it is illustrated by the following piece of code.

```
p(?,@,...).
p(1,a:0.3,...)  <-  ....
p(2,b:0.7,...)  <-  ....
```

The interpretation here is that in the goal `p(1,X,...)`, `X` can be abduced with an assumption cost (or hypothesis probability) of 0.3 whereas in the goal `p(2,X,...)`, `X`'s assumption cost is 0.7. The process `resolve_deadlock` can then use these attributes during the examination of the suspended state of the computation to decide which abducible(s) to abduce.

The extension of the model with heuristics such as the ones just described may help to alleviate a problem that most abductive models, including the one described in this paper, suffer from: the difficulty in computing minimal explanations and avoiding computing the same explanation multiple times.

4 Discussion. Related and further work

The relationship between abduction and constraint programming is well known and has been explored in a number of papers (eg. [2,10,11]). In constraint programming the computation interleaves between a constraint generation phase where a variable is instantiated to one of a set of possible values and a constraint satisfaction phase where it is checked whether the chosen value satisfies the imposed constraints; if it does not then another one must be tried. In abduction the computation interleaves between a hypotheses generation phase and a consistency phase that checks whether a newly generated hypothesis is consistent with the rest of the existing hypotheses or imposed constraints. In fact [10] claims that constraint solving is a subcase of abduction. Hence, many of the techniques devised for constraint satisfaction are applicable also to the case of abduction.

In particular, the extension of PARLOG with an abductive mechanism along the lines of the model proposed in [2] leads to a framework similar in many respects to the Pandora model ([1]) which combines stream AND-parallelism with OR-search (parallel or sequential) and is particularly suited to constraint satisfaction problems. However, there are some important differences between the two models especially with regard to the detection and resolving of deadlock. Pandora relies heavily on the use of PARLOG's (extended) metacall whereas our technique using short circuits is effectively independent of any particular underlying implementation mechanism; in fact our high-level transformation techniques could well be applied easily to programs written in other similar concurrent logic languages such as GHC or FCP ([15]). Nevertheless, Pandora could be used as a basis for developing an efficient sequential implementation of abductive PARLOG with minimal effort.

As we have stated already we were particularly interested in developing an abductive framework based on concurrent logic programming especially suited to distributed environments. The kind of applications we have in mind is multi agent systems, distributed expert and medical diagnosis systems ([3]) and diagnosis systems requiring reactive properties ([16]). We are currently extending the definition of `resolve_deadlock` to include the functionality described in the paragraph 3.5.

Regarding the *abductive policy* our model is adopting there are two issues involved ([10]): i) which relations are allowed to have abducible instances, and ii) which instances of an abducible relation can in fact be abducible. In abductive PARLOG a relation is abducible if it conatins at least one abducible mode declaration. Furthermore, all instances of that relation (along its abducible arguments) are abducible. If desired however, the model could be extended to restrict abduction to particular instances of a clause by dispensing with the abductive mode declaration and moving the abducible operator '@' to the head of particular clauses of the relation. In the following relation, for instance,

```
mode p(?,?,…).
p(@X,3,…) <- ….
p(X,4) <- ….
```

the first argument of p can be abduced only if the second one is instantiated to 3. The use of constraints in the top-level call abductive_parlog however effectively provides this functionality. Note here that the abductive mechanism we have described subsumes the (extended) four arguments abductive operator '@' proposed in [10].

5 Conclusions

We have presented a way to model abductive behaviour within the framework of concurrent logic programming by advocating the idea, first proposed for the general framework of concurrent constraint programming languages, of abducing arguments rather than predicates. The model has been tested by means of a meta interpreter written on top of the language PARLOG; however any other typical concurrent logic language could be used instead. The model inherits the high degree of concurrency enjoyed by concurrent logic languages and it is particularly suited to distributed environments.

Acknowledgements

I am grateful to Antonis Kakas for introducing me to the field of abductive reasoning and explaining its central concepts, and to Reem Bahgat for pointing out the similarities in concurrent logic languages between abductive behaviour and constraint satisfaction and explaining some key properties of the Pandora model.

References

[1] R. Bahgat (1993), *'Non-deterministic Concurrent Logic Programming in Pandora'*, World Scientific Series in Computer Science, Vol. 37, Singapore.

[2] P. Codogent and V. A. Saraswat (1992), 'Abduction in Concurrent Constraint Programming', Technical Report, Xerox Palo Alto Research Centre.

[3] A. Eliëns (1991), 'Distributed Logic Programming for Artificial Intelligence', *AI Communications*, Vol. 4, No. 1, pp. 11-21.

[4] K. Eshghi and R. A. Kowalski (1989), 'Abduction Compared With Negation By Failure', *Proc. 6th International Conference on Logic Programming,* Lisbon, Portugal, pp. 234-255.

[5] K. Furukawa (1993), contribution to 'The Fifth Generation Project: Personal Perspectives', eds. E. Shapiro and D. H. D. Warren, *Communications of the ACM*, March, pp. 48-101.

[6] S. Gregory (1987), 'Parallel Logic Programming in PARLOG', Addison-Wesley Publishing Company.

[7] A. C. Kakas, R. A. Kowalski and F. Toni (1992), 'Abductive Logic Programming', *Journal of Logic and Computation*, Vol. 2, No 6, pp. 719-770.

[8] A. C. Kakas and P. Mancarella (1990), 'Generalised Stable Models: a Semantics for Abduction', *Proc. 9th European Conference on Artificial Intelligence*, Stockholm, Sweden, pp. 385-391.

[9] A. C. Kakas and G. A. Papadopoulos (1994), 'Parallel Abduction in Logic Programming', *Proc. 1st International Symposium on Parallel Symbolic Computation*, Linz, Austria, (to appear).

[10] E. Maïm (1992), 'Abduction and Constraint Logic Programming', *Proc. 10th European Conference on Artificial Intelligence*, Vienna, Austria, pp. 149-153.

[11] A. Michael and A. C. Kakas (1994), 'Integrating Abduction and Constraint Logic Programming', Internal Report, Department of Computer Science, Univ. of Cyprus.

[12] A. Pool (1992), 'Logic Programming, Abduction and Probability', *Proc. International Conference on Fifth Generation Computer Systems*, Tokyo, Japan, pp. 530-538.

[13] V. A. Saraswat (1993), '*Concurrent Constraint Programming*', ACM Doctoral Dissertation Award, MIT Press series on Logic Programming.

[14] V. A. Saraswat, D. Weinbaum, K. Kahn and E. Y. Shapiro (1988), 'Detecting Stable Properties of Networks in Concurrent Logic Programming Languages', *Proc. 7th ACM Symposium on Principles of Distributed Computing*, pp. 210-222.

[15] E. Y. Shapiro (1989), 'The Family of Concurrent Logic Programming Languages', *Computing Surveys*, Vol. 21 (3), pp. 412-510.

[16] A. Wærn (1992), 'Reactive Abduction', *Proc. 10th European Conference on Artificial Intelligence*, Vienna, Austria, pp. 159-163.

Explanation of Designs through Heuristic Reasoning

B. D. Britt
Computer Science Department
Eastern Washington University
Cheney, WA 99004-2495
bbritt@ewu.edu

T. Glagowski
Electrical Engineering and Computer Science
Washington State University
Pullman, WA 99164-2752
glagowsk@spookey.spokane.wsu.edu

Abstract. This paper discusses an extension to derivational analogy. This extension, Reconstructive Derivational Analogy, recreates a possible design plan for a design. This design plan can then be used in a replay process, such as derivational analogy. This paper discusses the heuristic techniques used to reconstruct a design plan, given a design without a plan.

Keywords. Design, Rationales, Heuristic Classification, Explanation

1 Introduction

Derivational Analogy is one technique for improving the usefulness of a design [Carbonell, 1986]; essentially, a trace of the problem solving process that led to the design is saved, so that it can be replayed later on a new problem. Derivational Analogy has been used successfully in several automated design systems [Mostow, 1989; Mostow, *et al.*, 1989]. However, one limitation is that it can only be used when the problem solving that led to the existing design is explicitly available to the system. Many existing designs lack this type of recorded derivation. Industrial designers have large collections of tested designs that should be used as much as possible in any new design problem.

Our research has produced an algorithm, Reconstructive Derivational Analogy (RDA), that automatically reconstructs a design history for an existing design, thus allowing the power of derivational analogy to be extended to the large libraries of existing designs. This paper discusses the heuristic techniques used by the RDA system to reconstruct the design history.

2 Reconstructive Derivational Analogy

An overview of the RDA approach is presented in Figure 1, with the direction of the process given by the arrows (up-over-down). In RDA, an existing design is used to automatically reconstruct a design history that might have led to the creation of the design. This design history is then replayed on new requirements to create a new design meeting these

E. A. Yfantis (ed.), Intelligent Systems, 39–43.

requirements. The same rules are available for composition of the design as are used in the decomposition of this type of design problem. In Derivational Analogy the design history is stored as a decompositional hierarchy which is a trace of all problem decomposition steps made which transformed the Design Specification into the design. Since we are reconstructing this hierarchy, it is referred to as a "compositional hierarchy." Because the compositional hierarchy is a tree, each sub-module can only be used once by a composition step. The choice of a rule at one step can eliminate other rules which were available and will determine which rules may be applied at higher levels of the hierarchy.

It should be noted that the RDA algorithm discussed here is only responsible for the Reconstruction Stage of the process shown in Figure 1. The VEXED Derivational Analogy procedure is being used for the Replay Stage. VEXED was used as the forward design system; however, a different design system with comparable modules and data structures could be substituted. RDA produces a Design History which can be used by VEXED to produce a new design. For a further discussion of VEXED, the reader is referred to a set of previous publications [Mitchell et al, 1985; Mostow, 1989; Mostow, *et al.*, 1989; Steinberg, 1987].

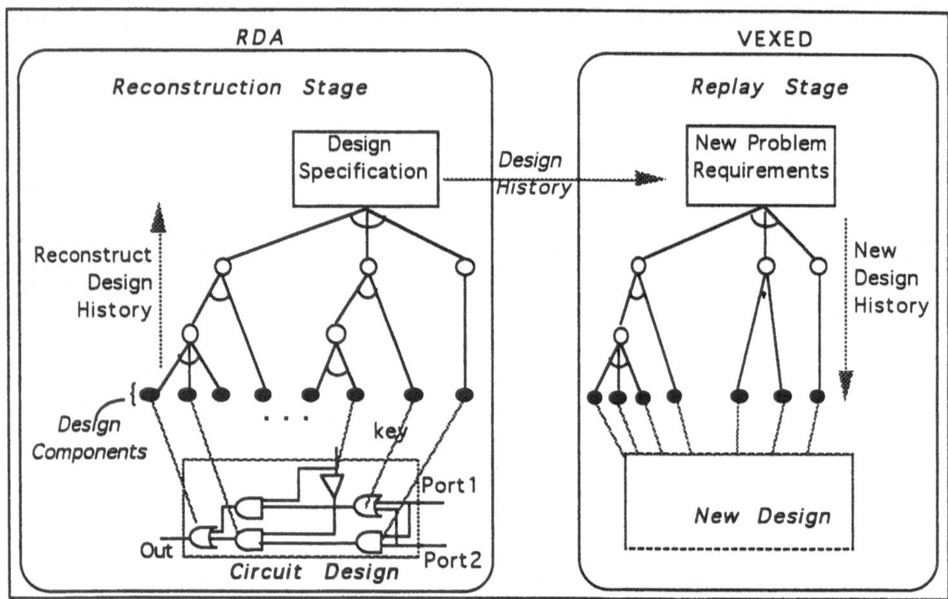

Figure 1: Overview of Reconstructive Derivational Analogy

There can be numerous design histories for a design, some of which will be more useful than others during replay. However, the only way to know for sure which design history would be best for replay is to create ALL possible design histories for this design, and then to compare these in the replay process. This makes this problem a good one for heuristic classification techniques which can navigate uncertain domains.

2.2.1 RDA ARCHITECTURE

The architecture for RDA/VEXED is given in Figure 2 below. Although our research is only responsible for producing the RDA algorithm, its use of the VEXED system requires that VEXED be included in this discussion. RDA consists primarily of the Design History Reconstruction Module (DHRM) and the Reconstruction Expert (RE). The RDA system rebuilds a design history using the available reconstruction rules, in this case the NMOS Circuit Rules. The DHRM is the engine which finds the applicable rules, applies the chosen rules to the circuit and rebuilds the design history. The RE is given the current circuit and the set of applicable rules, as well as any information about the new circuit problem, and uses this information to select the "best" rule for application at the current stage of reconstruction. A heuristic classifier type of expert system shell was used as a starting point for the RE. This approach is well suited for this problem because the reconstruction rules can be dynamically compared. The RE "gathers" supporting and opposing evidence for choosing the different rules. Certain rules may be discarded if they gather strong opposing evidence. Other rule solutions gather strength until enough evidence for choosing a particular rule is found. Sometimes, this is an easy process, when one rule is clearly better than the others. Other times, more evidence is requested, and the reconstruction rule base may need to be more closely evaluated. New evidence is dynamically requested by the RE. The heuristics used to confirm, oppose, discard or choose the reconstruction rules are discussed in more detail below.

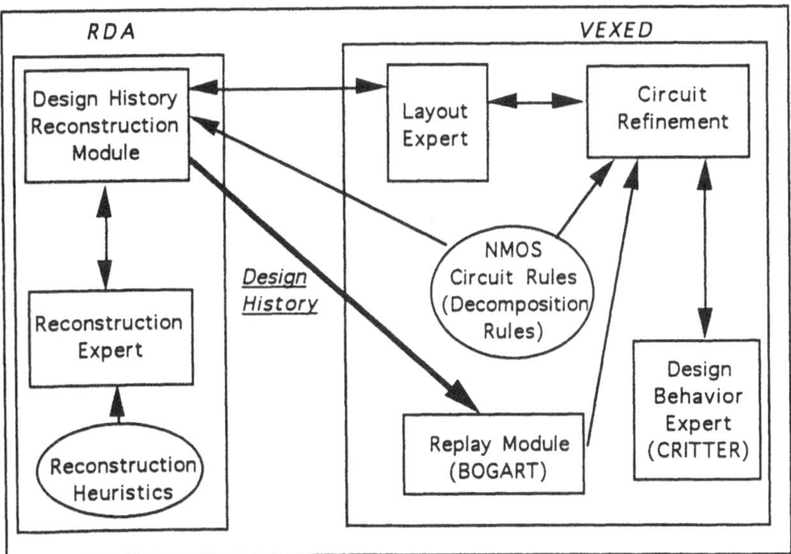

Figure 2: RDA Architecture

The rebuilt design history consists primarily of a rule tree and a replay plan. This replay plan refers to the rule tree and also includes additional information about the submodules created by the application of the rules. As mentioned previously, there are several points of interface between the RDA System and the design system, VEXED. RDA calls the VEXED Layout Expert to draw the circuit at each level of reconstruction. This enables the designer/user to see the new modules which are created by a rule application. Using the VEXED NMOS Circuit Rules as the reconstruction rules, RDA constructs a Design History which is given to the VEXED Replay Module and can then replayed on a new circuit design problem.

2.2.2 DESIGN KNOWLEDGE

The RDA algorithm is discussed in detail elsewhere. This paper focuses on the nature of the heuristic knowledge and techniques used to reconstruct a likely design plan. As discussed previously, heuristics are used to select from the set of potential rules at each stage of the reconstruction. The rules being selected from are the VEXED decomposition rules, although here they are being used more like "composition" rules.

The RDA heuristics are encoded as rules for the Reconstruction Expert (not to be confused with the VEXED decomposition rules). Several examples of these heuristics are given in Figure 3.

Pygmalion: Choose compositions that have known "improvements." An improvement is defined as anything which moves the design closer to the specifications given for the new design. This heuristic favors plans that can have easy substitutions made during replay.

Replay: Choose compositions that produce a derivational history which will have a high percentage of replay for the new design. Differences between the derivational history and the new design's solution should occur as low (close to the primitive modules) in the derivational history as possible. This heuristic strives to increase the usefulness of a plan during reuse.

Complexity: Choose compositions for more complex rules. This strategy will favor compositions that involve larger or more complex portions of the circuit design. There are two reasons for this heuristic. In general, more complex rules are more difficult to match: if a match is found for a complex rule then it is intuitive that this rule is more likely to be reflective of the original intent of the design. In addition, the use of more complex rules results in a simpler plan, and similer plans are preferable.

Output First: Choose compositions that will enable the output of the circuit to be established. This heuristic serves as a good starting point, directing the system to work backwards from the outputs to the inputs of the circuit.

Figure 3: Some heuristics for guiding rule selection

There are also *domain-dependent* heuristics which are associated with a particular type of circuit design rule. For instance, multiplexer rules have a heuristic associated with them which can be used to direct the further composition of a design. Both types of heuristics are discussed in more detail elsewhere [Britt, 1994].

3 Significance

Derivational analogy has been shown in the AI community to be an effective tool in automated case-based design. However, previous research has suffered from the real-world limitation that design histories are not always available for use in an automated system. Our research in digital circuit design has established the feasibility of reconstructing design histories for use by a derivational analogy system. We have achieved this reconstruction by applying current techniques of heuristic classification in an uncertain domain. This paper describes current work in the area of design and explanation and demonstrates that heuristic techniques can be used to produce a useful design history. RDA significantly extends the power of derivational analogy to a much larger real-world audience. Our results show that reuse rates of over 60% are possible with a completely automated system, thus establishing suggestive evidence that Reconstructive Derivational Analogy is not only feasible but desirable as well [Britt, 1994].

4 References

[Britt, 1994] Britt, B. D., "Reconstructive Derivational Analogy," Ph.D. Dissertation, Washington State University, July 1994.

[Carbonell, 1986] Carbonell, J. G., "Derivational Analogy: A Theory of Reconstructive Problem Solving and Expertise Acquisition," in Machine Learning: An Artificial Intelligence Approach, vol. 2, Morgan Kaufman, 1986.

[Mitchell et al, 1985] Mitchell, T.M., Mahadevan, S., and Steinberg, L., "LEAP: A Learning Apprentice for VLSI Design," *Proceedings of the Ninth International Joint Conference on Artificial Intelligence* (IJCAI), pp. 573-580, 1985.

[Mostow, *et al.*, 1989] Mostow, J., Barley, M., and Weinrich, T., "Automated Reuse of Design Plans," *Artificial Intelligence in Engineering*, 4(4);181-196, 1989.

[Mostow, 1989] Mostow, J., "Design by Derivational Analogy: Issues in the Automated Replay of Design Plans," *Artificial Intelligence*, 40(1):119-184, 1989.

[Steinberg, 1987] Steinberg, L., "Design as refinement plus constraint propagationplus what?," Proceedings 1987 IEEE International Conference on Systems, Man, and Cybernetics, Alexandria, VA 498-502, 1987.

SEARCH REDUCTION IN LINGUISTIC GEOMETRY

Boris Stilman
Department of Computer Science & Engineering
University of Colorado at Denver,
Campus Box 109, Denver, CO 80217-3364, USA.
Email: bstilman@gothic.cudenver.edu

Abstract. This paper reports new results on applications of Linguistic Geometry to optimization problems of space mission planning. The Linguistic Geometry is intended to discover the inner properties of human expert heuristics, which were successful in a certain class of complex control systems, and apply them to different systems. It relies on the formalization of search heuristics, which allow to decompose complex system into the hierarchy of subsystems, and thus solve intractable problems reducing the search. Currently we investigate heuristics extracted in the form of hierarchical networks of paths. The dynamic hierarchy of networks is represented as a hierarchy of formal attribute languages. This paper includes a brief formal survey of the Linguistic Geometry and example of solution of optimization problem for space robotic vehicles. This example presented informally demonstrates the dramatic reduction of search in comparison with conventional search algorithms (from billions to tens).

Keywords: Linguistic Geometry, Heuristic Search, Hierarchical Systems, Space Navigation, Networks of Paths, Formal Languages.

1 Introduction

It is well known that despite of the universal proliferation of computers there are many real-world problems where human expert skills in reasoning about complex systems are incomparably higher than the level of modern computing systems. At the same time there are even more areas where advances are required but human problem-solving skills can not be directly applied. For example, there are problems of planning and automatic control of autonomous agents such as space vehicles, stations and robots with cooperative and opposing interests functioning in a complex, hazardous environment. Reasoning about such complex systems should be done automatically, in a timely manner, and often in a real time. Moreover, there are no highly-skilled human experts in these fields ready to substitute for robots (on a virtual model) or transfer their knowledge to them. There is no grand-master in robot control, although, of course, the knowledge of existing experts in this field should not be neglected – it is even more valuable. It is very important to study human expert reasoning about similar complex systems in the areas where the results are successful, in order to discover the keys to success, and then apply and adopt these keys to the new, as yet, unsolved problems. The question then is what language tools do we have for the adequate representation of human expert skills? An application of such language to the area of successful results achieved by the human expert should yield a *formal, domain independent knowledge* ready to be transferred to different areas. Neither natural nor programming languages satisfy our goal. The first are informal and ambiguous, while the second are usually detailed, lower-level tools. Actually, we have to learn how we can formally represent, generate, and investigate a *mathematical model* based on the *abstract images* extracted from the expert vision of the problem.

E. A. Yfantis (ed.), Intelligent Systems, 45–54.

There have been many attempts to find the optimal (suboptimal) operation for real-world complex systems. One of the basic ideas is to decrease the dimension of the real-world system following the approach of a *human expert in a certain field*, by breaking the system into smaller subsystems. These ideas have been implemented for many problems with varying degrees of success [1, 14, 22]. Implementations based on the formal theories of linear and nonlinear planning meet hard efficiency problems [3, 11, 16, 20, 23]. An efficient planner requires an intensive use of heuristic knowledge. On the other hand, a pure heuristic implementation is unique. There is no general constructive approach to such implementations. Each new problem must be carefully studied and previous experience usually can not be applied. Basically, we can not answer the question: what are the formal properties of human heuristics which drove us to a successful hierarchy of subsystems for a given problem and how can we apply the same ideas in a different problem domain?

In the 1960's a formal syntactic approach to the investigation of properties of natural language resulted in the fast development of a theory of formal languages by Chomsky [4], Ginsburg [9], and others. This development provided an interesting opportunity for dissemination of this approach to different areas. In particular, there came an idea of analogous linguistic representation of images. This idea was successfully developed into syntactic methods of pattern recognition by Fu [7], Narasimhan [15], and Pavlidis [17], and picture description languages by Shaw [21], Feder [5], Rosenfeld [18].

Searching for the adequate mathematical tools formalizing human heuristics of dynamic hierarchy, we have transformed the idea of linguistic representation of complex real-world and artificial images into the idea of similar representation of complex hierarchical systems [24]. However, the appropriate languages should possess more sophisticated attributes than languages usually used for pattern description. The origin of such languages can be traced back to the research on programmed attribute grammars by Knuth [10],Rozenkrantz [19].

A mathematical environment (a "glue") for the formal implementation of this approach was developed following the theories of formal problem solving and planning by Nilsson [16], Fikes, Nilsson [6], Sacerdoti [20], McCarthy, Hayes [12, 13], and others based on first order predicate calculus.

To show the power of the linguistic approach it is important that the chosen model of the heuristic hierarchical system be sufficiently complex, poorly formalized, and have successful applications in different areas. Such a model was developed by Botvinnik, Stilman, and others, and successfully applied to scheduling, planning, and computer chess [1, 2].

In order to discover the inner properties of human expert heuristics, which were successful in a certain class of complex control systems, we develop a formal theory, the so-called *Linguistic Geometry* [25-29]. This research includes the development of syntactic tools for *knowledge representation* and *reasoning* about large-scale hierarchical complex systems. It relies on the formalization of *search heuristics*, which allow one to decompose complex system into a hierarchy of subsystems, and thus solve intractable problems, reducing the search. These *hierarchical images* were extracted from the expert vision of the problem. The hierarchy of subsystems is represented as a *hierarchy of formal attribute languages*.

2 Complex System

A ***Complex System*** is the following eight-tuple:

$$< X, P, R_p, \{ON\}, v, S_i, S_t, TR>,$$

where $X=\{x_i\}$ is a finite set of *points*; $P=\{p_i\}$ is a finite set of *elements*; P is a union of two non-intersecting subsets P_1 and P_2; $R_p(x, y)$ is a set of binary relations of *reachability* in X (x and y are from X, p from P); ON(p)=x, where ON is a partial function of *placement* from P into X; v is a function on P with positive integer values; it describes the *values* of elements. The Complex System searches the state space, which should have initial and target states; S_i

and S_t are the descriptions of the *initial* and *target* states in the language of the first order predicate calculus, which matches with each relation a certain Well-Formed Formula (WFF). Thus, each state from S_i or S_t is described by a certain set of WFF of the form $\{ON(p_j)=x_k\}$; TR is a set of operators, TRANSITION(p, x, y), of transition of the System from one state to another one. These operators describe the transition in terms of two lists of WFF (to be removed and added to the description of the state), and of WFF of applicability of the transition. Here,

Remove list: ON(p)=x, ON(q)=y;
Add list: ON(p)=y;
Applicability list: $(ON(p)=x) \wedge R_p(x, y)$,

where p belongs to P_1 and q belongs to P_2 or vice versa. The transitions are carried out with participation of one or many elements p from P_1 and P_2.

According to definition of the set P, the elements of the System are divided into two subsets P_1 and P_2. They might be considered as units moving along the reachable points. Element p can move from point x to point y if these points are reachable, i.e., $R_p(x, y)$ holds. The current location of each element is described by the equation ON(p)=x. Thus, the description of each state of the System $\{ON(p_j)=x_k\}$ is the set of descriptions of the locations of the elements. The operator TRANSITION(p, x, y) describes the change of the state of the System caused by the move of the element p from point x to point y. The element q from point y must be withdrawn (eliminated) if p and q belong to the different subsets P_1 and P_2.

The problem of the optimal operation of the System is considered as a search for the optimal sequence of transitions leading from one of the initial states of S_i to a target state S of S_t.

With such a problem statement for the search of the optimal sequence of transitions leading to the target state, we could use formal methods like those in the problem-solving system STRIPS, nonlinear planner NOAH, or in subsequent planning systems. However, the search would have to be made in a space of a huge dimension (for nontrivial examples). Thus, in practice no solution would be obtained.

We devote ourselves to the search for an approximate solution of a reformulated problem.

3 Trajectories

An element of the Complex System might follow a path to achieve the goal "connected with the ending point" of this path.

A ***trajectory*** for an element p of P with the beginning at x of X and the end at the y of X ($x \neq y$) with a length l is a following string of symbols with parameters, points of X: $t_0=\boldsymbol{a}(x)\boldsymbol{a}(x_1)\ldots\boldsymbol{a}(x_l)$, where $x_l = y$, each successive point x_{i+1} is reachable from the previous point x_i, i.e., $R_p(x_i, x_{i+1})$ holds for $i = 0, 1,\ldots, l-1$; element p stands at the point x: ON(p)=x. We denote $t_p(x, y, l)$ the set of trajectories in which p, x, y, and l coincide. $\boldsymbol{P}(t_0)=\{x, x_1, \ldots, x_l\}$ is the set of parameter values of the trajectory t_0.

Properties of the Complex System permit to define (in general form) and study formal grammars for generating different trajectories [25, 27].

A ***Language of Trajectories*** $L_t^H(S)$ for the Complex System in a state S is the set of all the trajectories of the length less than H. Different properties of this language and generating grammars were investigated in [25].

4 Trajectory Networks

A ***Language of Zones*** $L_Z(S)$ is a trajectory network language with strings of the form $Z=t(p_o,t_o,\tau_o)\ t(p_1,t_1,\tau_1)...t(p_k,t_k,\tau_k)$, where $t_o,t_1,...,t_k$ are the trajectories of elements $p_o,p_2,...,p_k$ respectively; $\tau_o,\tau_1,...,\tau_k$ are positive integer numbers (or 0) which "denote the time allotted for the motion along the trajectories" in a correspondence to the mutual goal of this Zone: to remove the target element – for one side, and to protect it – for the opposite side. Trajectory $t(p_o,t_o,\tau_o)$ is called the *main trajectory* of the Zone. The element q standing on the ending point of the main trajectory is called the *target*. The elements p_o and q belong to the opposite sides. A formal definition of this language is essentially constructive and requires showing explicitly a method for generating this language, i.e., a certain formal grammar, which is presented in [26-28]. To make it clearer let us show the Zone corresponding to the trajectory network in Fig. 1.

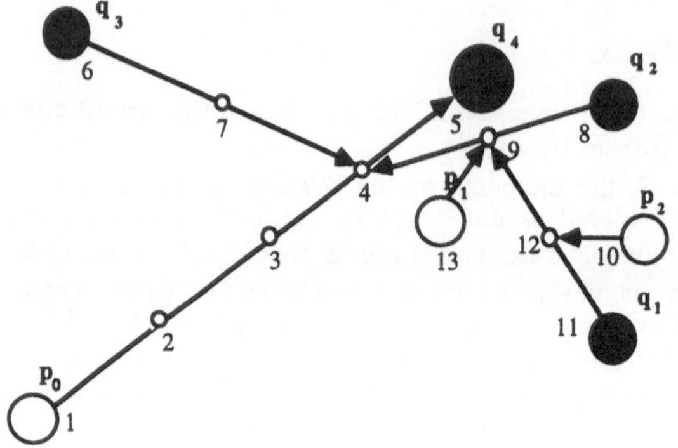

Figure 1: A network language interpretation.

$$Z=t(p_o,a(1)a(2)a(3)a(4)a(5), 4)t(q_3,a(6)a(7)a(4), 3)\ t(q_2, a(8)a(9)a(4), 3)t(p_1, a(13)a(9), 1)t(q_1, a(11)a(12)a(9), 2)\ t(p_2, a(10)a(12), 1)$$

Assume that the goal of the white side is to remove target q_4, while the goal of the black side is to protect it. According to these goals element p_o starts the motion to the target, while blacks start in its turn to move their elements q_2 or q_3 to intercept element p_o. Actually, only those black trajectories are to be included into the Zone where the motion of the element makes sense, i. e., the *length of the trajectory is less than the amount of time (third parameter* τ) *allocated to it.* For example, motion along the trajectories $a(6)a(7)a(4)$ and $a(8)a(9)a(4)$ makes sense, because they are of length 2 and time allocated equals 3: each of the elements has 3 time intervals to reach point 4 to intercept element p_o assuming one would go along the main trajectory without move omission. According to definition of Zone the trajectories of white elements (except p_o) could only be of the length 1, e.g., $a(13)a(9)$ or $a(10)a(12)$. As far as element p_1 can intercept motion of the element q_2 at the point 9, blacks include into the Zone the trajectory $a(11)a(12)a(9)$ of the element q_1, which has enough time for motion to prevent this interception. The total amount of time allocated to the whole bunch of black trajectories connected (directly or indirectly) with the given point of main trajectory is determined by the number of that point. For example, for the point 4 it equals 3 time intervals.

Languages of Trajectories and Zones describe the "statics", i.e., the states of the System. The Language of Translations [27, 28] is intended to describe the "dynamics" of the System, i.e., the transitions from one state to another.

5 Robotic Problem Example

These problems can be represented as Complex Systems naturally (Fig. 2). A set of **X** represents the operational district which could be the area of combat operation broken into smaller cubic areas, "points", e.g., in the form of the big cube of 8 x 8 x 8, n = 512. It could be a space operation, where X represents the set of different orbits, or an air force battlefield, etc. **P** is the set of robots or autonomous vehicles. It is broken into two subsets P_1 and P_2 with opposing interests; $\mathbf{R_p(x,y)}$ represent moving capabilities of different robots for different problem domains: robot p can move from point x to point y if $R_p(x, y)$ holds. These robots can move from one orbit to another. Some of them move fast and can reach point y (from x) in "one step", i.e., $R_p(x, y)$ holds, others can do that in *k* steps only, and many of them can not reach this point at all. **ON(p)=x,** if robot p is at the point x; **v(p)** is the value of robot p. This value might be determined by the technical parameters of the robot. It might include the immediate value of this robot for the given combat operation; $\mathbf{S_i}$ is an arbitrary initial state of operation for analysis, or the starting state; $\mathbf{S_t}$ is the set of target states. These might be the states where robots of each side reached specified points. On the other hand S_t can specify states where opposing robots of the highest value are destroyed. The set of WFF $\{ON(p_j) = x_k\}$ corresponds to the list of robots with their coordinates in each state. **TRANSITION(p, x, y)** represents the move of the robot p from the location x to location y; if a robot of the opposing side stands on y, a removal occurs, i.e., robot on y is destroyed and removed.

Space robotic vehicles with different moving capabilities are shown in Fig. 2. The operational district X is the cubic table of 8 x 8 x 8. Robot W-INTERCEPTOR (White Interceptor) located at 118 (x=1, y=1, z=8), can move to any next location, i.e., 117, 217, 218, 228, 227, 128, 127. The other robotic vehicle B-STATION (double-ring shape in Fig. 2) from 416 can move only straight ahead towards the goal area 816 (shaded in Fig. 2), one square at a time, e.g., from 416 to 516, from 516 to 616, etc. Robot B-INTERCEPTOR (Black Interceptor) located at 186, can move to any next square similarly to robot W-INTERCEPTOR. Robotic vehicle W-STATION located at 266 is analogous with the robotic B-STATION; it can move only straight ahead to the goal area 268 (shaded in Fig. 2). Thus, robot W-INTERCEPTOR on 118 can reach any of the points y ∈ {117, 217, 218, 228, 227, 128, 127} in one step, i.e., $R_{W\text{-}INTERCEPTOR}(118, y)$ holds, while W-STATION can reach only 267 in one step.

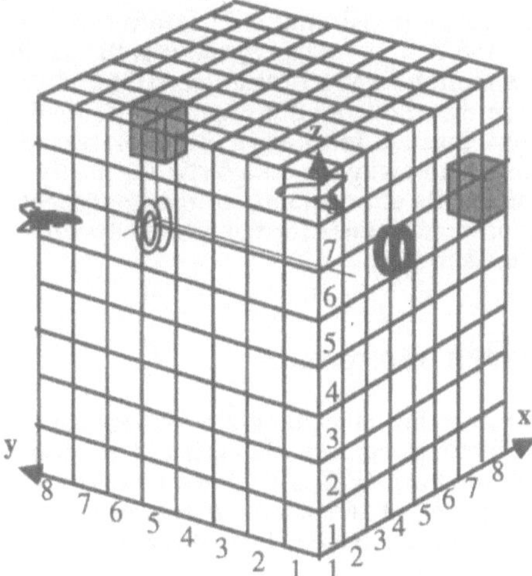

Figure 2: A space navigation problem for autonomous robotic vehicles.

Assume that robots W-INTERCEPTOR and W-STATION belong to one side, while B-INTERCEPTOR and B-STATION belong to the opposite side: W-INTERCEPTOR ∈ P_1, W-STATION ∈ P_1, B-INTERCEPTOR ∈ P_2, B-STATION ∈ P_2. Also assume that both goal areas, 816 and 268, are the safe areas for B-STATION and W-STATION, respectively, if station reached the area and stayed there for more than one time interval. Each of the STATIONs has powerful weapons capable to destroy opposing INTERCEPTORs at the next diagonal locations ahead of the course. For example W-STATION from 266 can destroy opposing INTERCEPTORs at 157, 257, 357, 367, 377, 277, 177, 167. Each of the INTERCEPTORs is capable to destroy an opposing STATION approaching its location from any direction, but it also capable to protect its friendly STATION approaching its prospective location. In the latter case the joint protective power of the combined weapons of the friendly STATION and INTERCEPTOR (from any next to the STATION area) can protect the STATION from interception. For example, W-INTERCEPTOR located at 156 can protect W-STATION on 266 and 267.

The battlefield considered can be broken into two local operations. The first operation is as follows: robot B-STATION should reach the strategic point 816 safely and stay there for at list one time interval, while W-INTERCEPTOR will try to intercept this motion. The second operation is similar: robot W-STATION should reach point 268, while B-INTERCEPTOR will try to intercept this motion. After reaching the designated strategic area the (attacking) side is considered as a winner of the local operation and the global battle. The only chance for the opposing side to revenge itself is to reach its own strategic area within the next time interval and this way end the battle in a draw. The conditions considered above give us S_t, the description of target states of the Complex System. The description of the initial state S_i is obvious and follows from Fig. 2.

Assume that motions of the opposing sides alternate and due to the shortage of resources (which is typical in a real combat operation) or some other reasons, each side can not participate in both operations simultaneously. It means that during the current time interval, in case of White turn, either W-STATION or W-INTERCEPTOR can move. Analogous condition holds for Black. Of course, it does not mean that if one side began participating in one of the operations it must complete it. Any time on its turn each side can switch from one operation to another, e.g., transferring resources (fuel, weapons, human resources, etc.), and later switch back.

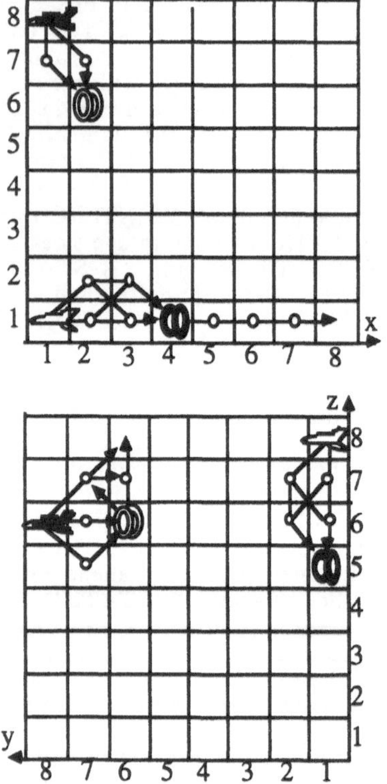

Figure 3: Interpretation of Zones in the initial state of the space system (2 projections)

It seems that local operations are independent, because they are located far from each other. Moreover, the operation of B-STATION from 418 looks like unconditionally winning operation, and, consequently, the global battle can be easily won by the Black side. Is there a strategy for the White side to make a draw? Of course, this question can be answered by the direct search employing, for example, minimax algorithm with alpha-beta cut-offs. Experiments with the computer chess programs showed that for the similar 2-D problem (in chess terms – the R.Reti endgame) the search tree includes about a million moves (transitions). Of course, in the 3-D case the search would require billions of moves. It is very interesting to observe the drastic reduction of search employing the Linguistic Geometry tools.

The details of generation of the Languages of Trajectories and Zones for the robotic systems are considered in [27].

6 Search Generation

Consider how the hierarchy of languages works for the optimal control of the space robotic system introduced above (Fig. 2). We generate the search of the Language of Translations representing it as a conventional search tree (Fig. 6) and comment on its generation.

First, the Language of Zones in the start state is generated. The targets for attack are determined within the limited number of steps which is called a horizon. In general, the value of the horizon is unknown. As a rule, this value can be determined from the experience of solving specific classes of problems employing Linguistic Geometry tools. In absence of such experience, first we have to consider the value of 1 as a horizon, and solve the problem within this value. If we still have resources available, i.e., computer time, memory, etc., we can increase the horizon by one. After each increase we have to regenerate the entire model. This increase means a new level of "vigilance" of the model, and, consequently, new greater need for resources.

In our case it is easy to show that within the horizons of 1, 2, 3, 4 all the models are "blind" and corresponding searches do not give a "reasonable" solution. But, again, after application of each of the consecutive values of the horizon we will have a *solution* which can be considered as an approximate solution within the available resources. Thus, let the horizon H of the language $L_Z(S)$ is equal to 5, i.e., the length of main trajectories of all Zones must not exceed 5 steps.

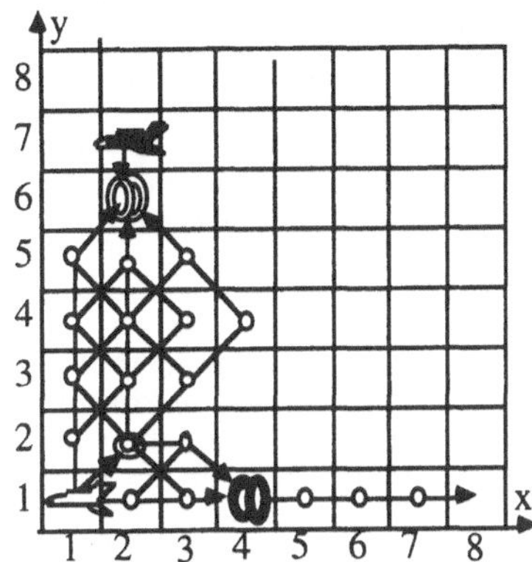

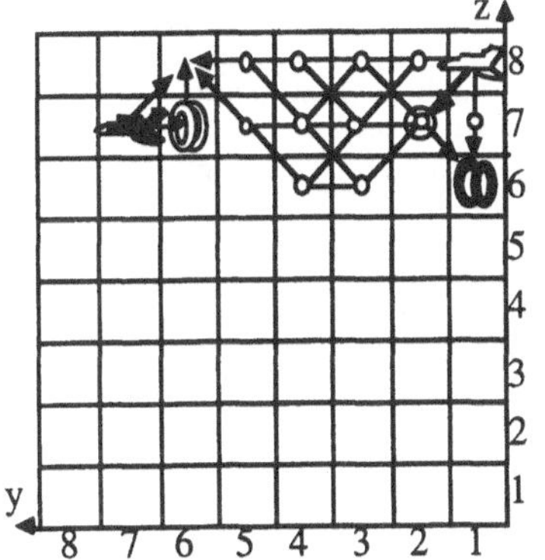

Figure 4: Interpretation of Zones in the state where control Zone from 118 to 268 was included first (2 projections).

All the Zones generated in the start state are shown in Fig. 3. Zones for INTERCEPTORSs and STATIONs as attacking elements are shown in Fig. 3. For example, one of the Zones for W-STATION, Z_{WS} is as follows:

Z_{WS}= *t*(W-STATION,***a***(266)***a***(267)***a***(268),3)
t(B-INTERCEPTOR,***a***(186)***a***(277)***a***(268),3)
t(B-INTERCEPTOR,***a***(186)***a***(276)***a***(267),2)
t(W-STATION, ***a***(266)***a***(277), 1)

The other trajectories of B-INTERCEPTOR, e.g., the second trajectory, ***a***(186)***a***(177)***a***(268), leading to the point 268 is included into different Zone; for each Zone only one trajectory from each bundle of trajectories with the same beginning and end is taken.

Generation begins with the move 1. 266-267 in the "white" Zone with the target of the highest value and the shortest main trajectory. The order of consideration of Zones and particular trajectories is determined by the grammar of translations. The computation of move-ordering constraints is the most sophisticated procedure in this grammar. It takes into account different parameters of Zones, trajectories, and the so-called chains of trajectories.

Next move, 1. ... 186-277, is in the same Zone along the first negation trajectory. The interception continues: 2. 267-268 277:268. Symbol ":" means the removal of element. Here the grammar terminates this branch with the value of -1 (as a win of the Black side). This value is given by the special procedure of "generalized square rules" built into the grammar. Then, the grammar initiates the backtracking climb. Each backtracking move is followed by the inspection procedure, the analysis of the subtree generated in the process of the earlier search. one branch (of two plies): 2. 267-268 277-268. The inspection procedure determined that the current minimax value (-1) can be "improved" by the improvement of the exchange in the area 268 (in favor of the White side). This can be achieved by participation of W-INTERCEPTOR from 118, i.e., by generation and inclusion of the new so-called "control" Zone with the main trajectory from 118 to 268. The set of different Zones from 118 to 268 (the bundle of Zones) is shown in Fig. 4. The move-ordering procedure picks the subset of Zones with main trajectories passing 227. These trajectories partly coincide with the main trajectory of another Zone attacking the opposing B-STATION on 516. The motion along such trajectories allows to "gain the time", i.e., to approach two goals simultaneously.

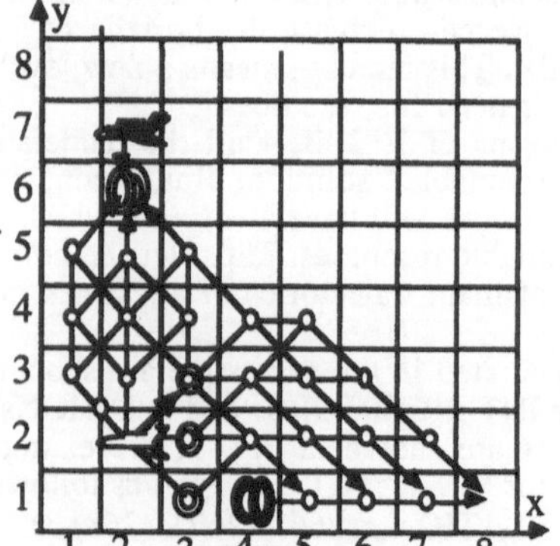

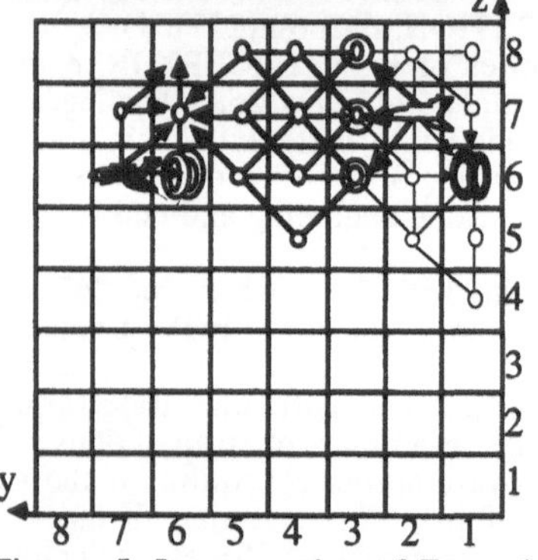

Figure 5: Interpretation of Zones in the state where the control Zone from 227 to 267 was included first (2 projections).

After the climb up to the move 1. ... 186-277, the tree to be analyzed consists of

The generation continues: 2. 118-227 277-267. Again, the procedure of "square rules" cuts the branch, evaluates it as a win of the black side, and the grammar initiates the climb. Move 2. 118-227 is changed for 2. 118-228. Analogously to the previous case, the inspection procedure determined that the current minimax value (-1) can be improved by the improvement of the exchange on 267. Again, this can be achieved by the inclusion of Zone from 118 to 267. Of course, the best "time-gaining" move in this Zone is 2. 118-227, but it was already included (as move in the Zone from 118 to 268). The other untested move in the Zone from 118 to 267 is 2. 118-228. Obviously the grammar does not have knowledge that trajectories to 267 and 268 are "almost" the same.

After the next cut and climb, the inspection procedure does not find new Zones to improve the current minimax value, and the climb continues up to the start state. The analysis of the subtree shows that inclusion of Zone from 118 to 268 in the start state can be useful: the minimax value can be improved. Similarly, the most promising "time-gaining" move is 1. 118-227. The Black side responded 1. ... 186-277 along the first negation trajectories ***a***(186)***a***(277)***a***(267) and ***a***(186)***a***(277)***a***(268) shown in Fig. 3 (better see yz-projection). Obviously, 2. 266:277, and the branch is terminated. The grammar initiates the climb and move 1. ... 186-277 is changed for 1. ... 186-276 along trajectory ***a***(186)***a***(276)***a***(266). Note, that grammar "knows" that trajectory ***a***(186)***a***(276)***a***(266) is active in this state, i.e., B-INTERCEPTOR has enough time for interception. The following moves are in the same Zone of W-STATION: 2. 266-267 276:267. The "square rule procedure" cuts this branch and evaluates it as a win of the Black side.

New climb up to the move 2. ... 186-276 and execution of the inspection procedure result in the inclusion of the new control Zone from 227 to 267 in order to improve the exchange in the area 267. The set of Zones with different main trajectories from 227 to 267 is shown in Fig. 5. Besides that, the trajectories from 227 to 516, 616, 716, and 816 are shown in the same Fig. 5. These are "potential" first negation trajectories. It means that beginning with the second symbol ***a***(336), ***a***(337), ***a***(338), or ***a***(326), ***a***(327), ***a***(328), or ***a***(316), ***a***(317), ***a***(318), these trajectories become first negation trajectories in the Zone of B-STATION h5. Speaking informally, from the areas listed above W-INTERCEPTOR can intercept B-STATION (in case of white move). The main trajectories of control Zones passing one of three points, 336, 337, or 338, partly coincide with the potential first negation trajectories. The motion along such trajectories allows to "gain the time", i.e., to approach two goals simultaneously. The move-ordering procedure picks the subset of Zones with the main trajectories passing 336. Thus, 2. 227-336.

This way proceeding with the search we will generate the tree that consists of 56 moves (Fig. 6). Obviously, this is a drastic reduction in comparison with a billion-move trees generated by conventional search procedures.

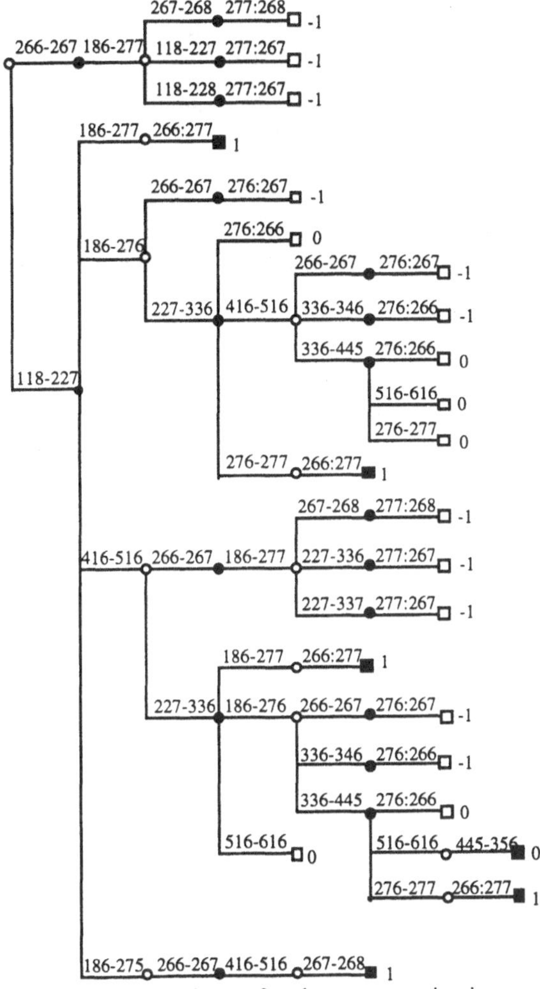

Figure 6: Search tree for the space navigation problem.

7 Discussion

The example considered in this paper demonstrates the power of the Linguistic Geometry tools that allowed to transfer heuristics discovered in one problem domain, specifically, in the game of chess, to another domain of simplified aerospace robotic vehicles. It is even more interesting that search reduction achieved in the original domain multiplied tremendously in the new domain. Indeed, the conventional approaches employing search algorithms with alpha-beta pruning require a billion move tree to solve 3D problem, while the tree presented in this paper (Fig. 6) consists of about 50 moves and is very similar with the tree from the 2D case [29]. Looking at the complexity of the hierarchy of languages which represents each state in the search process, it is very likely that the growth from the 2D case to 3D is linear with the factor close to one. This means that the complexity of the entire algorithm may be about linear with respect to the length of the input.

At the same time the simplified navigation problem considered here is still very close to the original chess domain. It is possible to predict that the power of Linguistic Geometry goes far beyond these limits. The definition of the Complex System is generic enough to cover a variety of different problem domains. The core component of these definition is the triple X, P, and R_p. Thus, looking at the new problem domain we have to define X, the finite set of points – locations of elements. We do not impose any constraints to this set while the space operational district X considered in this paper as well as the original chess board have different extra features, e.g., 3D space connectivity, which is totally unimportant for these problems. Thus, in the real world problems we can consider X as a set of orbits where the elements are in constant motion with respect to each other. The set of elements P, e.g., movable units, in our problem is quite small, while their moving capabilities, binary relations of R_p, are non-sophisticated. Indeed, during one time interval our robotic stations and space interceptors can move only to the next area. Even in the game of chess the moving capabilities of different pieces are much more advanced. This is exactly the place for introduction of the variable speed, the gravity impact, the engine impulse duration, etc.

Also, it should be noted that we introduced some extra constraints for the Complex System in examples considered in this paper. These are requirements of the motion alternation for the opposing sides and participation of the only element in each motion (in the form of the restriction for each side to participate in both local operations simultaneously). This introduction was done only for a transparent display of ideas and advantages the Linguistic Geometry. The generic definition of the Complex System (Section 2) to be applied to the real world problems does not include these constraints.

References

1. Botvinnik, M.M. (1984) Computers in Chess: Solving Inexact Search Problems. Springer Series in Symbolic Computation, Springer-Verlag, New York.
2. Botvinnik, M., Petriyev, E., Reznitskiy, A., et al. (1983) "Application of New Method for Solving Search Problems For Power Equipment Maintenance Scheduling", Economics and Mathematical Methods, 6, 1030-1041 (in Russian).
3. Chapman, D. (1987) "Planning for conjunctive goals" Artificial Intelligence , 32 (3).
4. Chomsky, N. (1963) "Formal Properties of Grammars", in Handbook of Mathematical Psychology, (ed. R.Luce, R.Bush, E. Galanter), Vol.2, J. Wiley, New York, pp. 323-418.
5. Feder, J. (1971) "Plex languages", Information Sciences, **3**, 225–241.
6. Fikes, R.E. & Nilsson, N.J. (1971) "STRIPS: A New Approach to the Application of Theorem Proving in Problem Solving", Artificial Intelligence, **2**, 189–208.
7. Fu, K.S. (1982) Syntactic Pattern Recognition and Applications, Prentice Hall, Englewood Cliffs.
8. Garey, M.R. and D.S.Johnson D.S. (1991) Computers and Intractability: A Guide to the Theory of NP-Completeness, W.H. Freeman and Co., San Francisco.
9. Ginsburg, S. (1966) The Math. Theory of Context-Free Languages, McGraw Hill, New York.
10. Knuth, D.E. (1968) "Semantics of Context-Free Languages", Mathem. Systems Theory, **2**, 127–146.
11. McAllester, D.& Rosenblitt, D. (1991) "Systematic Non-Linear Planning", *Proc. AAAI-91*, 634-639.
12. McCarthy, J. (1980) "Circumscription–A Form of Non-Monotonic Reasoning", Artificial Intelligence, **13**, 27-39.
13. McCarthy, J. and Hayes, P.J. (1969) "Some Philosophical Problems from the Standpoint of Artificial Intelligence", Machine Intelligence, **4**, 463–502.
14. Mesarovich, M.D., Macko, D., and Takahara Y. (1970) Theory of Hierarchical Multilevel Systems, Academic Press, New York.
15. Narasimhan, R.N. (1966) "Syntax–Directed Interpretation of Classes of Pictures", Comm. of ACM, **9**, 166–173.
16. Nilsson, N.J. (1980) Principles of Artificial Intelligence, Tioga Publ., Palo Alto, CA.
17. Pavlidis, T. (1977) Structural Pattern Recognition, Springer-Verlag, New York.
18. Rosenfeld, A. (1979) Picture Languages, Formal Models for Picture Recognition, Academic Press.
19. Rozenkrantz, D.J. (1969) "Programmed Grammars and Classes of Formal Languages", J. of ACM, **1**,107–131.
20. Sacerdoti, E.D. (1975) "The Nonlinear Nature of Plans", Proc. Int. Joint Conference on AI.
21. Shaw, A.C. (1969) "A Formal Picture Description Scheme as a Basis for Picture Processing System", Information and Control,**19**, 9-52.
22. Simon, H.A. (1980) The Sciences of the Artificial, 2-nd ed., The MIT Press, Cambridge.
23. Stefik, M. (1981)"Planning and meta-planning (MOLGEN:Part 2)",Artificial Intelligence, **2**,141-169.
24. Stilman, B. (1985) "Hierarchy of Formal Grammars for Solving Search Problems", in Artificial Intelligence. Results and Prospects, Proc. of the Int. Workshop, Moscow, 63–72, (in Russian).
25. Stilman, B. (1993) "A Linguistic Approach to Geometric Reasoning", Int. J. Computers and Mathematics with Applications, **26**(7), 29-57.
26. Stilman, B. (1993) "Network Languages for Complex Systems", Int. J. Computers and Mathematics with Applications, **26**(8), 51-79.
27. Stilman, B. (1993) "Syntactic Hierarchy for Robotic Systems", Integrated Computer-Aided Engineering, **1**(1), 57-82.
28. Stilman, B. (1994) "Translations of Network Languages", Int. J. Computers and Mathematics with Applications, **27**(2), 65-98.
29. Stilman, B. (1994) "A Formal Model for Heuristic Search", Proceedings of the 22nd Annual ACM Computer Science Conf., Phoenix, AZ, March, 380-389.

PROVIDING INTELLIGENT SUPPORT FOR PREDICTING COSTS OF WASTE WATER TREATMENT SYSTEMS

Michael Wood, Yan Wan, Michael Kaye, Bruce Bowhill, Tony Pe
University of Portsmouth, Locksway Road, Milton, Southsea, Hants, PO4 8JF, UK.

Abstract. This paper explores some of the issues raised by the initial phases of a project to develop an intelligent system to predict the capital and operating costs of waste water treatment systems. Five "paradigms" for intelligent support systems are distinguished - expert systems, induction systems, decision support systems, education systems, and the use of consultants. The paper discusses the appropriateness of these paradigms for the project, and the structure of the intelligent support system.

Keywords. Waste Water Treatment, Costing, Expert Systems, Decision Support Systems.

1 Introduction: the context of the project

The aim of the project discussed in this paper is to build an intelligent support system to predict the capital and operating costs of waste water treatment systems. Such forecasts are needed by a variety of parties: for example financial planners, designers, operations managers, government monitoring authorities, and so on.

The methods used to predict costs of treatment plants depend largely on the degree of accuracy which is required and the amount of information which is available. For long term financial planning, there are simple mathematical models for predicting both capital and operating costs from a few parameters such as the volume of the tanks in the proposed treatment system (Water Research Centre, 1977). More accurate estimates - as, for example, might be required by the designer of a specific plant, can be made at different levels of detail. These are all supported by a number of computer systems: for example STOM (Spearing, 1987), ICEMATE and ICEPAC (for the civil engineering costs). All these approaches are complicated by differences between plants - eg some are new plants while others are extensions, some are in urban areas whereas others are in rural areas; changing economic conditions; and factors depending on local circumstances.

The production of a reliable estimate needs expertise from a variety of different disciplines - accounting, quantity surveying, statistics, engineering and computer science - to name a few. In practice, of course, few people are experts in all of these areas - which means that an intelligent computer support system is likely to be useful to remedy any deficiencies.

We have interviewed a number of people in different roles (eg quantity surveyors, designers, managers, asset planners) in our case study organisation, both to learn about the problem domain and methods which are used to predict costs, and to explore some of the perceived difficulties in the current system. Our observations (in addition to the background outlined above) include:

1. The software support systems currently used in our case study company are all perceived as being difficult to use: a period of days or even weeks is seen as being necessary to achieve an adequate working knowledge of how to use these systems. Furthermore, they typically just cover a very small area of the general problem (eg civil engineering costs of particular components at a particular level of detail), and

E. A. Yfantis (ed.), Intelligent Systems, 55–59.

provide very limited guidance to the user or facilities for handling problems such as missing data.

2. Different individuals are expert in different areas, which means that arriving at overall cost predictions is likely to involve input from several different people.
3 Some estimates are presented in the form of intervals (ie "between X and Y"). The impression we obtained was that, on occasions, the statistical methods used to obtain the estimates were neither understood nor trusted. This is in line with a considerable body of research which explores the difficulties users have with statistics - for example Altman and Bland (1991) review the difficulties experienced by the medical profession and Hoadley and Kettenring (1990) consider the difficulties of engineers and physical scientists.
4. Financial concepts - such as net present values and cost index numbers - appear to be adequately understood but issues such as confusions over whether a discount rate includes or excludes inflation suggests this may not be so.

2 Paradigms for management support systems

We turn now to a discussion of some of the paradigms guiding the nature and development of support systems for management.

The expert system paradigm. The basic assumption here is that the aim is to mimic the expertise of human experts with a view to building a computer system which incorporates as much of this expertise as possible. The resulting computer system then gives advice, or takes action, in much the same way as an expert would. Most expert systems are probably advisory systems which give their advice by means of an interactive dialogue with the user. There is usually emphasis on the ability to *explain* advice in the same way that a human expert would - because only then will the system be credible and likely to be trusted and used in practice.

The induction system paradigm. Induction is the process of generalising from particular instances to general statements. The standard example in philosophy texts concerns the process of induction from observations of a number of white swans to the generalisation "all swans are white". The problem with this conclusion is that it is not correct: some swans are in fact black. Induction provides a useful method of reasoning but it is not - and cannot be - infallible.

Automating this process is an attractive possibility because this may provide a means of avoiding the problems of knowledge acquisition, and may indeed provide conclusions which go beyond the capabilities of human experts.

There are a number of approaches to this problem: case-based reasoning (Ketler, 1993), artificial neural networks, various "induction algorithms", and statistical techniques like multiple regression can be viewed as a means of generalising a general relationship from a sample of data.

The decision support system (DSS) paradigm. Another possibility is that the computer is used as tool to support the decision maker by "enabl[ing] the user to access data and/or models so that he or she may make better decisions" (Edwards, 1992, p. 115). The models in question are often mathematical ones for optimising certain variables. The critical difference between this and the expert system paradigm is that the "human queries the machine" (Turban, 1990, p. 22) and so controls the dialogue and the process of decision making. In practice the distinction between decision support and expert systems is inevitably hazy, but in principle in a decision support system the *user* would choose appropriate models and data - probably from a menu of possibilities, whereas in a expert advisory system, appropriate models would be suggested by the computer.

The training/education system paradigm. This would not normally be included in a list such as this. However, an obvious reaction to the question: "What support do people need to help them predict costs" is "They need training or education in appropriate areas." In principle the training requirements could be analysed, and appropriate training set up. The difficulties are fairly obvious: the range of disciplines needed (eg engineering, quantity surveying, accounting, mathematical modelling, statistics, at least); the time this would take; the fact that learning complex skills is a very time consuming and often ineffective process; and so on. However, this does represent an alternative support system to, for example, an expert advisory system.

The consultant paradigm. The final paradigm is the use of consultants in appropriate areas to support managers, designers and financial planners. This would involve employing the experts - accountants, statisticians, mathematical modellers, etc - directly, instead of eliciting their expertise and building it into a computer system. Again, from a practical point of view (as opposed to the artificial intelligence perspective), it is obviously important not to forget this possibility: some expertise may be so difficult to model, or be of such limited generality, that the effort involved in building a computer support system would not be justified.

These are all frameworks within which our task can be viewed. They are helpful in that they suggest an approach to follow, but the danger is that the assumptions of any one of these paradigms may be unnecessarily restrictive. For example, within the standard "educational" paradigm, a student's failure to use statistical methods correctly would be assumed to be due to either the student's incompetence or to inadequate pedagogy: this suggests that the remedy is to bring in more intelligent students or to improve teaching and learning methods. This ignores the possibility of improving computer support systems, or of redesigning the statistical concepts to make them more transparent (Wood, 1992).

The expert system paradigm assumes that experts are available, that their knowledge is the best source of expertise and can be elicited and implemented on a computer, and, perhaps most questionably, that mimicking human experts is an appropriate role for an expert system to try to fill. Even if this mimicking is perfect, users' interactions with a computer system are inevitably different from their interactions with a human expert, so the role of the expert system is inevitably different from that of the expert (Edwards, 1992; Collins, 1990). In practice, of course, this mimicking is often far from perfect: in order to make reasonable progress with a difficult problem there is a tendency for expert systems to restrict the boundary of their domain somewhat arbitrarily and assume a "closed-world" (Edwards, 1992) which ignores, for example, the wider organisational implications of the advice given. Both of these points: the fact that computer systems are inevitably treated differently from human experts, and the fact that they tend to have a restricted perspective, mean that the assumption that expert systems are a copy of a human expert is often very inaccurate.

Many authors have argued (eg Collins, 1990) that computers and human beings have different strengths; that humans can do things that computers cannot and vice versa, and so devising computer systems which copy human experts is not a sensible strategy. Human beings inevitably have more experience of the context of the decision and so are likely to take more relevant factors into account than the most sophisticated expert system. In addition, we have been able to improve on the expertise of our experts: as is often the case modelling expertise encourages and enables the knowledge engineer to improve that expertise.

King (1993, pp. 36-7) lists a number of problems which "stereotypical" expert systems are likely to encounter. These include problems of communication - both in terms of the language used for dialogue and recommendations (see also Wood, 1989), and the level of detail - and the fact that expert systems tend to be designed to achieve a single definite goal, whereas the needs of the user may be to "obtain contextual information, to focus attention on important topics, to help predict outcomes, and to criticise the solutions offered by others" (King, 1993, p. 37). This latter problem is particularly serious in the present project with its multiplicity of possible end-users (eg designer, financial planners, etc). King also mentions the possibility that a system which appears too prescriptive will be perceived as taking over control from the user, and may be resented or ignored for this reason. And, in addition, most current expert systems are considerably less impressive in practice than in initial concept: an expert system may not be capable of achieving its goal, and of explaining its reasoning sufficiently clearly to persuade the user to trust it.

There are also problems with the other paradigms, which cannot be reviewed in detail here. For example, there are strong arguments that successful generalisations, as for example in science, are often produced by inspired guesswork followed by systematically testing the resulting hypothesis (Popper, 1980), rather than by a systematic process of induction; some methods of induction make it difficult to provide explanations (eg artificial neural networks) - which are likely to be important for persuading users to trust conclusions; decision support systems may require too much expertise of their users to be helpful in a practical context; education and training schemes are notorious for being ineffective and failing to teach students what they "really" need to know; and consultants have a reputation for providing obvious answers at great expense which are not accepted because they come from outsiders.

Despite these points, all these paradigms may have something to offer, but probably not a complete solution. There would appear to be advantages in drawing from all of them, and severe disadvantages from remaining too firmly in one paradigm. According to Edwards (1992): "The majority of the literature treats DSSs and ESs as essentially different, and therefore separate, rather than dealing with them as parts of an integrated whole .... many of those in the ES 'camp' do not seem to be aware of the existence of DSSs." Despite this, attempts have been made to link the two paradigms of decision support systems and expert systems (Klein and Methlie, 1990; Edwards, 1992; Turban, 1990), and much of this work would probably incorporate induction systems too. We would also like to include aspects of the other two "paradigms" above: in practice this means that we do not rule out the possibility of using human expertise directly in parts of our system (some expertise may be too difficult or expensive to automate), and also some user education or training may be necessary if some concepts turn out to be too subtle to be "explained" by an advice system - possible examples are net present value and the discount rate, and some statistical concepts (Wood, 1992).

In this project we intend to combine aspects of all five paradigms, and also take account of the issues raised by the fact that our project does not have a single pre-defined customer or end-user, but rather a range of possible users.

3 The proposed structure for the cost prediction system

The proposed system system comprises a number of interconnected modules - for example there is a module for estimating the size of the required system, three modules for estimating the civil engineering costs at differing levels of detail, and so on. Most of these modules are likely to be expert systems or decision support systems. Some of the modules may also make use of induction systems: we are exploring the possibility of using case-based reasoning to draw conclusions from data on costs of past projects. It is also likely that we may decide that it is not cost effective to build a computer system for some of the modules. The module for predicting land prices may be one such example because these are likely to be dependent on the particular local circumstances and so any knowledge base would be of very limited generality; a human consultant would then be the preferred option for this part of the system. Finally we may have to borrow from the education paradigm if the final system requires, for example, any subtle statistical concepts with which users are not likely to be familiar. In this way it is likely that the final system will draw from all five of the paradigms discussed above.

There are various approaches to integrating an intelligent system comprising a collection of independent parts. At one extreme is distributed artificial intelligence: "agents are grouped together to form communities which cooperate to achieve the goals of the individuals and of the system as a whole" (Jennings, 1993, p. 224). This is an inappropriate tactic for this project because we are not trying to develop an entirely artificial intelligence: human beings will have a role to play. The notion of an intelligent user interface which helps users interact with the different modules to achieve their goals (Zhang, Nealon and Lindsay, 1993) is a more useful one. This interface will need to incorporate expertise on such matters as net present value and uncertainty management.

4 References

Altman, D. G., & Bland, M. J. (1991). Improving doctors' understanding of statistics. J. R. Statist. Soc. A, 154(2), 223- 267.

Collins, H. M. (1990). Artificial experts: social knowledge and intelligent machines. : MIT Press.

Edwards, J. S. (1992). Expert systems in management and administration - are they really different from decision support systems? European Journal of Operational Research, 61, 114-121.

Hoadley, A. B., & Kettenring, J. R. (1990, August). Communications between statisticians and engineers/physical scientists. Technometrics, 32(3), 243-247.

Jennings, N. (1993). Commitments and conventions: the foundation of coordination in multi-agent systems. The Knowledge Engineering Review, 8(3), 223-250.

Ketler, K. (1993). Case-based reasoning: an introduction. Expert Systems with Applications, 6, 3-8.

King, D. (1993). Building computerised financial advisors: the user model and human interface. in P. R. Watkins, & L. B. Eliot (eds), Expert systems in business and finance: issues and applications,Chichester: Wiley.

Klein, M., & Methlie, L. B. (1990). Expert systems: a decision support approach. Wokingham: Addison-Wesley.

Popper, K. R. (1980). The logic of scientific discovery. London: Hutchinson.

Spearing, B. W. (1987). Sewage treatment optimisation model - STOM - the sewage works in a personal computer. Proceedings of the Institution of Civil Engineers, 82(1),

Turban, E. (1990). Decision Support and Expert Systems: Management Support Systems. New York: MacMillan.

Wood, M. (1992). Using spreadsheets to make statistics easier for novices. Computers Educ., 19(3), 229-235.

Zhang, X., Nealon, J. L., & Lindsay, R. (1993). Intelligent user interface for multiple application systems. in M. A. Bramer, & R. W. Milne (eds), Research and development in Expert Systems IX, (pp. 301-316). Cambridge: CUP.

Planning Text for Interactive Plan Explanations *

Susan Haller
Computer Science and Engineering Dept.
University of Wisconsin – Parkside
Kenosha, WI 53141 USA

Abstract. The Interactive Discourse Planner (IDP) plans text to justify and describe domain plans interactively. A speaker uses plan justification to have a listener adopt a recommended plan; he uses plan description to enable a listener to execute a plan. The text plan that IDP formulates and executes incrementally is represented uniformly in the same knowledge base with the domain plans that are under discussion. In this way, the text plan and the domain plans are both accessible for analyzing the listener's feedback. IDP can interpret vaguely articulated feedback, generate concise replies and metacomments, and detect feedback that initiates digressions. This paper focuses on how IDP plans.

Keywords. Interactive Discourse, Text Planning

1 Introduction

In highly interactive settings, people often are called upon to analyze and respond to vaguely articulated questions like *Why?* and *What?*, and ill-formed queries like *Why take Maple?* and *What about Sheridan Drive?*. Listeners use these kinds of utterances to tell speakers efficiently what part of their communicative plan has failed, and how it can be replanned or expanded to succeed.

The Interactive Discourse Planner (IDP) plans text to justify and describe domain plans interactively. A speaker uses plan justification to have a listener adopt a recommended plan; he uses plan description to enable a listener to execute a plan. The text plan that IDP formulates and executes incrementally is represented uniformly in the same knowledge base with the domain plans that are under discussion. In this way, the text plan and the domain plans are both accessible for analyzing the listener's feedback. IDP can interpret vaguely articulated feedback, generate concise replies and metacomments, and detect feedback that initiates digressions. This paper focuses on how IDP plans.

As a testbed for the model, I have implemented a system that gives driving directions and route advice interactively. For giving driving route directions and advice, the system does not have a preset agenda. It describes routes and/or justifies a route choice until the

*I would like to thank my advisor, Stuart C. Shapiro, and the members of the SNePS Research Group in the Computer Science Department at the State University of New York at Buffalo. Their advice and comments are reflected in the research that this paper describes.

E. A. Yfantis (ed.), Intelligent Systems, 61–67.

listener's feedback indicates that she is satisfied. Therefore, IDP plans text in reaction to feedback that it uses to try to recognize an expansion of its text plan.

2 Motivation

Van Kuppevelt has developed a uniform conception of sentence topics (S-topics) and structurally higher-order discourse topics (D-topics) to define the latter in terms of the former [11]. He characterizes a ***topic*** T_p in terms of a contextually induced explicit or implicit question that he calls a ***topic-constituting question***. The *comment* on the topic, C_p, is the answer to the question.

Van Kuppevelt argues that an unsatisfactory answer to a question induces subquestions. When a question that constitutes topic T_p has been answered unsatisfactorily, it elicits subquestions thereby constituting subtopic T_{p_1} and so on. This process can continue recursively until the original, topic-constituting question is answered to the questioner's satisfaction. For example, in the following dialogue:

F_1 A: Tomorrow is Harry's birthday.
T_1 B: What would be a suitable present for him?
C_1 A: A suitable present for him would be a monkey-wrench.
  T_{1_1} B: What's that?
  C_{1_1} A: That's some kind of tool with which one can loosen or tighten bolts of various sizes.
  T_{1_2} B: Why would that be a suitable birthday present for him?
  C_{1_2} A: He recently came to borrow one from me.

a topic T_1 is introduced as a result of B's question at line 2. T_1 is that which is questioned, namely the undetermined present that is suitable for Harry. The answer to the question provides the comment C_1 on the topic T_1. It specifies the present. Question F_1 is a linguistic feeder. A feeder F_n is a linguistic or non-linguistic, topicless event that initiates the process of questioning in discourse.

Subquestions have a conjunctive property that is relevant to how IDP plans: the answers to all of them, in the order asked, constitute an appropriate and coherent answer to the original question. This point can be illustrated with the last example

F_1 A: Tommorrow is Harry's birthday.
T_1 B: What would be a suitable present for him?
C_1 A: A suitable present for him would be a monkey-wrench.
  A: That's some kind of tool with which one can loosen or tighten bolts of various sizes.
  A: He recently came to borrow one from me.

In this version of the interaction, A's answer anticipates B's questions thereby providing a satisfactory answer in the first pass. Because of the conjunctive property of subquestions, van Kuppevelt argues that topic-constituting questions can be implicit.

Topic-constituting questions restrict the development of a discourse. For van Kuppevelt, this restriction implies a program that must be followed for the discourse to come to a satisfactory end. This program is carried out by the question-answerer, and consists of providing an answer to the original topic-constituting question that is satisfactory. In the role of the question-answerer, IDP can carry out this program in one of two ways. One way is by anticipating the listener's subquestions and planning to answer them as part of the initial reply. The other way is by planning simple responses, and then planning additional responses in reaction to the listener's explicit subquestions.

3 Related Work in Text Planning

3.1 Text Planning and Hierarchical Planners

The STRIPS-style hierarchical expansion planner has the principal model for planning text [1, 3, 2, 6]. Shapiro notes that in hierarchical expansion planners, the plan operators and the procedural critics are written using a different formalism from the states of belief on which they operate [9].

The SNePS Actor is modeled as a cognitive agent operating in a single-agent world [4]. It integrates inference and acting by representing beliefs, plans, and acts as structured intensional entities in the SNePS semantic network knowledge representation and reasoning system [10, 8]. The agent has access to its concepts (represented as mental objects) and its beliefs about the world (stored as SNePS propositions). The agent's concepts include plans, acts, and goal states, and among the agent's propositional beliefs are ones about how these concepts relate to each other.

3.2 Related Work In Interactive Text Planning

There are two basic strategies for planning text for interaction. The first is to plan text to achieve a discourse goal, and then to formulate further text plans in reaction to the user's response. The Explainable Expert System (EES) Text Planner, is the primary example [6]. The second strategy involves formulating a single text plan in advance and then incrementally executing it. This is the approach used when discourse goals are known in advance. The tutorial system, the Explanatory Discourse Planner (EDGE), is an example of this approach [2].

IDP incorporates features of both approaches. Like the EES Text Planner, IDP plans only as needed in reaction to feedback. However, like EDGE, each plan that it formulates is incorporated into a single text plan that is used as the discourse context. This makes it possible for IDP to analyze feedback efficiently using techniques from plan recognition. The single text plan also contains the information that IDP needs to generate the kinds of concise, fragmentary replies that are appropriate in highly interactive discussions.

4 The Explanation Assumptions

IDP works in a collaborative, interactive mode in which the system is the primary speaker, the user is the primary listener, and the system is the uncontested expert. The formulation of text plans and the processing procedures for this mode of interaction rely on two simplifying assumptions that explanation systems typically make. I refer to these as the *explanation assumptions* (EAs):

EA-1: The explanation facility's knowledge is correct.
EA-2: The listener automatically believes what the explanation facility informs him of.

5 IDP's Goals and Plans

5.1 The Kinds of Goals

In IDP a *discourse goal* (DG) is a goal that has to do with the attitude or abilities of the listener. With respect to plan discussions, IDP can have one or both of the following DGs:

to have the listener adopt a plan ?p
to have the listener be able to execute a plan ?p

IDP may have to replan and expand its TP to achieve a DG.

A *content goal* (CG) specifies a system goal to have the listener know a proposition. By the Explanation Assumptions, the listener automatically believes what he is told. Therefore the effects of IDP's TPs are guaranteed. Since CGs match the effects of TPs, a CG is achieved immediately when IDP executes a TP that has the CG as one of its effects.

5.2 The Kinds of Text Plans

IDP's TP-operator's are based on Rhetorical Structure Theory (RST) [5]. In each rhetorical relation, there is one text unit which is essential; this is the *nucleus.* Other text units, which are not central, but which are related to the nucleus through a rhetorical relation, are called *satellites.* Given a nucleus, an IDP TP-operator selects a proposition for the satellite of a rhetorical relation.

Mann and Thompson describe a two-way division of the rhetorical relations. Each *presentational relation* relates two text spans for the purpose of increasing an inclination in the hearer. In contrast, a *subject-matter* relation is used by a speaker to inform the hearer of the rhetorical relation itself. I use the above classification of rhetorical relations to distinguish two kinds of TPs: *discourse text plans* (DTPs), and *content-selection text plans* (CTPs). 193z IDP uses DTPs to attempt to achieve its DGs. Since these goals have to do with affecting the attitudes and abilities of the listener, DTPs are based on speech acts and presentational rhetorical relations. The DTPs describe how to try to achieve DGs by selecting some minimal text content. IDP can augment this content with additional content by using one or more CTPs. CTPs are based on subject-matter rhetorical relations. TPs can be used to inform the user of additional propositions. IDP's high-level TP is always a DTP that can include other DTPs and CTPs. The TP always bottoms out in one primitive act, *say.* The argument to the say act is a text message that includes the proposition that is to be expressed.

6 The IDP Planner-Actor

To plan text interactively, I have modified the SNePS Actor in three ways. First, when presented with a choice of TPs, IDP prefers simple TPs to more complex ones. The only time that a complex TP is preferred, is when there are one or more active CGs that the complex TP achieves in addition to the DG that is being planned for. Secondly, IDP can use the effects of its text planning to constrain and focus further text planning. Thirdly, IDP can use its focusing mechanism to plan in two different modes.

6.1 Focusing Text Growth

Growth points are references to other TPs that are embedded in the body of each TP. Growth points suggest ways of adding content that augment the current TP. Hovy notes that treating growth points in a TP as *suggestions* for adding text rather than as *injunctions* makes the difference between a TP that is truly a plan, and a TP that is, in reality, a schema [3]. Hovy's RST planner currently treats growth points as injunctions.

In IDP focus is not solely a function of the current discourse entity. It is a function of the entity and aspects of it that are highlighted by the discourse so far. The entities that IDP is concerned with are propositions about plans and plan reasoning. When a proposition becomes known to the listener, it highlights other propositions as candidates for the subsequent expression and focus of attention. The candidates are screened using the TP that IDP has constructed and executed. To obtain the candidates, the IDP uses the *active path* of its TP and the associated *localized unknowns*.

6.2 Planning Text as Needed

In the IDP model, questions address the system's TP by making reference to what the system is doing to achieve its DG. Therefore, IDP uses feedback to try to recognize a TP to expand its current DTP. Following van Kuppevelt, the *local topic* is that which is questioned. IDP determines a local topic to search for a comment on it.

IDP processes questions in one of the following forms:

1. Why {not}?
2. Why {not} *plan*?
3. Why {not} *proposition*?

{not} indicates that the word "not" is an optional constituent of the input string. The local topic is a domain act or proposition. When there is a simple why-question (1.), the local topic is the last proposition that was expressed by the system with a say-act. If the question is in the form of 2. or 3., IDP makes the plan or proposition that is mentioned the local topic.

To select a TP-expansion, an important heuristic is the degree to which the proposed TP-expansion highlights the system's intent as realized by the DTP-level portion of its TP. Therefore, IDP considers the growth points for the DTPs on the active path in the following order:

1. DTPs that replan a DTP
2. CTPs that expand a DTP

IDP prefers DTPs that replan a DTP over CTPs that expand a DTP. This encodes a preference for plans that highlight the system's intent over plans that supply additional information.

Using its planning-replanning expansion procedure, IDP produces demonstration runs like the one that follows:

L:	`Should I take Maple or Sheridan to go to the Eastern-Hills Mall?`
IDP:	`take Maple.`
L:	`Why not take Sheridan?`
IDP:	`You could take Sheridan however,` `taking Maple avoids heavy traffic,`
L:	`Why?`
IDP:	`since taking Maple there are fewer businesses than taking Sheridan`
L:	`Ok.`

6.3 Anticipating Questions

In IDP's anticipatory planning mode, it anticipates the listener's implicit questions by collecting the candidates for answers to them. The potential answers are the localized unknowns. IDP finds the localized unknowns for each DTA before planning it, and sets CGs for the listener to know them.

In this mode, IDP produces recommendations like the one that follows:

L:	`Should I take Maple or Sheridan to go` `to the Eastern-Hills Mall?`

IDP:	`if now there was light traffic,`
	`you could take Sheridan however,`
	`now there is heavy traffic.`
	`you should take Maple.`
	`since taking Maple there are fewer`
	`businesses than taking Sheridan,`
	`taking Maple avoids heavy traffic.`

7 Status of Reported Work

IDP uses ten TP-operators to formulate TPs for justifying domain plan advice. After selecting and structuring content, IDP sends a text message to a generation grammar written in the Generalized Augmented Transition Network (GATN) formalism [7]. The same formalism has been used for the system's parser. IDP uses the SNePS knowledge representation and reasoning system as a platform running on Sun SPARC Workstations.

References

[1] D. E. Appelt. *Planning English Sentences.* Cambridge University Press, 1985.

[2] A. Cawsey. Generating explanatory discouse. In R. Dale, C. Mellish, and M. Zock, editors, *Current Research in Natural Language Generation.* Academic Press, 1990.

[3] E. H. Hovy. Unresolved issues in paragraph planning. In R. Dale, C. Mellish, and M. Zock, editors, *Current Research in Natural Language Generation.* Academic Press, 1990.

[4] D. Kumar, S. Ali, and S. C. Shapiro. Discussing, using and recognizing plans in SNePS preliminary report - SNACTor: An acting system. In *Proceedings of the Seventh Biennial Convention of South East Asia Regional Confederation.* Tata McGraw-Hill, 1988.

[5] W. C. Mann and S. A. Thompson. Rhetorical structure theory: A theory of text organization. Technical report, Information Sciences Institute, 1987.

[6] J. Moore and W. Swartout. A reactive approach to explanation: Taking the user's feedback into account. In C. Paris, W. Swartout, and W. Mann, editors, *Natural Language Generation in Artificial Intelligence and Computational Linguistics.* Kluwer Academic Publishers, 1991.

[7] S. C. Shapiro. Generalized augmented transition network grammars for generation from semantic networks. *American Association of Computational Linguistics*, 8, 1982.

[8] S. C. Shapiro and The SNePS Implementation Group. *SNePS-2.1 User's Manual.* Department of Computer Science, SUNY at Buffalo, 1992.

[9] S. C. Shapiro, D. Kumar, and S. Ali. A propositional network approach to plans and plan recognition. In *Proceedings of the 1988 Workshop on Plan Recognition*, Los Altos, CA, 1989. Morgan Kaufmann.

[10] S. C. Shapiro and W. J. Rapaport. The SNePS family. *Computers & Mathematics with Applications*, 23(2–5), 1992.

[11] J. van Kuppevelt. About a uniform conception of S- and D- topics. In *Proceedings of the Prague Conference on Functional Approaches to Language Description*, Prague, 1992.

KNOWLEDGE-BASED UNCERTAINTY MANAGEMENT: AN IDSS APPROACH

N. AYRE & T. J. ANDERSON
Department of Information Systems
University of Ulster at Jordanstown
Newtownabbey, Co Antrim
BT37 0QB, N. Ireland

e-mail: CBBF23@UJVAX.ULSTER.AC.UK
fax number: +44 232 362803

Abstract

This paper presents MATUM an intelligent decision support system for decision making under uncertainty. It supports the acquisition and manipulation of both quantitative and qualitative business data, thus allowing the decision maker to work in a more natural and intuitive manner. MATUM has been successfully applied to the domains of retail outlet site selection and reviewing computer hardware supplier tenders. The system architecture is discussed in some detail and a worked example is given.

Keywords

Intelligent Decision Support Systems, Uncertainty Management, MATUM.

1 Introduction

It is widely recognised that uncertainty pervades every aspect of our personal and business lives. In our personal lives we deal with many uncertainties but are often unaware that this is what we are doing. Yet we live quite comfortably with day-to-day uncertainties, making plans and decisions in their presence. Granger Morgan and Henrion (1992) link this ability to human cognitive heuristics, strategies, technologies and institutions to accommodate or compensate for the effects of uncertainty. They cite weather reports, pocket-sized raincoats and insurance as prime examples. Uncertainty does not play any lesser role in our business lives. Failing to deal with uncertainty in this environment can however have far reaching consequences, both for decision-makers as individuals and the organisation as a whole. The cost of failing to address these uncertainties will ultimately be a deterioration in the quality of plans produced and decisions taken. To paraphrase Mack (1971), dealing with uncertainty is not a secondary issue on the road to responsible business decisions, it is central to it.

1.2 Paper Overview

This paper consists of five main sections, beginning and ending with an introduction and conclusion respectively. In section two we discuss the concept of an intelligent decision support system and briefly outline one approach to uncertainty management in such systems.

E. A. Yfantis (ed.), Intelligent Systems, 69–76.

Section three presents MATUM a small prototype developed to test and evaluate the ideas discussed in section two. This is followed in section four by a worked example.

2 Decision Making & Uncertainty

Decision making has conventionally been considered an art or talent, gradually acquired through years of experience (learning by trial and error). Turban (1993) suggests that this is due to the variety of individual styles used in approaching and successfully solving similar problems. Styles often based on creativity, judgement, intuition, and experience, rather than on systematic quantitative methods founded on a scientific approach. In order to handle the complexity inherent in current business contexts, computer-based tools are increasingly available:- **computerised management support systems**, such as decision support systems (DSS), knowledge-based systems (KBS), executive information systems (EIS) and artificial neural networks (ANN).

Many man-years of research have been devoted to developing these support systems as stand alone tools. However in recent years the emphasis has moved to integrating these technologies creating the new more powerful paradigm of the **Intelligent Decision Support System**. This paper discusses the integration of two of these techniques, namely **decision support systems** and **knowledge-based systems** (Turban, 1993), with the aim of building an **intelligent decision support system** (IDSS) (Wilson,Wilson & Smith, 1993; Kuratani, 1986; Sebastian, 1993). This involves adding intelligence to a traditional DSS by merging with a KBS.

It has been suggested (Sebastian, 1993) that a further enhancement to intelligence is the addition of uncertainty handling mechanisms. Given the many application problems associated with the traditional approaches (Ayre & Hughes, 1993) we adopted a pseudo-uncertainty approach in the form of **Multiattribute Utility Technology** (MAUT) (Edwards & Newman, 1990; Edwards, von Winterfeldt & Moody, 1989). Business data tends to be neither purely quantitative nor qualitative in nature but a mixture of both. We suggest that any approach to manage uncertainty in this type of environment should cater equally for both types of data. (Ayre, Anderson & Glass, 1994). In response to this observation MAUT has been enhanced to cater for both qualitative and quantitative data. Section 3.3 gives a detailed discussion on how the IDSS (MATUM) supports this approach.

3 MATUM - Multi-Attribute Tool for Uncertainty Management

The system architecture (Figure 1) has five main components: a knowledge-base or model repository, a knowledge input tool, a model evaluation tool, a sensitivity analyses tool and an explanation generator.

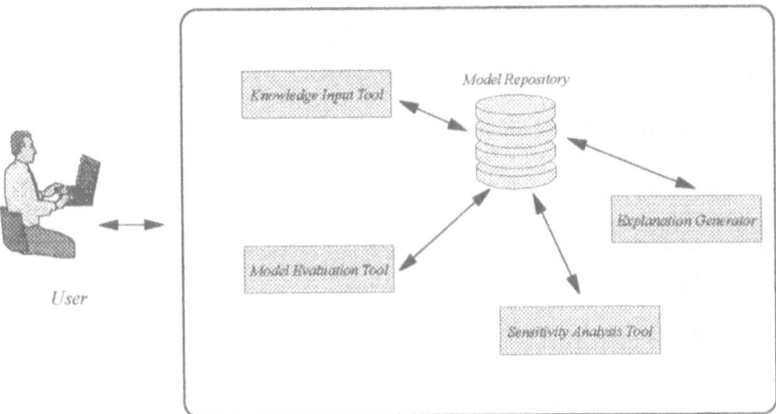

Figure 1 The Architecture of MATUM

3.1 The Model Repository

The Model Repository (knowledge-base) supports the development of new **Multiattribute decision models** (Edwards & Newman, 1990) and stores existing ones. A decision model consists of:

- A tree-like hierarchy of evaluation criteria or attributes, where attributes are related to each other in the traditional parent-child relationship. That is each decision has a number of major evaluation criteria or dimensions. Each of these 'parent' categories have a number of more specific areas of consideration.

- Appropriate weights associated with each attribute and a final normalised weight for each branch at twig-level. Some attributes are not as important as others, hence each attribute is assigned a weight which expresses its importance relative to the others. The weightings are obtained using rank sum weighting.

- A definition for every attribute included in the model. This description consists of one of four textual classifications coupled with a set of quantitative range variables. For example see Figure 2:

Attribute_Name : SIZE
Size_Definition : More Better Than Less
Size_MinPlaus : 60
Size_MaxPlaus : 160
...
Size_Rank : 10
Size_TotWeight : 0.055
Size_Weight : 0.333

Figure 2 An Attribute Definition

- One or more **decision options**, where a decision option represents one unique outcome of the decision making process. Each decision option holds the scores awarded to that option for each of the model attributes and a final overall score assigned to that option.

3.2 The Knowledge Input Tool

The knowledge-base is maintained through a Knowledge Input Tool (KIT). KIT allows the user to create and modify decision models, whilst still maintaining the integrity of the knowledge-base. Integrity is maintained through a series of validation and verification rules which interrogate all user input.

3.3 The Model Evaluation Tool

The Model Evaluation Tool (MET) 'runs' a model for any given set of decision options. In the early stages of a run MET is closely coupled with KIT. KIT prompts the user for actual values or scores for each twig-attribute for each decision option. MET contains a number of rules which use these actual values and the attribute's definition description to perform a numerical transformation on the values.

Given the nature of much business data, we recognise that it may only be possible to attach a realistic numerical score to some, but not all of these attributes. To cater for this the textual slot of the definition description allows for both subjective and objective attributes. If an attribute has a subjective textual definition (e.g. Judgmental) the user is given a verbal scale for example good, moderate, very poor and so forth. Once a selection has been made the system 'converts' this to a numerical score. If an objective definition (e.g. Less Better Than More) has been given the numerical transformation ensures that all scores relate to the same base line or have the same terms of reference. The choice of whether to provide a numerical score or use the semantic scale is left entirely with the decision maker, the idea being that they are the one with the domain expertise and knowledge.

The transformations, for those values with objective textual definitions, are achieved through the use of one of four equations: two linear and two bilinear.

Location of $L_A = 100\,(L_A - Min_{plaus}) / (Max_{plaus} - Min_{plaus})$
(More Better Than Less - Linear)

(i)

Location of $L_A = 100\,(Max_{plaus} - L_A) / (Max_{plaus} - Min_{plaus})$
(Less Better Than More - Linear)

(ii)

Location of $L_A = U_{max} + (100 - U_{max}) * (Max_{plaus} - L_A) / (Max_{plaus} - Min_{plaus})$
(Judgmental - Bilinear)

(iii)

Location of $L_A = U_{min} + (100 - U_{min}) * (L_A - Min_{plaus}) / (Max_{plaus} - Min_{plaus})$
(Judgmental - Bilinear)

(iv)

where:

L_A is the value to be transformed.
Min_{plaus} is the minimum plausible value (user defined).
Max_{plaus} is the maximum plausible value (user defined).
U_{min} is the minimum utility (user defined).
U_{max} is the maximum utility (user defined).

When the transformations are complete MET obtains the aggregate of the two sets of numbers held in the model: the importance weights, one for each attribute and the transformed scores awarded to each decision alternative on each attribute (Ayre & Hughes, 1993). This aggregate is achieved using a simple linear equation with the following mathematical description (Dawes, 1979):

$$Y = \sum_{i=1}^{n} W_i X_i$$

where:

Y is the final judgement score.
W_i is the weight attached to the ith attribute.
X_i is the score of the ith attribute.

Using these procedures the system forwards the 'best' decision option as the one with the highest Y value. Although more complex methods exist to achieve this aggregate research has shown that simple linear equations will produce as accurate results, making the use of more complex measures unwarranted for a system of this type.

3.4 The Sensitivity Analysis Tool

The Sensitivity Analysis Tool (SAT) supports all standard procedures relating to altering the values of individual attributes to determine the impact of that attribute. Its more attractive feature is the ability to 'split' models into mini-models which can be run as independent sub-decisions of the main decision to be taken.

3.5 The Explanation Generator

The Explanation Generator (EGO) is still under development but its task is to present the system findings in a more meaningful manner to the decision maker. It is envisaged that EGO will ultimately interface with domain specific knowledge-bases to assist with the interpretation of system answers.

4 A Simplified Site Location Example

Retailing involves selling goods and/or services to customers for personal or household use, and within these organisations store-planning is a key function. By store-planning we mean the process of identifying and explaining potential sites for locating new stores, which will provide the company with future growth opportunities. Based on an example in Mockler (1992), Figure 3 shows a simplified hierarchy for retail outlet site location.

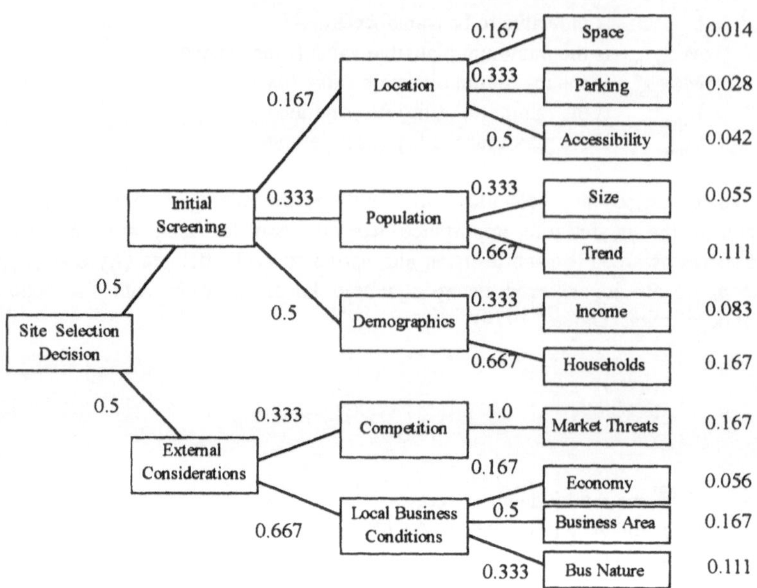

Figure 3 A Simplified Hierarchy for Retail Site Location

The figures represent the weights assigned to each attribute. For example Demographics has a weight of 0.5, whilst the final weight for Income is 0.083 (0.5 x 0.5 x 0.333). To assess each individual decision option the user scores each of the leaf nodes (Space, Parking etc.). Table 1 shows the results for one decision option for example to site the new store in 'Newtown'. Newtown has a population of 20,000, 12 miles from a city of 500,000.

ATTRIBUTE	WEIGHT	SCORE	W x S
Space	0.014	80	1.12
Parking	0.028	60	1.68
Accessibility	0.042	40	1.68
Size	0.055	70	3.85
Trend	0.111	95	10.55
Income	0.083	65	5.40
Households	0.167	90	15.03
Market Threats	0.167	50	8.35
Economy	0.056	80	4.48
Business Area	0.167	50	8.35
Bus Nature	0.111	100	11.10
		TOTAL =	**71.59**

Table 1 The results for Newtown

Newtown receives an overall score of 71.59 that is, the value for Y (the final judgement score) is 71.59. This score represents the aggregate of the two sets of numbers established through KIT: the importance weights, one for each attribute and the scores awarded to each decision alternative on each attribute. Using the procedures discussed above the 'best' decision is the one with the highest Y value. It should be noted that the decision to accept the system's findings or not, is up to the decision maker.

5 Conclusion

The system represents an intuitive approach to the problems of decision making under uncertainty and seeks to merge established techniques with the cognitive heuristics we use. This approach takes advantage of our natural ability to perceive relevant information and to generate creative options (Ezawa, 1993). In keeping with the traditional role of DSS and IDSS within business environments (Bidgoli, 1989), the system is not intended to take away or replace the decision maker's judgement, ability or experience, but rather to enhance it. MATUM covers each of Simon's (1977) three phases of decision making: Intelligence, Design and Choice. Decision making is an iterative process and very few decisions follow the sequence of intelligence, design and choice. MATUM fully supports decision making as an iterative process allowing the decision maker to express the flair, intuition, judgement and creativity at the heart of every good decision maker (Lucey, 1991). The underlying architecture has been generically designed and can be quickly applied to a variety of decision and planning problems (Ayre & Hughes, 1993). The system is currently being evaluated by a variety of personnel and whilst further development work will undoubtedly be required, early feedback is proving encouraging.

Acknowledgements

We gratefully acknowledge the financial support given by Digital Equipment Corporation and the Department of Education for Northern Ireland (DENI) throughout the course of this research project.

Bibliography

Ayre, N. & Hughes, J. G. (1993) "Integrating Uncertainty Management With An Automated Decision Support Tool", *Proceedings of the Higher Education Fundings Councils' Knowledge Based Systems Initiative*, (ed. E. S. Atwell), pp. 129-136, Cambridge, England.

Ayre, N., Anderson, T. J. & Glass, D. (1994) "Knowledge-Based Uncertainty Management", *Advances in Artificial Intelligence - Theory and Application II, Proceedings of the 7th International Conference on Systems Research, Informatics and Cybernetics*, (eds. J. W. Brahan & G. E. Lasker), vol. 2, pp. 43-48, Baden-Baden, Germany.

Bidgoli, H. (1989) *Decision Support Systems, Principles & Practice*, St. Paul:West Publishing Company.

Dawes, R. (1979) "The Robust Beauty of Improper Linear Models in Decision Making", *American Psychologist*, vol. 34, no. 7, pp. 571-582.

Edwards, W. & Newman, J.R. (1990) *Multiattribute Evaluation*, Sage University Paper series on Quantitative Applications in the Social Sciences, 07-026, Beverly Hills: Sage Publications.

Edwards, W., von Winterfeldt, D. & Moody, D. L. (1989) "Simplicity in Decision Analysis: An Example and a Discussion", in *Decision making, Descriptive, Normative, and Prescriptive interactions*, (eds. D. E. Bell, H. Raiffa & A. Tversky), pp. 443-464, Cambridge: University Press.

Ezawa, K. J. (1993) "A Normative Decision Support System", *Proceedings of AIEM'93*, Portland, Oregon.

Granger Morgan, M. & Henrion, M. (1992) *Uncertainty A guide to Dealing with Uncertainty in Quantitative Risk and Policy Analysis*, Cambridge: Cambridge University Press.

Kuratani, Y. (1986) "The Intelligent Decision Support System: Synthesis of a Decision Support System and an Expert System", *Proceedings of the 7th International Conference on Multiple Criteria Decision Making*, pp. 65-70.

Lucey, T. (1991) *Management Information Systems*, 6th ed., London: DP Publications.

Mack, R.P. (1971) *Planning on Uncertainty - Decision Making in Business and Government Administration*, New York: Wiley-Interscience.

Mockler, R. J. (1992) *Developing Knowledge-Based Systems Using An Expert System Shell*, New York: Macmillan Publishing Company.

Sebastian, H-J. (1993) "Intelligent Decision Support Systems", *Proceedings First European Congress on Fuzzy and Intelligent Technologies (EUFIT'93)*, vol. 1, pp. 99-307, Aachen, Germany.

Simon, H. A. (1977) *The New Science of Management and Decision*, Prentice Hall.

Turban, E. (1993) *Decision Support and Expert Systems: Management Support Systems*, 3rd ed., New York: Macmillan Publishing Company.

Wilson, F. A., Wilson, J. N. & Smith, A. M. (1993) "Computer-Based Systems: A Discussion of Their Application to Managerial Decision-Support", *Proceedings of The 1993 ACM SIGCPR Conference*, (ed. M. R. Tanniru), pp. 76-87.

Collective Learning as an Efficient Learning and Function Optimization Procedure

R. Salomon
International Computer Science Institute
1947 Center Sr., Suite 600
Berkeley, CA 94704-1198 USA

August 3, 1994

Abstract – Even though many strategies for function optimization have been developed, getting stuck in local minima is still a challenging problem. In this article we present a new procedure, *collective learning*, that based on genetic algorithms and steepest descent methods. We have tested our procedure on several neural networks and on several multimodal functions proposed by De Jong, Rastrigin and Schwefel. We attained an improved global convergence probability and a significant speedup compared to other procedures.

Keywords: Genetic Algorithm, Backpropagation, Multimodal Funcitons, Learning Procedure.

1 Introduction

Developing function optimization procedures is still an interesting area. Many methods have been developed, each with special advantages under certain circumstances [4]. However, getting stuck in local minima is a well-known problem of learning in neural networks and other kinds of optimization tasks. For the problems of interest one defines a fitness function (also called error or objective function) which is to be optimized. With respect to the fitness function, the term *local minimum* means that the fitness value is far from a known global value but no progress in the optimization process is attainable. A typical example on which the problem of getting stuck in local minima can be tested is the well-known exclusive-or. Here, a neural network has to learn the exclusive-or function $f(a,b) = \overline{a}b + a\overline{b}$. Using certain fixed network topologies, backpropagation or evolution strategies both will get stuck in an average of 20 % of trials.

When optimizing multimodal functions, one generally has to find a good tradeoff between fast local optimization with a high precision (e.g. hill climbing) and a high probability of finding the global optimum. For example, the steepest descent method may converge very fast but unfortunately merely to the next (local) optimum with respect to the (randomly) chosen initial point. Once getting stuck, there is almost no way for a gradient descent method to escape from such local optima. On the other hand, genetic algorithms are known to be able to come close to a global optimum but have trouble finding the exact objective value [1].

E. A. Yfantis (ed.), Intelligent Systems, 77–83.

Several strategies have been developed to deal with this problem of getting stuck in local optima. One approach is to use simulated annealing However, simulated annealing leads to notoriously slow convergence rates. Another common approach is to employ some heuristic. For example, if there is no further progress in convergence or the global minimum is not reached after a given number of iterations it is assumed that the process is stuck in a local minimum, in which case the procedure restarts with different initial conditions. The problem with this approach is that it is difficult to find a problem independent heuristic. Moreover, other approaches are trying to overcome local optima by multipopulations [3]. Here, several isolated populations perform a separate optimization process. At the end or after a certain number of time steps the best population is selected and the optimization process continues on the selected population. This approach suffers from the need of maintaining a large population of individuals, which results in large computing times. Moreover, the convergence speed attainable is slow. Mühlenbein [6] has developed the parallel genetic algorithm (PGA). This algorithm maintains a certain number of subpopulations each of which performs a local optimization for a given number of generations. At the end of such an isolated optimization process the best individuals are distributed into their vicinity. PGA runs well on parallel hardware and obtains good results in respect to global convergence. The drawbacks of this approach are the number of strategy parameters to be set a priori, and, in the sense of computational cost, the huge number of processors involved. Mühlenbein recently developed the Genetic Breeder Algorithm (BGA) [5] that is founded by a comprehensive theory. However, the BGA maintains very large populations for difficult problems.

2 Collective Learning Procedure

In [8] we proposed the *collective learning* procedure, a hybrid scheme which is loosely inspired by evolutionary considerations and provides a significant improvement in global convergence compared to other methods. Since then, we have extented the basic scheme by a genetic algorithm and it successfully applied to several function optimization problems. Collective learning is a hybrid method which is based on genetic algorithm or evolution strategy [7] respectively, and a self-adapting backpropagation algorithm [9] working as a local optimization procedure. The evolution strategy is very similar to genetic algorithms except that each object is specified by a set of floating-point values. In addition, the evolution strategy maintains a step size for each object. Each offspring inherits the step size from its parents, modifies it randomly (by a factor of 1.3, 1.0, or 0.7), and adds to each floating-point value a gauss random number multiplied by the step size.

The basic idea of collective learning is that the evolution strategy part is responsible to adapt learning parameters (eg. learning rate η, momentum α and perhaps others) as well as the initialization of the object variables and that backpropagation or any other local optimization procedure allows each individual to perform local adaptation steps.

To establish some biological components, collective learning maintains *one* population of size p. Then, in each generation it produces o offspring and determines their initial object variables / weights as well as local learning parameters by means of mutation and multiple-point crossing-over. Before collective learning makes the necessary selection, it gives each individual time to perform *exactly one local adaptation step*, which in our current approach is identical with one backpropagation step.

	Function	Limits
1	$f_1(\vec{x}) = \sum_{i=1}^{3} x_i^2$	$-5.12 \leq x_i \leq 5.12$
2	$f_2(\vec{x}) = 100\,(x_1^2 - x_2)^2 + (1 - x_i)^2$	$-2.048 \leq x_i \leq 2.048$
3	$f_3(\vec{x}) = \sum_{i=1}^{5} \text{integer}(x_i)$	$-5.12 \leq x_i \leq 5.12$
4	$f_4(\vec{x}) = \sum_{i=1}^{30} i\, x_i^4 + \text{Gauss}(0,1)$	$-1.28 \leq x_i \leq 1.28$
5	$f_5(\vec{x}) = 0.002 + \sum_{j=1}^{25} \frac{1}{c_j + \sum_{i=1}^{2}(x_i - a_{ij})^6}$	$-65.536 \leq x_i \leq 65.536$
6	$f_6(\vec{x}) = 20 + \sum_{i=1}^{20} x_i^2 - \cos(2\,\pi\, x_i)$	$-5.12 \leq x_i \leq 5.12$
7	$f_7(\vec{x}) = \sum_{i=1}^{10} -x_i\, \sin(\sqrt{\lvert x_i \rvert})$	$-500 \leq x_i \leq 500$

Table 1: Seven function test bed. The limits given above have been introduced by former authors and where used throughout this paper to give a better comparison.

In the subsequent *selection phase* the procedure selects the best p individuals according to the following scheme: For each individual the procedure maintains an *age count* and it calls an individual *young* iff the age count is below a given boundary. In the first stage, the procedure selects all young individuals and then, in the second stage, it uses the current fitness – with respect to the given task – to select further objects. Note that the individuals perform only *one step* and not, as in several other approaches, the total or a major part of the whole learning procedure. This sort of hybrid optimization process is known as Lamarckian learning, and the main idea behind this approach is that young and promising individuals with promising features have enough time adapting to the given task before they feel any selection pressure.

By means of these principles the procedure escapes from local minima when applied to the examples discussed below, and often converge to the global minimum.

3 Function Test Bed

This paper is about function optimization and particularly interested in multimodal functions. To this end, we used a wide range of test functions which are summarized in table 1. These test functions have been proposed to measure the performance of optimiziation procedures (especially genetic algorithms) and all functions differ in there degree of difficulty. The first five test functions have been introduced by De Jong [2]. The fifth De Jong function is known as pretty hard, because it contains 25 local optima each at $\vec{x} = (a_{1j}, a_{2j})^T$ with a corresponding function value of $f_5(\vec{x}) = 0.002 + 1/c_j$. De Jong originally defined $F_5 = 1/f_5$. However, most authors use f_5 (probably due to a typo). So do we, to allow a better comparison to others.

Function f_6 was proposed by Rastrigin which is highly multimodal and has its minimum at $\vec{x} = \vec{0}$ and $f_6(\vec{x}) = 0$. Function f_7 was proposed by Schwefel. Optimizing this function, one gets the problem that the global minimum is given by $\forall\, i\ x_i = 420.9687$ and that the second best minimum $x_j = -302.5253, \forall\, i\ i \neq j, x_i = 420.9687$ is far away from the global one. A one-dimensional version is printed in figure 1.

4 Results

We have applied the new collective learning procedure to the XOR problem mentioned above as well as to several other test functions summarized in table 1. Unless otherwise stated, we used in all trials a population size of $p = 30$, $o = 5$ offspring, and a bound of age = 5 for young objects.

Applied to the XOR problem the procedure converges in all of the 100 trials to the global minimum in an average of only **11** generations, which is very fast.

Table 2 presents a comparison between results obtained by De Jong's plan R1, Mühlenbein's PGA, and collective learning. Each row presents the number of generations needed to converge to the global minimum. One can see that our collective learning procedure is more than two orders of magnitude faster than De Jong's plans and we are surprised that it always found the global minimum at the fifth function, because this function contains 25 sharply local minima. If one compares PGA with collective learning then one should consider that PGA uses 80 instead of 30 individuals per generation respectively.

In table 3 we present a comparison of the number of function evaluations needed when applying PGA, BGA, and collective leaning to Rastrigin's function f_6. In this case collective learning (CL^{GA}) used a genetic algorithm insted of the evolution strategy. The genetic algorithm had the following parameters: each bit of the mantissa is toggled with a probability of 0.1, a crossover probability of 0.65, and a recombination probability of 0.5. For the cases $n \geq 50$ we chose $p = 50$, $o = 8$, and age = 5. One can see that collective learning is up to two orders of magnitude faster than PGA. In addition, our procedure needed approximately 25 minutes on a SUN SPARCstation 2 whereas PGA needed 81 minutes on a parallel hardware with 64 processors.

The results for Schwefel's function f_7 are not of that high quality. Mostly, after 500 generations all values but one have the desired value (420.9); the remaining one has a value of -302. It is very easy to explain this behavior. Collective learning performs its optimization implicitly on an envelope of the function of interest. Thus, it works very well on Rastrigin's function and has problems when a function, like Schwefels function f_7, has a chaotic distribution of the optima. The BGA overcomes this problem by maintaining a *huge* population of 500 individuals when merely optimizing 20 variables. Using such a large

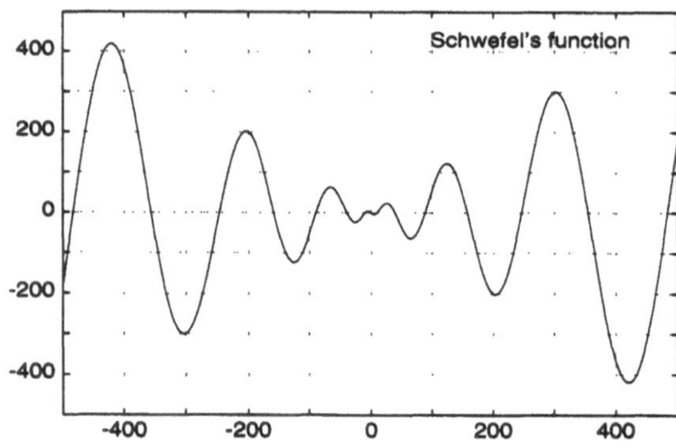

Figure 1: One-dimensional version of Schwefel's function $f_7(\vec{x}) = \sum_{i=1}^{1} -x_i \sin(\sqrt{|x_i|})$.

population, collective learning needs about 300 generations to find the global optimum. To attain a unimodal envelope of Schwefel's function we set the lower limit to 0 in further experiments; then collective learning needed about 360 generations ($p = 30$).

5 Enhancements

In this section, we describe a few enhancements for the evolution strategy variant of collective learning (collective learning^{ES}).

Due to its dynamic adaptation of the step size, the evolution strategy generally leads to a fast convergence. However, this adaptation can lead to getting stuck in a local optimum that is very close to the global one. For example, applying the collective learning^{ES} to Rastrigin's function f_6 it might be that a few object variable end up with a value of one instead of zero.

The use of a constant or minimal step size would overcome this problem; but this requires a priori knowledge about the function to be optimized. To avoid this, we developed the concept of *escaping individuals*. With a certain probability (e.g. 10 %) collective learning produces offspring which tend to go in the wrong direction. I.e., in case of finding the minimum they maximize their fitness value by approaching the next nearest maximum. This concept leads to a significant improvement. Applying to Rastrigin's function f_6 with $n = 20$, we obtained an average function value of 1.3 after 500 generations.

In a theoretical analysis [8] we have estimated the probability p_{cl} of attaining progress in the next generation with the following formula: $p_{\mathrm{cl}} = \int_{a_{\mathrm{cl}}}^{b_{\mathrm{cl}}} \mathrm{gauss}(x)$, whereas a_{cl} and b_{cl} are the coordinates of the next nearest maxima. This analysis suggests to use different random generators. Among others, we tried the following random generator:

$$y = -1^{\alpha} \begin{cases} \sqrt{2.0\,x} & \text{if } x < 0.5 \\ -\log\left(1.0 - 2\,(x - 0.5)\right) + 1.0 & \text{otherwise} \end{cases}$$

whereas x is a unique distributed random number [0..1), and α is randomly choosen to 0 or 1. In certain situations and especially in conjunction with escaping individuals, such random numbers result in a speedup of up to 25 %. However, these results are worse compared to the genetic algorithm variant. One reason might be that the recombination operator of the genetic algorithm is more adequate for linearly separable functions.

In addition, we introduced an upper age of individuals. Individuals that exceed that bound are removed from the population, which helps escaping from local minima.

	$f_1(x_i)$	$f_2(x_i)$	$f_3(x_i)$	$f_5(x_i)$
De Jong's R1	> 2000-8000	–	–	> 1500
PGA	15	16	33	16
Collective Learning	5	35	43	11

Table 2: Comparison of number of generations needed by De Jong's plan, PGA and collective learning. Note that PGA uses approximately three times more individuals per generation and, therefore, needs three times more function evaluations than collective learning.

	$n = 20$	$n = 50$	$n = 100$	$n = 200$	$n = 400$
PGA	9900	42753	109072	390768	7964400 (81 minutes)
BGA	3608	–	25040	52948	112634
CL^{GA}	4600	8650	12650	23750	31150 (25 minutes)

Table 3: Comparison of the number of function evaluations needed by PGA, BGA, and collective learning when applied to Rastrigin's function for several dimensions (20 to 400).

6 CONCLUSION

In this article we have presented the *collective learning* procedure, which is especially suitable for optimization of multimodal functions. We have applied it to several problems and the results are much better than those obtained by means of other procedures.

In the simulations it can be seen that once the procedure reaches a local minimum new offspring escape from it and reach the global one after a number of additional steps. In essence, we introduced the concept of interleaving learning / optimization with significant changes in the initialization process into one hybrid scheme. Each component can be substituted by any other problem adequate method. We discussed the disadvantages of single schemes and have shown that only the synthesis of different principles leads to such high performance.

Currently, we are working on evolving neural networks with a minimal topology. In this case, collective learning produces different topologies and each network performs exactly one backpropagation step in each generation. Although, we only used very simple topology modification operators, we obtained promising results. In our preliminary experiments, the procedure found networks with only *twelve connections* for the 4-x-4 encoder task within 6500 generations which is a reduction of almost 50 %. In addition, the procedure is simple to implement and it is inherently parallel.

Acknowledgements: We gratefully acknowledge Michael Luby and Rolf Pfeifer for helpful discussions, David Stoutamire for polishing up, and thanks in particular to Jerome Feldman for his support.

References

[1] R. Belew, J. McInery, and N. Schraudolph. Technical Report #CS90-174, University California at San Diego, June, 1990.

[2] Kenneth A. De Jong. *An Analysis of the Behavior of a Class of Genetic Adaptive Systems.* PhD thesis, University of Michigan, 1975.

[3] Reinhard Lohmann. In *Parallel Problem Solving.* Springer-Verlag, 1990.

[4] David G. Luenberger. *Linear and Nonlinear Programming.* Addison-Wesley Company, Menlo Park, California, 1984.

[5] Heinz Muehlenbein and Dirk Schlierkamp-Voosen. Predictive models for the breeder genetic algorithm. *Evolutionary Computation*, 1(1):25–50, 1993.

[6] H. Mühlenbein, M. Schomisch, and J. Born. In *Proceedings of the Fourth International Conference on Genetic Algorithms.* Morgan Kaufman Publisher, 1991.

[7] Ingo Rechenberg. *Evolutionsstrategie.* Problemata. Frommann-Holzboog, 1973.

[8] Ralf Salomon. In *The Third Annual Conference on Evolutionary Programming.* World Scientific Publishing, 1994.

[9] Ralf Salomon and J. Leo van Hemmen. In *International Conference on Artificial Neural Networks.* 1992.

SCIENTIFIC ANALYSIS LANGUAGE FOR RETRIEVAL AND ANALYSIS OF KNOWLEDGE

Pawan Lingras
Department of Computer Science, Algoma University College
Sault Ste. Marie, Ontario, P6A 2G4, Canada.

Abstract. Several researchers have emphasized the need for a natural language-like interactive query language for retrieval and updating of spatial and temporal information in geographic information systems. This paper focuses on the problem of knowledge retrieval from spatial and temporal knowledge bases. It is argued that a natural language processor may not be necessary for knowledge retrieval from spatial and temporal knowledge bases in many scientific and engineering applications. A scientific analysis language is proposed that can be used to simplify retrieval, analysis and display of results. The language design is similar to spreadsheet programs. However, the proposed language supports relatively large knowledge bases and facilitates use of statistical as well as artificial intelligence modeling tools. The language processor also provides spelling and grammar correction features.

1. Introduction

Data retrieval from spatial and temporal databases involves a certain amount of computer programming using query languages such as SQL or conventional procedural languages. The need for the computer programming restricts the database access to a few knowledgeable users. Simplification of the programming aspect in data retrieval will make the databases more accessible to experts in application areas who have limited knowledge of the database organization. The programming involved in data processing can be simplified using a menu driven or icon based system. Such a system will enable users to extract some of the frequently used quantities. However, such an approach lacks the flexibility that is generally required in scientific and engineering analysis. In some cases, the data obtained from such a system may have to be processed further using the conventional computer programming. Toledo and Davis (1992) listed various disadvantages of using menu driven or icon based systems. Development of a natural language interface for database retrieval is considered as a solution for making geographic information systems (GISs) more accessible to general users (Krzanowski, 1990; Lydiatt, 1992; Toledo and Davis; 1992). Toledo and Davis extended the natural language interface to all aspects of a GIS. A natural language interface that supports spatial operations on geographic information will be a major asset for a GIS. These natural language interfaces can successfully process simple queries. However, for more complex data retrieval, natural language processing may be difficult because of inherent ambiguities in natural languages. This paper proposes a scientific analysis language (SAL) which will be intuitively attractive for scientists and engineers.

E. A. Yfantis (ed.), Intelligent Systems, 85–94.

The design of SAL is based on the spreadsheet programs such as Lotus 1-2-3®. Spreadsheet programs are popular with business executives, scientists, and engineers because of their intuitive user interface. Memory restrictions make it impossible to use spreadsheets to manipulate large databases. Recent developments in spreadsheet technology provide better database management abilities. However, it is still difficult to manage large spatial and temporal databases using spreadsheets. Moreover, spreadsheet programs do not provide any support for knowledge bases. SAL provides spreadsheet-like data processing for large knowledge bases. Spreadsheets store data in a tabular format. Users can manipulate tables by easily specifying calculations involving various rows or columns. SAL uses similar view of the data. SAL programs produce tables. The first column of the table consists of the values of the independent variable. Subsequent columns consist of values of knowledge variables or values of an expression involving knowledge variables. The output from a SAL program can be directly supplied to a spreadsheet program or to other statistical and artificial intelligence analytical tools. Some of the natural language processing features such as spelling and grammar corrections are also used to minimize programming inconveniences.

2. An Informal Introduction to SAL Using Traffic Engineering Knowledge Base

This paper focuses on the problem of knowledge retrieval from spatial and temporal knowledge bases. It is argued that a spreadsheet-like interface which enables users to perform simple mathematical calculations may be more suitable for scientists and engineer. Such an interface should enable the user to retrieve knowledge from a knowledge base rather than restricting the user to a database. The organization of the knowledge base should be transparent to the user. The interface should also enable the user to activate various modeling procedures stored in the knowledge base. Some of the useful techniques from the area of natural language processing such as spelling and grammar corrections may make such an interface more user friendly. The remaining part of the paper describes scientific analysis language (SAL) which provides most of the facilities discussed above. It should be emphasized here that the language described in this paper focuses on knowledge retrieval from spatial and temporal knowledge bases in scientific and engineering applications. An interactive GIS query language for retrieval and updating of geographic information with natural language processing capabilities is still the best solution for GISs.

As the name suggests, SAL is designed to facilitate knowledge retrieval for scientific analysis. It is useful for scientists and engineers who deal with statistics. The design of SAL is similar to spreadsheet packages such as Lotus 1-2-3® and Quattro Pro®. Spreadsheets use a relatively small amount of data (roughly 10,000 records) in tabular format loaded in computer's main memory. Since the main memory can store a smaller amount of data than the secondary storage, it is not easily possible to process large databases or knowledge bases (several million records) using spreadsheet programs. A SAL program retrieves knowledge from a relatively large knowledge base stored on a disk. The knowledge base used by SAL consists of a database, a function base and a procedure base. The database stores factual information. The function base consists of several functions that compute some of the important parameters used in the analysis by manipulating data obtained from the database. The procedure base consists of procedures to perform statistical analysis, and apply artificial intelligence (AI) tools such as neural networks and expert systems. Design of the knowledge base allows for addition of new data, functions and procedures without making any changes to the SAL inference engine. The organization of knowledge base is transparent to the user. A user only specifies expression involving names of the knowledge variables. SAL communicates with the knowledge base to get the values of the knowledge variables. These values may be directly returned from the database or may be computed using the functions from the function base. The output from the SAL program is supplied to one of the procedures from the procedure base. The procedure base will generally support tabulation, graphical display, statistical analysis, and AI modeling.

Before a formal introduction to the scientific analysis language (SAL), we will look at some simple programs in SAL. Let us assume that we are designing a knowledge base to store traffic data (Lingras and Sharma, 1993a). The seasonal and permanent traffic counters scattered across the highway network are the major sources of traffic data. These traffic counters provide traffic volume, i.e., the number of vehicles that have passed through a particular section of a lane or highway in a given time period. Traffic volume can be expressed in terms of hourly or daily traffic. More sophisticated traffic counters provide additional information such as the speed, length and weight of the vehicle. Highway agencies generally have records from the traffic counters over a number of years. In addition to the data obtained from the traffic counters, traffic engineers also conduct occasional surveys of road users to get more information.

Table 1 shows a sample relation that can represent the information about traffic counters in Canadian province of Alberta. The Alberta Department of Transportation has 61 sites with permanent traffic counters. Each site is divided into three different types of traffic volume counts, traffic volume in each of the two directions and combined volume in both directions. Each row in the table represents the data for a particular hour for a particular traffic count. The notation AADT in the table stands for annual average daily traffic volume. The notation MADT stands for monthly average daily traffic volume. For each of the $61 \times 3 = 183$ traffic volume counts there will be $10 \times 365 \times 24 = 87600$ records for 10 years. That means totally there will be more than 16 million records in the table. It is not possible to store such a large amount of data in a single spreadsheet. Moreover, the table contains a large amount of redundant information. It is more appropriate to save such information in a relational (or an extended relational) database.

In addition to the database shown in table 1, traffic engineering knowledge base also consists of functions that calculate such quantities as level of service (LOS) for a given highway section. A traffic analyst also uses certain statistical tools for traffic analysis. Traffic researchers are also exploring the possibility of using AI tools for traffic analysis and modeling (Fwa and Chan, 1993; Sharma and Lingras, 1993b). The proposed knowledge base also contains these statistical and AI tools. SAL is used as an interface for this traffic engineering knowledge base.

Site	Year	AADT	Month	MADT	Date	Daily Volume	Hour	Hourly Volume
1	1980		Jan		1		1	
1	1980		Jan		1		2	
1	1980		Jan		1		....	
1	1980		Jan		1		24	
......	......	......	......	......	......	......	......	......

Table 1 Traffic Engineering Database using single Relation

A SAL program specifies the formulae to compute X (independent) and Y (dependent) values. This information is used to produce a table of X and Y values, or a graph of X versus Y values. The X and Y values can also be used for more sophisticated statistical, or artificial intelligence analysis. It is also possible to define multiple dependent variables such as Y1, Y2, Y3, etc. The following are some of the sample programs in SAL.

Example 1.

A program to produce a table of AADT values for years 1980 to 1990 at counter site 235.

Begin

X = Year **For** 1980 **to** 1990 Years at 235 Counter
Y = AADT
Table and Graph

End

Example 2.

A program to classify all the two-way counter sites according to AADT values for 1990. Information about the classification is in file class-det.inf.

Begin

X = Counter **For** 1990 Year at all both direction Counters
Y = AADT
Classify X class-det.inf

End

Example 3.

A program to produce a table of level of service values for hours between 7 am to 10 am and 3 pm to 6 pm in 1990 at counter site 240.

Begin

X = hour **For** 7 **to** 10 15 **to** 18 Hours on all days in 1990 Year at 240 Counter
Y = Level of Service
Table and Graph

End

Example 4.

A program to produce a table of MADT values and the ratio of MADT and AADT, for all months in year 1990 at counter site 235.

Begin

X = Month **For** all Months in 1990 Year at 235 Counter
Y = MADT
Y1 = MADT / AADT
Table and Graph

End

Example 5.

A program to produce a table of the ratio of daily volume on Sunday to average of daily volume on Tuesday to Thursday, for all weeks in May to August in year 1990 at counter site 235.

Begin

X = Week **For** May **to** Aug Months in 1990 Year at 235 Counter
Y = Month
Y1 = (Daily Volume **For** Sun Day) / (Average of Daily Volumes **For** Tue **to** Thu Days)
Table and Graph

End

Examples discussed in this section show that the SAL programs use column format similar to spreadsheets. SAL combines features from SQL and spreadsheets to provide a more intuitive interface for knowledge retrieval from spatial and temporal knowledge bases. Some of the programming inconveniences are reduced by using spelling and limited grammar correction facilities. The following section describes more formal specifications for the language.

3. Specifications for SAL

Any programming language consists of rules that specify the syntax and semantics of the language.

3.1. Syntax of SAL programs

The syntax of the scientific analysis language (SAL) consists of reserved words, special symbols, built-in functions, access variables, knowledge variables, and output variables. SAL is case insensitive, e.g., words *Begin, BEGIN, begin, BeGiN* are all treated identically.

a. Reserved words: These words have special meaning in SAL, and hence cannot be used for any other purpose. The reserved words are distinguished from other words in this paper using bold letters. The reserved words in SAL are: **Begin, End, For, to**.

b. Special symbols: Similar to reserved words, the special symbols also have special meaning. Special symbols in SAL are = , +, -, *, / , %, (,).

c. Built-in functions such as *sum of, average of, min of, max of.*

d. Access variables: The database is organized based on certain attributes or variables. These variables, called access variables, are used to restrict the scope of the analysis. Use of

Variable	Values
Year	all, 1900 to 2100
Month	all, Jan to Dec
Week	all, 1 to 52
Date	all, 1 to 31
Day	all, Sun to Sat
Hour	all, 1 to 24
Counter	all, both direction, first direction, second direction, a string identifying the counter number

Table 2. Allowable Values of Temporal/Spatial Variables

access variables results in faster retrieval. For the proposed traffic engineering database the access variables represent spatial and temporal attributes, and hence, they are also called temporal/spatial variables. The list of access variables may include Year, Month, Week, Date, Day, Hour, and Counter. Each of these variables could be proceeded by the letter 's'. For example, Year and Years refer to the same variable. Table 2 shows the possible values for each of these variables. A range of values is specified using reserved word to, e.g., Tue to Thu. A list of values can be specified by separating values and ranges by a space, e.g. Jan May to Aug Nov Dec.

e. Knowledge variables provide information stored in the knowledge base. The values of these variables may be stored in the database or may be obtained by executing one of the functions in the function base. In the traffic engineering knowledge base, the knowledge variables essentially provide traffic information and hence they are called traffic variables. Traffic variables may include Daily Volume, Hourly Volume, MADT, AADT, and Level of Service. All the variables in this list with the exception of level of service are stored in the database. Level of service is calculated using a function.

f. Output variables consist of one independent variable called X, and one or more dependent variables, named as Y, Y1, Y2, Y3, etc. In practice, there may be a limit on the number of dependent variables in a program.

Any sequence of characters other than the ones mentioned above is treated as a comment. For example, words such as *in, of, at* can be used to make programs more meaningful.

General format of a SAL program is as follows:

Begin

X = <access variable> **For** <qualifier>
Y = <expression>
[Y1 = <expression>
[Y2 = <expression>
......]]
<action> <parameter-list>

End

Quantities in square brackets are optional.

The statement X = <access variable> **For** <qualifier> provides definition of X variable. Similarly, the statement Y = <expression> provides definition of Y variable and so on. The variables X, Y, Y1, Y2, Y3, etc. are defined in that order, i.e. X is defined first, Y is defined next, Y1 is defined after Y, and so on.

The <qualifier> can be a list of values of access variables followed by the variable names.
Example. 1980 **to** 1990 Years at all two way counters.
Example. All weeks in May to Aug Months in 1990 Year at 235 Counter.

The expression can be

(i) <variable>, i.e., a knowledge, access or output variable such as Daily Volume, Hourly Volume, Day, Hour, Month, X, Y, Y1.

(ii) <variable> **For** <qualifier>.
<qualifier> is allowed to have only one value for each access variable (i.e. a list of values is not allowed). Example. Daily Volume **For** Sun Day.

(iii) <built-in function> <variable> **For** <qualifier>
Example. Average of Daily Volume **For** Tue **to** Thu Days.
Note that <qualifier> can have a list of values in this case.

(iv) infix expression involving the three types (i), (ii), (iii) and operators /, + , - , %, and *. If another output variable is used in an expression (e.g. Y2=Y1+Y), then it must be defined before its use. For example, Y can use X; Y1 can use Y and X; Y2 can use Y1, Y and X.
Example. MADT/AADT.
Example. Daily Volume **For** Sun Day / Average of Daily Volumes **For** Tue **to** Thu Days.

The <action> can be *Table and Graph* or any other executable procedure from the procedure base. The list of parameters will depend upon the procedure.

3.2. Semantics of SAL programs

Text included in the reserved words **Begin** and **End** is considered as the program in SAL. Anything before or after **Begin** and **End** will be treated as comments and ignored. The program must have the definitions of X and Y variables. The X definition is used to produce a column of values of the <access variable> for the time periods and locations (for a spatial and temporal knowledge base) specified in the <qualifier>. The Y definition is used to produce columns of Y values for all the values of X. The definitions of additional dependent variables Y1, Y2, Y3, etc. are optional.

If the <expression> is a <variable>, then Y column will contain the values of <variable> for all the values of X. The expression can also be a formula involving different knowledge variables and operators, e.g., MADT % AADT. The valid operators are + (addition), - (subtraction), * (multiplication), / (division). A special operator % is used to calculate the ratio of two variables expressed as percentage, i.e., MADT % AADT is equivalent to 100*(MADT / AADT). The operators *, / and % have a higher precedence than + and -. A set of parenthesis can be used to specify any other order of operations. The formula in the Y definition is evaluated for all values of X to create the Y column.

Sometimes, the expression can be a more complex formula involving the reserved word **For**. The reserved word **For** has two different purposes in an expression. The qualifier after **For** is used to restrict the scope of Y variable from a list of possible values. In Example 5, the X and Y definitions are:

X = Week **For** May to Aug Months in 1990 Year at 235 Counter
Y = Month

X values will be week numbers for months from May to Aug. Y values provide the month for each of those weeks. The expression for Y1 is of the form:

Y1 = (Daily Volume **For** Sun Day) / (Average of Daily Volumes **For** Tue **to** Thu Days)

Here, for a given week Daily Volume can take a list of values for days ranging from Sunday to Saturday. In the numerator, the qualifier followed by the reserved word **For** specifies exactly which value from this list should be used, namely the Daily volume for Sunday. The reserved word **For** can also be used to supply a list of values to functions such as *Average of*. In the denominator of the Y1 definition shown above, a list of Daily Volumes for Tue, Wed and Thu is supplied to the function *Average of*. The function then returns the average of the list of values.

One output variable can be used to define another output variable as long as the former is defined before its use. For example, Y3 = Y + (Y2 / Y1).

If <action> is *table and graph*, the table produced by SAL processor is supplied to a spreadsheet program. The spreadsheet program supports fancier input/output formatting as well as graphical output and statistical analysis. SAL also allows the user to run some of the predefined procedures using the resulting table as input. The SAL processor calls the procedure specified by the <action> and provides the necessary parameters. In Example 2,

Classify X class-det.inf

will classify all the X values according to the information supplied in the file class-det.inf.

The rigidity of programming language syntax is one of the unpleasant aspects of programming. The programmer has to provide the correct syntax before the program can be run successfully. The proposed system tries to minimize these inconveniences by being more tolerant towards spelling and grammatical mistakes in the program. For example, if the programmer misspells a word, the input module will use spell checker to look for a possible substitute. Similarly, in some cases, a programmer may not use the correct syntax, such as typing

For Tue **to** Thu

or

For Days Tue **to** Thu

instead of

For Tue **to** Thu Days

The SAL processor uses natural language processing features such as spelling and limited grammar corrections to correct the syntax as long as the program entered by the user is unambiguous.

4. Summary

A spreadsheet-like interface which enables users to perform simple mathematical calculations may be more suitable for scientists and engineers. Such an interface should also enable the users to retrieve the knowledge from a knowledge base rather than restricting the users to a database. The organization of the knowledge base should be transparent to the users. The interface should also enable the users to activate various modeling procedures stored in the knowledge base.

This paper describes scientific analysis language (SAL) as a possible substitute for natural language processing in spatial and temporal knowledge retrieval. The design of SAL is similar to spreadsheet packages that are used to manipulate tables. SAL provides the conveniences of spreadsheet programming for large knowledge bases (several million records). Generally, the output from SAL processor is redirected to a spreadsheet package such as Lotus 1-2-3® and Quattro Pro®. The results can then be displayed in graphical or tabular form similar to normal spreadsheet data. Spreadsheet packages also enable the users to perform simple statistical analysis. SAL also makes it possible to invoke other statistical and artificial intelligence tools from the knowledge base. The inconveniences of programming is further reduced by adding spelling and limited grammar correction features.

It should be noted that SAL combines features from SQL, spreadsheets, and natural language processing to provide an intuitive interface for knowledge retrieval in scientific and engineering applications. The requirements of GIS users can only be met by an interactive GIS query language.

Acknowledgment

This research is funded by a grant from NSERC, Canada.

5 References

[1] Armstrong, M.P., 1990. Database Organization Strategies for Decision Support Systems, *International Journal of Geographic Systems*, Vol. 4, No. 1, pp. 3-20.

[2] Egenhofer, M.J. 1992. Why not SQL!, *International Journal of Geographic Systems*, Vol. 6, No. 2, pp. 71-85.

[3] Krzanowski, R.M. 1990. Natural Language Interface to Geographic Databases: Experiments with Intelligent Assistant, *In the Proceedings of GIS'90*, Vancouver, British Columbia, pp. 513-518.

[4] Langran, G., 1989. A Review of Temporal Database Research and its Use in GIS applications, *International Journal of Geographic Systems*, Vol. 3, p. 215.

[5] Lingras, P.J. and Sharma, S.C., 1993a. A Knowledge-Based System for Traffic Engineering Analysis, *submitted to Canadian Journal of Civil Engineering*.

[6] Lingras, P.J. and Sharma, S.C., 1993b. Estimation of Traffic Parameters from Sample Traffic Counts using Neural Networks, *in preparation.*

[7] Lydiatt, G.W. 1992. Parser Design Strategies for GIS Natural Language Interfaces, *In the Proceedings of GIS'92*, Vancouver, British Columbia, pp. 513-518.

[8] Miller, H.J., 1991. Modelling Accessibility Using Space-time Prism Concepts within Geographical Information Systems, *International Journal of Geographic Systems*, Vol. 5, No. 3, pp. 287-301.

[9] Sharma, S.C., 1987. Driver Population Factor in New Highway Capacity Manual, Journal of Transportation Engineering, American Society of Civil Engineers, Vol. 113, No. 5, pp. 575-579.

[10] Sharma, S.C., Lingras, P.J., Hassan, M.U., Murthy, A.S., 1986. Road Classification and Driver Population, *Transportation Research Record 1090*, Transportation Research Board, National Academy of Sciences, Washington D.C.

Towards a Broader Theory for Abstract Interpretation

Kaninda MUSUMBU
LaBRI, CNRS/Université Bordeaux I
351, cours de la Libération,
33405 Talence Cedex FRANCE
e-mail: musumbu@labri.u-bordeaux.fr

Abstract

The aim of static analysis is to derive information about the actual operational behavior of programs. Introduced by P. and R. Cousot [6] for imperative languages, abstract interpretation is one of the technics used to analyze programs. Its application to logic programming has been broadly studied (see for example [2, 16, 10, 20, 4]). In [6], P. amd R. Cousot introduced some properties to assure the correctness of an abstract interpretation model. In this paper we try to enlighten the part of these properties. We focus our attention on some basic results that remain in the absence of one of these properties. On the other hand, many differents abstract interpretations are been proposed for what is essentially the same concept, we use the categorical language and approach to unify this theory in allowing simply to generality of concepts. That has help to formulate uniform definitions.

Keywords: Static Analysis, Abstract Interpretation, Category Theory

1 Introduction

The declarativity of logic programming (eg, pure Prolog) presents many optimization possibilities based on static program analysis methods. Abstract interpretation [6] is one of the most up-to-date and accurate method. Its application to logic programming has been proposed by many researchers (see for instance [3, 17, 10, 20, 5]).
In our framework, we call *abstract interpretation* the association of an abstract domain with a concretization function over this abstract domain. The abstract semantics of the program are defined as the least fixed point of a transformation $TSCT$ [12, 13, 21, 11]. Its existence is ensured if we assume that the sets $Asub_D$ are inductive and that the primitive operations are all monotonic. But, all these properties are not necessary to prove the *correctness* of the abstract semantics. The main goals of this paper is to analyze what happens when some of these properties are missing. To enlighten the *"essential part"* of each property and to a very general and powerful framework for abstract

E. A. Yfantis (ed.), Intelligent Systems, 95–108.

interpretation on logic programs [21, 12, 11].

Indeed, investigations into the abstract semantics of logic programming is generally restricted to domains which is being considered as complete lattice. This structure is very stronger for giving a general characterization of abstract interpretation theory. In the other hand, properties of abstract operations are important from the reusing point of view, to ensure first the existance of a fixpoint second the correctness of computed result. Yet the satisfaction by those operations of all properties are not necessary. Especially when we compare two abstract interpretations, there is a debate on what *"abstract interpretation is more expressive than other?"*
But there are some people, including me, who think that abstract interpretation can be view as a category, and we can built a category of abstract interpretations which might give a (good) general framework to study this theory.

The aim of our paper is to generalize the framework of abstract interpretation in using the *"category theory"* and to give a way to comparing the various abstract interpretations approach of programs logic.
Following [1, 6, 3], the *Abstract Interpretation* was originally designed by P. and R. Cousot as a *general methodology* for building static analyses of programs. The abstract interpretation of programs can be formalized as the effective computation of an approximation of collecting semantics of the programs. This collecting semantics is intended to model precisely SLD-resolution. It computes, given a predicat and a set of input substitutions, the set of output substitutions resulting from the application of the predicate to the set of input substitutions. From this semantics we can built a sequence of abstract semantics sound and less precise stage by stage.

The paper is organized in the following way. Section 2 contains some preliminary definitions needed for defining our abstract semantics. Section 3 presents a model of abstract semantics with its principal properties. Section 4 is the core of this paper. We translate in terms of category theory our abstract interpretation model. One advantage of this formalism is that each abstract interpretation model can be customized to the demands of particular application. Section 5 characterizes formally the comparison between two abstract semantics as two objets of a category of categories and abstract (concretization) functions as functors of this category. Finally, in section 6, we conclude on the mathematical structure really needed for general framework and how easy it is to implement the abstract algorithms.

2 A Collecting Semantics

The starting point of the present work is the definition of a fixpoint semantics for logic programs. This collecting semantics is intended to model precisely SLD-resolution with a left to right computation rule and a search. It computes, given a predicat and a set of input substitutions, the set of output substitutions. This collecting semantics will be abstracted in the next section to define its abstract version.

2.1 Normalized Programs and Substitutions

We assume the existence of sets F_i and P_i ($i \geq 0$) denoting sets of functors and predicate symbols of arity i and of an infinite set PV of program variables. Variables in PV are ordered and denoted by the $x_1, x_2, \ldots, x_i, \ldots$.

Normalized programs contain clauses with heads of the form $p(x_1, \ldots, x_n)$ where $n \geq 0$ and $p \in P_n$. Normalized clauses also contain bodies of the form $g_1, \ldots, g_n$ ($n \geq 0$) with g_i of the form $p(x_{i_1}, \ldots, x_{i_n})$ where $x_{i_1}, \ldots, x_{i_n}$ are all distinct variables and $p \in P_n$ or $x_i = x_j$ ($i \neq j$) or $x_i = f(x_{j_1}, \ldots, x_{j_n})$ where $x_i, x_{j_1}, \ldots, x_{j_n}$ are all distinct variables and $f \in F_n$.

The motivation behind these definitions is to allow the result of any predicate p/n to be expressed as a set of substitutions on program variables $x_1, \ldots, x_n$.

We assume the existence of another infinite set SV of standard variables. Terms and substitutions are constructed using program and standard variables. We distinguish two kinds of substitutions: *program substitutions* (*ps* for short) whose domain and codomain are subsets of PV and SV respectively, and *standard substitutions* (*ss* for short) whose domain and codomain are subsets of SV. In the following, PS denotes the set of *ps* and SS the set of *ss*. We use $var(o)$ to represent the set of variables in the syntactical object o and $dom(\theta)$ and $codom(\theta)$ to denote the domain and codomain of a substitution.

Definition 1 We call $Csub_D$ the set of all *program substitutions* whose domain is $D \subset PV$.

2.2 Concrete Primitive Operations

The operational semantics uses five primitive operations[1]. In the definitions, we make the assumptions that $var(c) = \{x_1, \ldots, x_m\}$, $var(head(c)) = \{x_1, \ldots, x_n\}$, and $x_{i_1}, \ldots, x_{i_n}$ is the sequence of variables occurring in g (from left to right).

$$\begin{aligned}\texttt{EXTC}(c,\theta) = \{ &\{x_1 \leftarrow t_1, \ldots, x_n \leftarrow t_n, \ldots, x_{n+1} \leftarrow y_1, \ldots, x_m \leftarrow y_{m-n}\} : t_i = x_i\theta \ (1 \leq i \leq n),\\ &y_1, \ldots, y_{m-n} \in SV,\ y_1, \ldots, y_{m-n} \text{ are all distinct, and}\\ &var(t_1, \ldots, t_n) \cap \{y_1, \ldots, y_{m-n}\} = \emptyset \}.\end{aligned}$$

$$\texttt{RESTRC}(c,\theta) = \{\theta_{/\{x_1,\ldots,x_n\}}\}.$$

$$\texttt{RESTRG}(g,\theta) = \{\{x_1 \leftarrow x_{i_1}\theta, \ldots, x_n \leftarrow x_{i_n}\theta\}\}.$$

[1]Operation RENAME is in fact used implicitly but it is important for the rest of the paper.

$\texttt{EXTG}(g,\theta_1,\theta_2) = \{\ \theta_1\sigma\ : \exists\theta_3$ such that $\theta_3 = \texttt{RESTRG}(g,\theta_1)$ and $\theta_2 = \theta_3\sigma$ where $dom(\sigma) \subseteq codom(\theta_3)$ and $(codom(\theta_1) - codom(\theta_3)) \cap codom(\sigma) = \emptyset\ \}$.

$\texttt{RENAME}(g,\theta)\ = \{\{x_{i_1} \leftarrow x_1\theta,\ldots,x_{i_n} \leftarrow x_n\theta\}\}$

The definition of substitution composition is slightly modified to take into account the special role held by program variables. The modification occurs when $\theta \in PS$ and $\sigma \in SS$ for which $\theta\sigma \in PS$ is defined by $dom(\theta\sigma) = dom(\theta)$ and $x(\theta\sigma) = (x\theta)\sigma$ for all $x \in dom(\theta)$. Unifiers are defined as usual but only belong to SS hereafter (using standard variables only). We use $mgu(t_1,t_2)$ to denote the set of most general unifiers of t_1 and t_2. We derive two built-in functions:

$$\begin{aligned}&\texttt{AI\_Var}(x_i = x_j,\Theta)\ =\ \Theta'\\ &\textbf{where}\quad \Theta' = mgu\{x_i\theta_{in}, x_j\theta_{in} :\ \forall\ \theta_{in}\ \in\ \Theta\}\end{aligned}$$

$$\begin{aligned}&\texttt{AI\_Func}(x_i = f(x_{j_i}\ \ldots,x_{j_n}),\Theta)\ =\ \Theta'\\ &\textbf{where}\quad \Theta' = mgu\{x_i\theta_{in}, f(x_{j_i}\ \ldots,x_{j_n})\theta_{in} :\ \forall\ \theta_in\ \in\ \Theta\}\end{aligned}$$

We denote $S = mgu\{x_i\theta_{in}, x_j\theta_{in} : \forall\theta_{in} \in \Theta\}$

2.3 Sets of Concrete tuples

The purpose of the semantics can be defined as follows: given a predicat p of arity n, a set of input substitutions Θ_{in}, whose domains are $\{x_1,\ldots,x_n\}$, define the set of output substitutions Θ_{out}, whose domains are also $\{x_1,\ldots,x_n\}$, representes the set of output substitutions we would obtained by executing $p(x_1,\ldots,x_n)$ on the set of input substitutions Θ_{in}. In the following we assume an underline program P.

Definition 2 A concrete tuple is a tuple of the form (Θ,p,Θ') where $\Theta,\Theta' \in Csub_D$, D is $\{x_1,\ldots,x_n\}$, n is the arity of p and p appears in the program P.
We use *cst* as an abbreviation of "set of concrete tuples".
CST denotes de set of all *csts*.

Property 3 Let $sct \in SCT$, sct is *functional* if and only if for all, (Θ,p), there exists at most one Θ' such that $(\Theta,p,\Theta') \in sct$. Then, $\Theta' = sct(\Theta,p)$.

Property 4 Let $sct \in SCT$, sct is *monotonic* if and only if for all, (Θ,p),(Θ',p), such that $\Theta \subseteq \Theta' \Rightarrow sct(\Theta,p) \subseteq sct(\Theta',p)$.

Property 5 Let $sct \in SCT$, sct is *continuous* if and only if for all, any increasing chain

$$\Theta_1 \subseteq \Theta_2 \subseteq \ldots \Theta_n \subseteq \ldots$$

satisfies

$$sect(\bigcup_{i=n}^{\infty}\{\Theta_i\}, p) = \bigcup_{i=n}^{\infty}\{sct(\Theta_i, p)\}$$

We denotes $SCCT$, the set of sct which are functional, monotonic and continuous.

2.4 A fixpoint Semantics

We are ready to define the semantics in terms of concrete operations and concrete tuples. The semantics is defined in terms of three functions and one transformation. Let b and c be a sequence of atoms and a clause using only predicat symbole from P. The three functions have the following signatures:

$$T_p(\bullet, p, \bullet) : Csub_D \times SCCT \longmapsto Csub_D,$$

$$T_c(\bullet, C, \bullet) : Csub_D \times SCCT \longmapsto Csub_D,$$

$$T_b(\bullet, SB, \bullet) : Csub_D \times SCCT \longmapsto Csub_D,$$

where D is defined respectively as the set of variables appearing in the head of clause p , in the body of clause c or in a call procedure b. The signature of the transformation is as follow:

$$TSCT : SCCT \longmapsto SCCT$$

The transformation and the functions are defined by the following equations:

$$TSCT(sct) = \{(\Theta, p, \Theta') : \Theta' = T_p(\Theta, p, Eta)\}$$

$$T_p(\Theta, p, sct) = \cup_{i=1}^{n}\{\Theta_i\},$$

where $\Theta_i = T_c(\Theta, C_i, sct)$,
and $C_i, \ldots, C_n$ are the clauses defining p.

$$T_c(\Theta, C, sct) = RestrC(C, \Theta')$$

where $\Theta' = T_b(ExtC(C, \Theta), SB, sct)$,
and SB is the body of C.

$$T_b(\Theta, (), stc) = \Theta,$$

$$T_b(\Theta, B.SB, sct) = T_b(\Theta_3, SB, sct)$$

$$\begin{array}{llll} \text{with} & \Theta_1 & = RestrB(B,\Theta), & \\ & \Theta_2 & = sct(\Theta_1,p) & \text{if } B \text{ is of form } p(\cdots), \\ & & = AI\_Var(\beta_1) & \text{if } B \text{ is of form } x_{i_1} = x_{i_2}, \\ & & = AI\_Func(\beta_1) & \text{if } B \text{ is of form } x_{i_1} = f(\cdots), \\ et & \beta_3 & = ExtB(B,\beta,\beta_2). & \end{array}$$

This recursive definition of $TSCT$ is not computable by usual recursive evaluation for two reasons at least:

1. S can be infinite;
2. sequences of transitions $\ldots, sct_i, sct_{i+1}, \ldots$ can be infinite.

3 The Basic Abstract Semantics

In this section, we recall the abstract semantics proposed in [21] and used subsequently in [12, 13, 14] to derive abstract interpretation algorithms. can then be approximated as follows. For each finite set D of program variables we assume the existence of a *cpo* $Asub_D$ whose elements are called abstract substitutions on domain D and denoted by β. Let $Csub_D$ be the set of program substitutions having domain D. The meaning of each abstract substitution is given through the concretization function: $Cc : Asub_D \rightarrow \mathcal{P}(Csub_D)$. We assume in the following that Cc is monotone. From this semantics one can derive several abstract fixpoint semantics, that "collect" different kinds of relevant information. The basic abstract semantics uses seven abstract primitive operations EXTC, RESTRG, EXTG, RESTRC, UNION, AI_VAR and AI_FUNC. The first four operations are safe abstractions of the corresponding "concrete" operations used by the operational semantics.

Property 6 An (abstract) operation

$$o_a : Asub_{D_1} \times \ldots \times Asub_{D_n} \rightarrow Asub_D$$

is safe abstraction of a (concrete) operation

$$o_c : CS_{D_1} \times \ldots \times CS_{D_n} \rightarrow \mathcal{P}(CS_D)$$

if and only if

$$(\forall i : 1 \leq i \leq n : \theta_i \in Cc(\beta_i)) \Rightarrow o_c(\theta_1, \ldots, \theta_n) \subseteq Cc(o_a(\beta_1, \ldots, \beta_n))$$

for all $\theta_i \in AS_{D_i}$ and $\beta_i \in CS_{D_i}$.

Property 7 An (abstract) operation

$$o_a : Asub_{D_1} \times \ldots \times Asub_{D_n} \rightarrow Asub_D$$

is monotonic if and only if

$$\forall \beta_i, \beta_j \in Asub_D : \beta_i \leq \beta_j \Rightarrow o_a(\ldots, \beta_i, \ldots) \leq o_a(\ldots, \beta_j, \ldots)$$

Definition 8 The abstract semantics of a program is defined as the least fixpoint of $TSAT$. whose signature is:

$$TSAT : SCAT \longrightarrow SCAT$$

where $SCAT$ is the abstract version of $SCCT$. To obtain is definition it suffices to replace all "concrete" notions by "abstract" one in the definition of $TSCT$.

4 Devoted duty

The abstract semantics of the program are defined as the least fixed point of a transformation $TSCT$ denotes $\mu(TSCT)$ [12, 13, 21, 11]. Its existence is ensured if we assume that the sets $Asub_D$ are inductive and that the primitive operations are all monotonic. But, all these properties are not necessary to prove the *correctness* of the abstract semantics. The main goal of this paper is to analyze what happens when some of these properties are missing. In this section, we will review the main properties required in general by the various abstract interpretation frameworks. In fact we will focus our attention on the monotonicity and consistency proprieties.

4.1 Monotonicity of primitive operations

These operations, which have been recalled in the above section, can be considered as functions of the form:

$$o_a : Asub_{D_1} \times \ldots \times Absub_{D_n} \rightarrow Asub_D.$$

In the theory, one requires that these operations be **monotonic**, i.e. such that

$$\forall \beta_1, \beta_1' \in Asub_{D_1}, \ldots, \beta_n, \beta_n' \in Asub_{D_n} :$$

$$(\beta_1 \leq \beta_1' \ \underline{\text{and}} \ \ldots \ \underline{\text{an d}} \ \beta_n \leq \beta_n') \Rightarrow o_a(\beta_1, \ldots, \beta_n) \leq o_a(\beta_1', \ldots, \beta_n')$$

4.2 Consistency of the primitive operations

To each *abstract* primitive operation, we associate a *concrete* operation defined on the set of *concrete* substitutions. This means that the *abstract* operation computes a correct approximation (superset) of the result returned by the *concrete* operation, which relates to the property of *"safe abstraction"*

4.3 Precision of the primitive operations

Saying that a primitive operation is *precise* is rather vague. On the contrary, one can precisely define two concepts: the *correctness* and the relation *"to be more precise than"*.

The abstract operation o_a is **correct** if:

$$\forall \beta_1, \ldots, \beta_n : o_c(Cc(\beta_1), \ldots, Cc(\beta_n)) = Cc(o_a(\beta_1, \ldots, \beta_n)).$$

Let o_1 and o_2 be two *abstract* versions of the same concrete operation. We say that o_1 *is more precise than* o_2 if:

$$\forall \beta_1, \ldots, \beta_n : Cc(o_1(\beta_1, \ldots, \beta_n)) \subseteq Cc(o_2(\beta_1, \ldots, \beta_n)).$$

4.4 What remains if missing one property

In this section, we will review the main properties required in general by the various abstract interpretation frameworks. In fact we will focus our attention on the monotonicity and Consistency properties.

4.4.1 Independence of the consistency and the monotonicity

First of all, the main conclusion of our abstract interpretation studies is the complete independence of the consistency and monotonicity properties.

In fact, the results for consistency remain valid in the case where all the primitive operations are nonmonotonic. As a matteroffact, it is not necessary to define an order on the abstract domain. What is important to ensure the correctness of the results, is that the operations are **consistent** and that the transformation $TSCT$ has a fixed point. $TSCT$ does not need to be monotonic, since there is no order on the abstract domain. This remark justifies the following proposition:

Proposition 9 If the primitive operations are consistent, then each fixed point of $TSCT$ is consistent.

On the other hand, the monotonicity of primitive operations is useful to ensure the existence of some **fixed point**. For this, we can chose any order on the abstract domain. There is no need for the function Cc to establish the least gap between this relation and the relation $\leq$ defined in the concrete domain. Different orders lead to different fixed points.

4.4.2 Monotonicity and Precision

When all requirements for the monotonicity are satisfied, it is possible to make some conclusions about the precision of the (abstract) results, that is to say about the least fixed point of $TSCT$ ($\mu(TSCT)$, for short).

Proposition 10 $\mu(TSCT)$ is the most precise fixed point of $TSCT$.

Proof Assume $sct = \mu(TSCT)$ and $Dr = dom(sct)$, and let sct', be another fixed point of $TSCT$.
Then we have:

$$sct \leq sct' \text{ since } sct \text{ is the least fixed point}$$

thus:

$$\forall(\beta, p) \in Dr : sct(\beta, p) \leq sct'(\beta, p),$$

thus:

$$\forall(\beta, p) \in Dr : Cc(sct(\beta, p)) \subseteq Cc(sct'(\beta, p)), \text{(since } Cc \text{ is monotonic).}$$

□
Indeed, one can easily prove, by induction on the sequence of approximations of $\mu(TSCT)$, that it is the least element of the set:

$$\{sct : TSCT(sct) \leq sct\}$$

Thus, if Cc is monotonic, all its SCT are consistent since they are greater than $\mu(TSCT)$ and thus less precise.

4.5 Remaining results, when there is a lack of monotonicity

In this section, we review three different cases of nonmonotonicity of the abstract interpretation. This can arise either when the primitive operations are not monotonic, or the concretization function or else the computed abstract substitutions.

4.5.1 Nonmonotonic primitive operations

Assume that all the objects needed for an abstract interpretation are given, but that one of the abstract operations is not monotonic for the corresponding order. The main trouble is that we can no longer ensure the monotonicity of the transformation $TSCT$, nor the existence of a fixed point.
Nevertheless, any existing fixed point will be consistent [21].

4.5.2 Nonmonotonic SCT

The notion of monotonic SCT (continuous, if the domain is infinite) permits the prove of the monotonicity of $TSCT$ ($TSCT$ is continuous) and then gives a natural process for building the least fixed point. This is absolutely needed to obtain these results. Nevertheless, a more general result holds whenever the primitive operations are monotonic.

5 Abstract Interpretation is a Category

The aim of static analysis is to derive information about the actual operational behavior of programs. This information is in fact completely determined by the standard operational semantics of the programming language. It is nevertheless convenient to use another, non standard semantics, as a basis for performing the analyses, because it is in practice impossible to "generate and analyse" all possible execution traces for a given program.

There are several approaches to the choice of the non standard semantics (see a discussion at paragraph 3.1). We basically follow the original approach of P. and R. Cousot. Their so-called *static* semantics provides a fixpoint characterization of the relevant information. We assume in the following for each abstract domain $Asub_D$ the existence of a preorder $\preceq$, and we consider the abstract substitutions as objets of category.
For all pairs of objets (β_i, β_j) , there exists an arrow $\beta_i \rightarrow \beta_j$ if, an only if, $\beta_i \preceq \beta_j$.

To complete this presentation, it only remains to resolve the problem of abstract operations.
Let β_i, β_o two abstract substitutions,
an operation o_a between this two objets such that $\beta_o = o_a(\beta_i)$, is an element of $\beta_o^{\beta_i}$ (i.e. collection of the continuous functions from β_i to β_j ordered pointwise). Then the proof of monotonicity of primitive operations, become a study of diagrams which are a typical way, in category theory, to describe equationnal reasoning.

Property 11 Let $Csub_D$ be the standard domain, $Asub_D$ is abstract version and Cc its concretization function. Then:

1. $\forall \beta_i, \beta_o \in Asub_D$, we have:

$$\exists f_a \in \beta_o^{\beta_i} \subseteq Asub_D \Rightarrow \exists f_c \in Cc(\beta_o)^{Cc(\beta_i)} \subseteq Csub_D$$

2. $\forall \beta_i, \beta_j \in Asub_D$, we have:

$$\beta_1 \preceq \beta_2 \Rightarrow Cc(\beta_1) \subseteq Cc(\beta_2) \text{ or } Cc(\beta_2) \subseteq Cc(\beta_1).$$

The difference between the old and new paradigms may be formally summarized as follows. For each pair of abstract interpretations $(Asub_D^1, Cc_1)$ and $(Asub_D^2, Cc_2)$ which satisfies (1), called *Galois connection*, forms an adjunction, i.e.

$$< Cc_1, Cc_2, \eta, \varepsilon >$$

where

1. Cc_1, Cc_2 are full functors,

2. η is a natural transformation of $Asub_d^1$ to $Asub_D^2$,

3. ε is a natural transformation of $Asub_d^2$ to $Asub_D^1$

such that:

$$Cc_2\varepsilon : Cc_2Cc_1Cc_2 \longrightarrow Cc_2$$

and

$$\eta Cc_2 : Cc_2 \longrightarrow Cc_2Cc_1Cc_2$$

This property whose importance seems to be *obvious*, allows one to deduce some conclusions about the precision of the results and especially, when we compare two abstract interpretations, there is a debate on what "abstract interpretation is more expressive than other?"

6 Discussion and Comparison with related works

Many authors have proposed frameworks for abstract interpretation(particularly, in logic programming). Cousot [7] declare: *The quest for unique general-purpose semantics for programming language has failed. A better approach is to establich correspondences between various semantics at level of abstraction".* We use the category theory to realize the unification of abstract theory.
The problem is characterized through the notion of functors:

Theorem 12 Let, $(Asub^1, \gamma_1)$ and $(Asub^2, \gamma_2)$ two abstract interpretations, exists a functor $\mathcal{F}$ such that the following diagram

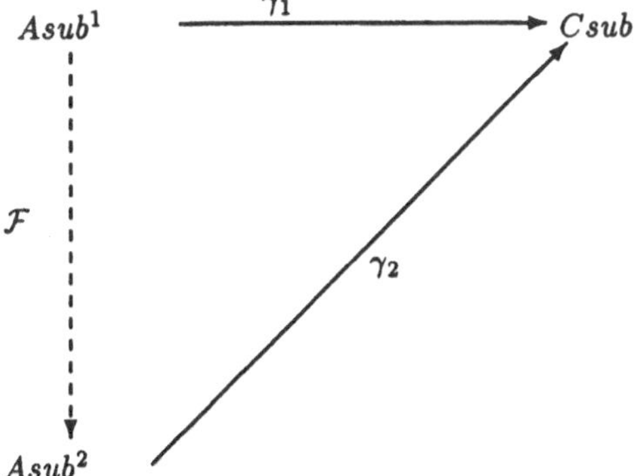

commutes. This diagram can be also viewed as a natural transformation, generaly expressed in terms of adjunction functions. We can construct a cartesian closed category with $Csub$ as the initial objet and Top=$\{\bot, \top\}$ as the terminal objet ($\bot$ means the failure or loop of program and $\top$ its success).

7 Conclusion

In this paper we have tried to enlighten the part of the properties imposed by P. and R. Cousot to ensure, in the theory, the correctness of any abstract interpretation model. The abstract semantics of the program are defined as the least fixed point of a transformation $TSCT$ [12, 13, 21, 11]. Its existence is ensured if we assume that the sets $Asub_D$ are inductive and that the primitive operations are all monotonic. But, all these properties are not necessary to prove the *correctness* of the abstract semantics. The main goal of this paper is to analyze what happens when some of these properties are missing. To enlighten the *"essential part"* of each property, we examine what happen when a property is omitted.

From the above, two questions arise: what structure leads to the most general theory? and what structure leads to the easiest implementation of efficient algorithms?
Our main conclusion is needlessness lattices are not necessary: the lowest bound is useless,and moreover in the general case this operation is not consistent.
For the $Asub_D$ sets, the more general structure is the inductive sets.
To generalize the notion of abstract interpretation we propose to use the category theory, which simplify the prove of *"correctness"*.

8 Acknowledgements

We acknowledge Baudouin Le Charlier for its helpful comments during his visit to the LaBRI in January 1993. Marc-Michel Corsini deserves special thanks for his thorough review of this paper.

References

[1] S. Abramski and C. Hankin, editors. *Abstract Interpretation of Declarative Languages.* Ellis Horwood Limited, West Sussex, England, 1987.

[2] M. Bruynooghe. A practical framework for the abstract interpretation of logic programs. *Journal of Logic Programming*, 10(2):91–124, February 1991.

[3] M. Bruynooghe, Janssens G., A. Callebaut, and B. Demoen. Abstract interpretation: Towards the global optimization of Prolog programs. In *Proceedings of the 1987 Symposium on Logic Programming*, pages 192–204, San Francisco, California, August 1987. Computer Society Press of the IEEE.

[4] P. Codognet and G. Filé. Computations, abstractions and constraints in logic programs. In *Proceedings of the fourth International Conference on Programming languages (ICCL'92)*, Oakland, U.S.A., April 1992.

[5] M-M. Corsini, G. Filè A complete framework for the abstract interpretation of logic programs: theory and application. *Research report* Università di Padova.

[6] P. Cousot and R. Cousot. Abstract interpretation: A unified lattice model for static analysis of programs by construction or approximation of fixpoints. In *Conference Record of Fourth ACM Symposium on Programming Languages (POPL'77)*, pages 238–252, Los Angeles, California, January 1977.

[7] P. Cousot et R. Cousot. Comparing the Galois connection and widening/narrowing approaches to abstract interpretation, (papier invité). In M. Bruynooghe et M. Wirsing, éditeurs, Programming Language Implementation and Logic Programming, Proceedings of the Fourth International Symposium, PLILP'92, Leuven, Belgique, 13–17 août 1992, Lecture Notes in Computer Science 631, pages 269–295. Springer-Verlag, Berlin, Germany, 1992.

[8] N.D. Jones and H. Søndergaard. A semantic-based framework for the abstract interpretation of Prolog. In S. Abramsky and C. Hankin, editors, *Abstract Interpretation of Declarative Languages*, chapter 6, pages 123–142. Ellis Horwood Limited, 1987.

[9] Kelly, G.M. Basic Concepts of Enriched Category Theory. Cambridge University Press 1882.

[10] T. Kanamori and T. Kawamura. Analysing success patterns of logic programs by abstract hybrid interpretation. Technical report, ICOT, 1987.

[11] B. Le Charlier and K. Musumbu. Une sémantique opérationnelle instrumentale pour prolog et son application à la preuve de consistance d'un modèle d'interprétation abstraite. In J.-P. Delahaye, editor, *Actes des Journées Francophones de Programmation Logique (JFPL'92)*, Lille, May 1992.

[12] B. Le Charlier, K. Musumbu, and P. Van Hentenryck. A generic abstract interpretation algorithm and its complexity analysis. In K. Furukawa, editor, *Proceedings of the Eighth International Conference on Logic Programming (ICLP'91)*, Paris, France, June 1991. MIT Press.

[13] B. Le Charlier and P. Van Hentenryck. Experimental evaluation of a generic abstract interpretation algorithm for Prolog. In J. Cordy, editor, *Proceedings of the IEEE fourth International Conference on Programming Languages (ICCL'92)*, Oakland, U.S.A., April 1992. IEEE Press.

[14] B. Le Charlier and P. Van Hentenryck. Reexecution in abstract interpretation of prolog. In K. Apt, editor, *Proceedings of the Join International Conference and Symposium on Logic Programming (JICSLP'92)*, Washington, U.S.A., October 1992. MIT Press.

[15] Longo G., and Moggi, E. Cartesian closed categories of enumerations and effective type structures Symposium on Semantics of Data Types, Khan, Mac-Queen,and Plotkin(eds), LNCS 173,(235-247) In Springer-Verlag, New York. 1984.

[16] K. Marriott and H. Søndergaard. Bottom-up abstract interpretation of logic programs. In R.A. Kowalski and K.A. Bowen, editors, *Proceeding of Fifth International Conference on Logic Programming (ICLP'88)*, pages 733–748, Seattle, Washington, August 1988. MIT Press.

[17] K. Marriott and H. Søndergaard. Notes for a tutorial on abstract interpretation of logic programs. In *North American Conference on Logic Programming (NACLP'89)*, Cleveland, Ohio, 1989.

[18] Mac Lane,S. Categories for the Working Mathematician. Springer-Verlag, New York. 1971.

[19] K. Marriott and H. Søndergaard. Semantics-based dataflow analysis of logic programs. In G. Ritter, editor, *Information Processing'89*, pages 601–606, San Fransisco, California, 1989.

[20] C.S. Mellish. Abstract interpretation of Prolog programs. In S. Abramski and C. Hankin, editors, *Abstract Interpretation of Declarative Languages*, chapter 8, pages 181–198. Ellis Horwood Limited, 1987.

[21] K. Musumbu. *Interprétation Abstraite de Programmes Prolog.* PhD thesis, Institute of Computer Science, University of Namur, Belgium, September 1990. In French.

[22] U. Nilsson. Systematic semantic approximations of logic programs. In P. Deransart and J. Małuszyński, editors, *Proceedings of the International Workshop on Programming Language Implementation and Logic Programming (PLILP'90)*, volume 456 of *Lecture Notes in Computer Science*, pages 293–306, Linköping, Sweden, August 1990. Springer-Verlag.

[23] R.A. O'Keefe. Finite fixed-point problems. In J-L. Lassez, editor, *Proceedings of the Fourth International Conference on Logic Programming (ICLP'87)*, pages 729–743, Melbourne, Australia, May 1987. MIT Press.

[24] H. Tamaki and T. Sato. OLD-resolution with tabulation. In E. Shapiro, editor, *Proceedings of the Third International Conference on Logic Programming (ICLP'86)*, volume 225 of *Lecture Notes in Computer Science*, pages 84–98, London, England, July 1986. Springer-Verlag.

[25] D. S. Warren. Memoization for logic programs. *Communications of the ACM*, 35(3), March 1992.

[26] W. Winsborough. Multiple specialization using minimal-function graph semantics. Technical report, Department of Computer Science, The Pennsylvania State University, August 1990. To appear in the Journal of Logic Programming.

A Tale of Three Tutoring Protocols: The Implications for Intelligent Tutoring Systems*

Ramzan A. Khuwaja[1], Allen A. Rovick[2], Joel A. Michael[2], Martha W. Evens[1]
[1] Computer Science Department
Illinois Institute of Technology, Chicago, Illinois 60616
[2] Department of Physiology
Rush Medical College, Chicago, Illinois 60612 USA

August 18, 1994

Abstract. This paper describes an analysis of human tutoring sessions that were conducted in the domain of cardiovascular (CV) physiology. Three different tutoring protocols have been used in these sessions. We have developed a framework under which we compare/contrast and analyze the characteristics of these tutoring protocols. This framework is based on the assumption that an effective tutor, in most cases, plays two roles simultaneously: an expert in the domain, and an expert in the process of tutoring. Over several years our human tutors have chosen increasingly to postpone intervention in favor of gathering more data about the source of the confusion underlying the student's pattern of errors. We discuss the implications of this analysis for the design of intelligent tutoring systems.

Keywords. Human Tutoring Protocols. Intelligent Tutoring Systems. Expertise.

1 Introduction

Tutoring a problem-solving task effectively in virtually any structured domain requires the tutor to perform two functions: create a problem-solving environment for the student, and teach flexibly in this environment in response to the needs of the student. The creation of a problem-solving environment requires the development of a set of rules that govern the interaction of the tutor and the student in the environment. This set of rules does not constrain the behavior of the tutor or the student at each step in the problem-solving process but rather emphasizes a generic way of proceeding in terms of higher level constraints on the tutoring process. We call this set of rules the tutoring protocol.

*This work was supported by the Cognitive Science Program, Office of Naval Research under Grant No. N00014-89-J-1952, Grant Authority Identification Number NR4422554, to Illinois Institute of Technology and Grant N00014-91-J-1622, Grant Authority Identification Number AA1711319 to Rush Medical College. The content does not reflect the position or policy of the government and no official endorsement should be inferred.

E. A. Yfantis (ed.), Intelligent Systems, 109–118.

2 Background

The knowledge domain in which our tutoring takes place is cardiovascular physiology, specifically the baroreceptor reflex, that part of the cardiovascular system responsible for maintaining a more or less constant blood pressure. This negative reflex system involves a multitude of anatomical components all over the body, but functionally one can think about the reflex in terms of the causal interaction between a limited number of parameters. Nevertheless, the number of parameters involved and the complexity of the causal interactions that occur (including the essential feature of negative feedback) make it difficult for students to master this system. Their greatest difficulties come in applying this knowledge in solving problems.

To assist students to achieve the desired degree of mastery we (AAR and JAM) produced a computer-assisted instructional program called CIRCSIM (Rovick & Michael, 1986). CIRCSIM presents the student with a defined perturbation to the system (something that will cause a change in blood pressure) and requires the student to make qualitative predictions (increase/decrease/no change) about the responses of seven cardiovascular parameters to that perturbation. CIRCSIM requires the students to begin by predicting the qualitative behavior of seven crucial variables immediately after the perturbation (in the Direct Response or DR phase). It then analyzes these predictions and displays a screenful of canned text to remediate misunderstandings uncovered in this analysis. Next the student is asked to predict the qualitative behavior of these same variables right after the baroreceptor reflex fires (in the Reflex Response or RR phase). Again the system analyzes error patterns in the predictions and provides a canned message. Finally, the student is asked to predict the behavior of these same variables after the system has stabilized again (during the Steady State or SS phase). CIRCSIM can display over 200 different paragraphs of text.

CIRCSIM has been shown to be effective (Rovick & Michael, 1992), but its authors decided that their students might be helped even more by a system supporting natural language interaction. We set out to build a machine tutor called CIRCSIM-Tutor. CIRCSIM-Tutor requires the students to make qualitative predictions, but it also requires that students be able to explain their predictions.

Since we belong to that part of the AI community that believes that the study of effective human tutors is the key to the development of effective machine tutoring systems (Galdes, 1990), we decided to record human tutoring sessions. We analyzed the transcripts of those sessions with regard to the process of tutoring, domain knowledge, student modeling, and language generation and understanding. These tutoring sessions all began by asking the student to make predictions about the response of CV parameters to a perturbation. The ensuing tutorial dialogue is aimed at correcting student misconceptions and assisting the student to explain the effects of the CV perturbation. The goal of the tutor was to help students internalize a mental model of the CV system and learn to use it in solving CV problems. Tutoring sessions always occurred shortly before the students were scheduled to use CIRCSIM in a computer laboratory setting.

The tutors in these sessions were AAR and JAM, both of whom are professors in the Department of Physiology and both of whom are teachers in the physiology course being taken by the student subjects. Both tutors are middle-aged white males.

The students were first year medical students at Rush Medical College. They ranged in age from 21 to 37 years with a mean age of 25 years; some were female and some were male. They were paid volunteers who understood that their participation in the experiment would have no bearing on their grade in the course, although they were told that they would learn something of relevance to them. (The advertisement used to recruit participants is headed "EARN WHILE YOU LEARN!") Students participating in the tutoring sessions were selected only on the basis of their availability, and no information about their performance in the physiology course was available when they were recruited to participate.

3 The Problem-Solving Environment of Our Keyboard-To-Keyboard Sessions

To directly capture the tutorial dialogue, we use two linked computers with the student and the tutor communicating by typing at the keyboard and reading the comments of the other on the computer screen. Before we analyze the transcripts of a keyboard-to-keyboard session, the dialogue captured is processed through a numbering program that tags each sentence or fragment with an identification number. AAR and JAM have, so far, conducted 47 tutoring sessions. These sessions are numbered K1, K2, ... K47. See Li et al. (1992) for a more detailed description of our Computer Dialogue System, which captures a dialogue between PCs.

When tutoring is carried out, whether face-to-face or employing a keyboard-to-keyboard communications channel, there are rules that define and constrain the interaction between the tutor and the student. These rules are usually implicitly understood by both parties to the interaction; some rules are generic to any tutoring interaction, while other rules are specific to the particular tutoring that is occurring. For example, the rules that govern the conduct of both parties in a session where a student in academic difficulty hires a tutor are not entirely the same as the rules that govern a session that occurs in the context of a course in which the tutor is a faculty member. In our situation it is clear that the tutoring session is controlled by the tutor/experimenter. In the same way, differences in the knowledge domain being tutored or in the kind of problems being solved will result in different tutoring rules.

One set of rules governs such aspects of the tutoring as how the communications medium is to be used (making entries at the keyboard, turn taking, how to interrupt, etc.), how the problem is defined for the student, what kind of problem solving behavior is expected of the student, what constitutes success in problem solving, how much time is available for the tutoring, etc. We will describe these constraints here, but they do not represent the focus of our interest. Rather, in the next section, we will address the rules that make up the tutoring protocol; these rules govern the interaction between the tutor and the student in the problem solving process. We call these rules the tutoring protocol.

The preceding description of our tutoring sessions applies to all 47 keyboard sessions that we have conducted. In the next section we will describe the changes that occurred in the tutoring protocols, the way in which the tutor interacted with the student, over the course of our experiments.

4 Protocols of the Keyboard-to-Keyboard Sessions

We have analyzed the transcripts of the keyboard-to-keyboard sessions to describe the differences in the tutoring protocols and the effect of these differences on the process of tutoring. The first author has also used a number of knowledge acquisition techniques (e.g., interviewing the tutors) to add to our understanding of these protocols.

A detailed investigation of the three protocols requires a framework on which the analysis can be based. The first author conducted a series of think-aloud sessions with one of the tutors (AAR) to develop a description of the problem-solving behavior of a domain expert. The next section describes the result of this analysis.

Every tutoring session alternates between two phases: a prediction collection phase (PCP) and a tutoring phase (TP). Variations in the ordering and arrangement of these two phases are the major differences between the three tutoring protocols; these differences determine the kind of information that the tutor collects about the student's cognitive state before the tutoring interaction begins.

4.1 Think-Aloud Sessions - A Method of Extracting the Problem-Solving Procedure of the Domain Expert

The prime purpose of these sessions was to determine how our tutors solve domain problems (from now on we will call this the "Expert Problem-Solving Procedure" - EPSP). In this experiment the first author (RAK) acted as the experimenter and tape-recorded the verbalization of the subject (AAR). In all, four think-aloud sessions were conducted. The experimenter selected the domain problems, from a large set of possible ones, for these sessions. Concurrent verbalizations (Ericsson & Simon, 1993) from these sessions were recorded on tape and then transcribed by hand.

The first author has analyzed these transcripts for a variety of purposes, one of which is to extract the general high level task structure of the EPSP. This task structure is shown in Figure 1 and is based upon a multi-layered qualitative causal model of our domain. A detailed discussion of the knowledge structures in our domain can be found in (Khuwaja, in preparation). The task structure of Figure 1 partitions the problem-solving behavior of the subject into the three phases used in CIRCSIM and in all the tutoring sessions (Direct Response - DR, Reflex Response - RR, and Steady State - SS).

Our analysis shows that the three most common operations performed by the subject in each phase are: identification of a physiology variable, prediction of the qualitative change in a variable, and propagation, in which the expert examines the causal relationships between the current variable and the next one and figures out which one is relevant. First among the variables that must be identified in solving a problem is the procedure variable; this is the first variable in our domain model to be affected by the perturbation in the problem description. The primary variable is the first prediction table variable to be affected. The regulated variable is the variable that the baroreceptor reflex system monitors. Neural or controlled variables are variables that are under direct neural control. The other variables are hemodynamic; they are under physical/chemical control. The correct execution of the three major operations (identification, prediction, and propagation) requires a detailed causal understanding of the domain. A detailed analysis of this causal knowledge led us to the development of a very sophisticated domain knowledge base for CIRCSIM-Tutor (v.3). A more detailed analysis of these think-aloud sessions can be found in (Khuwaja, in preparation).

Solve (CV perturbation problem)
    Solve (DR)
        Identify & Predict (Primary Variable)
            Identify & Predict (Procedural Variable)
            Propagate (Procedural Variable, Primary Variable)
        Identify & Predict (Regulated Variable)
            Propagate (Primary Variable, Regulated Variable)
        Predict (Rest of Prediction Table Variables)
            Propagate (Variable X, Variable Y)
    Solve (RR)
        Identify & Predict (Controlled Variables)
            Propagate (Regulated Variable (DR), Controlled Variables)
        Identify & Predict (Regulated Variable)
            Propagate (Controlled Variables, Regulated Variable)
        Predict (Rest of Prediction Table Variables)
            Propagate (Variable X, Variable Y)
    Solve (SS)
        Identify & Predict (Regulated & Controlled Variables)
            {Identify & Predict (Regulated Variable)
            Identify & Predict (Controlled Variables)} |
            {Identify & Predict (Controlled Variables)
            Identify & Predict (Regulated Variable)}
        Predict (Rest of Prediction Table Variables)
            {Propagate (Variable X, Variable Y)} |
            {Algebraic Addition (Variable X (DR), Variable X (RR))}

Figure 1: The Task Structure of the Expert-Problem-Solving Procedure

4.2 Three Tutoring Protocols: Common Characteristics

In this section we will describe the common characteristics of three tutoring protocols that we have used in our keyboard-to-keyboard tutoring sessions. In the following sections we will describe each protocol, independently, in detail.

All three of our tutoring protocols (Protocol 1, Protocol 2, and Protocol 3) break the tutoring process into three stages (i.e., DR, RR, and SS) just as in the Expert Problem-Solving Procedure. In each stage, two major (high level) operations are performed by the tutor. The first of these operations is "collect-student's-prediction." Here the tutor collects student's prediction for one or more physiology variables in the Prediction Table. The second operation is "tutor." These operations make up a part of the prediction collection and tutoring phases described at the beginning of Section 4. The chief difference between the protocols lies in the relationship between these two phases. All tutoring protocols for

DR require that the primary variable be predicted first and correctly, and the students are tutored until they have this correct. In DR the two groups of variables, neural and hemodynamic, are tutored differently. Tutoring in RR is driven by the two stereotypical response patterns that are possible, compensating for increased or decreased values of the regulated variable (the Mean Arterial Pressure). All three protocols rely on an "algebraic" method of making predictions in SS (which involves combining the values of the variable from DR and RR directly without considering causal propagation from one variable to the next). Most of the time, when the student used this algebraic method the tutor reinforced it. Only in a few sessions, was SS approached by the student and the tutor in a causal way. Both methods are listed in the SS portion of Figure 1.

4.3 Tutoring Protocol 1

This protocol was used in the first set (K1-K8) of transcripts and no formal specification was developed for it at the time it was used. The analysis presented in this section is solely based upon analysis of keyboard-to-keyboard sessions and interviewing and debriefing tutors.

Tutor (CV perturbation problem)
- Tutor (DR)
  - Collect & Tutor (Primary Variable)
    - Collect-student's-prediction (Primary Variable)
    - Tutor (Primary Variable)
  - Collect & Tutor (Rest of Prediction Table Variables)
    - Collect-student's-prediction (Variable X)
    - Tutor (Variable X)
- Tutor (RR)
  - Collect & Tutor (Prediction Table Variables)
    - Collect-student's-prediction (Variable X)
    - Tutor (Variable X)
- Tutor (SS)
  - Collect & Tutor (Prediction Table Variables)
    - Collect-student's-prediction (Variable X)
    - Tutor (Variable X)

Figure 2: The Task Structure of Tutoring Protocol 1

The task structure for this protocol is shown in Figure 2. The PCP and TP phases overlap greatly in this protocol. As a consequence the tutor provides immediate feedback for each student's prediction/response. If the student's prediction is correct then the tutor gives a positive acknowledgment and if needed provides additional (relevant) knowledge at that point. If the student's prediction is wrong the tutor tries to remedy the problem that caused the wrong prediction at that time before letting the student predict the remaining variables. The most obvious global plan in each stage (DR, RR, and SS) is to follow the operations in the respective stages of EPSP, but the tutor seems to avoid forcing the student to use this plan. Hence, the sequence in which variables are predicted/talked about is determined by the student. Our tutors seem to push the student to use the EPSP sequence only when the student is unable to progress in the process of problem-solving.

4.4 Tutoring Protocol 2

This is the second tutoring protocol used in our tutoring experiments (K9-K28). This time a formal specification was developed for the protocol. The task structure for this protocol is shown in Figure 3. In comparison with the first protocol, the PCP and the TP phases have a smaller overlap.

The prediction collection phase is quite complex compared to the first protocol. The tutor, besides collecting predictions for the Prediction Table variables, monitors and provides help (via a hinting process) for sequence errors. A sequence error occurs when a prediction is made at the wrong point compared to the order that the domain expert uses in the EPSP. The tutor often gives the student hints about the underlying problem-solving procedure, but this protocol does not force the student to predict variables in the EPSP sequence. The hinting process does not depend upon the problem. It reminds the student that a sequence violation has taken place and provides a general heuristic for dealing with this situation, "In order to predict a parameter you have to have predicted its determinants."

```
Solve (CV perturbation problem)
    Solve (DR)
        Collect & Tutor (Primary Variable)
            Collect-student's-prediction (Primary Variable)
            Tutor (Primary Variable)
        Collect (Rest of the prediction table variables)
            Collect-student's-prediction (Variable X)
                If "sequence" violation
                Then give a hint (but do not tutor)
        Tutor (Rest of Prediction Table Variables)
    Solve (RR)
        Collect (Prediction Table Variables)
            Collect-student's-prediction (Variable X)
                If "sequence" violation
                Then give a hint (but do not tutor)
        Tutor (Prediction Table Variables)
    Solve (SS)
        Collect-student's-prediction (Prediction Table Variables)
        Tutor (Prediction Table Variables)
```

Figure 3. The Task Structure of Tutoring Protocol 2

When tutoring begins, a complete column of predictions (either in DR, RR, or SS) is available to the tutor. Thus the tutor is much better informed about the student's knowledge of the domain, and is in a position to remedy the fundamental problems of the student based on patterns of errors (instead of individual errors). In the RR stage this protocol follows the pattern used in DR. The SS stage of this protocol is the same as RR except that no sequence checking is done because our analysis of the think-aloud sessions reveals that the experts have not established a definitive order.

4.5 Tutoring Protocol 3

This is the third protocol used in our human tutoring experiments (K20-K48). This protocol was carefully thought out and formalized by the authors at Rush before it was used in a tutoring situation. The high level task structure for this protocol is shown in Figure 4. The amount of overlap between the PCP and the TP phases is even smaller than in protocol 2 (see Figure 4). This small overlap is entirely due to the tutoring about the primary variable.

The prediction collection phase or PCP, is relatively simple compared to Protocol 2. The tutor does not provide any help in this phase and allows the student to predict variables in any order. After the tutoring about the primary variable in DR, the tutor is just a silent viewer. His job is restricted to prompting the student with questions like: "What variable do you want to predict next?" to obtain the next Prediction Table variable and "OK, how will it change?" to get the prediction for that variable.

Solve (CV perturbation problem)
    Solve (DR)
        Collect & Tutor (Primary Variable)
            Collect-student's-prediction (Primary Variable)
            Tutor (Primary Variable)
        Collect & Tutor (Rest of Prediction Table Variables)
            Collect-student's-prediction (Rest of Prediction Table Variables)
            Tutor (Rest of Prediction Table Variables)
    Solve (RR)
        Collect-student's-prediction (Prediction Table Variables)
        Tutor (Prediction Table Variables)
    Solve (SS)
        Collect-student's-prediction (Prediction Table Variables)
        Tutor (Prediction Table Variables)

Figure 4. The Task Structure of Tutoring Protocol 3

Before tutoring, the tutor has the student's complete solution for a stage, which means much more information is available than in the first protocol. Also, the tutor never interrupts the student during the PCP (unlike the second protocol), hence the record of the sequence of predictions and their values represent a true record of the student's performance. The tutoring phase is again, as in the second protocol, error-pattern driven; it lets the tutor start to develop a hypothesis immediately about the underlying cause of the student's problem. The sequence in which the student's prediction errors are remedied is not based upon the sequence of predictions from the EPSP. Instead, the tutor attacks fundamental student problems. It is the severity of the problem that determines the sequence in which errors are ordered and tutored.

5 Discussion and Conclusion

Comparison of the three tutoring protocols presented here shows that over a two-year period the tutors made significant changes in their tutoring protocols. During the first eight tutoring sessions that we recorded our tutors responded immediately to every student prediction. Their instincts told them to give each student as much immediate feedback as possible. As a result there is a large overlap between the prediction collection phase (labeled PCP in Figure 5) and the tutoring phase (labeled TP).

In the second set of sessions, in which the Protocol 2 was used, the tutor reduced the amount of feedback during problem-solving. The behavior of the tutor here is more like a coach. He watches the student's sequence of predictions and interrupts only if a violation takes place.

By the time the third set of tutoring sessions was carried out, the tutors had changed further. In Protocol 3 no help is provided in the PCP (see Figure 5). This protocol provides the student full freedom to use his/her knowledge to practice problem-solving under the watchful eyes of the tutor, who waits until the second phase to try to remedy the student's misconceptions. With a full column of predictions the tutor can make a very fine grained diagnosis (Michael et al., 1992) of the student's problem and tailor the tutoring accordingly.

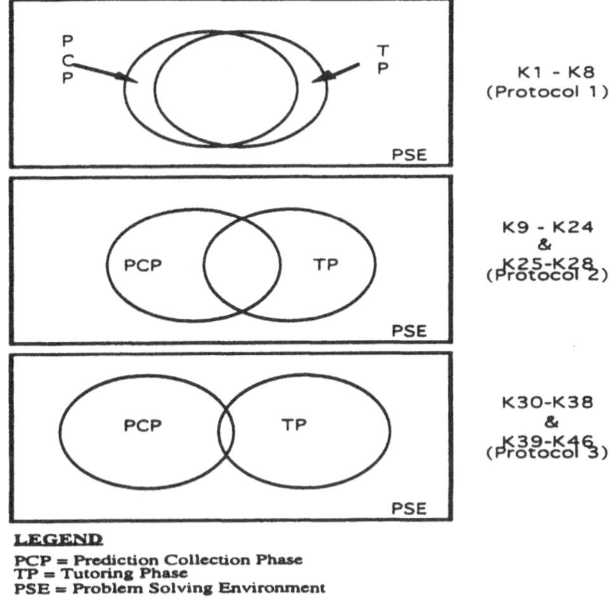

Figure 5: Interaction Between Prediction Collection and Tutoring Phases

The think-aloud sessions revealed that the tutors gradually became convinced that they tutor more effectively when they know more about the student's cognitive state before they begin. They have come to believe that they can provide much better remediation when they observe the student making a whole column of predictions. This waiting period gives them a chance to make inferences about the student's mental model of the cardiovascular system and to understand the problem-solving method that the student is trying to use.

They argue that the better the student model the more they can tailor the tutoring to the student and the more the student learns.

Can we present evidence for this argument? Not yet. There are too many other uncontrolled variables involved in the three different sets of tutoring sessions, which took place over more than two years. Also, we did not develop a satisfactory method of evaluating the effectiveness of a tutoring session until the last set of sessions was in progress. (Our pre- and post-test method is described by Michael et al. (1992)).

We believe that in order to develop effective Intelligent Tutoring Systems, one needs to observe the behavior of effective human tutors. We believe that an ITS should have an explicit "notion" of the tutoring protocol it uses. The tutoring protocol is a high level plan of the tutor. Hence in our view a machine tutor needs an explicit planning mechanism to handle the reasoning required for the tutoring protocol. We are currently developing an instructional planner for CIRCSIM-Tutor (v.3). This planner organizes the tutor's decision making into different levels. One level in our system is used to make explicit decisions about the tutoring protocol used. Handling the tutoring protocol explicitly not only enhances the flexibility of the system but also provides an experimental tool so that a better tutoring environment (and hence tutoring protocol) for a tutoring domain could be devised. Using CIRCSIM-Tutor (v.3) we hope to be able to study the effects of different tutoring protocols on the students.

6 References

Ericsson, K. A. & Simon, H. A. (1993) Protocol Analysis: Verbal Reports as Data, The MIT Press, Cambridge, MA.

Galdes, D. K. (1990) 'An empirical study of human tutors: The implications for intelligent tutoring systems', unpublished doctoral dissertation, Department of Industrial and Systems Engineering, Ohio State University, Columbus, OH.

Khuwaja, R. A. (in preparation) 'A model of tutoring: Facilitating knowledge integration using multiple models of the domain', doctoral dissertation, Illinois Institute of Technology, Chicago, IL.

Li, J., Seu, J., Evens, M., Michael, J. & Rovick, A. (1992) 'Computer dialogue system (CDS): A system for capturing computer-mediated dialogue', Behavior Research Methods, Instruments, & Computers, 24, 535-540.

Michael, J. A., Rovick, A. A., Evens, M. W., Shim, L., Woo, C., & Kim, N. (1992) 'The uses of multiple student inputs in modeling and lesson planning in CAI and ICAI programs'. Proceedings of the 4th ICCAL conference. Springer-Verlag, Berlin, 441-452.

Rovick, A. A. & Michael, J. A. (1986) 'CIRCSIM: An IBM PC computer teaching exercise on blood pressure regulation'. In Proceedings of XXX IUPS Congress. Vancouver, Canada.

Rovick, A. A. & Michael, J. A. (1992) 'The prediction table: a tool for assessing students' knowledge'. American Journal of Physiology, 263 (Advances in Physiology Education, 8), S33-S36.

QUANTUM MECHANICAL WAVE FUNCTION COLLAPSE AND THE PROBLEM OF FREE WILL IN PHYSICS

F. WINTERBERG
Desert Research Institute
University and Community College System of Nevada
Reno, Nevada 89506

Abstract

The physical universe is the universe of the countable belonging to the transfinite cardinal number $\aleph_0$ and, because of it, a fundamental Heisenberg-type nonlinear finite difference operator field equation, to explain all particles and fields, must with necessity be Galilei invariant. In such a theory, the fundamental field assumes the role of an aether as in pre-Einstein physics, with the system of galaxies at rest relative to this field. One solution of such a quantized fundamental field theory are collapsing wave packets as they occur in quantum mechanics. A finite difference equation belonging to $\aleph_0$ can be expressed as a differential equation of infinite order, thereby also belonging to $\aleph_1$, but because an infinite order differential equation requires an infinite number of initial conditions, the outcome of a quantum mechanical event must have its cause in the much larger universe of $\aleph_1$. It is then conjectured that the free will has likewise its cause in the much larger universe of $\aleph_1$ by its power to influence and violate the purely chaotic outcome of a quantum event.

The Brain as a Mechanism

At the root of all discussions about artificial intelligence is the controversy if the brain can be understood as a mechanism, ultimately to be reduced to the laws of quantum mechanics. Because quantum mechanics is statistical in its predictive power, the question has been raised if the statistical uncertainties of quantum mechanics permit the existence of what we know from our own immediate experience as "free will." It has sometimes been speculated that the "free will" comes from a few atoms in the brain, because a system consisting of a few atoms would be subject to large statistical fluctuations, described by the laws of quantum mechanics. It is argued that these fluctuations set off an amplification process within the brain eventually reaching macroscopic proportions and resulting in the phenomenon of the free will. This hypothesis, however, cannot work because the quantum mechanical fluctuations are chaotic, contradicting the rational behavior of our free will. It is mainly for these reasons that the hypothesis our brain is nothing more than a (admittedly very complex) machine (resp. computer) obtains its strongest support. This

E. A. Yfantis (ed.), Intelligent Systems, 119–133.

hypothesis is also known as the materialist view. It denies the existence of a hypothetical nonphysical entity, which somehow steers the brain and which is the seat of our being.

Our consciousness is the most fundamental experience we have about the universe, but the least understood. One thing should be certain: If a Super-Edison would be able to make an exact replica of a man - molecule-by-molecule, cell-by-cell - there can be little doubt that such a replica would experience the same kind of consciousness. Consciousness therefore rather appears as a not understood natural phenomenon. In the history of science, such a lack of knowledge is not new. The ancient Greeks for example, had no knowledge about chemistry and quantum mechanics and therefore were unable to understand the phenomenon of fire, even though they had the empirical knowledge for the conditions necessary to make fire. The question therefore is: Can consciousness be explained by the laws of physics as they are known to us, or are for this understanding unknown new laws necessary, transcending the laws of physics. The existence of such laws is suggested by mathematics, because mathematics gives us insight into laws unrelated to physical reality.

For a physical understanding of consciousness, the experience of the Cartesian "I think therefore I am" is insufficient. More important is the experience: "I can act according to my free will." It is this experience which seems to contradict the materialist view, because it makes us believe that we are different from a machine. One thing is certain: If behind this phenomenon of free will are new, still unknown, laws of nature, these laws must apply to all forms of matter. Their existence would become increasingly evident by going from lower to higher forms of life, very much as quantum phenomena become increasingly more evident by going from larger to smaller dimensions.

More specifically, our experience of a free will tells us that we determine the motions of our body. In this regard, it is of the utmost importance that these motions are subject to the conservation laws of energy momentum, etc., holding both in classical and quantum mechanics. According to Noether's theorem, these conservation laws are a consequence of the invariance under translations and rotations in space, and translations in time, and their inviolability in both classical and quantum mechanics cannot be doubted. But what distinguishes quantum mechanics from classical mechanics is that classical mechanics is completely deterministic, whereas quantum mechanics permits different outcomes as long as they do not violate the conservation laws. This particular property of quantum mechanics is reminiscent of our experience of a free will, giving us a choice of different outcomes within the limits set by the conservation laws. However, as noted above, the statistical nature of quantum mechanics can by itself not explain the phenomenon of the free will, because the different outcomes of quantum mechanics are chaotic, whereas the different possible outcomes motivated by our free will are ordered. It appears that this is not the only feature quantum mechanics and the free will have in common. According to the accepted quantum mechanical doctrine, there is no cause which determines the outcome actually taking place. The same seems to be true for actions done by a free will. By its very definition, the idea of a free will means complete freedom from all causes. In the context of quantum mechanics, Einstein expressed his disbelief in the violation of the law of causality by his saying "God does not play dice."

Because quantum mechanics and the free will seem to have in common a violation of the law of causality, we make the hypothesis that the stochastic behavior of quantum mechanics becomes gradually ordered by going from lower to higher forms of life, and is strictly true only in the asymptotic limit of a "lifeless" object. In this limit, the free will would become completely chaotic, as it is the case for the different possible outcomes in quantum mechanics. To explore the possible meaning of such a hypothesis, the still unresolved interpretation of quantum mechanics comes into focus.

The Interpretation of Quantum Mechanics

In predicting the outcome of an event in classical mechanics, one has to measure the positions and velocities of a mechanical system at some initial time. Inserted into the solutions of the Newtonian equations of motion, they determine the future of the system. The measurements are, of course, always somewhat inaccurate and, hence, the prediction derived from them, but this is not a principal obstacle to obtain a better prediction, because a better set of initial conditions through improved measurements can always be made in principle. In quantum mechanics, the situation is quite different. There the positions and velocities cannot be simultaneously measured with unlimited accuracy because of Heisenberg's uncertainty relations. The evolution in time of a quantum mechanical system is governed by Schrödinger's equation. It has the mathematical structure of a wave equation in the higher dimensional configuration space of all particles making up the system. A fundamental property of Schrödinger's equation is the spreading out of the wave function in space as time passes and, therefore, it appears that each particle gradually loses the localized character of a particle. In reality that is not what happens, because if a measurement is made at some later time, the particles are again found within a small volume, as if the wave function had collapsed into the volumes within which the particles are found. This strange collapse of the wave function can be demonstrated in the so-called double slit experiment. According to Feynman (1965), it shows all the paradoxes of quantum mechanics. In this experiment, a sufficiently intense beam of particles hits a wall containing two slits with a screen behind absorbing the particles. Because the particles are described by a Schrödinger wave, they generate on the screen an interference pattern. It therefore seems that the particles are not particles at all, but waves because only waves can explain the interference pattern. But if the intensity of the beam is reduced, one observes that the interference pattern of bright and dark fringes is generated by a larger or smaller number of localized flashes on the screen. The only rational explanation for the flashes is that the wave producing the interference pattern breaks up and "condenses," resp.

collapses, into the localized regions from where the flashes come. The impression for the existence of particles therefore appears as an illusion caused by this collapse of the wave into small volumes. If the beam is a light beam, this collapse would have to go with superluminal speed, contradicting the theory of relativity, outlawing velocities in excess of the velocity of light.

As a way out of this dilemma, two interpretations for this wave function collapse have been offered. In the so-called Copenhagen interpretation, the Schrödinger wave function is not seen as real, but only as a measure of our knowledge. In this interpretation, an event always results from the combined act of the mental and physical process recording the event (von Weizsäcker, 1977). And as our knowledge can abruptly change, so can the wave function through the result of a measurement. Different observers can, for this reason, have different wave functions and the appearance of superluminal events is not real. The Copenhagen interpretation is, for many physicists, very unsatisfactory because it introduces a subjective element. With few places in the physical universe where conscious observers are present, the Copenhagen interpretation contradicts our conviction that the universe is something absolute, independent from man.

The alternative De Broglie-Bohm pilot wave interpretation assigns the wave function the property of a new kind of force field, guiding a particle on a Brownian-motion-type stochastic trajectory (Bohm, 1961). This interpretation tries to give a "classical" explanation for quantum mechanics. However, it is hard to see how it can avoid the ultraviolet divergencies for this new field as they occur for the black body radiation field without quantum mechanics. Additional interpretational difficulties arise, if this model is extended to more than one particle, because the Schrödinger equation for more than one particle is a wave equation in an abstract higher dimensional configuration space. In addition, there seems to be experimental evidence contradicting the pilot wave hypothesis (Wang et al., 1991).

It appears the only way to evade these difficulties is to take the result of the double slit experiment at its face value, assuming that the wave function is real and that it can suffer a collapse with superluminal speed (Winterberg, 1991, 146). To avoid a contradiction with the theory of relativity, the experimentally verified results of the theory of relativity have to be sustained, at least in some asymptotic limit. In Einstein's theory of relativity, space and time form a continuum expressed by a Minkowskian pseudoeuclidean space-time metric. However, as it was already shown prior to Einstein by Lorentz and Poincaré, all the results of relativity can as well be understood by assuming an absolute space and absolute time, with space permeated by an aether, whereby rods in absolute motion against this aether are contracted and clocks go slower. Our knowledge that the universe has a distinguished frame of reference at rest with the very large system of galaxies, unknown at the time Einstein formulated the theory of relativity, gives this older pre-Einstein theory of relativity new credibility, because in it the system of galaxies can assumed to be at rest with regard to the aether.

Going beyond Lorentz and Poincaré, I make the hypothesis that the aether is nothing else than the fundamental field, conjectured by Heisenberg to explain all elementary particles and fields, as they are known to physics. It has been shown that a fundamental field equation proposed along these lines of reasoning admits wave modes propagating with superluminal velocities, just as they are required to explain the strange collapse of the wave function (Winterberg, 1991, 746). From there on, we may speculate what constitutes "physics" and what may lie beyond "physics," including the question if there is a hidden reality steering our mind. Light shed on this question may bring us closer to an answer of what constitutes the "free will."

The Fundamental Field Equation

To formulate a fundamental field equation in absolute space and time, we are guided by three postulates:

1. The field equation shall be Galilei invariant, with Lorentz invariance as a dynamic symmetry, valid in the asymptotic limit of low energies.
2. The only constants permitted to enter the field equations shall be Planck's constant h, Newton's gravitational constant G and the velocity of light c.
3. The field equation shall be a second order partial differential equation.

The first of these postulates is consistent with the assumption of an aether representing the fundamental field, with a distinguished reference system at rest with this aether. The second postulate follows from Planck's (1899) hypothesis that all physical constants should be derivable from the three universal constants h, G, and c. The third postulate has been made by Einstein in the context of his gravitational field equations. The reason given is that a field theory should be consistent with Newtonian mechanics making the same assumption.

The probably most simple field equation I have come up with, following these postulates, is the two-component operator field equation (Winterberg, 1990)

$$i\hbar\frac{\partial\psi_\pm}{\partial t} = \mp\frac{\hbar^2}{2m_p}\nabla^2\psi_\pm + 2\hbar c r_p^2\left(\psi_+^\dagger\psi_+ - \psi_-^\dagger\psi_-\right)\psi_\pm \tag{1}$$

where the field operators have to obey the canonical commutation relations

$$\left.\begin{aligned}\left[\psi_\pm(\underline{r})\psi_\pm^\dagger(\underline{r}')\right] &= \delta(\underline{r}-\underline{r}')\\ \left[\psi_\pm(\underline{r})\psi_\pm(\underline{r}')\right] = \left[\psi_\pm^\dagger(\underline{r})\psi_\pm^\dagger(\underline{r}')\right] &= 0\end{aligned}\right\} \tag{2}$$

In (1) $m_p = \sqrt{\hbar c/G} \simeq 2.2 \times 10^{-5}$ g and $r_p = \sqrt{\hbar G/c^3} \simeq 1.6 \times 10^{-33}$ cm, are the Planck mass and Planck length, derived from the two fundamental Planck relations $Gm_p^2 = \hbar c$ and $m_p r_p c = \hbar$.

Equation (1) resembles Heisenberg's nonlinear spinor wave equation, with the important difference that Heisenberg's equation is exactly relativistic, whereas equation (1) is exactly nonrelativistic. Equation (1) having the structure of a nonlinear Schrödinger equation, actually is the Heisenberg equation of motion for the field operators $\psi_{\pm}$.

In the framework of second quantization, equation (1) can be viewed to describe a medium composed of densely packed positive and negative Planck masses, interacting with each other via delta function contact type potentials. If the temperature of this Planck mass fluid is low, one can say that space is occupied by a two-component superfluid, one possessing a positive and the other one possessing a negative mass. This picture becomes more transparent by replacing the operator $\psi_{\pm}$, $\psi^{\dagger}_{\pm}$ with their expectation values $\phi_{\pm} = <\psi_{\pm}>$, $\phi^{*}_{\pm} = <\psi^{\dagger}_{\pm}>$, in what amounts making the Hartree approximation. (Because of the superfluid state for the positive and negative component, each described by a symmetric wave function, one actually has to use the better Hartree-Fock approximation.) It is this Planck aether model of the vacuum which hopefully can reproduce all fields and particles known to physics.

Because the Planck aether consists of both positive and negative masses it can, without the expenditure of energy, form pairs of vortex rings possessing opposite mass. Ultimately, it can form a lattice composed of such pairs of vortex rings, with the vortex core radius equal to r_p and the ring radius set equal to $r_o \gg r_p$. As Kelvin had already shown 100 years ago, a frictionless aether consisting of a lattice of vortex rings can transmit mechanical waves, which for small amplitudes have the same property as the waves derived from Maxwell's electromagnetic field equations. By an almost trivial generalization of Kelvin's derivation of electromagnetic waves, one can deduce a second transverse wave mode possessing the same property as small amplitude gravitational waves derived from Einstein's gravitational field equations (Winterberg, 1990). The model is even able to derive Dirac spinors, including the mass of the typical spinor in terms of the Planck mass. In addition, the model predicts the existence of not more than four particle families (Winterberg, 1991, 677). Finally, it can make plausible the existence of quarks (Winterberg, 1991, 551).

Hydrodynamic arguments suggest that the lattice of vortex rings is determined by the minimum drag quantum Reynolds number, whereby the ratio of the vortex ring radius and lattice constant to the vortex core radius would become equal to $\sqrt{Re} \sim 10^3$. This result is consistent with experimental results in high energy physics suggesting a unification of all interactions at a length about thousand times larger than the Planck length.

The zero point fluctuations of the Planck masses bound in the vortex filaments make them the source of longitudinal waves, leading to a scalar force coupling the vortex rings with the coupling constant turning out to be equal to Newton's constant. The Planck aether model therefore, not only can explain Maxwell- and Einstein-type waves as mechanical vortex waves, but can even provide a completely mechanistic interpretation of what is charge.

Because elementary particles appear as a kind of excitons held together by the gauge fields derived from this model, special relativity as a dynamic low energy approximation is fully recovered.

In the context of the conjectured wave function collapse as a real physical phenomenon, it is important that the Planck aether admits longitudinal wave modes involving both the positive and negative mass component of the Planck aether. These modes have a dispersion relation leading to a divergent phase velocity near the frequency*

$$\omega_o \sim c/r_o \tag{3}$$

with the phase velocity given by

$$v_{ph} \sim c \Big/ \sqrt{1 - (\omega_o/\omega)^2} \tag{4}$$

The Collapse of the Wave Function by Chaotic Entrapment

As Ehrenfest had shown, the motion of a wave packet under the influence of an outside force resembles the motion of a particle in classical physics (Ehrenfest, 1927). Only the wave packet as a whole would, for this reason, be subject to the relativistic law of motion. As long as the center of mass of a wave packet follows the laws of relativistic mechanics, there is no reason why its parts cannot assume superluminal velocities. The probability for the wave packet to collapse into a small volume would still be given by the Schrödinger equation, but the ψ-function would now present a real physical entity, not just a probability amplitude. The connection with the probability density $\psi^*\psi$ would simply mean that the probability for the collapse of the wave function to a particular point is proportional to the value of $\psi^*\psi$ at this point. It is, of course, very plausible that the collapse during which the density becomes very large, is enhanced at those locations where the density $\psi^*\psi$ prior to the collapse is large. The inner degrees of freedom of the wave packet describing the collapse would represent a kind of hidden variable, but because the collapse of the wave function is now considered real, taking place with superluminal speed in a finite time, von Neumann's nonexistence theorem for hidden variables does not apply.

We now propose a mechanism for the collapse. We first note that the particle, which would always be seen as a wave packet, is continuously exposed to the violent quantum fluctuations of the Planck aether all the way up to the Planck energy, but in particular to fluctuations of the longitudinal waves near the wavelength $\lambda_o \simeq r_o$, and where the phase velocity of these waves

* See appendix.

diverges. Because these fluctuating waves are longitudinal, they can entrap and accelerate the wave packet representing a particle to superluminal velocities, with the center of mass of the packet remaining subluminal. The region onto which the wave packet collapses must be occupied by a fermion absorbing the packet. The wavelengths of the fluctuating longitudinal waves are extremely short, and the wave pattern therefore changes very rapidly, generating within a short period of time a chaotic pattern of almost any possible form. Therefore, if from the many violently fluctuating configurations one representing a longitudinal wave field having superluminal phase velocity and converging onto a fermion is suddenly offered to the wave packet, the different internal parts of the packet are entrapped in this convergent longitudinal wave field and are accelerated with superluminal speed onto the absorbing fermion. Because the wavelengths of the fluctuating longitudinal waves represent quantum fluctuations at $\sim 10^{16}$ GeV, the collapse of the wave function can take place without appreciable disturbance even through dense matter.

For simplicity, we apply our hypothesis to the collapse of a photon (or another zero rest mass particle) wave function. A particle with zero rest mass is represented by a wave packet of wave frequency ω with a spread in time of the order $\Delta t \sim 1/\omega$. Near the frequency ω_o (with $\hbar\omega_o \simeq 10^{16}$ GeV), the wave packet of frequency ω is exposed during the time interval Δt to a very large number of chaotic oscillations of the zero point vacuum fluctuations, given by the ratio of the spherical surfaces in frequency space:

$$Z = (\omega_o/\omega)^2 \tag{5}$$

For a photon of ~ 1 eV, and with $\hbar\omega_o \simeq 10^{16}$ GeV, it follows that $Z \simeq 10^{50}$. If one of these longitudinal wave fluctuations has the form of a convergent spherical wave

$$\theta = \theta_o \frac{e^{i(kr+\omega t)}}{r}, \quad \frac{\omega}{k} = v_{ph}, \tag{6}$$

it would entrap the wave packet collapsing it with superluminal phase velocity towards $r = o$. For the collapse actually to take place, the total energy, $E = \hbar\omega$, of the wave packet must match the energy made available by the field of the convergent longitudinal wave. Since the energy of these aether waves is of the order $\hbar\omega_o \sim 10^{16}$ GeV, the matching would not be possible, except if these waves are a superposition of a positive and a negative energy wave. Because the phase velocity of these waves diverges at $\omega \sim \omega_o$, and because the collapse requires a very large phase velocity, one may assume that the positive energy wave has the energy $\hbar(\omega_o + \omega)$, and that the negative energy wave the energy $-\hbar\omega_o$, such that both combined give the energy $\hbar\omega$. The phase velocity of the negative energy wave at $\omega = \omega_o$ is infinite, but the phase velocity of the positive energy wave with a frequency $\omega_o + \omega$ with $\omega \ll \omega_o$ is according to (4) given by

$$v_{ph} \sim c\sqrt{\omega_o/\omega} \tag{7}$$

We now assume that after the wave packet has been offered a wave field having the structure (6) it together with this wave field forms a compound state. The lifetime of this compound state is $\tau \sim 1/\omega$, about equal the spread in time of the wave packet. It then follows that the largest length L, over which the wave packet can spread prior to its collapse is

$$L \simeq v_{ph}\tau = (c/\omega)\sqrt{\omega_o/\omega} \tag{8}$$

We then can show that for this correlation length L, the probability P that the wave packet is offered a longitudinal wave field causing its collapse, is of order unity. This surprising result emerges due to the immense magnitude Z of the possible chaotic wave patterns. The probability for a wave pattern to emerge out to these chaotic fluctuations, having the structure of a convergent wave field of wave number k within a volume L^3, and directed within an angle $\lambda/L \sim 1/Lk$ towards a center of convergence is

$$P = (Lk)^{-3}(Lk)^{-1} = (Lk)^{-4} = (\omega/\omega_o)^2, \tag{9}$$

It therefore follows that

$$ZP = 1 \tag{10}$$

The Origin of the Chaotic Fluctuations

In an attempt to explain the "free will", we make the hypothesis that it is an unphysical entity which can influence the quantum mechanical probabilities by influencing the value of the very large number Z of the fluctuations. To change the value of the product ZP by an appreciable amount, and thereby to change the quantum mechanical probabilities, would just need a small change in Z. One thing is certain: The hypothetical unknown entity would have to be totally different from what we believe is physics. Unlike our conscious feelings, physics consists of measurements. By their very nature, these measurements are always finite, and it is plausible to assume that this finiteness is a fundamental property of physics itself. If this is true, then physics is the world of the countable, assigned in transfinite set theory the cardinal number $\aleph_0$. The numbers belonging to $\aleph_0$ densely cover the continuum of space and time and, for this reason, are sufficient to describe physical reality. However, in mathematics there are infinitely more noncountable numbers than countable numbers. These are the transcendental numbers like π and e. They cover the continuum even more densely. According to Cantor's continuum hypothesis, they have the cardinal number $\aleph_1$, related to $\aleph_0$ by:

$$\aleph_1 = 2^{\aleph_0} \tag{11}$$

A differential equation as a model used to describe physical reality covers all numbers, including those belonging to $\aleph_1$, not only those belonging to $\aleph_0$. Therefore, if physical reality is restricted to those numbers having the cardinal number $\aleph_0$, a differential equation is, in principle, unsuitable to describe physical reality. It is for this reason that physical reality should be described by finite difference equations, and the same should be true for the fundamental field equation. With the Planck length r_p and Planck time $t_p = r_p/c$, it can be brought into the form (Winterberg, 1992):

$$it_p\Delta_1\psi_\pm = \mp \frac{1}{2}r_p^2 D_1^2\psi_\pm + 2r_p^3\left[\psi_+^\dagger\psi_+ - \psi_-^\dagger\psi_-\right]\psi_\pm \tag{12}$$

where Δ_1 and D_1 are finite difference operators. To be consistent, the field operators must obey commutation relations involving finite difference operations. For the commutator one has instead

$$\left[\psi(\underline{r})\psi^\dagger(\underline{r}')\right] = D(|\underline{r} - \underline{r}'|) \tag{13}$$

where D (|r|) is a generalized finite-distance, three-dimensional, delta function obeying the relation:

$$\frac{1}{D_1^3}D(|\underline{r} - \underline{r}'|) = 1 \tag{14}$$

The demand that physics should be described by finite difference equations, and the demand that the time sequence of cause and effect be sustained, seems to exclude the Minkowskian space-time structure of the special theory of relativity, because the property of the proximity of points in space is not relativistically invariant. It therefore follows that the introduction of a fundamental length in a relativistic quantum field theory destroys the invariance for the time sequence of cause and effect. The collapse of the wave function, going with superluminal velocity, and demanding the existence of an absolute space-time structure is therefore fully consistent with the demand that the laws of physics should be described by finite difference equations.

The remarkable feature of finite difference equations is that they are equivalent to differential equations of infinite order, suggesting a subtle relationship between $\aleph_0$ and $\aleph_1$. A differential equation of infinite order requires for its solution an infinite number of initial conditions, and it appears likely that the chaotic fluctuations have their origin in this very large number of initial conditions. The collapse of the wave function would therefore have its cause in the noncountable universe of $\aleph_1$. In a sense, this would imply the restoration of the law of causality for quantum mechanics, with the causes of the different chaotic outcomes located in the much larger universe of $\aleph_1$. But since even in this much larger universe, the laws would still be invariant under displacements in space and time, the conservation laws of physics would still apply, and the same would be true for quantum mechanics. The conservation laws resulting from the Euclidean-Galilean geometric property of space and time could, in general, not be sustained if the causes would have their origin in a higher dimensional space, a speculation which has been made from time to time. Therefore, if quantum mechanical events have their causes in the much larger universe of $\aleph_1$, the same would be true for the free will.

Under this hypothesis, artificial intelligence, understood to be realized by a machine consisting of a countable number of elements belonging to the universe $\aleph_0$ of physics, is impossible. The hypothesis can be supported by a thought experiment: An artificial intelligence machine (not a replica atom-by-atom - cell-by-cell of man), at some instance in time, having stored in it all the same information as it is stored at the same instance in the brain of a real person, with the information obtained from that brain by a set of physical measurements, is going to act deterministically. Now, if the artificial intelligence machine is built to work somewhat faster than the brain of the "real" person, and if it is brought into a sealed room with the real person, the real person can change his decision what to do against the prediction made by the artificial intelligence machine, simply because unlike the artificial intelligence machine, the real person has a free will. That this free will is connected to the much large universe of $\aleph_1$ also seems to follow from our ability to have an insight into the mathematical laws of this much larger universe.

According to Descartes, there are two entities of reality: "Res extensa" for the external world, and "res cogitans" for the inner world. With $\aleph_1$ infinitely larger than $\aleph_0$, such a dualism appears to be totally inadequate. Instead of Descartes' one-to-one dualism, it is rather a one-to-infinite relationship, like the relationship of $\aleph_0$ to $\aleph_1$.

APPENDIX

In the Hartree approximation, the operators in the nonlinear field equation are replaced by their expectation values and one obtains a nonlinear Schrödinger equation having the same form as the operator field equation. However, if the positive and negative mass components are superfluid, each component is described by a completely symmetric wave function and one has to make the Hartree-Fock approximation. The symmetric wave function of two identical Planck masses has the form

$$\psi(1,2) = \frac{1}{\sqrt{2}}\left(\varphi_1(\underline{r})\varphi_2(\underline{r}') + \varphi_1(\underline{r}')\varphi_2(\underline{r})\right) \qquad \text{(A.1)}$$

and their expectation value of a delta-function-type contact interaction is

$$< \psi(1,2)|\delta(\underline{r} - \underline{r}')|\psi(1,2) > = 2\varphi_1^2(\underline{r})\varphi_2^2(\underline{r}) \qquad \text{(A.2)}$$

with the direct and exchange integrals making an equal contribution. The Hartree-Fock approximation of (1), therefore, has a different form than (1), and is

$$i\hbar\frac{\partial\varphi_\pm}{\partial t} = \mp\frac{\hbar^2}{2m_p}\nabla^2\varphi_\pm \pm 2\hbar c r_p^2\left(2\varphi_\pm^*\varphi_\pm - \varphi_\mp^*\varphi_\mp\right)\varphi_\pm \qquad \text{(A.3)}$$

with

$$\begin{aligned} &< \psi_\pm > = \varphi_\pm \\ &< \psi_\pm^\dagger\psi_\pm\psi_\pm > \simeq 2\varphi_\pm^*\varphi_\pm^2 \\ &< \psi_\mp^\dagger\psi_\mp\psi_\pm > \simeq \varphi_\mp^*\varphi_\mp\varphi_\pm \end{aligned} \qquad \text{(A.4)}$$

Expressing (A.3) in its hydrodynamical form by putting

$$\begin{aligned} n_\pm &= \varphi_\pm^*\varphi_\pm \\ n_\pm v_\pm &= \mp\frac{i\hbar}{2m_p}\left[\varphi_\pm^*\nabla\varphi_\pm - \varphi_\pm\nabla\varphi_\pm^*\right] \end{aligned} \qquad \text{(A.5)}$$

one obtains from (A.3) (using $m_p r_p c = \hbar$)

$$\frac{\partial \mathbf{v}_\pm}{\partial t} + \nabla\left(\frac{v_\pm^2}{2}\right) = -2c^2r_p^3\nabla(2n_\pm - n_\mp) + \frac{1}{m_p}\nabla Q_\pm$$

$$\frac{\partial n_\pm}{\partial t} + \mathrm{div}(n_\pm \mathbf{v}_\pm) = 0 \qquad \text{(A.6b)}$$

$Q_\pm$ is the so-called quantum potential given by

$$Q_\pm = \frac{\hbar^2}{2m_p}\frac{\nabla^2\sqrt{n_\pm}}{\sqrt{n_\pm}} \tag{A.7}$$

For small amplitude disturbances, one can neglect the quadratic terms in (A.6a, b), and by adding and subtracting (A.6a) and neglecting the quantum potential, one has

$$\frac{\partial}{\partial t}(\underline{v}_+ + \underline{v}_-) = -2c^2r_p^3\nabla(n_+{}' + n_-{}') \tag{A.8a}$$

$$\frac{\partial}{\partial t}(\underline{v}_+ - \underline{v}_-) = -6c^2r_p^3\nabla(n_+{}' - n_-{}') \tag{A.8b}$$

where $n_\pm{}'$ are disturbances imposed on $n_\pm$, satisfying the linearized continuity equation

$$\frac{\partial n_\pm{}'}{\partial t} + n_\pm\nabla\underline{v}_\pm = o \tag{A.9}$$

Eliminating $n_\pm{}'$ from (A.8a, b) and (A.9), one obtains two wave equations

$$\frac{\partial^2}{\partial t^2}(\underline{v}_+ + \underline{v}_-) = c^2\nabla^2(\underline{v}_+ + \underline{v}_-) \tag{A.10a}$$

$$\frac{\partial^2}{\partial t^2}(\underline{v}_+ - \underline{v}_-) = 3c^2\nabla^2(\underline{v}_+ - \underline{v}_-) \tag{A.10b}$$

The first describes a wave propagating with the velocity of light. It has the characteristic feature of a compression wave with the positive and negative masses in phase, i.e., $n_+ = n_-$. In the second wave, which is faster than c by the factor $\sqrt{3}$, the total particle number is kept constant, that is $n_+ + n_- = r_p^{-3}$. The second wave does not propagate energy as long as $n_+ + n_- = r_p^{-3}$, but because it oscillates between positive to negative energies, it may entrap positive energy quasiparticles in its negative energy troughs.

If the waves propagate through the vortex lattice, the effect of the quantum potential can become important. Its influence can be estimated from the definition of the quantum potential and by expressing $n_\pm$ in terms of $v_\pm$ through the continuity equation:

$$\frac{1}{m_p}\nabla\dot{Q}_\pm = -\frac{\hbar^2}{4m_p^2}\nabla^4\underline{v}_\pm \tag{A.11}$$

Putting $\nabla^4 \sim 4/r_p^4$ (and with $m_pr_pc = \hbar$), (A.11) becomes

$$\frac{1}{m_p}\nabla\dot{Q}_\pm \sim -\frac{c^2}{r_p^2}\underline{v}_\pm \tag{A.12}$$

The average number density of Planck masses bound in the vortex filaments is of the order $(r_o/r_p)/r_o^3 = 1/r_o^2$. The average effect of the quantum potential therefore is obtained by multiplying (A.12) with the factor r_p^3/r_o^2:

$$\frac{1}{m_p}\nabla Q_\pm \sim -\left(\frac{c}{r_o}\right)^2 \underline{v}_\pm = -\omega_o^2 \underline{v}_\pm \tag{A.13}$$

The wave equations (A.10a, b) are therefore modified as follows

$$\frac{\partial^2}{\partial t^2}(\underline{v}_+ + \underline{v}_-) = c^2\nabla^2(v_+ + v_-) - \omega_o^2(\underline{v}_+ + \underline{v}_-) \tag{A.14a}$$

$$\frac{\partial^2}{\partial t^2}(\underline{v}_+ - \underline{v}_-) = 3c^2\nabla^2(v_+ - v_-) - \omega_o^2(\underline{v}_+ - \underline{v}_-) \tag{A.14b}$$

having the dispersion relations

$$\frac{\omega}{k} = \frac{c}{\sqrt{1 - (\omega_o/\omega)^2}}$$
$$\frac{\omega}{k} = \frac{\sqrt{3}\,c}{\sqrt{1 - (\omega_o/\omega)^2}} \tag{A.15}$$

and a common cut-off at $\omega = \omega_o$, where the phase velocity for both waves diverges.

REFERENCES

Bohm, D. (1961), Causality and Chance in Modern Physics, Harper & Brothers, New York.

Ehrenfest, P. (1927), Z.f. Physik. 45, 455.

Feynman, R.P. (1965), The Character of Physical Law, MIT Press, Cambridge, p. 127.

von Weizsäcker, C.F. (1977), Der Garten des Menschlichen, Carl Hanser Verlag, Munich, p. 169 ff.

Wang, L.J., X.Y. Zou and L. Mandel (1991), Phys. Rev. Lett. 66, 111.

Winterberg, F. (1988), Z.f. Naturforsch. 43a, 1131.

Winterberg, F. (1990), Z.f. Naturforsch. 45a, 1102.

Winterberg, F. (1991), Z.f. Naturforsch. 46a, 677.

Winterberg, F. (1991), Z.f. Naturforsch. 46a, 551.

Winterberg, F. (1991), Z.f. Naturforsch. 46a, 746.

Winterberg, F. (1992), Z.f. Naturforsch. 47a, 545.

EVALUATING LANGUAGE ACQUISITION MODELS — A COMMON FRAMEWORK

BRUCE S.N. CHEUNG
DEPARTMENT OF COMPUTER SCIENCE
THE UNIVERSITY OF HONG KONG
E-MAIL ADDRESS: bruce@csd.hku.hk

ROBERT C. UZGALIS
COMPUTER SCIENCE DEPARTMENT
UNIVERSITY OF AUCKLAND
E-MAIL ADDRESS: buz@cs.aukuni.ac.nz

Abstract

This paper provides a major contribution to knowledge. It develops a common framework to organize and classify the field of Automatic Language Acquisition. Such framework has been lacking. A taxonomy for classifying automatic language acquisition research is obtained according to the framework. The taxonomy obtained from the framework provides insight into the goals and approaches toward automatic language acquisition. This taxonomy also shows research possibilities which are less explored.

Keywords: *Artificial Intelligence, Automatic Langauge Acquisition, Machine Learning, Taxonomy.*

1. Introduction

Language acquisition has attracted researchers in Artificial Intelligence because nearly every child learns a language in the first five years of life despite the complexity of human languages, yet no one has formalize the process so a machine can do the same thing.

In the past two decades, automatic language acquisition has been investigated by many AI researchers. From this has come insight into possible mechanisms for language acquisition. However, evaluating an automatic language acquisition model is extremely difficult because of the vague definition of language acquisition and the complexity of the learning phenomenon. This paper develops a common framework to organize and classify the field of Automatic Language Acquisition. A taxonomy for classifying automatic language acquisition research is obtained according to the framework. The taxonomy obtained from the framework provides insight into the goals and approaches toward automatic language acquisition. This taxonomy also shows research possibilities which are less explored.

E. A. Yfantis (ed.), Intelligent Systems, 135–141.

2. A Common Framework — The Six-Questions Game

The framework here is expressed as a six-questions game and used to classify research in the field. In the six-questions game, there are a series of broad general questions about a language acquisition model, for each question, one should not provide any more precision about the model than what is required to answer the question. As questions proceed, the categorization of the model develops. The six questions are:

Question 1: What is the model for ?

The first question determines the goal of a model. Why does a certain language acquisition model exist? What is its goal? The expected answer to this question can either be ***simulation oriented*** or ***application oriented***.

Most of the language acquisition projects are simulations which try to mimic a child learning a language. They contributed much insight into the mechanisms of language acquisition. A computer model simulates a proposed psychological or cognitive model and serves to explain a set of phenomena. If simulation results match with human behavior, then the conclusion is drawn that the proposed model works.

Other projects have practical applications as their goals. They have carried out the initial exploration trying to discover a portable natural language processing system.

Question 2: What is be Learned?

Here the things-to-be-learned is an answer from one of the following classical paradigms, namely: ***Syntax-Inference***,

Syntax-semantic map, or the ***Developmental Approach***. Syntax-inference by omitting semantic distinctions provides an over-simplified model of language. It provides insight in the early stages of theory construction.

Syntax-semantic map approach recognizes syntactic structure and at the same time relates the syntactic structure to the input semantic structure. Similar to the syntax inference approach, the syntax-semantic mapping approach is used for convenience. Scholars in psychology and linguistics both find evidence that the formation of word classes and grammar rules interleave with the formation of concepts [Gleason 89, Katz 85, Levy & Schlesinger & Braine 88].

In the development approach, although some semantic entities may be assumed, most meanings must be learned in addition to the syntax and syntax-semantic map.

Question 3: What is the Input Data?

The input for a language acquisition model should reflect real situations. In application-oriented models, input comes from real situations. In simulation-oriented models, input should be an imitation of a child's language acquisition environment. For convenience, input to existing models usually consists of grammatical sentences (in the form of text) with or without meaning. Whether meaning is present depends on whether the model is syntax-inference, syntax-semantic map, or the developmental approach.

Question 4: How do Word-Class Distinctions Evolve?

In [Berman 88], Berman wrote: The way in which word-class distinctions evolve in children's emergent grammars is an issue which touches on very general concerns for language acquisition, such as whether the knowledge in question is innate or learned, and what developmental route it follows. However, Berman did not formulated the relation between the language developmental route and the way in which word-class distinctions evolve.

Word-class distinctions may either ***be assumed*** or ***not be assumed*** in the initial state. We discovered that models with assumed word-class distinctions usually assume some sophisticated linguistic theories and a specific language processing mechanism. Models without assumed word-class distinctions usually make use of a general cognitive approach — the co-occurrence statistics. This contrast reflects the contrast between *nativism* and *empiricism* of language modeling in psychology, linguistics, psycholinguistics and cognitive science. Nativism assumes innate knowledge and sometimes a specific mechanism for language acquisition, while empiricism assumes only a general information processing mechanism in human brains. The language development route for nativism usually involves determining the ordering of word classes in sentences and then more sophisticated phenomena. The language development route for empiricism usually follows the specific-to-general development sequence hypothesis: specific rules describing individual words are acquired first before they are generalized to more general rules describing word classes.

Question 5: What is the Target Grammar?

In every model, the acquired grammar is shaped by its assumptions on the grammar form and acquisition steps. In order to answer this question, one has to investigate the assumed ***grammar form*** and ***acquisition procedure***. Moreover, every language acquisition model involves ***generalization***. Overgeneralization during acquisition will result in an incorrect grammar. Therefore, each model must have criteria to stop generalization. To stop generalization too soon results in rules that are too specific. To stop generalization too late will result in a too permissive grammar. In general, there are two strategies to ensure proper degree of generalization: namely: prevention and preemption.

This answer to this question is descriptive and serve to describe details of individual models.

Question 6: What is the Metric of Success

What is "learning"? A learning model must have a definition of "learning" before one can determine whether an algorithm is learning, and one must discover the metric of success before one can evaluate how well it performs.

A general definition of learning is: a learner processes data to obtain improved performance. The definition is general because "performance" is not defined. It may be time efficiency, space efficiency, increase in the number of production rules, increase in dictionary size, decrease in error rate, etc. Different metrics of success come from different ***definitions of improved performance***.

Some learning models do not explicitly state the metric. Their implicit metric may include the size of lexicon, the number of rules, the number of word classes being formed, or the number of the mapping relationships, etc.

Although some metrics of success seem reasonable, they may lead to trouble. For example, many models increase the number of rules, or the number of sentences that can be parsed/generated. These criteria can easily be achieved by memorizing every input sentence in a separate rule (suppose the grammar form is the context free grammar). If only the latter metric is considered, the most general rule 'S $\rightarrow$ c^*' is able to accept every sentences (both grammatical or ungrammatical). Although these two trivial cases are not what we want, they satisfy the definition of learning based on the mentioned metrics of sucess. In fact there plenty of such metrics. If a model's metrics of success haven't been explicitly stated, we must determine some, and use them to measure how well the model learns.

3. The Language Acquisition Taxonomy

A taxonomy of language acquisition research can be formed based on classifications described from questions one, two and four. This taxonomy shows both milestones and unexplored areas of the field. Questions three, five and six are descriptive and serve to describe details of individual models. The taxonomy in Figure 1 shows that four areas: classes ***A, C, D*** and ***H*** have not been completely explored.

Examples for each class in the taxonomy of language acquisition research are shown:

- Application Oriented
  - Syntax-semantic Mapping

    Class B: *Word-class distinction assumed* — [Hirschman 86]

- Simulation Oriented
  - Syntax-semantic Mapping

    Class E: *Word-class distinction assumed* — [Harris 76, Wexler & Culicover 80, Pinker 84, Moulton & Robinson 85, Selfridge 86]

    Class F: *Word-class distinction NOT assumed* — [Anderson 83]

  - Developmental Approach

    Class G: *Word-class distinction assumed* — [MacWhinney 87]

  - Syntax Inference

    Class I: *Word-class distinction assumed* — [Berwick 85]

    Class J: *Word-class distinction NOT assumed* — [Berwick & Pilato 89, Wolff 87]

In [Cheung 94], the existing systems illustrating different categorizations of language acquisition research are summarized in terms of the six questions. They are: Sublanguage Acquisition [Hirschman 86] (***Class B***), Natural Language Acquisition by Robot [Harris 76] (***Class E***), Learnability Theory [Wexler&Culicover 80] (***Class E***), ACT* [Anderson 83] (***Class F***), the Competition Model [MacWhinney 87] (***Class G***), LPARSIFAL [Berwick 85] (***Class I***), SNPR [Wolff 87, Wolff 88] (***Class J***), and K-reversible Automaton Induction [Berwick& Pilato 87] (***Class J***),

Figure 1: A Taxonomy on Language Acquisition Research

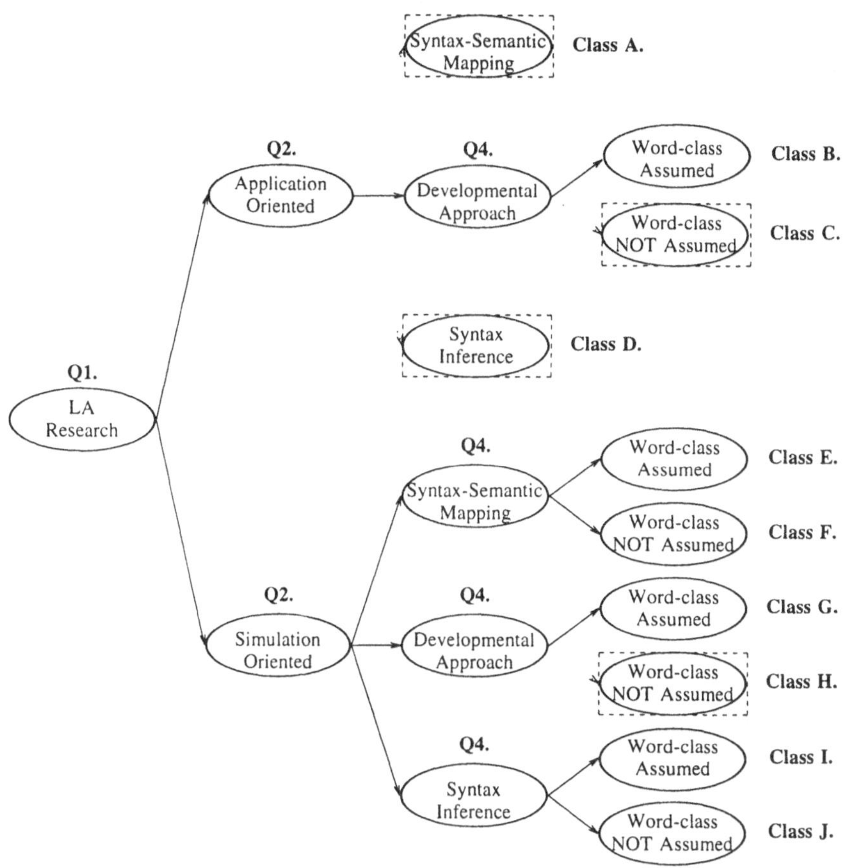

4. Conclusion

There are many different language acquisition models. This paper has described a common framework to organize them. The common framework is important so that different models can be compared and evaluated. The taxonomy implied by the framework exposes new areas which beg to be explored. They are: application oriented syntax-semantic map approach, application oriented developmental approach with word-class NOT assumed, application oriental syntax-inference approach, and simulation oriented developmental approach with word-class NOT assumed.

References

[1] John R. Anderson. 1983. *The Architecture of Cognition*. Cambridge, Mass.: Harvard University Press.

[2] Robert C. Berwick. 1985. *The Acquisition of Syntactic Knowledge*. Cambridge, Mass.: MIT Press.

[3] Robert C. Berwick, and Sam Pilato. 1987. Learning Syntax by Automata Induction. In *Machine Learning*, 2, 9-38

[4] Martin D.S. Braine. 1988. Modeling the Acquisition of Linguistic Structure. In Yonata Levy, Izchak M. Schlesinger, and Martin D.S. Braine (Eds), *Categories and Processes in Language Acquisition*. Hillsdale, New Jersey: Lawrence Erlbaum Associates.

[5] Bruce S.N. Cheung. 1994. *A Theory of Automatic Language Acquisition*. Ph.D. Thesis, University of Hong Kong.

[6] R.S. Cohen (Ed.), *The Form of Information in Science: Analysis of an Immunology Sublanguage*. Dordrecht, Boston, London: Kluwer Academic Publishers.

[7] Jean Berko Gleason (Ed). 1989. *The Development of Language*. Columbus, Ohio: Merrill Publishing Co.

[8] Larry R. Harris. 1976. Natural Language Acquisition by Robot. In J.C. Simon (Ed.), *Computer Oriented Learning*. Leyden: Noordhoff.

[9] L. Hirschman. 1986. "Discovering Sublanguage Structures". In R. Grishman, R. Kittredge (Fds), *Analyzing Language in Restricted Domains: Sublanguage Description and Processing*. Hillsdale, New Jersey: Lawrence Erlbaum Associates.

[10] Jerrold J. Katz (Ed.). 1985. *The Philosophy of Linguistics*. London, New York: Oxford University Press.

[11] Yonata Levy, Izchak M. Schlesinger, and Martin D.S. Braine (Eds). 1988. *Categories and Processes in Language Acquisition*. Hillsdale, New Jersey: Lawrence Erlbaum Associates.

[12] Pat Langley. 1982. Language Acquisition through Error Recovery. In *Cognition and Brain Theory*, 5, 211-255.

[13] Brian MacWhinney. 1987. The Competition Model. In Brian MacWhinney (Ed.), *Mechanisms of Language Acquisition*. New Jersey, London: Lawrence Erlbraum Associates.

[14] Janice Moulton, and George Robinson. 1981. *The Organization of Language*. Cambridge: Cambridge University Press.

[15] Janice Moulton, and George Robinson. 1985. Models for Language Cognition. In Andrew W. Ellis (Ed.), *Progress in the Psychology of Language*. London, Hillsdale, NewJersey: LEA.

[16] S. Pinker. 1984. *Language Learnability and Language Development*. Cambridge, Mass., London: Harvard University Press.

[17] S. Pinker. 1989. *Learnability and Cognition: the acquisition of argument structure*. Cambridge, Mass.: MIT Press.

[18] Mallory Selfridge. 1986. A Computer Model of Child Language Learning. In *Artificial Intelligence*, 29, 171-216.

[19] J. Slocum. 1986. ''How One Might Automatically Identify and Adapt to a Sublanguage: An Initial Exploration". In R. Grishman, R. Kittredge (Eds), *Analyzing Language in Restricted Domains: Sublanguage Description and Processing*. Hillsdale, New Jersey: Lawrence Erlbaum Associates.

[20] Kenneth Wexler, and Peter W. Culicover. 1980. *Formal Principles of Language Acquisition*. Cambridge, Mass.: MIT Press.

[21] J.G. Wolff. 1988. Learning Syntax and Meanings Through Optimization and Distributional Analysis. In Yonata Levy, Izchak M. Schlesinger, and Martin D.S. Braine (Eds), *Categories and Processes in Language Acquisition*. Hillsdale, New Jersey: Lawrence Erlbaum Associates.

Annotations, Signs, and Generally Paraconsistent Logics *

James J. Lu
Department of Computer Science
Bucknell University
Lewisburg, PA 17838
lu@sol.cs.bucknell.edu
(717) 524-1162

Erik Rosenthal
Department of Mathematics
University of New Haven
West Haven, CT 06516
brodsky@cs.newhaven.edu
(203) 932-7463

Abstract

Annotated logics are known to correspond to a special class of signed formulas called the *regular signed formulas*. In this paper, the fact that signed formulas can be embedded within annotated logics is proved. This embedding is used to demonstrate that annotated logics can be generalized in a manner that is paraconsistent with respect to both epistemic and ontological inconsistency while maintaining the distinction between the two types of inconsistency.

Keywords: *artificial intelligence, multiple-valued logics, signed formulas, annotated logics, inconsistency, paraconsistency*

*This research was supported in part by the National Science Foundation under grants CCR-9225037 and CCR-9202013.

E. A. Yfantis (ed.), Intelligent Systems, 143–157.

1 Introduction

Multiple-valued logics (mvl's) have been studied extensively in recent years, particularly by researchers interested in intelligent systems where knowledge representation and reasoning may require the ability to deal with uncertainty, inconsistency, and probability. One approach has been the incorporation of names for truth values directly into formulas. The language of signed formulas [23, 24, 25] employs sets of truth values called *signs* to enable the application of classical inference techniques to mvl's. Closely related notions were developed independently by Hähnle [10, 9, 11] and later in [2]. Other authors have used this approach for evidential reasoning [3] and for fuzzy operator logics [29, 18]. A number of researchers have *annotated* formulas with signs [4, 12, 13, 14, 15, 16, 6, 19, 27, 28].

Applications of annotated logics (also known as *paraconsistent* logics) have included reasoning in the presence of conflicting information [4, 13, 27, 6, 19], amalgamating databases [28, 22], and probablistic reasoning [26]. In [16], Kifer and Subrahmanian demonstrated how annotated logic programming can capture certain fragments of temporal reasoning as well as bilattice logic programming [8]. The language of signed formulas was developed to provide a classical framework for analyzing multiple-valued logics [23, 24, 25]. In [20], it was shown that annotated logics correspond to a class of signed formulas called the *regular signed formulas*. Consequently, proof theories developed for signed formulas specialize to annotated logics. This has allowed the development of improved proof procedures for several applications of annotated logics, including annotated logic programming [17], amalgamated knowledge bases [1], and theorem proving [20, 21].

One of the primary motivations for the development of annotated logics was to formalize the semantics of knowledge-based systems — for example, logic programs, deductive databases, and expert systems — that may contain inconsistent information[4, 13, 14]. In [13, 14], a distinction is made between *epistemic* inconsistency and *ontological* inconsistency of a system, each arising from a different type of negation. Intuitively, epistemic inconsistency arises from beliefs of reasoning agents, while ontological inconsistency arises from factual conflicts. It has been observed that both types of inconsistency often occur when the knowledge base of an expert system is acquired from more than one domain specialist. Current systems of annotated logics are *paraconsistent* (i.e., inconsistency tolerant) with respect to epistemic inconsistency, but these logics behave classically with respect to ontological inconsistency. Hence, when a system is ontologically inconsistent, all facts are trivially entailed. It is therefore desirable that both types of inconsistency be handled in an appropriate manner without rendering all conclusions possible.

In this paper, we further explore the relationship between annotated logics and signed formulas. We show that signed formulas may be embedded within annotated logics, thus establishing the fact that annotated logics and signed formulas are equally expressive. We use this relationship to demonstrate that annotated logics can be generalized in a manner that is paraconsistent with respect to both epistemic and ontological inconsistency while maintaining the distinction between the two.

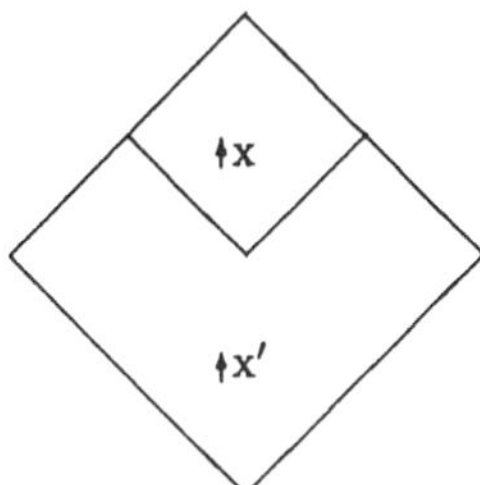

Figure 1: Regular Signs

2 Multiple-Valued Logics

We assume a language Λ consisting of logical formulas built in the usual way from a set $\mathcal{A}$ of atoms, a set of connectives, and a set of logical constants. To give a semantics to the language Λ, we associate with it a set Δ of truth values; an interpretation for Λ is then a function from $\mathcal{A}$ to Δ, i.e., an assignment of truth values to every atom in Λ. A connective Θ of arity n denotes a function $\Theta : \Delta^n \rightarrow \Delta$. Interpretations are extended in the usual way to mappings from formulas to Δ. Alternatively, a formula $\mathcal{F}$ of Λ can be regarded as denoting a mapping from interpretations to Δ.

In this paper, we assume that Λ has two binary connectives $\wedge$ and $\vee$ and several unary connectives (negation operators). We also assume that the set of truth values Δ is a complete lattice under some ordering $\leq$. The greatest and least elements of Δ are denoted $\top$ and $\perp$, respectively, and we use Sup and Inf to denote, respectively, the supremum (least upper bound) and infimum (greatest lower bound) of a subset of Δ. If $\langle P, \leq \rangle$ is any partially ordered set, and if $Q \subseteq P$, then $\uparrow Q = \{y \in P | (\exists x \in Q)\ x \leq y\}$. Note that $\uparrow Q$ is the smallest *upset* containing Q (see [7]). If Q is a singleton set $\{x\}$, then we simply write $\uparrow x$. We say that a subset Q of P is *regular* if for some $x \in P$, $Q = \uparrow x$ or $Q = (\uparrow x)'$ (the set complement of $\uparrow x$). In the former case, we call Q *positive*, and in the latter *negative*. The diagram in Figure 1 is a good way of thinking of regular signs. The larger diamond represents a lattice, and the point x is an element in the lattice. The inner diamond is the set $\uparrow x$, while the complement is the set $(\uparrow x)'$. Hähnle [10] has considered a class of multiple-valued logics, called *regular logics*, that appear to be closely related to regular signed formulas.[1]

We will be dealing with several logics in this paper. In a logic $\mathcal{L}$ in which there is a notion of satisfaction, we use the notation $I \models_{\mathcal{L}} \mathcal{F}$ to indicate that I is an interpretation that satisfies the formula $\mathcal{F}$. Similarly, If $\mathcal{F}_1$ and $\mathcal{F}_2$ are formulas in $\mathcal{L}$, we write $\mathcal{F}_1 \models_{\mathcal{L}} \mathcal{F}_2$ if whenever $I \models_{\mathcal{L}} \mathcal{F}_1$, then $I \models_{\mathcal{L}} \mathcal{F}_2$. Also, two formulas $\mathcal{F}_1$ and $\mathcal{F}_2$ in $\mathcal{L}$ are *$\mathcal{L}$-equivalent* if $I(\mathcal{F}_1) = I(\mathcal{F}_2)$ for any $\mathcal{L}$ interpretation I; we write $\mathcal{F}_1 \equiv_{\mathcal{L}} \mathcal{F}_2$.

[1] We are grateful to Reiner Hähnle for pointing out the similarity. The precise connection between the two is currently under investigation.

2.1 Signed Formulas

The development of signed formulas in [24] is motivated by the observation that there often exist two levels of reasoning when dealing with multiple-valued logics. On the one hand, at the object level, Δ represents the "truth values" over which propositions and formulas are associated. An example of an object level question is, "What is the truth value of A?" An answer denotes a value in Δ. On the other hand, at the meta-level, regardless of the choice of the underlying Δ, classical two-valued reasoning persists. The question, "Is δ the truth value of A?" has an answer that is either yes or no. The language of signed formulas attempts to capture this meta-level reasoning and is therefore cast within the framework of classical logic.

We use the term *sign* for any subset of Δ (and overload it by also using it for any expression that denotes a subset of Δ). We define a *signed formula* to be an expression of the form $S : \mathcal{F}$, where S is a sign and $\mathcal{F}$ is a formula in Λ; if $\mathcal{F}$ is an atom in Λ, we call $S : \mathcal{F}$ a *signed atom*. A signed formula is *regular* if every sign that occurs in it is regular. By Part 1 of Lemma 1 below, we may assume that no regular signed formulas have any signs of the form $(\uparrow x)'$, where $x = \bot$, since in that case $(\uparrow x)' = \emptyset$.

We are interested in signed formulas because they represent queries of the form, "Are there interpretations under which $\mathcal{F}$ evaluates to a truth value in S?" To answer such queries, we map formulas in Λ to formulas in a classical propositional logic Λ_S, called the *language of signed formulas*; it is defined as follows: The atoms are signed formulas and the connectives are (classical) conjunction and disjunction. We emphasize that a signed formula $S : \mathcal{F}$ is an atom in Λ_S regardless of the size or complexity of $\mathcal{F}$ and thus has no component parts in the language Λ_S. The set of truth values is of course $\{true, false\}$.

An arbitrary interpretation for Λ_S may make an assignment of *true* or $false$ to any signed formula (i.e., to any atom) in the usual way. Our goal is to focus attention only on those interpretations that relate to the sign in a signed formula. To accomplish this we restrict attention to *Λ-consistent interpretations.* An interpretation I over Λ assigns to each atom, and therefore to each formula $\mathcal{F}$, a truth value in Δ, and the corresponding Λ-consistent interpretation I_{Λ_S} is defined by $I_{\Lambda_S}(S : \mathcal{F}) = true$ if $I(\mathcal{F}) \in S$; $I_{\Lambda_S}(S : \mathcal{F}) = false$ if $I(\mathcal{F}) \notin S$. Note that this is a 1–1 correspondence between the set of all interpretations over Λ and the set of Λ-consistent interpretations over Λ_S. Intuitively, Λ-consistent means an assignment of *true* to all signed formulas whose signs are simultaneously achievable via some interpretation over the original language.

The following lemma from [25] is immediate.

Lemma 1. Let I_{Λ_S} be a Λ-consistent interpretation, let A be an atom and $\mathcal{F}$ a formula in Λ, and let S_1 and S_2 be signs. Then:

1. $I_{\Lambda_S}(\emptyset : \mathcal{F}) = false$;
2. $I_{\Lambda_S}(\Delta : \mathcal{F}) = true$;
3. $S_1 \subseteq S_2$ if and only if $S_1 : \mathcal{F} \models_{\Lambda_S} S_2 : \mathcal{F}$ for all formulas $\mathcal{F}$;
4. There is exactly one $\delta \in \Delta$ such that $I_{\Lambda_S}(\{\delta\} : A) = true$. □

In general, given an atom set, one can create logical formulas using any set of connectives.

Unless otherwise specified, the formulas from Λ_S that we will consider have only the two binary connectives $\wedge$ and $\vee$ and no unary connectives. There is no need for negations in the formulas: The restriction to Λ-consistent interpretations has the effect of building negations into the signs.

Many classical inference rules begin with links (complementary pairs of literals). Such rules typically deal only with formulas in which all negations are at the atomic level. Similarly, the negations in annotated logic formulas are at the "atomic level." A formula in Λ_S is called *Λ-atomic* if it has the property that whenever $S:A$ is an atom in the formula, then A is an atom in Λ; i.e., if the only atoms in the formula are signed atoms. There are many situations in which driving signs inward to the atomic level is straightforward. We will not consider them here; instead, we restrict attention to Λ-atomic formulas. Thus, for the remainder of the paper, we will use the notation Λ_S only for the logic of signed formulas restricted both to Λ-atomic formulas and to Λ-consistent interpretations.

2.2 Annotated Logics

We define the annotated logic $\mathbf{P}_\Delta$ as follows: If $A \in \mathcal{A}$ (the atom set of Λ), and if $\mu \in \Delta$ (the set of truth values), then $A : \mu$ is an *annotated atom*; an *annotated literal* is either an annotated atom or $\neg A$, where A is an annotated atom. Annotated formulas are constructed from annotated literals and the connectives $\wedge$ and $\vee$ in the standard way [5]. In this paper, we consider only formulas in *negation normal form* (NNF); i.e., formulas in which the negation symbol occurs only at the atomic level.

Interpretations for $\mathbf{P}_\Delta$ are interpretations for Λ; i.e., mappings from $\mathcal{A}$ (the set of atoms) to Δ. An interpretation I is said to satisfy

1. the annotated atom $A:\mu$ iff $I(A) \geq \mu$;
2. the annotated formula $\neg\mathcal{F}$ iff I does not satisfy $\mathcal{F}$;
3. the disjunction $\mathcal{F}_1 \vee \mathcal{F}_2$ if it satisfies either $\mathcal{F}_1$ or $\mathcal{F}_2$;
4. the conjunction $\mathcal{F}_1 \wedge \mathcal{F}_2$ if it satisfies both $\mathcal{F}_1$ and $\mathcal{F}_2$.

Annotated logics were introduced in [4, 13, 14] as a formalism for reasoning with inconsistent information. The lattice FOUR, displayed in Figure 2, which was used extensively in [4], the original study on paraconsistent logic programming, will frequently serve as an example in this paper. Intuitively, its elements represent, *inconsistent*, *true*, *false*, and *unknown*. A proposition $A:t$ over the lattice FOUR represents the assertion: "A is true." When presented with both $A:t$ and $A:f$ in different clauses in a theory T,[2] the semantics tell us that A is assigned the value $\top$, representing the assertion "A is inconsistent." Thus, the effect of contradictory information is localized to A and does not trivialize the semantics of T. See Section 4 for a more detailed discussion.

[2]A theory can be any fixed formula in the logic, but it can always be put into clause form, so in this paper we will restrict the term theory to mean a set of clauses.

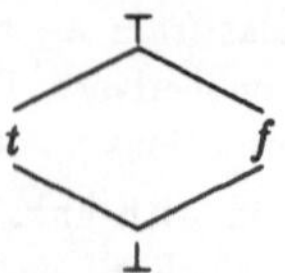

Figure 2: The Complete Lattice *FOUR*.

3 The Equivalence of Signed Formulas and Annotated Logics

In [20] it was shown that annotated logics can be captured by regular signed formulas.[3] In this section we describe that result and then show that the logic of signed formulas may be regarded as a sublogic of annotated logics.

3.1 Annotated Formulas as Regular Signed Formulas

Let Λ_S^* be the set of regular signed formulas, and define the mapping $Sgn : \mathbf{P}_\Delta \rightarrow \Lambda_S^*$ as follows. If $A : \mu$ is a positive literal, then $Sgn(A : \mu) = \uparrow\mu : A$, and if $\neg(A : \mu)$ is a negative literal, then $Sgn(\neg(A{:}\mu)) = (\uparrow\mu)' {:} A$. Since formulas in both $\mathbf{P}_\Delta$ and Λ_S^* are in NNF, the function Sgn may be extended to any formula in the obvious way. Observe that Sgn is a bijection between annotated formulas and regular signed formulas. The next lemma provides us with the key to the theorem that follows.

Lemma 2. Let I be an interpretation over Λ (and hence over $\mathbf{P}_\Delta$), and let I_{Λ_S} be the corresponding Λ-consistent interpretation. Then if $\mathcal{F}$ is any formula in $\mathbf{P}_\Delta$, $I \models_{\mathbf{P}_\Delta} \mathcal{F}$ iff $I_{\Lambda_S} \models_{\Lambda_S} Sgn(\mathcal{F})$.

Proof. Suppose first that $\mathcal{F}$ is the positive annotated literal $A : \mu$. Then I satisfies $A : \mu$ iff $I(A) \geq \mu$ iff $I(A) \in \uparrow\mu$ iff $I_{\Lambda_S}(\uparrow\mu{:}A) = \mathit{true}$. This argument extends to negative literals by taking complements and to arbitrary NNF formulas by the definitions of conjunction and disjunction. □

The theorem is now obvious.

Theorem 1. The mapping Sgn is a satisfiability preserving one-to-one correspondence between formulas in $\mathbf{P}_\Delta$ and formulas in Λ_S^*; i.e., if $\mathcal{F}$ and $\mathcal{G}$ are annotated formulas, then $\mathcal{F} \models_{\mathbf{P}_\Delta} \mathcal{G}$ iff $Sgn(\mathcal{F}) \models_{\Lambda_S} Sgn(\mathcal{G})$. □

3.2 Embedding Signed Formulas in Annotated Logics

The technique we use to embed signed formulas in annotated logics is to view each sign in a signed formula as a single annotation rather than as a collection of annotations. Since the sign is in fact

[3]This observation demonstrated that both inference rules of annotated logics are special cases of signed resolution, which, in turn, made possible a linear-style resolution proof procedure for annotated logics. This is quite useful for logic programming — see [21].

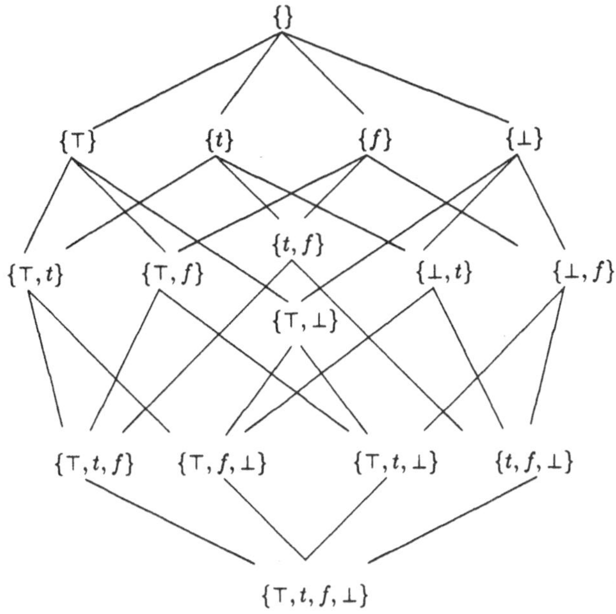

Figure 3: The Lattice $\aleph(FOUR)$.

a set, what we do is consider the annotated logic in which the underlying set of truth values is the power set (the set of all subsets) $\mathcal{P}(\Delta)$ of Δ. Let $\aleph(\Delta)$ denote the complete lattice $\langle\mathcal{P}(\Delta), \preceq\rangle$, where $a \preceq b$ if $b \subseteq a$. Note that $\aleph(\Delta)$ is the "upside-down" power set lattice of Δ. Note also that under the ordering $\preceq$, $\text{Sup}\{a, b\} = a \cap b$ while $\text{Inf}\{a, b\} = a \cup b$. For example, the sixteen element lattice $\aleph(FOUR)$ is pictured in Figure 3.

Let $\mathbf{P}_{\aleph(\Delta)}$ denote the annotated logic with $\aleph(\Delta)$ as the underlying set of truth values. Observe that if $S : A$ is any signed atom in Λ_S, then $A : S$ is an annotated atom in $\mathbf{P}_{\aleph(\Delta)}$ since $S \in \aleph(\Delta)$. Define the mapping $Ann : \Lambda_S \rightarrow \mathbf{P}_{\aleph(\Delta)}$ by $Ann(S : A) = A : S$ and extended it to arbitrary NNF formulas in the obvious way. Any annotated formula obtained under this translation is negation-free since the formulas in Λ_S are negation-free.[4] Thus Ann is a bijection between the set of all formulas in Λ_S and the positive formulas in $\mathbf{P}_{\aleph(\Delta)}$.

Interpretations for $\mathbf{P}_{\aleph(\Delta)}$ are mappings from atoms to arbitrary subsets of Δ, but Λ-consistent interpretations arise from mappings from atoms to *elements* of Δ. As a result, define an interpretation $I_\aleph$ for $\mathbf{P}_{\aleph(\Delta)}$ to be a *unit* interpretation if for every atom A, $I_\aleph(A) = \{\delta\}$ for some $\delta \in \Delta$. With each such interpretation, we associate the interpretation I over Λ defined by $I(A) = \delta$ if

[4] Recall that we are considering only negation-free Λ-atomic formulas in Λ_S.

$I_{\aleph}(A) = \{\delta\}$. This is obviously a one-to-one correspondence between unit interpretations and interpretations over Λ. Since we have already associated the Λ-consistent interpretation I_{Λ_S} with I in a one-to-one manner, this gives us a natural one-to-one correspondence between unit interpretations over $\mathbf{P}_{\aleph(\Delta)}$ and Λ-consistent interpretations over Λ_S. We write $I_{\aleph} \leftrightarrow I_{\Lambda_S}$ to denote this correspondence, which will be referred to several times in the remainder of this paper.

We will embed Λ_S in $\mathbf{P}_{\aleph(\Delta)}$ by mapping Λ_S to a sublogic of $\mathbf{P}_{\aleph(\Delta)}$. We denote the sublogic $\mathbf{P}^{+}_{\aleph(\Delta)}$ and define it to be the logic of $\mathbf{P}_{\aleph(\Delta)}$ with interpretations restricted to unit interpretations.

Lemma 3. Suppose $\mathcal{F}$ is a formula in Λ_S, and suppose that I_{Λ_S} and $I_{\aleph}$ are corresponding interpretations. Then $I_{\Lambda_S} \models_{\Lambda_S} \mathcal{F}$ iff $I_{\aleph} \models_{\mathbf{P}_{\aleph(\Delta)}} Ann(\mathcal{F})$.

Proof. It suffices to show that the lemma holds for atoms, so let $S : A$ be a signed atom, and let I be an interpretation with corresponding Λ-consistent interpretation I_{Λ_S} and corresponding interpretation $I_{\aleph}$ over $\mathbf{P}^{+}_{\aleph(\Delta)}$. Then I_{Λ_S} satisfies $S : A$ iff $I_{\Lambda_S}(S : A) = \mathit{true}$ iff $I(A) \in S$ iff $S \preceq \{I(A)\}$, where $\preceq$ is the ordering on $\aleph(\Delta)$, iff $I_{\aleph}$ satisfies $A{:}S$ in $\mathbf{P}^{+}_{\aleph(\Delta)}$. □

The next theorem follows immediately from the lemma.

Theorem 2. The mapping Ann is a satisfiability preserving one-to-one correspondence between formulas in Λ_S and formulas in $\mathbf{P}^{+}_{\aleph(\Delta)}$; i.e., if $\mathcal{F}$ and $\mathcal{G}$ are formulas in Λ_S, then $\mathcal{F} \models_{\Lambda_S} \mathcal{G}$ iff $Ann(\mathcal{F}) \models_{\mathbf{P}^{+}_{\aleph(\Delta)}} Ann(\mathcal{G})$. □

We should point out that the most general logic of signed formulas has not been embedded: The theorem applies to the logic with Λ-atomic formulas and interpretations restricted to Λ-consistent interpretations.

Composing the maps Sgn and Ann yields the following corollary to Theorems 1 and 2.

Corollary. If $\mathcal{F}$ and $\mathcal{G}$ are formulas in $\mathbf{P}_{\Delta}$, then $\mathcal{F} \models_{\mathbf{P}_{\Delta}} \mathcal{G}$ iff $Ann(Sgn(\mathcal{F})) \models_{\mathbf{P}^{+}_{\aleph(\Delta)}} Ann(Sgn(\mathcal{G}))$. □

The composite mapping $Ann \circ Sgn$ will play an important role in the next section, where we show that the non-equivalence of $\mathbf{P}_{\Delta}$ and $\mathbf{P}_{\aleph(\Delta)}$ is what makes the translation $Ann \circ Sgn$ interesting.

4 Inconsistency

Humans are able to reason in the presence of inconsistent or contradictory beliefs, and hence, in some sense, human reasoning does not take place within classical two-valued logic. Accordingly, a logic for modeling human reasoning should be *paraconsistent* — inconsistency tolerant. For example, as proposed by Kifer and Lozinskii [14], with t and $\top$ denoting, respectively, "true" and "inconsistent" in the belief lattice, the assertions $A{:}t$ and $B{:}\top$ indicate that the reasoning agent holds the belief that A is true and has an inconsistent belief regarding B. In other words, $B : \mu$ may be taken to mean: "The reasoner believes that B has a truth value at least μ." Such beliefs are held independent of reality, and hence are classified as *epistemic*. The important thing is that

$B:\top$ is satisfiable, and hence, although the reasoning agent's belief is inconsistent, inference is still possible. On the other hand, whether a reasoner's belief of a proposition is true, false, or inconsistent, there exists, independently, a real world (absolute) truth value for the proposition. Such information is *ontological*. In Kifer and Lozinskii's formulation ([13, 14]), such information must be consistent. Hence ontological inconsistency — for example, $\mathcal{F} \wedge \neg\mathcal{F}$ — is treated in a manner that is similar to classical logic: There are no models and all conclusions follow.

Suppose we choose to regard annotated logics as a formal framework on which knowledge bases may be constructed. Information stored in such a base, whether epistemic or ontological, may be gathered from multiple sources, and errors may be introduced even with respect to real world, factual information. Yet we would like to be able to use the information stored in the rest of the knowledge base — any information unrelated to the conflict — in a meaningful way. This is not possible with a standard annotated logic since ontological inconsistency is not "localized," and the logic is paraconsistent only with respect to inconsistent beliefs held by a reasoning agent. What we would like to achieve is a framework in which paraconsistency is retained with respect to both epistemic and ontological inconsistencies.

In a sense, one may view ontological information simply as beliefs at a "higher" or meta level, in which case it should be possible to tolerate ontological inconsistency without drawing arbitrary conclusions. In this section we show that if $\mathbf{P}_{\aleph(\Delta)}$ is the logic over which information is represented, a formalism is obtained in which epistemic and ontological inconsistency are both localized, yet the distinction between epistemic and ontological consequences is maintained. As a result, within such a knowledge base, one may query with respect to the beliefs of individual reasoning agents and with respect to information that is considered to consist of facts.

4.1 Negation

There are two types of negation in the annotated logics developed by Kifer and Lozinskii in [13, 14]. One is the negation symbol $\neg$ describe above, which they refer to as ontological negation. They call the second form epistemic negation, denoted $\sim$. The semantics of $\sim$ is defined by the following:

1. $\sim(A{:}\mu) \equiv_{\mathbf{P}_\Delta} A{:}(\sim\mu)$, where the latter occurrence of $\sim$ is a unary function on Δ;

2. $\sim\neg\mathcal{F} \equiv_{\mathbf{P}_\Delta} \neg\sim\mathcal{F}$, where $\mathcal{F}$ is an annotated formula;

3. $\sim(\mathcal{F}_1 \wedge \mathcal{F}_2) \equiv_{\mathbf{P}_\Delta} \sim\mathcal{F}_1 \vee \sim\mathcal{F}_2$, where $\mathcal{F}_1$ and $\mathcal{F}_2$ are annotated formulas;

4. $\sim(\mathcal{F}_1 \vee \mathcal{F}_2) \equiv_{\mathbf{P}_\Delta} \sim F_1 \wedge \sim F_2$, where $\mathcal{F}_1$ and $\mathcal{F}_2$ are annotated formulas.

For example, in the lattice FOUR, the commonly agreed upon function is the lattice isomorphism defined by $\sim t = f$, $\sim f = t$, $\sim\top = \top$, and $\sim\bot = \bot$.

Observe that repeated applications of the semantics of $\sim$ enumerated above allows the transformation of any formula in $\mathbf{P}_\Delta$ to an equivalent formula free of $\sim$. Hence, the results of the previous sections apply. What we wish to show here is that ontological negation $\neg$ in $\mathbf{P}_\Delta$ is equivalent to epistemic negation $\sim$ in $\mathbf{P}_{\aleph(\Delta)}$ in the sense described below. To accomplish this, we assume that all formulas in $\mathbf{P}_\Delta$ are free of $\sim$.

Consider the composite translation $Ann \circ Sgn$ that maps formulas in $\mathbf{P}_\Delta$ to formulas in $\mathbf{P}_{\aleph(\Delta)}$. It maps each positive annotated literal $A:\mu$ to $A:\uparrow\mu$ and each negative annotated literal $\neg A:\mu$ to $A:(\uparrow\mu)'$. Thus the range of the composite map contains no negative literals. We define a related function *Anno* that does map to negative literals. First, define $\sim$ to be any function on $\aleph(\Delta)$ that satisfies $\sim(S) = S'$ for regular subsets S of Δ.[5] We do not place any restrictions on how $\sim$ maps non-regular sets, although we could take $\sim$ to be the standard set complement function on Δ. Then define *Anno* on literals as follows: $Anno(A:\mu) = A:\uparrow\mu$ and $Anno(\neg A:\mu) = \sim A:\uparrow\mu$. The function *Anno* is extended to an arbitrary formula in $\mathbf{P}_\Delta$ by replacing each literal in the formula by its image under *Anno*.

Example. If $\mathcal{F}$ is the formula $(\neg A:t) \wedge (\neg B:t)$ in $\mathbf{P}_\Delta$, then $Anno(\mathcal{F})$ is the formula

$$(\sim A:\uparrow f) \wedge (\sim B:\uparrow t)$$

in $\mathbf{P}_{\aleph(\Delta)}$.

Observe that, given a formula $\mathcal{F}$ in $\mathbf{P}_\Delta$, the only difference between the formulas $Anno(\mathcal{F})$ and $Ann \circ Sgn(\mathcal{F})$ is that literals in $Anno(\mathcal{F})$ of the form $\sim A:\uparrow\mu$ occur in $Ann \circ Sgn(\mathcal{F})$ as $A:(\uparrow\mu)'$. Since $\uparrow\mu$ is a regular set, the definition of $\sim$ tells us that the two are equivalent in $\mathbf{P}_{\aleph(\Delta)}$. The next theorem is now immediate.

Theorem 3. Let $\mathcal{F}$ be a formula in $\mathbf{P}_\Delta$, and let I be an interpretation for $\mathbf{P}_\Delta$ with corresponding unit interpretation $I_\aleph$. Then $I \models_{\mathbf{P}_\Delta} \mathcal{F}$ iff $I_\aleph \models_{\mathbf{P}_{\aleph(\Delta)}} Anno(\mathcal{F})$ iff $I_\aleph \models_{\mathbf{P}_{\aleph(\Delta)}} Ann \circ Sgn(\mathcal{F})$. In particular, the logic $\mathbf{P}_\Delta$ is isomorphic to the sublogic $\mathbf{P}^+_{\aleph(\Delta)}$ of $\mathbf{P}_{\aleph(\Delta)}$. □

Observe that the conclusion of the theorem could have been stated (though more weakly) $I \models_{\mathbf{P}_\Delta} \mathcal{F}$ iff $I_\aleph \models_{\mathbf{P}^+_{\aleph(\Delta)}} Anno(\mathcal{F})$ iff $I_\aleph \models_{\mathbf{P}^+_{\aleph(\Delta)}} Ann \circ Sgn(\mathcal{F})$ since $\mathbf{P}^+_{\aleph(\Delta)}$ is the sublogic of $\mathbf{P}_{\aleph(\Delta)}$ formed by restricting attention to unit interpretations. Also, since $Anno(\neg A:\mu) = \sim A:\uparrow\mu$, this theorem tells us that $\neg$, the ontological negation in $\mathbf{P}_\Delta$, translates to $\sim$, the epistemic negation in $\mathbf{P}_{\aleph(\Delta)}$.[6]

4.2 Paraconsistency

We have already noted that one motivation for the development of annotated logics was *paraconsistency* — reasoning in the presence of inconsistency. Annotated logics that successfully deal with epistemic inconsistency but not with ontological inconsistency may be described as being *e-paraconsistent* but not *o-paraconsistent*. In this section we use the relationship between $\mathbf{P}_\Delta$ and $\mathbf{P}^+_{\aleph(\Delta)}$ to transform ontological inconsistency to epistemic inconsistency. This transformation can

[5] The unary function $\sim$ defined in this manner is not a lattice isomorphism, as required in [14].

[6] We have not included any discussion of implication In this paper, although Kifer and Lozinskii do make a distinction between ontological implication $\rightarrow$ and epistemic negation $\rightsquigarrow$. Here we make only the passing remark that ontological implication in $\mathbf{P}_\Delta$ translates to epistemic implication in $\mathbf{P}_{\aleph(\Delta)}$ since each implication is defined in terms of the corresponding negation operator: $\mathcal{F} \rightarrow \mathcal{G} \equiv_{\mathbf{P}_\Delta} \neg\mathcal{F} \vee \mathcal{G}$, and $\mathcal{F} \rightsquigarrow \mathcal{G} \equiv_{\mathbf{P}_\Delta} \sim\mathcal{F} \vee \mathcal{G}$.

be accomplished with either $Ann \circ Sgn$ or with $Anno$; we shall use $Ann \circ Sgn$ since the inclusion of $\sim$ into formulas does not appear to offer any conceptual advantage.

A theory T in the annotated logic $\mathbf{P}_\Delta$ is called *epistemically inconsistent* or *e-inconsistent* if $T \models_{\mathbf{P}_\Delta} A_1 : \top \vee \ldots \vee A_n : \top$ for atoms $A_1, \ldots, A_n$. Observe that this means that every model of T assigns $\top$ to at least one of the atoms $A_1, \ldots, A_n$. Kifer and Lozinskii [14] defined T to be e-inconsistent if there is a single atom A such that $T \models_{\mathbf{P}_\Delta} A : \top$. Thus, in their development, the theory consisting of the single formula $p : \top \vee q : \top$ is not e-inconsistent. We of course say that T is *epistemically consistent* or *e-consistent* if T is not e-inconsistent, and a clause of the form $p_1 : \top \vee \ldots \vee p_n : \top$ is called an *e-empty clause.* A theory T in $\mathbf{P}_\Delta$ is said to be *ontologically inconsistent* or *o-inconsistent* if T has no model (i.e., no satisfying interpretation) in $\mathbf{P}_\Delta$. Otherwise, we say that T is *ontologically consistent* or *o-consistent.*

Theorem 4. Let T be an o-consistent theory in $\mathbf{P}_\Delta$. Then $Ann(Sgn(T))$ is an e-consistent theory in $\mathbf{P}_{\aleph(\Delta)}$. Equivalently, if $Ann(Sgn(T))$ is e-inconsistent, then T is o-inconsistent.

Proof. Since the emptyset is the top element in $\aleph(\Delta)$, we must show that, for each atom A, $Ann(Sgn(T)) \not\models_{\mathbf{P}_{\aleph(\Delta)}} A:\emptyset$. So suppose there is an atom A for which $Ann(Sgn(T)) \models_{\mathbf{P}_{\aleph(\Delta)}} A:\emptyset$. Since every interpretation for $\mathbf{P}^+_{\aleph(\Delta)}$ is an interpretation for $\mathbf{P}_{\aleph(\Delta)}$ (but not the other way around), $Ann(Sgn(T)) \models_{\mathbf{P}^+_{\aleph(\Delta)}} A:\emptyset$. By Theorem 2, $Sgn(T) \models_{\Lambda^*_S} A:\emptyset$. Since no Λ-consistent interpretation may satisfy $A : \emptyset$ (Lemma 1), $Sgn(T)$ can have no model in Λ^*_S. By lemma 2, this implies that T has no model in $\mathbf{P}_\Delta$, contradicting our assumption that T is o-consistent. □

Define an interpretation I (for any annotated logic) to be *epistemically consistent* or *e-consistent* if it does not assign any atom to $\top$. Otherwise, I is *epistemically inconsistent* or *e-inconsistent.*

Lemma 4. Suppose T is a theory in $\mathbf{P}_{\aleph(\Delta)}$ that is not $\mathbf{P}^+_{\aleph(\Delta)}$-satisfiable and whose clauses contain only positive literals. Then each $\mathbf{P}_{\aleph(\Delta)}$-model of T is e-inconsistent.

Proof. Let I be a model of T in $\mathbf{P}_{\aleph(\Delta)}$. We must show that for some atom A, $I(A) = \emptyset$ since $\top = \emptyset$ in $\aleph(\Delta)$. Since I satisfies T, let $A_1 : \mu_1, ..., A_n : \mu_n$ be literals, one from each clause in T, satisfied by I. Thus $I(A_i) \geq \mu_i$ for $i = 1, ..., n$. Since each $I(A_i) \in \aleph(\Delta)$, each $I(A_i) \subseteq \Delta$. If for some i, $I(A_i)$ is empty, we are done. If not, let $\delta_i \in \mu_i$. Then $\{\delta_i\} \geq \mu_i$, so if $\tilde{I}$ is any unit interpretation such that $\tilde{I}(A_i) = \{\delta_i\}$, $\tilde{I}$ is a unit interpretation that satisfies T, contradicting the hypothesis. □

Lemma 5. A $\mathbf{P}_\Delta$-theory T is o-inconsistent iff there is an atom A and truth value $\mu \in \Delta$ such that $T \models_{\mathbf{P}_\Delta} (A:\mu \wedge \neg A:\mu)$.

Proof. Obviously, since the definition of o-inconsistent is having no model, if T is o-inconsistent then for any atom A and any truth value μ, every model of T satisfies $(A:\mu \wedge \neg A:\mu)$. Conversely, if $T \models_{\mathbf{P}_\Delta} (A:\mu \wedge \neg A:\mu)$ for some atom A and some truth value μ, suppose T has a model I. Then I satisfies $(A:\mu \wedge \neg A:\mu)$ in $\mathbf{P}_\Delta$. Thus, I satisfies both $A:\mu$ and $\neg A:\mu$, so $I(A) \geq \mu$ and $I(A) \not\geq \mu$, which is impossible. □

We are now able to prove the main result of this section.

Theorem 5. If T is an o-inconsistent theory in $\mathbf{P}_\Delta$, then $Ann(Sgn(T))$ is e-inconsistent but o-consistent in $\mathbf{P}_{\aleph(\Delta)}$. Conversely, if $Ann(Sgn(T))$ is e-inconsistent in $\mathbf{P}_{\aleph(\Delta)}$, then T is o-inconsistent in $\mathbf{P}_\Delta$.

Proof. We begin with the converse: If $Ann(Sgn(T))$ is e-inconsistent in $\mathbf{P}_{\aleph(\Delta)}$, then T is o-inconsistent in $\mathbf{P}_\Delta$ by Theorem 4. On the other hand, if T is o-inconsistent in $\mathbf{P}_\Delta$, observe first that any annotated formula containing only positive literals is o-consistent since the interpretation that assigns $\top$ to each atom satisfies the formula. Hence, since $Ann(Sgn(T))$ contains only positive literals, $Ann(Sgn(T))$ is satisfiable in $\mathbf{P}_{\aleph(\Delta)}$; i.e., $Ann(Sgn(T))$ is o-consistent in $\mathbf{P}_{\aleph(\Delta)}$.

To prove that $Ann(Sgn(T))$ is e-inconsistent in $\mathbf{P}_{\aleph(\Delta)}$, suppose it were e-consistent. Then it has a model J that does not assign any atom to the empty set (since $\top = \emptyset$ in $\mathbf{P}_{\aleph(\Delta)}$). Define the interpretation $I_\aleph$ as follows: For each atom A, choose $\delta \in J(A)$, and let $I_\aleph(A) = \{\delta\}$. Then, for every atom A, $J(A) \preceq I_\aleph(A)$ in $\aleph(\Delta)$, so, since $Ann(Sgn(T))$ contains only positive literals, $I_\aleph$ satisfies $Ann(Sgn(T))$. By Theorem 3, the corresponding interpretation I over $\mathbf{P}_\Delta$ satisfies T; this contradiction yields the desired result. □

4.3 An Example

This example illustrates Theorem 5; it comes from [14]. Consider the knowledge base T over the lattice FOUR:

$\neg bird(x){:}t \vee flies(x){:}t;$
$\neg penguin(x){:}t \vee \sim flies(x){:}t;$
$bird(tweety){:}t;$
$bird(fred){:}t;$
$penguin(fred){:}t.$

The first clause says that birds fly, and the second asserts the belief that penguins do not. We can see that T is e-inconsistent as follows: Resolving[7] on the first and fourth clauses produces the clause $flies(fred):t$. From the second and fifth clauses we infer $\sim flies(fred):t$, which is equivalent to $flies(fred){:}f$. From $flies(fred){:}t$ and $flies(fred){:}f$ we infer $flies(fred){:}\top$, so T is indeed e-inconsistent. Note that this inconsistency does not render all possible conclusions. In particular, $T \not\models_{\mathbf{P}_\Delta} flies(tweety){:}f$. Thus T is e-paraconsistent.

Suppose now that the fact $\neg penguin(fred):t$ is added to T, making the knowledge base o-inconsistent. In $\mathbf{P}_\Delta$, we may infer arbitrary conclusions from the new knowledge base, including, for example, $\neg bird(tweety):t$. But consider what happens in $\mathbf{P}_{\aleph(\Delta)}$. Applying $Ann \circ Sgn$ to T (including $\neg penguin(fred){:}t$) yields

$bird(x){:}\{t,\bot\} \vee flies(x){:}\{t,\top\};$
$penguin(x){:}\{t,\bot\} \vee flies(x){:}\{f,\top\};$
$bird(tweety){:}\{t,\top\};$
$bird(fred){:}\{t,\top\};$

[7] The reader unfamiliar with resolution in annotated logics should have no difficulty following this example simply by accepting the results of the resolution steps.

$penguin(fred){:}\{t, \top\};$
$penguin(fred){:}\{f, \bot\}.$

The last two clauses allow $penguin(fred):\emptyset$ to be inferred, so $Ann(Sgn(T))$ is e-inconsistent in $\mathbf{P}_{\aleph(\Delta)}$. On the other hand, $Ann(Sgn(\neg bird(tweety){:}t))$ is $bird(tweety){:}\{f, \bot\}$, and

$$Ann(Sgn(T)) \not\models_{\mathbf{P}_{\aleph(\Delta)}} bird(tweety) : \{f, \bot\}$$

since the $\mathbf{P}_{\aleph(\Delta)}$-interpretation that assigns $penguin(fred)$ to $\emptyset$ and $bird(tweety)$ to $\{t, \top\}$ satisfies $Ann \circ Sgn(T)$ but does not satisfy $bird(tweety){:}\{f, \bot\}$.

5 Conclusions

Annotated logics were originally developed to establish a framework for paraconsistent reasoning; i.e., for reasoning in the presence of epistemic — but not ontological — inconsistencies. The idea was that beliefs tend to be inconsistent but real world facts are incontrovertible. However, a data base may contain factual inconsistencies for any of a number of reasons, and it would seem to be desirable that such inconsistencies not render the entire data base useless. That is, an ontological inconsistency should not destroy the meaning of unrelated facts. Theorem 5 says that the move from $\mathbf{P}_{\Delta}$ to $\mathbf{P}_{\aleph(\Delta)}$ makes it possible to reason in the presence of both ontological and epistemic inconsistencies.

References

[1] Adalı, S., and Subrahmanian, V.S., Amalgamating Knowledge Bases II: Algorithms, Data Structures, and Query Processing, manuscript, 1993.

[2] Baaz, M., and Fermüller, C. G., Resolution for many-valued logics, *Proc. of Logic Programming and Automated Reasoning*, Lecture Notes in Artificial Intelligence 624, 107–118, 1992.

[3] Baldwin, J.F., Evidential Support Logic Programming, *Fuzzy Sets and Systems*, 24, 1–26, 1987.

[4] Blair, H.A. and Subrahmanian, V.S., Paraconsistent Logic Programming, *Theoretical Computer Science*, 68, 135–154, 1989.

[5] Chang, C.L, and Lee, R.C.T., *Symbolic Logic and Mechanical Theorem Proving*, Academic Press (1973).

[6] da Costa, N.C.A., Henschen, L.J., Lu, J.J., and Subrahmanian, V.S., Automatic Theorem Proving in Paraconsistent Logics: Theory and Implementation, *Proceedings of the 10th CADE*, 72–86, 1990.

[7] Davey, B.A., and Priestley, H.A., *Introduction to Lattices and Order*, Cambridge University Press (1990).

[8] Fitting, M., Bilattices and Logic Programming, in: *Journal of Logic Programming*, 11 (1991), 91-116.

[9] Hähnle, R., Towards an Efficient Tableau Rules for Multiple-valued Logics, *Proceedings of the Workshop on Computer Science Logic*, 248–260, 1990.

[10] Hähnle, R., Uniform Notation Tableau Rules for Multiple-valued Logics, *Proceedings of the International Symposium on Multiple-Valued Logic*, 26–29, 1991.

[11] Hähnle, R., *Automated Theorem Proving in Multiple-Valued Logics*, Oxford University Press, 1993.

[12] Kifer, M. and Li, A., On the Semantics of Rule-based Expert Systems with Uncertainty, *Proceedings of the 2nd International Conference on Database Theory*, 102–117, 1988.

[13] Kifer, M. and Lozinskii, E., RI: A Logic for Reasoning with Inconsistency, *IEEE Symposium on Logic in Computer Science*, 253–262, 1989.

[14] Kifer, M. and Lozinskii, E., A Logic for Reasoning with Inconsistency, *Journal of Automated Reasoning*, 9, 179–215, 1992.

[15] Kifer, M. and Subrahmanian, V.S., On the Expressive Power of Annotated Logics, *Proceedings of the North American Conference on Logic Programming*, 1069–1089, 1989. 1069-1089.

[16] Kifer, M. and Subrahmanian, V.S., Theory of Generalized Annotated Logic Programming and its Applications, *Journal of Logic Programming*, 12, 335–367, 1992.

[17] Leach, S.M. and Lu, J.J., Computing Annotated Logics, *proceedings of the International Logic Programming Conference*, 1994.

[18] Liu, X.H., Tsai, J.P., Weigert, T., Λ-resolution and the Interpretation of Λ-implication in Fuzzy Operator Logic, *Information Science*, 56, 259–278, 1991.

[19] Lu, J.J., Henschen, L.J., and Subrahmanian, V.S., N.C.A. da Costa, Reasoning in Paraconsistent Logics, *Automated Reasoning: Essays in Honor of Woody Bledsoe* (R. Boyer ed.), 181–210, 1991.

[20] Lu, J.J., Murray, N.V., and Rosenthal, E., Signed Formulas and Annotated Logics, *Proceedings of the 23rd International Symposium on Multiple-Valued Logics*, 48–53, 1993.

[21] Lu, J.J., Murray, N.V., and Rosenthal, E., A Framework for Reasoning in Multiple-Valued Annotated Logics, Computer Science Technical report 93-7, Bucknell University.

[22] Lu, J.J., Nerode, A., Remmel, J., and Subrahmanian, V.S., Towards a Theory of Hybrid Knowledge Bases, submitted. TR93-14, Cornell University, 1993.

[23] Murray, N.V. and Rosenthal, E., Resolution and Path Dissolution in Multiple-Valued Logics, *Proceedings of the International Symposium on Methodologies for Intelligent Systems*, 570–570, 1991.

[24] Murray, N.V. and Rosenthal, E., Signed Formulas: A Classical Approach to Multiple-Valued Logics, TR91-12, SUNY at Albany, 1991. Presented at the 1992 summer meeting of the ASL.

[25] Murray, N.V. and Rosenthal, E., Signed Formulas: A Liftable Meta-Logic for Multiple-Valued Logics, *Proceedings of International Symposium on Methodologies for Intelligent Systems*, 275–284, 1993.

[26] Ng, R. and Subrahmanian, V.S., A Semantical Framework for Supporting Subjective and Conditional Probabilities in Deductive Databases, *Journal of Automated Reasoning*, 10, 191–235, 1993.

[27] Subrahmanian, V.S., Paraconsistent Disjunctive Databases, *Theoretical Computer Science*, 93, 115–141, 1992.

[28] Subrahmanian, V.S., Amalgamating Knowledge Bases, *ACM Transactions on Database Systems*, (to appear).

[29] Weigert, T.J., Tsai, J-P., and Liu, X., Fuzzy Operator Logic and Fuzzy Resolution, *Journal of Automated Reasoning*, 10, 59–78, 1993.

A Type-Free System of Higher Order Logic for PROLOG[1]

Richard Butrick
Computer Science Department
Ohio University
Athens, Ohio 45701

Abstract. Ist order logic is used as a basis for the study of PROLOG. This is manifestly inappropriate. Syntactically, PROLOG permits self-predication and higher order predication (predicates of predicates). In this sense PROLOG is type-free whereas 1st order logic is not type-free and does not allow self-predication or higher order predication. Semantically, PROLOG allows for the truth of self-predication and higher order predication. The semantics of 1st order logic do not permit truth-value assignment to self-predication or higher order predications. Moreover, inferentially and semantically, PROLOG variables range over not just a domain of individuals, as is the case with 1st order logic, but also over the predicates defined over that domain. This paper develops a sound type-free system of logic with a syntax and semantics which permit self-predication and higher order prediction in general. The system introduced here, UHOL (universal higher-order logic), is much closer to the spirit of PROLOG than is 1st order logic or classical 2nd order or classical higher order logic - all of which are typed. This paper introduces a deductive apparatus for a system of logic which both syntactically and semantically allows higher order predications and proves that the deductive apparatus is sound. Matters pertaining to compactness, the Lowenheim-Skolem properties and completeness remain to be investigated.

Keywords. Type-Free Logic, Higher Order Logic, Soundness, PROLOG.

1 Prolog and Higher Order Predication

The following are legitimate PROLOG constructs:

1. self-predication: fun(fun).
2. higher order predication: good(being-cooperative). being-cooperative(Neitze).

The standard semantics of 1st order logic requires that constants in the subject position (individual constants) be assigned objects from the domain of interpretation D. Constants in the predicate position are then assigned subsets of D^n, for some n. Semantically, this precludes the same constant occupying both the subject and predicate position. In the semantics to be developed here all constants are assigned objects from the same universe (domain). Moreover, all variables, whether in the subject or predicate position, range over the same universe or domain of interpretation. This treatment of variables is much closer to the semantics of PROLOG implementations than the semantics of variables in 1st order logic. Granted, the official syntax (Edinburgh syntax) does not allow variables in the predicate position, but many implementations (LPA, Quintus) do allow variables in the predicate position.

[1]Some of the ideas central to this paper are based on the author's paper, *A Sound A-positional System of Higher Order Logic*, presented at the *Joint Conference of SEP and ASL*, York University, May 1993.

E. A. Yfantis (ed.), Intelligent Systems, 159–166.

2 Related Research

The meta-level programming capabilities of PROLOG have been extensively studied. Basically, the meta-level programming capabilities of PROLOG depend on the fact that terms (including predications and even rules) can occur in the predicate position. Moreover, Herbrand interpretations, though generally thought of in 1st order terms, allow for the truth of ground clauses of any syntactic form as permitted by the syntax of the system being considered. This potential for Herbrand interpretations allowing for the truth of higher-order predications has not, in general, been explored since 1st order Horn clause logic is generally used to study the logical properties of PROLOG. This paper develops a semantics which is seemingly very unlike Herbrand semantics. However, like Herbrand semantics, it allows for the truth of higher order predications and is in fact not as far removed from Herbrand semantics as it may at first seem.

3 UHOL (Universal Type-Free Logic)

SYNTAX

The vocabulary of the system consists of just constants and variables which may occupy any position in a predication. Thus a constant may occupy both the subject and predicate position in a predication. For example, pred2(pred1), pred1(obj), are legitimate constructs wherein pred1 occupies both the subject position and the predicate position in two different predications.

Vocabulary

1. Constants (finite): lower case letters.
2. Variables (finite): capital letters

Terms

1. Constants and variables are terms.

Atomic Sentences

1. If $\Pi, \Pi_1, \ldots , \Pi_n$ are terms then

   $\Pi\Pi_1, \ldots, \Pi_n$

is an atomic sentence.

Examples: *(a)* A *X; (b) X* A; *(c) X X; (d)* A *X,B,Y*

Sentences

1. Atomic sentences are sentences.

2. If Ψ, Φ are sentences and Δ is a variable then

$$(\Psi \,\&\, \Phi),\ (\Psi \rightarrow \Phi),\ (\Psi \vee \Phi),\ \neg\Psi,\ (\Delta)\Psi,\ (\exists\Delta)\Psi$$

are sentences.

4 Semantics

Interpretation $\mathfrak{I}$

A model M for UHOL is a structure, $< D, I, B >$. Unlike conventional Tarskian semantics, I is a function from the set of constants of UHOL into D, $I{:}C \rightarrow D$. In a conventional system a predicate constant, F^2, for example, is assigned a subset of D^2 and I is then a function from C into $D \bigcup D^2 \bigcup D^3 \ldots$. The domain D of interpretation for UHOL must contain everything used in the interpretation. Thus all the variables of UHOL range over the entire domain including what the interpretation assigns to predicates. Truth in an interpretation must also differ radically from the Tarskian basis for truth which is essentially set-based. Fa is true provided $I(a) \in I(F)$. This excludes the possibility of truth for a self-predication even if the syntax of 1st order logic permitted self-predication. Truth-value in the UHOL system is developed with reference to a basis set B which permits self-predication and closed chains of predication. Roughly, the basis set B plays the same role in the determination of truth in UHOL semantics that ground clauses play in PROLOG the determining truth-value of PROLOG clauses.

Let L_c be a set of constants for a UHOL language L and let D be a non-empty domain of objects. An interpretation $\mathfrak{I}$ is a structure
$\mathfrak{I} = < D, I, B >$
where
I: $L_c \rightarrow D$ and
B is a possibly empty set of n-tuples of D.

Satisfaction

$\mathfrak{I} \models_\alpha \Phi$ ($\mathfrak{I}$ models Φ at α)

Let α: $V \rightarrow D$, where V is the set of variables of L.

Alpha-variants: α' is said to be an alpha-variant of α at χ ($\alpha' =_\chi \alpha$) provided $\alpha' = \alpha$ or differs from α only with regard to what is assigned to χ.

Atomic Sentences

$\mathfrak{I} \models_\alpha \Pi\Pi_1 \ldots \Pi_n$ provided that $< I^*(\Pi), I^*(\Pi_1), \ldots, I^*(\Pi_n) > \in B$ where

(a) $\Pi, \Pi_1, \ldots, \Pi_n$ are constants or variables

(b) $I^*(\Delta)$ is $I(\Delta)$, if Δ is a constant and $I^*(\Delta)$ is $\alpha(\Delta)$, if Δ is a variable.

Compound Sentences

Satisfaction for compounds are treated as in standard Tarskian semantics.

Generalizations

$\mathfrak{I} \models_\alpha (\chi)\Phi$ provided that $\mathfrak{I} \models_{\alpha'} \Phi$, for all $\alpha' =_\chi \alpha$.

$\mathfrak{I} \models_\alpha (\exists\chi)\Phi$ provided that $\mathfrak{I} \models_{\alpha'} \Phi$ for some $\alpha' =_\chi \alpha$.

Examples:

1. $L_c = \{A, B\}$
(a) $D = \{1,2,3\}$ $B = \{<1,1>, <1,2>, <1,3>\}$ $I = \{<A,1>, <B,2>\}$

Then the following are true sentences:

(i) AA
(ii) $(X)AX$
(iii) -$(X)XX$
(iv) $(X)\neg BX$
(v) -AAA

(b) D = $\{1,2,3\}$ $B = \{<1,1,1>, <1,2,2>, <1,3,3>, <2,2,2>, <3,3,3>\}$ $I = \{<A,1>, <B,2>\}$

Then the following are true sentences:

(i) $(X)(\exists Y)AXY$
(ii) $\neg(A)(\exists Y)BXY$
(iii) $(X)(\exists Y)(\exists Z)XYZ$
(iv) $\neg[(X)(\exists Y)AXY \rightarrow (\exists Y)(X)AXY]$

2. Let $L_c = \{N, O, S, E\}$
$D = \{n, s, e\} \bigcup \mathbf{N}$ *(the natural numbers)*
$B = \{<n, x>: x \in \} \mathbf{N} \} \bigcup$
$\{<s, x, y>: y = x + 1, \textit{ for } x,y \in \mathbf{N} \} \bigcup$
$\{<e, x, x>: x \in \mathbf{N} \}$
$I = \{<N, n>, <O, 0>, <S, s>, <E, e>\}$

Then the following are true

(i) NO
(ii) $(X)(Y)(NX \wedge SXY \rightarrow NY)$
(iii) $\neg(\exists X)SXO$
(iv) $(X)(Y)(Z)(SXY \wedge SXZ \rightarrow EYZ)$
(v)} $[\Phi O \wedge (X)(\Phi X \wedge SXY \rightarrow \Phi Y)] \rightarrow (X)(NX \rightarrow \Phi X)$

3. Proof that, $(X)[(\exists Y)(LXY \wedge LYA) \rightarrow \neg LXA]$ $(= (X)\Phi X)$, is true in the following interpretation (M):

$D = \{l, a, b, c\}$
$B = \{< l, b, a >, < l, c, b >\}$
$I = \{< L, l >, < A, a >\}$

To prove: $M \models (X)\Phi X$. By definition, $M \models (X)\Phi X$ only if $M \models_d (X)\Phi X$, for all d. Assume to the contrary, $M \models_d (\exists X)\neg\Phi X$, for some d. Then $M \models_{d'} \neg\Phi X$, for $d' =_X d$ (d' is the same as d except perhaps at X). But $M \models_{d'} \neg\Phi X$
only if
$M \models_{d'} (\exists Y)(LXY \wedge LXA)$ and $M \models_{d'} LXA$
only if
$M \models_{d''} LXY$ and $M \models_{d''} LYA$ $(d'' =_Y d')$ and $d'(X) = b$
only if
$d''(Y) = b$ and $d''(X) = c$. But d'' can differ from d' only with regard to what it assigns to Y.

5 Some Comments Regarding UHOL and 1st Order Logic

In general, the meaning of the UHOL universal quantifier is not the same as the meaning of the universal quantifier in 1st order logic. The 'for all' in UHOL literally means 'for all'. In 1st order logic the 'for all' means for all objects of the lowest order as specified in the domain of interpretation. The variables of 1st order logic do not range over the sets assigned to predicates of the system. 1st order variables range over the 'individuals' in the domain. 2nd order variables range over the power set of the domain or Cartesian products thereof. UHOL is neither an extension nor a restriction of 1st order logic. With respect to 1st order logic UHOL constitutes a paradigm shift somewhat on the order of moving from one programming paradigm to another. Arguably, anything that can be done in 1st order logic can be done in UHOL. But not necessarily the same way.

The question of whether UHOL is as rich a paradigm as 1st order logic for representing knowledge and carrying out inferences could be tested by giving a UHOL axiomatization of set theory. This is like asking whether UHOL can be used to carry out mathematical reasoning. If not, then it is in some respects a weaker paradigm for knowledge representation and reasoning.

The further question of whether UHOL is a richer paradigm for knowledge representation and inference than 1st order logic could be based on the claim UHOL can be used to represent higher order reasoning in a direct way without a typing system. However, UHOL cannot be used to make the same statements as classical 2nd order logic. The special properties of classical 2nd order logic which allow complete formalizations of theories which are incompleteable by 1st order means depend on a "full semantics" (Shapiro [1]).

INFERENCE

A sound natural deduction inference system is developed for UHOL.

Derivation

A derivation is a series of lines of the form $P\ .\ \Phi$, Where Φ is a sentence and P is a set of sentences indicating the premise dependencies of Φ. Every line in a derivation is either a premise or an inferred line. The premise dependency of a premise is the premise itself. The premise dependencies of an inferred line are all the premise dependencies of the lines from which it was inferred except in the case of Conditional Proof as indicated below.

Rules of Inference

This is a standard Suppes type system of natural deduction.
1. Affirming the Antecedent $\{\ P\ .\ (\phi \rightarrow \Psi),\ P'\ .\ \Phi\ \} \vdash P \bigcup P'\ .\ \Psi$
2. Denying the Consequent $\{\ P\ .\ (\Phi \rightarrow \Psi),\ P'\ .\ \neg\Psi\} \vdash P \bigcup P'\ .\ \neg\Phi$
3. Double Negation $P\ .\ \neg\neg\Phi \vdash P\ .\ \Phi$
4. Conditional Proof
   $P \bigcup \{\Psi\}\ .\ \Phi \vdash P\ .\ (\Psi \rightarrow \Phi)$
5. Universal Specification $P\ .\ (\Delta)\Phi_\Delta \vdash P\ .\ \Phi_\Gamma$

where Φ_Γ is like Φ_Δ except for containing occurrences of Γ where Φ_Δ contains Δ. Γ is any constant or any variable which does not occur bound in Φ_Δ.

6. Universal Generalization $P\ .\ \Phi \vdash P\ .\ (\Delta)\Phi$,

where Δ is any unrestricted variable. A variable is restricted if it is free in a premise. It remains restricted in any subsequent line in which it is free and which depends upon that premise.

Soundness of the Rules of Inference

To prove: if $P_1\ .\ \Phi_1, \cdots, P_k\ .\ \Phi_k \vdash\ P\ .\ \Phi$ is a rule of inference then if
$I \models_\alpha P_1$ only if $I \models_\alpha \Phi_1, \cdots, I \models_\alpha P_k$ only if $I \models_\alpha \Phi_k$, for all α, then $I \models_\alpha P$ only if $I \models_\alpha \Phi$, for all α

1. Affirming the Antecedent (AA)

Assume $I \models_\alpha P \bigcup P'$. Then $I \models_\alpha P$ and $I \models_\alpha P'$. It then follows that $I \models_\alpha (\Phi \rightarrow \Psi)$, $I \models_\alpha \Phi$. If $I \models_\alpha (\Phi \rightarrow \Psi)$ then $I \models_\alpha \Psi$ or it is not the case $I \models_\alpha \Phi$. Since $I \models_\alpha \Phi$, $I \models_\alpha \Psi$. Then $I \models_\alpha P \bigcup P'$ only if $I \models_\alpha \Psi$, for all α.

2. Denying the Consequent (DC)

Assume $I \models_\alpha P \bigcup P'$. Then $I \models_\alpha P$ and $I \models_\alpha P'$. Then $I \models_\alpha (\Phi \rightarrow \Psi)$, $I \models_\alpha \neg\Psi$. If $I \models_\alpha (\Phi \rightarrow \Psi)$ then $I \models_\alpha \Psi$ or it is not the case $I \models_\alpha \Phi$. Since $I \models_\alpha \neg\Psi$, it is not the case $I \models_\alpha \Psi$. Hence it is not the case $I \models_\alpha \Phi$ and consequently $I \models_\alpha \neg\Phi$ and $I \models_\alpha P \bigcup P'$ only if $I \models_\alpha \neg\Phi$, for all α.

3. Double Negation (DN)

Assume $I \models_\alpha P$. Then $I \models_\alpha \neg\neg\Phi$. By definition, it is not the case that $I \models_\alpha \neg\Phi$. In which case $I \models_\alpha \Phi$.

4. Conditional Proof (CP)

To prove: If $I \models_\alpha P \bigcup \{\Psi\}$ only if $I \models_\alpha \Phi$ then $I \models_\alpha P$ only if $I \models_\alpha (\Psi \rightarrow \Phi)$

Assume $I \models_\alpha P \bigcup \{\Psi\}$ only if $I \models_\alpha \Phi$, $I \models_\alpha P$. Assume it is not the case that $I \models_\alpha \Psi$. Then $I \models_\alpha (\Psi \rightarrow \Phi)$. Assume $I \models_\alpha \Psi$. Then $I \models_\alpha P \bigcup \{\Psi\}$ and by the first assumption $I \models_\alpha (\Psi \rightarrow \Phi)$. As a corollary, CP is α_Δ-variant hereditary.

5. Universal Specification $[P \,.\, (\Delta)\Phi_\Delta \vdash P \,.\, \Phi_\Gamma]$

where Φ_Γ is like Φ_Δ except for containing occurrences of Γ where Φ_Δ contains Δ. Γ is any constant or any variable which does not occur bound in Φ_Δ.

Assume $I \models_\alpha (\Delta)\Phi_\Delta$. Φ_Γ is exactly like Φ_Δ except for containing occurrences of Γ where Φ_Δ contains free occurrences of Δ. Moreover if Γ is a variable, those occurrences are not bound. By the definition of satisfaction, $I \models_{\alpha'} \Phi_\Delta$ for every α' which at most differs from α in what it assigns to Δ. Assume Γ is a constant. No matter what I assigns to Γ there is some α' which assigns the same element to Δ. Then it must be that since $I \models_{\alpha'} \Phi_\Delta$, $I \models_\alpha \Phi_\Gamma$. Assume Γ is a variable. No matter what α assigns to Γ there is some α' which assigns the same element to Δ. Then it must be that since $I \models_{\alpha'} \Phi_\Delta$, $I \models_\alpha \Phi_\Gamma$.

6. Universal Generalization

$P \,.\, \Phi \vdash P \,.\, (\Delta)\Phi$
where Δ is any unrestricted variable. A variable is is restriced if it is free in a premise. It remains restricted in any subsequent line in which it is free and which depends upon that premise.

Assume: $(\alpha)(I \models_\alpha P$ only if $I \models_\alpha \Phi)$, where Δ does not occur free in P. To prove: $(\alpha)(I \models_\alpha P$ only if $I \models_\alpha (\Delta)\Phi)$

Assume $I_\alpha \models P$. Note that Δ does not occur free in P. If $I \models_\alpha P$ and Δ does not occur free in P, then $I \models_{\alpha'} P$ where α' differs from α only with regard to what it assigns to Δ. Then, using the first assumption, $I \models_{\alpha'} \Phi$ for any α'. But this proves that $I \models_\alpha (\Delta)\Phi$.

6 Closing Conjectures and Observations

It is conjectured that UHOL is complete, compact, and that it is subject to both the the upward and downward Lowenheim- Skolem results.

UHOL is not an alternative paradigm for (full) 2nd order logic. UHOL does not provide a categorical formalization of elementary arithmetic. Nor can UHOL be used to express certain essentialy 2nd order sentences (Geach-Kaplan sentence, finitude, etc.). However, pure 2nd order logic does not even allow second or higher order predication. In this regard UHOL is a more powerful paradigm than 2nd order logic. Moreover, traditional systems of higher order logic require a typing apparatus and preclude self-predication. UHOL like PROLOG allows unlimited orders of predication and self-predication. Moreover, PROLOG allows for a fuller concept of a term than UHOL. In PROLOG predicates and sentences can be values of variables. PROLOG in this sense is an even richer paradigm for representing knowledge and reasoning than UHOL. However, UHOL can be extended to include this feature of PROLOG.

SELECTED USEFUL REFERENCES

Andrews, P. B. [1986] *An Introduction to Mathematical Logic and Type Theory: to Truth Through Proof,* Academic Press.

Bohnert, H. G. [1977] *Logic: Its Use and Basis,* University Press of America.

Boolos, G. S. [1975] "On Second-Order Logic", *The Journal of Philosophy (vol. LXXII, no. 16, Sept. 18).*

Enderton, H. [1972] *A Mathematical Introduction to Logic,* Academic Press.

Feferman, S. [1984] "Toward Useful Type-Free Theories", *Journal of Symbolic Logic, (49).*

Gibbins, P. [1988] *Logic with PROLOG,* Oxford Applied Mathematics and Computing Science Series, Clarendon Press, Oxford.

Shapiro, S. [1991] *Foundations without Foundationalism,* Oxford Logic Guides, Clarendon Press, Oxford.

SEARCH REDUCTION TECHNIQUES IN JUMPING GAMES

Irena Pevac
Lazar Pevac
Computer Science Department
Central Conn. State University
New Britain, CT 06050 USA

Abstract. Different AI techniques to reduce the search space and the importance of problem representation will be discussed in several jumping game playing problems. State space reductions that do not affect the completeness, such as reducing the unproductive and unreachable states and avoiding expanding states that are functionally dependent will be introduced. In addition, domain oriented heuristics, such as introducing a metastrategy of playing the game by selective use of operators will also be discussed.

Keywords. Problem Solving, Search Reduction, Game playing, Heuristic Search

1 Introduction

The study of problem solving and game playing in artificial intelligence has attracted the attention of several researchers since the very first days. Shannon [9] worked in chess playing, Samuel's program [8] to play checkers reached the masters level, and Shoefield [10] developed a program for puzzle games.

This area remains challenging today. Games provide researchers with a clean domain for testing new techniques and ideas in AI. For instance, Reinefeld [7] has tested the efficiency of the IDA* algorithm on 3x3 and 4x4 puzzles. Epstein [3] used a metatheoretical approach by revising a collection of metarules to play and improve the strategy while playing tic tac toe and other two player games. Recent results in different fields of computer game playing including chess, go, draughts, othello, bridge etc. are presented in Levy & Beal [4].

AI techniques for problem solving, such as the carefully chosen representation of the problem, reduction techniques like deleting unsolvable states and omitting symmetric cases, will be used to reduce the search of two game playing problems. The search space reduction will include several types of functional relationships among different trees as well as parts of them and between trees and inverse trees. The investigation of functional relationships in other domains of problem solving could reduce the search space in heuristic as well as in the complete search approach. Also the importance of the selective use of operators and choosing the suitable order of their use will be demonstrated.

E. A. Yfantis (ed.), Intelligent Systems, 167–173.

2 Definition and Formal Representation of Games

The techniques will be explained on Hexagon and a Triangle game. The games are defined on graph configurations given in Fig. 1. At the beginning each node has a peg in it. To start we remove one peg. Further, moves are performed by jumping in any direction within the grid over another peg into the empty node beyond the jumped peg. The jumped peg is removed from the graph. The aim of the game is to reach a position with only one peg left. Game Hi-Q has different graph configuration but jumps are defined in the same way. Banjeri [1] and Coray [2] have introduced different search reduction techniques for Hi-Q.

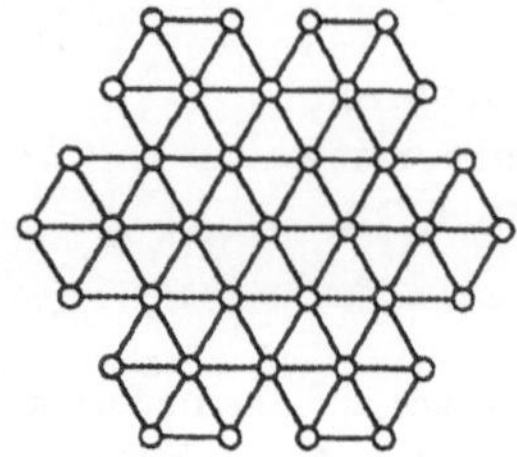

Fig. 1

The current computer programs do not accept graph configurations for further manipulation, so we need to introduce some other internal representation. When choosing among several possibilities we will select "the most suitable" one. Usually, the representation requiring less space would be preferable. The choice depends also on the heuristics introduced for the search reduction. Several representations can also be used, and then a conversion from one type to another is performed during the search. Of course, the cost of converting (time and space used to do it) should not overcome the advantages obtained by using different representations.

The current state configuration in Hexagon game can be presented as a set of those pairs (x, y) corresponding to nodes containing a peg satisfying the condition $(|x| \leq 3 \wedge |y| \leq 3 \wedge x * y < 0) \vee (|x| + |y| \leq 3 \wedge x * y \geq 0)$. The central point is a pair $(0, 0)$. This type of representation induces three scheme moves. This approach requires significant memory as every peg-filled node is represented as a pair of numbers.

A better way to represent a current state will be explained on a Triangle game to avoid technical details. Let us label the nodes as shown in Fig. 2. The current state is represented as a set of numbers that correspond to nodes that are filled with a peg. The operators are given in Fig. 3.

```
         9
      5     4
   6     0     3
8     1     2     7
```

Fig. 2

812	127	218	721	603	306
865	659	956	568	104	401
734	349	943	437	502	205

Fig. 3

The rule corresponding to triple 812 can be applied to a state S that contains 8 and 1 and does not contain 2. A new state produced is $S \setminus \{8\} \setminus \{1\} \cup \{2\}$.

In Hexagon game, a similar approach with labeled nodes such as in Fig. 4 would produce a list of 114 triples that define 114 operators for this game.

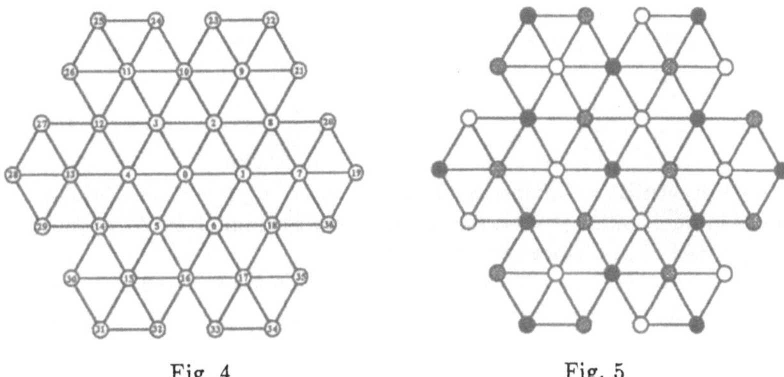

Fig. 4 Fig. 5

The search graph for the hexagon game is a forest consisting of 37 trees. The roots of these trees are initial states having 36 pegs. The leaves on level 35 are goal states having a single peg.

The inverse operator changes the current state with the triple laying on the same line in the graph configuration having two empty points and one boundary point filled with a peg to the new state in which the status of the previously empty and filled points of that triple is opposite. The status of the points outside the triple is not affected with the use of an inverse operator. The inverse search would contain 37 different trees. Each tree has the root configuration with one point filled with a peg. Instead of developing the search tree of depth 35 we may develop pairs of trees and inverse trees of depth k and 35-k respectively. The solution is obtained, for instance, if one element of the search tree of depth k matches the element of the depth 35-k in the inverse search tree.

Once the formal framework for these games (state space representation, rules and initial and goal states) has been specified, the search can be used to find one solution or a set of all solutions for each game.

3 State Space Reductions not affecting the completeness

The domain dependent knowledge for the games will be used to reduce the search space. In particular, we will prove that it is not possible to reach any goal state when we start from some initial states. Such states are unsolvable. Even if we are interested in finding the set of all possible solutions, omitting them does not eliminate any of the solutions. The technique will be explained on Hexagon game, but it can be applied to Hi-Q or the Triangle as well.

Let us introduce a node coloring into three different colors black B, white W, and shaded S as given in Fig. 5. There are 13 black points, 12 white points, and 12 shaded points in the graph for the hexagon game. The sums of black (white, shaded) nodes having a peg in it are denoted Σ_B, Σ_W, Σ_S, respectively. The initial states and final states ech are divided into three classes I_1, I_2, I_3 and F_1, F_2, F_3 according to the values of Σ_B, Σ_W, and Σ_S, respectively.

I_1	I_2	I_3	F_1	F_2	F_3
$\Sigma_B = 12$	$\Sigma_B = 13$	$\Sigma_B = 13$	$\Sigma_B = 1$	$\Sigma_B = 0$	$\Sigma_B = 0$
$\Sigma_W = 12$	$\Sigma_W = 11$	$\Sigma_W = 12$	$\Sigma_W = 0$	$\Sigma_W = 1$	$\Sigma_W = 0$
$\Sigma_S = 12$	$\Sigma_S = 12$	$\Sigma_S = 11$	$\Sigma_S = 0$	$\Sigma_S = 0$	$\Sigma_S = 1$

It will be proven that starting in an initial state of type I_1 none of the goal states can be reached. In addition, it will be shown that starting from an initial state of type I_2 (I_3) only F_3 (F_2) type of goal state can be reached.

Let us consider changing the color numbers when a move is applied to a current state. Any triple contains all three kind of colors. If we jump into a black (or white or shaded) colored point the corresponding color number Σ_B (Σ_W, Σ_S) will be increased by one, and the other two color numbers will decrease by one.

It is obvious that any type of a move changes the parity of all three color numbers. Previously odd color numbers become even and an even one becomes odd. Let us start with the initial state having 36 pegs, and apply 35 moves (odd number of moves) which lead to some final state with only one peg on the board. According to the discussion above, the color numbers in the final state must be of opposite parity to the color numbers in the initial state. Comparing initial and final types of color numbers I_1, I_2, I_3 and F_1, F_2, and F_3; it is evident that final states with color numbers of type F_2 (F_3) could be reached only starting from initial states with color numbers of type I_3 (I_2) respectively. If we start from the initial state with color numbers of type I_1 we cannot reach any final state. Also, no initial state would enable reaching the final state with color numbers of type F_1. We can delete all initial states of type I_1 since they are unsolvable and all final states of type F_1 since they are unreachable. This reduces the number of initial and final states from 37 to 24.

Further the rotation of 60° establishes 1-1 correspondence between initial configurations of type I_2 and those of type I_3. Further, it is easy to see that rotation of 60° establishes 1-1 correspondence between paths in search trees with initial states of type I_2 and paths in search trees with initial states of type I_3. Let S^c denote a complement of a state S. Let *op* denote an operator that will translate a state S_1 to state S_2 and let op^{-1} be an inverse operator as defined earlier. It can be shown that the diagram in Fig. 6 commutes.

Let us define $f(op)$ to be new operator which is applied to the triple $f(a1, a2, a3)$ whenever *op* is defined for the triple $a1, a2, a3$. For instance, let operator *op* be defined on triple $35, 18, 1$. It changes node occupancy *op*: 35 full, 18 full, 1 empty $\longrightarrow$ 35 empty, 18 empty, 1 full. Let f be rotation of 60°. Then $f(op)$ is defined on triple 20,8,2. It changes node occupancy $f(op)$: 20 full, 8 full, 2 empty $\longrightarrow$ 20 empty, 8 empty, 2 full.

If state S_1 is $S\backslash\{2\}\backslash\{1\}$ (all full except 2 and 10), S_2 would be a state $S\backslash\{2\}\backslash\{35\}\backslash\{18\}$ produced when *op* is applied to S_1. $f(S_1)$ is $S \setminus \{2\} \setminus \{3\}$ and $f(S_2)$ is $S \setminus \{3\} \setminus \{8\} \setminus \{20\}$.

Let $ROT = \{$ rotation of $k * 60^o \mid k \in \{1,2,3,4,5\}\}$ and let
$SYM = \{$symmetry with axis through nodes 0 and 1$\}$ and let f be any operator from $EQU = ROT \cup SYM$. It can be shown that the diagram in Fig. 7 commutes.

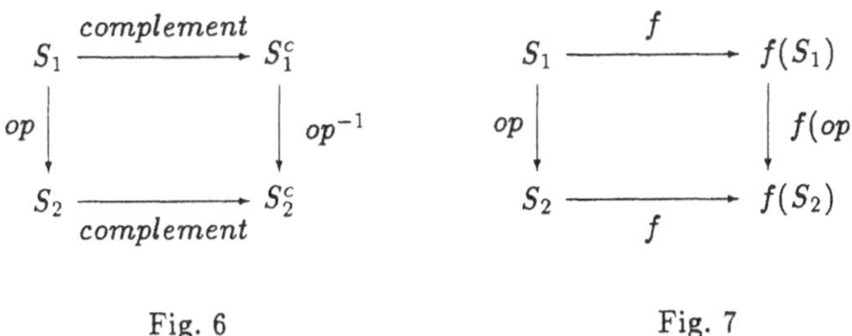

Fig. 6 Fig. 7

Let $Tree(k)$ denote a search tree whose root is a configuration that has all nodes filled with pegs and node k is empty. If we remove those trees that start with an empty black node (since they will never produce a goal state), there remain 24 trees. Let us introduce an equivalence relation $\sim$ to a set of states defined as $S_1 \sim S_2$ iff there exists $f \in EQU$ such that $f(S_1) = S_2$.

We may take only one representative from each equivalence class and omit expanding the other trees that are functionally dependent to the tree starting with a representative state. Let function $f \in SYM \cup ROT$ be used to obtain a root of unconsidered tree from one of the roots of representative trees. If we apply the same function f to any path in the tree, it will produce corresponding path in the tree not considered. In case of Hexagon this reduces the search to only three trees $Tree(1)$, $Tree(9)$ and $Tree(20)$.

Further, using composition of a complementing and $f \in SYM \cup ROT$, we can restore any inverse tree as well as any path in it. This cuts a search tree to half the depth of the tree. At each level we can eliminate all congruent configurations. If we find that complement of any configuration on level 17 is congruent to some configuration on level 18 this means that a solution path can be restored from it.

4 State Space Reductions Affecting the Completeness

Here we introduce a metastrategy of playing the game by selective use of operators. Instead of performing all possible moves in each current state, we introduce a metastrategy which will guide their selective application. These types of heuristics for state space reduction are domain oriented and rely on the possibility of performing certain subset of moves in one phase and applying the operators from the other subset later. If we find a solution in this reduced search space, we will have a solution for the initial search space. Since the solution depth for this particular game is fixed to 35 each solution is optimal. This would not be the case in other domains. The only danger in narrowing the search, which must be avoided, is to narrow the search to that extent that no solution exists in such subspace. In this case, of course, we would not have the solution of the initial problem either. As we have seen in section 4, black points in the graph configuration coloring in Fig. 5 cannot be filled with a peg in any final state. That means that those points must be empty before the end of the game. The strategy of performing the moves (i.e. operators will be divided into three phases).

4.1 Metastrategy

1) First we will perform, as soon as possible, the move that will empty the central black node labeled with 0 in Fig. 4. The introduced strategy of the game playing will not allow filling this node later.

2) Boundary black nodes labeled 19, 22, 25, 28, 31, 34 should also be cleared before the end of the game. The only type of move that enables this clearing is jumping with the black boundary peg over the corresponding shaded, or white node in front of it, if it is filled with a peg. The strategy is to clear those 6 black points in the last phase of the game without clearing the shaded or white points labeled 7, 9, 11, 13, 15 and 17. The pegs in those points will serve as bridges for jumping the pegs in the boundary black points in the last phase of the game.

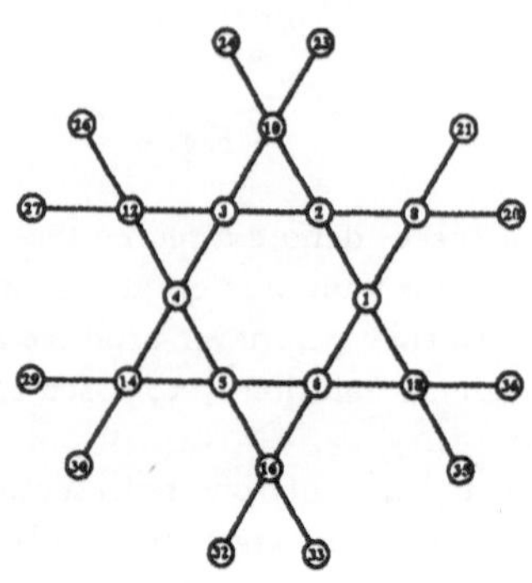

Fig. 8

With the restrictions mentioned above, in the second phase of the game playing we will apply moves that would clear black points which are neither central nor boundary. The other moves, except those affecting pegs in the forbidden parts of the graph configuration are also allowed. These produces a simpler configuration as shown in Fig. 8. The original problem is reduced to a simpler problem which has moves defined in the same way. The new graph configuration has 24 points, compared to 37 in the original game. In addition, there are 48 operators in the latter, as opposed to 114 in the former. This severely reduces the search tree branching factor. In organizing the new game search, the functional dependencies and coloring techniques mentioned in section 3 will be valid here too, since the graph configuration remains symmetric.

3) Clear the nodes 19, 22, 25, 28, 31, 34, 7, 9, 11, 13, 15 and 17.

Another strategy used for Hi-Q in Banerji [1] divides a graph configuration into four areas called submasks. One attempts the solution by solving the submasks in order, using only moves that do not cross the boundaries of the masks.

5 Applications of Functional Dependency in Other Areas

The functional dependency among subtrees of a search tree can be applied in different areas of AI. In robot planning the functional dependency among different plans of action can be established to reduce the search. In automated theorem proving this can be done for two formulas in the proof tree. If one is equivalent to the other after applying a substitution σ then subtree of the proof tree for the first formula can be obtained from the proof tree for the later one. Only one proof tree should be completed while the other can be derived from the first due to established functional relationship. Analogy based theorem proving is also based on the existence of functional dependency among formulas, but obtaining the analogous proof is not straight forward for nontrivial theorems. It usually requires some recovery.

6 References

[1] Banerji R. (1980) *Artificial Intelligence: A Theoretical Approach.* North Holland, New York.

[2] Coray G. (1976) *Additive Features in Positional Games. Computer Oriented Learning Processes,* edited by Simon. Noodhoff, Leyden.

[3] Epstein S. (1989) *The Intelligent Novice Learning to Play Better.* In Levy D. Beal D. edt. *Heuristic Programming in Artificial Intelligence, The First Computer Olympiad,* Ellis Horwood Ltd., Chichester.

[4] Levy D. Beal D. edt. (1989) *Heuristic Programming in Artificial Intelligence, The First Computer Olympiad,* Ellis Horwood Ltd., Chichester.

[5] Levy N., Newborn M., (1990) *How computers play chess,* W.H. Freeman & Co, New York.

[6] Rao V., Kumar V., Korf R. (1991) *Depth-first vs. best-first search.* AAAI-91, Anaheim, CA, 434-440.

[7] Reinefeld A. (1993) *Complete Solution of the Eight-Puzzle and the benefit of node ordering in IDA** Proc. 13th IJCAI 93 Shambery, France, 248-253.

[8] Samuel A. (1963) *Some Studies in Machine Learning Using the Game of Checkers.* In Computers and Thought ed. Feig., Feld., NY, McGraw Hill.

[9] Shannon E. (1950) *Programming a computer for playing chess.* Philosoph. Mag. 41(7), 256-275.

[10] Shoefield P. (1967) *Complete solution of the Eight-puzzle.* N.L. Collins, D. Michie (eds.), Machine Intell. 1, Amer. Elsevier, NY 125-133.

An Intelligent Terminal for Local and Remote Access to the Office Environment

V. Cappellini, A. Ferri, L. Lastrucci, A. Mecocci, A. Raggioli
Dipartimento di Elettronica
Università di Firenze
Via di Santa Marta 3 - 50139 Firenze - Italy

Abstract. To integrate telephone into the desk-top environment and to allow an efficient use of terminal resources even by unskilled people it has been developed *Julia*, an intelligent system to reduce user frustration and improve personal communications. It allows an human machine interaction by voice while the speech understanding system is based on a set of keywords.

The system is able to draw inference from a known domain and execute some reasoning and answering tasks in response to the user requests.

A large collection of sentences is used to address questions to the user when the reasoning process cannot go on. The sentences are selected in order to obtain a natural, not repetitive dialog between the user and the machine.

An intelligent graphical interface learns the user degree of skill and matches its help level to the user needs.

Keywords. Intelligent Kernel, Telephone, Desk-top Environment.

1 Introduction

Modern communication networks are called "integrated" because they allow the use of many different devices, like: fax, dial or touch-tone telephone and so on. Nevertheless, the growing number of services now available through modern communication networks has increased the complexity of the terminals. To allow an efficient use of terminal resources even by unskilled people it is important to simplify the man-machine interaction.

Another important feature toward an intelligent and personalized machine is to grant to the unsupervised terminal, a certain degree of interaction autonomy with remote users.

The use of remote terminals through the telephone system is particularly useful, since hundreds of millions of telephones are in use today. After decades of separate and parallel development in which no much has been done to exploit use of voice as a computer interface medium, the disciplines of voice communication and computing are finally integrated ([1], [2]). Equipped with speech recognition and synthesis equipment, a computing application can use telephones as input/output devices and it allows people to communicate directly with computers to perform simple tasks without needing operators making all telephone subscribers potential users.

In this paper an intelligent system based on a set of production rules is described. The system can manage a network terminal and allow machine interaction by voice.

E. A. Yfantis (ed.), Intelligent Systems, 175–179.

The system is designed to exploit some typical functionalities of an human secretary into an office environment.

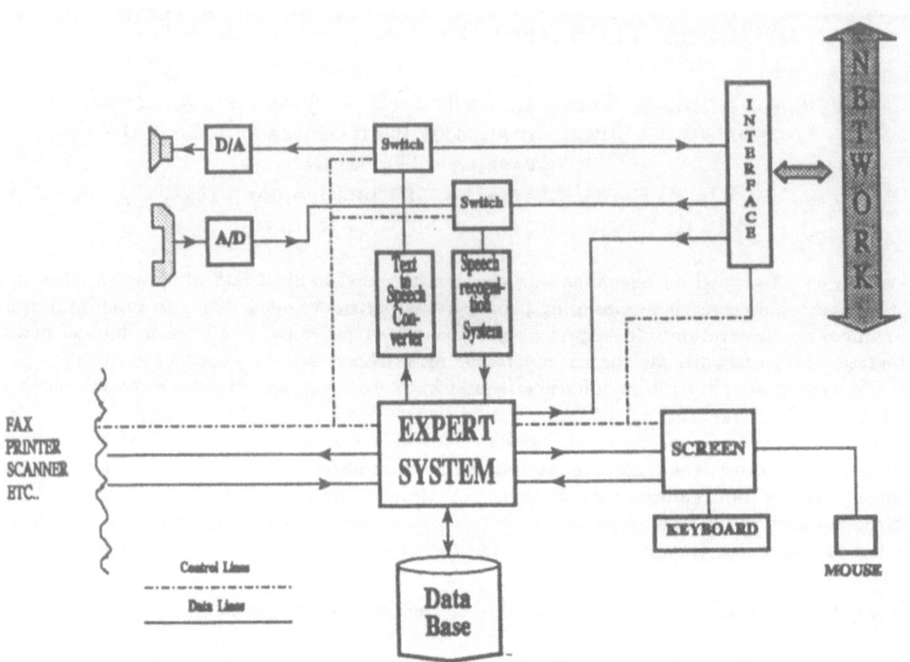

Figure 1: System architecture.

2 System Description

Most people never use telephone features since these features have complicated user interfaces. More important than assistance in making telephone calls is the assistance in answering them. To integrate telephone into the desk-top environment an intelligent system called *Julia* has been developed. It can also reduce user frustration and improve personal communications. Julia allows the possibility to receive audio telephone messages as ordinary electronic mail (email); an unified interface to handle voice mail and email provides the exchange of information between the computer and the telephone. Retrieving voice mail messages over the telephone, the answering machine application can convert regular email to speech, and read it to the user. If there is a fax machine available, the caller can instruct the answering machine application to fax you your regular email or documents. The computer-based answering assistant, by using a caller identification procedure, delivers personalized messages, routes calls to other numbers, or notifies urgent calls. Other information, such as personal diary, telephone number diary, appointment diary, the devices available to each user, the state device table, are stored in a database. An archive for messages, fax, images and data is available for each user. Time dependent personal data (if, where and when the owner is available, fax or messages to send, broadcast messages and so forth) are constrains for the system rules. Moreover, the owner is able to do a remote call by phone and ask the system about what happened during his absence. In the first phase of our project we interacted with the system through a multi-speaker discrete speech recognizer and a text-to-speech synthesizer. The speech understanding system was based on a set of keywords or tokens and it has been assumed the possibility to understand sentences only checking a keyword or certain chain of keywords within the phrase. Different keyword schemes are associated to different means. It has been pointed out before that the system should accept every input phrase belonging to natural language. If it finds a known keyword scheme in the phrase, it is able to understand the sentence. When a not complete keyword scheme is detected, the intelligent kernel makes context depended questions to gather the whole information (keywords) needed.

So doing, there are two possible situations: if no keyword is detected within the input sentence, the system replies with general questions like "what's your name", or "which is the required task?"; in the other case, when some keywords are detected, but they don't match any known scheme, the system makes specific questions to eliminate any ambiguity. The system is able to draw inference from known domain too, executing some simple reasoning and the answering to the user requests.

A large sentence collection is handled by the system to build questions when the reasoning process doesn't go on. The variety of this collection is determinant to have a natural machine behavior.

3 The Intelligent Kernel

System software architecture is similar to expert system architecture. It is articulated on three levels [4]: the first level is the *factual* or environment knowledge (i.e. known keywords). The second level is the *procedural* knowledge, necessary to perform tasks. The last level is the control strategy to select the necessary operator and find the solution to the input problem.

There are a lot of storing structures that can be used to represent knowledge. The basic system structure to store procedural knowledge is the production rule ([3], [4], [5]) modified adding two fields:

- AND_DO field, containing links to functions executed only if the rule hypothesis is verified.
- ASK field, where the information is inserted to allow the system to interact by questions with the user.

For the control strategy we modified the classic search in the state space called *hill climbing search* or gradient search [3]. When more than one operator (rule) can be applied to the actual state, the control strategy selects and applies the operator which conducts

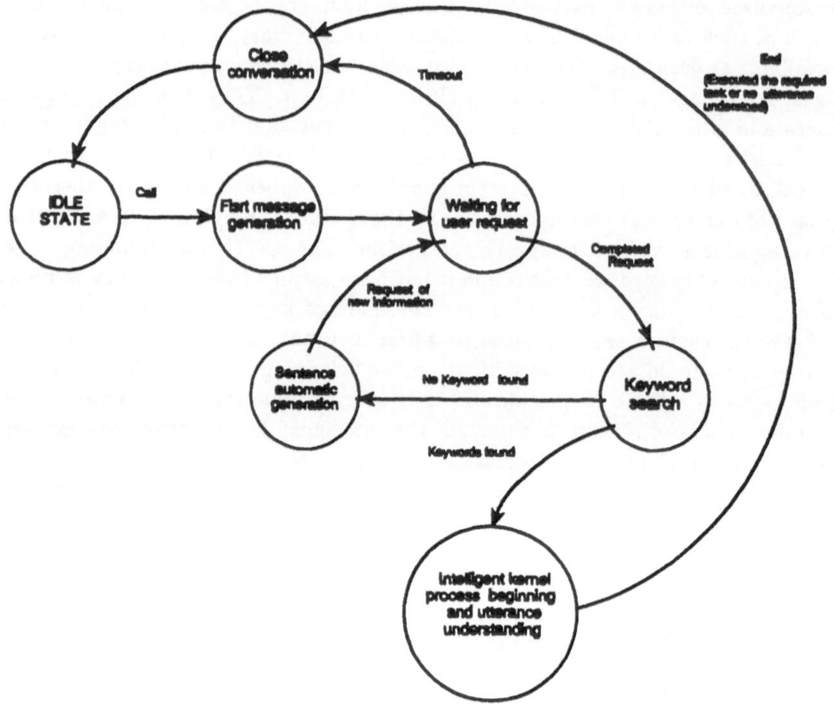

Figure 2: Intelligent kernel state diagram.

toward a state having the minimum distance from some solution. The selection is made using a slot, that every operator has, containing an heuristic number that indicates a "pseudo distance" from some solution or meta. The intelligent kernel moves its reasoning process over a graph instead than over a tree, allowing jumps between two different contexts (e.g. it is not necessary to undo an operation and restart a new one). The control strategy can operate in two ways depending on the system state. When a reasoning process starts, the control strategy operates in "NORMAL" mode and it remains in this mode as long as it is necessary to draw every possible inference from initial data (keywords found within the input sentence). If no solution is found during this step (e.g. sentence not understood), control strategy switches to the "ASKING" mode. While in normal mode reasoning, an operator is discarded if it is not applicable, in asking mode reasoning, the control strategy uses the asking field to build one or more questions about missing information. If the user doesn't furnish the required information, the system goes on producing new different questions with the same mean simulating a non repetitive natural conversation. A timeout mechanism stops the query session if the user doesn't supply missing information within a fixed time interval. If the user answers to the system questions furnishing the required information, the reasoning proceeds along a new path generated by this rule within the state space.

For the rule insertion into the procedural data base, a language was defined. In this way it was possible to type a standard ASCII files containing production rules and to obtain, through an appropriate compiler, a binary data base file.

Intelligent kernel can trace its reasoning steps, questions and user answers into an history file. These information are used to find the most frequently used path into the state space to speed up the reasoning process for the following user interactions. Moreover, expert users can use this file to update the rule database.

4 The Terminal Interface

To simplify the insertion and retrieval of information in local, a graphical interface was built using icons, windows, pop-up menus; it allows interactions by voice, mouse and keyboard.

The graphical interface is based on the system kernel previously depicted which guides the interaction between man and machine. The interface itself changes the user parameters finding the most frequent choices, the most frequent errors occurred and the user skill level.

The intelligent interface permits to adapt, in a learneable way, the help degree of the system depending on the user capability. When the user improves his ability, the interface automatically allows a faster interaction finding the most frequently used paths and furnishing a shortcut to them.

5 Conclusions

In this paper was described a system that realizes some simple functions of an human secretary. It allows a natural interaction with remote (linked by a telephone line) and local users, furnishing them the possibility to speak without any constrain and guiding the dialogue to obtain the desired responses, insert and retrieve informations from an electronic data base. The system is actually under test and preliminary results are very promising.

References

[1] R. Nakatsu, "Anser: An Application of Speech Technology to the Japanese Banking Industry", Computer, IEEE Computer Society, Vol 23, pp. 43-48, Aug. 1990.

[2] M. Lennig, "Putting Speech Recognition to Work in the Telephone Network", Computer, IEEE Computer Society, Vol. 23, pp. 35-41, Aug. 1990.

[3] E.Rich, "Intelligenza Artificiale," Mc Graw Hill libri Italia, first edition Oct. 1984.

[4] D.S.Nau, "Expert Computer Systems," Computer, IEEE Computer society, Vol 16, pp 63-85, Feb. 1983.

[5] N.J.Nilsson, "Principles of Artificial Intelligence," Springer-Verlang 1982.

The Terminal Interface

AN IMPLEMENTATION OF
THE FACETED CLASSIFICATION SYSTEM
FOR SOFTWARE REUSE

Ram Mareddy, Mirage Resorts, Inc., Las Vegas, NV 89109
M. Samadzadeh, Oklahoma State Univ., Computer Science Dept., Stillwater, OK 74078
E-mail: samad @ a.cs.okstate.edu

Abstract

This paper deals with the faceted classification system that organizes software artifacts with the help of predefined facet lists. Since the facet list or the vocabulary is standard for the librarian and the programmer, the chance of getting the right classification is improved. This system also provides flexibility by providing the ability to update the facet lists. The database for this system is organized using the relational model. This paper uses the faceted classification system with the help of the relational data model to arrive at a reuse system for the personal computer application development (or DOS application development) domain.

1. Introduction

Reusable software is widely believed to be the key to increased productivity in software development. While every "new" software system requires several components that have been developed before, these components frequently are re-developed for the current system. Developers will be able to concentrate on developing truly new components if they could easily integrate the past development into their current development. Even though the strategy of reusability offers great promise, it has been unfulfilled to a great extent [Biggerstaff87] [Hall92].

The beginnings of software reuse can be traced back to the 1968 NATO Software Engineering Conference which focused on building large, reliable software systems in a cost effective way [Krueger92]. Recent renewed interest in software reuse has a lot to do with the cost of developing software. With the cost of powerful hardware falling everyday, ambitious and more powerful software systems are being planned to provide solutions for various practical problems. Quality along with productivity should improve in building these systems, if well-tested and already developed components were to be used in constructing these systems.

The fundamental problem in reusing software is the storage, search strategy, and retrieval of the components that are required for building systems. Whenever a component is built, a library should be available for keeping this components for retrieval at a later time. For both these purposes, a classification system should be available which, when incorporated into a tool, will facilitate the storing and retrieval of software components.

E. A. Yfantis (ed.), Intelligent Systems, 181–194.

2. Previous Work

2.1 History and Past Research Efforts

As mentioned in the introduction, a historically important conference in the field of software reuse was the 1968 NATO Software Engineering Conference that was held in Garmisch, Germany. But the origins of software reuse can be traced back to as early as 1949 when the University of Cambridge proposed the first subroutine library on its stored program computer EDSAC [Tracz88a].

A more recent event in the field of reuse is applying it to alleviate the "Software Crisis" in the context of the development of the Ada programming language. This language was developed under the auspices of the US Department of Defense in the early 80's. Software reuse and the object-oriented approach are two of the aspects of program development that Ada's features were expected to support [Tracz87].

Ada has several language constructs to facilitate the development of reusable components. These constructs include a package construct that separates specification and implementation. There is an overload resolution construct that facilitates the semantic and syntactic reuse of functions. Ada also enforces strong typing. Added to this technical superiority, the Department of Defense directives on standardization, validation, and mandated use enhances the chances of software reuse making a significant impact on software crisis at DoD.

A large number of big corporations also contributed to the field of software reuse. These corporations include Boeing, Ford, General Dynamics, IBM, Lockheed, and Honeywell. They have projects addressing several aspects of reuse such as standardization of support tools, libraries,autoated catalog and library interaction, analysis and requirements definition, and code without language or implementation tricks [Tracz87].

2.2 Current Practice

As discussed in the previous section, the concept of software reuse has been in practice for some time. The following subsections detail the different ways this concept is practiced [Krueger92] [Hall87] [Zand94] [Zand93] [Zand92] [Swanson92].

2.2.1 Vertical Reuse or High-Level Languages Reuse in languages started with assembly languages. Assembly language routines provided an abstraction to the machine language routines for most of the hardware-level functions. In the newer languages, libraries of standard functions further extend this abstraction. Object-oriented languages such as C++ and Smalltalk provide powerful features for the user to extend such abstractions. When the concept of abstraction is further extended, we get Very High Level Languages (VHLL). These languages resemble application generators in the sense that the specification for a certain task is automatically transformed into executable systems. This property derives another name for VHLLs - executable specification languages.

2.2.2 Horizontal Reuse Within a level of abstraction, functions can be reused. Pipes and filters of UNIX come under this category. In this case, the output of one function is transferred to another function; in effect reusing the second function without having to rewrite it. Batch processing also comes under this category. Horizontal reuse is different from vertical reuse in the sense that the first category uses high-level languages whereas the second category uses the operating system functionality.

2.2.3 Design and Code Scavenging Sometimes programmers scavenge code and designs from a previously developed systems. Typically, the locations of this type of artifact and the concepts used reside primarily in a programmer's head. So, recalling from memory, a programmer will copy code fragments from an old system and integrate them into the current project. This type of reuse becomes more effective as the programmer gradually becomes more experienced.

2.2.4 Source Code Components Similar to the hardware components industry, this notion suggests an industry of off-the-shelf source code components. Ideally, functions with clearly defined input and output values can be manufactured for use in several different programs. Packaging of these components can use systematic techniques such as catalogs and libraries of components. To use these functions, the reuser can set the parameters rather than editing the source code directly, even with the availability of source code in such component libraries.

2.2.5 Application Generators For application generators, the input is the specification of the required task and the output is the executable program that implements the required task. These are similar to conventional programming language compilers, except that they are only higher level. Application generators typically focus on a narrow domain and generate code to solve the problems in that area. Application generators work at such a high level that they concentrate on what the system should do rather than how it is done. So, algorithms and data structures are automatically generated for the reuser. Application generators have been developed in several domains including database management, textual report generation, and graphical report generation [Burton87]. Lex and yacc come under this category. Lex addresses lexical analysis and yacc deals with parsing.

2.2.6 Software Schemas In software schemas, the notion is reusing algorithms and data structures rather than the code itself. An example of this schema is the PARIS system [Katz87]. In this system, the reuser starts the development by giving a problem statement, which is a formal set of computational requirements. Then, with its sophisticated retrieval system, PARIS supplies the schema that satisfies the problem statement.

2.3 Reuse Classification Criteria

Reusability has three important aspects: the abstraction level, customization methods, and reusability conditions [Lenz87] [Krueger92]. Classification systems used for software should facilitate all these three aspects of reuse.

2.3.1 Abstraction Level Abstraction is the most important feature of software reuse. If the implementation details are not hidden from the software artifact, a programmer will be forced to spend more time than necessary in understanding the internals. That may eventually force the programmer to redevelop that software artifact instead of reusing the already developed artifact. Reasonable amount of abstraction should be practiced in all units of software, be they specifications, designs, or code.

2.3.2 Customization Methods Once a software component is identified as a candidate to be used in the new system, it has to be tailored to meet the current needs. This can be done by simple parameterization or by changing of the internals. When the candidate artifact needs no tailoring at all, it would be the ideal form of software reuse.

2.3.3 Reusability Conditions When the domain in which the software artifacts are being developed reaches a certain degree of maturity, concepts used in that domain become apparent. This paves the way for higher software reuse. In a new domain, reusability becomes difficult because of the lack of software artifacts and the lack of domain knowledge by the programmer.

2.4 Economics

As mentioned in the introduction, the chief reason behind the increasing interest in software reuse is the enormous increase in the cost of developing complex software. A simple cost model proposed by John Gaffney of the Software Productivity Consortium is as follows [Tracz88b] [Barnes87] :

$$C = (1-R)*1 + b*R$$

where C is cost of developing software, R is the percentage of code reused, and b is the cost of reusing a line of code / cost to develop a new line of code.

As we can see from the above model, savings will be more when more code is reused or when the development of new code is higher than reusing the existing code. But all these gains that can be realized from the reuse of software are moderated by some organizational factors. Project managers in large organizations are not typically rewarded for the reuse of old software. Without organizational commitment, it becomes difficult to develop and maintain libraries with reuse in mind [Barnes87].

3. The Tool

A tool that facilitates the reuse of software artifacts is developed using the faceted classification scheme [Prieto-Diaz91] as the organizing mechanism. This tool has a graphical user interface (GUI) to make the reuser's interaction with the system as friendly as possible. This tool provides menu commands to store, search, and retrieve software artifacts.

This reuse tool applies the general faceted classification scheme towards the personal computer applications development domain. Hence part of the work was to come up with the facet lists and the standard vocabulary that is common to both the librarian and the user for developmental environments on personal computers.

The database that contains the names, properties, and descriptions of the software artifacts follows the relational database principles. Addition, search, and retrieval of the software artifacts are also done through this relational database. To support this, an implementation independent relational schema is also developed.

4. Implementation Platform

The tool is implemented on an IBM compatible personal computer running MS-DOS version 6.0 operating system with the MS-Windows graphical environment. IBM compatible personal computer is chosen because of its widespread availability, and MS-DOS is arguably the standard operating system for personal computers. MS-Windows environment is chosen to develop the Graphical User Interface (GUI) for this application.

The graphical interface for this application is developed in Microsoft Visual Basic version 3.0 and Microsoft Visual C++ version 1.0. The relational database which holds the information about reusable software artifacts is developed using the Microsoft Access version 1.1 database engine. For this database development, more sophisticated database environments such as the MS-SQL Server were considered. Since the SQL Server does not provide executables (i.e., the SQL Server should be present on the computer on which an application is running), Microsoft Access is given priority as the database engine. Microsoft Access 1.1 is also fully integrated into Microsoft Visual Basic 3.0 for all its database features. Other system related functions are developed using Microsoft Visual C++.

5. Data Model

5.1. Relational Databases

The relational model refers to a database system that contains only tables at the logical level for organizing data [Date91]. There might also be relations among these tables to connect the data among several tables. A user conceptualizes the data as tables and relations. Each table contains rows and columns. Each row, or a record from a table, is the complete description of one entity. This entity can be a person, a sale, etc. Each entity is described by several fields, characteristics, or attributes. These fields are represented by columns [MS-SQL93]

SQL (Structured Query Language) is a language (with various dialects) that implements the relational database model. This high-level language was originally developed at IBM in the mid-1970s. This language includes statements not only for querying and retrieving data from a database, but also for creating new databases and modifying and updating them. The American National Standards Institute (ANSI) recently came up with a standard SQL language called Transact-SQL [Transact-SQL93].

The first phase in the relational model is data modeling. In this phase, the specification of the system is collected in terms of tables, i.e., the rows and columns in each table. Here the primary and foreign keys for each table are also defined. A primary key is one or more columns that uniquely identify the rows or records in the table. A foreign key is a column in a table which is similar to a column in another table. With the help of foreign keys, two or more tables can be joined. The second phase is data definition. In this phase, storage is allocated to the database and then tables are created. So, this phase mainly creates the holders for data. The third phase is data manipulation. In this phase, the functions required for data retrieval and data modification are developed. In the SQL language, data retrieval is primarily done by the SELECT statement. And data modification is done by the INSERT, UPDATE, and DELETE statements. As the meanings suggest, to insert new data values into the tables, the INSERT statement is used, to modify the currently existing data, the UPDATE command is used, and to delete any data from the database, the DELETE statement is used.

The above three phases constitute the basic database operations. There are also some additional facilities provided for easy access to data and to keep the integrity of the data [Transact-SQL93]. Indexes can be defined on the tables to access data quickly. There are basically four types of indexes: composite, unique, non-clustered, and clustered [MS-SQL93]. A composite index is created on single or multiple columns. A unique index is created on single or multiple columns where these columns make a unique key. When a non-clustered index is created, the data is not ordered physically. With the clustered index the data is physically ordered. On any table, there can be as many composite, unique, and non-clustered indexes as needed. But, there can be only one clustered index, since this index physically orders the data [MS-SQL93]. It also should be noted that when multiple indexes are created, the clustered index should be created first.

Defaults, rules, and views are also part of the useful features of the relational model. Default values can be placed in the data holders when no value is specified. When a certain data is entered to be placed in the table, the rules check for the validity of that data for that column in terms of data types. Views help in presenting parts of a full table or database [MS-SQL93].

5.2 Entity-Relationship Diagram for the Tool

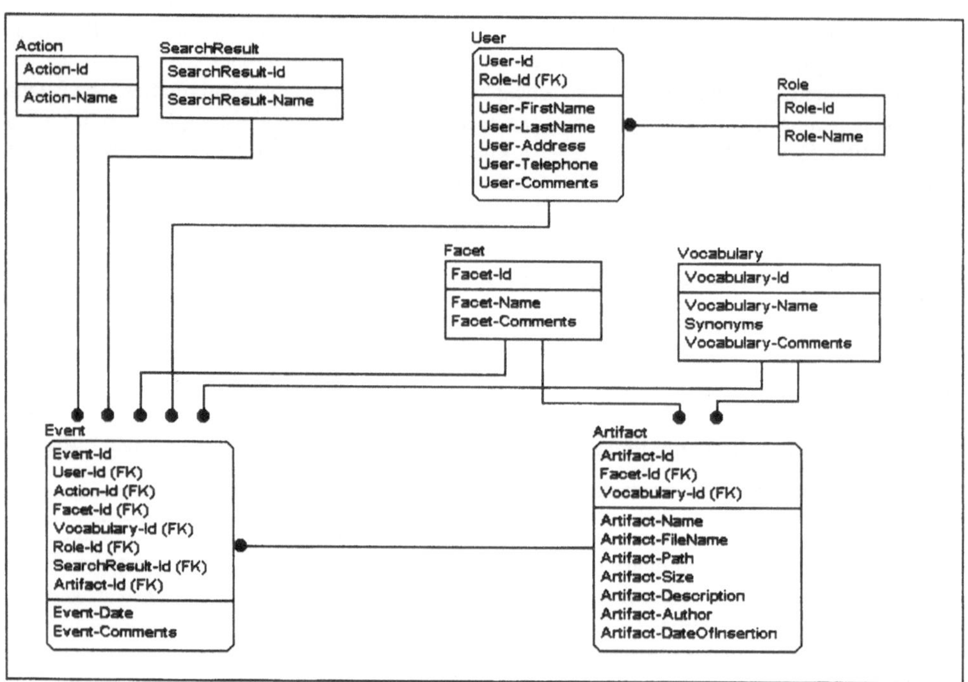

Diagram 1: Entity-Relationship diagram for the reuse tool.

All elements required for the reuse are stored using the relational model. Several interconnected tables are used in this model. Diagram 1 represents the entity-relationship diagram for this tool.

All the information related to the users is stored in the User table. This includes the name and address and the role of the user in this process.

The role of the user as far as this system is concerned can be one of the following: Developer, Manager, and Librarian. Developer is the one producing and reusing the software artifacts. Manager oversees the complete development process. Manager normally will not produce or reuse the software artifacts, but will be responsible for the development at a higher level. And finally, Librarian is responsible for the maintenance of this system. Librarian will constantly update and refine the vocabulary and monitor reuse in the system. This system provides the flexibility to add further roles.

Action tables are used in the description of the event. Three distinct actions are possible for the reuse events - inserting new software artifacts, retrieving the software artifacts for reuse, and general browsing. Developer and Librarian normally insert and retrieve and the Manager browses through the system.

Every event is tracked with the help of the Event table. This table stores which user is using the system, what is the purpose of the user's action, and the details of the facets selected. The event table also tracks the result of the search through the system.

The SearchResults table contains all the possible outcomes of the search. These are: Found, Not Found, and Aborted.

Facet and Vocabulary tables make up the necessary words required for the classification. Different facets and the vocabulary developed for the PC applications domain are discussed in detail under the section titled Faceted Classification System.

The Artifact table contains the classification and the location details of the software artifact. Even name, author, and date are made facets for this system, because, in short-term reuse, these facets are often used in the retrieval process. If there is any change to the facets, i.e. if the author, name, or date is removed from the list of facets, that information will still be available in this table.

6. User Interface

The tool is named Reuse With Facets, and, as mentioned in the previous section, it is developed for the Microsoft Windows Operating environment. The user interface conforms to the standards of the Windows operating environment.

Figure 1 shows a snapshot of the main window of the tool. There are four main menus available here: User, Librarian, Reports, and Help. The User menu contains three commands Retrieval, Insertion, and Exit. The Librarian menu facilitates the administrative functions of Facets, and Vocabulary. The Reports menu contains Classification and Usage commands that generate the reports on the current classification (facets and vocabulary) and the reuse related information in the system. The Help menu provides information to the user.

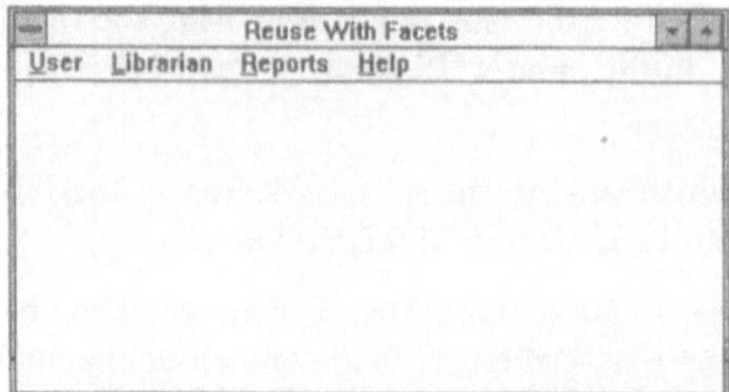

Figure 1: Main Window of the tool.

Figure 2 is a snapshot of the Search dialog box. A reuser provides the required vocabulary for various facets of the artifact. The empty facet or template takes all the vocabulary related to that facet into consideration. A user has the option to cancel the search or proceed for the search with the given information.

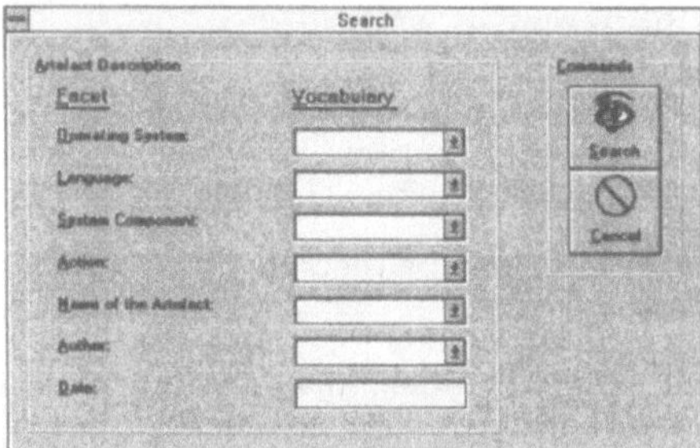

Figure 2 : Search dialog box of the tool.

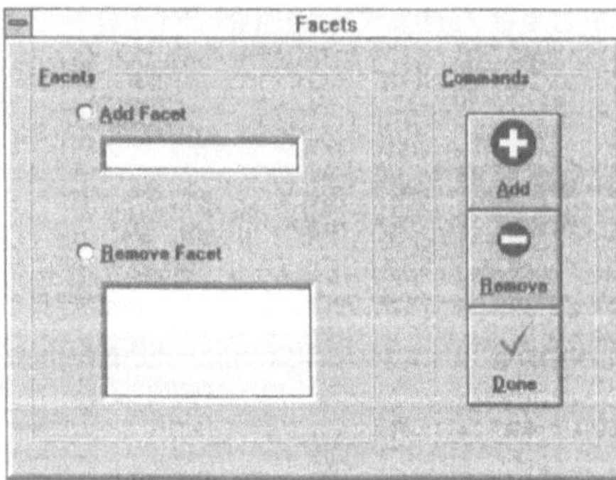

Figure 3 : Facet update dialog box for the tool.

Figure 3 is a snapshot of the dialog box that is displayed when the Facets command is chosen from the Librarian menu. This dialog box allows the librarian to add new facets to the system or remove the existing ones. Under the Remove Facet option button, all the existing facets are displayed and the librarian can choose the one that is to be removed and can click on the remove button. Usual precautionary messages are displayed when a facet is being removed from the system. All the artifacts containing this facet in their classification have this facet value removed from the corresponding tables.

Similar commands are available in the Librarian menu for deleting/adding vocabulary and artifacts. Also the Classification command in the Librarian menu provides the ability to change the classification of an existing artifact in the system.

7. Faceted Classification System

An important aspect of software reuse is the organization of the software artifacts so that they can be easily searched and retrieved. This organization methodology also should provide an easy way of inserting newly developed artifacts into the collection.

Three concepts of organization are widely used throughout libraries in the world. The first one is the Dewey Decimal System, which is used in a number of libraries in the United States. The second one is the Library of Congress Classification system, which is used by the library of congress and several other libraries in United States. The third is the Faceted Classification System, which is widely used in Europe and India [Prieto-Diaz91]. In the Dewey Classification System, possible classes are predefined. Hence, when a title needs to be classified under this system, the librarian should find a class that best fits the title. In order to do this, the librarian should have expertise in both the Dewey Classification System and the subject matter that the title represents. As expected, with several closely related classes, a title spanning across several classes becomes difficult to classify.

The Library of Congress (LC) system is similar to the Dewey Classification System in many aspects, especially in the predefined classes. But the final notation differs significantly from Dewey. The following example (taken from [Immorth71]) explains this difference. The third edition of Richard D. Altick and Andrew Wright's Selective Bibliography for the Study of English and American Literature has a call number Z2011.A4 1967 under the Library of Congress System and 016.82 A468s 1967 under the Dewey Decimal system. Each of these call numbers has three components: Class Number (Z2011 in LC and 016.82 in Dewey), Author Number (A4 in LC and A468s in Dewey), and Publication Date (1967). Z2011 and 016.82 are predefined classes in both of these systems.

With the Faceted Classification System, it becomes easy to decide on titles spanning across several classes. A faceted scheme contains as many facet lists as the domain to be classified requires. Each facet list contains as many keywords as needed to describe that facet. So, when a librarian wants to classify a particular title under this system, the librarian will select a term from each facet that suits the title. After all the terms are selected, the class that is arrived at will be the best fit for this title, hence offering flexibility and accuracy.

Another recent approach to information retrieval is by free text analysis [Prieto-Diaz91]. There are several reasons why free text analysis does not work for the analysis of source code functions. First of all, there might not be a lot of free text in the source code - unless the programmer decides to do a lot of commenting. Variable naming conventions can differ from programmer to programmer and also from function to function. In effect, by free text analysis of a function, it is not in general possible to find out what a function is doing or how it is doing it. In contrast, in the Faceted Classification System, we have predefined lists of key words. This list of key words or standard vocabulary can be updated as needed. Hence this classification not only provides the flexibility of having a reasonable number of keywords, but it also provides consistency among the programmers (or reusers) and the librarians of the reuse system [Prieto-Diaz91].

7.1 Facets and Vocabulary

Table 1 is the list of facets and vocabulary that is developed for the PC applications domain. For the purpose of the Faceted Classification System, a total of seven facets are suggested to characterize this domain (this system provides the flexibility of addition, deletion, and modification to these facets at any time). These seven facets are the Operating System for which the software artifact works, the Language in which the artifact is developed, the part of the system where this artifact works best, the action or service this artifact provides, a name given to this artifact by the developer, the name of the author, and the date of creation.

Each of these facets has a list of choices (or vocabulary) to further describe the software artifact. This vocabulary can also be modified as the repository of the artifact grows. The following paragraphs briefly describe the vocabulary for each of these facets.

In a large corporation, normally several operating systems are used on the PC domain. They include DOS, Windows, Windows NT, OS/2, and Unix. Others may be added as they become available. Similarly, several languages might be used in the development process, they include C/C++, Pascal, FORTRAN.

In the case of operating systems and languages, version also plays an important role. For example DOS can be further refined to DOS3, DOS4, DOS5, and DOS6. The Facets,Vocabulary, and files related to the older artifacts can be deleted to make room for the newer ones.

Operating System	Language	System Component	Action	Name	Author	Date
DOS	C	Specification	Average			
Windows	C++	Data Model	Count			
Windows	Pascal	User Interface	Color			
NT	FORTRAN	Editor	Shade			
OS/2	VB	Financial	Max			
UNIX	QB	Database	Min			
None	COBOL	Initialization	Sum			
All	Executable	Formatting	Lower			
	None	Date & Time	Upper			
	All	Math & Trig	Cos			
		Statistical	Sin			
		Text	Tan			
		Logical	Create			

Table 1: Facet and Vocabulary List.

The two facets called System Component and Action describe the functionality of the artifact. The System Component suggests the place where this artifact might be used. The Action suggests further specialization of the artifact that is under consideration.

The last three facets deal with the creation of the artifact. The Name given to it by the author, the name of the author, and the date of creation make up this set. Since these are important to identify the artifact, they are also stored in the artifact table of the database.

8. Example

Listing 1 has a function that changes the string of the color name to a hexadecimal color code. This function can be used to obtain a color name from the user and then generate the color code to actually use that color. So, this is mainly a user interface related function. And further, this is a color related function.

This function is written in Visual Basic and runs under the Windows operating system. The Name of this function is GetColorCode and is written by Mareddy on 01/01/94. So the class for this function according to the Faceted Classification System is Windows-VB-User Interface-Color-GetColorCode-Mareddy-010194. This name is obtained by the synthesis of all the facets for the function.

For the retrieval of this component the same class with most of the facets is inputted into the system. For example, if the Action facet for this artifact is selected as "Shade" instead of "Color", the system will miss this artifact. But in the system Shade is also defined as the synonym for color, so a further prompt will be given to the reuser, asking whether to look for classes for synonymous facets. Subsequently, the tool can retrieve the above artifact.

9. Summary and Conclusions

This paper and the tool briefly described mainly dealt with the implementation of the Faceted Classification System for reuse in Personal Computer applications development domain. As part of this effort a database schema, a list of facets, and a vocabulary pertinent to this domain were also developed.

This model can be effectively used in a multitude of development situations. Pure development corporations with development projects in several platforms will benefit the most. But, so will the Information Systems divisions of the corporations to which software development is not the main business. They will be able to save time and resources with the help of an effective classification system and a reuse tool.

Finally, the success of reuse depends on how strong the organizational commitment is to the reuse. In the case of the Faceted Classification System, it is equally important to develop and maintain a pertinent list of facets and vocabulary.

```
Function GetColorCode (sColorName As String) As Long
   '---------------
   'Declarations
   '---------------
   Dim sMessage As String
   Dim sTitle As String
   CONST G_BLACK=0
   CONST G_BLUE=1
   CONST G_GREEN=2
   CONST G_CYAN=3
   CONST G_RED=4
   CONST G_MAGENTA=5
   CONST G_BROWN=6
   CONST G_LIGHT_GRAY=7
   CONST G_DARK_GRAY=8
   CONST G_LIGHT_BLUE=9
   CONST G_LIGHT_GREEN=10
   CONST G_LIGHT_CYAN=11
   CONST G_LIGHT_RED=12
   CONST G_LIGHT_MAGENTA=13
   CONST G_YELLOW=14
   CONST G_WHITE=15
'-----------------------------------------
'Get the color code for color string
'-----------------------------------------
Select Case (sColorName)
   Case ("BLACK")
      GetColorCode = QBColor(G_BLACK)
   Case ("BLUE")
      GetColorCode = QBColor(G_BLUE)
   Case ("GREEN")
      GetColorCode = QBColor(G_GREEN)
   Case ("CYAN")
      GetColorCode = QBColor(G_CYAN)
   Case ("RED")
      GetColorCode = QBColor(G_RED)
   Case ("MAGENTA")
      GetColorCode = QBColor(G_MAGENTA)
   Case ("BROWN")
      GetColorCode = QBColor(G_BROWN)
   Case ("LIGHT GRAY")
      GetColorCode = QBColor(G_LIGHT_GRAY)
   Case ("DARK GRAY")
      GetColorCode = QBColor(G_DARK_GRAY)
   Case ("LIGHT BLUE")
      GetColorCode = QBColor(G_LIGHT_BLUE)
```

```
  Case ("LIGHT GREEN")
    GetColorCode = QBColor(G_LIGHT_GREEN)
  Case ("LIGHT CYAN")
    GetColorCode = QBColor(G_LIGHT_CYAN)
  Case ("LIGHT RED")
    GetColorCode = QBColor(G_LIGHT_RED)
  Case ("LIGHT MAGENTA")
    GetColorCode = QBColor(G_LIGHT_MAGENTA)
  Case ("YELLOW")
    GetColorCode = QBColor(G_YELLOW)
  Case ("WHITE")
    GetColorCode = QBColor(G_WHITE)
  Case Else
    'Error, ini file has wrong color name.
    sMessage = "Wrong Color code."
    sTitle = "InOut Board - GetColorName"
    MsgBox sMessage, MB_OK, sTitle
    GetColorCode = G_WRONG_COLOR
End Select
End Function
```

Listing 1: Color code function

References

[Barnes87] B. Barnes, T. Durek, J. Gaffney, and A. Pyster, "A Framework and Economic Foundation for Software Reuse", *Proceedings of the RMISE Workshop on Software Reuse*, Rocky Mountain Institute of Software Engineering, Boulder, CO, pp. 77-88, October 1987.

[Biggerstaff87] T. Biggerstaff and C. Richter, "Reusability: Framework, Assessment, and Directions", *IEEE Software*, pp. 41-49, July 1987.

[Burton87] B.A. Burton, R.W. Aragon, S.A. Bailey, K.D. Koehler, and L.A. Mayes, "The Reusable Software Library", *IEEE Software*, pp. 25-33, July 1987.

[Date91] C.J. Date and H. Darwen, *Relational Database Writings 1989-1991*, Addison-Wesley Publishing Company, MA, 1991.

[Hall87] P.A.V. Hall, "Software Components and Reuse - Getting More out of Your Code", *The International Journal of Information and Software Technology*, Vol. 29, No. 1, pp. 38-43, January/February 1987.

[Hall92] P.A.V. Hall, *Software Reuse and Reverse Engineering in Practice*, Chapman and Hall, New York, 1992.

[Immroth71] J.P. Immroth, *A Guide to the Library of Congress Classification*, Libraries Unlimited, Colorado, 1971.

[Katz87] S. Katz, C.A. Richter, and K.S. The, "PARIS: A System for Reusing Partially Interpreted Schemas", *Proceedings of the Ninth Annual International Conference on Software Engineering* , Washington, D.C., pp. 377-385, March/April 1987.

[Krueger92] C. Krueger, "Software Reuse", *ACM Computing Surveys*, Vol. 24, No. 2, pp. 131-183, June 1992.

[Lenz87] M. Lenz, H.A. Schmid, and P.W. Wolf, "Software Reuse through Building Blocks", *IEEE Software*, pp. 34-42, July 1987.

[MS-SQL93] *Microsoft SQL Server Implementation Notes*, Microsoft University, 1993.

[Prieto-Diaz91] R. Prieto-Diaz, "Implementing Faceted Classification for Software Reuse", *Communications of the ACM*, Vol. 34, No. 5, pp. 89-97, May 1991.

[Swanson92] J. E. Swanson and Mansur H. Samadzadeh, "A Reusable Software Catalog Interface", *Proceedings of the 1992 ACM/SIGAPP Symposium on Applied Computing (SAC'92)*, pp. 1076-1082, Kansas City, MO, March 1992.

[Tracz87] W. Tracz, "Ada Reusability Efforts: A Survey of the State of the Practice", *Proceedings of the Fifth Annual Joint Conference on Ada Technology and Washington Ada Symposium*, U.S. Army Communications-Electronics Command, Ft. Monmouth, NJ, pp. 35-44, March 1987.

[Tracz88a] W. Tracz, "Software Reuse Myths", *ACM SIGSOFT Software Engineering Notes*, Vol. 13, No. 1, pp. 17-21, January 1988.

[Tracz88b] W. Tracz, "RMISE Workshop on Software Reuse: Meeting Summary", *Tutorial on Software Reuse: Emerging Technology*, Boulder, CO, pp. 41-53, October 1988.

[Transact-SQL93] *Transact-SQL User's Guide for SQL Server*, Microsoft Corporation, 1993.

[Zand92] M. K. Zand, Mansur H. Samadzadeh, H. Saiedian, and H. Farat, "Classification and Identification of Software Components", *Proceedings of the Second Golden West International Conference on Intelligent Systems*, pp. 275-280, Reno, NV, June 1992.

[Zand93] M. K. Zand, K. M. George, Mansur H. Samadzadeh, and H. Saiedian, "An Interconnection Language for Reuse at the Template/Module Level", *The Journal of Systems and Software*, Vol. 23, No. 1, pp. 9-26, October 1993.

[Zand94] M. K. Zand, Mansur H. Samadzadeh, and H. Saiedian, "Version Management for ROPCO: A Micro-Incremental Reuse Environment", *The Journal of Information and Software Technology*, in print, scheduled to be published in 1994.

INVESTOR COUNSELING AND FINANCIAL MARKET SCREENING: AN EXPERIMENTAL STUDY TO EXTEND TRADITIONAL MEANS IN PORTFOLIO MANAGEMENT

JOHANNES GEIGER AND HUBERT F. HOFMANN
Department of Computer Science, University of Zurich
Winterthurerstr. 190, CH-8057 Zurich, Switzerland
email: {geiger,hofmann}@ifi.unizh.ch

Abstract

For investment advice, a framework of knowledge-based systems is yet to be developed and we attempt to make a step in this direction. While most financial institutions have already installed systems that support portfolio construction, they lack computer-support for investor counseling and market screening. To provide the foundation for such a framework, we suggest a conceptual model for investment advice. The Investor Counselor and the Market Screener are the co-operative tools derived from the suggested framework. To give the reader a feeling how the Investment Advisory prototype works, we will also present a counseling scenario to illustrate the prototype in action.

1 Introduction

An investment advisor must recognise an investor's needs and match these needs with appropriate investment vehicles. In this process, the advisor acquires investor data and interprets market signals. The investment advisor then allocates capital to different investment vehicles (bonds, stocks, etc.) through investment plans. The individual tuning of investment plans heavily influences the investor's satisfaction with given advice.

Investors often expect an advisor to be informed on all aspects of personal finance. Such expertise is impossible to attain for one person. Therefore, many financial institutions develop applications that assist their advisors in compiling appropriate investment plans [Pau 1991]. While most financial institutions have already installed portfolio management systems supporting portfolio construction, they lack computer-support for investor counseling and market screening [Gershman and Wolf 1985, Stansfield and Greenfield 1987, Chorafas and Steinmann 1991, Lebsaft 1991].

For investment advice, a framework of knowledge-based systems is yet to be developed and we attempt to make a step in this direction. In the next section, we will briefly describe the portfolio management process and propose an integrated view of portfolio management. The third section will present the foundation of our experimental study: a conceptual model for investment advice. Section four will propose an integrated view on the various tools which are available for portfolio management. In section five, we will outline the architecture of the investment advisory system and describe the main components of the laboratory prototype. We will especially focus on the description of the knowledge base. To give the reader a feeling how the Investment Advisory prototype works, the sixth section will contain a counseling session to illustrate the prototype in action.

E. A. Yfantis (ed.), Intelligent Systems, 195–207.

2 The Portfolio Management Process

In the portfolio management process, the information gathered from investor(s) and provided by financial analysts are the major types of information. They constitute two views of the portfolio management process (Fig. 1).

The *investor-oriented view* assumes that the satisfaction of a customer with a given advice on investments depends on the consideration of his or her individuality. From this viewpoint the construction of an investor profile, which reflects a customer's attitudes and attributes to investment, is at the centre of interest. The *market-oriented view* tackles the task in focusing on available investment vehicles. The risk-return payoffs that are associated with each investment vehicle provide the basis for specific combinations of securities from the market-oriented view. This portfolio reflects the actual profile of the observed investment market (market profile), i.e. the actual and the estimated market behaviour.

Considering these views, we can distinguish four elements of the portfolio management process: investor counseling, portfolio construction, market screening, and portfolio control. *Investor counseling* involves collecting and assessing investors and identifying investment objectives (for instance, long-term growth of capital and income without excessive fluctuations in market value). Investment objectives are necessary to evaluate given advice. For without deciding the final purpose of investing, it will be impossible to judge the success of investment decisions or the appropriateness of particular recommendation. Unfortunately, most individual and institutional investors cannot precisely articulate their investment objectives [Maginn and Tuttle 1983].

When *constructing the portfolio*, the investment advisor selects those specific securities in which to invest and determines the proportion of each security. The key issues of portfolio construction are selectivity, timing, and diversification [Sharpe 1987]. The aspect of selectivity means the examination of individual securities and entails forecasting the price movements of these securities. Timing involves forecasting the price movements of security classes. Diversification is the process of adding securities to a portfolio to minimise the portfolio's total risk.

Market screening is concerned which acquiring and analysing market signals that are essential in explaining environmental turbulence. Investment advisors focus on changes in the

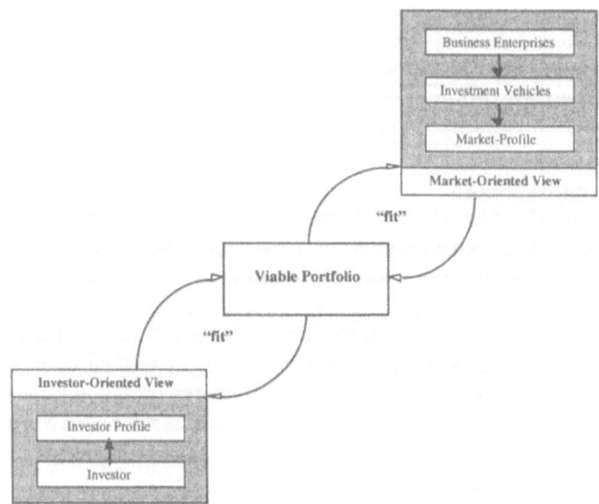

Fig. 1. Two Views of the Portfolio Management Process

investment market that do not correspond with their expectation of proper market behaviour. For example, an information inequilibrium of financial information that is observed by one of the market participants (investors, financial institutions, stock exchanges, and the state) where it was not anticipated (e.g. principal-agent problem [Arrow 1986]). The evaluation of such unexpected changes results in estimated risk factors of individual portfolios.

The fourth element concerns the *controlled, periodic repetition* of the previous phases. The currently held portfolio will often be viewed as suboptimal because of environmental change, for instance, modification of an investor's objectives or economic events. In response, the portfolio manager identifies a new "viable" portfolio and than tries to realise the necessary revisions to the current portfolio. The optimal structure of the new portfolio is a result of its balance between the investor-oriented view and the market-oriented view. A portfolio is "viable" in the sense that it can provide this double fit over a pre-defined period of time. That is, the selected portfolio is resistant against minor changes in the financial market. In other words, the portfolio is "optimal" with regard to both views of the portfolio management process.

3 A Conceptual System Model for Investment Advice

To integrate the investor-oriented and the market-oriented view, we propose the conceptual system model depicted in Fig. 2.
Data are represented as boxes, whereas the tasks are drawn as ellipses. Tasks and data are connected by arrows showing the direction of transfer. Such a conceptual model is enormously helpful in the development and maintenance of an investment advisory system. For example, the conceptual model enables all parties to commit their assumptions to a publicly examinable and executable form, and it supports early assessment of investment alternatives.

The starting point for investor counseling is a so-called *case description*. The investor's case description comprises *personal* and *financial data. Investment objectives* filter and specify, i.e. *map*, the investor-case description onto the *investor profile*, which represents an operational set of attributes defining the constraints for the selection of investment vehicles. For example, the investment objective "bond with high creditworthiness" is mapped onto the operational

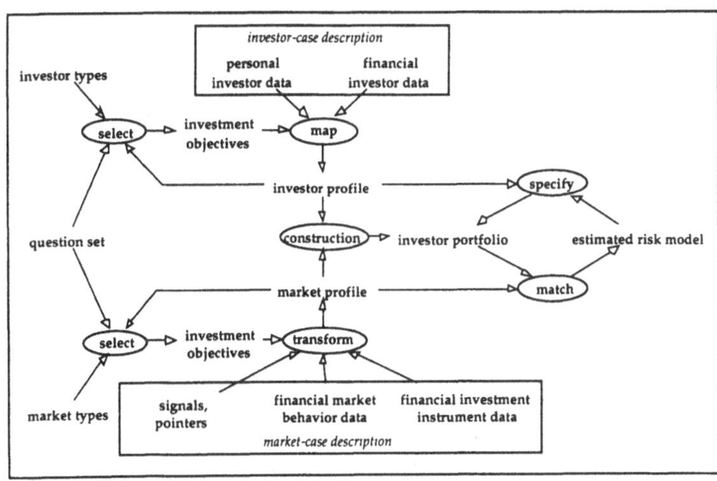

Fig. 2. Conceptual System Model for Investment Advice

attribute "AAA or AA rating is necessary to invest in a bond." Dynamically selected *question sets* construct the actual questionnaire used to acquire investor data. Thus, the questionnaire is neither pre-defined, nor static, but linked to investor types. *Investor types* are a coarse-grained view of an investor's individuality and therefore seldom met in pure form [Lewellen et al. 1977, Maginn and Tuttle 1983, Ruda 1988]. The investor advisor can both modify existing investor types and create new ones. In our current version of the Investment Advisory System we use four investor types: α (e.g. growth of income, long-term perspective, creditworthiness is restricted to the higher grades), β (e.g. guaranteed return, low risk tolerance, avoid foreign engagement), γ (e.g. highest return for a given level of risk, maximise the total return), and δ (e.g. highest level of risk, extraordinary gains in the shortest time possible). For example, the α-type restricts the bond-issuer and its creditworthiness to the higher grades. Therefore, a question in its set about creditworthiness is "Do you particularly avoid taking risks?"

While investor counseling results in an investor profile, market screening results in a *market profile*. The market profile combines the filtered market signals, the diagnosed market behaviour and the information about financial instruments in considering the investment objectives. The *market-case description* consists of those data and information that are used to estimate current and future market conditions. *Financial market behaviour data* describe the actual market behaviour, i.e. the sum of financial activities that market participants perform to achieve their investment goals. The interpretation of these data and the additional *signals and pointers* give a first picture of the actual situation at the investment market. Signals and pointers, which are often represented by internal and public index numbers, indicate future events that will change the current situation of the investment market. For example, prognosis of a partial illiquidity of a big company that currently is a market driver. Data about the *financial investment instruments* complete the market-case description. Financial investment instruments data are, For example, information about the international rating of a particular investment instruments (AAA rating of a bond).

To transform the market-case description into the market profile the investment objectives play a key role. From the market-oriented point of view, investment objectives are classified by so-called *market types*. The market types are categorised descriptions of essential properties of the investment market including, for example, interest rates or economic scenarios of particular branches. In other words, market types provide a set of stereotypic patters of environmental change that are anticipated to occur in the financial market. We distinguish four market types: The α type covers dynamic environmental conditions like turbulent market behaviour. It is conditioned to a weak form of efficient markets. While the β type fits reactive, but dynamic, market behaviour and a semistrong form of efficient markets, the γ type deals with clustered, mostly static, conditions of the financial market and a semistrong form of the efficient market hypothesis. The δ type focuses on randomised market behaviour and it assumes the strong form of efficient markets.

Different market types are linked to different investment strategies, which in turn are linked to norm-portfolios, i.e. pre-defined selections of types of investment vehicles that satisfy certain market situations. Depending on which market type or which combination of market types describes the investment market best, pre-defined question sets are selected to guide the investment advisor's specification of the market-oriented objectives. In other words, the selected question set depends on the market data acquired so far. The dynamically allocated question sets have to (1) point out market signals; (2) determine the economic situation of a particular branch; and (3) estimate the risk-return characteristic of investment vehicles. While investor and market types help to structure the counseling session in guiding the search for appropriate questions, they do not definitely classify the investor or the environmental behaviour of the financial market. Consequently, (new) information is permanently evaluated with respect to the identified type. This evaluation possibly identifies another type as more appropriate for the investor or the observed market and leads to the use of another question set.

Both the investor profile and the market profile are the input for the construction of the investor portfolio. Moreover, they are used to evaluate existing portfolios. The portfolio construction is largely based on optimisation models and thus can be automated, for example, by asset allocation tools. However, the controlling of existing portfolios cannot be automated completely because of the environmental dynamics in the financial market. For example, a change in the market profile has to be analysed with regard to the underlying causes and their resulting effects. Such a change analysis cannot be based solely on the resulting "numbers" representing the change. Thus, we describe the causes and effects of environmental change that are considered to be relevant in the *estimated risk model.* The estimated risk model provides the necessary information for adapting the structure of the viable portfolio. This control cycle has to be performed periodically in order to diagnose changes in the investor profile resp. the market profile adequately.

4 Information Infrastructure: An Integrated View

We propose an integrated view on the various tools of portfolio management to cope with the various tasks of portfolio management: portfolio planning, portfolio construction, and portfolio control. Investor Counselor, Portfolio Constructor and Financial-Market Screener are the different, though coupled, agents that should provide the necessary functionality for the bank staff. In combination, these agents constitute our idea of an investment advisory system. The three agents function as a front-end system that enhances and combines the traditional means of portfolio management (Fig. 3).

The data for portfolio management systems like indices or stock quotations are often provided by bank-internal data pools which are maintained by application specialists that are not directly involved in the maintenance of a specific investment advisory system. Therefore, an

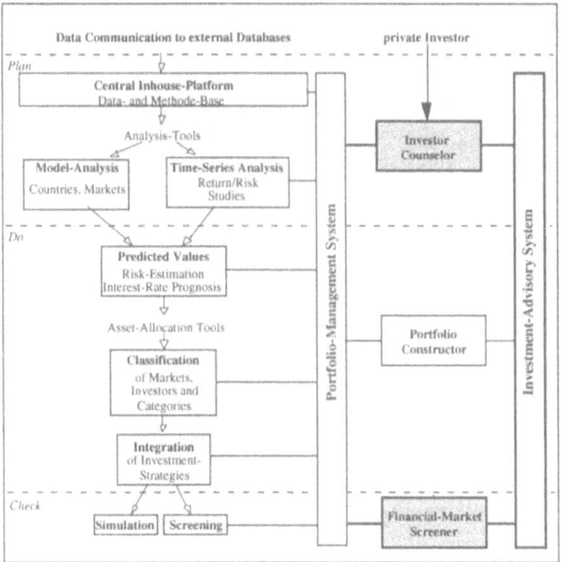

Fig. 3. Information Infrastructure of an Investment Advisory System

additional step is necessary to make use of the bank-internal investment data in a specific business area. The internal data are pre-processed in the business; for example, the analysis of data over time to perform various risk-return studies or through the use of economic models to estimate future market situations.

In general, historical data analysis and the prognosis of future trends are the input for portfolio construction. In many banks, mathematical optimisation is the primary means to construct viable portfolios. In a first step, bank-internal guidelines assign every investor a norm-portfolio. The portfolio manager then decides if the investment strategy should be an active or a passive one [Hofmann and Holbein 1994]. This decision heavily rests on the investor's financial goals. To keep the portfolio on track with the proposed investment goals, we can use two additional means: simulation and market-screening. For example, simulation based on interest scenarios or the analysis of changes in the financial market. Market screening focuses on signals which cause parameters of the economic models to change, e.g., interest rate or investment period.

In the literature almost all portfolio management systems operate on statistical data derived from mathematical models (for example [Hofmann 1993]). When using these systems, the portfolio manager can rely on quantitative (index) numbers, but often no qualitative measures support his or her decision making . To create a balance between qualitative and quantitative data we propose an infrastructure based on three agents which helps to interpret quantitative or qualitative investment data. The *Portfolio Constructor* performs the optimisation of portfolios' structures and thus it is not discussed further in this article (see, for example, [Sharpe 1987]). We will focus on the *Investor Counselor* and on the *Market Screener*. The Investor Counselor is a front-end application that deals with the identification and analysis of data provided by private investors. The Investor Counselor is implemented in Prolog and it is available in a prototype version. The Market Screener supports the interpretation of quantitative investment data that are available in the financial market. It guides the identification of market patterns that look promising for individual investments. This agent is still in its conceptual stage.

5 Investor Counselor and Market Screener: Agents' Structure

The Investment Advisory System consists of the two agents, Investor Counselor and Market Screener, and runs on a Macintosh computer. The Investment Advisory System assists an investment advisor in selecting appropriate question sets, in collecting and assessing current and estimated financial-market data, personal and financial investor data, and in identifying investment objectives. Finally, the Investment Advisory System produces an investor profile and a market profile. Although the Investment Advisory System's implementation is still in prototype stage, it has already shown that the established framework provides useful assistance to the software engineer in creating the knowledge base and specifying appropriate inference mechanisms.

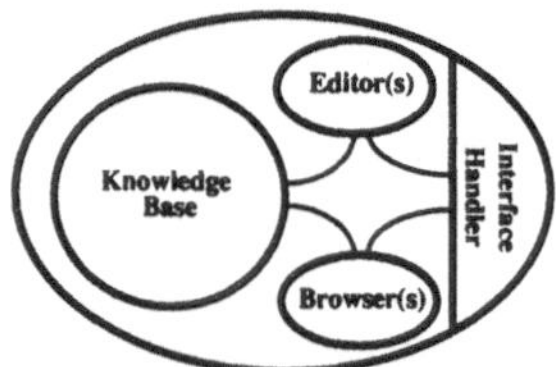

Fig. 4. The Structure of an Agent in the Investment Advisory System

The Investor Counselor and the Market Screener co-operate with the investment advisor, each solving a subtask using specific knowledge and specific inferences, and communicating with the other for information that is needed for their respective tasks. Fig. 4 shows the architecture of an agent in the Investment Advisory System. We separate the component handling the interaction from the components performing actual advisory. The *Interface Handler* shapes the interaction with the investment advisor. The Interface Handler distinguishes three types of questions. The first type constructs a standardised dialogue window to allow number input (e.g. when the investor defines the amount of money to invest). The second type supports questions where the investor can choose several alternatives (e.g. if the Investor Counselor asks about the intention of investment, the investor can either select emergency, purchase, speculation, or retirement planning). The third type defines yes-no questions. Moreover, a session history provides the chronological sequence of these questions and reports the connections between asked questions and established objectives. A session history primarily supports the advisor when evaluating a previous session. In this modular architecture, the notion of an explicit interface handler extends the functional kernel by different user interfaces that can be attuned to individual user needs and requirements. This obviously supports the idea of building several different prototype versions as design options.

During an investment session the prototype uses the different types of knowledge represented within the agents by means of the basic components: Knowledge Base, Browsers and (Objectives) Editors. For example, the *Counseling Browser* handles the investigation of investment objectives and establishes the investor profile. During the counseling session the hierarchy of objectives changes continuously. Previously established investment objectives are removed from the displayed hierarchy of objectives and added to the investor profile. The Counseling Browser allows switching between unanswered and answered questions of the predefined hierarchy, and it supports the investment advisor when changing given answers. The investment advisor can interrupt or finish a session at any point. The Investor Counselor then uses default values to establish the objectives necessary to construct a minimum investor profile. These default values are based on predetermined investor types.

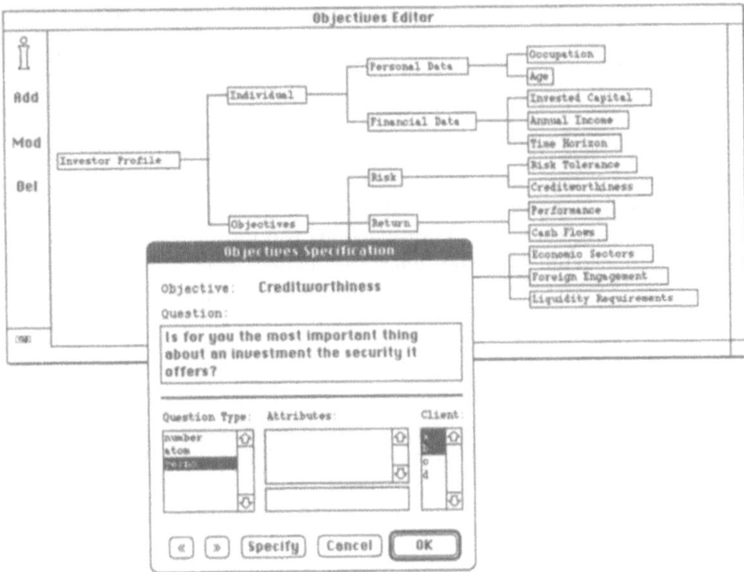

Fig. 5. The Objectives Editor

The flexibility of predefined investment objectives and question sets is a key issue, because financial institutions face a continually changing investor structure. As the underlying knowledge or data change, we must provide tools for adapting the counselor's knowledge base. The Investor Counselor provides an *Objectives Editor* (Fig. 5). With the Objectives Editor the investment advisor adds, deletes, and modifies question sets and investment objectives.

5.1 COUNSELING BASE

The Counseling Base contains predefined investment objectives, investor types, and question sets, and represents the primary instance on which problem-solving methods operate [Hofmann 1993]. As implemented currently, this component has its empirical basis in available investment surveys and relevant textbooks [Maginn and Tuttle 1983, Ruda 1988, Sharpe and Alexander 1990]. Investment objectives, investor types, and question sets are predefined in the sense that they are loaded before a session begins. However, the choice of which counseling base to load depends on the investment advisor. Therefore, the Investor Counselor not only focuses on the individuality of the investor, but also on the investment advisors needs and attitudes.

In the Counseling base, each concept (e.g. investment objective, investor type) is represented as a set of predicates in Prolog. This makes it easy to inspect, modify, adapt or extend them. The core of the counseling base is the objectives mapping process.

The collected answers of the investor are mapped to values representing the investor profile (Fig. 6). The mapping process distinguishes two kinds of mappings. Some investment objectives match the attributes of available investment vehicles directly, and therefore are applicable immediately in updating the investor profile (for example, invested capital and annual income).

However, many investment objectives require an explicit formalism to match distinctive investment attributes (for example, time horizon, creditworthiness, risk tolerance, performance objectives, and cash flows). In this case, we note that the mapping of investor answers depends on the state of counseling. For example, a speculator understands by a medium-term investment a period of four months, whereas a conservative investor expects such an investment plan to extend over four years. Therefore, the mapping of investment objectives is based on the actual investor type.

5.2 SCREENING BASE

The Screening Base consists of investment objectives resulting from market observations (e.g. risk/return goals), market types and question sets. A market type categorises the financial market to control the screening process of investment vehicles. Question sets guide the acquisition of market data, e.g. the interpretation of the financial market, the analysis of the economic situation of a particular branch, and the approximation of risk/return values for investment vehicles.

The degree to which the Market Screener supports the activities of the investment advisor is

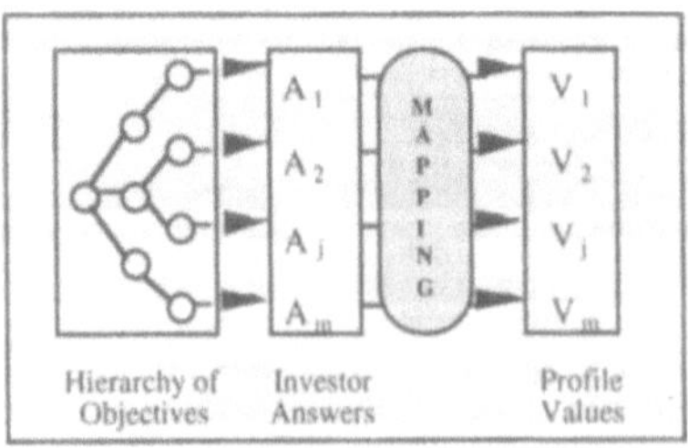

Fig. 6. Objectives Mapping Process

	Investor Type	α-Type	β-Type	γ-Type	δ-Type
Planning and Construction	Portfolio Structure	AAA, AA-Bonds	A, BBB-Bonds	BB, B-Bonds	CCC, CC, C-Bonds
	Management Strategy	passive	semi-active	semi-active	active
	Controlling Activity	few ←			→ frequent
Managing and Controlling	Decision Behaviour	programmed ←			→ non-programmed
	Management Task	strategic	strategic, tactic	tactic, operative	operative
	Short-Term Interest Rate	insignificant	significant	critical	survival
	Long-Term Interest Rate	survival	critical	significant	insignificant
	Interest-Rate Predictions	less important	useful	profitable	necessary

Fig. 7. Investmentstrategies

determined by the investor's classification (α, β, γ, δ types) and by the identified market type. Depending on the current investor type the Market Screener suggests in a first step different investment strategies (norm-portfolio) for the management of the investor's portfolio (Fig. 7). The suggested norm-portfolio is then adapted by the investment advisor to match the individual needs of the investor. However, the risk-return structure of the underlying norm portfolio determines the frequency of the control activities that are performed on the investment portfolio. Essential factors in this process are the performed portfolio management activity, e.g. a new investment or the controlling of existing portfolios, the investor type, the term structure of available investment vehicles, and the tendency of the interest rate.

In a next step the behaviour of the financial market — as it is perceived by the investment advisor — is classified according to the pre-defined market types (Fig. 8). Depending on the market type, certain characteristics of the investor's portfolio are suggested to fulfil his or her needs from the market-oriented point of view. For example, robustness of the portfolio and required variability of the portfolio structure.

	Market Type	α-Type	β-Type	γ-Type	δ-Type
Financial Market Characteristics	Market Behaviour	turbulent fields	disturbed, reactive	placid, clustered	placid, randomised
	Environmental Character	dynamic, complex	dynamic, simple	static, simple	static
	Information Efficiency	weak form	semi-strong form	semi-strong form	strong form
	Portfolio Variability	high ←			→ low
Design-Objectives	Portfolio Structure	viable ←			→ optimal
	Design Focus	Financial Market ←			→ Portfolio Structure
	Fin. Market Screening	essential ←			→ helpful

Fig. 8. Market Types

Finally, the Market Screener supports the matching process between the suggested classifications. Ideally, the actual market type and the current investor type match their predefined combinations (Fig. 9). These combinations are based on the investor's access to investment data. The symmetric access to investment data means that all investors have access to the same information, at least in principle. Whereas an access to investment data denotes the situation when one or more investors have access to more information than others.

If thc available investment pattern does not match the result of the investment advisor's analyses than additional information is gathered from the investor and additional market analysis are performed. This process is continued until one of the eight investment patterns matches the investor's needs.

	Market-Type	α-Type	β-Type	γ-Type	δ-Type	
Information-Flow	Asymmetrical Distribution of Information	α-Typ	β-Typ	γ-Typ	δ-Typ	Investor-Types
	Symmetrical Distribution of Information	δ-Typ	γ-Typ	β-Typ	α-Typ	

Fig. 9. Investment Patterns

6 An Application Scenario

This section provides an overview of a session with the Investor Counselor. We select characteristic parts of a counseling session to show the working of the Investor Counselor and its cooperation with the investment advisor.

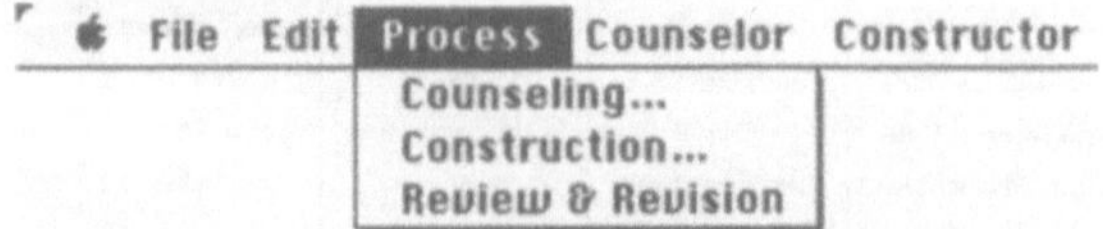

Fig. 10. Menu Bar of the Investor Counselor

The investment advisor starts a new session with the menu command "Counseling..." (Fig. 10). Right from the introductory phase of a session, the Investor Counselor associates the investor with a default type to make effective use of predefined question sets. However, this association is provisional and will be adapted to the investor's individuality as the counseling session proceeds.

A question from the introductory phase is, for example, Have you specific financial goals you wish to accomplish within a certain time period? This question results from the fact that certain investor types, i.e. the γ- or δ-type, define their investment objectives in advance, whereas other investor types expect the investment advisor to determine their objectives during the counseling session. Therefore, the investor's answer either strengthens the hypothesis that the investor is a γ- or δ-type, or in the case of a no answer suggests an α- or β -type.

After the introductory phase, the investment advisor mainly works with the Counseling Browser. The Counseling Browser is the core component of the prototype. It allows the investment advisor to follow the investor's flow of conversation easily and to record relevant data. The investment advisor can ask preselected questions or specify another topic of interest.

Fig. 11 shows a typical screen display during a counseling session. The Counseling Browser has four menus arranged in two rows. Each menu has one of its items selected. The leftmost menu of the top row displays the most general concepts of the objectives hierarchy, i.e. individual data and objectives data. The middle menu and the one on the right show the second and the third levels of the objectives hierarchy. The selections in the first three menus determine the text that is visible in the bottom menu, namely a list of questions to ask the investor. The actual investor type is used to determine the ordered sequence of displayed questions.

The hierarchy of investment objectives changes continuously, since established objectives are removed from the pre-defined hierarchy of objectives and are added to the investor profile. The Counseling Browser supports easy switching between objectives being asked for (session mode) and those already established (profile mode).

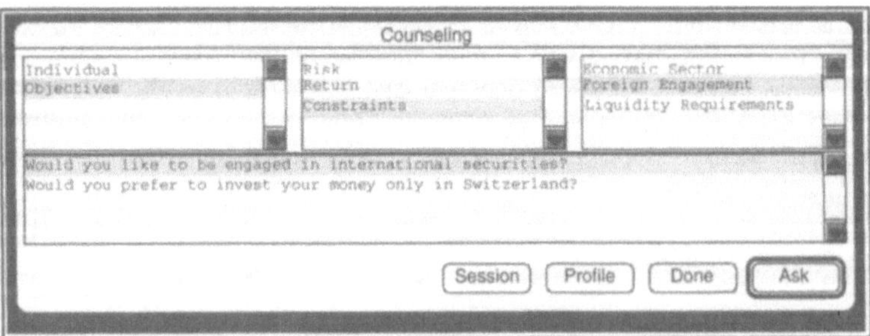

Fig. 11. Counseling Browser

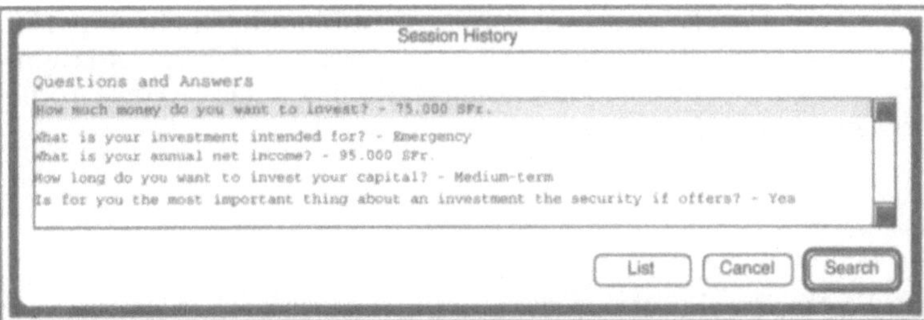

Fig. 12. Counseling History

In the *profile mode* the browser displays established investment objectives and supports the advisor when changing answered questions. If the investor changes his or her mind, for example, about the intention of investment then the advisor can easily select the questions about that topic. The investment advisor modifies the answer for the specified question, the Investor Counselor adapts the investor profile accordingly and proceeds with the counseling session.

In the *session mode*, the Counseling Browser displays all unanswered objectives of the hierarchy. Three situations can occur in this mode. First, the Counseling Browser examines the implications of the actual answer on the counseling process and considers the investor's need for more detail by starting a sub-dialogue. Second, the browser handles the situation where more than one question is necessary to clarify the investor's preferences for the actual investment objective. Third, the Counseling Browser removes an appropriately answered investment objective from the hierarchy and uses the predetermined counseling strategy for the current investor type to get the next investment objective.

To provide an overview of all questions asked and answers received, the Investor Counselor provides a session history (Fig. 12). The session history allows the investment advisor to work up a counseling session in chronological sequence.

Investor Profile

Investor	Charles Gambler
Occupation	Broker
Time Horizon	15 Years
Invested Capital	75.000 SFr.
Annual Income	95.000 SFr.
Performance	Investment
Risk Tolerance	60 Percent
Creditworthiness	AAA, AA, A
Cash Flows	None
Foreign Engagement	Yes
Liquidity	24 Hours

Economic Sectors

Prefer	Avoid
Banks	Aircraft
Building&Roads	Drapery&Stores
Communication	Electricity
Finance	Hotels&Caterers
Industrials	Mining

Cancel OK

Fig. 13. Investor Profile

Finally, the session results in a co-operatively defined investor profile (Fig. 13). The investor profile shows the established investment objectives mapped to attributes of investment vehicles. The investor profile constrains the subsequent construction of the investor's portfolio to ensure the consideration of his or her individuality.

7 Conclusions

We suggested a conceptual system model for investment advisory systems and used this model in developing a co-operative prototype. The prototype combines the framework and application-specific requirements. It uses the conceptual model for investment advice to guarantee transparency of inferences and to provide insight into the application domain.

The investment advisory system allows the investment advisor to use and maintain, for example, his own Counseling Base. This also supports the investment advisor's reflection on the application domain and thus supports the interpretation of derived solutions. Moreover, particular importance is given to a flexible system architecture to leave room for future changes.

In its current state, however, our investment advisory system is not embedded in the information infrastructure of a financial institution. Hence, necessary connections with external databases are simulated. In a next step of this research we should integrate the Investor Counselor and the Market Screener in an advisor's working process and validate the prototype's performance over extended periods of time. Also explanation facilities should be implemented to support co-operation between advisor and the investment advisory system.

We believe that the key to success in the area of investor advisory systems is careful management of user expectations, continuously guiding the user towards the areas of the system's competence.

Acknowledgements. We would like to thank Norbert E. Fuchs, Gustav Pomberger, and Gerald Quirchmayr for critical comments and constructive suggestions on this paper, and for their support of this work in general. We would also like to thank Ralph Holbein, Andrew Hutchison, Franz Lehner, and Patrick Scheidegger for discussing this manuscript and suggesting numerous improvements.

8 References

[Arrow 1986] K.J. Arrow, Agency and the Market. In: K.J. Arrow and M.D. Intrilligator (eds.), *Handbook of the Mathematical Economics.* Amsterdam, Elsevier Science Publisher, 1986, pp. 1183-1195.

[Chorafas and Steinmann 1991] D.N. Chorafas, H. Steinmann, *Expert Systems in Banking.* London, Macmillan, 1991.

[Gershman and Wolf 1985] A. Gershman, T.C. Wolf, Management of User Expectations in a Conversational Advisory System. *The Second Conference on Artificial Intelligence Applications*, Miami Beach, 1985, pp. 328-336.

[Hofmann 1993] H.F. Hofmann, The Investor Counselor: A Framework and Its Application. *Sixth International Symposium on Artificial Intelligence in Industry and Business*, Monterrey, 1993, pp. 46-53.

[Hofmann and Holbein 1994] H.F. Hofmann, R. Holbein, Reaching out for Quality: Considering Security Requirements in the Design of Information Systems. *6th Conference on Advanced Information Systems Engineering*, Utrecht, Netherlands, Springer, 1994, pp.

[Lebsaft 1991] E. Lebsaft, RAMSES — The Right Advice by means of Customer Servicing and Consulting Executed Step by Step: Experiences with an integrated Expert System fully operational in a Bank since 1988. *Expert System Integration*, Elsevier Science Publisher, Amsterdam, 1991, pp.

[Lewellen et al. 1977] W.G. Lewellen, R.C. Lease, G.G. Schlarbaum, Investment Strategy and Behavior among Individual Investors. *Journal of Business*, vol. 35, no. 3, 1977, pp. 296-333.

[Maginn and Tuttle 1983] J.L. Maginn, D.L. Tuttle, *Managing Investment Portfolios*. Boston, Gorham & Lamont, 1983.

[Pau 1991] L.F. Pau, Artificial Intelligence and FInancial Services. *IEEE Transactions on Knowledge and Data Engineering*, vol. 3, no. 2, 1991, pp. 137-148.

[Ruda 1988] W. Ruda, *Die Ziele privater Kaptalanleger (in German)*. Wiesbaden, Gabler, 1988.

[Sharpe 1987] W.F. Sharpe, *Asset Allocation Tools*. Redwood City, Scientific Press, 1987.

[Sharpe and Alexander 1990] W.F. Sharpe, G.J. Alexander, *Investments*. Prentice-Hall, Englewood Cliffs, 1990.

[Stansfield and Greenfield 1987] J.L. Stansfield, N.R. Greenfield, PlanPower: A Comprehensive Financial Planner. *IEEE Expert*, vol. 2, no. 2, 1987, pp. 51-56.

A First-Order Analogue of Higher-Order Unification

Wenchang Fang
Department of Business Administration
National Chung-Hsing University
No. 69, Sec. 2, Chien-Kuo N. Rd.
Taipei, Taiwan

Abstract. Unification is a basic process in theorem proving. Huet [10] makes a major contribution in this area, and his approach is used by most of the higher-order systems. Henschen [7] [8] suggests the use of an n-sorted first-order language for representing higher-order languages. The main idea is to translate a sentence A into an infinite set of T of axioms, then he proves the validity of A. Though differences existed in these two approaches, we try to find some rule to combine the two methods in which the explicitness of expression in higher-order logic is preserved and the convenience of first-order logic is offered. In particular, we try to combine the higher-order unification techniques of Huet into n-sorted unification. In [7] and [8], Henschen proposed a method in which the search for projections and imitations in higher-order logic was replaced by a search for renaming axioms in n-sorted logic. Unfortunately, no strategy was given to control this search or to coordinate the two inference rules, resolutions and naming. In this paper, we find normal first-order unification can use naming which result into the ordinary first-order resolution strategy.

Keywords. Higher-order logic, First-order logic, Theorem proving, Unification.

1 Introduction

Research in automated theorem proving has been concentrated on first-order logic. Higher-order logic[1] has been dealt with by only a few people. The $\lambda - calculus$ system was first introduced by Church [1]. Andrews [2] describes a resolution-like refutation system for type theory. Later he develops the TPS [3], which is an automated theorem proving system for proving theorems of both first and higher-order logic. Fang [5] [6] has discovered some heuristics for higher-order theorem proving.

One basic process in first-order theorem proving is unification, the search for two disagreements set in two terms to make them look alike. One who is familiar with automated theorem proving knows this process is quite straightforward and decidable [14]. Like the first-order unification, higher-order unification is also a fundamental tool for second-order logic [4] [12], general ω-order logic [9] [11] [13] and higher-order logic programming . Unfortunately, in languages of order three or higher, unification is undecidable. A unique most

[1] Also know as $\lambda - calculus$ or Type theory

E. A. Yfantis (ed.), Intelligent Systems, 209–220.

general unifier does not exist in general. In other words, two terms may have more than one unifier. It is Huet [10] who makes a major contribution in this area, and his approach is used by most of the higher-order systems. Henschen [7] [8] uses a different strategy. He suggests the use of an n-sorted first-order language for representing higher-order languages. The main idea is to take advantage of many existing first-order theorem proving programs with first-order unification to be used as a basis for a system which proves higher-order theorems. After translating a sentence A into an infinite set of T of axioms, he proves the validity of A. Observing the similarity between these two approaches, we try to find some rule to combine the two methods in which the explicitness of expression in higher-order logic is preserved and the convenience of first-order logic is offered. In particular, we try to embed the higher-order unification techniques of Huet into n-sorted unification. In [7] and [8], Henschen proposed a method in which the search for projections and imitations in higher-order logic was replaced by a search for renaming axioms in n- sorted logic. Unfortunately, no strategy was given to control this search or to coordinate the two inference rules, resolutions and naming. It is our thesis that the use of naming can be incorported into the normal first-order unification resulting in a name-if-needed-for-resolution strategy.

2 The Translation of $\lambda - calculus$

The following method is used by Henschen [7] to translate mathematical concepts from $\lambda - calculus$ to first-order n-sorted language.

Definition 1 *Let E be a subset of the set of types in a $\lambda - calculus$ wff. The* **type closure of** *E is the smallest set of types which contains E, the type TV and which contain α and β whenever it contains $(\alpha \rightarrow \beta)$*

Let S be a set of $\lambda - calculus$ sentences and $T(S)$ the type closure of S. Then **the associated n-sorted language K** contains the following symbols and types.

1. For a type α in $T(S)$, K contains a type, also denoted by α; note that although some types of $T(S)$ are complex, all types in K are simple.

2. For each type $\alpha = (\beta \rightarrow TV)$, one specialized binary predicate symbol of type $(\beta \rightarrow TV, \beta)$.

3. For each type $\alpha = (\beta \rightarrow \gamma)$ in $T(S)$, one specialized binary function symbol of type $(\beta \rightarrow \gamma, \beta, \gamma)$.

4. For each $\lambda - calculus$ identifier x of type α, a constant symbol c_x and a variable v_x both of type α.

5. For each type α in $T(S)$ the binary equality relation $EQUAL_\alpha$ of type (α, α).

6. A set of function symbols which are specified during the translation of S as described below.

7. The logical connectives, quantifiers and parentheses.

Definition 2 *Let A be a wff of $\lambda - calculus$ with a subexpression $(\lambda x B)$. An occurrence of the identifier x in B is called a* **bound occurrence** *of x. An occurrence of an identifier which is not bound is called a* **free occurrence.**

Note that $(\forall x A)$ and $(\exists x A)$ are abbreviations for formulas containing λx so that occurrence of x in A are also bound in these formulas.
Now we can define a map T from $\lambda - calculus$ to language K, which is almost the same as Henschen's approach. The minor modification is 5.

1. T maps the logical connectives and quantifiers onto the corresponding symbols of K.
2. If x is an identifier then $T(x)$ is v_x or c_x if the particular occurrence of x is bound or free, respectively.
3. If F is of type $(\alpha \rightarrow \beta)$ and B is of type TV, $T(FB)$ is $P(T(F), T(B))$, Where P is a predicate symbol of type $(\alpha \rightarrow \beta, \alpha)$.
4. If F is of type $(\alpha \rightarrow \beta)$ and B is of type $\beta(\beta \neq TV)$, $T(FB)$ is $f(T(F), T(B))$, where f is a function symbol of type $(\alpha \rightarrow \beta, \alpha, \beta)$
5. $T(\lambda x B)$ is recursively defined as $d(T(B))$. Here d is a new function symbol with appropriate type of K.
6. The naming axiom associated with d in step 5 is also need to include in K. The procedure of how to find the naming axioms will be showed in next section.

Example 1 *The definition of equality is:*

$$\forall x \forall y(x = y \leftrightarrow \forall f(f(x) \rightarrow f(y)))$$

can be translated into the following n-sorted clauses:

1. $\neg v_x = v_y \vee \neg P(v_f, v_x) \vee P(v_f, v_y)$

2. $P(v_f, v_x) \vee v_x = v_y$

3. $\neg P(v_f, v_y) \vee v_x = v_y$

Where P is of type $(\alpha \rightarrow TV, \alpha)$.

3 A rewriting Rule for Higher-Order Logic

We now give a presentation of how projection and imitation can be recaptured in n-sorted unification. Supposed we have two higher-order language terms e_1 and e_2, where e_1 is a flexible term and e_2 is a rigid term, both with the same type. They can be expressed in the following way:

$$\lambda u_1 u_2 \cdots u_{n_1} f(a_1, a_2, \cdots a_{p_1}) \quad n_1 \geq 0 \quad p_1 \geq 0$$

$$\lambda v_1 v_2\ cdotsu_{n_2} \mathcal{F}(b_1, b_2, \cdots b_{p_2}) \quad n_2 \geq 0 \quad p_2 \geq 0$$

1. $n_1 = 0$ and $n_2 = 0$, in other words, this is the situation of terms without $\lambda-$*headings*:

$$f(a_1, a_2, \cdots, a_{p_1})$$

$$\mathcal{F}(b_1, b_2, \cdots, b_{p_2})$$

The first thing we need to do is translate the two terms into first-order n-sorted language.

$$e_1' = f_{p_1}(\cdots f_1(v_f, a_1), a_2)\cdots), a_{p_1})$$

$$e_2' = f_{p_2}'(\cdots f_1'(c_{\mathcal{F}}, b_1), b_2)\cdots), b_{p_2})$$

Since e_2 is rigid, its headings can not be changed by substitution. We intend to let the heading of e_1' be adjusted to that of e_2' under certain situations. Therefore, two subcases are introduced: of rewriting rule

Notation :

f_i represents either function symbol or predicate symbol.
All upper-case letter represent constant.
All lower-case letter represent variable.

Imitation naming without $\lambda -$ *heading* :

From [9], f is $\lambda w_1 w_2 \cdots w_k \mathcal{F}(E_1, E_2, \cdots, E_{p_2-p_1+k})$,
where $E_i = h_i(w_1, w_2, \cdots, w_k)$
let $\mathcal{B}_1 = \mathcal{F}(E_1, E_2, \cdots, E_{p_2-p_1+k})$, and $\tau(\mathcal{B}_1) = \gamma$
Translate $\mathcal{B}_1$ into first-order n-sorted language:

$$T(\mathcal{B}_1) = f_{p_2-p_1+k}'(f_{p_2-p_1+k-1}'\cdots(f_1'(\mathcal{F}, E_1), \cdots), E_{p_2-p_1+k}) \qquad (1)$$

and

$$\begin{array}{ll} t_{k-1} = \lambda w_k \cdot \mathcal{B}_1 & \tau(t_{k-1}) = \alpha_k \rightarrow \gamma \\ t_{k-2} = \lambda w_{k-1} \cdot t_{k-1} & \tau(t_{k-2}) = \alpha_{k-1} \rightarrow (\alpha_k \rightarrow \gamma) \\ \quad\vdots & \quad\vdots \\ t_1 \;\; = \lambda w_2 \cdot t_2 & \tau(t_1) \;\; = \alpha_2 \rightarrow (\alpha_3 \rightarrow \cdots(\alpha_k \rightarrow \gamma)\cdots) \\ f \;\;\, = \lambda w_1 \cdot t_1 & \tau(f) \;\; = \alpha_1 \rightarrow (\alpha_2 \rightarrow (\alpha_3 \rightarrow \cdots(\alpha_k \rightarrow \gamma)\cdots) \end{array}$$

Now if we choose some suitable function symbol f_i and let its type be as following:

$$\tau(f_i) = (\alpha_i \rightarrow (\alpha_{i+1}\cdots(\alpha_k \rightarrow \gamma)\cdots)$$

then the Henschen's naming rule is formed as:

$$\begin{array}{rcl} T(\mathcal{B}_1) & = & f_k(T(t_{k-1}), w_k) \\ T(t_{k-1}) & = & f_{k-1}(T(t_{k-2}), w_{k-1}) \\ & & \vdots \\ T(t_2) & = & f_2(T(t_1), w_2) \\ T(t_1) & = & f_1(T(f), w_1) \end{array} \qquad (2)$$

From 2 we can get the following:

$$T(\mathcal{B}_1) = f_k(f_{k-1}(\cdots f_1(T(f), w_1), \cdots), w_k) \tag{3}$$

Note that $T(f)$ is translated to v_f, and abbreviated as f. Combine 1 and 3, the following result is obtained:

$$\begin{aligned} T(\mathcal{B}_1) &= f'_{p_2-p_1+k}(f'_{p_2-p_1+k-1}\cdots(f'_1(\mathcal{F}, E_1), \cdots), E_{p_2-p_1+k}) \\ &= f_k(f_{k-1}(\cdots f_1(f, w_1), \cdots), w_k) \end{aligned}$$

Do the same procedure for E_i, which can be obtained as:

$$E_i = g_k(\cdots g_1(h_i, w_1)\cdots, w_k)$$

The constraint of the rest paremater still remain the same. Finally, the rule can be expressed as:

$$f_k(f_{k-1}\cdots(f_1(f, w_1)\cdots), w_k)$$
$$= f'_{p_2-p_1+k}(f'_{p_2-p_1+k-1}(\cdots f'_1(\mathcal{F}, E_1)\cdots), E_{p_2-p_1+k})$$
$$E_i = g_k(\cdots g_1(h_i, w_1), w_2)\cdots w_k)$$
$$max(0, p_2 - p_1) \leq k \leq p_1$$
$$1 \leq i \leq p_2 - p_1 + k$$

Projection naming without $\lambda - heading$:

Also from [9], f is $\lambda w_1 w_2 \cdots w_k \cdot w_i(E_1, E_2, \cdots, E_m)$,
where $E_j = h_j(w_1, w_2, \cdots, w_k)$, $1 \leq j \leq m$, here $\tau(E_j) = r_j$
let $\mathcal{B}_2 = u_i(E_1, E_2, \cdots, E_m)$, and $\tau(\mathcal{B}_2) = \gamma$

$$\tau(w_i) = \alpha_i = (r_1, r_2, \cdots, r_m, \gamma) \tag{4}$$

Translate $\mathcal{B}_2$ into first-order n-sorted language:

$$T(\mathcal{B}_2) = f'_m(f'_{m-1}\cdots(f'_1(w_i, E_1), \cdots), E_m) \tag{5}$$

and

$$\begin{aligned} t_{k-1} &= \lambda w_k \cdot \mathcal{B}_2 & \tau(t_{k-1}) &= \alpha_k \to \gamma \\ t_{k-2} &= \lambda w_{k-1} \cdot t_{k-1} & \tau(t_{k-2}) &= \alpha_{k-1} \to (\alpha_k \to \gamma) \\ &\vdots & &\vdots \\ t_1 &= \lambda w_2 \cdot t_2 & \tau(t_1) &= \alpha_2 \to (\alpha_3 \to \cdots(\alpha_k \to \gamma)\cdots) \\ f &= \lambda w_1 \cdot t_1 & \tau(f) &= \alpha_1 \to (\alpha_2 \to (\alpha_3 \to \cdots(\alpha_k \to \gamma)\cdots) \end{aligned}$$

Now if we choose some suitable function symbol f_i and let its type be as following:

$$\tau(f_i) = (\alpha_i \rightarrow (\alpha_{i+1} \cdots (\alpha_k \rightarrow \gamma) \cdots)$$

then the Henschen's naming rule is formed as:

$$\begin{aligned} T(\mathcal{B}_2) &= f_k(T(t_{k-1}), w_k) \\ T(t_{k-1}) &= f_{k-1}(T(t_{k-2}), w_{k-1}) \\ &\vdots \\ T(t_2) &= f_2(T(t_1), w_2) \\ T(t_1) &= f_1(T(f), w_1) \end{aligned} \tag{6}$$

From 6 we can get the following:

$$T(\mathcal{B}_2) = f_k(f_{k-1}(\cdots f_1(f, w_1), \cdots), w_k) \tag{7}$$

Combine 5 and 7, the following result is obtained:

$$\begin{aligned} T(\mathcal{B}_2) &= f'_m(f'_{m-1} \cdots (f'_1(u_i, E_1), \cdots), E_m) \\ &= f_k(f_{k-1}(\cdots f_1(f, w_1), \cdots), w_{k-1}) \end{aligned}$$

Do the same procedure for E_j, which can be obtained as:

$$E_j = g_k(\cdots g_1(h_j, w_1) \cdots, w_k)$$

The constraint of the rest paremeters still remain the same. Finally, the rule can be expressed as:

$$f_k(f_{k-1} \cdots (f_1(f, w_1) \cdots), w_k)$$
$$= f'_m(f'_{m-1}(\cdots f'_1(w_i, E_1) \cdots), E_m)$$
$$E_j = g_k(\cdots g_1(h_j, w_1), w_2) \cdots w_k)$$
$$max(0, p_2 - p_1) \leq k \leq p_1$$
$$1 \leq i \leq p_2 - p_1 + k$$

Note here that m is decided from 4.

This conclude the rewriting rule for no $\lambda - heading$.

2. $n_1 \neq 0$ or $n_2 \neq 0$. Terms with $\lambda - heading$

$$\lambda u_1 u_2 \cdots u_{n_1} f(a_1, a_2, \cdots, a_{p_1})$$

$$\lambda v_1 v_2 \cdots v_{n_2} \mathcal{F}(b_1, b_2, \cdots, b_{p_2})$$

Note that either n_1 or n_2 could be zero, the λ terms may not exist at both terms. Translation to the first-order n-sorted language of the term with $\lambda u_i(T(e))$, where e is a $\lambda - calculus$ term.

$$\begin{aligned} T(e) &\rightarrow f_{p_1}(\cdots f_1(f, a_1), a_2)\cdots), a_{p_1}) \\ \lambda u_n \cdot (T(e)) &\rightarrow \lambda u_n(f_{p_1}(\cdots f_1(f, a_1), a_2)\cdots), a_{p_1}) \\ &\rightarrow d_n(f_{p_1}(\cdots f_1(f, a_1), a_2)\cdots), a_{p_1}) \end{aligned}$$

After n step:

$$\begin{aligned} \lambda u_1 \cdots u_n \cdot (T(e)) &\rightarrow \lambda u_1 \cdots u_n \cdot (f_{p_1}(\cdots(f, a_1), a_2)\cdots), a_{p_1}) \\ &\rightarrow d_1(d_2(\cdots d_n(f_{p_1}(\cdots f_1(f, a_1), a_2)\cdots), a_{p_1}\cdots) \end{aligned}$$

Of course, the naming axioms for the d_i functions would also be included. Still two subcases are provided here:

Imitation naming with $\lambda - heading$:

$$d_1(d_2(\cdots d_{n_1}(f_{p_1}(\cdots f_1(f, a_1), a_2)\cdots)a_{p_1})\cdots)$$
$$= d_1(d_2(\cdots(d_{n_2}(f'_{p_2}(\cdots f'_1(\mathcal{F}, E_1)\cdots), E_{p_2})\ cdots)$$
$$E_i = g_k(\cdots g_1(h_i, w_1), w_2)\cdots w_k)$$

Here h_i is a new variable with the appropriate type. g_j are new function symbols with appropriate type.

Projection naming with $\lambda - heading$:

$$d_1(d_2(\cdots d_{n_1}(f_{p_1}(\cdots f_1(f, a_1), a_2)\cdots)a_{p_1})\cdots)$$
$$= d_1(d_2(\cdots(d_{n_2}(f'_m(\cdots f'_1(w_i, E_1)\cdots), E_m)\cdots)$$
$$E_j = g_k(\cdots g_1(h_j, w_1), w_2)\cdots w_k)$$
$$1 \leq i \leq k \leq p_1$$
$$deside\ \ m\ \ from\ \ type$$
$$1 \leq j \leq m$$

In this case, i, j, k, m are the same as we discussed above.

The proof procedure for **naming with** $\lambda - heading$ is almost the same as the above.

4 Soundness and Completeness

Notation :

N_i is any disagreement node and has the form $< e_i^1, e_i^2 >$

r_i is the rewriting rule and has the form $wff_{a_i} \rightarrow wff_{b_i}$. After apply r_i to N_i, N_i can be unified.

c_i is the composite rule and is defined inductively as: $c_i = c_{i+1} \circ r_{i+1}$.

All the proof of the above lemma is trivial. After constructing the corresponding matching tree, three subcases exist:

Lemma 1 *If N_i is a* **TFN**, *then no r_i exists.*

Lemma 2 *If N_i is not a* **TFN** *or* **TSN**, *then there may exist more than one r_i.*

Lemma 3 *If N_i is a* **TSN**, *then there exists only one r_i.*

Soundness If the any corresponding matching tree for e_1 and e_2 possesses a **TSN**, then e_1 and e_2 are unifiable.

Proof: We use decreasing induction on i. Let us suppose that the corresponding matching tree for e_1 and e_2 has a **TSN** on some branch:

$$N_0 \xrightarrow{r_1} N_1 \xrightarrow{r_2} N_2 \ldots \xrightarrow{r_p} N_p = \textbf{TSN}$$

Base Step: The base step has been done in lemma 3.

Inductive Step: We assume that if c_{i+1} apply to e_{i+1}^1, then e_{i+1}^1 and e_{i+1}^2 can be unified.

Let

$$\begin{aligned} N_{i+1} &= < e_{i+1}^1, e_{i+1}^2 > \\ &= < r_{i+1} \circ e_i^1, r_{i+1} \circ e_i^2 > \end{aligned}$$

By inductive hypothesis that c_{i+1} can unify N_{i+1}, so

$$c_{i+1} \circ r_{i+1} \circ e_i^1 = c_{i+1} \circ r_{i+1} \circ e_i^2$$

that is, after apply c_i to e_i^1, e_i^1 and e_i^2 can be unified. **Q.E.D**

Completeness If e_1 and e_2 are unifiable, then the corresponding matching tree possesses a bf TSN.

Proof: The corresponding matching tree is the same as above:

$$N_0 \xrightarrow{r_1} N_1 \xrightarrow{r_2} \ldots \xrightarrow{r_p} N_p = \textbf{TSN}$$

Base Step: Since e_1 and e_2 are unifiable, from lemma 2 there exists at least one r_1, and **TSN** will be returned.

Inductive Step: We assume the assertion is true for i, there are two situations here:

1. N_i has one term $\longrightarrow$ **TSN** will be returned.
2. N_i has two term $\longrightarrow$ $N_i = < e_i^1, e_i^2 >$

From lemma 2, r_i exists and **TFN** does not exist.

$$\begin{aligned} N_{i+1} &= < e_{i+1}^1, e_{i+1}^2 > \\ &= < r_{i+1} \circ e_i^1, r_{i+1} \circ e_i^2 > \end{aligned}$$

By inductive hypothesis that c_{i+1} can unify N_{i+1}, so

$$c_{i+1} \circ r_{i+1} \circ e_i^1 = c_{i+1} \circ r_{i+1} \circ e_i^2$$

this repeat the first situation in the inductive step. And from the deriving procedure of section three, the second situation finally stop and return a **TSN**. Thus conclude our completeness. **Q.E.D**

Example 2 *(Cantor's theorem)*

$m, n \quad : i$
$S, f, g \quad : i \rightarrow i$
$N \quad : (i \rightarrow i) \rightarrow i$
$H \quad : i \rightarrow (i \rightarrow i)$

1. $H(N(f)) = f$

2. $\sim g = f, g(n) = f(n)$

3. $\sim n = S(n)$

The translation is as following:

$f_1 \quad : (i \rightarrow (i \rightarrow i), i, i \rightarrow i)$
$f_2 \quad : ((i \rightarrow i) \rightarrow i, i \rightarrow i, i)$
$f_3 \quad : (i \rightarrow i, i, i)$

1. $f_1(H, f_2(N, f)) = f$

2. $\sim g = f, f_3(g, n) = f_3(f, n)$

3. $\sim n = f_3(S, n)$
Start refutation:

(1,2) 4. $f_3(f_1(H, f_2(N, f)), n) = f_3(f, n)$

$\sigma = \ < g/f_1(H, f_2(N, f)) >$

(3,4) 5. $\square$
comment:
Rename n in clause 3, which become $\sim m = f_3(S, m)$.

$$\sigma = \ < m/f_3(f_1(H, f_2(N, f)), n), f_3(S, m)/f_3(f, n) >$$

Plug in the m of $f_3(f, m)$ with $f_3(f_1(H, f_2(N, f)), n)$, which is from the first m.

$$\begin{array}{ll} \sigma = & < f_3(S, f_3(f_1(H, f_2(N, f)), n)/f_3(f, n) > \\ \rightarrow & < f_3(S, f_3(f_1(H, f_2(N, f)), n))/f_3(S, f_3(h, n)) > \\ \rightarrow & < f_3(f_1(H, f_2(N, f)), n)/f_3(h, n) > \\ \rightarrow & < f_1(H, f_2(N, f))/h > \end{array}$$

Example 3 *(Pigeonhole principle)*

$$\begin{array}{ll} F & : \beta \rightarrow \alpha \\ G & : \alpha \rightarrow \beta \\ h & : \alpha \rightarrow \alpha \\ u, v & : \beta \\ E, x, y & : \alpha \\ X, Y & : (\alpha \rightarrow \alpha) \rightarrow \alpha \\ Z & : (\alpha \rightarrow \alpha) \rightarrow (\alpha \rightarrow \alpha) \end{array}$$

1. $u =_\alpha v, F(u) \neq_\alpha F(v)$

2. $F(u) \neq_\alpha E$

3. $X(h) \neq_\alpha Y(h), h(Z(h, x)) =_\alpha x$

4. $h(X(h)) =_\alpha h(Y(h)), h(Z(h, x)) =_\alpha x$

5. $x =_\alpha y, G(x) =_\beta G(y)$

The translation is as following:

$$\begin{array}{ll} f_1 & : (\beta \rightarrow \alpha, \beta, \alpha) \\ f_2 & : ((\alpha \rightarrow \alpha) \rightarrow \alpha, \alpha \rightarrow \alpha, \alpha) \\ f_3 & : (\alpha \rightarrow \alpha, \alpha, \alpha) \\ f_4 & : ((\alpha \rightarrow \alpha) \rightarrow (\alpha \rightarrow \alpha), \alpha \rightarrow \alpha, \alpha \rightarrow \alpha) \\ f_5 & : (\alpha \rightarrow \beta, \alpha, \beta) \end{array}$$

1. $u =_\alpha v, f_1(F, u) \neq_\alpha f_1(F, v)$

2. $f_1(F, u) \neq_\alpha E$

3. $f_2(X, h) \neq_\alpha f_2(Y, h), f_3(h, f_3(f_4(Z, h), x)) =_\alpha x$

4. $f_3(h, f_2(X, h)) =_\alpha f_3(h, f_2(Y, h)), f_3(h, f_3(f_4(Z, h), x)) =_\alpha x$

5. $x =_\alpha y, f_5(G, x) =_\beta f_5(G, y)$
   Start refutation:

(3,2) 6. $f_2(X, h) \neq_\alpha f_2(Y, h)$

$$\begin{array}{ll} \sigma = & < f_3(h, f_3(f_4(Z,h),x))/f_1(F,u) > \\ \rightarrow & < f_1(F, f_5(G,x))/f_1(F,u) > \\ \rightarrow & < u/f_5(G,x) > \end{array}$$

(4,2) 7. $f_3(h, f_2(X,h)) =_\alpha f_3(h, f_2(Y,h))$

$$\begin{array}{ll} \sigma = & < f_3(h, f_3(f_4(Z,h),x))/f_1(F,u) > \\ \rightarrow & < f_1(F, f_5(G,x))/f_1(F,u) > \\ \rightarrow & < u/f_5(G,x) > \end{array}$$

(7,1) 8. $u =_\alpha v$

$$\begin{array}{ll} \sigma = & < f_3(h, f_2(X,h)/f_1(F,u), f_3(h, f_2(Y,h)/f_1(F,v) > \\ \rightarrow & < f_1(F, f_5(G,x))/f_1(F,u), f_1(F, f_5(G,y))/f_1(F,v) > \\ \rightarrow & < u/f_5(G,x), v/f_5(G,y) > \end{array}$$

(8,5) 9. $x =_\alpha y$

$$\sigma = \; < u/f_5(G,x), v/f_5(G,y) >$$

(9,6) 10. $\square$

$$\sigma = \; < x/f_2(X,h), y/f_2(Y,h) >$$

5 Conclusion and Future Work

We have presented in this paper the results of general higher-order unification using n-sorted logic. As shown in our examples, the n-sorted logic approach allows us to use some modified first-order theorem proving programs. Besides Huet's higher-order unification, we offer another alternative to solve higher-order problems. Future work includes develop appropriate new first-order strategies for n-sorted logic to take advantage of the fact that much of the naming inferences will be done automatically within unification.

References

[1] Church A. A formulation of the simple theory of types. *Journal of symbolic logic*, 5:56–68, 1940.

[2] Peter B. Andrews. Resolution in type theory. *Journal of Symbolic logic*, 36(3):414–432, Sep. 1971.

[3] Peter B. Andrews, D. Miller, E. Cohen, and F. Pfenning. Automating higher-order logic. In *Contemptary Mathematics 29*, pages 169–192, 1984.

[4] J. L. Darlington. A partial mechanization of second-order logic. In B. Meltser and D. Michie, editors, *Machine Intelligence 6*, pages 91–100. American Elsevier Publishing Co., 1971.

[5] Wenchang Fang. *A Study of splitting in Higher-order theorem proving.* PhD thesis, Northwestern University, 1992.

[6] Wenchang Fang and Lawrence Henschen. Some heuristics for higher-order logic. In *Proceedings of the 1992 Golden West international conference on intelligence system*, pages 110–115, 1992.

[7] Lawrence J. Henschen. *A resolution style proof procedure for higher-order logic.* PhD thesis, University of Illinois at Urbana-Champaign, 1971.

[8] Lawrence J. Henschen. N-sorted logic for higher-order theorem proving. In *Proceedings of the ACM*, 1972.

[9] G. Huet. A mechanization of type theory. In *Proceedings of the third IJCAI*, pages 139–146, 1973.

[10] G. Huet. A unification algorithm for typed lambda-calculus. *Theoretical Computer Science*, pages 27–57, 1975.

[11] Lawrence C Paulson. Natural deduction as higher-order resolution. *Journal of logic programming*, 3:237–258, 1986.

[12] Tomasz Pietrzykowski. A complete mechanization of second-order type theory. *Journal of ACM*, 20(2):333–365, Apr. 1973.

[13] Tomasz Pietrzykowski and Jenson D. Mechanizing ω-order type theory through unification. *Theoretical Computer Science*, 3:123–171, 1976.

[14] J. A. Robinson. A machine oriented logic based on the resolution principle. *Journal of ACM*, pages 23–41, 1965.

Using String Logic Programming for Morphological Analysis: A Case Study on Indonesian Language

Hammam R. Yusuf*
Department of Computer Science
New Mexico State University
Las Cruces, NM 88003, USA
hyusuf@cs.nmsu.edu

Abstract

In this paper, we investigate a formalism based on string logic programming for analysing morphological structures in natural language processing systems. We will report the significant contribution of this research which can improve the effectiveness of morphological-parser development by enhancing the two-level morphology model. In particular, we describe a logical framework for analysing morphological structure and also present the application of the framework to a concatenative and generative language, such as Indonesian. A comparison with the two level model shows the improvement that our framework can achieve.

Keywords. Morphological Analysis, String Logic Programming, Two Level Morphology.

1 Introduction

Morphological analysis is a crucial step in natural language processing. Generally, morphological analysis is the first stage of analysis of a sequence of words forming a well-formed surface form, i.e. it produces trees of formatives constituting a correct parse of a compound word. Independent of the strategy being used, some information such as part-of-speech tag and case marker should be available to subsequent stages in a natural language understanding process.

One of the most influential approach to morphological analysis is the 'two-level' analyzer described by Koskenniemi [7]. This paradigm models a language's morphology and phonology as a set of finite-state transducers, each representing a phonological or morphophonological rule. These transducers operate on a parallel array, directly relating surface/lexical segment pairs.

*Supported by a graduate fellowship from The Agency for the Assessment and Application of Technology, Jakarta, INDONESIA. Many thanks to Dr. A. Rajasekar for introducing string logic programming and for a fruitful discussion on its applicability in natural language processing during the time I spent at the University of Kentucky.

E. A. Yfantis (ed.), Intelligent Systems, 221–230.

The description in two-level model proves to be unsuitable to describe generative language such as Indonesian. Koskenniemi notes that his system would require "extensions or revisions ... for an adequate descriptions of language possessing extensive infixation or reduplication" [7]. Gazdar [2] is less optimistic, pointing out that two-level model "has no straightforward way of talking about tree-structured objects". Rules in the two-level model are restricted to one-segment variations, while alternations which affect more than one segment must be lexicalized as suppletive.

In this paper, we use string logic programming [9] to develop a framework for morphological analysis by augmenting the two-level model with string resolution capabilities. String logic programs are extensions of logic program [13] which use strings as attributes for predicates instead of terms. We propose string-based framework which has proved useful for representing morphological rules in generative morphology paradigm. The main advantage of using the framework are improvement over the two-level model to cover more complex morphological process, a reduction in number of rules being used as well as an increase in the expressiveness of rule representation. To support the above argument, we demonstrate the application of our framework to Indonesian, a language possessing extensive affixation and reduplication.

2 Two-Level Morphology

The following is the description of two-level model as implemented in KIMMO system [5]. The model contains two type of components : automata and dictionary. The spelling-change rules that typically go along with affixation and inflection are encoded in a finite-state *automaton component*, while roots and affixes are listed with their co-occurence restrictions in a *dictionary component*. The two-level model is concerned with the representation of a word at two distinct levels, the surface level and the lexical level. At the surface level, words are represented as they may show up in text. At the lexical level, word consist of sequences of stems, affixes and boundary markers that have been pasted together without spelling changes. Thus we can represents the surface form **tries** as **try+s** at the lexical level. The lexical and surface levels are connected by several finite-state transducers that make up the automaton component.

A specific spelling change process in the two-level model amounts to a *constraint* on the correspondence between lexical and surface strings. For example, consider a simplified *y-change* process, as in **try+s** example, that changes "y" to "i" before adding "es". Interpreted as a constraint, y-change controls the occurence of the lexical/surface pairs **y/y** and **y/i**; lexical "y" must correspond to surface "i" when it occurs before lexical "+s", which will itself come out as surface "es" because of other constraints.

Table 1. Y-change State Table

	y	y	+	s	=	lexical characters
	i	y	=	s	=	surface characters
state1:	2	4	1	1	1	normal state
state2.	0	0	3	0	0	require +s
state3.	0	0	0	1	0	require s
state4:	2	4	5	1	1	forbid +s
state5:	2	4	1	0	1	forbid s

Each such constraint is implemented by a finite-state machine with two scanning heads that move together along the lexical and surface strings. The machine starts out in the state 1 and moves forward by changing state based on its current state and the pair of characters it is scanning.

For example, Table 1 taken from [5] describes the automaton that encodes the *y-change* process. In the notation, "=" is a certain kind of wildcard character. The use of ":" rather than "." after the state-number on some lines indicates that the ":" states are final states, which will accept end-of-input. In order to handle insertion or deletion, it is also possible to have a null character "0" on one side of a pair.

In processing the lexical/surface string pair **try+s/tries**, the automaton would run through the state sequence **1,1,1,2,3,1** and accept the correspondence. In contrast, with the string pair **try+s/tryes** it would block on **s/s** after the state sequence **1,1,1,4,5** because the entry for **s/s** in state 5 is zero. With the pair **try/tri** it would not block with any zero entries, but would still reject the pair because it would end up in state 2, which is not a final state.

3 String Logic Programs

In traditional first order logic and logic programming, *terms* form the basic building blocks which are used as arguments for predicate relationships. The main characteristics of terms is that they are precisely constructed and have a well-formed structure. The well-formed structure of the term allows for the use of a simple algorithm for unification of terms which forms the basis for the Robinson's resolution principle [10]. But this well-formed structure also disallows one to represent string information in an effective manner. String-based logic overcome this problem by defining an extension for the definition of terms in the form of strings.

Rajasekar [9] provided detailed syntax, semantics and proof procedures for string logic programs. Here we briefly review the syntax and the proof techniques for string logic programs. Strings are formed by concatenating string-variables and string-constants. String constants are made of basic elements in the language (eg. the alphabets in a language). A *string-variable* has two components, a name and a size given by an integer arithmetic expression (called the *bounds expression*). A string variable is denoted as follows: Let W be a string-variable name and let t be a bound expression then $\overset{t}{W}$ is a string-variable. The meaning of $\overset{t}{W}$ is that it denotes a string-variable which can take a string value whose size is equal to the value t. If S_1 and S_2 are strings then so is S_1S_2 (concatenation); the size of S_1S_2 is the sum of the sizes of S_1 and S_2.

From the above definitions it can be seen that we are restricting the value that can be taken by a string or a string variable and the restriction is provided by the bounds expression. When a string-variable name occurs more than once in a clause, it is considered to have the same size at each occurrence. This constraint is termed as the *bounds constraint.*

Definition 1 *If p is a n-ary predicate symbol and $S_1, \ldots, S_n$ are strings, then $p(S_1, \ldots, S_n)$ is a* (string) atom. *A string atom $p(S_1, \ldots, S_n)$ is a* ground string atom *if each S_i is a ground string.*

Example 1 *Some examples of string atoms and string logic program clauses:*
$name(john, las\_vegas)$

$concat(\overset{m}{X},\overset{n}{Y},\overset{m}{X}\overset{n}{Y})$
$inc\_by\_one(\overset{r}{X}0,\overset{r}{X}1).$
$inc\_by\_one(\overset{r}{X}1,\overset{m}{Y}0) \leftarrow inc\_by\_one(\overset{r}{X},\overset{m}{Y}).$
$add(\overset{r}{X}\overset{1}{A},\overset{m}{Y}0,\overset{p}{Z}\overset{1}{A}) \leftarrow add(\overset{r}{X},\overset{m}{Y},\overset{p}{Z}).$
$add(\overset{r}{X},\overset{m}{Y}1,\overset{p}{Z}\overset{1}{A}) \leftarrow inc\_by\_one(\overset{r}{X},\overset{n}{W}\overset{1}{A}), add(\overset{n}{W},\overset{m}{Y},\overset{p}{Z}).$

(The predicate $inc\_by\_one(X,Y)$ increments binary number X by 1 to produce Y, and the predicate $add(X,Y,Z)$ adds binary numbers X and Y to produce Z.)

Next we briefly discuss the proof procedure for answering queries in string logic programs. The proof procedure, called *SSLD-resolution* [9], is similar to SLD-resolution of Horn programs [13], except that instead of unifying predicates, the procedure matches two string atoms and converts them into a set of string equations and solves them incrementally as done in the case of constraint logic programming [3, 4, 14]. Hence instead of finding unifiers for two atomic expressions (as done in normal logic programs), we obtain a set of constraints in the form of string equations. We try to reduce them as much as possible in the process and also try to find if the set of constraints are inconsistent. A string equation is of the form $S_1 = S_2$ where S_1 and S_2 are strings.

The main difference between SLD-resolution and SSLD-resolution is in the way string matching is performed in SSLD-resolution as opposed to term unification in SLD-resolution. At any step in the SSLD-derivation, the Match algorithm takes the set of string constraints and the two atoms that need to be matched (A and A_m in the definition below) and forms a new set of string constraints. The new set of string constraints from the match algorithm augments the input set of string constraints along with constraints formed by the matching of the string atoms. The match algorithm returns a failure whenever the matched atoms is not consistent with the input set of string constraints. The operation of the match algorithm is similar to algebraic constraint solvers of the constraint logic programming systems [3, 4, 14]. More precise definition of the Match algorithm can be found in [9].

Definition 2 (SSLD-Derivation)
Let P be a string logic program and let G be a SSLD-goal. A String SLD-derivation *is a sequence of SSLD-goals $G_0 = G, G_1, \ldots,$ such that $\forall i \geq 0$, G_{i+1} is obtained from $G_i = \leftarrow (C_i, E_i), A_1, \ldots, A_n.$ as follows:*

1. *A_m is an atom in G_i and is called the selected atom.*
2. *$A \leftarrow B_1, \ldots B_r$ is a program clause in P standardized apart with respect to G_i*
3. *(C_{i+1}, E_{i+1}) is the output from the Match algorithm for the input $(C_i, E_i, A = A_m)$.*
4. *Let $\theta_{i+1} = \{e | e$ is a (variable substitution) equation of the form $\overset{n}{X} = S$ in $C_{i+1}\}$.*
5. *G_{i+1} is the goal $\leftarrow (C_{i+1}, E_{i+1}), (A_1, \ldots, A_{m-1}, B_1, \ldots B_r, A_{m+1},, \ldots, A_n)\theta_{i+1}$.*

Definition 3 *Let P be a string logic program and let G be a string goal. An SSLD-refutation is an SSLD-resolution which ends in a SSLD-goal with only a (possibly empty) constraint pair and no predicate goals.*

The final set of string constraint equations obtained from a SSLD-refutation is solved to obtain string substitutions for the string variables. Note that it is possible the final set of string constraints may not be solvable; in such a case the SSLD-resolution is considered to have ended in failure.

4 The Framework

A language will generally exhibit several different spelling-change processes; for example, Karttunen [5] mentions that Koskenniemi's analysis of Finnish uses 21 rules. By and large, these separate processes is encoded as separate automata, making the representation even more complex. In practice, the automata that express various constraint will inspect the lexical/surface correspondence in sequence. The correspondence will be accepted only if every automata accept it, that is if it satifies every constraint. In a string logic programming framework, this is analogous to satisfying a set of string equations in a procedural sequence of traversing a resolution proof tree, as decribed above.

Our framework begins with eliminating the redundancy of rules by combining several ordinary production rules into fewer general rule written in string logic program. For example, consider the following morphology rules/program:

Ordinary production rules
$singular(W, V) \rightarrow [say]$
$singular(W, V), ending(W, y), not\_abnormal(W) \rightarrow plural(says)$
$singular(W, N) \rightarrow [enemy]$
$singular(W, N), ending(W, y), abnormal(W) \rightarrow plural(enemies)$
$abnormal(enemy)$
$ending(enemy, y)$
$ending(say, y)$

String logic program
$plural(\overset{t}{W}\ s, V) \leftarrow singular(\overset{t}{W}, V), not\_abnormal(\overset{t}{W}).$
$plural(\overset{t-1}{W}\ ies, N) \leftarrow singular(\overset{t}{W}, N), abnormal(\overset{t}{W}).$
$abnormal(enemy).$

In the above program, N and V are ordinary variables, standing for noun and verb syntactic categories. W is a string variable standing for the word (or part of it in the case of the second *plural* clause). The program is just a fragment of a larger one which includes other details regarding *enemy* and *say*. By means of SSLD-resolution, string manipulation are hidden from the rewrite rule-set, whereas in ordinary rules, this manipulation must explicitly stated (eg. *ending*). Although the parsing mechanism is, therefore, still Prolog-like deductive mechanism, for linguistic purposes we can now visualize the parsing process as rewriting process where subsumption, in term of unification, is implicitly being performed.

One limitation of two-level model as addressed in [6] is due to lexical representation which tends to be arbitrary. Another limitation is that morphological categories are not directly encoded as part of the lexical form. Instead of morpheme like *Plural* or *Past*, we typically see suffix strings like +s and +ed, which do not by themselves indicate what morpheme they express. The string logic programming framework overcomes these limitations by enhancing the representation of the two-level model which includes all necessary syntactic and semantic information. as the arguments of a certain type of predicate (eg.*surface*, in the examples given below). These information are not readily available in the original model; and for many languages, analysis of morphological structure highly depends on the availability of these syntactic and semantic information.

Within our framework the y-change state table, as described in Table 1, is tranformed into the following string logic program:

$$surface(\overset{m}{W}\ s, plural) \leftarrow lexical(\overset{m}{W}, regular, Semantic)$$
$$surface(\overset{m-1}{W}\ ies, plural) \leftarrow lexical(\overset{m}{W}, irregular, Semantic)$$
$$lexical(say, regular, Semantic)$$
$$lexical(try, irregular, Semantic)$$

In comparison, we have shown that for the same morphological problem such as y-change, the number of rules decreases as well as improving the readability of rules. A more important issue is that information like *Syntatic* and *Semantic* information are embedded in the rule representation and provide additional restrictions for handling more complex mophophonological phenomena.

5 Application

A major motivation of developing the framework using string logic program is to enable effective application of the formalism to some derivational and concatenative languages. We start our description by outlining the morphological and phonological process in Indonesian. In a later section, we will propose an intelligent morphological analyzer for Indonesian.

5.1 Indonesian Morphology

In a brief outline, the major points of Indonesian morphology are as follows:

- the set of productive affixes is quite small (8 prefixes and 4 suffixes)
- circumfixation (combinational patterns of affixes and suffixes) is possible, in fact it yield to a large set of morphological structures
- there are few morphophonological rules
- affixing morphology is monosyllabic and morphological structure - not just the meaning of a word's component formatives- is crucial, and
- the same formatives can have different semantic effects depending on the value of the stem.

Circumfixation are used extensively in Indonesian word formation. The semantic resulted from affixing a word with prefix-suffix pattern usually is underivable from the compositional meaning of the affixes. For example, the prefix *ke* attaches to numeral *satu* (one) form *kesatu* (first) which is an adjective. The suffix *an* attach to the same numeral produce *satuan* (team) which is a concrete noun. However, when both *ke* and *an* appear as circumfixes, the resulting form *kesatuan* (unity) is an abstract nominal, a semantic effect which seems cannot be derived by the combination of semantic of individual affixes.

Reduplication is one of the unique characteristic of Indonesian since this phenomena have a tight interaction with phonological and morphological processes. Reduplication is accomplished by the parafixation of an empty affix synchronously with the base. The affix consists of a prosodic template, possibly dominating segmental slots (consonant Cs and verbal Vs). What appears to be segment-changing reduplication is argued to result from four mechanism [12]. They are: 1) lexical fossilization, 2) prespecification, 3) segment-filling default rules and 4) segment-changing rules. These mechanisms can be directly found in Indonesian, as follows:

1. Full reduplication, as in *kecil-kecil* (mostly small) from root *kecil* (small) and *berlari-lari* (running) from root *lari* (to run)

2. Partial reduplication, as in *lelaki* (male,men) from root *laki* (man)

3. Imitative reduplication, as in *bolak-balik* (to and from) from root *balik* (reverse) and *ombang-ambing* (drift to) from root *ombang* (float)

4. Circumfixial reduplication, as in *kekanak-kanakan* (childish) from root *kanak* (child)

Both transitive and intransitive verb roots can be reduplicated to indicate intensity. These reduplicated roots normally co-occur with prefixes as in the example which follows. Additionally, the verbal prefix meN participates in the morphological pattern shown subsequently:

1. $prefix + X - X : menulis - nulis \ (write : intensive)$

2. $X - prefix + X : tulis - menulis \ (write : reciprocal)$

In (1) the prefix meN is straightforwardly adjoined to a reduplicated transitive root (tulis 'write'), while in (2), the affix becomes a quasi-infix, appearing between the halves of the reduplication. The two prefixation and reduplication patterns in (1) and (2) have quite distinct semantic effects. A more complete discussion on this topic may be found in [8].

The above descriptions suggest the basic format of the analyzer, which assumes that identifying the word's root and its syntactic/semantic properties is the essential first step in processing a word in Indonesian. The regular monosyllabic character of affixes in the language provides the basis for an affix-stripping routine which divides the word into potential affixes and root. Before the semantic effects of any affixes can be determined, the position and identity of the root must be discovered.

5.2 Indonesian Morphological Analyzer System (IMAS)

Indonesian's regular syllable structure, coupled with the fact that all of its affixes are monosyllabic, makes it an extremely attractive candidate for string logic-based parsing. The identification of potential morpheme boundaries can be made much more efficient due to constrains established by the string logic program itself which then provide trees of syllable boundaries thru resolution.

We build an intelligent analyzer for Indonesian morphology IMAS which use the framework discuss above with SSLD-resolution as the inference machinery.

The proposed IMAS model for Indonesian correctly assign meaning to a sequence of formatives by identifying the root word and its morphological structure. The approach here conform to the word formation theory described in [1] with rule of the type:

$$[X]_{Cat} + (Case) \rightarrow [X + affix]_{Cat'} + (Case')$$

This schemata will analyze surface forms which do not exist as lexical entries.

For example, given an input word "membalikkan" (to reverse), we would have the following possible morphological structures for selection by the final stage of the analyzer :

```
[membalikkan, root]
[membalik, root] [kan, suffix]
[meN, prefix] [balikkan, root]
[meN, prefix] [balik, root] [kan, suffix]
```

Based on the regular morphological rule of Indonesian, the correct structure for the above example is the last one. In order to have this selection, the analyzer will make every possible dictionary lookup for the root word. It will match "balik" (reverse), but not "membalikkan".

The above possible morphological structure trees can be easily captured by the following string logic program :

$$surface(\overset{m}{X}\overset{n}{Y}\overset{p}{Z}) \leftarrow prefix(\overset{m}{X}), root(\overset{n}{Y}), suffix(\overset{p}{Z}).$$
$prefix(mem).$
$root(balik).$
$suffix(kan).$

The above program together with SSLD-refutation will produce the correct parse with variable binding: X=mem, Y=balik and Z=kan, corresponding to the last parse tree, while other possible bindings will be rejected by SSLD-refutation. Note that in the above logic program, the morphophonology process of meN and the first consonant for the root 'balik' (reverse) is implicitly carried out. The prefix meN mutate to *mem* when encountering the lower labial phoneme *b*.

For analysing the word "mempersoalkan" (to argue), which is a derived form of the root word "soal" (problem), we need to consider circumfixation "per-an" in place of prefixation-suffixation process. As pointed out in section 5.1 this phenomena can not be derived as prefix-suffix combination due to a semantic constraint being imposed of its derivational process. The word can be captured using the following recursive clause.

$$surface(\overset{m}{X}\overset{n}{Y}) \leftarrow prefix(\overset{m}{X}), surface(\overset{n}{Y}).$$
$$surface(\overset{m}{U}\overset{n}{V}\overset{p}{W}) \leftarrow root(\overset{n}{V}), circumfix(\overset{m}{U}, \overset{p}{W}).$$
$prefix(mem).$
$root(soal).$
$circumfix(per, kan).$

which will produce the following structure:

```
[meN, prefix] [[soal, root] [per-kan, circumfix]]
```

As described in section 2, the two-level model has a drawback in handling reduplication. We show how some types of reduplication found in Indonesian can be treated elegantly.

Given, input word "sekalisekali" (once in a while), the analyzer would use the following short program to parse it:

$$surface(\overset{m}{X}\overset{n}{Y}\overset{m}{X}\overset{n}{Y}) \leftarrow prefix(\overset{m}{X}), root(\overset{n}{Y}).$$
$$prefix(se).$$
$$root(kali).$$

which will produce the following bindings: X=se and Y=kali, whereas the parse trees with the answer substitution consisting other combination of X and Y will not have a successful SSLD-refutation.

The above example show a program for handling full reduplication. Next we consider imitative reduplication. From the example in 5.1 we want to derive the surface form of the word *ombang-ambing* (drift to) from root *ombang* (float). This is exemplified with the following string logic program:

$$surface(\overset{1}{X}\overset{m}{Y}\overset{1}{W}\overset{n}{Z}\overset{1}{U}\overset{m}{Y}\overset{1}{V}\overset{n}{Z}) \leftarrow root(\overset{1}{X}\overset{m}{Y}\overset{1}{W}\overset{n}{Z}), change_v(\overset{1}{X}, \overset{1}{U}), change_v(\overset{1}{W}, \overset{1}{V}).$$
$$root(ombang).$$
$$change_v(o, a).$$
$$change_v(a, i).$$

Another phenomena in Indonesian morphology, as noted in section 4, is the existence of *infix* such as "em", "el", "er". Forms which historically infixed appear to be lexicalized; no clear semantic effect can be assigned to these infixes, and speakers are not generally conscious of the infix as a distinct entity in the words. The following program simply process words with infix.

$$surface(\overset{p}{X}\overset{2}{Y}\overset{q}{Z}) \leftarrow lexical(\overset{p}{X}\overset{q}{Z}), infix(\overset{2}{Y}).$$
$$infix(el).$$
$$infix(em).$$
$$infix(er).$$

In our experimentation with this logical framework, IMAS shown a significant improvement over our previous implementation of morphological analyser in a multi-lingual machine translation system [15]. The resulting rule-set have 31% improvement in term of number of rules over system implemented with two-level model. In addition, our comparison with DCG implementation of morphological rules show 24% reduction in the number of defined rules. This suggests that result of applying the framework to morphological analysis is encouraging. One clear disadvantage of the current model is that subsumption is costly since SSLD-refutation requires time $\Omega(n^3)$ in the average case where n is the length of word. This result is due to the cost of string unification [11].

6 Conclusion

It has been shown that a logical framework based on string logic program proves to enhance the representation of two-level model and also add capabilities to handle a more complex morphological problem by attaching syntactic and semantic information to the rule representation. The framework works by satisfying a set of string equations which are resolved through SSLD-resolution. We have provided a case study to show the application of the framework to solve morphology problem in generative language. The framework is basically language independent, and can also be applied to other languages. We have also demonstrated how string logic program can be used to tackle problem that was originally difficult with two-level model, such as infixation and reduplication.

References

[1] M. Aronoff. (1976) *Word Formation in Generative Grammar.* MIT Press, Cambridge, MA.

[2] G. Gazdar, K. Pullum, and I . Sag. (1985) *Generalized Phrase-Structure Grammar.* Basil-Blackwell, Oxford, England.

[3] J. Jaffar and J.L. Lassez. (1987) Constraint Logic Programming. In *Proc. of POPL.*

[4] J. Jaffar and S. Michaylov.(1987) Methodology and Implementation of a CLP System. In *Proc. of Logic Programming Conference*, Melbourne.

[5] L. Karttunen and K. Wittenburg. (1983) Kimmo: A two-level Morphological Analysis of English. *Texas Linguistic Forum*, 22:217-228.

[6] L. Karttunen, R. Kaplan and A. Zaenen. (1992) Two-level Morphology with Composition. In *Proc. of COLING-92*, Nantes,France.

[7] K. Koskenniemi. (1983) Two level model for morphological analysis. In *Proc. IJCAI-83.*

[8] R. Macdonald and S. Darjowidjojo. (1967) *Indonesian Reference Grammar.* Georgetown University Press, Washington D.C.

[9] A. Rajasekar. (1993) Logic Programming Using Strings. Technical Report 227-93, Department of Computer Science, University of Kentucky.

[10] J.A. Robinson. (1965) A Machine-Oriented Logic Based on the Resolution Principle. *J.ACM*, 12(1).

[11] J. Siekmann. (1975) String Unification. Technical report, Essex University, Essex, England.

[12] Amy H. Urbach. (1987) Reduplication. Doctoral Dissertation, University of Texas, Austin, TX.

[13] M.H. van Emden and R.A. Kowalski. (1976) The Semantics of Predicate Logic as a Programming Language. *J.ACM*, 23(4):733–742.

[14] P. van Hentenryck. (1989) *Constraint Satisfaction in Logic Programming.* MIT Press.

[15] H. Yusuf. (1992) Analysis of Indonesian Language for Interlingual Machine Translation System. In *Proc. of COLING-92*, Nantes,France.

A Knowledge Organization Framework for Knowledge-Based Pattern Recognition

Yufeng F. Chen
Nazir A. Warsi
Army Center of Excellence in Information Sciences
Clark Atlanta University
Atlanta, GA 30314

Abstract. In this paper we present a knowledge organization framework for designing and using a knowledge-based system oriented to pattern recognition problems. The framework integrates a set of views for acquiring, structuring and using the necessary pattern recognition knowledge, with particular to the presence of noise and to the computational efficiency of the inference engine. Our approach differs from traditional problem-solving emphasis, but focusing on semantic abstraction as human does in mind for organizing his knowledge. In addition, we describe a very expressive representation language (KORL) to support the knowledge organization process. We consider those second-order predicate calculus features, such as self-describing capability, treatment of relations as first-class concepts, and the use of contexts, to be very useful.

Keywords. Knowledge Organization, Pattern Recognition, Ontology Building.

1 Introduction

Traditional approaches for handling pattern recognition problems have mostly relied upon procedural methods for the low-level analysis of signals as, for instance, segmentation and feature extraction. Recently, there is a need for applying knowledge-based methodologies to complex pattern recognition tasks, such as speech processing and waveform analysis. To cope with the complexity of individual concepts in the world, people mentally group them into categories. There are many natural categories, and they are often hierarchically organized: some categories are superordinate or subordinate to others. However, a classification scheme should not be arbitrary but should be based on the result of systematic study. The proposed organization scheme recognizes the importance of the different discrimination bases yet provides a uniform and principled methodology for knowledge organization.

1.1 Taxonomic Approach (Structure of Representation)

When classes are related by the specialization relationship, the resulting partial order (a tree or lattice) is known as a taxonomic hierarchy. A taxonomy is one of the most familiar knowledge structures for organizing information. In a taxonomy we can store information at the appropriate levels of generality and make that information available to more specific concepts

E. A. Yfantis (ed.), Intelligent Systems, 231–237.

by means of *inheritance* [Touretzky, 1986]. Another advantage of hierarchical structuring, besides representational compactness, is that it makes search more efficient. We can also use a relatively stable taxonomy as a basis for inferencing.

1.2 Ontological Engineering (Content of Representation)

The structural representation of knowledge is only half of the story. The question remaining is what to represent. Ontology is the study of the basic categories of existence: being. An ontology is a collection of abstract and concrete objects, and of relationships and transformations that represent the physical and cognitive entities necessary for accomplishing a task. Ontological analysis is a technique for the upper-level analysis of a domain. The issues include how to build the topmost levels of a large hierarchy of knowledge, how to integrate pieces of knowledge represented by different people, and how to control revision of knowledge. The semantic depth of the model of the domain depends on the richness of the ontology.

Two types of fundamental knowledge are needed to represent an ontology. *Primary concepts* are very general, top-level semantic categories. For example, there are many "natural-kind" concepts, such as *Person* and *Animal*, that are difficult to define or cannot be exhaustively described. On the other hand, what we call *concept primitives* are semantic elements that describe, distinguish, and classify other (nonprimary) concepts. Each concept primitive is a fundamental *property* and is represented by a slot in frame-based system. These primitives provide a basis for explaining the organization of concepts into families, as well as similarities among concepts within a single family, and similarities among the different families.

2 A Knowledge Organization Framework and Methodological Foundation

We present a knowledge organization framework for designing and using a knowledge-based system oriented to pattern recognition problems. The framework integrates a set of views for acquiring, structuring and using the necessary pattern recognition knowledge. We also describe the methodological foundation that supports our framework for knowledge-base organization.

2.1 System Analysis View

From the system analysis standpoint, we focus the knowledge-recovery effort on exploring the more general knowledge about the system from a real-world interpretation. There are two sub-levels of knowledge organization process in the analysis level: the system-general knowledge sub-level and the domain-specific knowledge sub-level (see Fig. 2.1).

At the system knowledge sub-level, the emphasis is on understanding the features of the system that distinguish it from other types of systems. For example, a knowledge-based pattern recognition system can be characterized by its motivation (to solve complexity and understanding problems), objective (to develop a general framework for all pattern recognition applications), approaches (statistical, structural, neural, or hybrid), significance (no other system has all these features together), language (KORL has both expressiveness and efficiency), scope (cover noisy environment), status (continue to expand the framework and language capability). The knowledge to be organized should be general for all pattern recognition modules, such as preprocessing, segmentation, classification, training, and evaluation.

At the domain-specific sub-level, we recognize a system by specifying the type of its application-specific knowledge, such as knowledge of speech processing or software reusability knowledge. This domain-specific knowledge can be used to guide a selection of suitable pattern recognition approaches. In addition, it is also important to decide how detailed the domain theory is to be represented; that is, we need to represent the boundary and the scope of the domain knowledge. An explicit description of the limitations of the system's knowledge is necessary for using the system properly. These limitations of the theory can be formalized as assumptions, described by our knowledge organization representation language (KORL), and added to the theory.

- System Sub-Level
  - Motivation
  - Objective
  - Approaches
  - Significance
  - Language
  - Scope
  - Status
- Domain Sub-Level
  - Type
    - Speech Recognition
    - Software Reusability, etc.
  - Domain Assumptions

Fig. 2.1 System Analysis View

2.2 Ontological Design View

From the standpoint of ontological design, the organization of knowledge emphasizes the methodology to generate a model for integration. An explicit design model for the ontology can prevent misunderstandings and improper use. We outline how to design an upper model for the ontological organization of a knowledge base. First, we make explicit the underlying principles of how and why this model has been developed. Then we must create for the upper model a level of abstraction that is sufficiently high that it generalizes across application domains, without being so high that the model has no semantics [Bateman *et al.*, 1990]. Next, a formal method is required to verify this model. It is important that both (1) the contents of such a level of abstraction be motivated on good theoretical grounds and (2) the mapping between that level and the specific domain can be specified. In addition, it is important to know not only why certain concepts are included in the upper model but also why others are excluded.

Furthermore, we need to set up guidelines to specialize the upper model for a domain application. The user constructs the domain model, defining the types of entities required, including actions, events, objects, and states. Domain-model entities must be linked to the upper model. For example, the top-level concept *Recognition* is partitioned into the concept *RecognitionProcess*, whose instances are algorithmic process, and *RecognitionObject*, whose instances are recognition subjects. In a software component recognition domain, the meaning of *ProgramConcept* will be derived from the general properties of *RecognitionObject*. Fig. 2.2. elaborates these issues.

- Ontological Model
  - Principles
    - Linguistics
    - Philosophy
  - Definitional Status
    - Dictionary Approach
    - Thesaurus Approach

Fig. 2.2 Ontological Design View

In our knowledge-based pattern recognition system, we intend to identify the different discrimination bases from the perspectives of knowledge-level and symbol-level uses [Newell, 1982]. The knowledge level specifies a distinct computer-system level, lying above the symbol level, that gives us the opportunity to analyze knowledge at a very high level of abstraction. For example, in software program recognition domain, a program concept classification ontology will cover programming concept, architectural concept, and domain concept [Kozaczynski *et al.*, 1992]. The symbol-level ontology includes the representational components for the pattern classification architectures. For example, the generic terms, *node* and *link*, can be used to express a multi-stage classifier or a network structure.

2.3 Semantic Implementation View

For a knowledge-based system, we must determine the meaning of a given term by its relationships with other represented terms. Therefore, our knowledge organization should reflect the architecture objects' functional roles within the representation language. Thus, the meaning of a term is given by the totality of the assertions in which it participates. In a random environment, the meaning of each *node (or neuron)* can be determined by a statistical decision rule (or activation function) and the relationship with other *nodes*.

2.4 Constraint Maintenance View

After being constructed, the knowledge organization much be maintained as new knowledge is added to the system. For example, when entering a new concept into an existing knowledge base, a user may have an idea of where the concept belongs in the taxonomy. For a knowledge-based system to have this capability, a detailed representation of the concept (in terms of slots and values) from the classification standpoint must be available. However, it is sometimes difficult to decide how much of this process should be automated and how much should be entrusted to the user.

Some system designers question the need for an automatic concept classifier and have decided to let the user make the judgment about where a concept should be located in the taxonomy. However, other designers are concerned about the ambiguity and inconsistent nature of human judgment and insist that knowledge-based systems should automatically maintain the taxonomy [Swartout and Neches, 1986]. However, the knowledge-based system can, for example, determine if user-entered taxonomic information is consistent with the (user-entered) slots and values.

Since most real-world concepts lack a complete definition, it is difficult to use the KL-ONE approach [Woods and Schmolze, 1992] to maintain a concept taxonomy. We adopt a constraint approach that permits partial definitions but provides for efficient consistency checking of the knowledge organization. We permit the use of an incomplete representation (in terms of primitives and associated values) for concepts. Thus, we allow a concept meaning based on necessary conditions if sufficient conditions are not available [Woods, 1991]. There are two varieties of constraint mechanisms for consistency checking. One is the restriction constraint, which enforces a legal value requirement directly to the specific property of a term. The other is the dependency constraint, which demands certain connections among terms.

3 Knowledge Organization Language

Most frame-based knowledge representation systems support only classification-based reasoning and simple *isA* inheritance. This section discusses features of the knowledge representation language that we consider they are useful for expressing the knowledge for a knowledge-based pattern recognition system.

3.1 Contextual Knowledge Representation

To express knowledge about domain scope for analysis-level knowledge organization, *context* representation capability can be very useful. Contexts provide a mechanism to describe several related assertions with a single frame unit. The second-order predicate takes as arguments a context and an assertion.

3.2 Inheritance through any Relation

While most other frame-based systems have only support for class specialization inheritance, KORL supports inheritance through any relation. For example, to express "all child nodes received a signal from their parent node," the *Signal* can be carried through the relation *childLink* on the frame *ParentNode*. This feature provides additional semantics to the ontological design.

3.3 Self-Describing Languages

If a representation language is self-describing, then it is easier for us to understand the language's features at the implementation-level. KORL explicitly represents the components of its representation languages as well as domain knowledge entered by the user. Thus, the language can encode its constructs within its own formalisms. If we desire a new type of inheritance, we may create a frame to implement it in terms of existing KORL features. For example, the slot *supralayerConnection* can be implemented by composing *InterlayerConnection* with itself in a recurrent neural network [Fiesler, 1994].

3.4 Relations as First-Class Objects

All constraint propagation involves slots. KORL can represent relations as full-fledged frames on equal footing with object frames. This allows the constraint features of the relations to be expressed declaratively and provides greater inferential power. For example, a slot may appear only on a frame that is an instance of the slot's domain class; the slot's value must be an instance of the range class; and the number of slot entries must agree with the entry format requirements. Without relations having first-order-object features, it will be very difficult to specify that one relation is tied to others.

4 Conclusions and Discussions

The field of pattern recognition has grown enormously in recent years and a variety of techniques have been developed for various applications. There is an increasing need to have a knowledge-based approach to guide the data preparation and offer the output interpretation for the complex pattern recognition systems. However, a principled organization of a knowledge base is essential for manipulating such invisible knowledge effectively. We have developed a methodology for the knowledge organization framework for its improvement in knowledge understanding and retrieval. In particular, we are concerned with the problem of how knowledge must be represented in a pattern recognition system and how representations can be transformed to different approaches accordingly.

Although their original purposes were not for knowledge-base organization, many researchers have implicitly applied reverse engineering [Cross II *et al.*, 1992] with semantic abstraction approaches. For example, in his heuristic classification paper, Clancey [1985] analyzed the inference structure underlying several expert systems, including MYCIN, BANNKER, and PUFF. However, the details of organization process for those expert systems were not described. The Ontolingua system [Gruber, 1992] represents an effort to represent the common features of different frame knowledge representation systems. However, the system is not a general representation language with automated reasoning capabilities. For our KORL language design, the self-describing features of the symbol-level view have been proposed in early knowledge representation languages, such as RLL [Greiner and Lenat, 1980].

Our framework for knowledge organization can be validated from two research domains: (1) the representation and reasoning of large scale knowledge-based systems [Lenat and Guha, 1990], and (2) the identification of fundamental terminological properties of semantic relationships by researchers in cognitive science and linguistics [Cruse, 1986]. We are continuing to investigate these underlying efforts for future research in the knowledge-based system organization.

5 References

[Bateman *et al.*, 1990] J. A. Bateman, R. T. Kasper, J. D. Moore, and R. A. Whitney, "A General Organization of Knowledge for Natural Language Processing: the Penman Upper Model," Technical report, USC/Information Science Institute, Marina del Rey, CA, 1990.

[Clancey, 1985] W. J. Clancey, "Heuristic Classification," *Artificial Intelligence Journal*, vol., 27, no. 3, 1985, pp. 289-350.

[Cross II *et al.*, 1992] J. H. Cross II, E. J. Chikofsky, and C. H. May, JR., "Reverse Engineering," *Advances in Computers*, vol. 35, Academic Press, 1992, pp. 199-254.

[Cruse, 1986] D. A. Cruse, *Lexical Semantics*. Cambridge: Cambridge University Press, 1986.

[Fiesler, 1994] E. Fiesler, "Neural Network Classification and Formalization," in *Computer Standard & Interfaces*, vol. 16, 1994.

[Greiner and Lenat, 1980] R. Greiner and D. Lenat, "RLL: A Representation Language Language," *Proceeding of the 1st AAAI Conference*, Stanford, CA, 1980, pp. 165-169.

[Gruber, 1992] T. R. Gruber, "Ontolingua: A mechanism to support portable ontologies," Stanford University, Knowledge Systems Laboratory, Technical Report KSL 91-66. Revision.

[Kozaczynski *et al.*, 1992] W. Kozaczynski, J. Ning,, and T. Sarver, "Program Concept Recognition," in *Proceedings of the 7th KBSE conference*, September, 1992, pp. 216-225.

[Lenat and Guha, 1990] D. B. Lenat and R. V. Guha, *Building Large Knowledge-Based Systems*, Addison-Wesley, Reading, Mass., 1990.

[Newell, 1982] A. Newell, "The Knowledge Level," *AI Journal*, 19(2): pp. 87-127.

[Swartout and Neches, 1986] W. R. Swartout and R. Neches, "The Shifting Terminological Space: An Impediment to Evolvability," *Proceedings of the Fifth National Conference on Artificial Intelligence, AAAI-86*, Philadelphia, Pa., August 1986, pp. 936-941.

[Touretzky, 1986] D. S. Touretzky, *The Mathematics of Inheritance Systems*, Morgan Kaufmann Publishers, Inc., Los Altos. CA., 1986.

[Woods, 1991] W. A. Woods, "Understanding Subsumption and Taxonomy: A Framework for Progress," in *Principles of Semantic Networks*, J. F. Sowa, ed., Morgan Kaufmann, San Mateo, CA, 1991, pp. 45-94.

[Woods and Schmolze, 1992] W. A. Woods and J. G. Schmolze, "The KL-ONE Family," *Computer & Mathematics*, vol. 23, no. 2-5, 1992, pp. 133-177.

An Interactive Front-End for Fuzzy Databases

Chengwen Liu
Department of Computer Science and Information Systems,
DePaul University, Chicago, Illinois

Qi Yang, Jing Wu and Clement Yu
Department of EECS, University of Illinois at Chicago

Abstract

Fuzzy databases have been introduced to deal with uncertain or incomplete information in many applications. We present the design and implementation of an interactive front end for fuzzy databases. Our system was implemented on top of existing relational DBMS and allows users to query, insert, update and delete data by using fuzzy SQL, which is an extension of standard SQL. The system also allows users to create, modify, delete and display fuzzy data definitions.

Keywords: fuzzy logic, fuzzy sets, fuzzy databases, fuzzy SQL.

1 Introduction

In order to extend the applicability of traditional databases, some new techniques have been proposed to deal with uncertain or incomplete information [3, 4, 12]. Fuzzy sets and fuzzy logic have been introduced into database systems for this purpose [10, 1, 5, 8, 13].

For example, the database in a police station has a relation containing descriptions of criminals / suspects. A suspect may be described as a "young" man with a weights of "about 190 lb". Then a person of 50 years old is unlikely to be the suspect, and a person of age 25 is more likely to be the suspect than a person aged 34 with respect to the age condition if "young" is considered to be between 20 and 35. A similar situation occurs with respect to weight. Fuzzy logic could play a significant role in such applications, since predicates in fuzzy logic have truth values between 0 and 1 and are suitable to capture degrees of uncertainty. In the example above, the truth value of "young" = 25 should be much higher than that of "young" = 50 which is 0, while that of "young" = 34 is in between.

Many useful proposals have been made for fuzzy relational databases[11, 5, 6, 13]. A fuzzy database library has been built by Omron Corporation [6], and a fuzzy SQL language has been proposed by extending standard relational SQL using fuzzy theory [7]. In this paper, we present the design and implementation of fuzzy SQL on top of relational DBMS.

E. A. Yfantis (ed.), Intelligent Systems, 239–247.

2 Fuzzy Databases

2.1 Data classification and Representation

In fuzzy databases, a value may be either a precise (or crisp) value or a fuzzy term. For numerical data, a fuzzy term may be a fuzzy number or a fuzzy label. For example, a relation about people may have an attribute, AGE, for people's age, and a value for AGE may be 28 or "about 35" or "young". The value 28 is a crisp value which is the same as ordinary databases. Both "about 35" and "young" are fuzzy terms; "about 35" is a fuzzy number and "young" is a fuzzy label.

Fuzzy data are described by membership functions, which characterize to what degree an individual element belongs to a fuzzy (imprecise) set. Membership functions of *fuzzy numbers* and *fuzzy labels* are defined based on their characterizations[6]. A fuzzy number is characterized by a width, a width specification method and the number appearing in the fuzzy number. All fuzzy numbers in the same column are characterized by the same width and the same width specification method. Widths and width specification methods are provided by users. Width specification method can be "Fixed Width" or "Ratio Width" as illustrated in Fig. 1.

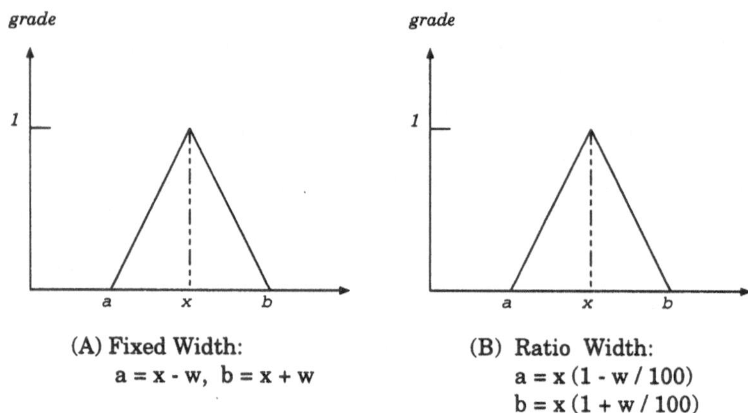

Figure 1: Membership functions of fuzzy numbers.

A fuzzy label is characterized by four numbers a, b, c and d, which are provided by users, such that $a \leq b \leq c \leq d$. The membership function of the fuzzy label characterized by a, b, c, d is illustrated in Fig. 2.

In fuzzy databases, *crisp data, fuzzy numbers* and *fuzzy labels* are allowed to be registered into the same column. For example, 28, "about 35", and "young" are registered into the same fuzzy column "age". Since our system is built on top of traditional relational database systems where data in the same column must be of the same type, a value for a fuzzy column is represented as a triplet (v, t, d), where v is a number, t indicates the type of the value and d is the reliability degree of the value, and stored in three columns. For crisp and fuzzy numbers, v is a number. For fuzzy labels, v is the numerical identifier of

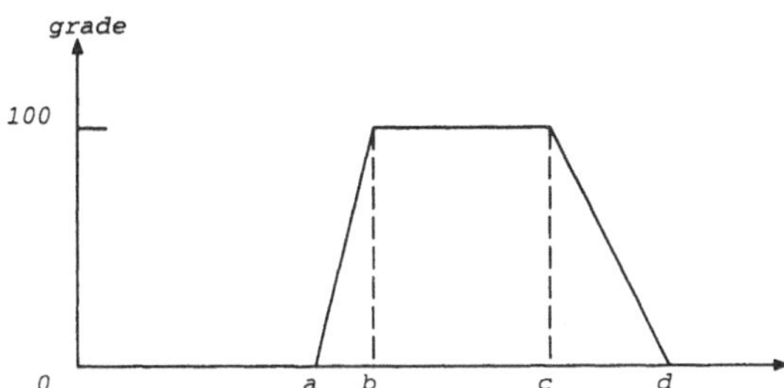

Figure 2: Membership function of a fuzzy label.

the fuzzy label.

2.2 Fuzzy Data Dictionary

Fuzzy data dictionary (FDD) consists of a set of tables that are used to store fuzzy schema and the membership functions of fuzzy numbers and fuzzy labels. Each fuzzy column is associated with one membership function for fuzzy numbers and one or more membership functions for fuzzy labels. Specifically, for each fuzzy column, the width and the width specification method are stored. For each fuzzy label of a fuzzy column, the system assigned identifier, the name of the label and the four numbers defining the membership function are stored.

3 Design and Implementation

The architecture of our system is shown in Fig. 3, where FDL is a fuzzy database library which consists of a set of functions to do fuzzy computation. Users are allowed to issue queries in Fuzzy SQL which is an extension of conventional SQL to support fuzzy logic. User queries will be translated by the front-end into conventional SQL so as to access the data stored in the relational database or the fuzzy data dictionary. Retrieved data are passed to FDL functions by the front-end to do fuzzy computation. For select queries, the tuples satisfying the search condition will be returned. Each output tuple is associated with a number d ($0 < d <= 1$) indicating to what degree that tuple satisfies the user query.

3.1 Fuzzy DDL

In this subsection, we describe the fuzzy data definition language (FDDL) supported by our system.

1. **Create relation schema**. In the fuzzy database system, we will allow data to be of types *type* or **F***type*, where *type* can be any data types used in SQL, such as *integer* or *float*. If a column is of type **F***type*, it indicates that the data type of the column is *type* and that the column may contain fuzzy data.

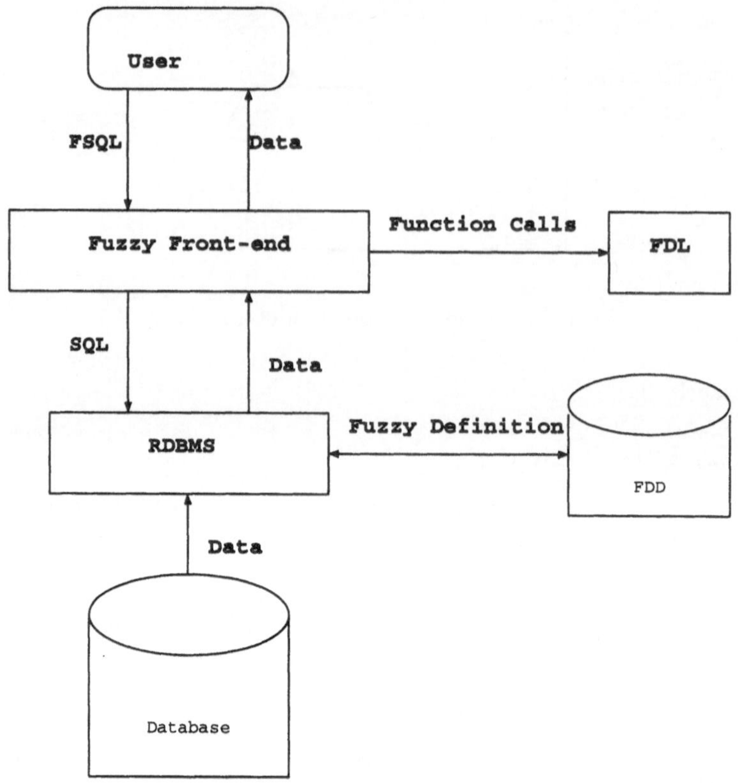

Figure 3: System Architecture.

Queries for creating relation schema are of the following form:

CREATE TABLE r_name
(a_1 $\quad$ type$_1$,
a_2 $\quad$ type$_2$,
... $\quad$ ...,
a_n $\quad$ type$_n$);

In the above, a_i, type$_i$ are the name and data type of the i-column in the relation r_name. If column a_i is of type **F***type*, the system will automatically create two columns associated with the column, one is named the status of the column, the other is named the reliability degree of the column. Both associated columns are of type *integer*. In this way, columns associated with columns containing fuzzy data are invisible to users, and users do not create such associated columns and input values for them.

For example, the following statement creates a relation SUSPECT with non-fuzzy columns S#, NAME and SEX, and fuzzy columns AGE, WEIGHT and HEIGHT.

```
CREATE TABLE SUSPECT
       ( S#      NUMBER NOT NULL PRIMARY KEY,
         NAME    VARCHAR(20),
         AGE     FNUMBER(3),
         HEIGHT  FNUMBER(3,2),
         WEIGHT  FNUMBER(3),
         SEX     CHAR(1)
       );
```

2. **Create fuzzy data dictionary**. FDD stores the characterizations of the membership functions for fuzzy numbers and fuzzy labels. The following statements are used to define fuzzy numbers and fuzzy labels.

DEFINE FUZZYNUMBER r_name
(a_1 w_1 m_1,
a_2 w_2 m_2,
......,
a_n w_n m_n);

where a_i is the name of a column of relation r_name which may contain fuzzy number. w_i and m_i are the *width* and *width specification method* of column a_i. m_i can be either "FIXED" or "RATIO". If the fuzzy number for a column a_i is defied as FIXED, then the range value for fuzzy number "about x" will be [x - w_i, x + w_i]. If it is defined as RATIO, then the range value for "about x" will be [x * (1 - w_i/100), x * (1 + w_i/100)].

For example, the following statement defines fuzzy numbers for columns AGE, HEIGHT and WEIGHT. Based on the definition, "about 40" years old represents 35 to 40 years old and "about 150" pounds represent 105 to 195 pounds. For the AGE column, the degree of matching between "about 40" and 40 is 1 while that between "about 40" and 38 is 0.6.

```
DEFINE FUZZYNUMBER SUSPECT
       ( AGE     5     FIXED,
         WEIGHT  30    RATIO,
         HEIGHT  20    RATIO
       );
```

DEFINE FUZZYLABEL r_name c_name
(l_1 l_min$_1$ l_max$_1$ r_max$_1$ r_min$_1$,
l_2 l_min$_2$ l_max$_2$ r_max$_2$ r_min$_2$,
......,
l_n l_min$_n$ l_max$_n$ r_max$_n$ r_min$_n$);

where l_i is the name of a fuzzy label in the column c_name of the relation r_name, while l_min$_i$, l_max$_i$, r_max$_i$ and r_min$_i$ are the four numbers to characterize the label. The fuzzy database system will automatically assign an identifier to each fuzzy label defined by the above statement.

For example,

```
DEFINE FUZZYLABEL SUSPECT AGE
        ( young      0, 0, 20, 30,
          middle-age 30, 35, 45, 50,
          old        50, 60, 200, 200
        );
```

3. **Modify fuzzy data dictionary**. Fuzzy data dictionary can be modified by the MODIFY FUZZYNUMBER and MODIFY FUZZYLABEL statements, which are defined in a manner similar to DEFINE FUZZYNUMBER and DEFINE FUZZYLABEL respectively. The statements MODIFY FUZZYNUMBER and MODIFY FUZZYLABEL allow users to re-characterize fuzzy data.

   For example, the following statement redefines the fuzzy number for column AGE of the SUSPECT table.

```
MODIFY FUZZYNUMBER SUSPECT
        ( AGE     10    RATIO,
          WEIGHT  20    FIXED
        );
```

   As a result, "about 40" years old becomes 36 to 44 and "about 150" pounds becomes 130 to 170. The degree of matching between "about 40" and 38 becomes 0.5.

4. **Delete from fuzzy data dictionary.** DELETE FUZZYNUMBER and DELETE FUZZYLABEL statements allow users to delete membership functions from the FDD.

5. **Display fuzzy data definitions.** DISPLAY FUZZYNUMBER and DISPLAY FUZZYLABEL statements allow users to display definitions of membership functions.

3.2 Fuzzy DML

In this subsection, we describe the fuzzy data manipulation language (FDML).

1. **Enter data into the database**. Data will be entered into database by statement INSERT INTO, which is of the following format:

   INSERT INTO r_name ($col_1, \ldots, col_n$)
   VALUES ($val_1, \ldots, val_n$);

where col_i is the name of a column in the relation *r_name*, val_i is the corresponding value for that column. If col_i is a fuzzy column, val_i will be either a crisp value or a fuzzy value (fuzzy number or fuzzy label).

For example, the following statements enter 3 tuples into the SUSPECT table.

```
INSERT INTO SUSPECT (S#, NAME, AGE, WEIGHT, HEIGHT)
       VALUES (1, NULL, about 40, about 130, tall);

INSERT INTO SUSPECT (S#, NAME, AGE, WEIGHT, HEIGHT)
       VALUES (2, NULL, young, about 180, about 6);

INSERT INTO SUSPECT (S#, NAME, AGE, WEIGHT, HEIGHT)
       VALUES (3, Smith, young, 150, 7);
```

2. **Update data in database.** Data in the database can be updated by the UPDATE statement which is of the following format:

   UPDATE   r_name
   SET $col_1 = Exp_1, \ldots, col_n = Exp_n$
   WHERE   qualification;

   where the *qualification* has same syntax as that in the standard SQL except that fuzzy data are allowed in the statement.

3. **Delete data from database.** Data can be deleted from database by statement DELETE, which is of the following format:

   DELETE FROM rel_name
   WHERE *qualification*;

   where the *qualification* has same syntax as that in the standard SQL except that fuzzy data are allowed in the statement.

4. **Retrieve data from database.** Data will be retrieved from the database by query SELECT, which is of the following format:

   SELECT   attr_list
   FROM $rel_1, rel_2, ..., rel_n$
   WHERE   *qualification*;

   The *attr_list* is the same as that in the standard SQL's SELECT query. The *qualification* is a single predicate or a set of predicates connected by "AND" and/or "OR". A predicate takes the following formats:

   (a) column_name < op > < constant_data > where constant_data could be a crisp value or a fuzzy value and column_name could be a fuzzy column or an ordinary column.

(b) $R.A < op > S.B$
where A is a column of table R and B is a column of table S.

(c) *val* $< op >$ *subquery*
where *val* is either a constant or a column name, $< op >$ is an operator such as **"in"** and **"> any"**.

(d) EXISTS subquery NOT EXISTS subquery.
The subquery can be a correlated or uncorrelated fuzzy query.
For example, "find all the young males whose weights are about 150 pounds" can be expressed as the following:

```
SELECT *
FROM SUSPECT
WHERE  AGE = young AND SEX = 'M'
       AND WEIGHT = about 150;
```

3.3 Query Processing

For a given user query, all the predicates in the where clause can be classified into fuzzy predicates and non-fuzzy predicates. For a predicate of format (a) or (b), if any of the operands is fuzzy (fuzzy value or fuzzy column), then the predicate is called fuzzy predicate; otherwise, it is called non-fuzzy predicate. For a predicate of format (c), if *val* is a fuzzy value or a fuzzy column or the subquery contains a fuzzy predicate, then it is called a fuzzy predicate; otherwise, it is called non-fuzzy predicate. For a predicate of format (d), if the subquery contains a fuzzy predicate, then it is called a fuzzy predicate; otherwise, it is called non-fuzzy predicate. Fuzzy predicates can not be passed to the relational DBMS. A non-fuzzy predicate may or may not be able to be passed to the relational DBMS depending on whether the query is in conjunctive or disjunctive form. If the query is in conjunctive form, i.e., all predicates are connected by AND, then the non-fuzzy predicates can be passed to the relational DBMS directly. In general, we convert the original fuzzy query into a standard SQL query and a set of fuzzy predicates. The select clause of the standard SQL query is derived from the original select list items by substituting fuzzy columns with internal columns. The from clause remains the same and the where clause contains as many non-fuzzy predicates of the original query as possible. The standard SQL query is passed to RDBMS for execution. For each retrieved tuple, the fuzzy predicates are evaluated by calling FDL functions to obtain a satisfactory degree. All the tuples having a degree greater than 0 are converted into fuzzy format and returned to the user.

Note that if the predicate is of format (c) and (d), the subquery is again converted into a standard SQL query which can be executed by RDBMS and a set of predicates that can be evaluated by calling FDL functions.

4 Summary

We implemented an interactive front-end for fuzzy relational databases which allows users to insert, delete, update and query fuzzy databases. Nested queries and fuzzy joins (join on fuzzy columns) are allowed in our system. However, the implementation for fuzzy join is inefficient. An efficient fuzzy join algorithm has been proposed[9]. We will include that algorithm in our system.

References

[1] P. Bosc, M. Galibourg and G. Hamon, *Fuzzy querying with SQL: Extensions and implementation aspects*, Fuzzy Sets and Systems, 1988.

[2] A. Cardena, I. Ieong et al., *The Knowledge-based Object-oriented Picture Query Language*, TKDE, 1993.

[3] E. F. Codd, *Extending the Database Relational Model to Capture More Meaning*, ACM TODS, Dec. 1979.

[4] T. Imielinski and W. Lipski, Jr, *Incomplete Information in Relational Databases*, JACM, 31(1984).

[5] D. Li and D. Liu, *A Fuzzy Prolog Database System*, Research Studies Press Ltd., 1990.

[6] H. Nakajima and T. Sogoh, *Fuzzy Database Library*, Proceedings of the 8th Fuzzy System Symposium, 1992.

[7] H. Nakajima, T. Sogoh and M. Arao, *Fuzzy Database Language and Library: Fuzzy Extension to SQL*, Proc. of the 2nd IEEE Conference on Fuzzy Systems, 1993.

[8] S. Shenoi and A. Melton, *An Extended Version of the Fuzzy Relational Database Model*, Info. Sci., 1990.

[9] Q. Yang, J. Wu, C. Yu and C. Liu, *Efficient Processing of Nested Fuzzy Queries in Fuzzy Databases*, TR, Dept of EECS, University of Illinois, Chicago, 1994.

[10] L. A. Zadeh, *Fuzzy Logic*, IEEE Computer, April, 1988.

[11] M. Zemankova and A. Kandel, *Fuzzy Relational Databases: a key to expert systems*, Verlag TUV, Rheinland, 1984.

[12] M. Zemankova and A. Kandel, *Implementing Imprecision in Information Systems*, Info. Sci., 1985.

[13] W. Zhang, C. Yu, G. Wang, T. Pham, and H. Nakajima, *A Relational Model for Imprecise Queries*, International Symposium on Methodologies in Intelligent Systems, Trondheim, Norway, 1993.

Extending Occam's Razor

Kevin S. Van Horn & Tony R. Martinez
Computer Science Department
Brigham Young University
Provo, UT 84602 USA

Abstract: Occam's Razor states that, all other things being equal, the simpler of two possible hypotheses is to be preferred. A quantified version of Occam's Razor has been proven for the PAC model of learning, giving sample-complexity bounds for learning using what Blumer et al. call an Occam algorithm [1]. We prove an analog of this result for Haussler's more general learning model, which encompasses learning in stochastic situations, learning real-valued functions, etc.

1. Introduction

Work in machine learning often makes use of the Occam's Razor principle: given two explanations of the data, all other things being equal, the simpler of the two is preferable. Occam's Razor can be applied once we have a measure of the complexity of a hypothesis. Such a measure is obtained by choosing some means of representing hypotheses, and defining the complexity of a hypothesis in terms of the size of its smallest representation.

Blumer et al. [1] have formalized this intuition for a restricted model of learning. They show that the problem of learning under the PAC model, with a hypothesis space of large or infinite VC dimension, can be solved with an *Occam algorithm*: an approximate minimization algorithm that finds a near-simplest hypothesis correctly classifying all the training examples. The PAC model, however, has a number of limitations. It assumes that the problem is to learn the correct classification of instances, and that some member of the given hypothesis space correctly classifies all instances. This rules out problems such as learning real-valued functions, probability distributions, class probability distributions as a function of the instance to be classified, and the Bayes-optimal classifier in a stochastic setting.

Haussler [3] has generalized the PAC model to deal with these situations and others. In this paper we prove, for Haussler's model, an analog of the Occam algorithm result. The approach we analyze is essentially the "hold-out" method often used in applied statistics. That is, one tries to minimize the empirical risk (a measure of error on the training sample) with different bounds on the complexity of the hypotheses to be considered, then uses a separate hold-out set to choose the best complexity bound. As in [1], attention is paid to avoiding exact minimization and its attendant intractability — a limited increase in hypothesis complexity is allowed over what is strictly needed to attain a given level of empirical risk. We obtain sample complexity bounds similar to those in [1], but for Haussler's more general learning model.

2. Background

2.1. The PAC model and Occam's Razor

E. A. Yfantis (ed.), Intelligent Systems, 249–259.

The PAC (probably approximately correct) model of learning [1, 4–6] is a simplified learning model first introduced by Valiant [6]. The elements of the model are an *instance space* X and a "stratified" *hypothesis space* $H = (H_i)_{i \geq 1}$, where $H_i \subseteq H_{i+1}$ and we write H for $\bigcup_i H_i$. The elements of H are functions from X to $\{0,1\}$. H_n can be thought of as the set of all hypotheses of size at most n, where size is defined in terms of some means of representing the elements of H.

In the PAC model it is assumed that there is some unknown *target* $f \in H$ which determines the classification of instances, and some unknown distribution D governing the frequency of occurrence of instances. The *error* of a hypothesis $h \in H$ is the probability that $h(x) \neq f(x)$ (h misclassifies x) when x is drawn at random from the distribution D. The goal of learning is to find a hypothesis with small error.

An algorithm for learning H takes as input a sequence of *training examples* $(x_i, f(x_i))$, where the x_i are randomly and independently chosen according to D, and outputs a hypothesis $h \in H$. Its *sample complexity* $m(\epsilon, \delta, n)$ is the worst-case number of examples needed to have confidence $1 - \delta$ of returning a hypothesis with error at most ϵ, when the target f may be any element of H_n and D may be any distribution over X.

Blumer et al. [1] discuss the Occam's Razor approach to learning: find a near-smallest *consistent* hypothesis (a consistent hypothesis correctly classifies all m training examples). Their result is framed in terms of the *VC dimension* of H_i, a combinatorial parameter that is generally related to the number of bits or parameters required to specify an arbitrary hypothesis in H_i (see [1]). Suppose we have an algorithm A, functions $p(s)$ and $b(s,m)$, and constant a satisfying the following:

1. $0 \leq a < 1$;
2. $p(s) \geq 1$ and $p(s)$ is bounded by a polynomial in s;
3. the VC dimension of $H_{b(s,m)}$ is at most $p(s)m^a$;
4. given m training examples as input, A outputs a consistent hypothesis $h \in H$ of size at most $b(s,m)$, where s is the size of the smallest consistent hypothesis.

Then A is called an *Occam algorithm.* Blumer et al. show that any Occam algorithm serves as a learning algorithm with sample complexity

$$O\left(\epsilon^{-1} \log \delta^{-1} + (p(s)\epsilon^{-1} \log \epsilon^{-1})^{\frac{1}{1-a}}\right).$$

Note that exact minimization of the hypothesis size is not needed — even a very weak approximation algorithm will do. This is important, because exact minimization is often NP-hard.

2.2. Haussler's Extension of the PAC Model

Haussler [3] has extended the PAC model to handle more general learning situations. The elements of his model are an *instance space* X, an *outcome space* Y, a *decision space* Y', a *hypothesis space* H, an *assumption space* $\mathcal{A}$, and a *loss function* $l : Y' \times Y \to [0, M]$ (for some $M > 0$). The elements of H are functions $h : X \to Y'$, training examples are from $X \times Y$, and the elements of $\mathcal{A}$ are probability distributions over $X \times Y$. $l(y', y)$ is the loss incurred when a hypothesis outputs y' for an instance x whose outcome is y. The PAC model can be considered a special case of this model, with $Y = Y' = \{0,1\}$, $l(y',y) = 1$ if $y' \neq y$ and 0 otherwise, and $\mathcal{A}$ being the set of all distributions over $X \times Y$ satisfying $\mathbf{Pr}[y = f(x)] = 1$ for some $f \in H$.

As with the PAC model, it is assumed that there is some unknown *target* $D \in \mathcal{A}$ which determines the frequency of occurrence of instances $x \in X$ and their outcomes $y \in Y$. The *risk* $\mathbf{r}(h)$ of a hypothesis h is the expected value of $l(h(x),y)$ when (x,y) is drawn at random from the distribution D. A learning algorithm takes training examples drawn from this same distribution. The goal of learning is to find a hypothesis whose error is close to the minimum possible for hypotheses from H.

Haussler defines sample complexity in terms of the following family of metrics on the real numbers: for any real $\nu > 0$, d_ν is a measure of relative distance defined by

$$d_\nu(r,s) = \frac{|r-s|}{\nu + r + s}$$

for any $r, s \geq 0$. The condition $d_\nu(r,s) \leq \alpha$ is equivalent to

$$\frac{1-\alpha}{1+\alpha}r - \frac{\alpha\nu}{1+\alpha} \leq s \leq \frac{1+\alpha}{1-\alpha}r + \frac{\alpha\nu}{1-\alpha};$$

for small α this says that s differs from r by at most a multiplicative factor of about $1+2\alpha$ and an additive term of $\alpha\nu$. The sample complexity $m(\alpha,\nu,\delta)$ of a learning algorithm is then the worst-case number of examples needed to have confidence $1-\delta$ of returning a hypothesis h satisfying

$$d_\nu(\mathbf{r}(h), \inf\{\mathbf{r}(h') : h' \in H\}) \leq \alpha,$$

when the target distribution may be any $D \in \mathcal{A}$.

To our knowledge the literature contains no analog, for Haussler's model, of Blumer et al.'s Occam algorithm result. (Nor does it contain such an analog for the simpler PAC extension of just allowing stochastic targets.) However, Haussler gives useful sample complexity results for the case that H has a finite *pseudodimension*. The pseudodimension of H, denoted $\mathrm{pdim}(H)$, depends on the loss function l and may be considered a generalization of the VC dimension; in fact, $\mathrm{pdim}(H) = \mathrm{VCdim}(H)$ if the hypotheses of H are Boolean-valued and $l(y',y) = (y' \neq y)$. (Note: in this paper we abuse terminology by writing $\mathrm{pdim}(H)$ when we really mean $\mathrm{pdim}\{f_h : h \in H\}$, where $f_h(x,y) = l(h(x),y)$.) Details on the pseudodimension may be found in [3].

Given a sequence of examples $\bar{z} = (x_1,y_1),\ldots,(x_m,y_m)$, the *empirical risk* $\hat{\mathbf{r}}(h;\bar{z})$ of h is the average loss of h on $\bar{z}$, i.e. $\hat{\mathbf{r}}(h;\bar{z}) = \frac{1}{m}\sum_{i=1}^{m} l(h(x_i),y_i)$. We write $\hat{\mathbf{r}}(h)$ when $\bar{z}$ is understood. A learning algorithm is said to use *empirical risk minimization* [3,7,8] if it returns a hypothesis whose empirical risk is the minimum possible for hypotheses from H.

Combining Lemma 1 and Theorem 7 of Haussler [3] gives this result: any learning algorithm that uses empirical risk minimization has a sample complexity that is

$$O((\alpha^2\nu)^{-1}(\mathrm{pdim}(H)\log(\alpha\nu)^{-1} + \log\delta^{-1})). \tag{1}$$

3. Summary of Results

We begin with some definitions.

Definition 1 *We say that hypothesis class H is* stratified by dimension *if $H = (H_i)_{i \geq 1}$, where $H_i \subseteq H_{i+1}$ and $\mathrm{pdim}(H_i) \leq i$ for all i. We define the* size *of $h \in H$, denoted $\mathrm{siz}(h)$, to be $\min\{i : h \in H_i\}$.*

Definition 2 *Let $p(s) \geq 1$ be monotonic and polynomially bounded in s, and let $0 \leq a < 1$. We say that A is a* loose ERM (empirical risk minimization) algorithm *for $H = (H_i)_{i \geq 1}$* with bound $p(s)m^a$ *if*

- *A takes as input a sequence of examples $\bar{z}$ and an integer* size bound *i satisfying $1 \leq i \leq m \stackrel{\text{def}}{=} |\bar{z}|$;*
- *A outputs a hypothesis $h \in H_i$ satisfying $\hat{\mathbf{r}}(h; \bar{z}) \leq \hat{\mathbf{r}}(h'; \bar{z})$ for all $h' \in H$ s.t. $p(\mathrm{siz}(h'))m^a < i + 1$.*

Thus a loose ERM algorithm loosens the requirements of empirical risk minimization in hopes of making the problem tractable. In essence, we achieve the minimum risk attainable with hypotheses of size at most s using a hypothesis of size at most $i \approx p(s)m^a$. Associated with each loose ERM algorithm for H is a learning algorithm for H:

Definition 3 *Let A be a loose ERM algorithm for H, $0 < c < 1$, and $0 < \rho \leq 1$; then $HO[A, c, \rho]$ is the following algorithm:*

1. *Input a sequence of examples $\bar{z}$.*
2. *Split $\bar{z}$ into two sequences: $\bar{z}_1$ (the* training sample*), containing the first $\lfloor |\bar{z}|/(1+\rho) \rfloor$ elements of $\bar{z}$, and $\bar{z}_2$ (the* test sample*), containing the remaining elements of $\bar{z}$. Let $m \stackrel{\text{def}}{=} |\bar{z}_1|$.*
3. *For all $1 \leq i \leq \lfloor \log m / \log(1/c) \rfloor$, compute $h_i \stackrel{\text{def}}{=} A(\bar{z}_1, \lfloor mc^i \rfloor)$.*
4. *Output that h_i minimizing $\hat{\mathbf{r}}(h_i; \bar{z}_2)$.*

The above is essentially the "hold-out" method often used in applied statistics (see [2], for example). Our contribution is to analyze the sample complexity of $\mathrm{HO}[A, c, \rho]$ given the weakened requirement that A be a loose ERM algorithm (as opposed to doing strict empirical risk minimization), and show that we get sample complexity bounds analogous to those for Blumer et al.'s Occam algorithms.

We modify Haussler's definition of sample complexity to take into account hypothesis complexity by adding a size parameter n. We also need to take into account the possibility that the hypothesis returned has size greater than n, and hence might have a risk less than the minimum achievable by hypotheses from H_n.

Definition 4 $d'_\nu(r, s) \stackrel{\text{def}}{=} d_\nu(r, \max\{r, s\})$.

Definition 5 *The sample complexity $m(\alpha, \nu, \delta, n)$ of a learning algorithm for $H = (H_i)_{i \geq 1}$ is the worst-case number of examples needed to have confidence $1 - \delta$ of returning a hypothesis h satisfying*

$$d'_\nu(\inf\{\mathbf{r}(h') : h' \in H_n\}, \mathbf{r}(h)) \leq \alpha$$

when the target distribution may be any $D \in \mathcal{D}$.

Our result is that if H is stratified by dimension and A is a loose ERM algorithm for H with bound $p(s)m^a$, then HO$[A, c, \rho]$ is a learning algorithm for H with sample complexity

$$O((\alpha^2\nu)^{-1}\ln\delta^{-1} + ((\alpha^2\nu)^{-1}p(n)\ln(\alpha\nu)^{-1})^{\frac{1}{1-a}}).$$

4. Proof of Result

Our proof will require some theorems from Haussler and two lemmas. We will assume throughout that M is a bound on the loss function for whatever hypothesis class H is being discussed. The theorems from Haussler give sample complexity bounds for hypothesis classes of finite cardinality or finite pseudodimension.

Theorem 1 *(Theorem 1 of [3].) Let H be a finite set of hypotheses; let a sequence $\bar{z}$ of m examples be drawn randomly and independently from the distribution D; and let $\nu > 0$ and $0 < \alpha < 1$. Then*

$$\mathbf{Pr}[\exists h \in H.\, d_\nu(\hat{\mathbf{r}}(h;\bar{z}), \mathbf{r}(h)) > \alpha] \le 2|H|\exp(-\alpha^2\nu m/M).$$

For $\delta > 0$ and $m \ge M(\alpha^2\nu)^{-1}(\ln|H| + \ln(2/\delta))$ this probability is at most δ.

Theorem 2 *(Theorem 7 of [3].) Let H be a set of hypotheses with* $\mathrm{pdim}(H) = n$ *for some $1 \le n < \infty$; let a sequence $\bar{z}$ of $m \ge 1$ examples be drawn randomly and independently from the distribution D; and let $0 < \nu \le 8M$ and $0 < \alpha < 1$. Then*

$$\mathbf{Pr}[\exists h \in H.\, d_\nu(\hat{\mathbf{r}}(h;\bar{z}), \mathbf{r}(h)) > \alpha] \le 8\left(\frac{16eM}{\alpha\nu}\ln\frac{16eM}{\alpha\nu}\right)^n \exp\left(-\frac{\alpha^2\nu m}{8M}\right).$$

The following lemma is used to bound the risk of the hypothesis output by a loose ERM algorithm. In what follows we write $\mathrm{lln}\,(x)$ for $\ln(x\ln x)$. Note that $\mathrm{lln}\,(x) = O(\ln x)$.

Lemma 3 *Let m be a positive integer and $0 \le a < 1$; let H be a set of hypotheses with $\mathrm{pdim}(H) \le nm^a$; let a sequence $\bar{z}$ of m examples be drawn randomly and independently from the distribution D; and let $0 < \alpha < 1$, $\delta > 0$ and $0 < \nu \le 8M$. Then*

$$\mathbf{Pr}[\exists h \in H.\, d_\nu(\hat{\mathbf{r}}(h;\bar{z}), \mathbf{r}(h)) > \alpha] \le \delta$$

whenever

$$m \ge \max\left\{\left(\frac{16M}{\alpha^2\nu}n\,\mathrm{lln}\,\frac{16eM}{\alpha\nu}\right)^{\frac{1}{1-a}},\ \frac{16M}{\alpha^2\nu}\ln\frac{8}{\delta}\right\}. \tag{2}$$

Proof. By Theorem 2 it suffices to show that

$$8\left(\frac{16eM}{\alpha\nu}\ln\frac{16eM}{\alpha\nu}\right)^{nm^a}\exp\left(-\frac{\alpha^2\nu m}{8M}\right) \le \delta$$

when the bound (2) on m holds. Taking logarithms of both sides of the above inequality and rearranging yields

$$m \ge \frac{8M}{\alpha^2\nu}\left(nm^a\,\mathrm{lln}\,\frac{16eM}{\alpha\nu} + \ln\frac{8}{\delta}\right). \tag{3}$$

From the bound (2) on m we have that

$$\frac{m}{2} \geq \frac{8M}{\alpha^2\nu} \ln \frac{8}{\delta};$$

thus (3) will be satisfied if

$$\frac{m}{2} \geq Bm^a \quad \text{where} \quad B \stackrel{\text{def}}{=} \frac{8M}{\alpha^2\nu} n \,\text{lln} \frac{16eM}{\alpha\nu},$$

which can be rewritten as $m^{1-a} \geq 2B$ and then as $m \geq (2B)^{\frac{1}{1-a}}$. This latter inequality follows directly from (2). □

The next lemma is used to bound the error incurred in step 4 of HO$[A, c, \rho]$.

Lemma 4 *Let $\rho, C, m > 0$; let H be a set of $\lfloor C \ln m \rfloor$ hypotheses; let a sequence $\bar{z}$ of at least ρm examples be drawn randomly and independently from the distribution D; and let $\alpha^2\nu \leq M/(3\rho)$; then*

$$\mathbf{Pr}[\exists h \in H.\ d_\nu(\hat{\mathbf{r}}(h; \bar{z}), \mathbf{r}(h)) > \alpha] \leq \delta$$

whenever

$$m \geq \frac{2M}{\alpha^2\nu\rho} \max\left\{0.44 + \ln\ln \frac{2M}{\alpha^2\nu\rho}, \ln \frac{2C}{\delta}\right\} \tag{4}$$

Proof. By Theorem 1, the above-mentioned probability will be at most δ if

$$\rho m \geq \frac{M}{\alpha^2\nu}(\ln(C \ln m) + \ln(2/\delta)),$$

which can be rewritten as

$$m \geq \frac{M}{\alpha^2\nu\rho}(\ln\ln m + \ln(2C/\delta)).$$

This inequality in turn will hold if

$$m \geq \frac{2M}{\alpha^2\nu\rho} \ln \frac{2C}{\delta} \tag{5}$$

and

$$m \geq B \ln\ln m \quad \text{where} \quad B \stackrel{\text{def}}{=} \frac{2M}{\alpha^2\nu\rho}. \tag{6}$$

From the bound (4) on m we see that (5) holds, so it remains only to show that (6) holds. We shall use the fact that $B \geq 6$, since $\alpha^2\nu \leq M/(3\rho)$ from the statement of the lemma.

Let $f(b) \stackrel{\text{def}}{=} b\ln(1.55 \ln b)$. Since $0.44 > \ln 1.55$, we have by (4) that $m > f(B)$. In fact, $m = f(b')$ for some $b' > B$, since $f(b)$ is continuous and increasing for $b > 1$, and $f(b) \to \infty$ as $b \to \infty$. Thus (6) holds if $f(b) \geq B \ln\ln f(b)$ for all $b \geq B$; this in turn holds if

$$f(b) \geq b \ln\ln f(b) \quad \text{for all} \quad b \geq 6 \tag{7}$$

Defining $g(b) \stackrel{\text{def}}{=} b^{0.55}/\ln(1.55\ln b)$, for all $b \geq 6$ we have

$$\begin{aligned} & f(b) \geq b\ln\ln f(b) \;\Leftrightarrow\; 1.55\ln b \geq \ln(b\ln(1.55\ln b)) \\ & \Leftrightarrow\; b^{1.55} \geq b\ln(1.55\ln b) \;\Leftrightarrow\; g(b) \geq 1. \end{aligned}$$

Writing $\text{sgn}(x)$ for the sign of x and noting that $1.55\ln b > 1$ for $b \geq 6$, we have that

$$\begin{aligned} \text{sgn}(dg/db) &= \text{sgn}(\ln(1.55\ln b)\cdot 0.55b^{-0.45} - b^{0.55}\cdot(1.55\ln b)^{-1}1.55b^{-1}) \\ &= \text{sgn}(s(b)) \text{ where } s(b) \stackrel{\text{def}}{=} 0.55\ln(1.55\ln b) - (\ln b)^{-1}. \end{aligned}$$

But $s(6) > 0$ and $s(b)$ is an increasing function of b, so $s(b) > 0$ for all $b \geq 6$. Hence $dg/db > 0$ for all $b \geq 6$. But $g(6) \simeq 2.6 > 1$, hence $g(b) > 1$ for all $b \geq 6$. Thus (7) holds. □

We now mention some properties of d_ν and d'_ν. Haussler proves the following properties of d_ν:

1. d_ν is a metric on the nonnegative reals, i.e. $d_\nu(r,s) \geq 0$, $d_\nu(r,s) = 0$ iff $r = s$, $d_\nu(r,s) = d_\nu(s,r)$, and $d_\nu(r,t) \leq d_\nu(r,s) + d_\nu(s,t)$ (triangle inequality.)
2. d_ν is compatible with the ordering on the reals, i.e. if $r \leq s \leq t$ then $d_\nu(r,s) \leq d_\nu(r,t)$ and $d_\nu(s,t) \leq d_\nu(r,t)$.

We state without proof the following easily-verified properties of d'_ν:

1. if $r_2 \leq r_1$ then $d'_\nu(r_1,r_2) = 0$;
2. $0 \leq d'_\nu(r_1,r_2) \leq d_\nu(r_1,r_2)$;
3. $d'_\nu(r_1,r_3) \leq d'_\nu(r_1,r_2) + d'_\nu(r_2,r_3)$ (triangle inequality.)

Finally, we are ready for the main theorem.

Theorem 5 *Let $H = (H_i)_{i\geq 1}$ be a hypothesis space stratified by dimension; let n be a positive integer and $r_0 = \inf\{\mathbf{r}(h) : h \in H_n\}$; let A be a loose ERM algorithm for H with bound $p(s)m^a$; let $0.05 \leq c < 1$ and $0 < \rho \leq 1$; let $\bar{z}$ be a sequence of m examples drawn randomly and independently from the distribution D; let $h_{out} = HO[A,c,\rho](\bar{z})$; and let $0 < \delta \leq 1$, $0 < \alpha < 1$, and $0 < \nu \leq 8M$. Then*

$$\mathbf{Pr}[d'_\nu(r_0, \mathbf{r}(h_{out})) > \alpha] \leq \delta \tag{8}$$

whenever $m \geq (1+\rho)(B+1)$, where

$$B = \max\left\{\frac{K}{\rho}\left(0.44 + \ln\ln\frac{K}{\rho}\right), \frac{K}{\rho}\ln\frac{2(C+6)}{\delta}, \left(\frac{K}{c}p(n)\,\text{lln}\,\frac{81M}{\alpha\nu}\right)^{\frac{1}{1-a}}\right\},$$

$K = 55M/(\alpha^2\nu)$, and $C = 1/\ln(1/c)$.

Before giving the proof, we have a few comments. Recall that HO$[A,c,\rho]$ runs A with various size bounds of the form $\lfloor mc^i \rfloor$, where i goes from 1 to a maximum value. The requirement $0.05 \le c < 1$ merely says that the size bound decreases as i increases, but not by more than a factor of 20 at each step. Recall also that A is run on the first $\lfloor |\bar{z}|/(1+\rho) \rfloor$ examples, with the remaining examples used to choose the best size bound. Thus the requirement $0 < \rho \le 1$ says that at least (about) half — but not all — of the examples are used as input to A. Finally, note that a, c, ρ, and M are constants, $\operatorname{lln}(x) = O(\ln x)$, and

$$\ln\ln \frac{55M}{\alpha^2 \nu \rho} = O(\ln \frac{1}{\alpha\nu}),$$

so that $B = O((\alpha^2\nu)^{-1} \ln \delta^{-1} + ((\alpha^2\nu)^{-1} p(n) \ln(\alpha\nu)^{-1})^{\frac{1}{1-a}})$.

Proof. HO$[A,c,\rho]$ will be run with $m_* = \lfloor (1+\rho)^{-1} m \rfloor$ training examples and $m_{**} = m - m_*$ test examples. Since $m \ge (1+\rho)(B+1)$, we have that

$$m_* \ge \lfloor (1+\rho)^{-1}(1+\rho)(B+1) \rfloor = \lfloor B+1 \rfloor > B$$

and

$$m_{**} = m - m_* \ge m - \frac{m}{1+\rho} = \frac{\rho}{1+\rho} m \ge \rho \lfloor (1+\rho)^{-1} m \rfloor = \rho m_*.$$

Thus it will suffice to show that (8) holds whenever steps 3 and 4 of HO$[A,c,\rho]$ are run with $m_* > B$ training examples and $m_{**} \ge \rho m_*$ test examples. We write $\bar{z}_1$ for the training examples and $\bar{z}_2$ for the test examples.

Viewing B as a function of α, we have that $B(\alpha)$ is continuous and decreasing in α, and $B(\alpha) \to \infty$ as $\alpha \to 0$; hence $m_* > B(\alpha)$ implies that $m_* = B(\alpha_*)$ for some $\alpha_* < \alpha$. By the definition of r_0 as an infimum, there are hypotheses $h \in H_n$ with risk arbitrarily close to r_0; in particular, there exists some $h_* \in H_n$ with $d'_\nu(r_0, \mathbf{r}(h_*)) \le \alpha - \alpha_*$. By the triangle inequality for d'_ν, it then suffices to show that

$$\mathbf{Pr}[d'_\nu(\mathbf{r}(h_*), \mathbf{r}(h_{out})) > \alpha_*] \le \delta \tag{9}$$

whenever steps 3 and 4 of HO$[A,c,\rho]$ are run with $m_* = B(\alpha_*)$ training examples and $m_{**} \ge \rho m_*$ test examples.

It will be useful to have a simpler lower bound on m_*. By the requirements that $\nu \le 8M$ and $\alpha, c < 1$ we have that $55M/(\alpha_*^2 \nu c) > 55/8$ and $\operatorname{lln}(81M/(\alpha\nu)) > \operatorname{lln}(81/8)$; hence using $m_* = B(\alpha_*)$ we have

$$m_* > \left(\frac{55}{8} p(n) \operatorname{lln} \frac{81}{8}\right)^{\frac{1}{1-a}} > (21.6\, p(n))^{\frac{1}{1-a}} \tag{10}$$

Let us define the following:

- $h_i = A(\bar{z}_1, \lfloor m_* c^i \rfloor)$ for $1 \le i \le \lfloor C \ln m_* \rfloor$, as in the definition of HO$[A,c,\rho]$.
- $j = \max\{i \in \mathcal{Z} : m_* c^i \ge p(n) m_*^a\}$.
- $n' = \lfloor m_* c^j \rfloor$.

We wish to ensure that $1 \leq j \leq \lfloor C \ln m_* \rfloor$, so that $h_j = A(\bar{z}_1, n')$ will be one of the hypotheses considered in step 4 of HO$[A, c, \rho]$. We will have $j \geq 1$ if $m_* c \geq p(n) m_*^a$, i.e. $c \geq p(n) m_*^{a-1}$. From (10) and the fact that $a < 1$ we have that

$$p(n) m_*^{a-1} < p(n)(21.6\, p(n))^{\frac{a-1}{1-a}} = 21.6^{-1};$$

but $c \geq 0.05$ (from the statement of the theorem) $> 21.6^{-1} > p(n) m_*^{a-1}$, so $j \geq 1$. For the upper bound on j, note that since $m_* c^j \geq p(n) m_*^a$, we have

$$j \leq \ln(p(n) m_*^{a-1}) / \ln c = \ln(p(n)^{-1} m_*^{1-a}) / \ln(1/c) \leq \ln m_* / \ln(1/c)$$

(the last inequality uses $p(n) \geq 1$, $m_* > 1$, and $a \geq 0$); but since j is an integer we have

$$j \leq \lfloor \ln m_* / \ln(1/c) \rfloor = \lfloor C \ln m_* \rfloor.$$

By the triangle inequality for d'_ν, we will have $d'_\nu(\mathbf{r}(h_*), \mathbf{r}(h_{out})) \leq \alpha_*$ if the following hold for appropriate positive values a_i summing to 1:

1. $d'_\nu(\mathbf{r}(h_*), \hat{\mathbf{r}}(h_*; \bar{z}_1)) \leq a_1 \alpha_*$;
2. $d'_\nu(\hat{\mathbf{r}}(h_*; \bar{z}_1), \hat{\mathbf{r}}(h_j; \bar{z}_1)) = 0$;
3. $d'_\nu(\hat{\mathbf{r}}(h_j; \bar{z}_1), \mathbf{r}(h_j)) \leq a_2 \alpha_*$;
4. $d'_\nu(\mathbf{r}(h_j), \hat{\mathbf{r}}(h_j; \bar{z}_2)) \leq a_3 \alpha_*$;
5. $d'_\nu(\hat{\mathbf{r}}(h_j; \bar{z}_2), \hat{\mathbf{r}}(h_{out}; \bar{z}_2)) = 0$;
6. $d'_\nu(\hat{\mathbf{r}}(h_{out}; \bar{z}_2), \mathbf{r}(h_{out})) \leq a_4 \alpha_*$.

Thus we prove (9) by showing that conditions 2 and 5 always hold, and that the following hold for appropriate positive values f_i summing to 1:

$$\begin{array}{rcl} \mathbf{Pr}[\text{Condition 1 doesn't hold}] & \leq & f_1 \delta \\ \mathbf{Pr}[\text{Condition 3 doesn't hold}] & \leq & f_2 \delta \\ \mathbf{Pr}[\text{Condition 4 doesn't hold}] & \leq & f_3 \delta \\ \mathbf{Pr}[\text{Condition 6 doesn't hold}] & \leq & f_4 \delta. \end{array} \tag{11}$$

By definition, $\hat{\mathbf{r}}(h_{out}; \bar{z}_2) \leq \hat{\mathbf{r}}(h_j; \bar{z}_2)$, so condition 5 holds. Since $h_* \in H_n$ we have $\text{siz}(h_*) \leq n$, hence using the monotonicity of p and the definition of j,

$$p(\text{siz}(h_*)) m_*^a \leq p(n) m_*^a \leq m_* c^j < \lfloor m_* c^j \rfloor + 1 = n' + 1.$$

But $h_j = A(\bar{z}_1, n')$ and A is a loose ERM algorithm, so $\hat{\mathbf{r}}(h_j; \bar{z}_1) \leq \hat{\mathbf{r}}(h_*; \bar{z}_1)$, and condition 2 holds.

We now specify the values f_i summing to 1 and a_i summing to 1:

- $f_1 = f_3 = (C+6)^{-1}$, $f_2 = 4(C+6)^{-1}$, and $f_4 = C(C+6)^{-1}$.
- $a_1 = a_3 = (6+\sqrt{2})^{-1}$, $a_2 = 4(6+\sqrt{2})^{-1}$, and $a_4 = \sqrt{2}(6+\sqrt{2})^{-1}$.

Using $m_* = B(\alpha_*)$ and $55 > (6+\sqrt{2})^2$ we obtain

$$m_* \geq \frac{M}{a_1^2\alpha_*^2\nu\rho}\ln\frac{2}{f_1\delta}.$$

Now apply Theorem 1 with a singleton hypothesis set, $a_1\alpha_*$ for α and $f_1\delta$ for δ; then using the facts that $\rho \leq 1$ and $d'_\nu(r,s) \leq d_\nu(r,s)$ we obtain

$$\mathbf{Pr}[\text{Condition 1 doesn't hold}] \leq f_1\delta.$$

Using $a_1 = a_3$ and $f_1 = f_3$ we also obtain

$$m_* \geq \frac{M}{a_3^2\alpha_*^2\nu\rho}\ln\frac{2}{f_3\delta}.$$

Since the choice of h_j is independent of $\bar{z}_2$, and $|\bar{z}_2| = m_{**} \geq \rho m_*$, we can again apply Theorem 1 to obtain

$$\mathbf{Pr}[\text{Condition 4 doesn't hold}] \leq f_3\delta.$$

Using $m_* = B(\alpha_*)$, $55 > (6+\sqrt{2})^2$, $81 > 4e(6+\sqrt{2})$, and $\rho \leq 1$, we obtain

$$m_* \geq \max\left\{\left(\frac{16M}{a_2^2\alpha_*^2\nu}c^{-1}p(n)\,\mathrm{lln}\frac{16eM}{a_2\alpha_*\nu}\right)^{\frac{1}{1-a}}, \frac{16M}{a_2^2\alpha_*^2\nu}\ln\frac{8}{f_2\delta}\right\}$$

Referring back to the definitions of j and n', we note that $m_*c^{j+1} < p(n)m_*^a$ (recall that $0 < c < 1$), hence

$$n' = \lfloor m_*c^j \rfloor \leq m_*c^j < c^{-1}p(n)m_*^a.$$

Then applying Lemma 3 with this upper bound on n' and the preceding lower bound on m_* we obtain

$$\mathbf{Pr}[\exists h \in H_{n'}.\ d_\nu(\hat{\mathbf{r}}(h;\bar{z}_1), \mathbf{r}(h)) > a_2\alpha_*] \leq f_2\delta.$$

Since $h_j = A(\bar{z}_1, n')$ and hence $h \in H_{n'}$, we then obtain

$$\mathbf{Pr}[\text{Condition 3 doesn't hold}] \leq f_2\delta.$$

Finally, we look at condition 6. In step 4 of the algorithm we look at $k \stackrel{\text{def}}{=} \lfloor C\ln m_* \rfloor$ hypotheses, which are tested on $m_{**} \geq \rho m_*$ examples. Using the facts that $\nu \leq 8M$ and $\alpha_* < 1$, we have

$$(a_4\alpha_*)^2\nu < \frac{2}{(6+\sqrt{2})^2}8M < M/3 \leq M/(3\rho).$$

Furthermore, using $m_* = B(\alpha_*)$ and $55 > (6+\sqrt{2})^2$ we obtain

$$m_* \geq \frac{2M}{a_4^2\alpha_*^2\nu\rho}\max\left\{0.44 + \ln\ln\frac{2M}{a_4^2\alpha_*^2\nu\rho}, \ln\frac{2C}{f_4\delta}\right\}.$$

Thus the conditions of Lemma 4 hold; applying this lemma, we obtain

$$\mathbf{Pr}[\exists 1 \leq i \leq k.\ d_\nu(\hat{\mathbf{r}}(h_i;\bar{z}_2), \mathbf{r}(h_i)) > a_4\alpha_*] \leq f_4\delta.$$

and hence $\mathbf{Pr}[\text{Condition 6 doesn't hold}] \leq f_4\delta$.

Thus we have proven that the inequalities (11) hold, which completes the proof of (9), and hence of the theorem itself. □

References

[1] Blumer, A., et al. (1989.) Learnability and the Vapnik-Chervonenkis dimension. *Journal of the ACM* 36, 929-965.

[2] Devroye, L. (1988.) Automatic pattern recognition: a study of the probability of error. *IEEE Transactions on Pattern Analysis and Machine Intelligence* 10, 530–543.

[3] Haussler, D. (1992.) Decision theoretic generalizations of the PAC model for neural net and other learning applications. *Information and Computation* 100, 78–150.

[4] Kearns, M. (1990.) *The Computational Complexity of Machine Learning.* Cambridge, MA: MIT Press.

[5] Natarajan, B. K. (1991.) *Machine Learning: A Theoretical Approach.* San Mateo, CA: Morgan Kaufmann.

[6] Valiant, L. G. (1984.) A theory of the learnable. *Communications of the ACM* 27, 1134–1142.

[7] Vapnik, V. N. (1982.) *Estimation of Dependences Based on Empirical Data.* New York: Springer-Verlag.

[8] Vapnik, V. N. (1989.) Inductive principles of the search for empirical dependences. In *Proceedings of the 2nd Annual Workshop on Computational Learning Theory.* San Mateo, CA: Morgan Kaufmann.

MODELLING OCCAM2 PROGRAMS WITH META LOGIC PROGRAMMING

Kang Zhang
Department of Computing, Macquarie University
Sydney, NSW 2109, Australia

Abstract. Several message−passing parallel programming languages have become available on various MIMD machines (e.g. Occam on the transputer and CMMD on the CM−5) in recent years. Yet run−time characteristics of this class of languages has not been widely studied, and the programming support is much needed for more productive and efficient program development. This paper provides an overview of a modelling tool for Occam2, implemented as two levels of meta interpreters. The general features of a message−passing programming paradigm is modelled through an extended Prolog with the notion of dynamically created, time−dependent, communication processes. The extended communication and process creation and termination mechanisms, and their interpretations are described. The second level interpreter interprets the primitive processes and constructions of the Occam2 language. Given a performance metrics of the target machine, e.g. a network of transputers, the interpreter can predict the program speed in a fashion of event−driven simulation. Communication−related errors, such as deadlock, can be detected with an extended version of the high level interpreter.

Keywords. Meta Logic Programming, Program Modlling, Occam, Prolog

1 Introduction

Several message–passing parallel programming languages have become available on various MIMD machines, e.g. Occam on the transputer and CMMD on the CM–5, in recent years. Generally, parallel programming introduces an extra dimension of complexity over conventional sequential programming. This is particularly true with message–passing parallel programming. When designing a message–passing program, one must consider the allocation of multiple tasks to be run simultaneously and the coordination of these tasks through inter–task communication. The ideal parallelism is usually achieved through concurrent execution of multiple tasks by maximising the ratio of computations verses communication among the tasks. However, concurrency and task interaction introduce new problems into programs, such as the management of shared resources and the possibility of deadlocks (when several tasks mutually wait on each other to advance). Another problem is that when several tasks are running concurrently, it is not always possible to predict their relative speeds. Therefore, running the same program with the same data may not always give the same results. Such programs are said to be non–deterministic and are apparently very difficult to debug.

E. A. Yfantis (ed.), Intelligent Systems, 261–273.

This paper introduces a tool which models message–passing parallel programs written in Occam2, predicts the program performance, assists the program analysis, and detects communication deadlocks. Occam2 is a concurrent programming language, based on the concepts of CSP [1] and designed to express parallel algorithms and their implementation on a network of processing components, typically transputers.

Tools for monitoring or modelling Occam2 programs have been developed in recent years. Among these, graphical approach for visualising Occam2 program structure and execution behaviour have been most widely used [2,3]. Program visualisation exploits the power of computer graphics to convey the static or run–time information and thus is particularly suitable for concurrent and parallel programs [4,5,6]. It is, however, not always easy to view graphically the misbehaviour, such as mutual waiting and deadlocking, of a concurrent program. Therefore, some researchers have developed software analysing packages or hardware monitoring tools integrated with visualisation front–ends or as stand–alone systems [7,8]. The usefulness of a software analysing package can be restricted by the the set of information recorded during the program execution, or by the complexity of the analysing program. Hardware monitoring approach can overcome these problems but is more costly, and the portability is usually limited by the actual hardware implementations. Logic programming languages, however, typically Prolog, have shown to be powerful and higher level specification and verification tools [9,10].

We demonstrate, in this paper, the power of meta logic programming in modelling and analysing message–passing programs written in Occam2. We start by introducing, in Section 2, a simulation model which characterises the state transitions caused by various program events, followed by a description of the modelling language known as ETDLP in Section 3. Section 4 describes an Occam2 interpreter written in ETDLP. We then present a scheme for detecting communication deadlocks in Section 5, and conclude the paper in Section 6.

2 Program Events

In order to model a message–passing program, we need to define certain program events that characterise the concurrent behaviour of the program, such as communication events. In general, an *event* is a change in the state of a program. Independent tasks, each consisting of a sequential set of instructions, are usually referred to as *threads* in a message–passing program (or *processes* as in Occam2). Multiple simultaneously executed threads may communicate with each other through message–passing. The characteristics and communication pattern of a message–passing program are typically shown through the events that affect the state of threads and that are externally observable. In Occam2, these events include thread creation, thread termination, channel input (including racing inputs), channel output and hanging. Figure 1 is a state transition diagram, which illustrates the way in which the state of a thread is altered due to various events.

A thread creation event is triggered by the activation of a child process within a PAR (for parallel) construction. A newly created thread is suspended when there is no processor available for its execution. When a child process of PAR terminates, i.e. after the execution of the last statement in the child process, a thread termination event occurs. Obviously, a processor becomes available after this event. Therefore, a currently suspended thread can be unsuspended for execution.

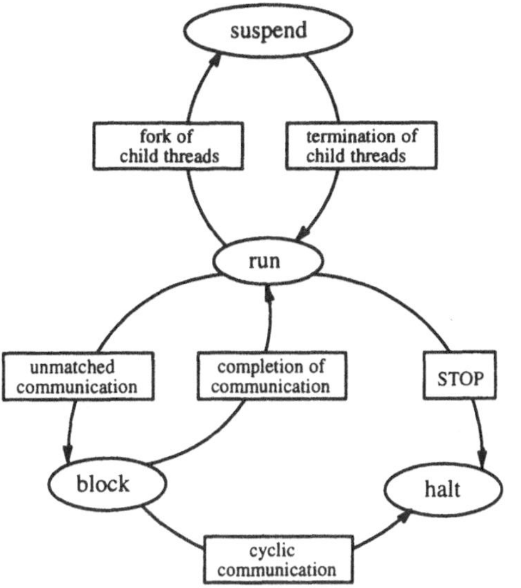

Figure 1 State transition diagram

Communication in Occam2 is synchronised and unidirectional via statically declared channels. A communication can proceed only when both input and output events involved become ready. It is also assumed, as dictated by Occam2 language definition, that no more than one parallel thread can access a single variable. It is shown in Figure 1 that a channel input (output) event in a thread causes the whole thread to be blocked if the corresponding output (input) is not ready. Conversely, a thread blocked due to such an unmatched input/output event can be released for execution when one matching input/output event occurs so that the communication can complete. When several channels are involved in a cyclic chain of attempted communications, they will be waiting on each other and thus blocked forever. This situation is known as *deadlock*, which causes the whole program to hang. The program also hangs upon the execution of a STOP process which, according to its Occam2 semantics, performs no action and never terminates.

After identifying the program events and the state transition rules, we now present a high level, communication–driven simulation tool using meta logic programming.

3 Communication–Driven Simulation

3.1 EXTENDED TDLP

The communication–driven simulation tool to be presented in this section is a modified and extended version of TDLP (for Time–Dependent Logic Programming), previously proposed by Hickey *et al* [Hickey 92]. TDLP is an extension of Horn clause logic programming. The notion of dynamically created, time–dependent communicating threads are introduced in TDLP such that a deterministic message–passing system can be modelled in an event–driven fashion. We have enhanced TDLP for the support of modelling a form of non–determinism, and of synchronising timing for parallel threads. The extended version is called ETDLP. The non–deterministic feature is reflected in the racing condition of the guarded alternative processes seen in CSP. This is handled in Occam2 by the ALT construction. Consider the following example program:

```
ALT
  chan_1? var_1
   process_1
  chan_2? var_2
   process_2
     ...
```

The first channel, say `chan_i`, that receives an input will instantiate its target variable `var_i`; and the corresponding process `process_i` can then proceed while all other processes are discarded. If more than two inputs become available at the same time, a random choice will be made on the available inputs.

ETDLP defines a set of simulation primitives which model the channel communication, thread creation and termination activities, and the execution times of program components. The simulation primitives are described below.

send(C, M) – that sends the message M through the channel C.

wait(C, M) – that suspends the current thread until a message is received and assigned to M.

fork(F, A) – that starts a new thread whose arguments are specified in the list A.

race(L, win(M, V, P)) – that determines, from the list L of suspended waiting channels, a winning channel which is the first to receive a message M. The message M is then assigned to the variable V and the corresponding process P is selected for execution.

join – that terminates all the parallel threads forked from the same point of their parent thread. The execution returns to the remaining part of the parent thread.

hold(X) – that suspends the current thread for T time units, associated with the time–variable X. The execution of the thread components following the *hold* are resumed after such a suspension.

This last simulation primitive models the execution time of a system component. A thread can be suspended for a given amount of time t_i which is specified symbolically by a time–variable x_i. This is given as a fact

```
time_variable(xi, ti).
```

It is assumed that the system to be modelled includes a set of performance measurements expressed as such facts. *Race* and *join* are the new simulation primitives added to the original set of TDLP primitives and are introduced to handle racing conditions and time synchronisation.

3.2 THREAD SCHEDULING AND TIMING

The operational semantics of Prolog is modified for ETDLP in which Prolog goals are replaced by time–associated threads. Each thread is a tuple of four arguments

(Id, T, S, G)

where a goal is associated with a thread identifier Id, the starting time T and a tag S identifying the thread's seniority, or the position in the parallel spawning tree. The thread identifier is dynamically generated at run–time to distinguish individual threads for correct timing and profiling. The thread's seniority is expressed by a unique sequence of integers and is also generated at run–time to identify the position of the thread in the spawning tree.

Threads are stored in a stack operated in a similar way as the Prolog goal stack [12]. Two queues are used and implemented as lists for recording the current message–passing states. The *send–list* stores all the sending messages and their associated channels, specified by *send(C, M)*, in the order of simulated time. The *wait–list* records the information of all suspended threads. The information also includes a message and a channel, specified by *wait(C, M)*.

When parallel threads are forked, they are marked by seniority tags indicating their positions on the spawning tree. When sibling threads belonging to (i.e. forked from) the same parent thread complete their execution, the completion time of the parent thread is recorded according to the time when the slowest child thread completes. The thread identifier has the form `id(N, PosIndex)`, where `N` is incremented for each fork of a new thread and its sum with `N` of the parent thread becomes the head of the position index list `PosIndex`. Therefore, `N` indicates the total number of the child threads, while `PosIndex` indicates the depth at which the thread was forked. This tagging scheme is illustrated by a simple example in Figure 2. Threads belonging to the same sibling group and thus having a common parent can be easily identified.

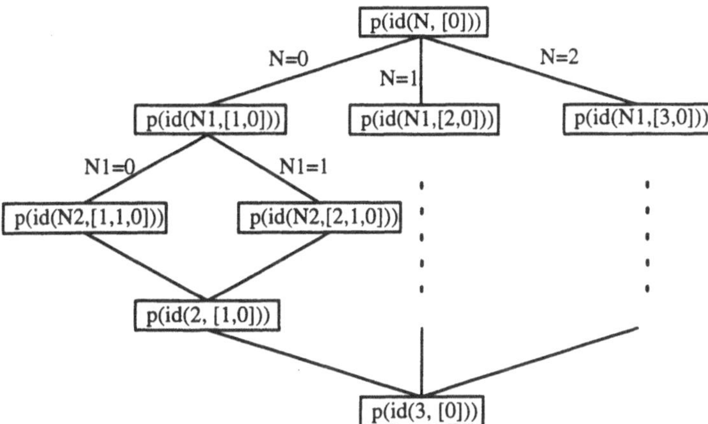

Figure 2 Tagging scheme for synchronising parallel threads

The thread scheduling policies in the simulation are communication–driven and are briefly summarised below.

* A thread is suspended whenever a channel communication primitive (say, an input) can not find a matching primitive (output) for the same channel.
* A suspended thread resumes its execution when a matching primitive event becomes ready and the communication is performed.

* Threads suspended due to outputs are stored in the send–list in the order of simulated time.
* Threads suspended due to inputs are stored in the wait–list in an FIFO order.
* Multiple alternative input choices of an ALT construction, when suspended, are kept in a racing list in the wait–list. The racing list is released from the wait–list when one of the alternative inputs has been activated by a corresponding output.

3.3 EVENT–TIME GRAPH

It has been demonstrated by Hickey *et al* that event graphs can be generated for deterministically parallel programs written in Ada interpreted in TDLP [11]. We choose a graphical notation, which we will call *event–time graphs*, to visualise the communication pattern and elapsed times of multiple parallel Occam2 threads in a more intuitive manner. ETDLP is able to generate the event–time graph of a given program based on the performance metrics specified by time–variables. Figure 3(b) illustrates an event–time graph showing the communication pattern and efficiency of a program. The program has a skeleton form shown in Figure 3(a).

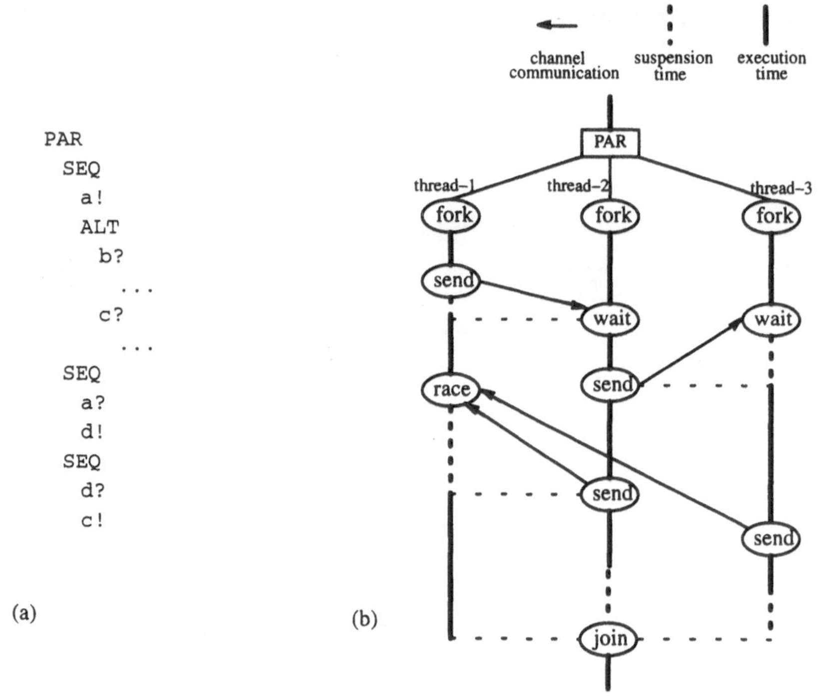

Figure 3 Event graph of an Occam2 program

Different variations of event–time graphs have previously been proposed in the literature [13,14]. In the event–time graph we have defined, the vertical direction represents time and later times are lower than the earlier times. The horizontal direction represents space. A PAR construct is explicitly drawn in a rectangular node and always paired with a *join* node, which joins together all the child threads of the PAR construction. All the communication events defined as simulation primitives are represented as oval nodes. A SEQ is implicitly shown as a vertical sequence of events. An ALT is simply shown as a *race* node. In the example shown above, three threads are forked and executed in parallel. The arrowed lines indicate inter–thread communications. The thick solid lines indicate the execution times of computations which are not shown in the program skeleton. The dotted thick lines indicate the wasted times due to the suspension of unmatched communication events or due to the waiting of the last thread in the same PAR construction to finish. The lengths of these lines reflect proportionally the execution times. Therefore, the notation of event–time graphs can express the program efficiency through the comparative lengths of thick solid lines and dotted lines. Various communication patterns including anomalous ones may be visually identified on this type of event–time graphs. For instance, a deadlocked communication can easily be seen when there are at least two communication lines crossed on each other. The following section describes an ETDLP implementation of an interpreter for a main subset of Occam2 language.

4 An Occam2 Interpreter

Occam2 is the derivative of CSP which is widely known and used [1]. Although Occam2 is a language in its own right, it has always been closely associated with the transputer, a building block for multiprocessor systems. Occam2 contains a small number of operation mechanisms and has a unique syntax. The most basic statements, i.e. assignment and communication, are themselves regarded as primitive processes. An Occam2 program contains a static number of processes and static communication paths.

Based on the Occam2 language syntax, we have defined a syntax tree format in terms of Prolog structures. A parser for generating the syntax tree has been implemented in Prolog using the conventional compilation techniques. This section describes the Occam2 interpreter, interpreted by ETDLP. The execution of the two nested interpreters is invoked by a goal

```
etdlp(p(id(0,[0]),0,[int_occam(Tree,Declarations,Output)])).
```

The ETDLP interpreter starts from the root level thread identified as `id(0,[0])`, where the first zero indicates zero number of child thread initially, and `[0]` is the position index of the root level thread. The second argument of `p` is also a zero, indicating that the simulated time starts from zero. The third argument is a list including the top level call to the Occam2 interpreter.

In the following description of the interpreter, the Occam2 syntax convention is used when referring to program components. A particular attention needs to be paid to the use of "process" in this context. According to the Occam2 definition, a process refers to a basic primitive or a construction of primitives. Compared with the "thread" in ETDLP, an Occam2 "process" is a lower level concept. This can easily be understood when considering that the Occam2 interpreter is the second level meta–interpreter on top of the ETDLP interpreter.

We describe the interpreter from the top level interpretation to the interpretation of various constructions. Some of these constructions, such as parallel construction PAR, are defined in terms of the ETDLP simulation primitives.

4.1 TOP LEVEL INTERPRETATION

At the highest level of the Occam2 interpreter, the syntax tree representation of a program, and a set of initial variables and channel declarations form the input to the interpreter. The output is a list of instantiated output variables. This is expressed by

```
int_occam(GrossTree, InitDeclarations, OutVarList) :-
    initial_env(InitDeclarations, GrossTree, Tree, Env1),
    emulate(Tree, Env1, Env),
    return(OutVarList, Env).
```

A list of all the variable bindings are kept and updated in an environment throughout the interpretation. The initial variable declarations are collected and stored into the environment. After the interpretation of the whole program by `emulate/3`, the output variables are instantiated according to the final version of the environment.

4.2 PROCESSES

The simplest process in Occam2 is either an assignment, an input, an output, SKIP or STOP. An assignment process is simply an assignment statement in the form of

variable := expression.

Input and output processes are used to achieve communication, which is an essential part of Occam2 programming. Values are passed between concurrent processes by communication on channels. Each channel provides unbuffered, unidirectional point–to–point communication (each channel must have exactly one sender and one receiver). The primitive process SKIP starts, performs no action and terminates, while the primitive process STOP starts, performs no action and never terminates. Therefore, these primitive processes can be interpreted as the following

```
process(skip, _, _).
   process(stop, _, _) :- hang.
process(assign(Var, Expr), Env, NewEnv) :-
   expression(Expr, Env, Value),
   store_in_env(Var, Value, Env, NewEnv),
   hold(assignment).

process(input(Chan, Var), Env, NewEnv) :-
   wait(Chan, Message),
   store_in_env(Var, Message, Env, NewEnv),
   hold(input).
process(output(Chan, Expr), Env, Env) :-
   expression(Expr, Env, Value),
   send(Chan, Value),
   hold(output).
```

Neither of SKIP and STOP produces any effect on the variable environment. When interpreting an assignment, the result of the expression is used to update the environment by the `store_in_env` predicate. An input process is suspended until a message is received on the appropriate channel, and then the input variable is bound to the message to be stored in the environment. Conversely, an output process is interpreted as evaluating the output expression before sending the expression result down the channel. The *hold* primitive simulates the execution time (e.g. duration in machine clock cycles) of the process concerned.

4.3 CONSTRUCTIONS

Primitive processes can be combined using a construction. A sequential construction is built by combining processes in a sequence using a SEQ, and a condition construction using an IF. A loop is built by combining processes in a WHILE construction. These constructions can be interpreted as

```
process(seq([]), Env, Env) :- hold(seq).
process(seq([P|Rest]), Env, NewEnv) :-
   process(P, Env, Env1),
   process(seq(Rest), Env1, NewEnv).
process(if([choice(else, P)|_]), Env, NewEnv) :-
   process(P, Env, NewEnv), hold(if).
process(if([choice(Bool, P)|_]), Env, NewEnv) :-
   expression(Bool, Env, true),
   process(P, Env, NewEnv), hold(if).
process(if([choice(Bool, _)|RestGC]), Env, NewEnv) :-
   expression(Bool, Env, false),
   process(if(RestGC), Env, NewEnv).
```

An IF construction can have a number of alternative branches, rather than just two for THEN and ELSE. This is why the third "if" clause is needed. Alternative choices are expressed in a list of `choice(Ci, Pi)`s, where C_i is a condition in Boolean expression and P_i is the corresponding branch process.

Parallel processes can be built with a PAR construction, and they start simultaneously. A PAR construction terminates when all the parallel processes terminate. Therefore, its ETDLP interpretation can be expressed as

```
process(par([]), Env, Env) :- join, hold(par).
process(par([P|Rest]), Env, NewEnv) :-
   fork(process, [P, Env, Env1]),
   process(par(Rest), Env1, NewEnv).
```

where all the parallel processes are spawned as independent and simultaneously started processes by the `fork` primitive. The termination of all the parallel processes is marked by the `join` primitive. Each parallel process is interpreted as a thread in ETDLP.

An alternation construction ALT combines a number of processes guarded by inputs. It performs the process associated with an input guard which is the first to be ready. When more than one inputs become ready at the same time, an arbitrary choice is made on the available inputs. The simulator decides the "arbitrary" choice based on the ordering of the available inputs in the wait–list. The ALT construction is interpreted as the following

```
process(alt([]), Env, Env) :- hold(alt).
process(alt(AltList), Env, NewEnv) :-
   collect(AltList, RaceList, Env),
   race(RaceList, win(Mess, Var, P)),
   store_in_env(Var, Mess, Env, Env1),
   process(P, Env1, NewEnv).
```

The interpreter first collects the alternatives whose Boolean guards are evaluated to be true and those which do not have a Boolean guard. The collected alternatives are qualified for the race, which will produce a winning process. The input message of the winning process is stored in the environment and the process itself is then evaluated.

Both IF and ALT constructions may be nested. The interpretations of nested IFs and ALTs are straightforward and not shown here.

4.4 REPLICATORS

Occam2 allows the construction of arrays of processes by using a device, called *replicator*, together with one of the constructs SEQ, PAR, ALT and IF. A replicator construction has the format

```
Rep name=base FOR count
  Process
```

where "Rep" is either SEQ, PAR, ALT or IF. "Process" is executed for "count" times, given that the replicator variable "name" starts from "base". Both "base" and "count" are expressions. The replicated construction can be expressed as

```
process(rep(Construct,Name,Base,Count,Process),Env,NewEnv) :-
   expression(Base, Env, BaseVal),
   expression(Count, Env, CountVal),
   rep_process(Construct,Name,BaseVal,CountVal,Process,Env,Ne
wEnv).
```

where `rep_process` interprets different constructions in a similar fashion. The interpretation of SEQ is given as an example below

```
rep_process(seq, _, _, 0, _, Env, Env) :- hold(seq_rep).
rep_process(seq, Name, Base, Count, P, Env, NewEnv) :-
   Count > 0,
   store_in_env(Name, Base, Env, Env1),
   process(P, Env1, Env2),
   Count1 is Count - 1,
   Base1 is Base + 1,
   rep_process(seq, Name, Base1, Count1, P, Env2, NewEnv).
```

5 Deadlock Detection

A deadlock occurs when several threads mutually wait on each other to advance. As described in Section 3.3, the synchronous communication scheme of Occam2 is modelled by scheduling threads involved in communications through two suspending lists, i.e. the wait–list and send–list. Based on the scheduling policy, any waiting communication event should be stored in either wait–list or send–list depending on whether the event is an input or output event. Therefore, a deadlock situation can be determined by examining the communication pattern of the events stored in the two lists. The question then is: at which point of the execution should we examine the existence of a deadlock? Making the best use of the information provided by the two lists, we find that the most appropriate point for checking possible deadlocks is when the deadlocks have just occurred. Consider a situation where a deadlocked communication occurs within a conditional loop:

```
WHILE run
  PAR
    SEQ
      chan1 ? run
      chan2 ! a
    SEQ
      chan2 ? X
      chan1 ! FALSE
```

The deadlock situation will never be detected if the detecting point is chosen outside of the WHILE loop. It is also inappropriate to choose the point between the two communication events to detect the deadlock. So a most natural choice of a point is at the end of a PAR construction which consists of multiple parallel threads (e.g. within the WHILE loop in the above example). This point is marked in ETDLP by a simulation primitive *join* (Section 3.1). An advantage of detecting deadlocks at the end of a PAR is that a local deadlock can be detected as soon as it occurs and before it is involved in a larger scale of anomalous communication. This assists an effective debugging of the program. Another advantage is that the threads, which are not involved in the deadlock, are not checked at all as they are not recorded in either the send–list or the wait–list.

Our deadlock detection scheme consists of three main steps as described below.

Step1: remove any non–communication events from all the threads stored in the wait–list and send–list, and tag each communication–related event (i.e. *wait, send* or *race*) with a unique natural number.

Step2: record the connectivity and ordering of any two communication events in an arc `a(ei,ej)`, where e_i and e_j are the tags for two separate events (we will simply call them event e_i and event e_j). Two recording actions are performed:

(a) An arc `a(ei,ej)` is recorded if both events e_i and e_j belong to a same thread and the event e_j is the immediate successor of the event e_i.

(b) Two arcs, which we will call a *pair* `a(ei,ej)` and `a(ej,ei)`, are recorded if the events e_i and e_j belong to two different threads and they form a matching pair of input/output events. In Occam2, such a pair represents two ends of a communication channel. The recorded arcs form a directed graph representing the communication pattern.

Step3: check whether the graph includes any cyclic path(s), where more than two events are involved. If such a cyclic path is found, the existence of deadlock and all the events involved in the deadlock are reported.

The ordering rule (b) in Step 2 is due to the fact that the communication in Occam2 is synchronised so that a communication can proceed only when both input and output events involved become ready. In the top level deadlock detection procedure listed below, `filter/2`, `connector/2` and `cyclic/1` correspond to the above three steps respectively. The definition of `filter/2` is not listed here because it is relatively long and mainly involves nested list processing.

```
lock_detect(SendList, WaitList) :-
    append(SendList, WaitList, AllEvents),
```

```
        filter(AllEvents, CommEvents),
        connector(CommEvents, Graph),
        not(cyclic(Graph)).

    connector([], []).
        connector(CommEvents, Graph) :-
        mark(CommEvents, MarkedEvents, GraphPart1, 1),
        make_graph(MarkedEvents, GraphPart2),
        append(GraphPart1, GraphPart2, Graph).

    cyclic(Graph) :-
        connected(X, X, Graph, [A,B,C|D]),
        write('Events involved in deadlock: '),
        write([A,B,C|D]).
    connected(X, Y, Graph, [Y]) :-
        member(a(X, Y), Graph), !.
    connected(X, Y, Graph, [Z|Path]) :-
        member(a(X, Z), Graph),
        connected(Z, Y, Graph, Path).
```

Some aspects of this program, and indeed other programs shown throughout this paper, can be improved in their execution efficiency. For example, we may use difference–lists to make the program shorter and also more efficient. But by doing this, we sacrifice the readability. The deadlock detection algorithm itself can be no doubtably be improved.

6 Conclusion

The modelling system described in this paper has been implemented in Sicstus Prolog. We have tested several Occam2 programs by automatically translating them into syntax trees and then running through the two levels of meta–interpreters. The performance metrics for individual Occam2 primitive operations were defined according to the transputer engineering data. As a result, the simulated execution times, measured in clock cycles, are close (less than 10% difference) to the real–time measurements when run on the transputer.

It has been shown that the Prolog implementation of the two levels of interpreters is concise and is of practical use for the Occam2 program analysis. Apart from the the syntax tree generator, i.e. the parser, the "rule" format of the Occam2 interpreter, required by the ETDLP interpreter, is also generated by a small Prolog program. The development of the entire system is, therefore, a cost–effective process by using Prolog. A standard exercise in Prolog in converting the ETDLP interpreter to a tracer should provide another useful debugging tool. With the availability of the graphical and X Window interfaces for the Sicstus Prolog, the dynamic execution of multiple parallel threads can be traced graphically, and the powerful capability of graphical reasoning can enhance the quality of such execution animation. These will be our future work.

7 References

[1] Hoare C.A.R. (1985) Communicating Sequential Processes. Prentice–Hall, Englewood Cliffs, NJ.

[2] Abdennadher N. *et al* (1992) 'A Graphical Environment for Debugging Occam Programs on Transputer Network', in S. Tzafestas *et al* (eds.), Parallel and Distributed Computing in Engineering Systems, Elsevier Science Publishers B.V., 65–69.

[3] Marwaha G. and K. Zhang K. (1994) 'Parallel Program Visualisation for a Message–Passing System', Proc. 13th Ann. IEEE Int. Phoenix Conf. on Computers and Communications, Phoenix, USA, 200–205.

[4] Heath M.T. and Etheridge J.A. (1991) 'Visualising the Performance of Parallel Programs', IEEE Software, 29–39.

[5] Pancake C. (1990) 'Visualising the Behaviour of Parallel Programs', Supercomputer, 39, VII–5, 31–37.

[6] D. Zernik D. *et al.* (1992) 'Using Visualisation Tools to Understand Concurrency', IEEE Software, 87–92.

[7] Arvind D.K. (1992) 'On the Detection of Communication–Related Errors in Concurrent Programs', Parallel Computing, 18, 1381–1392.

[8] Knowles A.E. and Illiev M.S. (1988) 'Monitoring Facilities on the ParSiFal T–Rack', Proc. CONPAR 88, Cambridge University Press, 399–406.

[9] Yalcinalp L.U. and Sterling L. (1989) 'An Integrated Interpreter for Explaining Prolog's Successes and Failures', in H. Abramson and M.H. Rogers (eds.), Meta–Programming in Logic Programming, The MIT Press,191–203.

[10] Rondogiannis R. and Cheng M.H.M. (1992) 'A Prolog System for Detecting Deadlocks in Concurrent Programs', Proc 1st Int. Conf. on Practical Application of Prolog, London, UK.

[11] Hickey T.J. *et al* (1992) 'Computer–Assisted Microanalysis of Parallel Programs', ACM Tran. Programming Languages and Systems, Vol.14, No.1, 54–106.

[12] Warren D.H.D. (1983) 'An Abstract Prolog Instruction Set', Technical Note 309, AI Centre, SRI International.

[13] Lamport L. (1978) 'Time, Clocks, and the Ordering of Events in a Distributed System', Communications of the ACM, Vol.21, No.7, 558–565.

[14] Netzer R.H.B. and Miller B.P. (1992) 'Optimal Tracing and Replay for Debugging Message–Passing Parallel Programs', Proc. Supercomputing'92, Minneapolis, MN, USA.

PERFORMANCE COMPARISON OF DIFFERENT LEARNING METHODS FOR WEATHER FORECASTING OPERATIONS

V. R. Kumar
P. Guignard
C. Y. C. Chung
Knowledge-Based Systems Laboratory
CSIRO, Division of Information Technology
Locked Bag 17, North Ryde, NSW 2113, Australia
E-Mail:{kumar, paulg, chung}@syd.dit.csiro.au

Abstract. In this paper, we present the comparative results of applying three different machine learning methods namely, induction, neural networks and k-nearest neighbour on a ten year historical weather data set for weather forecasting operations. A ten year weather data set was assembled from the archives of the Australian Commonwealth Bureau of Meteorology for our experiments. The results indicate that the Neural Network technique performs better than other two techniques for this probme domain.

Keywords. Machine Learning, Neural Networks, Inductive Learning, k-Nearest Neighbour, Weather Forecasting, Knowledge-Based Systems

1 Introduction

Machine Learning, a branch of AI has evolved through several stages in the past four decades. Much of the research in machine learning has been devoted to defining various paradigms, establish relationships among them, and elaborate the algorithms that characterise them. Recently, researchers have focussed more on applying machine learning techniques to real-world problems. But despite the increasing use of machine learning techniques in real-world applications, they have not yet found widespread application in weather forecasting operations. One reason for this delay has been the difficulty in obtaining suitable weather forecasting data sets.

In this paper, we present our comparative results of applying three different machine learning techniques namely, neural networks, induction, and *k*-nearest neighbour on a ten year weather data set assembled from the achieves of the Commonwealth Bureau of Meteorology of Australia. The aim is to predict the occurrence of rainfall over Melbourne city and its suburbs of Australia during a 24 hour period beginning at 9am local time using the learning techniques. We presents some results of our experimental studies on the data set. We present the results of the three techniques by carrying out some experimental studies on a data set. We take into account one data set namely, the ten year data set out of the two (ten year data set and thirty year data set) that we acquired for our experimental study. Performances of neural nertworks and induction tecniques by directly applying them on these two data sets have already been reported seperately in [1] and [5], and their comparative study in [6]. In this paper, apart from neural network and induction, we take into account another technique namely, the *k*-nearest neighbour for this experimental study on the ten year data set. Details about our experimental study are given in section 7.1.

E. A. Yfantis (ed.), Intelligent Systems, 275–281.

2 Weather Forecasting By Forecasters

The accuracy of forecasts produced by a given forecasting procedure typically varies with factors such as geographic location, season, categories of weather, quality of input data, lead time and validity time [3]. Forecasting of the weather for Melbourne and its suburbs of Australia is carried out by the Victorian Regional Forecasting Centre (VRFC). The forecasts are currently made routinely by the forecasters at about 8pm local time or thereabouts on the previous day, and is a prediction of the occurrence and accumulated rain depth over 24 hours beginning at 9am next day.

The current method used in the VRFC to produce the 24 hour rain forecasts studied involves both judgmental and "objective" computer-based information processing. Specifically, observational data and objectively generated data such as numerical weather prediction output, are examined, evaluated and judgmentally combined in order to arrive at the issued forecast. The judgmental activity is described by forecasters in terms of rules based on atmospheric physics, descriptive meteorology and professional experience. It should be emphasised here that the forecasts made by the forecasters are the most likely "category of rainfall" judged on the basis of weather information available at the time.

3 Need For A Better Forecasting Procedure

Forecasting experts expressed a need for a decision support system to assist them in weather forecasting operations. The forecasters gave several reasons for the need of such a system and the reasons are explained by Kumar [5]. In short, decision support systems assist in providing guidance to the users in decision making by facilitating access to relevant information, and by facilitating necessary information processing. The decision support system incorporating expert/knowledge based system technology can help in analysing information and operate in the way in which forecasters process information. They can help to search for a better forecasting procedures in a systematic way. They can explain their decision making process to the users.

4 Need for Machine Learning in Weather Forecasting Operations

Based on a study conducted by the CSIRO, Division of Information technology on the possibility of building an expert/knowledge-based system for weather forecasting operations using the manual knowledge acquisition process, it was found out that this technique does not work well due to several factors [5]. Hence, it was decided to try machine learning techniques to acquire the necessary weather forecasting knowledge automatically from the weather data set.

5 Learning Methods for Weather Forecasting Operations

There are almost many paradigms for machine learning as there are systems. Methods have been reported which claim to learn by analogy, being told, cases, debugging, discovery, doing, examples, experimentation, explanations, imitation, instruction, observation, rote, and taking advice. In this study, we concentrate on three types of classification techniques namely, induction, neural networks and k- nearest neighbour method.

5.1 Induction based Learning

Induction is the process of discovering general laws by the observation and analysis of specific examples. The application that induction addresses all involve classification. The members of this family are sharply characterised by their representation of acquired knowledge as decision trees. The decision trees are developed for classification task. The underlying strategy is non-incremental learning from examples. For our experiments we use C4.5 [7] and CN2 [2]. Both these algorithms produce rules automatically from the raw data sets and can handle continuous and discrete type data sets effectively.

5.2 Neural Network based Learning

Among different machine learning techniques, neural networks have emerged as a very successful learning technique and has been applied widely to a variety of real world problems. Neural computing is one of the fastest growing area of artificial intelligence There are several neural network paradigms. Among them, the back propagation paradigm is one of the most popular and widely used [1]. This paradigm is based on an iterative gradient algorithm designed to minimise the mean square error between actual output of a multi-layer feed forward perceptron and the desired output [8]. We decided to use the fully connected feed forward back propagation neural network tool implemented by the Oregon Graduate School for this experimental study.

5.3 k-Nearest Neighbour based Learning

One example of a good general-purpose heuristic that is useful for a variety of combinatorial problems is the nearest neighbour algorithm, which works by selecting the locally superior alternative at each step. The advantage of this method is that it is less combinatorially explosive compared to other methods with no heuristic search. k-nearest neighbour algorithm is an enhanced version of the nearest neighbour algorithm with k referring to the number of nearest neighbours considered for evaluating the most likely outcome. The procedure to evaluate directly the effectiveness of the k-nearest neighbour classifier on the furnished data set is to consider various values for k and then pick the optimum k value. We decided to use the k-nearest neighbour algorithm implemented by the CSIRO Division of Information technology.

6 The Weather Data Set

A ten year and a thirty year weather data sets were assembled from the archives of the Australian Commonwealth Bureau of Meteorology. For this experimental study of the data set , we concentrated on the ten year data set. The data set had a total of 2663 cases. Each case corresponds to the weather data of a day at a particular time. Each case had 129 weather predictor items (attributes) and one outcome classification. The outcome classification had the mean depth of rainfall in millimetres. If the depth was greater than 0.0, then rain fell on that day, else no rain was recorded on that day. The attributes of the data set did not have any discrete values and were all of type continuous (real numbers).

7 Experimental Study of the Data Set

7.1 Experimental Procedure

Same training and test cases from the ten year data set were used for all the three learning methods. The best results that evolved out of the experimental studies are summarised in this paper. In all cases the objective was to predict whether rain would occur in the subsequent time interval. The time interval here is 24 hours.

Two sets of experiments were carried out. In the first, equal number of training and test cases were used and overlap between the sets in the experiment was avoided. The data used for the first experiment is shown in Table I. The figures in brackets refer to the case numbers in the data set (eg.1-100) and to the fraction of positive and negative cases (eg. 49 positive (rain) and 51 negative (no rain) in 1-100 cases) respectively.

In the second experiment, the same test set was used (cases 1001-2000) while the number of training cases was increased from 50 to 400. This experiment was carried out to analyse the performance of the three methods when the number of training cases are increased. This is shown in Table II.

Total number of Cases	Training cases	Test cases
200	100 (1-100, 49, 51)	100 (101-200, 62, 38)
400	200 (1601-1800, 92, 108)	200 (1801-2000, 121, 79)
600	300 (401-700, 229, 71)	300 (701-1000, 167, 133)
800	400 (301-700, 223,1 77)	400 (701-1100, 210, 190)

Table I: Experiment I. Equal Number of Training and Test Cases

Training cases	Test cases
50 (1-50)	1000 - 2000
100 (1-100)	1000 - 2000
150 (1-150)	1000 - 2000
200 (1-200)	1000 - 2000
300 (1-300)	1000 - 2000
400 (1-400)	1000 - 2000
500 (1-500)	1000 - 2000

Table II: Experiment II. Varying Training Cases and Same Test Cases

With C4.5, the performance results were evaluated on pruned and unpruned decision trees and also on the production rules generated by C4.5. CN2 does not provide facility to evaluate the decision trees. Evaluation of the production rules generated by CN2 was carried out. The results summarised using these two techniques are the best performance results obtained as a result of rigorous tuning of the alogorithms. Experiments for the fully connected feedforward backpropagation neural network were conducted by varying the hidden layers and the hidden units. The number of hidden layers were varied from 0 to 5. The number of hidden units were varied from from 10 to 70. Iterations up to 100 were performed. The nearest neighbour algorithm relies on a set of previously classified cases as a set of reference (or training) cases in the solution space. Classification is based on finding the k cases with the shortest squared distances to the unseen case and on voting with the majority. The value of k was varied from 1 to 99 in steps of one. Unlike induction and neural networks, k-nearest neighbour does not require processing or tuning based on the training cases to develop the classification algorithm.

7.2 Experimental Results

The Best Configuration for Neural Network is shown in brackets in row 4 of Tables III and IV. These represent the peak performance obtained by the neural network for the configuration (eg. [126-24-2] (31) represents 126 input units, 24 hidden units, 2 output units and 31 iterations). The peak performance for k-nearest neighbour was obtained by varying the k value from 1 to 99 in steps of 1. The k value for which the peak performance was obtained is shown in brackets in row 5 of Tables III and IV. Tables III and IV are shown graphically in figures 1 and 2 respectively.

Training -Test Cases / Tools	100-100	200-200	300-300	400-400
C4.5	0.750	0.685	0.647	0.670
CN2	0.720	0.680	0.697	0.655
NN	**0.780** [126-24-2] (31)	**0.820** [126-58-2] (24)	**0.723** [126-24-2] (24)	**0.735** [126-17-2] (32)
K-NN	0.750 (5)	0.745 (39)	0.703 (15)	0.707 (19)

Table III: Best Performance Results of Experiment I

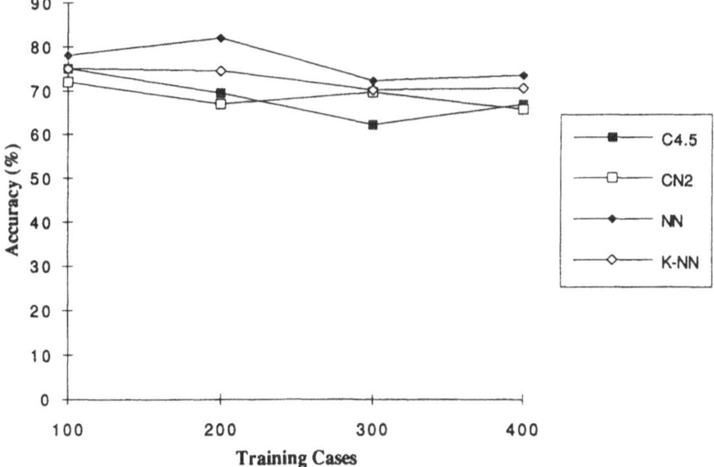

Figure 1: Comparative Results (Graph) for Experiment I

Training Cases / Tools	50	100	150	200	300	400	500
C4.5	0.616	0.606	0.656	0.627	0.621	0.653	0.661
CN2	0.636	0.648	0.615	0.655	0.652	0.671	0.651
NN	**0.734** [126-27-2] (20)	**0.725** [126-55-2] (9)	**0.746** [126-60-2] (10)	**0.749** [126-40-2] (12)	**0.749** [126-19-2] (26)	**0.756** [126-45-2] (11)	**0.755** [126-10-2] (39)
K-NN	0.611 (5)	0.664 (3)	0.646 (11)	0.698 (39)	0.698 (65)	0.704 (73)	0.707 (81)

Table IV: Best Performance Results of Experiment II

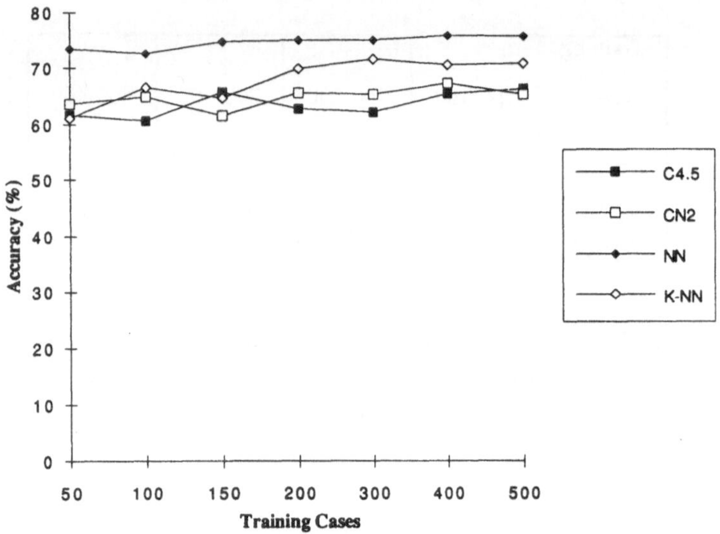

Figure 2: Comparative Results (Graph) for Experiment II

8 Conclusions And Future Work

In this paper, we have conducted an experimental study on the ten year weather data set using induction, neural network, and k-nearest neighbour techniques. We conducted two experiments namely, varying the sixe of training and test cases, and varying the size of training cases while keeping the test cases constant. In the first experiment the accuracy for induction varies between 64.7% and 75%, the accuracy for the neural network varies from 72.3% and 82% and that for the k-Nearest Neighbour between 70.3% and 75%. Despite the increase or decrease in the number of training and test case sizes, there is no significant variation to the accuracy. There does not appear to exist a correlation between the number of cases in the data set, the number of positive and negative cases in the test and training set on the one side and the accuracy on the other side. The fluctuation in the accuracy appears to be caused by the characteristics and distribution of the cases in the hyperspace. The second experiment shows that by varying the number of training cases the accuracy for induction varies between 60.6% and 67.1%, for the neural network between 72.5% and 75.6%, and for the k-Nearest neighbour between 61.1% and 70.7%. Therefore varying the size of the training cases by keeping the test cases constant does not help to improve the accuracy of classification significantly on unseen cases. With the neural network, there is no significant difference in accuracy with respect to various configurations, eg. increasing or decreasing the hidden layers or number of units in the hidden layer. Neural network performed better than induction and k-Nearest neighbour for this problem domain.

Guignard [4] presents a formalism that is suitable for the evaluation of decision performance that is free of bias and subjectivity. He argues that this approach is necessary for the application of machine learning to real-world problems because machine learning techniques are increasingly linked to decisions where cost-benefit considerations must take place. It is therefore important to consider whether machine learning provides a suitable environment for cost-benefit analysis and decision making. Guignard has successfully tested his formalism on the *k*-nearest neighbour algorithm [4]. Our future work will investigate the application of Guignard's approach to induction and neural networks using the weather data sets.

9 Acknowledgments

The authors are thankful to the Bureau of Meteorology Research Centre, Melbourne, for providing the necessary weather data sets, Ross Quinlan for providing C4.5, and the Turing Institute for providing CN2.

10 References

[1] Chung, C.Y.C., Kumar V.R. (1993) 'Knowledge Acquisition using a Neural Network for a Weather Forecasting Knowledge-Based System' International Journal on Neural Computing and Applications Vol.1 No.3 pp 215-223.

[2] Clarke P., Niblett T. (1989) 'The CN2 Induction Algorithm' Machine Learning 3 pp 261-283.

[3] Fraedrich K., Leslie L.M. (1987) 'Combining Predictive Schemes in Short-term Forecasting' Monthly Weather Review 115 pp 1640-1644.

[4] Guignard P. (1994) 'Quantitative Evaluation of Classification Performance for Machine Learning : ROC Analysis and Sensitivity/Specificity Plots' 3rd Golden West International Conference on Intelligent Systems, Las Vegas, June.

[5] Kumar V.R., Chung C.Y.C., Lindley C.A. (1994) 'Towards Building an Expert System for Weather Forecasting Operations' International Journal on Expert Systems with Applications Vol.7 No.2 pp 373-381.

[6] Kumar V.R., Chung C.Y.C., Lindley C.A. (1993) 'Learning to Perform Weather Forecasting Operations' Proceedings of the Fourth Scandinavian Conference on Artificial Intelligence, Stockholm, Sweden, pp 147-156.

[7] Quinlan J.R. (1993) C4.5: Programs for Machine Learning, Morgan Kauffman Publishers, California, USA.[8] Rumelhart D.F., Hinton G.E., Williams R.J. (1986) 'Learning Internal Representations by Error Propagation' Parallel Distributed Processing, MIT Press pp 318-362.

[8] Rumelhart D.F., Hinton G.E., Williams, R.J. (1986) 'Learning Internal Representations by Error Propagation' Parallel Distributed Processing', MIT Press pp 318-362.

[illegible] ranging [illegible] the [illegible] between [illegible] for the [illegible] between [illegible] and [illegible]. Therefore [illegible] the size of the [illegible] category [illegible] the [illegible] does not [illegible] the [illegible] cases. All [illegible] where [illegible] different [illegible] with respect to [illegible] configuration [illegible] [illegible] better than [illegible] neighbour [illegible].

Dietterich [4] [illegible] support for the evaluation of [illegible] only [illegible] that is the [illegible] We expect [illegible] this approach [illegible] for the application [illegible] machine learning to real world problems because meaning [illegible] learning [illegible] than [illegible] [illegible] specifically [illegible] in [illegible] our [illegible] will investigate the [illegible] approach [illegible] instance [illegible] learning.

9 Acknowledgements

[illegible]

10 References

[illegible]

QUANTITATIVE EVALUATION OF CLASSIFICATION PERFORMANCE FOR MACHINE LEARNING ROC ANALYSIS AND SENSITIVITY/SPECIFICITY PLOTS

Paul A Guignard, Ph.D.
Knowledge-Based Systems Laboratory
CSIRO Division of Information Technology
Locked Bag 17, North Ryde, NSW 2113, Australia
email:paul.guignard@syd.dit.csiro.au

Abstract. Machine learning algorithms, such as induction and neural nets, are being increasingly applied to real-world situations with important economic consequences. In this paper we apply the ROC (Relative Operating Characteristics) formalism to the evaluate decision performance in a way that is free of bias and subjectivity. We argue that this approach is necessary for the application of machine learning to real-world applications. We support our argument by applying this approach to a classification system for weather forecasting based on the nearest neighbour algorithm, using real-world data supplied by the Australian Bureau of Meteorology.

Keywords: Machine Learning, Classification Performance, Relative Operating Characteristics, ROC analysis, Nearest Neighbour Algorithm.

1 INTRODUCTION

Applications for Machine Learning exist in knowledge-based systems, expert systems, computer vision, speech understanding, intelligent tutoring systems, etc. Researchers and engineers are now focusing on applying ML techniques to real-world problems [1] and are therefore extending the range of applications to environments with economic (and sometimes health-related) consequences. As such ML becomes increasingly linked to decisions where cost-benefits considerations must take place. It is therefore important to consider whether ML provides a suitable environment for cost-benefit analysis and decision making.

In this paper we present a formalism based on Signal Detection Theory that provides a framework for evaluating decision performance that is free of bias and subjectivity. This formalism (referred to as Relative Operating Characteristics formalism) has been applied to evaluate the "effectiveness" of tests in medicine, weather forecasting by human observers and in other situations where human operators need to weigh options [2, 3, 4, 5, 6] . It has also been used to evaluate classification performance of discriminant analysis [7, 9] We believe that this formalism is needed for evaluating the performance of machine learning algorithms such as classifiers, and for their appropriate application to real-world situations. We apply it to a classification system for weather

E. A. Yfantis (ed.), Intelligent Systems, 283–290.

forecasting based on the nearest neighbour algorithm and using real-world data by the Australian Bureau of Meteorology.

2 USUAL EVALUATION OF CLASSIFICATION SYSTEMS

A perfect (binary) classification system would be able to discriminate between two populations (ie positive and negative) without error. That is, it would be able to identify all events belonging to the positive and negative populations without misclassification (that, is without classifying a positive event with the negative population or vice versa). Most classification systems are not perfect in that some misclassifications occur. It is therefore customary to describe performance using a classification matrix which shows both the cases correctly classified and those incorrectly classified.

$P(S/s)$	$P(N/s)$
$P(S/n)$	$P(N/n)$

Where $P(S/s)$, $P(S/n)$, $P(N/s)$ and $P(N/n)$ are the true positive, false negative, false positive and true negative fractions respectively. This matrix has the drawback that it is difficult to compare classifiers where both P(S/s) and P(N/n) are different.

A moment's reflection shows that real-world decision making covers situations where one can ill afford to miss a positive case (for example the consequence of not detecting fatigue in an aircraft frame can be catastrophic). Alternatively, calling a negative case positive can be equally damaging (for example in medicine where treatment for an ailment can have serious side effects and can be economically costly to both the patient and society). In the first case it would be better to be very conservative in maintaining the aircraft than to miss a true positive case (this is a low threshold decision). In the second example, physicians would wish to avoid false positive diagnosis (high threshold decision). These two examples illustrate that the decision process need to be able to adjust the positive threshold to reach the most appropriate compromise. This is equally valid whether the classification is carried out by a human or by a machine. The "most appropriate" compromise is usually reached by bringing into the decision making process considerations that lie outside the mechanisms embodied in a test or classifier.

3 SIGNAL DETECTION THEORY - ROC ANALYSIS

ROC analysis, based on signal detection theory, addresses this problem by providing a framework for modelling the positive and negative populations and the decision process itself [2].

Let us have two populations, one positive and one negative, which, when plotted against a criterion for "positiveness", overlap, as illustrated in Figure 1 with normal distributions. A good discrimination criterion is one that, when the negative and positive populations are plotted against the criterion, minimises their overlap. When two populations have negligible overlap, the criterion produces a near perfect discrimination between positives and negatives. Different classification algorithms use different criteria and result is varying separation between the populations.

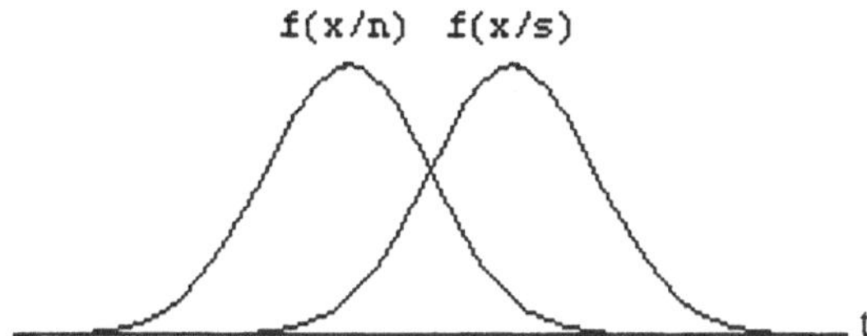

Figure 1: Positive and negative populations plotted against "positiveness"

With respect to Figure 1, $f(x/n)$ and $f(x/s)$ represent the normal negative and positive populations. These two distributions do not have to be Gaussian but it is assumed that they are precisely known, that is, the status (positive or negative) of every case can be ascertained objectively.

For each threshold t one can calculate the true positive fraction (TPF) and the false positive fraction (FPF) by taking the normalised integral between t and ∞ for both distributions. In medicine, TPF and FPF are usually referred to as sensitivity and specificity respectively and are equal to $P(S/s)$ and $P(N/s)$. TPF versus FPF is the called the Relative Operating Characteristics (ROC) plot.

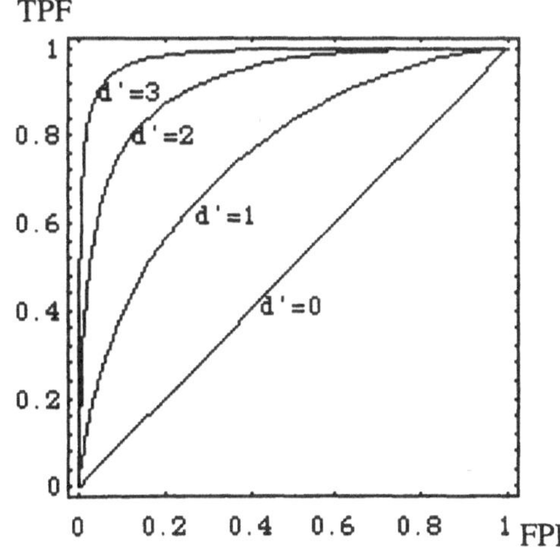

Figure 2: ROC curves for normal distributions, equal variance cases

When the positive and negative distributions are normal, that is when the Gaussian Assumption (GA) for the distributions is applicable, the value $(m_s - m_n)/\sigma_n$ (where m_n, m_s are the means of the positive and negative populations and σ_s, σ_n their respective standard deviations) is usually referred to as d' and is a measure of the separation between the two Gaussian curves. It is frequently used as an index of detectability. Figure 2 shows ROC curves for the equal variance case and several value for d'. In the unequal variance Gaussian cases, the ROC curves do not necessarily have the properties of symmetry shown in Figures 2. In this case, the use of either a single index such as d' without knowledge of the underlying shape of the probabilities can lead to gross misinterpretation of performance and erroneous conclusions as to the comparative values of different tests [2, 8].

4 DATASET AND CLASSIFICATION ALGORITHM

The dataset is a 10-year weather dataset assembled from the archives of the Commonwealth Bureau of Meteorology of Australia. It has a total of 2663 cases, with each case corresponding to the weather data for a specific day in a geographical area around Melbourne. Each case has 129 predictors or attributes (meteorological measurements) and one outcome classification, the amount of rain the following day over Melbourne airport. The dataset has been described more fully in [10,11].

The classification algorithm used is a nearest neighbour algorithm ($k-NN$ classifier) developed by the CSIRO Division of information technology. Based on a previous study [11], k was varied as a function of the number of cases in the reference data set, as it was found that it can have a significant impact on classification accuracy. In each case the classification was obtained by voting, that is by taking the classification given by the majority. Additional information on the algorithm can be found in the above paper.

5 EXPERIMENTAL PROCEDURE

A set of 500 successive cases was selected as reference set against which 1000 cases (testing set) were tested for classification using the $k-NN$ algorithm. In order to produce several operating points for plotting the ROC curve, a varying decision threshold given by the number of nearest neighbours giving positive answers was adopted. For example, a low threshold positive decision was taken to be when only a few (say 10%) out of the nearest cases voted positive. Similarly a high threshold positive decision was taken to be when most of the nearest neighbours voted positive. The threshold was varied in steps of $k/10$, thus giving a total of 10 levels for the classification criterion. The operating points were calculated as follows: for TPF it is the sum (over 1000 test cases) of all the cases $C(S/y)$ that are really positive and that have at least y

positive nearest neighbours, divided by T^{pos}_{test}, which is the total number of positive cases in the testing set. Similarly for FPF.

equation V.1

$$TPF_{t=y} = \sum_{test\_cases} C(S/y) \Big/ T^{pos}_{test}$$

equation V.2

$$FPF_{t=y} = \sum_{test\_cases} C(N/y) \Big/ T^{neg}_{test}$$

Four data sets were used as shown in Table 1 below. The accuracies quoted correspond to the k value giving maximum accuracy [11]. TPF, FPF and the accuracy were calculated using the equations above.

Reference cases	Testing cases	*k*	TPF	FPF	accuracy
150	1000	11	0.554	0.251	0.65
300	1000	65	0.575	0.127	0.72
400	1000	73	0.554	0.127	0.70
500	1000	81	0.594	0.166	0.71

Table 1: Experimental datasets with k, TPF and FPF for maximum accuracy

6 RESULTS AND DISCUSSION

Figure 3 shows the ROC plots for the four experiments described in Table 1. In each case, the experimental points were obtained by varying the threshold t_i from 0 to k in steps of k /10 and the TPF and FPF calculated (using equation V.1 and V.2 applied the entire testing set) and then plotted on the graph, as described in Section 5. The markers on each curve (one marker per curve) are the operating points corresponding to the accuracies stated in Table 1. These points are frequently given when the performance of a classifier is described using one measure only.

Looking at Figure 3, it is clear that the ROC performance of curves for nref=150 is inferior to those for nref=300, 400 and 500. In this case, inferior refers to the fact that the classifier with 150 reference cases does not achieve as good a separation between the positive and negative populations, as explained in section 2 (ie, for each choice of FPF, the TPF is lower). Note here that it has not been established that the positive and negative populations can be described using Gaussian distributions (as in the illustrations of Section 2).

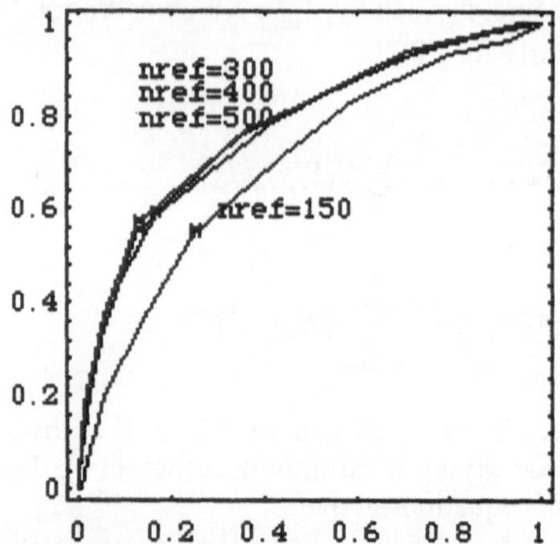

Figure 3: ROC plots for the four experimental datasets

The markers on Figure 3 clearly show the TPF performance of the classifiers, with FPF between 0.15 and 0.25, to be between 0.55 and 0.6. While this may be acceptable for some applications (such as the prediction of mild rain over an airport), it might not be acceptable if the performance referred to the prediction of ice on an airport runway. In the later case the FPF would presumably be too high and an operating point with a lower FPF might be selected. The impact of this requirement on the TPF is clearly shown on the plot and the impact on the efficiency of the airport operations can be studied. This illustrates the practical importance of the ROC approach.

Using the curve corresponding to nref=500, the experimental operating points produced using equations V.1 and V.2 are shown in Table 2 below.

TPF	FPF	t_i/k
0	0	<0.1
0.026	0	<0.2
0.178	0.012	<0.3
0.387	0.062	<0.4
0.594	0.166	<0.5
0.783	0.412	<0.6
0.938	0.703	<0.7
0.996	0.955	<0.8
1	1	--> 1

Table 2: Experimental operating points for the experimental dataset with nref=500

When a requirement (as in the example of the airport above) is expressed in terms of acceptable TPF or FPF, one can select the threshold t_i corresponding to the desired operating point. For example, if an application can accept a FPF of up to 0.4, then a possible operating threshold, given by t_i/k, could be 0.6 (Table 2). The corresponding TPF would be 0.783. This is a clearly higher than the TPF given by the marker discussed previously (0.594). To achieve this relative performance, the designer of the classifier, in our example, would tune the algorithm so that a case is identified as positive when only 40% of the nearest neighbours are positive. This is different from the performances reported in Table 1 which were calculated using a threshold of 50% (re by voting with the majority). This illustrates that, using the *k--NN* algorithm, ROC curves can be produced and used practically to tune classifiers based on selectable TPF versus FPF trade-offs. These trade-offs are, often, selected by considering the cost-effectiveness of the intended application.

For the ROC curves for nref= 300, 400 and 500, the curve of best fit calculated using the Gaussian assumption, that is, by assuming that both the positive and negative populations can be represented by Gaussian distributions, is given by:

equation VI.1

$$\sigma_s/\sigma_n = 0.900 \qquad m_s - m_n = 1.070$$

That is, the standard deviation of the negative cases distribution is 0.900 that of the positive distribution and the distance between the two distributions is 1.070 times the standard deviation of the negative population. That is, the characteristics of the distributions can be investigated with the ROC formalism and some aspects of the datasets can be studied. This is in addition to the decision options that are offered by the ROC formalism.

7 CONCLUSION

Relative Operating Characteristics (ROC) analysis is a formalism is based on Signal Detection Theory. It provides a framework for decision performance analysis that is free of bias and subjectivity, by supporting the representation of classification performance over a wide range of decision criteria. This formalism has been applied to medicine, weather forecasting and to other situations where human observers need to weigh options. Our interest was to apply the ROC analysis to machine learning (ML). This task as urgent because ML is increasingly linked to decisions where cost-benefits considerations must take place such as in health, the operation of power plants, etc.

In this paper we have applied the ROC formalism to a machine learning classifier applied to a real-world dataset supplied by the Commonwealth Bureau of Meteorology of Australia. The machine learning algorithm used was the *k*-Nearest Neighbour (*k-NN*) described in a previous work [11]. We selected the *k-NN* algorithm because the voting mechanism associated with this algorithm (to decide the outcome of the classification experiment) can be simply mapped into a varying decision threshold criterion. ROC curves were obtained for 4 different weather datasets. We showed how the usual way of reporting performance in ML corresponds to one operating point on the ROC curve. Using the ROC formalism, we showed how different operating points can be more appropriate to different situations, depending on the type of trade-offs associated with the decision. These trade-offs usually involve considerations that are outside of the ML algorithm and the dataset. We also showed how different operating points can be selected and the *k-NN* algorithm tuned to the reflect this choice. Finally we applied the ROC formalism to the study of the characteristics of the underlying negative and positive distributions that comprise the dataset.

We conclude that ROC formalism may have an important practical role in the application of machine learning to real-world situations, that it can be used practically with *k-NN* algorithms to both study the performance of the algorithm over a wide range of decision criteria and to decide which operating point to select, and finally that this formalism supports the investigation of the properties of the datasets themselves. In the future, we propose to investigate the use of the ROC formalism to other ML algorithms and classifiers.

8 REFERENCES

[1] Segre, A.M. (1992), Applications of machine learning, *IEEE Expert,* pp.30-34.

[2] Green D. M., Swets J.A., Signal detection theory and psychophysics, J Wiley & Sons, 1966.

[3] Metz C.E., Starr S.J., Lusted L.B., Quantitative evaluation of visual detection performance in medicine: ROC analysis and determination of diagnostic benefit; in Medical Images: formation, perception, measurement, 220-241, 1976.

[4] Swets J.A., ROC analysis applied to the evaluation of medical imaging techniques; Investigative Radiology, 14, 109-121, 1979.

[5] Swets J.A., Measuring the accuracy of diagnostic systems. Science, 240, 1285-1293, 1988.

[6] Allen G., Le Marshall J.F. An evaluation of neural networks and discriminant analysis methods for application in operational rain forecasting. 1993.

[7] Mason I. A model for the assessment of weather forecasts. Australian Meteorological Magazine, 30, 291-303, 1982.

[8] Goodenough D.J., Metz C.E., Lusted L.B.; Cavaet on use of the the parameter d' for evaluation of observer performance; Radiology, 106, 565-566, 1973.

[9] Murphy A.H.; Scalar and vector partitions of the probability score: Part I. Two-state situation. Journal of Applied Meteorology, 11, 253-282, 1972

[10] Kumar V.R., Chung C.Y.Cl, Lindley C. A., Learning to perform weather forecasting operations. Proceedings of the Fourth Scandinavian Conference on Artificial Intelligence, Stockholm, Sweden, 174-156, 1993.

[11] Kumar V. R., Guignard P.A., Chung C.Y.C., Performance comparison of different learning methods for weather forecasting operations. Third Golden West International Conference on Intelligent Systems. Las Vegas. 1994.

SEMANTIC TREES FOR DISJUNCTIVE LOGIC PROGRAMS

John R. Fisher
Computer Science Department,
California State Polytechnic University
Pomona, CA 91768

Abstract: This paper presents a unique tree-based method for formally specifying the semantics of disjunctive logic programs. The trees themselves are a kind of 'and-tree' determined by the clauses in the given program. Some connections between this tree-based specification and well-founded semantics are given; in particular, the tree-based specification coincides with the well-founded semantics for nondisjunctive logic programs without function symbols. The tree-based specification is important because it also serves (in part) as a formal design specification for a top-down interpreter which computes answers based upon the semantics.

Keywords: logic programming, disjunctive logic programs, indefinite and definite programs, negation as failure, formal negation, program trees, full trees, semantic trees, well-founded semantics, bounded trail property.

1 Motivation and Background

Consider the familiar Prolog meta-interpreter M written in Prolog:

```
prove(true) :- !.
prove((G,Goals)) :- !, prove(G), prove(Goals).
prove(G) :- clause(G,B), prove(B).
```

Which tree is searched by M? For example, consider the abstract logic program P (or its Prolog equivalent):

```
p <- q(1), q(2)
q(x) <- r(x)
q(2) <- q(1)
r(1) <-
```


E. A. Yfantis (ed.), Intelligent Systems, 291–305.

For the goal <- p, an SLD-derivation (or the Prolog search) searches (or grows) the following *derivation tree* **D**, pictured in Figure 1 on the left, using the clauses of P (adapted from Bol [2], p. 117):

Figure 1. SLD tree on the left, semantic 'and' tree on the right

However, M actually searches (or grows) the tree **T**, pictured in Figure 1 on the right.

So M does not (exactly) simulate SLD deduction. To emphasize this point further, suppose that M were to be augmented with a loop-check, such as in

```
prove(true, Trail) :- !.
prove((G,Goals), Trail) :- !, prove(G, Trail), prove(Goals, Trail).
prove(G, Trail) :- loop_check(G,trail), !, fail.
prove(G, Trail) :- clause(G,B), prove(B,[G|Trail]).

loop_check(G,[F|R]) :- G == F, !.
loop_check(G,[F|R]) :- loop_check(R).
```

Then the new M appropriately finds **NO** loops in T. But, an augmented SLD procedure with a loop check on lead (or selected) goals could prune D at the second occurrence of 'q(1)', marked with the asterisk (*), and fail to answer correctly for the goal <-p.

We call a tree like T a *semantic program tree*. For positive logic programs, the general definition of a semantic program tree, or a *P-tree* for short, requires the trees themselves to be finite (a finite data structure), to have unordered branchings determined by ground instances of clauses of the program P, and to allow repeated (but separately identifiable) nodes (because the clauses of P could sometimes lead to repeated occurrences).

We say that a ground positive literal of the program is a *tree-consequence* of the program P provided that there is some (finite) P-tree rooted at the literal having all 'true' leaves. Then

Proposition. *A ground literal L is a tree-consequence of the positive logic program P if, and only if, L belongs to the least (positive) model of P.*

Thus, we see that M directly implements the tree-based "semantics" defined above (which is equivalent to the standard least model semantics). Or, to exaggerate a little, we could say that M is an executable version of the tree-based semantics for positive programs. It is interesting that the tree-based specification is both a *requirements* specification (because it is equivalent to least-model semantics) and a *design* specification (because of its direct relationship to the meta-interpreter M).

(The reader is invited to write a Prolog meta-interpreter that *does* simulate SLD-deduction. It is not M!)

Now, if we turn our attention to logic programs with negation as failure, we will see that the distinction between derivation trees and semantic trees is more important. Both SLDNF-resolution and SLS-resolution modify the SL-resolution engine in attempts to compute reasonable answers for goals for (non disjunctive) logic programs with negation as failure. SLDNF is based upon Clark's completion semantics (Clark [4]) and is appropriately suited to programs without loops (and finite failure). SLS resolution (Przymusinski[10]) attempts to accommodate programs that have some loops for positive literals only, but is restricted to stratified programs (so no "loops through negation"). Also, SLS resolution, being based on the SL-derivation trees, is prone to unsound loop checks like in the previous example (the tree D).

The problem here is that SLDNF and SLS are searching, or growing, the wrong trees! We take a considerably different approach in this paper. The class of disjunctive programs with negation as failure considered here contain disjunctive clauses of the form

$$A_1,A_2,\ldots,A_k <- B_1,\ldots,B_m,not(C_1),\ldots,not(C_n)$$

where each A_i, each B_i, and each C_i is a positive literal (possibly containing variables), $k >= 1$, $m,n >= 0$. The sequence of literals $A_1,A_2,\ldots,A_k$ constitutes the *head* of the clause and this sequence is a *disjunction*. The sequence of literals $B_1,\ldots,B_m,not(C_1),\ldots,not(C_n)$ is a *conjunction* and is the *body* of the clause. In particular, there is no stratification assumed. We design a constructive "specification of semantics" based upon semantics program trees (**NOT** SL-type procedural trees), in a fashion similar to that discussed above for positive programs. This specification uses a (finite) tree data structure to determine (or support) meanings. The clauses of the program, together with well specified contrapositive clause forms associated with the program, are used to specify the semantic trees. The tree-based specification must accommodate looping along trails in the trees. The trails can stay in a tree or leave at a negative leaf. Recursion, or looping, may be positive (within a single tree), or through negation (involving nodes in more than one tree). The tree-based specification must specify generally when a P-tree is "full" enough: For non disjunctive logic programs, this happens when all leaves are 'true', or a repeat node (from the current or some previous tree), or negative of the form not(L), in which case P-trees rooted at L need to be considered. (For disjunctive logic programs, the characterization is somewhat more complicated.) The specification uses three truth values, with positive looping counting as failure, and looping through negation counting as indeterminacy (undefined truth value). Lastly, but closely related to the previous requirements, we require that the tree-based semantics should correspond -- as much as we can guarantee -- to the well-founded semantics for non disjunctive logic programs with negation as failure (VanGelder, Ross, and Schlipf [12]).

The resulting constructive tree-based specification of semantics for disjunctive logic programs then serves as the design (and requirements) specification for a meta-interpreter that computes well-founded semantics, just as for positive programs the tree-consequence semantics introduced earlier could serve as a design (and requirements) specification for the meta-interpreter M. For "reasonable" programs, the tree-based specification properly and significantly subsumes what

SLDNF and SLS can compute, and, in addition, the tree-based specification gives an extension to disjunctive logic programs with negation as failure.

The paper provides the formal definitions for the tree-based specifications and characterizes the basic propositions regarding its properties. In particular, for nondisjunctive logic programs with negation as failure, the tree-based semantics is provably equivalent to the well-founded semantics if B_P is finite. Proofs of the technical theorems will be published elsewhere (Fisher [5]).

The literature has references to similar trees for non-disjunctive logic programs, referred to as "clause trees" (Pereira, Alferes, and Aparacio [9]), or sometimes as "proof trees" (Bruffaerts and Henin [3]). We believe the formal specification approach taken by this paper is unique. There is the possibility that other formal software specification methodologies would be applicable.

2 Disjunctive Logic Programs

Let us assume that P is a disjunctive logic program whose clauses may have negation-as-failure literals in the bodies of its clauses. Thus, the clauses of P can be described as having the form

$$A_1,A_2,\ldots,A_k \leftarrow B_1,\ldots,B_m,not(C_1),\ldots,not(C_n)$$

where A_i, each B_i, and each C_i is a positive literal, $k >= 1$, $m,n >= 0$. The sequence of literals $A_1,A_2,\ldots,A_k$ constitutes the *head* of the clause and this sequence is a *disjunction*. The sequence of literals $B_1,\ldots,B_m,not(C_1),\ldots,not(C_n)$ is a *conjunction* and is the *body* of the clause. The sequence $B_1,\ldots,B_m$ is the *positive* part of the body and the sequence $not(C_1),\ldots,not(C_n)$ is the *negative part* of the body. If k=1 then the clause is said to be *definite*, otherwise it is *indefinite*. An *indefinite* program must have at least one indefinite clause, otherwise the program is *definite*.

In what follows, we will need to refer to contrapositive forms of a clause. A *primary alternative* of the clause

$$A_1,A_2,\ldots,A_k \leftarrow B_1,\ldots,B_m,not(C_1),\ldots,not(C_n)$$

has the form

$$A_j \leftarrow alt(\sim A_1), \ldots,alt(\sim A_{j-1}),alt(\sim A_{j+1}),\ldots,alt(\sim A_k),B_1,\ldots,B_m,not(C_1),\ldots,not(C_n)$$

where $1<= j <= k$. There are k primary alternatives if k>=2. If k=1 then the clause is definite and does not have any primary alternatives. The '~' denotes *formal negation*. The 'alt' forms are special markers for the alternatives. Note that there are now two kinds of negation that could be referred to: 'not' is negation as failure, and '~' is formal negation. We will need to maintain a careful distinction between these two negations. For a primary alternative the sequence $alt(\sim A_1),\ldots,alt(\sim A_{j-1}),alt(\sim A_{j+1}),\ldots,alt(\sim A_k)$ is called the *alternative* part of the body.

There are other contrapositive forms of clauses of an indefinite program that could be useful. These are called *backlinks* (for reasons that will be apparent later). They are formed as follows. Suppose that

$$A \leftarrow \alpha, B, \beta$$

is either a definite clause of P or a primary alternative whose head is the positive literal A and B is a literal in the positive part of the body; α, β are (possibly empty) sequences of the other literals of the body. Then

$$\sim B \leftarrow \alpha,\ \sim A,\ \beta$$

is a backlink clause, where '~' is formal negation. Later in this paper (Section 5), we will characterize which backlinks are *potentially useful*.

Working Example. Consider the indefinite program P (X is a variable)

p(X), q(X) <- r(X), not(s(X))
s(a) <- p(a)
r(a) <-   r(b) <-
v(a) <-   v(b) <-   v(c) <-

d(X) <- p(X), w(X)
d(X) <- q(X), v(X)
w(a) <-   w(b) <-
k(X) <- not(d(X))

The primary alternatives of the indefinite clause are

p(X) <- alt(~q(X)), r(X), not(s(X))
q(X) <- alt(~p(X)), r(X), not(s(X))

Here are all of the possible backlinks:

~r(X) <- alt(~q(X)), ~p(X), not(s(X))
~r(X) <- alt(~q(X)), ~q(X), not(s(X))
~p(X) <- ~d(X), w(X)

~w(X) <- p(X), ~d(X)
~q(X) <- ~d(X), v(X)
~v(X) <- q(X), ~d(X)

Given a disjunctive logic program P, we say that the *usable* clauses of P are the definite clauses of P together with the primary alternatives of P and the backlink clauses of P. Note that the only usable clauses of P that actually belong to P are the definite clauses of P. The other usable clauses are contrapositive forms of clauses of P.

3 Program Trees

P-trees are constructed using the usable clauses of P. Let B_P be the Herbrand base of P (set of ground positive literals of P), and let

$$\sim B_P = \{ \sim b \mid b \ \varepsilon\ Bp \}.$$

The *branchings* for P-trees are formed using the usable clauses of P. If

$$a \leftarrow c_1,\ldots,c_s$$

is a ground instance of a usable clause of P, then the corresponding branching node is

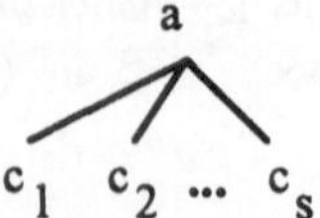

P-trees of height 0 are just elements of Bp ∪ $\sim B_P$. P-trees of height 1 are those just described using a single branching node, rooted at some a ε Bp ∪ $\sim B_P$. If T is a P-tree and c is a leaf not of the form 'alt(-)' or 'not(-)' then T may be extended using another branching at that leaf, as described above for P-trees of height 1. Negation-as-failure nodes 'not(-)' and alternatives 'alt(-)' *must be* leaves in the P-trees. Inductively, A *P-tree* is any *finite* tree that can be constructed in this fashion. The height of such a tree is, in general, the length of the longest branch from the root of the P-tree to its deepest leaf. If c <- is a ground instance of a unit clause of P, then we write the corresponding branching P-tree node as

For the Working Example. Figure 2 shows some P-trees that will be referred to later.

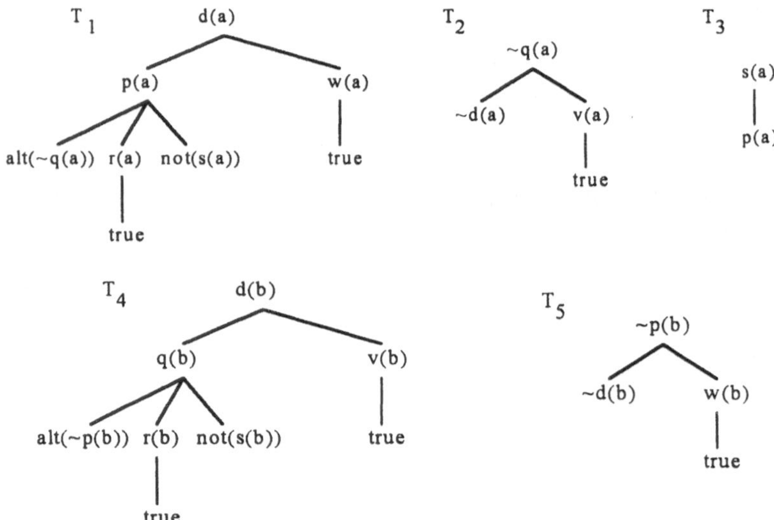

Figure 2. Some semantic trees for the working example

Definition 1. Suppose that a ε Bp ∪ $\sim B_P$ and that S is a subset of Bp ∪ $\sim B_P$. Then T is an *S-full P-tree rooted at a* if T is a P-tree rooted at a each of whose leaf nodes is either

1) an element of S, or
2) of the form ~b where b ε S
3) a literal in Bp ∪ $\sim B_P$ which does not unify with the head of any clause of P, or

4) a literal in Bp U ~B_p which has itself as an ancestor in T, or
5) the true leaf, true, or
6) a negation-as-failure node of the form not(b),or
7) an alternative form alt(b).

If P-tree T is {}-full then we simply say that T is *full*.

Pictorially, Figure 3 portrays such a tree (all leaves having one of the displayed forms)

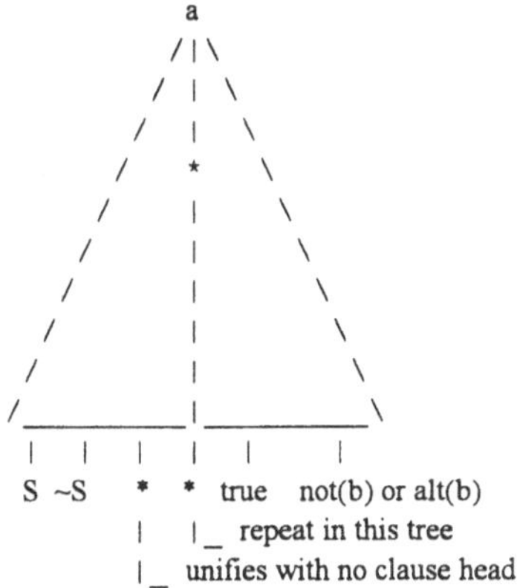

Figure 3. Leaves in a full tree

An *ancestor trail* is a sequence $a_0,a_1,\ldots,a_n$ of nodes in P-trees such that a_{i+1} is either a positive node which is a child of a_i or else a_{i+1}=b where 'not(b)' or 'alt(b)' is a child of a_i. Note that ancestor trails can wind through several trees. Trails can leave a particular tree at a negative leaf.

For the Working Example. Consider the ancestor trail {d(a),p(a),~q(a),~d(a)} which winds from T_1 through T_2. T_1 is full, and T_2 is {d(a),p(a)}-full. Similarly, the trail {d(a),p(a),s(a),p(a)} leads through T_1 and into T_3. T_3 is {d(a),p(a)}-full. One *could*, of course, extend T_3 some more, to produce a full P-tree. This is a general phenomenon, characterized in the following proposition.

Proposition 1. *Suppose that 'sw' refers to either of the predicates 'not' or 'alt'. Suppose that $T_1, \ldots, T_n$ is a sequence of P-trees rooted at $a_1, \ldots, a_n$, respectively, that $sw(a_{i+1})$ is a leaf of T_i, $i = 1,\ldots,n$, that S_i is the ancestor trail in T_i of $sw(a_{i+1})$, and that T_{i+1} is an $(S_1 \cup \ldots \cup S_i)$-full P-tree rooted at a_{i+1}, $i = 1,\ldots,n-1$. (In particular, T_1 is a full P-tree.) Pictorially, we would have*

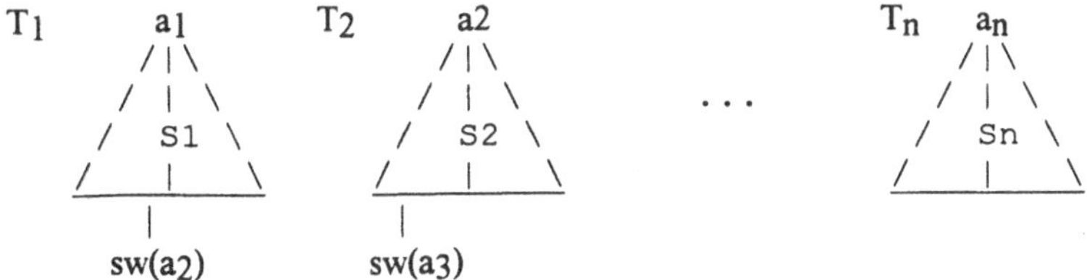

Then, some extension of T_n is a full P-tree rooted at a_n.

We have used 'sw' to represent either of 'not' or 'alt', suggesting that 'sw' means "switch to another tree". In the conclusion of this proposition the term *extension* refers to a P-tree formed using clauses of P in the natural way to create a bigger tree containing the given tree as a top portion (the extended branches being below some of the leaves of the given tree).

S-full P-trees, where S is a trail of ancestors, are important because of their use in forcing a termination to the computations of truth values. Here are the technical definitions.

4 Tree-based Semantics

Define a mathematical relation R on the set

$$(B_P \cup \sim B_P) \times \{t,f,u\} \times 2^{(B_P \cup \sim B_P)}$$

where t, f, u stand for 'true', 'false', 'undetermined', respectively. We will use the notation "...=...#..." to describe this relation. That is, write a = v # S provided that (a,v,S) is in R, where a is a ground atom in $B_P \cup \sim B_P$, v is a truth value in {t,f,u} and S is a subset of $B_P \cup \sim B_P$. The definition is recursive, and given in three parts, as follows.

Definition 2(part 1). Define *a = t # S* to mean that there is some S-full P-tree T rooted at a such that every leaf node of T is either

(i) true,
(ii) of the form ~b where b ε S (*ancestor resolution* rule),
(iii) of the form not(c), and c = f # (S ∪ S') where S' is the set of positive literals which are ancestors of not(c) in T, or
(iv) of the form alt(~d), where ~d = t # (S ∪ S') where S' is the set of positive literals which are ancestors of alt(~d) in T.

Pictorially, there must be a P-tree having *every* leaf of the following allowed forms:

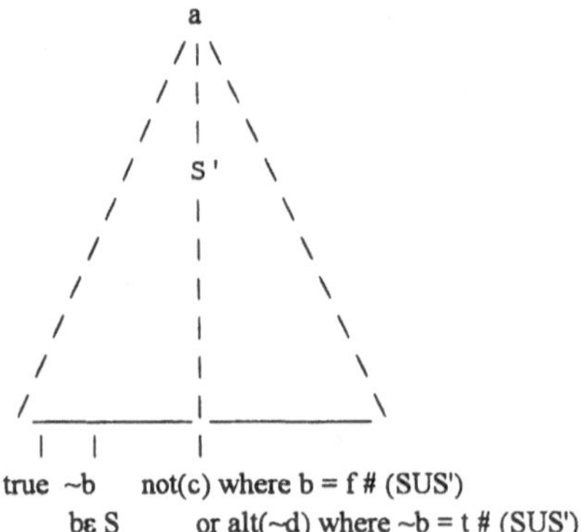

Figure 4. Full tree for truth value 't'

Definition 2(part 2). Define $a = f \# S$ to mean that every S-full P-tree T rooted at a has at least one leaf which has one of the following forms:

(i) a literal b ε Bp U ~B_P which does not unify with the head of any usable clause of P,

(ii) a literal c ε Bp U ~B_P which has itself as an ancestor in T

(iii) not(b) where b = t # (S U S'), and S' is the set of positive literals which are ancestors of the leaf not(b) in T.

Pictorially, every S-full P-tree rooted at a must have at least one leaf of the following forms:

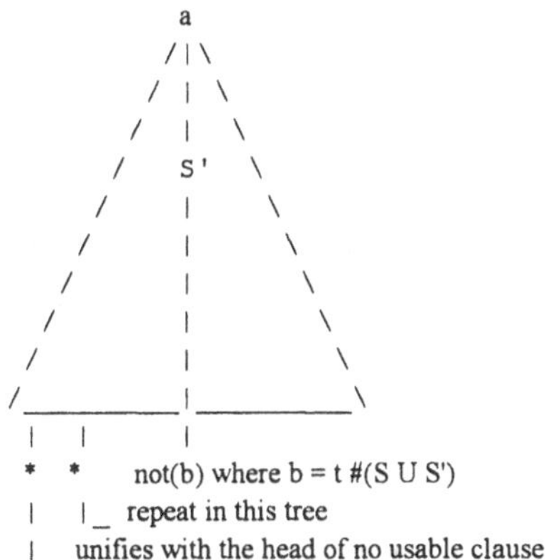

Figure 5. Full tree for truth value 'f'

Note that a leaf of the form 'alt(-)' never can contribute to failure (f truth value) of the root of the tree. Alt-leaves can contribute to "truth" by allowing resolution with an ancestor, but otherwise their appearance contributes to "indeterminacy".

For the Working Example. We have

(a) ~q(a) = t # {d(a),p(a)} is established using T_2,

(b) ~p(b) = t # {d(a),p(a)} is established using T_5,

(c) s(a) is neither t nor f #{d(a),p(a)} (neither definition part 1 nor 2 applies to s(a)),

(d) s(b) = f # {} since s(b) unifies with the head of no usable clause,

(e) d(a) is neither t nor f #{}, because of (c),

(f) d(b) = t # {}, using (b) and (d) in T_4.

Note that we did not explicitly examine all P-trees rooted at d(a) in order to confirm (e). We leave this to the reader.

The *bounded trail property* (BTP) states that every ancestor trail (if sufficiently extended) through a forest of P-trees eventually stops at a node having no descendants or else the trail eventually repeats an element previously encountered on the trail. Note that the working example satisfies the BTP.

Proposition 2. *Disjunctive programs without function symbols satisfy the bounded trail property.*

Two example programs which do not have the BTP are P_1 = {p(x) <- p(f(x))} and P_2 = {p(x) <- not(p(f(x))}. Many programs with lots of function symbols do have the BTP. We believe that the BTP is interesting (and convenient for characterizations) but make no claims here regarding its *general* decidability, since it requires a suspicious sounding "halting" requirement.

Proposition 3. *If P satisfies the bounded trail property, then for every literal a, at least one of a = t # {} or a = f # {} does not hold.*

Thus, for programs with the bounded trail property, we may finish the truth-value definition, as follows.

Definition 2(part 3). *a = u # {}* means that neither a = t # {} nor a = f # {} holds.

For the Working Example. We can now say that d(a) = p(a) = u # {}; i.e., that d(a) and p(a) are undetermined by the program. Of particular interest are the truth values determined for k(a), k(b) and k(c). Recall that the definition of k in P was

k(x) <- not(d(x))

We have

k(a) = f # {}since d(a) = t # {}
k(b) = u # {} since d(b) = u # {}
k(c) = t # {} since d(c) = f # {}

Now, of course, the definitions of truth value based on trees must be used with care. For example, in the program

a <- not(b)
a <- not(a)
a <- c

we have that a = t # {} based upon the first clause (or corresponding tree), whereas if the first clause were ignored, then we would have had a = u # {}, and we would have had a = f # {} if only the last clause were available. As for well founded semantics, t supersedes u, which in turn supersedes f; that is, t > u > f. For the tree-based semantics, this is a consequence of the three parts of the definition for truth values. A rough characterization of this would be: a literal is true if at least one tree supports with all "truthful" leaves, or the literal is false if all trees trying to support have at least one "failing" leaf, otherwise the literal is indeterminate.

An example can be used to motivate the use of 'alt' literals in the alternative clauses. Consider the program

a, b <-
c <- not(a).

Now we have a = u # {}. To emphasize why this is the case, observe that

a

|

alt(~b)

is the only full P-tree rooted at a, and clearly ~b = f # {}, but this last fact does not "falsify" the alt(~b) leaf, as previously noted. Thus c = u # {}.

A *bottom-up* characterization for the semantics corresponding to the semantic trees specification can be formulated as follows. Let us assume that the program P itself is already grounded, and let us also assume that for any ground positive literal L, L only occurs (as one of the disjuncts) in the head of finitely many clauses of P. A sequence of programs P_i and sequences of truth sets T_i, false sets F_i, and undetermined sets U_i are define by induction.

$P_0 = P$
T_0 = the set of heads of body-less clauses of P_0. These can be disjuncts.
F_0 = the set of literals occurring in the head of no clause of P_0.
$U_0 = Bp \setminus (T_0 \cup F_0)$.

Now, assuming that P_i, T_i, F_i, and U_i have been defined for $i < k$,

P_k is obtained from P_{k-1} by modifying or deleting clauses of P_{k-1}:
Erase body literals L from clauses of P_{k-1} when $L \varepsilon T_{k-1}$.
Erase body literals not(L) from clauses of P_{k-1} when $L \varepsilon F_{k-1}$.
Erase a clause of P_{k-1} when the clause has a body literal not(L) and $L \varepsilon T_{k-1}$.
(Erase clauses D <- . . . where $D \varepsilon T_{k-1}$.)
$T_k = T_{k-1}$ U {heads of body-less clauses of P_k}
U {stretch and factor disjuncts in T_{k-1} using clauses from P_k}.
$F_k = F_{k-1}$ U {positive literals of U_{k-1} now occurring in the head of no clause of P_k}.
$U_k = Bp \setminus (T_k \cup F_k)$.

The definition of the T_k is suggested by the approach of Rajasekar, et. al. [11], where disjuncts or states are used. *Stretching* and *factoring* can be understood using an example. Suppose that disjunct 'a v b' is in T_{k-1} and that clauses 'c v d <- b' and 'c <- a' are in P_k. Then a v b can be stretched using the two clauses, obtaining 'c v c v d', and then factoring produces 'c v d' in T_k. These operations correspond to clausal resolution on the body literals of the clauses ('a' and 'b' in the example), followed by the elimination of repeated factors produced in the resolvent; this is a traditional theorem-proving technique. Stretching can only be performed using clauses with a single positive body literal (but the corresponding head may be disjunctive).

Finally, let the *net* truth, false, and undefined sets be given as follows:

$T = \bigcup T_k \quad k = 1..\infty$
$F = \bigcup F_k \quad k = 1..\infty$
$U = Bp \setminus (T \cup F)$

For the Working Example. If one does the relevant calculations, one gets
$T = T_2$ = {r(a), r(b), w(a), w(b), v(a), v(b), v(c), p(b) v q(b), d(b), k(c)}

$F = F_2 = \{s(b), s(c), r(c), w(c), p(c), q(c), d(c)\}$
$U = \{p(a), q(a), s(a), d(a), k(a)\}$

For the working example, the bottom-up characterization of semantics and the tree-based specification give the same truth values to positive literals. We conjecture that this is true more generally. For nondisjunctive programs we have the following proposition, proved in Fisher [5].

Proposition 4. *For nondisjunctive logic programs with negation as failure, if Bp is finite, then the tree semantics is the same as well-founded semantics characterized using the bottom-up definition. That is,*

$$T = \textit{true positive literals} = \{a \in Bp \mid a = t \# \{\}\}$$
$$U = \textit{undefined literals} = \{a \in Bp \mid a = u \# \{\}\}$$
$$F = \textit{false positive literals} = \{a \in Bp \mid a = f \# \{\}\}$$

A stronger version of this bottom-up characterization would interpret disjunction exclusively (if it could): If positive literal A has been added to T_k, if each of A and $B_1, B_2, \ldots, B_n \in U_{k-1}$ and disjunct $D = A \vee B_1 \vee B_2 \vee \ldots B_n \in T_{k-1}$ then remove D from T_k and add each of $B_1, B_2, \ldots, B_n$ to F_k. In addition, one must insist that F_k be purged of positive literals that appear in T_k (literals which used to be false because of exclusive disjunction, but now have become true because they are separately supported). The stronger approach would be in adherence to the *generalized closed world assumption* (GCWA). The version presented above corresponds more to the *weak generalized closed world assumption* (WGCA) suggested for positive programs in Rajasekar, et. al. [11]. The GCWA approach forces more "disjunctive literals" to be *false* because disjunction is being interpreted exclusively.

For example, consider the logic program

```
a , b <-
b <-
c <-  not(a).
```

Both the tree-based semantics and the bottom-up characterization conclude that a = u # {}, whereas a semantics using the GCWA would insist that "a is false".

In Fisher [7], the ancestor resolution rule was given a procedural characterization and used as an enhancement for SLD resolution, resulting in what was called SLD/AER deduction (SLD + *ancestor erasure rule*). The corresponding set-theoretical semantics used sets of disjuncts, or states, as presented and characterized in (Rajasekar [12]). The resulting characterizations in Fisher [7] of soundness and completeness for procedural SLD/AER were for disjunctive programs *without* negation as failure (so-called positive programs). The tree-based semantics presented in this paper provides a generalization of the previous concepts to disjunctive programs *with negation as failure*, using an extension of well founded semantics.

We are not claiming to have *the correct approach* to semantics for disjunctive programs. Rather, our purpose is to explain the top-down, semantic tree specification approach. We cannot offer at this time a tree-based specification that would correspond to the stronger version imposing the GWCA. An excellent discussion of semantics issues for disjunctive logic programs is in the paper by Apt and Bol [1].

In the truth value definition, negation-as-failure nodes and alternative nodes were not allowed to be the roots of P-trees, and no truth value was independently ascribed to 'not(...)' nor to 'alt(...)' literals. Informally, we do so as follows:

not(b) = t # S if b = f # S	alt(~b) = t # S if ~b = t # S
not(b) = f # S if b = t # S	alt(~b) = u # S if ~b = f or u # S
not(b) = u # S if b = u # S	

Using this informal notation, we have

Proposition 5 (Tabulation). *Suppose that*

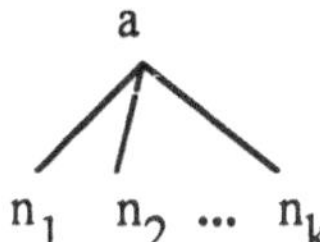

is a P-tree branching based upon a usable clause of P. Suppose that a = v # {}, and that n_i = v_i *# {a} for i = 1,...,k, where v,* v_i *ε {t,u,f}. If this is the only P-tree rooted at a then*

$$a = min\{v_i \mid i = 1, \ldots, k\} \# \{\}$$

where the ordering is the usual t > u > f. On the other hand, if $T_1,\ldots, T_m$ *are all of the P-trees rooted at a, and if a =* v_j *# {} when only the subprogram growing* T_j *is considered, then, for a net result*

$$a = max\{v_j \mid j = 1,\ldots, m\}\# \{\}.$$

This shows that tree semantics (and well founded semantics, insofar as the two coincide) is a sort of "maxi-min computation". Using a metaphor of deliberation, one seeks the strongest overall argument, where each individual argument is only supported by (or is as strong as) its weakest evidence.

5 Useful Clauses

Usable clauses were for an indefinite logic program were characterized in the first section. It is probably apparent that not all of the usable clauses would actually be needed to grow P-trees in order to determine truth values. The following proposition shows that formally negative literals can never sustain a 't' truth value on their own.

Proposition 5. *Suppose that P is a disjunctive logic program with no formally negative literals in any clause. Then, for any formally negative ground literal ~a ε ~Bp we have*

$$\sim a = f \# \{\}$$

The proposition may seem surprising at first, but recall that ~a = t # S has occurred in the examples only when S contained sufficient ancestors for ancestor resolution.

Definition 3. Suppose that P is a disjunctive logic program and that H is a positive literal (which can contain variables). H is said to be an *indefinite literal* (with respect to P) provided either that h unifies with some literal in the head of some indefinite clause of P, or else there is some definite clause A <- $B_1,\ldots,B_r$,not(C_1),...,not(C_s) of P such that H and A have most general unifier σ and for some j=1,...,r, $\sigma(B_j)$ is an indefinite literal.

In the Working Example of section 1, the literals p(x), q(x), s(a), d(x) are all indefinite literals (as would be any variants or instances of any of these literals).

Definition 4. Suppose that, as before,

$$\sim B <- \alpha, \; \sim A, \beta$$

is a backlink clause of P. This backlink is said to be *potentially useful* provided that the positive literal B is an indefinite literal.

Here is another small example, showing how indefinite information depending upon other indefinite information can be accessed through usable clauses.

```
a, b <-                 e <- c
c, d <- b               e <- d
e <- a
```

Figure 6 shows the program's logical dependencies graphically

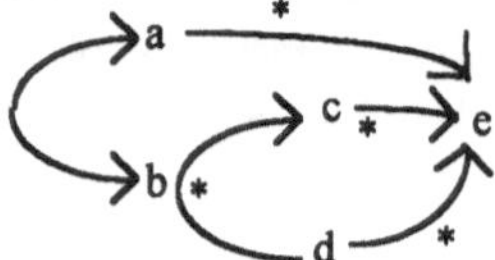

Figure 6. Potentially useful backlinks

We have marked the clauses that generate potentially useful backlinks with an asterisk (*); for example, for the clause e<-a, the actual backlink would be ~a<-~e. Note that these potentially useful backlinks "link back from indefinite literals". Let the reader write down all of the usable clauses for this indefinite program. Here are some P-trees that establish e = t # {}:

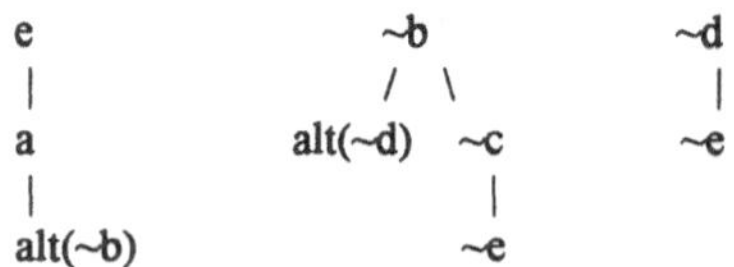

Figure 7. Forest showing e = t # {}

Note that the middle tree shows a use of a backlink of a primary alternative clause. The last tree shows ~d = t # {e,a,~b}, the next to last shows that ~b = t # {e,a}, and the first that e = t # {}. Note how the backlinks allow reasoning "around the arrow diagram".

Proposition 6. *Suppose that P is a disjunctive logic program, that aε B_P, that a is an indefinite literal of P, and that a = t # {}, using a supporting forest of program trees F. Then any backlink clause actually used to grow a branch of some tree in F must be a potentially useful backlink.*

Let the reader provide an example where a potentially useful backlink cannot actually be used to support a truth value of t for any literal. This explains the "potential" in the terminology "potentially useful".

For an indefinite logic program, the ratio of potentially useful to usable backlinks could be called the *backlink utility ratio*. For the Working Example of section 1, this ratio is 2/6=1/3. It would be interesting to establish some mathematical relationships for this ratio.

6 References

[1] Apt, K.R., and R. Bol (1994), Logic programming and negation: a survey, Preprint.

[2] Bol, R.N. (1991). *Loop Checking in Logic Programming*, Thesis, Centre for Mathematics and Computer Science, Amsterdam.

[3] Bruffaerts,A., and Henin,E. (1988), Proof trees for negation-as-failure: yet another Prolog meta-interpreter. *Proceedings of the Workshop on Meta-Programming in Logic Programming*, University of Bristol, Bristol, England, June, pp.133-146.

[4] Clark, K. (1978). Negation as Failure, in H.Gallaire and J.Minker, eds., *Logic and Databases*, Plenum Press, pp.293-322.

[5] Fisher, J.R. (1994), Top-down tree specification of semantics for logic programs with negation as failure, in preparation. (*Tech. Report #1993-06*, Computer Science Department, California State Polytechnic University, Pomona.)

[6] Fisher, J.R. (1992), Tree specification of semantics for logic programs with negation as failure, *Proc. Third Annual California State University Artificial Intelligence Symposium*, California State University, San Luis Obispo, June 1992, pp.158-66.

[7] Fisher, J.R. (1989), *GPL Notes, Generalized Prolog, Tech. Report #1989-01*, California State Polytechnic University, Pomona.

[8] Lloyd, J.W. (1987), *Foundations of Logic Programming*, 2nd ed., Springer-Verlag.

[9] Pereira, L.M., J. Alferes, and J.N. Aparacio (1990), Top-down procedures for well-founded semantics, *Technical Report*, AI Centre, Uninova.

[10] Przymusinski, T.C. (1989), On the declarative and procedural semantics of logic programs, *J. Automated Reasoning*, 5, pp.167-205.

[11] Rajasekar, A., J. Lobo, and J. Minker (1989). Weak generalized closed world assumption, *J. Automated Reasoning*, Vol. 5, No. 3, pp. 293-307.

[12] Van Gelder, A., A. Ross, and J.S. Schlipf (1991), The well-founded semantics for general logic programs, *J. ACM*, Vol. 38, No. 3, pp. 620-650.

THE OK BDI ARCHITECTURE

Deepak Kumar
Department of Mathematics
Bryn Mawr College
Bryn Mawr, PA 19010
(610) 526-7485
dkumar@cc.brynmawr.edu

Stuart C. Shapiro
Department of Computer Science
State University of New York at Buffalo
Buffalo, NY 14260
(716) 645-3181
shapiro@cs.buffalo.edu

Abstract: The design of a belief-desire-intention (BDI) architecture is presented. The architecture is defined using a unified object-based knowledge representation formalism, called the OK formalism, and a unified reasoning and acting module, called the OK rational engine. Together they form the OK BDI architecture for modeling rational agents endowed with beliefs, desires, and intentions.

Keywords. BDI Architectures, Knowledge Representation and Reasoning, Acting, and Planning.

1 Introduction

A survey of AI systems would reveal that it is somewhat awkward to do acting in reasoning (or logic-based) systems (but it is convenient to talk about representational and reasoning issues using them), and it is awkward to study reasoning and representational issues in systems designed for acting/planning. Thus, most "good" planning/acting systems are "bad" knowledge representation and reasoning (KRR) systems and vice versa. For example, in a recent symposium on "Implemented KRR Systems" [21] out of a total of 22 KRR systems presented only 4 systems had capabilities for representation and reasoning about actions/plans (RHET [1], CYC [16], CAKE [20] and SNePS [25]). The work presented in this paper presents an approach that bridges this "representational/behavioral gap." We extend the ontology of an intensional KRR system to facilitate representation and reasoning about acts and plans. I.e. a computational cognitive agent modeled using the extended ontology has representations for beliefs, acts, and plans, and is able to reason about them. I.e., the agent is able to represent its beliefs and desires (the 'B' and the 'D' of 'BDI').

In most current AI architectures reasoning is performed by an inference engine and acting is done under the control of some acting executive (or a plan/act interpreter). Our approach is based on the viewpoint that logical reasoning rules implicitly specify the act of believing. Reasoning is the process of forming new beliefs from other beliefs. The connectives and quantifiers of the inference rules govern the derivation of new beliefs. Reasoning can be looked at as a sequence of *actions* performed in applying inference rules

E. A. Yfantis (ed.), Intelligent Systems, 307–317.

to derive beliefs from other beliefs. Thus a reasoning rule can be viewed as a rule specifying an act—that of believing some previously non-believed proposition—but the believe action is already included in the semantics of the propositional connective. Thus, the inference engine can be viewed as a *mental actor*. This enables us to establish a closer relationship between rules of inference and rules of acting (or planning). This suggests that we can integrate our models of inference and acting by eliminating the acting executive (plan/act interpreter). These ideas are used in developing a computational model— called a *Rational Engine*, that is a unified model of acting and inference and can be used for modeling rational cognitive agents and their behavior. Acting and reasoning about beliefs, actions, and plans is performed by a single component— the Rational Engine. The rational engine implements the underlying logic as well as notions of intentionality (the 'I' of 'BDI').

The work presented here has evolved from research involved in extending a semantic network-based KRR system, SNePS (whose rational engine called SNeRE is described in [9, 10, 11, 12]), into a BDI architecture. In this paper we use an object-oriented approach to describe the architecture. The resulting architecture is independent of, yet isomorphic to, the SNePS formalism. The resulting architecture enjoys all the advantages of object-oriented design—the ontology is easily extendible, as is the underlying logic, and amenable to a concurrent implementation [8, 11].

2 The OK Architecture

We have defined the OK[1] architecture to have the following constituents:

$$\text{OK BDI Architecture} = \text{OK Formalism} + \text{OK Rational Engine}$$

The OK formalism is a conceptual, extendible, object-oriented hierarchy of classes that correspond to the various ontological components of a knowledge representation system—individuals, propositions, and acts. While several object-based AI systems are already in existence, this is the first one that makes an object-oriented commitment to the entities of a KR formalism itself. The OK rational engine is defined using methods inherited or specialized by the classes of the OK formalism.

2.1 The OK Formalism

The representational formalism is described as a conceptual object-oriented hierarchy. This is depicted in Figure 1. In an intensional representational framework, anything a cognitive agent can think about is termed a "mental concept" or a conceptual entity. More specifically these can be individuals, beliefs (propositions), or acts. In addition to standard beliefs that an agent is able to represent, we also define a special class of beliefs called *transformers*. A *transformer* is a propositional representation that subsumes various notions of inference and acting. Being propositions, transformers can be asserted in the agent's belief space; they are also beliefs. In general, a transformer is a pair of entities—$(\langle\alpha\rangle, \langle\beta\rangle)$, where both $\langle\alpha\rangle$ and $\langle\beta\rangle$ can specify beliefs or acts. Thus, when both parts of a transformer specify beliefs, it represents a reasoning rule. When one of its parts specifies beliefs and the other

[1] OK stands for *O*bject-based *K*nowledge.

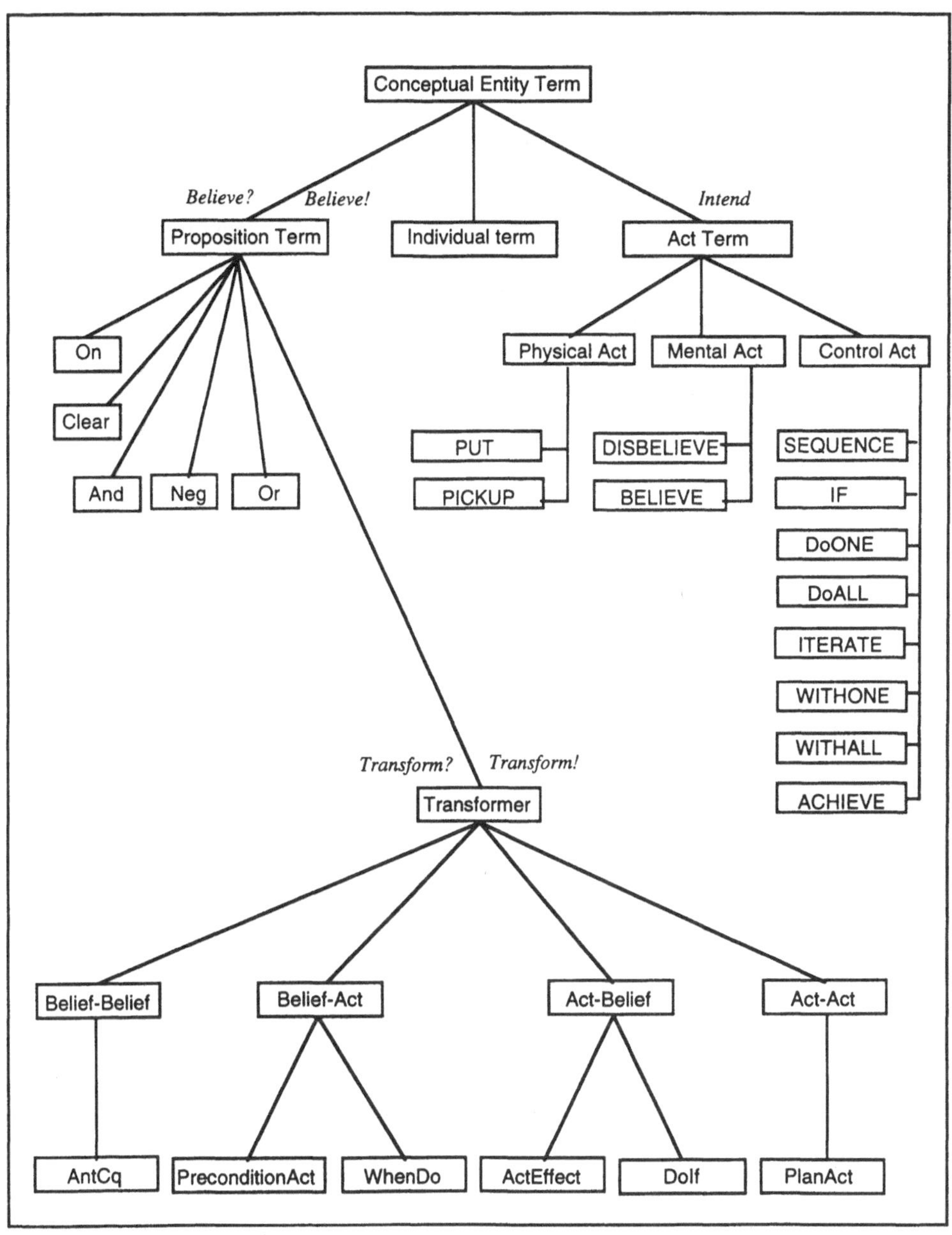

Figure 1: An Object-Oriented KR Formalism

acts, it can represent either an act's preconditions, or its effects, or a reaction to some beliefs, and so on. What a transformer represents is made explicit by specifying its parts. When believed, transformers can be used during the acting/inference process, which is where they derive their name: they transform acts or beliefs into other beliefs or acts and vice versa. Transformations can be applied in forward and/or backward chaining fashion. Using a transformer in forward chaining is equivalent to the interpretation "after the agent believes (or intends to perform) $\langle\alpha\rangle$, it believes (or intends to perform) $\langle\beta\rangle$." The backward chaining interpretation of a transformer is, "if the agent wants to believe (or know if it believes) or perform $\langle\beta\rangle$, it must first believe (or see if it believes) or perform $\langle\alpha\rangle$." There are some transformers that can be used in forward as well as backward chaining, while others may be used only in one of those directions. This depends upon the specific proposition represented by the transformer and whether it has any meaning when used in the chaining process. Since both $\langle\alpha\rangle$ and $\langle\beta\rangle$ can be sets of beliefs or an act, we have four types of transformers— *belief-belief, belief-act, act-belief,* and *act-act.*

Belief-Belief Transformers: These are standard reasoning rules (where $\langle\alpha\rangle$ is a set of antecedent belief(s) and $\langle\beta\rangle$ is a set of consequent belief(s)). Such rules can be used in forward, backward, as well as bidirectional inference to derive new beliefs. For example, a class of transformers that represent antecedent-consequent rules is called AntCq transformers. We will use the notation

$$\langle\alpha\rangle \rightarrow \langle\beta\rangle$$

to write them. For example "All blocks are supports" is represented as

$$\forall x[\mathsf{Isa}(x, \mathsf{BLOCK}) \rightarrow \mathsf{Isa}(x, \mathsf{SUPPORT})]$$

In addition to the connective above (which is also called an or-entailment), our current vocabulary of connectives includes and-entailment, numerical-entailment, and-or, thresh, and non-derivable. Other quantifiers include the existential, and the numerical quantifiers (see [23, 9]).

Belief-Act Transformers: These are transformers where $\langle\alpha\rangle$ is a set of belief(s) and $\langle\beta\rangle$ is a set of acts. Used during backward chaining, these can be propositions specifying preconditions of actions, i.e. $\langle\alpha\rangle$ is a precondition of some act $\langle\beta\rangle$. For example, the sentence "Before picking up A it must be clear" may be represented as

$$\mathsf{PreconditionAct}(\mathsf{Clear}(\mathsf{A}), \mathsf{PICKUP}(\mathsf{A}))$$

Used during forward chaining, these transformers can be propositions specifying the agent's desires to react to certain situations, i.e. the agent, upon coming to believe $\langle\alpha\rangle$ will form an intention to perform $\langle\beta\rangle$. For example, a general desire like "Whenever something is broken, fix it" can be represented as

$$\forall x[\mathsf{WhenDo}(\mathsf{Broken}(x), \mathsf{FIX}(x))]$$

Act-Belief Transformers: These are the propositions specifying effects of actions as well as those specifying plans for achieving goals. They will be denoted ActEffect and PlanGoal transformers respectively. The ActEffect transformer will be used in forward chaining to accomplish believing the effects of act $\langle\alpha\rangle$. For example, the sentence, "After picking up A it is no longer clear" is represented as

$$\mathsf{ActEffect(PICKUP(A), \neg Clear(A))}$$

It can also be used in backward chaining during the plan generation process (classical planning). The PlanGoal transformer is used during backward chaining to decompose the achieving of a goal $\langle\beta\rangle$ into a plan $\langle\alpha\rangle$. For example, "A plan to achieve that A is held is to pick it up" is represented as

$$\mathsf{PlanGoal(PICKUP(A), Held(A))}$$

Another backward chaining interpretation that can be derived from this transformer is, "if the agent wants to know if it believes $\langle\beta\rangle$, it must perform $\langle\alpha\rangle$," which is represented as a DoIf transformer. For example, "Look at A to find out its color" can be represented as

$$\mathsf{DoIf(LOOKAT(A), Color(A, ?color))}$$

Act-Act Transformers: These are propositions specifying plan decompositions for complex actions (called PlanAct transformers), where $\langle\beta\rangle$ is a complex act and $\langle\alpha\rangle$ is a plan that decomposes it into simpler acts. For example, in the sentence, "To pile A on B first put B on the table and then put A on B" (where piling involves creating a pile of two blocks on a table), piling is a complex act and the plan that decomposes it is expressed in the following proposition:

$$\mathsf{PlanAct(SEQUENCE(PUT(B, TABLE), PUT(A, B)), PILE(A, B))}$$

Our present model of acting is based upon a state-change model (see [13]). We identify three types of states —external world states, mental states (belief space), and intentional states (agent's current intentions). Accordingly, we identify three classes of actions — physical actions, mental actions, and control actions that bring about changes in their respective states. Thus PICKUP is a physical action, we have BELIEVE and DISBELIEVE as mental actions whose objects are beliefs, and control actions are described below. Acts can be *primitive* or *complex* (not shown in the figure). A primitive act has an effectory procedural component which is executed when the act is performed. Complex acts have to be decomposed into plans.

Plans, in our ontology, are also conceptual entities. However, like acts, we do not define a separate class for them as they are also acts— albeit control acts. Control acts, when performed, change the agent's intentions about carrying out acts. Our repertoire of control actions includes *sequencing* (for representing linear plans), *conditional, iterative, disjunctive* (equivalent to the OR-splits of the Procedural Net formalism [22, 26]), *conjunctive* (AND-splits), *selective*, and *achieve* acts (for goal-based plan invocation).

Sequencing Act: SEQUENCE(α_1, α_2)
The acts α_1 and α_2 are performed in sequence. For example:

$$\mathsf{SEQUENCE(PICKUP(A), PUT(A, TABLE))}$$

is the act of first picking up A and then putting it on the table.

Disjunctive Act: DoONE($\alpha_1, \ldots, \alpha_n$)
This act represents a nondeterministic choice. One of the acts $\alpha_1, \ldots, \alpha_n$ is performed. For example:

$$\text{DoONE(PICKUP(A), PICKUP(B))}$$

is the act of picking up A or picking up B.

Conjunctive Act: DoALL($\alpha_1, \ldots, \alpha_n$)
All of the acts $\alpha_1, \ldots, \alpha_n$ are performed in some order. For example:

$$\text{DoALL(PICKUP(A), PICKUP(B))}$$

is the act of picking up A and picking up B.

Conditional Act: IF((π_1, α_1), ..., (π_n, α_n))
Some act α_i whose π_i is believed is performed. For example:

$$\text{IF((Clear(A), PICKUP(A)), (Clear(B), PICKUP(B)))}$$

is the act of either picking up A (if A is clear) or picking up B (if B is clear).

Iterative Act: ITERATE((π_1, α_1), ..., (π_n, α_n))
Some act α_i whose corresponding π_i is believed is performed and the act is repeated. For example:

$$\text{ITERATE((Clear(A), PICKUP(A)), (Clear(B), PICKUP(B)))}$$

is the act of picking up A (if A is clear) and picking up B (if B is clear).

Achieve Act: ACHIEVE(π)
The act of achieving the proposition π. For example:

$$\text{ACHIEVE(Clear(A))}$$

is the act of achieving that A is clear.

Single-object Qualifier Act: WITHONE(x, y, ...)(π(x, y, ...), α(x, y, ...))
Find some x, y, etc. that satisfy π and perform act α on them For example:

$$\text{WITHONE(x)(Held(x), PUT(x, TABLE))}$$

is the act of putting on the table something that is being held.

Multiple-object Qualifier Act: WITHALL(x, y, ...)(π(x, y, ...), α(x, y, ...))
Find all x, y, etc that satisfy π and perform the act α on them. For Example,

$$\text{WITHALL(x)(Held(x), PUT(x, TABLE))}$$

is the act of putting on the table everything that is being held.

These control acts are capable of representing most of the existing plan structures found in traditional planning systems (and more). We should emphasize, once again, that since plans are also conceptual entities (and represented in the same formalism) they can be represented, reasoned about, discussed, as well as followed by an agent modeled in this architecture.

2.2 The OK Rational Engine

Next, we will outline details of the integrated reasoning and acting module—called a *Rational Engine* (as opposed to an inference engine that only performs inference). A Rational Engine is the 'operational' component of the architecture (the interpreter) that is responsible for producing the modeled agent's reasoning and acting (and reacting) behavior. It is specified by three types of methods (or messages) (see Figure 1)—

Believe– A method that can be applied to beliefs for assertional or querying purposes. Consequently there are two versions—

Believe!(π)– where π is a proposition, the method denotes the process of asserting the proposition, π, in the agent's belief space. It returns all the propositions that can be derived via forward chaining inference/acting.

Believe?(π)– where π is a proposition, it denotes the process of querying the assertional status of π. It returns all the propositions that unify with π and are believed by the modeled agent either explicitly or via backward chaining inference/acting.

Intend– that takes an act as its argument (*Intend*(α)) and denotes the modeled agent's intention to perform the act, α.

Transform– These methods enable various transformations when applied to transformers. Corresponding to backward and forward chaining interpretations there are two versions— *Transform*? and *Transform*!, respectively.

Notice that the first two also correspond to the propositional attitudes of belief and intention. The methods *Believe* and *Intend* can be invoked by a user interacting with the agent. New beliefs about the external world can be added to the agent's belief space by using *Believe*! and queries regarding agent's beliefs are generated using *Believe*?. These methods, when used in conjunction with transformers lead to chaining via the semantics of the transformers defined above. The architecture also inherently provides capabilities for consistency maintenance. Each specific object that is a belief can have slots for its underlying support. The support is updated and maintained by the *Believe* methods as well as the mental actions BELIEVE and DISBELIEVE (together they form the TMS). The effectory procedures for BELIEVE and DISBELIEVE are implemented as belief revision procedures. We have found that such an integrated TMS facility simplifies several action and plan representations (see [14] for details). The *Intend* method is used to specify the fulfillment of agent's intentions by performing acts. All these methods can be specified (and specialized) for the hierarchy as well as inherited. Thus, domain specific acts (physical acts) will inherit the standard method for the agent to accomplish its intentions (i.e. the specific theory of intentionality employed), where as specializations of the *Intend* method can be defined for mental and control acts (to implement the semantics of respective acts).

Thus, an object-oriented design not only provides a uniform representational formalism, it also facilitates an extendible ontology. The semantics of representations is described by reasoning and acting methods that can be either individually specified or inherited and further specialized, as the case may be. Further, we would also like to claim that the representational formalism is 'canonical' in that its user interface (which is mainly defined via 'print methods') is also extendible. For example, the same object (say, a belief proposition) can be displayed as a frame, a predicate, a semantic network, or some other communicational entity (*ala* KIF) (see [9]). At present, these ideas are implemented using SNePS (for *S*emantic *N*etwork *P*rocessing *S*ystem) [24, 23] —an intensional, propositional, semantic network system used for modeling cognitive agents. SNePS-based cognitive agents have network representations for individuals, propositions, deduction rules, actions, acts, and plans. Acting and reasoning about beliefs, actions, and plans, is performed by a single component, SNeRE— the *SNe*PS *R*ational *E*ngine.

3 Related Work

Our use of the term 'BDI Architectures' comes from Georgeff who mentions the challenges of designing rational agents capable of goal-directed as well as reactive behavior based on the attitudes of beliefs, desires, and intentions [6]. Georgeff specifically mentions that, 'the problem that then arises is specifying properties we expect of these attitudes, the way they interrelate, and the ways they determine rational behavior in a situated agent.' As explained in Section 1, we have taken the task of designing BDI architectures by defining a unified intensional representational formalism; identifying the semantic interrelationships between beliefs, desires, and intentions; capturing these into the idea of transformers; and finally designing a rational engine that brings about rational behavior based on these entities.

There has been work describing formal BDI models [3, 18]. There are also architectures that have been proposed that address various issues relating to rational agency. For instance [2, 17] describes a high-level BDI architecture that specifically focuses on issues relating to resource boundedness of rational agent behavior. Their work explores the hypothesis that plans, once committed, in addition to guiding the agent's actions, also constrain the agent's reasoning behavior. Rao and Georgeff have also studied formally the nature of intention and commitment in the context of rational agent behavior [18, 19]. The architecture reported in [19] provides a very simplistic representation of beliefs (thus suffering from some of the concerns mentioned in Section 1) together with a transition network-like formalism for plans. It is a (limited, though successful) attempt towards bridging the their earlier work on PRS [5, 7] and their later work on formal foundations of rational agents [18]. The work presented here complements these models. It provides a general representational framework which these models lack. At the same time, it can facilitate easy incorporation of their ideas by virtue of the extendibility of the design.

We have taken a unified approach to representations. Drummond expresses the need for a single unified formalism for representing beliefs, acts, and plans [4]. This facilitates a single reasoning module to be able to reason about beliefs, acts, and plans. We have taken this approach a step further by explicitly identifying the semantic relationship between inference and acting so that a single module, a rational engine, in addition to reasoning, is also responsible for carrying out physical acts and plans (see [15] for examples). In our formalism, act representations are different from standard operator-based representations of classical planning/acting systems. Elsewhere, we have also shown how even simple act representations can benefit from an integrated TMS [14]. In the presence of a TMS even the simplest acting model (that of adding and deleting the act's effects) implements the extended STRIPS assumption. As a result, ours is a deductive approach to acting. While this leads to tractability concerns, we feel that it provides consistency in the modeled agent's belief space and forms the basis for rational behavior. This also facilitates a deductive approach to hierarchical plan decomposition (specific PlanAct and PlanGoal propositions can be deduced in order to find plan decompositions). Search during reasoning/acting/plan decomposition is focused by means of some KR principles, the Uniqueness Principle being one (there is a one-to-one correspondence between instances and intensional entities) [10]. The Uniqueness Principle helps focus the chaining (method/message propagation) through a restricted set of entities.

The object-oriented approach provides a promising approach to building BDI architectures. It can be used to implement a unified representational formalism that bridges the gap between classical approaches to representation/acting/planning and the emerging paradigms for designing and implementing integrated intelligent architectures.

References

[1] James Allen. The RHET System. In Charles Rich (Guest Editor), editor, *SIGART BULLETIN Special Issue on Implemented KRR Systems*, pages 1–7, June 1991.

[2] Michael E. Bratman, David J. Israel, and Martha E. Pollack. Plans and Resource-Bounded Practical Reasoning. *Computational Intelligence*, 4(4), 1988.

[3] P. R. Cohen and H. J. Levesque. Intention is choice with commitment. *Artificial Intelligence*, 42(3), 1990.

[4] Mark E. Drummond. A representation of action and belief for automatic planning systems. In Michael P. Georgeff and Amy L. Lansky, editors, *Reasoning about Actions and Plans - Proceedings of the 1986 Workshop*, pages 189–212, Los Altos, CA, 1987. AAAI and CSLI, Morgan Kauffmann.

[5] M. P. Georgeff, A. Lansky, and P. Bessiere. A procedural logic. In *Proceedings of the 9th IJCAI*, 1985.

[6] Michael P. Georgeff. Planning. In *Annual Reviews of Computer Science Volume 2*, pages 359–400. Annual Reviews Inc., Palo Alto, CA, 1987.

[7] Michael. P. Georgeff and Amy. Lansky. Procedural knowledge. Technical Note 411, AI Center, SRI International, 1987.

[8] Deepak Kumar. An AI Architecture Based on Message Passing. In James Geller, editor, *Proceedings of The 1993 AAAI Spring Symposium on Innovative Applications of Massively Parallel Architectures*. AAAI Press, March 1993.

[9] Deepak Kumar. *From Beliefs and Goals to Intentions and Actions— An Amalgamated Model of Acting and Inference*. PhD thesis, State University of New York at Buffalo, 1993.

[10] Deepak Kumar. A unified model of acting and inference. In *Proceedings of the Twenty-Sixth Hawaii International Conference on System Sciences*. IEEE Computer Society Press, Los Alamitos, CA, 1993.

[11] Deepak Kumar. The SNePS BDI architecture. *Journal of Decision Support Systems— Special Issue on Logic Modeling*, 1994. Forthcoming.

[12] Deepak Kumar, Susan Haller, and Syed S. Ali. Towards a Unified AI Formalism. In *Proceedings of the Twenty-Seventh Hawaii International Conference on System Sciences*. IEEE Computer Society Press, Los Alamitos, CA, 1994.

[13] Deepak Kumar and Stuart C. Shapiro. Architecture of an intelligent agent in SNePS. *SIGART Bulletin*, 2(4):89–92, August 1991.

[14] Deepak Kumar and Stuart C. Shapiro. Deductive efficiency, belief revision and acting. *Journal of Experimental and Theoretical Artificial Intelligence (JETAI)*, 5(2), 1993.

[15] Deepak Kumar and Stuart C. Shapiro. Acting in Service of Inference (and *vice versa*). In Douglas D. Dankel II, editor, *Proceedings of The Seventh Florida AI Research Symposium (FLAIRS 93)*. The Florida AI Research Society, May 1994.

[16] Douglas B. Lenat and Ramanathan V. Guha. The Evolution of CYCL, The CYC Representation Language. In Charles Rich (Guest Editor), editor, *SIGART BULLETIN Special Issue on Implemented KRR Systems*, pages 84–87, June 1991.

[17] Martha E. Pollack. Overloading Intentions for Efficient Practical Reasoning. *Noûs*, XXV(4):513–536, September 1991.

[18] Anand S. Rao and Michael P. Georgeff. Modeling Rational Agents within a BDI-Architecture. In *Principles of Knowledge Representation and Reasoning— Proceedings of the Second International Conference(KR91*, pages 473–485. AAAI, IJCAI, CSCSI, April 1991.

[19] Anand S. Rao and Michael P. Georgeff. An Abstract Architecture for Rational Agents. In Bernhard Nebel, Charles Rich, and William Swartout, editors, *Proceedings of the 2nd Conference on Principles of Knowledge Representation and Reasoning*, pages 439–449, San Mateo, CA, 1992. Morgan Kaufmann Publishers.

[20] Charles Rich. CAKE: An Implemented Hybrid KR and Limited Reasoning System. In Charles Rich (Guest Editor), editor, *SIGART BULLETIN Special Issue on Implemented KRR Systems*, pages 120–127, June 1991.

[21] Charles Rich. Special Issue on Implemented Knowledge Representation and Reasoning Systems—Letter from the Guest Editor. *SIGART Bulletin*, 2(3), June 1991.

[22] Earl D. Sacerdoti. *A Structure for Plans and Behavior.* Elsevier North Holland, New York, NY, 1977.

[23] S. C. Shapiro and The SNePS Implementation Group. *SNePS-2 User's Manual.* Department of Computer Science, SUNY at Buffalo, 1989.

[24] S. C. Shapiro and W. J. Rapaport. SNePS considered as a fully intensional propositional semantic network. In N. Cercone and G. McCalla, editors, *The Knowledge Frontier*, pages 263–315. Springer–Verlag, New York, 1987.

[25] Stuart C. Shapiro. Case studies of SNePS. *SIGART Bulletin*, 2(3):128–134, June 1991.

[26] David E. Wilkins. *Practical Planning-Extending the Classical AI Planning Paradigm.* Morgan Kaufmann, Palo Alto, CA, 1988.

Evolving the Size of Rule-Based Fuzzy Systems

Mark G. Cooper & Jacques J. Vidal

University of California, Los Angeles
4531 Boelter Hall
Los Angeles, California 90024

Abstract. Our research has focused on the development of a methodology which employs a genetic algorithm to automatically generate fuzzy process controllers. Unique to this approach is that each controller is represented as an unordered list of an arbitrary number of rules. This paper focuses on a critical aspect of this methodology: the emergence of an appropriately-sized rule base as a result of the genetic search. Specifically, we establish the efficacy of specialized "add rule" and "delete rule" mutation operators, and propose a mechanism by which rule base size evolves.

Introduction

We have recently introduced a novel method for evolving rule-based fuzzy controllers (Cooper & Vidal, 1993; Cooper & Vidal, 1994; Cooper 1994). This approach differs from prior genetic-fuzzy techniques in that each controller is represented as an unordered sequence of an arbitrary number of fuzzy rules. This produces a more parsimonious representation of the underlying fuzzy structure than prior genetic efforts in which lists of parameters evolve under a fixed set of membership functions, or tables of output values evolve to a fixed set of physical system configurations. In addition, while an appropriate action may be strongly conditioned by a subset of the control variables, techniques based upon solution space partitioning require that every output be dependent upon every input. To overcome this limitation, we devised a mechanism through which the consequent of a rule may be calculated solely from relevant variables. We have successfully applied our method to several nonlinear control problems, such as the benchmark cart-pole problem, boat steering, and aircraft landing.

In this paper we focus on an essential feature of the algorithm—the emergence of an appropriately-sized rule base as a result of the genetic search—to determine the mechanism by which rule base size evolves. We continue this first section by briefly introducing fuzzy controllers and genetic algorithms. In Section 2 we present our methodology for automatically generating rule-based fuzzy controllers. In Section 3, we describe the series of experiments designed to illuminate the sizing mechanism. Finally, in Section 4 we discuss the implications for generating fuzzy rule bases and present our conclusions regarding the sizing mechanism.

FUZZY CONTROL

A fuzzy controller consists of a number of rules (the *rule base*) generally expressed in the form "If a_1 is u_1 and a_2 is u_2 ... and a_n is u_n then c is v" where a_n correspond to the control variables, c is the output variable, and u_n and v are descriptors such as "large positive," "small negative," "near zero," etc. Associated with each descriptor is a *membership function* which specifies the degree to which a given input satisfies the descriptor.

Fuzzy rules are evaluated and combined according to the Fuzzy Associative Memory Model described by Kosko (1992). Instantiations of the input variables are presented to the rule base in parallel, with the membership function for each antecedent clause applied to its corresponding input to produce a *fit value* for the clause. If a rule has more than one antecedent,

E. A. Yfantis (ed.), Intelligent Systems, 319–326.

the minimum fit value is taken as the fit value for the entire rule. This process determines the applicability of a rule. Next, the consequent of each rule is calculated as the region under its membership function below the antecedent fit value. This collection of consequent regions constitutes the output of the fuzzy system; however, in most cases a single value (i.e. a control action) is required. The combination of consequent regions to produce a crisp output value is called defuzzification, and is generally accomplished by overlaying the regions and computing the centroid.

GENETIC ALGORITHMS

Genetic algorithms refer to a class of probabilistic search techniques that emulate the mechanics of evolution (Goldberg, 1989). They are capable of globally exploring a solution space; pursuing potentially fruitful paths while simultaneously examining random points to reduce the likelihood of settling for a local optimum. Recently, genetic algorithms have been applied to the automatic generation and tuning of fuzzy rule bases.

The evolving system must first be converted into a string of bits which encodes the information necessary to reconstruct it. Initially, a random set, or *population*, of strings is generated. Each string is then evaluated according to a given performance criterion and assigned a *fitness score*. The strings with the best scores are used in the reproduction phase to produce the next generation. The cycle of evaluation and reproduction continues for a predetermined number of generations or until an acceptable performance level is achieved.

The reproduction of a pair of strings proceeds by copying bits from one parent string until a randomly-triggered *crossover* point, after which bits are copied from the other. As each bit is copied, the possibility exists that the bit will be inverted, or *mutated*. While crossover and mutation comprise the operator set used in nearly all genetic algorithms, in our method two additional mutation operators for randomly adding and deleting rules are included. These are demonstrated to be factors in evolving rule base size.

The Compact Rule Base Approach

In this section we summarize our methodology for evolving rule-based fuzzy controllers. A more detailed description of the algorithm appears in Cooper (1994). The selection and encoding of the rule base is the central issue in evolving fuzzy controllers. Sufficient information must be encoded in the string to reconstruct the fuzzy system, and the representation must be amenable to evaluation, and within the genetic framework, reproduction.

The membership function for each variable in a fuzzy system is chosen as a triangle characterized by the location of its center and the half-length of its base. A single fuzzy rule, therefore, consists of the concatenation of one-byte unsigned characters (assuming values from 0 to 255) which specify the centers and half-lengths of the membership functions for each input and output variable (Figure 1). The rule descriptions for a single fuzzy system are concatenated into a single bit string where the number of rules is not restricted. Although each position within a rule has specific semantics, the rules themselves are unordered.

One drawback of prior attempts at fuzzy system learning is the assumption that each rule is dependent upon combinations of all of the input variables. Without the ability to ignore irrelevant variables, rules must evolve which account for every combination of spurious variables with each required value for the relevant ones. We consider this situation in our algorithm by ignoring a variable when its half-length falls outside of a specified range ([40, 215] out of an overall variable range of [0, 255] in our experiments).

The most significant problem in genetically combining strings consisting of unordered lists of rules is that while two rule bases may share several specific rules, these rules may appear in different orders. To be meaningful, the genetic paradigm requires that the rules in two reproducing strings be aligned so that similar rules are combined with each other. Therefore, before reproduction, substrings must be shifted to bring similar rules in registration. This is accomplished by aligning the rules according to the similarity of their antecedents—those most closely matching are aligned first, followed by the best match among those that remain, and so on (Figure 2). Any remaining rules are added to the end of the string. While this matching scheme is suboptimal, it requires considerably less computation than the optimal solution, and has been found in most cases to perform nearly as well. The remainder of the genetic algorithm proceeds normally—a random initial population is generated, each system is evaluated with respect to its performance on a problem, and the strings that perform the best are selected for reproduction to produce the next generation.

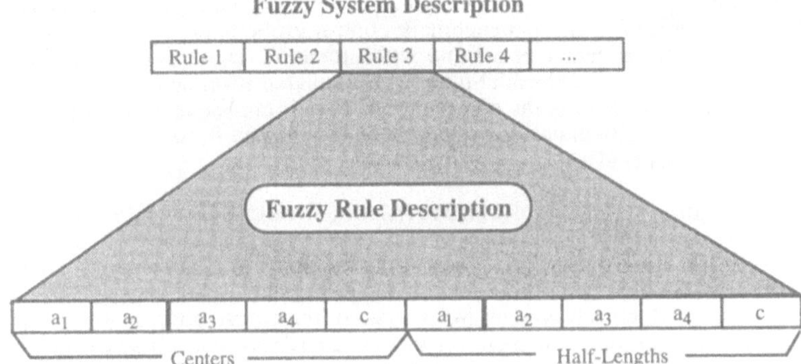

Figure 1. Organization of the Fuzzy System Population

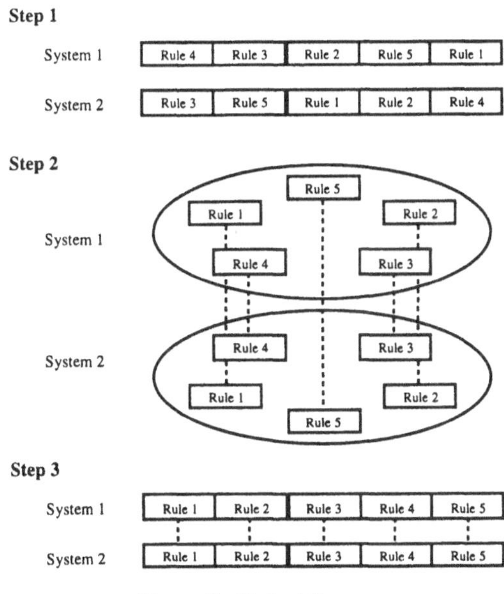

Figure 2. Rule Alignment

Evolving Rule Base Size

In this section we present three experiments whose combined goal is to suggest a mechanism by which our method evolves fuzzy rule base size. In the first experiment, we observed the effect of add and delete rule mutations on controller quality and rule base size. In the second, we tracked the distribution of rule base sizes throughout evolution to determine whether or not they tend toward homogeneity. Finally, we examined individual add and delete-rule mutations to characterize their effects on generated offspring.

All of these experiments were performed by evolving controllers for the single-pole cart-pole system. The objective of this benchmark control problem is to specify translational forces that position a cart at the center of a finite width track while simultaneously balancing a pole hinged on the cart's top. The physical plant was simulated using a set of well known nonlinear differential equations describing the cart and pole dynamics (Wieland, 1991). The applicability of our method to this problem has previously been established in Cooper & Vidal (1993), and is expanded upon in Cooper (1994).

THE EFFECT OF ADD AND DELETE RULE MUTATIONS ON RULE BASE QUALITY AND SIZE

In the first experiment we determined whether the inclusion of add and delete rule mutations has an effect on controller quality and rule base size. Two variables were considered—the number of rules initially assigned to each random fuzzy system (three possibilities—too few, adequate, and too many) and the set of mutation operators that were enabled (four possibilities—add and delete, add only, delete only, and neither). Each of the twelve combinations was applied to the evolution of four rule bases, with each trial characterized by the size of the highest-scoring rule base that it produced. The results of these trials, presented in Table 1, suggest that the evolutionary process takes advantage of these mutations to produce more "appropriately" sized rule bases.

Table 1. Sizes of the Rule Bases Evolved

Small Initial Rule Base (initial population size = 3)				
Include Both	4	4	11	4
Add Only	4	7	6	3
Delete Only	3	3	3	3
Neither	3	3	3	3
Intermediate Initial Rule Base (initial population size = 20)				
Include Both	6	9	5	8
Add Only	25	24	25	29
Delete Only	7	7	7	6
Neither	20	20	20	20
Large Initial Rule Base (initial population size = 40)				
Include Both	7	16	8	22
Add Only	45	42	45	44
Delete Only	8	8	11	10
Neither	40	40	40	40

To confirm this hypothesis, we performed a series of two-way Analyses of Variance (ANOVAs) on this set of values, and found an interaction between the rule base size and the included mutation set ($F(6, 36) = 50.09$, $p < 0.05$), indicating that they influence each other with respect to rule base size. In addition, we found significant main effects for both the size of the rule base ($F(2, 36) = 307.74$, $p < 0.05$) and the set of mutation operators included ($F(3, 36) = 152.3$, $p < 0.05$). Therefore, the effect that the initial rule base size has on the number of rules in the evolved rule base depends upon the mutation set enabled. When there are too few or too many rules, the add and delete rule mutations attempt to augment or diminish the rule base respectively. Hence, we can conclude that these mutations do affect the sizes of the evolved rule bases.

It is interesting to note, however, that the scores of the evolved controllers were similar regardless of the set of mutations enabled (Table 2). We examined the effects of the size of the initial rule base and the set of included mutation operators with respect to score in a two-way ANOVA and found no interaction between them ($F(6, 36) = 0.80$, n.s.). Similarly, no main effect was found for the included mutation set ($F(3, 36) = 0.40$, n.s.). In other words, overall system performance did not depend upon the set of mutations included. This result is not surprising in light of our prior research where we were able to generate successful controllers having as few as three and as many as fifty rules.

Table 2. Scores of the Best-Performing Systems

Small Initial Rule Base (initial population size = 3)				
Include Both	127.1	117.5	158.1	138.4
Add Only	133.3	156.3	133.6	40.0
Delete Only	137.2	80.0	140.6	24.9
Neither	120.3	76.8	107.1	139.2
Intermediate Initial Rule Base (initial population size = 20)				
Include Both	160.3	163.1	120.1	159.9
Add Only	161.8	163.4	163.2	162.4
Delete Only	159.2	161.3	164.1	153.0
Neither	160.4	162.5	161.1	161.5
Large Initial Rule Base (initial population size = 40)				
Include Both	164.1	163.7	163.2	159.2
Add Only	164.3	165.3	164.8	163.6
Delete Only	163.1	165.0	159.6	164.5
Neither	164.3	164.9	164.7	163.0

However, we did find a significant main effect for the rule base size ($F(2, 36) = 19.75$, $p < 0.05$), where systems initialized with a small number of rules performed considerably worse overall than the other two groups. This indicates that the add rule mutation is less effective at augmenting small rule bases than the delete rule mutation is at diminishing large ones. However, initializing with 40 rules did not significantly improve performance over initializing with 20 rules.

Rule Base Size Convergence

In the second experiment, we tracked the distribution of rule base sizes both from the entire population and from the subset of the population chosen for reproduction to determine whether population size becomes homogeneous during evolution, and whether variations in rule base size are selected against. We hypothesized that a core of highly-scoring systems is perpetuated throughout the genetic search, and that successful systems are typically variations on these systems. If this is true, then we would expect that systems that differ significantly in size from this core (as a result of add and delete rule mutations) would reproduce very infrequently.

The density graphs in Figure 3 plot the relative number of controllers having a particular rule base size as a function of generation. It is clear that the systems chosen for reproduction were drawn from the entire range of rule base sizes. This experiment also demonstrates the beneficial effect of the add-rule mutation on a population. While each first-generation system was initialized with between three and thirty rules, many successful, reproducing systems had more than thirty rules. In fact, the best-scoring system in the second trial had thirty-seven rules. These results indicate that the populations do not become homogeneous with respect to rule base size. Rather, it appears that sub-populations of controllers with varying sizes evolve in parallel.

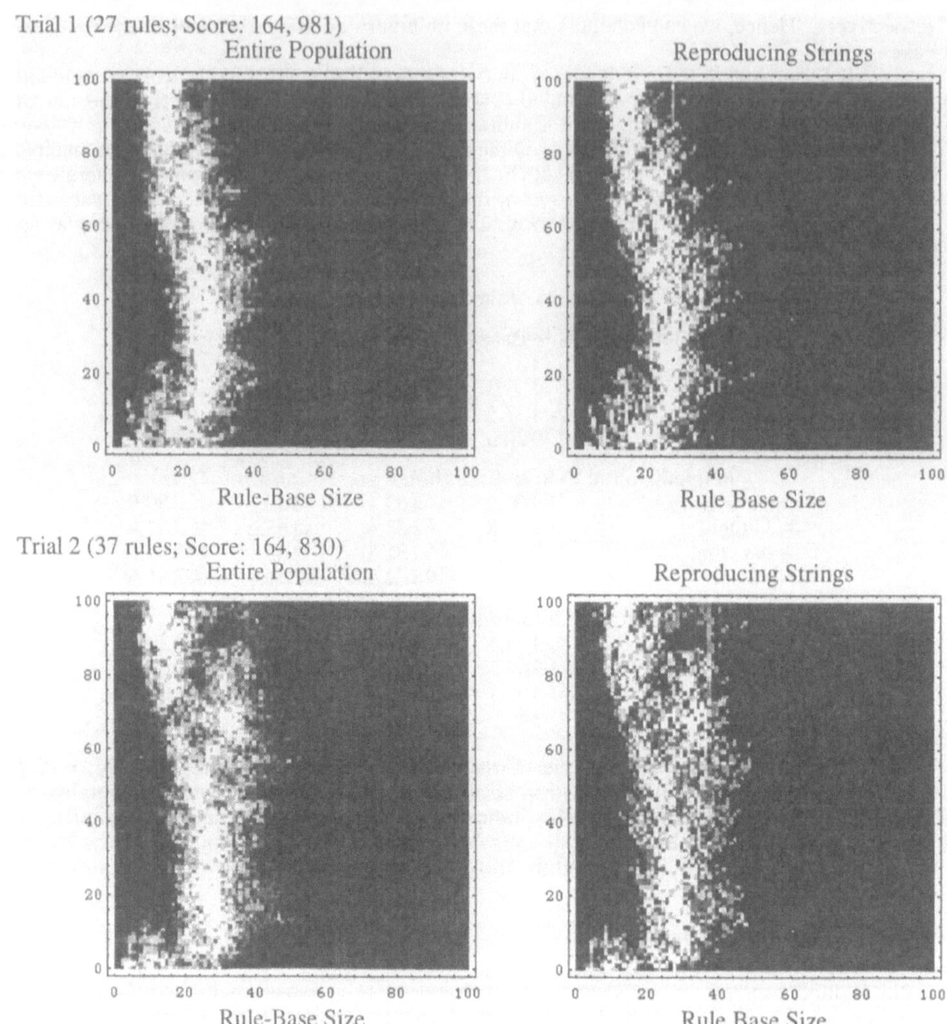

Figure 3. Relative Number of Controllers of Each Rule Base Size as a Function of Generation

THE EFFICACY OF INDIVIDUAL ADD AND DELETE RULE MUTATIONS

The final experiment examined the effect of adding and deleting rules on individual controllers. A population of random controllers initialized with between three and thirty rules was generated and evolved through fifty generations. As each new rule base was created, the number of additions and deletions triggered was recorded, along with the scores of the resulting systems relative to their parents and peers. We expected that these mutations would usually degrade rule base performance, and only occasionally improve it.

We found that in the first few generations, the add and delete rule mutations had a significant beneficial effect. Nearly all reproducing systems exhibited at least one such mutation.

Surprisingly, there was no discernible evolutionary preference for either mutation. At about the tenth generation, the number of reproducing controllers exhibiting these mutations diminished, and after the twenty-third generation, no such system reproduced. These observations support the conclusions of the previous experiment—the sub-populations converged to a point where adding or deleting rules (or mating between sub-populations) disrupted the interactions between the existing rules, degrading performance such that these controllers were not reproduced.

A similar pattern emerged when comparing the performance of systems that were produced using the add and delete mutations with that of siblings that were not. In the first few generations, the controllers produced using the mutations performed better and reproduced more frequently, but after the tenth generation the siblings having no such mutations dominated. Thus, from this experiment we can conclude that the add and delete rule mutations have their most significant effect early in evolution when the population is broadly searching for an adequate solution, and has a lesser effect once the population has converged.

Conclusions

The results of this investigation suggest several conclusions regarding the add and delete mutations and rule base size:

The delete-rule mutation has a more profound effect on rule base size than does the add-rule mutation. Fuzzy controllers initialized with a large number of rules made liberal use of the delete-rule mutation to reduce the rule base size. When the rule bases were initialized with a large number of rules and the delete-rule mutation was enabled, the resulting best-scoring rule base in all cases had many fewer rules than in the original population.

In general, the add-rule mutation alone is not sufficient to build rule bases. Even with the add-rule mutation enabled, populations initialized with very small rule bases could not consistently evolve successful controllers. This is because the probability that adding a random rule will actually improve the performance of an existing rule base is small. We looked at several additional systems initialized with only three rules that evolved through one-thousand generations. We found that the performance of these rule bases was generally better than that of those evolved through only one hundred generations, but that the rule base sizes remained roughly the same. It appears that it is easier to delete superfluous rules than to build effective new rules.

The add-rule mutation does have a positive effect on the evolution of rule base size. The presence of the add-rule mutation permits flexibility to the evolutionary process. As evident in Figure 3, rule base sizes *can* increase, with the best-scoring system often having more rules than any initial system. This is because it is much more likely that a rule added randomly to a large rule base will produce an incremental improvement since its effect relative to the rest of the rule base will be small. However, a random rule added to a small rule base will have a much greater (typically negative) effect.

For practical purposes, it is preferable to start with a rule base that is too large than too small. Since population size increases only linearly with rule base size (as opposed to exponentially in several prior genetic-fuzzy approaches), it is advantageous to allow the genetic search to reduce the size of the rule bases rather than attempt to build them up. Additionally, a large rule base can be consolidated after evolution using non-genetic techniques (Song, et.al., 1993).

These results also suggest a mechanism by which rule base sizes may evolve: *As long as there are enough rules to effectively control the problem, the rule bases will settle into sub-populations of controllers of "acceptable" size. Continued training may then improve performance within each sub-population, but usually does not change the size of the rule base.*

As training progresses, sub-populations converged upon rule bases of various size. The genetic search then essentially becomes an optimization over rule bases of each size, with neither the add or delete rule mutations improving the performance of the system. While incremental improvement typically continues throughout evolution, adding or deleting a rule has a much greater impact on the performance of a rule base and is much less likely to improve it than the small change effected by a bit mutation.

References

Cooper, M.G. (1994) *Genetic Design of Rule-Based Fuzzy Controllers.* Ph.D. Dissertation. Department of Computer Science, University of California, Los Angeles.

Cooper, M.G. & J.J. Vidal (1993) 'Genetic Design of Fuzzy Controllers.' Presented at the Second International Conference on Fuzzy Theory and Technology; Durham, NC.

Cooper, M.G. & J.J. Vidal (1994) 'Genetic Design of Fuzzy Controllers: The Cart and Jointed Pole Problem.' Presented at the Third IEEE International Conference on Fuzzy Systems; Orlando, FL.

Goldberg, D.E. *Genetic Algorithms in Search, Optimization, and Machine Learning.* Reading, MA: Addison-Wesley.

Kosko, B. (1992) *Neural Networks and Fuzzy Systems: A Dynamical Systems Approach to Machine Intelligence.* Englewood Cliffs, NJ: Prentice-Hall, Inc.

Song, B.G., R.J. Marks II, S. Oh, P. Arabshahi, T.P. Caudell, & J.J. Choi (1993) 'Adaptive Membership Function Fusion and Annihilation in Fuzzy If-Then Rules.' in *Proceedings of the Second IEEE International Conference on Fuzzy Systems.* New York, NY: Institute of Electrical and Electronics Engineers. pp 961-967.

Wieland, A.P. 'Evolving Controls for Unstable Systems,' in S. Touretzky et.al. (ed.), *Connectionist Models: Proceedings of the 1990 Summer School.* San Mateo, CA: Morgan Kaufmann Publishers, Inc. pp 91-102.

ANALOG: A Knowledge Representation System for Natural Language Processing

Syed S. Ali
Department of Computer Science
Southwest Missouri State University
901 South National Avenue
Springfield, MO 65804
ssa231f@cnas.smsu.edu

Background

The general task of work in knowledge representation and reasoning (KRR) is, trivially, the representation of knowledge. The domain for which the representation task has been undertaken is, often, secondary to the description of the overall task as an exercise in knowledge representation. Thus, the task of representing and using mathematical, common sense, visual, language-based, and logical knowledge is "lumped" together under the rubric of KRR (for examples of the diversity of goals of KRR see [12]). I will argue that in natural language processing and understanding there is a is a strong argument for the language structure and use mediating any attempt at representation. The task of knowledge representation for natural language processing and understanding is a knowledge-intensive one. To understand a natural language sentence a NLP system must, minimally, be able to represent the content of the sentence in the language of representation. These representation languages have been largely unmotivated with respect to the natural language they may be representing. In [12] only six of the twenty-two KRR systems presented are driven by natural language processing concerns.

Logic

Logic is a popular choice of representation languages because of is prefabricated syntax, and model-thoretic semantics. However, the syntax and semantics of first-order predicate logic is most definitely not that of natural language. I will argue that if the domain of representation is natural language, then the (minimally, syntactic) form of the natural language should strongly influence the representation.

If we contrast the various goals of researchers in the domains for which these kinds of knowledge are to be used, we may come to the conclusion that the goals differ significantly and, indeed, may conflict. Consider the traditional use of representations based on first-order logic. Here, we have a system that has a powerful, well-understood, inferential machinery but weak expressive power. Collections of logical formulas do not seem to capture the intuitive use of concepts by people. This representation for knowledge is unstructured

E. A. Yfantis (ed.), Intelligent Systems, 327–332.

and disorganized. What's missing is that first-order predicate does not provide any special assistance in the problem of what Brachman called "knowledge structuring" [6]. That is, the specification of the internal structure of concepts in terms of roles and interrelations between them and the inheritance relationships between concepts Attempts to incorporate knowledge-structuring into the representation language are typified by the use of frames, or frame description languages (FDLs). These are structured slot-filler structures representing concepts that stand in various relationships (i.e., taxonomic) to each other. However, this shift into slot-filler structures where the expressive power is greater but the inferential machinery is underspecified, perhaps, goes too far. The cost of knowledge-structuring is a weaking of the inferential machinery available.

Logic is a traditional representational medium for attempts to understand natural language. As a formal language logic has the desirable property that expressions of a formal language can be paired with interpretations. Thus a "correct" mapping from natural language sentences into logic allows the original sentences to be interpreted. However, the mapping from natural language structures into the standard syntax of logic (and from logic to natural language) is unnatural in that the structure of the natural language (which can be significant) is usually lost. This is typified by the "de-structuring" of noun phrases that occurs when it is translated into logic. This is a consequence of the atomic nature of variables in logic and results in the separation of constraints on variables from the variables themselves. This most often occurs when these constraints are moved to the antecedents of rules that type the variables. This results in the loss of the simple predicate-argument structure of the original sentence being represented.

Goals

This work is based on the assumption that natural language determines the design of a knowledge representation and reasoning (KRR) system. In particular, I present a KRR system for the representation of knowledge associated with natural language dialog. I will argue that this may done with minimal loss of inferential power and will result in an enriched representation language capable of supporting complex natural language descriptions, support for some discourse phenomena, standard first-order inference, inheritance, and terminological subsumption. Some of the goals of a more natural logic and its computational implementation in a knowledge representation and reasoning system are discussed below.

It should possess as much of the machinery of traditional FOPL as possible. It is not an accident that logical form representations are popular; it is a consequence of their power and generality.

The mapping from natural language sentences into logical form sentences should be as direct as possible, and the representation should reflect the structure of the natural language (NL) sentence it purports to represent. This is particularly well illustrated by rule-type sentences, such as "small dogs bite harder than big dogs," where their representation takes the form of an implication whose antecedent constraints specify what types of dog bite harder than another type. This representation, as a logical rule, contrasts with the predicate-argument structure of the original sentence, as below:

$$\forall x, y((\text{small}(x) \wedge \text{dog}(x) \wedge \text{large}(y) \wedge \text{dog}(y)) \\ \Rightarrow \text{bites-harder}(x, y))$$

By comparison, the representation of *Fido bites harder than Rover* is more consistent with the original sentence,

bites-harder(Fido, Rover)

This is so, despite the intutive observation that the two sentences have nearly identical syntactic structure, and similar meaning.

The subunits of the representation should be conceptually complete in the sense that any component of the representation of a sentence should have a meaningful interpretation independent of the interpretation of the entire sentence representation. For example, the sentence "dogs bite" is typically translated as:

$$\forall x[\text{dog}(x) \Rightarrow \text{bites}(x)].$$

With this translation, we might ask what is the meaning of x? Presumably, some thing in the world, or a set denoting the non-empty universe. Note that the original sentence mentions only dogs. We suggest that a better translation might be:

$$\text{bites}(\forall\, x \text{ such that } \text{dog}(x))$$

where the variable, x, has its own internal structure that reflects its conceptualization. Note that I am suggesting something stronger than just restricted quantification (the above could also be represented as $(\forall x : \text{dog}(x))\,[\text{bites}(x)]$ using restricted quantifiers). Complex internalized constraints (that is, other than simple type) and internalized quantifier structures characterize this proposal for the representation of variables. Thus representation of the sentence: *Every big dog that is owned by a bad-tempered person bites* should mimic the representation of *dogs bite.*

A high degree of structure sharing should be possible, as multi-sentence connected discourse often uses reduced forms of previously used terms in subsequent reference to those terms. This corresponds to the use of pronouns and and some forms of ellipsis in discourse. An example of this phenomena is the representation of intersentential pronominal reference to scoped terms, e.g.,

Every apartment had *a dishwasher.* In some of them *it* had just been installed.
Every chess set comes with *a spare pawn. It* is taped to the top of the box.

(examples from [10]). The structures that are being shared in these sentences are the variables corresponding to the italicized noun phrases. Logical representations can only model this "sharing" by combining multiple sentences of natural language into one sentence of logic. This method unnatural for at least two reasons. Firstly, when several sentences must be so combined into one sentence the resulting logical sentence is overly complex as a conjunction of several potentially disparate sentences. Secondly, this approach is counter-intuitive in that a language user can re-articulate the original sentences that he/she represents which argues for some form of separate representations of the original sentences. The problem with logic in this task is that logic requires the complete specification of a variable, corresponding to a noun phrase, and its constraints in the scope of some quantifier. This difficulty is not restricted to noun phrases, indeed it is frequently the case that entire subclauses of sentences are referred to using reduced forms such as "too" e.g.,

John *went to the party.* Mary did, *too.*

A language-motivated knowledge representation formalism should model this sort of reference, minimally by structure sharing.

Any computational theory must incorporate knowledge-structuring mechanisms. In particular, subsumption and inheritance of the sort supported in frame-based and semantic network based systems. A taxonomy provides "links" that relate more general concepts to more specific concepts. This allows information about more specific concepts to be associated with their most general concept, and information filters down to more specific concepts in the taxonomy via inheritance. More general concepts in such a taxonomy *subsume* more specific concepts, the subsumee inheriting information from its subsumers. For atomic concepts, subsumption relations between concepts is specified by the links of the taxonomy. A clear example of subsumption in natural language is the use of descriptions such as *person that has children* subsuming *person that has a son.* If one were told: *People that have children are happy,* then it follows that *People that have a son are happy.* The intuitive idea is that more general descriptions should subsume more specific descriptions of the same sort, which in turn inherit attributes from their more general subsumers.

ANALOG

I have presented some general arguments for considering the use of a more "natural" (with respect to language) logic for the repesentation of natural language sentences. I have also presented some characteristics of natural language that a knowledge representation and reasoning system should support. In the full paper I will clarify the motivations for this work with specific examples, present an alternative representation for simple unstructured variables which involves according variables potentially complex internal structure, and reify the logic of these variables. This involves providing a syntax and semantics of the logic of structured variables. The syntax of the logic is specified by a complete definition of a propositional semantic network representation formalism (an augmentation of [14]). The implemented system is called ANALOG (A NAtural LOGic) [4, 5, 2, 1, 3].

The representations that that result from using these structured variables have the following advantage:

- The representation of numerous types of quantifying expressions, using structured variables, is more "natural" than typical logics, because the mapping of natural language sentences is direct. Thus parsing and generation of natural language sentences (particularly those involving restrictive relative clauses) is simplified.

- With structured variables that contain their own binding structures (quantifiers), no open sentences are possible. This is consistent with natural language where few, if any, open sentences occur (minimally, noun phrases are, by definition, typed).

- Sentences involving non-linear quantifier scopings, such as branching quantifiers and the donkey sentences can be represented [11, 9] because quantifier scope specification is internal to the structured variable.

- Subsumption can be defined syntactically it terms of the structure of structured variables. For example, from a rule representing *Every boy loves a girl* all rules involving more restricted boys follow directly by variable subsumption (e. g., *Every boy that owns a red car loves a girl*). This sort of subsumption can be defined without the traditional T-Box and A-Box distinction (as in, for example, [7, 8]) and has a great deal of utility.

- Since the underlying representation is semantic network-based the representation of sentences that share structures (e. g., *Every farmer that owns a donkey beats it. His wife does too.*) is both possible and reasonable.

An example of the capabilities of the ANALOG system for NLP is given in Figure 1. ANALOG includes a generalized augmented transition network (GATN) natural language parser and generation component linked up to the knowledge base (based on [13]). A GATN grammar specifies the translation/generation of sentences involving complex noun phrases into/from ANALOG structured variable representations.

```
: Every man is mortal
I understand that every man is mortal.
: Who is mortal
Every man is mortal.
: Is any rich man mortal
Yes, every rich man is mortal.
: John is a man
I understand that John is a man.
: Is John mortal
Yes, John is mortal.
: Who is mortal
John is mortal and every rich man is mortal and every man
is mortal.
: Are all rich young men that own some car mortal
Yes, every young rich man that owns some car is mortal.
: Any rich young man that owns any car is happy
I understand that every young rich man that owns any car is
happy.
: Is John happy
I don't know.
: Young rich John owns a car
I understand that mortal rich young John owns some car.
: Who owns a car
Mortal rich young John owns some car.
: Is John happy
Yes, mortal rich young John is happy.
```

Figure 1: Example of NLP in ANALOG.

References

[1] Syed S. Ali. A Propositional Semantic Network with Structured Variables for Natural Language Processing. In *Proceedings of the Sixth Australian Joint Conference on Artificial Intelligence.*, November 17-19 1993.

[2] Syed S. Ali. A Structured Representation for Noun Phrases and Anaphora. In *Proceedings of the Fifteenth Annual Conference of the Cognitive Science Society*, pages 197–202, Hillsdale, NJ, June 1993. Lawrence Erlbaum.

[3] Syed S. Ali. Node Subsumption in a Propositional Semantic Network with Structured Variables. In *Proceedings of the Sixth Australian Joint Conference on Artificial Intelligence.*, November 17-19 1993.

[4] Syed S. Ali. *A "Natural Logic" for Natural Language Processing and Knowledge Representation.* PhD thesis, State University of New York at Buffalo, Computer Science, January 1994.

[5] Syed S. Ali and Stuart C. Shapiro. Natural Language Processing Using a Propositional Semantic Network with Structured Variables. *Minds and Machines*, 3(4):421–451, November 1993. Special Issue on Knowledge Representation for Natural Language Processing.

[6] Ronald J. Brachman. On the Epistemological Status of Semantic Networks. In N. V. Findler, editor, *Associative Networks: Representation and Use of Knowledge in Computers.* Academic Press, New York, 1979.

[7] Ronald J. Brachman, Richard E. Fikes, and Hector J. Levesque. KRYPTON: a Functional Approach to Knowledge Representation. *IEEE Computer*, 16(10):67–73, 1983.

[8] Ronald J. Brachman, Victoria Pigman Gilbert, and Hector J. Levesque. An Essential Hybrid Reasoning System: Knowledge and Symbol Level Accounts of KRYPTON. *Proceedings IJCAI-85*, 1:532–539, 1985.

[9] Peter Thomas Geach. *Reference and Generality.* Cornell University Press, Ithaca, New York, 1962.

[10] Irene Heim. Discourse Representation Theory, 1990. Tutorial material from ACL-90.

[11] W. V. Quine. *Philosophy of Logic.* Prentice-Hall, Englewood Cliffs, NJ, 1970.

[12] Charles Rich, editor. *Special Issue on Inplemented Knowledge Representation and Reasoning Systems*, volume 2. ACM Press, June 1991. SIGART Bulletin.

[13] S. C. Shapiro. Generalized augmented transition network grammars for generation from semantic networks. *The American Journal of Computational Linguistics*, 8(1):12–25, 1982.

[14] S. C. Shapiro. Cables, Paths, and "Subconscious" Reasoning in Propositional Semantic Networks. In John F. Sowa, editor, *Principles of Semantic Networks*, pages 137–156. Morgan Kaufmann, 1991.

Conversion of Complete Logic Programs to Double Defined Logic Programs

Faye Liu
Douglas Huntington Moore
Department of Computer Science & Computer Engineering
La Trobe University
Bundoora Vic 3083 Australia

Abstract. The double defined logic programming system has been introduced to overcome the incompleteness and limitations caused by applying the "Negation as Failure" rule and to build a constructive negation in logic programming. A detailed procedure of conversion of complete logic programs to double defined logic programs is introduced in this paper. The conversion is a one-to-one mapping between complete logic programs and double defined logic programs. The proposition that contradictions do not theoretically exist in a double defined logic program is proven.

Keywords. Double defined logic programming, Complete logic programs, Conversion, Contradiction.

1. Introduction

To implement constructive negation in logic programming without destroying the property of monotonic reasoning of Horn clause logic programs, a new logic programming system - double defined logic programming system[3] - has been introduced. A double defined logic program converted from a general logic program is a Horn clause logic program, hence the reasoning on a double defined logic program with SLD-Resolution is sound and complete[1][2]. Meanwhile constructive negation is built in a double defined logic program.

The basic idea of double defined logic programming is to define a logic program from both positive and negative prospectives. The negative subgoals are independent from positive subgoals. This means that negative subgoals are derived directly from the program, but not from failing to prove the positive subgoals.

The concept P-domain[3] has been introduced. Let $\boldsymbol{P}$ be a logic program and $q(X)$ be a predicate defined in $\boldsymbol{P}$, the P-domain of the predicate $q(X)$ is defined as $D[q]=\{s \in U_l \mid P \models q(s)\}$ ($\boldsymbol{U_l}$ is the Herbrand universe of $\boldsymbol{P}$). The P-domain of a predicate is a subset of the Herbrand universe of the program.

The neg-predicate[3] of a predicate is defined as the negation of the predicate. Based on neg-predicates, concepts neg-atoms and neg-clauses have been defined. Negative information exists in a logic program in forms of neg-facts and neg-rules, therefore the fact set and the rule set of the

E. A. Yfantis (ed.), Intelligent Systems, 333–340.

program become a double defined fact set and a double defined rule set. The logic program with a double defined fact set and a double defined rule set is called a double defined logic program.

Since a neg-predicate which exists independently in a logic program is defined as the negation of a predicate, negation is, therefore, deleted from the program. For example, let $\boldsymbol{P}$ = *{ p(a). p(b). t(b). t(c). q(X,Y) :- p(X), not(t(Y)).}* be a general logic program. *p(X), t(X)* and *q(X,Y)* are the predicates defined in $\boldsymbol{P}$. To convert $\boldsymbol{P}$ to the double defined logic program $\boldsymbol{P}_d$, we define neg-predicates *np(X), nt(X)* and *nq(X,Y)* as the negations of the predicates. Neg-predicates *np(X), nt(X)* and *nq(X,Y)* occur in the converted program $\boldsymbol{P}_d$ as independent predicates.

General logic programs were divided into two categories -- Complete logic programs and Incomplete logic programs[4]. A complete logic program is the program in which the P-domain of each predicate equals the union of the P-domains of the premise of the predicate. A complete logic program is equivalent to its completion[4]. Therefore the "Negation as Failure" rule can be applied directly to a complete logic program and the reasoning is sound and complete.

A general logic program which is not complete is called an incomplete logic program. There is a predicate defined in the program, which satisfies the condition that the P-domain of the predicate includes the union of the P-domains of its premise, but not equals. An incomplete logic program is a part of its completion[5]. This means that the program is not equivalent to its completion. "Negation as Failure" can only be applied to the completion of the program, but not the program itself.

The negation implemented by "Negation as Failure" is not constructive. In this paper, we will discuss the procedure of the conversion of a complete logic program to a double defined logic program so that a constructive negation is built in the program.

2. Conversion of Complete Logic Programs to Double Defined Logic Programs

Let $\boldsymbol{P}_g$ be a complete logic program and let $\boldsymbol{F}$ and $\boldsymbol{R}$ be the fact set and rule set of $\boldsymbol{P}_g$. $\boldsymbol{P}_d$ is denoted as the double defined logic program converted from $\boldsymbol{P}_g$. $\boldsymbol{F}_d$ and $\boldsymbol{R}_d$ are the double defined fact set and the double defined rule set of $\boldsymbol{P}_d$. The conversion of a complete logic program to a double defined logic program involves 4 steps.

(1) Defining the neg-predicates and a variable restrictor

Define neg-predicates for all the predicates which occurred in the original program $\boldsymbol{P}_g$.

One of the problems in converting a complete logic program to a double defined logic program is that the converted program may not satisfy weak covering[1]. For example, the neg-rules such as *nq(X,Y) :- p(Y).* and *nt(X,Y,Z) :- nr(X), p(Y), q(X,W)..* To solve the problem, an unary predicate *var_retor(X)* has to be defined as a variable restrictor. *var_retor(X)* is defined as $\forall s \in U_l$ *var_retor(s)* is a clause of the converted program. Whenever a converted rule does not satisfy weak covering, *var_retor(X)* can be added to the body of the rule to restrict the variables without changing the ranges of the predicates in the rule. After adding variable restrictor, the neg-rules above become *nq(X,Y) :- p(Y), var_retor(X).* and *nt(X,Y,Z) :- nr(X), p(Y), q(X, W), var_retor(Z).*

---

1 If every rule in a logic program **P** satisfies the criterion that all the variables which occur in the head of the rule also occur in the body of the rule, then **P** is called a logic program satisfying weak covering.

(2) Converting the fact set $\boldsymbol{F}$ to a double defined fact set $\boldsymbol{F_d}$

Firstly, add the neg-facts into the fact set according to the problem statements. Second, find all the predicates which do not have definition rules. If *q(X)* is one of them, according to the Lemma 1 in [3], add set $\{nq(s) \mid s \in U_l \text{ and } s \notin D[q]\}$ to the fact set. Finally, add the set $\{\mathrm{var\_}retor(s) \mid s \in U_l\}$ to the fact set to define the variable restrictor.

The fact set $\boldsymbol{F}$ is, therefore, converted to a double defined fact set $\boldsymbol{F_d}$.

(3) Converting the rule set $\boldsymbol{R}$ to a double defined rule set $\boldsymbol{R_d}$

Within the rule set $\boldsymbol{R}$, replace the negations of atoms with neg-atoms and replace the negations of neg-atoms with the atoms themselves. For example, change *not(q(X))* to *nq(X)* and change *not(nq(X))* to *q(X)*. Then, according to the Theorem 1 in [3], derive all the neg-rules and append variable restrictors when it is necessary. Finally, add the neg-rules into the rule set.

The rule set $\boldsymbol{R}$ is converted to a double defined rule set $\boldsymbol{R_d}$.

(4) Creating contradiction check C

A contradiction is defined as that both an atom and its neg-atom can be derived from a double defined logic program $\boldsymbol{P_d}$ (ie. $\exists s \in U_l$ such that $\boldsymbol{P_d} \models q(s)$ and $\boldsymbol{P_d} \models nq(s)$ ($q(X)$ is a predicate of $\mathbf{P_g}$)).

A contradiction check is a set of rules which is used to check if there is any contradiction in the program. A contradiction check can be built in different ways. The simplest way is to define

contradiction :- q(X), nq(X). for all of the predicates in the program.

If detailed information about contradiction is demanded, contradiction check can also be designed like

contradiction :- q(X), nq(x), write(' Contradiction is detected on predicate q when X= '), write(X), nl.

A contradiction does not theoretically exist in a double defined logic program. This proposition will be proven in the following section.

On the conclusion of this four-step procedure, we arrive at the converted program

$$\boldsymbol{P_d} = \boldsymbol{F_d} \cup \boldsymbol{R_d} \cup \boldsymbol{C}$$

The following example converts a complete logic program to a double defined logic program.

Example

A complete logic program $\boldsymbol{P}$ with

Fact set $\boldsymbol{F}$: *r(a). r(b). s(b). s(c). t(d,b).*

Rule set $\boldsymbol{R}$: R_1 *t(X,Y) :- not(r(X)), s(Y).*

R_2 *q(X,Y) :- not(t(Y,X)), r(Y).*

R_3 *p(X,Y,Z) :- t(X,Y), not(q(X,X)), s(Z).*

R_4 *p(X,Y,Z) :- t(X,Z), r(Y).*

We are going to convert the program to a double defined logic program.

(1) Define all the neg-predicates and a variable restrictor.

The neg-predicates are *nr(X)*, *ns(X)*, *nt(X,Y)*, *nq(X,Y)*, and *np(X,Y,Z)*. The variable restrictor is *var_retor(X)*.

(2) Convert the fact set to a double defined fact set.

The Herbrand universe U_l = { *a, b, c, d* }. Since *r(X)* and *s(X)* have no definition rules, according to the Lemma 1 in [3], *D[r]* = { *a, b* }, *D[s]* = { *b, c* }, *D[nr]* = { *c, d* }, and *D[ns]* = { *a, d* }. Therefore neg-facts *nr(c), nr(d), ns(a), ns(d)* can be added to the fact set. Set $\{\mathrm{var\_}retor(s) \mid s \in U_l\}$ should be added to the fact set to define the variable restrictor.

The double defined fact set becomes F_d :

r(a). r(b). s(b). s(c). t(d,b). nr(c). nr(d). ns(a). ns(d).
var_retor(a). var_retor(b). var_retor(c). var_retor(d).

(3) Convert the rule set to a double defined rule set.

Replace all the negative atoms with neg-atoms in all the rules.

According to the Theorem 1 in [3], the neg-rules derived from R_1 are *nt(X,Y) :- r(X).* and *nt(X,Y) :- ns(Y).*. After appending variable restrictor, they become *nt(X,Y) :- r(X), var_retor(Y).* and *nt(X,Y) :- ns(Y), var_retor(X).*

In the same way, neg-rules *nq(X,Y) :- t(Y,X).* and *nq(X,Y) :- nr(Y), var_retor(X).* are derived from rule R_2.

Neg-rules *np(X,Y,Z) :- nt(X,Y), nt(X,Z). np(X,Y,Z) :- nt(X,Y), nr(Y), var_retor(Z). np(X,Y,Z) :- q(X,X), nt(X,Z), var_retor(Y). np(X,Y,Z) :- q(X,X), nr(Y), var_retor(Z). np(X,Y,Z) :- ns(Z), nt(X,Z), var_retor(Y). and np(X,Y,Z) :- ns(Z), nr(Y), var_retor(X).* are derived from rules R_3 and R_4.

The double defined rule set becomes R_d :

R_1 *t(X,Y) :- nr(X), s(Y).*
R_2 *q(X,Y) :- nt(Y,X), r(Y).*
R_3 *p(X,Y,Z) :- t(X,Y), nq(X,X), s(Z).*
R_4 *p(X,Y,Z) :- t(X,Z), r(Y).*
R_5 *nt(X,Y) :- r(X), var_retor(Y).*
R_6 *nt(X,Y) :- ns(Y), var_retor(X).*
R_7 *nq(X,Y) :- t(Y,X).*
R_8 *nq(X,Y) :- nr(Y), var_retor(X).*
R_9 *np(X,Y,Z) :- nt(X,Y), nt(X,Z).*
R_{10} *np(X,Y,Z) :- nt(X,Y), nr(Y), var_retor(Z).*
R_{11} *np(X,Y,Z) :- q(X,X), nt(X,Z), var_retor(Y).*
R_{12} *np(X,Y,Z) :- q(X,X), nr(Y), var_retor(Z).*
R_{13} *np(X,Y,Z) :- ns(Z), nt(X,Z), var_retor(Y).*
R_{14} *np(X,Y,Z) :- ns(Z), nr(Y), var_retor(X).*

(4) Create a contradiction check C

A contradiction check C is defined as :

contradiction :- r(X), nr(X).
contradiction :- s(X), ns(X).
contradiction :- t(X,Y), nt(X,Y).
contradiction :- q(X,Y), nq(X,Y).
contradiction :- p(X,Y,Z), np(X,Y,Z).

We end up with the converted program $P_d = F_d \cup R_d \cup C$.

3. The Complete Double Defined Logic Programs

The procedure of the conversion of a complete logic program to a double defined logic program has been discussed in the previous section. We will now go on to consider a case of the double defined logic program: complete double defined logic programs. A double defined logic program is said to be complete if it can be converted from a complete logic program. This gives rise the question "Is the conversion one-to-one?" In fact, two questions are involved. (a) Can a complete logic program be converted to more than one double defined logic program? (b) Can a double defined logic program be converted from more than one complete logic program? Theorem 1 will answer these questions.

Theorem 1

The conversion from complete logic programs to double defined logic programs by the procedure in Section 2 is a one-to-one mapping on the condition that regardless of the contradiction check .

Theorem 1 can be explained as (a) the double defined logic program converted from a complete logic program by the procedure in Section 2 is unique on the condition that regardless of the contradiction check and (b) a double defined logic program can only be converted from unique complete logic program through the procedure in Section 2.

Proof (a)

Suppose P^1_d and P^2_d are two double defined logic programs converted from a complete logic program P_g with the procedure in Section 2.

∵ the conversion dose not change the Herbrand universe

∴ P^1_d and P^2_d have the same Herbrand universe

∵ the conversion dose not change the definitions of predicates

∴ the definitions of predicates in P^1_d and P^2_d are identical

∵ the neg-predicate of a predicate is defined as the negation of the predicate

∴ the definitions of neg-predicates in P^1_d and P^2_d are identical

∵ P^1_d and P^2_d have the same Herbrand universe

∴ the variable restrictors defined in the two program are identical

∵ P^1_d and P^2_d have the same Herbrand universe and the same definitions of predicates

∴ the fact sets of P^1_d and P^2_d created with the procedure in Section 2 are identical

∵ the definition of variable restrictors are identical in P^1_d and P^2_d

∴ the rule sets of P^1_d and P^2_d created with the procedure in Section 2 are identical

∴ programs P^1_d and P^2_d are identical

Proof (b)

Suppose P_d is a double defined logic program converted from tow complete logic programs P^1_g and P^2_g by the procedure in Section 2.

∵ P_d can be converted by both P^1_g and P^2_g and the conversion dose not change the Herbrand universe and the definitions of predicates of the programs

∴ the Herbrand universes of P^1_g and P^2_g are identical and the definitions of the predicates in both programs are identical.

∵ the double defined fact set F_d can be converted from both F^1 and F^2 and the conversion does not change the positive information

∴ the fact sets F^1 and F^2 are identical

∵ the double defined rule set R_d can be converted from both R^1 and R^2 and the conversion does not change the positive information

∴ the rule sets R^1 and R^2 are identical

∴ the complete logic programs F^1 and F^2 are identical //

Theorem 1 indicates that the conversion from complete logic programs to double defined logic programs is, in fact, a one-to-one mapping. Based on this theorem, we can state the following definition.

Definition (Complete Double Defined Logic Program)

A double defined logic program converted from a complete logic program is called a complete double defined logic program.

Since a neg-predicate is a predicate, a neg-clause is a Horn clause. Therefore a double defined logic program is a Horn clause logic program. The following theorem indicates a nature of a complete double defined logic program as a Horn clause logic program.

Theorem 2

A complete double defined logic program is a complete logic program.

Proof

Let P_d be a double defined logic program which is converted from a complete logic program P_g.

Let U_l be the Herbrand universe of P_g.

∵ the conversion from P_g to P_d does not change the Herbrand universe

∴ U_l is also the Herbrand universe of P_d

For every predicate $q(X)$ defined in P_g.

∵ P_g is complete

∴ $$D[q] = \bigcup_{i=1}^{m} \bigcap_{j=1}^{k_i} D[p_{ij}] \tag{1}$$

where $R_i \; q(X) \leftarrow p_{i1}(X) \wedge p_{i2}(X) \wedge \ldots \wedge p_{ik_i}(X)$ $(\, i = 1, 2, \ldots, m \,)$ are all the definition rules of $q(X)$.

∵ the conversion does not change the P-domains of predicates in P_g (Definition 7 in [3])

∴ formula (1) is also true in P_d

According to the Theorem 1 in [3], all the definition rules of $nq(X)$ are

$$nq(X) \leftarrow np_{1l_1}(X) \wedge np_{2l_2}(X) \wedge \ldots \wedge np_{ml_m}(X)$$

We want to prove that

$$D[nq] = (\bigcup_{j_1=1}^{k_1} \ldots \bigcup_{j_m=1}^{k_m}) \bigcap_{i=1}^{m} D[np_{ij_i}] \tag{2}$$

∵ P_g is complete

∴ $$\begin{aligned} D[nq] &= U_l - D[q] \\ &= U_l - \bigcup_{i=1}^{m} \bigcap_{j=1}^{k_i} D[p_{ij}] \\ &= \bigcap_{i=1}^{m} \bigcup_{j=1}^{k_i} (U_l - D[p_{ij}]) \end{aligned}$$

$$= \bigcap_{i=1}^{m} \bigcup_{j=1}^{k_i} D[np_{ij}]$$

$$= (\bigcup_{j_1=1}^{k_1} \cdots \bigcup_{j_m=1}^{k_m}) \bigcap_{i=1}^{m} D[np_{ij_i}]$$

(2) is proved //

As having been mentioned in Section 2 that contradiction does not theoretically exist in a complete double defined logic program. The following theorem will prove it.

Theorem 3

Let P_d be a complete double defined logic program. For every predicate in P_d, the intersection of the P-domains of the predicate and its neg-predicate is empty.

Theorem 3 can be explained as follows: if P_d is a complete double defined logic program and $q(X)$ is a predicate of P_d, then there is no such an element $a \in U_l$ satisfying $P_d \models q(a)$ and $P_d \models nq(a)$.

Proof

Let F_d and R_d be the fact set and rule set of P_d. We assume that the definition of the logic program is not contradictory (ie. there do not exist an element $a \in U_l$ and a predicate $t(X)$ in P_d, such that $t(a) \in F_d$ and $nt(a) \in F_d$).

Let $q(X)$ be a predicate of P_d. R_i $q(X) \leftarrow p_{i1}(X) \wedge p_{i2}(X) \wedge \ldots \wedge p_{ik_i}(X)$ ($i = 1, 2, \ldots, m$) are all of the definition rules of $q(X)$.

According to the Theorem 1 in [3], all the neg-rules with heads $nq(X)$ are R_l' $nq(X) \leftarrow np_{1l_1}(X) \wedge np_{2l_2}(X) \wedge \ldots \wedge np_{ml_m}(X)$ (where l_1, l_2, ... , l_m are all the combinations of $1, \ldots, k_1$; $1, \ldots, k_2$; ... ; $1, \ldots, k_m$).

Assume $a \in U_l$ and q(a) is a clause or logical consequence of P_d (ie. $a \in D[q]$).

$\because$ $D[q] = \bigcup_{i=1}^{m} \bigcap_{j=1}^{k_i} D[p_{ij}]$

$\therefore$ $\exists i$ $(1 \le i \le m)$ such that $a \in \bigcap_{j=1}^{k_i} D[p_{ij}]$ (ie. $a \in D[p_{i1} \wedge p_{i2} \wedge \ldots \wedge p_{ik_i}]$)

$\therefore$ $a \in D[p_{ij}]$ $(j = 1, 2, \ldots, k_i)$

We can assume that predicates $p_{i1}, p_{i2}, \ldots, p_{ik_i}$ do not have definition rules.

Otherwise, for example, $p_{ij}(X) \leftarrow t_1(X) \wedge \ldots \wedge t_s(X)$ is a rule of P_d,

then $a \in D[p_{i1} \wedge p_{i2} \wedge \ldots \wedge t_1 \wedge \ldots \wedge t_s \wedge \ldots p_{ik_i}]$.

$\because$ p_{ij} does not have definition rules and P_d is a complete logic program

$\therefore$ according to the Lemma 1 in [3], $D[p_{ij}] = \{ x \mid x \in U_l$ and $p_{ij}(x) \in F_d \}$ and

$D[np_{ij}] = U_l - D[p_{ij}]$ ($j = 1, 2, \ldots, k_i$)

$\therefore$ $a \notin D[np_{ij}]$ ($j = 1, 2, \ldots, k_i$)

$\therefore$ $a \notin \bigcap_{i=1}^{m} D[np_{ij}]$ ($j = 1, 2, \ldots, k_i$)

$\therefore$ $a \notin (\bigcup_{j_1=1}^{k_1} \cdots \bigcup_{j_m=1}^{k_m}) \bigcap_{i=1}^{m} D[np_{ij_i}]$

$\therefore$ $a \notin D[nq]$

On the other hand, assume $a \in U_l$ and $nq(a)$ is a clause or a logical consequence of P_d, (ie. $a \in D[nq]$)

$\because \quad D[nq] = (\bigcup_{j_1=1}^{k_1} \dots \bigcup_{j_m=1}^{k_m}) \bigcap_{i=1}^{m} D[np_{ij_i}]$

$\therefore \quad \exists j_1, \dots, j_m$ such that $a \in \bigcap_{i=1}^{m} D[np_{ij_i}]$ (ie. $a \in D[np_{1j_1} \wedge np_{2j_2} \wedge \dots \wedge np_{mj_m}]$)

We assume that predicates $np_{1j_1}, np_{2j_2}, \dots, np_{mj_m}$ do not have definition rules

$\therefore \quad a \notin D[p_{ij_i}]$ ($i = 1, 2, \dots, m$)

$\therefore \quad a \notin \bigcap_{j=1}^{k_i} D[p_{ij}]$ ($i = 1, 2, \dots, m$)

$\therefore \quad a \notin D[q]$ (ie. $q(a)$ is not a clause or a logical consequence of $\mathbf{P_d}$) //

4. Conclusions

In this paper we have introduced the procedure of the conversion of a complete logic program to a double defined logic program as a means of eliminating negation based on the "Negation as Failure" rule and implementing constructive negation in the program. A double defined logic program is a Horn clause logic program, hence deductions on a double defined logic program with SLD-resolution are sound and complete.

Contradiction and contradiction checks have been introduced. Theoretically speaking, contradiction does not exist in the double defined logic programming.

The conversion of complete logic programs to double defined logic programs is an one-to-one mapping. This means that a complete logic program can be converted to unique double defined logic program; a double defined logic program can be converted from unique complete logic program. Based on Theorem 1, complete double defined logic programs have been defined. A complete double defined logic program is a complete logic program.

5. References

[1] Clark K.L. (1977) *Negation as Failure*, Symposium on Logic & Databases, Plenum Press, pp. 293 - 322.

[2] Jaffar J. (1983) *Completeness of the Negation as Failure Rule*, Proceedings of The Eighth International Joint Conference on Artificial Intelligence, Volume 1, pp. 500 - 506

[3] Liu F. and Moore D.H. (1993) *Double Defined Logic Programming*, Proceedings of the 6th Australian Joint Conference on Artificial Intelligence, pp. 27 - 32.

[4] Liu F. and Moore D.H. (1993) *Implementing Negation in Incomplete Logic Programs through Double Defined Logic Programming*, Proceedings of IEEE Australian and New Zealand Conference on Intelligent Information Systems, pp. 709 - 713.

[5] Lloyd J.W. (1987) *Foundations of Logic Programming*, Springer-Verlag, Berlin.

AN EFFICIENT METRIC FOR HETEROGENEOUS INDUCTIVE LEARNING APPLICATIONS IN THE ATTRIBUTE-VALUE LANGUAGE[1]

Christophe Giraud-Carrier and Tony Martinez
Brigham Young University, Department of Computer Science, Provo, UT 84602

Abstract. Many inductive learning problems can be expressed in the classical attribute-value language. In order to learn and to generalize, learning systems often rely on some measure of similarity between their current knowledge base and new information. The attribute-value language defines a heterogeneous multi-dimensional input space, where some attributes are nominal and others linear. Defining similarity, or proximity, of two points in such input spaces is non trivial. We discuss two representative homogeneous metrics and show examples of why they are limited to their own domains. We then address the issues raised by the design of a heterogeneous metric for inductive learning systems. In particular, we discuss the need for normalization and the impact of don't-care values. We propose a heterogeneous metric and evaluate it empirically on a simplified version of ILA.

1. Introduction

Many inductive learning systems use the classical attribute-value language as their representation language. In the attribute-value language, a training example is a vector whose entries are pairs consisting of an attribute name and its value. In most cases, attribute names are omitted as the context unambiguously determines which entry corresponds to which attribute. Some of the attributes are identified as being input attributes, and some (often a single one) are identified as output attributes. Each input attribute's value ranges over some domain, and the cross product of all the domains defines the input space of the application. Hence, training examples are points in the input space, labeled with some output value.

In many cases, inductive learners are concerned with the classification of points into disjoint categories. The system uses training examples to derive a classification that is then generalized to previously unseen points of the input space. Generalization typically consists of either one (or a combination) of two processes: 1) the discovery of subsets of points (or regions) of the input space that map to a given concept [4, 5], and 2) the use of proximity (or similarity) to known points [2, 6, 7, 12]. When a new point is shown, its classification is determined either by its belonging to one of the identified regions, or by the classification of the "closest" known points. Classification by closeness is the essence of most nearest-neighbor algorithms (see [3] for a survey) and memory-based reasoning (see for example [11]).

To find the "closest" known points to a new point in the input space, the system needs some measure of proximity or similarity, i.e., a distance metric defined on the input space. Because the domains of attributes vary in nature, the input space is often heterogeneous, in the sense that topological properties such as distance do not have the same definition in all dimensions. The notion of distance between two points in such heterogeneous spaces is non trivial. This paper shows the limitations of homogeneous metrics, and proposes a (combined) heterogeneous distance measure that accounts for heterogeneous spaces, in the context of inductive learning. Empirical results demonstrate the superiority of the proposed measure over homogeneous metrics.

---

[1]This work was supported in part by grants from Novell Inc., and WorPerfect Corp.

E. A. Yfantis (ed.), Intelligent Systems, 341–350.

Section 2 addresses the issue of similarity, discusses two representative homogeneous metrics, and gives examples of the inadequacy of these metrics in heterogeneous domains. Section 3 discusses two issues raised by the design of a metric for heterogeneous inductive learning applications. Section 4 proposes a heterogeneous metric that extends the similarity function of [2]. Section 5 overviews a simplified version of ILA [6] that serves as a basis for empirical evaluation. Finally, Section 6 concludes the paper.

2. Considerations on the Notion of Similarity

We consider here two types of attributes: nominal and linear. Nominal attributes have discrete values that are not related in any way other than the fact that they belong to the same set. Linear attributes have values that can be ordered in the usual sense, and whose ordering is relevant to the context in which they are used. Linear attributes may be discrete and ordered or continuous. The attribute *blood group*, for example, is a nominal attribute, while the attribute *weight* is linear. Each kind of attribute gives rise to a distinct notion of similarity or distance. We discuss two of them here, and show that they are mutually incompatible. The selected metrics are in no wise unique, only representative. They serve as illustration, are commonly used [2, 5, 12], and have been found to give good empirical results on their respective domains of applications.

2.1. Distance for Nominal Spaces

For nominal data, the notion of how far apart two values are reduces to a simple binary relation. Two values are either the same, or they are different. If the values are the same, then the distance is 0; otherwise, the distance is 1. The one-dimensional definition extends to the multi-dimensional case in a straight forward way. Formally,

$$\text{Let } x = (x_i) \text{ and } y = (y_i) \text{ be two } n\text{-dimensional nominal vectors}$$

$$\text{Let } dn(x_i, y_i) = \begin{cases} 1 & \text{if } x_i \neq y_i \\ 0 & \text{otherwise} \end{cases}$$

$$\text{Then } DN(x, y) = \sum_{i=1}^{n} dn(x_i, y_i)$$

DN (Distance for Nominal spaces) conveys the intuitive idea that the farther away two vectors are, the less similar they are. So it can be used directly to choose a closest match.

However, *DN* is inadequate on linear spaces. Consider the following example, where each attribute is linear and ranges over [0, 10] in the natural numbers.

$$x = (2,1,3)$$
$$y = (3,1,4) \Rightarrow DN(x,y) = 2$$
$$z = (2,6,0) \Rightarrow DN(x,z) = 2$$

Though $DN(x,y)=DN(x,z)$, it appears that (given linear attributes) y is closer to x than z is. The problem is that the ordering implies varied magnitudes in the differences between values, while *DN* only accounts for equality or inequality. It takes any magnitude greater than 0 to 1, and leaves 0 magnitudes at 0. In some sense, it forces a step function on a linear domain. Things become even worse in continuous spaces where the probability of any two values being equal is extremely low.

2.2. Distance for Linear Spaces

Linear attributes are ordered (e.g., continuous values). The ordering gives rise to a natural measure of distance, namely, the farther away things are in the ordering, the larger their distance should be. This is consistent with our everyday notion of distance. The distance between two values can be defined as the absolute value of their difference. Since every ordered set is in one-to-one correspondence with a subset of the natural numbers, this distance is well-defined on all domains. A common generalization of this one-dimensional definition to the multi-dimensional case is the classical Euclidean distance.

However, this definition implicitly assumes that all attributes range over the same domain (e.g., reals, integers). If the domains are different, then some attribute distances may dominate others in the overall distance. For example, if x and y are linear attributes ranging over [0, 1] in the real numbers and [0, 100] in the natural numbers, respectively, then distances along y are likely to dominate distances along x. The smallest distance in y is equal to the largest distance in x. The problem is one of scale. We thus modify the definition as follows.

First, let us argue that in most practical learning applications, linear attributes are bounded, that is they have a smallest and a largest possible value. These can easily be obtained from the associated dataset. To eliminate the effects of statistical outliers, the dataset must be ridden of examples whose attributes have such "irregular" values. Alternatively, the linear attributes (especially continuous ones) can be discretized into finitely many classes.

Then, let *range(i)* denote the range of values for attribute i, i.e., the difference between the maximum and minimum values of attribute i. We now define *NDL* (Normalized Distance for Linear spaces) by:

$$\textit{Let } \mathbf{x} = (x_i) \textit{ and } \mathbf{y} = (y_i) \textit{ be two } n\textit{-dimensional linear vectors}$$

$$\textit{Let } ndl(x_i, y_i) = \frac{|x_i - y_i|}{range(i)}$$

$$\textit{Then } NDL(\mathbf{x}, \mathbf{y}) = \sqrt{\sum_{i=1}^{n} [ndl(x_i, y_i)]^2}$$

The division by *range(i)* causes all attribute distances to fall within the range [0, 1]. Hence, all attributes make a normalized contribution to *NDL*. Again, *NDL* conveys the intuitive idea that the farther away two vectors are, the less similar they are. So it can be used directly to choose a closest match.

However, *NDL* is inadequate on nominal spaces. Consider the following example, where each attribute ranges over the discrete set {0, 1, 2, 3, 4, 5}. For nominal domains, we let *range(i)* be the number of possible values in the domain minus 1.

$$x = (3,1,2)$$

$$y = (2,1,1) \Rightarrow NDL(x, y) = \sqrt{2}/5$$

$$z = (3,4,2) \Rightarrow NDL(x, z) = 3/5$$

Though $NDL(x,y)<NDL(x,z)$, it appears that (given nominal attributes) z is closer to x than y is. The problem is that *NDL* treats nominal values as if they had inherent magnitudes. Instead of making the distance between any two distinct nominal values the same, it assigns different distances based on some non-existent, but assumed, intrinsic ordering.

One may be tempted to claim that the above example is a result of an artifact of encoding. If all nominal values are "one-hot" encoded (i.e., as many bits as there are values, one set for each distinct value), then *NDL* would make z closer to x as expected. However, such encodings cause a multiplication of the number of attributes. Moreover, they reduce nominal spaces to Boolean spaces, and it is straight forward to show that on Boolean spaces *NDL=DN*.

2.3. Inadequacy of Homogeneous Metrics on Heterogeneous Spaces

The above examples show that if a learning system uses a homogeneous metric and applies it in a non-discriminatory way to all applications, then it is likely that its accuracy will suffer on some applications. It is clear that similar results hold when applying either of the homogeneous metrics to a heterogeneous space. Consider the following example, where the first attribute is nominal and ranges over the discrete set {0, 1, 2, 3, 4, 5}, and the other two attributes are linear and range over [0, 10] in the natural numbers.

$$x = (4,1,0)$$
$$y = (4,7,0) \Rightarrow NDL(x,y) = {}^{6}\!/_{10} \wedge DN(x,y) = 1$$
$$z = (3,0,2) \Rightarrow NDL(x,z) = {}^{\sqrt{6}}\!/_{10} \wedge DN(x,z) = 3$$

The two distances naturally give inconsistent results. *DN* leads to the conclusion that y is closer to x than z is, while *NDL* leads to the opposite conclusion. This inconsistency is a direct consequence of the problems discussed above. It follows that to correctly handle heterogeneous spaces, a heterogeneous metric must be employed.

3. Designing a Metric for Heterogeneous Learning Applications

We address two important issues raised by the design of a metric for heterogeneous learning applications. One has to do with the heterogeneity of the learning space, and the other with the common use of rules containing *don't-care* values.

3.1. Heterogeneity of the Learning Space

In designing a metric for heterogeneous spaces, it is desirable not to loose accuracy on homogeneous spaces. One simple solution consists of combining the homogeneous metrics into a single metric that applies each homogeneous metric to its corresponding attributes. Such a combination may suffer from the scale problem.

In this case, the problem is no longer with each attribute's range, but with the ranges of values covered by the homogeneous metrics. If these are different, then one of the metrics may indeed dominate the overall result. Hence, it is not sufficient that each metric be adequate on its own space. They must be somehow normalized with respect to each other. The normalization in *ndl* causes *ndl* to have exactly the same range of values as *dn*, namely [0, 1]. Hence, not only is *ndl* a more accurate linear metric (for linear spaces), it is also "compatible" with *dn*.

There is a possible problem with such a normalization however. Indeed, by normalizing, we artificially force complete mismatches (i.e., distance of 1) to the extrema. It would be reasonable instead to assume that when two values are far enough apart (based on some threshold value), then there is a complete mismatch between them. In other word, a non-linear normalization technique may be more appropriate. Another alternative consists of using multiplicative scale factors on nominal attribute distances. We do not address this problem in this paper.

3.2. Using Rules with Don't-Care Values

A common and useful extension, made to the classical attribute-value language, is the addition of a special *don't-care* symbol for input attributes. The interpretation for a vector containing input attributes set to don't-care is as follows. An attribute whose value is not don't-care is said to be *asserted.* A vector whose input attributes are all asserted represents a single point of the input space. A vector that has input attributes set to don't-care represents all of the points in the input space that would be obtained by replacing all of the don't-care symbols by any one of the possible values of the corresponding input attributes. Consider the following example where each attribute ranges over the discrete set {1, 2, 3, 4}.

Let $v = (1,*,3,2)$

Then v represents the following points

(1,1,3,2), (1,2,3,2), (1,3,3,2), *and* (1,4,3,2)

Hence, such vectors represent subsets of points, or regions, in the input space. We discuss two situations in which they arise, and show the impact they have on the definitions of distance.

3.2.1. Origins of Don't-Cares. Vectors with input attributes set to don't-care represent subsets of points in the input space. Hence, they can be viewed as *rules* in the sense that they map a set of points, rather than a single point, to a given output value (or concept). The number of points represented by a rule is a measure of that rule's *generality.* The larger the set of points, the greater the generality. Such rules arise in at least two situations that are both relevant to learning systems.

The purpose of inductive learning systems is to extract, from a set of training examples, the critical features that govern an application. Some of these critical features, or set of attribute-value pairs that are good predictors of the output, may be represented as rules. Such rules are the result of the learning system's ability to generalize. A commonly used generalization rule in learning systems is the *dropping-the-condition* rule [8]. In the attribute-value language, it translates into the setting of one (or more) of the input attributes of a given vector to don't-care. The input attributes that remain asserted can then be viewed as critical features (those that are don't-care do not affect the output value).

Another (possibly complementary) situation in which rules arise is by being provided *a priori* (by some teacher). Indeed, such rules may also be used as an encoding of prior knowledge, or as a particular instantiation of domain knowledge or commonsense. They then serve as learning biases for the system (see, for example [5]). For example, when learning about the flying abilities of birds, one may bias the system with a rule encoding the fact that birds typically fly. The system then only needs to focus on exceptions.

3.2.2. Effects of Don't-Cares. Since they do not have any intrinsic value, don't-cares are not comparable with other values. To find a closest vector in the current knowledge base, distance is measured from vectors in the knowledge base to new vectors. If x and y are two vectors such that all of their attribute values are equal except for some attributes that are don't-care in x and asserted in y, and if d is any well-defined distance, then we claim that the following should hold.

$$d(x,y)=0 \quad (1)$$
$$d(y,x)\neq 0 \quad (2)$$

That is, the distance measure is not symmetric. In (1), x is in the current knowledge base, y is a new vector, and the equality is an immediate (and natural) consequence of the definition of don't-cares. In (2), y is in the current knowledge base, x is a new vector, and the inequality requires some explanation.

If reasonable at all, inequality (2) is only useful when unseen vectors (either in training or in testing) are allowed to have don't-care input attributes. We argue that such situations may arise and claim that in such cases, the inequality should hold. As seen in Section 3.2.1., rules may be the expression of domain knowledge. There is no reason to assume that all of these are available *a priori*. It is plausible that the learning system (much like a human in this case) is exposed to examples and rules in an unpredetermined order. Hence, certain rules may come in later in the learning process. For example, if you live in Australia, most of the birds you see are ostriches that do not fly. Thus, you are exposed to exceptions before you may be taught the more general rule about birds flying ability. In testing, allowing don't-cares is a way to potentially discover default rules for situations in which much information is missing. This is particularly useful in bottom-up inheritance, where, based on the current knowledge base, one can obtain the default classification of a very large set of points. In this latter case, a don't-care can also be thought of as a don't-know or missing value.

If (1) holds and the distance is symmetric, then the generality of the new vector may cause several vectors in the current knowledge base to be equidistant to it, as in the following example, where all attributes are Boolean, and the last one is the output. The knowledge base is *KB*, the new vector is x, and * stands for don't-care.

$$KB=\{(1,1,0,1,1),(1,1,*,1,0),(*,0,*,1,1)\}$$
$$x=(1,*,*,0,?)$$

All of the vectors in *KB* have the same distance to x. However, in this case, where a default classification is expected, it seems reasonable to expect the more general of the vectors in *KB* to be closest. Inequality (2) increases the distance to more specific existing vectors, thus causing the system to favor the more general vectors.

Another impact of the variation in generality of different vectors and (1) is the need for some normalization of the overall distance, so as to distinguish a perfect match from a don't-care. Since attributewise distances range over [0, 1], this can be done by dividing the overall distance by the number of asserted input attributes in the vector from which distance is computed. For example, it is better to have 2 mismatches out of 5 asserted input attributes than 1 mismatch out of only 2 asserted input attributes.

4. DHET - A Metric for Heterogeneous Inductive Learning

Based on the above discussion, we propose the following modified set of definitions. Let *num_asserted(x)* be the number of asserted input attributes in vector x.

Let $x=(x_i)$ *and* $y=(y_i)$ *be two* n*-dimensional vectors*

$$Let\ dn'(x_i, y_i) = \begin{cases} 1 & if\ x_i \neq * \wedge x_i \neq y_i \\ 0 & otherwise \end{cases}$$

$$Then\ DN'(\boldsymbol{x}, y) = \frac{\sum_{i=1}^{n} dn'(x_i, y_i)}{num\_asserted(\boldsymbol{x})}$$

$$Let\ ndl'(x_i, y_i) = \begin{cases} \frac{|x_i - y_i|}{range(i)} & if\ x_i \neq * \wedge y_i \neq * \\ 1 & if\ x_i \neq * \wedge y_i = * \\ 0 & otherwise \end{cases}$$

$$Then\ NDL'(\boldsymbol{x}, y) = \frac{\sqrt{\sum_{i=1}^{n} [ndl'(x_i, y_i)]^2}}{num\_asserted(\boldsymbol{x})}$$

$$Finally,\ DHET(\boldsymbol{x}, y) = \frac{\sqrt{\sum_{i=1}^{n} \begin{cases} dn'(x_i, y_i) & if\ attribute\ i\ is\ nominal \\ [ndl'(x_i, y_i)]^2 & if\ attribute\ i\ is\ linear \end{cases}}}{num\_asserted(\boldsymbol{x})}$$

The value of 1 for *dn'* and *ndl'* when x_i is asserted and y_i is not reflects the fact that we consider such a case to be a "complete" mismatch. *DN'* is the reciprocal of the conceptual similarity measure of [12]. *DHET* (Distance for HETerogeneous spaces) is an extension of IBL's similarity function [2]. The two metrics are equivalent when a purely instance-based learning algorithm is used, and *DHET* extends IBL's similarity function to inductive learning algorithms that use and/or create general rules. *DHET* is also similar to the distance metric used by NGE [10]. It is somewhat simpler since rules containing don't-cares represent hyperplanes rather than hyperrectangles. It is also inherently heterogeneous whereas NGE's distance is designed for purely continuous domains. Extensions to heterogeneous spaces have been proposed (see for example [13]). NGE's distance is also a weighted sum, where each attributewise distance is assigned its own weight in the computation of the final distance. Mechanisms to assign weights to attributes in distance computation may be found in [10, 11, 13]. *DHET* could be similarly extended.

5. Evaluation

In this section, we give empirical evidence of the adequacy of *DHET*. The results were obtained by executing the same algorithm, only varying the metric it uses. ILA (Incremental Learning Algorithm) was chosen because of its simplicity, execution speed, and the direct impact the metric has over its accuracy. ILA bears resemblance with IBL algorithms [2] and NGE [10]. It can be viewed as a nearest-hyperplane algorithm. We first overview a simplified version of ILA. Details of the full algorithm are in [6]. We then present results of simulations on a variety of applications.

5.1. ILA - Algorithmic Overview

ILA's representation language is the classical attribute-value language, with the addition of the special *don't-care* symbol. For convenience, examples and rules are viewed as *input-output pairs*.

Examples have all of their input attributes asserted to some value. ILA discovers new rules by generalizing from examples. The generalization rule it uses is the classical drop the condition rule [8], which causes the value of an input attribute to be set to don't-care.

ILA implements a network of simple nodes that adapts to newly acquired knowledge by dynamically adding and/or removing nodes in the network. The network's structure is a balanced binary tree. No parameters need to be set *a priori*. Parallelism at the node level is inherent.

For our purposes, each node stores a pair, together with counters for each output value (vector *counters*), and a counter for the number of pairs covered (*num_covers*). The counters are used to handle noise. For each output value, they keep track of the number of training pairs whose inputs match exactly the input of the node's stored pair. The output whose count is highest is the node's output. The counter num_covers keeps track of the number of pairs that have the same output as the stored pair, and are covered by it. We say that a pair *p covers* a pair *q* if and only if the set of points represented by *q* is a proper subset of the set of points represented by *p*. The value of num_covers can serve as the degree of confidence in the node's stored rule.

Training pairs are incrementally presented to the system. ILA's learning consists of two phases: *execution* and *adaptation*. In the execution phase, the training pair is presented to the network and a winning node is identified. Identifying the winner is a simple broadcast (of the training pair) and gather operation, that first minimizes the distance (i.e., node with smallest distance wins), then resolves ties by giving priority to the more specific node (i.e., larger number of asserted inputs), then to the node with the largest value of num_covers. Winning distances are passed up and nodes are tagged as self, right or left, based on which node wins the competition. A winner is then easily identified by going down the tree until a node marked as self is found. If the network is empty, a new node is automatically added for the training pair. Giving priority to more specific nodes helps in handling exceptions, and giving priority to nodes whose num_covers value is higher increases the chances of selecting a "most likely to be correct" winner.

In the adaptation phase, counters and num_covers are updated, and a new node is created for the training pair, if 1) the training pair is an exception to, 2) covers, or 3) is too different from the winning node's stored pair. If it covers it and has the same output, then the winning node is also deleted. An attribute is dropped if the training pair and the winning node's stored pair have the same output, differ in the value of exactly one of their asserted input attributes, and their relative numbers of asserted input attributes differ by at most 1. Precautions are taken to avoid the creation of a rule that covers all of the input space (i.e., drop all attributes).

Evaluating ILA for predictive accuracy on a test set consists of presenting every pair in the test set and only running the execution phase. The output of the winning node is the system's output and prediction of the new pair's output.

5.2. SIMULATION RESULTS

A variety of datasets from the Irvine repository of machine learning datasets [9] were chosen, representing a mixture of homogeneous and heterogeneous applications. Two modifications are made however. All continuous attributes are made discrete (using a K-means algorithm on each input), and all don't-know values are arbitrarily treated as don't-care values. Other (possibly more appropriate) ways of handling don't-know values are the topic of future research (three of them are discussed in [1]). Also, even though the attributes of the application lenses are all claimed to be nominal, the first one (i.e., age) appears to have a definite natural ordering to it. Hence, the dataset lenses is considered to be heterogeneous.

ILA's predictive accuracy (in %) on a test set was gathered for each application. Each of the reported results is an average over 10 runs of ILA for that application. In each run, the training set and the test set are randomly regenerated. Each set contains one half of the examples in the complete dataset. The datasets used and corresponding results of simulations with *DN'*, *NDL'*, and *DHET* are given in Table 1. Table 2 summarizes the averages on each kind of datasets and shows the overall averages for all the datasets.

Table 1 - Datasets Used and ILA's Simulation Results

Dataset	#instances	#inputs	#out. values	DN'	NDL'	DHET
zoo	90	16N	7	95.3	94.7	95.3
mushroom	8124	22N	2	100	100	100
chess-end	3196	36N	2	90.5	90.1	90.5
tic-tac-toe	958	9N	2	80.4	75.8	80.4
LED	1000	7N	10	70.2	70.2	70.2
molecular-bio	106	57N	2	77.9	66.8	77.9
iris	150	4L	3	90.5	94.0	94.0
glass	214	9L	7	66.4	67.9	67.9
breast-cancer	699	9L	2	94.4	94.7	94.7
pima-indians	768	8L	2	63.4	67.2	67.2
vowel	528	10L	11	78.2	91.9	91.9
sonar	208	60L	2	79.8	85.4	85.4
lenses	24	3N, 1L	3	65.0	80.0	80.0
hepatitis	155	13N, 6L	2	77.3	78.6	78.6
soybean-sm.	47	31N, 4L	4	98.8	98.8	98.8
credit-screen.	690	8N, 6L	2	79.5	80.3	80.5
australian-crx	690	8N, 6L	2	79.9	82.2	82.0

Table 2 - ILA - Summary Results

Space	DN'	NDL'	DHET
Nominal	85.7	82.9	85.7
Linear	78.8	83.5	83.5
Heterogeneous	80.1	83.9	84.0
Overall	81.6	83.4	84.4

As expected, *DHET=DN'* on nominal datasets, and *DHET=NDL'* on linear ones. Also, each homogeneous metric outperforms the other on its respective spaces. The above results suggest that *DHET* is an adequate metric for inductive learning in the attribute-value language. Over the selected datasets, representing a variety of input spaces, the average predictive accuracy obtained with *DHET* exceeds that obtained with either of the homogeneous metrics discussed here.

6. Conclusion

In this paper, we have addressed the problem of designing an efficient metric for heterogeneous inductive learning applications in the attribute-value language. We gave examples of the inadequacy of homogeneous metrics, and proposed a metric, *DHET*, that effectively handles homogeneous and heterogeneous spaces. *DHET* also accounts for inductive learning algorithms that deal with rules containing don't-cares. Results of simulations with ILA were given, that confirm these findings. On average, the results with *DHET* exceed those obtained by homogeneous metrics on all spaces.

Future research includes the following extensions to *DHET*:

- Possible normalization based on standard deviations (rather than ranges)
- Non-linear normalization for linear attributes
- Individual weighting of features (likely depend upon the learning algorithm)
- Adequate treatment of *don't-know* (or missing) values.

References

[1] Aha, D.W. A Study of Instance-Based Algorithms for Supervised Learning Tasks. Technical Report, University of California, Irvine, 1991.

[2] Aha, D.W., Kibler, D., and Albert, M.K. Instance-Based Learning Algorithms. *Machine Learning*, 6, 1991, 37-66.

[3] Dasarathy, B.V. *Nearest Neighbor (NN) Norms: NN Pattern Classification Techniques*. IEEE Computer Society Press, 1991.

[4] Dietterich, T.G., and Michalski, R.S. A Comparative Review of Selected Methods for Learning from Examples. In Michalski, R.S., Carbonell, J.G., and Mitchell, T.M., (Eds.). *Machine Learning: An Artificial Intelligence Approach*. Tioga Publishing Company, Palo Alto, CA, 1983, Chapter 3.

[5] Giraud-Carrier, C., and Martinez, T.R. Using Precepts to Augment Training Set Learning. In *Proceedings of the 1993 International Conference on Artificial Neural Networks and Expert Systems (ANNES'93)*, 1993, 46-51.

[6] Giraud-Carrier, C., and Martinez, T.R. ILA: Combining Inductive Learning with Prior Knowledge and Reasoning. Submitted.

[7] Kibler, D., and Aha, D.W. Learning Representative Exemplars of Concepts: An Initial Case Study. In *Proceedings of the Fourth International Workshop on Machine Learning*, 1987, 24-30.

[8] Michalski, R.S. A Theory and Methodology of Inductive Learning. *Artificial Intelligence*, 20, 1983, 111-161.

[9] Murphy, P.M., and Aha, D.W. *UCI Repository of machine learning databases*. Irvine, CA: University of California, Department of Information and Computer Science, 1992.

[10] Salzberg, S. A Nearest Hyperrectangle Learning Method. *Machine Learning*, 6, 1991, 277-309.

[11] Stanfill, C., and Waltz, D. Toward Memory-Based Reasoning. Communications of the ACM, Vol. 29, No. 12, December 1986, 1213-1228.

[12] Sun, R. A Connectionist Model for Commonsense Reasoning Incorporating Rules and Similarities. *Knowledge Acquisition*, 4, 1992, 293-321.

[13] Wettschereck, D., and Dietterich, T.G. An Experimental Comparison of the Nearest-Neighbor and Nearest-Hyperrectangle Algorithms. To appear in *Machine Learning*.

IMPROVING DECISION SUPPORT THROUGH HYPERMEDIA

V. R. Kumar
C. A. Lindley
Knowledge-Based Systems Laboratory
CSIRO, Division of Information Technology
Locked Bag 17, North Ryde, NSW 2113, Australia
E-Mail : {kumar, lindley}@syd.dit.csiro.au

Abstract. Traditional knowledge-based systems have not been very successful in solving highly complex or poorly structured problems or problems that require "common-sense" knowledge or judgement. One way of dealing with such problems is to incorporate formalised aspects of the decision process into a knowledge-based system, while using human judgement for those aspects of the problem that cannot be easily automated. In this case hypermedia technology can be used to provide rich, complex, and loosely structured information in support of human problem solving: the user interprets the current situation to provide context-dependent and ongoing inputs to automated decision processing. However, using manually constructed hyper-links compounds the knowledge engineering problem, since a human author must create the hyperstructure between media comonents, and the links between the hypermedia system and the knowledge base. This paper describes how statistical information retrieval methods can successfully be used for automatic and dynamic link generation both within the hypermedia system and between a hypermedia system and a knowledge base, eliminating the need for manual link creation and thereby greatly simplifying the knowledge engineering process. The techniques have been demonstrated in the Lambda decision support environment developed by the CSIRO Division of Information Technology. We describe the architecture of Lambda, and describe how automated decision processing is integrated with the Hyperbrowser, a generic software tool for automated hypertext browsing.

Keywords. Knowledge-Based Systems, Decision Support Systems, Hypermedia, Information Retrieval.

1 Introduction

In recent years, the decision support system (DSS) field has come to encompass such paradigms as expert systems (ESs), intelligent decision support systems (IDSS), active decision support systems (ADSSs), and systems that seek the advantage of integrating DSSs with ESs (Integrated DS-ESs). Due to the complexity of decision making in real environments, better and more efficient decision making capabilities are needed, along with improved support and justification for decisions. Integrating Hypermedia technology with ES-DS technologies has great potential to address this need [8]. The so called "Knowledge Engineering Bottleneck" in knowledge-based system methodologies has long been perceived as the major difficulty to be overcome in developing knowledge-based systems. Integrating Knowledge-Based System/Decision Support System techniques with Hypermedia compounds this problem if it is necessary to manually construct links within Hypermedia Systems and between Hypermedia and Knowledge-Based Systems. This paper presents the *LAMBDA* System, an integrated knowledge-base and hypermedia decision support environment, that avoids the authoring problem by using statistical information retrieval methods for automatic Hypermedia and Hypermedia-to-Knowledge-Base System link generation. Using statistical information retrieval techniques for automated link

E. A. Yfantis (ed.), Intelligent Systems, 351–361.

generation allows the benefits of combined Hypermedia and Knowledge-Based System technology to be realised without increasing the knowledge engineering effort.

2 Hypermedia, Knowledge-Based Systems, Decision Systems, and Text Retrieval

2.1 Hypermedia Systems

Hypermedia technology is based upon recent advances in information technology that provide computer access to very large databases, high-bandwidth data communications networks, on-line manuals, and CD-ROM. These advances, and particularly the rapid increase in the number of available on-line databases, have accentuated the importance of computer based information retrieval methods. Most of the available information is in the form of text, although the amount of digitised image, audio and video data is growing. Hypermedia is a powerful new information representation technology that brings new perspectives to accessing information from large on-line databases by providing superstructural and infrastructural networks of links between media components. Hypermedia has the potential to become a significant application area, encompassing and extending word processing, spreadsheets and general database applications. Hypermedia differs from conventional representations of information in three significant ways: network representation of information, reader participation in the information accessing process, and machine support for accessing information [1]. Readers of hypermedia choose their own path through structured information, exploring related topics at will. Appropriately designed hypermedia systems can record the user's path, save and restore interupted sessions, and provide access to recorded paths that different "experts" would have suggested for particular purposes.

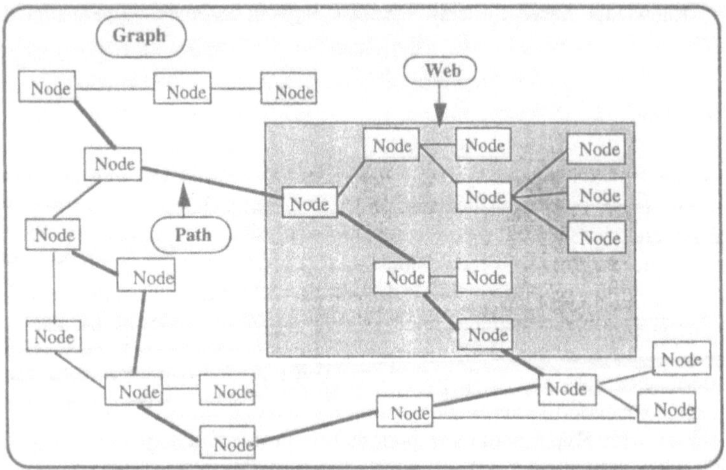

Figure 1: Structure of Hypermedia Information

Figure 1 represents the structure of hypermedia information. A node can consist of a discrete block of *editable data*. This data may be text, graphics, animation, sound, music, or any other type of information that a computer can control or present. Links are defined to interrelate nodes, and are typically represented as data within the hypermedia system that are stored and accessed separately from the nodes that they refer to. Links can therefore be traced automatically, providing networks of potential data presentation sequences for system users. The links between the nodes in any hypermedia system are normally constructed manually by subject experts and/or authors, and hence is a time-consuming process. This is a serious drawback in any hypermedia system. However, a more fundamental problem with manually defined link structures concerns the fundamental concept of what a link is supposed to be.

A distinction may be made between *structural links* that define specific structures within a hypermedia system (eg. defining the chapters, sections, and subsections of a conventional book), specific *referential links* that provide links from explicit references to the objects referred to (eg. other documents, subsections within a document, figures, etc.) and *semantic links* that define connections between points of interest. There is no general view of the nature of semantic links other than that they interconnect "related" information, where relatedness is informally understood in terms of the meaning of the linked items. In practice this means that a hypermedia system author believes that two items of media are related, and so creates a link between them. However, relatedness is completely dependent upon the context: why the user is following links between data units, what problem they are trying to solve, what satisfaction they are seeking (eg. simple entertainment, or the solution to a specific problem), how their own knowledge and understanding relates to the information present in the system, etc.. The hypermedia author must severely limit the number of links and link types available to users, thereby severely limiting the potential usefulness and usability of the system, or is faced with an impossible knowledge engineering task.

2.2 Knowledge-Based/Decision Support Systems

Knowledge-Based/Decision Support Systems that emerged from applied artificial intelligence in the 1970's have demonstrated their ability to solve difficult problems in specific domains. Any system that possesses some form of knowledge about the problem domain can be called a knowledge-based system. However, Knowledge-Based Systems are more typically understood as systems in which an explicit (declarative) model of knowledge about the problem domain is expressed in a formal language and used with general-purpose inference procedures to solve problems. Expert Systems are a subset of Knowledge-Based Systems, typically understood as systems that address problems difficult enough to require significant human expertise for their solution. Knowledge-based systems possess the capability to provide explanations about decision making processes. They have the ability to explain *why* a decision was made and also *how* the decision was made. The extent of the explanation capabilities typically depend on the knowledge base; the richer the knowledge base, the richer the explanation may be.

The difficult process of acquiring the knowledge of a problem domain during system building, often called the "knowledge engineering bottleneck", has emerged as a key issue underpinning the creation of Knowledge-Based Systems. Unfortunately there is no all-encompassing unified theory of how knowledge is acquired, and probably never will be. The success of any given knowledge-base application depends heavily upon how easy it is to develop an appropriate representation of knowledge adequate for solving the problem at hand. The more that a problem depends upon the excercise of judgement, the application of common-sense or general knowledge, or an appreciation of dynamic aspects of the situation in which the system is used, the more difficult it is to develop an appropriate knowledge model. The difficulty of representing sufficient contextual knowledge can be identified as the source of the typical brittleness and inflexibility of knowledge-based systems.

One approach that is being pursued to address this problem is to actually represent large amounts of common-sense knowledge to supplement more problem-specific knowledge bases [5]. This approach has the drawbacks of, firstly, assuming at least implicitly that "common-sense knowledge" is somewhat static, and, secondly, of being unable by itself to maintain dynamic situational awareness. An alternative approach is to represent well-understood aspects of a problem-solving process or a problem domain, and to rely upon user inputs to provide the mapping from the represented knowledge models to the current situation. The drawback here is in the system relying more heavily upon users, where the initial aim was for the system to help and support users. Hypermedia technology can therefore be used to supplement knowledge-base technology to provide additional support beyond declarative knowledge models.

2.3 Integrating Knowledge-Based/Decision Support Systems with Hypermedia

Hypermedia technology can be used to provide rich, complex, and loosely structured information in support of human problem solving: the user interprets and applies this knowledge to the current situation to provide inputs to ongoing automated decision processing. Hence, integrated hypermedia and knowledge base technologies can help to redress the limitations of knowledge systems while still supporting, extending, and complementing human decision processes (this development extends initial work in using links to documents that represent the *source* of knowledge used to develop a rule or structure in a knowledge system, used for purposes of explanation, as demonstrated by Jansen [3].

In addition to text, the user is able to selectively see information and explanations about conceptually difficult problem attributes through the use of pictures, full-motion video, animation or graphics, and hear information with stereo sound, voice recordings, or music. Research has shown that when a system is presented with relevant audio/visual capabilities, retention increases by 10 percent and persuasiveness by 43 percent. For example, if a physician is unfamiliar with terminology, procedures, or methodologies associated with a newly identified disorder, greater insight could be obtained by requesting indepth definitions, explanations, or a visual example of the application of the recommended treatment. This would be particularly useful for demonstrating a deviation from standard procedures in a step-by-step manner via full-motion narrated video with text or graphics superimposed to highlight critical points. Ragusa [7] describe the reality and promise of the synergistic integration of these two technologies namely, the knowledge-based system and Hypermedia. Efficient retrieval of relevant information to support the decision making process of a knowledge-based system can be very crucial to the success of such a system.

In addition to enhancing automated decision functions, the integration of hypermedia and knowledge-based system technology addresses the particular problem of users becoming “lost in hyperspace”, ie. knowledge-based system structures can provide a situationally dependent organisational framework for accessing and navigating through hypermedia networks.

The major outstanding challenge for integrated hypermedia/knowledge-based systems is to avoid compounding the “knowledge engineering bottleneck” by extending the structuring activities of knowledge acquisition to include the creation of hypermedia link networks, and hypermedia-to-knowledge-based system links. Here it is proposed that the creation of such links can be fully automated by using statistical information retrieval methods for dynamic link generation.

2.4 Information Retrieval

The rapid increase in the number of available on-line databases has accentuated the importance of computer based information retrieval methods. Most of the information in current databases is in the form of text; although the amount of digitized image and video data is growing, text will continue to be critical due to its unique role as a medium for communication. Much of the research in information retrieval uses a statistical approach. The retrieval strategies and indexing capabilities used in this approach are simple, easily implemented, and effective. The last few years have seen considerable interest within the information community in the development of intelligent information retrieval systems, concentrating on the syntactic and semantic analysis of text and the representation of text in a suitable form [4]. Artificial intelligence based and statistical approaches are not competing alternatives, but may be used together to achieve more effective information retrieval. Ongoing research is addressing the development of unified theoretical frameworks for combined IR/AI systems.

3 Lambda

A considerable amount of work has been done on intelligent information retrieval systems that integrate information retrieval with knowledge-based system techniques to improve retrieval, and to integrate key-word and expression searching within hypermedia systems. However, very little work has been done to integrate statistical information retrieval methods with knowledge-based systems to improve decision support during decision making processes, or as automated dynamic link-generation facilities within hypermedia. *Lambda* is a decision support environment that uses statistical information retrieval to integrate automated, dynamic Hypermedia technology with Decision Support/Knowledge-based systems technology in order to improve the decision support process. Lambda has been developed by the CSIRO Division of Information Technology. The architecture of Lambda is shown in figure 2.

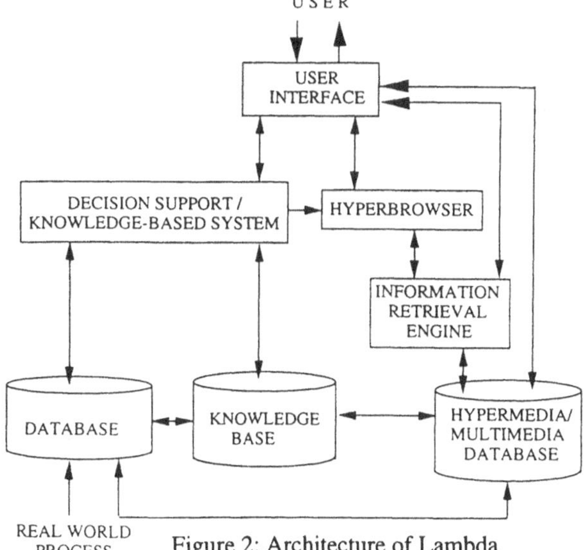

Figure 2: Architecture of Lambda

3.1 Knowledge-based Decision Support Functions

The Lambda decision support/knowledge-based system uses conventional rule-based decision processes. It consists of a knowledge base, inference engine and explanation facility. The knowledge base possesses knowledge about the problem domain in the form of rules. The inference engine fires rules based upon the current state of system knowledge. Conventional knowledge-base dialogue facilities are provided allowing the system to:
- present message information to the user
- present message information along with a number of selection options, and accept a selection specified by the user
- present message information along with an unconstrained data field, and accept arbitrary text inputs from the user

The input functions allow users to provide information to the system, including problem details and inputs that may require the exercise of common-sense, situational, or judgemental knowledge.

3.2 Enhanced Knowledge Base Explanation and User Support Functions

In addition to providing conventional explanation functions (eg. traces of the rule firing sequence for a forward-chaining inference process), the dialogue funtions also allow the user to request supporting information. When selected, the supporting information facility invokes a general-purpose hypermedia interface, referred to as the Hyperbrowser.

The invokation of the Hyperbrowser by the knowledge base involves the knowledge process setting the search context by specifying a particular index identifying the document set that search is to be carried out within. This allows dynamic and automatic modification of the search context during inferencing. Searched items may be located locally, or distributed across a network. The hyperbrowser is invoked with an initial search string set by the knowledge system, which may be purpose specific, may be the content of message information presented to the user by the knowledge system, or may be the textual content of currently active or firing rules or other knowledge structures. If knowledge structures are used, the search mechanism is capable of identifying knowledge sources if they are present within the underlying media base (implementing the kind of knowledge source link described by Jansen [3]. If user message material is used, the initially retrieved information is likely to be related to the problem or question being asked of the user, and thereby supports the user in providing relevant information.

3.3 The Hyperbrowser

The Hyperbrowser is a generic software tool for hypertext browsing via semantic links that are created automatically and dynamically by hidden statistical information retrieval algorithms. This tool was built as a separable part of the Lambda system, and it runs across multiple platforms.The Hyperbrowser retrieval system consists of a set of C++ software routines for ranked information retrieval. The routines have been used for standard information retrieval, and also for automated link generation in hypertext and multimedia applications. Retrieval takes place within the scope of a number of *source objects* that may be individual text documents, or other media types with associated text that is used for search and indexing. The source text is subdivided into *logical text units* (or LTUs), that represent the individual targets of retrieval

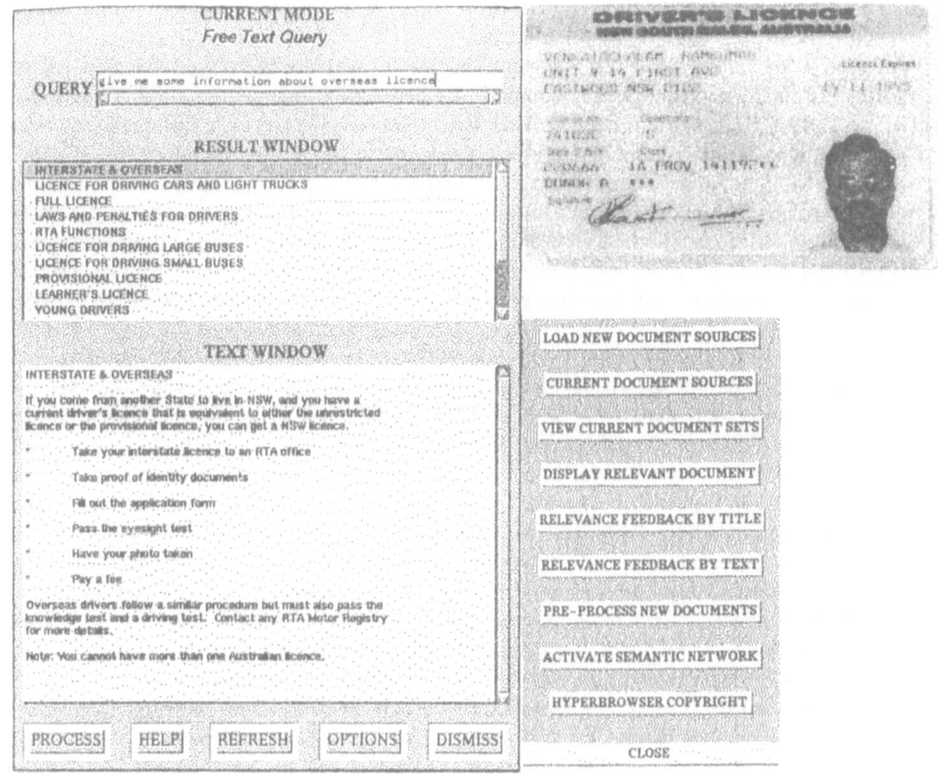

Figure 3: Output of the Hyperbrowser to a free text query

(ie. the items that are retrieved). LTUs are separated within source documents by appropriate markers. The user of the Hyperbrowser can provide an arbitrary text string as the search string. Alternatively, the user can select any subset of the currently displayed text, or the whole complete displayed item, as the search string (using associated text for non-text items), effectively defining hyperlinks from the current objects to other objects in the database. This varied and flexible source of initial search strings makes the system highly interactive, and overcomes the limitations imposed by static and predefined semantic links in the hyperstructure. An example of the Hyperbrowser output is shown in figure 3. Once a list of items has been returned by the Hyperbrowser retrieval engine, a particular item can be selected (eg. by the user) and retrieved from the storage system.

By using the information retrieval routines, the Hyperbrowser creates links between different nodes automatically during user interaction and thus avoids manual structuring of the information. Linking can occur from a complete text unit, or from an arbitrary chunk of subtext selected by the user (the text in each case being used as the search string input to the information retrieval system). The selection and retrieval of text (and associated media types) is a continuing process by which the user can navigate through the database until he/she is satisfied with what they have accessed. Users can also provide an arbitrary search string, and so are not limited to searches based upon currently displayed data.

Lindley [6] conducted a series of experiments to evaluate the effectiveness of statistical information retrieval techniques in automatically generating links within a hypermedia system. The results of the experiements showed that statistical retrieval techniques can be very successful for automated link generation. Users have far more freedom than in manually constructed hypermedia systems, since they can select their own arbitrary anchor points for links. Anchor points therefore require no presuppositions to be made about what may or may not be interesting in the currently displayed text. Moreover, the ability of users to provide their own search string (by a free text entry field) allows more specific expressions of the target subject, and results in an interface that is much more interactive than typical hypertext or hypermedia systems.

When Lambda decision processes invoke the Hyperbrowser, they do not simply invoke the system at some general or arbitrary starting point. Rather, they can invoke the system at a starting point specified within the decision process, and thereby at a point appropriate to the decision-making context. In this way contextuality is accounted for in both directions: users provide general (world-oriented) contextual information to the decision process, and support for supplementary decision-making by users is invoked in a way that is sensitive to the internal context of the automated decision process. Control over context is achieved in two ways:

1. the decision process can specify the *index* used for search, and therefore the *context* of all search operations. The context can in principle be any arbitrary subset of all media objects available to the system. The context specification can limit the context of all subsequent browsing operations by the user, until reset by the automated decision process.
2. the decision process can invoke the Hyperbrowser with an *initial search string*. That string can be derived from messages associated with current rules, from the rules themselves, or have the form of more general text specified within the decision process (eg. comments attached to knowledge structures). The effect of specifying the initial search string is that the Hyperbrowser will be invoked displaying an initial list of items relevant to the current point in the decision process.

The information retrieved by the Hyperbrowser can in principle include any combination of text, video, audio, graphics, or animation, although in all cases the search and indexing is only carried out on text associated with the other media forms (which may be titles or content descriptions). The Hyperbrowser provides users with access to hypermedia information that can provide information in a variety of media forms to support user decision-making that is required to provide inputs to the knowledge-based decision process. In this way, a very large amount of heterogeneous and loosely structured information can used in the solution of a problem.

The Lambda interfaces provide an additional option to use semantic network processing (if a network is present). In this mode of operation, a thesaurus is used for query expansion in order to take into account semantic information that is missed by purely lexical IR techniques. The thesaurus can be general-purpose or problem-specific, depending upon user needs. The use of the thesaurus is described in more detail in [6].

3.4 The Indexing and Retrieval Process

The Hyperbrowser retrieval process requires preprocessing of the material that is to be retrieved. Input parameters for the preprocessor include:

- the name of a file (the *source file*) that lists all of the source documents to be referenced by a particular retrieval index. The source documents may be located at any accessible position within a network environment.
- the name of the *index file* that is to be generated.

Preprocessing involves the application of:

1. *conflation algorithms*, which carry out word stemming on the input text files to overcome problems with different word endings, variations in spelling, punctuation, and hyphenation.
2. *stop lists*, to filter out terms within the input text files that add little to the mapping operation, such as conjuncts, articles, etc..
3. *synonynm* lists to ensure that similarity matching is not limited to identical word forms, but captures identity of meaning.
4. information trace generation algorithms, to produce information traces for the whole text base, and for each LTU.

The information traces are fundamental to the retrieval process. To generate an information trace, a pre-filtered document is parsed into character substrings, or *ngrams*, of length n (where a default value of n = 3 has been found to provide the best results). The number of occurrences of each unique ngram are counted. The information trace for an LTU contains a list of each ngram that the LTU contains, along with the frequency of occurrence of each ngram within that LTU. The information trace for a data set (or database) contains the total number of occurrences of each unique ngram within the data set. The information traces for an LTU or a data set also include the *relative frequency* of each ngram within the LTU or data set, respectively.

The output of the preprocessor is an index file that contains:

- an information trace for the whole data set covered by the index
- an information trace for each LTU, along with a reference to the LTU (eg. path name and offset within the file that an LTU belongs to)

The index produced by the preprocessor is used for information retrieval. Input parameters used by the retrieval process are:

- the name of the index file
- a threshold for minimum item similarities

The algorithm used by the retrieval process is:

- accept an input search string
- apply conflation algorithms, stop-list filtering, and synonym expansion to the input string
- generate an information trace for the input string
- generate similarity measures that represent the similarity of the input string to the LTUs within the target data set, as a function of ngram frequencies within the search string, the LTUs, and the overall data set
- return a list of LTUs in decreasing order of similarity to the input string, for those LTUs with a similarity above the specified similarity threshold. The returned list includes the path name to the associated item in the data set.

The Hyperbrowser uses document ngram statistics to create links automatically. An ngram is a character substring of length n, where a value of n = 3 has been found to work well in practice. After filtering, conflation, and synonym expansion, each LTU and each search string is parsed to create an *information trace*. An information trace contains a list of each ngram in its parent object, together with the absolute and relative frequencies of occurrence of the ngram in the parent. The similarity calculation is then made by comparing the information trace of a search string with those of the target objects. Detailed descriptions of this technique can be found in [9] and [10].

4 Ongoing Work

Ongoing work by the CSIRO Division of Information Technology is aimed at increasing the usefulness and flexibility of initial system functions. Current project activity involves extensions to the initial Lambda system that focus upon distribution of the information retrieval algorithms. This will allow searching to be conducted within an open network environment, as long as appropriate infrastructural support is provided. The aim is to allow the searching process to access relevant information on the network without itself having to know anything about the underlying network structure. Further extensions will include the encapsulation of system processes within agent shells, allowing concurrent search and access to collective data resources by multiple users and/or automated decision processes.

5 Conclusions

In this paper we have demonstrated successfully the possibility ot using statistical information retrieval methods for automatic and dynamic link generation both within the hypermedia system and between a hypermedia system and a knowledge base, thus eliminating the need for manual link creation and thereby greatly simplifying the knowledge engineering process. This approach goes a step ahead of the traditional approach to explanation in any knowledge-based system where the system is limited to responding with the list of rules it fired during its reasoning process.

6 Acknowledgments

The authors are thankful to Kai Foong for her contribution towards building an interface for the Hyperbrowser Tool, and to Rosemary Irrgang for her work on the information retrieval engine. The authors also thank Anne-Marie Vercoustre for extensive comments on the paper. This project was supported by a research grant from the CSIRO Institute of Information Sciences and Engineering.

7 References

[1] Conklin J. (1986) 'A survey of hypertext' MCC Technical Report STP-356-86, MCC Software technology program, Austin, TX.

[2] Holsapple C.W., Pakath, R., Jacob, V.S., Zaveri, J.S. (1993) 'Learning by Problem Processors' Decision Support Systems 10 pp 85-108.

[3] Jansen B., Robertson J. (1989) 'Management of Wool Dark Fibre Risk Knowledge Using Hypertext' CSIRO Division of Information Technology Technical Report TR-FD-89-05.

[4] Kumar V.R. (1994) 'Intelligent Information Retrieval Systems: A Survey' CSIRO Division of Information Technology Technical Report (to be published).

[5] Lenat D. B., Feigenbaum E. A. (1991), 'On the thresholds of knowledge Artificial Intelligence' 47(1-3) pp 185-230.

[6] Lindley C.A., Kumar V.R., Irrgang R., Robertson J.R. (1994) 'An Evaluation of Information Retrieval Methods and Semantic Network Processing for Automatic Link

Generation in Hypermedia Systems' Proceedings of the Second International Interactive Multimedia Symposium, Perth, Australia pp 290-297.

[7] Ragusa J.M., Orwig G.W. (1991) 'Integrating Expert Systems to Multimedia: The Reality and Promise' Proceedings of the First World Congress on Expert Systems pp 2919-2930.

[8] Sipior J.C., Garrity E.J. (1992) 'Merging Expert Systems with Multimedia Technology' Data Base Winter pp 45-64.

[9] Teufel B. (1989) 'Informationsspuren zum numerischen und graphischen Vergleich von reduzierten naturlichsprachlichen Texten.' Informatik-Dissertationen ETH Zurich NR.13. Verlag der Fachvereine Zurich.

[10] Teufel B., Schmidt S. (1988) 'Full text retrieval based on syntactic similarities' Information Systems Vol.31 No.1.

The Design of Object-Oriented Meta-Architectures For Programming Languages*

Guruduth Banavar and Gary Lindstrom
Department of Computer Science
University of Utah, Salt Lake City

Abstract. This paper is a survey of the design of four object-oriented meta-level architectures for programming languages. We present overviews and compare the salient features of the meta-architectures of Smalltalk, Common Lisp Object System (CLOS), a Scheme Compiler, and Etyma, our framework for modular systems. This comparison clarifies important architectural aspects of the surveyed systems, such as the space of concepts captured by the architectures, and the abstractions that embody similar language concepts across the architectures. We find that there are considerable differences in the goals and conceptions of these architectures, yet they can all be used for similar applications. Finally, we point out some strengths and weaknesses of the architectures surveyed.

1 Introduction

Object-orientation is a popular design technique that has been used to model application domains of all varieties [1]. A recently emerging trend is to apply the object-oriented (O-O) method to the design of O-O language processors themselves, thereby harnessing the much touted advantages of abstraction and reuse in this domain also. For example, the Meta Objects of the CLOS MetaObject Protocol (MOP) [2] is an O-O model of certain useful concepts in CLOS. A recently proposed O-O model for a compiler for a non O-O language, Scheme, [3] is another example. One of the earliest O-O programming systems, Smalltalk-80 [4], was itself built upon an intricately interconnected group of meta-classes.

The above languages embody O-O meta-level architectures, or meta-architectures for short, in the sense that they model the fundamental concepts in the language, such as *class* and *method*, as interacting meta-classes. This has resulted in reflective, flexible, and extensible language designs. Many of these advantages stem from the fact that reified meta-classes are candidates for systematic reflective access. That is, a system that has a well-designed meta-architecture can essentially provide users not only with its standard

*This research was sponsored by the Advanced Research Projects Agency (DOD), monitored by the Department of the Navy, Office of the Chief of Naval Research, under Grant number N00014-91-J-4046. The views and conclusions contained in this document are those of the authors and should not be interpreted as representing official policies, either expressed or implied, of the Advanced Research Projects Agency or the US Government. Contact author: G. Banavar, Computer Science - 3190 MEB, University of Utah, Salt Lake City, UT 84112 USA, e-mail *banavar@cs.utah.edu*, phone +1-801-581-8378, fax +1-801-581-5843.

E. A. Yfantis (ed.), Intelligent Systems, 363–374.

interface, but with an alternative interface — a "side door" to the internal architecture, which is typically a subset of the meta-architecture interface. Information access and refinement via this alternative interface can enable applications to fine-tune a language implementation to suit its particular needs. Meta-classes can be specialized to suit specific tasks using standard O-O techniques such as inheritance. In a compilation setting, meta-classes can even be specialized to statically optimize run-time data layout or generate optimized code for particular special cases.

It is important to clarify the relationship between the concepts of *meta-architecture, reflection*, and *metaobject protocol* (MOP). A meta-architecture models, systematically implements, and documents the fundamental concepts of a system. A meta-architecture is O-O if the concepts are modeled as collaborating classes. A system is reflective if its users have introspective (i.e. read) and/or intercessory (i.e. modification) access to the internal implementation of the system. Finally, a MOP documents and illustrates a disciplined method of reflective access to a carefully chosen subset of a system's O-O meta-architecture.

The semantics of a programming language is rarely made formally explicit in the language implementation, let alone made usefully accessible from within the language itself. This may be due to the fact that the design of programming languages is generally considered to be an amorphous creative activity, carried out by the best experts in the field. However, this need not necessarily be the case — an important point of this paper is that the design discipline encouraged by object-oriented methods can be fruitfully applied to the design of programming languages themselves. Furthermore, a significant part of the semantics of a language can be reified as an O-O meta-architecture. A well-designed meta-architecture enables reuse, reflection, and the design of a suitable MOP, and thus brings the advantages mentioned above.

The traditional architecture for processing high level languages involves the following, usually separate, languages: the source language, the target language (this is not significant in the case of direct interpreters), and the processor description (implementation) language. An important observation is that the meta-architecture of a language is expressed in its processor description language. From this it follows that a language can have multiple meta-architectures corresponding to multiple processor descriptions, each designed with different requirements. It also follows that a language's meta-architecture need not necessarily be meta-circular, i.e. expressed in the source language itself. In fact, in order to have an O-O meta-architecture, a language need not even be object-oriented. However, its implementation language must. The meta-architectures of dynamic languages such as Smalltalk are meta-circular since a large part of the language is implemented using the language itself, as is its meta-architecture.

In this paper, we contrast the three meta-architectures mentioned above, Smalltalk, CLOS MOP, and the Scheme compiler MOP, along side our own meta-architecture based on a language called Jigsaw [5]. Smalltalk and CLOS are general-purpose O-O languages that enjoy significant followings. Smalltalk was built ground-up based on a remarkably coherent meta-architecture, while the meta-architecture for CLOS was retrofitted onto the language. The Scheme compiler MOP shares the goals of the CLOS MOP, but is significantly different from the above two since Scheme is not O-O, and its MOP is compile-time oriented. Finally, our framework, Etyma, attempts to generalize and bring together many of the concepts from the above meta-architectures.

The paper continues by discussing the design issues investigated in this survey, followed by detailed descriptions of the four meta-architectures under consideration. Finally, we provide a summary of architectures and conclude.

2 Design Issues

The design of meta-architectures for languages is driven by various considerations. In this section, we outline some of the issues that govern the design of meta-architectures, the main categories being (i) the pre-stated design goals of the language as governed by the requirements of applications, and (ii) the requirements of semantic models of languages, also driven by applications — i.e. how the abstractions of meta-architectures must capture the crucial semantics of their languages.

Goals and Application Requirements

Consider the competing goals of *generality vs. backward compatibility.* A primary requirement in the design of the CLOS MOP was backward compatibility with several existing LISP Object systems which were pairwise incompatible. The refineability of the object model enabled by the MOP essentially brought about the backward compatibility. Hence, it was sufficient to model just the object system in the meta-architecture, in such a manner that backward compatibility can be achieved. One can imagine that the meta-architecture in such a scenario could be markedly constrained by the existing object models. On the other hand, the Smalltalk and Etyma meta-architectures were built from scratch based upon a uniform model, i.e. every concept in these systems is modeled in the meta-architecture. Moreover, Etyma was designed from the start with the explicit purpose of abstracting semantic commonalities in module-based languages and systems. As a result, the abstractions provide a clean module and inheritance model, leaving the rest of the language design open.

An important application of meta-architectures is the support for flexibility and extensibility of a language via *reflection and MOPs.* In addition to introspective access, a MOP typically provides intercessory access to meta-objects, making it possible to refine them incrementally using standard O-O techniques and hence amend the existing language design itself. The user can define new meta-object classes as specializations (subclasses) of the standard meta-object classes, refining their behavior as necessary. The most important goal of the CLOS and Scheme compiler meta-architectures is to provide a metaobject protocol for users. This has led to the pragmatic design goals of (i) controlled and carefully documented user extensibility, (ii) interoperability of separately designed extensions, (iii) efficiency via user specialization, and (iv) ease of use. In the context of MOP design, there is a tradeoff between implementor freedom and user extensibility. The CLOS MOP designers deal with this tradeoff by explicitly specifying restrictions on the usage of the MOP.

Given reflective access to its meta-architecture, a language's source programs are made up of base language code interspersed with meta-code, i.e. the code that accesses the meta-architecture. One design issue in such systems is the *execution time of meta-code.* Dynamic environments like Smalltalk and CLOS require meta-code to execute at run-time, while the Scheme MOP runs meta-code at compile-time. Dynamic architectures exhibit meta-circularity, and hence a tight coupling between the language and the meta-architecture. While this coupling enhances application development flexibility, it causes the meta-level architecture to become difficult to disentangle from their base languages for separate reuse.

The need for self-applicability (*meta-circularity*) is an important design issue. It is a fundamental requirement in the dynamic environments of Smalltalk and CLOS MOP. It is not even possible in Scheme since it is a non O-O language with an O-O meta-architecture. The requirements of static typing and separate compilation make it impossible to express the Jigsaw language (on which Etyma is based) using itself. The details of this are beyond the scope of this paper [6].

A language meta-architecture supports *reuse* of design and code just as a domain specific O-O framework does. O-O frameworks for several domains have been constructed [1]. Similarly, an O-O framework can be used as a reusable domain model for O-O languages and systems — indeed, this is a primary goal of Etyma.

An important use of meta-architectures is as an *aid in understanding/maintaining* the system. Another use is the *construction of program analysis tools* such as browsers and debuggers. Specific meta-architectures have also been used for other applications such as *persistence*.

Requirements of Semantic Models

In addition to general goals such as the above, meta-architecture designs have several semantic requirements. For instance, type-checking is an important issue that must be taken into early consideration when building a meta-architecture. Although meta-architectures are considered extensible, the design decisions in the area of typing built into existing architectures pervade the entire model, and make it hard to retrofit significant static semantics. Another example of a fundamental requirement is the semantics of inheritance.

Feature	Smalltalk-80	CLOS MOP	Etyma
Inheritance	Single	Multiple	Unbundled
Encapsulation	Object-level	(none)	Object-level
Method dispatch	Single	Generic/Multi	Single
Static Typing	No	No	Structural

Figure 1: Selected O-O semantics of surveyed languages

By its very nature, the meta-architecture of a language captures and constrains the semantics of the base language. The space of high-level language semantics is broad and a subspace of it must necessarily be carved out by a particular meta-architecture. The larger the subspace, the more complex and potentially less useful the meta-architecture. Once the subspace is chosen, the particular way in which semantic concepts within this subspace are modeled is also significant. Furthermore, the location of the point representing the base semantics of the language must be chosen within this subspace. Figure 1 tabularizes a sample of O-O semantic choice points of the languages surveyed here. In the following sections, we describe the specific semantics captured by meta-architectures for O-O languages in detail, including support for inheritance, encapsulation, method dispatch, instantiation, static typing, and abstract classes.

We now turn to a more detailed treatment of specific meta-architectures.

3 Smalltalk

Smalltalk is based on a uniform model of communicating objects. It has a small number of concepts — object, class, instance, message, and method. Every concept in the system is modeled as an object, either instantiable (class object) or not (instance object). The most primitive low-level operations in the system are delegated to a virtual machine. Objects communicate via messages; the semantics of messages are implemented by receivers as methods.

Smalltalk's notion of objects is captured by class `Object` which provides the basic semantics, including message handling, of all objects in the system. The semantics of classes is captured by class `Class` along with its superclass `Behavior` which defines the state required by classes, such as for instance variables and a method dictionary. Further, the class `CompiledMethod` embodies the notion of a class' method; this class defines a method `valueWithReceiver:` to evaluate itself. The O-O semantics of Smalltalk captured by the meta-architecture is given in Figure 2.

Inheritance	The class `Class` implements a message `subclass:...` which accepts one class parameter, the superclass, thus implementing single inheritance. The subclasses of `Class`, which are the meta-classes of individual classes, inherit the method `subclass:...`, thus individual class objects also respond to this message. The assumption that classes have a single superclass permeates the system. Inheritance of state and methods is captured by the superclass of `Class`, class `Behavior`, which implements methods to compute the set of instance variables and methods available to instances of a class.
Message handling	Class `Object` defines a method `perform:withArguments:` that handles message dispatch using a primitive method of the Smalltalk virtual machine. There is also a class `MessageSend` that captures the notion of a message-send. However, for efficiency, an instance of this class is not created for every message-send in the system.
Encapsulation	Instance variables are encapsulated in Smalltalk. Method handling code searches only the method dictionary of a class, but not the instance variables. Method objects have access to instance variables since they refer to a scope object that records the variable objects accessible within that scope.
Instance creation	Class `Behavior`, the superclass of class `Class`, defines a message `new:`, which calls a primitive message to create an instance of the receiver (which must be a class).

Figure 2: Smalltalk's O-O semantics

Smalltalk is a "dual hierarchy" language, as are most object-oriented languages. That is, it has a cleanly articulated class-subclass hierarchy as well as a class-instance hierarchy. In most languages, however, the class-instance hierarchy is not interesting since it comprises only two levels — that of all classes and all instances. In Smalltalk, this hierarchy is deeper, and is recursive, as described below.

Every object in Smalltalk is an instance of some class. Since classes themselves are objects, each class object is an instance of yet another class, usually referred to as a *metaclass* object. For example, a class `Foo` is an instance of its metaclass, given by the expression `Foo class`. Such metaclass objects are themselves instances of an ordinary class called `Metaclass`. The metaclass of class `Metaclass` itself is given by `Metaclass class`, which is also an instance of class `Metaclass`, just as `Foo class` is. The above recursion puts an end to the infinite regression of metaclasses.

Consider the class-subclass hierarchy of metaclasses. Every class in Smalltalk inherits from class `Object`; hence the subclass hierarchy is a singly rooted tree. The class `Metaclass` mentioned above is also a subclass of `Object`. The instances of class `Metaclass`, such as `Foo class`, `Metaclass class`, and even `Object class`, are all (meta)classes. These metaclasses are subclasses of class `Class`, which is a subclass of class `Object`[1].

In Smalltalk, the meta-architecture is really "infinitely open," in the sense that every single concept in the system (except some primitive operations that are performed directly by the virtual machine) is captured as an object which users can not only specialize, but also browse and access, i.e. directly edit and modify, which of course is strongly discouraged.

4 CLOS MOP

As mentioned earlier, a primary requirement in the design of CLOS was backward compatibility with existing LISP systems. It was recognized that these incompatibilities could be reconciled if a family of languages, rather than a single one, were defined. Thus, the CLOS meta-architecture was designed to facilitate modeling an entire *space* of language designs, with the default CLOS design being a distinguished point. Furthermore, a protocol (MOP) has been carefully designed and documented to access this meta-architecture usefully.

The CLOS object system supports the standard concept of *classes*, which can be instantiated into *instances* [7]. Class attributes are called *slots*. A distinguishing feature of the CLOS model is the notion of *generic functions* which are defined independent of any class, and can be specialized into *methods* that are applicable to specific classes. Generic functions can be dispatched based on *multiple* arguments (multi-methods).

The CLOS meta-architecture specifies the following basic meta-object classes corresponding to the basic concepts of the language: `class`, `slot-definition`, `generic-function`, and `method`. All user-defined metaobjects must be designed to be subclasses of one of the above meta-object classes. The specified default semantics of the CLOS language are embodied by specializations of the above classes; with names beginning with `standard-..`, e.g. `standard-class`, and `standard-method`. The manner in which the meta-architecture captures basic O-O semantics in CLOS is given in Figure 3.

The class-subclass hierarchy of the CLOS meta-architecture is as follows. At the root is class `t` which has one subclass `standard-object` capturing the semantics of all objects in the system. Every class created in the system must have `standard-object` as its superclass. One subclass of `standard-object` is the class `metaobject`, of which the basic meta-object classes mentioned above are subclasses.

The class-instance hierarchy of CLOS essentially has four levels. Individual CLOS classes are instances of class `class` or one of its subclasses. Class `class` is an instance of (its own subclass) `standard-class`, as are most other meta-object classes.

The CLOS MOP has functions for systematic *introspective access* to its meta-objects. For example, the programmer can access the class meta-object of a given object, the

[1] The actual subclass hierarchy of Smalltalk is slightly more involved than what is described here, due to the desirability of symmetric class and metaclass hierarchies, but the given description will suffice for this discussion.

Inheritance	Generic functions specialized on the `class` metaobject class implement the semantics of multiple inheritance. A *class precedence list*, i.e. a total ordering on a class' superclasses, is computed by the generic `compute-class-precedence-list`. The generic function `compute-slots` computes the full set of slots accessible from instances of the class. The semantics of slot property union is implemented by `compute-effective-slot-definition` which is called by `compute-slots`. The generic function `class-default-initargs` computes the full set of initialization arguments required by the class.
Generic invocation	A *discriminating function* associated with a `generic-function` metaobject provides the semantics of (multi-)method dispatch. The discriminating function is computed by a generic function `compute--discriminating-function`. Dispatch proceeds by first finding the set of applicable methods for the given set of arguments from the set of all methods associated with the generic function metaobject, via `compute-applicable-methods`, and computing an *effective method* via `compute-effective-method`.
Slot access	The function `slot-value`, a wrapper for the generic function `slot-value-using-class`, is used for slot access. The generic function itself is specialized to class and slot metaobjects implementing the semantics of slot access in CLOS objects.
Instance creation	The generic function `make-instance` and `allocate-instance`, both specialized to `class` metaobject class, implement instance creation. Prior to creating an instance of a class, it is *finalized* by computing the actual structure of the class as described under "inheritance" above.

Figure 3: CLOS's O-O Semantics

class meta-object's name, superclasses, slots (each of which is a meta-object on its own), subclasses and methods. The details of each slot meta-object, generic function, and method can also be accessed. Using these functions, it is possible to, for example, reconstruct a textual description of an object's class.

CLOS MOP is a *layered* protocol, i.e. the protocol specifies meta-architecture functionality at various levels of detail, with higher levels delegating work to lower levels, so that user-refinement can be made at various granularities of semantics. For instance, a top layer protocol concerned with inheritance is the generic function `finalize-inheritance`, which delegates to the next layer, `compute-class-precedence-list` and `compute-slots`, which further delegates to `compute-effective-slot-definition`.

A large number of applications that the CLOS MOP can be put to are illustrated in [2]. These include specialized classes such as counted classes and encapsulated classes. CLOS MOP has also been utilized to provide a significant persistence facility [8].

5 Etyma

Etyma is a general meta-level architecture for O-O languages realized as a C++ framework (in the sense of [1]). The primary abstractions of Etyma are based on a language called Jigsaw, a module manipulation language designed to model the semantic foundations of object-orientation, especially *inheritance*, in all its forms. A basic premise of this work is that O-O concepts, properly formulated, can be applied not only to traditional programming language design, but for the broader design and implementation of O-O programming systems, such as linkers/loaders, library management tools, configuration management systems, type checkers, etc.. We name our framework Etyma (the plural of "etymon," taken from the etymology of "etymology") since it is a collection of root concepts from which other concepts are formed by composition or derivation.

In Etyma, as in Jigsaw, the central concept is that of a *module*, akin to a *class*, which can be informally defined here as any software unit that provides a set of services as specified by its interface. A module consists of a set of labels (identifiers) each associated with either (i) a *value* (e.g. an integer, a function) in a language's value domain, or a *type* in the language's type domain, or (ii) a *location*, in which case the label corresponds to a mutable instance variable, or (iii) a (nested) module or its *interface*. The key characteristic of this model is that modules can be combined using a suite of unbundled and general module *combinators* to achieve various effects of inheritance, sharing, and encapsulation. A summary of the key semantics captured by the meta-architecture is given in Figure 4.

Inheritance	The semantics of the usual kinds of inheritance is supported by some combination of the primitive operators `merge`, `override`, and `copy_as`, with various other effects achieved via the operators `rename`, `restrict`, and `freeze`. All of these operators are implemented as methods of class `Module`.
Typing	A static type system with subtypes and inherited types is supported. The structural type of modules is captured by class `Interface`. Etyma also has a hierarchy of type meta-classes to capture the type space of programming languages.
Encapsulation	Supported by the `hide` operator of `Module`. Hidden attributes are removed from a module's interface, and are accessible only by a class' own methods.
Method dispatch	Supported by the `select` method of class `Instance`. In the default case, `select` dynamically dispatches on attribute name.
Abstract classes	Modules can have attributes whose types are declared but which are not defined. Such modules correspond to abstract classes, and cannot be instantiated.
Instantiation	Supported by the method `instantiate` of class `Module`, which returns an object of class `Instance`.

Figure 4: Etyma/Jigsaw O-O semantics

The class-subclass hierarchy of Etyma has class `Etymon` at the root. A subclass `TypedValue` embodies the typed value domain of languages, and another subclass `Type` embodies the corresponding type domain. Class `Module` is a subclass of `TypedValue`, as is class `Instance`, but via class `Record`. There is a parallel type hierarchy with classes `Interface`, `InstanceType`, and `RecordType`.

In [9], we have described a preliminary C++ prototype implementation of Etyma. The design of abstractions in Etyma has been guided mostly by semantic concerns, with ideas based on a denotational description of the Jigsaw language. Etyma can be used to describe and build processors for many systems that can be construed to be module-based. Languages that are derived from the framework are called *client* languages, and processors for them are constructed by *extending* the framework. The client language is in general unrelated to the framework implementation language — an extension of Modula-3, Modula-π, is presented in [5], and examples of simple languages based on C++ are given in [9]. Etyma is being used as the meta-architectural framework for a larger initiative for evolutionary support for modular architectures, in which a module-based server-style linker/loader is being designed as an extension. In this extension, UNIX ".o" and ".so" object files are regarded as specializations of class `Module`, thus enabling the use of comprehensive inheritance semantics [10] and type checking [11] in their composition.

6 A MOP for Scheme Compilers

Like the CLOS MOP, this MOP carefully chooses a useful portion of the internal functionality of a scheme compiler in order to provide the Scheme programmer with the desirable attributes of flexibility and control over layout and access over run-time data. Many of the details of this MOP are still under development [3, 12], so we only give a general description of it.

Unlike the CLOS MOP, this is a compile-time MOP, i.e. the accessible meta-architecture is specializable to control the *static* behavior of the compiler. Such static specialization is utilized at run-time. However, meta-code (the code that accesses the MOP) is not executed at run-time. This architecture decouples the language of the static processor (compiler), and hence the meta-architecture, from the source language itself — thus it is not meta-circular. The meta-architecture is expressed in an O-O extension of LISP called Traces [13]. This meta-architecture attempts to capture certain aspects, such as procedures and pairs, of a non O-O base language, Scheme.

The primary abstraction in this meta-architecture is what is termed a "contract." A *contract metaobject* represents a group of interrelated source program fragments that must agree on the layout of run-time data. For example, a `lambda` abstraction and all applications of (i.e. calls to) it would be such a group. Contracts essentially capture the notion that an abstraction and its uses must *statically* agree on conventions such as run-time layout. Other such static "contracts" include `cons` pairs along with its accessors `car` and `cdr`, and `let` environments along with their variable accesses.

Dependencies between abstractions and their uses are traced by flow analysis on source program fragments captured as *program graph metaobjects.* Source programs are translated into a register transfer language captured as *RTL metaobjects.* For instance, when a function application is required to generate code, it delegates the job to its contract metaobject, which further requests the appropriate program graph metaobjects to generate RTL metaobjects.

A few applications of the Scheme MOP are illustrated in [12]. These include extending the base Scheme language to support procedures with extra data attached to them, immutable data structures, and procedures that are dispatched based on the number of input parameters.

7 Summary and Conclusions

In this section, we attempt a summary of the meta-architectures surveyed. Of course, it is impossible to provide a comprehensive summary of the depths of the meta-architectures; instead we give a broad comparison of some essential aspects.

Figure 5 shows the abstractions, both O-O (e.g. class) and non-O-O (e.g. function), captured as meta-classes by the architectures. Figure 6 gives a comparative summary of the architectures in the areas of inheritance, method dispatch, encapsulation, and static typing.

	Smalltalk	CLOS MOP	Etyma	Scheme
Meta	`Metaclass`	(none)	(none)	N/A
Class	`Class`	`standard-class`	`Module/Interface`	N/A
Instance	`Object`	`standard-object`	`Instance/InstanceType`	N/A
Function	`BlockClosure`	(none)	`Function/FunctionType`	*lambda contract*
Variable	`Variable`	(none)	`Location/LocationType`	*let contract*
Primitive Value	`Magnitude`, etc.	(none)	`PrimValue/PrimType`	*[primitive] contract*

Figure 5: Summary of selected abstractions

Inheritance	In Smalltalk, the class `Behavior` and its subclass `Class` together model single inheritance semantics. In CLOS MOP, generic functions specialized to class `class` model multiple inheritance semantics. In Etyma, class `Module` implements unbundled inheritance operators. The default semantics of inheritance is significantly broader in Etyma/Jigsaw compared with the defaults in either Smalltalk or CLOS MOP.
Method dispatch	In Smalltalk, method dispatch is done by the `perform:...` method of class `Object`. In CLOS MOP, a discriminating function associated with class `generic-function` performs method dispatch. In Etyma, it is done by the method `select` of class `Instance`. CLOS MOP, by virtue of its very general model of generic functions and multi-methods, provides the most sophisticated method dispatch semantics.
Encapsulation	Smalltalk supports strong encapsulation of instance variables, and Etyma/Jigsaw encapsulates module attributes subjected to `hide` operations. CLOS MOP's default encapsulation semantics is weak, although metaobjects could be specialized using the MOP to support better encapsulation.
Static typing	Static typing and separate processing of modules are highly desirable attributes for languages. The Jigsaw language supports static type rules which have been incorporated into Etyma's `Interface` abstraction. Etyma also incorporates a comprehensive model of type meta-classes. Such a model is practically absent in the other meta-architectures.

Figure 6: Summary of O-O semantics

Although Smalltalk cannot boast generality in the area of inheritance, it still provides the most uniform and comprehensive model of concepts as objects in its meta-architecture.

It is also the most comprehensively designed meta-architecture, considering the complexity of the interacting dual hierarchies of meta-classes. The CLOS MOP, on the other hand, provides a pragmatic and systematically documented MOP, making it the most useful to applications. Etyma/Jigsaw provides significant generality compared with the other architectures, but its utility is yet to be demonstrated. The Scheme MOP is as yet experimental and in the process of being developed, but is unique and very promising.

Smalltalk is the clear winner in the area of abstractions for non O-O concepts in the language. The abstractions are general, broadly conceived, and uniform. The Scheme MOP attempts to capture only those basic concepts in the language which are important from a compilation standpoint. Etyma is currently attempting to design general abstractions covering the space of basic values and types in some commonly found languages.

In conclusion, meta-architectures are powerful, flexible, and extensible by their very nature. There are considerable similarities and differences in the goals that the architectures surveyed here are trying to achieve, as well as in their conceptions. Each was designed with a different set of requirements, yet they can be used for similar applications. The space of meta-architectures span from the pragmatic to the very general. In this paper, we have surveyed the goals, semantic models, and applications of four meta-architectures — Smalltalk, CLOS MOP, Scheme compiler and Etyma — and highlighted their salient features.

Acknowledgments

We gratefully acknowledge useful discussions and input from Gregor Kiczales, Bob Kessler, and Tim Moore.

References

[1] Johnson, R. E. and Russo, V. F., (1991), "Reusing object-oriented designs," Tech. Rep. UIUCDCS 91-1696, University of Illinois at Urbana-Champagne.

[2] Kiczales, G., des Rivières, J., and Bobrow, D. G., (1991), *The Art of the Metaobject Protocol.* Cambridge, MA: The MIT Press.

[3] Lamping, J., Kiczales, G., Rodriquez, L., and Ruf, E., (1992), "An architecture for an open compiler," in *Proc. of the IMSA '92 Workshop on Reflection and Meta-level Architectures.*

[4] Goldberg, A. and Robson, D., (1983), *Smalltalk-80: The Language and its Implementation.* Addison-Wesley.

[5] Bracha, G. and Lindstrom, G., (1992), "Modularity meets inheritance," in *Proc. International Conference on Computer Languages*, (San Francisco, CA), IEEE Computer Society, pp. 282–290. Also available as Technical Report UUCS-91-017.

[6] Bracha, G., (1992), "The programming language *jigsaw*: Mixins, modularity and multiple inheritance,". PhD thesis, University of Utah. Technical report UUCS-92-007; 143 pp.

[7] Keene, S. E., (1989), *Object-Oriented Programming in Common Lisp.* Reading, MA: Addison-Wesley.

[8] Lee, A. H., (1992), "The persistent object system MetaStore: Persistence via metaprogramming,". PhD thesis, University of Utah. Technical report UUCS-92-027; 171 pp.

[9] Banavar, G. and Lindstrom, G., (1993), "A framework for module-based language processors," Computer Science Department Technical Report UUCS-93-006, University of Utah.

[10] Orr, D. B. and Mecklenburg, R. W., (1992), "OMOS — An object server for program execution," in *Proc. International Workshop on Object Oriented Operating Systems*, (Paris), IEEE Computer Society, pp. 200–209. Also available as technical report UUCS-92-033.

[11] Banavar, G., Lindstrom, G., and Orr, D., (1994), "Type-safe composition of object modules," in *Computer Systems and Education: In honour of Prof. V. Rajaraman*, pp. 188–200, Bangalore, India: Tata McGraw Hill Publishing Company, Limited. ISBN 0-07-462044-4. Also available as Technical Report UUCS-94-001.

[12] Kiczales, G., Lamping, J., and Mendhekar, A., (1994), "What a metaobject protocol based compiler can do for lisp." Unpublished report. A modified version to be presented at the OOPSLA '94 workshop on O-O Compilation.

[13] Kiczales, G., (1993), "Traces (a cut at the "make isn't generic" problem).," in *Proc. of Int'l Symposium on Object Technologies for Advanced Software*, vol. 742 of *Lecture Notes in Computer Science*, Springer Verlag.

Non-Well-Founded Set Theory and the Circular Semantics of Semantic Networks

Robin K. Hill
Computer Science Department
226 Bell Hall
State University of New York at Buffalo
Buffalo, New York 14260
hill@cs.buffalo.edu

Abstract

The theory of non-well-founded sets–that is, sets that are potentially circular in hereditary membership–can be applied to a semantic network knowledge/belief representation, SNePS. It provides a particular type of node, the base node, which represents a primitive concept, with a semantics that is both influenced by and influences its dominating compound nodes. A semantic function is defined that interprets each node in the network as a non-well-founded set over the sensory input nodes (words, visual stimula, etc.).

Under certain axioms governing SNePS structure, results show that the semantics does not allow meanings that are circular to the point of vacuity, and that the semantics is inherent in the graphical structure of the SNePS network itself. The semantics supports SNePS principles such as that the meaning of each node (concept) is distinguished from all others, and that the meaning of a node is internal, dependent on its location with regard to other nodes in the network (rather than on external phenomena). An enhanced semantics, which incorporates into the hyperset semantics the relations used to label the arcs, is also developed.

1 Circularity in Semantics

Several researchers in artificial intelligence and knowledge representation note that some real-world phenomena seem inherently circular [Barwise, 1989, pages 194–198], [Nebel, 1991], [Smith, 1991, pp. 265 ff.]. The semantics of SNePS is intended to be circular, notwithstanding its acyclic graph representation, in the sense that the meanings of certain directly-connected nodes influence each other. We address the provision of circularity in semantics using the theory of non-well-founded sets as the tool.

2 Non-Well-Founded Set Theory

Set theory provides a rigorous environment in which to reason about objects. Set theorists adopt a homogeneous typing under which all members of sets are regarded

E. A. Yfantis (ed.), Intelligent Systems, 375–386.

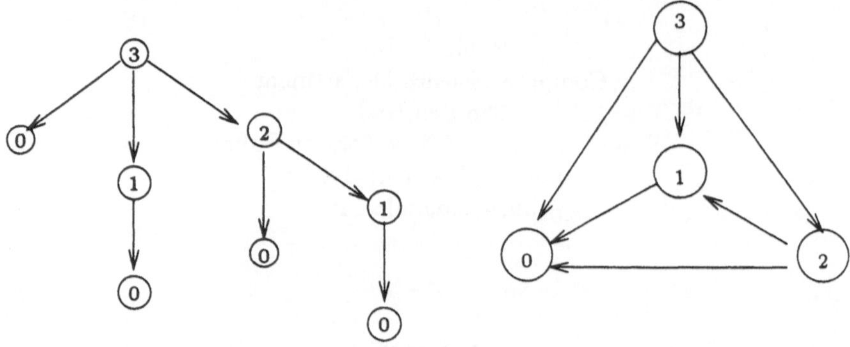

Figure 1: Graphical pictures of 3 as a (hereditary) set

as sets themselves; sets are ***hereditary***. A member of a set S might also be a subset of S. When we need concrete objects that are not sets, we will call them 'atoms'. This simple mechanism permits the standard set-theoretic definition of the natural numbers $\mathcal{N}$ as $\emptyset, \{\emptyset\}, \{\emptyset, \{\emptyset\}\}, \{\emptyset, \{\emptyset\}, \{\emptyset, \{\emptyset\}\}\}, \ldots$. Each $n \in \mathcal{N}$ is represented by the set $\{m \in \mathcal{N} \mid m < n\}$, or, equivalently, $n+1 = n \cup \{n\}$, where $\emptyset$ is identified with zero. For example, $\emptyset \subset \{\emptyset, \{\emptyset\}\}$ and $\emptyset \in \{\emptyset, \{\emptyset\}\}$; 0 is both a member and a subset of 2. The hereditary nature of these sets allows them to be meaningfully depicted as rooted directed graphs, with the arrows showing membership. Figure 1 shows the set we call '3' in two ways, on the right with unique occurrences of the member nodes.

Various axiomatic systems of set theory have been developed; one standard is that called *Zermelo-Frankel Set Theory with the Axiom of Choice*, abbreviated ZFC. It has nine axioms, with the Axiom of Foundation, disallowing any sets that contain themselves, either directly or indirectly, like the sets below:

$$a = \{a\} \tag{1}$$
$$b = \{s, t\}, \text{ where } s = \{t\} \text{ and } t = \{s\} \tag{2}$$

In his theory of non-well-founded sets, Peter Aczel negates the Axiom of Foundation, retaining the others [Aczel, 1988]. The statement of his axiom relies on the graphical depiction of sets, where an ***accessible pointed graph*** or ***apg*** of a set is a directed graph with a distinguished node called the ***point*** from which all other nodes are reachable, which has a ***decoration***, an assignment of sets to each node such that the children are members of their parents' sets. (A node with no children is assigned the empty set.) Non-well-founded set theory is ZFC with the Axiom of Foundation replaced by a negation of it called the Anti-Foundation Axiom.

The Anti-Foundation Axiom: Every graph has a unique decoration.

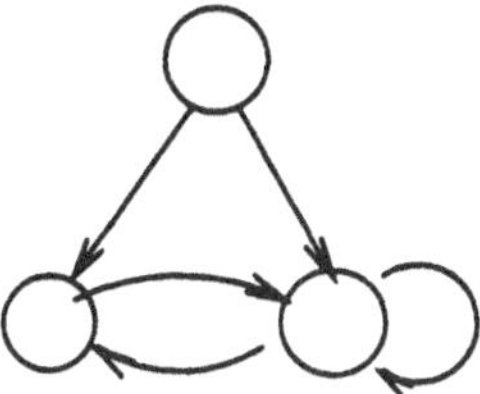

Figure 2: Other pictures of Ω

In other words, even cyclic graphs represent sets. All acyclic graphs, such as that of Figure 1, still represent sets, too. A simple example of one of the new non-well-founded sets, or *hypersets*, is given by the picture below.

Aczel calls the hyperset depicted Ω. Like other hypersets, this one has many pictures—including those in Figure 2. Note that writing down its contents in standard set notation is problematic. Because it is its own child, the solitary node would have to be decorated with something like this: $\{\{\{\ldots\}\}\}$.

We now bring in a set of atoms $\mathcal{A}$; a *decoration* then assigns to each childless node either an atom or the empty set, with the decoration of other nodes made up of the sets assigned to their children, as before. The notation $V_{\mathcal{A}}$ stands for all hypsersets over the set of atoms $\mathcal{A}$, all hypersets that may (but are not required to) include elements from $\mathcal{A}$ or other sets that include elements from $\mathcal{A}$. The range of the decoration is $V_{\mathcal{A}}$. The thrust of the Anti-Foundation Axiom is that every picture, even if it has cycles, has a membership-relative assignment of sets to each node.

To put it another way, an equivalent of the Anti-Foundation Axiom called the Solution Lemma that any system of membership relations can be satisfied by hypersets. The notions of a system of equations and its solution are built up as in albegra, with the equations providing a definition for each "indeterminate", expressed as a hyperset over atoms and indeterminates (and so possibly circular), and the solution being an assignment to each indeterminate of a hyperset over *atoms only*, not other indeterminates, such that all of the original relationships given by the equations still hold when its assignment is substituted for the indeterminate.

> **The Solution Lemma:** Every system of equations in a collection $\mathcal{X}$ of indeterminates over $\mathcal{V}_{\mathcal{A}}$ has a unique solution.

This gets us non-well-founded set theory without reliance on graph notation or concepts. We will use both forms.

3 SNePS and its Principles

The particular knowledge representation used is SNePS, developed by Shapiro et alia. SNePS is a semantic network representation, with nodes and directed arcs, and it adheres to principles that promote intensionality, such as relative belief spaces; it has other significant aspects, such as inference systems, that do not bear directly on this work. SNePS networks are acyclic and can be defined compositionally, as new nodes and arcs are added to networks in certain patterns, thereby creating new larger networks.

A SNePS network is a propositional semantic network, that is, one in which every proposition represented in the network is represented by a node (rather than an arc). Arcs are best regarded as punctuation, having no conceptual semantics. For this reason, it is forbidden to add an arc between two existing nodes. Certain arc labels come with SNePS, while others necessary for a particular implementation are to be defined by the user. Together they form the set of *relations* of the implementation.

Arcs are directed; the node at the origin is called the 'tail' node and the node at the arrowhead, the 'head' node. There would be no point in connecting two nodes with multiple instances of the same arc (arcs with the same label), but there may well be multiple arcs with the same label emanating from the same tail node but terminating at different head nodes, or multiple arcs with different labels connecting two nodes. Nodes with no arcs emanating from them are called 'atomic'. They include *sensory nodes*, which represent the real-world interface, being associated with printed text strings, other visual or tactile or aural data, etc., and *base nodes*, which represent individual concepts, and hence have in-arcs from other nodes.[1] Nodes that do have arcs emanating from them, i.e., that dominate others through their out-arcs (and may themselves be dominated with in-arcs), are called *molecular nodes*. They are defined to be sets of *wires*, structures consisting of a labeled arc and the node at its head.

> A **wire** is an ordered pair $\langle r, n \rangle$, where r is a SNePS relation, and n is a SNePS node. [Shapiro, 1991, page 145]

A guiding force in the theory behind SNePS is the Uniqueness Principle, which states that each concept in the modeled "mind" of a cognitive agent is represented by a unique node in the SNePS network. In SNePS, the FIND/BUILD mechanism creates networks, performing one of two operations when some new concept is submitted to it: (1) if the concept already exists (a node with exactly the right connections is already in the network), then the new information is added to it, or (2) if such a node does not exist, it is created and assigned an unused unique identifier.

4 Circularity through Non-well-founded Sets

Along with the degree and nature of the circularity postulated for SNePS networks, another outstanding question of current SNePS research is the semantics of base nodes, which seem heavy with meaning that doesn't "go" anywhere. Non-well-founded set theory is the stone that can kill both birds. Let the meaning of a node, informally, be the set of meanings of subordinate nodes, except that *base nodes have circular meaning*, participating also in the meaning of the parent node.

[1] *Variable nodes*, which represent arbitrary concepts (individuals or propositions) are not in the scope of this work.

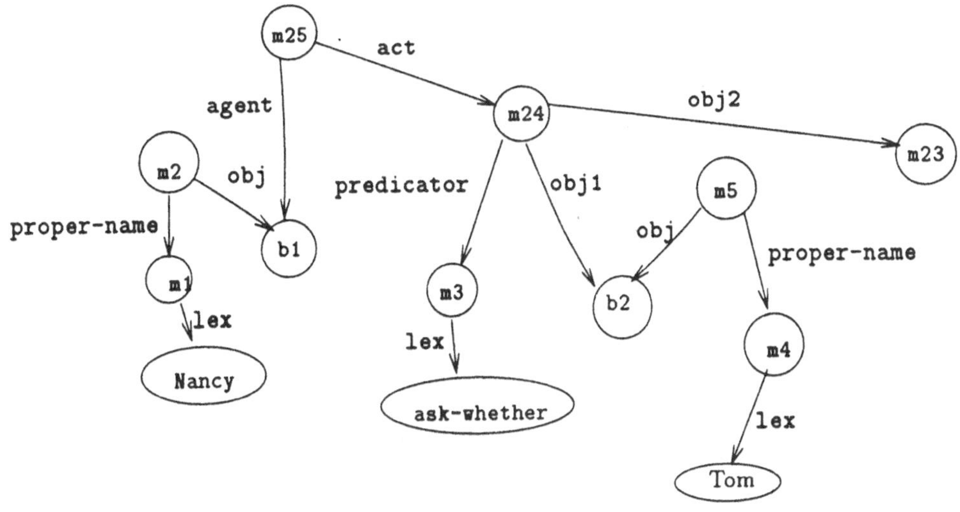

Figure 3: The SNePS network S

Sensory nodes will be the atoms $\mathcal{A}$. To focus on the graphical structure, arc labels will be stripped off for now. Ergo, construed as hereditary sets, base nodes will be members of their parent nodes, and vice versa, which can be shown graphically by an additional edge from a base node to each of its parents. The semantics $\mu(n)$ of a node n will then be the hyperset, over the sensory atoms, that is depicted by the subgraph with n as its point.

An example of a SNePS network (taken from a larger SNePS network) is shown in Figure 3 and called S. S is the SNePS representation of the sentence fragment "Nancy asks Tom whether (`m23`)"—the rest of the structure, rooted at `m23`, is omitted. Base nodes include `b1`, representing the concept of Nancy, and `b2`, Tom; `m25`, `m2`, `m24`, etc., are molecular nodes; **`Nancy`**, **`ask-whether`**, and **`Tom`** sensory nodes representing the respective written words. The relations are { **`proper-name`**, `obj`, **`lex`**, **`agent`**, **`act`**, **`predicator`**, `obj1`, `obj2` }. The new graphical structure that includes circularity is shown as the form called S^* in Figure 4.

To formalize the suggested semantic function from nodes to hypersets, we partition the set of nodes in S into the subsets BASE, SENSORY, and two types of molecular nodes—those being treated as atoms due to circumscription of the network, the subset MOLATOM (in the example, {`m23`}), and those treated as molecular, MOLFULL (including `m1`, `m2`, `m25`, `m24`, and so forth). Each sensory node $s \in$ SENSORY is tagged with its lexeme or other sensory datum, and each "atomic" molecular node $a \in$ MOLATOM with its label. The decoration and the semantic function μ are formally derived to allow for circularity by making the meaning of a node the hyperset assigned to it, respecting the meanings of surrounding nodes. No semantics is assigned to nodes from SENSORY or MOLATOM, on the principle that they are best regarded only as sources of input (actual or po-

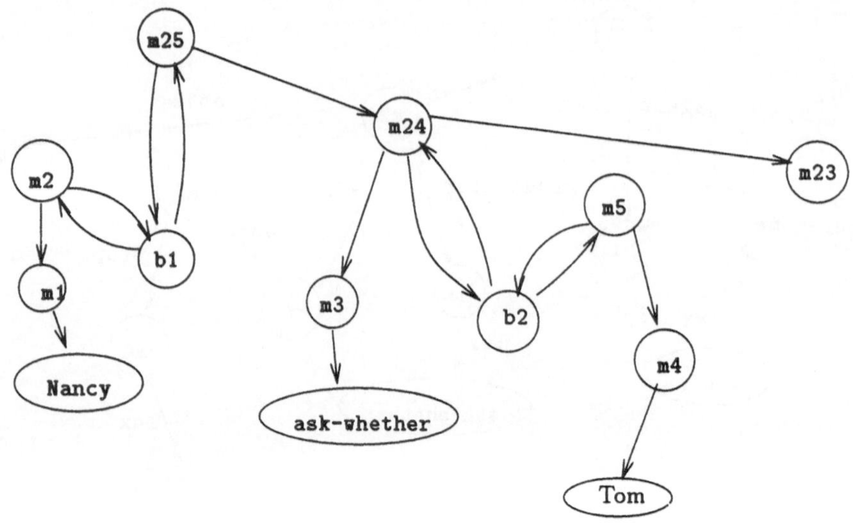

Figure 4: The derived graph S^*

tential), not as meaningful in their own right. (Of course, any molecular node in MOLATOM may be interpreted fully under μ whenever desired.) Under this definition, the semantic value of a base node b is influenced by its parent nodes, as well as vice versa, as desired.

For example, the meanings under μ of the base nodes b1 and b2 in the given S (Figure 3) are shown in Figure 5 as pictures, where the function f turns out to be the hypserset assignments given by μ. We can develop this solution rigorously using the Solution Lemma. Since they comprise the indeterminates,

$$\mathcal{X} = \{\mathtt{b1}, \mathtt{b2}\}, \tag{3}$$

and

$$\mathcal{A} \quad = \quad \{\mathtt{Nancy}, \mathtt{ask\text{-}whether}, \mathtt{m23}, \mathtt{Tom}\}. \tag{4}$$

As noted, the node m23 is taken as an atom, rather than as a set with its own subordinates/members, for convenience. The universe of hypersets, then, is

$$V_{\mathcal{A}} = V_{\{}\mathtt{Nancy}, \mathtt{ask\text{-}whether}, \mathtt{m23}, \mathtt{Tom}\} \tag{5}$$

The solution sought will be an assignment f of sets from $\mathcal{V}_{\mathcal{A}}$ to b1 and b2. The hyperuniverse $\mathcal{V}_{\mathcal{A}'}$ is all hypersets over $\mathcal{A} \cup \mathcal{X}$. We need a system of equations defining b1 and b2, where each set on the right-hand side is in $\mathcal{V}_{\mathcal{A}'}$:

$$\mathtt{b1} \quad = \quad \{\mathtt{m2}, \mathtt{m25}\} = \{\{\{\mathtt{Nancy}\}, \mathtt{b1}\}, \{\mathtt{b1}, \{\{\mathtt{ask\text{-}whether}\}, \mathtt{b2}, \mathtt{m23}\}\}\} \tag{6}$$
$$\mathtt{b2} \quad = \quad \{\mathtt{m24}, \mathtt{m5}\} = \{\{\{\mathtt{ask\text{-}whether}\}, \mathtt{m23}, \mathtt{b2}\}, \{\mathtt{b2}, \{\mathtt{Tom}\}\}\} \tag{7}$$

The function μ is an assignment of hypersets to b1 and b2 such the equations that defined them in terms of their membership still hold when those hypersets are

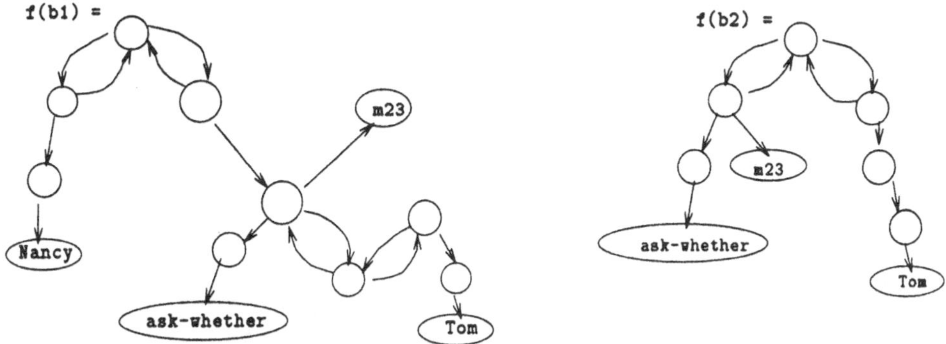

Figure 5: Assignments to **b1** and **b2** under μ

used in place of the indeterminates **b1** and **b2**. Indeed, that is the case when the hypersets pictured in Figure 5 are substituted for occurrences of the indeterminates in Equations 6 and 7. For the definition of μ as the solution to the system of equations given by the network, the computation of the sets, and verification that they are indeed the solution to the defining equations, see [Hill, 1994]. We have a function μ with the desired scope:

$$\mu : \text{MOLFULL} \cup \text{BASE} \longrightarrow V_{\text{SENSORY} \cup \text{MOLATOM}} \tag{8}$$

In a full network with no molecular nodes treated as atoms, all molecular and base nodes would be assigned hypersets over sensory nodes only, that is, hypersets from V_{SENSORY}. The crucial point is that the semantics given to nodes are hypersets over the atoms, not over other ungrounded hypersets. No references to intermediate nodes, such as **m24** or **b2**, are left.

5 Results and Implications of μ

Results derived from the foregoing bolster the respectability of μ as a semantics for SNePS. For proofs, see [Hill, 1994].

Theorem 1 states that no node is circular to the point of "vacuity".

Theorem 1 $\forall$ *nodes* $n \in S, \mu(n) \neq \Omega$

Shapiro employs a definition of *domination* of one node by another that is analogous to hereditary set definition, providing Theorem 2 to the effect that, for the meaning of a molecular node, what you see is what you get. The apg rooted at a node $n \in S^*$ is $\mu(n)$.

Theorem 2 (Node-Picture Principle)

$$\forall m \in \ MOLFULL, \mu(m) = \{\mu(c) \mid m \ dominates \ c\}$$

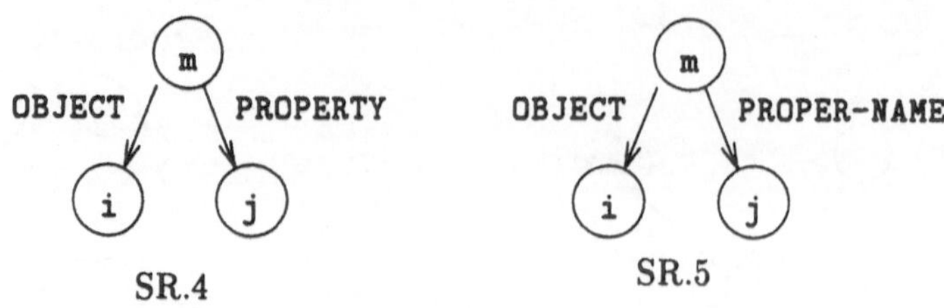

An analog to the Uniqueness Principle holds for the semantic function μ as Theorem 3. Nodes with different meanings under μ are different nodes and therefore represent different concepts (except in the case where uniqueness depended on distinction of arc labels).

Theorem 3 (Uniqueness Principle under μ) *In any $C' \in SNets'$ derived from a full network C, unless nodes n and m dominated exactly the same subordinate nodes (in which case the arc labels differed),*

$$n = m \textit{ if and only if } \mu(n) = \mu(m)$$

6 A Richer Semantics, with Relation

So far in this development, the arc labels, taken from a set of relations $\mathcal{R}$ supplied by SNePS and by the user, have not been considered, and the semantic function μ has been defined over the skeletal structure of a SNePS network only. What is the role of arc labels, and what exactly is the nature of their "punctuation" function [Shapiro and Rapaport, 1991, pages 221-222]? In [Shapiro and Rapaport, 1987], the two networks given by syntactic rules SR.4 and SR.5 differ only in that one has an arc labeled PROPERTY and the other has an arc in the same position labeled PROPER-NAME; the two networks have different semantics, as expressed in the semantic rules SI.4 and SI.5.

SI.4 m is the Meinongian objective corresponding to the proposition that i has the property j.

SI.5 m is the Meinongian objective corresponding to the proposition that Meinongian objectum i's proper name is j. (j is the Meinongian objectum that is i's proper name; its expression in English is represented by a node at the head of a LEX-arc emanating from j.)

If arcs have no semantic import, but the meaning of a node is the entire network in which it is embedded, in what principled way can the "structural" contribution of an arc be distinguished from the "semantic" contribution of a node? If arcs make fixed contributions to the meanings of molecular nodes, they should be involved in the semantic function μ. For an exploration of several alternative solutions, see [Hill, 1994, Chpater 7].

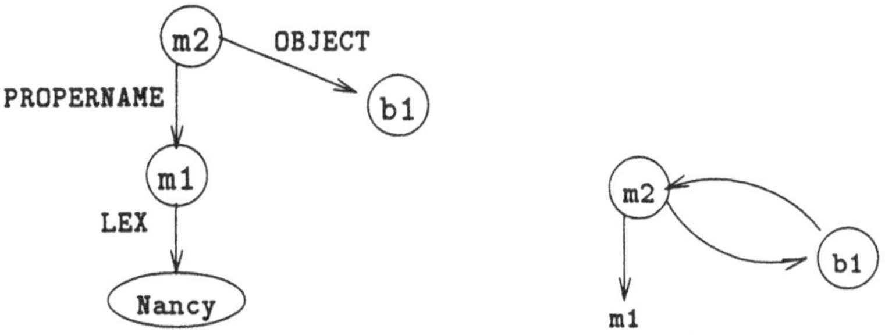

Figure 6: Z Figure 7: Z^*

Here we present a solution that respects nicely both the definition of a molecular node and the treatment of atoms as the sole grounding of decorations. Recall that a molecular node is a set of a wires. Since wires are ordered pairs $\langle$relation, head-node$\rangle$, they can be expressed as sets in the usual way:

$$\langle a, b\rangle = \{\{a\}, \{a, b\}\}$$

These properties can be used to enhance the original μ semantics to provide a much richer value for the semantics $\mu(n)$ of a node n, which treats the set of relations $\mathcal{R}$ as *atoms*, along with the sensory data.

Let us return to the previous example, S and its S^* of Figures 3 and 4, and (because the computations become quite complex) define an even smaller network context, the very restricted Z, shown in Figure 6, from which Z^* is derived. Instead of the semantically sterile labels for molecular nodes, we use their definitions as sets of wires, which are ordered pairs, and convert the ordered pairs to sets.

$$
\begin{aligned}
\mathtt{m1} &= \{\langle \mathtt{LEX}, \mathtt{Nancy}\rangle\} && (9)\\
&= \{\{\{\mathtt{LEX}\}, \{\mathtt{LEX}, \mathtt{Nancy}\}\}\} && (10)\\
\mathtt{m2} &= \{\langle \mathtt{PROPERNAME}, \mathtt{m1}\rangle, \langle \mathtt{OBJECT}, \mathtt{b1}\rangle\} && (11)\\
&= \{\{\{\mathtt{PROPERNAME}\}, \{\mathtt{PROPERNAME}, \mathtt{m1}\}\}, \{\{\mathtt{OBJECT}\}, \{\mathtt{OBJECT}, \mathtt{b1}\}\}\} && (12)
\end{aligned}
$$

For this small example, the set of atoms, extended to include the relations that participate in the semantics, is $\mathcal{A} = \{\mathtt{Nancy}, \mathtt{LEX}, \mathtt{PROPERNAME}, \mathtt{OBJECT}\}$. To apply the Solution Lemma to the same task as before—finding assignments to the selected set of indeterminates, $\mathcal{X} = \{\mathtt{m2}, \mathtt{b1}\}$—we need a system of equations expressing them as hypersets over $\mathcal{A} \cup \mathcal{X}$.

$$
\begin{aligned}
\mathtt{m2} &= \{\{\{\mathtt{PROPERNAME}\}, \{\mathtt{PROPERNAME}, \{\{\{\mathtt{LEX}\}, \{\mathtt{LEX}, \mathtt{Nancy}\}\}\}\}\},\\
&\qquad \{\{\mathtt{OBJECT}\}, \{\mathtt{OBJECT}, \mathtt{b1}\}\}\} && (13)\\
\mathtt{b1} &= \{\mathtt{m2}\} && (14)
\end{aligned}
$$

Compare these equations to the system that would have been used under the original μ, where the set of atoms was $\mathcal{A} = \{\mathtt{m1}\}$:

$$
\begin{aligned}
\mathtt{m2} &= \{\mathtt{m1}, \mathtt{b1}\} && (15)\\
\mathtt{b1} &= \{\mathtt{m2}\} && (16)
\end{aligned}
$$

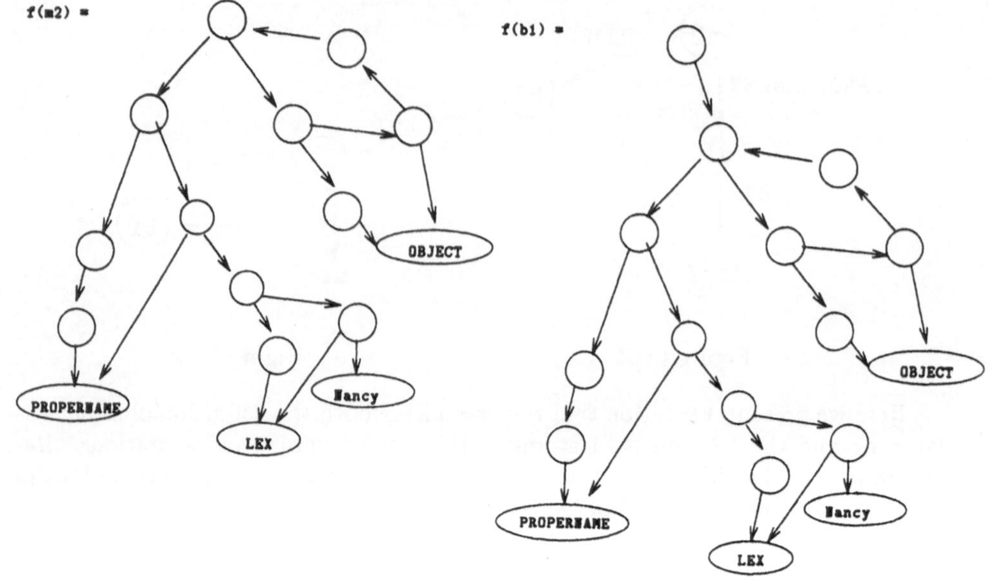

Figure 8: Assignments to b1 and b2 under μ_r

In Equations 13 and 14, m1 no longer exists as an object, having been superseded by its definition as a set of wires.

As the solution f, of course, we want hypersets over $\mathcal{A}$, that is, hypersets from the universe $\mathcal{V}_{\mathcal{A}}$, such that the following relationships are maintained:

$$f(\texttt{m2}) = \{\{\{\texttt{PROPERNAME}\}, \{\texttt{PROPERNAME}, \{\{\{\texttt{LEX}\}, \{\texttt{LEX}, \texttt{Nancy}\}\}\}\}, \{\{\texttt{OBJECT}\}, \{\texttt{OBJECT}, f(\texttt{b1})\}\}\} \quad (17)$$

$$f(\texttt{b1}) = \{f(\texttt{m2})\} \quad (18)$$

The solution is shown graphically in Figure 8. Compare the hypersets assigned here (via f, which is the enhanced semantic function μ_r made manifest), to those given under the plain μ, shown in Figure 5. For the definition of μ_r, computation of the solution, and its verification, see [Hill, 1994, Chapter 7].

We have computed the semantics of nodes in a standard SNePS network Z, incorporating rather than neglecting the arc labels (and even allowing for multiple arcs between the same pair of head and tail nodes, a configuration that can exist in SNePS where the arcs are labeled with distinct relations). Given a SNePS network S, the set of molecular nodes MOLFULL is not unstructured primitives, but consists of wires, sets of ordered pairs $\langle r, n\rangle$, and the relations are included in the range of the semantic function.

$$\mu_r : \text{MOLFULL} \cup \text{BASE} \longrightarrow V_{\text{SENSORY} \cup \text{MOLATOM} \cup \mathcal{R}} \qquad (19)$$

7 Results and Implications of μ_r

Though complex, μ_r is significant for reasons of the integrity of the semantics. It seems obvious that the two arcs labeled **PROPERTY** and **PROPERNAME** in SR.4 and SR.5 above have something to do with the establishment of distinct meanings for their respective dominating molecular nodes. In fact, they could both occur in the same cognitive agent, as discussed above, since the BUILD command of SNEPSUL would not judge them to violate the Uniqueness Principle. In other words, the new version of Theorem 3 would state that $n = m \Leftrightarrow \mu_r(n) = \mu_r(m)$, no longer qualified by the exclusion of the case where n and m dominate exactly the same structure but have different arc labels. Theorem 2 does not hold for μ_r, however, since the SNePS network itself does not show the hyperset structure rooted at nodes if relations from the arc labels are to be atoms along with the sensory nodes.

The treatment given above distinguishes SNePS from other semantic network approaches that have explicitly-named relations between nodes, but no way to build them into nodes themselves at a fundamental level. The definition of the SNePS object "wire" as a node and relation is the key here. (Of course, any semantic network treatment could have such a definition added to it.) On the other hand, the original semantic function μ, which ignores arc labels in favor of node identifiers and connectivity, shows *what* participates in the meaning of a node (that is, what other nodes) without making a commitment as to *how*, and could be applied (with its handling of circularity) to any graphically structured knowledge/belief representation—even those that do not allow propositions modelled to have multiple arguments in a single position (see [Shapiro, 1991, page 138ff.] for comparison).

8 Contributions to Semantics of Representations

The semantic function μ expresses the meaning of a concept, such as **grandmother**, in terms *not* of other concepts, such as **grandparents' house**, **lilac eau de toilette**, and **blood being thicker than water**, but in terms of the discrete sensory stimula involved—the voice, the sight of the house, the scent of lilacs, and the myriad other components of feeling that contribute to the cognitive agent's notion of grandmother. Motivations and principles of SNePS are also supported. The meaning of a node is highly dependent on its location within the surrounding network, rather than on some external property of the concept itself.

We have seen, as an improvement on the basic idea, an enhanced semantics μ_r incorporates the relations used as arc labels into the semantics as atoms, treating them also as primitives. It is also straightforward to define a "measured" μ^δ that provides semantics to some degree of elaboration δ, useful for computational and algorithmic analysis [Hill, 1994]. A further use of μ would be to provide the semantics of the entire "mind" of a cognitive agent, the union of meanings of all nodes constituting the *point basis*, those nodes from one of which all other nodes are reachable.

Since non-well-founded sets must rely on apgs for their finite depiction, so a knowledge representation that embraces circularity will naturally be graphical, providing an advantage exclusive to graphical representations.

References

[Aczel, 1988] Peter Aczel. *Non-Well-Founded Sets.* CSLI Lecture Notes; Number 14. Center for the Study of Language and Information, Stanford, California, 1988.

[Barwise, 1989] Jon Barwise. *The Situation in Logic.* Center for the Study of Language and Information, Leland Stanford Junior University; Stanford, California, 1989.

[Hill, 1994] Robin K. Hill. Issues of semantics in a semantic-network representation of belief. Technical Report 94–11, Department of Computer Science, SUNY at Buffalo, 1994.

[Nebel, 1991] Bernhard Nebel. Terminological cycles: Semantics and computational properties. In John F. Sowa, editor, *Principles of Semantic Networks*, chapter 11, pages 331–361. Morgan Kaufmann Publishers, Inc., San Mateo, California, 1991.

[Shapiro and Rapaport, 1987] S. C. Shapiro and W. J. Rapaport. SNePS considered as a fully intensional propositional semantic network. In N. Cercone and G. McCalla, editors, *The Knowledge Frontier*, pages 263–315. Springer–Verlag, New York, 1987.

[Shapiro and Rapaport, 1991] Stuart C. Shapiro and William J. Rapaport. Models and minds: Knowledge representation for natural-language competence. In Robert Cummins and John Pollock, editors, *Philosophy and AI: Essays at the Interface*, pages 215–259. MIT Press, Cambridge, MA, 1991.

[Shapiro, 1991] Stuart C. Shapiro. Cables, paths and "subconsious" reasoning in propositional semantic networks. In John F. Sowa, editor, *Principles of Semantic Networks*, chapter 4, pages 137–156. Morgan Kaufmann, San Mateo, CA, 1991.

[Smith, 1991] Brian Cantwell Smith. The owl and the electric encyclopedia. *Artificial Intelligence*, 47:251–258, 1991.

A NEW SYSTEM APPROACH TO STUDY PROGRAM PROPERTIES

P. A. VENKATACHALAM[1], P. V. RAJA[2]

[1] *School of Electrical and Electronic Engineering, Universiti Sains Malaysia, Tronoh, MalaysiaTel: 60 5 3676901 ; Fax : 60 5 3677443 ; E-mail:pav@kcp.usm.my*

[2] *Department of Computer Science, University of Oakland, Rochester , Michigan*

Abstract

Syntactic errors can be easily detected in a program. But semantic errors are not easily detectable. So proving program properties still poses a problem to researchers. Even though many approaches have been reported these could not yield fruitful results. We apply here a new sytem approach which converts a program into set of recurrence equations on an abstract program schema which is independent of a language or computer. Once a program schema is given an interpretation and assigned an initial state vector it becomes an executable program.

Introduction

Given a program the syntactic errors can be easily detected. But there are no easy ways to identify the semantic errors. So the program properties viz. Termination, Correctness, Equivalence and Efficiency of a program still remain unsolved in complete sense. There are many methods [1,2] reported to prove the program properties. But they have been found to be not so effective. In this work an attempt is made to formulate a new discrete system approach to prove programs. The program schema is defined as an abstract computational model which not only makes the computation machine independent but also allows direct interpretation of results into real life programming language. Then the concept of representing a Program Schema by a system of recurrence equations with number of steps of computation as the independent variable is presented.

Program Schema

Program schema, or Simple Program Schema, consists of a finite ordered sequence of statements such as START (**s**), END (**e**), ASSIGN (**a**), TRANSFER (**p**), constant and variable symbols without assigning any specific meaning to them. For example a program schema may be represented as shown in Fig.1. It is possible, given a program schema, it is wasy to define a program graph [3] to represent the schema which will give a partial meaning but this is not necessarily be unique. For the above schema the program graph is shown in Fig.2.

Interpretation

The program schema described above only constitute the syntax of a program without having any semantic content for the statement, function or a symbol. If interpretations are given to them from a specific domain, the schema will become a program.

E. A. Yfantis (ed.), Intelligent Systems, 387–392.

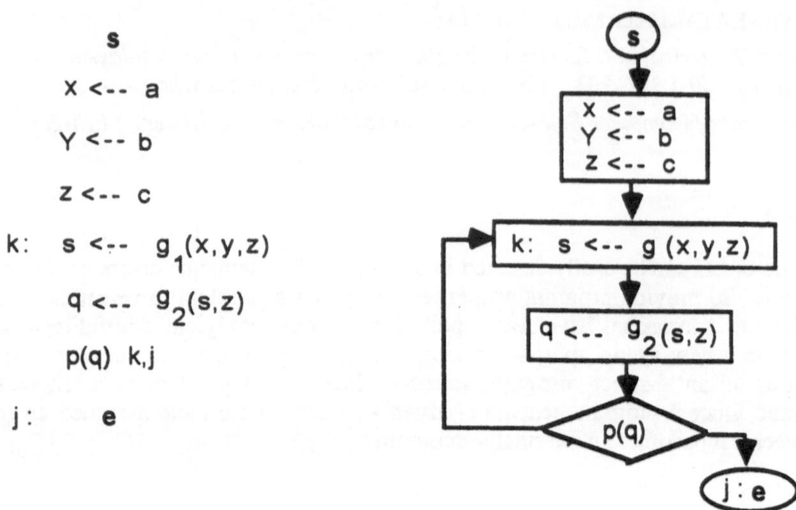

Fig.1 Program Schema

Fig.2 Program Graph

An interpretation (total interpretation) I, consists of a non empty set of elements D_I, called domain of interpretation.

Each variable X_i is assigned some element $I(X_i) \in D_I$.

Each constant symbol K_i is assigned some element $I(K_i) \in D_I$.

Each f_i is assigned some n-adic function $I(f_i) : D_I^n \rightarrow D_I$.

Each p_i^n is assigned a function $I(p_i^n) : D_I^n \rightarrow \{T,F\}$.

Under different interpretations a program schema may represent different programs. We say an interpretation *partial* if assignments to the constant, function and predicate symbols are made as above but the program variables are left unassigned.

State Vector

The values assigned to the variables in a schema are ordered in an arbitrary fashion and grouped in a vector called the state vector. If x is a state vector, then x specifies mapping of the variables $X_1, X_2, ..., X_m$ to elements in D_I. Let $x[X_i]$ denote the value assigned to X_i. Then $\overline{X}_o = (x_o[X_1], x_o[X_2], ..., x_o[X_m])$ where x_o is the initial state vector.

Execution Sequence and the Value of a Schema

Once a partially interpreted schema (P, I) is assigned an initial state vector x_0, the execution starts at s-node and follows the arrows of the directed labelled graph from node to node residing in only one node at any one time. If the execution is at a a-node, then in the next step the execution will move to its immediate descendent in the graph. Suppose if the execution is at a p-node. Then there are two paths along which it could move. The path chosen by the execution depends upon the truth value of the predicate.

We first recursively define the value of the function and predicate terms with respect to a partial interpretation I and a state vector x.

(i) Let t be a function term. Then its value is given by

$$\begin{aligned} \text{Val}\ (t, I, x) = \underline{\text{if}}\ \ t = x_i\ \ \underline{\text{then}}\ \ x[x_i] \\ \underline{\text{else}}\ \underline{\text{if}}\ \ t = k_i\ \ \underline{\text{then}}\ \ I(k_i) \\ \underline{\text{else}}\ \underline{\text{if}}\ \ t = f(t_1, t_2, \ldots, t_n) \\ \text{then } I(f)\ [\text{Val}(t_1, I, t), \text{Val}(t_2, I, x), \ldots, \text{Val}(t_n, I, x)], \end{aligned}$$

where $t_1, t_2, \ldots, t_n$ are function terms.

(ii) Let $P_Z = p_i(t_1, t_2, \ldots, t_n)$ be a simple predicate-term. Then its value is given by

$$\text{Val}\ (P_Z, I, x) = \underline{\text{if}}\ I(p_i)\ [\text{Val}\ (t_1, I, x), \ldots, \text{Val}(t_n, I, x)] = T\ \underline{\text{then}}\ T\ \underline{\text{else}}\ F.$$

iii) Let $P_T = B(t_1, t_2, \ldots, t_n)$ be a predicate-term where B is a boolean function and $P_1, P_2, \ldots, P_k$ are simple predicate terms. Then its value is given by

$$\text{Val}\ (P_T, I, x) = \underline{\text{if}}\ B\ [\text{Val}\ (P_1, I, x), \ldots, \text{Val}(P_k, I, x)] = T\ \underline{\text{then}}\ T\ \underline{\text{else}}\ F.$$

Let S be the set of statements in a schema P and let G = (C, y) be the directed labeled graph of P where C denotes the set of nodes of G and y is a mapping $C \leftarrow C \cup C^2$. Let [c] denote the statement at node c. Then if [c] is an assignment statement y(c) is a single node q and if [c] is a test statement, y(c) is (q, r), where c, q, r ε C.

Suppose [d] is an assignment statement, $x_i \leftarrow \tau$, where t is a function term. Then [d](I, x) is the value of the state vector after execution of this statement with respect to state vector x and is given by

$$\begin{aligned} [d]\ (I, \xi)(j) &= \xi[j] \text{ for } i \neq j \\ &= \text{Val}\ (\tau, I, \xi) \text{ for } i = j \end{aligned}$$

Suppose [d] is a test or conditional statement (CS). Then, [d] (I, x) = x. Let a schema P be assigned a partial interpretation I and an initial state vector x_o. Then we represent the value of P as Val (P, I, x_o) = E(P, I, x_o, s), where E is called the partial execution function and is defined as follows:

E(P, I, x, d) = if ([d] AS and j(d) = q), then E[P, I, [d] (I,x), q]
else if ([d] CS and j (d) = (q, r)), then if Val ([d], I, x) = T, then E (P, I,x, q)
else E (P, I, x, r)
else if [d] UT(unconditional transfer) and j(d) = q, then E (P, I, x, q)
else if [d] = 0(null), then x.

From the above we see how the execution proceeds and the Val (P, I, x) is evaluated. A particular execution leads to a particular path through the program graph either from s to e or to an infinite loop. In the latter case the execution is said to divergent. If e is reached, the execution is convergent. If execution is convergent, there exists Val (P, I, x). If execution is divergent, Val (P, I, x) is undefined.

The execution sequence (EXS) of a schema with a partial interpretation I and initial state vector x_o is the sequence

$$(x_o, d_o), (x_1, d_1), \ldots, (x_n, d_n), (x_{n+1}, d_{n+1}), \ldots$$

where E (P, I,x_i, d_i) --> E(P,I,x_{i+1}, d_{i+1}) for i = 0,1,2, ...

The execution sequence is finite if for some k, d_k = C and is infinite otherwise. We would normally consider d_o to be the start node S and hence EXS (x_o) implies EXS (P,I,x_o,s) unless otherwise specified.

A step of computation is the transition (x_i, d_i) --> (x_{i+1}, d_{i+1}) and corresponds to the execution of the statement occurring as node label of d_i. There may exist a different execution sequence for every interpretation I on a schema.

Recurrence Relations of a Program Schema:

We number the nodes of the program graph of a schema P in some sequence and call these numbers as node numbers. Let P be an arbitrary schema with variables = {x_1, x_2, ... x_m}. Let Y = {1, 2, ... , y, , v} be the node numbers of the program graph of P. We then write [y] to represent the statement that appears at node numbered with y in the graph.

Let R be the set of reals. In what follows the domain of an interpretation will always be taken R. Consider a schema P with partial interpretation I. Let x_o be the state vector that assigns the value $\overline{x}_0 = \left\{x_1^0, x_2^0, \ldots, x_m^0\right\}$ to the variables of P. Then we define the following functions.

(i) $p: N \times R^m \rightarrow Y$ is a mapping such that $p(n, \overline{x}_0)$ denotes the node number of the program graph of P at which the control resides at n-th step of computation starting with initial state vector $\overline{x}_0$. A variable X_i of P is said to be defined at node y if X_i is assigned a value in the statement at node y.

(ii) $d: \{1, 2, \ldots, m\} \times Y \times N \times R^m \rightarrow \{0,1\}$ is a mapping such that
$d(i,y,n, \overline{x}_0) = 1$, if $p(n, \overline{x}_0) = y$ and X_i is defined at y ; = 0, otherwise.

Consider now a partially interpreted schema (P, I) with input assignment $\overline{x}_0$. If the execution sequence of P for this input assignment is infinite, then $x_{i,n}$ represents the value of the variable X_i after n steps of computation for all n. Suppose the execution sequence is finite and of length k. Then $x_{i,n}$ has the same meaning as above for $n \leq k$, and we define $x_{i,n} = x_{i,k}$ for n>k.

Let $X_{i,n}$ denote the value of the variable X_i after n-th step of computation and let us represent $d(i, y, n, \overline{x}_0)$ as d (i,y,n) for simplicity ($\overline{x}_0$ being implicit).

Let $p(n, \overline{x}_0) = y$ and $[y] = (x_i \leftarrow f(x_1, x_2 \ldots, x_m))$.

Then we have the recurrence relation

$$x_{i,n} = [1 - d(i,y,n)]\, x_{i,n-1} + d(i,y,n)\; f(x_{1,n-1}, \ldots, x_{m,n-1})$$

Thus we get a recurrence equation for every assignment statement and with n as the parameter, assuming that any variable X_i is assigned a value only once in the schema. In general there is a possibility that X_i may have multiple assignments in a schema.

Let $v = \{v_1, v_2, \ldots, v_r\}$ be the set of <u>node-numbers</u> at which x_i has been assigned values as follows:

$V_1 = (x_i \leftarrow g_1\ (x_1, x_2, \ldots, x_m))$.
$V_2 = (x_i \leftarrow g_2\ (x_1, x_2, \ldots, x_m)))$.
. . .
$V_r = (x_i \leftarrow g_r\ (x_1, x_2, \ldots, x_m)))$.

We note that if $d(i,v_j,n) = 1$ then $d(i,v_k,n) = 0$ for all $k \neq j$. Hence for any arbitrary step of computations, we have

$$X_{i,n} = \left[1 - \sum_{j=1}^{r} \delta\left(i, v_j, n\right)\right] X_{i,n-1}$$

$$- \sum_{j=1}^{r} \left[\delta\left(i, v_j, n\right) g_j\left(X_{1,n-1}, X_{2,n-1}, \ldots, X_{m,n-1}\right)\right]$$

The above procedure yields a system of recurrence relations R(P,I) corresponding to the set of assignment statements in the schema P for a partial interpretation I.

We now show that the system of equations R(P,I) is equivalent to P in some sense and state the following Lemma and Theorem.

Lemma:

Let $x_1, x_2, \ldots, x_m$ be the program variables, n be the step of computation and R(P,I) be the system of recurrence equations of the schema P for the partial interpretation I. Then $x_{1,n}, x_{2,n}, \ldots, x_{m,n}$ are uniquely defined in R(P,I) for every n and every x_0.

Theorem

Let (P,I) be a partially interpretted program for any variable x_i of P. The value of x_i after n steps of computation is equal to that of $x_{i,n}$ in R(P, I).with respect to an initial assignment $\overline{x}_0$

Corollary:

If for initial assignment $\overline{x}_0$ a partially interpreted program (P, I) terminates in p steps, then Val (P, I, $\overline{x}_0$, \$) = $(x_{1,p'}, x_{2,p'}, \ldots, x_{m,p'})$ for $p' \geq p$.

Conclusion

Program schema which is independent of a particular computer or a programming language has been presented. The interpretation of the statements, symbols and functions of the schema has been described. The schema with interpretation can be transformed into a set of recurrence relations. Using the generated recurrence relations to study to the properties of program may be found [3] as an extension to this work..

References

[1] Manna, Z, "*Computation of Recursive Programs - Theory vs Practice*", AFIAS conference proceedings, vol. 40, 1972.

[2] Cooper D.C., "*Some transformations and standard forms of graphs with application to computer programs*" Machine Intelligence Vol. 2, (Ed.).Ella Dale and D.Mitchie, 1968.

[3] Venkatachalam P.A., "*Convergence of Programs by Discrete System Techniques*", Ph.D thesis, I.I.T., Kanpur, 1973.

EVOLUTIONARY PROCESSING[1]

PIERRE A. I. WIJKMAN, Royal Institute of Technology & University of Stockholm, Department of Computer and Systems Sciences, Electrum 230, 164 40 Kista, Sweden, e-mail: pierre@dsv.su.se

Abstract - We present in this paper a framework for evolutionary computation. We reformulate the foundation of evolutionary computation by introducing a number of new concepts that are centred around the general concept of system. We derive from this foundation the functionality of the, so called, developmental system. This functionality is equal to a maximally effective search system. Finally, we propose, as a hypothesis, a form that should implement this functionality. We arrive at the conclusion that the form of the developmental system should be equal to a parallel generate-and-test search employing a relative generator and a relative tester.

1 INTRODUCTION

The idea to look in a new direction, in a different area, for the solution to some problem is an example of an often very fruitful methodology for solving problems called reasoning by analogy. In this paper we look at the area of evolutionary biology to solve problems within the area of machine learning. Earlier approaches in the same direction are Frasers work on "Genetic System Simulation" 1957 [2], Fogel, Owen and Walshs work on "Evolutionary Programming" 1966 [1], Rechenbergs work on "Evolution Strategies" 1973 [5], and Hollands work on "Genetic Algorithms" 1975 [4]. Recently a more general approach was taken by Spears, De Jong, Bäck, Fogel and Garis in their work on "Evolutionary Computation" [6] which compared the most important approaches. The approach presented in this paper distinguish itself from previous work in that it is looking at the problems involved from a new viewpoint by introducing a general and abstract framework as its basis. The results from this "foundation reconstruction" are a number of general conclusions regarding the nature of the process of evolution. These conclusions are hopefully of interest both for machine learning and evolutionary biology.

2 FOUNDATION

The foundation of the framework presented in this paper consists a number of concept definitions. The cornerstone in this foundation is the concept of system.

Definition of system. A system is described by the two complementary concepts of *form* and *functionality*:

- **Form.** The systems form, or static properties, are described by the concept of *composition* and the concept of *structure*. The composition is composed of a finite number of *sub systems* and the structure is composed of a finite number of *connec-*

[1] Currently there exist two established names for general approaches trying to formalise the process of evolution; "Evolutionary Algorithms" and "Evolutionary Computation". The name "Evolutionary Algorithms" seems to focus on the form, or static properties, of the process of evolution while the name "Evolutionary Computation" seems to focus on the functionality, or dynamic properties, of the process of evolution. We therefore propose the more embracing name "Evolutionary Processing" as an alternative to these names.

E. A. Yfantis (ed.), Intelligent Systems, 393–403.

tions. The sub systems determine how the signals carried by the connections are transformed. The connections determine how the sub systems are connected.

- **Functionality.** The systems functionality, or dynamic properties, are described by the concept of *stimuli-response unit*. A system has some specific finite number of stimuli-response unit's. A system can in this case be seen as a whole, or black box, that, within specific time limits, *transforms signals* from and to the environment. The *output signals* from the system are dependent on the *input signals* given to the system and the *internal state* of the system.

Definition of general form description. The concept of general form description is a type of description that does not specify a system form in every detail. It is a general description that only describes the most essential part of a specific system form.

Definition of form space. The set of sub systems in the composition of a system and the set of connections in the structure of a system can, in general, be combined in very many different ways. Assume that we have a distance function for system form d_{form}. We will refer to this function as the *form distance function* and to the calculated result of this function as the *form distance value*. We can, with the help of this form distance function *order* the set of all possible system forms into, what is called, a *metric space*. We will refer to this metric space as the *form space*. The size of the form space grows *polynomially with the size of the composition of a system*, i.e. with the number of sub systems in the composition of the system and *exponentially with the size of the structure of a system*, i.e. with the number of connections in the structure of the system.

Definition of general functionality description. The concept of general functionality description is a type of description that does not specify a system functionality in every detail. It is a general description that only describes the most essential part of a specific system functionality.

Definition of functionality space. The set of stimuli-response unit's in the functionality of a system can, in general, be combined in very many different ways. Assume that we have a distance function for system functionality d_{func}. We will refer to this function as the *functionality distance function* and to the calculated result of this function as the *functionality distance value*. We can, with the help of this functionality distance function *order* the set of all possible system functionality's into, what is called, a *metric space*. We will refer to this metric space as the *functionality space*.

Definition of form-functionality topography. The form-functionality topography is the shape, or landscape, that takes form when we plot the functionality distance function, given some general functionality description as one of its arguments, over the whole form space.

Definition of form-functionality effectiveness. A system is said to be transforming a signal in a form-functionality effective way if both of the following two conditions hold for the system:

- **Condition 1.** The system is form effective. We say that a system is form effective when the size of the composition of the system and the size of the structure of the system is small, i.e. if the size of the composition of the system + the size of the structure of the system is growing *polynomially* with the size of the input signal of the system + the size of the output signal of the system.

- **Condition 2.** The system is functionality effective. We say that a system is functionality effective when the response time for all of its stimuli-response unit's is short, i.e. if the response time for all of the systems stimuli-response unit's is growing *polynomially* with the size of the input signal of the system + the size of the output signal of the system.

Definition of main question. The main question is the question that framework presented in this paper address. We will distinguish two kinds of systems; the *developmental system* and the *performance system*. A developmental system can *construct* performance systems and a performance system can perform some arbitrary task.

Main question: What are the specific form and the specific functionality of a developmental system that have the following general form and functionality description:

- **Form.** The specific form description of the developmental system should be consistent with the general form description for a system given in the definition of a system.
- **Functionality.** Given the following as input;
  - **input 1;** a general form description for a performance system, and
  - **input 2;** a general functionality description for a performance system,

  the developmental system should output;
  - **output 1;** a specific form description for a performance system that is consistent with input 1, and
  - **output 2;** a specific functionality description for a performance system that is consistent with input 2,

  in a maximally form-functionality effective way.

From this definition the following two observations follows:

- **Observation 1.** Since one, in general, can construct several different performance systems from a general form description, the problem for the developmental system is in essence a *search problem*. The task for the developmental system is to *find* a specific performance system that is consistent with the general form and functionality descriptions given to the developmental system.
- **Observation 2.** This search task should be performed in a *maximally form-functionality effective way*.

From these two observations the following conclusion regarding the functionality of the developmental system follows:

- **Conclusion.** The developmental system is in essence equal to a *maximally form-functionality effective search system*.

3 FUNCTIONALITY OF THE DEVELOPMENTAL SYSTEM

In this section will we first describe why the search problem defined by the main question is a difficult search problem. We will thereafter define the functionality of the maximally form-functionality effective search system.

3.1 Difficulties of Search

The search problem that the developmental system should solve is a very difficult search problem. The reason for this is described in the following:

- **Size.** The form space is, in general, very large. The number of performance systems that can be constructed grows exponentially with the size of the structure of the performance system, i.e. with the number of connections in the structure of the performance system.

- **Shape.** The form-functionality topography is, in general, is arbitrarily shaped. Since the form space, in general, is very large the only alternative left is to use information to constrain the form space. The only source for this information is the information that is implicit in the form-functionality topography. The search system must accordingly use the form-functionality topography to guide the search process. If, for example, the form-functionality topography was formed as one single large valley it would be very simple to find the bottom of this valley by an ordinary local information hill down climbing search method. This method does, however, only work in strictly monotonical topographies and will accordingly not perform well in arbitrary fitness topographies. The generality of the main question implies that the search method must be able to deal with arbitrary fitness topographies.

3.2 Functionality of the Maximally Form-Functionality Effective Search System

The search system has not information a priori of the whole form-functionality topography. The search system must *explore* the form-functionality topography *in a number of steps*. The search system must, in essence, use information that is gained during the search in order to constrain the form space. A search system that uses information that has been gained during the search is called a *cumulative search system*. A cumulative search system keeps some or all the information from previous search trials and use this information to guide the next search trials. The maximally form-functionality effective search system is, accordingly, a type of cumulative search system that use the information that is implicit in the form-functionality topography in an optimal way. Before we can give the definition of the functionality of the maximally form-functionality effective search system we have to define the concept of *neutral-positive form change* and the concept of *connected form cluster*:

Definition of degree of neutral-positive form change. If we make a number c of form changes to a system s_i, i.e. a number of additions and/or deletions of connections in the structure of the system s_i, and the resulting new system s_j is a system with equal or lower functionality distance value, in relation to the a priori given general functionality description implicitly given in the system s_{gen}, than the original system s_i has we say that we have made a *degree c of neutral-positive form change* to the original system s_i.

We will denote the degree c of neutral-positive form change between (1) two systems s_i and s_j and (2) between two sets of systems S_i and S_j with;

$$(1)\quad s_i \underset{\text{FormChange}(c,=+)}{\rightarrow} s_j \qquad (2)\quad S_i \underset{\text{FormChange}(c,=+)}{\Rightarrow} S_j$$

In case (2), *all* systems in the set of systems S_i can be transformed to *all* systems in the set of systems S_j by a degree c of neutral-positive form change.

Definition of degree of connected form cluster. Assume that we have a set of systems such that the following holds;

$$S = S_1 \cup \ldots \cup S_n, \quad \text{where;} \quad S_1 \underset{\text{FormChange}(c,=+)}{\Rightarrow} \cdots \underset{\text{FormChange}(c,=+)}{\Rightarrow} S_n.$$

We then say that the set of systems S forms a *degree c connected form cluster*. We will denote a set of systems that form a degree c connected form cluster with $S^{\text{FormChange}(c,=+)}$.

Definition of maximally form-functionality effective search functionality. Assume that we are searching for a performance system with the functionality value v. A maximally form-functionality effective search proceeds in a number n of steps. During each step the maximally form-functionality effective search tries to transform some performance system into another performance system by a degree of neutral-positive form change in the following manner;

$$s_1 \underset{\text{FormChange}(\_,=+)}{\rightarrow} \cdots \underset{\text{FormChange}(\_,=+)}{\rightarrow} s_n, \quad \text{where;} \quad d_{\text{func}}(s_n, s_{\text{gen}}) = v,$$

and the *path* that this series of transformations takes is the path that minimise the following sum;

$$\text{trialsInAverage} = \frac{\left|S_1^{\text{FormChange}(c_1,=+)}\right|}{\left|S_2^{\text{FormChange}(c_2,=+)}\right|} + \ldots + \frac{\left|S_{n-1}^{\text{FormChange}(c_{n-1},=+)}\right|}{\left|S_n^{\text{FormChange}(c_n,=+)}\right|}, \quad \text{where;}$$

$$s_1 \in S_1^{\text{FormChange}(c_1,=+)}, \ldots, s_n \in S_n^{\text{FormChange}(c_n,=+)}.$$

The parameters $c_1, \ldots, c_n$, is set in an optimal way, i.e. in such a way that the trials in average sum is minimised.

The form-functionality topography can be thought of as partitioned into several form clusters with appropriate connectedness values set a priori. The maximally form-functionality effective search system proceeds by climbing down from one form cluster to another, lower, form cluster. There exist though, in general, several alternative paths that a search system can take. The maximally form-functionality effective search system takes the path that requires the least total search effort. In this way, the maximally form-functionality effective search system makes an optimal use of the information that is implicit in the form-functionality topography.

4 FORM OF THE DEVELOPMENTAL SYSTEM

In this section will we give a definition of the *form of the developmental system*. This form should implement the, in section 3 defined, functionality of the developmental system (equal to the functionality of the maximally form-functionality effective search system). In addition to this, the form of the developmental system should, according to the main question, be maximally form effective. The, in this section defined, form of the developmental system is *hypothesised* to be a *maximally form effective* and *arbitrary close approximative implementation* of the functionality of the developmental system.

4.1 Difficulties of Implementation

In this section will we describe the difficulties that are involved in finding the form associated to the functionality of the developmental system.

Problem 1. The form-functionality topography is partitioned in the form dimension into many different form clusters where each form cluster has a specific degree of connectedness. What is the form of the stimuli-response unit that can differentiate between different form clusters in the form-functionality topography.

Problem 2. The form-functionality topography is partitioned in the functionality dimension into many different levels of distance functionality value. What is the form of the stimuli-response unit that specifies the differentiation between different functionality distance value levels in the form-functionality topography?

Problem 3. The form-functionality topography is partitioned both in the form and functionality dimension. What is the form of the stimuli-response unit that specifies which *path* the optimal search system should take among all different form clusters in its way to finding performance systems with lower and lower functionality difference value in the form-functionality topography?

4.2 Form of the Maximally Form-Functionality EffectiveSearch System

The form of the maximally form-functionality effective search system is based on the form of a general search system.

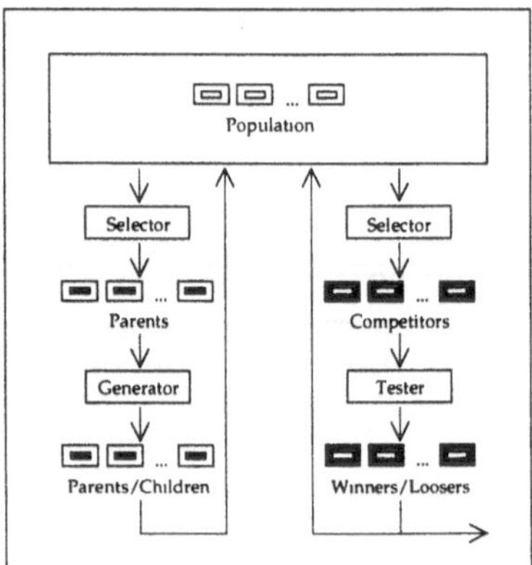

Figure 2. The form of the developmental system.

The form of the maximally form-functionality effective search system consists of (1) the population, (2) the parent selector, (3) the generator, (4) the competitor selector and (5) the tester. These sub systems are structured as is visualised in figure 2.

4.2.1 *The Population*

The population is a sub system that consists of a finite number of performance systems. The reason for having *several* performance systems in the population is to enable a better approximation of the form-functionality topography. With several performance systems situated on different locations in the form-functionality topography the search system can get an approximate picture of the form-functionality topography at some specific moment. This information could then be used to cumulatively guide the movement of the performance systems in the population in some direction in the form functionality topography.

The population consists initially of performance systems that have a form (and a functionality) that is, within the limits defined by the general form description, random. These performance systems changes over time from this initial random state towards a more ordered state. This change takes place because the generator adds performance systems to the population and because the tester deletes performance systems from the population. By having a population, a generator that can differentiate between different forms in the form-functionality topography and a tester that can differentiate between different functionality distance values in the form-functionality topography the proposed developmental system form can *in parallel* explore several alternative paths among all different form clusters in the way to finding performance systems with lower and lower functionality difference values in the form-functionality topography. This solves approximately the problem with *form-functionality differentiation*, i.e. the problem of path finding. The number of performance systems in the population determines how closely the form of the developmental system approximates the functionality of the developmental system.

4.2.2 *The Parent Selector*

The parent selector is a sub system that selects performance systems from the population. The selected parent performance systems is then given to the generator. The parent selector is selecting performance systems from the population in a random way. The performance systems in the population can, however, be more or less ordered. The result of this is that the parent selector can appear to be more or less biased, or informed, in its selection of parent performance systems.

4.2.3 *The Generator*

The generator is a sub system that constructs, or generates, child performance systems. The information for this generation comes from (1) the general form description, and, (2) the parent performance systems that is given to the generator by the parent selector. The generated child performance systems are added to the population. The generator is biased in its generation of child performance systems by the *form* of the parent performance systems given to the generator by the parent selector. The generator considers thus only the form of the parent performance systems in generating child performance system forms. The generator can belong to one of the following three categories:

Absolute. An absolute generator is not taking any parent performance systems as input. This type of generator can not be part of a cumulative search system and will not be considered further.

Semi-relative. A semi-relative generator is taking exactly one parent performance system as input. To generate a child performance system from a single parent performance system involves the following two steps:

1. Make a copy of the parent performance system.
2. Make a number of random minimal form changes to the form of the copied performance system.

Relative. A relative generator is taking more than one parent performance system as input. To generate a child performance system from several parent performance systems involves the following two steps:

1. Make copies of the parent performance systems.
2. Chose the form for the child performance system by using some procedure that in some random way chooses partial forms (see below), i.e. connections, from either of the copied performance systems.

The purpose for the generator is to *continually direct* the search process towards the bottom in the form-functionality topography. To be able to do this the generator must be able to differentiate between different form clusters in the form-functionality topography. The generator is, in essence, working in a purely random way. The performance systems in the population can, however, be more or less ordered. The result of this impact from the population on the generator is that the generator can appear to be more or less biased, or informed, in its generating of child performance systems.

4.2.3.1 Pro and Cons with the Semi-Relative and the Relative Generator

The following is a discussion of the pros and cons with the semi-relative and the relative generator. In the following discussion a *partial form* of a system is some part of the systems form; i.e. a sub set of the structure of the system and a *positive partial form* of a system is a partial form that, in some way, contributes to the system having a lower functionality distance value. There is, in general, a *difference* between the form of the parent performance system(s) and the form of the generated child performance system(s). This fact means that the generator in effect makes a *change* to the form of the generated child performance system(s) relative to the form of the parent performance system(s).

Semi-relative. A semi-relative generator employing a small degree of random form change is very sensitive to the problem of local minima. A semi-relative generator employing a large degree of random form change will be less sensitive to the problem of local minima but is, however, working in a *less cumulative* way, i.e. a lesser amount of the information given by the parent performance system is used. The semi-relative generator can in this case reach a larger number of performance systems but the focus of generation is less sharp. The number of systems that the semi-relative generator can reach grows, in general, exponentially with the degree of random form change made by the semi-relative generator.

There is thus an unsolvable conflict for the semi-relative generator; the semi-relative generator must at the same time be both (1) *non sensitive to the problem of local minima* and (2) *cumulative*. This conflict becomes larger the more developed the parent performance system is. The reason for this is that a more developed parent performance system is more sensitive to random form change. The conclusion from this is that the semi-relative generator not can adaptively bias the *degree of random form change*, the *location of the random form change* and the *range of the random form change* to be made to the form of the generated child performance system. When the generator is semi-relative then are the performance systems in the population, in generating child performance systems, *isolated* form each other in the form-functionality topography. This means, in turn, that the semi-relative generator not can differentiate between any form clusters in the form functionality topography.

Relative. The relative generator is not sensitive to the problem of local minima. The relative generator could jump over arbitrary large hills in the form-functionality topography. In more connected form clusters the relative generator can make both a small and a large degree of random change. In more unconnected form clusters the relative generator can only make a large degree of random form change. The relative generator is, non the less, working in a *cumulative* way, i.e. the relative generator use the information given by the parent performance systems. The more similar the forms of the parent performance systems are, i.e. the lower form distance value they have, the less degree of random form change is made, and, vice versa. In this way can the relative generator adaptively set the degree of random form change.

There is thus no conflict for the relative generator; the semi-relative generator is at the same time both (1) *non sensitive to the problem of local minima* and (2) *cumulative*. Add to this that the relative generator will not have a performance that decreases with the developmental level of the parent performance systems. The reason for this is that the relative generator can bias the *location* and the *range* of the random form change. The relative generator makes no change at those places where the partial forms are identical. The relative generator makes a change at those places where the partial forms are different. This range of this change is limited to the partial forms of the parent performance systems. The relative generator acts thus both as a "smart stabiliser" that does not change good partial forms of the parent performance systems and as a "smart combiner" that do change unequal parts by combining possibly already good partial forms.

The performance systems in the population determine co-operatively the bias of the *degree* of the random form change, the *location* of the random form change and the *range* of the random form change. When the generator is relative then are the performance systems in the population, in generating child performance systems, *not isolated* form each other in the form-functionality topography. This means, in turn, that the relative generator adaptively can differentiate between any form clusters in the form functionality topography.

Conclusion. The conclusion from this discussion is thus that the generator should be relative (to the form of the parent performance systems). A population together with a form relative generator solves the problem with *form differentiation* approximately. The larger number of performance systems there are in the population, the better this approximation is.

4.2.4 The Competitor Selector

The competitor selector is a sub system that selects competitor systems from the population. The selected competitor performance systems is then given to the tester. The competitor selector is selecting performance systems from the population in a random way. The performance systems in the population can, however, be more or less ordered. The result of this is that the competitor selector can appear to be more or less biased, or informed, in its selection of competitor performance systems.

4.2.5 The Tester

The tester is a sub system that validates, or tests, competitor performance systems. The information for this testing comes from (1) the general functionality description, and, (2) the competitor performance systems that is given to the tester by the competitor selector. The tested competitor performance systems with the highest functionality distance valued is deleted from the population. The tester is biased in its testing of performance systems by the *functionality* of the competitor performance systems given to the tester by the competitor selector. The tester considers thus only the functionality of the competitor performance systems in testing the competitor performance systems. The tester can belong to one of the following three categories:

Absolute. An absolute tester is not taking any competitor performance systems as input. This type of tester can not be part of a cumulative search system and will not be considered further.

Semi-relative. A semi-relative tester is taking exactly one competitor performance system as input. To test a single competitor performance system involves the following;

- Calculate the functionality distance value of the input competitor performance system. If this value is above some threshold value t then the competitor performance system should be deleted from the population. The threshold value t could, for example, be probabilistic.

Relative. A relative tester is taking more than one competitor performance system as input. To test a number of competitor performance systems involves the following:

- Set up a *functionality competition* between the competitor performance systems:
    - Calculate regularly the functionality distance value for these competitor performance systems.
    - Delete the competitor performance systems that have functionality distance values higher than the competitor performance systems that have the lowest functionality distance value from the population.
    - Continue the functionality competition until some specific number of competitor performance systems remains.

The purpose for the tester is to *continually guide* the search process towards the bottom in the form-functionality topography. To be able to do this the tester must be able to differentiate between different functionality value levels in the form-functionality topography. The tester is, in essence, working in a purely random way. The performance systems in the population can, however, be more or less ordered. The result of this impact from the population on the tester is that the tester can appear to be more or less biased, or informed, in its testing of competitor performance systems.

4.2.5.1 Pro and Cons with the Semi-Relative and the Relative Tester

The following is a discussion of the pros and cons with the semi-relative and the relative tester.

Semi-relative. The semi-relative tester is, in its working, directly dependent on the general functionality description. If the general functionality description is very general, then will the semi-relative tester provide only a very rough guidance in the form-functionality topography, i.e. it will partition the form-functionality topography in the functionality dimension in very few partitions. This will, in turn, make the search system form-functionality ineffective. If, on the other hand, the general functionality description is more specific, then will the semi-relative tester provide a smooth guidance in the form-functionality topography. This will, in turn, make the search system form-functionality effective. The semi-relative tester is, however, in this case dependent on that the general functionality description is specific, i.e. highly informed a priori.

Relative. The relative tester can, in contrast to the semi-relative tester, serve as a good continuos guide in the form-functionality topography without the constrain that the general functionality description must be highly specific. The reason for this is that the relative tester encourages the performance systems in the population to *develop stimuli-response units that not are defined in the general functionality description.* The reason for this is that the systems will try to push each other to the general limits defined by the general functionality description.

Conclusion. The conclusion from this discussion is accordingly that the tester should be relative (to the functionality of the competitor performance systems). A population together with a functionality relative tester solves the problem with *functionality differentiation* approximately. The larger number of performance systems there are in the population, the better this approximation is.

5 CONCLUSION

We first reformulated the foundation of evolutionary computation. We showed that the functionality of the developmental system was equal to a maximally effective search system. We then proposed, as a hypothesis, a form that should implement this functionality. We arrived at the following general conclusion regarding the nature of the form of the developmental system; the form should be equal to a parallel generate-and-test search with a relative generator and a relative tester.

REFERENCES

1. Fogel, L. J., Owens, A. J., Walsh, M. J., Artificial Intelligence Through Simulated Evolution, Wiley Publishing, New York 1966.
2. Fraser, A. S., Simulation of genetic systems by automatic digital computers, Australian Journal of Biological Science, Australia 1957.
3. Futuyama, D. J., Evolutionary Biology, Sinauer Associates, Inc, Massachusetts, 1986.
4. Holland, J. H., "Adaptation in Natural and Artificial Systems", Ann Arbor, Michigan: The University of Michigan Press, 1975.
5. Rechenberg, I., "Evolutionsstrategie: Optimierung Technisher Systeme nach Prinzipien der Biologischen Evolution", Frommann-Holzboog, Stuttgart, 1973.
6. Spears, W. M., De Jong, K. A., Bäck, T., Fogel, D. B., Garis, H, "An Overview of Evolutionary Computation", Machine Learning ECML-93, Springer-Verlag, Berlin Heidelberg, 1993.

MODEL-BASED DIAGNOSIS OF THE HUMAN BODY

JAN ERIC LARSSON
Knowledge Systems Laboratory, Stanford University
701 Welch Road, Building C, Palo Alto, CA 94304, USA
Phone: +1 (415) 723–0948, E-mail: Larsson@KSL.Stanford.Edu

This paper describes the use of explicit models of goals and functions for monitoring and diagnosis of intensive-care unit patients. The method is based on multilevel flow models, (MFM), and used in the Guardian system. The use of MFM makes it possible to provide the system with alarm analysis, fault diagnosis, and generation of explanations, subject to strict demands of meeting hard real-time deadlines.

1. Introduction

The possibility of automating monitoring and diagnosis of intensive-care unit patients is an important objective, since it will increase both safety and efficiency of treatments, especially in difficult, unexpected, and time-critical situations. The Guardian system, [6], based on the BB1 architecture, [5], aims at supporting hospital personnel under such circumstances.

The Guardian project addresses a wide range of problems in real-time intelligent monitoring and control, such as continuous acquisition and interpretation of data, construction of therapy plans and monitoring the execution of such plans, diagnosis and alarming of unexpected disorders and complications, agility in reaction to problems requiring immediate attention, explanation of physiologic and pathophysiologic phenomena, and closed-loop control.

Expert systems have had a large impact on automated diagnosis. However, rule-based expert systems have several weaknesses, mainly because of the difficulty of building and maintaing large rule bases and also since the inference algorithms are not well-suited for real-time action or fast responses. A current trend is to use models instead of rules, and since monitoring and diagnosis largely deal with the *goals* and *functions* of the target system, it seems to be a good idea to use *explicit* representations of these concepts in diagnostic algorithms.

This paper describes algorithms based on one type of explicit means-end models, *multilevel flow models*, (MFM). MFM was developed by Lind, who has suggested a syntax for a formal language and given general ideas on its use, [13].

The programs presented in this paper provide three algorithms. The *alarm analysis* algorithm takes as input a set of alarm states such as *normal*, *low flow*, *high flow*, *low volume*, and *high volume*. Each alarm is associated with a corresponding MFM object, and the method separates primary from consequential alarms. The *fault diagnosis* algorithm uses an MFM model to produce a "backward chaining" style of diagnosis. The algorithm will find faults and give remedies. The *explanation generation* algorithm uses the fault state produced by the fault diagnosis algorithm to generate causal explanations in restricted natural language.

E. A. Yfantis (ed.), Intelligent Systems, 405–412.

2. Related Work

The main contributions of MFM have been made by Morten Lind and his group. Lind describes the basics of MFM, [13], and gives suggestions for a diagnostic system, [14]. He has also treated real-time diagnosis, [15], and design of operator interfaces, [12]. Lind's group has developed a graphical interface, [18], a STRIPS planning system, [17], and a system for alarm analysis and fault diagnosis, [1].

Larsson has developed diagnostic MFM algorithms for measurement validation, [8], alarm analysis, [7], and fault diagnosis, [9], implemented in the real-time expert system shell G2, see also [10–11]. Sassen and Jaspers have constructed an MFM fault diagnosis algorithm for COGSYS, [19], and Walseth has developed a measurement validation and fault diagnosis algorithm based on Lind's graphical interface, [21–22].

The area of model-based reasoning and diagnosis is overviewed in for example [2, 4].

3. Basic ideas of MFM

In multilevel flow modeling, a system is modeled as an artifact, i.e., a man-made system constructed with some specific *purpose* in mind. The purposes of a system are modeled with the MFM concept of *goals*, i.e., objectives of running the system. The physical *components* of a system are used to provide one or several *functions*. These functions are the means with which the goals are achieved. The concepts of goals and functions are explicitly represented as graphical objects in the MFM models.

Just as important are the different *relations* between goals, functions, and components. In MFM, these relations are explicitly described. A set of functions used to fulfill a goal are grouped together and connected to that goal via an *achieve relation*. If a subgoal is a necessary condition for a function to be working, it will be connected to the function via a *condition relation*. If a physical component is used for a certain function, the component object is connected to the function object via a *realize relation*. These relations connect the objects into a graph, i.e., the MFM model proper. Algorithms can then traverse this graph in order to perform different reasoning tasks.

The following example will help to explain the use of MFM, see Figure 1.

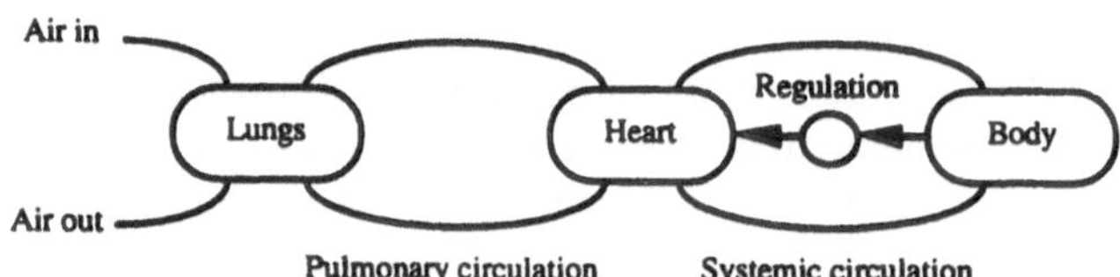

Figure 1. A simplified description of a part of the human physiology, including the intake of oxygen, the circulation of blood, and regulation of the heart rate. Oxygen-rich air flows through the lungs, driven by respiration, and delivers the oxygen into the blood. The circulation of blood consists of two parts, the pulmonary circulation through the lungs, where the blood is oxygenated, and the systemic circulation through the rest of the human body, where the oxygen is

used in the metabolism. The heart rate is regulated in several ways; the one shown here is based on hydrogen ion receptors in the lower brain-stem. When the hydrogen ion concentration rises, (a sign of high carbon dioxide and, indirectly, of low oxygen concentration), receptors cause the heart rate to increase via nervous stimulation.

This is a simplistic example, but it will serve well to explain the concepts of MFM. There are two *goals* that this system fulfills:

- Maintain the circulation
- Provide oxygen for the metabolism

The given example is small, but still several *functions* are present:

- The *pumping* of blood performed by the heart
- The *transport* and *storage* of blood in the lungs and the body
- The *providing* of oxygen from the environment
- The *transport* of oxygen in the blood
- The *consumption* of oxygen in the body tissues
- The (indirect) *observation* of oxygen concentration
- The nervous feedback *control* of the nervous system
- The *control action* of the cardiac muscle and ennervation
- Etc.

The third type of MFM objects are the physical *components*, i.e., the organs:

- The heart
- The lungs
- The body tissues
- The blood vessels
- Etc.

The relations between these objects are as important as the objects themselves. For example, the heart *realizes* the function of transporting blood through the circulation, and the latter function *achieves* the goal of maintaining the circulation.

In the MFM model, goals, functions, and relations are described in a graphical language, see Figure 2.

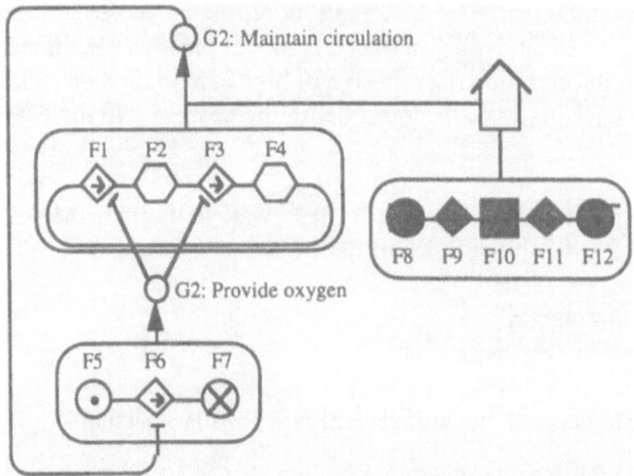

Figure 2. An MFM model of the circulation and oxygen flow in the human body. The toplevel goal of maintaining circulation is achieved by a network of mass flow functions, describing the right and left parts of the heart, (F1 and F3), as transport functions and the pulmonary and systemic circulations as storages, (F2 and F4). The circulation goal is achieved by control via the manager function, (to the right), which in turn depends on the information flow below it. The receptors are described by an observer function, (F8), the feedback control by a decision function, (F10), and the control action by an actor, (F12). The transport functions of the heart are conditioned by the goal of providing oxygen, since the heart muscles need oxygen to work. The goal of providing oxygen is achieved by a network of flow functions, here modeled simply as a source, (F5), a transport, (F6), and a sink, (F7).

4. An MFM Model of the Human Body

For the Guardian project, a large MFM model of the human body has been developed. It covers all systems needed for intensive-care unit montoring, such as the heart, circulation, the body fluid volume, the nutrition, respiration, body temperature, acid-base balance, the concentrations of sodium, potassium, calcium, and magnesium, and the, often multiple, regulatory mechanisms for all these systems.

A first version of this MFM model was built during the autumn of 1994, with the author acting as both domain expert and knowledge engineer. The knowledge in the model was compiled from medical textbooks such as [3] and [20]. An early verification, performed by a medical student, showed that the model was generally reasonable and covered all major systems needed in intensive-care monitoring and diagnosis. Currently, the model is being revised by a team of professional physicians, with the author acting as the knowledge engineer. This effort is a part of the current phase of the Guardian project, which aims at increasing and verifying the knowledge base of the Guardian system. This phase is planned to last for three years, but the MFM model should be ready after the first.

It is worth mentioning that MFM provides a relatively easy knowledge acquisition and knowledge engineering task. Its graphical nature and hard syntax make model building effective, and it also excludes the possibility of some internal inconsistencies. The high level of abstraction in the description of goals and functions makes it possible to give crisp, correct, and still highly useful models of complicated, real-world systems like the human body.

The three algorithms, (alarm analysis, fault diagnosis, and explanation generation), have been verified and found to work without any problems on the Guardian model, which is probably the largest MFM model in the world. Judging from these experiences, it is clear that MFM scales favorably, without changing any qualitative properties in behavior or results. This is true both for model building activities and for the execution of the diagnostic algorithms.

Reasonable assessments of the efforts needed to construct three MFM models are shown in Table 1. The knowledge acquisition includes the time needed to study the system and to find out the physical facts of how it works, while the model construction time was used for identifying the goals and functions and constructing the actual MFM model. Since it is difficult in practise to compartmentalize the different phases from each other, the figures in Table 1 are approximations. Although no comparative studies are available, it is the author's experience that the MFM modeling effort is much smaller than the effort needed to turn the same knowledge into a rule base. Furthermore, the model-based and graphical nature of MFM makes it considerably easier to handle MFM models than to use knowledge encoded in rules.

Process	Size	Knowledge Acquisition	Model Construction
Lab Tanks	"toy"	0	1 hour
Steritherm	medium	2 months	1 week
Guardian	large	4 months	1 week

Table 1. Three modeling tasks of different size. The times are fairly reasonable estimates of the man-hours used. Note, though, that the Guardian effort is not yet finished.

5. Implementation

The diagnostic algorithms have been implemented in the expert system tool G2, [16], and tested on two systems: a simple lab process with two cylindrical tanks, and Steritherm, a process for ultra-high temperature treatment of dairy products. The G2 implementation is commercially available and has been sold to CERN, the European high-energy research facility in Switzerland.

The methods have also been implemented in C for Macintosh systems, and in Common Lisp for use in the Guardian project. These implementations have been tested on several target systems, among other the large MFM model of the human body.

The diagnostic reasoning tasks described above were implemented as searches in MFM graphs. The fault diagnosis method uses a depth-first search, and since the size of every subtree is known in advance, it is simple to obtain a worst-case estimate of the computational effort needed to diagnose any goal. The ability of giving worst-case estimates of the time needed for a diagnostic subtask makes MFM well-suited for diagnosis under hard real-time constraints. Whenever a subtask must be finished in a fixed time, the known worst-case can be used to test whether it is indeed possible to comply with the deadline.

The *hierarchical structure* of MFM models is also valuable. The fault diagnosis algorithm starts from the topmost goals, i.e., the most general goals, that concern production and safety demands of the entire system. Then the search moves downwards in the MFM graph and attacks more specific subproblems. This means that the diagnostic resolution will be finer and finer as more time is available. Whenever the time runs out, a more or less coarse diagnosis is available. Thus, MFM allows *graceful degradation* under hard real-time constraints.

MFM's graphical structure makes the described algorithms behave favorably when the problems are scaled up. All three methods increase at worst linearly in effort needed when the size of the MFM model goes up. However, most incoming data cause only local changes in the diagnosis, and very often, the increase is less than linear. For example, the cost of performing a fault diagnosis of one subpart of a model is not affected at all by adding another, (separate), subpart.

The C implementation highlights the efficiency of the diagnostic methods. Table 2 shows some data for the C implementation of the fault diagnosis algorithm, when executed on a SPARC station 10. The Lisp implementation provides an automatic rule generator for the fault diagnosis algorithm, and the number of rules needed is given in the table, to give a general idea of the descriptive power of the three MFM models. Given the number of rules, an apparent speed in rules per second can be computed. Note that this speed goes up with increasing model size, due to less influence of start-up and bookkeeping times. It should also be noted that a state-of-the-art rule-based system like G2 triggers some 1300 rules per second. The fastest system availabe, RTworks from Talarian Corporation, reaches 10 000 rps. Most model-based systems are considerably slower.

Process	Objects	Rules	Worst-Case	RPS
Lab Tanks	27	39	480 μs	81 000
Steritherm	99	150	630 μs	240 000
Guardian	331	544	1100 μs	500 000

Table 2. Sizes and execution speeds for three MFM models. The columns list the number of MFM objects in the model, the number of rules needed for a rule-based system to perform an equivalent fault diagnosis, the worst-case execution time for a diagnosis, and the apparent speed in rules per second, should the diagnosis have been performed by a rule-based system.

The algorithms were all tested in realistic fault situations and the conclusion was that they worked accurately and gave correct and useful information.

6. Conclusions

The Guardian system needs capabilities for monitoring, diagnosis, and explanation to solve its task in the intensive-care unit enviroment. Doing this, it must be able to react quickly in real-time and meet hard real-time deadlines. Furthermore, its knowledge must be reliable and easy to understand and explain. Our experiences so far show that MFM can provide capabilities that are very useful for Guardian. The hard structure and high-level concepts of MFM support the verification of the model built, and the good real-time properties and speed of the algorithms satisfy the computational demands. It is clear that MFM can provide models and algorithms that are both useful and reliable under difficult, real-world circumstances.

7. Acknowledgements

I would like to thank Barbara Hayes-Roth and the staff at the Knowledge Systems Laboratory for valuable help and support. I would also like to thank the originator of MFM, professor Morten Lind, and my Ph. D. supervisor, professor Karl Johan Åström, and Doctor Karl-Erik Årzén for inspiration and support during earlier phases of my MFM research.

The project is supported by grants from the Swedish Research Council for Engineering Sciences, the Swedish Institute, the Royal Physiographic Society in Lund, and the Nils Hörjel Research Fund at Lund Institute of Technology.

8. References

1. Creutzfeldt, J., "Sensor Validation, Alarm Analysis, and Fault Diagnosis in Large Processes," Technical report, Institute of Automatic Control Systems, Technical University of Denmark, Lyngby, Denmark, 1990. In Danish.
2. Davis, R. and W. Hamscher, ''Model-Based Reasoning: Troubleshooting,'' in H. Shrobe, (Ed.): *Exploring Artificial Intelligence*, Morgan Kaufmann Publishers, Inc., San Mateo, California, pp. 297–346, 1988.
3. Guyton, A. C., *Textbook of Medical Physiology*, Seventh edition, W. B. Saunders Company, Philadelphia, 1981.
4. Hamscher, W., L. Console, and J. de Kleer, (Eds.), *Readings in Model-Based Diagnosis*, Morgan-Kaufmann Publishers, Inc., San Mateo, California, 1992.
5. Hayes-Roth, B., "A Blackboard Architecture for Control," *Artificial Intelligence*, vol. 26, no. 3, pp. 251–321, 1985.
6. Hayes-Roth, B., R. Washington, D. Ash, R. Hewett, A. Collinot, A. Viña, and A. Seiver, "Guardian: A Prototype Intelligent Agent for Intensive-Care Monitoring," *Artificial Intelligence in Medicine*, vol. 4, no. 2, pp. 165–185, 1992.
7. Larsson, J. E., "Model-Based Alarm Analysis Using MFM," *Proceedings of the 3rd IFAC International Workshop on Artificial Intelligence in Real-Time Control*, Rhonert Park, Sonoma, California, 1991.
8. Larsson, J. E., "Model-Based Measurement Validation Using MFM," *Proceedings of the IFAC Symposium on On-Line Fault Detection and Supervision in the*

Chemical Process Industries, University of Delaware, Newark, Delaware, 1992.

9. Larsson, J. E., "Model-Based Fault Diagnosis Using MFM," *Proceedings of the IFAC Symposium on On-Line Fault Detection and Supervision in the Chemical Process Industries*, University of Delaware, Newark, Delaware, 1992.
10. Larsson, J. E., *Knowledge-Based Methods for Control Systems*, Doctor's thesis, TFRT–1040, Department of Automatic Control, Lund Institute of Technology, Lund, 1992.
11. Larsson, J. E., "Diagnostic Reasoning Strategies for Means-End Models," *Automatica*, vol. 30, no. 5, 1994.
12. Lind, M., "Human-Machine Interface for Diagnosis Based on Multilevel Flow Modeling," *Proceedings of the 2nd European Meeting on Cognitive Science Approaches to Process Control*, Siena, Italy, 1989.
13. Lind, M., "Representing Goals and Functions of Complex Systems—An Introduction to Multilevel Flow Modeling," Technical report, Institute of Automatic Control Systems, Technical University of Denmark, Lyngby, Denmark, 1990.
14. Lind, M., "Abstractions Version 1.0—Descriptions of Classes and Their Use," Technical report, Institute of Automatic Control Systems, Technical University of Denmark, Lyngby, Denmark, 1990.
15. Lind, M., "An Architecture for Real-Time MFM Diagnosis," Technical report, Institute of Automatic Control Systems, Technical University of Denmark, Lyngby, Denmark, 1990.
16. Moore, R. L., H. Rosenof, and G. Stanley, "Process Control Using a Real-Time Expert System," *Proceedings of the 11th Triennial IFAC World Congress 1990*, Tallinn, Estonia, pp. 241–245, 1991.
17. Norby Larsen, M., "Strips as a Planning Method within Abstractions and MFM Modeling," Technical report, Institute of Automatic Control Systems, Technical University of Denmark, Lyngby, Denmark, 1990.
18. Osman, A., "The Interface Substrate," Technical report, Institute of Automatic Control Systems, Technical University of Denmark, Lyngby, Denmark, 1990.
19. Sassen, J. M. A., *Design Issues of Human Operator Support Systems*, Doctor's thesis, Delft University of Technology, Faculty of Mechanical Engineering and Marine Technology, Delft, 1993.
20. Taverner, D., *Taverner's Physiology*, Fourth edition, Hodder and Stoughton, London, 1983.
21. Walseth, J., *Diagnostic Reasoning in Continuous Systems*, Ph. D. thesis, ITK-rapport 1993: 164–W, Division of Engineering Cybernetics, Norwegian Institute of Technology, Trondheim, Norway, 1993.
22. Walseth, J. Å, B.¯A. Foss, M. Lind, and O, Ögaard, "Models for Diagnosis—Application to a Fertilizer Plant," *Proceedings of the IFAC Symposium on On-Line Fault Detection and Supervision in the Chemical Process Industries*, University of Delaware, Newark, Delaware, 1992.

INTELLIGENT DECISION MAKING IN TWO-LEVEL ACTIVE SYSTEMS

RYSZARD A. WASNIOWSKI
New Mexico Highlands University, Las Vegas, NM 87701

Many control applications require cooperation of two or more intelligent subsystems. Besides the good performance of each subsystem, effective coordination of these subsystems is very important to achieve the desired general system performance. In this paper the optimization of such systems is considered. At the high level there is a controller (CO) who exerts control on N subsystems. At the low level each subsystem has its own local controller. The objective function of each local controller assesses the performance of its corresponding subsystem. The objective function of the high level coordinator assesses the performance of the overall system. The game based approach and neural networks based approach are considered for the optimization of such systems.

INTRODUCTION

It is common modern practice to consider the overall system as separate process units and the overall control scheme as a hierarchical structure containing some levels. The lowest level contains local feedback control loops with the objective of maintaining stable control of basic process parameters. The intermediate levels specify the set points of the local feedback controllers according to a desired operational strategy. The highest level specifies economic constraints on the intermediate levels optimization problem and contains management decisions. Although some basic principles in coordinating multiple systems were developed in early eighties many practical problems are still open. The main difficulty in coordinating multiple systems comes from the lack of precise system models and parameters as well as the lack of efficient tools for system analysis, design, and real-time computation of optimal solutions. New methods for analysis and design are thus required for the closed-loop coordination of multiple systems. It should be noted that the operation of hierarchical systems draws heavily on the information transfer between adjacent levels, the explicit specification of the submodels and subobjecives, and the proper adjustment of the subsystem activities. The situation becomes more complicated when we consider human factor in a system. In such systems it is possible to observe a phenomenon called the lack of interests agreement. For example, the interest of one of the systems elements does not always coincide with the interest of the system as a whole. Such optimization problems arise in practice in many computer-controlling engineering processes, economic systems and networks. Since the decision-making problem depends on goal function of subsystem a game approach is well suited for dealing with this problem.

E. A. Yfantis (ed.), Intelligent Systems, 413–419.

PROBLEM FORMULATION

The problem of decision making in a two-level system consisting of a controller (CO) and a certain number of subsystems A_i, $i \in I = \{1,2,\ldots,N\}$ (see Fig.1) is formulated as follows: does there exist a decision of CO at which certain permissible rigors of subsystems work provides the required functioning of a system as a whole?

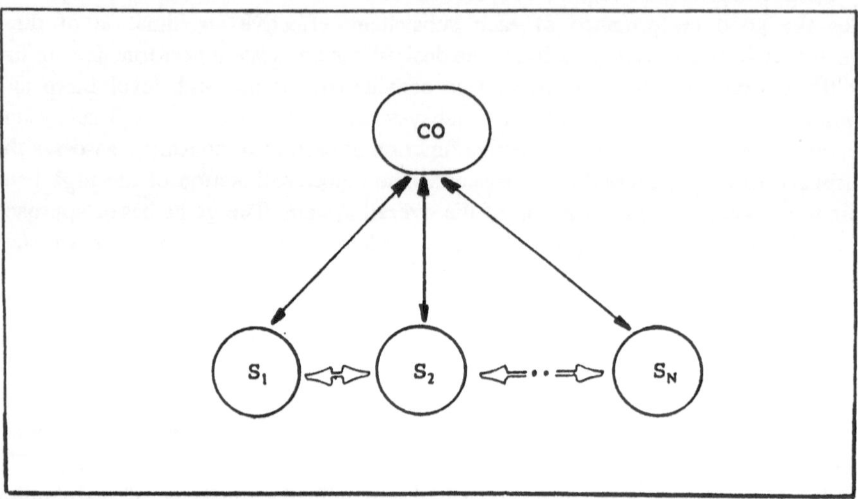

Figure 1. The structure of the two-level system.

Formally the problem is denoted by

$$\Phi(x, y) \to \max_{x \in X}$$

$$\Phi: X \times Y \to R$$

where: Φ is the objective function of a system;

$$x=(x_1,x_2,\ldots,x_N),\ x_i \epsilon X_i \subset R,\ x \epsilon \underset{i \epsilon I}{X} X_i \subset R$$

denote decision made by CO;

and

$$y=(y_1,y_2,\ldots,y_N),\ y_i \epsilon Y_i \subset R,\ y \epsilon \underset{i \epsilon I}{X} Y_i \subset R$$

denote decisions of subsystems,

Considering active subsystems behavior it is assumed that the subsystems have their own objectives defined by the functions $\varphi_i(x_i,y_i)$, $i \epsilon I$ and the above problem take the following form:

$$\Phi(x,y) \rightarrow \max_{x \epsilon X}$$

$$\forall_{i \epsilon I} \quad \varphi_i(x_i,y_i) \rightarrow \max_{y_i \epsilon Y_i}$$

It is assumed that CO has priority in decision making and makes decisions based on the principle of the maximal warranted income. The formulation of this problem is idealized. In real situations we deal with decision making under conditions of indeterminacy caused by insufficient definition of limitations, decision area, objective of decision making, and information errors. In such cases we often have at our disposal only the estimations (given, e.g., by experts) of certain quantities. Fuzzy sets have been proposed as mathematical models of such classes of objects. In [9] an analysis has been carried out of problems that occur in doing so, according to the level of information about subsystems: decision making by CO when subsystems models are known; and, decision making when subsystems models are not known, but there is a possibility of obtaining information about the subsystems while they are functioning; It has been shown that, in several cases, the first problem may be solved by means of a mathematical programming and the other problem by means of a game theory.

GAME BASED DECISION MAKING

In this case the process of decision making consists in the choice of strategy $x \in X$ by CO that informs the subsystems about his choice and then each subsystem chooses its own strategy $y_i \in Y_i$, $i \in I$.

If we assume that the CO knows the functions of subsystems $\varphi_i(x_i, y_i)$, $i \in I$ and sets Y_i and first makes a decision then we receive Germeier game in which every decision $x \in X$ of the CO is concerned as the warranting strategy in Germeier sense.

Determining optimal warranting strategies becomes complicated when CO does not know accurately the goal functions of subsystems and the sets Y_i or does not know them at all. Then, there appears the problem of decision making optimization in the conditions of incomplete information. In this case when decisions are made repeatedly, there is a possibility of using the information from the past. This problem has been depicted in [1,10]. A particular classes of games useful here are so-called Stackelberg games. We now consider a two-level system where CO control $x \in X$ is common for all subsystems. If the CO knows the objective functions of the subsystems, then it knows their optimizing controls $y_i = T_i x$ for every $x \in X$. Then he can determine a control x^S such that along with the corresponding optimal local controls y_i^S. The objective function of the overall system is maximized, x^S is then the Stackelberg control of the system. In addition to the exchange of information by means of introducing an extra function or collecting information from the subsystems which was discussed above it is possible to introduce a dialogue between CO and the subsystems aided by technological means e.g. ,computer conference. Many practical problems, (especially where it is difficult or impossible to introduce formalization, e.g., some planning systems) require an approach based on the dialogue between the levels.

The dialogue scheme of decision making in the two level systems is conceived as the scheme in which CO and the subsystems frequently present proposals of their decisions, estimations of CO decisions, preferences, bargaining, changes etc. until the proposals of both sides are settled. An essential feature of the dialogue schemes of decision making is the fact that the stage of the realization of decisions is not performed until the decisions CO and the subsystems are definitely determined. One should emphasize that the dialogue scheme of decision making does not need to be and usually is not an adaptation scheme. It signifies that both CO and the subsystems do not use an algorithm of determining their proposals for defining subsequent proposals in a dialogue. Generally, the rule of decision making CO in iterated k step dialogue may be presented in the following way:

$$x_i^k = \{x^D_i(x^{D,k-1}_i, y^{D,k-1}_i)\}$$

Where $x^{D,k}_i$ and $y^{D,k}_i$ is an information about favored decisions exchanged between CO and the subsystems during the dialogue, and $k=1,2,\dots,K$, K is number of steps in the dialogue. The number of steps of a dialogue may be given or defined by stop rules. CO and the subsystem in the dialogue are continuously directed by their own goals i.e. maximization or minimization of their goals functions. Accepting the idea of a dialogue system of decision making involves the following problems: Whether the dialogue system of decision making will correspond to practice and whether the decision developed in the dialogue will constitute the basis for practical applications (CO and subsystems are still directed by their own goals). Practical experiments [10] prove that where both sides i.e. CO and the subsystems and whenever such compromises are possible, introducing the dialogue system of decision making is effective.

DECISION MAKING WITH NEURAL NETWORKS BASED PREDICTOR

In this part we will discuss the possibility of using neural nets based predictor. A coordinator that combines the techniques of intelligent control and neural networks, and forms the high-level coordinator in a hierarchical structure, is proposed. The basic idea is to estimate the effects of the control commands to subsystems using neural networks based predictor and modify these commands to achieve the desired performance. (see Fig.2)

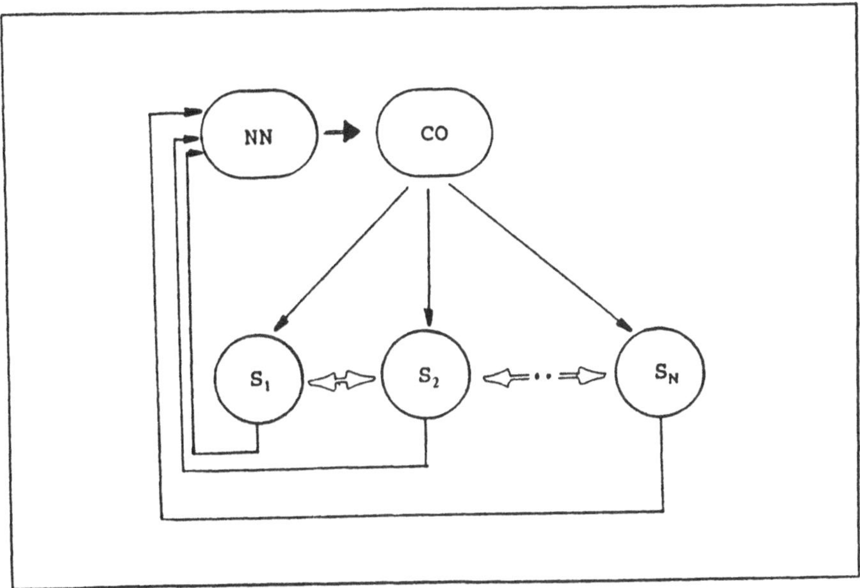

Figure 2. Two-level system with neural networks (NN) based predictor.

Because the internal structure and parameters of the low level are not affected by using the proposed method, some commercially designed controllers for single systems can be coordinated to perform more sophisticated tasks for multiple systems than originally intended.

The development of dynamical mappings

$$(x^0{}_1,y^0{}_1),(x^1{}_1,y^1{}_1),\dots,(x^k{}_1,y^k{}_1)$$
$$(x^0{}_2,y^0{}_2),(x^1{}_2,y^1{}_2),\dots,(x^k{}_2,y^k{}_2)$$
$$\dots$$
$$(x^0{}_N,y^0{}_N),(x^1{}_N,y^1{}_N),\dots,(x^k{}_N,y^k{}_N)$$

where k denotes decisions of subsystems and CO in the current step,

can be approximated by a multilayer perceptron. In other words neural networks can be trained to learn the dynamical properties of the system from examples. Note that the interactions among subsystems are implicitly included in the training data. One of the most popular neural networks structures is the multilayer perceptron with the back propagation algorithm [4,8]. The computation of back propagation algorithm includes two steps: compute the neural network's output forward from its input to output layers, and modify the connection weights backward from its output to input layers. A three-layer (with one hidden layer) perceptron was designed to learn the relationship of the above dynamical mappings, and to form the neural networks based predictor. After the neural network is "well trained," its weights are no longer modified. However, for time-varying mappings, it is meaningless to say that neural network was "well trained." In practice it is also desirable to operate the system in closed-loop. This means that the neural network based predictor should be updated to be able to track a time-varying system. In order to update the neural networks based predictor "on-line," the basic idea is to the weights of the neural network are modified using the a posteriori prediction error.

APPLICATION AND CONCLUSION

A special interactive simulation system for conducting experiments of decision making in two level system has been developed [10]. In this system, the decision process is conducted through the set of terminals. Having developed a concrete algorithm of decision making one may begin an experiment in the interactive way. Conducting an experiment consists of decision making by the players simulating a real system under consideration.

In this paper an approach based on the Germeier and Stackelberg solution of de cisionmaking in two-level system as well as neural networks based predictor have been discussed. It is assumed that the nsystem under consideration consists of two levels. At upper level there is a coordinator whose decisions are transmitted to a lower level. The lower level consists of N subsystems, each equipped with it own local controller and its own individual objective function. The objective function of the coordinator assesses the global performance of the overall system. Algorithms and programs for experimenting with problems of decision making for resource allocation and planning were developed.

REFERENCES

[1] Cruz, J.B. Jr, Stackelberg Strategies for Hierarchical Control of Large Scale Systems in: Second Workshop on Hierarchical Control W. Findaisen Ed. Warsaw, 1979.

[2] Germaier, Ju.B. Games with antagonistic interests , Nauka, Moscou,1976

[3] Khanna T., Foundations of Neural Networks, Addison-Wesley Pub. Co., 1990

[4] Kosko B., Neural Networks for Signal Processing, Prentice-Hall, Englewood Cliffs, New Jersey, 1992

[5] Mesarovic, M.D., Macko, D., Takahara, Y., Theory of Hierarchical Multilevel Systems, Academic Press, 1970

[6] Siman, M., Cruz J.B. A Stackelberg Strategy for Games with Many Players, IEEE Trans. Automatic Control, AC-18, 1973

[7] Stackelberg, von H. The Theory of the Market Economy, Press, Oxford University Press, 1952

[8] Vemuri V., Artificial neural networks: Theoretical concepts, IEEE Computer Society, 1988

[9] Wasniowski, R., Active Systems and Their Applications to Management, Systems Science, Vol. 4, No2, 1987

[10] Wasniowski, R., Problems of Decision-making in two-levels Systems, Technical Report, FRC, Technical University of Wroclaw, 1987

AUTOMATIC GENERATION OF C PROGRAM CODE FROM DATA FLOW DIAGRAMS

Young K. Nam, Jinsam Kim, Kongseon Lee
Systems Engineering Research Institute/KIST, Taejon, Korea

Lawrence J. Henschen
Dept. of EECS, Northwestern University, Evanston, IL60208

Abstract

In this paper we propose a method for generating C code from a specification of requirements written in DFDs (Data Flow Diagram) and DD (Data Dictionary) by using Prolog as an intermediate language. The automatic generation of C code takes two transformation steps. The specification is transformed into a Prolog program and the transformed Prolog program is compiled into procedural languages such as C or Pascal so that the specification is directly executable in the conventional programming environment. The transformation results in two kinds of code; one is directly executable C programs which mimics Prolog behavior and the other is modules with procedure calls for its lower diagram and skeleton code for functional primitives. This work makes it possible to rapidly have a prototype and remove the design stage from the software development life cycle. The execution speed of Prolog programs will be enhanced by using conventional optimized compilers.

Keywords : Requirements specification, Compiling Prolog, Structured Analysis Techniques, Language translation.

1 Introduction

Over the past two decades, structured specification techniques, such as the Structured Analysis Technique [3, 20], have drawn a lot of attention from software developers to researchers due to their use of intuitively appealing notation and concepts. But like many other specification techniques, the lack of formality of descriptions for the concepts and notation used in such techniques prohibits their wide spread of use in the software development industries. There are usually trade-offs between rigid specifications and user friendly specifications. The mathematically rigorous specifications may provide a facility for reasoning, verifying the specifications, or checking consistency between requirements and design specifications. Furthermore, the specifications written in the rigorous formal style can be

E. A. Yfantis (ed.), Intelligent Systems, 421–436.

directly implemented without further transformations. But it may be very difficult to learn or to analyze requirements so formally from the beginning. On the other hand, the structured techniques provide a pictorial description of the specifications so that it is very intuitively appealing and easy to use for analysts. However, it can not be transformed into design specs or implemented directly without further work. Those techniques can not represent all the necessary information such as control or time-dependent information.

DFDs are one of the most widespread formalisms for requirements specification for data-processing problems, and a variety of software-development methodologies and tools having DFDs as their underlying model is in use. The popularity of DFDs is largely due to the simplicity and intuitive appeal of their graphical notation: bubbles are used to represent functions resulting from system decomposition, arcs connecting them represent functional dependencies among their input and output data, and suitable visual conventions are provided to represent permanent data stores and data exchange with the external environment.

The Prolog language is the best known example of a logic programming implementation, with the uniqueness that it can be viewed declaratively as well as procedurally. Therefore, Prolog may provide an attractive tool in software engineering, particularly in the areas of *executable* specification and rapid prototyping [2, 9, 10, 11, 12]. The standard software development cycle of analysis, design, and implementation provides many benefits, but it also has serious drawbacks. One problem is that the mapping from a declarative specification to its procedural implementation is complex. There is no automated procedure to transform analysis into design, and verifying program correctness is an unsolved problem. The nonexplicit representation of requirements reveals itself in maintenance problems. The lack of executable specifications (due to informal syntax and semantics) leads to long delays in evaluation whether the direction chosen for development is right.

Prolog provides a direction that narrows the gap between analysis and implementation. Prolog's declarative nature is ideally suited for the analysis phase, while Prolog's procedural capabilities are well suited for software implementation by making a specification *executable*. The specification becomes a program that can be run and debugged. A user can see whether his requirements as stated are what he actually wants, and the programmer can verify the correspondence of the program to the user's requirements. Unfortunately, these advantages are obscured by the pragmatic problem of efficiency. Logic programs can often be surprisingly inefficient. Many commercial compilers for Prolog are available, but they still suffer from a speed problem.

We have already suggested an idea compiling Prolog programs with list structures into procedural languages such as C or Pascal in order to use Prolog as a software requirements analysis and programming tool [14, 15, 16, 17]. In this paper we propose a system called START (STructured Analyzing and Reporting Tool), for transforming a specification written in DFDs into Prolog programs and compiling the transformed Prolog programs into procedural programs such as C or Pascal so that the following advantages can be obtained;

Figure 1: Proposed System

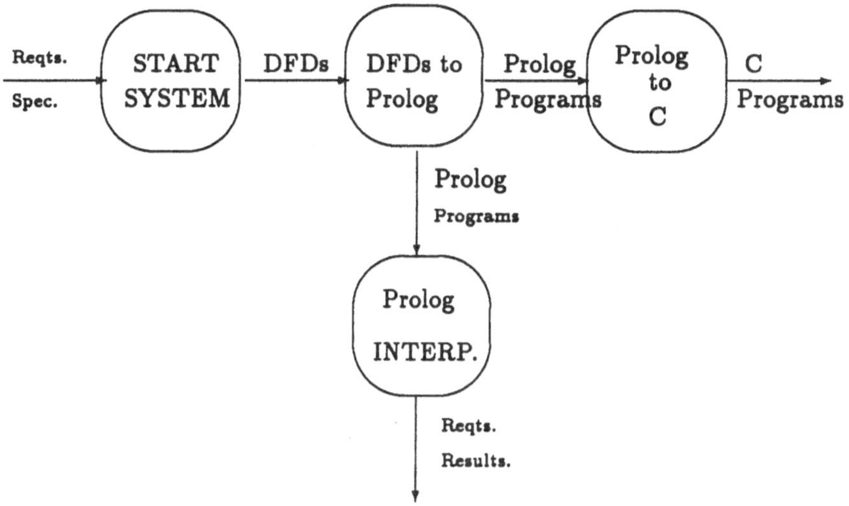

- A specification written in DFDs can be executable so that a rapid prototyping is possible.
- The enhancement of execution speed of Prolog programs by using conventional optimized compilers can be obtained.
- Once an application is conformed by executing the requirements specifications, programmers can directly turn the programming stage so that the gap between design and implementation can be narrowed.

The description of mapping from DFDs into C code is shown in Figure 1. We assume that readers have some acquaintance with logic programming languages, in particular Prolog, and Structured Analysis Techniques. In Section 2, the work related to executable data flow diagrams is sketched. Sections 3 and 4 represent mapping from DFDs to Prolog and from Prolog to C.

2 The Related Work

Most of research related to DFDs started by pointing out the flaws and inaccuracies and introduced formalisms as an extension to DFD [6, 7, 19, 18]. Based on the formalisms, some of them make DFDs executable [7, 13, 4].

The approach of Ward [19] extends the DFD notation to represent control and timing by the definition of the so-called transformation schema. Control and data-processing are separated to reflect the typical architecture on which the systems being modeled are implemented. In [6], France suggested a formal specification tool, which is based on C-DFDs [19] for syntactics and ASTS (Algebraic state transition systems) for semantically. This model supports synchronous and asynchronous data flows between processes. In [18], a formal specification language, which is called FDFD (Formal Data Flow Diagrams), based on Petri nets has been developed by using an advantage that Petri nets can be represented both graphically and algebraically. The method can analyze both structural and behavioral consistency. Those formal languages are strict and rigorous and able to analyze specifications for an given application but quite difficult to understand for analysts and do not have tools for supporting its corresponding model.

Among the executable specifications, the newest one is by Fuggetta et al. [7]. Fuggetta et al. introduced VLP, an executable visual language for formal specifications and prototyping which integrates entity-relationship (ER) and data flow diagrams in a semantically rigorous and clear way. The language can define data flow synchronously and asynchronously. Data types and functions are defined in a way similar to what is done in Pascal-like languages. The primitives for manipulating aggregations and relations are also provided in the language. SAME, developed by Docker [4], is a CASE prototyping tool and an environment in which DFDs can be executed as a two-dimensional program graph. The tool is implemented in Prolog; a predicate name is one of the types of data flows such as importer, exporter, or type-checking, etc., and the terms of the predicate consist of an object name and the kind of object. The inference is based on a tree derived from a hierarchy of data flows in the diagram. The system is different from ours in the point of an inference model; in SAME it is based on data flows in a diagram, but START is based on process. Meeson [13] developed a compiler for an extension of Ernest, which is a functional programming language, for executing DFDs represented in the functional programming language. The compiler analyzes the connectivity of a data flow diagram and constructs an abstract syntax tree for the equivalent Ernest function definition.

Our system does not change or extend the Structured Analysis Technique so that Prolog code is transparent for users and is directly executable. Furthermore, both executable code and skeleton code are readable by users, and incremental programming development is possible based on those code.

3 Mappings DFDs into Prolog

A DFD models an application in terms of the following elements:

- Process (Data Transform) : Abstractions of an application's processing elements.
- Data Stores : Abstractions of repositories of data.
- External Entities : Abstractions of external objects interacting with the application.

Figure 2: The relationship between DFDs and corresponding Prolog program in an application

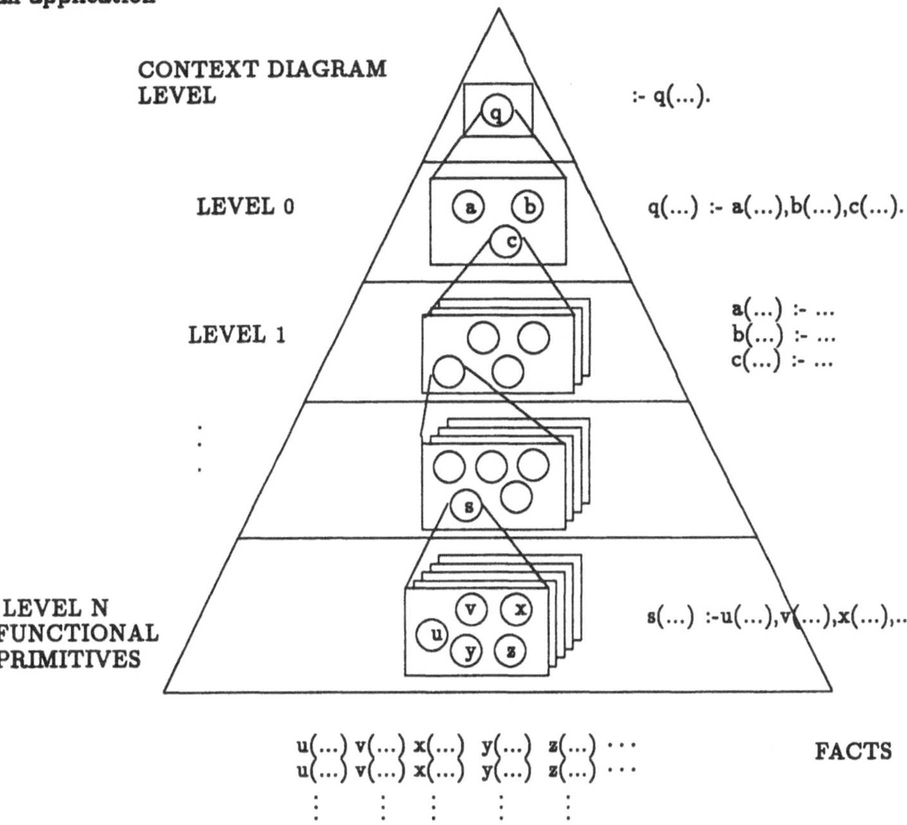

- Data flows : Abstractions of data communication between the above elements

We use a numbering mechanism similar to the method in [3]. An idea for transforming DFDs into Prolog is presented in [11] by Lazarev. Given DFDs for an application, a Prolog program is generated. A basic idea is that each process is named as a predicate with its process name, whose terms consist of input and output data flows. A context diagram serves as a query. A process and its corresponding lower diagram make a horn clause. Many facts may exist for a functional primitive process. Figure 2 shows a relationship between DFDs and a Prolog program. We recap the method briefly and present some modification to the method of [11].

1. Describe process name as a predicate name. All predicates have two arguments; an input argument represented as structure $i(I_1,I_2,\ldots,I_n)$ where $I_1,I_2,\ldots,I_n$ are input data flows; and an output argument represented as structure $o(O_1,O_2,\ldots,O_n)$ where $O_1,O_2,\ldots,O_n$ are output data flows.

2. If a data flow has mutually exclusive data (OR relation), then represent a predicate including the data flow with several predicates with the same name. In this case only one predicate can be supplied for the process at a time.

3. For any process $i_1.i_2., \cdots, .i_j$ in any level diagram except for the lowest one, make a horn clause by assigning process $i_1.i_2., \cdots, .i_j$ as its head predicate and its lower diagram as its body. The processes indexed from $i_1.i_2., \cdots, .i_j.i_{k1}$ to $i_1.i_2., \cdots, i_j.i_{kn}$ in the lower diagram are represented as subgoals and the processes in the lowest diagram are represented as facts. The clause must satisfy the following conditions;

   (a) Each input of each subgoal must be matched with some input of the head predicate or with an output of some previous subgoal (a balanced input).

   (b) Each output of each subgoal must be matched with some output of the head predicate or with an input of some later subgoal (a balanced output).

   (c) Unmatched inputs and outputs of a predicate including a head predicate are replaced by *nil*. At least one input and one output of a subgoal must be non-nil.

In [11], a cyclic DFD is transformed into a non-cyclic DFD by assigning a functional process which handles data flows getting into a process making the DFD cyclic. This method requires that an analyst be aware that the process is split into two processes such that each data flow is handled separately. If this transforming process is done manually by an analyst himself, there would be no problems in splitting a process into two or more processes with different inputs and outputs. But a purpose of the method in this paper is that a Prolog program is automatically generated from DFDs in the START system so that it is impossible to draw the DFDs without cyclic data flows and no semantical differences. Another problem is that if a DFD has cyclic data flows, then it may not correct in displine because cyclic data flows can be handled by iterations or recursive calls in a process.

For OR-correlations [11], Lazarev asks analyst to mark whether more than two processes in a DFD have OR correlation, i.e., only one process is called in a clause but not both predicates in one clause. Then there should be several clauses with the same heads for predicates having OR-correlations. Since all mutually exclusive predicates can be represented in a clause and there are no information losses for getting answers from the clause if all bubbles in a DFD are represented in a clause.

With those considerations, for a given application, clauses generated from the above algorithm have the following properties;

- Except for a context diagram, there is a clause for each DFD.
- For each functional primitive in the lowest level DFDs of a system, there may be several predicates.

- Terms that a predicate can have are only i(I_1,I_2,...,I_n) and o(O_1,O_2,...,O_m).
- A recursive call may not exist in the application.
- Modes that a predicate can have are of variable or constant.

4 From Prolog to C code

Two kinds of code are generated; one is an executable C program and the other is skeleton C code which consists of module stubs with function names and its parameters. The first one is used for directly executing modules with real data, and the second code allows programmers to get into programming stage with no more design work such as drawing structured charts or decomposition charts. The parameter types are automatically generated and inferred from the Data Dictionary. In what follows, how those programs are generated and what kind of functions are necessary will be described.

4.1 Prolog Execution Behavior

A Prolog program consists of a set of predicate definitions. A predicate definition consists of a sequence of clauses having the same predicate symbol. Conceptually, a clause corresponds to a procedure definition, where the head gives the formal parameters and the literals in the body correspond to the sequence of procedure calls defining the procedure body. A predicate definition corresponds to a procedure definition, each clause for the predicate corresponding to an alternative body for the procedure. Thus, a predicate definition can be thought of as a simple generalization of procedure definitions in traditional languages, in that multiple alternative bodies, not necessarily mutually exclusive, are permitted. In this view, the terms in the head of the clause correspond to the formal parameters, and each literal in the body of a clause corresponds to a procedure call.

Parameter passing in procedure calls is via a generalized pattern matching procedure called *unification* that finds a most general unifier for two literals, namely, the calling predicate and the head predicate of the corresponding clause.

Operationally, the execution of a Prolog program follows the textual order of clauses and literals. Execution begins with a query from the user, which is a sequence of literals processed from left to right. The processing of a literal proceeds as follows: the clauses for its predicate are tried in order, until one unifies with it. If there are any remaining clauses for that predicate whose heads might unify with that literal, a backtrack point is created to remember this fact. After unification with head, the literals in the body are processed from left to right. If unification fails at any point, execution backtracks to the most recent backtrack point, and the next clause is tried and so on. Execution terminates when all the literals have been processed completely.

```
main()
{

/* Type Declaration */
system_name(i(I1,...,In),o(O1,...Om));
printf(output_variable_name);

}
```

Figure 3: Structure of executable code of main program in C

4.2 Executable C code

As we noted in the previous section, a clause corresponds to a procedure definition, where the head gives the formal parameters and the literals in the body correspond to the sequence of procedure calls defining the procedure body. In a general Prolog program, there may be several clauses with the same head, which belong to a procedure, but in our proposed system there is only one clause with the same head except for facts. Therefore a procedure is generated in C for each clause, whose name is the head predicate name and whose body consists of a sequence of procedure call statements corresponding to the subgoals in the clause. Since a query that corresponds to a context diagram is recognized as the application system name, its name is included in the *main*() program as a call for its lower diagram. A subprocedure for a clause includes three parts as follows;

- Instructions for restoring information for backtracking process.
- Operation codes for unification process.
- Procedure calls for the subgoals in the body of the clause.

4.2.1 Routine for backtracking process

The most important thing that should be considered in transforming from Prolog to C is how to reflect backtracking behavior and recursive calls. As indicated previously, recursive calls do not exist and only one clause may exist under the same head name. In the normal Prolog program's execution, the information that should be saved before unification and restored on backtracking is the address of the alternative clauses and values of registers or variables in the clause

Before firing a unification against a clause or a fact, the values of terms in the clause should be saved in the local variables which is allocated by the system. Saving of the addresses of the alternative clauses are handled in two ways according to objects to be unified. If the current goal is unified against a clause, the address of the alternative clause is not necessary since there is no other clause with the same head. If it is unified against a fact, the address of the fact to be executed next is maintained by a pointer, where facts with the same predicate name are connected by linked lists.

4.2.2 Operation codes for Unification

Since the terms a predicate can have are only functional terms i and o with data flows for its arguments and there are no recursive calls unification process will be easily converted into procedural statements such as assignment or if-then-else statements according to the conditions of variables in the predicate. Thus, once the mode of a term in the predicate has been determined, i.e., it is a variable or constant type, then automatically the kind of statements will be decided. There are two approaches for deciding types of variables in the predicate. The first one is that, in the declaration part, a mode of a query which may occur in a program is defined. Then the query should be used in the given type. Otherwise, an error message will be issued. The second method is that the translation process is postponed until the mode of the query is known. Then the types of the remaining predicates are computed as computation continues. In this paper we take the latter method.

4.2.3 Procedure calls for subgoals

It is simple to generate code for calling subgoals in a clause as a subprocedure call in the C language; the subprocedure name corresponds to the head name in the clause and subgoals appear in the body of the subprocedure. But, when there is a subgoal which returns with failure, it is not like a normal return in C. In Prolog, there are two kinds of returns; success or failure. If the return is of success, then the next subgoal in the clause will be executed, otherwise the alternative clause will be executed. In order to represent the return status, $flag$ and $state$ variables are used. If $flag$ is false, then reduce the value of $state$ variable in order to execute the previous subgoal, otherwise the next subgoal is continued.

4.3 Skeleton C code

Once a prototype for the given application has been confirmed by execution, the next step would be the programming stage. Since the procedural information is already included in the sequence of clauses and subgoals in the clause, the remaining work is only the programming job for each module which is represented in the lowest diagrams. In our system, the calling sequence for each diagram and the module skeleton are generated. In another words, for clauses in the Prolog program its procedure name and procedure calls for subgoals are generated and for functional primitives corresponding to facts, only the procedure name and its parameters are declared.

```
for(;;)
switch(state) {
    case -1 ; restore arguments values;
              return FALSE;

    case 0 ; flag = subquery_0(arguments);
             if (flag = FALSE) {state = -1; break;}
             .
             .
             .

    case N ; flag = subquery_N(arguments);
             if (flag = FALSE) {state = N-1; break;}
}
```

Figure 4: Procedure calls for subgoal in a clause

5 An Application Example

For an execution example, a three level diagram of the model of the project life cycle, which originally designed by Yourdon for simplicity of presentation, is used. From the diagram executable C code and skeleton C code are generated automatically. The system has been developed on UNIX-based workstations using X Window X11R5 and C-ISAM.

6 Conclusions

We have briefly described a method for generating C code from DFDs; executable C code and skeleton C code for an application. The advantages and relevances are also discussed. The method consists of mapping DFDs into a Prolog program by analyzing the properties of the relationship between the Structured Analysis Technique and Prolog. The translated Prolog program is compiled into procedural code according to its execution mechanism. A call to a unifiable clause is coded into a subprogram call as C or Pascal programs do. The unification processes are translated into procedural statements by analyzing the conditions of variables in the clause. An application example with the project life cyle for getting C code from Data flow diagrams is shown. Getting C code from DFDs including timing and control information remains as a further study.

References

[1] W. Clocksin and C. Melish, "Programming in PROLOG," *Springer-Verlag*,

```
main()
{
    #include type.h;

    system_name(I1,...,In,O1,...Om);
    printf(output_variable_name);

}

process_i(formal parameters);
{
    process_i_1(actual parameters);
    process_i_2(actual parameters);
         .
         .
         .
    process_i_n(actual parameters);
}

fact_name(formal parameters);
{

}
```

Figure 5: C code skeleton for a system

1981.

[2] R. Davis, "Runnable Specification as Design Tool," in *Logic Programming* (Clark and Tarlund, eds.), Academic Press, London, 1982, pp. 141-149.

[3] T. DeMarco, *Structured Analysis and System Specification*. Englewood Cliffs, NJ: Prentice-Hall, 1978.

[4] T. W. G. Docker and G. Tate, "Executable data flow diagrams," in D. Barnes and P. Brown (eds), *Proceedings of the BCS/IEE conference Software Engineering '86*, Peter Oeregrinus Lts, 1986, pp. 352-370.

[5] T. W. G. Docker, "SAME-A structured analysis and its implementation in Prolog," in *Proceedings of 5-th International Conference on Logic Programming*, Cambridge, MA: MIT Press 1988.

[6] R. France, "Semantically Extended Data Flow Diagrams: A Formal Specification Tool," *IEEE Trans. on Software Engineering*, Vol 18, No. 8, 1992, pp 329-346

[7] A. Fuggetta, et al., "Executable Specifications with Data-flow Diagrams," *Software-Practice and Experience*, vol. 23(6), June 1993, pp. 629-653.

[8] C. Gane and T. Sarson, *Structured Systems Analysis: Tools and Techniques.* Englewood cliffs, NJ: Prentice-Hall, 1978.

[9] R. A. Kowalski, "AI and Software Engineering," *Datamation*, Nov., 1984, pp. 92-102.

[10] G. L. Lazarev, "Solving Problems with Prolog," *AI EXPERT*, July, 1987, pp. 59-68.

[11] G. L. Lazarev, "Executable Specifications w'th Prolog," *Dr. Dobb's Journal*, Oct., 1989, pp. 61-68

[12] U. Leibrandt and P. Schnupp, "An Evaluation of Prolog as a Prototyping System," in *Rapid Prototyping* (R. Budde ed.), Springer-Verlag, Berlin, 1984, pp 424-433.

[13] R. N. Meeson, M. B. Dillencount and A. M. Rogerson, "Executable data flow diagrams," in *Proceedings of First International Workshop on Computer Aided Software Engineering*, Cambridge, MA, 87.

[14] Y. Nam and L. J. Henschen, "Compiling Linear Recursive Programs with List Structure in Prolog into Procedural Languages," in *Proc. of COMPSAC90*, Chicago, IL, Nov. 1990.

[15] Y. Nam and L. J. Henschen, "Compiling Recursive Programs with List Structure in Prolog into Procedural Languages," in *Proc. of COMPSAC91*, Tokyo, Japan, Sept. 1990.

[16] Y. Nam and L. J. Henschen, "On Generating Efficient Procedural Code from Linear Recursive Prolog Progam with List Structure," *Proc. of IPCCC91*, Scottsdale, Arizona, Mar. 1991, pp. 806-812.

[17] Y. Nam and L. J. Henschen, "Compiling Recursive Functional Prolog Programs with List Structure into Procedural Langaues," to be published in *Festschrift for Woody Bledsoe*, Kluwer Academic Publishers, Texas, Nov. 1991.

[18] T. H. Tse and L. Pong, "Towards a formal foundation for DeMarco data flow diagrams," *The Computer Journal*, vol 32(1) 1-12 (1989).

[19] P. T. Ward, "The transformation schema: an extension of the data flow diagram to represent control and timing," *IEEE Trans. Software Eng.*, vol. SE-12, p 22, 1986.

[20] E. Yourdon, *Modern Systems Analysis*, Englewood Cliffs, NJ: Prentice-Hall, 1989.

```
/* DATA FLOW DIAGRAMS FOR SOFTWARE LIFE CYCLE */
```

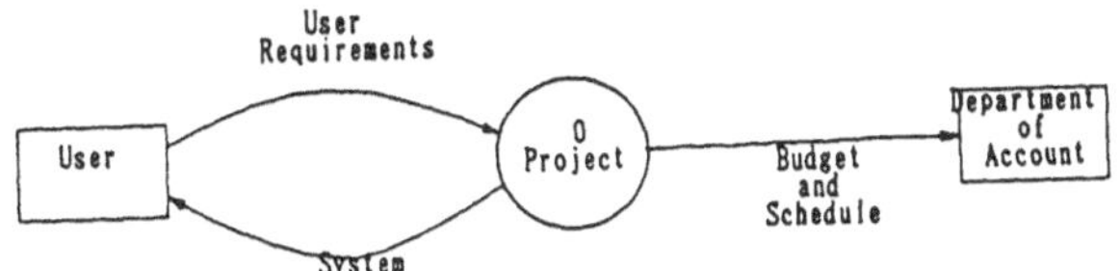

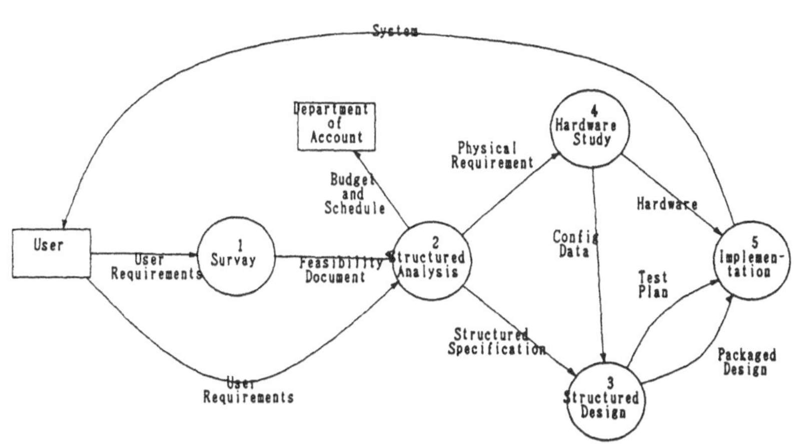

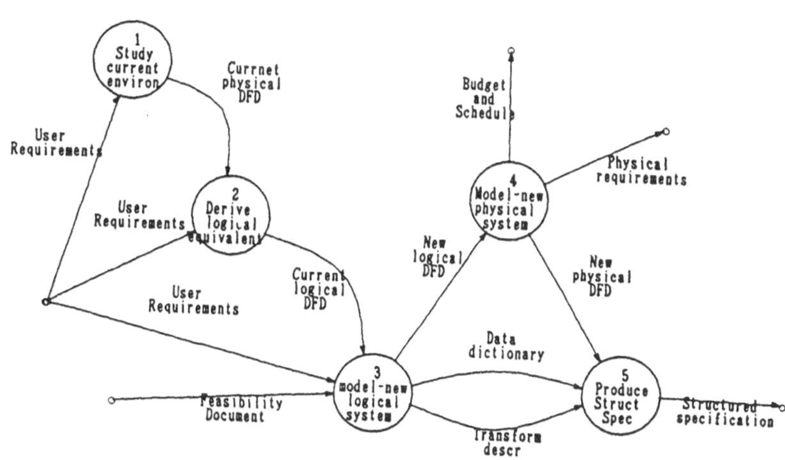

```
/* PROLOG PROGRAMS GENERATED FROM DATA FLOW DIAGRAMS */


  project( i( UserRequirements  ) , o( System ,BudgetandSchedule  ) ):-

      survay( i( UserRequirements  ) , o( FeasibilityDocument  ) ),
      structuredAnalysis( i( FeasibilityDocument ,UserRequirements  ) , o( BudgetandSchedule ,PhysicalRequirement ,StructuredSpecification  ) ),
      structuredDesign( i( ConfigData ,StructuredSpecification  ) , o( TestPlan ,PackagedDesign ) ),
      hardwareStudy( i( PhysicalRequirement  ) , o( Hardware ,ConfigData  ) ),
      implementation( i( Hardware ,TestPlan ,PackagedDesign  ) , o( System  ) ).

  structuredAnalysis( i( FeasibilityDocument ,UserRequirements  ) , o( BudgetandSchedule ,PhysicalRequirement ,StructuredSpecification  ) ):-

      studycurrentenviron( i( UserRequirements  ) , o( CurrnetphysicalDFD  ) ),
      derivelogicalequivalent( i( UserRequirements ,CurrnetphysicalDFD  ) , o( CurrentlogicalDFD ) ),
      modelnewlogicalsystem( i( UserRequirements ,FeasibilityDocument ,CurrentlogicalDFD  ) , o( NewlogicalDFD ,Datadictionary ,Transformdescr  ) ),
      modelnewphysicalsystem( i( NewlogicalDFD  ) , o( NewphysicalDFD ,BudgetandSchedule ,Physicalrequirements  ) ),
      produceStructSpec( i( NewphysicalDFD ,Datadictionary ,Transformdescr  ) , o( Structuredspecification  ) ).

  survay( i( userrequirements  ) , o( feasibilityDocument  ) ).
  structuredDesign( i( configData ,structuredSpecification  ) , o( testPlan ,packagedDesign  ) ).
  hardwareStudy( i( physicalRequirement  ) , o( hardware ,configData  ) ).
  implementation( i( hardware ,testPlan ,packagedDesign  ) , o( system  ) ).
  studycurrentenviron( i( userRequirements  ) , o( currnetphysicalDFD  ) ).
  derivelogicalequivalent( i( userRequirements ,currnetphysicalDFD  ) , o( currentlogicalDFD ) ).
  modelnewlogicalsystem( i( userRequirements ,feasibilityDocument ,currentlogicalDFD  ) , o( newlogicalDFD ,datadictionary ,transformdescr  ) ).
  modelnewphysicalsystem( i( newlogicalDFD  ) , o( newphysicalDFD ,budgetandSchedule ,physicalrequirements  ) ).
  produceStructSpec( i( newphysicalDFD ,datadictionary ,transformdescr  ) , o( structuredspecification  ) ).
```

```
/* SKELETON C CODE GENERATED FROM THE PROLOG PROGRAM  */

/*
 *  program produced by pl2c..
 */


#include "STARTtypes.h"


int
project(UserRequirements_type       UserRequirements,
    System_type     *System,
    BudgetandSchedule_type  *BudgetandSchedule)
{
    /*
     *      local variable declaration
     */
    FeasibilityDocument_type        FeasibilityDocument;
    PhysicalRequirement_type        PhysicalRequirement;
    StructuredSpecification_type    StructuredSpecification;
    ConfigData_type ConfigData;
    TestPlan_type   TestPlan;
    PackagedDesign_type     PackagedDesign;
    Hardware_type   Hardware;

    /*
     *      functions call
     */
    survay(UserRequirements,
            &FeasibilityDocument);

    structuredAnalysis(FeasibilityDocument, UserRequirements,
            BudgetandSchedule, &PhysicalRequirement, &StructuredSpecification);

    structuredDesign(ConfigData, StructuredSpecification,
            &TestPlan, &PackagedDesign);

    hardwareStudy(PhysicalRequirement,
            &Hardware, &ConfigData);

    implementation(Hardware, TestPlan, PackagedDesign,
            System);

}

int
structuredAnalysis(FeasibilityDocument_type FeasibilityDocument,
    UserRequirements_type   UserRequirements,
    BudgetandSchedule_type  *BudgetandSchedule,
    PhysicalRequirement_type        *PhysicalRequirement,
    StructuredSpecification_type    *StructuredSpecification)
{
    /*
     *      local variable declaration
     */
    CurrnetphysicalDFD_type CurrnetphysicalDFD;
    CurrentlogicalDFD_type  CurrentlogicalDFD;
    NewlogicalDFD_type      NewlogicalDFD;
    Datadictionary_type     Datadictionary;
    Transformdescr_type     Transformdescr;
    NewphysicalDFD_type     NewphysicalDFD;
    Physicalrequirements_type       Physicalrequirements;
    Structuredspecification_type    Structuredspecification;

    /*
     *      functions call
     */
    studycurrentenviron(UserRequirements,
            &CurrnetphysicalDFD);

    derivelogicalequivalent(UserRequirements, CurrnetphysicalDFD,
            &CurrentlogicalDFD);

    modelnewlogicalsystem(UserRequirements, FeasibilityDocument, CurrentlogicalDFD,
            &NewlogicalDFD, &Datadictionary, &Transformdescr);

    modelnewphysicalsystem(NewlogicalDFD,
            &NewphysicalDFD, BudgetandSchedule, &Physicalrequirements);

    produceStructSpec(NewphysicalDFD, Datadictionary, Transformdescr,
            &Structuredspecification);

}
```

```
/* EXECUTABLE C CODE GENERATED FROM THE PROLOG PROGRAM */

/*
 *  program produced by pl2c..
 */


#include <stdio.h>

#include "trace.h"
#include "constant.h"

/*
project( i(UserRequirements), o(System, BudgetandSchedule) )        :-
    survay( i(UserRequirements), o(FeasibilityDocument) ),
    structuredAnalysis( i(FeasibilityDocument, UserRequirements), o(BudgetandSchedule, PhysicalRe
quirement, StructuredSpecification) ),
    structuredDesign( i(ConfigData, StructuredSpecification), o(TestPlan, PackagedDesign) ),
    hardwareStudy( i(PhysicalRequirement), o(Hardware, ConfigData) ),
    implementation( i(Hardware, TestPlan, PackagedDesign), o(System) ).
 */
int
project(Arg* arg, Trace** ptrace)
{
    int     i, flag, state = 0;
    Trace*  trace = *ptrace;
    Arg     arg0, arg1, arg2, arg3, arg4;
    if (trace == NULL)
    {
            trace = (Trace*) malloc(sizeof(Trace));
            *ptrace = trace;
            for (i = 0; i < arg->iArity; i++)
                    trace->orig.i[i] = *arg->i[i];
            for (i = 0; i < arg->oArity; i++)
                    trace->orig.o[i] = *arg->o[i];

            for (i = 0; i < MAX_SUBTRACE; i++)
                    trace->subTrace[i] = NULL;
            for (i = 0; i < MAX_LOCAL; i++)
                    trace->l[i] = NULL;
    }
    else    state = 4;

    arg0.i[0] = arg->i[0];
    arg0.o[0] = &trace->l[0];
    arg0.iArity = 1;        arg0.oArity = 1;

    arg1.i[0] = &trace->l[0];
    arg1.i[1] = arg->i[0];
    arg1.o[0] = arg->o[1];
    arg1.o[1] = &trace->l[1];
    arg1.o[2] = &trace->l[2];
    arg1.iArity = 2;        arg1.oArity = 3;

    arg2.i[0] = &trace->l[3];
    arg2.i[1] = &trace->l[2];
    arg2.o[0] = &trace->l[4];
    arg2.o[1] = &trace->l[5];
    arg2.iArity = 2;        arg2.oArity = 2;

    arg3.i[0] = &trace->l[1];
    arg3.o[0] = &trace->l[6];
    arg3.o[1] = &trace->l[3];
    arg3.iArity = 1;        arg3.oArity = 2;

    arg4.i[0] = &trace->l[6];
    arg4.i[1] = &trace->l[4];
    arg4.i[2] = &trace->l[5];
    arg4.o[0] = arg->o[0];
    arg4.iArity = 3;        arg4.oArity = 1;
    for(;;)
    switch(state) {

    case    -1:
            for (i = 0; i < arg->iArity; i++)
                    *arg->i[i] = trace->orig.i[i];
            for (i = 0; i < arg->oArity; i++)
                    *arg->o[i] = trace->orig.o[i];
            free(trace);
            *ptrace = NULL;
            return  FALSE;

    case    0:
            flag = survay(&arg0, &trace->subTrace[0]);
```

Approach $C - C_d$
A New Contradiction Handling Strategy & Extended Logic Programs

Suryanil Ghosh
Department of Computing Science
University of Alberta
Edmonton, AB T6G-2H1
Canada

Abstract. In this paper we expand the language of extended logic programs (ELPs) to allow for *reasoning in the presence of inconsistent information* . This extension increases the expressive power of this nonmonotonic reasoning framework and allows one to explicitly reason in presence of contradictions without trivialization approach of classical logic. Furthermore, we extend the language of ELPs allowing *reasoning beyond paraconsistency*, i.e. reasoning in the presence and from inconsistent information, and the propagation of the information that, in the line of reasoning, inference have been made based on inconsistent information. This extension expands the horizon of reasoning.

Keywords. Inconsistency Handling, Paraconsistency, Extended Logic Programs, Knowledge Representation.

1 Background & Motivation

In [4, 7, 6] we proposed a paraconsistent system (an extension of positive logic programs) based on the new inconsistency handling strategy, *Approach C*. The new strategy captures a contradiction by introspecting its conclusions during the reasoning process, looking for the mutually complementing inconsistent pairs of conclusions and replacing them by a new conclusion which represents the contradictory pairs. In this way we avoid the existence of complementing inconsistent pairs in our conclusions and disallow their formation in the final set of conclusions. Otherwise, either the theory is trivialized as by classical logic approach or unintuitive conclusions derived from the contradictory pairs leads to more confusion [7, 6].

Extended logic programs [1] allow for negative conclusions in rules. So, the question arises, of how to deal with contradictions in the database. The language of extended logic programs generates both *ontological* and *epistemic inconsistency.* Ontological inconsistency owes its origin directly to *explicit negation.* Epistemic inconsistency owes its origin directly to explicit negation and indirectly to *negation by default*, which is an added feature of extended logic programs. (For more on the two types of inconsistency refer to [5]).

E. A. Yfantis (ed.), Intelligent Systems, 437–442.

But the difference in their origin does not express them differently in the theory. Hence, the way they are treated are similar. We apply our inconsistency handling strategy to enhance the language of ELP to handle both ontological and epistemic inconsistency in a more satisfactory way. We extend the answer set semantics [1] attached to extended logic programs to embed in it our inconsistency handling strategy *Approach C* and give it the ability to deal with contradictions in a more intuitive way.

In applying *Approach C* to handle inconsistency we introduce the new connective C to the existing language of extended logic programs [1]. $C\alpha$ means α *is in contradiction)*, i.e. there is contradictory information available about α in a theory (here α denotes an atom). By our semantics if α and $\sim \alpha$ are in the same theory, we replace them by $C\alpha$. We allow formulas of the form $C\alpha$ in the bodies of the rules of an ELP.

For the purpose of reasoning beyond paraconsistency, we apply *Approach* C_d to ELP and introduce a new connective C_d. A formula $C_d\alpha$ means that α *is the consequence of a premise of a rule (of the extended language of ELP) influenced by inconsistent information.* This relationship of the premise of the rule to inconsistent information can be in three distinct possible ways (see [6]). *Approach* C_d enables us to propagate the information throughout the chain of reasoning, that in its process, inference have been made from inconsistent information. *Approach* C_d increases the expressive power of the language of ELP, allows for explicit reasoning from contradictions (which are captured in a paraconsistent form) and enables us to distinguish between conclusions inferred from inconsistent information and that from consistent information. Hence expanding the horizon of reasoning.

We call this modified language of ELP, *Paraconsistent Assumptive Specification* (PAS). Existing answer set semantics [1] for extended logic programs lacks the ability to handle inconsistency. By it, an inconsistent answer set is trivialized to give the set of all literals of the language. What we would like our extended answer set semantics to achieve compared to the existing one is made clear below:

Let us consider the following set of rules in our extended language of ELP:

$$\{a \leftarrow not\, g;\ \sim a \leftarrow not\, f;\ e \leftarrow b;\ b \leftarrow\}$$

Answer set semantics of ELP [1] does not have an answer set for the above set of rules. The contradictory information a and $\sim a$ sets off the explosive approach by which we get the set of all literals of the above theory. Hence, trivializing the theory. By applying *Approach C* we preserve the theory, and get as a conclusion the following set:

$$\{Ca, b, e\}$$

By just applying *Approach C*, we recover from the problem posed by the classical explosive approach, but we still have a problem. The problem which arises here is that if we infer $\sim e$ from some other part of the program, say by appending a rule:

$$\sim e \leftarrow Ca$$

to the given set of rules, we get another inconsistency Ce. Getting this inconsistency is not intuitive as the inference of e is from a consistent piece of information b, whereas $\sim e$ is inferred from Ca, an inconsistent information. To our understanding it is a rational necessity to differentiate between conclusions derived from inconsistent information and consistent information, when reasoning in presence of inconsistency. As, in the inconsistent information there can be an intrinsic error. We consider the information derived from inconsistent information to have a lower epistemicity than information derived from consistent information. So instead of arriving at the inconsistent information Ce, we would prefer the conclusion of e from the set of rules. This we will achieve by inferring Ca (which we do by applying *Approach C*) and e from the first set of rules. $C_d \sim e$ is concluded from the rule:

$$e \leftarrow Ca$$

based on the idea that if the premise of a rule is influenced by a contradiction (here Ca), the head of the rule, which we conclude, is tagged with the information that it is inferred from a contradictory premise or a premise influenced by contradictory information. Thus we are able to differentiate between conclusions inferred from consistent and inconsistent information. Now from all the above rules we get the conclusion set $\{Ca, b, C_d \sim e, e\}$. We do not get the conclusion Ce anymore.

2 Formalizing Our Framework

Let us consider a language $\mathcal{L}$ consisting of atomic sentences a, b, c, ..., p, q, ..., logical connectives $\wedge$ (and), $\sim$ (explicit negation), *not* (negation by default) and the new ones C (in-contradiction) and C_d (contradiction-influenced). Formulas of $\mathcal{L}$ will be defined in the following way. Formula of the form p will be called *atoms*. A *literal* is an atom α or a negated atom of the form $\sim \alpha$. We use Φ to denote the set of all literals. *Extended literals* are of the form $not\,\phi$, where $\phi \in \Phi$. We denote the set of all extended literals by Φ^x. *Paraconsistent literals* are formulas of the form $C\phi$ and *Extended Paraconsistent literals* are formulas of the form $C_d\phi$, where $\phi \in \Phi$. We use Ψ and Ψ^x to denote the set of all paraconsistent and extended paraconsistent literals respectively.

Definition 2.1 (Paraconsistent Assumptive Specification (PAS))
By a paraconsistent assumptive specification P we will mean a collection of clauses of the form

$$F \leftarrow G_1 \wedge \ldots \wedge G_m$$

where, either $F \in \Psi^x$ and $m = 0$, or, $F \in \Phi$, $G_i s$ (for $i = 0, \ldots, m$) are formulae belonging to the set of formulae $\Phi \cup \Phi^x \cup \Psi \cup \Psi^x$ and $m \geq 0$. □

Our semantics is not "contrapositive" with respect to $\leftarrow$ and $\sim$; it assigns different meaning to the rules $p \leftarrow \sim q$ and $q \leftarrow \sim p$. The reason is that it interprets expressions like these as *inference rules*, rather than conditionals.

The basics of the semantics which we are proposing here is the same as that of PS (paraconsistent specification) as proposed in [6]. Literals α and $\sim \alpha$ are not related to each other in the usual classical logic sense, i.e. if α is $true$ we should not conclude that $\sim \alpha$ is $false$ or vice-versa. Truth valuation of α and $\sim \alpha$ in an interpretation are independent of each other.

We will give the truth-theoretic evaluation of sentences in our language in two steps. First we will define the truth-valuation in an inconsistent context, i.e. when we allow both sentences α and $\sim \alpha$ to be simultaneously true. Here we will initially just deal with simple sentences belonging to the set of literals Φ. In the second step we will define the truth-valuation of sentences in our language in a paraconsistent context when the inconsistent sentences are no more true but a new formula capturing the inconsistency is true. In this second step of truth-valuation of sentences in a paraconsistent context we deal with literals, extended literals, paraconsistent literals and extended paraconsistent literals. Finally we will complement the definition for truth-valuation of literals in an inconsistent context by paraconsistent and extended paraconsistent literals.

We will now define truth of sentences in an inconsistent context. Let us consider a set W of literals, paraconsistent literals and extended paraconsistent literals. W represents the current (working) inconsistent set of beliefs of a reasoner. We will inductively define the notion of truth ($\models_{in}$) in an inconsistent context of formulae of $\mathcal{L}$ w.r.t. W.

Definition 2.2 (Truth-valuation in an inconsistent context)

$$W \models_{in} \alpha \textit{ (resp. } \sim \alpha\textit{) iff } \alpha \textit{ (resp. } \sim \alpha\textit{)} \in W.$$

□

We will now define truth-valuation of sentences in a paraconsistent context. Let us consider a set W^p of literals, paraconsistent and extended paraconsistent literals and W a set of the same. W adheres to the truth-valuation $\models_{in}$. W^p represents the current (working) paraconsistent set of beliefs, while W represents its current (working) inconsistent set of beliefs. We will inductively define the notion of truth ($\models_{C-C_d}$) in a paraconsistent context, of formulae of $\mathcal{L}$ w.r.t. a pair $M^p =< W^p, W >$.

Definition 2.3 (Truth-valuation in a paraconsistent context)

(1) $M^p \models_{C-C_d} C\phi$ *iff* $M \models_{in} \phi$, $M \models_{in} \sim \phi$ *and* $C\phi \in W^p$
(2) $M^p \models_{C-C_d} \phi$ *iff* $M^p \not\models_{C-C_d} C\phi$, $M \models_{in} \phi$ *and* $\phi \in W^p$
(3) $M^p \models_{C-C_d} not\phi$ *iff* $M^p \not\models_{C-C_d} \phi$ *(resp.* $C\phi$*)*
(4) $M^p \models_{C-C_d} G_1 \wedge \ldots \wedge G_n$ *iff* $M^p \models_{C-C_d} G_i$ *for* $i = 1, \ldots, n$
(5) $M^p \models_{C-C_d} C_d\phi$ *iff* $C_d\phi \in W^p$

where ϕ*s are literals and* G_i*s are in the set* $\Phi \cup \Phi^x \cup \Psi \cup \Psi^x$. □

Definition 2.4 *Any formula G of our language $\mathcal{L}$ true by the truth-valuation in paraconsistent context is true by the truth-valuation in inconsistent context and the literals, paraconsistent literals and extended paraconsistent literals in W^p are also in W.* □

Proposition 2.1 *Any formula G of our language $\mathcal{L}$ true by the truth-valuation in inconsistent context is not necessarily true by the truth-valuation in paraconsistent context.* □

Now we will define a set W^p of literals, paraconsistent and extended paraconsistent literals satisfying a *paraconsistent assumptive specification* (PAS) P. We will call such a set an *answer set* of P. Answer sets give the intended semantics of a PAS. The precise definition of these notions will be given in the following steps.

Definition 2.5 (Belief set) *Let P_0 be a paraconsistent assumptive specification consisting of rules of the form*

$$F \leftarrow$$

where $F \in \Phi$ *or* $F \in \Psi^x$.
Let W be a set of literals, paraconsistent literals and extended paraconsistent literals and W^p be the same. Let $M^p =< W, W^p >$. W^p is called a belief set *of P_0 iff it is a minimal set with the property $M^p \models_{C-C_d} F$ (or CF) for every rule from P_0.* □

Example 2.1 *Let PAS P_0 consist of clauses:*

$$\{q \leftarrow;\ C_d p \leftarrow;\ C_d\neg q \leftarrow;\ r \leftarrow;\ s \leftarrow\}$$

Let $W = \{q, \sim q, Cq, C_d \sim q, C_d p, r, s\}$ and thus the corresponding $W^p = \{Cq, C_d p, C_d \sim q, r, s\}$. We can see that P_0 has one belief set W^p. □

Definition 2.6 (Answer set) *Let P be an arbitrary paraconsistent assumptive specification, W be a set of literals, paraconsistent literals and extended paraconsistent literals of P and W^p be a set of the same. For every $M^p = < W^p, W >$, by P_{M^p} we will denote the paraconsistent assumptive specification obtained from P by:*

1. *Replacing all rules of the form:*

   $$F \leftarrow G_1 \wedge \ldots \wedge G_m$$

   where $F \in \Phi$, G_is in $\Phi \cup \Phi^x \cup \Psi \cup \Psi^x$ and $m \geq 0$,
   by rules of the form:

   $$C_d F \leftarrow$$

   iff

   (a) (i) $M^p \models_{C-C_d} G_1 \wedge \ldots \wedge G_m$ and
   (ii) there exists formulas of the form $C\phi_c$ (resp. $C_d\phi_c$)[1] $\in \{G_1, \ldots, G_m\}$;

   OR

   (b) (i) $M^p \models_{C-C_d} C\phi_c$ such that ϕ_c is any of the formulas G_i, and
   (ii) $M^p \models_{C-C_d} G_i$ for the rest of the G_is.

2. *Removing from the premises of the remaining rules of P all formulae G such that $M^p \models_{C-C_d} G$.*

3. *Removing all remaining rules with non-empty premises.*

Then W^p will be called the answer set of P iff W^p is a belief set *of P_{M^p}.* □

Example 2.2 *Let us consider the PAS P given as a motivating example in Section 1:*

$$\{a \leftarrow not\, g;\ \sim a \leftarrow not\, f;\ \sim e \leftarrow Ca;\ e \leftarrow b;\ b \leftarrow\}$$

By the program transformation of Definition 2.6 we get the program P_{M^p}:

$$\{a \leftarrow;\ \sim a \leftarrow;\ C_d \sim e \leftarrow;\ \sim e \leftarrow;\ b \leftarrow\}$$

from P by considering W^p to be the following set:

$$\{Ca, C_d e, b, \sim e\}$$

Now again, W^p is the belief set *(refer to Definition 2.5) of P_{M^p}, hence it is the* answer set *of P.* □

Example 2.3 *Consider the PAS P given below:*

$$\{q \leftarrow;\ p \leftarrow q;\ \sim q \leftarrow p;\ r \leftarrow s;\ s \leftarrow\}$$

By considering $W^p = \{Cq, C_d p, C_d \sim q, r, s\}$ for P, we get the transformed program P_{M^p} same as the PAS P_0 in Example 2.1, which has W^p as the belief set. Hence W^p is the answer set of P. □

[1] $C\phi_c$ (resp. $C_d\phi_c$) belonging to the set of paraconsistent literals Ψ (resp. extended paraconsistent literals Ψ^x.

3 Conclusions

In [6] we have applied *Approach* $C - C_d$ to positive logic programs. So this paper complements that work by its extension to extended logic programs. In [7, 6] we have also extensively discussed the merits of our inconsistency handling strategies over the existing approaches handling inconsistency, so we do not discuss it here. Our semantics collapses to that of answer set semantics for extended logic programs [1] and to that of stable model semantics for general logic programs [3]. We have showed the relationships in the full paper [5], which we are unable to do here because of space constraint. This work has been further extended (in [7]) by applying *Approach* $C - C_d$ to disjunctive databases [2].

References

[1] M. Gelfond and V. Lifschitz. Logic programs with classical negation. In *Proceedings of the 7th. International Conference on Logic Programming*, pages 579–597. MIT Press, 1990.

[2] M. Gelfond and V. Lifschitz. Classical negation in logic programs and disjunctive databases. *New Generation Computing*, 9 (Nos. 3 and 4), 1991. Selected papers from the 7th. ICLP, 1990.

[3] M. Gelfond and V.Lifschitz. The stable model semantics for logic programming. In R. Kowalski and K. Bowen, editors, *Proceedings of the 5th. Logic programming symposium*, pages 1070–1080. MIT Press, Cambridge, Massachussets, 1988.

[4] Suryanil Ghosh. A new strategy to handle inconsistency: Extended abstract. In *Proceedings of* Reflections on the Future Conference, Edmonton, Alberta, Canada, September 29-30 1993. Dept. of Computing Science, University of Alberta.

[5] Suryanil Ghosh. Applying Approach $C - C_d$ to Extended Logic Programs. manuscript, 1994. Dept. of Computing Science, University of Alberta.

[6] Suryanil Ghosh. Paraconsistency and beyond: A new approach to inconsistency handling. In *Proceedings of the Eighth International Symposium on Methodologies for Intelligent Systems, (to be published as Lecture Notes in Computer Science /LNAI series).* Springer-Verlag, October 16-19 1994.

[7] Suryanil Ghosh. Paraconsistency and beyond: Issues and approaches in reasoning. manuscript of Ph.D. thesis (in preparation), 1994. Dept. of Computing Science, University of Alberta, Canada.

NEURAL NETWORKS AND GENETIC ALGORITHMS

STABILIZING TECHNIQUES IN TRAINING FEEDFORWARD NEURAL NETWORKS

Carl. G. Looney
Computer Science Department/171
University of Nevada
Reno, NV 89557, USA
email: Looney@cs.unr.edu

Abstract. We review methods and techniques for training feedforward neural networks that avoid problematic behavior, accelerate the convergence, and verify the training.

Keywords. feedforward neural networks, training, acceleration, stabilization, testing

1 Introduction

Figure 1 shows a *feedforward neural network* (FNN) with N inputs, M *neurodes* (neural nodes) in a single hidden layer, and J output neurodes. Any input feature vector $\mathbf{x} = (x_1,...x_N)$ is mapped via hidden neurodes to activated outputs $y_1,...,y_M$, which in turn are mapped by output *neurodes* to activated outputs $z_1,...,z_J$. The weights $\{w_{nm}\}_{NM}$ at the hidden neurodes and $\{u_{mj}\}_{MJ}$ at the output neurodes, respectively, combine linearly with inputs to those neurodes that are activated as outputs by *sigmoidal* functions h(-) and g(-), respectively, per Figure 2.

$$r_m = \Sigma_{(n=1,N)} u_{nm}x_n, \quad y_m = h(r_m) = 1/[1 + \exp(-\alpha r_m + b)] \quad \text{(unipolar)} \tag{1}$$

$$s_j = \Sigma_{(m=1,M)} u_{mj}y_m, \quad z_j = g(s_j) = 1/[1 + \exp(-\alpha s_j + b)] \quad \text{(unipolar)} \tag{2}$$

The *stretching* factor is α. The *bias* b in the unipolar sigmoids offsets the sigmoid ("S" shaped curve) so as to squash the sums r_m and s_j into [0,1], while bipolar activations squash the activated outputs into [-1,1]. The weights assume values in [-d,d] for some d, e.g., d = 4.0. The training of FNNs consists of adjusting the weights so as to force the FNN black box to map a sample set of Q exemplar input feature vectors $\{\mathbf{x}^{(q)}\}$ into the set of output identifier training vectors $\{\mathbf{t}^{k(q)}\}$, with each $\mathbf{x}^{(q)}$ mapping to $\mathbf{t}^{k(q)}$, where k = k(q) denotes the class to which $\mathbf{x}^{(q)}$ belongs. The map is many-to-one. The *backpropagation* (BP) algorithm does *supervised training* on the Q exemplar pairs $\{(\mathbf{x}^{(q)},\mathbf{t}^{k(q)})\}$. The FNN maps each $\mathbf{x}^{(q)} = (x_1^{(q)},...,x_N^{(q)})$ into $\mathbf{z}^{(q)} = (z_1^{(q)},...,z_J^{(q)})$, compares it to the output $\mathbf{t}^{(q)} = (t_1^{(q)},...,t_J^{(q)})$, and adjusts all weights to minimize the *partial sum squared error* SSE function $E^{(q)}$ for exemplar pair $\{\mathbf{x}^{(q)},\mathbf{t}^{(q)}\}$, where

$$E^{(q)} = \Sigma_{(j=1,J)} [t_j^{(q)} - z_j^{(q)}]^2, \ ..., \ E^{(Q)} = \Sigma_{(j=1,J)} [t_j^{(Q)} - z_j^{(Q)}]^2 \tag{3}$$

BP uses the gradient of each $E^{(q)}$ with respect to the weights to make an adjustment to all weights in the direction of steepest descent of $E^{(q)}$ in the weight space. It does this for each q = 1,...,Q, which constitutes one *epoch*. Epochs are repeated until the training is done. When all SSEs $E^{(q)}$ are small and each output $\mathbf{z}^{(q)}$ is closer to $\mathbf{t}^{(q)}$ than to any other $\mathbf{t}^{(p)}$, for $p \neq q$, the training is complete. The *batch* mode is sometimes used where the weight increments are kept

E. A. Yfantis (ed.), Intelligent Systems, 445–450.

as running totals, averaged, and then added to the weights at the end of each epoch. It is well known that the batch mode is slower (see Parlos et al [16]). We note that batch training is ***not*** the same as steepest descent on the *total* SSE $E = E^{(1)} + \ldots + E^{(Q)}$, where each single weight adjustment is for all partial SSEs at once and is equal to an entire epoch (it avoids thrashing that increases and decreases a weight over an epoch under the various gradients of $E^{(q)}$. Figure 3 gives the flow chart for the backpropagation process for I epochs. We call the algorithm that uses the total SSE the *fullpropagation* algorithm (see Looney [12]).

Figure 1 - A Feedforward Neural Network

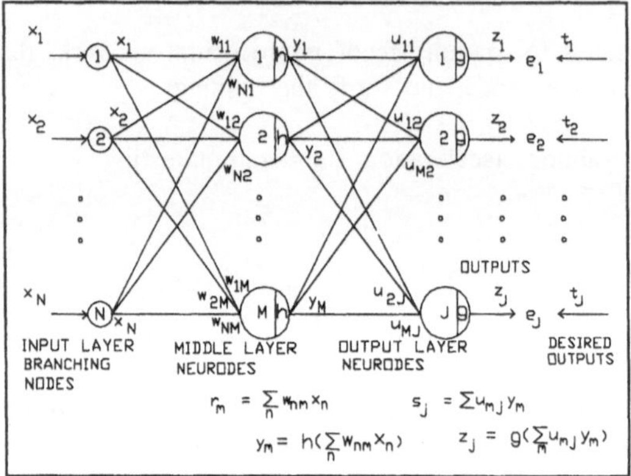

Figure 2 - Sigmoid Functions

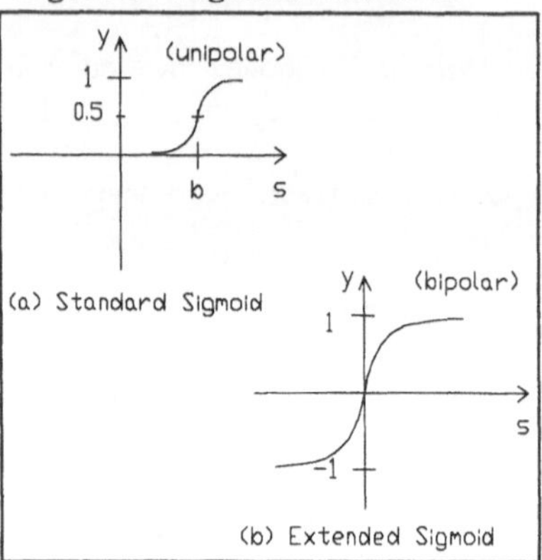

2 Gradient Training and Stability

For training purposes, the total SSE is a function $E(\mathbf{w},\mathbf{u})$ of $\mathbf{w} = \{w_{nm}\}$ and $\mathbf{u} = \{u_{mj}\}$. Thus

$$E(\mathbf{w},\mathbf{u}) = \Sigma_{(q=1,Q)} \| \mathbf{t}^{(q)} - \mathbf{z}^{(q)} \|^2 = \Sigma_{(q=1,Q)}\Sigma_{(j=1,J)}[t_j^{(q)} - z_j^{(q)}]^2 =$$
$$\Sigma_{(q=1,Q)}\Sigma_{(j=1,J)}[t_j^{(q)} - g(\Sigma_{(m=1,M)}u_{mj}h(\Sigma_{(n=1,N)}w_{nm}x_n^{(q)}))]^2 \tag{4}$$

Iterations on $E^{(q)}$ for $\mathbf{w}$ and $\mathbf{u}$ (α is absorbed into the step gain η) use

$$\mathbf{w}^{(r+1)} = \mathbf{w}^{(r)} - \eta[\nabla E] = \mathbf{w}^{(r)} + \Delta\mathbf{w}, \quad \mathbf{u}^{(r+1)} = \mathbf{u}^{(r)} - \eta[\nabla E] = \mathbf{u}^{(r)} + \Delta\mathbf{u} \tag{5}$$
$$w_{nm}^{(r+1)} = w_{nm}^{(r)} - \eta(\partial E(\mathbf{w}^{(r)},\mathbf{u}^{(r)})/\partial w_{nm}), \quad u_{mj}^{(r+1)} = u_{mj}^{(r)} - \eta(\partial E(\mathbf{w}^{(r)},\mathbf{u}^{(r)})/\partial u_{mj}) \tag{6}$$

The increments used in BP, a process shown in Figure 3, are derived via chain rules to be

$$u_{mj}^{(r+1)} = u_{mj}^{(r)} + \eta(t_j - z_j)z_j(1-z_j)y_m \tag{7}$$
$$w_{nm}^{(r+1)} = w_{nm}^{(r)} + \eta\{\Sigma_{(j=1,J)}(t_j - z_j)[z_j(1-z_j)]u_{mj}[y_m(1-y_m)][x_n]\} \tag{8}$$

Some destabilizing phenomena are: i) *weight drift*, observed by Rumelhart et al [18], Schmidt et al [19] and others, where the SSE decreases with continued training but the learning becomes more *specialized* (the FNN recognizes all input exemplar vectors but misclassifies some input vectors close to them - *generalized* learning is where new input vectors close to any exemplar are recognized as belonging to the exemplar's class); ii) *premature saturation* (Lee et al [10]), which occurs in the weight region where the sigmoid has slope near zero so that a large number of steps reduces the error very little; iii) *thrashing* (for BP), where a weight learns and then unlearns when adjustments are made for different partial SSEs; and iv) *saturation* in the vicinity of a local minimum (Parlos et al [16]), caused by the gradient approaching zero so that a large number of steps does not yet reach the minimum.

Figure 3 - The Backpropagation Flow Chart

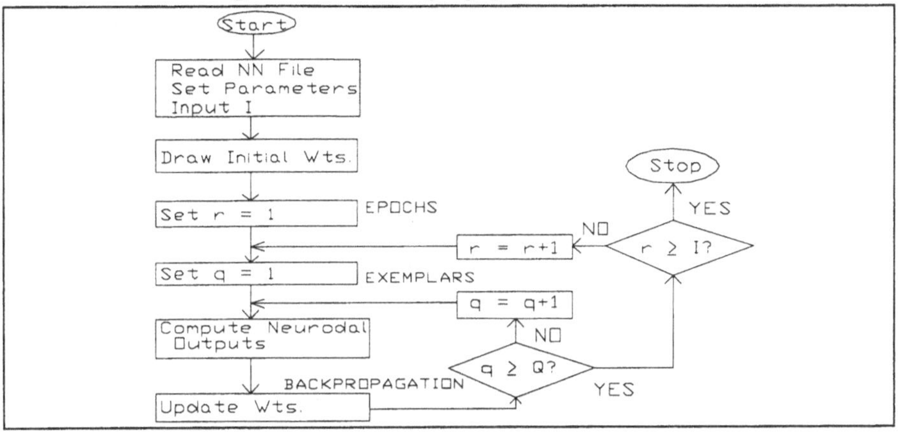

Start
Read NN File
Set Parameters
Input I
Draw Initial Wts.
Set r = 1
EPOCHS
Set q = 1
EXEMPLARS
Compute Neurodal
Outputs
BACKPROPAGATION
Update Wts.
q ≥ Q?
NO
q = q+1
YES
r ≥ I?
NO
r = r+1
YES
Stop

According to some (Lee et al [10], Li et al [11], Schmidt et al [19], Looney [12]), the convergence behavior of gradient training varies greatly with the choice of the initial weight set. Thus any randomly drawn initial weight set should be adjusted into an area of the weight domain with low lying SSE values. Zurada [22] trains on the rise rate α also. Adjustment of the step gain η speeds up convergence. We invert the output layer sigmoids, which are monotonic increasing. We invert the training outputs $\{t_j^{(q)}\}$ by putting $\tau_j^{(q)} = g^{-1}(t_j^{(q)})$ for the jth output on the kth exemplar by means of $s = \{\ln[z/(1-z)] + b\}/\alpha = g^{-1}(z)$ (from $z = g(s) = 1/[1 + \exp(-\alpha s+b)]$, that is, $(1-z)/z = \exp(-\alpha s+b)$. Instead of minimizing $(t_j^{(q)} - z_j^{(q)})^2$ (Looney [14]), we use

$$(g^{-1}(t_j^{(q)}) - g^{-1}(z_j^{(q)}))^2 = [\tau_j^{(q)} - g^{-1}(g(s_j^{(q)}))]^2 = (\tau_j^{(q)} - s_j^{(q)})^2 \qquad (9)$$

$$E\sim = \Sigma_{(q=1,Q)}\Sigma_{(j=1,J)}[\tau_j^{(q)} - s_j^{(q)}]^2, \qquad s_j^{(q)} = \Sigma_{(m=1,M)}u_{mj}y_m^{(q)}) \qquad (10)$$

3 Training Ability and Architecture

The training ability of FNNs is intimately related to their architectures. It was proved by Hornik et al [7] that a FNN with a single hidden layer of neurodes can approximate any Borel measurable mapping of an n-dimensional space into an m-dimensional space as closely as desired, given sufficiently many neural nodes in the middle layer. For large FNNs, an additional hidden layer may be used to reduce the total number of neurodes, and thus the number of weights. Chester [4] concluded that a single hidden layer is inferior to two hidden layers, but Baum and Haussler [3] used a single hidden layer successfully on his counterexample functions.

The work of de Villiers [5] showed that there is no statistically significant difference between the performance of FNNs with one or two hidden layers. Training on certain exemplar data sets is sometimes quicker with two hidden layers, but not always (see Johansson et al [8]). It is more sensitive to initial weight sets and to local minima, with more twisting ridges and canyons of the error function above the weight space. With too few neurodes, a FNN will not train, while too many allow overtraining and specialization when generalization is the goal.

4 Some Stable Training Methods and Techniques

First, FNNs should always use bipolar sigmoid activation functions. Stornetta and Huberman [20] have found that this increases learning by 30% to 50%. It also reduces the errors associated with estimating and adjusting the biases. *Second*, fullpropagation (on E rather than sequentially on $E^{(q)}$) should be used if the FNN is not so large as to exceed computer memory. *Third*, the gradient must be normalized so the gradient steps do not approach zero near a local minimum. Parlos et al [16] showed a speed increase of ten-fold over BP, and it is faster than *quickprop* (Fahlman [6]). *Fourth*, always add 0.1 to the sigmoid derivatives in Equations 7 and 8 to prevent saturation (Fahlman [6]). *Fifth*, adjust the step gain (learning rate) η up or down on each iteration (we assume fullpropagation here so a different step gain is not necessary for each $E^{(q)}$ and each weight). Sixth, the Hestenes-Stiefel form of conjugate gradient training is the most effective overall, and no momentum nor step gain are necessary (see Johansson et al, [8]). It is not quite as fast as some of the projection methods or quasi-Newton methods (Watrous [21]), but is much less complex than quasi-Newton and less greedy than projection methods (Kung et al, [9]).

Some simple methods that provide better training on the average than BP are random optimization methods where no gradients are computed (see Baba [1]). After drawing an initial weight set, one way to proceed is to draw a random direction, and then evaluate the SSE function E on a slice in that direction by a cubic polynomial (a quartic polynomial could also be used - the slices of E take a quartic or quadratic shape on the weight space). Purely random draws may be used, but without a strategy, they take considerable computing time. Another strategy is a sequence of univariate random draws (Looney, [13]) where all weights are fixed except one at a time, and a sample of 6 to 12 points is drawn in the weight region [-d,d], with the one that yields the lowest SSE value being retained. This converges quickly at first, but becomes progressively slower and does not overtrain. It produces a generalized solution when there are not too many neurodes. A hybrid approach is to use the direction of steepest descent from the gradient, then use a quadratic, cubic, or quartic polynomial to approximate the SSE function E along a slice in that direction, and take the minimum of the polynomial as the new weight point. These methods can also adjust the initial weights before gradient descent is started.

5 Understanding the FNN Black Box

It is known (Minsky and Papert [15]) that a single neurode represents a separating hyperplane in the feature space. A sum of incoming values and weights $s_m = w_{1m}x_1 + ... + w_{Nm}x_N = \tau_m$ at the mth hidden neurode, describes a hyperplane. If $\tau_m = 0.5$, then if $s_m < 0.5$, the feature vector is on one side of the hyperplane, while $s_m > 0.5$ means that it is on the other side. A layer of

M neurodes partitions the feature vector space into convex regions that are the intersections of 2M halfspaces. The output of the mth neurode is high if the input feature vector **x** is on the correct side of each of the M hyperplanes, and thus designates the convex region as a class. Note that each convex region is separated multilinearly.

A second layer of neurodes is necessary to combine multiple convex regions together into a nonconvex, and thus nonlinearly separable, class (Bose and Garga [2]). The output **y** of the first layer of neurodes is put into a second layer neurode. The combination of highs and lows of $\mathbf{y} = (y_1,...,y_M)$ represent a convex region for each high. The second layer neurode performs and OR function essentially, and outputs a high for the correct combination. Thus the training identifier **t** for that class must have this component high and all others low, or some workable variation of that. Every class in a feature space can be partitioned into linearly separable subclasses, which can be separated by a single layer of neurodes. A second layer of neurodes is necessary to lump the subclasses into a class. Thus the number M of hidden neurodes in a FNN with a single hidden layer is the number of linearly separable subclasses. Without apriori knowledge, the usual case, we may take M = 2K, where K is the number of classes. For two hidden layers, we may take $M_1 = M_2 = \sqrt{(2K)}$ because $(M_1)(M_2) = 2K$ (each neurode in the second hidden layer should split each halfspace from the first layer into two parts). If needed, we adjoin more hidden neurodes. If possible after training, we prune neurodes and train (tune) for greater generalization.

6 Testing the Training

Poh [17] uses a method from Rumelhart that validates and verifies the training. Given a set of exemplar input/output vector pairs $\{\mathbf{x}^{(q)},\mathbf{t}^{(q)}\}_Q$, he selects 65% of these at random for the training. He selects another 25% for the final testing (verification), and the remaining 10% for validation during training. After a segment of training, he puts the validation set through the FNN to obtain the SSE(10%). When SSE(10%) stops decreasing and/or begins to increase, the (validated) weights are ready for testing. The test set is then put through the FNN to obtain the SSE(25%), which is divided by the number of exemplar pairs in that set, the SSE(65%) is divided by the number of exemplars in the training set, and the resulting MSE(25%) and MSE(65%) (mean square error) values are compared. If MSE(25%) < 2xMSE(65%) then the training is accepted, or else it is started over with another initial weight set.

7 References

[1] N. Baba (1989), A new approach for finding the global minimum of error function of neural networks, *Neural Networks*, vol. 2, no. 5, 367-373.

[2] Bose, E. B., and Garga, A. K. (1993), Neural network design using Voronoi diagrams, *IEEE Trans. Neural Networks*, vol. 4, no. 5, 778-787.

[3] Baum, E. B., and Hausler, D. (1989), What size net gives valid generalization?, *Neural Computation* 1, 151-160.

[4] Chester, D. (1990), Why two hidden layers are better than one, *Proc. 1990 IEEE Joint Int'l Conf. Neural Networks*, Washington, D.C., 265-268.

[5] de Villiers, J. (1993), Backpropagation neural nets with one and two hidden layers, *IEEE Trans. Neural Networks*, vol. 4, no. 1, 136-141.

[6] Fahlman, S. E. (1988), *An Empirical Study of Learning Speed in Backpropagation*, Technical Rpt. CMU-CS-88-162, Carnegie-Mellon University.

[7] Hornik, K., Stinchcombe, M., and White, H. (1989), Multilayer feedforward networks are universal approximators, *Neural Networks*, vol. 2, no. 5, 1989, 359-366.

[8] Johansson, E. M., Dowla, F. U., and Goodman, D. M. (1992), Backpropagation learning for multilayer feedforward neural networks using the conjugate gradient method, *Int'l J. Neural Systems*, vol. 2, no. 4, 291-301.

[9] Kung, S. Y., Diamantaras, K., Mao, W. D., and Taur, J. S. (1991), Generalized perceptron networks with nonlinear discriminant functions, appeared in *Neural Networks: Theory and Applications*, edited by R. J. Mammone and Y. Zeevi, Academic Press, San Diego, 245-279.

[10] Lee, Y., Oh, S. H., and Kim, M. (1991), The effect of initial weights on premature saturation in backpropagation learning, *Proc. 1991 Int'l Joint Conf. Neural Networks*, Seattle, vol. 1, 765-770.

[11] Li, G., Alnuweiri, H., and Wu, W, (1993), Acceleration of backpropagation through initial weight pre-training with delta rule, *Proc. 1993 IEEE Int'l Conf. NNs, San Francisco*, vol. 1, 580-585.

[12] Looney, C. G. (1994), Stabilization and Speedup of Convergence in Training Feedforward Neural Networks, *Neurocomputing* (in press).

[13] Looney, C. G. (1993), Neural networks as expert systems, *J. Expert Systems with Applications*, vol. 6, no. 2, 129-136.

[14] Looney, C. G. (1992), An inverse map analysis of neural networks, *Proc. 1992 Golden West Int'l Conf. Intelligent Systems*, Reno, June, 1992.

[15] Minsky, M. L., and Papert, S. A. (1988), *Perceptrons*, MIT Press, Cambridge, MA.

[16] Parlos, A. G., Fernandez, B., Atiya, A., Muthusami, J., and Tsai, W. K. (1994), An accelerated learning algorithm for multilayer perceptron networks, *IEEE Trans. Neural Networks*, vol.5, no. 3, 493-497.

[17] Poh, H. L. (1991), *A Neural Network Approach for Marketing Strategies Research and Decision Support*, Ph.D. Thesis, Stanford University.

[18] Rumelhart, D. E., Hinton, G. E., and Williams, R. J. (1986), Learning internal representations by error propagation, in *Parallel Distributed Processing: Explorations in the Microstructures of Cognition*, Edited by Rumelhart and McClelland, MIT Press, 318-362.

[19] Schmidt, W., Raudys, S., Kraaijveld, M., Skurikhina, M., and Duin, R. (1993), Initializations, backpropagation, and generalization of feedforward classifiers, *Proc. 1993 IEEE Int'l Conf. Neural Networks*, San Francisco, vol. 1, 1993, 598-604.

[20] Stornetta, W. S., and Huberman, B. A. (1987), An improved three-layer backpropagation algorithm, *Proc. First IEEE Int'l Conf. Neural Networks*, San Diego, vol. 2, 645-651.

[21] Watrous, R. L. (1987), Learning algorithms for connectionist networks: applied gradient methods of nonlinear optimization, *Proc. First IEEE Int'l Conf. Neural Networks*, San Diego, vol. 2, 619-627.

[22] Zurada, M. (1993), Lambda learning rule for feedforward neural networks, *Proc. 1993 IEEE Joint Int'l Conf. Neural Networks*, Seattle, vol. 3, 1808-1811.

TWO GENETIC ALGORITHM ENHANCEMENTS

John Paxton
John Evans
Computer Science Department
Montana State University
Bozeman, MT 59717 USA

June 6, 1994

Abstract. Genetic algorithms [5] have been successfully applied to several real world problems and hence have attracted much attention in the field of machine learning. This paper discusses two enhancements to genetic algorithms, one which greatly reduces the run-time of genetic algorithms by replacing an explicit diversity calculation with an implicit one, and the second which yields a superior learning curve to the traditional genetic algorithm by utilising the idea of recessive and dominant genes.

Keywords. Genetic Algorithms.

1 Introduction

Genetic algorithms combine elements of bottom up approaches to learning (e.g. neural networks) with elements of top down approaches (e.g. decision trees [6]). The two standard implementations of genetic algorithms are called the Michigan approach and the Pitt approach [1]. The research that this paper reports upon has been conducted using the Pitt approach. Thus, many rule sets compete with one another to perform a prespecified task and the fittest rule sets have a higher probability of having progeny in the next generation. In our research, we have developed two enhancements to the Pitt approach that yield superior performance of the genetic algorithm under certain circumstances.

2 Improvement One: Outside-Crossover

The first enhancement relates to the diversity criterion built into many genetic algorithms [7]. This criterion selects the next generation of rules based on both a fitness and a diversity rating. Including diversity increases the probability of populating local maxima, but does so at the price of a high computational cost. Some previous research has investigated alternative approaches to maintaining diversity[4]. We decided to throw out the explicit diversity computation completely and investigate what would happen if diversity were maintained implicitly by occasionally producing new rules by mating a rule from the previous generation with a randomly-created rule.

E. A. Yfantis (ed.), Intelligent Systems, 451–456.

Our standard experiment consisted of a population of 40 rules. Each rule has 40 binary digits. The rules are initialized randomly. The fitness of each rule is computed by counting the number of 1's [2] in it. Succeeding generations are computed from the previous generation by selecting each rule with probability 0.60 for two point crossover. A certain percentage of those rules are then probabilistically selected for "outside-crossover" with a randomly-generated rule. For those rules not selected for outside-crossover, their mates are probabilistically chosen, based on the mate's fitness.

For example, if the population size is 40, then 24 rules are selected for crossover on average. If the rate of outside-crossover is 60%, then 14 of these 24 rules can be expected to be selected for outside-crossover with a randomly-generated rule. The other 10 rules have their mate chosen from the current rule set based on the mate's fitness rating. In this example, 48 rules would now be in contention for the next generation; 40 of the rules would be selected based on their fitness. There is then a 0.01% chance that one of these rules will have one of its genes mutated.

When we conducted experiments, we used a modified version of Winston's method [7] as a baseline for comparison. Rules are selected for crossover as in our method, but only internal crossovers occur. After crossovers are finished, 40 rules are retained based on a combination of their fitness and their diversity. Each of our experiments encompassed 50 generations and each trial was repeated ten times. Our investigations so far have shown three major phenomena.

(1) The amount of time needed to conduct experiments is drastically reduced. If there are k rule sets generated per generation, m rule sets kept per generation, and n genes per rule, then the time complexity of Winston's method [7], a method that maintains diversity explicitly, is $O(m*k \log k + m*k*n)$ while the complexity of our method is only $O(m*k + k*n + m*n)$. To illustrate this improvement concretely, we timed both Winston's method and our method for 50 generations and took the average of 10 trials to produce our results. When m and n were both 40, we found that Winston's method took on average 630 time units to run while outside-crossover varied smoothly between 23 in the case of 0% probability of outside crossover to 57 in the case of 100% probability of outside crossover, as can be seen in Figure 1. Other choices for m, k, and n produced results that similarly showed large disparities in the average run-time.

(2) At the standard parameter settings, the average population fitness shows a substantial improvement over the average population fitness of Winston's method (Figure 2), where the explicit diversity check can hold back overall fitness averages. The real test of fitness, however, lies in the fitness of the most-fit rule at the end of each generation. These results are displayed in Figure 3. Winston's method fares much better in this comparison, with its most fit rule rivaling that of the most fit rule encountered by our method.

(3) One unfortunate consequence of our method is that the diversity of the rules is degraded. Table 1 shows diversity measurements at the end of 5, 10, and 50 generations. At the end of 5 generations, Winston's method shows a diversity average of 20.11, meaning that any given rule differs on average from any other given rule by 20.11 genes out of a possible 40. All cases of outside-crossover except for 100% outside-crossover fared badly in terms of diversity, with any given rule differing on average by less than one gene from any other. However, the case of $p = 1.00$ ended with an average diversity of 5.5 genes. This

Figure 1

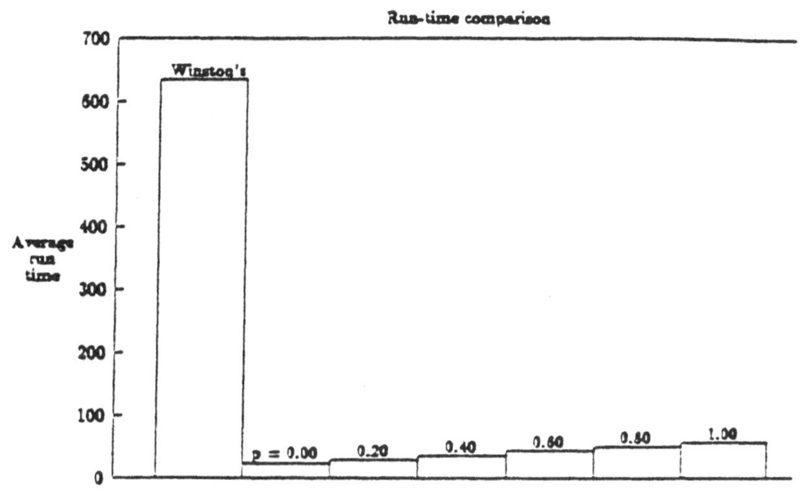

Figure 2

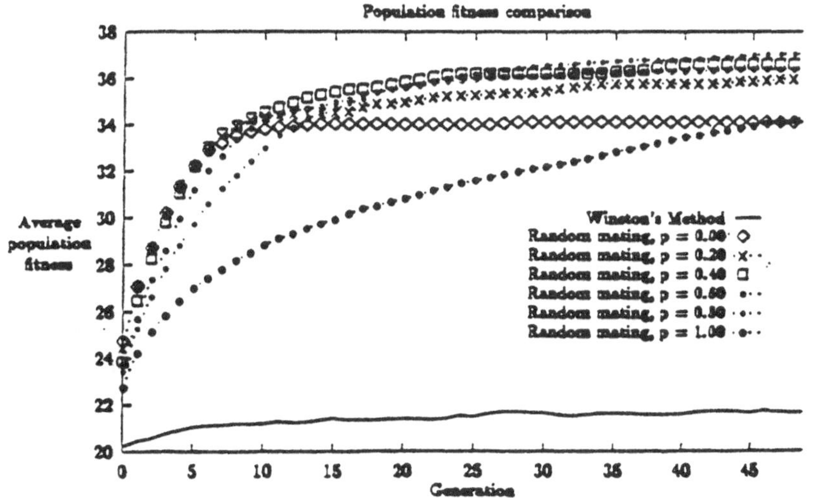

Table 1

Average Diversity Results				
Method	5 generations	10 generations	30 generations	50 generations
Winston's	20.11	20.15	20.10	20.15
p = 0.00	2.44	0.21	0.01	0.00
0.20	4.98	0.56	0.06	0.00
0.40	4.49	1.14	0.01	0.10
0.60	6.13	3.13	0.11	0.02
0.80	12.13	4.99	0.26	0.12
1.00	17.01	15.57	9.17	5.50

may account for the somewhat steadier rate of increase in the average population fitness.

3 Improvement Two: Recessive and Dominant Genes

The second topic of this paper utilizes the idea of recessive and dominant genes to investigate the possibility of accelerating the rate of learning in a genetic algorithm. We used two neighboring genes to code for a phenotype of 1 or 0, with 1's being dominant and 0's being recessive. Thus, genotypes 01, 10, or 11 constituted a phenotype 1, and genotype 00 constituted a phenotype 0. Fitness is determined by counting the number of phenotype 1's present in a rule.

The population size was held at 40, and each rule consisted of 80 genes, keeping a perfect fitness score at 40. Since a rule initialized randomly with 80 binary digits has an expected initial fitness of 30 (as opposed to an expected initial fitness of 20 for methods that don't used recessive and dominant genes), we compensated by forcing the gene pair 00 to appear with probability 0.5 at each gene locus. The other gene pairs of 01, 10, and 11 appear with approximate probability 0.167. This results in the recessive-dominant populations having the same starting average fitness as the populations that don't use recessive and dominant genes. However, randomly generated outside-crossover mates that have recessive and dominant genes will typically have a higher fitness rating than randomly generated outside-crossover mates that do not.

When recessive and dominant genes were added to the genetic algorithm, Figure 4 shows that both our method and Winston's had a faster learning curve. It is not surprising that these populations learn faster, since the randomly-generated rules probabilistically have the higher fitness scores to pass along to their progeny through outside-crossover. While helpful for this concept, there are other concepts where recessive and dominant genes could decelerate the learning curve. More investigation is needed to determine when recessive and dominant genes are useful.

Figure 3

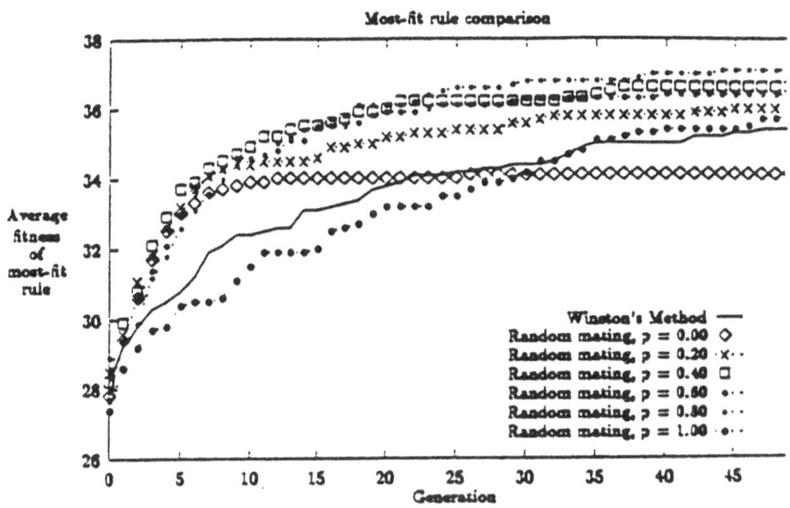

Figure 4

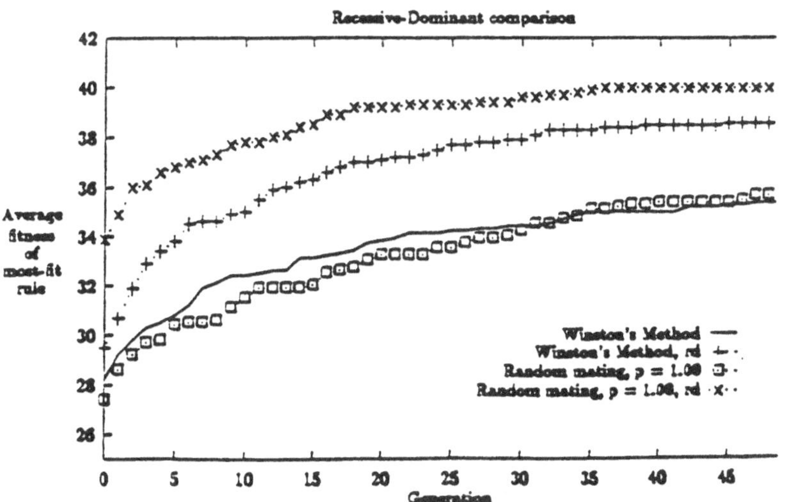

4 Conclusion

The largest contribution of outside-crossover appears to be that run-time measurements are drastically reduced due to the elimination of the computationally expensive explicit diversity check, and that fitness is still not penalized accordingly. Diversity does suffer, but not catastrophically in the case of 100% outside-crossover. By adding recessive and dominant genes into the scheme, learning can be accelerated in certain situations.

Future directions for this research include a more detailed analysis of the most advantageous parameter setting of the probability for outside-crossover, as well as investigating the applicability of introducing recessive and dominant genes. It may also be advantageous to hybridize Winston's method with outside-crossover. If one had a predetermined lower bound on the overall population diversity, a genetic algorithm could begin with outside-crossover and then switch to Winston's method when the diversity dropped below this level.

References

[1] K. DeJong, *Genetic-Algorithm-based Learning*. In Michalski, Carbonell, Mitchell (editors) **Machine Learning II: An Artificial Intelligence Approach.**, Morgan Kaufmann, 611-638, 1990.

[2] K. DeJong, W. Spears & D. Gordon, *Using Genetic Algorithms for Concept Learning*. In **Machine Learning**, Volume 13, 161-188, Kluwer Academic Publishers, Boston, 1993.

[3] S. Forrest & M. Mitchell, *What Makes a Problem Hard for a Genetic Algorithm? Some Anomalous Results*. In **Machine Learning**, Volume 13, 285-319, Kluwer Academic Publishers, Boston, 1993.

[4] D. Greene & S. Smith, *Competition-Based Induction of Decision Models from Examples*. In **Machine Learning**, Volume 13, 229-258, Kluwer Academic Publishers, Boston, 1993.

[5] J. Holland, *Escaping Brittleness: The Possibilities of General-Purpose Learning Algorithms Applied to Parallel Rule-Based Systems*. In Michalski, Carbonell, Mitchell (editors) **Machine Learning II: An Artificial Intelligence Approach.**, Morgan Kaufmann, 593-624, 1986.

[6] J. Paxton, *Adapting Decision Trees to Produce Continuous Classifications*. In Proceedings of the 1992 Golden West International Conference on Intelligent Systems, 166-171, 1992.

[7] P. Winston, **Artificial Intelligence.**, Third Edition, 505-528, 1992.

Chaos, Neural Networks And Gaming

Steven Walczak

University of South Florida
4202 East Fowler Ave., CIS 1040
Tampa, FL 33620-7800

James Krause

University of Tampa
401 W. Kennedy Blvd.
Tampa, FL 33606-1490

Abstract

Games such as roulette and black jack are statistically measurable, but games (e.g., football and horse racing) which involve human/animal performance are much more difficult to predict. The theory of chaos claims that all physical actions are perfectly predictable if sufficient information concerning the action is known prior to its occurrence. Using the theory of chaos as a motivation, a neural network system for forecasting the outcome of college football games has been developed. The design methodology for the football outcome prediction network is discussed in detail. Difficulties with using neural networks in dynamic athletic domains are discussed and methods for overcoming those difficulties are presented. Results of applying the football outcome prediction network are given for several top 25 (using the Associated Press team rankings) collegiate football teams during the 1993-4 season.

Introduction

Games of chance, such as roulette, blackjack or craps, are measurable through the use of probability. This is how casinos and other gaming institutions make a profit, by setting the return rates below the probability for a win. For example, a roulette wheel has 38 locations into which the ball may fall, hence the probability of winning is 1 in 38. However, casinos payoff roulette wins on a single number at the rate of 35 to 1. If every number were to be covered by a five dollar chip, the house would still make a fifteen dollar profit on every turn of the wheel.

If an individual could consistently choose the winning number, the casinos would soon be bankrupt. Unfortunately, knowing the probability of winning does not enable a game player to choose the winning numbers. However, according to chaos theory, the outcome of each roulette wheel turn and each toss of dice is perfectly predictable. Chaos theory claims that even the most random (chaotic) events have an underlying order (see Gleick, 1987), or chance events are actually predictable if the underlying equations of order can be discovered.

To test this theory, a brief ad hoc experiment was conducted. In the experiment, the position of a fair die (height from surface and angle of impact and starting position of the die) was controlled to eliminate variables in the underlying equation and the die was dropped onto a surface ten times. The *a priori* hypothesis is that the same number will appear on the die each time it is dropped. The probability of this event occurring is 1 in 6^{10} or approximately 1 in 60 million (i.e. worse odds than winning most state run lotteries). In this experiment, the die is

E. A. Yfantis (ed.), Intelligent Systems, 457–466.

positioned with the number three appearing on the uppermost surface and dropped from a height of seven inches onto a uniform hard smooth surface which did not contain any obstacles to impede the path of the die. The results of this experiment support the claims of chaos theory with the number five coming face up ten times in a row.

Chaos theory claims that every physical occurrence is predictable, including the outcome of a dice throw on the craps table, if the right information about the event is available (e.g., a chaotic system equation). Why hasn't chaos theory been used by game players? A problem with chaotic systems is that the underlying mathematical equations, which can be used to accurately predict all future events in the system, are extremely complex. An example of a chaotic equation is the Julia set equation of $z^2 - \lambda z$, where λ is the imaginary number 0.7373688 + 0.6754903i - cycle of Siegel disks (Keen, 1989). Furthermore, the chaotic system equations for games of chance have not been discovered yet and certain gaming industry events (e.g., horse races and football games) do not lend themselves to chaotic system solutions due to the influence of non-physical events such as motivation on the outcome. Approximations to the chaos equations for determining the outcome of a dice roll are not acceptable either, since even very small errors or changes in the equations can produce very large differences in the outcome rendering the system useless to game players.

Although chaos theory itself cannot provide a solution for improving the probability of winning, chaos theory implies that knowing a sufficient amount of information about the particular gaming domain can improve the outcome for game players. An example of this improvement is the system of counting cards in blackjack which removes the advantage of the house and turns the game into one which has an equal chance of winning as losing. In other gaming domains such as dog and horse racing and setting the point spread for football games, knowledge of historical information concerning the contestants in a specific gaming event is used by gamblers to improve their probability for a winning outcome.

Background

Artificial intelligence provides a tool for categorizing large quantities of information and inferring outcomes from the information. This tool is neural networks. Neural networks have already been used for several years in the financial industry to perform financial forecasting and manage portfolios (Widrow et al., 1994). Viewed from the proper perspective, the financial industry is analogous to the gaming industry. Both attempt to enable an individual to gain monetary wealth through the assumption of risk. In the financial industry, neural networks process significant amounts of historical information concerning stocks, bonds, etc... to determine relationships or patterns amongst the input criteria which can predict the future value of the financial medium analyzed by the neural network. Applying neural networks in the gaming industry can produce similar effects.

Neural networks are given an input vector of numeric information from which one or more output values are determined. An advantage of neural networks is that they learn to produce accurate output by being trained on a large collection of historic examples from the domain (e.g., previous horse races). The learning performed by the neural networks is done automatically through the training process, so the complex mathematical formulae of chaos theory need not be incorporated into the system by the user. Furthermore, neural networks are capable of recognizing extremely complex relationships among the input values and the desired output values. Once trained, neural networks provide instantaneous results.

Neural network applications for the domain of gaming are virtually non-existent. Only one company, RaceCom located in Jacksonville, Florida, has produced neural network tools for predicting the outcomes of dog and horse races. Why aren't these neural networks in large use by game players? There are many difficulties with using a neural network to predict the outcome of a race. The network must be trained with a large quantity of historical information about the prior performance of each contestant in the race. The number of training examples required is a function of the number of nodes in the network (Hertz et al., 1991), including input criteria. For example, a neural network for predicting the outcomes of horse races could have seventeen input criteria (e.g., position at the starting gate, record of the jockey riding the horse today, ...) which would require over 1200 unique training examples to produce accurate learning. Acquiring this database of training examples is a time consuming process for the game player and a reliable and accurate source of this historic information must be found. The network must be constantly re-trained with new information as new races are run for the output to remain accurate. Neural networks require that all input data be in numeric form and characteristics, such as the blood lines of a dog or horse, which affect the betting strategies of game players are difficult or impossible to encode numerically. Finally, the learning and re-learning being performed by the neural network is being done using the game players money and this may be too risky. Although neural networks are usually very accurate, the desired accuracy may not be achieved immediately, but over time and this results in an initial capital loss to the game player.

Since RaceCom has already produced systems for dog and horse races, this research concentrates on producing a system for determining the outcome of football games, more specifically the final point spread which is used by a gamester for placing wagers. The Football Outcome Prediction Network (FOPN) neural network has been designed to predict the point spread of football games and is being prototyped on NCAA Division I football games. The development of FOPN addresses the various issues which have hindered the usage of neural networks in the gaming industry.

Developing A Neural Network To Predict Football Game Results

As stated previously, two specific reasons exist for the lack of use of neural networks in gaming. First, the number of training examples available to enable the neural network to learn to predict the outcome of the games is extremely limited. This problem is particularly serious in the domains of collegiate and professional football due to the changing characteristics of each team (e.g., players) from one season to the next. In the domain of horse racing, all previous races run by the same horse can serve as training examples to a neural network, but in the domain of football, only games from the current season can serve as accurate training examples. Second, many intangible factors such as home field advantage, motivation, and leadership are difficult to encode numerically and have a definite effect on the outcome of the game.

The FOPN neural network models described below use standard back propagation with supervised learning. All input values are normalized between negative one and positive one. Each neural network attempts to predict the point spread of a game with respect to a specific team. For example, Penn State played Iowa in the third game of the 1993 season and an FOPN network trained to predict Penn State games would produce a point spread for the Penn State team. A separate neural network could be trained to predict Iowa games and could be used to predict the point spread with respect to Iowa.

The solution to the problem of too few training examples is to minimize the size of the neural network. The number of training examples required to produce accurate training is proportional to the number of nodes in the neural network, so smaller networks will require fewer training examples. This presents a paradox to the neural network designer, since the network designer, following the principles of chaos theory, desires to include as much information as possible about the two football teams into the network which increases the number of nodes in the overall neural network. To simultaneously minimize the number of nodes in the FOPN neural network while maximizing the information content of the input vector, input nodes are chosen to correlate to or cover a large number of domain factors. For example, the original FOPN neural network had three input nodes, a hidden layer with three nodes, a second hidden layer with two nodes and one output node. The input nodes in this network represented the AP (Associated Press) rank of each team in the game to be predicted and the professionally established point spread. The AP rank of a team indicates the offensive and defensive strength of the team and marginally implies the prior schedule strength (the quality of the opponents) of the team.

Additional architectures for the FOPN neural network schema have been designed to increase the knowledge available to the neural network. As shown in Figure 1, the first alternative design uses four input values and also increases the first hidden layer to contain four nodes. The additional input node represents the motivation of the football team. Another network architecture utilizes six input and six first hidden layer nodes with the two additional input nodes representing the win and loss ratio as a percentage of games won for each of the two teams. Both of these "increased knowledge" architectures outperform the original three input model. Results for the FOPN neural networks given in the next section are for these two alternative network models.

Many intangible factors (e.g., home field advantage and motivation) will affect the outcome of a football game. These intangibles are difficult to encode numerically. Home field advantage can easily be encoded as a probable point advantage and is assumed to be correlated to the professional point spread which is already used as an input value. Motivation or the inspiration a team feels when playing against the number one rated team in the nation or an inter-state rival is much more difficult to encode. These additional fuzzy input nodes, although difficult to quantify, greatly enhance the accuracy of the FOPN neural networks. Numeric values for the motivation input node, used in the four and six input versions of FOPN, are determined heuristically. Heuristics which increase the base value for motivation include: first game of the season, losing the previous week, playing a team ranked in the AP top 25, playing a higher ranked team, or playing a conference or other traditional school rival. Conversely, heuristics which decrease the value for motivation include: playing a lower ranked team and playing after three or more consecutive losses.

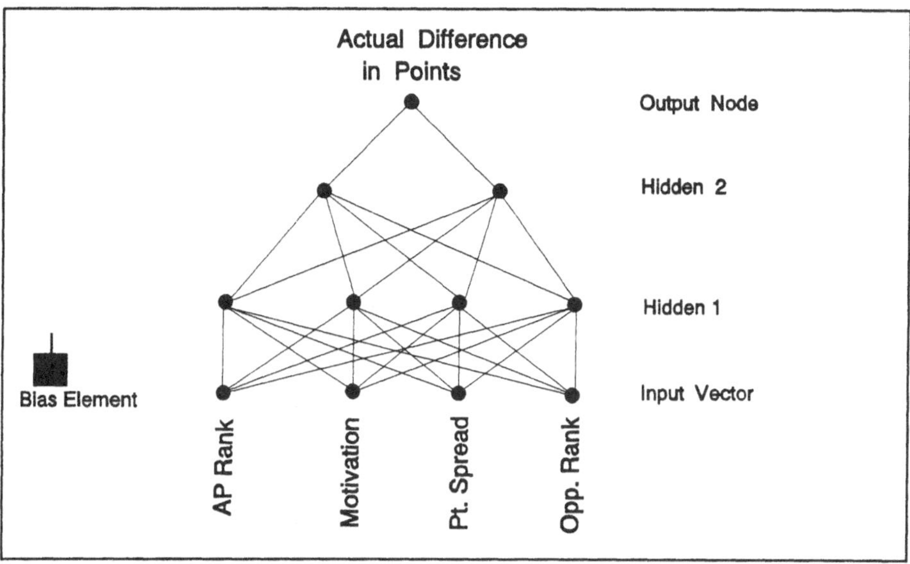

Figure 1: FOPN neural network configuration.

Results Of Applying The FOPN To Collegiate Football

The FOPN has been trained with information on various collegiate football teams from the 1993-94 season. Due to the training requirement of the network, predictions for specific football game outcomes could not be made for any games during the first week of the season. The teams modeled by the FOPN neural network are: Penn State (PSU), Florida (UFL), Florida State (FSU), Miami (UM), and Notre Dame (UND). Except where specified, all results are from the four input node FOPN neural network model.

Research was conducted to determine the effectiveness of the FOPN on a week-by-week basis throughout the football season. The methodology for this phase of the research was to train the FOPN from the input values and results from the first week of the season, then require the network to produce an output for the next (second) game to be played. After the FOPN had predicted an outcome for a specific game, the game results were included in the training example set and the FOPN neural network was re-trained. The process continues iteratively until predictions for the entire football season had been made. The team modeled by FOPN in this phase of the research is Penn State (PSU), which had AP rankings ranging from 7th to 19th place with an average ranking for the season of 13th place. Results of this research are presented in Table 1. Remember that the FOPN for PSU is predicting the number of points PSU will win the game by, with negative results indicating a loss. The evaluation column states if the PSU FOPN was correct from a gamester perspective. In other words, the FOPN is incorrect if the professional odds value is between the actual outcome and the FOPN value, meaning that the professional odds predicted the actual outcome better than FOPN. Otherwise, the professional odds are incorrect and the FOPN value indicates the direction of the error. When FOPN is correct, gamblers placing a wager on the side of the professional odds indicated by FOPN would win their wager.

Table 1: FOPN weekly results for Penn State's 1993-94 Season

Professional Odds	Actual Outcome	FOPN Predicted Outcome	Evaluation
5.5	1	17	Incorrect
5.5	31	5	Incorrect*
14	24	9	Incorrect
17.5	63	34.5	CORRECT
5.5	-8	69.5	Incorrect
-3.5	-18	-26.5	CORRECT
11	7	-11.5	CORRECT
10	14	27	CORRECT
17.5	22	17	Incorrect*
4.5	1	3	CORRECT

From Table 1, it can be seen that the FOPN network performs poorly for the first three predictions, but this corresponds to the problem of having a sufficient number of training examples to accurately learn the prediction patterns. Additional research with the other football teams verified that the FOPN starts to perform accurately with a minimum of four training examples. In the final seven games of the season, the FOPN accurately predicts over 71 percent (5 out of 7) of the games. For two of the incorrect FOPN predictions (marked with an asterisk * in Table 1), the difference between the prediction value of the FOPN and the professional odds is less than one point. Even though the FOPN was incorrect for these two games, the fact that the neural network prediction is nearly identical to the professional odds maker prediction should be interpreted to mean that not enough knowledge is available concerning the game to be predicted or the professional odds are exactly correct. In either case, a prudent gamester would not place a wager under these circumstances.

Another interesting statistical measure of the FOPN performance for gamesters is that four of the last seven games played by Penn State were against other teams ranked in the AP top 25 at the time the game was played. Wagers are made more frequently on games involving the top ranked football teams. For these four games (the 5th-7th and 10th row in Table 1), the FOPN neural network was correct 75 percent, or three out of four times.

The next phase of research with the FOPN neural network models reveals that using two FOPN neural networks simultaneously, one for each team in the game to be forecast, improves the prediction accuracy of the FOPN. Each FOPN neural network predicts the point difference between a specific team and an arbitrary opponent (the point spread). A partial explanation of the improvement derived from two FOPNs is that the final score difference from previous games used in the training set can bias the output of an FOPN. For example, Florida State outscored their opponents by more than 35 points per game, including their seven point defeat from Notre Dame.

The methodology for the research using multiple FOPNs is to design separate FOPNs for each team competing in games to be forecast. In this research, an FOPN was developed for FSU, UFL, UM and UND, with FSU being the only opponent faced in common by each of the other three college football teams. Each FOPN is trained from examples of all games played previously in the season to the game being forecast. This means that the FSU FOPN was trained three times on different example sets. After each FOPN is trained, it is stored in a knowledge-base of team FOPNs. Re-training of all FOPNs is performed when new examples become available. A front-end processor (see Figure 2) acquires the FOPN input values, rank of each team, motivation value for each team and the professional odds (point spread), then activates the appropriate FOPN for each team in the game to be forecast. The individual FOPN forecasts are combined using the following equation:

$$[1]\ (FOPN_1 \text{ value - professional odds}) - (FOPN_2 \text{ value + professional odds})$$

The professional odds are assumed to be given to the front-end processor with respect to the first FOPN's team and $FOPN_1$ and $FOPN_2$ represent the output values of the FOPN for the respective teams. Equation [1] normalizes the FOPN output values and determines the relative difference to the professional odds. A positive value from equation [1] indicates that the first team will win by more than the professional odds, while a negative value indicates that team 1 will lose or possibly win by less than the professional odds. The results of the combined FOPN evaluation tests are shown in Table 2. The equation [1] value is given in the DIFF column. Team one is always FSU, while the opponent for the team 2 FOPN is given in the corresponding rows of Table 2.

For the combined FOPN system to give a correct prediction, if the DIFF value is positive, the game score difference must be greater than the professional odds. Otherwise, for a negative DIFF value, the game score difference must be less than the professional odds. The professional odds and game score difference displayed in Table 2 are with respect to the FSU team.

While the FOPN for all four teams predicted wins, the equation [1] difference value yields a more accurate result. The combined FOPN accurately predicts that Miami will lose by more than the professional odds and that Notre Dame will upset FSU and win. Unfortunately, the combined FOPN is incorrect for the UFL versus FSU game, indicating that UFL will beat the point spread. This artifact may be caused by the fact that UFL also outscores its opponents by a large margin, on average by more than 23 points. As can be seen in Table 2, the FOPN predictions for both FSU and UFL are large.

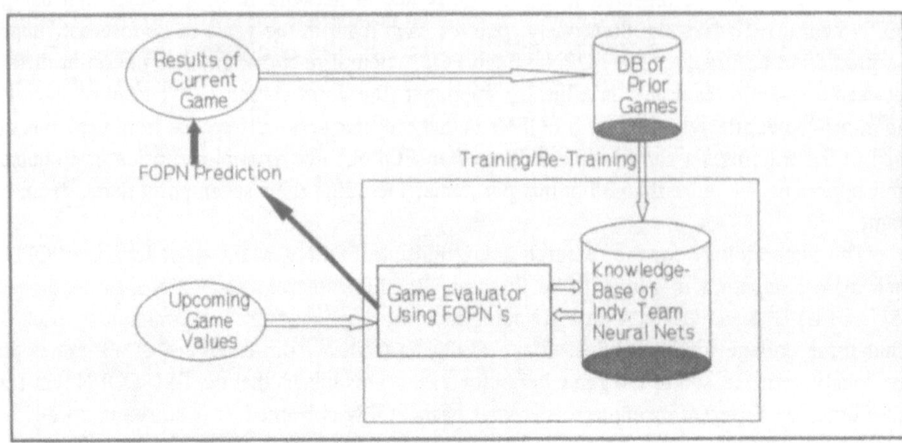

Figure 2: Combined FOPN system architecture.

Table 2: Combined FOPN results.

Professional Odds	Game Score Difference	FSU-FOPN$_1$ Prediction	FOPN$_2$ Prediction	Team modeled by FOPN$_2$	DIFF
13	18	39.5	5.3	UM	8.2
7	-7	11	4.7	UND	-7.7
10.5	12	27	14	UFL	-8

Finally, research has been conducted to examine an increase in the size of the FOPN neural networks, to six input nodes, by including knowledge about the win percentages of both teams. The six input node FOPNs are evaluated using the combined FOPN system approach described above. Results for the larger FOPN neural networks are shown in Table 3.

Table 3: Combined six input node FOPN results.

Professional Odds	Game Score Difference	FSU-FOPN$_1$ Prediction	FOPN$_2$ Prediction	Team modeled by FOPN$_2$	DIFF
13	18	31	-4	UM	9
7	-7	9.26	1.3	UND	-6.04
10.5	12	6.7	2.8	UFL	-17.1

The results shown in Table 3 are practically speaking identical to the results from the smaller network results shown in Table 2. An interesting product of the larger FOPN network architecture is that the UM FOPN prediction shown in Table 3 indicates that UM will lose the game by four points, even though UM had been undefeated in the season up to this point. The undefeated season by UM means that all the prior game examples used to train the UM FOPN had positive results, but the six input node FOPN was still able to predict a defeat.

Another interesting result not shown in Table 3 is a dramatic decrease in the training time of the expanded FOPN neural networks. Training an FOPN consists of presenting the prior game examples in random order repeatedly until a minimum threshold for error is achieved. The four input node FSU FOPN required 10000 training iterations for the UM game, 16000 training iterations for the UND game and 75000 training iterations for the UFL game. The increase in the number of iterations corresponds to the point in the season or number of training examples available. The six input node FSU FOPN only required 5000, 10000 and 16000 training iterations respectively. Producing identical results, or better in the case of the UM FOPN, and decreasing the number of iterations required to enable the FOPN neural network to learn accurate prediction patterns represents a significant time savings to gamblers.

Summary And Future Research

Traditionally, many game players utilize historic information, but the information is frequently incomplete or sometimes ignored by the player. Neural networks learn to recognize complex patterns which may escape notice by even experienced gaming industry experts. To be used in the gaming domains of football or other athletic contests, neural networks must overcome two problems. These problems are an insufficient quantity of training examples and transforming intangible factors into numeric input values. This research shows that minimizing the size or number of nodes in a neural network reduces the requirement for a large database of historic training examples. For the domain of collegiate football, a minimum of four training examples is sufficient to produce accurate predictions of the outcome of future games. Heuristic methods are applied to produce numeric values for intangible factors, such as team motivation, included in networks.

The FOPN neural network is a standard back-propagation neural network which uses supervised learning to predict the point difference between two teams in a football game. Over 70 percent of the games analyzed by FOPN are correctly predicted. Improved prediction accuracy is achieved by using a system of two FOPN neural networks which calculates the normalized difference between the two FOPN predictions.

As with any knowledge-based system, an increase in the quantity and quality of knowledge available to a system will improve the performance of the system. Neural networks applied in the football domain are restricted in size due to the small quantity of training examples available. Future research will investigate the use of training examples from previous seasons and heuristics for determining when previous season training examples may or may not be appropriate due to a change in personnel or significant changes to other team factors.

Additional future research will look at applying the FOPN system to the domain of professional football games. To apply FOPN in other sport domains such as professional football, a universal ranking system similar to the AP ranking system for collegiate football must be identified. Also, the motivations of professional teams may be more complex than collegiate teams, necessitating the use of additional input nodes to capture the relevant knowledge.

Neural network technology has not been widely used in the gaming industry. The FOPN neural network system has demonstrated that the use of neural networks can greatly improve the probability of winning in the gaming industry.

References

Gleick, J. (1987). Chaos: Making A New Science, Penguin Books, New York.

Hertz, J., Krough, A. & Palmer, R. G. (1991). Introduction To The Theory Of Neural Computation, Addison-Wesley, Redwood City, CA.

Keen, L. (1989). Julia Sets. In Chaos and Fractals: The Mathematics Behind the Computer Graphics, Devaney & Keen (eds.), American Mathematical Society, Providence, RI.

Widrow, B., Rumelhart, D. E. & Lehr, M. A. (1994). Neural Networks: Applications in Industry, Business and Science. Communications of the ACM, 37(3), 93-105.

Application of Genetic Algorithms to 2D Velocity Inversion of Seismic Refraction Data

Li Li
Seismo. Lab. MS 174
University of Nevada, Reno
Reno, NV 89557
email: lili@seismo.unr.edu

Sushil J. Louis
Dept. of Computer Science
University of Nevada
Reno, NV 89557
email: sushil@cs.unr.edu

James N. Brune
Seismo. Lab. MS 174
University of Nevada, Reno
Reno, NV 89557
email: brune@seismo.unr.edu

Abstract: This paper explores the possibility of using genetic algorithms to invert seismic refraction travel time data for a two-dimensional velocity structure. The problem is usually high-dimensional, non-linear and multi-modal. Our chromosome is encoded as a two-dimensional real number array. Binary tournament selection, two-point crossover and block mutation directed by prior geological knowledge are used in this paper. Preliminary results indicate that the genetic algorithm efficiently fits the data. Given sufficient data sampling, the genetic algorithm is capable of finding geologically plausible two-dimensional velocity structures.

Keywords Genetic Algorithms, Seismic Refraction Survey, Two-Dimensional Seismic Velocity Inversion, Travel Time Inversion.

1 Introduction

Use of seismic refraction data collected at the surface to find the underlying velocity structure of the rocks has been one of the main tasks of seismology. The data set is composed of the travel times of the first arriving seismic waves from a source to receivers on the surface. The seismic sources can be earthquakes or manmade sources e.g. chemical blasts or nuclear explosions . The technique used to infer the subsurface velocity structure is called travel time inversion. These problems are usually highly-dimensioned, multi-modal and non-linear.

There are many methods to carry out travel time inversion. Among them trial and error, matrix inversion and tomographic inversion are commonly used. The trial and error methods need a lot of experience and related geological knowledge of the working area, and are laborious. The matrix inversion and simple tomographic inversion methods all involve some kind of linearization of the problem, therefore a good starting velocity model

E. A. Yfantis (ed.), Intelligent Systems, 467–473.

is essential for these methods. It is also easy to get trapped at a local optimum. Due to the large number of unknowns, these methods all need to use iteration, and sometimes have difficulty in converging.

Genetic algorithms (GAs) were designed to work on non-linear, multi-modal and poorly understood problems [1]. They are stochastic, parallel search algorithms, based on natural selection and evolution. After an initial population is created, the GA starts to evolve good solutions moving towards global optima. Three basic genetic operators, selection, crossover and mutation are used to carry out the search. At each generation, every individual in the population is assigned a fitness value according to its performance, and fitter individuals have more offspring in subsequent generations.

Travel time inversion can be considered as a search problem, that is, from the model space we want to find a model such that the calculated travel times from this model fit the observed travel times the best. From this point view, we may use GAs to solve our inversion problem.

Though GAs have found recent application in seismic waveform inversion for a 1D velocity structure [3, 2], to our knowledge the GA has not been used in travel time inversion for 2D velocity structures. In this study, we explore the possibility of using GAs to invert a set of synthetic travel time data for the velocities. The travel times in forward problem are calculated by solving the Eikonal equation with a finite difference method [4].

2 GAs for 2D Velocity Inversion

The unknown parameter in our problem is a 2D velocity array which forms a cross section of velocity as shown in Figure 1. We could encode each element in the 2D velocity array either as a six bit binary string or as a real number between 1.8 and 7.2 km/sec. A GA with binary string chromosome is called a Low Level Language GA (LGA) and a GA with real number chromosome is called High Level Language GA (HGA) following [5]. In both GAs, the velocity is in the range from 1.8 km/sec to 7.2 km/sec, a full range of reasonable velocities of the earth's upper crust. We use a 2D chromosome to represent the 2D velocity model and the velocity at any point is defined either by six bits in a LGA or by one real number in a HGA.

The adoption of HGA is mainly suggested by the really large number of unknowns (usually in the order of 120*20 or more) and the fact that the performance of an individual is not sensitive to a single velocity parameter but to a block of velocities. The preliminary results show that the high level language GAs give better results and are easier to handle than the binary GA. Therefore we only describe the operators used in our real number GAs. We use selection, crossover and mutation operators to act on the old generation to produce the new generation.

2.1 Initial Population

Seismologists usually assume that the velocity distribution of upper crust is smoother horizontally than vertically. Thus, individuals in the initial population are formed in

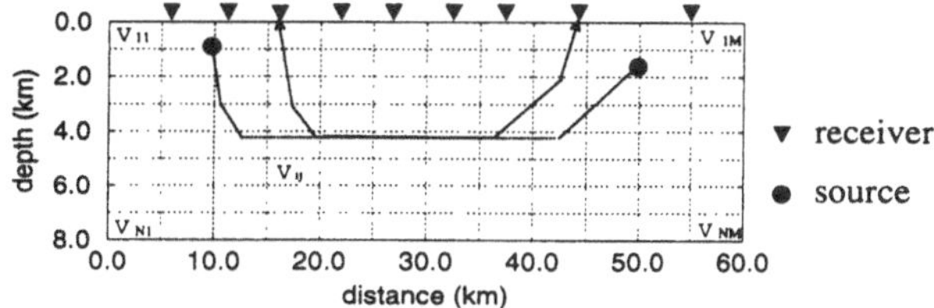

Figure 1: Parameterization of velocity cross section

the following way: We divide the velocity cross section into N layers (N=3 here). The thickness and velocity of each layer is randomly chosen.

2.2 Selection

Selection, crossover and mutation act on the old generation to produce the new generation.

We use a variation of binary tournament selection where every individual is assured a chance of mating. Each pair of parents undergo either crossover (with probability P_c) *or* mutation (with probability $1 - P_c$) to produce two children. From the two parents and two children, we select the two fittest individuals to enter the next generation. The fitness of a chromosome is defined as the inverse of the average differences between observed and calculated travel times (called residuals).

To increase diversity in the population during later generations, we vary the probability of crossover, and thus mutation, over time. We decrease the probability of crossover and therefore increase the probability of mutation when the average fitness of the population is close to the highest fitness of the population.

2.3 Crossover and Mutation

The crossover operator randomly chooses two points in the 2D chromosome, then flips a coin to decide whether to do crossover vertically (crossover 1) or horizontally (crossover 2) as shown in Figure 2.

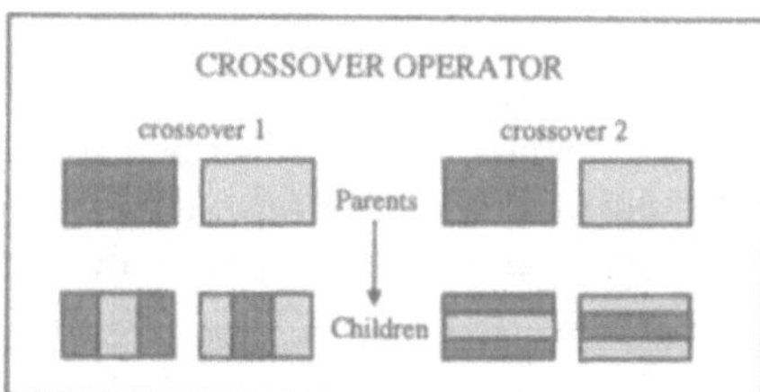

Figure 2: Our two dimensional crossover operators

Our mutation operator uses geological and geophysical knowledge to increase the efficiency of the search process. We provide some background on seismic refraction surveying before describing our mutation operator.

In multi-source refraction survey, the travel time data set can be divided into several subsets according to the geometry of the survey (the location of sources and receivers). A subset of data may be more sensitive to some area of the velocity cross section, and less sensitive to other areas. For example, the travel times to the receivers that are very close to a source are much more sensitive to the velocities at shallow depths around that source, while deeper velocities are more responsible for travel times to distant receivers.

We divide the velocity cross section into M segments horizontally and N layers vertically. M and N can be determined by using knowledge of surface geology and analyzing the travel time curves. The travel time data of receivers at each segment are organized into N subsets according to the source to receiver distance and the direction of the ray (the propagation path of seismic energy from source to receiver). Each subset of data has its own (approximately) corresponding area in the chromosome and its own velocity limit.

We flip an unbiased coin to decide whether to do mutationr at shallow depths or at deeper depths. Once this is decided, we analyze the residuals of each subset of data in all segments and find the area of the chromosome responsible for the worst fit (maximum residual). Within that area, we mutate a randomly chosen block by reinitializing that block with a random velocity in the velocity limit of this area.

3 Results and Analysis

We generated a synthetic data set from a one-dimensional velocity structure with four layers, plus a high velocity block in the center as shown in the lower left part of Figure 3. The velocities for the four layers are 2.5, 3.5, 5.5, 6.3 km/sec respectively. The 2.5 km/sec layer ends just below the topographic line. The high velocity block has a velocity of 6.8 km/sec, representing a possible cooled magma body. The colors that represent the velocities are also shown in the lower right part of this figure. The synthetic data sets (the small circles at the top of Figure 3) are composed of nine seismic refraction profiles with shot locations at $S01$, $S02$, $S03$, $S04$, $S05$, $S15$, $S16$, $S17$, $S18$. The circles with different color represent travel times from different source to the same receiver. The travel time displayed is called reduced travel time which equals the travel time minus distance divided by reduction velocity 6.0 km/sec. The numbers above the topography show the receiver locations. The first arrivals from every shot to all the receivers are recorded as our data. The geometry of the source and receiver distribution and the topography are taken after the 1982 USGS refraction survey in the area around Yucca Mountain, Nevada.

We have tested different probabilities of crossover and mutation and different population sizes for this synthetic data set. The best model is found when using $P_c = 0.8$, $P_m = 0.5$ (the probability of doing mutation at shallow depth when GA decides to do mutation for the next pair of parents), and a population size of 200 for 150 generations. Figure 4 shows the residuals versus generation curve for this good GA run.

The GA starts to converge at about generation 60, but the directed block mutation operator is still making small improvements up to the end. The final smallest residual is about 0.037 sec, a fairly good fit to the data.

The best velocity model from this GA run is shown in Figure 5. The upper part of the figure displays the fitting of the calculated travel time (the curves) using this velocity

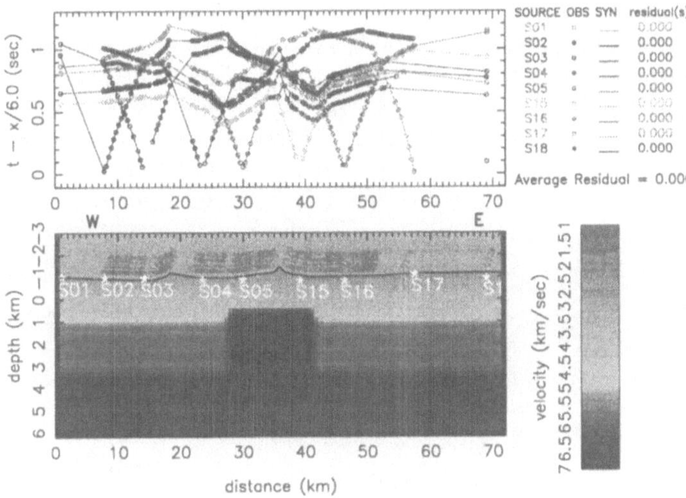

Figure 3: Top: Synthetic data set; Bottom: Original velocity model used to generate the synthetic data

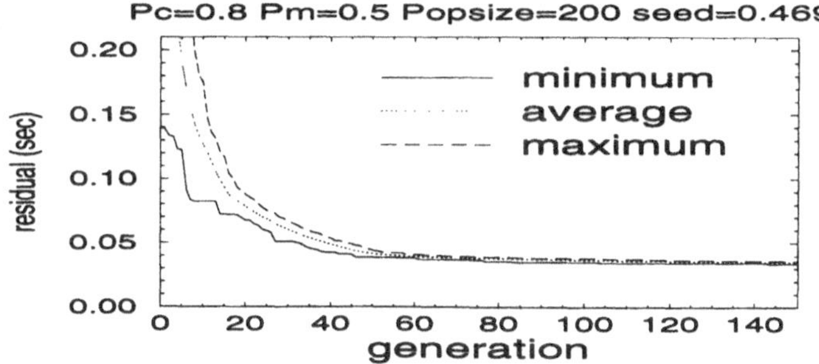

Figure 4: Convergence behavior for a good GA run

model to the data (the circles). Although the velocity contrasts have been smoothed, and velocity at every grid point is not exactly the same as in the theoretical model (Figure 3), the general pattern of the velocity distribution is well resolved at shallow and intermediate depth. The high velocity block is found by the GA without any specific prior knowledge put into the genetic algorithm. The deviation of the velocities at deeper depth from the theoretical model are mainly due to the poor sampling of the data. To better resolve this part of the velocities, we need to extend the source receiver distance.

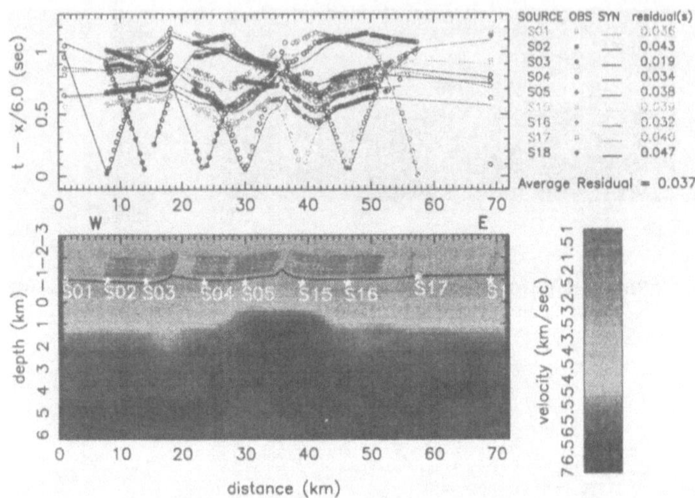

Figure 5: Top: fit to the data; Bottom: resulting velocity structure from the good GA run

The results indicate that the GA can fit seismic refraction data very well and can find a reasonably good velocity structure from the given travel time data. The use of block mutation helps to generate a geologically plausible velocity model. The dynamically changing crossover rate reduces the probability of stalling the GA at a local optimum. Currently, we are applying the technique to some real data sets, and trying to put more geological and geophysical constraints into our genetic algorithm.

References

[1] J. Holland. *Adaptation In Natural and Artificial Systems*. The University of Michigan Press, Ann Arbour, 1975.

[2] M. Sambridge and G. Drijkoningen. Genetic algorithms in seismic waveform inversion. *Geophysical Journal Inernational*, 109:323–342, 1992.

[3] M. K. Sen and P. L. Stoffa. Rapid sampling of model space using genetic algorithms: Example for seismic waveform inversion. *Geophysical Journal Inernational*, 108:281–292, 1992.

[4] J. Vidale. Finite-difference calculation of travel times. *BSSA*, 78(6):2062–2076, 1988.

[5] W. G. Wilson and K. Vasudevan. Application of the genetic algorithm to residual statics estimation. *Geophysical Research Letters*, 18(12):2181–2184, 1992.

REVERSE ENGINEERING: A CASE STUDY ON NEURAL NETWORK SOFTWARE

Y. B. Reddy
Dept. of Math and Computer Science
Grambling State University
Grambling ,LA 71245, USA
August 5, 1994

Abstract. An experiment was conducted using CADRE/teamwork Ensemble package on neural network software written in C-Language. System design was extracted using structure charts, control flow diagrams, coverage reports, function understanding, metrics reports, test case verification, and parameter reference in functions. Dead code was eliminated and increase in efficiency was identified using metric reports. The system then redesigned to C++ (using object-oriented design of Ensemble package).

Keywords. Reverse Engineering, Re-Engineering, Program Understanding, Design Recovery, Object-Oriended Design,

1. Introduction

Reverse Engineering for software is the process of analyzing a program in one effort to create a representation of the program at a higher level of abstraction than source code [1]. It is a process of design recovery. Reverse engineering tools extract data, architectural, procedural design, and information from an existing program. Re-engineering, also called renovation or reclamation, not only recovers design information from existing software, but also uses this information to alter or reconstitute the existing system in an effort to improve its overall quality [2]. In most cases, re-engineering software implements the function of the existing system. But at the same time, the software developer also adds new functions and/or improves overall performances. Re-engineering combines the analysis and design extraction features of reverse engineering with a restructuring capability for program data, architecture, and logic [3].

The reverse engineering tools can be categorized as static or dynamic. A static reverse engineering tool (by far the most common) uses program source code as input and analyses and extracts program architecture, control structure, logic flow, data structure, and data flow. Other tools in this category apply a technique called program slicing. The software engineer specifies the types of program structures (data declaration, loops, other logic) that are of interest and the reverse engineering tool removes extraneous code, enabling only code of interest to be represented. Dependency analysis tools perform most of the functions (above) but in-addition, tools in this sub-category build graphical dependency maps that show the links between data structures, program components, and other user specified program characteristics. Static reverse engineering tools are called "code visualization" tools[4]. In fact, by enabling the software engineer to "visualize" the program, such tools greatly improve the quality of changes that are made and productivity of the people making them. Dynamic reverse engineering tools monitor the software as it executes and uses information obtained during monitoring to build behavioral model of the program. Although such tools are relatively rare, they provide important information for software engineers who must maintain real-time software or embedded systems.

Although re-engineering tools offer significant promise, relatively few industry quality tools are in use today. Existing re-engineering tools can be divided into two sub-categories: code re-structuring tools and data re-engineering tools. Code restructuring tools accept unstructured source code as input, perform the reverse engineering analysis, and then restructure the code to conform to modern structured

E. A. Yfantis (ed.), Intelligent Systems, 475–482.

programming concepts. Although such tools can be useful, they focus solely on the procedural design of a program. Data re-engineering tools work at other end of the design spectrum. Such tools assess data definitions or database described in a programming language (usually COBOL) or database description language. They then translate the data description into graphical notation that can then be analyzed by a software engineer. Working interactively with the re-engineering tool, the software engineer can modify the logical structure of the database, normalize the resultant files, and then automatically regenerate a new database physical design. The tools may use an expert system and knowledge-base to optimize the re-engineered software for improved performance.

2. Backpropagation model

Backpropagation is a widely used neural network model during recent years for most of the pattern recognition problems [5]. Multilayer Backpropagation paradigm is a feedforward neural network having more than one hidden layer. In the present problem, the program written for backpropagation in C-language is selected to reverse engineering case study [6]. In the present research, the neural network program is used to identify a fault component in a hierarchically connected sensor output using the two gates "and" and "or" [7]. The 'and' gate takes the inputs and outputs the minimum value. The 'or' gate takes the inputs and outputs the maximum value[1]. The experiment for neural network model was conducted with six inputs, two hidden layers, and seven projected test outputs[2]. The main idea of conducting an experiment is to cover most of the lines of code and branches with test input values. Test coverage reports are also generated to find the behavior of the program. It is clear that the program [6] is well written but it misleads the user while executing the program particularly when calling the functions *'dread'* and *'dwrite'* in 'output_generation' (user needs to remember the previous data file name without extension after the period). The two modules adds the extensions as: *dread* adds *_v* to data file and (b) *wtread* adds *_w* to the data file to separate these from others in the directory. The program can be used to train various data files and generate outputs for any trained pattern. The execution of the program and the necessary modifications are discussed below. The program never keeps track of previous weights if a user wants to train with more than one input file or further trains the system with the same data again if the system does not reach the minimum required error. The design extraction using Ensemble documentation is shown in Figure 1. Each time the program executes as if it was started for the first time (weights are initialized each time). With little modifications the program can be further trained for more than one data file at different times so that we can save previous trained time. The design modifications with added functions are shown in Figure 1 (with dotted lines) and metric reports with existing design and modified design are shown in Appendix A.

3. Program Understanding

It is required to understand actual representation of the system [3]. To understand actual representation of the software system, one can read the documentation, read the source code, run the program using test data, understand the storage allocation, and the cross references. Simple documentation, that is reading the source code, may not be helpful many times. The dynamic behavior of program is very important. In the present study, if a user executes the program with various examples, the result will be very clear and it will be difficult to find an error. While working closely, it is difficult to think that the system must be trained from previous point without wasting the time that we have spent already. After training the system with one data file, the weights are adjusted to minimize the error. Training the system with more data, the weights will be adjusted to our satisfaction. If we do not care for this important point, i.e. further training of the system, we are repeating the whole process again and again

---

[1] Reference [7] provides the syntax diagrams.

[2] References [5] and [6] provides the standard neural network diagrams.

and wasting time. This is one of the main difference, in understanding, between neural network software and other software.

To understand the proposed neural network software [6], Teamwork/Ensemble software was used. It included metric reports, cross reference of functions, reference of library functions, test criteria, coverage reports, and many important points to understand a software system. Function call (figure 1) map was extracted from cross references and coverage and metric reports (Appendix A see run1). Due to unclear output statement in the function output_generation, line coverage and branch coverage of 'dread' and 'wtread' were zero. Further understanding of the functions 'output_generation', 'dread', 'wtread' metric reports in Appendix A(see run2) were produced. More and more executions were performed with less iterations and more iterations in a continuous process. After performing 5000 iterations to train the system, we get a tolerant error 'e1=0.032.' We continue to perform 1000 iterations using the same or different data file until we get the system tolerant error 'e2=0.12'. Then it was found that e2 greater than e1. This was not accepted because further training would reduce the error. The error was corrected with introduction of two simple functions 'ybr1' and 'ybr2' (dotted line extension in Figure 1) and few statements in the main function without changing the original system and with improved performance.

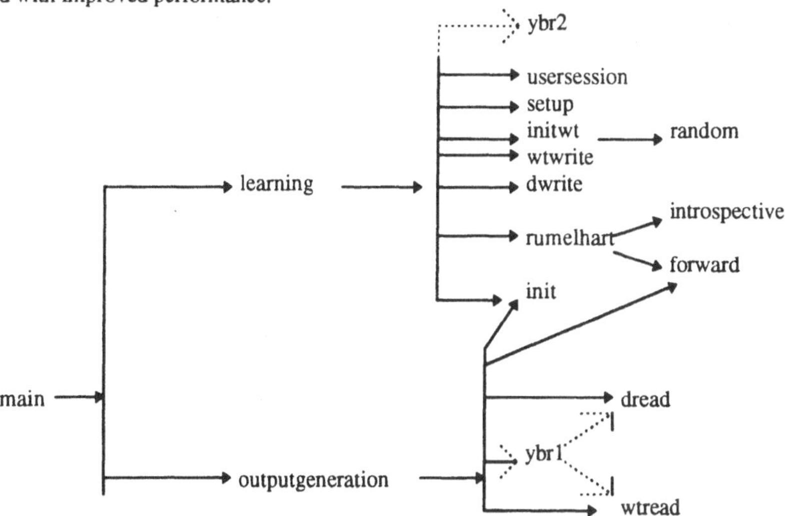

Figure 1: Design extraction using Ensemble program
(--> shows the design modification)

The modification in main(), learning, output_generationand the two functions ybr1 and ybr2 are given below (the bold case shows introduction of new lines or modifications):

```
ybr1()
{
printf("\nIf you did not train the system and you want to use pre-trained system\n");
printf("enter the data file name used previously for training without extension\n");
printf("\nif you just train the system and use, please enter the same name\n");
printf("\nIf you do not follow the instructions program terminates or you get bad results\n");
printf("\n\tType the task name:");  scanf("%s", task_name);
dread(task_name); init();             wtread(task_name);
}
/*=====================================================*/
ybr2()
```

```
{
 printf("\nTo further train the system with another data file with same\n");
 printf(" number of inputs, outputs then enter the data file name\n");
 printf(" \nenter the name of the data file:");
 scanf(" %s", task_name);
 printf("\nEnter total number of input samples in this data file:");  scanf(" %d", &ninput);
 printf("\nMax number of iterations?: ");             scanf (" %d", &cnt_num);
printf("\nexecution starts....");
}
learning( )
{
  int result;
 if (ln == 0)  {  user_session();    set_up();     init(); } else ybr2();
do{ -----
/*-------- see reference [6] for complete program ------*/
}
/* modifications in output_generation()  */
output_generation( )
{
  ------
scanf("%s", ans);
  if ((ans[0]=='y') || (ans[0]=='Y'))  ybr1();
/*  statements given below are removed from output_generation
  {   printf("\n\t Type the task name : ");
      scanf("%s", task_name);
     dread(task_name);   init();   wtread(task_name);   }     */

/* modified and required part of main function  is shown below  */
main()
{ -----
strcpy(task_name, "*********");
printf("you want to use prelearned system or train first time: enter p or f: ");
scanf(" %s",yb);
if  (yb[0]=='f') { ln = 0;       printf("\nselect learning\n");
    else  { ln = 1; printf("\nselect output generation \n"); }

  do { - - - -
  ---------
scanf("%s",cont);  ln += 1;
  } while ((cont[0] =='y') || (cont[0] =='Y'));
  ------
}
```

The main() function modified to verify either user was training the system first time or further training. The output_generation() was modified by deleting five statements and including one if statement to transfer the control. The learning() function was modified with an if statement. The modifications were shown above in bold font. With this simple improvement, the total performance of the system was changed. If we train the system completely at one time, we can still use the system for further training or testing even after we restart the system. The simple modification helps no loss of training state, can train to required minimum error from previous point and can test any time.

4. Design Recovery

Design recovery is a subset of reverse engineering in which domain knowledge, external information, and documentation or fuzzy reasoning are added to the observations of the subset system to identify meaningful higher level abstractions beyond those obtained directly by examining the system itself [1 -3]. Design recovery recreates the design abstractions from a combination of code, documentation (existing or extracted from code), personal experience, and general knowledge about the problem and application domains. Design recovery must reproduce all of the information required to understand the functionality of the program The primary purpose of reverse engineering of a software system is to increase the overall comprehensibility of the system for both maintenance and new development.

The objective of the design recovery [8] is to understand the software system or program with the help of related information. The related information may be structured charts, control flow diagrams, coverage reports, metric reports, cross references, and object design information (local variables, global variables, code, and parameter transformation) for all modules. To build complete documentation using ensemble package, we use one icon instruction called 'Build Documentation.' The outline of the system design will be extracted from cross references. The design of each object can be obtained from the following information generated by Ensemble documentation report:

structured charts	-	provides global variables, library functions, and reference direction
control flow diagrams	-	control flow with internal logic
coverage reports	-	control flow and source coverage
metric reports	-	metric information about functions in the model
cross references	-	function reference, library functions, global variables
object design	-	local variables, global variables, code, parameter transformation, and copy of object (code)

With the help of above main headings and other information, the module design and the system design [9] were recovered (complete information was printed by using Ensemble documentation) and shown in Figure 1. In the metrics report, the line and branch coverage for dread and wtread were '0' (see Appendix A: run1 part). It shows that 'dread' and 'wtread' were never executed if we do not understand the functionality of the program. After recovering the design and understanding the modules and the relation between the modules, the experiment was conducted again. The coverage for 'dread' and 'wtread' were generated and given in Appendix A (run 2 part). Some of the functions have less than 100% of line or branch coverage due to the presence of 'if' statements. The design was modified with little code variation by adding two functions. The modified design works to train the system with initial values and any number of training files. The adjusted recent weights are used for further training. Continuous training of neural network system saves time to reach minimum error.

5. Object-oriented design

Object-oriented design (OOD) is another part of the teamwork package. The documentation reports generated by Ensemble is very useful to design the new object-oriented system. Programmer has options to create private or public objects. The global variables, local variables, and function references generated by the documentation are used as part of the structured charts and class diagrams. Friend operations can be done using foreign modules. Private function declaration will be done by using the functions in structured charts without referring in class diagrams. The global variables are referred by using the 'Wok' diagram. The direction of the data flow between the functions is done using 'Couple' notation. There are other notations for complete design of the object model. Once the object model is created, the simple execution step using the icon from class diagram or structured diagram helps to generate the C++ code. The code of C-working program (algorithm part) will be introduced as part of the generated code to complete the C++ program. To complete the object-oriented design, there are four types of diagrams: (a) Dependency diagram (road map of design), (b) Inheritance Diagram(to derive new classes from existing class), (c) Class Diagram(implementation of classes), and Class Structure Chart (implementation of classes). We get warnings if we do not create the first two types of diagrams, but we must draw correctly the class and structured diagrams. While generating C++ code, the user chooses to develop header and class files in a single file or two separate files. If the user assumes separate files the class part will be created as .hxx file and The sample code generated by the class and structured diagrams is given in Appendix B. The user can introduce the algorithm written in C-language and compile and execute the C++ programs. The OOD gives the idea of design of total system and will be helpful for future modifications and maintenance.

6. Observations

Understanding code is an intelligent activity. None of the tools give the complete functionality of the system and what is missing to a specific problem. But the tools help to narrow the problem and speed the process. In the present problem, if the programmer (author) does not have the knowledge of neural network basics, it is very difficult to identify the error. The program is not a commercial package and the error was not intentionally created by the programmer in the book[6]. While converting to object-oriented design, we find that there are many differences between scientific packages and business packages (the author worked on bank account package to understand the object-oriented design). In scientific packages, private and protected has a minor role and business has a major role. In scientific packages, we have class dependency and inheritance is less compared to business software. The object has a more execution part in a scientific program. The object-oriented design is required in scientific packages because of technology changes particularly with graphics outputs and interfaces.

7. Conclusions

Ensemble structured charts are useful to find the global variables and library functions used in a module. Structured charts also help us to use the variables in library functions. Metric reports helps the line coverage, branch coverage, data complexity, and cyclomatic complexity. Static variables and macros are printed clearly. We can come to a conclusion that Ensemble documentation helps in design recovery and OOD tool helps to design to C++. Experience shows that tools have a major role in today's world to perform our work faster and accurately. The Ensemble tool from CADRE Technology has documentation with self explanatory tutorial examples. They have excellent email and telephone support.

8. References

[1] Chikofsky, E. J. and J. H. Cross III. (1990) "Reverse engineering and design recovery: A taxonomy" IEEE software, Jan. pp 13 - 17.

[2] CASE Tools for Reverse Engineering. CASE outlook (1988) Vol. 2, no. 2, pp1-15.

[3] T.A.Corbi. (1989) " Program understanding: Challenge for the 1990s" IBM Systems Journal, Vol. 28, No. 2, pp 294-306.

[4] Oman, P. (1990) " Maintenance Tools" IEEE software, May, p 21-23.

[5] Rumelhart, D.E., Hinton, G.E., and Williams, R.J. (1987) Parallel Distributed processing, Vol. I, MIT press, pp318-364

[6] You-Han Pao. (1989) Adaptive Pattern Recognition and Neural Networks Addition-Wesley Publishing company, Inc.

[7] Piotr Gmytrasiewicz, Jere A. Hassberger, and John C. Lee. (1990) "Fault Tree Based Diagnostics Using Fuzzy Logic" IEEE Trans. On Pat. Anal. and Mach. Int. Vol. 12, No. 11, Nov. pp 1115-1119.

[8] Ted J. Biggerstaff. (1989) Design recovery for maintenance and reuse, IEEE Computer, July, pp 36-49.

[9] Cadre Tech. Inc.; 222 Richmond Street, Providence, RI 02903; Ph - 1-800-548-7645.

[10] Robert S. Arnold (ed) (1993) "Software Re-engineering", IEEE Computer Society Press.

Appendix A

Function summary Report

Ensemble Report:
Metrics

This report contains metrics information about function in the model.
Model='bj1_user'

	run1	run2	run3
Average cyclometric complexity =	6.13	6.0	5.47
Average data complexity =	6.30	6.3	5.91
Number of total lines =	396	390	405
Number of lines executed =	316	347	362
Percent lines executed =	79.80%	88.97	89.38
Number of total branches =	151	147	149
Number of branches executed =	114	127	123
Percent branches executed =	75.50%	86.39%	82.55

WARNING: Metrics that could not be calculated for a function show up with ** for their metric value. For data complexity, check to make sure the function is part of your model. For coverage metrics, check to make sure the function is part of your coverage set. For branch coverage, this could mean there were no branches in the function.

Function Summary Report

file	function	cyclo			data			%line			%branch		
		run1	*run2*	run3	run1	*run2*	run3	run1	*run2*	run3	run1	*run2*	run3
bj.c	dread	4	4	4	11.0	11.0	11.0	0	70	70	0	66	66
bj.c	dwrite	6	6	6	14.0	14.0	14.0	80	80	80	80	80	80
bj.c	forward	6	6	6	8.0	8.0	8.0	100	100	100	100	100	100
bj.c	init	5	5	5	7.0	7.0	7.0	100	100	100	100	100	100
bj.c	initwt	3	3	3	2.50	2.5	2.5	100	100	100	100	100	100
bj.c	introspective	6	6	6	10.0	10.0	10.0	92	92	92	70	80	70
bj.c	learning	3	3	**4**	0.38	0.38	0.44	72	100	100	50	75	66
bj.c	main	7	7	7	0.67	0.67	1.0	82	82	82	55	55	55
bj.c	output_generation	10	*8*	**8**	1.60	1.60	2.67	73	87	84	83	85	85
bj.c	random	1	1	1	2.0	2.0	2.0	100	100	100	**	**	**
bj.c	rumelhart	20	20	20	7.33	7.33	7.33	94	94	94	92	97	92
bj.c	set_up	2	2	2	11.00	11.0	11.0	100	100	100	100	100	100
bj.c	user_session	8	8	8	7.0	7.0	7.0	80	80	80	71	85	71
bj.c	wtread	5	5	5	6.0	6.0	6.0	0	72	72	0	75	75
bj.c	wtwrite	6	6	6	6.0	6.0	6.0	78	78	78	80	80	80
bj.c	ybr1			**1**			**0.5**			**100**			**
bj.c	ybr2			**1**			**4.0**			**100**			**

Appendix B

```
/* .cxx file generated by OOD package */
#include "level1.hxx"
//$$:begin_note code_body
void    level1::learning()
{

}
void    level1::outputgeneration()
{

}
void    level1::ybr2()
{

}
void    level1::user_session()
{

}
void    level1::set_up()
{

}
void    level1::wtwrite()
{

}
void    level1::init()
{

}
void    level1::ybr1()
{

}
void    level1::random()
{

}
void    level1::rumelhart()
{

}
void    level1::initwt()
{

}
void    level1::dread()
{

}
void    level1::usersession()
{

}
void    level1::dwrite()
{

}
void    level1::introspective()
{

}
void    level1::forward()
{

}
void        level1::wtread()
{
}
```

```
/* .hxx file generated by OOD package */
#ifndef level1_hxx_included
#define level1_hxx_included

        class level1
        {
        public:
//$$:begin_note public_decls
        //$$:end_note c

        void    learning();
        void    outputgeneration();
        void    ybr2();
        void    user_session();
        void    set_up();
        void    wtwrite();
        void    init();
        void    ybr1();
        void    random();
        void    rumelhart();
        void    initwt();
        void    dread();
        protected:

        //$$:begin_note protected_decls
        //$$:end_note c


        public:

        //$$:begin_note private_decls
//$$:end_note c

        void    usersession();
        void    dwrite();
void    introspective();
        void    forward();
        void    wtread();
        //The following logical components are declared:
//nteger m;
        //integer i;
        //integer nsample;
        //char ans;
//char dfile;
        };

        #endif          //ifdef level1_hxx_included
//$$:end_pu level1.CD
```

A GENETIC ALGORITHM APPROACH TO SOLVING THE BATTLEFIELD COMMUNICATION NETWORK CONFIGURATION PROBLEM

Tony F. Chang
VerTec Solutions, Inc.
5601 Roanne Way, Suite 606
Greensboro, NC 27409
or
P O Box 49811
Greensboro, NC 27419
tchang@ai.uga.edu

Walter D. Potter
Computer Science Dept and
Artificial Intelligence Programs
University of Georgia
Athens, GA 30602
potter@pollux.cs.uga.edu

Abstract

The problem of finding a low-cost set of communication network components is a very difficult and time-consuming task. This is because of the large number of possible configurations derivable from the numerous types and quantities of components. We developed a genetic algorithm approach to solving the battlefield communications network configuration problem[Chan93, Chan94, Pott92-1, Pott93-1]. A prototype expert module (named IDA-NET) has been developed to perform this task. To further prove that our genetic algorithm approach is an appropriate method towards solving this problem, the result of IDA-NET is compared with the results generated by a general heuristic algorithm (i.e., the hill-climbing search algorithm). The comparison shows that the genetic algorithm approach performed *better* than the other heuristic algorithm.

1. Introduction

The Intelligent Decision Aiding Representation System (IDARS) is a frame-based object-oriented platform that is intended to support the intelligent tutoring of U.S. Army planning personnel in battlefield communications network configuration. U.S. Army personnel configure communications networks composed of Mobile Subscriber Equipment (MSE), which provide communications support for a typical five-division corps in an area of up to 15,000 square miles.

The need for reliable battlefield communications support is essential for a military operation. The communications aspect of a mission is a particularly difficult task because of the numerous dimensions involved such as: types of components needed, quantities of the components, and where the components should be located. The main objective of the Intelligent Training and Decision Aiding for MSE project is to develop a prototype Intelligent Tutoring System (ITS) and Intelligent Decision Aid (IDA) for a Network Planning Facility.

Usually, network planning is done by the personnel who have received intensive training and have some amount of network planning and maintenance experience. Planning an optimal network is the type of problem that is affected by combinational explosion. As the number of network components increases, the number of possible network configurations increases as well. Besides, as a battlefield communications system is a key part of a mission, the system may need

E. A. Yfantis (ed.), Intelligent Systems, 483–496.

to be reconfigured frequently due to the mission dynamics. Thus, there is a need to develop an intelligent computer system that can aid the network planners in configuring the communications network.

In order to achieve these goals, the single most important module of an ITS/IDA, the Expert Module, must be fully developed. The Expert Module provides expertise in the target domain. It solves the problem that the student is learning to solve. Student remediation as well as remediation strategies are based on the variance between the solution provided by the expert module and the solution developed by the student.

A research prototype expert system called IDA-NET (the Intelligent Decision Aid Network Expert Module) has been developed to support intelligent MSE network configuration [Chan93, Chan94, Pott92-1, Pott93-1]. Based on the user-defined mission requirements, IDA-NET determines the types of components needed to support the mission, the quantity of each component type needed, and the locations (i.e., the peaks in the supplied terrain information) where backbone components should be placed in the battlefield in order to support the communication network in the most efficient way. In IDA-NET, we use a two-phase genetic algorithm approach. A genetic algorithm, in short, is a heuristic search mechanism based on natural selection and natural genetics. The first-phase of IDA-NET takes the input of mission requirements (i.e., the number of Mobile Subscriber Radio Telephones (MSRTs), and the number of Direct Secured/Nonsecured Voice Telephones (DSVTs and DNVTs)) to determine the types and quantities of components needed; the second-phase, then, takes the result of the first-phase, along with battlefield terrain data and determines the best locations to place the backbone components. In the first-phase, a "hybrid genetic algorithm" [Davi91], which has an integer-based representation rather than a binary-based representation has been used. Every individual in a chromosome corresponding to the quantities of a network component is needed to support the defined mission. The typical binary representation genetic algorithm is used in the second-phase, but incorporates the use of variable length binary strings. Here, the length is a function of the number of peak locations (i.e., the high ground locations) identified in the battlefield.

In this paper, basic problem of network configuration is discussed in section 2. Section 3 includes an introduction to the Genetic Algorithms (GAs). Section 4 contains the brief status and introduction of the IDA-NET v1.0. The rest of the paper is devoted to a discussion of the development and enhancement of the IDA-NET v2.0 and its results. Moreover, to show that our genetic algorithm approach is an appropriate method towards solving this problem, the result of the IDA-NET v2.0 is compared with the results generated by a hill-climbing search algorithm. Finally, the last section contains a short summary and an indication of future directions for this research project.

2. Problem Complexity

Designing an efficient communication network involves specifying those factors that together produce optimum network performance. This type of problem is affected by combinational explosion. As the number of network components increases, the number of possible network configurations increases as well [Mitt89]. This situation will get worse if more factors must be taken into consideration in the process of network planning, such as network size, device locations, and link speed. These goals may conflict and vary dynamically over time, depending on user needs. Given the fact that network configuration is a type of problem affected by combinational explosion, it is clear that an exhaustive search method which is guaranteed to find an optimal configuration set is impractical. In our application, since a typical MSE five-division corps contains 42 Node Centers and other numerous types of components, the search space size is greater than 3 trillion. Thus, we need a good heuristic algorithm to solve the MSE configuration problem.

Besides, there are two essential factors in planning a network, namely, the cost and the performance. First, we need to select a low-cost, acceptable configuration while satisfying the customer's criteria. Then, we need to select and place (or set up) the components at locations where we can fully utilize their features and achieve the best performance.

The complexity of network planning and configuration arises essentially from the large number of possible solutions that attempt to satisfy all the requirements at the lowest cost. It is extremely difficult to design a network that satisfies the customer's performance criteria while minimizing network cost. The job is handled by human network planners. Network planners, however, need to be highly trained, and have extensive experience with the network components and planning.

The use of expert systems to solve this configuration problem is appropriate due to the fact that designing a complex communication network in most cases relies on expertise more than theoretical knowledge. There are a number of reasons for this [Ceri90]:

- Computer network theory is of little help in the matter of configuration; computer architecture experts are driven by "rules-of-thumb" during their job; these rules can be represented by rule-based systems.
- No standard procedures for network configuration are available. Most of the approaches found in the literature deal with subproblems of limited scope.
- A configuration system should be able to explain the partial and final results when required; expert systems can handle this job.
- Expert systems also can provide some partial or generic solutions; even if not complete, these results are often useful advice for the users.

A pure expert system, however, is not guaranteed to come up with the best configuration. On the basis of the above criteria, it is impractical to use the exhaustive search method to find the best solution. We should instead use a heuristic approach. Methods for solving this problem are not new; good heuristic algorithms have been widely used in many communication network problems. In the development of a heuristic approach, there are two ways: the rule-based approach, and the functional approach. The former uses typical if-then rules; the latter is driven by the knowledge about the structure and behavior of the network and its components (i.e., the "rules-of-thumb"). The major points of emphasis are the heuristic synthesis or design of a satisfactory network. IDA-NET, however, could be considered as a combination of the two approaches.

Many expert systems have been developed for network configuration, planning, management, and troubleshooting in past years. In [Chan93, Chan94-1, Pott93-1], there are more detail discussion of other expert systems which are desired for network configuration. Among them, DESIGNet is the one that employ genetic algorithm to optimize the solution to network linking. More details on Davis's research and other survey papers that deal with network management, design, and planning can be found in [Cron88, Davi91, Eric89, Hieb88].

3. The Genetic Algorithm

Genetic algorithms [Davi91, Gold89, Holl75] are *heuristic* search routines based on natural selection and natural genetics. Genetic algorithms are able to converge on the optimum solution in complex search spaces. The typical representations of the solutions are bit strings -- lists of 0's and 1's. The fitness of an individual is measured by the objective function. The genetic algorithm uses this fitness value to compare an individual with other individuals in the same population.

The basic operations involved in a genetic algorithm are: 1) mate selection, 2) crossover, and 3) mutation. The purpose of mate selection in a genetic algorithm is to give more reproductive chances to those population members that are the most fit [Davi91]. During this stage, strings with a higher fitness value have a higher probability of contributing one or more offspring in the next generation. The function of the crossover operator is to make the next generation of strings differ from those of their parents. A simple crossover operation is to split the parent strings at the same randomly chosen position and swap their right-hand sections. Mutation has an associated probability parameter that is typically quite low. It is the occasional random alteration of the value of a string position. In the binary based representation, that means changing a 1 to a 0 or vice versa. However, there are many variations on those components which enhance the performance of the genetic algorithm [Davi91, Gold89, Jog89, Liep90, Pott90, Pott92-2]. In IDA-NET, we also deployed variations of the operators which will be described in later sections.

4. Development of IDA-NET: The Network Expert Module

In IDA-NET, we use two GA modules (namely GA-1 and GA-2) to generate the optimal or near optimal solution. GA-1 takes the input of mission requirements to determine the types and quantities of components needed to satisfy both the mission requirements and the terrain data restrictions. The consideration of terrain data restrictions is critical in our case. By giving the same set of mission requirements, GA-1 should produce similar answers. However, not all the terrain have the same situations and environments. Thus, without taking terrain data into consideration in mission planning, GA-1 may produce a result which is not suitable to use in certain terrain. The GA-1 uses a hybrid representation scheme [Davi91] where each individual in the population is an integer sequence instead of a binary representation. Each position in the sequence corresponds to a particular component type, and the values of each position represent the quantities of the component types. The range of values is between the minimum and maximum allowable for each component. Because of the fact that we use an integer representation in GA-1, we cannot use a typical crossover technique, such as two-point crossover. In the case of an integer-based string, we can use either uniform crossover or average crossover [Davi91]. Uniform crossover creates new children by randomly selecting sequence elements from either one of the two parent sequences. These random selections are positional so that the child's elements will be guaranteed to fall in the proper range for the corresponding components. On the other hand, average crossover produces a child that is the result of averaging the corresponding positions of two parents. This method also ensures positional range constraint satisfaction; however, after extensive experiments, we found uniform crossover superior to average crossover.

The goal of GA-2 is to determine the best possible locations for the backbone network components in the battlefield in order to support the required mission. This includes Node Center positioning to support multiple radio subscribers, called Mobile Subscriber Radio Telephones (MSRTs), and wire subscribers called Directed Secured/Nonsecured Voice Telephones (DSVTs/DNVTs), within the battlefield. Our objective function in this phase emphasizes ncde center positioning with respect to the terrain, component capabilities, and connective features (i.e., distances) between nodes.

In IDA-NET, we use the modified selection and mutation schemes rather than the standard schemes [Davi91, Rawl91]. The performance of our system has been improved greatly by these two new processes. Under the standard mate selection process (i.e., the simple roulette wheel selection scheme), the search for the location of the chosen slot is performed via a linear search from the beginning of the list. In this case, each selection takes $O(n)$ steps, because on average half of the list will be searched. The run time for mate selection is $O(n^2)$. This is because we need n spins in a generation to fill the population. We were able to revise our approach using a selection method that takes $O(n \log n)$ time [Rawl91]. Instead of using linear search to find the location of the chosen slot, a binary search is used. This method requires $O(\log n)$ steps per spin and n spins. Thus, the overall run time is reduced to $O(n \log n)$.

In a normal mutation process, we would go through every new individual, testing each component to determine whether that value would be changed by mutation. We found a method in [Davi91] which allows us to calculate the location of the components to be mutated; thus we can skip all of those components that we were sure would not be mutated. Figure 4 shows the pseudo code for such a method (modified from the GENESIS package). This method is much more efficient because mutation is so rare (usually, only about 0.03% components are mutated). By applying this method, we were able to speed up the mutation process also, but only by a constant speedup factor in the neighborhood of 97% (based on the mutation probability).

5. Technical Information about GA-1 and GA-2

The objective function of GA-1 [Chan93, Chan94, Pott92-1, Pott93-1] determines the fitness of an individual based on the following issues: (1) satisfying mission requirements; (2) minimizing total components; (3) maintaining proper connectivity support; (4) avoiding any violations of the minimum or maximum constraints of the components; (5) including required components; and (6) optimizing component mix. In short, we attempt to use the minimum available (cheaper) components to achieve the mission requirements. When determining the best solution, we also avoid deploying the more expensive types of components. In addition, our objective function considers the ability to support the entire list of components using the available UHF antenna links. We want to use the exact number of links to support the proposed components. That means the best combination is to provide the exact number of UHF antenna node center support. The more excess links that are available, the lower the fitness, and vice versa. We also follow a 3-to-1 ratio for the small extension node types (SEN-V2:SEN-V1 = 3:1). Thus, any solutions that have ratios larger or smaller than 3:1 will receive lower fitness. The closer to the 3-1 ratio, the higher the fitness.

As mentioned above, the goal of GA-2 [Chan93, Chan94, Pott92-1, Pott93-1] is to determine the best possible locations for the backbone network components in the battlefield. In GA-2, we use a standard binary representation scheme. Based on the given terrain data, we select the major peaks. Thus, the length of our GA population bit string varies in our particular case. Each bit represents whether or not a node center is placed on a corresponding peak (Figure 1). The GA-2 objective function uses the number of node centers produced by GA-1 and the user supplied terrain data to evaluate locations of the node centers. Here, we considered the following factors: (1) the number of available node centers; (2) the number of peaks identified in the terrain; (3) whether or not a peak is in enemy territory; (4) the line of sight condition between peaks; (5) the transmission constraints of the components, and (6) the linkage for multiple node centers (i.e., connecting the backbones). The final GA-2 objective function value, like that in GA-1, is a product of terms relating to the above issues. For more detail information on GA-1 and GA-2, please refer to [Chan93, Chan94, Pott92-1, Pott93-1].

In a 10 peak battlefield where we want to deploy 4 node centers, a possible solution would look like:

0 0 1 0 0 1 1 0 0 1

This means that we should place our 4 node centers on peaks 3, 6, 7, and 10.

Figure 1

An Example Chromosome in the GA-2

6. Hybrid Modifications Of GA & Fine-Tuning Of The Internal GA Operating Parameters

Incorporating various hybrid modifications such as post-evaluation [Pott90] or engineered conditioning [Pott92-2] in order to improve the accuracy and reliability of the solutions produced by GA-based search schemes has been shown to be effective in certain cases. Engineered Conditioning (EC) is a Genetic Algorithm operator that works together with the typical genetic algorithm operators (mate selection, crossover, and mutation) in order to improve convergence toward an optimal solution. To a certain extent, the EC operator is very similar to the hill-climbing mutation scheme [Bram91, Davi91], except that only the dominant individual is conditioned in the population at each generation, and the other GA operators continue to be used. When the EC operator is incorporated with the genetic algorithm, the resulting hybrid scheme produces improved reliability by exploiting the global nature of the genetic algorithm, as well as the 'local' improvement capabilities of the EC operator. In the current implementation of the EC operator, three local search or 'conditioning' tests are performed. The pseudo code for these operators is described in Figure 3. We incorporated this scheme into GA-2. More information on this operator can be found in [Pott92-2].

```
for each individual's element do {                              /* test 1 */
        If it is TRUE, then change it to FALSE;
        test the fitness of the new string, and keep the better one }
for each individual's element do {                              /* test 2 */
        If it is FALSE, then change it to TRUE
        test the fitness of the new string, and keep the better one  }
for each individual's element do (let the counter variable as i) {          /* test 3 */
        If it is TRUE, then change it to FALSE
        for each individual do (let the counter variable as j) {
```

```
            If (i not equal to j) and (individual j == FALSE)
            then test the fitness of the new string, and keep the better one. }
    }
    keep the string with best overall fitness value;
```

Figure 3

Pseudo code for the EC operator (binary representation version)

However, the original EC operator only works with binary representation chromosomes. In GA-1, we thus developed an integer version of the engineered conditioning scheme called "Integineered Conditioning(IC)," which is similar in nature to the binary string conditioning scheme. In our integer conditioning scheme, we make two passes through the best individual in each generation. During the first pass, we increment each component's quantity and test the fitness of the resulting solution. On the second pass, we decrease each component's quantity instead. Then, we keep the best overall solution. In the IDA-NET's integineered conditioning scheme, the component's quantity are increased or decreased by 1; however, the increment or the decrement value can be changed in a more intelligent fashion. For instance, the changing value can be bigger if the objective value is low and vice vera. Figure 4 shows the pseudo-code of the "Integineered Conditioning" scheme:

The control parameters of a genetic algorithm can have a significant impact on its performance. The performance was seen to be sensitive to the combined influences of population size, crossover, and mutation rates and to the number of crossover points used in each mating [Scha89]. In addition to fine-tuning the processing of IDA-NET, various mutation and crossover probabilities are tested. This leads to the conclusion that, generally, in order to get the best performance, 0.7 crossover and 0.03 mutation probabilities should be used for GA-1, and 0.7 crossover and 0.035 mutation probabilities for GA-2. By using these parameters and 1000 as the population size, the results of the prototype are or are very near the best solutions we found through exhaustive search. Larger population sizes gave a higher likelihood of obtaining optimal solutions, but at the expense of longer run time.

```
for each individual do {                                /* first pass */
        increase each component's quantity by 1;
        test the fitness of the new string, and keep the better one }
for each individual do {                                /* second pass */
        decrease each component's quantity by 1;
        test the fitness of the new string, and keep the better one }
keep the string with best overall fitness value
```

Figure 4

Pseudo code for the Integineered Conditioning(IC) Scheme

7. Hill-Climbing Heuristic Approach To This Problem

In order to further evaluate the efficiency of the Genetic Algorithm, the hill-climbing heuristic search algorithm was employed to solve the same problem as in IDA-NET's GA-1. The same objective function is used to ensure an accurate comparison. A Hill-climbing based prototype (named Battlefield Network Configuration--BNC) was developed. As in the GA-1, it takes the input of mission requirements to determine the types and quantities of components needed to satisfy the mission requirements.

The following figure (Figure 5) shows the algorithm of our search mechanism:

*{** modified hill-climbing algorithm **}*
1. set the acceptance_level = 10;
set the stable_level = 10;
2. generate an initial configuration and evaluate it with our objective function until it meets the acceptance_level;
3. go through all the rules and modify the configurations using different operators
4. evaluate the new configurations
5. select the better configurations
6. go to step 3 until no improvements in stable_level

Figure 5
Modified Hill-climbing algorithm

The algorithm is basically a modified hill-climbing algorithm. As there is a huge search space involved, the best-first search algorithm might be inefficient because it takes large amounts of memory and time. The single most important step in the algorithm is step 3. Now let us go through the current operators in our prototype.

The current prototype involved only 2 operators:

1. adder: randomly add values to each component;
2. subtracter: randomly subtract values from each component;

Usually, a normal hill-climbing scheme would encounter several pitfalls, such as a local maximum, plateau, or ridge. In order to avoid those types of problems, when applying the operators to change the solution set, (1) use a big jump if the mission requirement of the objective value is too low; and (2) use random ranges which minimize the chance of the solution being locked in the case of local maximum or ridge.

8. Results Comparison

To get better view on our results, 30 runs with same mission requirements (154 MSRTs and 459 DNVTs) have been conducted on IDA-NET with IC, IDA-NET without IC, and BNC. To ease the discussion, SYSTEM I, II, and III has been used to refer to IDA-NET with IC, IDA-NET without IC, and BNC respectively.

From the result, SYSTEM I has the highest average fitness value of 373.5, while SYSTEM II and III has an average of 306, 244.1, respectively (the optimal solution is 437.7 in this case). Clearly, this lead us to the conclusion that the performance of SYSTEM I is better than SYSTEM II, and SYSTEM II is better than SYSTEM III.

To further prove that there are significant differences between the three SYSTEMs, two ANOVA tables has been computed (Table 1 and Table 2). Table 1 is the ANOVA table based on SYSTEM I, II, and III. It tests whether SYSTEM I, II, and III are equally well designed. From Table 1, the F-test value is 15 while the F(0.95, 2, 87) is 3.1. Thus, we will reject the null hypothesis at the 0.05 level of significance; and conclude that there are significant differences between them. Table 2 shows the ANOVA table conducted on STSTEM II and SYSTEM III. It tests whether SYSTEM II, and III are equally well designed. The F-test value is 6 while the F(0.95, 1, 58) is 4. Again, there are significant difference between SYSTEM II and III.

Since SYSTEM III (BNC) employed the hill-climbing heuristic search scheme, it may run into the local maximum problem, and plateau problem. This explained the results of SYSTEM III that it always has unsatisfied results. Equipped with the reproduction operators, the SYSTEM II (Genetic Algorithm without IC) performed better over the SYSTEM III. As we have mentioned in the previous chapter, IC (Integineered Conditioning) enabled our system to produce improved reliability by performing additional local search. Thus it always leads us to near or very near optimal solutions.

Table 1
ANOVA table for SYSTEM I, II, and III

Anova: Single Factor

SUMMARY

Groups	*Count*	*Sum*	*Average*	*Variance*
GA-1 w/IC	30	11204.8	373.4933333	6254.406851
GA-1 w/o IC	30	9177.8	305.9266667	8242.704782
BNC	30	7323.43	244.1143333	10588.24669

ANOVA

Source of Variation	*SS*	*df*	*MS*	*F*	*P-value*	*F crit*
Between Groups	251249.4464	2	125624.7232	15.02367097	2.48519E-06	3.101291668
Within Groups	727475.3915	87	8361.786109			
Total	978724.8378	89				

Table 2
ANOVA table for SYSTEM II, and III

Anova: Single Factor

SUMMARY

Groups	*Count*	*Sum*	*Average*	*Variance*
GA-1 w/o IC	30	9177.8	305.9266667	8242.704782
BNC	30	7323.43	244.1143333	10588.24669

ANOVA

Source of Variation	*SS*	*df*	*MS*	*F*	*P-value*	*F crit*
Between Groups	57311.46828	1	57311.46828	6.08694344	0.016592272	4.006864174
Within Groups	546097.5928	58	9415.475738			
Total	603409.0611	59				

9. Enhancements and new features of the IDA-NET v2.0

The IDA-NET v2.0 has two major enhancements over the previous version. They are: (1) it is equipped with an explanation engine called Simple Explanation Engine (SEE); and (2) it utilizes real terrain data supplied by the user instead of random-generated terrain data.

IDA-NET's Simple Explanation Engine (SEE)

One of the important features of an expert system is the ability of explaining the decisions of the system. People will not accept results unless they have been convinced of the accuracy of the reasoning process that produced those results. In IDA-NET v2.0, a simple explanation engine (named SEE) was implemented. With this feature, the system is able to explain the results

generated by GA-1 and GA-2. All this information is stored in the .SEE file. The user is allowed to view this file after the operations of GA-1 and GA-2. To read the SEE file, the user can select the VIEW SEE option under the VIEW menu. Figure 6 shows a sample segment of the explanation.

GENERATION # 11
OBJ. VALUE : 331.137557
CONFIGURATIONS:
NC-2 LEN-1 SEN1-9 SEN2-3 RAU-1 NAI-7 SCC-1
CARDINALITY: 476.309469
Each component is scaled based on its component type.
For those components with fewer quantities available,
we give them a higher cardinality term.
This term is calculated based on the following formula:
cardinality-term=(constant)/(component/corps)*percent));
It sums the percentage weights of the total components;
Thus, more components relates to more penalty on an equal basis.
(......)
DNVTVIOLS : 0.790017
PASSED! This configuration is able supported the mission requirements,
However, it supported more than the mission requirements needed (i.e. some penalty).
MSRTVIOLS : 0.88
PASSED! This configuration is able supported the mission requirements,
However, it supported more than the mission requirements needed (i.e. some penalty).

Figure 6

Incorporation of Real Terrain Data

In IDA-NET v1.0, the terrain data used by the system is randomly generated. To make the system more realistic, IDA-NET v2.0 incorporates real terrain data. In order to utilize this new feature, a user needs to provide the map's filename, map scale, number of pixels on X-coordinates and Y-coordinates, and the elevations. After that, the system will load the map data into memory. From now on, the GA-1 and the GA-2 will utilize the terrain data supplied as above to assist its decision making. By clicking the MAP menu, VIEW MAP item, the defined map will be drawn on the screen. After the processes of the GA-1 and the GA-2, the VIEW MAP item will also show the locations of the node centers suggested by the system. Without this feature, IDA-NET may generate results that are not suitable to apply in certain fields. Moreover, real terrain data is an important element in the knowledge base. Figure 7 shows the new flowchart of IDA-NET.

INPUT: ***Mission requirements, and map data***
OUTPUT: *Component list and backbone network*
PROCESS:-Phase 1: ***Analyze the map data, and use mission requirements to determine the ecessary component types and quantities to support the mission and the supplied terrain data.***
-Phase 2: *Use results from phase 1 and terrain data to determine the locations for the backbone network.*

Figure 7

By employing this feature, the important step is to identify the locations of peaks. Technically, to identify peaks on a map is to locate all the local maximums on the map since peaks are always surrounded by other peaks; the highest points are not the peaks. For example, in the following picture (figure 8), the highest three points are: point 3 and 2 other points *right below* point 3. However, the actual peaks should be point 1, point 2 and point 3. All points between line 1 and line 2 cannot be consider as peaks, same as points between line 2 and line 3. Since the purpose of this research project is not to find the correct peaks, a simple algorithm is to used to find the peaks.

1. divide the map into several pieces (a rectangle), in this case 25 rectangles;
2. find the highest point in each rectangle;
3. compare these 25 points and delete the point(s) if they are too close;
4. use the remain points as peaks.

Again, this algorithm is *not* a good method to find the peaks. There are some other algorithms desired solely for this purpose.

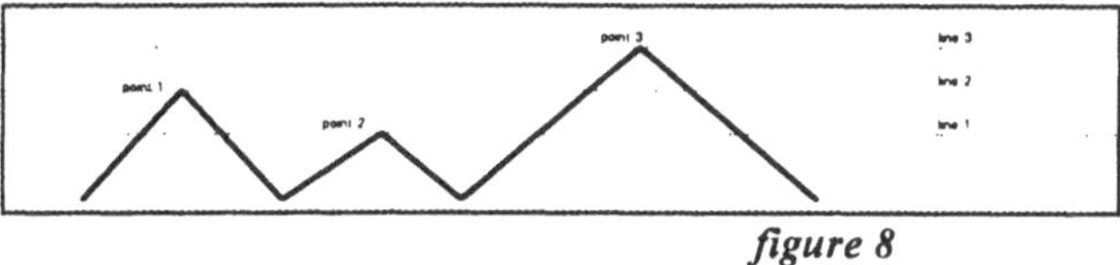

figure 8

10. Conclusion And Future Directions

In this paper, it has been shown that the nature of the network planning and configuration problem can be effectively solved via heuristic search algorithms in a network expert system. The prototype IDA-NET which was developed to aid battlefield communication network configuration has been discussed. IDA-NET uses a two-phase genetic algorithm approach to target this issue. The first phase determines the types and quantities of components needed to satisfy the mission requirements. The second phase takes the results of the first phase and the terrain data to determine the best possible locations for the node centers. After extensive experiments, IDA-NET has produced optimal or near-optimal solutions in the majority of cases. To further verify the result of IDA-NET (GA approach), the result is compared with the hill-climbing algorithm. The results show that the genetic approaches lead us to better solutions than hill-climbing. However, there are still several items that would improve our current prototype system that we feel need to be addressed.

(1) The current IDA-NET system does not have the ability to connect all the components. To evolve IDA-NET into a more complete intelligent decision aid system, the mechanism to connect all the different components is necessary. This would be a very challenging task because of the many (combinatorially explosive) arrangements of links between node centers and other network components. Hugh amounts of knowledge are also required in this case.

(2) The current version of IDA-NET used a simple algorithm to identify the peaks. In order to find the real peaks, another intelligent search algorithm is needed.

(3) Our current prototype only has the capability for node placement of the node centers. One enhancement of IDA-NET is to incorporate a scheme for placement of the other network components. This feature will involve a lot of knowledge on the network components.

(4) Another item is to evolve IDA-NET into a stand-alone intelligent decision-aid system. Thus, the network manager would use IDA-NET to assist in determining the network component compliment and the backbone deployment locations based on current mission requirements without the presence of IDARS. In other words, by enhancing IDA-NET into a stand-alone intelligent decision aid system; IDA-NET can be used in real world situations rather than just in a classroom environment.

References

[Ande88]
Anderson, J.R. (1988). "The Expert Module," *Foundations of Intelligent Tutoring Systems*, (M.C. Polson and J.J. Richardson, eds.) LEA Publishers, Hillsdale, N.J.

[Bach84]
Bachant, J., and J. McDermott (1984). "R1 Revisited: Four Years in the Trenches," *AI Magazine*, Vol. 5, No. 3.

[Ceri90]
Ceri, S., and L. Tanc (1990) "Expert Design of Local Area Networks," *IEEE Expert*, Vol. 5, No. 5, pp. 23-33, October.

[Chan93]
Chang, Tony F, W. D. Potter, P. Gillis, and M. Sanders (1993) "Genetic Algorithms in Battlefield Communications Network Configuration", *Proceedings of the 31st Annual Southeast Conference*, pp. 96-102, April.

[Chan94]
Chang, Tony F., (1994) "Genetic Algorithms in Battlefield Communications Network Configuration", *Masters Thesis*, Artificial Intelligence Programs, University of Georgia, Athens, GA.

[Davi87-1]
Davis L., and M. Steenstrup (1987). "Genetic Algorithms and Simulated Annealing: An Overview," *Genetic Algorithms and Simulated Annealing*, pp. 1-11, Morgan Kaufmann Publishers, Inc.

[Davi87-2]
Davis L., and S. Coombs (1987). "Genetic Algorithms and Communication Link Speed Design: Theoretical Considerations," Proceedings of the Second International Conference on Genetic Algorithms, pp. 252-256.

[Davi89]
Davis L., and S. Coombs (1989). "Optimizing Network Link Sizes with Genetic Algorithms," *Modeling and Simulation Methodology*, pp. 317-331, North-Holland Publishers.

[Davi91]
Davis, L., (ed.) (1991). *Handbook of Genetic Algorithms*, Van Nostrand Reinhold, New York.

[Ferg88]
Ferguson I. A., and D. R. Zlatin (1988) "Knowledge Structures for Communications Networks Design and Sales," *IEEE Network*, Vol. 5, No. 2, September.

[Gill91]

Gillis, P.D., and R.W. Pitts (1991). "A C++ Frame System for Decision Aiding and Training for MSE," *AI Exchange*, Vol. V, No. 2, pp. 10-12, Office of Artificial Intelligence Analysis and Evaluation.

[Gold89]

Goldberg, D.E. (1989). *Genetic Algorithms in Search, Optimization, and Machine Learning*, Addison-Wesley Publishing Co.

[Holl75]

Holland, J.H. (1975). *Adaptation in Natural and Artificial Systems*, Ann Arbor: The University of Michigan Press.

[Jog89]

Jog, P., J. Y. Suh, and D. V. Gucht (1989). "The Effects of Population Size, Heuristic Crossover and Local Improvement on a Genetic Algorithm for the Traveling Salesman Problem," *Proceedings of the 3rd International Conference on Genetic Algorithms*, Morgan Kaufmann Publishing, San Mateo, CA.

[Liep90]

Liepins, G.E., M.R. Hilliard, J. Richardson, and M. Palmer (1990). "Genetic Algorithm Applications to Set Covering and Traveling Salesman Problems," *OR/AI: The Integration of Problem Solving Strategies*, (Brown, ed.).

[Mant86]

Mantelan, L. (1986). "AI Carves Inroads: Network Design, Testing, and Management," *Data Communications*, pp. 106-123, July.

[McDe81]

McDermott, J. (1981). "R1: The Formative Years," *AI Magazine*, Vol. 2, No. 2.

[Payn93]

Payne, Linda W., and W. D. Potter (1993), "Student Modeling in a Network Configuration Intelligent Tutoring System, *Proceedings of the 31st Annual Southeast Conference*, pp. 119-126, April.

[Pott90]

Potter, W.D., J.A. Miller, and O.R. Weyrich (1990). "A Comparison of Methods for Diagnostic Decision Making," *Expert Systems with Applications: An International Journal*, vol. 1, pp. 425-436, 1990.

[Pott92-1]

Potter, W.D., P. Gillis, R. Pitts, J. Young, and J. Caramadre, "IDA-NET: An Intelligent Decision Aid for Battlefield Communications Network Configuration," *Proceedings of the Eighth IEEE Conf. on Artificial Intelligence for Applications (CAIA'92)*, pp. 247-253, March, 1992.

[Pott92-2]

Potter, W.D., J.A. Miller, B.E. Tonn, R.V. Gandham, and C.N. Lapena (1992). "Improving the Reliability of Heuristic Multiple Fault Diagnosis Via The Environmental Conditioning Operator," *International Journal of Applied Intelligence*.

[Pott93]

Potter, W. D., F. L. Chang, P. Gillis, and M. Sanders, "An ITS Expert System Module to Aid Battlefield Communication Network Configuration", *Proceedings of the Sixth International Conference*, Edinburgh, Scotland, June, 1993.

[Rawl91]

Rawlins, G.J.E., (ed.) (1991). *Foundations of Genetic Algorithms*, Morgan Kaufmann Publishers, San Mateo, CA.

[Zhan88]

Zhan, W., S. Thanawastien, and L. M. L. Delcambre (1988). "SIMNETMAN: An Expert Workstation for Designing Rule-Based Network Management Systems," *IEEE Network*, Vol. 5, No. 2, pp. 35-42, September.

EVOLVING CASCADE-CORRELATION NETWORKS FOR TIME-SERIES FORECASTING

J.R. McDONNELL and D.E. WAAGEN†
NCCOSC, RDT&E Div., San Diego, CA 92152
†Systems Integration Group, TRW, Ogden, UT 84403
mcdonn@cod.nosc.mil dwaagen@ow009.bmd.trw.com

Abstract. This investigations applies evolutionary search to the cascade-correlation learning network. Evolutionary search is used to find both the input weights and input connectivity of candidate hidden units. A time-series prediction example is used to demonstrate the capabilities of the proposed approach.

Keywords. Evolutionary Search, Cascade-Correlation Learning Architecture, Time-Series Models, Neural Networks.

1 Introduction

Time-series forecasting is concerned with making future predictions about a process based on its past behavior. Real applications exist in many fields including economics, engineering, and science. Since linear time-series models are not always appropriate when attempting to model an unknown system, a multitude of nonlinear modeling techniques have been proposed (e.g., Casdagli and Eubank, 1992). For example, artificial neural networks are one common class of nonlinear time-series models as discussed by Cichocki and Unbehauen (1994). However, one drawback with applying a purely nonlinear modeling approach to an unknown system is that a linear process may not be easily represented in a nonlinear model (Weigend and Gershenfeld, 1993). The cascade-correlation learning architecture (CCLA) developed by Fahlman and Lebiere (1990) represents a modeling approach which accommodates both linear and nonlinear structure within a single model. The presented work discusses how evolutionary search can be used for generating traditional CCLAs as well as pruned CCLAs. The benefits of pruning the CCLA during construction are realized in a parsimonious model with potentially better generalization capabilities.

The present investigation is concerned only with auto-regressive (AR) type models. A linear AR model for next-step predictions is given by

$$\hat{x}(k) = \sum_{i=1}^{M} a_i x(k-i) + e(k)$$

where M is the order of the AR model, a represents the coefficient associated with each tapped-delay, and e is the forcing function or a white noise term. Evolutionary search has been previously applied by D. Fogel (1991) for determining both the model order and coefficients of linear time-series models. A predictive, nonlinear AR model is described by

E. A. Yfantis (ed.), Intelligent Systems, 497–506.

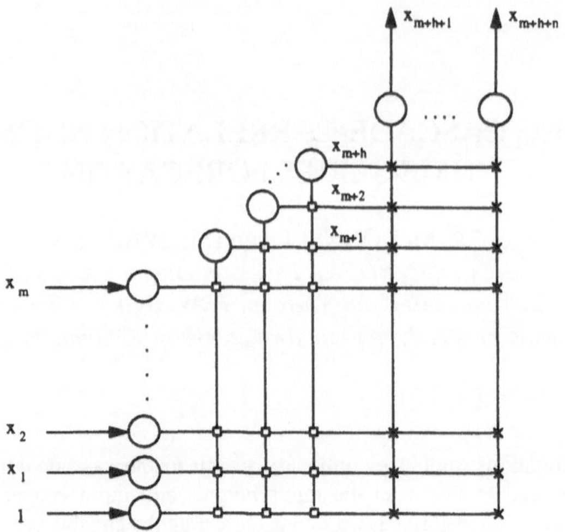

Figure 1. The cascade-correlation network structure. The boxes □ indicate weights which are frozen when the unit is incorporated into the network. The crosses × indicate weights which are modified after a new hidden unit is incorporated.

$$\hat{x}(k) = f(x(k-1),\dots,x(k-M)) + e(k)$$

where $f(\ )$ is the nonlinear mapping. The present investigation is concerned with the combined linear+nonlinear AR model

$$\hat{x}(k) = \sum_{i=1}^{M} a_i x(k-i) + \sum_{j=1}^{N} b_j f_j(x(k-1),\dots,x(k-M)) + e(k)$$

which may incorporate more than one nonlinear component. This is the essence of the CCLA.

2 The Cascade-Correlation Learning Architecture

The cascade-correlation learning algorithm was introduced by Fahlman and Lebiere (1990) as a means of automatically constructing feedforward networks by adding hidden units until a desired mapping is achieved. The network is initialized with only the input units mapped directly to the output units. New hidden units are added on an individual basis (one-by-one) into the network until a termination criterion is met. Each new hidden unit is selected from a pool of candidate units and is incorporated into the network with its input weights frozen. Using a pool of candidate units increases the likelihood of finding a good hidden unit by reducing the susceptibility to poor initial conditions. The goal of each candidate unit in the population pool is to maximize the correlation between its output and the residual output error of the network. This is the correlation aspect of the cascade-correlation algorithm. Each newly incorporated hidden unit is connected to *all* input and previously generated hidden units in the network as shown in Figure 1. Likewise, each newly incorporated hidden unit is also fully-connected to all of the output units. After each hidden unit is inserted into the network, training takes place on the output layer weights with the hidden unit weights remaining fixed.

3 Evolutionary search

Evolutionary search algorithms are based on the paradigm of natural evolution as an optimization technique. Evolutionary algorithms (EAs) encompass a variety of multi-agent stochastic search techniques including evolutionary programming (L. Fogel et al., 1966; D. Fogel, 1991; D. Fogel, 1992), evolution strategies (Schwefel, 1981; Bäck and Schwefel, 1993), and genetic algorithms (Goldberg, 1989; Holland, 1992). The presented work employs a modified version of evolutionary programming (EP) by adding a recombination operator to the EP paradigm. The resultant algorithm is very similar to an evolution strategy (ES). A general evolutionary search algorithm is given in Figure 2.

In 1958, Brooks described a *creeping random* method where k points were generated via Gaussian perturbations about a search point. The best point was kept and the process repeated. Brooks (1958) observed that "there are some rather intriguing analogies that can be made between the creeping random method and evolution." This analogy was also apparent to L. Fogel et al. (1966) who applied a random search strategy termed *evolutionary programming* (EP) to the optimization of finite state machines. More recently, D. Fogel (1991; 1992) has extended the EP paradigm to address combinatorial and real-valued function optimization problems as well as applied it to a variety of problems in system identification and control.

```
evolutionary search procedure
begin
k=0
initialize parent population P(k)
evaluate P(k)
do {
        generate offspring O(k) from P(k)
        evaluate O(k)
        select P(k+1) from {P(k) ∪ O(k)}
        k=k+1
} while (terminate condition not met)
end
```

Figure 2. *An evolutionary search algorithm.*

4 Neural network complexity

Parsimonious networks are less susceptible to overfitting a sample data set and thereby usually result in neural networks with better generalization capabilities. Techniques employed for generating parsimonious structures include increasing network size by the addition of hidden units/layers or pruning an oversized, trained network to yield a smaller network. The CCLA falls into the former class of approaches. Since it is desirable to construct parsimonious architectures, a cost or objective function is formulated which incorporates both system performance and model complexity.

The common form of such a risk function is given by Haykin (1994) as $R(w)=E_s(w)+\lambda E_c(w)$ where E_s is the standard performance measure, E_c is the complexity penalty, and λ is the regularization parameter. The standard performance measure is usually the sum-squared error. As noted by Haykin (1994), there exists a similarity between the risk function $R(w)$ and the composition of statistically derived complexity terms like the minimum description length (MDL) complexity criterion formulated by Rissanen (1986). Initially, $\lambda=0$ so that performance is not sacrificed at the expense of complexity. The regularization parameter λ may be adapted during the training process using a technique described by Weigend et al. (1992).

5 Evolving cascade-correlation architectures

Evolving the weights and connectivity structure of a population of candidate hidden units in a cascade-correlation network is computationally less demanding than evolving the weights and connectivity of a population of neural networks. A valid criticism of this approach is that in exchange for these computational savings, one is solving only small parts of a large problem and thereby reducing the benefits associated with using global search methods on large scale optimization problems. However, the combinatorial optimization capabilities of evolutionary algorithms make them an appropriate search for determining the input weights and connections of the candidate nodes.

A population of candidate hidden units is randomly initialized with full connectivity as in the standard CCLA. The same optimization objective (the absolute value of the covariance between the hidden unit and the residual error) used to train CCLA units is employed to represent the fitness of each candidate node

$$S(a_i) = \sum_o \left| \sum_p (z_{i,p} - \bar{z}_i)(e_{p,o} - \bar{e}_o) \right|$$

where z_p represents the output of each candidate node for pattern p and e is the residual error as measured at the network's output unit o. The weight vector for each candidate unit is modified using standard EP (D. Fogel, 1991)

$$w'_{i,j} = w_{i,j} + \sqrt{Sf / S(a_i)} \cdot N(0,1)$$

where Sf represents the arbitrarily selected scaling factor. Recombination takes place according to

$$\mathbf{w}_k = \mathbf{w}_i + \alpha(\mathbf{w}_j - \mathbf{w}_i)$$

where the indices and scaling coefficient are selected at random such that $i,j \in \{1,...,\mu\}$, $i \neq j$, and $\alpha \sim U(0,1)$. Selection is deterministic so that a new parent set is formed from the best μ members of the set comprised of the original parents, the mutated offspring, and the recombined offspring. After an arbitrary number of generations, the best candidate unit is inserted into the network.

The objective function which incorporates the unit's complexity is formulated using the standard CCLA performance measure $S(a_i)$ in the risk function $R(\mathbf{w})$. Thus the complexity cost retains the general form of the risk function and is given by

$$\Phi(a_i) = -S(a_i) + \lambda E_{c,N}(\mathbf{w},k)$$

where $E_{c,N}(\mathbf{w},k)$ represents the complexity cost for N samples and k parameters. The MDL complexity term $E_{c,N}(\mathbf{w},k)=0.5 \cdot k \log N$ was employed since the connections are strongly specified. The variable complexity regularization parameter λ is tied to the risk Φ of the best candidate node in the population at each generation. No degree of optimality is implied by the given formulation of $\Phi(a_i)$. Although not yet implemented, a reasonable stopping criterion would be to test the current model performance and the newly generated model performance on a validation set. When the current model has superior performance to the newly created model, training is stopped and no more units are added.

All candidate nodes are initially fully connected. Diversity is introduced in the mutation step of the algorithm by randomly selecting an input connection to a candidate node and flipping its bit (i.e., 1→0 or 0→1). Recombination of the connectivity arrays is limited to the bitwise logical 'AND' and 'OR' operators as illustrated in Figure 3. The AND operation reinforces common connectivity structures while the OR operation retains any connectivity which exists between two randomly selected candidate nodes. Structural modifications occur less frequently than weight perturbations as suggested by Yao (1993) who advises that different time-scales should be applied at different levels of evolution. That is, variable connectivity should be considered a larger evolutionary step and occur less frequently than a smaller evolutionary step such as weight perturbations.

The output weights are found deterministically. The optimal (in a least-squares sense) output weight set is determined using the pseudoinverse. Iterative deterministic methods such as the LMS rule are also appropriate for determining the weights from the hidden units to the output units.

Parent Node 1	Operation	Parent Node 2		Offspring
[1111 0010]	AND	[1110 0110]	→	[1110 0010]
[1111 0010]	OR	[1110 0110]	→	[1111 0110]

Figure 3. *Bitwise logical operations applied to the candidate node connectivity strings.*

6 A Time-Series Modeling Example

Evolutionary learning was applied to the CCLA in an effort to model the sunspot time-series data set. This data set has served as a benchmark for a variety of statistical and neural network models. The average relative sunspot number represents a daily mean value taken from up to fifty observing stations throughout the world. The sunspot data set is typically broken down into a training set (years 1700-1920) and two test sets (years 1921-1955 and 1956-1979). This distinction results from the different statistical characteristics between the two test sets. The normalized MSE (NMSE) is used to evaluate the performance of the model which generates next step predictions. The NMSE is given by

$$NMSE = \frac{\sum_{i=1}^{N} (y_i - \hat{y}_i)^2}{\sum_{i=1}^{N} (y_i - \bar{y})^2}$$

where $\hat{y}$ is the single-step prediction and $\bar{y}$ is the mean of the target values. An *NMSE=1* implies that the estimate is just the average of the target values. For the sunspot data set, the NMSE is referenced as the MSE of each data segment scaled by the variance of the full data set

$$NMSE = \frac{1}{N \cdot \hat{\sigma}_{all}^2} \sum_{i=1}^{N} (y_i - \hat{y}_i)^2$$

Pruning was not employed in the initial CCLA-EP experiment which consisted of 100 parents (candidate nodes) and 100 evolutionary training iterations. The experiment was arbitrarily stopped after the addition of three hidden units. We note that as additional hidden units are added, the performance of each model on a validation set could be used as a termination criterion. This idea is easily implemented since a new model is built on the existent one. Table 1 shows that the performance of the relatively more complex model (as demonstrated by the higher number of parameters) is comparable to results generated using regularization techniques in other investigations.

The next set of CCLA-EP experiments incorporated pruning during network construction and consisted of 50 parents with 100 evolutionary training cycles. Training was arbitrarily stopped after four hidden units were added. Again, a validation set could have been used as a stopping criterion. For the pruned network, good generalization occurs on both test sets with roughly half as many parameters as the unpruned CCLA-EP network as given in Table 1. Figure 4 shows the test and training results for the CCLA with complexity regularization.

Table 1. NMSE Performance and complexity of various sunspot models.

	NMSE Train: 1700-1920	NMSE Test: 1921-1955	NMSE Test: 1956-1979	No. of parameters
Tong & Lim (1980)	0.097	0.097	0.28	16
Weigend et al. (1992)	0.082	0.086	0.35	43
Svarer et al. (1993)	0.090	0.082	0.35	12-16
Deco et al. (1994)	0.091	0.087	0.32	n/a
CCLA-EP (3 hid. nodes, not pruned)	0.084	0.082	0.36	58
CCLA-EP (4 hid. nodes, pruned)	0.094	0.083	0.25	27*

** 25 if output units are pruned. See text.*

The linear (AR) portion of the model describes most of the dynamics of the sunspot series. The linear model is augmented by the nonlinear (sigmoidal) components of the CCLA as shown in Figure 5. The weights corresponding to the pruned architecture are given in Table 2. The 10 year difference between the $x(k-2)$ and $x(k-12)$ inputs to the first hidden unit corresponds with the cycle of maximum power spectral content for the first 280 years (1700-1979). More recent observations (1980-1987) of sunspot activity indicates that the maximum power spectral content corresponds to an 11 year cycle for the first 288 years (1700-1987).

Additional pruning may take place on the output vector. For example, zeroing out the two smallest output weights listed in Table 2 yields the same performance values given in Table 1 and reduces the model complexity to 25 parameters.

7 Conclusion

Pruned cascade-correlation learning architectures represent a combined linear+nonlinear time-series modeling technique with potentially better generalization capabilities. Evolutionary search is appropriate for simultaneously determining both hidden unit input structure and weights. The reader is cautioned not to draw any statistical conclusions for the presented example since Marple (1987) points out that time-series models "from short data records is a difficult problem in general." Additionally, the shift in the maximum power spectral energy content cycle observed for the example data set implies a time-invariance which is not addressed with the proposed modeling approach.

8 Acknowledgments

The authors gratefully acknowledge the support of Dr. Al Gordon, NRaD Deputy for Science. Special thanks to David Fogel, Tom English and the anonymous reviewers for their comments and suggestions.

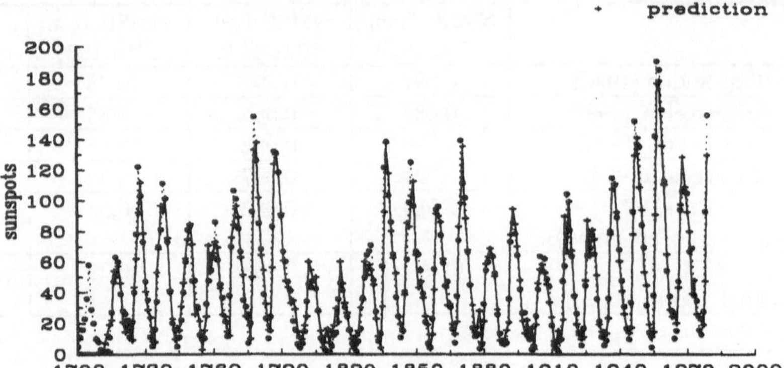

(a)

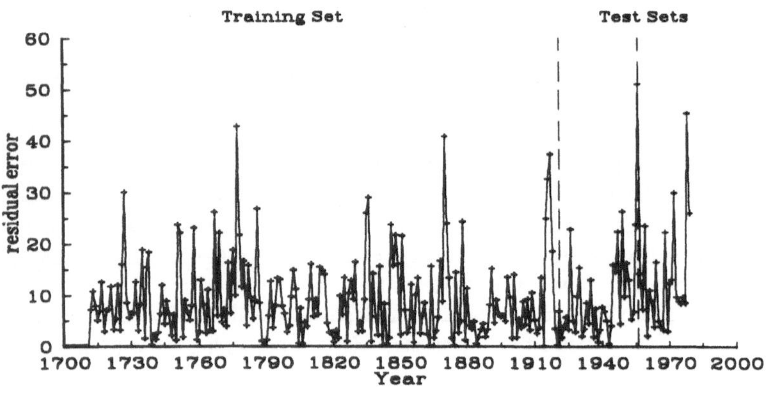

(b)

Figure 4. *Results of the pruned, 4 hidden node CCLA with evolutionary learning. (a) An approximation of the sunspot data over the years 1712-1979 and (b) the magnitude of the residual error for each year.*

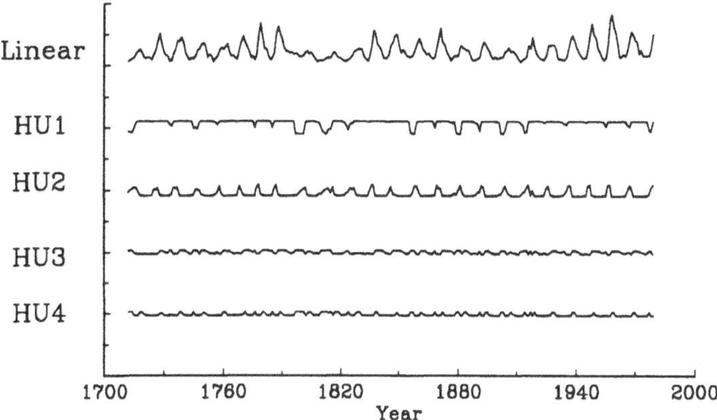

Figure 5. *The relative contributions of the linear portion and each nonlinear component of the pruned CCLA. The hidden units are designated by HU and the order in which they were incorporated.*

Table 2. The hidden unit and output weights for the pruned CCLA sunspot model.

HU Wts	*x(k-2)*	*x(k-8)*	*x(k-11)*	*x(k-12)*	HU1	HU2	HU3
HU1	11.318	-	-	-34.448	-	-	-
HU2	-	-	25.289	-13.869	-	-	-
HU3	12.231	-	-	-		8.719	-
HU4	-	22.981	-	-	-8.998	-11.681	-

Output Wts	bias	*x(k-1)*	*x(k-2)*	*x(k-3)*	*x(k-4)*	*x(k-5)*	*x(k-6)*	*x(k-7)*
	0.0524	-0.086	0.171	-0.062	0.129	-0.001	0.028	0.001
x(k-8)	*x(k-9)*	*x(k-10)*	*x(k-11)*	*x(k-12)*	HU1	HU2	HU3	HU4
-0.052	0.143	-0.159	-0.247	0.878	-0.103	-0.091	0.027	-0.029

9 References

Bäck, T., and Schwefel, H.-P. (1993). "An overview of evolutionary algorithms for parameter optimization," *Evolutionary Computation*, Vol. 1, No. 1, pp. 1-24.

Brooks, S.H. (1958), "A discussion of random methods for seeking maxima," *Operations Research*, 6, pp. 244-251.

Casdagli, M., and Eubank, S. (eds.) (1992), *Nonlinear Modeling and Forecasting*, Santa Fe Institute XII, Addison-Wesley.

Cichocki, A., and Unbehauen, R. (1993), *Neural Networks for Optimization and Signal Processing*, John Wiley & Sons.

Deco, G., Finnoff, W., and Zimmermann, H.G. (1994). "Unsupervised mutual information criterion for elimination of overtraining in supervised multilayer networks," submitted to *Neural Computation*.

Fahlman, S.E., and Lebiere, C.L. (1990). "The cascade-correlation learning architecture," Technical Report CMU-CS-90-100, Carnegie Mellon University, Pittsburgh, PA.

Fogel, D.B. (1991). *System Identification through Simulated Evolution: A Machine Learning Approach to Modeling*, Ginn Press.

Fogel, D.B. (1992). Evolving Artificial Intelligence, Ph.D. Dissertation, University of California, San Diego.

Fogel, L.J., Owens, A.J., and Walsh, M.J. (1966). *Artificial Intelligence through Simulated Evolution*, John Wiley & Sons.

Goldberg, D.E. (1989). *Genetic Algorithms in Search, Optimization and Machine Learning*, Addison-Wesley.

Haykin, S. (1994). *Neural Networks: A Comprehensive Foundation*, MacMillan.

Holland, J. H. (1992). *Adaptation in Natural and Artificial Systems*, MIT Press.

Marple, S.L. (1987). *Digital Spectral Analysis with Applications*, Prentice-Hall.

Rissanen, J. (1986). "Stochastic complexity and modeling," *The Annals of Statistics*, Vol. 14, No. 3, pp. 1080-1100.

Schwefel, H.-P. (1981). *Numerical Optimization of Computer Models*, John Wiley & Sons.

Svarer, C. , Hansen, L.K., and Larsen, J. (1993). "On design and evaluation of tapped-delay neural network architectures," *IEEE Int. Conf. on Neural Networks*, San Francisco.

Tong, H. and Lim, K.S. (1980). "Threshold autoregression, limit cycles and cyclical data," *Journal Royal Statistical Society B*, Vol. 42.

Weigend, A.S., Huberman, B.A., and Rumelhart, D.E. (1992). "Predicting sunspots and exchange rates," in M. Casdagli and S. Eubank (eds.), *Nonlinear Modeling and Forecasting*, Addison-Wesley.

Weigend, A.S., and Gershenfeld, N.A. (1993). "The future of time-series: learning and understanding," in A.S. Weigend and N.A. Gershenfeld (eds.), *Time-Series Prediction: Forecasting the Future and Understanding the Past*, pp. 1-70, Addison-Wesley.

Yao, X. (1993). "A review of evolutionary artificial neural networks," *International Journal of Intelligent Systems*, Vol. 8, No. 4, pp. 539-567.

ADAPTIVE USER MODELS FOR INTELLIGENT INFORMATION FILTERING

KENRICK J. MOCK
Department of Computer Science
University of California at Davis, California, 95616
mock@cs.ucdavis.edu

V. RAO VEMURI
Department of Applied Science
University of California at Davis, Livermore, California, 94550
vemuri@icdc.llnl.gov

Abstract. As networked systems grow in size, the amount of data available to users has increased dramatically. The result is an information overload for the user. In this project, an intelligent information filtering system reduced the user's search burden by automatically eliminating incoming data predicted to be irrelevant. These predictions are learned by adapting an internal user model which is based upon user interactions. This report describes the information filtering problem and examines three techniques for filtering information: global hill climbing, genetic algorithms, and preliminary work with neural networks using radial basis functions.

1 The Information Overload Problem

With the advent of networked systems, computer users are inundated with information that they cannot efficiently utilize. Tools are urgently needed to assist the user with information filtering devices in order to reduce the user's search burden. This project examined the Usenet News system as a testbed for the filtering algorithm. In the Usenet system, users throughout the world intermittently post articles to a common bulletin board. The number of articles posted may be very large; e.g., newsgroups may receive hundreds of articles daily. The goal is to predict whether new articles are likely to be of interest, or not of interest, based upon the prior behavior of the user. This is an extremely fuzzy and difficult problem to define because users are notorious for their inconsistency in their behavior patterns and changing interests.

One of the difficult constraints imposed by this type of problem is the necessity for incrementality. Many learning algorithms, such as those based on neural networks, require repeated training epochs over a fixed data set. In the Usenet News problem, the data set is constantly changing as incoming messages are posted. To ensure consistency, the method would need to store all messages that were ever posted. When new messages arrive, the system would need to retain the old as well as the new messages. This is clearly undesirable due to the time requirements for training and the space required to store all messages. Many approaches to the information filtering problem bypass this problem by typically forcing the user to explicitly define what should be filtered, e.g. via a keyword-based database language [1].

E. A. Yfantis (ed.), Intelligent Systems, 507–516.

2 Previous Work

In addition to the keyword approaches, neural networks may also be used to classify incoming articles. Eberts examined the approach of feeding words into a feedforward backprop network [2]. In his tests, the article headers were sufficient to correctly classify articles in most instances. This approach did not address the problem of a dynamically changing data set described in the previous section, but only tested upon a static store of messages. Another neural net implementation was examined by Jennings and Higuchi [3] in an associative network which associated weights with other words to arrive at an overall interest "activation" for each article.

A hill climbing and genetic algorithm approach through competitive agents for information filtering has also been investigated by Baclace, Sheth and Maes, and Stevens [4,5,6]. In their approaches, agents classify news messages depending on terms from the header or body fields. As the system is used over time, the agents will model the frequency of textual patterns which appear in the articles which are read. Although effective, exploration to other messages which may be of interest are limited by using the same user model on other newsgroups, unless the user explicitly directs the system to explore new messages and topics.

The genetic system described here is similar to the work in [4,5,6], but learning is automatic, user-modifiable, and actively explores different topics within a newsgroup.

3 Usenet Data

The data structure for a typical Usenet post is shown below. Entries exist for the subject, author, keywords, and other references within the header of each message. Some of these fields are shown below:

Newsgroup: comp.ai **Subject**: Genetic algorithms
Author: mock@ucd **Body**: ... (body of message)

Throughout this project, the words in the **Subject** and **Author** fields were extracted from each article as features for classification. Future work will include using text extracted from the body of each article.

4 Global Hill Climbing

A simple method to automate the filtering process is to memorize features extracted from each article, and then perform hill climbing on those features by counting the number of times the feature is accepted and rejected. As the user reads messages, she indicates whether or not each message read was accepted or rejected. This outcome is used to increment the weights accordingly. An example is shown in Table 1. Here, the feature "genetic" has appeared in five accepted articles, the feature of the author "hoyle@ucdavis" has appeared in three accepted articles and one rejected article, etc. This data indicates an interest in articles posted by hoyle or containing the word "genetic", and a disinterest in articles containing the word "flames". The table grows or is updated as new articles are read.

In addition to using words from the articles as features, the system is also capable of using feedback from other users as features. These other users are local users running the same news system who are willing to share their own "accepted" and "rejected" reviews with others. As the user reads messages, the outcome is compared to the reviews made by others, and the table is updated as before. In Table 1, the other user "Michelle" has accepted four articles the current user has accepted, and Michelle has rejected one article the current user has accepted. Similarly, Michelle has rejected two articles the current user has accepted, and rejected three articles the current user has rejected. This table indicates that the current user's accepted messages strongly correspond with Michelle's accepted messages, while the current user's rejected messages slightly correspond with Michelle's rejected messages. In this fashion, filtering is possible in a collaborative fashion as in [1].

Word	Accepted	Rejected
genetic	5	0
algorithm	3	3
flames	2	7
hoyle@ucdavis	3	1
Michelle Accepted	4	1
Michelle Rejected	2	3

Table 1

Given such a table, classification of new messages is performed by extracting the features from the new article and then computing the sum of all the Accepted and Rejected values from matching features in the table. If the percentage of accepted values exceeds A, the message is classified as being of interest. If the percentage is less than B, the message is classified as being of no interest. Messages in between are marked unknown. In this project, A was set to 0.7 and B to 0.3 so that the range of percentages could be divided fairly evenly.

5 Local Genetic Hill Climbing

The local genetic hill climbing method employed in this project is similar to the global hill-climbing method except instead of a single dynamic table there is a population of many tables, where each table constitutes an individual, and each individual performs its own hill climbing as well as genetic crossover [7]. In this project, the table size was set to 20, allowing each individual to identify combinations up to 20 features. Most articles consist of only 4 or 5 features; consequently, each individual has the capacity to represent many different types of messages. These feature tables comprise the chromosome of each individual. The entries in each table are initialized to features selected randomly from a set of 100 previously stored messages, and the accepted and rejected values initialized to small random numbers between 0 and 5. In addition to the table, each chromosome also consists of a history variable which counts the number of times a particular individual has made a correct, or incorrect, prediction. This variable is initialized to zero.

5.1 HILL CLIMBING COMPONENT

The classification procedure first finds the single individual who best matches the input. This is determined by computing how many features from the input are present in the individual. The individual with the most matching features is selected as most representative of the input article. If at least C percent of the features match, then the accepted and rejected values from the matching features of this single individual are added and used to compute a prediction in a manner identical to that of the global hill climbing approach. If the best individual doesn't match at least $C\%$, it is deemed too distant from the input to make an accurate prediction, and the prediction is set to Unknown. In this project, C was set to 50%.

The hill-climbing learning procedure is outlined next. Each new article read by the user is classified as accepted or rejected. If rejected, then the rejected values of all features matching the input are incremented in the tables of all individuals. Similarly, if accepted, then the accepted values of all features matching the input are incremented in the tables of all individuals. This is a global update similar to the global hill climbing approach.

The next step is to update the individuals with greater than $C\%$ of its features matching the features of the message. The prediction of all of these individuals is computed; if the prediction equals the actual value, then the history variable is incremented. If it is incorrect, the history is decremented. This history variable stores the number of times a particular individual has been correct, and will be used in the genetic algorithm fitness function.

The final step is to create new individuals if necessary. If the best matching individual has a match percentage less than C, then no individual exists which matches the input and one must be created. The individual with the smallest fitness and lowest match percentage is selected and its features set to those of the input. In addition to exactly matching new messages which are far away from existing individuals, this process also adds new features into the gene pool for the genetic algorithm component of the system.

5.2 GENETIC ALGORITHM COMPONENT

After the user finishes reading messages, the genetic algorithm component goes into effect. As discussed in section two, it is not practical to store all previously read messages for training. Instead, the system trains in batches, using only the set of messages read in the previous session. With a fitness function based upon these messages alone, it is possible that individuals which classified well on old messages will classify poorly on the new message set. Consequently, it is possible that potentially good individuals which performed well on old messages will be given an unfair fitness. These individuals may then be left out of mating, or may be destroyed during the crossover process.

To help make the fitness more fair to these old individuals, the history variable is included into the fitness calculations. This will bias the fitness by including the previous performance of each individual in the past. For each message in the set of articles read, a prediction is computed using the prediction procedure given in section 5.1. The fitness of each individual is given by the history variable plus the sum of the correct predictions minus the sum of the incorrect predictions.

The crossover operation uses Fitness Proportionate Reproduction [7] to select two parent individuals from the population. The actual crossover operation is two point crossover; two indices are selected at random from one of the parents, and all features and values within this selection are switched among parents. The history variables for both parents are then set to the average of the history value of the two parents.

The mutation operation works by selecting a single individual at random and then either adding a positive or negative random value to one of the feature values, or by selecting a new feature at random from the set of new messages and randomly replace an old feature.

6 Comparison of Global Hill Climbing to Local Hill Climbing/Genetic Method

The global hill climbing method's main strength lies in its simplicity and predictive abilities for features which have been previously encountered. In this case, the system will build a compact representation with no need to worry about destroying potentially good individuals, as is the case in the genetic method. However, the genetic method is actually more powerful than the global method. The global method is unable to discern fine differences in features because it linearly combines all input features; e.g., if we are not interested in messages with the features "case-based" and "algorithms" but we are interested in messages with the features "genetic" and "algorithms", then the global method will be using the same accepted and rejected values for the word "algorithms" and may be unable to correctly classify one or both classes of messages. On the other hand, the genetic approach can simply create two separate individuals which can classify both classes of messages, resulting in a non-linear mapping from input to classification. Furthermore, via crossover, the genetic approach will also tend to explore new areas of the search space which would not normally be examined through a strict hill climbing approach. Due to the greater complexity and generality of the genetic approach over the global hill climbing method, training for this scheme should also be more difficult.

7 Experimental Results

The global hill climbing algorithm and the local hill climbing genetic algorithm were both tested on a set of 100 messages extracted from the comp.ai artificial intelligence newsgroup. A user read all of these messages and marked each as accepted or rejected. The first 50 messages were used for training, and the system predicted the users' choices for the rest of the unread messages. These results do not incorporate collaborative data which would be collected when multiple users share reviews with each other; i.e. the system was run with only a single user.

Both the global and local hill climbing and genetic algorithm schemes perform global updates after each message is read. When using the genetic algorithm, crossover was performed using the batch of previously read messages; i.e. there are three executions of the genetic component, one using the first 15 messages to determine fitness, the next operating upon the next batch of 15, and finally the last operating upon the next batch of 20 messages. The number of generations varied from 0 to 5, the probability of mutation was set to 2%, and the population size was set to 50. Note that the data set is small and the data space is extremely large; an experiment with larger data sets to more accurately represent the data space will likely produce more valid results.

7.1 GLOBAL HILL CLIMBING

A summary of the results for the global hill climbing method is shown in figure 1 below. As expected, the percentage of correctly classified unread messages increases as more messages are read and more information is known about the interests of the user, ending at 52% correctly classified after 50 messages are read. The percentage of incorrect messages remains fairly low (including both false positives and false negatives) at about 5%, while the percentage of unknown messages decreases quickly from a large value to approximately 46%. Global hill climbing ended up giving better overall results than the genetic scheme.

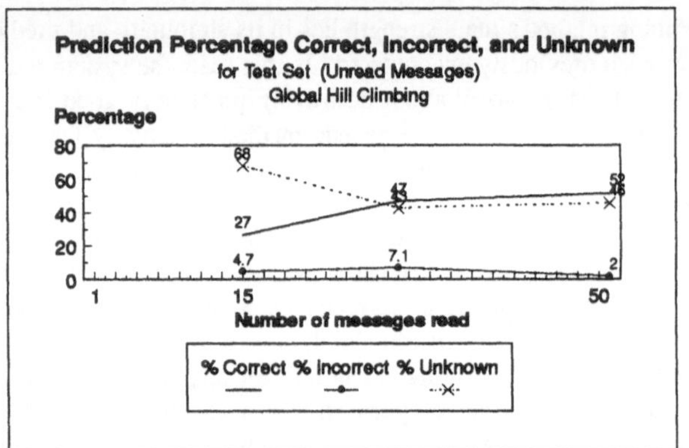

Figure 1. Prediction percentages for Global Hill Climbing Scheme.

7.2 LOCAL HILL CLIMBING WITH GENETIC COMPONENT

Figure 2 depicts the local hill climbing scheme with a single generation per batch. After the first fifteen messages are processed, a single crossover/mutation operation is performed across all individuals, and then once again after the next fifteen and finally the next twenty messages are read, for a total of three crossover sessions. The percent of correct classifications is slightly lower than the global hill climbing scheme at 44% opposed to 52%, probably due to the slower training time than the global scheme, and the relatively small amount of training performed (only 50 messages).

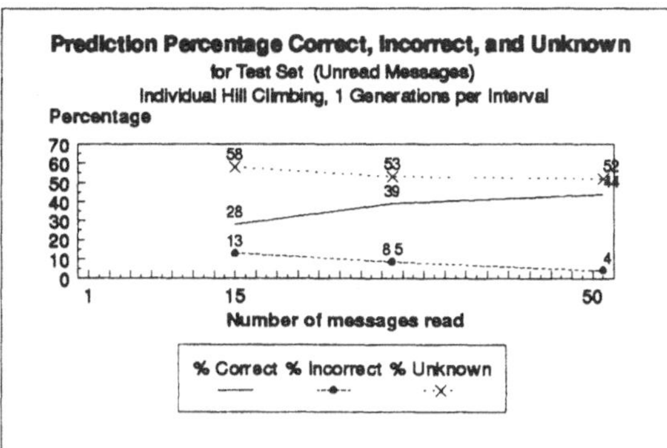

Figure 2. Prediction percentages for Local Hill Climbing / Genetic Scheme, 1 Generations

Figure 3 depicts the prediction percentages for 3 generations per interval. The percent correct matches (42%) is about the same result as the previous case with 1 generation, although the error is slightly less in this run.

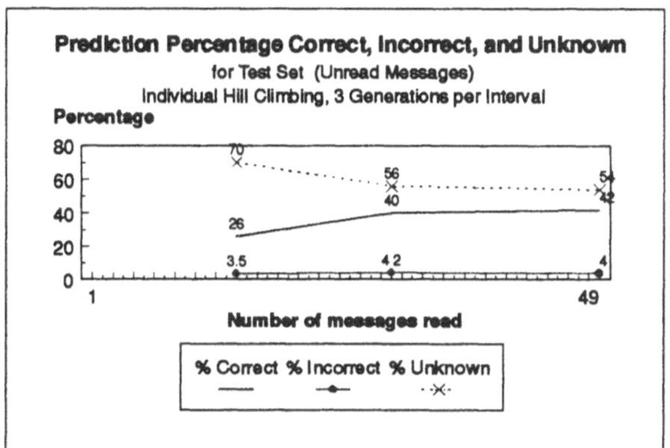

Figure 3. Prediction percentages for Local Hill Climbing / Genetic Scheme, 3 Generations

A final run with 5 generations produced a marked decrease in percent correct classifications. At this point, the amount of crossover is destroying too many old individuals which are necessary to perform well, and the accuracy of correct predictions decreases down to 36%. The error stays approximately constant throughout all runs, but the number of unknown messages increases and the percentage of correctly classified messages decreases. Runs with more generations resulted in worse behavior and are not shown here.

7.3 ANALYSIS OF RESULTS

While these results indicate that the global scheme performs better on this set of data, further testing is necessary before conclusions may be drawn and generalized. In particular, a larger data set and testing with larger batch sizes may result in significantly different performance.

Furthermore, the genetic scheme does perform fairly well on its own, correctly classifying a large portion of the data set and making relatively few errors.

Further insight may also be gained by examining which messages are being classified by the genetic scheme and comparing these to the answers classified by the global scheme. Comparison of these answers indicates that the genetic scheme often makes correct classifications of messages which the global scheme marked as unknown. In other words, the genetic scheme has effectively enlarged the search space to include new areas which the global scheme would not consider, as shown in Table 2. In the run with 1 generation per batch, 18% of the correctly classified answers were classified as unknown by the global scheme. This percentage reached 30% in the run with 3 generations per batch. Since the genetic scheme is capable of classification in cases where the global scheme is not, the two systems could feasibly be combined together to produce a single system with higher overall precision and breadth than either alone.

Percent Correct Answers for Local Genetic Hill Climbing Not Given by Global Hill Climbing Method

1 Generation	18
3 Generations	30

Table 2: Genetic scheme correctly classifies messages unknown by Global Hill Climbing Method

7 Neural Network Approach

Preliminary experiments were also performed using a radial-basis function neural network to classify messages. In this approach, the extracted features are presented to the input units and the output units determine whether or not the article is accepted or rejected. The neural network algorithm, developed by Platt, combines gradient descent with memorization so that new cases may be learned quickly but network size is controlled through gradient descent [8]. Moreover, the network is more powerful than the global hill climbing approach since it is able to non-linearly combine any of the input features in making a classification.

7.1 NEURAL NETWORK ARCHITECTURE

A major problem in using a traditional backpropagation neural network in the setting of information filtering is that a large amount of training data and many epochs are neeessary to create a trained network. This poses a problem to users who want the system to learn and quickly adapt to their interests. A solution for this problem is to combine gradient descent with memorization in a radial basis function network. If the network poorly classifies an input, then a new hidden unit is created which matches the input pattern (memorization). On the other hand, if the network is fairly close in classification, standard LMS gradient descent is performed to adjust the weights even closer to the desired output. In information filtering, memorization occurs initially and with new cases, resulting in quick learning, while gradient descent is used to fine-tune the network.

The network employed in this project consists of three layers. The first layer consists of nodes which correspond to features present in the input. The same input features were used as in the previously described schemes. If the feature is present in the input article then the node is 1, otherwise the node is 0. New input nodes are allocated when novel features are presented. The second layer combines the active input units and computes a radial Gaussian distance metric. Finally, the results of the middle layer are linearly summed in the final layer, which consists of a single node indicating whether the article is accepted or rejected. If the input space is two dimensional, one interpretation of the network is that each second layer unit is a Gaussian shaped figure which is either "accepted" or "rejected." The network then attempts to create units to accurately cover the space (memorization), or change the shape of the units to cover the space (gradient descent).

The criteria for memorizing new inputs is twofold. If the input vector is far away from the nearest second-layer unit ($> \delta$)and the error between the desired output and the network's output is large ($> \varepsilon$), then the current input is memorized. Both criteria are necessary to control the amount of memorization which is performed when gradient descent would be more effective. Memorization is performed by allocating a new second-layer unit with input weights matching the input vector and output weights set to the desired output. If the memorization criteria are not met, then standard gradient descent is used to modify the weight values.

A more detailed description of the neural network may be found in [8]. In this project, δ varied from 0.7 down to 0.07 and ε was set to 0.05. These are the same parameter values used in the simulations described by Pratt [8].

7.2 NEURAL NETWORK RESULTS

Although still under development, the neural network approach appears promising. In the test cases used with the other schemes, preliminary results after reading 50 messages have very low error (2%) and good correct classification (40%) but do not yet generalize well in classifying messages with unseen features, resulting in a large percentage of unknown classifications (58%). Ongoing work with modifying the network configuration and testing with larger data sets may correct these problems.

8 Conclusion

This project has examined a hybrid hill climbing/genetic approach, a strictly global hill climbing approach, and a neural network approach to the problem of automated information filtering. All filters adapt to user interests without the need for explicit user programming. The first two schemes result in user models which are easily modifiable by the user. One drawback of the neural network scheme is that the network is difficult for users to change directly. While all techniques result in fairly good filters, the global hill climbing method appears to give slightly more accurate results, suggesting that user interests may be modeled fairly well with a linear model of the input features. The more complicated non-linear models may require additional data to adapt to the user's interests. However, the genetic scheme does appear promising in exploring alternate areas of the data space not examined by the hill climbing method alone. Further testing and the exploration of other classification schemes is necessary to better understand the scope and limitations of the approaches presented here.

9 Acknowledgements

Work reported in this paper is supported in part by a grant from UC Micro and Apple Computer, Inc. The authors wish to thank Dr. Rao Machiraju, Dr. Mike Graves, and Rick Borovoy for several useful discussions.

10 References

[1] Goldberg, D., Nichols, D., Oki, B., Terry, D. (1992). "Using Collaborative Filtering to Weave an Information Tapestry," *Communications of the ACM*, 35 (12), 61-70.

[2] Eberts, R. (1991). "Knowledge Acquisition Using Neural Networks for Intelligent Interface Design," *Proceedings of the 1991 IEEE International Conference on Systems, Man, and Cybernetics*, 1331-1335.

[3] Jennings, A. & Higuchi, H. (1992). "A Personal News Service Based on a User Model Neural Network," *IEICE Transactions Inf. & Systems*, E75-D(2), 198-209.

[4] Baclace, P.E. (1992). "Competitive Agents for Information Filtering," *Communications of the ACM*, 35 (12), 50.

[5] Stevens, C. (1992). "Automating the Creation of Information Filters," *Communications of the ACM*, 35 (12), 48.

[6] Sheth, B. and Maes, P. (1993). "Evolving Agents For Personalized Information Filtering," *Proceedings of the Ninth IEEE Conference on Artificial Intelligence for Applications*, 1993.

[7] Holland, J. (1975). *Adaptation in natural and artificial systems.* Ann Arbor, MI: University of Michigan Press.

[8] Platt, J.(1992). "Learning by Combining Memorization and Gradient Descent," Neural Processing, 1992.

USING GENETICALLY EVOLVING MULTI-LAYER CELLULAR AUTOMATA FOR IMAGE PROCESSING

P Sahota, M F Daemi and D G Elliman
Department of Computer Science
University of Nottingham
Nottingham NG7 2RD, UK.

Abstract This paper describes the use of a genetically controlled multi-layered automata model to tackle image processing problems. It extends the work carried out previously to provide a generalised system for evolving multi-layered cellular automata (CA) transition rules. The transition rules are located with the help of a genetic algorithm, and once trained the system is able to process unseen images. Using a multi-layered automata model in place of a single layered one allows far greater processing to be achieved. The work carried out here demonstrates the successful evolution of transition rules for a two-dimensional CA to perform edge detection in noisy images.

1. Introduction

Cellular automata are often used for the simulation of complex dynamical physical systems. They have a high degree of parallelism which makes them suitable for fabrication by chemical structures consisting of many interacting molecules [1]. The uncertainty in predicting how local CA rules affect the overall system, makes the design of rules to perform a required task very difficult. One encouraging solution to the design has been to employ genetic search techniques to locate CA transition rules [2,3,4]. The work carried out previously in developing genetically evolved CA rules for image processing [5] has been extended in this paper to incorporate multi-layered CA models. The results show that a slightly modified genetic algorithm is able to find the correct transition rules to perform edge detection on noisy images using a double layered automata.

2. Cellular Automata

Cellular automata consist of a regular lattice of sites (cells), each site existing in one of n states. Each cell is connected to a number of neighbouring cells, collectively known as the Neighbourhood Set. In general, for a two-dimensional CA the rule can be given as

$$S_{i,j}^{t+1} = R(\underline{N}_{i,j}^{t}),$$

where $S_{i,j}^{t+1}$ is the new state at site (i,j) with a neighbourhood set of states $\underline{N}_{i,j}^{t}$, using the transition rule R.

E. A. Yfantis (ed.), Intelligent Systems, 517–524.

The CA used in this paper employs a 5-cell neighbourhood set (the cell itself and its 4 adjacent neighbours), and binary state cells, $S_{i,j} \in \{0,1\}$. This configuration requires a state transition rule table of length 32 entries to completely define every possible rule. Our model uses two separate CA layers, each with its own rule table, and so the overall system employs a rule table of length 64. There is thus $2^{64} = 1.8 \times 10^{19}$ different possible rules that can exist.

3. Genetic Algorithms

Genetic algorithms (GAs) were first described by J Holland [6], and provide an efficient means of scanning huge search spaces for good solutions. They maintain a population of candidates. The success (fitness value) of each individual is evaluated and is used to selectively reproduce the fruitful ones for the next generation. Two genetic operators are applied to the individuals in an effort to increase their fitness values. The main operator in Holland's GA is crossover, which takes two individuals and exchanges part of their genetic code. The second operator mutation, randomly changes the value of a gene according to a pre-defined mutation rate. Reproducing the individuals with crossover and mutation provides an efficient searching technique. In this paper, GAs are used to search automata rules that can solve a given problem.

4. Image Processing Model

The model used operates in two distinct stages - the training stage and the execution stage.

4.1 Training Stage

Figure 1 shows the model during the training stage. The user must provide pairs of images as examples of the type of processing required. These are indicated in the diagram as the input and ideal images. The input images are in turn mapped to the automata cells in the first layer, a black pixel being represented by a cell state of 1. The first layer CA uses the transitions rules provided by the GA to process the image. A fixed processing time is allowed for each layer, which should be chosen to allow the cells to reach stable states. Our model selects the number of processing cycles to be twice the image width so that cells at one end of the image can propagate and affect cells at the other end. Each layer passes its resulting image to the next layer for processing.

Initially, the function selector provides families of random CA rule tables. Each family provides one rule table per layer, and so uniquely defines the processing task for each layer. Each family thus produces its own group of output images from the input images. The matching process compares these output images to the ideal images and assesses how well each family is able to perform the desired transformations. A fitness value is assigned to each family as a measure of its success. Using information from the fitness value, the genetically controlled function selector is able to suggest further improved rule tables to test. The whole process is repeated until a family of rule tables have evolved with the desired level of accuracy.

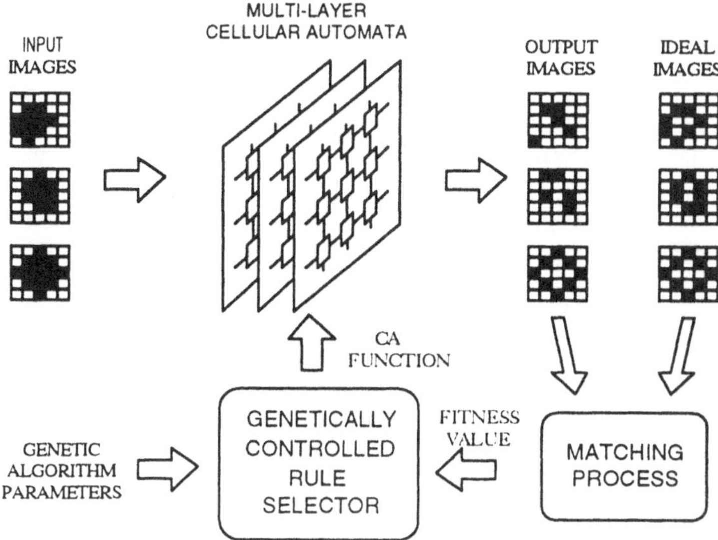

Figure 1 - Training Model

4.1.1 Genetically Controlled Rule Selector

The rule selector attempts to discover rapidly the set of rules necessary to perform the desired transformations. At the heart of this module lies a genetic algorithm. The controller maintains a population of 40 chromosomes. The individuals are binary strings of length 64, corresponding to the entries in the rule transition table for both layers. The mapping of the rules onto the chromosome is thus straight forward, with the search space of automata rules corresponding exactly to the search space of the GA. The GA used employs a fitness-proportionate reproduction scheme with mutation and two-point crossover.

4.1.2 Matching Process

One of the key issues in the success of the system hinges on the matching process being able to translate the image differences into a single accurate fitness value suitable for the genetic controller. For the problem of edge detection, a simple bit-by-bit count was performed yielding a bit error for each image. The overall fitness value, F, was chosen to be the root mean square of the fitness values produced by each image pair.

$$F = \sqrt{\frac{1}{i}\sum_{i} m(O_i, I_i)^2}$$

where $m(O_i, I_i)$ returns the pixel match between the ith output and ideal images.

4.2 Improved Training Schemes

The model described using the standard GA was unable to correctly locate the optimum rules in the time allotted. Three separate schemes were employed in an attempt to enhance the search technique.

4.2.1 Collapsed Image

The first was based on the observation that a significant proportion of the transition rules give rise to an automata collapsed image i.e. an output image where the cells turn either all white or all black. At this stage any information content of the image is lost. To avoid saturation of the population with these undesirable rules, their assigned fitness values should remain low. However, to increase efficiency, some information should be sought from these rules, and so the model incorporates a modification of the penalty method as described in [7]. For a collapsed image, the fitness of the rule is adjusted to

$$F_{modified} = F.\frac{C_{col}}{C_{max}}$$

where C_{max} is the maximum allotted number of automata cycles and C_{col} is the cycle number at which the automata image collapses. This penalty function scales the fitness value according to when the image collapses, inflicting a high punishment to rules with a rapid collapsing rate.

4.2.2 Bonus Scheme

The second improvement was to adapt a *bonus* scheme. Since the ideal images comprise of only a small proportion of black images, then obtaining a black pixel match is much more difficult than obtaining a white pixel match. Scoring one point for each match, regardless of the colour of the match does not reflect this quality. The improved method allocated a higher score for a successful match of a minority shade pixel. A more accurate assessment of the rules are thus provided to guide the GA.

4.2.3 Fitness Scaling

The final improvement used a fitness scaling type of strategy [7]. At fixed generation intervals a baseline is chosen to be 90% of the average fitness value. All individuals are subsequently assessed relative to the baseline. This technique was adopted to jump start the stagnated evolution that occurred. For the basic GA it was noted that after many generations the population contained substantially different strings with similar rated fitness values. These rules were being reproduced at approximately the same rate. The effect of including a scaled fitness in the evaluation procedure enforced fiercer competition between closely matched rivals, emphasising small differences in strength.

4.3 Execution Stage

Once an optimal set of rules have been found for the problem at hand (or the maximum assigned number of generations reached) the system is ready to begin processing. Unseen images are shown to the automata individually, which subsequently updates its cells by the specified number of cycles using the optimal rule set found previously. The input image is thereby transformed through the layers producing the processed output image.

5. Training Images

Figure 2 shows three sample image pairs that were provided as examples of edge detection in noisy images for the double-layered automata training.

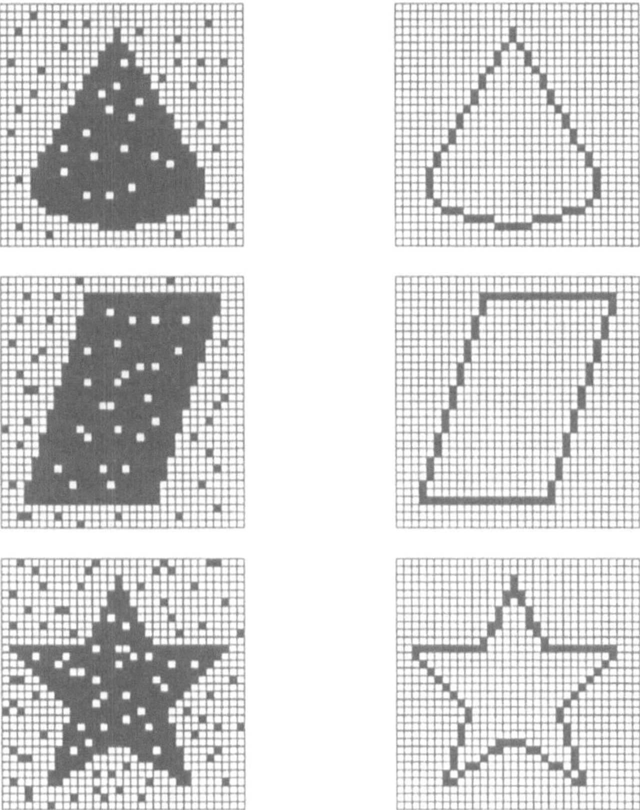

6. Results

6.1 Training

Figure 3 shows the average performance of 5 runs of the entire training process using the two-layered automata. The genetic algorithm maintained a population of 40 candidates, using a crossover rate of 0.7 and a mutation rate of 0.01. The process was allowed to run for 1000 generations. For the bonus scheme, a black pixel match was given a score that was 3 times higher than the white pixel match.

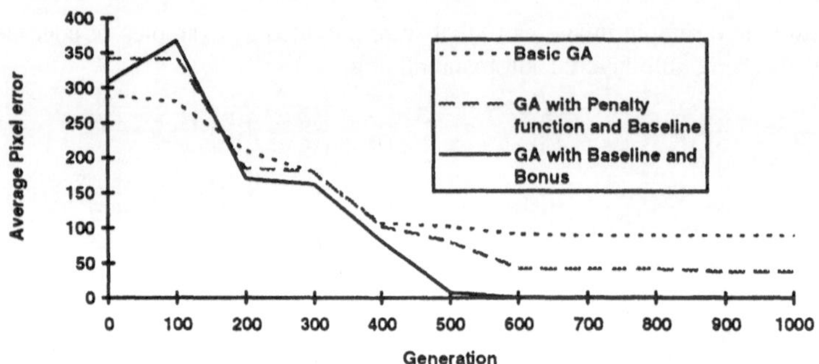

Figure 3 Average Population Fitness During Training

6.2 Execution Stage

An optimum set of transition rules were usually found for the runs using both the baseline and bonus schemes. These rules were used to process unseen images that were of a higher resolution than the test images. The transition rules produced successful edge detection on the noisy images. One such example is given in figure 4.

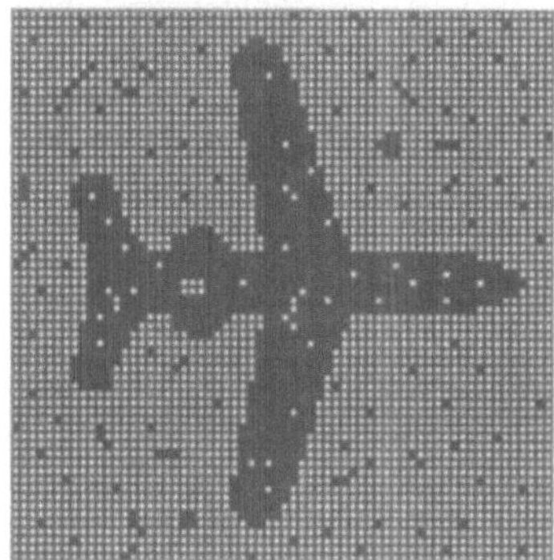

Figure 4.1(a) - Noisy Unseen Image

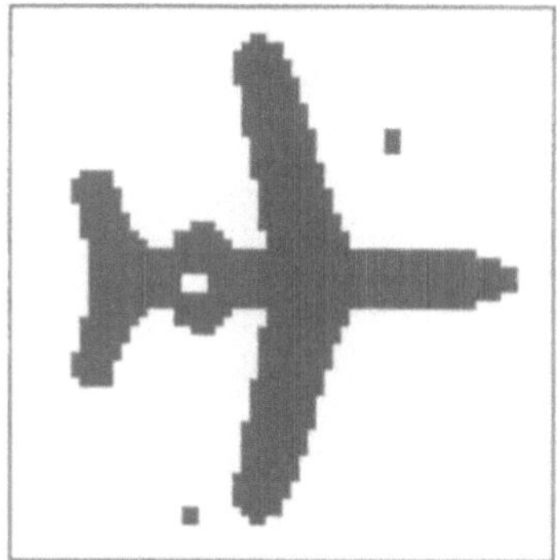

Figure 4.1(b)
Image produced by first CA layer

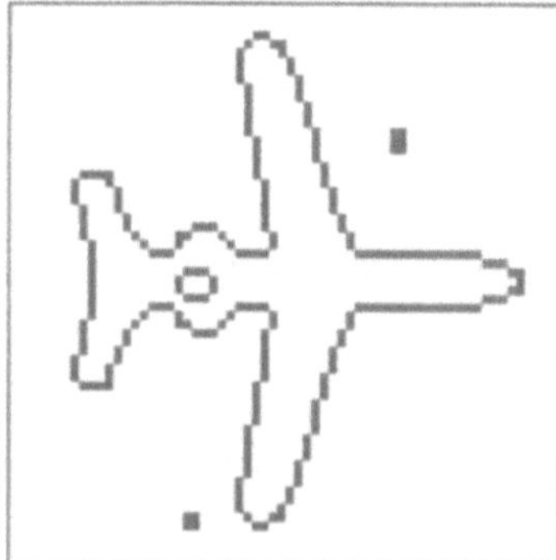

Figure 4.1(c)
Image produced by second CA layer

7. Discussion and Conclusions

The results show that a genetic algorithm can be adapted to evolve the rule tables for two-layered CAs to perform edge detection on noisy images. For the first 1000 generations the basic GA gave a relatively poor performance, not even arriving at a close solution. At this stage the three schemes were adopted in an attempt to improve the searching procedure. Using the penalty and baseline schemes together gave an improves performance, but the best results were achieved by the combination of the baseline and bonus schemes. Adding the penalty scheme to this combination did not improve the speed of convergence, and so was not shown on the graph.

For the processing of the unseen image (figure 4), the trained automata model attempted to find the outline of the noisy plane image that had twice the resolution of the training images. Since the automata rules are local by nature, they can perform computation on any size of image. The rule table that evolved provided each layer with a well defined specific task. The first layer filtered out the noise, and the second layer provided edge detection on the noiseless image. As can be seen clearly, the noise removal process was not perfect. It was unable to remove the larger blocks of noise. This was not totally surprising, however, since the system was trained with images having only small specks of noise. The trained system had treated the larger areas of black noise as small objects, and the group of white noise pixels as a hole in the plane.

This generalised technique can be used for a greater number of layers to yield more complex degrees of processing. The multi-layer automata model provides a viable architecture for molecular implementation; on the microscopic level the rules are very simple. On the macroscopic level, useful image processing problems may be solved.

8. References

[1] Babloyantz A. Molecules, *Dynamics and Life: An Introduction to the Self Organisation of Matter*, Wiley Interscience, New York, 1986.

[2] Richards, Fred C, Meyer Thomas P, Packard Norman H. *Extracting Cellular Automaton Rules Directly from Experimental Data*, Physica D 45, pp189-202, 1990.

[3] Sims Karl, *Interactive Evolution of Dynamical Systems*, 1992.

[4] Mitchell Melanie, Crutchfield James P, Hraber Peter T, *Evolving Cellular Automata to Perform Computations: Mechanisms and Impediments*, submitted to Physica D, Oct 1993

[5] Sahota P, Daemi MF, Elliman DG, *Training Genetically Evolving Cellular Automata for Image Processing*, International Symposium on Speech, Image Processing and Neural Networks, pp753-756, 1994

[6] Holland, John H. *Adaptation in natural and artificial systems*, The University of Michigan Press, Ann Arbor, 1975.

[7] Goldberg, David. *Genetic Algorithms in Search, Optimization, and Machine Learning*, Addison-Wesley, 1989.

Speech Recognition Using Neural Networks(HearNet): Mapping From A Set Of Phoneme Strings To Character/Word Recognition

Byoung Jik Lee

Department of Computer Science
University of Iowa
Iowa City, IA 52240 USA

August 9, 1994

Abstract. An approach to part of speech recognition using neural networks(HearNet) is described. Speech recognition is a complex problem, because it is a speaker-dependent problem and speech is continuous. HearNet proposes an intermediate step, which maps from a set of phoneme strings to character/word recognition. This can function as a filtering step for many different speakers. The system gradually learns to convert the phoneme strings to English words. After 100 epochs, 92.2% of training data characters are recognized correctly and 87.5% of test data characters are recognized correctly. Recognition results perfectly on entire words with a 62% on training data set and 50% on test data set. Including one mistake in an entire word, the performance is increased from 62% to 92.5% and from 50% to 82%. The system shows how it learns to recognize from phoneme strings to words.

Keywords. Neural Networks, Speech Recognition, Phoneme Recognition.

1 Introduction

Speech recognition has been one of the most studied subjects using neural networks. The problem of speech recognition is a speaker-dependent and context-dependent problem. In addition to this, the speed of speech varies and speech is continuous [2]. Various groups have tried to solve this problem. By providing some representation of spoken words, the system learns to recognize them. It works for a small number of vocabularly and isolated words. But the performance is not good on speaker-independent continuous speech recognition. If we have a step which maps from speech wave form to phonemes, then we can map from phonemes to characters/words effectively. There are fewer phonemes than words in any languages. This intermediate step can function as a filtering step to recognize speech for many different speakers. Sejnowski proposed mapping from letters to phonemes in his work NETtalk [4] for speech generation. My work is a mapping from phonemes to letters for

E. A. Yfantis (ed.), Intelligent Systems, 525–530.

part of speech recognition. The system shows the process of learning to recognize words from phoneme strings.

2 Architecture

2.1 Network Architecture

As showing in figure 1, I utilize a window size of three for recognizing words. The input layer has three groups, each with 57 units. The phonemes are represented locally in 52 units. Five additional units are for stress and word boundaries. The total input units are 171 units. One window of strings of a word comes to this 171 units at a time. In each group, only two units are activated for an input. The hidden layer has 40 units. The output layer has 26 units. In the output layer, the letters are represented locally in 26 units, one letter for each unit. Between adjacent layers, the units are fully connected to each other. In the network, there are 5,580 weight parameters. In the output layer, the most active unit is selected as the output letter. A logistic function is used as an output function.

$$F(x_j) = \frac{1}{1+e^{-Net_j}}$$

A simulator was written in C and it took about two hours to train for 30 epochs on an IBM RS6000-530 system.

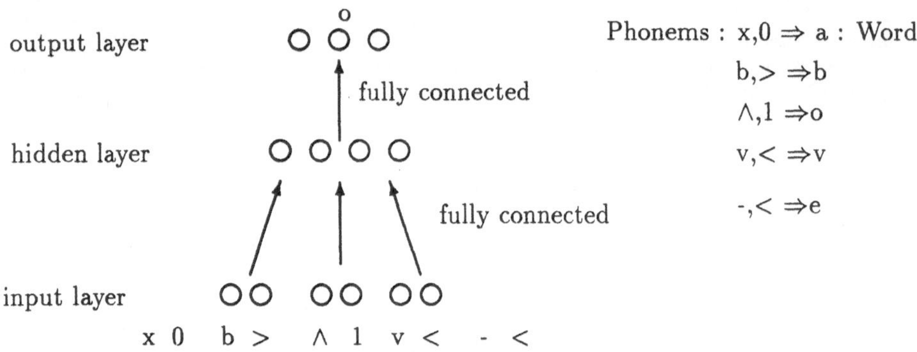

Figure 1: HearNet Architecture

2.2 Data Set

The data set is from Sejnowski [4]. From the Miriam Webster's Pocket Dictionary which contains 20,012 words, a subset of the most commonly used English words was selected by the Brown corpus [1]. This data set contains 934 words. Among them, 748 words are used as a training data set and 186 words are used as a test data set.

2.3 Learning Algorithm

In this work, the standard backpropagation algorithm of Rumelhart et. al. [3] is used as the training algorithm. By using a training data set, the backpropagation algorithm minimizes the error between the target output and the actual output. The weights are adjusted by a gradient descent process in weight space. Each weight is updated by an amount propotional to the first derivative of the error with respect to the weight.

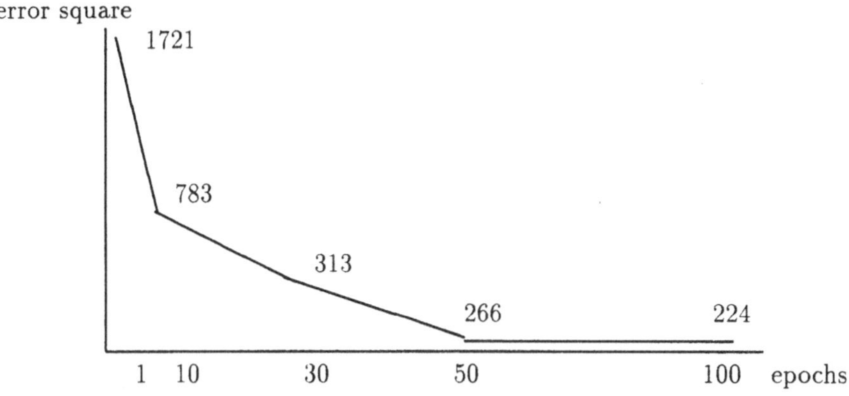

Figure 2: Learning Curve

$$Error\ Square = \frac{1}{2}\sum_{p,k}(target\ output_{p,k} - actual\ output_{p,k})^2$$

where p refers pth training vector and k refers kth output unit. At the beginning of learning, i.e., during early epochs, the system seems to have a childlike hearing ability. For example, for the word 'nation', the system recognizes it as 'natiyn' at the beginning. But finally the system recognizes it correctly. There are more examples, it recognizes 'clear' as 'cliar', 'color' as 'coler', 'complex' as 'comples', 'day' as 'dai', 'during' as 'doring' at the beginning stage of learning, but eventually HearNet recognizes all of them correctly.

3 Result

3.1 Accuracy Measurement

By looking at three phoneme strings at a time, the system classifies the center letter of three phoneme strings. The accuracy measurements [6] for characters are as follows.

$$A_a = \frac{Cor_a}{Num_a}$$

$$Accuracy = \frac{\sum_{i=a}^{z} Cor_i}{\sum_{i=a}^{z} Num_i}$$

where Num_a is the total number of phonems for the letter a in the data sets, Cor_a is the number of phonemes that are classified correctly for the character a, and A_a is the accuracy of the corresponding character a. The avereage accuracy of total characters is 91% after 100 epochs.

In the consecutive two same letters in a word, the latter has phoneme representation '-' which does not have a proper target output. Since the target output is always varing, it can not be learned. 'Dinner' and 'difficult' are examples. For this case, if the confidence of letter decision for the latter is below .25, then I decide the first output letter of these two consecutive letters as the output letter of the latter. This increases the performance in all training and test data sets. The accuracy performance with this rule is 92.2% for training data and 87.5% for test data. The accuracy measurements for words is as follows.

$$WordAccuracy = \frac{C_w}{N_w}$$

where N_w is the total number of sets of phoneme strings, i.e., the number of words in data sets. C_w is the number of words which are recognized correctly, i.e., all characters in the word are recognized correctly. The word accuarcy is 65%. When we accept a word eventhough there is one mistaken letter in the word, the word accuracy increases up to 92.5% in training data and 82% in test data.

Table 1. The accuracy of mapping for characters.

Epochs	*Character Accuracy*		*Word Accuracy*	
	Training data total 4091 chs	Test data 1037 chs	Training data 748 words	Test data 186 words
30	3544(87%)	890(86%)	405(54%)	83(45%)
60	3602(88%)	893(86%)	434(58%)	90(48%)
100	3639(91%)	900(87%)	453(61%)	90(48%)
100 with rule	3680(92%)	908(88%)	486(65%)	93(50%)
Considering a mistake in the entire word			692(93%)	153(82%)

In general as showing in table 1 and figure 2, until epochs 30, the system learns rapidly(87% for characters and 54% for words) and then learns slowly. When we check the correctness for each character, there are three cases. One is the correctly recognized case(correctness). The others are error cases. Over-rated case is that though there is not such a character, the system decides it. Under-rated case is that for a character, the system classifies it as a different character. The accuracy is high as showing in table 1 and table 2. Most of the results are similar as at the output of epochs 30 except in the case of j, q, and x. The total occurrences of them was too low to be learned from this data until 30 epochs, but they are recognized perfectly by epochs 100 because they are trained sufficiently. The performance of e is a little degraded in epochs 100, but not much. As showing in table 2, most of the errors come from vowels. For example, target letters 'a', 'e', and 'i', are much over-rated.

Table 2. The accuracy and confusion matrix for training data set

target	actual recognized output
ch	a b c d e f g h i j k l m n o p q r s t u v w x y z
a(309)	253(82%) e-28 i-9 l-1 m-2 o-7 q-4 u-4 z-1
b(56)	56(100%)
c(152)	132(87%) s-16 t-2 v-1 x-1
d(147)	140(95%) e-3 i-1 t-2 u-1
e(511)	a-10 459(90%) h-1 i-10 k-3 l-5 o-2 q-6 t-1 u-11 w-1 y-1 z-1
f(80)	e-1 77(96%) h-1 v-1
g(90)	88(96%) i-2
h(141)	c-1 140(98%)
i(283)	a-5 c-1 e-8 g-2 h-1 249(88%) q-1 t-1 u-5 y-9 z-1
j(7)	7(100%)
k(32)	c-6 e-1 23(72%) n-1 w-1
l(189)	a-2 184(97%) u-3
m(120)	120(100%)
n(280)	280(100%)
o(286)	a-1 c-1 e-3 f-2 i-4 255(89%) q-2 u-15 w-2 y-1
p(107)	105(98%) r-2
q(6)	6(100%)
r(302)	301(99.7%) u-1
s(242)	c-6 e-12 h-2 222(92%)
t(316)	i-1 315(99.7%)
u(138)	a-3 e-2 g-1 h-2 n-1 o-16 t-1 112(81%)
v(41)	41(100%)
w(69)	a-2 e-2 h-1 o-9 s-1 t-1 u-8 45(65%)
x(12)	12(100%)
y(73)	e-2 i-6 q-3 t-1 u-3 58(78%)
z(2)	e-1 s-1 o(0%)
over-rated	a-23 b-0 c-15 d-0 e-63 f-2 g-3 h-8 i-33 j-0 k-3 l-6 m-2 n-2 o-34 p-0 q-16 r-2 s-18 t-9 u-51 v-2 w-4 x-1 y-11 z-3
under-rated	a-56 b-0 c-20 d-7 e-52 f-3 g-2 h-1 i-34 j-0 k-9 l-5 m-0 n-0 o-31 p-2 q-0 r-1 s-20 t-1 u-26 v-0 w-24 x-0 y-15 z-2

3.2 Generalization

As showing in table 1, the character accuracy and word accuracy between the training data set and the test data set are not much different. The difference of character accuracy between the training data set and the test data set is only 4%. They show that the system works well on generalization.

4 Conclusions and Future Works

An intermediate step of speech recognition is proposed in which sets of phoneme strings are mapped to letters/words. It can function as filtering step in speech recognition for many different speakers. This work shows a process of learning by correcting mistakes. By providing more examples, the system learns how to map from phoneme strings to letter/word recognition. This work can be used in any other languages. As described, most learning has been done by 30 epochs and by 100 epochs there is still some progress, but not much. Here is a limitation of backpropagation. By using paradigms other than neural networks, we can build a hybrid system to have better performance. Much works has been done on English orthography [5]. By utilizing this, we can bulid a rule based system which is different from the neural network paradigm. Hybrid systems which combine neural networks and rule based system can perform better than each seperate system.

5 Acknowledgements

The data set was from Terry Sejnowski[4] whom I wish to thank and is provided by the CMU. I want to thank Gregg C. Oden for his encouragement and advice.

6 References

[1] Kuchera, H., Francis W. N. (1967) Computational Analysis of Modern-Day American English. Brown University Press. Providence, Rhode Island.

[2] Lippman Richard P. (1989) "Review of Neural Networks for Speech Recognition" Neural Computation Vol. 1 pp 1-38 MIT Press

[3] Rumelhart D. E., Hinton G. E. Williams R. J. (1986) "Learning Internal Representations by Backpropagation" Parallel Distributed Processing Vol. 1 pp 318-362 Cambridge MA MIT Press

[4] Sejnowski T. J. and Rosenberg C. R. (1987) "Parallel Networks that Learn to Pronounce English Text" Complex Systems Vol. 1 No. 1 pp 145-168.

[5] Venezky R. L. (1970) The Structure of English Orthography. The Hague.

[6] Zhang X., Mesirov J. P. and Waltz D. L. (1992) "Hybrid System for Protein Secodary Structure Prediction" Journal of Molecular Biology. Vol. 225 pp 1049-1063.

An Adaptive Training Method of Back-Propagation Algorithm

Jang-Hee Yoo and Jae-Woo Kim
Artificial Intelligence Division
Systems Engineering Research Institute
P.O. Box 1, Yoosung, Taejeon, 305-600, Korea

Jong-Uk Choi
Dept. of Information Engineering
Sangmyung Women's University
7 Hongji-Dong, Jongro-Gu, Seoul, 110-743, Korea

Abstract

Currently, the back-propagation is the most widely applied neural network algorithm at present. However, its slow learning speed and local minima problem are often cited as the major weakness of the algorithm. In this paper, described are an adaptive training algorithm based on selective retraining of patterns through error analysis, and dynamic adaptation of learning rate and momentum through oscillation detection for improving the performance of back-propagation algorithm. The usefulness of proposed algorithms was demonstrated in experiments with the XOR and Encode problems.

Keywords : Back-Propagation, Convergence Speed, Degree of Generalization

1 Introduction

Back-propagation algorithm has been known to be useful in training multi-layered neural networks, and thus has been effectively applied to various fields such as pattern recognition, signal and image processing, forecasting, robot control, etc. However, as the number of iterated updates of weight vectors is n^2 when the number of nodes in each layer is n, the required computation time becomes enormous with the increase of the number of nodes. In addition, it is quite possible for the algorithm to converge into local minimum. To overcome these problems, many researchs on improving convergence speed have been done [2, 10, 12, 14]. The research efforts can be classified into the following three categories.

First, heuristic knowledge obtained through repeated experiments can be embedded into algorithm improvement. For example, many iterations are required when the learning rate is small and the error curve is declining very slowly, because differential value becomes very small. In contrast, when the learning rate is a large, overshooting problem arises which is caused by the large curvature in the differential values. Therefore, dynamic

E. A. Yfantis (ed.), Intelligent Systems, 531–536.

adaptation of learning rate or reuse of gradient depending on the size of variations are frequently employed [5, 6]. Second, as the learning algorithms of neural networks are nothing but solving non-linear optimization problems [13], attempts have been made to improve algorithm performance based on numerical methods. As differential values of high-order equations includes more information of the search space, when calculating weights of next step, frequently employed are well-formed mathematical theories: Newton method, Quasi-Newton method, and differential values with the values of gradient [1, 4, 9]. Third, in contrast to modifying learning algorithms, external factors such as training set and training order are modified: in the review technique the patterns identified difficult to train in the trace of learning degrees of each category will have more chances of training than others [7], while in the preparation technique the number of training samples gradually increases to prevent overfitting [8].

In this paper an adaptive back-propagation algorithm is proposed, based on a selective retraining of patterns through analyzing error curve, and a dynamic adaptation of learning rate and momentum through detections of oscillation. Usefulness of the proposed algorithm was tested in XOR and Encoder problems.

2 Error Analysis and Selective Retraining

Back-propagation algorithm is a gradient descent algorithm in which MSE(mean square error) is employed [3, 12, 15] for minimizing error in weight-error space. An error measure, RMS(root-mean square) [12, 15] is derived which normalizes MSE as equation-(1).

$$E_{rms} = \frac{1}{PK}\sqrt{\sum_{p=1}^{P}\sum_{k=1}^{K}(d_{pk} - o_{pk})^2} \tag{1}$$

In the equation, P is the number of training patterns, and K denotes the number of nodes in output layer. The d_{pk} is value of $k-th$ output node for $p-th$ input pattern, and o_{pk} is the actual output. The value of E_{rms} is more descriptive than MSE in comparing training results of algorithms and thus is more effective in measuring the accuracy of mapping and association[6].

E_{rms} can be used as an error measure in back-propagation algorithm which continue training processes until the value of E_{rms} becomes less than the predetermined tolerance value. The algorithm which uses E_{rms} and a fixed value of predetermined tolerance has two serious problems. First, some of input patterns, even though they are not responsible for error, should undergo training processes because of the error caused by other patterns, especially when the size of input pattern is large. Second, as the value of E_{rms} is used as an error measurement, degree of learning obtained for each pattern is not accurately reflected. One of the solutions to the problem might be to calculate average RMS for all training patterns and individual RMS for each pattern, and then to train specific patterns which have large values of RMS than average RMS.

In many cases, the weights incorrectly fit actual output of specific patterns in backpropagation algorithm. The incorrect fitting can be detected by identifying the output node of k which has the maximum value of error for pattern p, defined as following:

$$E_{pk_{max}} = max_{k=1}^{K}(|d_{pk} - o_{pk}|) \tag{2}$$

As a conclusion, retraining which reflect characteristic of each pattern can be done by detecting incorrect fittings and by utilizing error measurements of $E_{pk_{max}}$(max_out) and E_{rms}(average_rms). Figure-1 describes the selective retraining algorithm proposed in this research.

```
current_tss = max_output = 0.0;
Loop number_of_pattern for a epoch
      compute_actual_output();
      compute_error();
      current_tss = current_tss + current_error;
      If number_of_output > 1 Then
            current_rms = current_error/ number_of_output;
      else current_rms = 0.0;
      If (max_output - current_rms) > average_rms Then
            adjust_weights();
end_of_loop
average_rms = current_tss / (no_of_pattern * no_of_output);
```

Figure 1: Selective Retraining

Applications of the algorithm may not only reduce training time, but also increase recognition rate by selective retraining.

3 Dynamic Adaptation of Learning Rate and Momentum

In the back-propagation algorithm, weights are recursively adjusted with a set of pairs (input values and corresponding output values) until the value of difference between desired output and actual output is less the predetermined tolerance value. Weight adjustment is done based on the generalized equation-(3) [3, 11, 12]:

$$\Delta W_{ji}(t) = \eta(\delta_j o_i) + \alpha \Delta W_{ji}(t-1) \tag{3}$$

In the equation, t is time sequence and η denotes the learning rate. As the learning rate becomes larger, the change of weight is becoming larger. Therefore, training with larger learning rate might be finished earlier. However, in that case convergence is not guaranteed, because oscillation can arise. It is desirable that the learning rate be maximized for speedy convergence within a range to prevent oscillations. The variable α is a momentum term introduced to provide speedy training while preventing oscillations, indicating the size of weight adjustment based on previous changes of weights. Current research on improving training speed mainly focuses on modifications of terms included in the equation-(3) [10, 12, 15].

In the back-propagation algorithm of equation-(3), oscillation can be detected by analyzing error curves. Oscillation is indicated by irregular fluctuations of decrease and increase of the error measurement term E_{rms}, and should be detected within a predetermined interval(number of epochs) to be applicable to dynamic adaptation of learning rate and momentum. In the analysis of error curve, the error curve is likely to converge into a global minimum or escape from local minimum, when the error monotonously decreases or increases for the predetermined interval. Therefore, whenever the frequency of error decreases falls below minimum value but jumps above maximum value predetermined in the error analysis, learning rate and momentum are modified. Otherwise, initial learning rate and initial momentum are assigned. Figure-2 describes the algorithm of dynamic adaptation of learning rate and momentum.

```
delta_error = average_rms_{t-1} - average_rms_t;
If delta_error < 0.0 Then
      oscillation = oscillation + 1;
  If reference_interval = TRUE Then
      If (oscillation > max_freq) || (oscillation < min_freq) Then
            oscillation = 0;
      learning_rate = initial_learning_rate
                  * (interval_size - oscillation) / interval_size;
      momentum = initial_momentum * ((1.0 - initial_learning_rate)
                + learning_rate);
      oscillation = 0;
end_if
```

Figure 2: Dynamic Adaptation of Learning Rate and Momentum

To apply the algorithm of Figure-2, the initial learning rate and initial momentum, in addition to the interval(reference_interval) for error analysis and error decreasing frequency(min_freq, max_freq) for detecting oscillations, should be determined. The proposed algorithm may effective in training relatively complex patterns by detecting oscillations and quickly adapting to them.

4 Experimental Results

In this research, the proposed algorithm was tested in XOR problem and 8-3-8 Encoder problem. The algorithms of Figure-1 and Figure-2 were programmed in C-language running on SUN SPARCstation. Patterns tested were of uniform distribution, obtained by the random number generator. Initial learning rate was set 0.5 and initial momentum was set 0.9. The interval for oscillation detection was set the number of training patterns.

Table-1 shows comparison results of standard algorithm and proposed algorithm. In the test, after training 100 patterns, investigated are average RMS values when the training iterations reach 500, 1000, and 2000 epochs, average recognition rate measured after 2000 epochs, and recognition rate for 1000 new patterns.

Table 1: Experimental Result of Algorithm Performance

NeuralNets Algorithm	*Task*	*Neural Topology*	*AverageRMS*			*Average Correct*	*Gen. Test*
			500	1000	2000		
Standard BP	*XOR*	2x3x1	0.0174	0.0145	0.0133	99%	92.0%
	Encode	8x3x8	0.0133	0.0100	0.0093	98%	72.4%
Proposed BP	*XOR*	2x3x1	0.0083	0.0040	0.0024	100%	95.3%
	Encode	8x3x8	0.0150	0.0138	0.0120	97%	74.3%

Performance results such as the number of iterations and convergence speed are sensitive to the initial weights. Therefore, the same set of initial weights was employed for comparing two algorithms and the performance tests were repeated in multiple times for preventing statistical biases. In the test, as is shown in Table-1, the proposed back-propagation algorithm demonstrated better performance than standard algorithm in solving XOR problem which has only a single output node. In solving Encode problem, convergence speed was a little slow, but the degree of generalization was increased. Modification of weights through selective retraining reduced computation complexity and eventually decreased training time. Intersting result obtained was that the standard algorithm showed much better performance where the proposed algorithm was poor. The opposite was true. The result shows that the initial weights might be a critical factor in determining convergence speed.

5 Conclusions

In this paper, a back-propagation algorithm is proposed, combining a dynamic adaptation of learning rate and momentum, and selective retraining method. The proposed algorithm may contribute to improving convergence speed and enhancing degree of generalization, by decreasing computation complexity through selectively training patterns and effectively preventing oscillation through dynamical adaptation of learning rate and momentum.

Future research should be done on determining the interval and the frequency in error analysis for detecting oscillations, in addition to determining initial learning rate and initial momentum. Also, more research should be done on assignment of appropriate initial weights and on improving degree of generalization, which is one of the most important goals of neural network learning.

References

[1] Becker, S., and Le Cun, Y. (1989) "Improving the Convergence of Back-Propagation Learning with Second Order Methods," *in Proceeding of the 1988 Connectionist Models Summer School*, Morgan Kaufmann Publishers, pp.29-37.

[2] Fahlman, Scott E. (1989) "Faster-Learning Variations on Back-Propagation : An Empirical Study," *in Proceeding of the 1988 Connectionist Models Summer School*, Morgan Kaufmann Publishers, pp.38-51.

[3] Hecht-Nielsen, R. (June 1989) "Theory of the Backpropagation Neural Network," *in Proceedings of 1989 International Conference on Neural Networks*, Washington D.C., Vol.I, pp.593-601.

[4] Himmelblau, D. M. (Jan. 1990) "Introducing Efficient Second Order Effects into Back Propagation Learning," *in Proceedings of the International Joint Conference on Neural Networks*, Washington D.C., Vol.I, pp.631-634.

[5] Hush, D. R., and Salas, J. M. (July 1988) "Improving the Rate of Back-Propagation with the Gradient Algorithm," *in Proceedings of the IEEE International Conference on Neural Networks*, San Diego, Vol.I, pp.441-447.

[6] Jacobs, R. A. (1988) "Increased Rates of Convergence Through Learning Rate Adaption," *Journal of Neural Networks*, World Scientific, Vol.1, pp.295-307.

[7] Mori, Y., and Yokosawa, K. (1989) "Neural Networks that Learn to Discriminate Similar Kanji Characters," *Advances in Neural Information Processing Systems I*, pp.332-347.

[8] Ohnishi, N., Okamoto, A., and Sugie, N. (Jan. 1990) "Selective Presentation of Learning Samples for Efficient Learning in Multi-Layer Perceptron," *in Proceedings of the International Joint Conference on Neural Networks*, Washington D.C., Vol.I, pp.688-691.

[9] Parker, D. B. (June 1987) "Optimal Algorithms for Adaptive Networks: Second Order Back-Propagation, Second Order Direct Propagation, and Second Order Hebbian Learning," *in Proceedings of the IEEE International Conference on Neural Networks*, San Diego, Vol.II, pp.593-600.

[10] Pfister, M. and Rojas, R. (Oct. 1993) "Speeding-up Backpropagation - A Comparison of Orthogonal Techniques," *in Proceedings of 1993 International Joint Conference on Neural Networks*, Nagoya, Vol.I, pp.517-523.

[11] Rumelhart D. E., Hinton G. E., and Willians R. I. (1986) "Learning Internal Representations by Error Propagation," in Rumelhart, D. E, and McClelland, J. L., and PDP Research Group(Eds.) *in Parallel Distributed Processing*, Vol.1, MIT Press, pp.318-362.

[12] Smith, M. (1993) *Neural Networks for Statistical Modeling*, Van Nostrand Reinhold.

[13] Watrous, R. L. (June 1987) "Learning Algorithm for Connectionist Networks: Applied Gradient Methods of Nonlinear Optimization," *in Proceedings of the IEEE International Conference on Neural Networks*, San Diego, Vol.II, pp.619-627.

[14] Werbos, Pau J. (Oct. 1990) "Backpropagation Through Time : What It Does and How to Do It," *in Proceedings of the IEEE*, Vol.78, No.40, pp.1550-1560.

[15] Zurada, Jacek M. (1992) *Introduction to Artificial Neural Systems*, West Publishing Company.

MULTIVERSION INFORMATION RETRIEVAL: PERFORMANCE EVALUATION OF NEURAL NETWORKS VS. DEMPSTER-SHAFER MODEL

G.V. Meghabghab(1) and D.B. Meghabghab(2)
(1)Department of Mathematics and Computer Science
VALDOSTA STATE UNIVERSITY
VALDOSTA, GA 31698 USA
e-mail:gmeghab@grits.valdosta.peachnet.edu
(2)Instructional Technology Program
VALDOSTA STATE UNIVERSITY
VALDOSTA, GA 31698 USA
e-mail:dmeghab@grits.valdosta.peachnet.edu

Abstract. A new Neural network paradigm called "selection paradigm" (SP) is proposed to solve the problem of algorithm selection in multiversion information retrieval. SP learns to choose a preferred pattern for a given situation from a set of n alternatives based on examples of a human expert's preferences. The testing results after training are compared to Dempster-Shafer's (DS) model. The agreement between a document belief and two or more reformulated query beliefs by two or more different algorithms is computed by DS. Final testing results indicate that SP can produce consistent rank orderings and perform better in certain information retrieval situations than DS's model.

Keywords. Singleversion Information Retrieval, Multiversion Information Retrieval, Dempster-Shafer Model, Performance Evaluation, Connectionist Selection Paradigm.

1 Introduction

The term Multiversion Information Retrieval (MIR) refers to the existence of several different algorithms for realizing one information retrieval process and the use of the best algorithm for a given situation. The search quality in "single version systems," (i.e., systems with only one algorithm for each process) varies to a great extent due to difference in users' search strategies, search requests, subject of queries, etc. Given any particular algorithm, there are situations where the use of this algorithm would result in lower relevant document retrieval relative to other algorithms. Therefore, it seems appropriate for functioning Information Retrieval (IR) systems to have more than one algorithm for each process and use the best for a given situation. Such systems are called "multiversion" IR (Frants et. al. [1993]). To make an IR system more flexible and to increase the quality of service it provides to users, it should provide for more than one algorithm version for

E. A. Yfantis (ed.), Intelligent Systems, 537–545.

each particular search case and select the most appropriate algorithm for the case (i.e., user search, user request, user knowledge, etc..). The selection of the best algorithm allows the system to function to its optimal state, and, hence, to provide optimal information retrieval to users.

Query reformulation is best suited for the application of "multiversion" schema because it yields an efficient evaluation of the relevant documents retrieved. An additional reason for using multiversion construction of query formulation is its use in the feedback process. As a rule, the process of feedback assumes the change of query formulation to retrieve highly relevant documents. The information system uses a set of different algorithms for constructing query formulation. Clearly, it is necessary to develop a selection mechanism which chooses the best algorithm (i.e., from a set of available algorithms). The results of two different techniques on the multiversion information retrieval systems are evaluated: Connectionist selection paradigm and Dempster-Shafer evidential reasoning (Shafer [1976]).

2 Connectionist Selection Paradigm

There is a new widespread interest in the use of connectionism for real-world practical problem solving. Its principal areas of application involved relatively low- level signal processing and pattern recognition tasks (Meghabghab and Nasr [1993]; Dagli et. al. [1991],[1992],[1993]). Connectionist networks may also be useful in higher- level tasks that are currently tackled by expert systems and knowledge engineering approaches (Gallant [1988]; Meghabghab and Meghabghab [1994]).

MIR can be modeled by connectionist systems as follows: the expert is given a set of n alternatives as input (n may be either small or large) for a given problem domain, and must select the most desirable or preferable alternative. How can a learning system such as a connectionist network be designed to learn different alternatives or choices from a human expert examples? The answer lies in using a network which is trained to produce a numerical output "score" for each input alternative. Since the system learns by examples, it is trained on a data base of queries which are assigned a numerical score by a human expert. To make a choice, the network scores each alternative and selects the answer with the highest score. The pitfalls of this approach are:

1. In many domains where n is large, the expert spends a lot of time creating a data base with evaluated alternatives.

2. In many domains, human experts do not think in terms of absolute scoring functions; this would make it extremely difficult to create training data containing absolute scores, since scoring is alien to the expert's subjective thinking about the problem.

A better method for creating training data is by recording the expert in action (i.e., each of the alternatives the expert chooses from are recorded for each problem along with the best selected alternative). The network is taught to compare more than two alternatives, rather than scoring individual alternatives. The input consists of two sets of alternatives or more; the output is set at 1 or 0, depending on which alternative is better. From a set of recorded human expert preferences, one can then teach the network the expert's choice as the alternative. One concern with this approach is that during performance mode or after the network is trained, it may be necessary to make $n * m$ (where m is the number of input alternatives and n is the number of examples to train the network on) comparisons to select the best alternative, whereas only individual scores are needed in the other approach. However, the network can select the best alternative with n comparisons in order by checking the list of alternatives and comparing the current alternative with the best alternative. If the current alternative is better, it becomes the new best alternative; if it is worse, it is rejected. Another potential concern is that the network which only knows how to compare choices may not produce a consistent rank-ordering (i.e., it may indicate that alternative a is better than alternative b, b is better than c, and c is better than a). This makes it difficult to determine which query must be selected. It is possible to guarantee consistency, however, with a constrained architecture, which forces the network to come up with absolute numerical scores for individual queries.

3 Dempster-Shafer belief function model for multiversion information retrieval

From the single version information retrieval point of view, Dempster-Shafer belief model (Shafer and Logan [1987]; Shafer et. al. [1987]) is a model that measures the agreement or relevance of a document d to a query q. From the multiversion information retrieval view, Dempster-Shafer model will measure the relevance of a document d to different formulations of a query q by using different search techniques. It will evaluate the relevance of a document d to the same query formulated differently by measuring the relevance of the documents selected that are relevant to document d.

3.1 Document representation

Let θ denote a finite set of propositions. Suppose that only one proposition is true, but we do not know which one it is. Let A be a subset of θ. We define belief (Bel) by: $Bel : 2^{\theta} \rightarrow [0,1]$. Bel distributes degrees of belief (Shafer and Logan, 1987) for all subsets of θ. Bel is said to be a belief function iff there is a random non empty set S of θ such that: $Bel(A) = Pr[S \subset A], \forall A \subset \theta$. If Bel is a belief function and S is its associated random set, then there is only one function $m : 2^{\theta} \rightarrow [0,1]$, called basic probability assignment (bpa), such that : $m(A) = Pr[S = A], \forall A \subset \theta$. The quantity $Bel(A)$ measures the total belief committed to A. The set $F = [A|A \subset \theta, m(A) > 0]$ is called a focal set of Bel and its elements are called focal elements of Bel.

Another function that conveys the same information that Bel does is the function $Pl : 2^{\theta} \rightarrow [0,1]$ which distributes degrees of plausibility for all subsets of θ. Function Pl is called plausibility function and is defined by (2), where $\neg A$ is defined by $\neg A = \theta - A$. It can be shown that $Bel(A) \leq Pl(A)$, for all subsets of θ, and that :

$$\begin{aligned} Pl(A) &= 1 - Bel(\neg A) && (1) \\ &= \sum_{B \cap A \neq \emptyset} m(B) \qquad \forall A \subset \theta && (2) \end{aligned}$$

The following definitions are applicable to Dempster-Shafer model for information retrieval. Let D denotes the set of documents and T denotes the set of controlled terms. T(d) then denotes the subset of terms in T that appear in document d. S(t) also denotes the subset of terms in T that are synonyms to term t. Assume that t $\in$ S(t) and that S(t)=S(x) $\forall x \in$ S(t). (S(t)) designates a function that returns the descriptor of S(t). Let $C= [\alpha | \alpha \in T, \alpha =$ Descriptor $(S(\alpha))]$ denotes the subset of terms in T that are descriptors. Assume that each set of synonyms, S(t) $\forall t \in$ T, has only one descriptor which belongs to S(t). Let $\alpha \in C$ be a descriptor, Since $\alpha \in S(\alpha)$, and exactly one term in $S(\alpha)$ is a descriptor, it must be α. We know that any term in $S(\alpha)$ can appear in document d, but only descriptor α represents the elements of $S(\alpha)$ and the semantic concept that they convey. Therefore, the relative frequency of α as a representative of $S(\alpha)$ in each document $d \in D$ is given by:

$$m_d(\alpha) = \frac{\sum_{t \in S(\alpha)} f(t,d)}{\sum_{t \in T(d)} f(t,d)} \tag{3}$$

$m_d(\alpha) = 0$ only when f(t,d)=0, $\forall$ terms t such that $t \in S(\alpha)$.

Suppose we were to describe the belief function representation of documents, and suppose that $\theta = [\alpha | \alpha \in C, N(\alpha) = \emptyset]$, and θ^* is the power set of θ. If $x \in C$ and $N(x) \neq \emptyset$, then x labels a subset of θ. The term x is given by the union of the atomic descriptors in N(x). If $x \in C$ and $N(x) = \emptyset$, then x is an atomic descriptor and $x \in \theta$.

As the set of descriptors, C, is a subset of θ^* we can take $m_d : \theta^* \rightarrow [0,1]$ as defined in (3), to be the belief function Bel_d. We can define for any $S \in \theta$, Bel_d and Pl_d:

$$Bel_d(S) = \sum_{t \subset S} m_d(t) \tag{4}$$

$$Pl_d(S) = \sum_{S \cap t \neq \emptyset} m_d(t) \tag{5}$$

3.2 Query representation

Each user's query is given by (F_q, W). F_q is a subset of descriptors contained in D and $W : F_q \rightarrow [0, \infty]$ is a weight function that expresses the user's belief in each descriptor $t \in F_q$ as a representative of the semantic content of the documents to be retrieved. Then, each user's query is transformed by the retrieval system into a belief function query Bel_q with focal set F_q and bpa m_q. In this case, the bpa m_q, $\forall\ \alpha \in F_q$, is given by:

$$m_{q(\alpha)} = \frac{w(\alpha)}{\sum_{t \in F_q} w(t)} \tag{6}$$

$Bel_q(\alpha)$ is the user's belief in α as being the best representative of the semantic content of the documents that he or she would like to retrieve.

3.3 Agreement between a query and a document

Let Bel_q and Bel_d be belief functions representing the user's query d and the document q, respectively. The agreement between these two functions is an estimate of the degree of relevance of document d to the query q. This estimate is a function of the semantic relations among the descriptors in query q and document d. Let S_d and S_q be the random sets of Bel_q and Bel_d, respectively. Then we define the agreement between d and q:

$$A(d,q) = Pr(S_d \cap S_q \neq \emptyset)$$

The ranking of documents in D with respect to a query q is achieved by estimating the degree of relevance of each document d to the given user's query. The ordering of documents according to decreasing values of these agreements will achieve the appropriate ranking.

3.4 Agreement between many multiple search queries and a document

Let Bel_{q1} and Bel_{q2} be belief functions representing two reformulated user queries by two different query reformulation algorithms of the same original query q. They can be combined when their associated random sets are probabilistically independent and $Pr(S_{q1} \cap S_{q2} \neq \emptyset)$ > 0. Let Bel denote the combined belief function, then its bpa m is given by:

$$m(A) \quad = \quad Pr(S_{q1} \cap S_{q2} \equiv A | S_{q1} \cap S_{q2} \neq \emptyset) \qquad \forall A \subset \theta.$$

The rule that permits to combine two or more belief functions is known as the Dempster rule, or orthogonal sum, which is presented as follows:

$$K^{-1} \quad = \quad \sum_{A_1 \cap A_2} m_1(A) m_2(A) > 0$$

$$m(A) \quad = \quad K. \sum_{A_1 \cap A_2 \neq \emptyset} m_1(A_1) m_2(A_2)$$

Where K is a constant of normalization. Its inverse can also be expressed in either of the following equations:

$$\begin{aligned} K^{-1} &= \sum_{A_1 \in F_1} m_1(A_1) \sum_{A_1 \cap A_2 \neq 0} m_2(A_2) > 0 \\ &= \sum_{A_1 \in F_1} m_1(A_1) Pl_2(A_1) \end{aligned}$$

and

$$K^{-1} \quad = \quad 1 - \sum_{A_1 \in F_1} m_1(A_1) Bel_2(\neg A_1)$$

Let Bel_d be the belief function representing the document d. The agreement between Bel_{q1} and Bel_{q2} and the function Bel_d is an estimate of the degree of relevance of document d to query q, which are reformulated in two different queries: q_1 and q_2. This estimate is a function of the semantic relations among the descriptors in both query q_1 and query q_2 and document d. Let S_{q1}, S_{q2}, and S_d be the random sets of Bel_{q1} , Bel_{q2}, and Bel_d, respectively. Then we define the agreement between d, q_1, and q_2 by:

$$A(d, q_1, q_2) = Pr(S_d \cap S_{q1} \cap S_{q2} \neq \emptyset)$$

The ranking of documents in D with respect to queries q_1 and q_2 is achieved by estimating the degree of relevance of each document d to both queries q_1 and q_2 combined. The ordering of documents according to the decreasing values of these agreements will achieve the appropriate ranking.

4 Experimentation, results, and conclusion

The experimental collection consisted of approximately 2500 documents, which were randomly selected from the journal Computer Science VINITI (journal of abstracts), 1982-1987. The experiment was based on 35 search requests. For each search request, the users selected relevant documents based on specific criteria. As a result, each search request had a list of relevant documents contained in the experimental collection. Two different methods for constructing query formulation were used in the experiment, and for each search request there were two query formulations. For each search request, query reformulation was used in the search, and in each output the relevant documents were identified. Then, recall (R) and precision (P) levels were computed for each output. The best query formulation was identified by using both Connectionist Selection Paradigm and Dempster-Shafer's technique:

1. Following the standard practice of back propagation (Rumelhart and McClelland, [1986]), a selection paradigm network has one input layer, one or more layers of hidden units, and an output layer, with full connectivity between adjacent layers. The input layer represents two final query formulations of the same search $query_1$ and $query_2$ (i.e, subject, type of publication, name of publication, and date) and the output layer has only a single unit to represent the best query formulation. The "teacher signal" for the output unit would be 1 if $query_1$ was better than $query_2$ from the relevant documents point of view. The proposed selection paradigm overcomes the limitation only by being able to consider queries in isolation (i.e., without knowledge of other existing alternative queries). In addition, the sophisticated encoding scheme that was developed to encode transition information was not needed, since comparisons can be based solely on the final query negotiation. The selection paradigm approach offers greater sensibility in distinguishing between close alternatives as it is the closest to a human expert's knowledge. A major problem with this approach is the consistency of the network's comparisons. Two properties are needed for complete consistency:

   (a) The comparison between any two queries must be unambiguous (i.e., if the network indicates that $query_a$ is better than $query_b$ when $query_a$ position is $query_1$ and $query_b$ position is $query_2$, the selection network will be consistent if it still

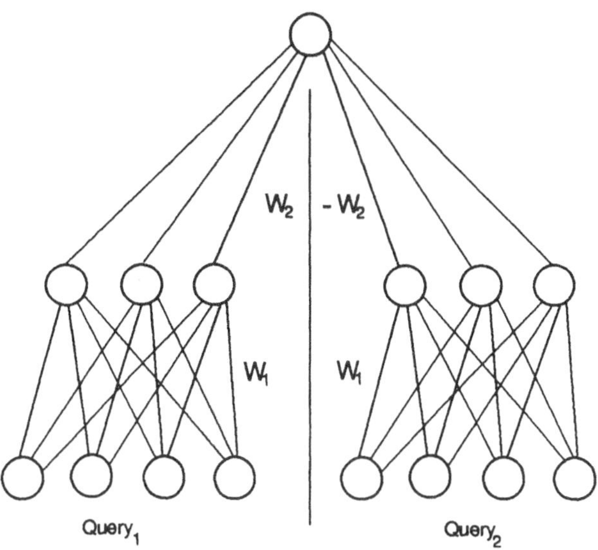

Figure 1 : Selection Paradigm Neural Network applied to $Query_1$ and $Query_2$

retrieves $query_a$, which is better than $query_b$ when $query_a$ position is $query_2$ and $query_b$ position is $query_1$.

(b) The comparisons must be transitive (i.e., if $query_a$ is judged better than $query_b$ and $query_b$ is judged better than $query_c$, the network retrieves $query_a$, which is better than $query_c$).

For this particular application, the network has two different queries as input. Thus, the symmetry among weights at the hidden layer enforces the output inversion symmetry. The transitivity and rank-order consistency can be guaranteed by the inversion shown in Figure 1. As seen in Figure 1, the selection paradigm network is made out of two halves with each working on one particular query. As long as that the halves are not cross-coupled, then the evaluation of each of the input queries will be measured by a single real number. Because real numbers rank-order consistently, the network's performance is always consistent. The training procedure consisted of 20 queries out of 35 queries with a varying number of hidden layers and one single output unit. The queries in the training set were randomly selected from the whole set of 35 queries. Testing performance of this approach is compared to that of Dempster-Shafer's, which is summarized in Table 1.

2. Dempster-Shafer's approach was applied to the same set of queries. A relative frequency table was built on the whole set of documents and measurements of Bel_d and Pl_d were derived from m_d, which, in turn, was derived from the frequency table. In addition, a relative frequency of descriptors, and degrees of agreement of the documents for the 25 set of queries were established by both algorithms. The results indicate that ANN is closer to the best theoretical value established by Frants et. al.

Characteristics Used in the Averaging Process	*Values of the Averages of Search Characteristics as a result of search based on query formulation by:*		
	$Algorithm_1$	$Algorithm_2$	Theoretical Result[a]
R	0.423	0.473	0.476
P	0.494	0.343	0.491
$\sqrt{R.P}$	0.457	0.403	0.46
$R+P$	0.917	0.816	0.967
F(R,P)	0.827	0.858	0.810
DS	0.42	0.39	0.46
ANN	0.459	0.455	0.46

R is recall, P is precision, DS stands for Dempster-Shafer, ANN artificial neural networks, and F(R,P) is defined by: $F(R,P) = 1 - \frac{1}{\frac{1}{R}+\frac{1}{P}-3}$

[a]The Theoretical result of the best query formulation was determined by Frants et.al.[1993]

Table 1 : Values of search characteristics after averaging all search requests

[1993] than that of the Belief function model. Because more queries are needed to confirm that ANN retrieves more relevant documents than Dempster-Shafer's belief function model, other subject domains such as information science is being considered for evaluation by both approaches. Table 1 summarizes the whole performance evaluation schema of ANN, Dempster- Shafer model, and few results obtained from Frants et.al.[1993]. ANN is by far, and in many cases, the best approach for the specific set of queries.

5 References

Dagli, C., Burke, L.I., Fernandez, B.R. and Gosh, J.(Eds) [1991,1992,1993]. Intelligent Engineering Systems through Artficial Neural Networks. ASME Press, New York, NY, Vols. **1,2,3**.

Frants, V.I., Shapiro, J., and Voiskunskii, G. [1993] "Multiversion Information Retrieval Systems with feedback mechanism selection" *Journal of the American Society for Information Science*, Vol 4, *No 1*, pp 19-27.

Gallant, S.I. [1988], "Connectionist Expert Systems" , *Comm. ACM* , Vol **31**, 152-169.

Meghabghab, G.V. and Nasr, G.E. [1993] "Segmentation of MRI's of the brain for the detection of Multiple Sclerosis by Mathematical Morphology and Neural Networks" in Dagli, C., Burke, L.I., Fernandez, B.R. and Gosh, J.(Eds) [1993], Intelligent Engineering Systems through Artficial Neural Networks, ASME Press,Vol. **3**, New York, NY, pp 387-396.

Meghabghab, G.V. and Meghabghab, D.B. [1994] "INN: An Intelligent Negotiating Neural Network for Information Systems: A design Model" *Information Processing & Management*, **Vol. 30**, *No 5*, pp 663-685.

Rumelhart, D.E. and McClelland, J.L. (Eds.) [1986]. Explorations in Parallel Distributed Processing. MIT Press,Cambridge, MA, **Vol. 1**.

Shafer, G. [1976]. A Mathematical Theory of Evidence. Princeton University Press, Princeton, NJ.

Shafer, G. [1987], "Belief Function and Possibility Measures" in Bezdek, J. (Ed.), The Analysis of Fuzzy Information. CRC Press, **Vol. 2.**, Boca Raton, FL.

Shafer, G. and Logan, R. [1987] "Implementing Dempster's Rule for Hierarchical Evidence" *Artificial Intelligence*,**Vol. 33**, pp 271-298.

Shafer, G., Shenoy, P.P., and Mellouli, K. [1987] "Propagating Belief Functions in Qualitative Markov Trees" *International Journal of Approximate Reasoning*, **Vol. 1**, pp 349-400.

A Study of Genetic Algorithms to Find Approximate Solutions to Hard 3-SAT Problems

Jeremy Frank
frank@cs.ucdavis.edu
(916) 758-5925
Division of Computer Science
University of California at Davis
Davis, CA. 95616

Abstract

Genetic algorithms have been used to solve hard optimization problems ranging from the Travelling Salesman problem to the Quadratic Assignment problem. We show that a Simple Genetic Algorithm can be used to solve an optimization problem derived from the 3-Conjunctive Normal Form problem. By separating the populations into small sub-populations, parallel genetic algorithms exploit the inherent parallelism in genetic algorithms and prevent premature convergence. Genetic algorithms using hill-climbing conduct genetic search in the space of local optima, and hill-climbing can be less computationally expensive than genetic search. We examine the effectiveness of these techniques in improving the quality of solutions to 3-SAT problems.

1 Introduction

The Genetic Algorithm proposed by Holland [Ho] has become the basis for a new approach to optimization problems. In a genetic algorithm, sets of solutions called populations are succeeded by new solutions which are closer to the desired optima. Successive generations are created by selecting solutions from the previous generation using a fitness function related to the objective function of the optimization problem. The new population is then subject to genetic operators such as crossover and mutation to extend the search to other parts of the search space [Go]. Genetic algorithms (hereafter called GAs) have been tested on a wide variety of problems, including the combinatorially difficult problems in the class NP. A typical NP-Hard problem is the Boolean Satisfaction Problem, where we desire an assignment of boolean variables which satisfies a boolean formula. One version of the Boolean Satisfaction problem is called the 3-Conjunctive Normal Form problem (3-SAT) in which disjunctions of 3 distinct variables are conjoined in a formula. The optimization problem for 3-SAT, called MAX 3-SAT, requires maximizing the number of true clauses (disjunctions) in the formula, and is NP-complete.

Figure 1 shows a sample problem of this type.

E. A. Yfantis (ed.), Intelligent Systems, 547–554.

$$(\neg X \vee \neg Y \vee \neg Z) \wedge (X \vee Y \vee \neg V) \wedge (X \vee \neg Y \vee \neg Z) \wedge (U \vee V \vee W)$$

Figure 1: A 3-SAT problem with 4 clauses and 6 variables.

Backtracking procedures are used to find exact solutions to 3-SAT problems. Recent research has shown that, for randomly generated formulas, if the ratio of clauses to variables is close to 4.3 then these problems are the most difficult to solve [MiSe]. Greedy local techniques, used to solve MAX 3-SAT and 3-SAT show improvement over even the best backtracking methods [Se], [SeKa]. However, there are problems which can fool greedy algorithms into finding local optima unless the algorithms are specially modified [SeKa]. This brings to mind the so-called "deceptive" functions which are shown to fool GAs into finding suboptimal solutions [Mu].

To improve the performance of GAs on deceptive problems researchers have introduced parallel genetic algorithms, (referred to generally as PGAs). PGAs are designed to enhance genetic diversity by separating sub-populations. In some PGAs individuals are separated and only allowed to mate with certain neighboring individuals [Mu2], while in other algorithms sub-populations are allowed to exchange individuals after a certain time period has elapsed [Ta]. To improve some parallel algorithms, hill-climbing is introduced. Performing crossover on local optima gives better results than crossover on other points, and in such cases hill-climbing improves individuals more quickly than the genetic operators [MuSc].

In this paper we construct several GAs to solve the optimization version of the 3-SAT problem. We analyze these GAs and investigate the settings of parameters such as crossover probability, mutation probability, migration rate, migration interval and population parameters. We use a variety of performance measures to analyze the results of our algorithms. For the test suite, we propose to use randomly generated "hard" 3-SAT problems of different sizes.

2 The Genetic Algorithm and 3-SAT

To build a simple genetic algorithm (SGA) to solve the 3-SAT problem we first investigate the encoding. A natural encoding is a string of 0s and 1s indicating the truth assignment to each boolean variable indicated by the position of the bit in the string. The fitness function we use in this case is exactly the function to be optimized, the number of true clauses for a particular assignment. Crossover and mutation naturally follow from such an encoding and objective function. For the problem in Figure 1 a sample chromosome for $UVWXYZ$ is 100110; all clauses in the problem evaluate to true using this assignment.

2.1 The Hill-climbing Genetic Algorithm

In the Hill-climbing Genetic Algorithm (HGA), hill-climbing is done before crossover and mutation; thus the HGA can be considered to be a genetic algorithm operating in the space of local optima for a given problem. It should be pointed out that there may still be an exponential number of local optima in the search space. In cases where it is important to find as many local optima as possible, this method guarantees finding only local optima and attempts to find better local optima using the genetic operations. It may also be less expensive to use hill-climbing to do some of the work of the GA, thereby reducing the population size and/or the number of generations required to converge [Mu2]. For our experiments we shall use steepest ascent hill-climbing; among the options to be explored later are tabu search [Gl], simulated annealing and greedy local search [Se].

2.2 The Parallel Genetic Algorithm

Genetic algorithms are just as susceptible to premature convergence as other search procedures. To address this problem, genetic algorithms induce diversification throughout the lifetime of the algorithm using parallel populations which exchange individuals for genetic crossover [Mu]. These algorithms are known as parallel genetic algorithms or PGAs. Wright argues that for PGAs to avoid local optima a PGA should be divided into isolated sub-populations, one of which will eventually take over the others [Mu]. We take this view in designing a PGA to solve 3-SAT problems. Accordingly, the PGA is divided into sub-populations arranged in a ring [Mu]. Each subpopulation sends a small number of randomly picked individuals to it's neighbors after a number of generations [Ta]. These individuals randomly replace members of the populations they are sent to. By keeping the rate of migration small and the intervals of migration reasonably large, we maintain diversity in the sub-populations, but allow opportunities for improvement by exchanging individuals. Other schemes include elitism in choosing migrants and reverse elitism in choosing the individuals to replace in the new population. This algorithm can be modified by allowing hill-climbing.

3 Analysis

Our goal is to find the best GA configuration to solve 3-SAT problems. We ran a variety of experiments varying each of the GA parameters available and measured the value and variance of the local optima found on randomly generated problems. We tested problems of 25 variables. We tested the GAs on 100 randomly generated "hard" CNF problems [ChKa].

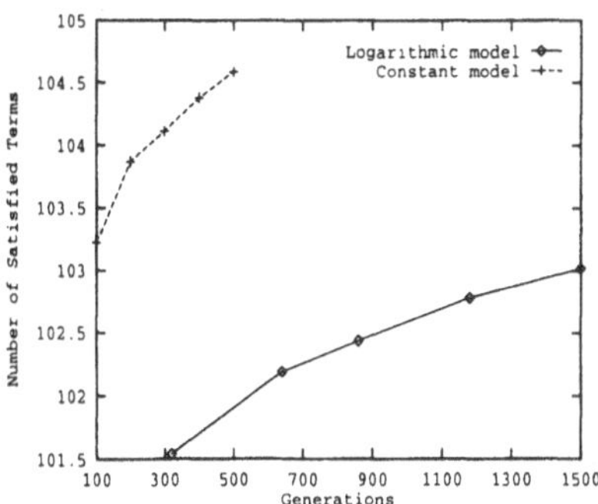

Figure 2: Performance of GAs under constant and log models.

In order to analyze our results we measured the average number of clauses solved by each algorithm. In many cases this average is skewed by problems which have satisfying assignments, so we also recorded the number of problems solved by each algorithm.

3.1 Population Size and Generations SGA

Goldberg [Go2] presents a method of sizing populations and determining the number of generations to convergence for serial and parallel GAs. We used this work to find the optimum number of generations for SGA using both the model of constant time convergence and logarithmic time convergence. For these experiments we fixed the probability of crossover at 20% and the probability of mutation at 2%. Figure 2 shows the results of this experiment. The logarithmic model requires only 4 members, and convergence time is logarithmic in the number of variables. In the constant model, the population size is 40, with the time to convergence constant. We see that the performance of the constant model is superior to that of the logarithmic model. For the remainder of the experiments we used 300 generations and 40 population size. It is interesting to note that the performance difference may be due to the population disparity, and should be the subject of future work.

3.2 Crossover and Mutation for SGA and HGA

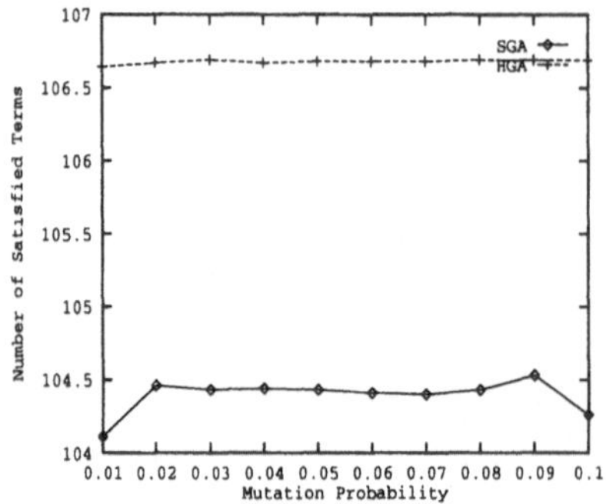

Figure 3: The Effect of Mutation on SGA and HGA.

We examined the effects of crossover and mutation on the performance of SGA and HGA. We varied the crossover probability from 10% to 100% by 10%, and varied the mutation probability from 1% to 10% by 1%. Figure 3 shows the effects of the mutation rate on the average number of clauses satisfied for SGA and HGA. We see that there is little overall effect in altering the mutation rate for either algorithm, although the data suggests that high rates below 0.1 are best. The data also suggests that increased mutation rates improve the performance of HGA, however slightly.

Figure 4 shows the effects of crossover on SGA and HGA. Here we see that higher rates of crossover improve the performance of SGA and HGA. This is most interesting in the result for HGA, since it indicates that hill-climbing is solely responsible for the increase in performance of HGA over SGA.

Figures 3 and 4 indicate that HGA outperforms SGA by a significant margin. This is more pronounced when the number of problems solved is analyzed. SGA failed to solve more than 3 of the instances in either the crossover or the mutation experiments, while HGA typically solved 67-69 of the problems.

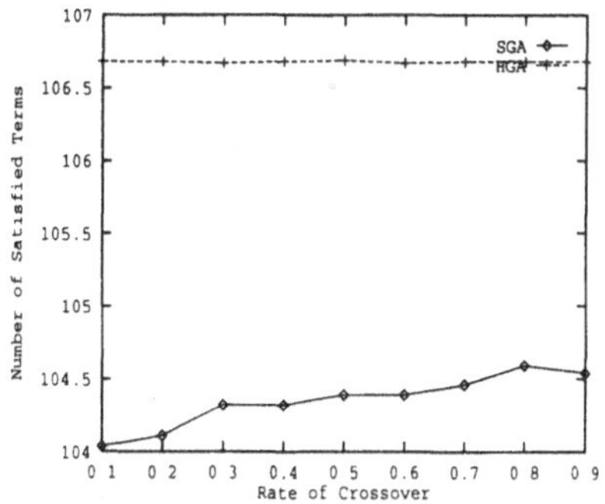

Figure 4: The Effect of Crossover on SGA and HGA.

3.3 Migration Rate and Interval for PGA and HPGA

Due to constraints imposed by the crossover mechanism, the size of any population must be both even and ≥ 4. In order to prevent larger population sizes from influencing the effects of parallelism on solving problems we divided the original population into different numbers of sub-populations. We required that the migration interval evenly divide the number of generations, and we also limited the number of migrations so that in all cases at least a few generations without migration would pass. Finally, while it is plausible to allow the rate of migration to be as high as 100% we bounded the number of migrants to 50%. Under these limitations we varied the following parameters: the number of subpopulations, the migration rate, and the migration interval. For these experiments we fixed the probability of crossover at 70% and the probability of mutation at 5%.

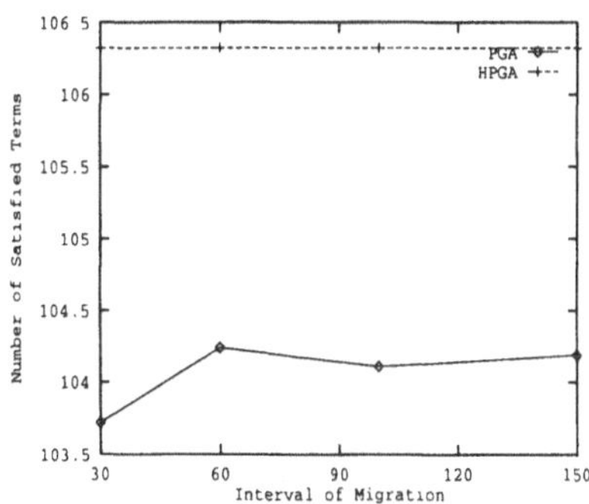

Figure 5: The Effect of Migration Interval on PGA and HPGA.

Figure 5 shows the effects of varying the migration interval on PGA and HPGA. To do this we kept the same maximum number of generations (300) and allowed migration after different length intervals. We held the number of subpopulations to 2 and the migration rate at 10%. If we recall Figure 2 we see that performance increases as the maximum number of generations increases, so we expect to see the same result here. Indeed, the figure shows that longer migration intervals improve the performance.

Figure 6 shows the effects of migration rate on PGA and HPGA. Here we held the migration interval at 100 and the number of sub-populations at 2 and tested migration rates

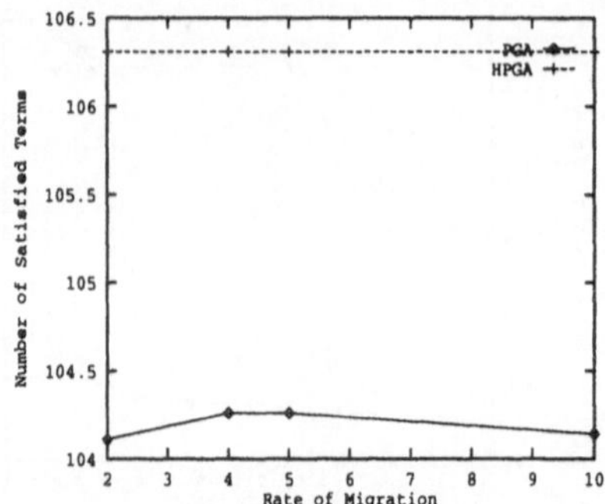

Figure 6: The Effect of Migration Rate on PGA and HPGA.

of 10%, 20%, 25% and 50%. We see that migration rates of 20% and 25% performed slightly better than the others. It is our opinion that using an elitist scheme would improve this result. Since we select migrants and replace at random, we do not guarantee introducing better material into subpopulations, merely new material.

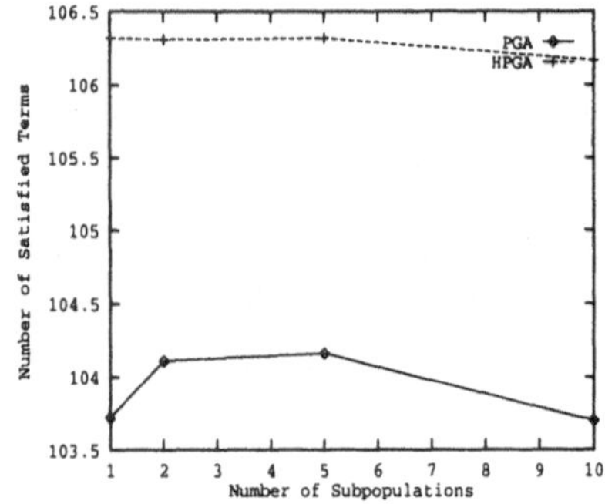

Figure 7: The Effect of Migration Rate on PGA and HPGA.

Figure 7 shows the effects of varying the number of subpopulations on performance. When the number of migrants, and the interval are held constant, this has the effect of varying the rate of migration and the rate at which information propagates from one sub-population to another. For this experiment we held the number of migrants to 2 and the interval of migration at 100. The figure indicates that the optimal number of subpopulations is 5, which corresponds to a migration rate of 25%. This result matches the previous result on the effects of migration rate.

Figures 5, 6, and 7 indicate that hill-climbing improves the effectiveness of PGA. Again, this is most apparent if we consider the number of problems solved by each algorithm. PGA never solved more than 1 problem, which HPGA solved as many as 32 problems. We can therefore conclude that adding hill-climbing can significantly enhance the power of GAs. However, more work must be done to analyze the cost of adding hill-climbing to GAs.

Algorithm	Maximum Problems Solved
SGA	3
HGA	69
PGA	1
HPGA	32

Figure 8: Maximum Problems Solved.

3.4 Does Parallelism Improve GA Performance?

It is interesting to note that parallelism as implemented in this work does not enhance the performance of GAs in solving CNF problems. The average number of clauses solved by SGA reached a maximum of 104.5 as shown in Figure 3, while for PGA the average was only 104.25 as shown in Figure 5. Again, the story is best told by examining the number of problems solved. We recall from the previous sections that HGA solved as many as 69 problems, while HPGA never solved more than 32.

4 Analysis

The data clearly shows that the addition of hill-climbing enhances the performance of GAs to solve 3-SAT problems. When the GA operates in the space of local optima it automatically improves the quality of solutions since any local optima is clearly a better solution than any of it's neighbors. Where there is an inexpensive algorithm to find local optima, this technique can enhance the performance of GAs at little cost.

Parallelism resulted in a degradation of GA performance, as clearly shown by Figure 8 and as seen in the graphs. It is interesting to note that the average number of clauses satisfied did not fall significantly when parallelism was added to HGA. However, the number of problems solved did fall by at least half. This implies that HPGA almost finds satisfying assignments for many solveable problems. Our explanation is that, by cutting the population sizes in the PGA, we reduce the quality of local optima that are found in each sub-population, thereby degrading overall performance. If this is true, then migration using random selection and replacement is not effective enough at recovering the ground lost by using smaller population sizes. Elitism may help to solve this problem, but it is possible that the only solution is to increase the population sizes so that each subpopulation of the PGA has enough members to guarantee that good local optima are found.

5 Future Work

A great deal of future work remains to be done in this area. We should test other hill-climbing methods such as tabu search and GSAT in HGA, and analyze the CPU performance of GAs compared with GAS employing hill-climbing. In PGA we currently select migrants and replace elements of the new populations at random. A wide variety of selection schemes should be tested and compared, including elitism in choosing migrants and reverse elitism in replacing members of the host populations. In addition, it would be useful to consider either annealing schedules or adaptive schemes for migration. Such schemes would either send out migrants or request neighboring populations to send migrants when the subpopulation converges. We should also consider new neighborhood topologies such as grids or hypercubes. Finally, the ultimate goal of a parallel algorithm is to save time as well as improve performance. We should investigate distributed or massively parallel implementations of GAs and analyze real time performance gains.

References

[ChKa] P. Cheeseman, Bob Kanefsky, William Taylor. "Where the *Really* Hard Problems Are." *Proceedings of the* 12th *IJCAI,* 1991.

[Gl] F. Glover. "Tabu Search for Nonlinear and Parametric Optimizations", *Discrete Applied Mathematics*, 1991

[Go] D. Goldberg. *Genetic Algorithms in Search, Optimization and Machine Learning.* Menlo Park, Wiley and Sons, 1989.

[Go2] D. Goldberg. "Sizing Populations for Serial and Parallel Genetic Algorithms." *Proceedings, 3d International Conference on Genetic Algorithms*, 1989.

[Ho] J. Holland. *Adaptation in Natural and Artificial Systems.* Univ. of Michigan Press, 1975.

[LiBa] G. Liepens, S. Baluja. "apGA: An Adaptive Parallel Genetic Algorithm." In *Computer Science and Operations Research, New Developments in Their Interfaces,* Balci, Sharda, Zeinos, ed., Pergamon Press, 1992.

[MiSe] D. Mitchell, B. Selman, H. Levesque. "Hard and Easy distribution of SAT Problems." *Proceedings, AAAI,* 1992, pp. 459-465

[MuSc] H. Mülenbien, M. Schomisch, J. Born. "The Parallel Genetic Algorithm as a Function Optimizer." *Proceedings, 4th International Conference on Genetic Algorithms,* 1991, pp. 271-278.

[Mu] H. Mülenbien. "Evolution in Time and Space: The Parallel Genetic Algorithm." In *Foundations of Genetic Algorithms*, G. Rawlins, ed., Morgan Kaufman, 1991.

[Mu2] H. Mülenbien. "Asynchronous Parallel Search by the Parallel Genetic Algorithm." *Proceedings, 3d IEEE Symposium on Parallel and Distributed Systems,* 1992, pp. 526-533.

[Se] B. Selman. "A New Method for Solving Hard Satisfiability Problems." *Proceedings, AAAI,* 1992, pp. 440-446.

[SeKa] B. Selman, H. Kautz. "An Empirical Study of Greedy Local Search for Satisfiability Testing." *Proceedings, AAAI Spring Symposium on Artificial Intelligence and NP-Hard Problems,* 1993, pp. 149-155.

[Ta] R. Tanese. "Distributed Genetic Algorithms." *Proceedings, 3d International Conference on Genetic Algorithms,* 1989.

FINGERPRINT IMAGE CODING BY A CLUSTERING LEARNING NETWORK

W. Chang, H. S. Soliman & A. H. Sung
Department of Computer Science
New Mexico Institute of Mining & Technology
Socorro, New Mexico 87801-4682 USA
Internet E-mail changw@nmt.edu

Abstract. A self-organizing neural network performing Learning Vector Quantization (LVQ) is proposed in this paper to compress image data in still pictures. The advantages of our model are its low training time complexity, high utilization of neurons, robust clustering capability, and simple computation; further, a VLSI implementation is highly feasible. By unsupervised learning, our LVQ neural model finds near-optimal clustering from image data and builds a compression codebook in the synaptic weights. The compression results are competitive comparing with the currently popular transform codings such as JPEG and wavelet methods. The neural codebook trained by a few pictures can be used to compress other pictures efficiently. Special image types such as the fingerprints exhibit this property in our experiments. Other experiments involve some filtering effects and techniques to enhance the neural codebook learning to yield higher picture quality.

Keywords. Image Compression, Neural Networks, Learning Vector Quantization.

1 Introduction

The research of better image compression schemes have become important and several *de facto* standards have been developed [1]-[4]. Conversion of relatively high rate data to lower rate data virtually always entails a *loss* of fidelity or an increase in distortion. Hence, the goal of data compression is to obtain the best possible fidelity for the given compression, or, equivalently, to maximize the compression for a given fidelity [5]. We introduce a neural network approach which is based on a *compression-through-classification* scheme and produces visually lossless results with high compression.

Image compression using neural networks started with an experiment of training a *backpropagation network* (BPN) to perform transform learning [6]. Our approach, on the other hand, addresses a self-organizing, vector-clustering technique to learn the principal components (features) in the pictorial data by *memorizing* them in the form of synaptic

E. A. Yfantis (ed.), Intelligent Systems, 555–564.

weights. Such a learning is to "embody" the compression coding into the adaptive weights for a simple and fast operation. The merit of neural networks is their capability of approximating some function which may not be known *a priori*, or to find a near-optimal solution for a hard computation problem (such as the vector clustering problem). The advantage of using a neural codebook instead of some complex transform is that once the neural codebook has been built by training with one image, it can be used to compress other pictures of the same type in a speedy operation. An example is a neural codebook trained by one fingerprint image and used to "cross-compress" other fingerprint images.

Classical vector quantization as an image compression technique can produce low data rates but there are issues of complex computation and the *blockiness* or *staircase* artifacts in the decompressed picture [4], [7], [8], [9]. The artifacts are caused by the small errors in the decompressed pixels which line up on the boundaries between subimage blocks. Our approach of addressing these issues will be described.

Implementation of *Counterpropagation Networks* (CPN) and *Kohonen Feature Maps* (KFM) for image compression is suggested in [10]-[13]; and in our past work, we have experimented with various networks [14]-[17]. Self-organization in CPN and KFM is a statistical *natural-clustering* process in which the network performs *competitive learning* to *perceive* pattern classes based on data *similarity* or *principal component analysis* [18], [19]. The coding of subimages (input data patterns) from the target picture is a *compression-by-classification* process which generates the class indices (neuron indices) as code words [5]. We implement *Learning Vector Quantization* (LVQ) of [20] into a neural model and name it *Sampling Frequency-Sensitive Network* (SFSN) which clusters most efficiently and yields the least *Normalized Mean-Square Error* (NMSE) among the various networks we have experimented. The learning speeds of SFSN and three other major self-organizing networks, namely, CPN, KFM, and a *Winning-frequency Self-organizing Network* (WFSN, [21]), are compared in Figure 1 by the following method: Upon the finishing of each training cycle (when all training vectors are presented once), the NMSE is accumulated by the differences of the winning weight vectors and the training vectors. The SFSN has the fastest diminishing NMSE in the training cycles.

In section 2, the vector clustering problem and SFSN which incorporates a *Neuron Replenishment* (NR) technique and a *Centroid Adaptation at Class Stabilization* (CACS) method are explained. Section 3 describes the compression process and compares the results with JPEG and the Wavelet/Scalar Quantization (WSQ) method developed by Los Alamos National Laboratory (LANL) and FBI. Section 4 concludes the SFSN approach.

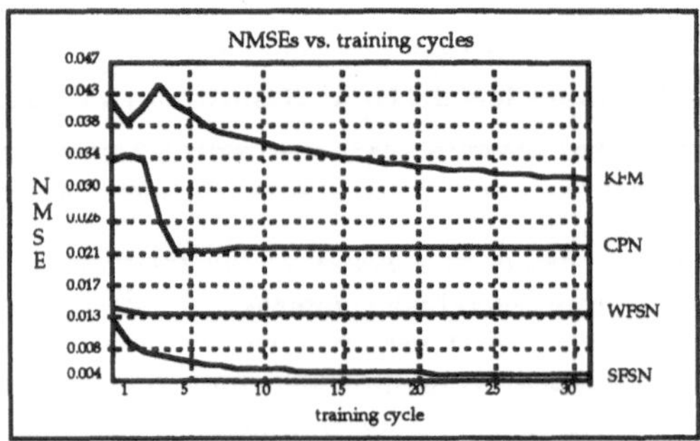

Figure 1. Four major natural-clustering self-organizing networks adapt their weight vectors to the centroids of training vector clusters. SFSN has the fastest approximation (with the smallest error).

2 A Clustering Neural Model

The clustering problem is to approximate the cluster (class) centroids in the training vectors by the weight vectors such that the accumulated distance between each training vector and its closest weight vector is within a tolerance. Such an accumulated distance is called an error or distortion in the quantization process. Since the training vectors can be partitioned into an exponential number of formations it is computationally hard to search for a formation having a cluster number equal or less than a given weight number, and to yield a distortion smaller than a given tolerance [22], [23]. The weight vectors arranged in the clustering learning process can be used as a classification codebook. Self-organizing networks such as KFM and CPN adapt weight vectors to the centroids of the pattern clusters in a faster convergence rate than the traditional vector quantization method [24]. However, the *nearest-neighbor* method of KFM and CPN leaves a large amount of neurons underutilized after training. Our model resolves the issue by applying LVQ which employs no neighborhood neurons.

2.1 FINDING NATURAL CLUSTERS

Data clustering is an important methodology in exploratory data analysis. We assume the data constitute a set of N points, $\boldsymbol{D} = \{\boldsymbol{x}_i: i = 1, ..., N\}$ each of which is an ordered m-tuple, where m is the number of *features*, $\boldsymbol{x}_i = \langle x_{i1}, ..., x_{im}\rangle$. The features themselves will be named $P^1, ..., P^m$. The data can thus be viewed as a collection of N points in a m-dimensional space, the *feature space*. The objective of clustering is to partition the data set into some non-empty subsets such that *alike* data are grouped together and data in different subsets or clusters are not *alike* [25].

Usually, blocks of subimages are treated as vectors and the similarity is indicated by the Euclidean distance between these vectors [5], [7]-[9], [25]. A clustering of a vector set $\boldsymbol{D}$ of size N is a partition of $\boldsymbol{D}$ into some number J of subsets, clusters $C_1, ..., C_J$, which are mutually exclusive and jointly exhaustive of $\boldsymbol{D}$. Each vector $\boldsymbol{x}_i \in \boldsymbol{D}$ is assigned a number $j \in \{1, ..., J\}$, the cluster index to which $\boldsymbol{x}_i$ belongs, and has no ordinal significance. Let $\boldsymbol{y}_i$ be the centroids of C_i, so

$$C_i = \left\{\boldsymbol{x} \in \boldsymbol{D} \mid (\forall j)\left[j \neq i \Rightarrow \|\boldsymbol{x} - \boldsymbol{y}_i\| < \|\boldsymbol{x} - \boldsymbol{y}_j\|\right]\right\}$$

where $\|\cdot\|$ is the Euclidean norm. Each vector in C_i is closer to $\boldsymbol{y}_i$ than to any other centroid $\boldsymbol{y}_j$. The partition is a *natural-clustering* process, and C_i's are *natural clusters* since each region of each C_i tends to be convex or isotropic [10].

When the dimension of the vector space v (the training subimage size) and the number of clusters J (the codebook size) are both chosen *a priori*, the iterative algorithm for finding the set C_i ($i = 1, 2, ..., I$) which produces the minimum accumulated error such that

$$E_{min} = \min\left\{\Sigma_i \Sigma_j \sqrt{\left(\boldsymbol{x}_j - \boldsymbol{y}_i\right)^2}\right\}$$

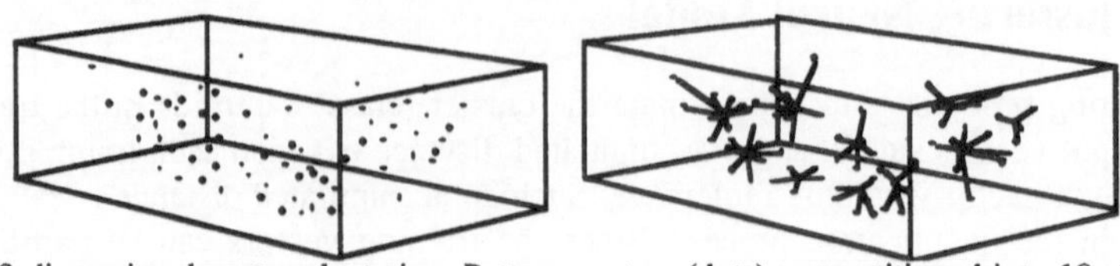

Figure 2. A 3-dimensional vector clustering. Pattern vectors (dots) are partitioned into 10 clusters by the nearest-neighbor rule (nearest to a centroid vector). The small lines are the distances between each pattern vector and its centroid vector.

where y_i is the centroid of C_i and nearest to x_j, has a deterministic polynomial time complexity if the vector space is discrete [23]. That is, given N positions and I learning vectors, the time requires to go through all the different positions for the J learning vectors has the complexity $O(N^J)$.

Figure 2 illustrates the clustering of data patterns in a 3-dimensional vector (feature) space. Giving vectors at N different positions, the formation of 10 clusters has roughly N^{10} different combinations. The search for one formation having a minimum number of clusters and come up with a satisfactorily small distortion, has an exponenetial time complexity. The clusters are generalized classes and obtaining fewer clusters having a smaller accumulated error guarantees higher compression for the coded image and better fidelity for the decompressed image.

2.2 SAMPLING FREQUENCY-SENSITIVE NETWORK (SFSN)

The SFSN (Fig. 3) has the advantages of lower time and space complexities since the SFSN learning rules require less training iteration than other networks and the number of neurons required in SFSN is far less than that in CPN or KFM to form the same number of classes. The basic SFSN learning rules are as follows:

1) Initialize weight vector W_i of the ith neuron in the C-neuron network such that

$$W_i(0) = R(i),\ i = 1, 2, ..., C \ \text{ and } \ W_i(0) = [w_{i1}(0), w_{i2}(0), ..., w_{im}(0)]$$

where $R(\cdot)$ is a random vector generation function, m is the vector dimension (training subimage size), (0) is the training time index at beginning. The initial weights can also be randomly selected from the training subimages [20], [21].

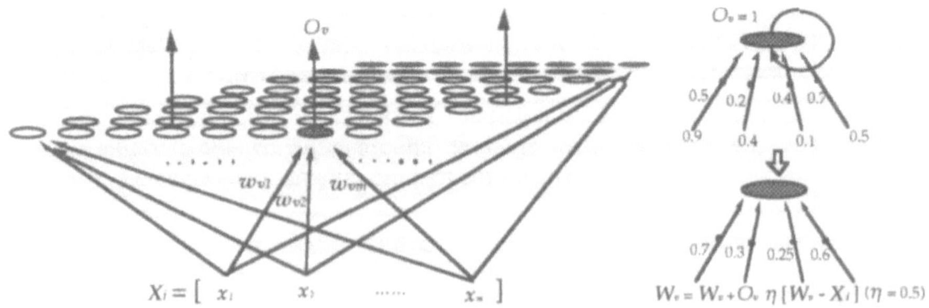

Figure 3. SFSN structure and weight adaptation (The learning rate is given as 0.5, as an example.)

2) Select an initial learning rate η such that $0 < \eta < 1$.
3) Present a training vector $X(t) = [x_1(t), x_2(t), ..., x_m(t)]$.
4) Compute the distance $D_i(t)$ by

$$D_i(t) = \sum_{j=1}^{m} [x_j(t) - w_{ij}(t)]^2,.$$

5) Set output O_i of the ith neuron by

$$O_i = \begin{cases} 1, & D_i(t) < D_j(t), \quad \text{for all } j, \; i \neq j \\ 0, & \text{otherwise} \end{cases}$$

6) Each W_i adapts by

$$W_i(t+1) = W_i(t) + \eta(t) \cdot O_i(t) \cdot [X(t) - W_i(t)].$$

7) Update η by $\eta(t+1) = \eta(t) + \eta_\Delta$ where η_Δ is a small negative term.
8) Repeat steps 3 through 7 until $t = N \cdot K$ where N is the number of training vectors, K is the *sampling frequency* such that in each iteration all training vectors are presented once.

Preliminary experiments exhibit a constant $K = 32$ is sufficient for the learning process of SFSN. This is due to the class stabilization initiates in only a few training iterations and the centroid approximation afterward is relatively fast to approach the the convergence property. The class stabilization can be observed in the simulation and can be used an indicator for initiating a learning enhancement technique to shorten the prolonged centroid approximation stage and to improve the neural codebook described.

2.3 CENTROID ADAPTATION AT CLASS STABILIZATION (CACS)

The SFSN training process consists of two stages: *class forming* and centroid approximation. The class forming stage completes as the training vectors reach the *class-stable* condition. That is, on input X_j, the class index C_j is the index argument of W_c, the nearest weight in the vector space from X_j [20], so

$$C_j = \arg\min_i \{\|X_j - W_i\|\}$$

and the classes stabilize when

$$C_j^t = C_j^{t+1} \Rightarrow C_j^t = C_j^u \quad \text{for all } u > t \qquad \text{(Eq. 1).}$$

Figure 4 illustrates the class forming stage in the following: After presenting the 6 patterns (in the same numbering sequence) in the feature space for 3 cycles, the 3 weight vectors (squares) have moved into to the corresponding cluster and will not change the clusters they occupy (no matter how many extra training cycles are provided). That is, each pattern is measured closer to *a fixed* weight vector. The classes are stabilized. Note in the first cycle pattern 1 attracts two different weight vectors but attracts only one weight vector in the latter cycles.

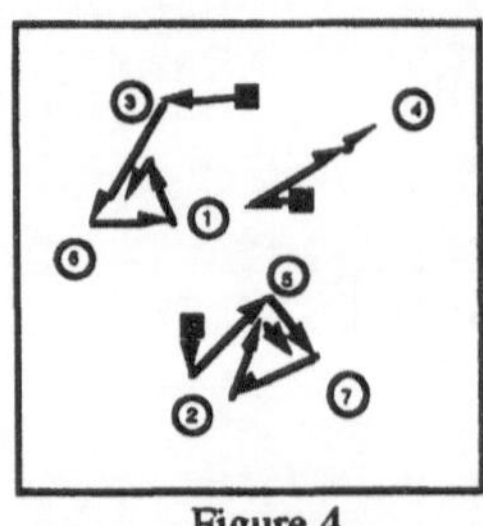

Figure 4

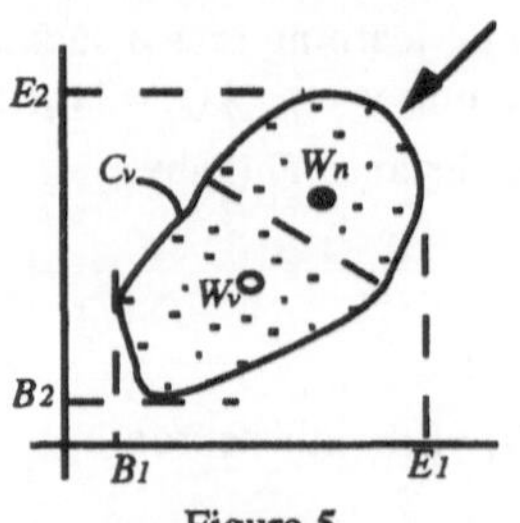

Figure 5

Figure 4. Class stabilization. Weight vectors (squares) move to centroids of training vector (circles) clusters. The numbers indicate the presenting sequence of the training vectors (during training).
Figure 5. The NR technique. A new adaptive weight vector W_n moves into the old region of W_v, at away from W_v on the long axis (the arrow) of the region. The region is then split into two smaller regions (clusters).

The class forming stage takes a few training cycles (within 32 iterations of target data in our experiments) and this conforms with [20] (classes are formed with training patterns of the size 30 to 50 times of the number of weights or neurons.) In our experiments, $K = 32$ provides a sufficient number of iterations since $N \cdot K >> D \cdot 50$ where D is the network (codebook) size. Hence, the codebook and the training time can be improved by the following CACS procedures:
(1) Terminate the training process after the class forming stage (satisfying Eq. 1).
(2) Compute centroids in each cluster to substitute the corresponding weight vector.
Thus, the prolonged centroid approximation stage can be eliminated.

2.4 NEURON REPLENISHMENT (NR)

By adopting a codebook replenishment vector quantization technique reviewed in [8] into SFSN, we propose the NR technique. The idea is to *perturb* the *idling* or the *least-used* neuron W_n (the neuron winning the fewest training patterns or none) by a weight adaptation so that the neuron moves into the class region C_v having the largest distortion, and thus it splits C_v with the original winning neuron W_v in training (Fig. 5).

Let B_i and E_i indicate the beginning and the ending of C_v on the *i-th* dimensional axis in the m-dimensional vector space ($1 \leq i \leq m$) and V_i be the position of W_v on the *i-th* axis, the new position N_i of W_n (on the *i-th* axis) is defined as follows:

$$N_i = \begin{cases} V_i + s\cdot(E_i - V_i) & \text{if } (E_i - V_i) > (V_i - B_i) \\ V_i + s\cdot(V_i - B_i) & \text{if } (E_i - V_i) < (V_i - B_i) \\ V_i + s\cdot r\cdot(E_i - V_i) & \text{if } (E_i - V_i) = (V_i - B_i) \end{cases}$$

where s is a small random real ($0 < s < 1$) and r is a random bipolar binary ($r \in \{1, -1\}$). Thus, W_n is placed away from W_v by a random distance, in general, on the longer axes of C_v (Fig. 5). The NR method reduces distortions by splitting classes having larger distortions into smaller classes with smaller distortions. The codebook representing the classes is hence refined.

Neither CACS nor NR causes a higher training time complexity in the original learning rules since:

• the class index C_j of each input X_j can be found and recorded in step 4 of SFSN training rules, and the number of input patterns won by W_c can also be counted by a S_c,

• each B_i and E_i of a class region can also be updated by X_j, and

• a distortion e_c in a class C can be accumulated with the distance computed in step 3 of SFSN learning rules.

These properties allow us to identify the class-stable condition (Eq. 1) by C_j, find the classes having larger distortions by e_c, select the idle or less-used neurons by S_c, and move the selected neurons by B_i and E_i.

3 Measurements and Results

The comparison of processing fingerprints by the classical entropy coding method of Lempel-Ziv & Welch (LZW), transform coding methods of JPEG and WSQ, and our SFSN model is in Table 1, and the fingerprints are in Figure 6. The compression and distortion measurements are given in the following subsections.

3.1 DISTORTION AND FIDELITY MEASUREMENTS

The NMSE and the Normalized SNR (NSNR) are standard distortion and fidelity measurements [26]:

$$NMSE = \frac{\Sigma[x_i - x_i']^2}{\Sigma x_i^2} \quad \text{and} \quad NSNR_{dB} = 10 \cdot \log_{10} \frac{\Sigma x_i^2}{\Sigma[x_i - x_i']^2} = -10 \cdot \log_{10}(NMSE)$$

where x_i is a pixel in the original picture and x'_i is the corresponding pixel in the recalled picture. Fidelity is a quantitative measurement while quality is a subjective matter. However, high fidelity usually indicates high quality.

Table 1. Compression Comparison (Fig 6)

Method	NMSE	NSNR (dB)	Bit Rate	Ratio *Q:1*
Thumb Fingerprint f1 (768², 256-gray, Fig. 6A*)*				
LZW	N/A (Lossless)	N/A	6.84	1.17
JPEG	0.00069	31.63	0.67	12.00
WSQ	0.00119	29.24	0.60	13.33
SFSN (Fig. 6B)	0.00115	29.38	0.56 (0.36)†	11.88 (22.16)†
Thumb Fingerprint f6 (768², 256-gray, Fig. 6C*)*				
LZW	N/A (Lossless)	N/A	6.50	1.23
JPEG	0.00061	32.17	0.67	12.00
WSQ	0.00070	31.56	0.60	13.33
SFSN (Fig. 6D)	0.00049	33.12	0.56 (0.41)†	11.88 (19.48)†

† The values parenthesized in the bit rates and *Q's* are optimized (class data are further LZW-coded) since JPEG and WSQ use entropy coding. The 0.56 bit rate of SFSN is the conventional VQ bit rate derived from 512 classes (neurons) and 4x4 subimages. LZW, JPEG and WSQ bit rates are estimated from their optimized *Q* ratios, i.e., 8 bits divided by *Q*.

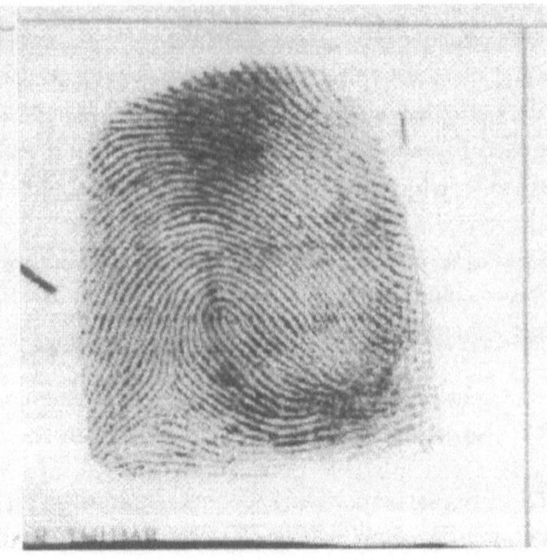

(A) Thumb fingerprint f1, 768x768, 256 grays.

(B) SFSN compressed-decompressed result of f1:
511 classes, trained with 4x4 subimages,
Bit rate = 0.56 bits/pixel (0.36, optimized),
Q compression ratio = 11.88 (22.16, optimized),
NSNR = 29.38 dB, NMSE = 0.00115.

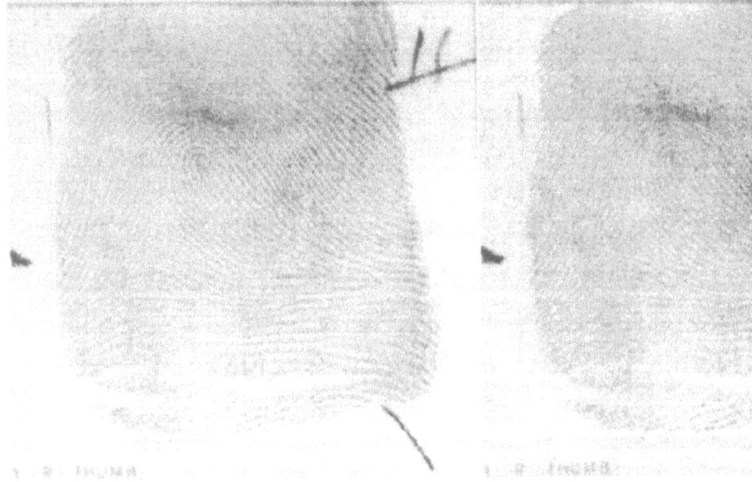

(C) Thumb fingerprint f2, 768x768, 256 grays,
a typical lightly imprinted thumb in the FBI's files.

(D) SFSN compressed-decompressed result of f2:
511 classes, trained with 4x4 subimages,
Bit rate = 0.56 bits/pixel (0.41, optimized),
Q compress ratio = 11.88 (19.48, optimized),
NSNR = 33.12 dB, NMSE = 0.00049.

Figure 6. Fingerprints and the SFSN's compressed-decompressed results.

3.2 COMPRESSION RATIO MEASUREMENTS

The image data I to be compressed is split into N training vectors of m dimensions, i.e., $I = (m, N)$. The coded class data requires $N \cdot \log_2 C$ bits where C is the number of codebook entries (winning neurons). The *critical* compression ratio is the storage size of the original image to the storage size of all the necessary information needed (the fixed-length coded class data and the codebook) to reconstruct the replica [27]:

$$Q = \frac{N \cdot m \cdot b}{N \cdot \lceil \log_2 C \rceil + C \cdot m \cdot b}$$

where b is the bit planes in the picture. When a *general* codebook is built to code a large number of images or a specific type of pictures (e.g., fingerprints), the shared-codebook compression ratio Q_g becomes:

$$Q_g = \frac{N \cdot m \cdot b}{N \cdot \lceil \log_2 C \rceil + \frac{C \cdot m \cdot b}{P}} \approx \frac{N \cdot m \quad b}{N \cdot \lceil \log_2 C \rceil} \qquad \text{(Eq. 2)}$$

for a large P, the number of coded pictures. Hence, the VQ image coding employs the more generally adopted *bit rate* [5], [7]-[9], the number of bits used to code a single pixel:

$$\frac{b}{Q_g} = \frac{\lceil \log_2 C \rceil}{m} \qquad \text{(Eq. 3).}$$

Both Q and *bit rate* are reasonable compression measurements providing that the subimage is large and the codebook is kept small. Coding 4x4 subimages from a 128x128 picture with a codebook having 1024 or more entries would not be practical since there are only 1024 subimages. The target picture should generate a sufficient number of subimages to justify the compression bit rate according to the codebook size (Eq. 2 and 3). Generally, a codebook of 1024 or less entries is used to encode 3x3, 4x4, or 5x5 *non-overlapping* subimages from a 512x512 or larger picture. Overlapped subimage coding (for low-spatial frequency filtering [17]) has smaller effective compression subimage sizes in determining the bit rate, e.g., 4x4 subimages overlapping by 1 pixel on each side actually have 3x3 effective compression areas. Since the number of partitioned subimages to encode is increased in the calculation of Q_g (In Eq. 2, the lower N in the divider term is increased while the original picture size in the denominator remains.)

4 Concluding Remarks

The robust generalization and classification capabilities of neural networks make them strong candidates for codebook building mechanisms. The advantage of building an image compression codebook is that it can be used to encode and decode other pictures on the fly afterwards [7]. We call this independent adaptivity a *cross compression* capability. We use the neural codebook trained by one picture to process other pictures. The compression bit rate remains the same but the NSNR differs. Figure 7 is the statistics chart of such "cross compressions" obtained by training SFSN with one of the five selected pictures and then compress all fourteen pictures. Only through such a cross compression process can we know which pictures have richer pattern classes than others (or which two are similar). As expected, the codebook trained with one fingerprint picture (Fingerprint f1) can process all other fingerprints with good fidelity. Also, the f1-trained codebook works better on Boy and House than the codebook trained by Lena or Tiffany. Hence, Boy and House, Lena and Tiffany, may be identified as two different groups. The compression coding learned by SFSN is well demonstrated as an independent adaptivity to images excluded from the training set.

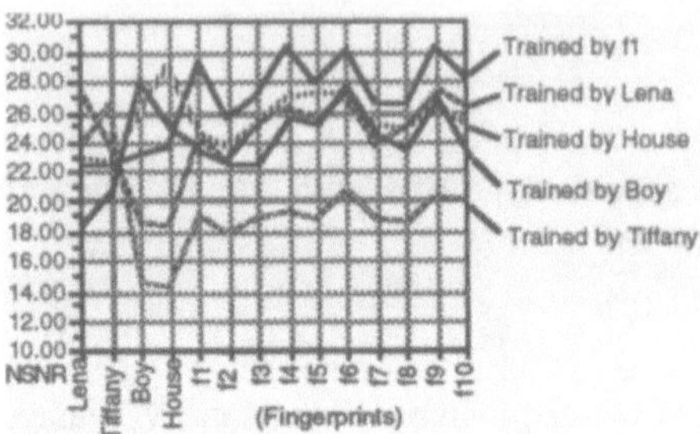

Figure 7. Each line is the NSNR's of the 14 images compressed-decompressed result with the codebook trained by one of the five images (Lena, Tiffany, Boy, House, and f1. f1 - f10 are fingerprints).

5 References

[1] Bradley, J. N., Brislawn, C. M., and Hopper, T. (1993) "The FBI wavelet/scalar quantization standard for gray-scale fingerprint image compression", SPIE Proceedings: Visual Information Processing II, vol. 1961, pp. 128-137.

[2] Daubechies, I. (1992), Ten Lectures on Wavelets, Society for Industrial and Applied Mathematics, Philadelphia, Pennsylvania.

[3] Hudson, G. P., Yasuda, H., and Sebestyen, I. (1988), "The international standardization of a still picture compression technique", Proceedings of IEEE Global Telecommunications, pp. 1016-1021.

[4] Wallace, G. K. (1991), "The JPEG still picture compression standard", IEEE Transaction on Consumer Electronics.

[5] Gray, R. M. (1984), "Vector quantization", IEEE Acoustics, Speech, and Signal Processing, vol.1, no. 2, pp. 4-29.

[6] Cottrell, G. W., Munro, P., and Zipser, D. (1988) "Image compression by back propagation: an example of extensional programming", in Model of Cognition: A Review of Cognitive Science, N. E. Sharkey (Ed.), Ablex Publishing, Norwood, New Jersey, vol. 1, pp. 208-240.

[7] Gersho, A. and Ramamurthi, B. (1982), "Image coding using vector quantization", Proceedings of IEEE Int'l Conf. on Acoustics, Speech, and Signal Processing., pp. 428-431, Paris.

[8] Nasrabadi, N. N. and King, R. A. (1988), "Image coding using vector quantization: a review", IEEE Trans. on Communications, vol. 36, no. 8, pp. 957-971.

[9] Ramamurthi, B. and Gersho, A. (1986), "Classified vector quantization of images", IEEE Trans. on Communications, vol. 34, no. 11, pp. 1105-1115.

[10] Hecht-Nielsen, R. (1987), "Counterpropagation networks", Applied Optics, vol. 26, pp. 4979-4984.

[11] Hecht-Nielsen, R. (1988), "Applications of counter-propagation networks", IEEE Trans. on Neural Networks, vol. 1, pp. 131-141.

[12] Kohonen, T. (1982), "Self-organized formation of topologically correct feature maps", Biological Cybernetics, vol. 43, pp. 59-69.

[13] Kohonen, T. (1988), Self-Organization and Associative Memory, Springer-Verlag, New York, New York.

[14] Chang, W., Soliman, H. S., and Sung, A. H. (1992) "Image data compression using counterpropagation network", Proceedings of IEEE Int'l Conf. on Systems, Man, and Cybernetics, vol. 1, pp. 405-409.

[15] Chang, W., Soliman, H. S., and Sung, A. H. (1993) "Preserving visual perception by learning natural clustering", Proceedings of IEEE Int'l Conf. on Neural Networks, vol. 2, pp. 661-666.

[16] Chang, W., Soliman, H. S., and Sung, A. H. (1994) "A learning vector quantization neural model for image data compression," Proceedings of IEEE Data Compression Conf., pp. 493.

[17] Chang, W., Soliman, H. S., and Sung, A. H. (1994) "A vector quantization neural model to compress still monochromatic images", Proceedings of IEEE Int'l Conf. on Neural Networks, vol. 6, pp. 4163-4168.

[18] Oja, E. (1982), "A simplified neuron model as a principal component analyzer", Journal of Mathematical Biology, vol. 15, pp. 267-273.

[19] Linsker, R. (1988), "Self-organization in a perceptual network", Computer, vol. 21, pp. 105-117.

[20] Kohonen, T., Kangas, J., Laaksonen, J., and Torkkola, K. (1992), "LVQ_PAK: a program package for the correct application of learning vector quantization algorithms", Proceedings of IEEE Int'l Joint Conf. on Neural Networks, vol. 1, pp. 725-730.

[21] Fang, W.-C., Sheu, B. J., Chen, O. T.-C., and Choi, J. (1992), "A VLSI neural processor for image data compression using self-organization networks", IEEE Trans. on Neural Networks, vol. 3, pp. 506-518.

[22] Davis, H. F. and Snider, A. D. (1988), Introduction to Vector Analysis, Wm. C. Brown Publishers, Dubuque, Iowa.

[23] Jolion, J.-M., Meer, P., and Bataouche, S. (1991), "Robust clustering with applications in computer vision", IEEE Trans. on Pattern Analysis and Machine Intelligence, vol. 13, no. 8, pp. 881-895.

[24] Kosko, B. (1992), Neural Networks and Fuzzy Systems, Prentice-Hall, Englewood Cliffs, New Jersey.

[25] Wu, Z. and Leahy, R. (1993), "An optimal graph theoretic approach to data clustering: theory and its application to image segmentation", IEEE Trans. on Pattern Analysis and Machine Intelligence, vol. 15, no. 11, pp. 1101-1113.

[26] Pratt, W. K. (1978), Digital Image Processing, Wiley Publishing, New York, New York.

[27] Mougeot, M., Azencott, R., and Angeniol, B. (1991), "Image compression with back propagation: improvement of the visual restoration using different cost functions", IEEE Trans. on Neural Networks, vol. 4, pp. 467-476.

PARALLEL GENETIC PROCESSES

Klaus P. Kratzer
Rainer A. Scholze
Fachhochschule Ulm
Prittwitzstr 10
89075 Ulm, FRG

Abstract. A distribution strategy for genetic algorithms on MIMD parallel systems is investigated. Based on a generic software component, the control flow and specific genetic operators are mapped to a complex of individual processors (transputers). The traveling salesman problem serves as exemplary application for a distributed genetic optimizer. Finally, the benefits and drawbacks of this approach are discussed.

Keywords. Genetic Algorithms, Parallel Computing, MIMD Configuration, Transputer, Distributed Optimization.

1 The Appeal of Distributed Genetic Search

Genetic algorithms ([4]) are a powerful generalized paradigm for searches in large and possibly multi-dimensional solution spaces. Therefore, one of their principal applications is the optimization of solutions for selection and permutation problems of a complexity which precludes exhaustive search strategies. There are significant advantages of genetic algorithms over conventional heuristic approaches.

- You do not need extensive expertise in the application domain; even the naïve use of a genetic search strategy may prove surprisingly powerful.
- There is no definite termination criterion for a genetic search algorithm. The quality of solutions improves stepwise which opens up graceful tradeoffs of time against quality.
- A number of parameters and the genetic operators serve as handles to tune a genetic optimizer to a specific application environment.

However, with the advent of massively parallel systems the most important advantage seems to be that

- Genetic algorithms display an extremely high potential for parallelization ([3]). This raises the question whether to work on a population in common storage or to distribute the population among several processors with local storage facilities.

The focus of this paper is on the development of a concept to support the distribution and parallelization of genetic algorithms on an MIMD (multiple instructions / multiple data) parallel system with distributed storage. This concept leads to a re-usable, generalized, and distributed genetic optimizer which is operational within single-processor and multi-processor environments. Finally, some empirical results and the merits and drawbacks of this approach are discussed.

E. A. Yfantis (ed.), Intelligent Systems, 565–571.

2 The Distribution Strategy

Basically, a genetic algorithm alternates between the selection among a given set of possible solutions and the recombination of a decimated population applying genetic operators, specifically one ore several versions of

- *mutations,* deriving a new member of the population from a surviving member and/or
- *crossovers*, the combination of traits taken from several members of the population.

Parallelization with distributed storage requires a somewhat extended concept as shown in figure 1.

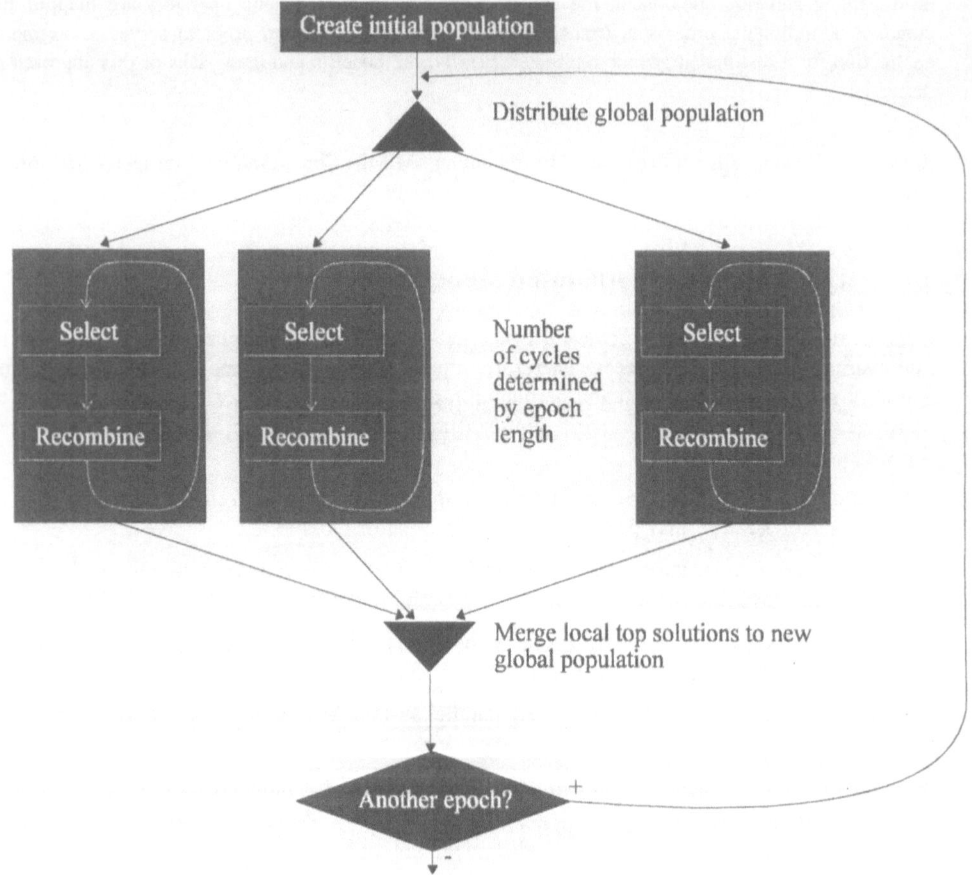

Figure 1: Parallel Genetic Processes

Each of the processors in a distributed MIMD configuration receives a copy of the global population. These populations are developed *individually* by each node which is somewhat at variance with the theory underlying the genetic approach. However, there is a periodical synchronization point to merge the various "tribes" of solutions and reset each participating processor to a new global population which, hopefully, includes a variety of promising approaches to a global optimum. This merge operation is the major limitation of the overall system under Amdahl's law stating that the speedup gained by truly concurrent processes has an upper bound depending from the ratio between the serial and the parallel components of an algorithm. Therefore, it is desirable to increase the number of selection/recombination cycles performed individually by each processor *(epoch)*. The length of an epoch is a major system parameter; while an increasing epoch length adversely affects the performance of the overall system, the synchronization overhead inherent to every distributed process system is dramatically reduced.

3 The Target Hardware

The target system is a complex of loosely interconnected 32 bit transputers with RISC architecture and four bitserial DMA channels each. These processor components are perfectly suitable for an MIMD complex with distributed memory. A wide range of processors with varying performance characteristics is available. In this case, a number of T805 processors with 30 MIPS and a 4.3 MFLOPS floating point unit were employed.

Four bitserial and bi-directional DMA channels *(links)* for each processor serve as communication devices. The maximum transfer speed to adjacent nodes is 20 Mbit/s. This data transfer is concurrent to the CPU's processing of instructions. Barely 0.6 μs are needed in the CPU to set up a transfer.

Concurrent segments of an algorithm are organized as processes which, in turn, are statically mapped to the hardware. The processes follow Hoare's CSP model ([5]). They are controlled by a hardware-resident scheduler. Synchronization and communication between processes is effected by means of software channels. Within a transputer these channels are implemented by transfers of words. Externally, the channels are mapped to links. Due to the concurrent nature of the link operation one rather transfers one big chunk of information than smaller items.

The big advantage of such an MIMD system is its scalability which allows a stepwise buildup of the system with hardly any modification on software level. The topology of the system is somewhat restricted by the limited number of channels. There are three characteristic configurations, the transputer farm as two-dimensional layout, the linear pipeline and the tree structure (see Figure 2) which ensures the least number of data transfers via links needed for any communication from the root down to the workers.

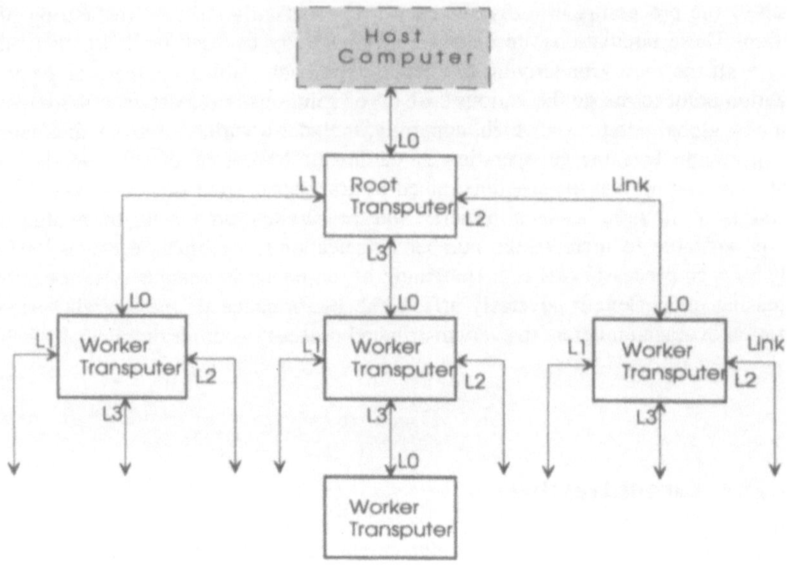

Figure 2: An MIMD Transputer System With Tree Configuration

4 The Software Architecture

The implementation relies on an all-purpose generic software component *(PGO for "parallel genetic optimizer")* encapsulating all static and procedural knowledge of genetic algorithms and their distribution. An executable instantiation only needs some further application-specific declarations (e.g., the nature of a solution, sensible mutation and crossover operators plus some system parameters) to operate as full-fledged genetic optimizer.

After an early implementation in Occam2 the programming language Ada was adopted as language of choice for this purpose. Carefully written Ada code ports with hardly any effort and excellent Ada development systems are available on all conceivable hardware platforms, in particular on parallel hardware ([1]). The distribution among the transputers in the system requires a additional Ada concept, the partition, as definition of an executable image. Partitions have been added to the language by several vendors individually in recent years and are now standardized in the Distributed Software Systems Annex of the Ada 9X standard ([6]). An instantiation of the generic component takes in a full descrption of the optimization problem and possibly specialized genetic operators and parameters to generate a customized optimization tool for a specific purpose.

The process structure resulting from an instantiation is shown in figure 3. There is a central PGO *(parallel genetic optimizer)* process in the master unit which farms out work to the several worker units which, in turn, supply an EW *(epoch worker)* process to perform the actual selection/recombination cycles. To achieve node-to-node communication there are MMX *(master multiplexer)* processes to handle all communication via links and WMX *(worker multiplexer)* processes in leaf transputers to accept and return data.

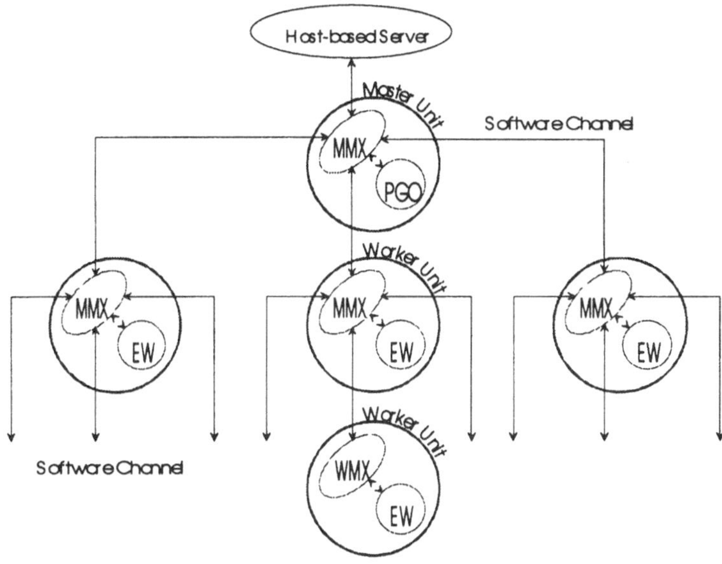

Figure 3: Process Structure for PGO

5 Tests & Results

One of the most interesting and hardest NP-problem classes to be tackled by any heuristic or probabilistic optimizer is the class of permutation problems, with the traveling salesman problem as its canonical incarnation. By adding mutation and crossover operators specifically designed to preserve the integrity constraint inherent to permutations, *PGO* was transformed into a dedicated TSP optimizer (see also [2]).

Numerous experiments investigated the behavior of this optimizer with different settings of the parameters. It turns out that the most important parameters are

- *EPOCH* which determines the length of independent operation of each *EW* process and
- *FAN-OUT* which gives the number of processors operating in parallel.

A suite of hard TSP problems was used to measure the effects of different settings of these parameters on the number of recombinations needed to achieve a certain and pretty high level of convergence a solution. This level was chosen after a number of trial runs for 50 stations with the genetic optimizer and some heuristic optimizers.

The aim of this investigation was to find out to which extent the behavior of the distributed algorithm deviates form a centralized version and which benefits are to be gained. All benchmarks were conducted with an even probability of a surviving solution to take part in the recombination process and an even distribution of crossovers and mutations.

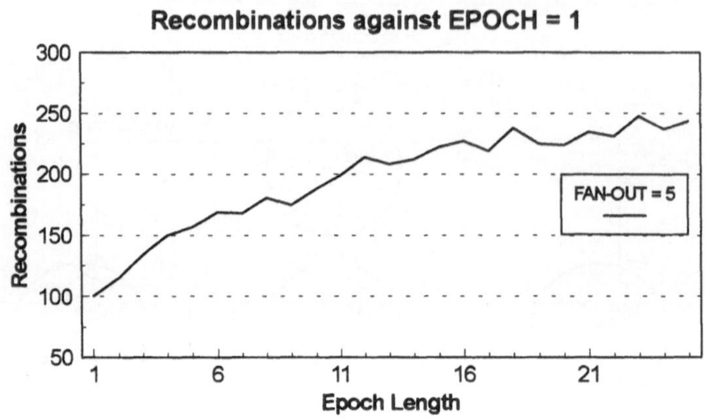

Figure 4: Recombinations (in percent) Against Epoch Length

Figure 4 shows the first example of a gradual increase of the *EPOCH* parameter with five worker processes. It is evident that an unlimited increase leads to duplicate work being performed. Usually, only low settings of this parameter (e.g., 5) are required to offset dominant system overhead.

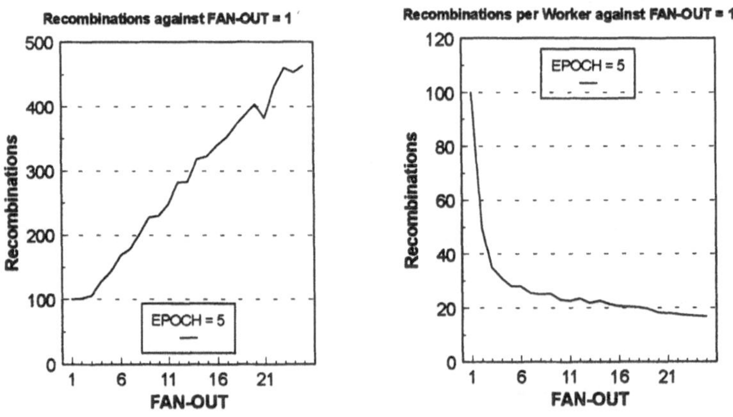

Figure 5: Recombinations (in percent) Against Fan-out

Figure 5 gives an idea of the effectivity of an increased fan-out in total recombinations and in recombinations per worker process. As it turns out there is a lower saturation level which depends on the epoch length and the population size. Part of the problem is again duplicate work which becomes dominant with higher numbers of workers, apart from the need to increase the epoch length to offset the higher cost of synchronization and communication. First improvements with special parametrizations for genetic operators to give each worker a special work profile avoid this problem and lead to even better performance.

6 Summary & Current Activity

The distribution concept presented in this paper is surely workable and efficient for the operation of genetic algorithms on MIMD configurations. Most important is that it is easy to handle, easily portable and ideal for the investigation of system behavior. The cost in terms of investment and system overhead is surprisingly low.

Further investigation to achieve an optimal time/space/quality ratio is underway. Presently, the implementations are refined and the suitability of different hardware and software configurations and layouts is evaluated.

This work is funded by the State of Baden-Württemberg under grant KA-411-0 (93).

7 References

[1] Alsys: *The Ada Development Environment for the Transputer*, Reston 1992.

[2 Buckles, B.P., Petry, F.E., Kuester, R.L.: *Schema Survival Rates and Heuristic Search in Genetic Algorithms*, Proc. Tools for AI, IEEE 1990.

[3] Goldberg, D.E.: *Sizing Populations for Serial and Parallel Genetic Algorithms*, Proc. 3rd Int'l Conf. on Genetic Algorithms, San Mateo 1989.

[4] Goldberg, D.E.: *Genetic and Evolutionary Algorithms Come of Age*, CACM 37 (1994) 3.

[5] Hoare, C.A.R.: *Communicating Sequential Processes*, CACM, 21 (1978) 6.

[6] Intermetrics: *Ada 9X Reference Manual*, Cambridge 1993.

An Intelligent Neural Network Based System for 3-D Motion Analysis

Ting Chen
Argonne National Laboratory
9700 S. Cass Ave.
Argonne, IL 60439
U. S. A.

ABSTRACT

A new intelligent neural network based system for 3-D motion analysis is proposed in this paper. Experiments are conducted to corroborate the proposed techniques.

1 The Proposed System

1.1 2-D Hopfield Network for Matching

Let p_i and p'_i be vectors in 3-D space and represent a unique point in frame 1 and frame 2, respectively. The goal here is to find reliable and fundamental descriptors relevant to a 3-D point and define the constraints of the mapping from p_i to p'_i under the 3-D rigid motion assumption.

For pointwise matching, the following two measures are used to represent the attributes of each point: (1) the Euclidian distance between the point and the centroid of the object, and (2) the angles of polygons adjacent to the point. The former is explicitly defined as a global feature and the latter a local feature. All the extracted features should be 3-D translation, rotation and scaling invariant under the *rigidity constraint* commonly used in motion analysis.

Under the assumption of one-to-one matching, the *problem constraint* amounts to the uniqueness of matching, i.e., each point in one frame eventually matches only one point in the other frame, or no more than two points in one frame should match the same point in the other frame.

In the 2-D array neural network, each column corresponds to a feature of the data in frame 1, while each row corresponds to a feature of the data in frame 2. If there are N_1 data points in frame 1 and N_2 data points in frame 2, the neural network is an $N_2 \times N_1$ 2-D array as shown in Figure 1. The final output of the neuron at position (i, k) in the

E. A. Yfantis (ed.), Intelligent Systems, 573–577.

network reflects the support for the hypothesis of mapping the *kth* neuron in frame 1 to the *ith* neuron in frame 2. Matching between two frames can be formulated as minimizing an energy function. In other words, minimization of this energy function results in stable pairings that are optimal with respect to the constraints defined in the previous section. The energy is expressed as follows:

$$E = \frac{A}{2}\sum_{i}\sum_{k}\sum_{l\neq k} V_{ik}V_{il} + \frac{B}{2}\sum_{i}\sum_{k}\sum_{j\neq i} V_{ik}V_{jk} + \frac{C}{2}\sum_{i}\sum_{k}\sum_{j\neq i}\sum_{l\neq k} M_{ikjl}V_{ik}V_{jl} + \frac{D}{2}\sum_{i}\sum_{k} N_{ik}V_{ik} \tag{1}$$

where A, B, C, D are constants, M_{ikjl} is the compatibility measurement, N_{ik} is the magnitude of the point-to-point constraint, and V_{ik} is the output of neuron at position (i,k) which is equal to "1" if the mapping is established between the *ith* point in frame 2 and the *kth* point in frame 1, or equal to "0" otherwise.

1.2 Learning Network for Motion Estimation

A precise mathematical model of the 3-D spatial rigid motion involving translation and rotation over time is

$$p_i' = Rot \cdot p_i + Tran \tag{2}$$

where *Rot* denotes a 3-D rotation matrix and *Tran* denotes a 3-D translation vector.

The transformation in Equation 2 can be decomposed into a rotation followed by a translation in 3-D space, i.e., $p_i'' = Rot \cdot p_i$ and $p_i' = p_i'' + Tran$.

The transformation vector is simply the displacement between the object centroids in two time frames, i.e.,

$$\begin{aligned} Tran &= [Tran_x, Tran_y, Tran_z]^T \\ &= [C_{2x}, C_{2y}, C_{2z}]^T - [C_{1x}, C_{1y}, C_{1z}]^T \end{aligned} \tag{3}$$

where

$$C_1 = [C_{1x}, C_{1y}, C_{1z}]^T \quad and$$
$$C_2 = [C_{2x}, C_{2y}, C_{2z}]^T.$$

The rotation matrix *Rot* used here is in a very general form. Given any arbitrary rotational axis k with any rotational angle θ, the rotation matrix can be defined as follows [1]:

$$\begin{aligned} Rot &= [n, o, a] \\ &= \begin{bmatrix} n_x & o_x & a_x \\ n_y & o_y & a_y \\ n_z & o_z & a_z \end{bmatrix} \end{aligned}$$

$$= \begin{bmatrix} k_x^2 vers\theta + cos\theta & k_x k_y vers\theta - k_z sin\theta & k_x k_z vers\theta + k_y sin\theta \\ k_x k_y vers\theta + k_z sin\theta & k_y^2 vers\theta + cos\theta & k_y k_z vers\theta - k_x sin\theta \\ k_x k_z vers\theta - k_y sin\theta & k_y k_z vers\theta + k_x sin\theta & k_z^2 vers\theta + cos\theta \end{bmatrix} \quad (4)$$

where $vers\theta = 1 - cos\theta$, $k = [k_x, k_y, k_z]^T$, and k is an normalized vector.

Extracting motion parameters between two frames is, in fact, a process of deriving the translation vector *Tran* and the rotation matrix *Rot* based on point correspondences obtained at the matching stage. As described in Equation 3, the translation vector *Tran* can be easily obtained by calculating the displacement of the object geometrical centroid positions in two frames. The remaining problem then is how to derive the rotation matrix *Rot* based on the available information.

In the proposed three-layered neural network, the inputs to the first (or input) layer are the components of the point vector $p_i = [p_{ix}, p_{iy}, p_{iz}]^T$. A 3×3 weight matrix is constructed between layer 1 and layer 2:

$$W = [W_1, W_2, W_3]^T$$

where

$$W_1 = [w_{11}, w_{12}, w_{13}]$$
$$W_2 = [w_{21}, w_{22}, w_{23}]$$
$$W_3 = [w_{31}, w_{32}, w_{33}]$$

and w_{ij} represents the sense and strength of the connection between neuron j of layer 1 and neuron i of layer 2.

Let $y_i = [y_{ix}, y_{iy}, y_{iz}]^T$ denote the actual outputs of layer 2. Then, we can obtain

$$\begin{aligned} y_i &= [W_1 \cdot p_i, W_2 \cdot p_i, W_3 \cdot p_i]^T \\ &= \begin{bmatrix} w_{11}p_{ix} + w_{12}p_{iy} + w_{13}p_{iz} \\ w_{21}p_{ix} + w_{22}p_{iy} + w_{23}p_{iz} \\ w_{31}p_{ix} + w_{32}p_{iy} + w_{33}p_{iz} \end{bmatrix}. \end{aligned} \quad (5)$$

The weight vector matrix V between layer 2 and layer 3 is defined as

$$V = [V_1, V_2, V_3]^T$$

where

$$V_1 = [v_{11}, v_{12}, v_{13}],$$
$$V_2 = [v_{21}, v_{22}, v_{23}],$$
$$V_3 = [v_{31}, v_{32}, v_{33}],$$

and, v_{ij} denotes the sense and strength of the connection between neuron j of layer 2 and neuron i of layer 3.

By applying the least-mean-square algorithm and taking ε^2 itself as an estimate of ε^2, the estimated gradients can be obtained:

$$\hat{\nabla}_{ix} = \frac{\partial \varepsilon_{ix}^2}{\partial W_1} = -2\varepsilon_{ix} p_i \tag{6}$$

$$\hat{\nabla}_{iy} = \frac{\partial \varepsilon_{iy}^2}{\partial W_2} = -2\varepsilon_{iy} p_i \tag{7}$$

$$\hat{\nabla}_{iz} = \frac{\partial \varepsilon_{iz}^2}{\partial W_3} = -2\varepsilon_{iz} p_i. \tag{8}$$

The updating rule for the weight matrix W can be described as follows:

$$\begin{aligned} W^{(n+1)} &= [W_1^{(n+1)}, W_2^{(n+1)}, W_3^{(n+1)}]^T \\ &= \begin{bmatrix} W_1^{(n)} + 2\mu\varepsilon_{ix}^{(n)} p_i \\ W_2^{(n)} + 2\mu\varepsilon_{iy}^{(n)} p_i \\ W_3^{(n)} + 2\mu\varepsilon_{iz}^{(n)} p_i \end{bmatrix} \end{aligned} \tag{9}$$

where μ is the learning rate and n represents the number of iterations.

2 Experimental Results

A sequence of two frames was created for a synthetic object. The second frame was obtained from the first one based on a rotational axis $k = [1, 1, 1]^T$ and a rotational angle $\theta = 10°$. The computed rotation matrix was as follows:

$$Rot = \begin{bmatrix} 0.989151 & -0.089821 & 0.110822 \\ 0.104456 & 0.984225 & -0.90169 \\ -0.94043 & 0.105438 & 0.981052 \end{bmatrix}.$$

By comparing our results with the ideal answer

$$Rot = \begin{bmatrix} 0.9899 & -0.0952 & 0.1053 \\ 0.1053 & 0.9899 & -0.0952 \\ -0.0952 & 0.1053 & 0.9899 \end{bmatrix},$$

the result from the proposed system is obviously very accurate.

3 Conclusions

Although the application of artificial neural network techniques to conventional computer vision problems is a relatively new development, it has received increasing attention in recent years. The research presented in this paper contributes to this area by proposing novel approaches to solve classical 3-D rigid motion analysis problems. The system is much simpler than the commonly used quaternion-based approach. Moreover, the structure of the proposed learning neural network suggests a very fast implementation scheme.

References

[1] R. P. Paul. *Robot Manipulators: Mathematics, Programming, and Control.* The MIT Press, 1986.

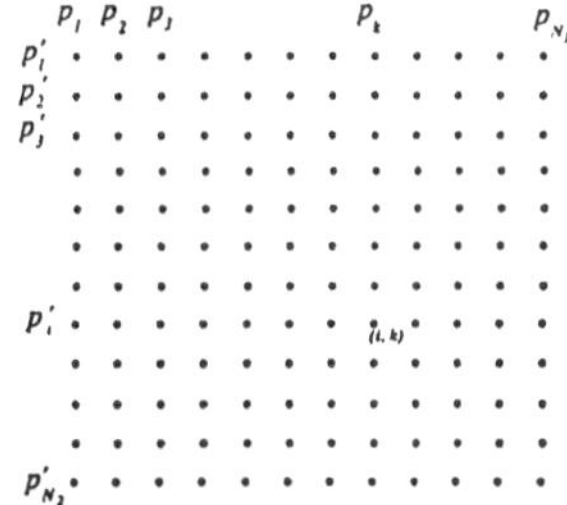

Figure 1: 2-D Hopfield network model for matching

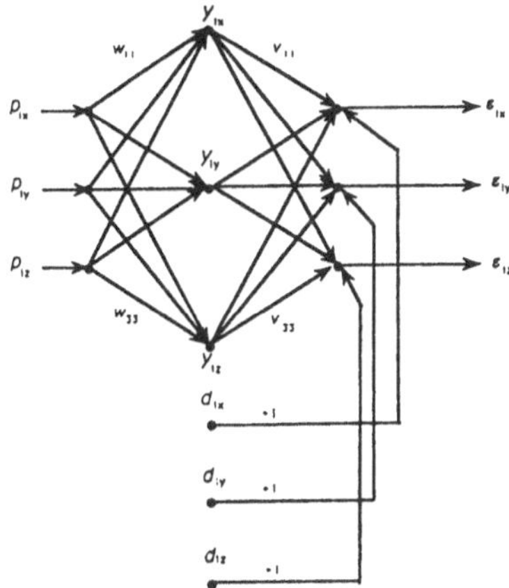

Figure 2: Learning network for motion estimation

Neural Networks for Rainfall Forecasting

L. Lastrucci
M. Maggini
Dipartimento di Sistemi e Informatica
Università di Firenze
Via di Santa Marta 3 - 50139 Firenze - Italy

Abstract. Neural networks can be used as pattern recognition diagnostic techniques with some short-term predictive skill. In this paper we describe some preliminary results for rainfall forecasting with neural networks. Performance measures were computed for 3 and 6 hours lead time forecasts on the training and the validation data of Pisa's area comparing feedforward and recurrent networks. Feedforward nets showed good capability for short time rainfall forecast (3 hours) while recurrent nets seem to be more appropriated for longer forecast period. The models appear to perform quite well even if no height data were used. The observational errors are the main source of errors of this forecast approach, specially if a significant number of conflicting examples is used to train the net.

Keywords. Recurrent Neural Networks, Rainfall Forecasting.

1 Introduction

Rainfall is one of the most difficult elements of the hydrologic cycle to forecast. The prediction with time leads of the order of hours or, at most, days is meteorologists' and hydrologists' dream.

Rainfall forecasting models in use are typically valid on large scales, but a more accurate prediction on smaller space scales is desired. The availability of more powerful computers has allowed a better modeling of the atmosphere and much of the emphasis on forecasts has been placed on the machine. Artificial Intellingence (AI) techniques are being developed to enhance meteorologists's pattern recognition prolificiency. Two successful AI techniques are expert systems and neural networks (NNs). For expert systems a knowledged engineer tries to extract the knowledge from one or more experts who know how to solve the problem in question. Neural networks, on the other hand, do not require any prior knowledge of a solution. A NN "learns" what it needs to know about a particular problem.
Our work's objectives were:

- The model should predict the weather on Pisa's area
- It should be able to predict rainfall at least several hours lead time.
- It should be simple enough to realistically operate in real time and with reasonable computational needs.

E. A. Yfantis (ed.), Intelligent Systems, 579–582.

- The model should allow updating, given new precipitation observations.
- Inputs to the model should be easily measured.

The complexity and non-linearity of the problem makes it attractive to try the neural network approach, which is inherently suited to problems that are mathematically difficult to describe. Even if there was little research done with NNs in meteorology they have the potential to solve pattern recognition problems that other methods have not yet been able to solve.

Moreover, in an adaptive forecasting environment, users may desire to continuously improve the capability of the model. The use of NNs can support this request. Additional training of the NN could be approached in two possible techniques: real-time directed and off-line learning. Real-time learning is desirable to guide the NN toward the recognition of the specific storm event behavior at hand. It is possible that the NN retains most strongly the transformation relationship for the latest training set. However, the underlying transformations learned by the NN in all previous trainings are still present. This approach would provide a forecast model which recognizes the general transformations characteristic of all storm events and it adapts to specific details of an event being observed. The off-line learning process is implemented as a maintenance procedure, performed on the NN as new rain events are added to the local database.

In the following sections we'll describe the architectures and some preliminary results obtained with these models.

2 Data Processing

To predict rain and clear weather we used ground station observation data at each hour in Pisa's area. Because precipitation is a phenomenon non linearly related to a series of meteorological variables, as inputs we had: wind direction and intensity, visibility, phenomenon, ceiling, quantity and type of clouds, temperature, dewpoint temperature and ground-level pressure.

Only meteorological situations with significant (for our purpose) events (phenomena 17, 60-69, 80-89, 90-99 for the international encoded system) were considered. A preprocess was carried out on some of these variables before presenting them to the input of the NN. For the wind direction, the main four sectors were binary encoded with 4 bits; the visibility range was non-linearly encoded with 4 bits, too; the quantity of the most significant clouds, if below 4000Ft, was considered and weighted according to an empirical criterion of relationship with the probability of rain (Tab. 1). The phenomenon was binary encoded, too. The NN was trained on 2 years of data and we tested the model on 1 year of data never seen before.

3 NN Architecture and Experiment Results

Some experiments were carried out to compare feedforward and recurrent neural networks on the same tasks. While a feedforward network is a static system where the output value at time T depends only by the input vector at time T, a recurrent network is a dynamic

significant cloud types	*cloud weights*
CB	1
NS	0.9
TC	0.8
CU	0.7
SC	0.5
ST	0.1

Table 1: Cloud weights.

system with an internal state used to compress the significant history of the input temporal sequence. During the training the gradient of the error function is computed using the BPTT (Back Propagation Through Time [1]) an extension of the famous BP algorithm for recurrent NNs. So doing the network learns the proper state transition function in order to predict the given output targets. The system learned by a RNN is described by the following set of equations:

$$H(t) = f(H(t-1), I(t))O(t) = g(H(t))$$

where t is the time, H(t) is the internal state of the network (recurrent layer activation), I(t) is the input vector, O(t) is the output vector, f(.) and g(.) are the transition and the output function respectively.
In our case the net had two output neurons and the targets were the binary encoding of the phenomena: rain, clear weather.

In the first case the nets were trained to produce 3 hours lead time forecast of rainfall using a series of real examples. An independent series of events were used as a verification set for evaluating NN performance on independent data. We used a tree layers (16-20-2) fully connected net with squashed neurons for the feedforward net and a four layers (16-12-4-2) architecture for the recurrent net. Results are reported in Tab. 2.

Some experiments were carried out with the same architectures to produce 6 hours lead time forecasts (Tab. 3).

4 Conclusions

Performance measures were computed for 3 and 6 hours lead time forecasts on the training and the validation data of Pisa's area, comparing feedforward and recurrent neural networks. Feedforward nets showed good capability for short time rainfall forecast (3 hours) while recurrent nets seem to be more appropriated for longer forecast period. The observational errors are the main source of errors of this method, specially if a significant number of wrong examples is used to train the net. Nevertheless, we must consider that no height data were used for the model. That in order to have easily measured inputs, but of course the task became harder.

		rain	*c. weather*	*n. of frames*	*recog. %*
Feedforward n.	*rain*	350	133	483	72
	clear weather	277	1065	1342	79
	Accuracy %				78
Recurrent n.	*rain*	340	143	483	70
	clear weather	309	1033	1342	77
	Accuracy %				75

Table 2: Forecast results for the 3 hours lead time nets.

		rain	*c. weather*	*n. of frames*	*recog. %*
Feedforward n.	*rain*	314	169	483	65
	clear weather	403	939	1342	70
	Accuracy %				69
Recurrent n.	*rain*	333	150	483	69
	clear weather	349	993	1342	74
	Accuracy %				73

Table 3: Forecast results for the 6 hours lead time nets.

NNs are nothing more than pattern recognition diagnostic techniques with some short-term predictive skill. Our idea is that NNs, as any AI technique, can not replace meteorologists. Nevertheless, their patter mapping capability can help meteorologists, enhance numerical model outputs and analysis giving new insights into old forecasting problems.

References

[1] Pearlmutter, B. "Learning state space trajectories in recurrent neural networks", Neural Computation, 1(2): 263-269, 1989.

[2] Lee, J. Wegener, R.C. Sengupta, S.K. and Welch, R.H. "A neural network approach to cloud classification", IEEE trans. Geosci. Remote Sensing, 28 (5): 846-855, 1990.

[3] French, M.N., Krajewski, W.F, and Cuykendall, R.R. "Rainfall forecasting in space and time using a neural network", Journal of Hydrology, 137 (1992) pp. 1-31, 1992.

[4] Mc Cann, D.W. "A neural network short-term forecast of significant thunderstorms", Weather and Forecasting, Vol. 7, pp. 525-534, 1992.

[5] Hewitson, Crane "Large scale atmospheric controls on local precipitation in Mexico", Geophysical Research Letters, Vol. 19, no. 18, pp. 1835-1838.

RECEIVER FUNCTION INVERSION USING GENETIC ALGORITHMS

Serdar Ozalaybey
Martha K. Savage
Seismological Laboratory
University of Nevada
Reno – 89557
email: serdar@quake.unr.edu

Sushil J. Louis
Dept. of Computer Science
Mackay School of Mines
University of Nevada
Reno – 89557
email: sushil@cs.unr.edu

Abstract. We use a genetic algorithm to attack the receiver function inversion problem where the objective is to retrieve shear wave velocities of rock layers as a function of depth. The receiver function inversion is a non–linear, multi–modal, and multi–parameter optimization problem suited for genetic algorithms. We implement a three operator genetic algorithm consisting of selection, single–point crossover, and mutation. We find that a modified selection algorithm based on "elitism" works better than simple roulette–wheel selection for this problem. The results show that a genetic algorithm based parallel search method can find solutions that can fit the data well. The method is powerful in finding a good subset of optimal models from a randomly chosen population of models sampling various portions of the entire model space. We include a priori information in the inversion to direct the search towards a desired set of solutions. This helps to eliminate many multi–modal solutions.

Keywords. Genetic Algorithms, Inversion, Receiver Functions, Seismic Waveform.

1 Introduction

A receiver function (RF) is simply defined as a time series consisting of arrivals of earthquake shear waves generated at the interfaces of the earth's layered crust where each layer's velocity differs significantly beneath a recording receiver. The aim of RF inversion is to retrieve shear wave velocities of rock layers as a function of depth from this time series. RF inversion, like many geophysical inverse problems, can be cast as a multi–modal, multiparameter, non–linear optimization problem. Existing methods for RF inversion rely on using the gradient of the objective function to improve upon some initial model in an iterative linearized fashion [1]. Consequently, these linearized methods depend strongly on the initial model and can be trapped in local minima. Further, RF inversions are known to be non–unique [1]. The model space contains more than one set of unknown parameters corresponding to the global maximum of the objective function. The non–uniqueness problem is overcome by using a priori constraints obtained from independent geophysical and/or geological observations.

A genetic algorithm (GA) is a randomized parallel search technique based on natural selection and the process of evolution. The GA works with a finite distributed set of model parameters

E. A. Yfantis (ed.), Intelligent Systems, 583–588.

(binary strings) from the solution space to generate a new set of model parameters. The genetic search process iteratively seeks to evolve good solutions from this initial population. An objective function is used to evaluate the "fitness" value of each string. Further generations of the binary strings are created by favouring the survival and recombination of fitter strings. GAs require only the fitness value of each string to move toward an optimal solution in the search space.

GAs have found recent applications in seismic waveform inversion as an alternative approach to linearized methods [6, 7]. The RF inversion problem, which belongs to the class of seismic waveform inversion, has not been previously attempted using GAs. In this study, we explore the feasibility of using GAs for solving this multi–modal, multi–parameter, non–linear optimization problem.

2 Inversion and Model Parametrization

The optimization problem we will consider involves the determination of the shear wave velocity as a function of depth (vertical shear velocity structure) for a flat layered 1D earth model. Our objective is to translate the information contained in the RFs into a local vertical shear velocity structure. We parametrize the velocity structure as M dimensional vectors of unknown shear wave velocities. We choose a total of 23 flat layers, each with unknown shear velocity, to represent a layered earth crust from surface to a depth of 40 km. Layer thicknesses are kept fixed; 1 km at shallow depths (down to 6 km), and 2 km for deeper layers. Figure 1 shows a schematic of the layered earth crust model.

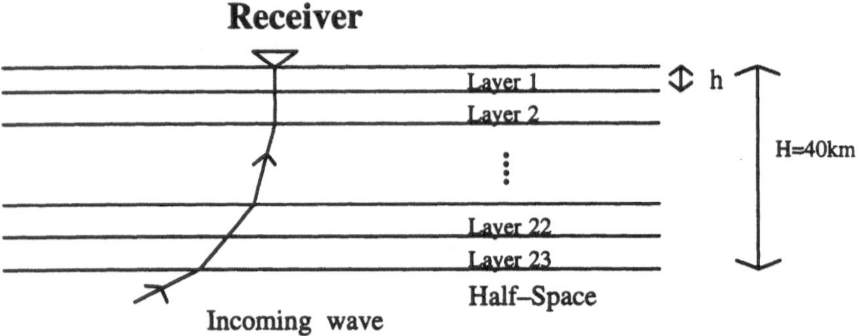

Figure 1. Schematic of 23 flat layered 1D earth model. Shear waves generated at the layer interfaces beneath the receiver are recorded at the surface.

The forward problem is to find the modeled RF from a given set of shear velocities and is solved by a propagator matrix synthetic seismogram method developed by Kennett [4].

A time domain zero–lag normalized cross–correlation coefficient will be used as the objective or fitness function [7]:

$$C(m) = d * g(m) \;/\; [(d * d)^{1/2} \; (g(m) * g(m))^{1/2}] \qquad (1)$$

where * represents cross–correlation, **d** is the observed receiver function (data), **g** is the forward modeling operator, **m** is the unknown shear velocity vector, and **g(m)** represents the modeled receiver function. We attempt to find values of **m** that maximize C(**m**).

3 Parameter Coding and Genetic Algorithm

We choose an unsigned integer binary string representation for each model parameter. The search space of each model parameter is defined by its pre–specified minimum and maximum values and the string length. To test the algorithm, we use a 5 bit string length with the following parametrization; we divide the model into 3 regions: for layers from surface to 6 km depth where rapid variation of shear velocity is expected, we let velocity vary from 1.5 km/s to 4.0 km/s. This range results in a resolution of 0.11 km/s for these parameters. For layers between 6 and 26 km, velocity is allowed to vary between 2.5 km/s and 4.5 km/s with a resolution of 0.064km/s, and for layers between 26 and 40 km, a range of 3.0 to 4.7 km/s is used with a resolution of 0.055 km/s. This forms a model parameter search space with 23 unknowns each having 2^5 possible values making up a search space size of 2^{115}. Each model parameter is coded as a 5 bit long binary string. The coded model parameters are then concatenated to form a 'chromosome'.

4 Application and Discussion

Initially, we use a simple three operator GA with selection (roulette–wheel), single point crossover, and mutation [3]. We have performed a series of GA runs using different values for initial population size Ps, cross–over probability Pc, and mutation probability Pm to seek the maximum value of the objective function. Our goal is to find well fit models with Ps rather small (30 to 50) and within a small number of generations because of the computational cost of the evaluation of equation (1). We found greatly different solutions corresponding to the multi–modal peaks in the search space from different GA runs. To reduce the effect of non–uniqueness we introduce a priori information into the inversion by constraining the velocity of the surface layer as 1.8 km/s [5]. This was done by fixing the value of the first 5 bits in the chromosome. We use Ps=48 and let crossover probability Pc vary between 0.6–0.7 and mutation probability Pm vary between 0.01–0.005 for different GA runs. In all cases, convergence and performance of the algorithm was poor; maximum correlation value did not exceed 0.86 within 60 generations.

To improve the convergence and performance of the algorithm we employed a two–step selection process called 'selection with elitism'. In this processes, when a new generation population has been produced the next generation population is created from this generation and previous generation population by ranking them according to their fitness. This doubles the population size [2]. We then continue with roulette–wheel selection, crossover and mutation operations using only the better half of this population. This process guarantees the better fit models will not be lost and directs the search towards highly fit models. This is expected to cause quick convergence and a higher performance. One disadvantage is that elitism may result in premature convergence. This is overcome by increasing crossover and mutation rates. A maximum correlation of 0.926 was obtained using Ps=48, Pc=0.8, and Pm=0.03 for 80 generations. Figure 2 shows two of the GA runs that resulted in the highest fitness values with and without elitist selection.

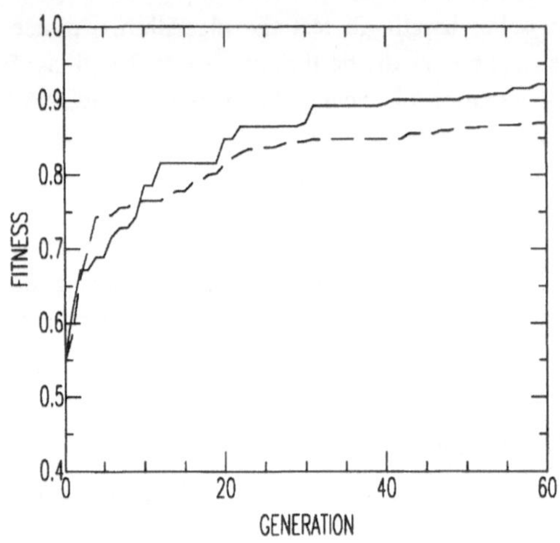

Figure 2. Two sample GA runs with (solid) and without (dashed) elitism.

The best fitting model obtained from elitist selection GA run is shown in Figure 3. The waveform is fit quite well and the resulting velocity model contains geologically reasonable velocity variations. We have obtained a good subset of optimal models from a randomly chosen population of models sampling various portions of the entire model space. However, some of the models obtained by other GA runs contained significant amount of velocity variations between adjacent layers, a problem which is common to seismic waveform inversions. This is typically solved by introducing smoothness constraints into the inversion to avoid oscillatory solutions. In this study, we do not attempt to find what is the optimum way to include model smoothness constrains in a GA because this could be a research area itself depending on the nature of the problem. We leave this as a potential future study. We note that the models that have high correlation values (> 0.9) would be appropriate for the initial velocity model required for iterative linearized inversion schemes.

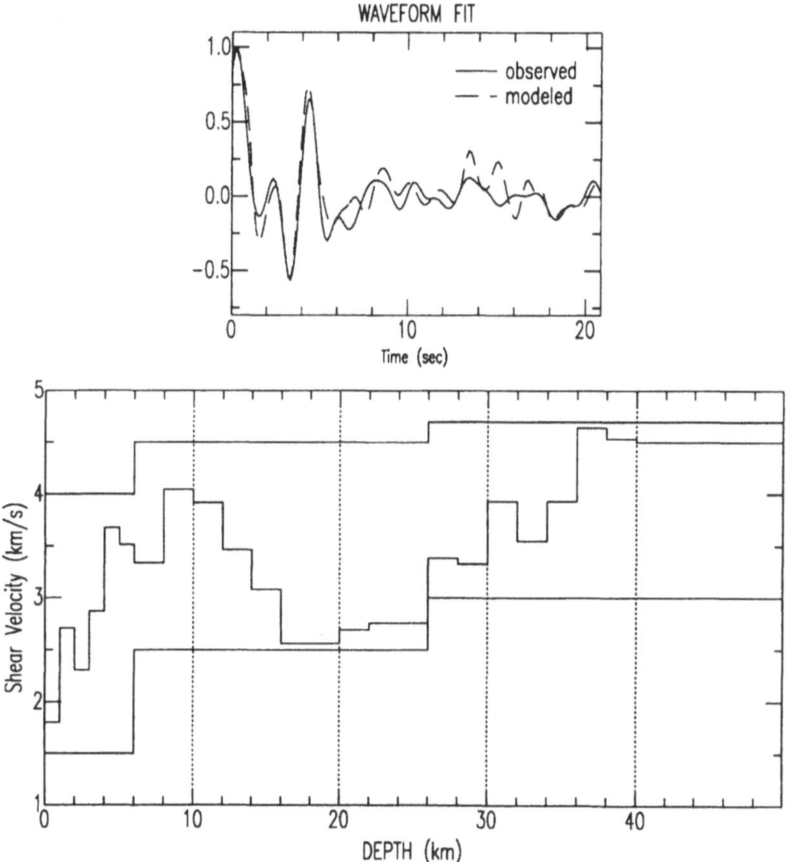

Figure 3. Waveform fit: observed (solid) modeled (dashed) receiver functions (top). Corresponding velocity model obtained from the best GA run. Upper and lower bounds of the model search space are indicated (bottom).

5 References

[1] Ammon, C. J., G. E. Randall, and G. Zandt, On the nonuniqueness of receiver function inversions, *J. Geophys. Res.*, **95**, 15,303–15,318, 1990.

[2] Eshelman, L. J., *Foundations of Genetic Algorithms–1: The CHC Adaptive Search Algorithm: How to have Safe Search When Engaging in Nontraditional Genetic Recombination*, 265–283, Morgan Kauffman Publishing Company, **1991**.

[3] Goldberg, D. E., *Genetic Algorithms in Search, Optimization, and Machine Learning*, 442 pp., Addison Wesley Publishing Company, **1989**.

[4] Kennett, B. L. N., G., Seismic wave propagation in stratified media, 342 pp., Cambridge University Press, New York, **1983**.

[5] Mangino, G. S., G. Zandt, and C. J. Ammon, The receiver structure beneath Mina, Nevada, *Bull. Seism. Soc. Ame.*, **83**, 542–559, 1993.

[6] Sambridge, M., and G. Drijkoningen, Genetic algorithms in seismic waveform inversion, *Geophys. J. Int.*, ***109**, 323–342*, 1992.

[7] Stoffa, P. L., and M. K. Sen, Nonlinear multiparameter optimization using genetic algorithms: Inversion of plane wave seismograms, *Geophysics*, **56**, 1794–1810, 1991.

AN AUTOMATIC TRANSCRIPT EVALUATION SYSTEM

Robert Burke* Bin Cong* Yueming Li**
***South Dakota State University, Brookings, SD 57007**
****Louisiana State University, Baton Rouge, LA 70803**

Abstract. Universities face an important challenge in evaluating the academic transcripts of transfer students. Accurate translation of a student's academic history into the course numbering and naming system of the receiving institution is important to the student's degree progress. Often the timeliness and outcome of the evaluation process is also a key determinate of the student's choice of transfer institutions. Unfortunately the evaluation task is time consuming, requires expert judgement based on a wide knowledge base, is somewhat idiosyncratic to the rater, and may vary across departments. In order to solve some of these problems we developed an intelligent system based on neural network models which will learn from an expert evaluator in order to efficiently analyze and convert transcripts of transfer students. In this paper, we will discuss our models and their performances.

Keywords. Neural Networks, Transcript Evaluation, Recognition.

1 Introduction

Over the past few years, South Dakota State University has accepted increasing numbers of transfer students. While the increase in enrollment is welcome, it taxes resources allocated to the process of transfer evaluation. Currently, all the accompanying transfer transcript evaluation work is done by college deans and department heads without the benefit of an automated system. One result is a major time impact on administrators, often at the expense of other important tasks. Another result is a sense of frustration at the tedious nature of the task and the ensuing question, "Isn't there a way to automate this?" This paper gives a positive answer.

Simply stated, the problem we are considering in this paper can be defined as follows: the input to the system is a transfer student's list of courses stated in the language of originating university, and the output is the decision set, i.e., a list of equivalent courses at SDSU with associated grades and credits. When stated in these terms the task seems oversimplified, as if it could be solved by a simple look up table with one-to-one linkages. However, the environment of the task makes it more complex than that. We have identified the following environmental features:

1. The problem is a global task and very time consuming.

2. Current automated approaches require large maintenance efforts to update rule bases.

E. A. Yfantis (ed.), Intelligent Systems, 589–597.

SSU/JE/BIOL/100/1/A/F 89**BIOL/157/0.66/A/F 89/C
SSU/JE/BIOL/100/1/A/F 89**BIOL/157/0.66/A/F 89/C
SSU/JE/BIOL/100/1/A/F 89**BIOL/157/0.66/A/F 89/C
SSU/JE/BIOL/100/1/B/F 89**BIOL/157/0.66/B/F 89/C
SSU/JE/BIOL/100/1/B/F 89**BIOL/157/0.66/B/F 89/C
SSU/JE/BIOL/100/1/B/F 89**BIOL/157/0.66/B/F 89/C
SSU/JE/BIOL/100/1/N/F 89**BIOL/157/0.66/N/F 89/NC
SSU/JE/BIOL/100/1/N/F 89**BIOL/157/0.66/F/F 89/NC
SSU/JE/BIOL/100/1/W/F 89**BIOL/157/0.00/W/F 89/NC
SSU/JE/BIOL/100/3/A/F 89**BIOL/157/2.00/A/F 89/C

Table 1: Examples of expert evaluated transcript.

3. Course lists and catalogs at both the originating and receiving institutions change continuously, as do, to a lesser extent, policy rules.

4. Within SDSU there are individual and college differences between expert evaluators.

5. The evaluation process is a heuristic process.

Given this environment, it seemed most appropriate to design a system that could emulate an expert evaluator and in fact could be trained by that evaluator to generate an appropriate set of course list outcomes. Such a system would have the benefit of not requiring table changes or rule system changes, but would be "trained" by the expert evaluator during the normal course of decision making.

Based on this notion of a model system, and given the environment of the problem, we decided to develop a system based on neural networks to perform transcript evaluations. Two neural network models, a Backpropagation (or simply Backprop) model and a Monte Carlo model, were used to implement the automatic transcript evaluation systems. Both performed well on our testing data. Interestingly, the two models exhibit a trade off between speed and success rate.

The data for our training and testing data sets are data from actual transcript decisions in the College of Arts and Science at SDSU. This data set allows us to compare a wide range of course work from a variety of originating institutions and the consistent record of a single evaluator. Table 1 is an example.

Each entry in Table 1 has 13 fields separated by a slash. The input fact, or the course list from the originating institutions, is separated by "**" from the output fact, or equivalent course list in SDSU. Field 1 is the university name; field 2 is the evaluator's initial; field 3 and 8 are department names; field 4 and 9 are course numbers before and after translation respectively; field 5 and 10 are number of credits before and after translation respectively; field 6 and 11 are grades before and after translation respectively; field 7 and 12 are the semesters when the course was taken; field 13 is the decision made by the evaluator, accept or not accept.

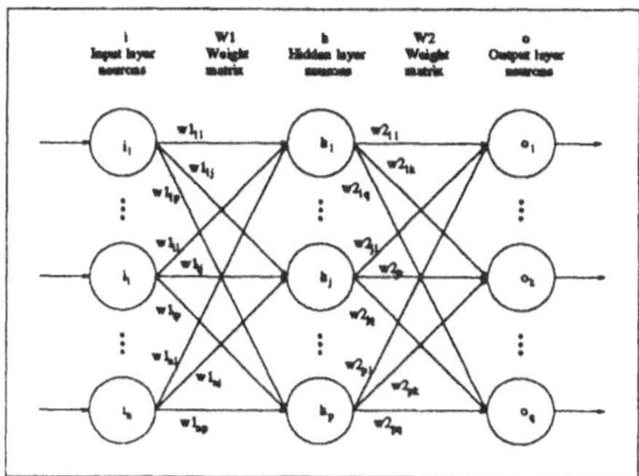

Figure 1: A neural network configuration with three layers.

2 The Neural Network Models

Backpropagation network (BPN) was developed independently by a number of researchers [3, 9, 10, 11, 13]. It was first formalized by Werbos [13]. BPN is designed to be a multilayer, feedforward and error- Backpropagation network.

The Backprop model used in this project has three layers of neurons: an input layer, a hidden layer, and an output layer, as illustrated in Figure 1. There are two layers of synaptic weights.

The neural network configuration of the Monte Carlo model used in our research is the same as that of the Backprop model.

The *Monte-Carlo* approach trains a network by adjusting the value of a randomly chosen element in one of the weight matrices. If the change has an improving effect, the change is accepted. Otherwise, the change may be or may not be accepted depending on a probability function. This approach is related to simulated annealing or statistical cooling. The claimed advantage of simulated annealing is that it reduces the chance of falling into a local optimum [6]. Aarts and Van Laarhoven [1] presented a general theoretical framework of the statistical cooling algorithm. The framework was applied to the traveling salesman problem to obtain an optimal result. Faigle and Schrader [5] observed that the best results can be achieved by iterating at a fixed temperature 'long enough' before lowering the control parameter to the next level. Generally the probability of accepting the weight adjustment remains the same or varies infrequently. They also provided theoretical support to this practical observation. Hashemi et.al [7] used a Monte-Carlo paradigm to study the toxicological effect of a toxicant on 14 different animals and humans. For their application, the predicting ability of the Monte- Carlo paradigm is better than the Backpropagation training paradigm. The optimal triangulation problem was investigated by Sen and Zheng [12] using Simulated Annealing. The simulated annealing was able to produce near-optimal or optimal solutions for the triangulation problem.

0	1 0 0	1	0,	0	1 5 7	0 6 6	0	1,
0	1 0 0	1	0,	0	1 5 7	0 6 6	0	1,
0	1 0 0	1	0,	0	1 5 7	0 6 6	0	1,
0	1 0 0	1	1,	0	1 5 7	0 6 6	1	1,
0	1 0 0	1	1,	0	1 5 7	0 6 6	1	1,
0	1 0 0	1	1,	0	1 5 7	0 6 6	1	1,
0	1 0 0	1	5,	0	1 5 7	0 6 6	5	0,
0	1 0 0	1	5,	0	1 5 7	0 6 6	4	0,
0	1 0 0	1	6,	0	1 5 7	0 0 0	6	0,
0	1 0 0	3	0,	0	1 5 7	2 0 0	0	1,

Table 2: The numeric version of the transcript shown in Table 1.

3 Implementation

Neural networks usually manipulate numeric values. However because a transcript file contains textual data, the transcript must be converted to a numeric representation. The numeric output of networks must also be mapped back to the text version for humans to understand. Field 1 and Field 2 are university name and the evaluator respectively, as described in Section 1. Since there are no counterparts after translation, they are removed during the conversion. Field 7 and field 12 do not change during the transcript translation. They are also removed during the conversion. The mapping functions of other fields are as follows:

Field 3 and field 8:

Dept:	BIOL	ENG	ENGL	MATH
Value:	0	1	2	3

Field 6 and field 11:

Grade:	A	B	C	D	F	N	W
Value:	0	1	2	3	4	5	6

Field 13:

Decision:	C	NC
Value:	1	0

Field 4, Field 5, field 9 and field 10: Since these fields are already or can be consider as numerical values, no mapping is necessary for these fields except that the period in the credit hour after translation are dropped out. For example, 2.66 will become 2 6 6 after conversion.

Table 2 is the numeric version of Table 1.

The algorithm used in the Backprop program is as follows [2]:

1. Compute the hidden-layer neuron activations:

   $h = F(W1 \cdot i + thresh1)$

   where h is the vector of hidden-layer neurons, i is the vector of input-layer neurons, $W1$ is the weight matrix of synapses connecting the input and hidden layers. $thresh1$

is the vector of thresholds for the hidden layer: $thresh1 = thresh1 + \alpha \cdot e$ where α is the learning rate and e is the hidden layer error computed in step (4), and $F()$ is a sigmoid activation function: $F(x) = 1/(1 + e^{-x})$

2. Compute the output-layer activations:

   $o = F(W2 \cdot h + thresh2)$

   where o is the vector of output layer, $W2$ is the weight matrix of synapses connecting hidden and output layers, and $thresh2$ is the vector of thresholds for the output layer: $thresh2 = thresh2 + \alpha \cdot d$ where d is the output layer error computed in step (3).

3. Compute the output-layer error (The difference between the target and the observed output):

   $d = o(1 - o)(o - t)$

   where d is the vector of errors for each output neuron and t is the target activation of the output layer.

4. Compute the hidden-layer error:

   $e = h(1 - h)W2 \cdot d$

   where e is the vector of errors for each hidden-layer neuron.

5. Adjust the weights for the second layer of synapses:

   $W2 = W2 + \Delta W2$

   where $\Delta W2$ is the matrix representing the change in matrix. It is computed as follows: $\Delta W2_t = \alpha \cdot h \cdot d + \theta \cdot \Delta W2_{t-1}$

6. Adjust the weights for the first layer of synapses:

   $W1 = W1 + \Delta W1_t$

   where $\Delta W1_t = \alpha \cdot i \cdot e + \theta \cdot \Delta W1_{t-1}$

7. Repeat steps 1 to 6 on all pattern pairs until the output- layer error (vector d) is within the specified tolerance for each pattern and for each neuron.

In the above algorithm, the learning rate term, α, is the indication of the effect of the weight change on each pass. This is typically a number between 0 and 1. The momentum term, θ, indicates how much a previous weight change should influence the current weight change. Initially, random values between -1 and +1 are assigned to the weights between the input and hidden layers, the weights between the hidden and output layers, and the thresholds for the hidden-layer and output-layer neurons.

When a network is trained using the Monte Carlo approach, like the Backpropagation approach, the weight matrices are initialized to random values between -1 and 1. The training process is an iterative procedure. Each time a forward pass is performed, that is, the result of layer i will be fed to layer $i + 1$. The difference, say $D1$, between the computed output of the output layer and the target output of the same layer is then computed. After one forward pass is done, a randomly chosen element from a randomly chosen weight matrix is adjusted by a small value. The output vector is computed again using the new matrices. A new difference, say $D2$, between the computed output and the target output is computed. If $D2$ is smaller than $D1$, the weight adjustment is accepted. Otherwise, the acceptance of the weight adjustment depends on the probability function as introduced below. The steps needed for training the network are as follows [4]:

1. For the given input data, complete the forward pass.

2. Compute the squared deviation (D) of the difference between the target output and the computed output using the following formula:

$$\sum_{i=1}^{m}\sum_{j=1}^{q}\frac{T_{ij}-O_{ij}{}^{2}}{m \cdot q}$$

   where t_{ij} is the target output for the jth node of the ith training pair, o_{ij} is the computed output for the jth node of the ith training pair, m is the number of training pairs in the training set, and q is the number of output nodes in the output layer.

3. Adjust one element of one of the two weight matrices and regenerate a new output vector of the net by repeating the forward pass.

4. Compute a new squared deviation D' using the new output vector in the formula in step (2).

5. Compute the objective function $x = D - D'$.

6. If the objective function x is greater or equal to zero (the new output is better), the weight adjustment in step (3) is accepted; otherwise, the weight adjustment is accepted according to Boltzmann law [7]:

$$P(x) = e^{\frac{A \cdot x}{K \cdot T}}$$

   where $P(x)$ is the probability of accepting x, A is some positive constant with appropriate unit, K is the Boltzmann's constant. A/K can be chosen as $(npq)2 * 103$[7]. T is some artificial temperature. T can be computed using: $T = T_0/(1+t)$

   where T_0 is the initial temperature, and t is an artificial time which is related to some step. For example, One step can be one cycle of complete manipulation of all the weights in both weight matrices.

4 Results

The transcript data file has 136 entries or input-output pairs. Of these 123 were used to train the networks and 13 of them (every 10th line) were used to test the networks.

The statistics of the predicting error of Backprop model is shown in Figure 2. The error indicates the difference between a field of the computing output and the corresponding field of the target output. From Figure 2, one can see that 35 percent of the fields have an error of 0.0, 43 percent have an error of 0.1, and 20 percent have an error of bigger than 0.1. If an error of 0.1 is acceptable, 78 percent of the fields are acceptable.

The predicting error of Monte Carlo model is shown Figure 3. Figure 3 shows that 58 percent of the fields have an error of 0.0, 33 percent have an error of 0.1, and 8 percent have an error of bigger than 0.1. If an error of 0.1 is acceptable, 91 percent of the fields are acceptable.

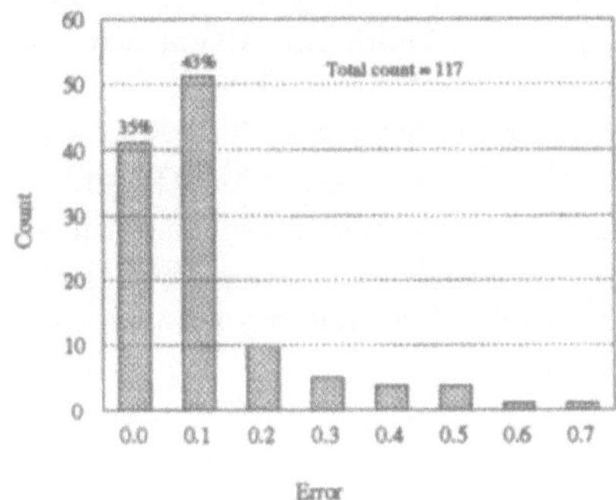

Figure 2: Histogram of errors of Backprop model.

5 Discussion

By observing Figure 2 and Figure 3 one can see that the Monte-Carlo model behaves better than the Backprop model. There are many factors that affect the behavior of a neural network. For example, for Monte-Carlo method, the values of the constant A and the Boltzmann's constant K, the cooling speed T, and the weight adjustment all affect the behavior of the Monte-Carlo. Some factors may affect the net's predicting precision. Others may affect the converging speed during the training of the net.

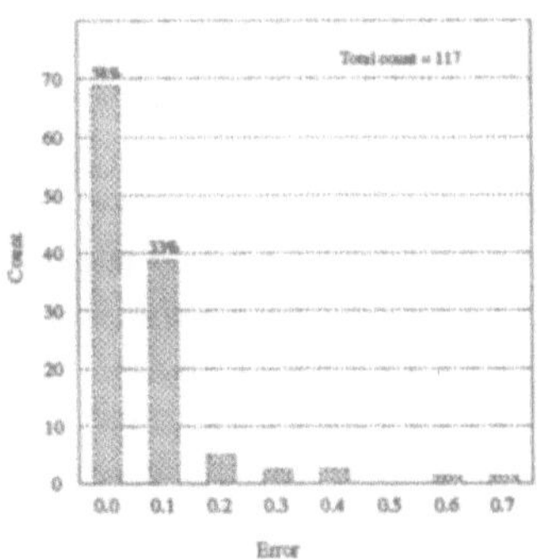

Figure 3: Histogram of errors of Monte Carlo model.

There are also many factors that affect the behavior of the Backprop net. These factors include MOMENTUM and TOLERANCE. There are no specific rules to determining these factors. They can only be decided by trying different values.

There are some additional characteristics that are unique to each method that bear further discussion. For example, although the predicting precision of the Monte-Carlo net is higher than the Backprop net, its converging speed is much slower. The reason is that adjustment of weight matrix elements is done randomly in Monte-Carlo paradigm. It is something like blind behavior. For example, some adjustments may have no effect and simply waste time. On the other hand, the weight adjustment in the Backprop model is directed by the error.

Another difference between the models is that there is a bigger chance for the Backprop model to be trapped in a local optimum. The Backprop model uses error to direct the weight adjustment. If a local optimum is reached, the corresponding error will be the minimum. Any adjustment of weight will give a bigger error. Therefore, training will stop here. On the other hand, the Monte-Carlo model uses random adjustment of weight. It is easier to escape a local optimum.

Some factors affect both models. For instance, the number of hidden layers and the nodes in each hidden layer may affect both predicting precision and converging speed. Another factor is the quality of the data. Two training facts are conflicting if they have the same inputs but different outputs. Some measures may be taken to remove conflicting data from the data file to improve the behavior of a net. However, in reality there will be naturally occurring conflicting cases in transcript data.

The predicting results of the two models shows that it is feasible to evaluate the transcript using a neural network if the net is well trained. For example, the Monte-Carlo model predicts 58 percent of outputs correctly and 33 percent with one grade of difference. If one grade of difference is acceptable, the net would predict 91 percent with acceptable error. Recall that the field 13 in the training fact represents whether a course transfer is acceptable or not. The Monte- Carlo model predicts this field with 85 percent accuracy.

One immediate application of our system is to let it serve as a preprocessor for the human expert. The expert starts working on the results of the net. This way, a great amount of time will be saved.

6 Conclusions

From the experimental results, it can be concluded that it is possible to use neural networks to do the automatic transcript translation. For the given set of test data, the test results also show that the Monte Carlo model works better than a Backprop model.

Further improvement should focus on the training speed. A good solution is to use a parallel computer system.

References

[1] Aarts, E.H.L. and P.J.M. Van Laarhoven, (1985) 'Statistical cooling: A general approach to combinatorial optimization problems', Philips Journal of Research, Vol.40,

No.4.

[2] Blum, A. (1992) Neural Networks in C++, Wiley Professional Computing.

[3] Bryson, A.E. and Y.-C. Ho, Applied Optimal Control, New York: Blaisdell.

[4] Burke, R. and B. Cong. (1993) 'An Intelligent System for Transcript Evaluation', Proceedings of 26th Annual Small College Computing Symposium, University of Northern Iowa. Cedar Falls, Iowa. April 16-17.

[5] Faigle, U. and R. Scherader, (1988) 'On the convergence of stationary distributions in simulated annealing algorithms', Information Processing Letters, Vol.27, No.4, pp189-194.

[6] Freeman, J.A. and D. M. Skapura, (1992) Neural Networks algorithms, applications, and Programming Techniques, Addison Wesley.

[7] Hashemi, R.R., A.H. Chowdhury, N.L. Stafford and J.R. Talburt, (1993) 'Prediction Capability of Neural Networks trained by Monte-Carlo Paradigm', Proc. of ACM Symposium on Applied Computing, pp9-13,

[8] Li, Y. (1994) Transcript Evaluation Using Neural Networks, Master Degree Research Paper, Computer Science Department, South Dakota State University.

[9] Parker, D.B. Learning Logic, Technical Report TR-47, Center for computational Research in Economics and Management Science, Massachusetts Institute of Technology, Cambridge, MA.

[10] Rumelhart, D.E., Ge.E. Hintorn, and R.J. Williams(1986a), 'Learning Representations by Back-Propagating Errors', Nature 323, 533-536, Reprinted in Anderson and Rosenfeld[1988].

[11] Rumelhart, D.E., Ge.E. Hintorn, and R.J. Williams(1986b). 'Learning internal Representations by Error-Propagating Errors', In Parallel Distributed Processing, vol.1, chap. 8, Reprinted in Anderson and Rosenfeld[1988].

[12] Sen, S. and S.Q. Zheng,(1993) 'Near-Optimal Triangulation of a Point Set by Simulated Annealing', Proc. of ACM Symposium on applied computing, pp1000-1008.

[13] Werbos, P. Beyond Regression: New Tools for Prediction and Analysis in the Behavioral Sciences, PhD thesis, Harvard, Cambridge, MA, August 1974

EXTRAPOLATION OF VIBRATION DATA USING NEURAL NETWORKS

MARC KARAM and ANDRZEJ M. TRZYNADLOWSKI
Department of Electrical Engineering
University of Nevada, Reno
Reno, NV 89557-0153, USA

Abstract - Vibration data obtained from systems monitoring large electromachine systems are often distorted because of high sensitivity of the sensors. Clipped peaks represent one of the most common distortions of vibration waveforms. In the existing practice, clipped waveforms are rejected from further analysis. The paper describes an endeavor of reconstruction of the missing peaks. Commercial neural network software was used to extrapolate the undistorted portion of a waveform into the region of the clipped peak. The developed neural networks, training process, and experimental results are presented.

1 Introduction

Analysis of vibration data in large electromachine systems, such as turbine-generator sets and electric motor drives, is an important means of machine diagnostics and preventive maintenance [1-4]. In the subsequent considerations, references are made to specific vibration monitoring equipment and diagnostic procedures, namely these of the Bently Nevada Corporation (BNC), a major U.S. manufacturer of such equipment and provider of diagnostic services. BNC's database was used as a source of material for the described study.

As illustrated in Fig. 1, vibration signals that represent the lateral displacement of a rotating shaft are obtained from two proximity transducers (probes) located in the direct (*d*) and quadrature (*q*) axes of the shaft. A synchronizing signal, allowing space orientation of the vibration and used for measurement of the rotating speed of the shaft, is generated by the so-called keyphasor probe mounted over a small notch machined in the shaft. Several such probe sets are placed along the shafts of the monitored machinery. Analog electric signals from the probes are transmitted through cables to a data acquisition system and converted to a digital format [5].

High sensitivity of the monitoring system, necessary for high accuracy of the measurements, backfires in the form of frequent distortions of the vibration data. Particularly troublesome are saturation effects that result in clipped peaks of vibration waveforms. Therefore, tedious visual inspection of the waveforms must be done before further processing. In the existing practice,

E. A. Yfantis (ed.), Intelligent Systems, 599–608.

clipped waveforms are rejected in order to avoid incorrect diagnostic conclusions [6,7]. If the amount of the remaining data is insufficient, the measurements must be repeated, which is expensive and time consuming. Examples of undistorted and clipped vibration waveforms are shown in Fig.2.

The study described in this paper is based on the assumption that if the loss of information due to the clipped peak is not excessive, an attempt can be made to recover the missing bytes of data by extrapolating the undistorted portion of a waveform into the region of the clipped peak. Neural network software was used for this purpose since the classic numerical extrapolation has been found to yield unsatisfactory results. The developed neural networks, their training, and experimental results are described.

2 Neural Networks

In the BNC's systems, a single cycle of vibration is represented by 32 bytes. A limit of up to 5 bytes to be recovered has been assumed. Fig. 3 shows an example of a clipped waveform with 5 bytes of the missing positive peak.

The neural network software package DynaMind 3.0 from NeuroDynamX Inc., along with the associated NeuroLink 2.0 program, set up on an IBM PC 486 computer, were used for implementation of the extrapolating neural networks [8]. The structure of such a network, of the common, backpropagation type, is shown in Fig. 4. Experiments involved waveforms with exactly 5 bytes missing from the positive peak. Therefore, the network has 27 inputs, for the 27 bytes of the undistorted part of the waveform, and 5 outputs, for extrapolation of the missing 5 bytes of the clipped peak. There are two hidden layers, with 13 and 6 neurons in the first and second layer, respectively. Similar neural networks have already successfully been used in preceding studies devoted to recognition and classification of distorted data [9,10].

As explained later, a single neural network has been proven insufficient for the required task, and several networks had to be developed. However, all of them have the same number of layers and neurons in the first three layers.

3 Training

To focus the network's attention on the relevant features of the processed waveforms, the original, raw data were subjected to pre-processing. The waveforms were normalized with respect to their amplitude and initial phase. Assuming that a vibration waveform constitutes a coarse sinewave, the normalized samples (a single cycle of vibration) were made to have the negative peak of 0, the positive peak of 1, and the first byte closest to 0.5, as shown in Fig.

5.

The training was performed using undistorted data, i.e., the extrapolating network was trained to predict bytes no. 6 through 10 from the remaining 27 bytes of a sample. Initially, a single network was experimented with, but the results were unsatisfactory. An analysis of the data has shown that the typical peaks can be classified into four basic categories shown in Fig. 6. Interestingly, it turned out that the 27 bytes of the input information allow to predict the type of missing peak. This prediction was performed by three so-called classifying neural networks, similar in structure to the network in Fig. 4 but with two outputs only. Classifying network C12 was trained to distinguish between types 1 or 3 and 2 or 4, network C13 between types 1 and 3, and network C24 between types 2 and 4. In the training process, the two outputs were set to 01 or 10, and when testing was conducted an output of less than 0.33 was interpreted as a 0 and that of more than 0.67 as a 1. A high, almost 100%, success rate was achieved.

Following the peak classification, data samples were applied to appropriate four extrapolating networks, E1 through E4. The data flow is illustrated in Fig. 7 that shows all the seven neural networks used and the reference output values for the classifying networks. The whole computing process was automatized utilizing the NeuroLink 2.0 software.

4 Experimental Results

A comparison between an example actual peak and the predicted one is shown in Fig. 8. For an additional evaluation, numerical extrapolation has been employed using the cubic spline [11]. As exemplified in Fig. 9, the results were significantly less accurate than these obtained from the neural networks. It is typical for numerical extrapolation procedures that they tend to yield the least accurate results in the areas around the minima and maxima of a function.

For the purpose of machine diagnostics, vibration data are subjected to various signal processing operations, the Fast Fourier Transform (FFT) being most widely used. Therefore, it is important that the FFT of a reconstructed waveform be possibly close to the FFT of the original waveform. The high quality of operation of the developed neural networks has been confirmed by comparing these FFTs. Example FFTs are shown in Fig. 10. It can be seen that the individual coefficients of the transform of the original waveform are practically identical with these of the same waveform but with the peak extrapolated by a neural network.

Performance of a neural network is usually evaluated using the Mean Square Error (MSE) of the network output values. A summary of the experimental results, involving trained networks in the configuration shown in Fig. 7 and 200 representative samples of vibration data selected from the used database of over 1600 samples, is provided in Table I.

Table I. Experimental Results

No. of Samples	MSE	Minimum	Average	Maximum
200	5 Bytes of a Peak	0.0022	0.0207	0.0490
	32 FFT Coefficients	0.0092	0.0733	0.1884

5 Conclusion

Feasibility of application of neural network software to recovery of clipped peaks of vibration waveforms has been demonstrated. Neural networks can significantly improve efficiency of processing vibration data in machine diagnostics procedures. The study opens the way for incorporation of neural networks into expert systems for preventive maintenance of large electromachine systems.

Acknowledgments

The project described was supported by the IEEE Industry Applications Society's Myron Zucker Student-Faculty Grant. Bently Nevada Corporation provided the vibration data and invaluable technical advice.

References

[1] J.T. Renwick, "Condition monitoring of machinery using computerized vibration signature analysis," *IEEE Trans. Ind. Appl.*, vol. IA-20, no. 3, 1984, pp. 519-527.

[2] J.T. Renwick and P.E. Babson, "Vibration analysis - a proven technique as a predictive maintenance tool," *IEEE Trans. Ind. Appl.*, vol. IA-21, no. 2, 1985, pp. 324-332.

[3] B.C. Bhaoval, "Vibration analysis and monitoring technique," *J. Inst. Eng.*, vol. 68, March 1988, pp. 132-137.

[4] J. Matthew, "Monitoring the vibrations of rotating machine elements," *Conf. Rec. 12th Bienn. Conf. on Mechanical Vibration and Noise*, 1989, pp. 15-22.

[5] "Rotating Machinery Information Systems and Services, *Bently Nevada Corp.*, 1990.

[6] "Vibration measurement - basic parameters for predictive maintenance of rotating machinery," Application Note, *Bently Nevada Corp.*, 1986.

[7] A. Muszynska and D. Bently, "Fundamentals of rotating machine diagnostics," *Bently Nevada Corp.*, 1992.

[8] "DynaMind Developer User's Guide," *Neuro-DynamX Inc.*, 1992.

[9] M. Evans and A.M. Trzynadlowski, "Verification of vibration data in electromachine systems using neural-network software," *Conf. Rec. 1992 IEEE-IAS Ann. Mtg.*, pp. 60-65.

[10] M. Karam, M. Ghassemzadeh, N. Dai, M. Gandikota, and A.M. Trzynadlowski, "Validation and recovery of vibration data in electromachine systems using neural network software," *Conf. Rec. 1993 IEEE-IAS Ann. Mtg.*, pp. 225-232.

[11] C. De Boor, "A Practical Guide to Splines," *Springer Verlag*, 1978.

Figures

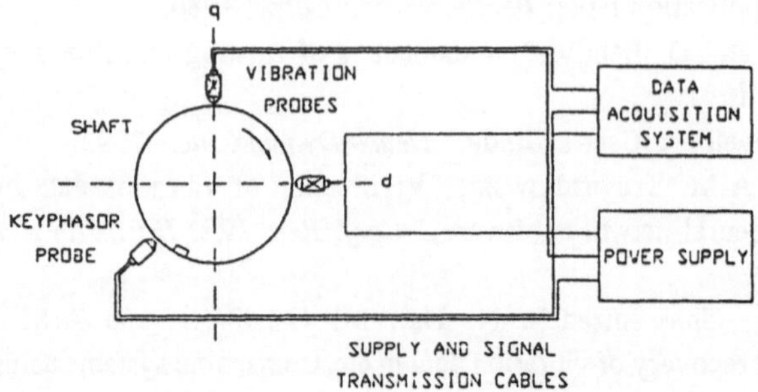

Fig. 1. The vibration monitoring system

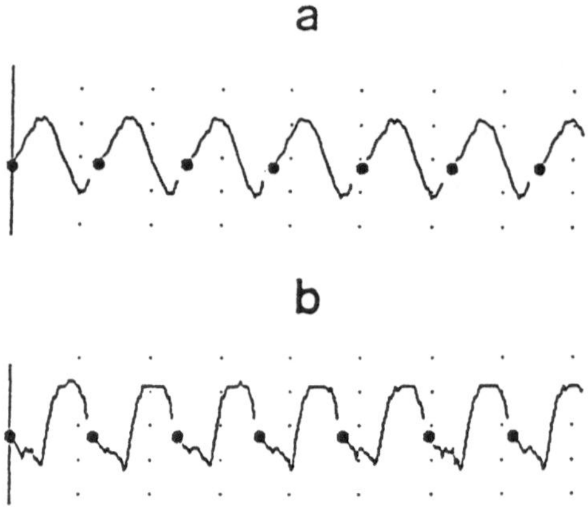

Fig. 2. Example vibration waveforms :
(a) undistorted, (b) clipped

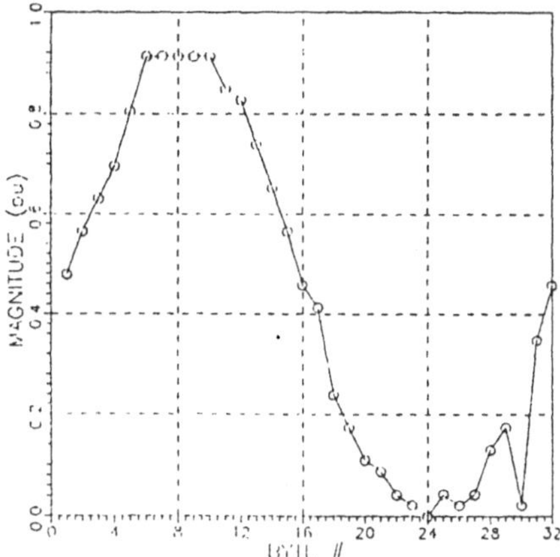

Fig. 3. An example clipped waveform with 5 bytes of the missing positive peak

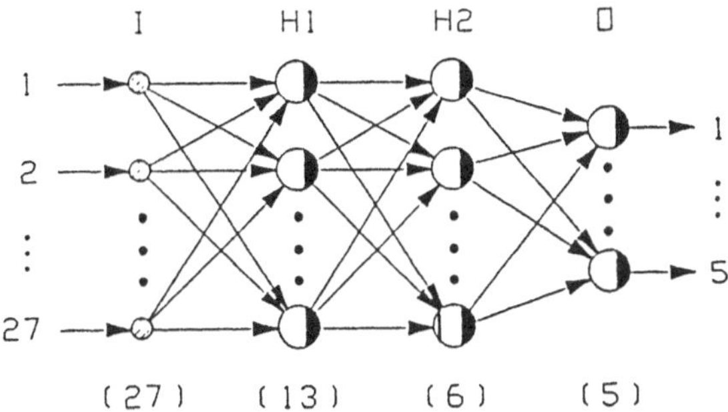

Fig. 4. The structure of an extrapolating Neural network

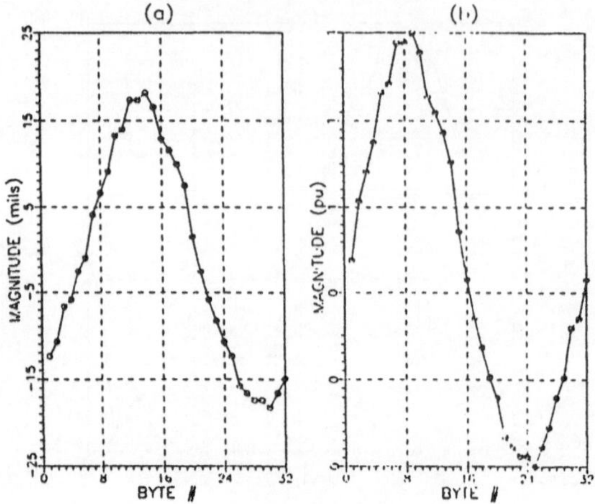

Fig. 5. Example data sample:
(a) original, (b) normalized

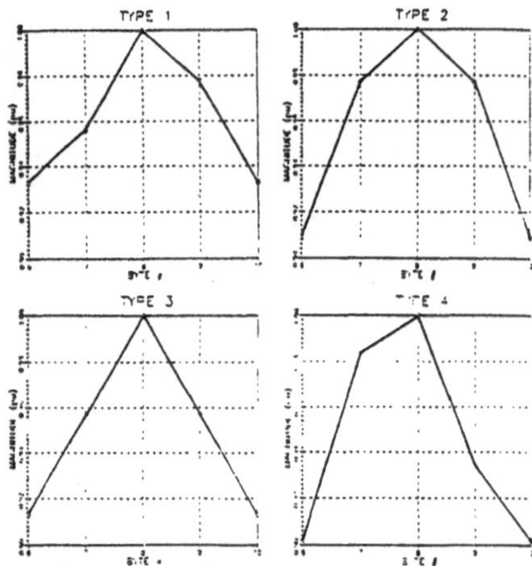

Fig. 6. The four basic types of 5-byte peaks

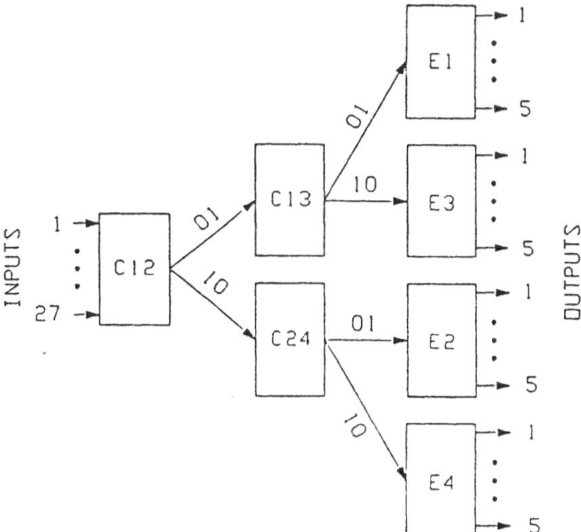

Fig. 7. The data flow block diagram

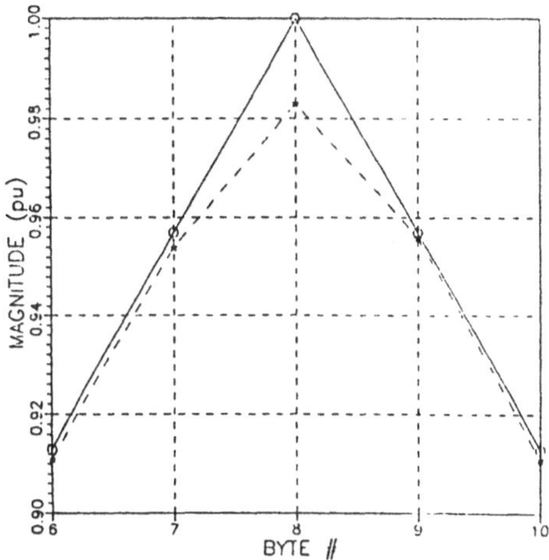

Fig. 8. Comparison between an example original peak (solid) and that extrapolated by a neural network (dashed)

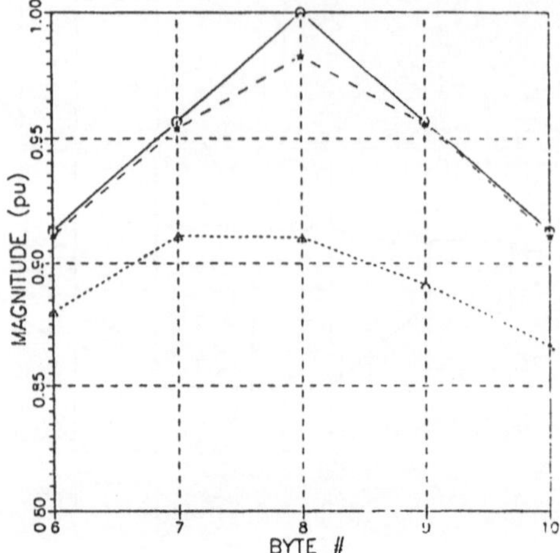

Fig. 9. Comparison between an example original peak (solid) and those obtained from a neural network (long dash) and by numerical extrapolation (short dash)

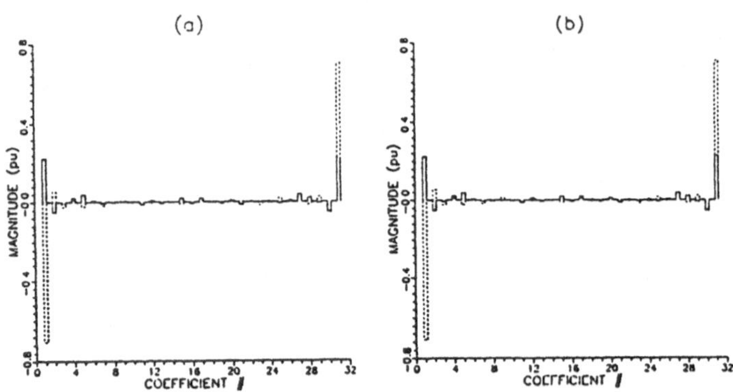

Fig. 10 Comparison between real (solid) and imaginary (dashed) FFT coefficients of (a) and example original waveform and (b) the same waveform but with a recovered peak

Design Strategies for Evolutionary Robotics

Andrew Murray
Dept. of Computer Science
University of Nevada
Reno - 89557
email: murray@cs.unr.edu

Sushil J. Louis
Dept. of Computer Science
University of Nevada
Reno - 89557
email: sushil@cs.unr.edu

Abstract: This paper deals with the question of how to balance evolutionary design and human expertise in order to best design robots which can learn specific tasks. We study two behavioral tasks, approach and avoidance, and provide some preliminary results.

Keywords Genetic Algorithms, Computational Design, Autonomous Agents, Robot.

1 Introduction

Our primary objective is to investigate the balance between human design and computational design as a foundation for designing autonomous agents. In this research two fundamental design tradeoffs are addressed. First, the design and configuration of external sensors may be under evolutionary control in order to simulate real animals and their adaptability. This is the structure design task. Second, the control strategies which map sensory input to effector output may also be under genetic control and need not be explicitly defined. This is the control design task. We explore the tradeoff between human design of structure and control and computational design of structure and control. Nature indicates that natural selection, the process of evolution, is a powerful design paradigm, thus we use evolutionary computation algorithms, specifically genetic algorithms to computationally design the structure and control strategies of our autonomous agents. In other words, we search through possible structure designs while searching for co-adapted control strategy designs to maximize performance at a particular task. We believe sensible human design must be balanced with evolutionary design to reduce human design time and maximize performance and flexibility.

Previous work in this field has concentrated more on the learning of control strategies although some recent work has explored the evolution of structure and control [2, 3, 6]. We use a genetic algorithm (GA) to design both the structure and to find a control-strategy for our agent [5, 4].

We use a simulated autonomous robot or SIMBOT, that learns two basic types of behavior, approach and avoidance. For approach, the simbot has "ears" which "hear" food sources placed in a simulated environment. The simbot also possesses four touch whiskers in order to detect and avoid environment boundaries and other obstacles that are present in the environment. A Genetic Algorithm guides the placement and sensitivity of

E. A. Yfantis (ed.), Intelligent Systems, 609–616.

sensors and evolves control mappings between sensor input and motor output that allows the simbot to perform the two tasks.

2 The Genetic Algorithm

Genetic algorithms (GAs) are stochastic, parallel search algorithms based on the mechanics of natural selection, the process of evolution. GAs were designed to efficiently search large, non-linear search spaces where expert knowledge is lacking or difficult to encode and where traditional optimization techniques fail. They are flexible and robust, exibiting the adaptiveness and degradation of biological systems. However, like other robust search algorithms, they often require a large number of iterations before converging on a candidate solution.

Conceptually, GAs use the mechanisms of natuaral selection in evolving individuals that, over time, adapt to an environment. In our case, the simulated robots which performed the target behaviors the best were more likely to survive and propagate their designs to future generations. This weeding out procedure is performed with a population of candidate designs known as phenotypes, and iterated over many generations.

Mathamatically, each individual is comprised of bit strings (genotype) which encode sensor parameters and control mapping strategies. The initial population consists of random bit strings. The only requirement for the GA to work properly is to be able to compare the relative fitnesses for each individual, and thereby select individuals for mating proportionately to there fitnesses.

The mating process involves swapping portions of the bit strings in each of two parent individuals. In our experiments, simple one point crossover was used. This procedure involves choosing a random crossover point and then swapping the contents of the parent bit strings beyond that point. In addition to crossover, bit mutation is also performed in order to prevent loss of genetic material and to promote exploration of the search space. Mutation is generally performed at a much lower probability rate than crossover. By changing crossover and mutation rates, a wide range of searching techniques can be achieved.

3 The Environment

The environment was a square area of 300 units on each edge (figure 1). Obstacles consisted of random rectangular shapes with maximum edge lengths of 50 units. Up to ten rectangular obstacles of random dimensions and location could be placed in the environment.

Food sources were represented by small red circles and would disappear when "eaten". These entities were also placed at random locations and could be removed if they appeared inside obstacles, or otherwise unreachable to the simbot. The eating process was simulated by a simple proximity function between food sources and the center of the modeled simbot.

For control and comparison purposes, the experiments were run in groups of identical environments. Two different environments were used in each of the three stages of experimentation. Touch experiments used 4 and 10 obstacles, hearing experiments used 5 and

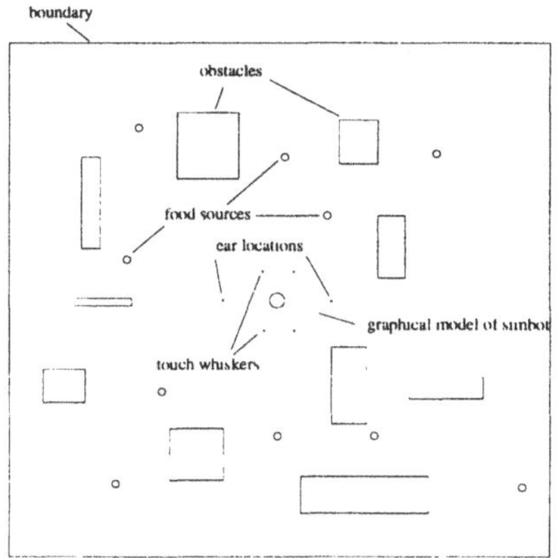

Figure 1: Simulated environment.

10 food sources and experiments for both simply paired these two quantities.

4 Simbot Design

A simbot was modeled by 7 points: 1 for the center of the body, 4 for the locations of the tips of the 4 touch whiskers, and 2 for the sound receptors or "ears" (figure 1). The whiskers protruded from the center of the simbot at an approximate angle of 30 degrees to either side of the direction of motion, both forwards and backwards. Unlike the whiskers, the locations and range of the ears were not fixed, but were under genetic control.

The motion of the simbots was generated by two locomotive tracks, left and right. Each track had four speeds: one reverse speed, one stop, and two forward speeds. Turning was accomplished by differences in track velocities.

5 Experimental Setup

In the first step of experimentation, we only considered obstacle avoidance. We started with a direct mapping scheme to map touch sensors to motor outputs. In this scheme the left touch sensor had a higher probability of connecting to the left motor and vice-versa. This mapping made it more difficult to find an effective control strategy because the only way to turn in the opposite direction when sensing an obstacle is to decrease the velocity of the motor opposite to the side of the touch sensor, assuming that full speed ahead is the normal operating state [1]. Switching to a cross mapping, where the left sensors had a higher probability of influencing the right motor was more effective (see figure 2). The gates were connected in two 2 x 2 grids such that both sensor outputs were input to the first level of logic gates and the outputs of these gates were input to the second level of gates as shown in figure 2. The final output of these gates was applied to the motor control. The logic gates themselves consisted of standard AND, OR and XOR gates as well as a few custom definitions which allowed the 16 possible truth tables for the four possible inputs (figure 2). Thus, 4 bits were needed to encode each gate, and the entire chromosome consisted of 32 bits for eight gates giving rise to a search space of 2^{32}.

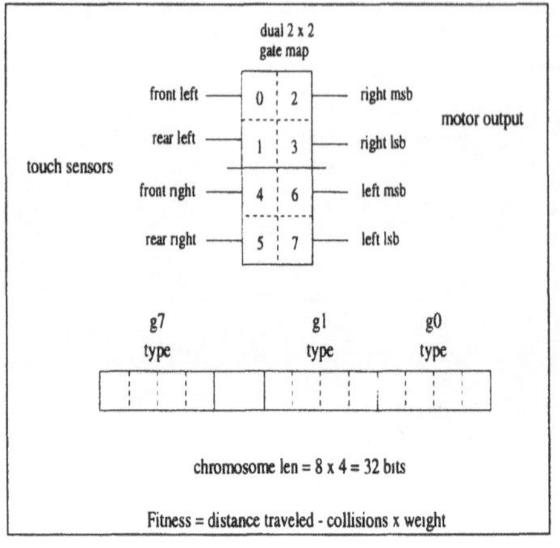

a — gate — out
b

1. a AND b
2. ~a AND b
3. a AND ~b
4. ~(a AND b)
5. a OR b
6. ~a OR b
7. a OR ~b
8. ~(a OR b)
9. a XOR b
10. ~(a XOR b)
11. 0
12. 1
13. a
14. b
15. ~a
16. ~b

Figure 2: Left: Touch mapping. Right: Logic gate definitions.

The fitness function was defined as the distance traveled per simulated run minus the number of collisions times a weighting factor which could be changed to maximize performance.

$$\text{Fitness} = \sum_{i=1}^{n} d_i - \sum_{j=1}^{m} c_j$$

d = distance traveled + straight motion bonus.
n = length of simulation.
c = penalty for crashing.
m = number of collisions.

In early experiments, the simbots developed circular motion because of two reasons: First, circular patterns were somewhat effective in avoiding obstacles and second, circular motion was being rewarded almost as much as straight motion. In order to promote straight motion, the final version of the fitness function added a bonus to the distance when both motors were on full.

The genetic algorithm used single point crossover and bit mutation, while the selection strategy used was elitist.

6 Preliminary Results

The length of simulation was set at 1500 moves so that simbots would have to evolve avoidance behavior in order to obtain a high fitness. A weighting factor of 10 was experimentally chosen as it produced good results. In all tests, a population size of 30 was chosen mainly due to time constraints and the fact that in our experiments larger population did not significantly improve results. The simulations were run for 40 generations. The two parameters common to all experiments were the crossover and mutation rate.

6.1 Touch

In addition to changing crossover and mutation rates, two distinct environments were created, one with 4 obstacles and the other with 10. Results comparing the two environments are shown in figure 3. As expected, it took longer for the simbots to elvolve in the environment with more obstacles. Also the final performance of these simbots was slightly less than those in the four obstacle environment since there was less opportunity to achieve long straight runs, which is the highest rated bevhavior.

In general, higher crossover rates degrade the performance of the GA in later generations. This is mainly due to the greater probability of destroying highly fit schema corresponding to desirable simbot behavior.

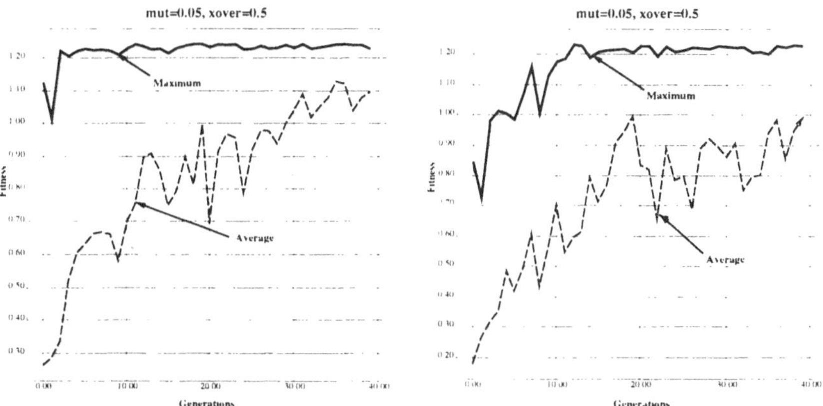

Figure 3: Left: Four obstacle environment. Right: Ten obstacle environment.

6.2 Hearing

The next step was to evolve approach behavior. Food sources gave off a sound that could be used by the simbot to locate and consume food. In this case, as expected, crossed mapping was less effective because in order to turn towards a detected object, the motor track on the same side as the hearing sensor had to be decreased, again assuming forward motion in free space. The mapping for hearing was almost identical to touch except that the inputs were taken from differences in sound level input between left and right ears (figure 4). Using absolute sound levels as inputs didn't work as the resolution of 4 bits wasn't nearly enough to establish desired behavior. In addition to the 32 bit control chromosome, another chromosome with 6 bits was added to evolve other hearing parameters, 3 bits for range and 3 bits for separation between the ears.

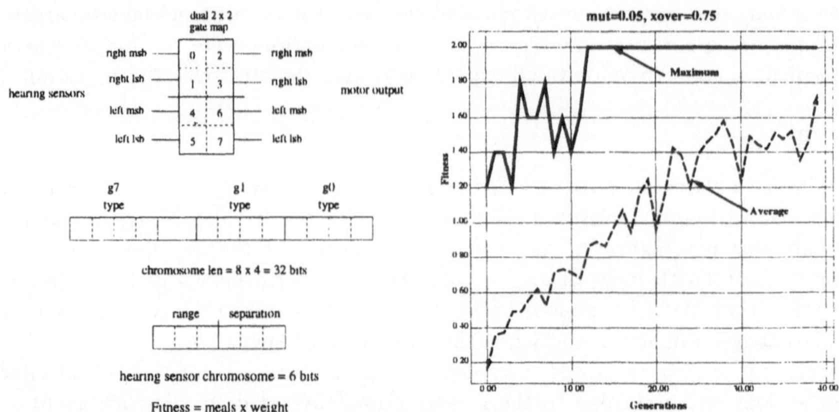

Figure 4: Left: Sound mapping. Right: 10 food sources

The fitness for these experiments was simply the number of food sources eaten times a weighting factor. The weighting factor was set to 200 to achieve fitnesses which approximated those in the touch experiments. Although it made no difference in the hearing experiments it was important to balance touch and hearing fitnesses in later experiments which combined both avoidance and approach behaviors. Crossover and mutation was performed as before but was also performed on the 6 bit sensor chromosome independently.

Figure 4 also shows a typical result for an environment with ten food sources. Individuals which could track down and eat all ten sources evolved after only 12 generations. On average, the simbots would consume between 70% to 80% of the available food sources. As in the touch experiments, higher crossover and mutation rates degraded the average performance.

Another interesting result was that the sensor parameters evolved differently for different environments. In experiments with 5 food sources, the ears tended to be far apart and had medium to long ranges. For 10 food sources, the ears were close together and had short to medium range. This makes perfect sense if you consider that simbots with wide separation and long ranges would not be able to discriminate between several food sources in close proximity, and would become confused. Likewise, simbots with short ranges and ears close together would not be able to detect food sources from long distances, and would have to stumble upon them randomly before control mapping could take over.

6.3 Touch and Hearing Simultaneously

The first attempt at combining the two behaviors involved switching between the two controls, depending on if a touch whisker was active. We reasoned that if no whiskers were active, the simbot would be in free space and should be allowed to search for food. When a touch whisker became active, the simbot would have to switch over to touch control in order to avoid the obstacle. However, this reasoning is only valid *after* the simbot has learned both types of behavior. At first, the simbots would simply hit an obstacle, switch to touch mapping and back away from it, and then go straight back into the obstacle, since it would be in the exact same situation it was in before. In addition, they were being rewarded for this behavior since the simbot was indeed moving in an oscillating manner, but of course not getting anywhere.

To solve this problem, we included a training period in which the simbots were allowed to evolve each type of behavior independently, then the two were combined. The length of training became a fourth variable along with crossover, mutation, and number of obstacles/food sources in the environment. This method produced good results, but a timeout counter had to be added which forced the simbots to stay in touch control mode for a specified number of moves after detecting an obstacle. This gave the simbots enough time to completely turn away from obstacles.

Figure 5 compares having 12 generations of training (6 for each type of behavior), and 24 generations of training in an environment with 10 obstacles and 10 food sources. We chose a low crossover rate, 0.5, and mutation rates, 0.05 to maximize performance. As expected, the simbots with more training achived maximum performance sooner and the learning curve was a little steeper than those with less training. However, average performance was comparable after 40 generations.

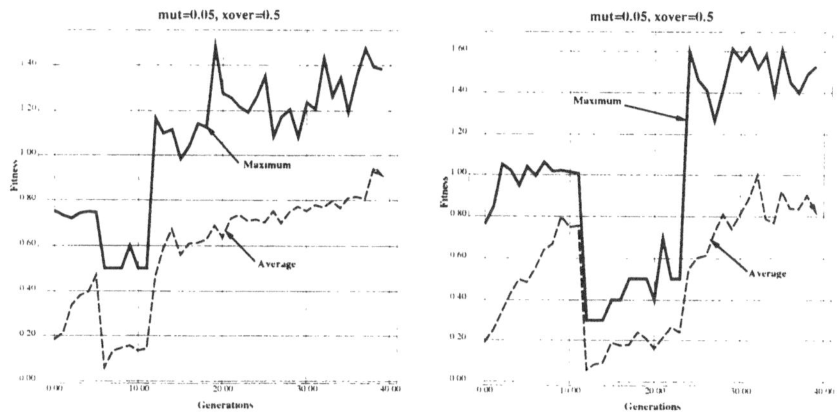

Figure 5: Left: 12 generation training period, Right: 24 generation training period.

The effects of crossover and mutation were the same as before. In fact, in one experiment with high crossover and mutation rates, the simbots did not develop avoidance behavior in a limited number of generations. As a result, they displayed the same oscillating behavior described earlier.

7 Conclusions

In answering the question posed in the introduction, human design played a major role in affecting the performance of the simbots in these particular experiments. In fact, due to the simplistic nature of the environment, basic heuristic rules could have been implemented to achieve higher performance in the simbots. However, as the complexity of tasks and environment increases, genetic control is likely to be more effective. Where heuristic control is not obvious, genetic algorithms would be more effective in finding solutions to problems, since it would be difficult to a priori specify the relation of sensor input to motor output. In addition, control could evolve over time to better adapt to an environment, overcoming shortcomings in initial designs. The idea is to start with some human designed control but provide enough flexibility so that if the environment changes or if there are bugs in the control algorithm, the GA can modify the mapping from sensors to effectors and thus better adapt a robot to its current environment.

In these experiments, the control mapping evolved to mimic the heuristics that were already known to be effective for the types of behavior being studied. Future work includes modifying the simulator in order to give more control to the GAs by combining the individual mappings into one input/output map or letting the simbots figure out for themselves when to switch between touch and hearing control. If successful, more senses such as vision might be added to increase complexity of the simulations. Also, more interesting behaviors or combinations of behaviors could be studied. Finally, we plan to develop a hardware platform to test the efficacy of simulation results on a real robot.

References

[1] Valentino Braitenberg. *Vehicles.* The MIT Press, Cambridge, MA, 1986.

[2] D. Cliff, P. Husbands, and I. Harvey. "evolving visually guided robots". Technical Report CSRP 220, School of Cognitive and Computing Science, University of Sussex, 1992.

[3] Marco Dorigo and Marco Colombetti. "robot shaping: Developing situated agents through learning". Technical Report TR-92-040 Revised, International Computer Science Institute, University of California, Berkeley, 1993.

[4] D. E. Goldberg. *Genetic Algorithms in Search, Optimization, and Machine Learning.* Addison-Wesley, Reading, MA, 1989.

[5] J. Holland. *Adaptation In Natural and Artificial Systems.* The University of Michigan Press, Ann Arbour, 1975.

[6] John R. Koza. *Genetic Programming.* The MIT Press, Cambridge, MA, 1992.

Efficient Construction of Networks for Learned Representations with General to Specific Relationships

Cory Barker and Tony Martinez
Brigham Young University, Provo, Utah 84602
cory@axon.cs.byu.edu, martinez@cs.byu.edu

Abstract

Machine learning systems often represent concepts or rules as sets of attribute-value pairs. Many learning algorithms generalize or specialize these concept representations by removing or adding pairs. Thus concepts are created that have general to specific relationships. This paper presents algorithms to connect concepts into a network based on their general to specific relationships. Since any concept can access related concepts quickly, the resulting structure allows increased efficiency in learning and reasoning. The time complexity of one set of learning models improves from O(*n* log *n*) to O(log *n*) (where *n* is the number of nodes) when using the general to specific structure.

1. Introduction

Many machine learning algorithms learn by either generalizing or specializing existing representations [Michalski 1983, Mitchell 1982, Hayes-Roth 1978, Vere 1975, Barker 1993]. The learning process typically involves a search for the best new representation. The search is potentially time consuming and can be benefited by using parallel architectures for implementation. This paper presents a set of algorithms to create a structure that connects rules or concepts based on their general to specific relationships. The resulting natural structure has potential to increase performance of learning algorithms.

Learning systems often represent concepts, rules, or features as sets of attribute-value pairs or input-value pairs. Such rules cover subsets of the input space. Each input-value pair (i, x) contained in the representation of a rule restricts the rule to match only those input points where input $i = x$. Thus, rules with fewer input-value pairs are more general while rules containing more input-value pairs are more specific.

If a rule A has input-value pair (i, x) and rule B has input-value pair (i, y) such that $x \neq y$, then A and B are said to be *discriminated.* The sets of input points covered by A and B are disjoint. If two rules are not discriminated they are said to *overlap.*

Rules can be generalized by removing input-value pairs and specialized by adding input-value pairs. The specialization S obtained by adding one or more input-value pairs to a rule R matches a subset of the points in the input space that are matched by R. S is said to be *more-specific-than* R [Mitchell 1982].

The more-specific-than relation can be computed by comparing the input-value pairs that define two rules. Given two rules A and B with input-value sets p and q, respectively, A is more-specific-than B if-and-only-if $p \supset q$.

The more-specific-than relation defines a *partial order* on the set of rules. A partial order gives rise to a *semi-lattice.* The algorithms described in this work create this lattice structure. The structure is a digraph whose vertices or nodes are rules and which has an edge or connection from rule r to rule s if and only if s is more-specific-than r. The resulting network structure can be implemented using either software or hardware.

E. A. Yfantis (ed.), Intelligent Systems, 617–625.

Using a structure that exploits the general-specific relationship between rules allows increased efficiency in learning and reasoning. One family of learning models [Barker 1994] incorporates a new training example in time O(*n* log *n*) (where *n* is the number of nodes) without using the structure. The complexity improves to O(log *n*) when the structure is used. The contribution of this paper is to present efficient algorithms for construction of the network.

2. General-Specific (GS) architecture

A network using the GS topology is a set of nodes *N* and a set of edges *E*. Let *A*, *B*, and *C* be nodes from *N*. Let r_a, r_b, and r_c be the rules for nodes *A*, *B*, and *C*, respectively. The set of nodes *N* always contains a node *R* that is the root node. The root node contains the rule that is fully general; it matches all points in the input space. Suppose the network is defined such that there is an edge from *A* to *B* if-and-only-if r_a is a generalization of r_b.

An example of such a network is shown in Figure 1. The example has three inputs {*Wind Speed, Wind Direction, and Visibility*} (abbreviated S, D, and V, respectively). *Wind Speed* has three possible values {*Low, Medium, High*} (abbreviated L, M, and H, respectively). *Wind Direction* has two values {*Head, Tail*} (abbreviated H and T, respectively). *Visibility* has two values {*Clear, Cloudy*} (abbreviated Clr and Cld, respectively). The root node is connected to all other nodes since the root is a generalization of every possible rule. Node 1 is a generalization of both Nodes 2 and 3. Note that the connections between the root and Nodes 2 and 3 are essentially redundant. These connections can be obtained by computing the transitive closure of other connections. The number of connections required in the network can be reduced by removing these redundant connections.

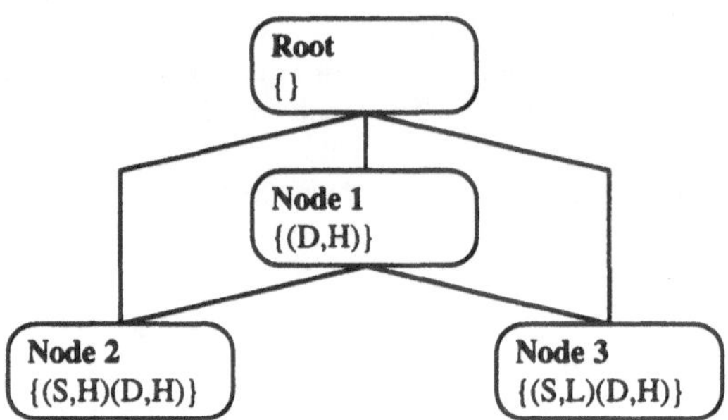

Figure 1. General/Specific Architecture

The definition for edges is modified to remove edges that are redundant. An edge connects nodes *A* and *B* if-and-only-if r_a is a generalization of r_b and there is not some other node *C*, such that r_c is a generalization of r_b and r_c is a specialization of r_a. In other words, two nodes are connected if the rule in one node is either the most specific generalization or the most general specialization of the rule in the other node.

An example of a network without redundant connections is shown in Figure 2. Note that a node can still have multiple generalizations. Node 3 is connected to Nodes 1 and 2 since both nodes are generalizations of Node 3 but Nodes 1 and 2 are not generalizations or specializations of each other. The resulting network is thus a directed acyclic graph (DAG) rather than a tree.

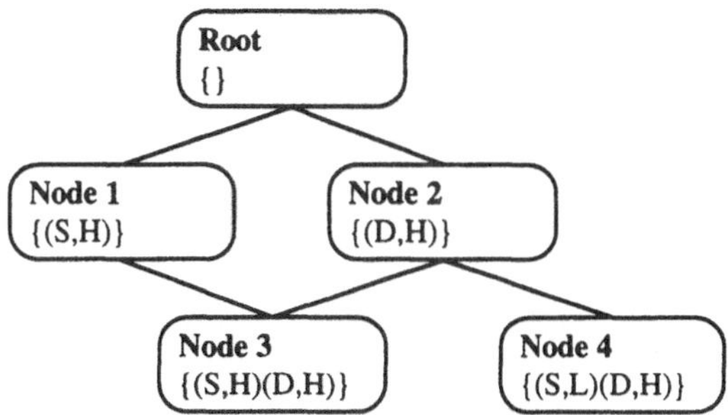

Figure 2. General/Specific Architecture

3. Adding Nodes

When adding a new node to a GS network the new node must be connected to existing nodes that are the most specific generalizations of the new node. The new node must also be connected to existing nodes that are the most general specializations of the new node. It is desirable to reduce the number of existing nodes that must be searched in order to connect a new node. By using information that is already known about a new node the new node can be connected in a more efficient manner.

New nodes that have no known relationship with other nodes in the network are linked to their generalizations by starting at the root node. The procedure is described in Section 3.1. Then, using one of the generalizations *N* that was linked to the new node, specializations of the new node can be found efficiently by starting at node *N* in the network. This process is described in Section 3.2. When a node is known to be a generalization of an existing network node *N*, generalizations of the new node can be found efficiently by starting at node *N* in the network. This process is not described in this work but is essentially the opposite of that described in Section 3.2. Once a node is linked into the network, other links previously existing in the network may have become redundant. Section 3.3 shows how these links are removed.

3.1. Finding Generalization Links

The routine *LinkRoot* in Figure 3 finds the most specific generalizations of a new node without knowing any relationship between the new node and any existing nodes. The routine is called starting at the root node. The root node is known to be a generalization of all possible other nodes. However, there may be some specialization *S* of the root that is also a generalization of the new node. In such a case the new node should not be linked to the root because the root is not the most specific generalization.

The routine checks all specializations of the root node. If no specialization is found to be a generalization of the new node then the root is known to be the most specific generalization and the root is linked to the new node.

If any specialization *S* of the root node is found to be a generalization of the new node then LinkRoot is called recursively starting at *S*. A flag is set to indicate that the root node should not be linked to the new node. LinkRoot must be called recursively because node *S* may have a specialization that is a generalization of the new node. In such a case the new node should not be linked to node *S* because *S* is not the most specific generalization.

The specializations of node *S* are compared to the new node. If no specializations are found to be generalizations of the new node then node *S* is linked to the new node. Otherwise the process is repeated with the specialization by again calling LinkRoot recursively and node *S* is not linked.

Note that the search is limited to nodes that are generalizations of the new node. Thus the number of nodes visited is relatively small.

```
LinkRoot(G(search node), N(new node)).
    Link = TRUE.
    For all nodes S that are specializations connected to node G.
        If S is a generalization of N then
            LinkRoot(S, N).
            Link = FALSE.
        End.
    End.
    If Link and a link between G and N is not a redundant link then
        Link G and N.
    End.
End.
```

Figure 3. Link Generalizations

Figure 4 shows an example of adding a new node (Node 8) using the LinkRoot procedure. LinkRoot begins at the root node. Specializations connected to the root node, Nodes 1 and 2, are compared to the new node. Node 1 is a generalization of the new node, so (1) a flag is set indicating that the root node should not be linked to the new node and (2) LinkRoot is called recursively starting at Node 1.

Specializations of Node 1, Nodes 5 and 6, are compared with the new node. Neither node is a generalization of the new node, so neither node is processed further. The *Link* flag remains set to true since no specializations of Node 1 were found to be generalizations of the new node. Therefore, Node 1 is linked to the new node as a generalization. The recursion terminates and returns to the LinkRoot procedure at the root node.

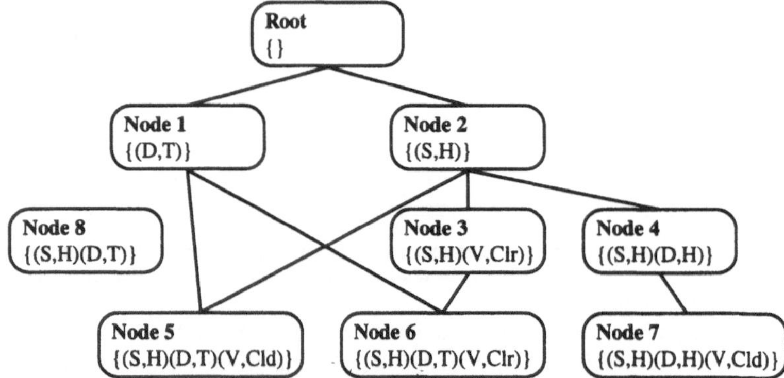

Figure 4. Finding Generalizations using LinkRoot

Node 2, the remaining specialization of the root node, is a generalization of the new node, so LinkRoot is called recursively starting at Node 2. Nodes 3, 4, and 5 are compared to the new node.

None of the nodes are generalizations of the new node, so they are not processed further. The *Link* flag remains set to true since no specializations of Node 2 were found to be generalizations of the new node. Therefore, Node 2 is linked to the new node as a generalization. The recursion terminates and returns to the LinkRoot procedure at the root node. No specializations of the root node remain to be processed, so the linking procedure has completed. The resulting network is shown in Figure 5.

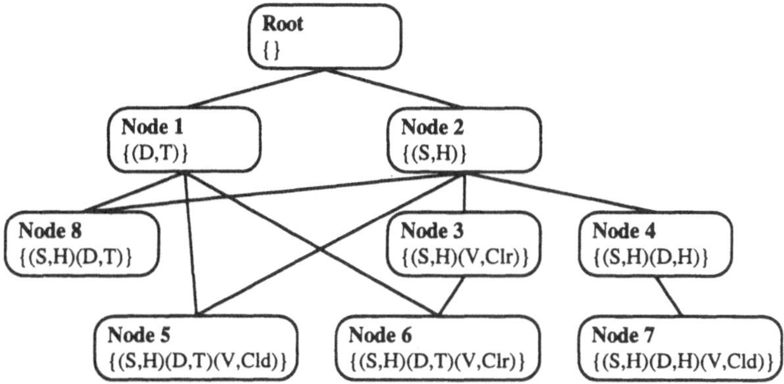

Figure 5. Network After Linking Generalizations

3.2. Specialization Links

This section explains how a new node is linked to its specializations. When linking to specializations, we assume that a generalization of the node is always available, since the node can be linked to its generalizations using the method described in the previous section.

The *specialization DAG* of a node *N* is the set of nodes and connections starting at *N* and going in the direction of specialization.

The routine *LinkSpec* shown in Figure 6 finds the most general specializations of a new node *N* given an existing node *G* in the network that is a generalization of *N*. By using node *G* that is already connected in the network the search is narrowed to only those nodes in the specialization DAG starting at node *G*. Since *G* is a generalization of *N*, *N* is a specialization of *G*, and all specializations of *N* are guaranteed to be contained in the specialization DAG of *G*.

The routine searches the specialization DAG of *G* starting at *G* and proceeding in a general to specific direction. Let *T* be a node in the specialization DAG of *G* that is also found to be a specialization of *N*. Since the search is general to specific, *T* is guaranteed to be a most general specialization. Therefore, Node *T* is linked to node *N*.

Node *T* is not guaranteed to be a most general specialization if the network allows redundant links. If redundant links are allowed, *G* can be linked to two specializations, S_1 and S_2, where S_1 is a specialization of S_2. The algorithm would link both nodes to *N* but the link to S_1 would be redundant.

All nodes in the specialization DAG beyond *T* are specializations of *T* so any links between those nodes and node *N* are redundant links. Therefore, the search of the DAG does not continue beyond node *T*.

Let *F* be a specialization of *G* that is not a specialization of *N*. If node *F* overlaps node *N* it is possible that some specialization of *F* is a specialization of *N*. Therefore, the search of the DAG is continued to the next level by calling the routine recursively starting with node *F*. If node *F* does not overlap (is discriminated with) node N then there is no specialization of F that can ever be a specialization of N, so the search of the DAG does not continue.

LinkSpec(G, N).

1. For all nodes S that are specializations connected to node G.
    2. If S is a specialization of N then
        3. If a link between S and N is not a redundant link then
            4. Link S and N.
        5. End.
    6. else If S overlaps N then
        7. LinkSpec(S, N).
    8. End.
9. End.

End.

Figure 6. Link a Node to its Specializations

As an example of LinkSpec, Node 8 in Figure 5 from the last section will be linked to its specializations. Since Node 8 is already linked to its generalizations, one of the generalizations can be selected and used as the starting point for LinkSpec. Suppose Node 2 is selected.

The routine starts at Node 2. Nodes 3, 4, and 5 are compared to Node 8 since they are specializations connected to Node 2. Node 5 is found to be a specialization of Node 8, so Node 5 is linked to Node 8 as a specialization. Node 3 is not a specialization of Node 8. However, Node 3 does overlap with Node 8. Therefore, LinkSpec is called recursively starting at Node 3.

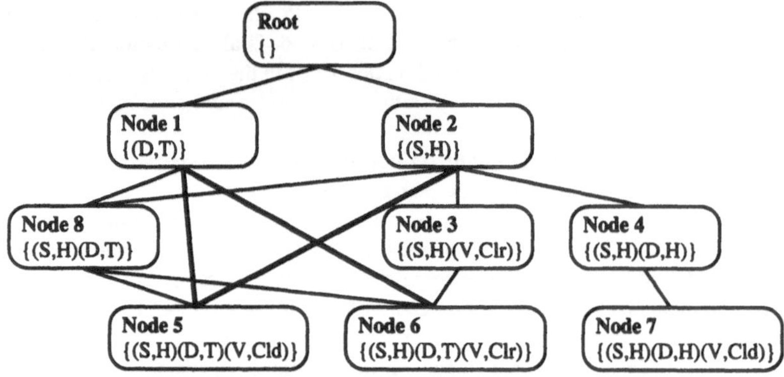

Figure 7. Completed Specialization Links

Specializations of Node 3 are compared with Node 8. Node 6 is found to be a specialization of Node 8, so Node 6 is linked to Node 8. No other specializations of Node 3 exist, so the recursion terminates and processing continues back with specializations of Node 2.

The remaining specialization of Node 2, Node 4, is discriminated with Node 8 by the *Wind Direction* input. Therefore LinkSpec is not called recursively on Node 4 and the linking process is complete. The resulting network is shown in Figure 7.

3.3. Removing Redundant Links

Figure 7 shows Node 8 after it has been linked to both specializations and generalizations as described in the previous sections. Note that linking Node 8 into the network has created three redundant links (Node 1 to Node 5, Node 1 to Node 6, and Node 2 to Node 5). The routine *RemoveRedundant* must be run on Node 8 to remove the extra links.

RemoveRedundant considers all pairs of nodes (*G*, *S*), where *G* is a generalization directly connected to *N*, *S* is a specialization directly connected to *N*, and *N* is a new node that has just been linked into the network. If a link exists between *G* and *S*, the link is redundant and is removed. This simple routine keeps the network free from redundant links because it is run whenever a new node is added to the network. Thus, when the routine is run, the only redundant links in the network are those that were created by adding the new node. The routine is not intended to remove arbitrary redundant links.

RemoveRedundant(*N*).

1. For all nodes *G* that are generalizations connected to node *N*.
    2. For all nodes *S* that are specializations connected to node *N*.
        3. If *G* and *S* are linked then
            4. Unlink *G* and *S*.
        5. End.
    6. End.
7. End.

End.

Figure 8. Remove Redundant Links After Linking a New Node

The example shown in Figure 7 has four general/specific pairs for Node 8, (Node 1, Node 5), (Node 1, Node 6), (Node 2, Node 5), and (Node 2, Node 6). The first three pairs of nodes are linked so the links are removed giving the network shown in Figure 9.

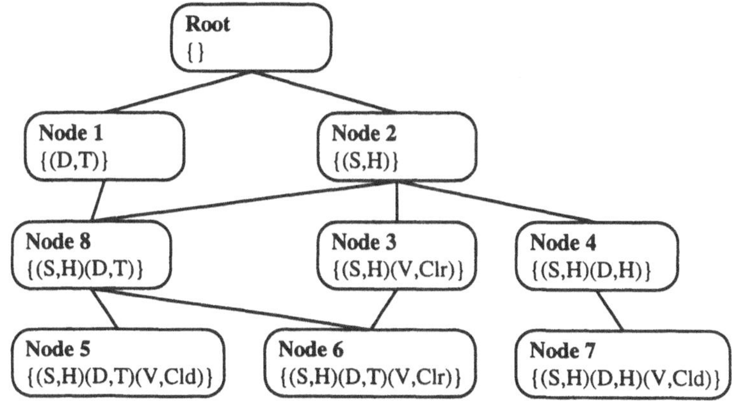

Figure 9. Network After Removing Redundant Links

4. Deleting Nodes

When a node *N* is deleted, all pairs of nodes (*G*, *S*), where *G* is a generalization directly connected to *N* and *S* is a specialization directly connected to *N* must be considered for linking. The process is essentially the reverse of removing redundant links when a node is added. If a link between *G* and *S* is not redundant (treating the node *N* as though it has already been removed from the network) then *G* and *S* are linked. If the node *N* is not treated as though is has already been removed from the network, then *N* will cause all links between all pairs (*G*, *S*) to appear redundant.

After the nodes around the node being deleted are linked together, then the node being deleted is unlinked from the network.

RemoveLinks(*N*).

1. For all nodes *G* that are generalizations connected to node *N*.
    2. For all nodes *S* that are specializations connected to node *N*.
        3. If a link between *G* and *S* is not a redundant link (assuming *N* is deleted) then
            4. Link *G* and *S*.
        5. End.
    6. End.
7. End.
8. For all nodes *G* that are generalizations connected to node *N*.
    9. Unlink *G* and *N*.
10. End.
11. For all nodes *S* that are specializations connected to node *N*.
    12. Unlink *S* and *N*.
13. End.

End.

Figure 10. Removing Links when Deleting a Node

As an example of removing a node from a network, Node 2 will be removed from the network in Figure 9. Three new links must be considered for creation (Root, Node 3), (Root, Node 4), and (Root, Node 8). Only the last link (Root, Node 8) is redundant so the other two are created. Then Node 2 is removed from the network giving the result shown in Figure 11.

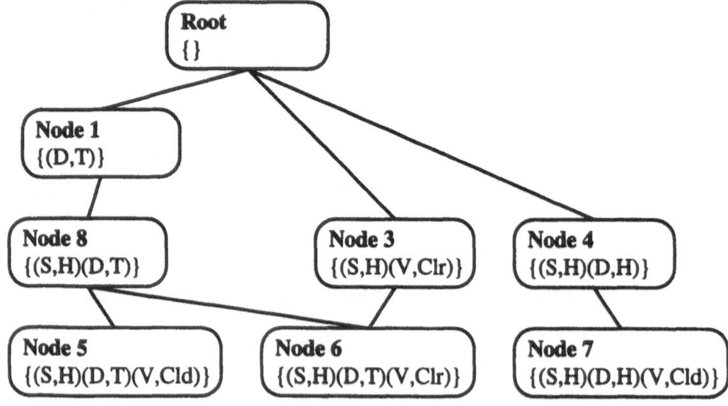

Figure 11. Network After Deleting Node 2

5. Conclusion

Learning systems often use representations for rules, features, or concepts that give rise to a more-specific-than relation. Using the more-specific-than relation, rules can be connected into a lattice structure such that rules with general-to-specific relationships are connected. The resulting network structure can be used to efficiently link new nodes into the structure as well as to delete existing nodes from the network. The connections between related nodes can be used to increase performance of learning and reasoning.

6. References

Barker, J. C. and Martinez, T. R. (1993a). GS: A Network that Learns Important Features. *Proceedings of The World Congress on Neural Networks*, Portland, Oregon, July 11-15, 1993.

Barker, J. C. and Martinez, T. R. (1993b). Generalization by Controlled Intersection of Examples. *Proceedings of The Sixth Australian Joint Conference on Artificial Intelligence*, Melbourne, November, 1993.

Barker, J. C. (1994). *Eclectic Machine Learning*. Doctoral Dissertation, Computer Science Department, Brigham Young University, Provo, Utah.

Hayes-Roth, F. and McDermott, J. (1978). An Interference Matching Technique for Inducing Abstractions. *Communications of the ACM, 21, 5*, 401-410.

Michalski, R. S. (1983). A Theory and Methodology of Inductive Learning. *Artificial Intelligence, 20*, 111-161.

Mitchell, T. M. (1982). Generalization as Search. *Artificial Intelligence, 18*, 203-226.

Vere, S. A. (1975). Induction of Concepts in the Predicate Calculus. *Proceedings of the Fourth International Joint Conference on Artificial Intelligence*, IJCAI, Tbilisi, USSR, 281-287.

AIR POLLUTION SOURCE APPORTIONMENT USING GENETIC ALGORITHMS

Lyle C. Pritchett
Govardhan Mekala
Department of Computer Science
University of Nevada, Reno
Reno, NV 89557 USA

David N. Wittorff
Department of Physics
University of Nevada, Reno
Reno, NV 89557 USA

Abstract: Effective control of air pollution levels in a given airshed depends upon the accurate determination of the sources of those pollutants. An effective modeling technique to identify and quantify those sources is the receptor-oriented approach, particularly the Chemical Mass Balance (CMB) model. However, CMB modeling often has difficulty resolving collinear sources (sources with similar chemical profiles). In addition, the user must be familiar with the airshed in question and knowledgeable in atmospheric chemistry to manually select the chemical species and sources to be used in the fitting process. Genetic algorithms (GAs) provide alternate approaches which avoid extensive user interaction and potentially improve reproducibility. An implementation of a GA-based CMB modeling program is described in detail, and results of CMB modeling on real world data sets are compared to the results from traditional CMB software.

Keywords: chemical mass balance, receptor model, source apportionment, air pollution, genetic algorithms.

Introduction

The relatively recent emphasis on air quality issues is based not only on increased public awareness of perceptible visibility problems, but also on growing evidence of health and global environmental impacts. Any effective approach to emissions control, however, requires accurate identification of the major pollution sources in an airshed. This insures not only that the major sources are targeted, but also that control strategies specific and effective for those sources are implemented.

E. A. Yfantis (ed.), Intelligent Systems, 627–636.

A proven modeling technique to identify these major sources is the receptor-oriented approach, whereby chemical fingerprints of suspected sources are matched to chemical profiles of ambient receptor samples. This matching process allows quantitative estimations of the relative impacts of those sources on ambient air quality to be made, and it is used extensively by local, state, and federal agencies.

However, the most commonly used modeling software often has difficulty in handling collinear sources (sources with similar chemical profiles) because it is based on linear least-squares regressions. In addition, the software is highly interactive with the user, who must be familiar with the airshed in question, knowledgeable in atmospheric chemistry, and responsible for manually selecting the chemical species and sources to be fit to the ambient data. This user dependence leads to obvious problems with reproducibility and accuracy of the model results. Genetic algorithms (GAs) provide an alternate approach to traditional modeling software which reduces this need for extensive user interaction and improves the reproducibility of the results.

Background

Receptor models are approved by the US Environmental Protection Agency (EPA) for state and local planning purposes. In fact, receptor modeling is a crucial element in the development of State Implementation Plans (SIPs) required for areas violating federal standards for particulate and key gaseous pollutant concentrations. However, despite this widespread use, the results of such modeling may be heavily dependent upon the skill of the modeler, a less-than-desirable basis for potential legal action against an emission source and for large amounts of money spent on control equipment.

The CMB approach is the most commonly used receptor model (Gordon, 1988). CMB modeling involves calculating fractional contributions of sources to best explain the aerosol material collected at the receptor site. It depends on the assumption that the absolute levels of the chemical species are diluted by means of transportation and dispersion, but that the relative chemical composition of a plume does not change. At the receptor site, such diluted source plumes combine linearly to comprise the ambient material collected, according to

$$\mathbf{M} = \sum_{j=1}^{J}(\mathbf{F}_j\mathbf{M}) \qquad [1]$$

where **M** is the total mass of material collected on an ambient sample, **J** is the total number of contributing sources, and $\mathbf{F}_j$ is the fraction of the ambient mass contributed by source **j**. Similarly, the calculated mass of each species **i** is given by

$$\mathbf{M}_i = \sum_{j=1}^{J}(\mathbf{F}_j\,\mathbf{M}\,\mathbf{S}_i) \qquad [2]$$

where $\mathbf{S}_i$ is the fractional contribution of species **i** to source **j**.

The CMB approach most commonly used is one developed for the US EPA by the Desert Research Institute (Reno, NV). This modeling technique allows the user to try different combinations of sources and species to achieve a maximum agreement between observed and predicted data (Watson *et al.*, 1990). This agreement is measured in terms of high regression coefficients, low χ^2 (chi-square) values, and high explained-mass values.

The accuracy and relevancy of the modeling results are highly dependent upon the species and sources selected to be included in the fit. Omitting a key fitting species may have dramatic effects on the calculated source fractions. Likewise, manually removing a source affects the relative proportions of the remaining sources due to the interactions of their common species. The results from the DRI CMB modeling are therefore dependent upon the user's skill and knowledge to make good modeling decisions; some *a priori* knowledge of the airshed is also assumed.

The traditional CMB model depends on the convergence of linear solutions to a single optimum. With the inclusion of two or more sources with similar chemical profiles, convergence is not possible. The presence of source collinearity can be detected by the software, but the operator must manually eliminate one or more of the sources to allow a final fit to be determined, again leading to operator dependence of the final CMB results.

Genetic Algorithm

The genetic algorithm (GA) provides a solution to these traditional CMB limitations. Instead of the user selecting the species and elements to be used in the fitting process, a GA-based program can examine a large number of possible solutions in an efficient manner. This "survival of the fittest" algorithm works on numerical strings which represent potential solutions to the problem (Goldberg, 1989). The GA is a stochastic search operation which can effectively test and refine many potential solutions in parallel.

A previous attempt to use GAs to implement CMB modeling was tested with synthetic data sets (Cartwright and Harris, 1993). This present work is applied to actual data sets for which traditional CMB modeling results are available for comparison.

Coding

CMB modeling results are the fractions (usually expressed as percentages) of the receptor sample attributable to each source. Therefore, each chromosome is a linear list of source contributions as percentages. Each source fraction value is encoded in binary representation in 22 bits as integer variables. During decoding these values are divided by 10,000 to yield percentages with accuracies of four decimal places; this is same resolution available in version 7.0 of the DRI/EPA CMB program.

Selection

The selection of parents from one generation to be included (either directly or via crossover) in the next generation is based upon the relative fitness of those parents. Chromosomes which represent better intermediate solutions are selected more frequently to participate in building the new generation.

The selection scheme used here is similar to that described in previous work (Cartwright and Harris, 1993). The actual algorithm is derived directly from Goldberg's textbook under the name "stochastic remainder without replacement" (Goldberg, 1989).

Another refinement to the selection scheme is the use of elitism. Elitism guarantees that at least one copy of the individual with the highest fitness value in a given population will appear unchanged in the subsequent generation (bypassing the crossover and mutation operators) which in turn provides prime genetic material for ensuing generations. In some situations elitism may lead to premature convergence on a local optimum, but in this application speed of convergence is also a concern.

Crossover And Mutation

The crossover operator in this work implements two-point crossover, whereby chromosome segments between two randomly selected points are swapped between the parent strings. A high crossover rate allows faster exploration of the search space, but an excessively high rate toward the end of the search will disrupt chromosomes with high fitness values. Therefore, the crossover rate in this current work decreases dynamically as the search progresses to shorten the time required to find a solution.

The mutation operator is applied after the parent chromosomes are selected and crossover is performed; it merely reverses individual bits from 1 to 0 or 0 to 1. A high mutation rate, like a high crossover rate, is desirable at the beginning of the GA search, but will cause instabilities toward the end of the search. The mutation rate is also dynamically adjusted.

The Fitness Function

The key to any GA is the fitness function, which directs the search. The EPA/DRI CMB approach is based on manually striving toward three goals: a low χ^2 value, a high fraction of available species used as fitting species, and a percentage of explained ambient mass close to 100%. The GA's fitness values comprise these same three measures, accounting for the relative importance of the three goals (χ^2 values and number of fitting species take precedence over mass percentages).

To simplify conversion of these three fitness measures into a single fitness value, they are expressed as "percents of ideal". For example, the explained mass percentage is converted to a 0 to 100% scale, where 100% represents a perfect fit (100% of the ambient mass explained) and any deviation (positive or negative) from 100% mass explained results in a lower value on the ideal scale. The formula for this conversion is

$$\mathbf{I}_m = 100 / \left[1 + \left(\frac{\left| 100 - \sum_{j=1}^{J} (100\, F_j) \right|}{100} \right) \right] \qquad [3]$$

where $\mathbf{I}_m$ is the ideal mass value.

The χ^2 value is a modified version of the standard statistical definition:

$$\chi^2 = \frac{1}{DF}\sum_{i=1}^{I}\left[\frac{\left(M_i - \sum_{j=1}^{J} F_j MS_{ij}\right)^2}{V_i}\right] \qquad [4]$$

where **DF** is the degrees of freedom and V_i is the variance. The **DF** term is calculated as

$$DF = I - J \qquad [5]$$

and V_i is calculated as

$$V_i = \sigma^2_{M_i} + \sum_{j=1}^{J}\left(F_j M\, \sigma_{S_{ij}}\right)^2 \qquad [6]$$

where σ_{M_i} is the uncertainty of the mass of ambient species **i** and $\sigma_{S_{ij}}$ is the uncertainty of species **i** in source **j**. Note that the chemical data used in the CMB model include uncertainties which represent the accuracy and precision of analytical and sampling methods. These uncertainty values are typically 1σ (one sigma) values (68% confidence limits). Equation 4 incorporates the uncertainties of the observed and calculated values and is called the "reduced chi-square" value (Watson *et al.*, 1990).

Again, to make the calculation of a single fitness value from χ^2 easier, the χ^2 value is mapped onto an "ideal" scale, where 100% represents a perfect fit ($\chi^2 = 0.0$), by

$$I_{\chi^2} = \frac{100}{\left(1 + \frac{\chi^2\, DF}{100}\right)} \qquad [7]$$

The percentage of chemical species which are included in the fitting process as compared to the total number of fitting species is a straightforward calculation:

$$I_s = 100\left(\frac{N_f}{N}\right) \qquad [8]$$

where N_f is the number of fitting species and **N** is the total number of species measured.

The three goodness-of-fit measures, all in a range of 0 to 100%, are multiplied by empirically derived weighting factors and linearly combined to yield a single fitness value for each chromosome in the GA population:

$$F = I_m W_m + I_{\chi^2} W_{\chi^2} + I_s W_s \qquad [9]$$

where W_m, W_{χ^2}, and W_s represent the fractional weights applied to the explained mass, χ^2, and fitting species terms, respectively. These raw fitness values are further subjected to a scaling function (Goldberg, 1989). This scaling function depresses the relative importance of a few high fitness individuals at the beginning of the run to prevent premature convergence and magnifies small differences toward the end of the run to encourage additional evolution.

Test Data Sets

Data from actual CMB studies performed at DRI were used to validate the GA approach described here. Two recent studies provided simple data sets: an ambient pollution study conducted in Reno, Nevada (Wittorff, 1994), and a diesel emissions study performed at a mid-town sampling site in Manhattan, New York (Wittorff *et al.*, 1994a; Wittorff *et al.*, 1994b).

The Reno project included 48 chemical species for 33 source samples. Of these source sample profiles, five dissimilar profiles (little or no collinearity) were selected to construct simulated receptor samples. These samples were then used to validate the GA source apportionment approach. No random errors were introduced to simulate analytical "noise".

The Manhattan data set included 48 chemical species, 26 source sample types, and 13 ambient samples (including one sample which represents the average of the other 12). This latter data set was used as a real-world test of the GA approach; CMB results using the EPA/DRI approach were available for comparison.

Note that many of the sources in both projects represent variations on a single source type. For example, several motor vehicle exhaust profiles represent various relative mixes of gasoline and diesel emissions. Normally the user would manually determine the one single variation for each source type which best explains the ambient data to avoid the collinearity problem discussed earlier. In the GA program, all sources participate in the fitting process, and final source contributions which were less than 3% were disregarded as insignificant. To avoid excessively long calculation times in this present work, the fitting species were determined in advance and not allowed to evolve.

Results and Discussion

Because of the GA's stochastic nature and its heavy dependence upon the random numbers generated by the computer, ten runs were performed on each CMB fit, each with a different random number seed value.

Results of multiple GA runs were compared for variations with each other and for spread in the final population around the optimal individuals. The effects of dynamically changing crossover and mutation rates were tabulated and evaluated, and optimal crossover and mutation rates were determined. The results of this tuning process are presented in Table 1.

The results of the simulated data set created from known combinations of five Reno source profiles are presented in Table 2. The data in the table clearly show that the GA approach can apportion source contributions accurately.

Parameter	Value
Population Size	100
# of Generations	250
Starting Crossover Probability	0.90
Ending Crossover Probability	0.66
Starting Mutation Probability	0.01
Ending Mutation Probability	0.001
Linear fitness scaling factor	2.00
Fraction of fitness, explained mass	0.1
Fraction of fitness, chi-squared	0.8
Fraction of fitness, fitting elements	0.1

Table 1. GA Program Parameters

Sample	Results	% Explained Mass	χ^2	% MV*	% RD*	% NACL*	% SUL*	% AMNIT*
SIM1	model	100.0	0.00	75.0	5.0	10.0	5.0	5.0
	GA	100.0	0.00	75.2	5.0	9.8	5.0	4.9
SIM2	model	100.0	0.00	80.0	10.0	10.0	0.0	0.0
	GA	100.0	0.00	79.7	10.1	10.2	< 3.0	< 3.0
SIM3	model	100.0	0.00	50.0	20.0	10.0	10.0	10.0
	GA	100.0	0.00	50.4	19.7	10.0	10.2	9.7
SIM4	model	100.0	0.00	40.0	40.0	0.0	10.0	10.0
	GA	100.0	0.00	39.9	40.2	< 3.0	10.0	9.8
SIM5	model	100.0	0.00	20.0	60.0	0.0	10.0	10.0
	GA	100.0	0.00	21.1	59.0	< 3.0	10.0	9.9

* MV = Motor Vehicle Exhaust
RD = Road Dust
NACL = Sodium Chloride
SUL = Sulfate, Ammonium Sulfate
AMNIT = Ammonium Nitrate

Table 2. Summary of Simulated CMB Results: Simulated Contributions and Best of Ten GA Runs

The results of GA CMB calculations are compared with CMB calculations for selected Manhattan samples in Table 3. Included in the comparison are percent-mass-explained values, χ^2, and fractional contributions attributed to each source category. Sources allowed to participate in this fitting process were preselected based on the original CMB results to purposefully avoid collinearities. The contributions for each of the six sources calculated by the GA are generally well within 5% (absolute) of the values calculated by the traditional

CMB model. Percent-mass-explained and χ^2 values also compare favorably. (Note: the large mass-explained values are due to sampling artifacts).

Sample	CMB	% Explained Mass	χ^2	% MV*	% RD*	% MAR*	% SUL*	% AMNIT*	% FEORE*
001	DRI	167.3	0.11	115.2	19.9	0.9	7.2	18.0	5.3
	GA	165.3	0.12	113.9	19.5	< 3.0	6.6	18.3	5.4
	Δ	-2.0	0.01	-1.3	-0.4	--	-0.6	0.3	0.1
007	DRI	137.9	0.07	78.8	14.1	5.8	18.9	16.3	3.3
	GA	136.7	0.11	78.7	13.1	< 3.0	19.0	15.9	3.5
	Δ	-1.2	0.04	-0.1	-1.0	--	0.1	-0.4	0.2
008	DRI	115.8	0.30	66.4	7.2	6.3	16.7	12.8	5.6
	GA	116.2	0.32	65.5	9.4	6.2	16.4	12.8	4.9
	Δ	0.4	0.2	-0.9	2.2	-0.1	-0.3	0.0	-0.7
009	DRI	111.1	0.21	56.8	5.5	13.7	15.3	13.2	5.3
	GA	106.7	0.23	51.6	6.8	13.6	15.4	13.2	4.9
	Δ	-4.4	0.02	-5.2	1.3	-0.1	0.1	0.0	-0.4

* MV = Motor Vehicle Exhaust
RD = Road Dust
MAR = Marine Aerosol
SUL = Sulfate, Ammonium Sulfate
AMNIT = Ammonium Nitrate
FEORE = Iron Ore

Table 3. Summary of Manhattan CMB Results: DRI Modeling and Best of Ten GA Runs with Preselected Sources

The second run of the Manhattan samples included all 33 source samples to test the GA's ability to handle collinearities. Table 4 presents comparisons of multiple GA runs (each started with a different random number seed) for four collinear road dust sources and four ambient samples. As expected, the attribution between the four sources varies from one run to another, reflecting the GA's stochastic nature; however, in general the total contributions from all four sources (a generic category of "road dust") are fairly constant. The notable exception to this is sample 004, which shows two distinct sum values. For this one sample apparently two distinct maxima exist in the solution search space.

Conclusions

The GA approach to CMB receptor modeling described here is used to locate a global maximum in the search space composed of chemical species and individual source

Sample	Run	% Explained Mass	χ^2	% PRD1*	% PRD2*	% PRD3*	% URD1*	% Sum
002	1	115.0	0.72	< 3.0	6.6	78.8	5.9	91.3
	2	106.5	0.67	4.3	< 3.0	66.2	6.2	76.7
	3	102.2	0.79	4.7	< 3.0	58.9	6.5	70.1
	4	116.7	0.82	< 3.0	< 3.0	79.4	8.1	87.5
003	1	111.4	0.63	3.0	< 3.0	66.6	6.5	76.1
	2	115.2	0.64	< 3.0	< 3.0	65.7	6.4	72.1
	3	125.6	0.86	< 3.0	3.4	53.5	6.2	63.1
	4	109.4	0.65	5.6	< 3.0	62.0	6.5	74.1
004	1	138.8	0.89	< 3.0	< 3.0	104.9	4.2	109.1
	2	142.9	0.86	4.8	< 3.0	105.0	< 3.0	109.8
	3	106.0	0.39	< 3.0	< 3.0	52.5	4.9	57.4
	4	107.0	0.41	< 3.0	< 3.0	52.4	< 3.0	52.4
005	1	110.1	0.60	< 3.0	< 3.0	79.8	< 3.0	79.8
	2	101.0	0.59	< 3.0	< 3.0	65.1	3.1	68.2
	3	114.2	0.54	< 3.0	< 3.0	79.7	< 3.0	79.7
	4	111.8	0.58	< 3.0	< 3.0	79.3	< 3.0	79.3

* PRD1 = Paved Road Dust Source 1
PRD2 = Paved Road Dust Source 2
PRD3 = Paved Road Dust Source 3
URD1 = Unpaved Road Dust Source 1

Table 4. Summary of Manhattan CMB Results: Variations in Collinear Sources

contributions. The search space is large enough that exhaustive search techniques are impractical. GAs are well-suited for parallel searches for solutions in such spaces.

The results depicted in Tables 2 and 3 indicate that the GA results are comparable to results obtained from the traditional CMB modeling. Additional tuning of the fitness function and other parameters should allow the differences between the two methods to be further minimized. In this feasibility study, the results obtained so far are encouraging.

The GA approach also gracefully handles the problem of source collinearities. As mentioned previously, the EPA/DRI software depends on the convergence of linear solutions. The generic GA tends towards only one maximum fitness peak, but over the process of many runs with different random number seeds, two or more peaks representing multiple optimal solutions are detectable. The data in Table 4 suggest that large local maxima may be problematic. A further refinement of the GA in the form of niche formation (Deb and Goldberg, 1989) might allow multiple optima to be collapsed into a single source category.

References

Cartwright, H.M., and S.P. Harris (1993). "Analysis of the Distribution of Airborne Pollution Using Genetic Algorithms", **Atmospheric Environment, 27A:12**, 1783-1791.

Deb, K., and D.E. Goldberg (1989). "An Investigation of Niche and Species Formation in Genetic Function Optimization", **ICGA Proceedings**, 42-50.

Goldberg, D.E. (1989). **Genetic Algorithms in Search, Optimization, and Machine Learning**, Addison-Wesley Publishing Company, Reading, MA.

Gordon, G.E. (1988). "Receptor Models", **Envir. Sci. & Technol., 22:10**, 1132-1142.

Watson, J.G., N.F. Robinson, J.C. Chow, R.C. Henry, B.M. Kim, T.G. Pace, E.L. Meyer, and Q. Nguyen (1990). "The USEPA/DRI Chemical Mass Balance Receptor Model, CMB 7.0", **Environ. Software, 5:1**, 38-49.

Wittorff, D.N. (1994). The Contributions of Diesel Emissions and Road Sanding and Salting Material to Ambient PM_{10} Concentrations. Masters Thesis, University of Nevada, Reno, NV. University Microfilms, Intl., Ann Arbor, MI.

Wittorff, D.N., A.W. Gertler, J.C. Chow, W.R. Barnard, H.A. Jongedyk (1994a). The Impact of Diesel Particulate Emissions on Ambient Particulate Loadings. Report in preparation, Desert Research Institute, Reno, NV.

Wittorff, D.N., A.W. Gertler, J.C. Chow, W.R. Barnard, H.A. Jongedyk (1994b). The Impact of Road Sanding and Salting on Ambient PM_{10} Concentrations. Report in preparation, Desert Research Institute, Reno, NV.

A TRANSFORMATION FOR IMPLEMENTING NEURAL NETWORKS WITH LOCALIST PROPERTIES

George L. Rudolph
Tony R. Martinez
Computer Science Department
Brigham Young University Provo, Utah 84602
e-mail: george@axon.cs.byu.edu, martinez@cs.byu.edu

Abstract. Most Artificial Neural Networks (ANNs) have a fixed topology during learning, and typically suffer from a number of short-comings as a result. Variations of ANNs that use dynamic topologies have shown ability to overcome many of these problems. This paper introduces Location-Independent Transformations (LITs) as a general strategy for implementing feedforward networks that use dynamic topologies. A LIT creates a set of location-independent nodes, where each node computes its part of the network output independent of other nodes, using local information. This type of transformation allows efficient support for adding and deleting nodes dynamically during learning. In particular, this paper presents LITs for the single-layer competitve learning network, and the counterpropagation network, which combines elements of supervised learning with competitive learning. These two networks are *localist* in the sense that ultimately one node is responsible for each output. LITs for other models are presented in other papers.

Keywords. Artificial Neural Networks, Hardware Implementation Design, Dynamic Topologies

1. INTRODUCTION

Artificial Neural Networks (ANNs) use a different computational paradigm than conventional von Neumann mechanisms. ANNs are composed of nodes and weighted connections between nodes, where each node computes its output based on a function of its weighted inputs. The overall function that a network computes is typically changed by altering the values of the weights between nodes, until the desired result is achieved. The main features of ANNs are learning ability, generalization, parallelism, self-organization and fault-tolerance. These features allow ANNs to solve various applications not handled well by current conventional computational mechanisms. Application areas include, but are not limited to, problems requiring learning, such as pattern recognition, control and decision systems, speech, and signal analysis [2].

Hardware support for ANNs is important for handling large, complex problems in real time. Learning times can exceed tolerable limits for complex applications with conventional computing schemes. Furthermore, hardware is becoming cheaper and easier to design. The *Location-Independent Transformation* (LIT) is a general implementation strategy for ANNs that overcomes several weaknesses of current hardware implementation methods.

Most ANNs use only static topologies—the topology is fixed initially, and remains the same throughout learning. ANNs with static topologies typically suffer from the following short-comings:

- sensitivity to user-supplied parameters
- local error minima during learning
- no *a priori* mechanism for deciding on an effective initial topology (number of nodes, number of layers, etc.)

Current research is demonstrating the use of dynamic topologies in overcoming these problems [3], [9], [11], [13]. A *dynamic topology* is one which allows adding and deleting entire nodes and individual weighted connections during learning.

Early ANN hardware implementations are model-specific, and are intended to support only static topologies [4-5], [10]. More recent *neurocomputer* systems have specialized neural hardware, and seek to support more general classes of ANNs [6], [12], [17]. Although some neurocomputers could potentially support dynamic topologies more directly in hardware, rather

† This research is funded by grants from Novell Inc. and Word Perfect Corp.

E. A. Yfantis (ed.), Intelligent Systems, 637–645.

than in software, they currently do not Of course, general parallel machines, like the Connection Machine [8] and the CRAY [1], can simulate the desired dynamics in software, but these machines are not optimized for neural computation. LIT supports general classes of ANNs *and* dynamic topologies in an efficient parallel hardware implementation.

A LIT transformation redesigns the network so that each node contains enough information locally to compute its part of the network output, independent of any other node. A network whose nodes have this property is said to be *location-independent.* The nodes also are location-independent—regardless of the physical location of any node in the network, the relative order in which they compute results, or the order in which those results are gathered, *the individual computations are the same, and the network output is the same.* Furthermore, because a node's information is local, adding or deleting nodes from the network can be done without affecting any other nodes. Thus, location-independence allows efficient support for dynamic topologies.

In this paper, the term *Control Unit* refers to a mechanism that broadcasts inputs to a network, and gathers results from it. The term *original* refers to a network before it is transformed, and the term *transformed* refers to a network after it has been transformed. These terms apply similarly to the nodes as well. The number of layers refers to the number of *weight layers.*

LIT is a two-step process:

1. Construct a set of LI-nodes based on the original model.
2. Embed the nodes in a tree.

This process is outlined in figure 1. Based on an original ANN model (left), a set of LI-nodes is constructed (middle), and the nodes are embedded in a tree (right).

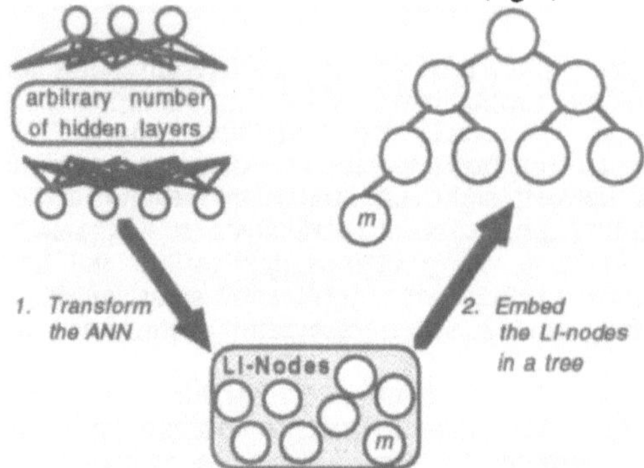

Figure 1. General LIT Transformation

LIT is not simply a reorganization of the original network, but rather involves redesigning the basic structure. The heart of a transformation is the construction step: It defines the mapping between the original network and the set of LI-nodes. The mapping, in turn, affects how the behavior of the original is modeled in the transformed network. Thus, a construction typically also involves reformulating the network equations, in order to describe precisely the behavior of the transformed network. Although the original and transformed networks compute equivalent functions, correspondences between the two are not always direct, nor obvious. Constructions vary across different ANNs. LI-nodes for CL, which has only a single weight layer, do not have the same structure as LI-nodes for multilayer backpropagation.

A communication topology can be chosen as a matter of speed, cost and reliability. The same topology can be used with many models. A binary tree topology is specified in the second step in figure 1.. It is a simple, fast, regular topology that naturally supports broadcast and gather operations. It provides a basis for applying the transformation and explaining the associated execution and learning algorithms, without obscuring the details with a complex interconnect. However, a hypercube, or some other more fault-tolerant topology, could also be used.

An example illustrates the support for dynamic topologies. When a node is added to (deleted from) an original network, both the node and its connections to other nodes are added (deleted). If a hidden node 6 were added to the original network in figure 2, four connections to the input nodes and three connections to the output nodes would also be added. This is accomplished in the transformed network by allocating a free node, such as the left child of node 3, and initializing it as a 4|3 node with the corresponding weights. The deletion of a hidden node in the original network, such as node 5, is accomplished in the transformed network by marking the corresponding node as "free". No other nodes are affected.

LITs have been developed for backpropagation [15], and Adaptive Self-Organizing Concurrent Systems' Adaptive Algorithm 2 [9], [14]. Transformations for other important ANNs are also being developed. LITs can potentially support a broad set of ANNs, thus allowing one efficient implementation strategy to support inherently dynamic ANNs and dynamic variations of many static ANNs.

This paper presents the transformations for two networks, the competitive learning (CL) network with a single weight layer, and the counterpropagation network (CPN). CL and CPN are examples of *localist* networks, i.e. for each possible output, there is a single node that is ultimately responsible for that output. CL uses unsupervised learning, while CPN combines elements of competitive and supervised learning. Both the original CL and CPN models use static topologies, but this paper also discusses how LIT can support extended dynamic versions of these models. While support for dynamic versions of static ANNs is demonstrated, LIT is not concerned with devising such extensions, but rather seeks to support extensions being devised by others. The intent of this paper is to introduce the LIT strategy for localist ANNs— the descriptions given here should be seen as representative of the general class of localist ANNs, not exclusive of other, more complex models. Furthermore, a complete transformation description requires formal definitions of the equations and algorithms of each model. For the purposes of this paper, and for brevity, only informal descriptions are given here, and it is assumed that the reader is familiar with the original models.

Section 2 describes the LIT for CL. Section 3 descibes the LIT for CPN. Section 4 is the conclusion.

2. Transformation Of A Competitive Learning Network

The goal of the CL model is to spontaneously classify sets of similar inputs to the same class, and sets of different inputs into different classes, according to critical features discovered by the network [16]. The original CL has one output node for each output class, and weights from each input node to every output node. The output nodes are classical sum-of-products nodes— the node with the highest activation for a particular input is chosen as the output class for that input. (The input nodes are place-holders for the input values.) Thus, the output of the network is the class of the input. During learning, only the winning node adjusts its input weights—the other nodes make no changes. A node causes its input weights to change so that the weights are more similar to the current input vector.

The original CL model reveals the localist nature of computation in the original CL model:

- The activation value for each class is computed locally at each output node.
- A single, global winner node determines the output of the network.
- Determination of a global winner can be accomplished using localized comparisons of subsets of the output nodes.

The LIT construction given here is based on these ideas.

Figure 2 shows the transformation of an 8x3 network into a LIT network with three 8x2 nodes. There is one node in the transformed network for each original output node. A transformed CL node stores two vectors. One vector contains weights on inputs $\mathbf{w}_j$, whose elements w_{ij} correspond to the weight connected to the respective original input node. The other vector contains the activation value net_j, and the class $class_j$ (or class number j) as a pair.

The original node 2, in figure 2, has eight weights on inputs, and outputs its class number 2. The transformed node 2 has eight weights in w_j and outputs the tuple (*net*$_j$, *class*$_j$).

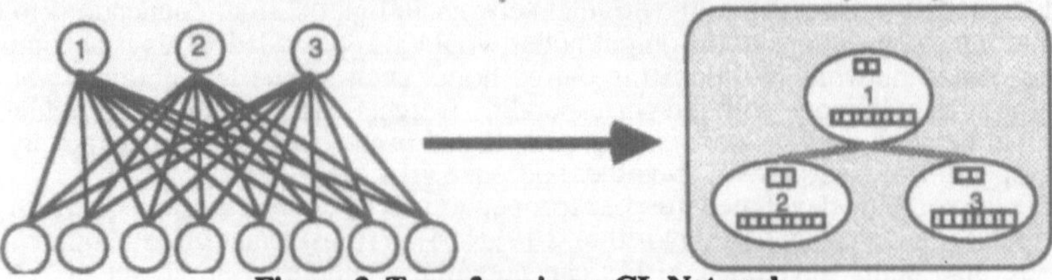

Figure 2. Transforming a CL Network

Figure 3 shows the structure of a new node in more detail. Each transformed node outputs its activation because determining a global winner requires non-local access to the activation values of each node. In the original network, the output nodes compete in some fashion, the node with the highest net activation being the winner, and all others being "quenched" somehow. The mechanisms for quenching are usually assumed, not explicitly given, but typically require that each node has access to the activation of every other node. The transformed network localizes this competition by means of a gather operation. Each node compares its own activation with the activations in the tuples received from its children. The tuple with the highest activation is sent to the parent, the others are discarded. In case of a tie, the tuple with the lowest class number is chosen. Thus, at any given transformed node, at most three tuples are compared. The output of the root node is the tuple which contains the class that has the highest activation value.

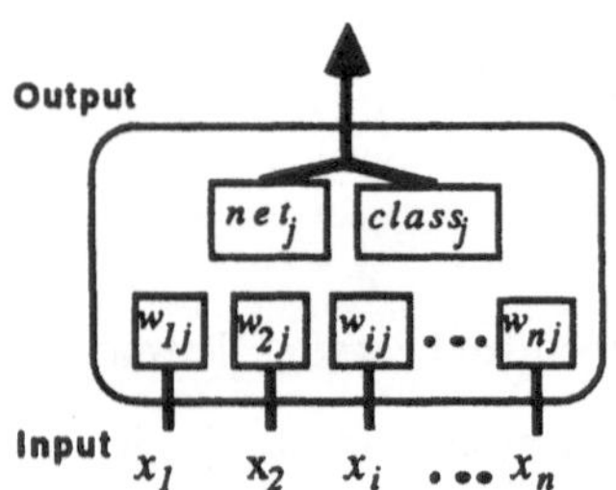

Figure 3. A CL Node

It may appear at first glance that the amount of parallelism in the network is degraded by the transformation—the transformed network has fewer nodes, and therefore each node does more work. This impression is false. The input nodes in an original CL network don't really *do* anything—conceptually, they are input buffers that hold the values of input variables. The real work is done at the output nodes (as there are no hidden nodes). The amount of real parallelism in computing the sum-of products is ultimately implementation-dependent, and LIT does not constrain implementation. Hence, the amount of parallelism is actually constrained by the number of output nodes, and by the physical implementatation of a node. In any case, if there is not a large number of nodes, then there cannot be a lot of parallelism.

During execution, each node receives all the inputs by broadcast, and performs a *sum-of-products* function on the inputs and input weights. The result is each node's activation. Each node then sends its class and activation value up to its parent. Each parent compares its activation with those of its two children, and sends the class with the highest activation up to its parent. The others are ignored. The output at the root of the network (tree) is the output class to which the input vector belongs, along with the activation of the node corresponding to that class. The root node this single tuple to the Control Unit.

During learning, each node receives the input by broadcast, computes an activation, and sends its class and activation up to its parent (as above for execution mode). The Control Unit receives the class activation of the node with the highest activation. The output class is broadcast back to the network, in order to select the winning node, and "quench" all others. The selected node then alters its weights appropriately; no other nodes make changes.

The learning algorithm guarantees that only one node can win a competition. The learning equation shows that a winning node will change its input weights to respond more strongly to the current input.

The learning equation

$$\Delta w_{ij} = g(x_i / v - w_{ij})$$

for the standard CL model [16] gives a very non-robust type of learning. Both the zero vector and the vector with all ones will always be put in the class with the lowest number, because in each case, the activations of all nodes is the same. This helps to illustrate the generality of LIT—a different, more robust learning equation (algorithm) could be substituted in the original model, and the overall transformation would still be the same.

The original CL model above only supports a static topology—it describes how to change the weights, but provides no information about adding or deleting nodes and connections. An example illustrates the support for dynamic topologies. Assume that appropriate extended dynamic equations and algorithms exist: Typical schemes for adding nodes involve using a distance metric computed at each node. If the "distance" of the input pattern from the *winning* node is too far, a new output node is added to the network, and the original winning node is penalized in some fashion. A node can be deleted from the network if the node never wins a competition or if its average activation value is below some threshold over some period of learning. When a node is added to (deleted from) an original network, both the node and its connections to other nodes are added (deleted). If a new output node is added to the original network in figure 2, eight connections to the input nodes would also be added. Additional connections among the output nodes may be required also, depending upon how the global competition is handled. Adding a new output node is accomplished in the transformed network by allocating a free node, such as the left child of node 3, and initializing it as a 8x2 node with the corresponding weights and output values. The deletion of an output node in the original network, such as node 3, is accomplished in the transformed network by marking the corresponding node as "free". No other nodes are affected.

3. Transformation Of A Counterpropagation Network

The goal of CPN is to learn approximations to input-output pairs presented to the network [7]. The original network is presented as a four-layer "counterpropagation" flow between five sets of nodes (figure 4). However, , without loss of generality, CPN can be viewed as single-layer competitive learner with adjustable output weights.

Throughout section 3, it is necessary to refer to node layers and weight layers both. To avoid confusion, *layers* refers to weight layers, and *sets* refers to node layers.

The organization of the CPN can be simplified as follows:

1. Node sets 1 and 5 are combined into a single input set, divided into a (x,y) vector pair as desired, where n is the size of x, and k is the size of y.
2. Node sets 2 and 4 are combined into a single output set, divided into a (x', y') vector pair corresponding to (x,y).
3. Node set 3 remains as in the original description.
4. There is a component-wise weight layer from sets 1 and 5 to sets 2 and 4 respectively, used only during learning. This is a pass-through layer whose weight values (set to 1) never change. After the transformation, it will not be needed.
5. The typical competitive method for determining the winner among the nodes of set 3 involves connections from each node in the set to every other node in the set. The transformed CPN network performs the selection in the same fashion as the CL network (see section 2). Hence, these connections can be ignored in the present discussion.

At this point, the network can be seen as a 2-layer (three, counting the pass-through layer) feed forward network with one set of hidden nodes (see figure 5), without loss of functionality. However, one more reorganizing step will show that CPN can be considered a competitive learner with the addition of adjustable weights as output. Several additional notational simplifications, along with relabelling (noted in figure 5), done for the purposes of this paper, are given to clarify this:

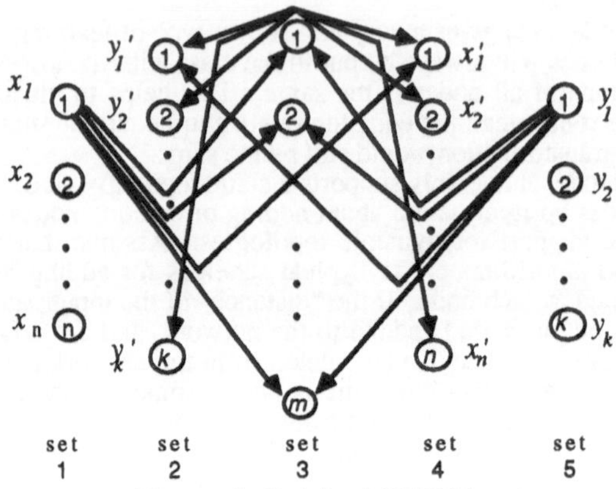

Figure 4. Original CPN Structure

- The input vector pair $(\boldsymbol{x},\boldsymbol{y})$ is treated as a single vector $\boldsymbol{X}$, with components x_i, where $1 \leq i \leq k+n$.
- Similarly, the output vector pair $(\boldsymbol{x}', \boldsymbol{y}')$ is treated as a single vector $\boldsymbol{Y}$, with components y_i, where $1 \leq i \leq k+n$.
- The input weight vectors $\boldsymbol{u}$ and $\boldsymbol{v}$ are treated as a single weight vector $\boldsymbol{U}$, with components u_{ij}, where $1 \leq i \leq k+n$, $1 \leq j \leq m$
- Node set 3 is labeled $\boldsymbol{H}$, where h_i is the ith node in that set.
- The output weight vector $\boldsymbol{W}$ (for $\boldsymbol{x}'$ and $\boldsymbol{y}'$) is treated as a single vector $\boldsymbol{W}$, where w_{ij} is the weight from h_i to y_j, and $1 \leq i \leq m$, $1 \leq j \leq k+n$.

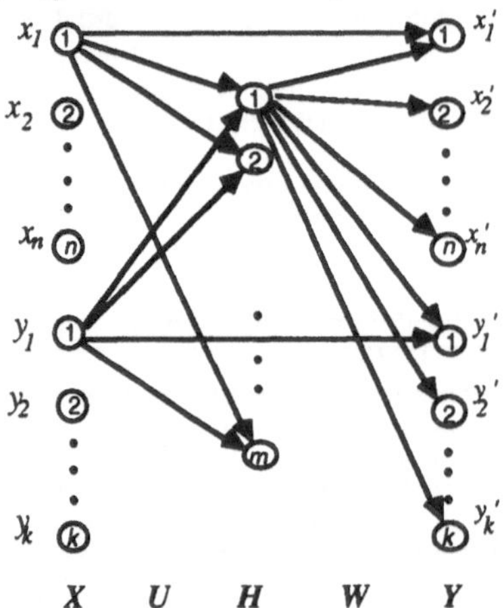

Figure 5. Rearranged CPN Network

The nodes are classical sum-of-products nodes. The nodes in set $\boldsymbol{H}$ compete in winner-take-all fashion, hence this hidden set acts as a selector for the output vector. Because $\boldsymbol{H}$ is a selector, there is no recombining of the weights in layer **W** to determine the output of the network, in either execution or learning mode. Only the winning node's output weights are (effectively) transmitted to the output layer **Y**, because all other results are zero. This suggests that the real work of determining the output involves the weights in **U**, the first weight layer, and the nodes in **H**. Hence, CPN behaves as a spontaneous classifier like the original CL model of section 2: **H** corresponds to the output nodes, **U** corresponds to the single weight layer, and **X** corresponds to the input nodes. What CPN has, that CL does not, is **W**, which are adjustable output values associated with each output class.

In transforming the CPN network, there are *m* nodes in the transformed network, one corresponding to each node in **H.** The structure of a transformed CPN node is shown in figure 6. Each node has a fixed class number, which is also its priority and index, from 1 to *m*. In case of a tie, the node with the lowest index wins. A node stores $2(k+n)$ weights: $k+n$ weights on inputs u_j, and $k+n$ weights on output w_j. The weights that each transformed node stores are the input and output weights of the corresponding original node.

There are four ways in which the CPN network behaves differently from the CL network:

1. Each node outputs w_j in addition to net_j and $class_j$.
2. The winning node adjusts both its w_j and its u_j.
3. The input and output vectors can contain real values.
4. The learning equations are different.

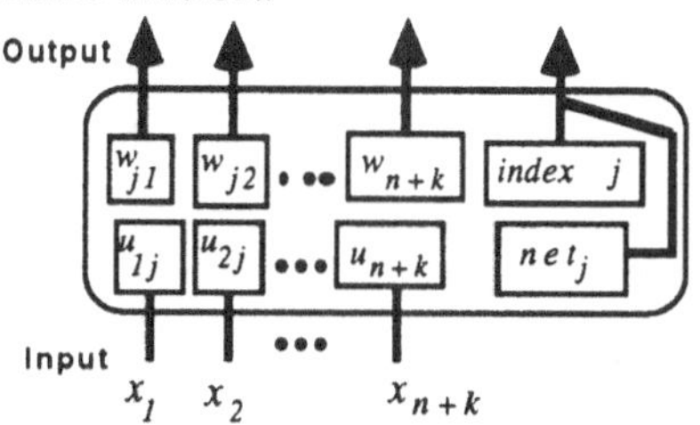

Figure 6. Structure of a Transformed CPN node

During execution, a node receives all the inputs by broadcast and performs a *sum-of-products* function on the inputs and input weights. The result is the node's activation. The node then sends its activation value and output class up to its parent.

In learning, only the winning node's weights are adjusted, just as in the standard CL model. Hecht-Nielsen's original equations allow for different learning constants for the **x** and **y** weights, but otherwise the equations for the two vectors are identical [7]. Each node receives the input, computes an activation (as above), and sends its activation and class up to its parent. The node with the highest activation is chosen, and its class is broadcast back to the network. The winning node then alters its weights on input *and* output appropriately. No other nodes make changes.

The original CPN model above only supports a static topology—it describes how to change the weights, but provides no information about adding or deleting nodes and connections. The idea used here to support dynamic topologies is essentially the same as for CL in section 3. Adding nodes involves using a distance metric computed at each node. If the "distance" of the input pattern from the *winning* node is too far, a new hidden node is added to the network, and the original winning node is penalized in some fashion. A hidden node can be deleted from the network if the node never wins a competition or if its average activation value is below some threshold over some period of learning.

Assume there is an original CPN network, like the one shown in figure 5, with $n=20$, $k=10$, and $m=10{,}000$. When a node is added to (deleted from) the original network, both the node and its connections to other nodes are added (deleted). If a new hidden node is added to the original network, 30 (i.e. $n+k$) connections to the input nodes and 30 connections to the output nodes would also be added. Additional connections among the hidden nodes may be required also, depending upon how the global competition is handled. In the transformed network, adding a new hidden node is accomplished by allocating a free node in the tree and initializing it as a 30x(30+2) node with the corresponding weights and index values. The deletion of a hidden node in the original network is accomplished in the transformed network by marking the corresponding node as "free". No other nodes are affected.

4. Conclusion

ANNs that use a static topology, i.e. a topology that remains fixed throughout learning, suffer from a number of short-comings, such as parameter problems, local minima, and less than general optimization ability. Current research is demonstrating the use of dynamic topologies in overcoming some of these problems. The Location-Independent Transformation (LIT) is a general strategy for implementing variations of ANNs that use dynamic topologies during learning.

This paper defined LITs for the competitive learning and counterpropagation networks. These ANNs are examples of the more general class of *localist* ANNs. The LIT descriptions for these two models have three main parts:

1. Construct a transformation (mapping) between the original ANN and the LIT uniform tree topology.
2. Demonstrate the equivalence of the original and transformed networks.
3. Since the original ANNs use static topologies, demonstrate how LIT supports extended versions which use dynamic topologies.

For ANNs that are inherently dynamic, and do not need to be extended, step 3 is subsumed by step 2.

Current work includes the following:

- transformations for other feedforward networks, recurrent and relaxation networks, and
- VLSI design and fabrication of LIT models.

5. References

[1] Almasi, G., A. Gottlieb. *Highly Parallel Computing*. Redwood City, CA: The Benjamin/Cummings Publishing Company, Inc. 1989.

[2] DARPA. *Neural Network Study*. AFCEA International Press, 1988.

[3] Fahlmann, Scott, C. Lebiere. The Cascade-Correlation Learning Architecture. in *Advances in Neural Information Processing* 2. pp. 524-532. Morgan Kaufmann Publishers: Los Altos, CA.

[4] Farhat, N., D. Psaltis, A. Prata, and E. Paek. Optical Implementation of the Hopfield Model. *Applied Optics, Vol. 24, #10.* pp.1469-1475 . 1985.

[5] Graf, H., L. Jackel, W. Hubbard. VLSI Implementation of a Neural Network Model. In *Artificial Neural Networks: Electronic Implementations*, Nelson Morgan, Ed. pp. 34-42. 1990.

[6] Hammerstrom, D., W. Henry, M. Kuhn. Neurocomputer System for Neural-Network Applications. In *Parallel Digital Implementations of Neural Networks*. K. Przytula, V. Prasanna, Eds. Prentice-Hall, Inc. 1991.

[7] Hecht-Nielsen, R. Counterpropagation Networks. *Applied Optics, Vol. 26, #23.* pp. 4979-4984. December, 1987.

[8] Hillis, W. Daniel. *The Connection Machine*. Cambridge, Mass.: MIT Press, 1985.

[9] Martinez, T.R., D.M. Campbell. A Self-Adjusting Dynamic Logic Module. *Journal of Parallel and Distributed Computing, Vol. 11, #4*. pp. 303-313. 1991.

[10] Mead, Carver. *Analog VLSI and Neural Systems*. Addison-Wesley Publishing Company, Inc., 1989.

[11] Odri, S.V., D.P. Petrovacki, G.A. Krstonosic. Evolutional Development of a Multilevel Neural Network. *Neural Networks, Vol. 6, #4*. pp. 583-595. Pergamon Press Ltd.: New York. 1993.

[12] Ramacher, U., W. Raab, J. Anlauf, U. Hachmann, J. Beichter, N. Brüls, M. Weißling, E. Schneider, R. Männer, J. Gläß. Multiprocessor and Memory Architecture of the Neurocomputer SYNAPSE-1. *Proceedings, World Congress on Neural Networks 1993, Vol. 4*. pp. 775-778. INNS Press, 1993.

[13] Reilly, D.L., L.N. Cooper, C. Elbaum. Learning Systems Based on Multiple Neural Networks. (Internal paper). Nestor, Inc. 1988.

[14] Rudolph G., and T.R. Martinez. An Efficient Static Topology for Modeling ASOCS. *International Conference on Artificial Neural Networks*, Helsinki, Finland. In *Artificial Neural Networks*, Kohonen et al, pp. 729-734. North Holland: Elsevier Publishers, 1991.

[15] Rudolph G., Martinez, T. R. An Efficient Transformation for Implementing Dynamic Backpropagation Networks. *Technical Report BYU-CS-94-1.*

[16] Rumelhart, D., J. McClelland, et. al. *Parallel Distributed Processing: Explorations in the Microstructure of Cognition, Vol. 1*. MIT Press, 1986

[17] Shams, S. Dream Machine—A Platform for Efficient Implementation of Neural Networks with Arbitrarily Complex Interconnect Structures. *Technical Report CENG 92-23*. PhD Dissertation, USC, 1992.

THE APPLICATION OF NEURAL NETWORK TO FAULT DIAGNOSIS

JIANN-LIANG CHEN, HUNG-FA SUN, and RONLON TSAI
Advanced Technology Center
Computer and Communication Research Laboratories
Industrial Technology Research Institute
Hsinchu, Taiwan, R.O.C.

1. Introduction

For the last two decades, due to the hardware cost in computer systems has been dropped tremendous and the variety communication techniques has been developed, the demand of distributed processing is terribly increasing in our society. Many distributed applications have pervaded our lives, such as automated teller machine networks, air-line reservation systems, and on-site validation of credit card [1]. The well developed network technology enable the large enterprises to use networks of various types to provide integrated communication services. To manage such large and multi-vendor based networks, Integrated Network Management System (INMS) with efficient management procedures is necessary [2]. However, because it is so complicated and time-consuming to handle and scrutinize the network fault status, the fault management, especially on diagnosing faults, is the first step in the integrated network management. To get through the work, a research project aimed at developing an effective tool for diagnosing faults was conducted recently by the authors.

Fault diagnosis is for determining network problems by analyzing the alarm and event reports on managed objects (devices) and by pinpointing the problems with various diagnostic tests [3]. And the activity of fault diagnosis often requires experienced network managers and sophisticated mechanisms [4]. In large networks, a network failure may result enormous number of alarm/event reports. However, the processes of determining the problem and performing diagnostic tests to narrow down the problem to certain network elements are not transparent. Therefore, the activity of fault diagnosis has to invoke various diagnostic tools among different elements and resolve the problems as a whole. And better approaches to consolidate existing tools in an integrated way that presents to the managers a complete and consistent view of the network information are required. In addition, the INMS always has to cover up the real-time nature [5].

With these considerations in mind, an approach of connectionist expert systems to solve fault diagnosis problems is proposed in this paper [6]. The basic design structure is composed of three stages including input stage, counter-propagation network, and decision table. These stages will be discussed in Section 3 (Proposed INMS).

Continually, we introduce the management information. We also talk about the

E. A. Yfantis (ed.), Intelligent Systems, 647–653.

implementation issue in Section 4 (Implementation Issue). Finally, the conclusions are outlined in Section 5 (Conclusions).

2. Management Information

To achieve the interoperability of network information, many activities, such as ITU-T, IETF, ..., and special Forums, devote to define the network management standards. Among those standards, CMIP (Common Management Information Protocol) of ISO and SNMP (Simple Network Management Protocol) of Internet are the two most popular standard protocols in current network market [7]. Considering the "simple and implementable" feature, we choice SNMP as our first implementation [8]. Following SNMP-based management standard (see Figure 1), SNMP manager can issue a "GetRequest" or "GetNextRequest" to access the network information by receiving "GetResponse" from the specified SNMP agent.

From Figure 1, we can find that the Management Information Base (MIB) is the source of network information. So far, a lot of management groups have been defined in MIB-II (see Figure 2) which is a network management de-factor standard now [9]. Therefore, based on the standard management information, called as object values, we build an INMS to analyze these object values derived from various managed elements and release the fault diagnosis problems.

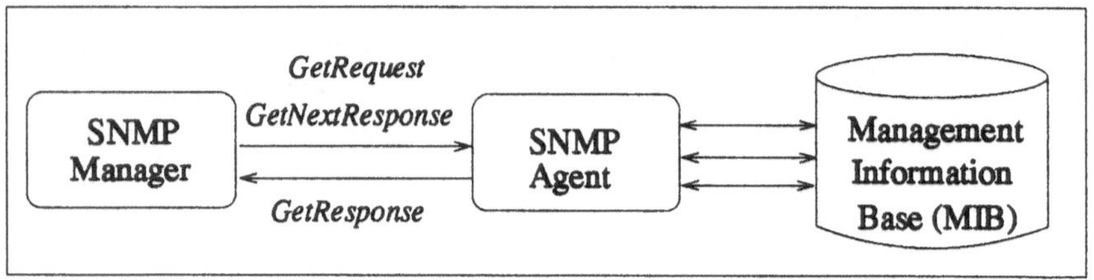

Figure 1. The brief SNMP-based architecture

3. Proposed INMS

The proposed fault diagnosis mechanism consists of a set of diagnostic tools and a connectionist expert system. The tools are presented by a window-based GUI (Graphical User Interface) which cooperates with some specific configuration management functions to allow users to select network elements for various tests. The connectionist expert system applies neural networks technique in cooperation with a rule-based expert system to provide automatic fault diagnosis on certain levels [10]. The connectionist expert system paralyzed to perform an expert's procedure in analyzing the event reports from Element Management Systems (EMSs; see Figure 3) and to determine the potential network problems in the sense of real-time. In the following, three stages shown in Figure 4 under design are described

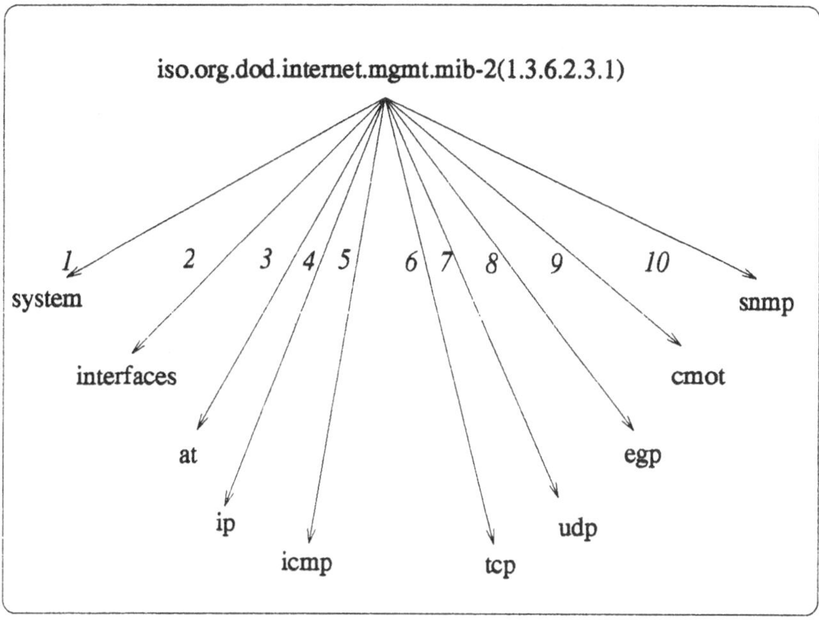

Figure 2. The management standard MIB-II

in detail.

1. **Input Stage.** In a multi-vendor based network environment, event reports are notified to INMS from various EMSs. Hence, it is important to provide event reports with a standardized style which uses a common set of notification types with standardized parameters. In our work, SNMP-based object values are adopted. These notifications provide information that INMS can act upon relating to fault diagnosis activities. Event reports are transmitted from EMSs to INMS through communication services and protocols defined in SNMP shown as Figure 1 and referred to as the input patterns of our research.

2. **Counter-Propagation Network.** Although many types of connectionist expert system and machine learning techniques have been developed, the research described here focuses on a three-layer feedforward network with counter-propagation learning rule [11]. The proposed counter-propagation network consists of two phases including learning and diagnosis phase as shown on Figure 5. In the learning phase, the network generates its knowledge by learning from the past experiences. The learning is achieved through clamping the known input patterns and the output classes (training sets) in the states that represents the relations between a set of event reports and their corresponding network problems, and adjusting the interconnection

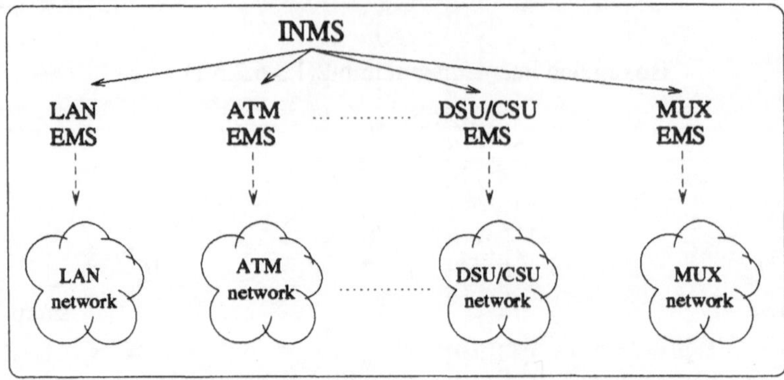

Figure 3. Connection between INMS and EMSs

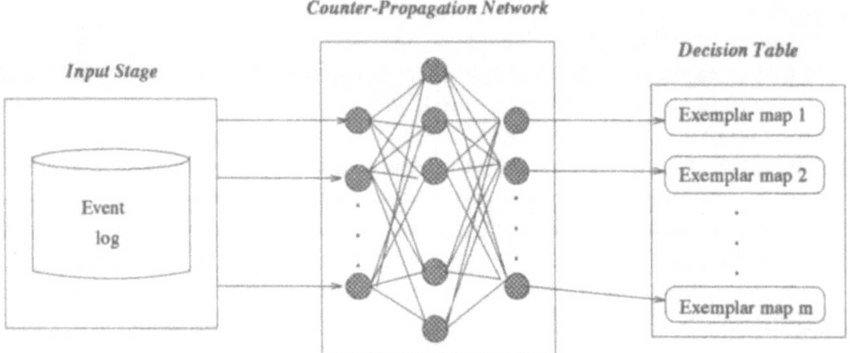

Figure 4. The stages under design

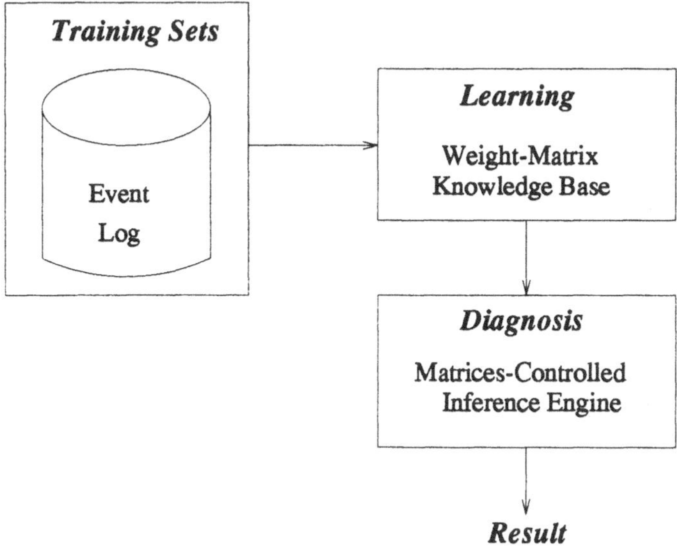

Figure 5. The counter-propagation network

weights among processing elements until an equilibrium state is approached. The final weight matrices are able to support the inference engine responsible for determining network problems from new event reports. In the diagnosis phase, possible network problems are determined based on the weight matrices, and associated diagnostic tests are applied according to the knowledge base of the expert system (see Figure 6).

3. **Decision Table.** In real world, the diagnostic processes of network problems are highly dependent on network configuration. Different network configurations may require different diagnostic expertise stored in the knowledge base of the expert system. In the proposed diagnosis mechanism, diagnostic experiences can be generated through diagnostic tools and saved into the connectionist expert system to build a decision table. Therefore, the table can indicate that each possible output exemplar map will correspond to one of network fault status.

4. Implementation Issue

The INMS is currently implemented by using HP OpenView SNMP software platform on SUN-SPARC series workstations with Solaris 2.x OS. In addition, Motif, the GUI of OpenView applications, gets high regard in the implementation. Presently, we use "xnmgraph" command line to get the management information

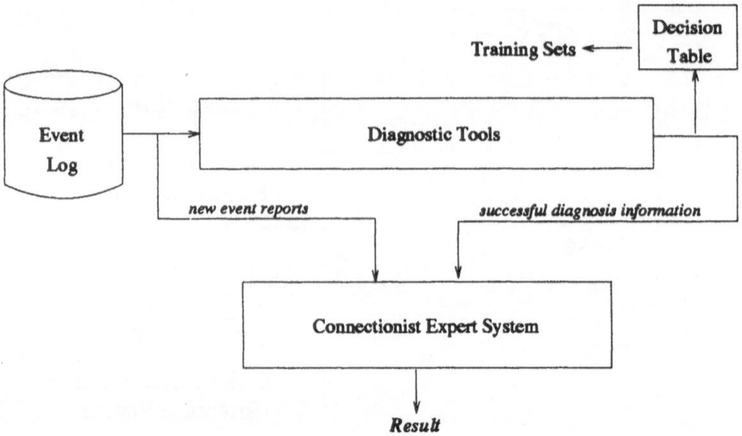

Figure 6. The connectionist expert system

and then feed into connectionist expert system (background process). Upon receiving the results, we then use "xnmgraph" command to illustrate the line chart or issue an emergency message. Unfortunately, the aforementioned processing may weaken the real-time nature. Therefore, strengthening the AP function under OpenView platform is our future mission.

5. Conclusions

An approach of integrating diagnostic tools and connectionist expert systems to solve fault diagnosis problems is proposed here. The proposed system, INMS, that utilizes artificial neural network technique is applied to determine network problems in real-time. The basic structure under design is composed of three stages including input stage, counter-propagation network, and decision table. The INMS is implemented by using HP OpenView software platform on SUN-SPARC series computer to manage a heterogeneous network environment. In view of the effectiveness and efficiency of the proposed approach, the connectionist expert system may be a valuable tool to assist network managers in real-time solving fault diagnosis problems.

6. Acknowledgements

This paper is a partial result of the project no. 37H1200 conducted by the ITRI under sponsorship of the Ministry of Economic Affairs, R.O.C.

7. References

[1] R. Popescu-Zeletin, V. Tschammer, and M. Tschichholtz, "Y' Distributed Application Platform", Computer Communication, Vol. 14, No. 6, pp. 366-374, 1991.

[2] H. W. D. Chang, B. S. P. Lin, and J. L. Chen, "Fault Diagnosis for Integrated Network Management Systems", Proc. of 7th International Joint Workshop on Computer Communications, pp. 365-372, 1992.

[3] "OSI-Based Network Management", DataPro, C05-010-651, June 1989.

[4] A. Leinwand and K. Fang, "Network Management: A Practical Perspective", Addison-Wesley Publishing Company, Inc., 1993.

[5] Special Issue on Real-Time Systems, Proceedings of the IEEE, Vol. 82, No. 1, 1994.

[6] S. I. Gallant, "Connectionist Expert System", Communications of ACM, pp. 152-169, 1988.

[7] B. A. Amatzia, C. Asheem, and W. Unni, "Network Management of TCP/IP Networks: Present and Future", IEE Network Magazine, July 1990.

[8] J. Galvin and K. McCloghrie, "Administrative Model for Version 2 of the Simple Network Management Protocol (SNMPu2)", Network Working Group, Request for Comments 1445, Trusted Information Systems, Hughes LAN Systems, April 1993.

[9] K. McCloghrie and M. T. Rose, "Management Information Base for Network Management of TCP/IP-based Internets, Internet Working Group Request for Comments 1213, March 1991.

[10] T. Sorsa, H. N. Koivo, and H. Koivisto, "Neural Networks in Process Fault Diagnosis", IEEE Transactions on System, Man and Cybernetics, Vol. 21, No. 4, pp. 815-825, 1991.

[11] R. H. Nielsen, "Counter-Propagation Networks", Appl. Opt., Vol. 26, pp. 4979-4984, 1987.

EXPERT SYTEMS

RSCL3: A LEARNING EXPERT SYSTEM FOR INTELLIGENT TUTORING

S. H. Rubin*
Department of Computer Science
Central Michigan University
Mt. Pleasant MI 48859 USA

Abstract. One of the central problems facing contemporary researchers in artificial intelligence is that of educating man and machine. This paper describes a methodology, RSCL3, which learns and teaches how to multiply matrices. Here, planning is involved in the path from the knowledge engineer to the machine to the user. That is, the RSCL3 algorithm is applied to learning how to teach what it has learned. The RSCL3 algorithm is a program, which was first conceived as any other rule-based system -- except that it contains a knowledge amplifier. This effective amplifier extends and extrapolates basis knowledge through the use of domain symmetries. It is an inductive technique, which creates new knowledge that is open under deduction. It is expected that this technology will ultimately reduce the cost of expert system applications.

Keywords. Expert Systems, Hybrid Learning Systems, Knowledge Amplifier, Random Seeded Crystal Learning.

1 Introduction

A central aspect of human intelligence is the ability to plan, that is, to generate action sequences that achieve one's goals. The pivotal role played by human planning is shared by the machine. It too must generate action sequences to achieve a specified goal. Sometimes these action sequences are pre-defined as is the case for an algorithm. However, the most interesting, and potentially most productive, cases occur in an expert system. Here, the machine is free to generate its own action sequences by way of an agenda, or task-planning mechanism.

This paper describes a method for planning the induction of new rules where they are most likely to be needed in the short-term. Perhaps this process can be likened unto post-tetanic potentiation in the brain by way of analogy. This is a process for automatically inducing new rules from old ones. Such a process is potentially intractable. It is only natural that the inductive process be constrained to the set of rules most likely to be needed in the immediate future. This working set of rules is defined by way of analogy with their basis -- that is, the set of most-recently-fired rules.

---

* The research described herein originated during the tenure of the author's three year ASEE/ONT sponsored postdoctoral fellowship (1990-93) with NCCOSC in San Diego, CA.

E. A. Yfantis (ed.), Intelligent Systems, 657–663.

2 Etiology

First among the novel machine learning technologies developed by the author, during the course of a three-year ASEE-sponsored sabbatical, was a methodology termed, "constrained-set generalization." This technology was successfully applied to an automatic message correction system. It had the capability to capture user knowledge in the form of sets -- a form suitable for subsequent automatic generalization. This system was suggestive of a tutoring system because it had the side effect of teaching the operator message corrections. Thus, the operator learned from the machine what not to do, or at least what to be on the watch for.

Next among the novel machine learning technologies developed was the predecessor to the one described herein, namely "random seeded crystal learning," or RSCL for short. The basic idea of this technology was to amplify a rule base using a knowledge of domain symmetries to effect the amplification.

The first version of RSCL -- RSCL1 -- ran far too slow to be of practical use -- even on a super computer. The next version, RSCL2, was implemented on a DAP-610 super computer in FORTRAN PLUS. It was used to inductively create naval AEGIS command and control rules. Measured tests revealed that the system exceeded break even. That is, it was able to learn more knowledge than it was programmed for. These impressive results can be found in the literature [8, 9, 10, 11].

The current version of RSCL -- RSCL3 -- carried much of the methodology of its predecessor with two exceptions. First, the algorithm for finding rule symmetries was rendered more efficient. Second, this improved algorithm allowed for the porting of the code to a Mac IIfx after being rewritten in Think Lightspeed C.

2.1 PLANNING IN A TUTOR

Tutoring is a linguistic exchange whose goal, in general, is to clarify a body of knowledge to which the student has already been exposed -- for example, knowledge obtained through lectures or reading [12]. Intelligent tutors enable one-on-one individual attention.

An intelligent tutor must be able to learn as well as teach. It should learn from the student any improvements to existing techniques or perhaps novel techniques not previously known to the tutor.

An intelligent tutor tries to model the student in an attempt to teach more effectively. It also needs knowledge about mistakes the student is likely to make, manifestations of these mistakes, and methods for constructing them [12].

2.2 INTELLIGENT TUTORING SYSTEMS

Any intelligent tutoring system must contain knowledge of the subject area. It may not be able to solve the problems it presents to the student, but it must at least be able to learn new knowledge from the teacher. The development of artificial intelligence technologies gave tutoring systems the capability to generate, and a limited capability to understand, natural language. It also gave them more flexibility to interpret the answers of a nontraditional student -- that is, the student who has the correct answer, but expresses it in a fundamentally different way [12]. Finally, the development of intelligent heuristics enabled tutoring systems to express their own knowledge, enabling them, for example, to explain their problem-solving reasoning.

A calculus tutor, developed at Stanford in the LISP programming language, and a tutor called SPADE-O, developed at the MIT artificial intelligence labs and used to teach program design methods have two things in common [6]. First, in both the tutor and the user communicate in natural language. Second, both are able to solve problems in their respective problem domains on their own.

The well-documented problem with expert systems is that it is time consuming and expensive to program in all of the human expert's knowledge. Thus, the concept of a knowledge amplifier was born. The idea for an intelligent tutoring system had its inception in the need for an improved expert system [7].

2.3 THE KNOWLEDGE AMPLIFIER

The knowledge amplifier is the critical component of a computer tutor. It is defined through the use of expert systems targeted to the domains of knowledge acquisition. Observe that the definition of a knowledge amplifier allows for self-reference -- implying the potential for non-linear amplification! Much amplification occurs by way of design automation, through a process that we term, *conversational learning*.

In essence, while expert systems can not directly assist with their own design, tutoring systems can indirectly assist with the design of tutoring systems through the medium of expert systems. In effect, this provides an example of setting the task of one expert system to the design of another. In particular, such assistance is necessary for maintaining scripting languages, handlers, and daemons.

Knowledge amplification improves the performance of an expert system. Thus, it designates an endeavor, which will contribute to the overall quality of a tutoring system. Automating knowledge acquisition will allow for the creation of tutoring systems of greater practical utility for less cost. An example of the workings of a knowledge amplifier is provided in section 4.

3 Methodology

An expert system can be amplified or extended through the extraction of symmetry and the introduction of randomness. These Yin and Yang of nature have been applied to the creation of random seeded crystal learning (RSCL) -- a program run on a DAP-610 SIMD super computer having 4,096 nodes.

3.1 PERTINENT MODELS

A key model, underpinning this project, is the use of induction as opposed to deduction in the generation of rules for ultimate use in a programming tutor. For example, programming is said to be an art in part because one can never be quite sure as to how any non-trivial program will behave prior to testing multiple-execution paths. That is, programming involves the use of heuristics, which are by definition inductive.

If programming were not inductive, then one would never need to test a program -- the performance of any program would be guaranteed prior to its execution. Obviously, this is far from the case.

Deductive tools have been addressed by other researchers [1, 5]. Here, the knowledge engineer can specify rules for performing peephole optimization to rules for generating corner-point software test cases. Deductive tools can be specified using rules, since rules are universal. It follows that a deductive methodology can be subsumed by an inductive one, and not vice versa. Clearly, the inductive methodologies have an important place in the overall plan for tutoring systems.

Suppose that all of the rules in the rule set are what Gregory Chaitin [2] calls random. This refers to the fact that no pair of rules in the set can be compressed -- just as any infinite sequence of truly random numbers can not be compressed into any algorithmic generator. However, it turns out that every non-trivial domain has some degree of symmetry inherent in it. Similarly, every non-trivial domain has some degree of randomness associated with it. The relative balance of randomness and symmetry, or Yin and Yang, varies with domain.

3.2 PSYCHOLOGY OF ANALOGY

Analogical reasoning is a mechanism for using experience in solving a problem to solve a new, similar problem [4]. Analogy may be a powerful new paradigm with which to exploit reusable specifications, but it has received little attention in the relevant literature. The perceived power of analogy is its potential to retrieve knowledge from one domain and apply it to a different domain. The capability for transferring knowledge across similar domains makes for true learning systems.

Transformational analogs are similar rules. An example based upon a psychological experiment conducted by Gick and Holyoak [3] will serve to clarify the concept. They show how small groups of soldiers can converge from many directions upon a fortress occupied by terrorists and successfully capture it. They go on to show how this scenario is analogous to radiating a tumor from several directions with weak rays, which are collectively strong enough to kill the tumor while preserving the surrounding tissue. Furthermore, they note the importance of similarity in encoding to successful "spontaneous analogical recall."

3.3 TRANSFORMATIONAL ANALOGY

Software represents a class of domains that are quite symmetric. That is, for a simple example, if an isomorphism can be established between Program A and Program B, then the two programs are said to be symmetric as formally defined. If A represents a program to compute the annual rate of return on a principal invested at 10 percent interest say and B represents a program to compute the proliferation rate of an anaerobic bacterial culture, then the two programs are obviously symmetric. However, if instead B' represents a program to extrapolate a sentence for a person having an aphasia, then the two programs are relatively random.

Suppose that you, an intelligent person, were given Program A and asked to generate B. Would this not be trivial? It would because of the symmetry property. However, given the task to generate B' instead, it would not be at all trivial in view of the randomness property. The central thesis of this project is that intelligence is necessary to high-level software automation. That is to say, should it not be easier for an expert system to induce the proper structure for B than for B'? Notice that the choice of B vs. B' will not matter much to any deductive-based translation methodology. Nevertheless, this raises the question that given that it is possible to induce B from A using some mechanization of nature's symmetry -- how do we do it?

Clearly, search is involved here. In fact, it turns out that search is inherent to any inductive methodology; although, knowledge provides the necessary hedge to constrain the search.

Object-oriented expert systems contain rule predicates consisting of functions and procedures -- i.e., software. Thus, rule pairs may express different degrees of symmetry. The RSCL methodology exploits these symmetries to induce candidate new rules. These rules may or may not be correct. They can not be assigned an absolute correctness metric. Rather, they can be assigned a relative (i.e., in comparison to each other) metric. Then, it follows that a rule can not contradict any rule, in the rule base, having a relatively greater probability of correctness. If no such contradiction occurs then the symmetric rule is saved and is indistinguishable from those that were manually generated. Otherwise, the new rule, and any of its progeny, are expunged.

Notice that this system allows for the creation of new rules through many successive transformations. These rules, while symmetric, may not look anything like their parents. It turns out that this amplification process can provide up to n^2 symmetric rules for a basis of n rules, in the limit [8]. Thus, an expert compiler consisting of perhaps 1,000 rules could be written with the effort formerly required to write say 50 rules. Consistency is guaranteed -- saving an enormous amount of additional manual labor.

Many ~~involved~~ details such as temporal matching of fired rules, queue management techniques, and similar implementation concerns have been omitted from this descriptive overview for the sake of brevity. These details can depend on the available platform.

4 Results and Conclusion

The following matrix multiplication tutor, while not necessarily indicative of the quality or sophistication of a knowledge amplifier, is representative of the embedded technique we call, *transformational programming.* It also hopefully shows that computer tutors are potentially of great utility. That, simply put, is our conclusion.

Monday, Feb. 28, 1994: 12:30

Good Afternoon CPS 510 student. How can I help you today
?My problem is matrix multiplication.

Searching... found.

To SOLVE MATRIX MULTIPLICATION, I believe you mean:
  READ DIMENSIONS (M, N), and then
  READ ENTRIES ($u_{i,j}$), and then
  MULTIPLY MATRICES.

Correct
?Y

To READ DIMENSIONS (M, N), I believe you mean:
  READ A.M, and then
  READ A.N, and then
  READ B.M, and then
  READ B.N.

Correct
?Y

To READ ENTRIES ($u_{i,j}$), specify matrices:
?A
?B
To READ ENTRIES ($u_{i,j}$), do you mean:
/* A system for hypertext help may be inserted here. */
  A row major read, or
  A column major read?

Please make your selection:
?A row major read

Working:
  The following rules were invoked in sequence to instantiate object module f8' in the CASE repository.
    DR1.
To MULTIPLY MATRICES, do you mean:
/* A system for hypertext help may be inserted here. */
  A row major multiplication, or
  A column major multiplication?

Please make your selection:
?A column major multiplication

Working:
The following rules were invoked in sequence to instantiate object module f10' in the CASE repository.
DR4.

The functional problem description:
SOLVE MATRIX MULTIPLICATION has been reduced to primitive recursives. List the synthesized pseudocode
?Y

Monday, Feb. 28, 1994: 12:36

```
Procedure: SOLVE MATRIX MULTIPLICATION;

        READ A.M;
        READ A.N;
        READ B.M;
        READ B.N;

        /* Note: Documentation generated automatically using trace */

        /* Read in row major order: */
        FOR U=A..B
                FOR J=1,U.N
                        FOR I=1,U.M
                                READ u_i,j;

        /* Multiply in column major order: */
        FOR I=1,M
                FOR J=1,N
                {
                        C_i,j = 0;
                        FOR K=1,N
                                C_i,j = C_i,j + A_i,k * B_k,j;
                }

        Print C;

End: SOLVE MATRIX MULTIPLICATION.
```

Specify target language:
?C

Terminate
?Y

End of run.
Elapsed time 11 minutes, 43 seconds.
Logoff at 12:41 on Monday the 28th of Feb., 1994.

5 References

[1] Blackburn, M.R. (1989) "Using expert systems to construct formal specifications," IEEE Expert, Spring, 4, 62-74.

[2] Chaitin, G.J. (1975) "Randomness and mathematical proof," Scientific American, May, 232, 47-52.

[3] Gick, M.L and Holyoak, K.J. (1983) "Schema induction and analogical transfer," Cognitive Psychology, 15, 1-38.

[4] Harandi, M.T. and Bhansali, S. (1989) "Program derivation using analogy," Proceedings of the Eleventh International Joint Conference on Artificial Intelligence, Detroit, MI, pp. 389-394.

[5] Liu, H. (1993) "Software engineering practice in an undergraduate compiler course," IEEE Transactions on Education, 36, 104-108.

[6] McClure, C. (1989) CASE is software automation, Prentice-Hall, Inc., Englewood Cliffs, CA.

[7] Rubin, S.H. (1991) "Learning in the large: case-based software systems design," Proceedings of the 1991 IEEE International Conference on Systems, Man, and Cybernetics, Charlottesville, VA, pp. 1833-1838.

[8] Rubin, S.H. (1992a) "Case-based learning: a new paradigm for automated knowledge acquisition," ISA Transactions: Special issue on artificial intelligence and competitive manufacturing, 31, 181-209.

[9] Rubin, S.H. (1992b) "Intelligent compilation: bootstrapping case-based learning," Heuristics: The Journal of Knowledge Engineering, 5, 13-43.

[10] Rubin, S.H. (1993a) "Machine learning and expert systems," AI Expert, 8, 32-37.

[11] Rubin, S.H. (1993b) "New knowledge for old using the crystal learning lamp," Proceedings of the 1993 IEEE International Conference on System, Man, and Cybernetics on Systems Engineering in the Service of Humans, Le Touquet, France, pp. 119-124.

[12] Woolf, B. and McDonald, D.D. (1984) "Building a Computer Tutor: Design Issues," Computer, Sept., 17, 61-73.

NO CAUSALITY IN FUNCTION: BUILDING A FUNCTION-CENTERED KNOWLEDGE BASE

David J. Russomanno
Department of Electrical Engineering
The University of Memphis, Memphis, TN 38152 USA

Ronald D. Bonnell
Department of Electrical and Computer Engineering
The University of South Carolina, Columbia, SC 29208 USA

Abstract. The search for a methodology for representing knowledge about conceptual blocks and how devices work in an Expert System for Failure Modes and Effects Analysis (XFMEA) has led to considering how a functional knowledge representation can be exploited and generalized in the domain of reliability analyses. This paper describes the essence of a functional representation, whose foundations were first proposed by Sembugamoorthy and Chandrasekaran, that has been adapted for XFMEA. The paper proposes the NCIF (No-Causality-In-Function) principle which is viewed as a generalization necessary for migration to a function-centered ontology base. In XFMEA, a simulation subsystem organized around a function-centered knowledge base provides the inference procedure with a focus of attention directed toward expected goals and guides the reasoning process in determining the effects of a system's failure modes.

1 Introduction

The engineering reliability technique referred to as Failure Modes and Effects Analysis (FMEA) [1] can be thought of as a detailed cause and effect analysis, with failure effects being categorized according to severity. The analysis considers the effects of a single-item failure mode on overall system operation and spans a myriad of applications, including electrical, mechanical, and software domains.

FMEA methodology is based on a hierarchical approach to analysis. The indenture level of the analysis (i.e., the item levels that describe the relative complexity of assembly or function) coincides with the life-cycle of the overall system design. Often, FMEA is incorporated into a system's conceptual design, before commitments are made to a specific technology or implementation.

Intelligent FMEA systems must explore architectures that capture the information-processing activity of the entire analysis process. Traditionally, FMEA knowledge tends to be restricted to small groups of specialists and an analysis can only be generated by a team of engineers having a thorough knowledge of the system's design and application. Hence, a cooperative effort covering diverse areas of expertise is required. Russomanno [2] has been involved in research to develop an Expert System for Failure Modes and Effects Analysis (XFMEA) based upon the blackboard paradigm. In the blackboard model of XFMEA, the

E. A. Yfantis (ed.), Intelligent Systems, 665–679.

requisite problem-solving knowledge is partitioned into a set of knowledge sources (KSs) that embody the knowledge of a particular task or viewpoint of the domain. The top-level KSs are the *User*, *Design Expert*, *Mission Expert*, *Simulation Expert*, *Analysis Generator*, and the *Discourse Manager*, as shown in Figure 1. In particular, the *Simulation Expert* generates a simulation of the system under study. A significant challenge in the design of this KS is choosing a set of representations to describe systems and a strategy for integrating these representations at multiple levels of abstraction. The three chief components of this KS are the *Black Box CAD Tools*, the *Behavioral Modeler*, and the *Functional Modeler*.

The goal of this paper is to convey the essence of the functional representation[1] and its implications on the organization of the *Simulation Expert's Functional Modeler*. In addition, several research issues of more universal interest in the design of a function-centered knowledge base are discussed.

2 Evolution of Model-Based Reasoning

Initially, system representation emphasized device structure (i.e., the static physical characteristics of the device based on its topology) and deriving numerical relationships among structural constituents to produce a simulation. The next step focused on utilizing a methodology for qualitatively simulating and deriving the system's behaviors (i.e., its dynamic characteristics). Causal reasoning, which builds on representations of structure and behavior,

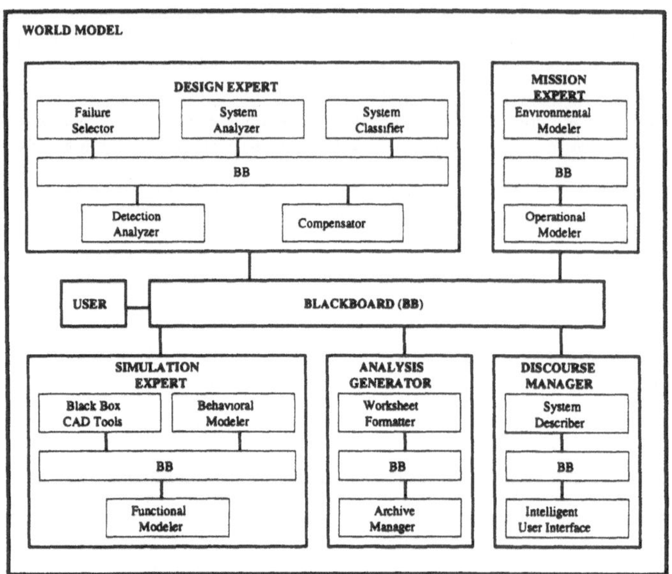

Figure 1. XFMEA Blackboard Model

[1]In this paper, the phrase *functional representation* is not intended to imply that there exists a single representation; instead, it is being used to imply the representation first proposed in [3], and its extensions [4-6], as well as the adaptations and applications proposed in this paper.

addresses the issues of how systems achieve their behavior. Neither structural nor behavioral models, including causal models, explicitly describe the functionality of a device.

The development of representations that focus on the function of devices is a natural evolution in model-based reasoning. For example, to perform an FMEA, human experts must develop an accurate mental model of how systems or subsystems work. Humans reduce the cognitive load of understanding systems by hierarchically decomposing an overall device into subsystems [7]. This requires the expert to understand, on a qualitative level, how the system is designed to perform certain functions. For example, consider the following human expert's description of the function of a hot water heater's temperature subsystem:

The function of the temperature subsystem is to control the water temperature at the outlet faucet by maintaining the temperature in the tank through the controlled addition of gas to the burner and tap water to the tank.

The description indicates a function (to control), and a method by which the function is obtained (by maintaining). Notice that when stating a method by which the function is obtained, a theory of causality is implied.[2] Such a knowledge organization, representation, and acquisition scheme is needed to build expert systems for FMEA that use descriptions of function in device simulation of item failure effects. Moreover, the representation can be used to explain a system's goals, that is, functions, in terms of the behaviors accomplished, which are explicitly represented in the acquired knowledge base.

2.1 Functional Reasoning Overview

When performing an FMEA, a numerical or qualitative model simulating the behavior of a system may provide unnecessary detail, and a more abstract model--a functional model is more efficient and appropriate. The function of a system is its intended purpose and its behavior is how that purpose is achieved [4]. A system is made of devices and accomplishes its overall functionality via the functions of its devices. A device's function is achieved by a detailed set of causal relations that constitute the behavior of the device. As a design becomes more specific, detail is added to the behavioral segments that make up an overall function. Sembugamoorthy and Chandrasekaran [3] view a behavioral segment as a compiled relation such as $A \Rightarrow B$; and this segment may actually be made up of several behavioral segments (e.g., $A \Rightarrow A' \Rightarrow A'' \Rightarrow B$).

As early as 1971, Freeman and Newell [8] identified the need and utility of a functional-reasoning methodology. The functional modeling scheme described is based upon the research of how model-based reasoning is used rather than how it can be derived, and is exemplified by the work of Chandrasekaran [9], Keuneke [10], Franke [11], and Gero [12], among others. The functional modeling scheme is expressed as a set of primitives that describe functionality at an abstract level. The scheme is top-down, that is, the functionality is explicitly described by primitives, and if known, detailed behaviors manifested by qualitative simulations are added to the representation. This approach to explicitly representing functionality is a paradigm shift from earlier work by de Kleer [13] in teleological reasoning in which a device's function is derived from the system's structure and behavior. In the teleological approach, an expert

[2] A proposal of this paper is that only when a function is linked to a behavior is causality implied.

recognizes structural configurations to infer purpose. For example, when analyzing a circuit an engineer could recognize a common collector configuration and infer that the circuit is being used for the purpose of increasing the bandwidth, or recognize a common emitter configuration and infer that the circuit's purpose is to increase the gain.

Keuneke [10] has delineated a set of functional primitives based upon the work of Sembugamoorthy and Chandrasekaran [4]:[3]

- *ToMake*, places an entity in a certain state
  e.g., *ToMake* (Engine Start)
- *ToMaintain*, maintains an entity state once it has been reached
  e.g., *ToMaintain* (Engine Temperature)
- *ToPrevent*, disallows an entity to achieve a certain state
  e.g., *ToPrevent* (Engine Overspeed)
- *ToControl*, regulates state changes via a relationship
  e.g., *ToControl* (Engine Speed)

The primitives essentially describe the function's purpose by relating an entity and a state and can be viewed as *verbs* that dictate certain simulation procedures and semantics.[4]

A primary goal of employing a functional representation in XFMEA is that by explicitly representing purpose by primitives, the system can be simulated at a more abstract level, that is, the effects of system failures can be analyzed before the detailed design is known. This approach eliminates the extensive detail that results when describing a system either numerically or qualitatively as a long sequence of causal states. Specifically, while performing a functional FMEA early in the system's conceptual development or proposal stage, the designer or analyst uses building blocks, which are tied to a functional knowledge representation, to achieve the design purpose. Then as the design progresses, physical components that participate in the detailed behavior are configured and the causal network is built. Hence, the knowledge of the design and its failure modes and effects have been organized hierarchically from function down to detailed behaviors.

3 Reasoning About Function in XFMEA

3.1 Specification of Goals, Functions, and Causality

From the most general level of abstraction, the placement of a functional building block in a conceptual design does not imply a specific behavior, only a top-level goal is specified. At the most, the designer implies some abstract notion of I/O pairs. Moreover, a causal theory of how the function is achieved is not implied. A function, such as *Transformer*, can be achieved via many different causal theories and even I/O pairs are not needed at the most abstract level of functionality--only a description of the purpose or the goal of the function needs to be captured.

Only when a designer commits to a specific technology to implement the function is causality imposed. This view, which is being adopted in XFMEA, contrasts with Bonnet's [14]

[3]Bonnet [14] has identified an additional primitive, *ToAllow*, which captures the purpose of passive functions. An example of a passive function is the purpose of the tank in a hot water heater system. Passive functions, which were also proposed by Keuneke, generally do not require state changes in the behaviors that accomplish their functions.

[4]The Japan Value Engineering Society [15] has formalized approximately 140 functional verbs.

perspective in which functionality and a theory of causality are inseparable. The contrast can be attributed to the level of abstraction upon which Bonnet's perspective is based. That is, most, if not all functional representations assume that a function's causality is packaged with the functional declaration. However, this view assumes that the function represented has sufficient implementation details such that its behavior has semantic identity (i.e., behavioral meaning has been ascribed to the function).

A function can be specified without prescribing a causal theory or a behavioral semantic. For example, the function *control the traffic at the street intersection* implies no means by which the function is achieved. A police officer or an automated traffic light can be used to accomplish the same goal. Hence, the specification of this goal has no behavioral identity until the designer commits to some mechanism to accomplish the function.

The proposal here is that a function-laden knowledge base must support elliptic, functional teleological statements, that is, declarations of functions that omit the behavior or causal details of how the function is achieved. These statements emphasize only the goal that has been ascribed to the functional declaration.

The design of a function-laden knowledge base must include general semantic categories of functions that are causality-free, that is, intensional declarations of functions that adhere to a NCIF (No-Causality-In-Function)[5] principle. The motivation is that at very abstract levels functions must be represented in an achievement-free manner. Functions can be achieved via many different causal theories and only when the designer makes a commitment to a particular device is the notion of causality specified. Only when a function is linked to a behavior (e.g., filling in Chandrasekaran's *By* clause) is causality implied by a function. By adhering to the NCIF principle for representing functional semantic categories, a top-level function does not apply to a particular object and its set of causal implications; instead, a methodology for capturing a *standardized* functional representation is proposed.

3.2 Ontological Distinctions

The design of any knowledge-based system must address primitive notions from which the knowledge acquisition task can express captured conceptualizations. The problem of capturing knowledge at a level which is too specific has been identified by numerous researchers and Lenat [16] has coined the problem as the *Representation Trap*.

Building a function-laden knowledge base raises important issues concerning generality, knowledge reuse, sharing, and the intended underlying semantic interpretation of the base. A methodology for ontological design is needed to build a knowledge base for XFMEA's *Functional Modeler.*[6] Fink [17] has described functional primitives, such as: the *Transformer*, the *Regulator*, the *Reservoir*, the *Conduit,* etc. A set of primitives, including the *Transformer*, are viewed in XFMEA as semantic categories that form a function-centered ontology base, and are very general categories that can be useful in virtually any subject. The primitives (i.e., the suite of *to do something* verbs) of Sembugamoorthy and Chandrasekaren [3], and Keuneke [10], among others, are viewed as linguistic elements that convey the goals of the general semantic categories, as well as the purposes of other functional specializations of the top-level

[5]Russomanno has coined the acronym NCIF in the spirit of NFIS (No-Function-In-Structure) principle.

[6]Here, a function-centered ontology refers to general semantic categories, relations, and assumptions that make naturally occurring functions explicit.

categories. That is, when capturing the purpose of general, functional semantic categories, the primitives of Keuneke [10] are being used in this work in a way that implies no causal means. A goal applies to thresholds and quantities only when specifying the causal details of its behavior, and by specifying the goal as an instance of a semantic category (i.e., specifying extensional knowledge about a functional instance of a top-level semantic category). For example, the function *to make hair dry* describes the purpose of a hair dryer, but only when a commitment is made to a technology to implement the hair dryer, can quantities, such as 120 V electric outlet, be specified as predicating conditions for the goal to be achieved.

Russomanno [2] is currently involved in research to build a computational construct for XFMEA's *Functional Modeler* that utilizes a linguistic element (primitive)[7] in a function's declaration (e.g., *ToMake*(Entity State)), and a semantic category to provide general concept inheritance (e.g., *InstanceOf*(regulator)). With this approach, it is possible for two particular functions to be an instance of the same semantic category, but they can still be described by two different linguistic elements (primitives). For example, both a power supply and a hot water heater's tank are instances of the semantic category *Reservoir*.[8] However, the power supply's goal is *ToMake*(Device Energized), while the tank has a passive function, that is, it achieves its function by existence rather than accomplishing its behavior by a set of state transitions. This approach is viewed as useful in moving toward a more formal methodology in function-centered ontology utilization. By attempting to adhere to the NCIF principle, and employing an initial set of semantic categories (e.g., Fink's primitives) and the primitives denoted in Sembugamoorthy [3] and extended by Keuneke [10] as the basis for computational constructs for XFMEA's *Functional Modeler*, new concepts are only introduced as a specialization of these categories. Additional research is needed to develop the function-centered ontology base and to select a set of semantic categories and primitives that cover the conceptualizations of many domains. This research is beyond the scope of this paper; nevertheless, illuminating these issues suggests that the recognition of a function-centered ontology, as well as its associated methodology and terminology, will foster the continued investigation of these issues and the prevention of the increasing proliferation of *roll your own* knowledge bases.

4 Knowledge Engineering Issues in XFMEA

4.1 Functional Ontology/Functional Block Diagram Mismatch

Traditionally, when designing a system with a rigorous reliability requirement, the design and reliability evaluation begins with a functional and reliability block diagram. However, in some instances, when an FMEA expert analyzes such a system, a different functional knowledge hierarchy from the one depicted in the functional block diagram, is used to mentally simulate the effects of item failure modes. Traditional functional block diagrams tend to correspond to

---

[7] In the remainder of the paper, functional primitives refer to the linguistic elements that are used to describe a functional semantic category. Functional semantic categories (e.g., the regulator, the transformer, the conduit, etc.) are those top-level functions which are common to many domains. These semantic categories make up a linguistic stock that is achievement free.

[8] Fink [17] describes the *Reservoir's* purpose as: to store a substance for later output.

the engineer's physical taxonomic decomposition of the system.[9] For example, when modeling a communication system's receiving functions, Figure 3 [7], the engineer depicts the functional blocks *RCVR front end* and *FTN CKTS*. Notice the loop from the *FTN CKTS* to the *RCVR front end* has a higher-level purpose: *ToMake*(input_noise null). In addition, the loop is an instance of the general semantic category *regulator*. The human expert, perhaps unknowingly, guides his or her inference process based on a function-centered ontology. A functional ontology/functional block diagram mismatch may exist and becomes evident when observing a novice engineer performing an FMEA. As the engineer matures, he or she develops a knowledge organization of the system that is not directly tied to the functional blocks depicted on the diagram. Instead, a set of basic functional primitives are used to describe the purposes of a device's subsystems.

In Figure 2, there is an implicit functional block of feedback with a purpose of nulling the input noise. The feedback loop is an important functional organizer for representing the device at the block diagram level, and provides the expert with a higher-level concept for predicting the effects of a failure mode; hence, this higher-level concept must be acquired in the functional knowledge base. Since there is a small probability that a functional entity in a block diagram will have a one-to-one correspondence with the functional knowledge organization of an expert or with a physical system component,[10] resolving the mismatch among the functional block diagram, the functional knowledge hierarchy used for analysis, and the system's physical organization is a key to becoming an FMEA expert. No better example exists to illustrate this observation than that of feedback in control systems. An FMEA expert has knowledge that feedback is occurring which includes the knowledge and inference procedures that accompany a higher-level abstraction. At the lowest level of abstraction, the feedback process is not explicit, but rather a detailed behavior manifested by components such as operational amplifiers; while at the most abstract level, feedback is a higher-level concept such as *a loop to cancel noise*.

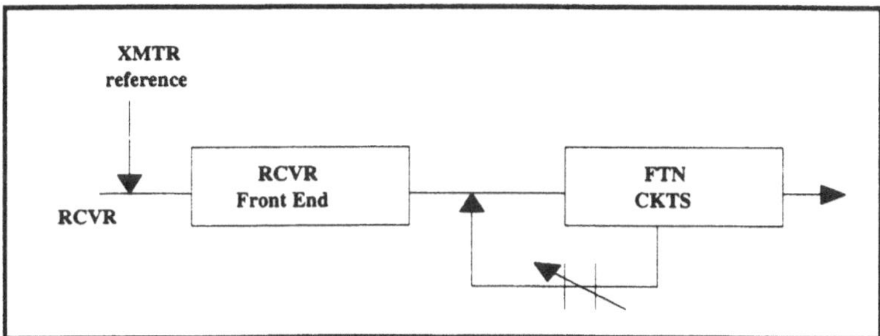

Figure 2. Functional Block Diagram of a Receiver's Front End

[9]There is a difference between the traditional functional block diagram and functional decomposition. A functional decomposition is based on a functional ontology rather than a design engineer's traditional conceptualization of a functional taxonomic hierarchy, i.e., a functional block diagram, which is based on a structural hierarchy.

[10]Gero [12] refers to a direct mapping between function and structure as a *catalog lookup*. *Catalog lookups* are not a common occurrence and cannot be regarded as functions in the mathematical sense.

The traditional functional block diagram used in reliability analysis tends to be the classification of the system types that constitute an overall system, that is, it is a taxonomic organization based on structure, rather than a function-centered ontological knowledge hierarchy for describing the effects of functional failure modes. Based on the detailed study of how experts determine system failure mode effects [7], the functional block diagram does not produce the same hierarchy that an expert utilizes to mentally simulate failure effects. Traditional functional blocks cut across physical units, rather than being organized around a function-centered ontology useful for simulating failure and reasoning. The functional hierarchy used by an expert to mentally simulate an implemented system is organized around higher-level functions such as "to cancel noise," and can have ties to causal knowledge that manifests the function. Concepts such as noise, power, information, loops, and feedback are relevant for mentally simulating the effects of item failures. The following sections discuss implications of a knowledge representation that makes semantic distinctions among, and ascribes conceptual identity to, signals, power, and feedback loops.

4.2 Signals and Power

Functional primitives and semantic categories can be used to make distinctions explicit between entities. The function of an information signal and a power signal can be differentiated with a functional primitive. *ToMake*(Device Energized) is used to represent the functionality of a power signal, whereas, *ToControl* (Device State) is used to describe how an information signal is employed for the purpose of controlling a device's operating state. For example, consider the eight-channel multiplexer schematic shown in Figure 3. The purpose of inputs 11, 12, and 13 are *ToControl* (Multiplexer Output), that is, these input bits are used to select which of the inputs (1-7, 9) are routed to the output (14). In addition, inputs 8 and 16 have the function *ToMake* (Multiplexer Energized), while input 10 has the function *ToMake*(Multiplexer Enabled) and input 15 has the function *ToMake*(Multiplexer.Output Tristated).[11] Distinguishing between information and power signals is important in providing an FMEA analyst with high-level knowledge about what functions a designer intended to accomplish with each building block in the system design.

Failure to energize a device, in other words, to not make a device energized, that is, ¬*ToMake*(Device Energized), prevents the device from operating. If the *ToMake* (Device Energized) function is achieved, but due to a component failure, the function *ToControl*(Device State) fails, the device still operates, but can produce an additional failure effect on the overall system due to a degraded information signal. In the multiplexer example, the device can be energized, but an information signal failure can result in the wrong input being routed to the output. By explicitly representing the function of each constituent in the design, reasoning about the failure implications of energy signals can be differentiated from information signals. The knowledge engineer provides the XFMEA system with knowledge structures to reason, explain, and simulate at a more abstract level by utilizing a functional representation.

[11] The multiplexer example is not intended to imply that a device's structure is necessary to describe functionality. The multiplexer provides a good example for differentiating between control and power signals.

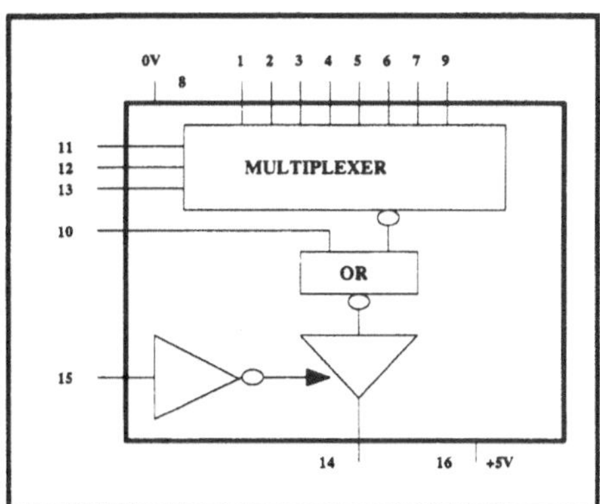

Figure 3. Eight-Channel Multiplexer

4.3 Loops and Feedback

The described functional representation provides the basis for an organizational kernel for conceptualizing loops and feedback in system analysis.[12] Feedback can be characterized by its goal: to allow some *controlled* process parameter to be *monitored* then *compared* to some reference parameter so that an appropriate control *action* may be invoked based on the *monitored* and *referenced* parameters, respectively. Although feedback loops in an electronic design are implemented by the behaviors exhibited by a sequence of local events made up of operational amplifiers, resistors, and other signal conditioning circuits, there exists a macro-level conceptual feedback loop with a teleology. To perform a computer-aided functional FMEA, knowledge about loops and their purpose must be captured and then recognized in system analysis.

Consider the water tank leveling system (WTLS) illustrated in Figure 4 [18]. This system is an elementary example of a feedback control system that achieves its goal via the behaviors of its constituent components, that is, each constituent of the system, other than the tank and the stopcock, combines to form one functional block--a control system for a water tank leveling system. The function of the WTLS is to maintain the water level in the tank (i.e., *ToMaintain*(Tank H_20.Level)).

Just as a mathematical model, which is an abstraction of the characteristics of a *feedback system*, can be derived for a system, an abstract notion of higher-level functions can be generated. In Figure 5, the WTLS can be conceptualized as a control system consisting of a controller and feedback elements operating on a plant. The mapping from physical components to building blocks, such as a controller and feedback element, requires a *cognitive clustering and leap* by the analyst. The motivation for representing the functionality of a feedback system is to analogously capture higher-level abstractions that have roles in performing a functional FMEA on design concepts. That is, the analyst who possesses a knowledge of feedback, and its accompanying inference strategies, can reify[13] a building block and consider the effects of failing to achieve the function without considering design details.

[12]Keuneke [10] provides an abstract schema for feedback function and behavior.
[13]Reify refers to the process of conceptualizing an abstraction as an object.

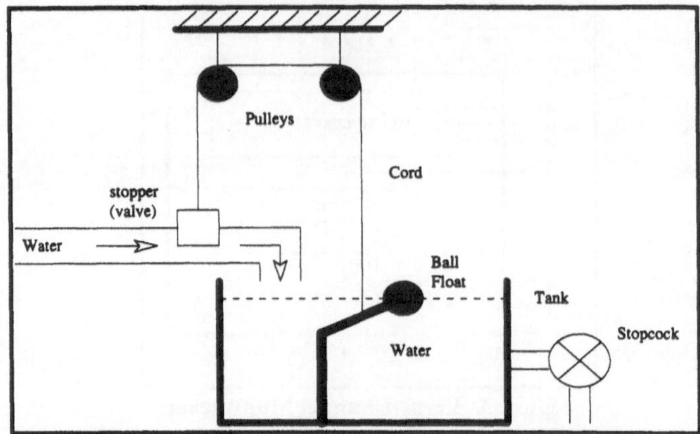

Figure 4. Water Tank Leveling System (WTLS)

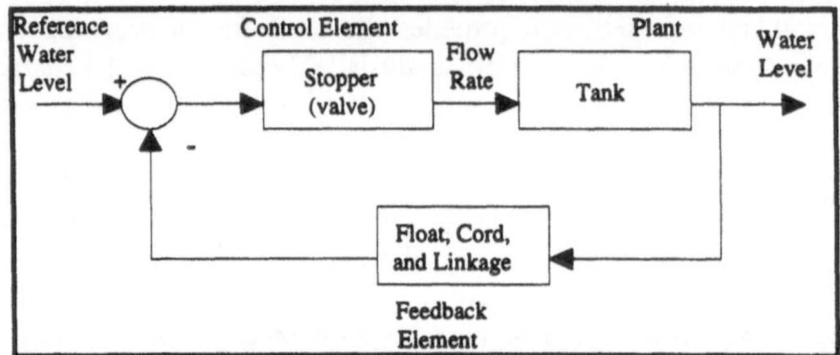

Figure 5. Control Diagram for the Tank Leveling System

In the WTLS example, the maintenance function, whose purpose is conveyed by the linguistic element *ToMaintain*, is an instance of the semantic category *regulator*; however, there are feedback systems (e.g., servomechanisms), whose goals are to track a system's output based on a varying input, which are best described with the primitive *ToControl*.[14] Hence, in XFMEA, control system functions are instances of either the *servo* or *regulator* top-level semantic categories.

5 Formalization of Knowledge-Based Functional FMEA

The following sections explore how a functional representation methodology is applied in the design of an expert system for FMEA that can reason at the functional level. The functional representation is employed to overcome the mismatch between the functional ontology and the functional block diagram and to provide a formalism for representing the results of the FMEA *couched* in the same syntax and semantics as used in the inference process to derive the analysis conclusions.

[14]Bonnell has suggested adding the primitive *ToTrack* to XFMEA's linguistic stock to convey the purpose of the *servo* category.

5.1 Functional FMEA Overview

To appreciate the organizational benefits of the functional approach, consider the simple example of analyzing a portable communications gear (PCG) system. The PCG consists of two major physical units: a chassis and a headset. The user communicates by pressing a button on the chassis and speaking into the headset's microphone and receiving information via the speakers in the headset.[15] Typically, an FMEA considers each physical component of the PCG based on a structural taxonomic decomposition, Figure 6a. Hence, any model of the system bases its relationships on physical parts (i.e., the system's topology). However, a functional decomposition reveals more about a system's operation and has implications for a higher-level simulation, Figure 6b.

In Figure 6a, each element in the decomposition is a physical entity and no conceptualization of purpose is conveyed. Moreover, the decomposition is based on the perspective that a physical component is built up from the assembly of its parts. Although the decomposition *bottoms out* with the same physical components as depicted in Figure 6b, the goals in which the components participate are not explicit. The structural decomposition can be considered to utilize a physical *partOf* relation.

In Figure 6b, the decomposition is based on the perspective that the PCG's conceptual parts (i.e., the transmit, receive, buzz, and energize functions) contribute to the whole, not as a structural part, but as an essential constituent in the accomplishment of the PCG's top-level goal. As used in Figure 6b, the relationship between the PCG and the function transmit can be characterized as a functional *partOf* relation. In the functional decomposition, the *partOf* relation is emphasizing an abstract conceptualization of participation in purpose, whereas in the physical decomposition, the *partOf* relation is stressing the structural parts that make up a whole.[16]

By acquiring a knowledge base organized around functions, reasoning about higher-level concepts, such as failure to transmit, can now be considered. Organizing the device around functions does not limit the knowledge base to a physical-functional match (i.e., a function coincides with a physical component or subsystem), therefore, high-level concepts can be considered. Most systems (e.g., mechanical, electrical, hydraulic, etc.) are built on a set of functional primitives and categories that can be used for simulation. A functional organization represents an ontological choice and can be thought of as a conceptual bias that limits the search space when simulating a system. Here, a conceptual bias refers to restricting the vocabulary that is used to describe systems. When simulating the PCG at the functional level, the functional semantic categories are limited to a set such as: *Transmit, Receive, Buzz,* and *Energize*. With this bias, the simulation of the PCG is expressed entirely in terms of these functions along with their associated primitives and extensions. The question of how to select a suitable bias set is difficult to answer and probably depends upon the problem type and the

---

[15] In this example, the chassis has the following subcomponents: repeating coil, buzzer, power supply, antenna, etc., and the headset has these subcomponents: speaker, microphone, etc.

[16] It should be noted that Iris [19] proposes four models of the part-whole relation. However, both Figures 6a and 6b would be classified as employing the function-component, part-whole relation. In Iris [19] the instance in which an artifact or concept has a functional decomposition not based on a physical taxonomy is not differentiated from the instance in which only a sophisticated structural organization is specified. For example, a police officer standing in an intersection is *partOf* the function *to control the traffic at the intersection*. However, he/she has no structural identity in the goal, only an abstract notion of participation.

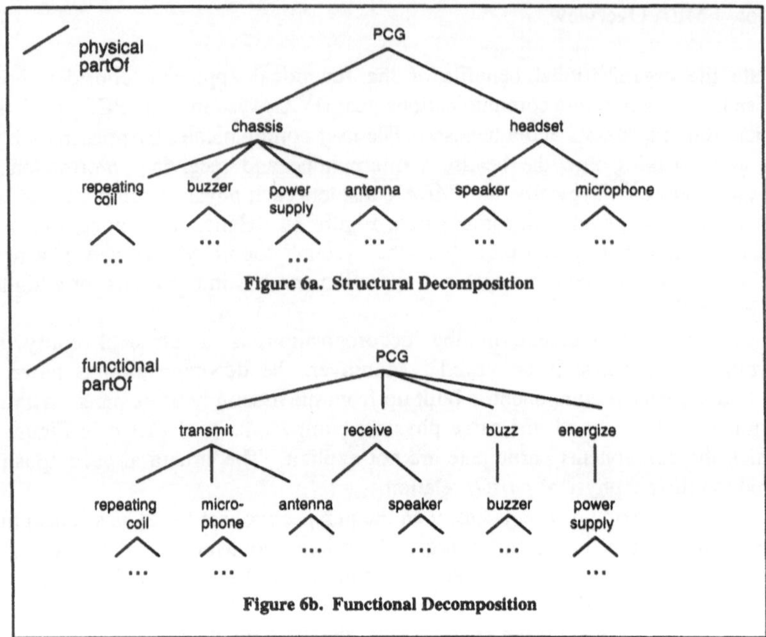

Figure 6a. Structural Decomposition

Figure 6b. Functional Decomposition

Figure 6. Decompositions of the PCG

domain.[17] There may be instances in which new primitives and semantic categories must be hypothesized to provide satisfactory simulations of failure mode effects. Naturally, the choice to simulate systems at the functional level is determined by the stage of the design process and is most useful early in the conceptual development of systems.

The functional representation described concurs with recent literature on alternative methods for preparing FMEAs [20] in which functions are treated as virtual components at a more abstract level in analysis. The partial, top-level functional representation of *PCGTransmit*[18] is shown in Figure 7 utilizing the *ToMake* primitive of Keuneke [4]. Notice that the *PCGTransmit* function's purpose is described via the primitive *ToMake*. In addition, the functional dependency of the *PCGTransmit* function on the function *PCGEnergize* is captured via the *hasPrerequisite* relation, and its relationship to the semantic category *Transmit* is captured via the *instanceOf* relation.

The function *PCGTransmit* requires the initial state *pressed MircophoneSwitch* for its detailed behavior ***pcgTransmitBehavior***[19] to be accomplished. *PCGTransmit* is made up of constituents that participate in the behavior ***pcgTransmitBehavior***. Hence, if a component fails

[17]A functional organization is not unique. Analysts and designers can view a system's purposes differently based on their background knowledge and experience.

[18]It should be noted that *PCGTransmit* is an extension of the top-level semantic category *Transmit*; hence, at this level of abstraction the *PCGTransmit* function's *By* clause has behavioral semantic identity, while the functional semantic category *Transmit* is represented in an achievement-free context.

[19]***pcgTransmitBehavior*** denotes a set of partial states, which when traversed, accomplish the function *PCGTransmit*. The *Provided* relation is the prerequisite for the detailed behavior to be accomplished, while the ***hasPrerequisite*** relation is the predicating condition for the goal *PCGTransmit* to be satisfied.

(e.g., the repeating coil) and it participates in the behavior ***pcgTransmitBehavior*** that fails as a result of the component failure, then the function *PCGTransmit* will also fail. This failure will propagate up the functional hierarchy (via the link to behavior and the *SubFunctionOf* relation) and result in the function of the PCG not being achieved. In addition, the PCG functions such as *PCGBuzz, PCGReceive,* and *PCGEnergize,* are instances of the semantic categories: *Buzz, Receive,* and *Energize,* respectively. These PCG functions are virtual components at a given level of abstraction and their failure effects can be simulated analogously.

FUNCTION: *PCG Transmit*

ToMake:	(rf signal present $Receiver)
instanceOf:	(*Transmit*)
hasPrerequisite:	(*PCGEnergize*)
SubFunctionOf:	(PCG)
If:	(pressed MicrophoneSwitch)
By:	(pcgTransmitBehavior)
Provided:	<Predicate: Energized State Achieved ...>

Figure 7. *PCGTransmit* Functional Representation Augmented with Pointer to Behavior

5.2 Expressing the FMEA Worksheet with the Functional Representation

Table 1 illustrates how components affect functionality for the PCG and can be used to synthesize the FMEA worksheet at a more abstract level [20]. This table conveys how components ultimately affect functionality without expressing the detailed inference mechanisms and knowledge representation relationships needed to propagate a component failure through behavior and ultimately to function. The table assumes that the PCG is a physical device, that is, it has been implemented down to the component level. In addition, the table can be viewed as an intermediate step in conveying knowledge to the user about how component failures affect functionality. For example, if the antenna fails, the function transmit and receive will fail, or if the microphone fails, the transmit function will fail.

The functional representation is not only useful in organizing the goals and purposes of a system's top-level functional blocks, but it also provides the framework for linking to the behavior of the system as detail is added to the design. That is, by exploiting the knowledge base's relationships between function and behavior, an organization is provided for linking goals to the causal means by which they are accomplished. In addition, the functional representation is utilized in formalizing the FMEA worksheet in terms of primitives and semantic categories which have direct relationships to simulation methods.

In Table 2, the traditional FMEA worksheet is retained; however, more precise semantics are imposed on the relationships among components, functions, and their failure effects. Moreover, the interpretation of the table is expressed in the same formalisms that are used in the expert system's knowledge base. The table specifies that the power supply failure, which has the function of making a device energized and the failure mode *loss of power*, results in the overall system failing to transmit a signal, receive a signal, energize a device, or buzz a device. The cause of the failure can be attributed to the state *battery is dead*. The state *battery is charged* is required to achieve the behavior that accomplishes the goal of the power supply.

Table 1. Functional FMEA Matrix

DESIGN ENGINEER ________ RELIABILITY ENGINEER ________ SYSTEMS ENGINEER ________			FUNCTION AFFECTED BY FAILURE			
REF.	FAILURE RATE λ	COMPONENT	TRANSMIT	RECEIVE	BUZZ	ENERGIZE
A1	2.01	ANTENNA	√	√		
M1	6.06	MICROPHONE	√			
P1	14.1	POWER SUPP.	√	√	√	√
C1	1.36	CLAPPER			√	
C2	1.93	COIL	√			
S1	1.76	SWITCH			√	
...	...	...	...	...	...	...

Table 2. FMEA Table Entry for the Power Supply

ITEM	FUNCTION	FAILURE MODE	CAUSE	EFFECT ON SYSTEM
Power Supply	ToMake ($Device Energized)	Loss of power	Battery is in state dead	¬Transmit($Signal) ∨ ¬Receive($Signal) ∨ ¬Energize($Device) ∨ ¬Buzz($Device)

6 Conclusion

Organizing an expert system's knowledge base around a function-centered ontology provides the capability for incorporating computer-aided FMEA early on in the conceptual design of an engineering system and serves as the basis for generating a high-level simulation and explanation of a system's failure modes and effects. In particular, XFMEA's *Simulation Expert* adopts a knowledge organization that provides the framework for a total simulation package integrating functional, as well as behavioral and numerical methods.

This paper proposes the NCIF principle with the motivation that a function-laden knowledge base must specify general, functional-semantic categories that are stated in an achievement-free manner. Therefore, a causal reasoning system (i.e., the means by which the detailed behaviors, which make up a function, are realized from structure) can be viewed as a subcase of the functional representation only when sufficient detail has been specified by an engineer or analyst to impose a behavioral semantic upon a functional block in a design.

The acquisition of a set of functional primitives and semantic categories facilitates knowledge reuse and sharing of reasoning and representation methods. In addition, such a formalism enhances the knowledge acquisition process and improves XFMEA's operational phase (i.e., reliability, maintainability, and extensibility) by imposing a structure on the FMEA results that is consistent with the knowledge base.

REFERENCES

[1] MIL-STD-1629A. (1980) *Procedures for Performing a Failure Mode Effects Criticality Analysis*, Springfield, VA, NTIS.

[2] Russomanno, D.J., Bonnell, R.D., and Bowles, J.B. (1994) "Viewing Computer-Aided Failure Modes and Effects Analysis from an Artificial Intelligence Perspective," *Integrated Computer-Aided Engineering*, John Wiley & Sons, 1(3): 209-228.

[3] Sembugamoorthy, V. and Chandrasekaran, B. (1986) "Functional Representation of Devices and Compilation of Diagnostic Problem Solving Systems," in J. Kolodner and C. Reisbeck (eds.), *Experience, Memory, and Reasoning*, Lawrence Erlbaum Associates, pp. 47-73.

[4] Keuneke, A.M. (1991) "Device Representation: The Significance of Functional Knowledge," *IEEE Expert*, April, pp. 22-25.

[5] Babin, B. and Loganantharaj, R. (1991) "Representing Functional Knowledge," *Proceedings of the Fourth International Conference on Industrial and Engineering Applications of Artificial Intelligence and Expert Systems*, Kauai, HI.

[6] Bonnet, J. (1991) "Functional Representations: A Support for Enriched Reasoning Capabilities," *Knowledge Systems Laboratory Report No. KSL 91-58*, Stanford University, Stanford, California.

[7] Kurland, L.C. and Tenney, Y.J. (1988) "Issues in Developing an Intelligent Tutor for a Real-World Domain: Training in Radar Mechanics," in J. Psotka, L.D. Massey, and S. A. Mutter (eds.), *Intelligent Tutoring Systems Lessons Learned*, Lawrence Erlbaum Associates.

[8] Freeman, P. and Newell, A. (1971) "A Model for Functional Reasoning in Design," *Second International Joint Conference on Artificial Intelligence*, September, pp. 621-640.

[9] Chandrasekaran, B., et al. (1986) "Functional Representations as a Basis for Generating Explanations," *Proc. IEEE International Conf. Systems, Man, and Cybernetics*, pp. 726-731.

[10] Keuneke, A.M. (1989) *Machine Understanding of Devices: Causal Explanation of Diagnostic Conclusions*, doctoral dissertation, Dept. of Computer and Information Science, Ohio State University, Columbus, Ohio.

[11] Franke, W. (1991) "Deriving and Using Descriptions of Purpose," *IEEE Expert*, April, pp. 41-47.

[12] Gero, J.S. (1990) "Design Prototypes: A Knowledge Representation Schema for Design," *AI Magazine*, Winter, pp. 27-36.

[13] de Kleer, J. (1985) "How Circuits Work," in D. G. Bobrow (ed.), *Qualitative Reasoning about Physical Systems*, MIT Press, pp. 205-280.

[14] Bonnet, C. (1992) "Towards Formal Representation of Device Functionality," *Tech. Report 92-54*, Knowledge Systems Laboratory, Stanford University.

[15] Katai, O., Kawakami, H., Sawaragi, T., and Iwai, S. (1991) "A Knowledge Acquisition System for Conceptual Design Based on Functional and Rational Explanations of Designed Objects," in J.S. Gero (ed.), *Artificial Intelligence in Engineering Design*, Springer-Verlag, pp. 281-300.

[16] Lenat, D.B. and Guha, R.V. (1990) *Building Large Knowledge-Based Systems Representation and Inference in the CYC Project*, Addision-Wesley.

[17] Fink, P.K. and Lusth, J.L. (1987) "Expert Systems and Diagnostic Expertise in the Mechanical and Electrical Domains," *IEEE Transactions on Systems, Man, and Cybernetics*, vol. SMC-17, no. 3, May/June, pp. 340-349.

[18] DiStefano, J., Stubberud, A.R., and Williams, I.J. (1967) "Theory and Problems of Feedback Control Systems," *Schaum's Outline Series*, McGraw-Hill.

[19] Iris, A., Litowitz, B.E., and Evens, M.W. (1988) "Problems of the part-whole relation," in M. W. Evens (ed.), *Relational Models of the Lexicon: Representing Knowledge in Semantic Networks*, Cambridge University Press, pp. 261-288.

[20] Sexton, R.D. (1991) "An Alternative Method for Preparing FMECAs," *Proc. Ann. Reliability and Maintainability Symp*., IEEE Press, pp. 222-225.

Methodology for Expandable Expert System Development

Wei Dai
Artificial Intelligence Systems
Telecom Australia Research Laboratories
770 Blackburn Road, Clayton, Victoria 3168, Australia
Email : w.dai@trl.oz.au

Abstract. This paper presents a system for expandable expert system development (SEED). SEED has an open architecture that supports an arbitrary number of application platforms for various domain problems. Flexibility and applicability have been ensured by vertical and horizontal expansions of environment capabilities. Each application platform of SEED consists of its knowledge and control. Selection of an application platform at a specific level is supported by adaptive abstraction techniques that manipulate the knowledge and control components throughout the system. An important objective of this research is to identify a methodology to effectively use available expert system resources at various application platforms. There are several application products generated from such an approach including an expert system for version control, an intelligent diagnostic tool for telecommunications, and cooperative expert systems for multimedia support.

Keywords. Expert System Specifications, Adaptive Abstraction, Software Architecture, Application Platform.

1 Introduction

The study of the potential contributions of an expert system (ES) development environment towards effective expert systems production needs a conceptual framework within which to analyze their scope and feasibility. SEED, an expandable system for ES development, has been designed to uncover opportunities for incremental progress in ES techniques used at various levels of problem-solving platforms. SEED employs available resources such as ES primitive facilities from a conventional software environment [1] and ES specification concepts [2,3] as the fundamental support for the environment. It represents an experimental activity in which various ES components/capabilities have been generated and used in terms of conventional software context. Environment capabilities can be expanded horizontally and vertically to support flexibility and applicability.

2 Background

Recent research [2] has viewed ES development in terms of problem and solution specifications. Problem specifications deal with the goal structures required to solve a problem. Solution specifications describe various activities, levels of interactions etc., involved in achieving a conclusion. These concepts have further been addressed in [3] as model specifications and process specifications to describe the static and dynamic aspects of expert system development processes. Model specification is formed by a domain model and a state model that respectively specify domain facts and problem-solving states. Process specification has been described by functional and behavioural specifications that characterize state transformations and interactions between problem-solving levels. The methodology used in building SEED has been influenced by these distinctions in order to integrate domain, states, functional and solution specifications into a common ES development framework.

E. A. Yfantis (ed.), Intelligent Systems, 681–689.

3 Conceptual Framework

SEED aims to provide an open and expandable architecture for ES development. The fundamental support of SEED comes from its primitive layer which contains the previously implemented ES primitives [DW90]. The primitives represent various low-level activities for ES development. They have been classified into several categories: production rule base (PRB), inference control (ICS), knowledge data store (KDS), domain dependent (DDP) information module. The primitive layer plays an important role in further expansion of the system capabilities. Concepts and principles regarding how the expansion may occur therefore need to be identified. Based on the existing ES specification concepts, key issues are raised to guide the formation of ES skeletons residing at a level higher than the primitives. These issues are described in terms of domain, state, behavioural and functional specifications involved in ES development. Domain and state specifications deal with the static aspects of the ES development process. Behavioural and functional specifications determine the reasoning behaviours and problem-solving capabilities in an expert system. Domain specific expert systems are built upon specific inference strategies and specific knowledge sources and knowledge data stores. Inference strategies are described through behavioural and functional specifications, knowledge sources and knowledge data stores through domain and state specifications. We shall distinguish four types of specifications :

Domain Specification

Domain information refers to each specific domain area with which the system problem solver is confronted. It contains the expert knowledge, e.g. rules in solving various domain problems as well as descriptions of domain entities. "A specification is intended to describe all the required properties of the system while leaving all other properties unconstrained..." [4]. Domain specification describes domain knowledge in terms of primitive domain entities (e.g. concepts)and their relationships. It focuses on how domain entities and their patterns are fitted within some representation templates and how these templates can be used.

State Specification

State information aims to facilitate level interactions and provide various data views for users or problems-solvers working at different levels of the environment. The hierarchical organization of state information will help define levels of interaction and enable problem-solvers to access the relevant information efficiently. State information includes the existing domain data, current problem-solving states, task information (goals to be achieved, etc) at various levels of interaction.State specification provides effective operations in describing state information at each (horizontal) level and the overall structural (vertical) descriptions.

Behavioural Specification

"A specification is a plan for a solution or a generator of behaviours which satisfies the predicates..." [5]. Behaviour specification describes and controls the problem-solving activities of the inference program. We have implemented a goal-directed inference mechanism where behavioural specification is responsible in specifying control structures and generating goals and their associated information. The mechanism is influenced by the subgoaling techniques used in Soar [6] and intelligent control from blackboard control architecture [7].

Functional Specification

Functional specifications aim to realize various identified tasks by dynamic invocation of the primitive ES routines. There is no fixed order in which these primitives are applied. However, the sequence of invocation of these primitives normally complies with the logical intentions. Input/output parameters of the primitive routines can also be used as a clue for the invocation sequence of some primitives, e.g. values of input parameters should be known before the primitives are used.

The above specification concepts have dominated the development of system skeletons from which higher levels of environment capabilities such as ES tools or other application routines, can be efficiently constructed. Figure I presents such a conceptual framework.

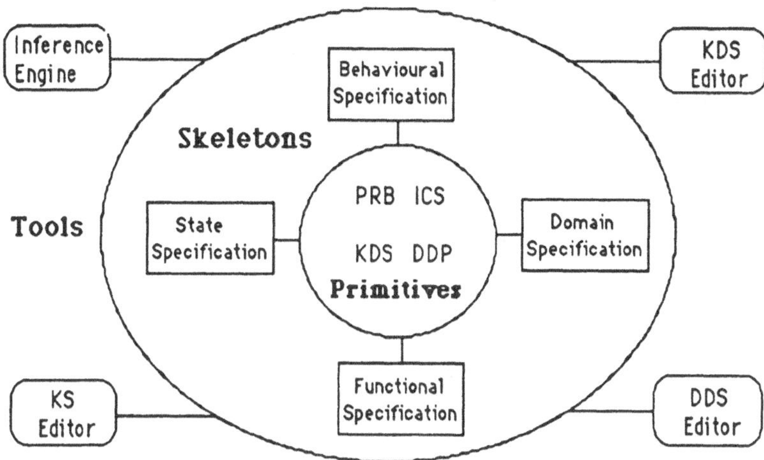

Fig. I : Conceptual Framework for Expert Systems Development

4 The System

SEED is characterized with its capabilities for expansion and for abilities of binding ES components by adaptive abstraction. These features will be discussed in the following sections.

4.1 Expandability

SEED is a system for expandable ES development environment where new ES capabilities can be developed in two directions, i.e. vertical expansion and horizontal expansion, to enhance environment flexibility and applicability. SEED expandability comes from the users working at various levels of problem-solving platforms. New application platforms can be built to enhance its applicability (vertical expansion). Generating new functionalities can be based on the existing facilities at the same platform (horizontal expansion).

4.1.1 Flexibility Flexibility addresses the horizontal expansion of the environment by which a new representation mechanism and relevant operations can be developed based on the existing resources at that level. It focuses on the development of the additional-level capabilities based on the resources of current level, e.g. knowledge representation capabilities and operations required in performing some specific inference tasks. In the current SEED environment, a production rule representation has been implemented which is supported by the PRB module. In order to experiment how horizontal expansion can be achieved in such an environment, we use the existing PRB primitives given in Table I (a) to develop a new set of frame-based (FRB) primitives that are described in Table I (b), to expand knowledge representation capabilities of the primitive level. Many of the generated FRB primitives from Table I (b) are implemented using the existing mechanisms from the counterpart PRB primitives. However in practice, the horizontal expansion may not be on a one to one transformation basis (e.g. not all the available resources are used).

```
prb.newk(in rbname, out rbd)            - - create a new rulebase with descriptor rbd
prb.delk (in rbname)                    - - delete a rulebase
prb.openk(in rbname, in mode, out rbd)  - - open a rulebase
prb.closk (in rbd)                      - - close a rulebase
prb.newr (in rbd, in rulename, out rd)  - - create a new rule with descriptor rd
prb.delr ( in rbd, in rulename)         - - delete a rule
prb.lsize(in rbd, in rd, out sizelhs)   - - get size (number of premises) of lhs
prb.asize(in rbd, in rd, out sizeaction) - - get size (number of actions) of the rule
prb.objod(in rbd, in object, out rod)   - - obtain object descriptor
prb.nrule(in rbd, in rod, in type, out nrule)           - - number of relevant rules
prb.openr (in rbd, in rulename, in mode, out rd)        - - open a rule
prb.putl (in rbd,in rd,in premise, in obj,in pattern)   - - build lhs premise
prb.putr (in rbd,in rd, in obj,in conclusion)           - - build rhs
prb.puta (in rbd,in rd,in ith, in action)               - - build ith action
prb.lhs(in rbd,in rd,in premise,out obj,out pattern)    - - get lhs information
prb.rhs(in rbd,in rd, out obj,out conclusion)           - - get rhs information
prb.action (in rbd,in rd,in ith, out action)            - - get ith action
prb.rule(in rbd, in rod, in type, in ith, out rd)       - - obtain the ith rule
```

Table I (a) : Current PRB Primitives

```
frb.newk(in fbname, out fbd)            - - create a new framebase with descriptor fbd
frb.delk(in fbname)                     - - delete a framebase
frb.openk(in kbname, in mode, out fbd)  - - open an framebase
frb.closk(in fbd)                       - - close a framebase
frb.newf(in fbd, in fname, out fd)      - - create a frame node with descriptor fd
frb.delf( in fd, in framename)          - - delete a frame node
frb.ssize(in fbd, in fd, out nslot)     - -  get the number of slots
frb.objod(in fbd, in object, out fod)   - - obtain frame node descriptor
frb.openf(in fbd, in fname, in mode, out fd)           - - open a frame-node
frb.putslot(in fbd,in fd, in islot, in obj, in stype)  - - build a slot
frb.putact(in fbd,in fd, in islot, in action)          - - build associated action
frb.slot(in fbd, in fd, in ith, out fod, out stype)    - - get ith slot
frb.getact(in fbd, in fd, in ith, out action)          - - get action from ith slot
frb.sizeframe(in fbd, in fod, in type, out nframe)     - - obtain number of relevant nodes
frb.frame(in fbd, in fod, in type, in ith, out fd)     - - get ith frame node
```

Table I (b) : FRB Primitives through Horizontal Expansion at Primitives Level

4.1.2 Applicability Applicability refers to the vertical expansion of the environment by which higher levels of software modules such as skeletons, tools can be built from the lower-level. SEED supports an arbitrary number of application levels. Each level is characterized by its knowledge and control. These levels are generated by vertical expansions of the existing application platforms. There is no explicit link built among these levels. It is up to the user to choose which level he or she is willing to work from, although it is possible to build a new level of applications (in this case, user takes the role of a system developer) if desired. An example of designing a goal directed inference program at a new application level is given in

Table II. Users are expected to put together the relevant skeletons and fill in the details into these skeletons. Inference behaviours of the program are guided by a list of goals identified during inference processes. Major goals have been classified as the followings : task identification, plan generation, and plan execution. SKE routines are implemented under the functional and behavioural specifications using lower-level facilities, e.g. ES primitives. Generation of this new application level is described by its own functional and behavioural specifications. The functional specification describes the set of available facilities to be used in the work, e.g. SKE routines, while the behavioural specification combines these facilities into a specific form (see Table II).

```
procedure Goal_Directed_Inference (goal)          - - new inference facility
begin
    moregoal : boolean; goal : string;            - - types declaration
    SKE_generate_plans(goal);                     - - generate plans for the goal
    SKE_execute_plans(goal,moregoal);             - - execute plans
    if moregoal then                              - - task identification
    Goal_Directed_Inference (goal);               - - recursion
end
```

Table II : A New Application Level

4.2 Glueing ES components together

We use the technique of Adaptive Abstraction to glue the ES components from SEED together (based on the assumption that an application level has been pre-selected by the user) to form various ES application programs. Adaptive Abstraction is a way of organizing knowledge and control strategy that facilitates level-adaptation in a multi-layered expert system development environment. This will enable the system developer (user) to switch to a different application level when the current abstraction level is judged to be either too difficult to work at, e.g. a special skill is required; or too rigid to change. In the former case, the user will go to the higher level for the sake of simplicity of application programs. In the latter case, the user will come to the lower level for greater control. Switching among these levels is not always caused by the above factors. For example, the processing result (success, failure, etc.) at a specific application level may also influence the adoption of a new application level of the environment. The roles of the environment facilities, e.g. primitives, skeletons, tools, have been described by the ES specifications. The use of adaptive abstraction on these specification modules will lead to different types of expert systems. A general picture has been given in Figure II.

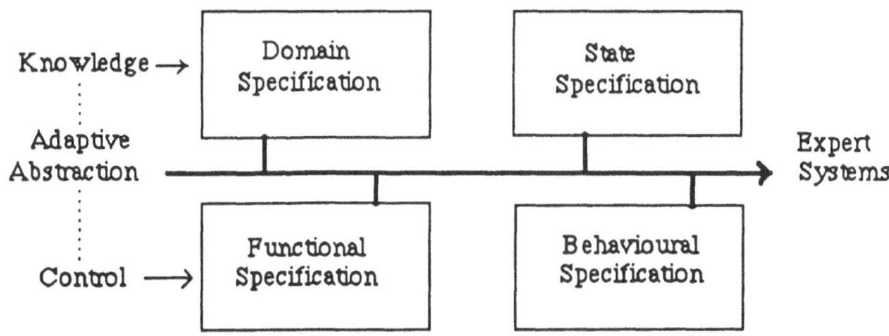

Fig. II : Adaptive Abstraction

Knowledge and control (see Fig. II) integration is addressed through Adaptive Abstraction whose semantics is defined according to each specific application, e.g. a solution architecture. In practice, only the facilitiesof each module which have the shared objectives with the others (facilities from other modules) are linked with the major trunk (Adaptive Abstraction). Identification of the required facilities to perform a task for Adaptive Abstraction is helped with Functional Specification. It is up to Behavioural Specification to put relevant facilities into various forms. Adaptive Abstraction supports glueing of the system facilities by allowing them to move to the application platform rather than having the call sent to those facilities. This reduces communication overhead, and allows flexible component reconfigurability.

At each application level, we emphasize the need to isolate the functionalities of the relevant programs from the physical nature of the objects by defining various abstract data types. Abstract data types attempt to describe the external properties of the desired software objects rather than their data representations. They are also used for interface definition between levels of interaction to support an open architecture organization. Software components generated at each application level from specification skeletons are modular and independent to support our emphasis on software reconfigurability. Expert systems generated from this approach covered a comprehensive range of architectures including cooperative expert systems.

5 Implementation

SEED has been implemented through the adoption of software engineering techniques as solutions. We view each software component in SEED as a data object. A data object is characterized by a Name, Properties and Values. Each object can have any number of properties each of which is defined by a Type that restricts the range of its values. A Type can be either a primitive type such as integer, real, boolean, string etc., or a structured type which is an arbitrary collection of other data objects (called the components of the structure). Each environmental object is associated with a relevant set of operations. It is quite similar to the operand-operator relationship in a programming language. The objects and their operations are used to implement domain, state, behavioural and functional specifications at each level of the environment. These objects are used for different purposes, e.g. control, representation, and construction of new objects. For instance, typical objects in the environment include *plans* used for inference process, *rules* for building KS, goals and domain entities (also called subjects) for building KDS. KS and KDS are themselves SEED data objects. Any number of KS or KDS can be stored in the SEED data store, which can be open simultaneously at any time within the problem-solving process of expert systems. Information can thus be stored at a very fine granularity stressing the modularity and re-use of knowledge.

6 Applications

Several applications have been implemented using SEED resources at various levels. These applications are briefly described in the following sections followed by their evaluation.

6.1 An expert system for version control

The task is to maintain an up-to-date version of a program within a complex software system (traditionally it is done by the Unix Make tool) by performing only the minimum amount of work (re-compilation and re-linking) necessary to rebuild a program after a modification has been made. Relevant ES primitives from different groups such as *prb, kds, ics* and *ddp* (see Figure I) have been invoked from a conventional language to build inference program and knowledge editor for the version control expert system. The inference program generates a list of instructions for maintaining the correct version of the file using backward-chaining strategy.

6.2 An intelligent tool for telecommunication network maintenance (INTENEMA)

INTENEMA has been designed as a help desk to assist telecommunication customer service operators to improve customer service in fault reporting situations. The aim of INTENEMA is to efficiently distribute technical expertise to customer service operators in various business divisions associated with total service delivery. Capabilities of INTENEMA include local diagnostic problem solving and communication with external software systems when the repair of a fault requires the joint attention of other business divisions. More detailed descriptions about INTENEMA can be found in [8].

6.3 Cooperative expert systems for multimedia support

The multimedia support expert systems constitute a group of cooperative expert systems to maintain a multimedia communication network. These expert systems use existing application programs for controlling individual parts of the telecommunication facilities such as the video processing sub-system. The participating expert systems are generated from a cooperative framework [9] using part of the SEED resources. Their external behaviours (ways of dealing with external systems) can be modified in a flexible way to suit various situations under such a framework. These include the description of the situation under which an expert system is engaged in a collaborative problem solving process. Communications among the participating expert systems are carried out by the low-level communication primitives responsible for message passing, synchronization, etc.

6.4 Evaluation

Here we provide some information regarding development efforts of SEED resources. It took about three months for building the ES primitives, additional one month and two months respectively for ES skeletons and tools. The development of MultiMedia (MM) and INTENEMA expert systems took about eight and nine months respectively including the time spent on the SEED software components, when SEED resources were not available for the applications. This was very similar to the total effort spent in building the same applications without using SEED resources. However, by using the available SEED resources, supposing that MM and INTENEMA were new applications, it would take about three and four months respectively for the above applications at the tools level. For some simple applications, SEED resources need not be built before hand. For example, it would take about seven months to build the Make expert system by implementing the SEED resources first (when SEED is not available). This would be a disadvantage as it normally takes about four months starting from scratch. However, if SEED is already available due to previous applications, it will take about one month working at the tools application platform.

7 Related Work

SEER [10] addressed the effective use and representation of shallow and deep knowledge required in an expert system, and the updating of the knowledge base guided by its knowledge of principle. Levels of interaction are focused on the domain models rather than on the environment itself.

ABE [11] is a new generation of software architecture which addresses the problem of combining conventional computing functions with knowledge processing capabilities to build intelligent systems. It attempts to solve the major problems of AI tools, such as information re-use, large-scale application problems etc., by the importation of existing software, including conventional and knowledge processing tools. The major research effort of ABE is to provide levels of capabilities, such as the virtual machine level for which a special operating system has been designed, the system framework level and the tools level including skeletal systems where the final applications can be developed. These levels form a hierarchical architecture that isolates application developers from lower-level details. ABE aims mainly at providing an environment built on a new operating system where system developers can assemble solutions by connecting the existing software components. There is a lack of interfaces with conventional programs where ES components can be directly used at different levels of ABE.

8 Conclusions

SEED evolved from a generic architecture INDEX [12] for integrating AI and SE techniques within a conventional software environment. The power of SEED comes from its capacity to build larger and larger systems, and support for software reconfiguration and arbitrary selection of application levels. The choice of one form of knowledge (representation and control) can have a great impact on the performance of an expert system. There is no single mechanism that is best for all problems. In many cases, sharing and re-using knowledge will involve translating from one representation to another [13]. Horizontal expansion of a SEED application level will fulfill this requirement (see example from Table I). Creation of a new application level can be achieved through vertical expansion of the system which is supported by reusing the available SEED resources located from different application levels. Software re-use for SEED can take place in three different modes which are consistent to the issues raised in [13]. The first mode is to use the approach behind SEED without directly using its software modules. This can be achieved within most of the conventional software environments. The second mode is through the inclusion of source specifications. This will require a common implementation language with SEED. There are currently two versions of the system which are written in Fortran77 and C++ respectively. The third mode is through the run-time invocation of external facilities. SEED primitive facilities, i.e. object libraries of ES primitives, can be linked with a conventional application program as the primitives are only interfaced with primitive data types, e.g. integer, real, character, which can be directly translated into most of the conventional languages.

Acknowledgments

The author would like to thank Dr Sid Wright for his contributions in building INDEX system on which this work is based. Thanks are also due to Dr Jacques Guy and Chris Rowles for their useful comments on an earlier draft of the paper. Permission of Managing Director of Research and Information Technology, Telecom Australia, to publish the above paper is hereby gratefully acknowledged.

References

[1] Dai,W. and Wright,S.L. (1990) 'Primitives for constructing rule-based expert systems', Proceedings of the 4th Australian Joint Conference on Artificial Intelligence, World Scientific, Singapore, pp 432-447.

[2] Zualkernan, I.A. and Tsai, W.T. (1988), 'Are Knowledge Representations the Answer to Requirement Analysis?', Proceedings of IEEE Int'l Conf. in Computer Languages, IEEE Computer Society Press, Los Alamitos, Calif., pp 437-443.

[3] Yen, J. and Lee, J. (1993) 'A Task-Based Methodology for Specifying Expert Systems', IEEE EXPERT, February, 1993, pp 8-15.

[4] Zave, P. (1988) 'Assessment', in Report on the Fourth International Workshop on Software Specification and Design, Software Engineering Notes, Vol 13, No. 1. ACM Press, January, 1988. pp 40-41.

[5] Arango, G. and Freeman, P. (1988) 'Application of Artificial Intelligence', in Report on the Fourth International Workshop on Software Specification and Design, Software Engineering Notes, Vol 13, No. 1. ACM Press, January, pp 32-38.

[6] Laid,J., Rosenbloom,P., Newell,A. (1986) Universal Subgoaling and Chunking, Kluwer Academic Publishers.

[7] Hayes-Roth,B. (1993) 'Intelligent Control', Artificial Intelligence, 59, Elsevier, pp 213-219.

[8] Dai,W. Beler,M. and Rowles,C.D. (1994) 'INTENEMA: An Intelligent Tool for Telecommunication Network Maintenance', Proceedings of The Seventh International Conference on Industrial & Engineering Applications of Artificial Intelligence & Expert Systems, IEA/AIE'94, Gordon and Breach Science Publishers, USA, pp 353-360.

[9] Dai,W. and Wright,S.L. (1993) 'Framework for Building Intelligent Cooperative Systems', Proceedings of the 6th Australian Joint Conference on Artificial Intelligence, World Scientific, Singapore, pp 203-208.

[10] Kim,J.J. (1993) 'Preliminary Results on a Self-Expanding Expert System Environment. Proceedings of the 2nd Pacific Rim International Conference on Artificial Intelligence, Seoul. pp 456-462.

[11] Erman,L.D., Lark,J.S. and Hayes-Roth,F. (1988) 'ABE: An environment for engineering intelligent systems', IEEE transactions on software engineering, vol. 14, no. 12, December 1988. pp1758-1770.

[12] Dai,W. and Wright,S.L. (1993) 'INDEX : An Architecture that Integrates Expert System Techniques within Conventional Software Environment', In Niku-Lari, A. (ed.), Expert Systems Applications & Artificial Intelligence. EXPERSYS-93, IITT-International Technology Transfer Series, Gournay sur Marne, France, pp 13-18.

[13] Neches,R., Fikes,R., Finin,T., Gruber,T., Patil, R., Senator, T. and Swartout, W.R. (1991) 'Enabling Technology for Knowledge Sharing.',AI Magazine, Fall, 1991, pp 36-56.

Intelligent Telecommunication Services: Adaptive And Demonstrational User Interfaces

Marc Yvon
Frédéric Lefèvre and Philippe Piernot

Laboratoire d'Intelligence Artificielle de Paris 5
UFR de Mathématique et d'Informatique
Universite René Descartes
45, Rue des Saints-Pères
75006 Paris - France

yvon, lefevre, piernot@math-info.univ-paris5.fr

Abstract. In collaboration with France Telecom (Centre National d'Etudes des Télécommunications) we have been applying adaptive and demonstrational techniques to the Djinn [1] user interface, a state of the art software application that integrates the major telecommunication services (phone, fax, answering machine, French Minitel [2], file transfer, etc.) and bundles the services together in a single application running on a PC or Macintosh platform. Adaptive and Demonstrational interfaces aim at helping the user work with a customized environment. Adaptive interfaces adapt themselves according to the user's needs; and demonstrational systems create parameterized procedures by demonstration in order to extend application capabilities and eliminate repetitive tasks ([2] and [9]). Many adaptive and demonstrational research systems have been implemented ([1], [3], [5], [6], [7], [8], and [12]), but the telecommunication world has never been the application domain of any of them. We first present Djinn and the characteristics of the telecommunication services, then we describe a telecommunication prototype with a separate adaptive module and a demonstrational module.

Keywords. Adaptive Interfaces, Demonstrational Interfaces, Telecommunication Services, Irreversible Events, Event Inhibition, Journal, Macro Commands, Scripts, Direct Manipulation.

1 Djinn

Djinn is a product that integrates the major telecommunication services by bundling them together into a single application. The user communicates with the external world through

[1] Trademark of France Telecom.

[2] The Minitel allows the connection from the regular phone line to many videotext services. Most connections start by dialing 3615, 3616 or 3617 ¡service's name¿.

E. A. Yfantis (ed.), Intelligent Systems, 691–698.

a modem to place a phone call, to send a fax to one or more recipients, to record incoming voice messages, to transfer files to another computer, and to access French Minitel online services. The user maintains his work environment by updating (writing, modifying, deleting) information associated with his correspondents (e.g. home/work phone number, an address, etc.) in a repertoire, metaphor of a address book. This repertoire is a relational database whose representation is alphabetically indexed. The user keeps track of all performed operations in two distinct data structures, the outgoing journal and the incoming journal. The first one summarizes all operations performed by the user (e.g. sent a fax to Mrs. Smith at 8:30am on 11-03-93); the second one summarizes all operations originating from the outside (e.g. Mrs. Smith called at 10:00am on 11-03-93). Djinn plays an important role in a desktop office automation environment [13]. Let us also just mention that the user can access an electronic mail with the Minitel. The purpose of this study is to enhance Djinn into an "intelligent" communications agent through adaptation and end-user programming by demonstration.

2 Characteristics of Telecommunication Services

A software application that integrates telecommunication services will likely have daily use, because the telephone and the fax machine are invoked in an office almost as often as the pencil. Furthermore, people or companies dealing with many external communications definitely will have their work habits defined over a certain period of time. For example, a user may send a fax to the same set of people at the end of each month for billings.

2.1 Time

An important characteristic of the application domain is the time consideration, a strong parameter of the working context. For instance, in a word processor, the time when actions are performed is not relevant, whereas the time that a fax is received may be quite important. The user frequently manipulates services and performs tasks according to a given time or date; he can therefore work in a time dimension not restricted to the present time. Actions can be planned, allowing the user to pilot his application.

2.2 Irreversible Events

Once a service has been invoked by the user, it is impossible to undo because it takes place in the external world. The physical action of sending a fax cannot be undone:Êit is an irreversible action. Actions taking place in this environment are very different from the ones we usually find in adaptive and demonstrational systems. A text editor easily tolerates mistakes by undoing errors; whereas irreversible actions must be manipulated carefully.

3 Telecommunication Prototype

3.1 New Journal

As we mentioned, the time handling in a telecommunication software, in the past as well as now and in the future, is definitely a parameter to consider in a system where actions can be programmed and anticipated (e.g. telecommunications expected within a time interval). The user could organize his future work by defining and planning a set of actions in a time scheduler, hence introducing an order in task invocation. We have merged the outgoing and incoming journals into a single journal not restricted to the past events, i.e. it also contains the future events. The user can include in his journal an event that has not yet been sent or received; once it has been invoked it becomes an event of the past. We have defined several views of the journal with some filters in order to visualize subsets of it, such as all events past and/or future, related to phone calls, to faxes that will be sent, etc. The user can also define his own filters to perform specific operations. Many criteria could be used for building a filter, such as a correspondent identification, a date, a time interval, etc.

The user provides to the system through the journal, an image of his working methodology,with the past events stored and the future events inserted. He demonstrates by giving hints about how he behaves, guiding the system in its inferences. The journal is therefore a means of expression for the user and a source of observation for the system. The user can inhibit an event from the journal, and as a result the system will not take it into account in its inferring process. This inhibition feature is important to treat exceptions, an event that the user does not consider as relevant for the system. The event is then marked in the journal with a special feedback. The inferences made by the system are added to the journal with a different notation than the other events. Figure 1. illustrates an instance of the journal:

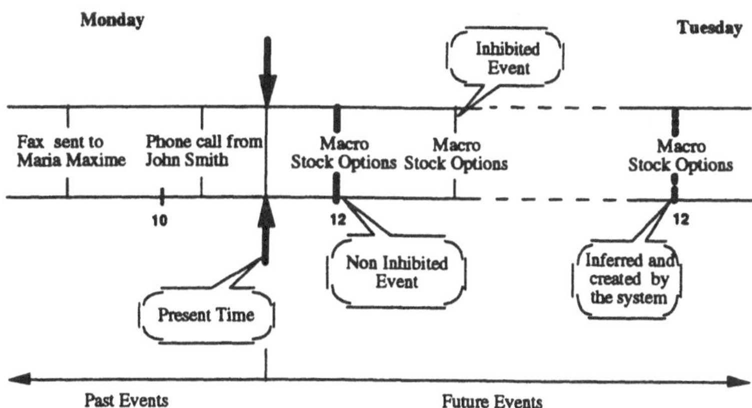

Figure 1. The Journal: a Time-Based programming Environment.

The journal is not restricted to the past events, it is more than a trace, it contains actions scheduled in the future. In our example Figure 1., we have a journal representation for Monday and Tuesday.

These actions are sometimes inferred by the system when regularities are found . The macro command "Stock Options" inserted in the journal for Tuesday at 12 has been inferred because of the one inserted by the user for Monday at the same time.
The user can guide the system in its inferences, he can also inhibit an action for it not to be taking into account by the system. The macro command "Stock Options" inserted in the journal for Monday around 13 was explicitly inhibited by the user. Exceptions are taking care of by inhibiting events.

3.2 Adaptive Module

Our adaptive module optimizes the application's use in modifying the user interface to better fit the user's needs. The main goal of adaptive techniques is to adapt the user's working environment according to his requirements and his working habits (in one working session and/or several sessions). It is intended to provide a customized user interface. The module is always active as a background task looking at the user's actions and the information gathering is a continuous recording process. Its inferences and its adaptations are made from the collected information. Unlike the demonstrational module the user does not explictly activate nor deactivate the adaptive behaviors in the middle of a work session, but he could turn the behaviors on or off at the application startup. Some side effects of this continuous recording process have to be taken into account though, such as dealing with a large amount of data being collected as well as the diversity of data. As a matter of fact, the user is not necessarily paying much attention to his actions and how he manipulates the data while interacting with the user interface. The user's working methodology will therefore not be clearily presented to the system, making the adaptations not so obvious to detect and introducing probably a lot of noisy information.

The user's intention is anticipated by the system, performing personalized operations in the user interface, such as saving the window position upon quitting so that the user will see the same layout next time. Unlike the demonstrational module it does not need to infer from a generalized trace to get significant information; its adaptation could be made from a single action. We have seen that the adaptive module permanently proposes adaptations with respect to previous user's manipulations. The user's implicit validation to these propositions also changes the frequency with which the adaptive system interacts with the user (e.g. the more the user accepts adaptations, the more the system proposes them). One way for the user to express his intention to the system is to perform tasks, keeping in mind that the system is always observing him.

The module uses a set of specific metrics to filter the collected information. These metrics are based on different kinds of observations, such as the use and manipulation frequencies, the response time (keyboard or mouse), the absolute objects' coordinates on the screen, the relative objects' coordinates to other objects, the time (current and periods), and the error and help message rates. Significant adaptations found by the module for a given user are saved in an external resource file called the user model. This user-specific adaptation file is loaded at the beginning of a work session and updated whenever a new adaptation is validated for this particular user. The user could consult his adaptation model, by editing the user model associated with him, and eventually modify it. On the one hand, such an approach would allow the explicit communication of all kinds of information and adaptations to the user, and to share common adaptations by several users; on the other hand the user would have to understand the knowledge representation language.

3.3 Demonstrational Module

Our demonstrational module enhances the application's performances in providing macro commands on top of the regular application's functionalities. Its principal goals are to reduce the time spent by the user in performing a task (e.g. avoiding repetitive tasks) and to extend the existing application by supplying a set of new tools (e.g. accomplishing new tasks). The module is not active most of the time, as long as the user has not explicitly requested it. We differentiate two modes: a recording mode and an executing mode. In the first one, the system builds macro commands by recording user's actions in a trace using a begin and an end recording session. In the second one, the system instantiates a generalized macro in order to execute it (this mechanism is not present in the adaptive module). Those macro commands are categorized by their potential to be applied and executed in contexts slightly different from the one in which they were created.

This module uses the new journal to schedule macro command invocations (cf Chapter 3.1).

Unlike the adaptive module, the information gathering is not a continuous recording process. The system records a subset of the user's actions without taking into account all the events generated in the interface. The user has to invoke the demonstrational module's services before any kind of inferences can be realized. The user demonstrates while being in the recording mode what he wants to do. Additionally, he will likely have to guide the system in its inference in order to get the most suitable generalization. This last point is one of the most critical in Programming-By-Demonstration systems, especially how to keep track of the execution context in which the examples were shown and how to present the most appropriate feedback to the user.

3.4 Comparisons between these two modules

We distinguish between our adaptive module and our demonstrational module to better understand their integration in the prototype. In this area of user interface research, no clear distinction has so far been made. Figure 2. summarizes the main differences:

	Adaptive	Demonstrational
Purpose	Customizing the user's working environment	Creating new tasks, macro commands, in the user interface
User control	Adaptation active or deactive by user preferences	Demonstration alternatively active and deactive within the same session, with the recording keys
Activity	Continuous	Partial
Information processed	All actions in the user interface	The user's actions specified in the macro commands and manipulated in the journal
User model	Based on information extracted (metrics) from the user's action trace	
User validation	Depends on the adaptation's consequences	Depends on new demonstrations by examples
Additional user	Validation operations	Recording (Begin...End) and demonstrations of intentions

Figure 2. Comparisons between Adaptive and Demonstrational.

We have been developing a prototype in SmallTalk that includes all of the ideas and the features presented in this abstract. First, we have designed a new interface supporting the objects' direct manipulation in the user interface (telecommunication services, repertoire, new journal, etc.) and have defined a list of events for each object. Secondly we have been integrating the AIDE architecture [11] to our prototype (defined as an AIDE-based application) for handling the events (AideEvent Manager). The AIDE architecture is in charge of storing, merging and generalizing events, and also detects loops. The dialogue between the AIDE-based application and the AideEvent Manager is different in the teaching mode and the execution mode. The teaching mode tries to generalize an event; whereas the execution mode gets a generalized event from the macro definition.

4 Conclusions

We have emphasized in this study that adaptive and demonstrational techniques require a specific user interface environment in order for an application to support them. The bottom line for an 'intelligent' system is to be able to look at the user's activity and the interface's status. Without these observations, no inferences can be made. The telecommunication domain has appeared to be an excellent one for applying these techniques. The user could manipulate services in his interface, and create macro commands and organize them in a time-based programming environment. Our system infers generalizations from the user's working methodology and proposes adaptations to the user's interface. Special care in dealing with irreversible events has been emphasized, especially for undoing actions.

After the programming phase of our research project, a second phase will evaluate the user's behaviors in front of adaptive and demonstrational techniques applied to telecommunication services. The user's reactions study, for both beginners and experts, will help us measure the system and will play a key role in improving the user system interaction (e.g. defining appropriate feedback). Two other research teams, consisting of psychologists and ergonomists, are also involved in this phase of the project.

5 Acknowledgments

Major support for this work described in this abstract was provided by the Centre National de la Recherche Scientifique and the Centre National d'Etudes des Tlcomunications (project CNET/CNRS/COGNISCIENCE with the LIAP5 laboratory). For help with this abstract, we would like to thank our thesis advisor Professor Norbert Cot.

6 References

[1] Cypher (1991) EAGER: programming repetitive tasks by example, Proceedings of CHI '91, ACM, New Orleans, May 1991, pp.33 - 40.

[2] Cypher (1993), Watch What I Do: Programming by Demonstration, MIT Press, Cambridge, 1993.

[3] Finzer, Finzer W. and Gould L. (1984), Programming by rehearsal, Report No. SCL-84-1, Xerox PARC, May 1984.

[4] Gaines B. and Monk A. (1990), Adaptive User Interfaces, Academic Press, San Diego, 1990.

[5] Halbert D. (1984), Programming by Example, Ph.D. Thesis, Department of Electrical Engineering and Computer Science, University of California Berkeley, 1984. Also published as: Halbert D., Programming by example, Technical Report OSD- T8402, Office Systems Division, Xerox PARC, December 1984.

[6] Lieberman H. (1980), "Tinker: Example-Based Programming for Artificial Intelligence," Proceeding of seventh International Joint Conference on Artificial Intelligence, IJCAI, Vancouver, August 1981, p. 1060.

[7] Maulsby D., Kittlitz K. and Witten I. (1989), Metamouse: specifying graphical procedures by example, Proceeding of SIGGRAPH 89, Vol. 23, No. 3, ACM, Boston, August 1989, pp. 127 - 136.

[8] Myers B. (1988), Creating User Interfaces by Demonstration, Academic Press, San Diego, 1988.

[9] Myers B. (1992), Demonstrational Interfaces: A Step Beyond Direct Manipulation, IEEE Computer, Vol. 25, No. 8, IEEE, August 1992, pp. 61 - 73.

[10] Piernot P. and Yvon M. (1991), "Interfaces Graphiques Intelligentes: Le Cas des Interfaces Démonstrationnelles," Internal Document, Ecole des Hautes Etudes en Informatique of Paris, December 1991.

[11] Piernot P. and Yvon M. (1993), The AIDE Project: An Application-Independent Demonstrational Environment, Watch What I Do, Cypher A. ed, MIT Press,Cambridge, 1993, pp 382 - 401.

[12] Smith D. (1975), Pygmalion: A Creative Programming Environment, Report No. STAN-CS-75-499, Department of Computer Science, Stanford University, June 1975. Also published as: Smith D., Pygmalion: A Computer Program to Model and Stimulate Creative Thought, Birkhauser, Basel, 1977.

[13] Yvon M. and Lefevre F. (1994), "Interfaces Adaptatives et Démonstrationnelles au Service des Télécommunications: Etude de Djinn," Technical Report No. LIAP5-1-94, UFR de Mathématiques et d'Informatique, Université Paris 5, January 1994.

Multi-Layered Hybrid Architecture to Solve Complex Tasks of an Autonomous Mobile Robot

François Tièche, Claudio Facchinetti, Heinz Hügli
Institute of Microtechnology
University of Neuchâtel
Rue de Tivoli 28, CH-2003 Neuchâtel, Switzerland
(tieche@imt.unine.ch)

Abstract. In this paper, we present the implementation of an autonomous mobile robot controller developed according to the principle of a multi-layered hybrid architecture. This architecture is composed of four layers: sensori-motor, behavioural, sequencing, and strategic. The paper describes its general structure and the function of its main elements. It further analyses the development of an example task presenting the advantages of the hybrid architecture.

1 Introduction

The ability of a mobile robot to achieve reliably tasks in a real environment depends essentially on the architecture of its controller. We use a multi-layered hybrid architecture that combines the advantages of both the behavioural and the centralised architectures. This architecture distributes distinct competence levels on several layers: the top layer is responsible for symbolic planning, the intermediate layers are behavioural-based, and, finally, the bottom layer controls the robot. Our architecture extends the behavioural approach discussed in [1] to more complex tasks, by offering the possibility to define and execute the goals as sequences of simple behaviours. In order to evaluate our architecture, we choose a task where the robot has to tidy up chairs in a room, by pushing and aligning them, using sequences of simple vision-based behaviours. The architecture is realised in the form of a development environment called MANO (Mobile Autonomous robot system NOmad200). Its main features are (i) the possibility to control either a real robot or a simulated one, (ii) a set of concurrent processes implementing the various levels of the architecture, and (iii) a blackboard, handling information exchange between elements of the architecture.

2 Related work

Traditional architectures — such as centralised or hierarchical — split up the robot control in three modules responsible for: sensing, planning and acting. These architectures are

E. A. Yfantis (ed.), Intelligent Systems, 699–704.

convenient for high-level planning tasks. The sensing module builds a high-level representation from sensed data. Using this information, the planning module generates the robot actions executed in the acting module. Such architectures are not time-efficient and have difficulties to take into account some of the uncertainties issued from the real world. The subsumption architecture [1] separates the robot control in several layers of modules. Each module is responsible for the complete processing from sensing to control and interacts directly with the environment. These modules are organised hierarchically: the upper ones activate or deactivate the modules of the underlying levels. For real applications, it is often difficult to partition a global task in a set of elementary modules because (i) the decision element is distributed over several modules, and (ii) there is no model of the robot's world.

Both architectures share advantages and disadvantages depending on complexity of the task they have to realise. The hybrid architecture [2] [5] [6] integrates and organises them to take advantage of each, resulting in a multi-layered hierarchical architecture. The lowest layers are organised according to the behavioural architecture. The topmost layer is a module responsible for the high-level planning based on a map representation of the world. At each level the sensor interpretation is used both for control at the same level and to feed the upper level. The layers are structured according to response time (quick reaction at lower level versus slower reaction at higher level), data abstraction (signal versus symbols), and locality of spatial information (local measures versus global map).

3 Architecture

Our architecture is composed of four layers operating asynchronously with respect to each other: sensori-motor, behavioural, sequencing and strategic. The lowest one called sensori-motor is based on control theory and on signal processing. It is responsible for the elementary movements of the robot and processes data acquired by the sensors. The second layer is behavioural-based and controls the robot with respect to the environmental characteristics. Next, the sequencing layer implements tasks described as a sequence of behaviours. It acts by selecting the elementary behaviours that form the tasks. The strategic level has both global and symbolic knowledge of the world. It is used to define the long term strategy to reach a given goal.

The **sensori-motor layer** is characterised by fast interactions and is usually hard-wired. The movements of the robot are controlled by servo loops, both for velocities and position. This layer is also used for the processing of sensor data which — at this level — are essentially local measures of the world. The sensori-motor layer interacts with the environment sending command signals to the actuators and receiving signal from the sensors. Measures of the world are send to the behavioural level.

The **behavioural layer** (Figure 1) is made of a set of concurrent behaviours, reacting with the environment. We call the closed loop formed by the world, the sensor, the behaviour module, and the actuator *external behaviour*. By analogy, we call the module responsible for processing of measures of the world, used to update the internal database, *internal behaviour*. The set of external behaviours defines the capability of the robot to interact with its environment. Each behaviour extracts specific world characteristics from the measures provided by the sensori-motor layer: we call them *sign patterns*. Each time an expected sign pattern appears, the behaviour is stimulated. It then controls the robot so that the sign pattern remains present. Each behaviour informs the sequencing layer on its internal state using so-called *stimuli* signals. We distinguish two kind of behaviours: (i) simple behaviours with a two-values stimulus (not stimulated and stimulated), and (ii) goal-driven behaviours that stop when a expected configuration of sign patterns appears. Their stimuli take the values: not stimulated, stimulated, satisfied, and failed.

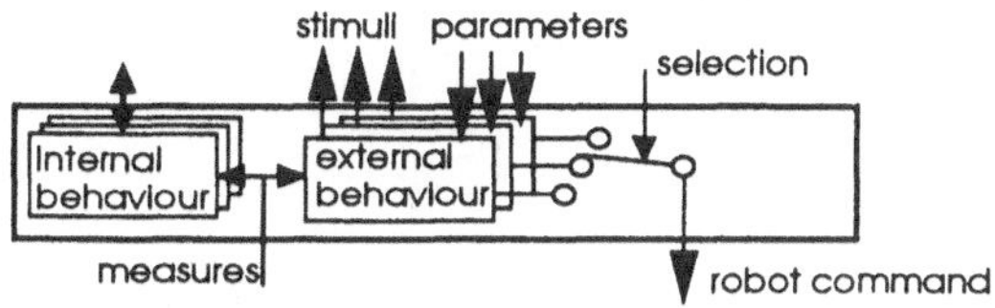

Figure 1: *Behavioural layer.*

While each behaviour solves a small part of a robot task, the **sequencing layer** composes them to achieve a more complex one. According to a given strategy, this task parametrises and selects one-by-one the suitable behaviour, depending also on the stimuli it receives. At this level, the world representation reflects the current states of the behaviours and consists of the set of stimuli at a given time.

The aim of the **strategic layer** is to achieve a task using knowledge-based reasoning. It solves tasks such as map building, map validation, navigation. It needs a global world representation providing spatial relationship between objects. Long term goals are achieved by scheduling individual tasks according to information provided by the map and, retroactively, by the tasks.

4 Development environment MANO

MANO is the development environment [4] used to test the mobile robot task execution. It implements the principle of our hybrid architecture (Figure 2). The core of this environment is composed of a virtual robot unit and of a blackboard handling the communication between the different layers. The four layers of the hybrid architecture are connected to these central elements. The sensori-motor layer is implemented on dedicated hardware located in the robot itself and on additional external units. The three other layers together with the blackboard and the virtual unit are distributed over a network of SUN work stations.

The virtual robot unit links the robot and the blackboard. It offers unified access to either the real or the simulated robot and provides extended capabilities to monitor sensor data, commands, position etc. The blackboard is the communication channel between the virtual robot and the upper layers. It acts as a server, other devices can connect as clients from any point of the Sun network.

The **sensori-motor** layer is implemented on a number of PC-boards: the servo loops controlling the robot are on board while some vision processing is currently performed remotely. Nomad 200 — from Nomadic Technologies, Mountain View CA — is a one-meter-tall robot moved by a three wheel motion system. It provides sensors of different

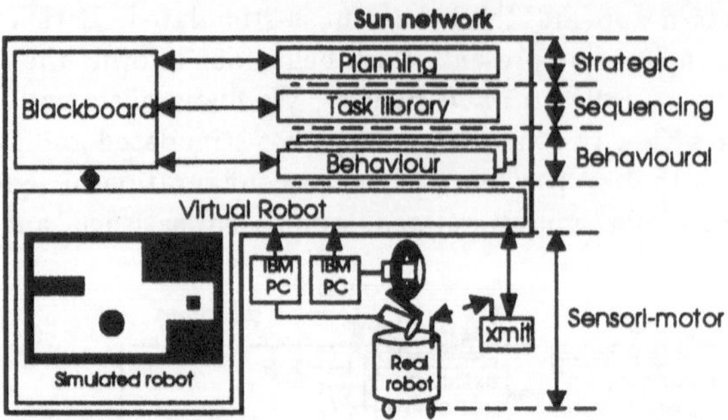

Figure 2: *Development architecture MANO.*

types: sonars, infrared range-sensors and tactile sensors. Two vision-based sensors have been added: a system observing landmarks and a second one using structured light [3]. The former uses a camera and active illumination to enhance the contrast of landmarks made of reflective material, and performs their detection and tracking. The latter uses a camera and a laser in order to deliver a 3D profile of the environment in front of the robot. The **behaviours** are fully independent and run concurrently as individual Unix processes. They are client of the blackboard server and read from it (i) the sensed data provided by the sensori-motor level and (ii) the parameters provided by the sequencing layer. The behaviours write their robot commands and their stimuli on the blackboard. The **sequencing** tasks are implemented in form of state machines and are placed in a library. They are connected to the blackboard in order to exchange the selected behaviour and the stimuli with the behavioural layer. The sequencing tasks use function parameters to exchange information with the strategic layer. Currently, the **strategic** level is realised as a single Unix process calling the sequencing tasks. It receives informations from the sequencing layer, and calls then the adequate tasks to reach a given goal.

5 Application: sequencing task for tidying up chairs

As an example of a task implemented on MANO, we describe here *TidyUpChairs* in a room. It illustrates how a specific task is ported onto our multi-layered hybrid architecture. The robot has to detect chairs which are located arbitrarily in a room, and to push them up to a tidying up area which is a virtual line defined with respect to a fixed position of the environment. This fixed position, called *home*, is defined by two landmarks. Two vision sensors are used: *vision by landmark* detects chairs marked with reflective material and homing landmarks, while *vision by structured light* detects obstacles in front of the robot. Odometry is used to move the robot to the virtual line and to bring the robot back to the homing area. Figure 3 shows the *TidyUpChairs* task decomposed in a sequence of simple behaviours. First, the robot performs a homing (HO behaviour), then searches chairs by looking around (SC). If a chair is found, it goes towards the selected chair (GC). Now, the robot turns around the chair until it is positioned on the side of the chair opposite to the virtual line (AC) and pushes the chair (PC) until the line is reached. Finally it returns to the homing area (RH) and adjusts its position (HO).

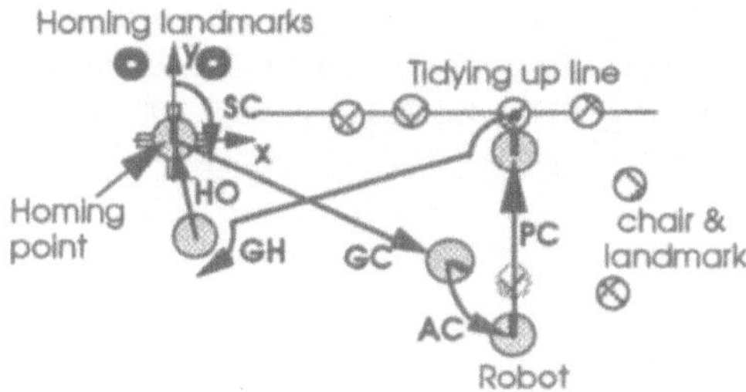

Figure 3: *TidyUpChairs decomposed by behaviours.*

The behaviours needed to tidy up chairs are describe below:

• Wandering around (WA): this behaviour moves the robot forward. If an obstacle is detected, the robot turns away and starts moving forward again.

• Homing (HO): based on the vision by landmarks system, the homing behaviour brings the robot in a fixed configuration with respect to two landmarks.

• Searching a chair (SC): searching a chairs is a behaviour which is stimulated when a chair landmark appears. It then turns the robot in the direction of the nearest landmark.

• Going to a chair (GC): this behaviour moves the robot forward and servoes its orientation by centring the centermost landmark in the image.

• Aligning on the chair (AC): this behaviour moves the robots around a chair, until it is oriented perpendicular to the virtual line.

• Pushing the chair (PC): this behaviour moves the robot forward. Using the odometry, it stops the robot when the tidy up line is reached.

• Returning home (RH): Using the odometry, this behaviour brings the robot back to its home position.

• Getting position (GP): this internal behaviour returns the current robot position.

Figure 4: *TidyUpChairs seen as an state automata.*

At the sequencing layer, this task has a pre-programmed structure described by a state automata (figure 4). The circle indicates the selected behaviour, and the arrows show the result of the executed behaviour. The concentric circles are final states: success or failure. All the behaviours used are goal-driven. Only the *satisfied* stimulus provides a "success" output. The three other values of stimuli or a stimulus providing by an obstacle detection behaviour give a "failed" output. The *TidyUpChairs* task performs as desired. Many tests have been run with various chairs and homing landmark configurations: the programmed sequence invariably leads to the kind of path shown in figure 3 (path: HO-SC-GC-AC-PC-RH-HO).

6 Conclusion

We presented the implementation of a multi-layered hybrid architecture used to control a mobile robot. The levels considered are: sensori-motor, behavioural, sequencing, and strategic. The architecture runs on a network of workstations, a Nomad200 mobile robot and dedicated vision hardware. To illustrate the architecture functionality, we presented the *TidyUpChairs* task as well as its decomposition with respect to the hybrid architecture. Formally, the goal was expressed as an finite automaton that triggers behaviours. MANO — the development environment we realised — is now fully operational.

Acknowledgements

This work is bound to project 4023-027037 of the Swiss National Research Program "Artificial Intelligence and Robotics" (NRP23), conducted in collaboration with the Institute of Informatics and AI of the University of Neuchâtel, Switzerland.

References

[1] Brooks R.A. (1986) "A Robust Layered Control System for a Mobile Robot System", IEEE Journal of Robotics and Automation, Vol. RA-2, No 1, pp. 14-13.

[2] Connell J. H. (1992) "SSS: A Hybrid Architecture Applied to Robot Navigation", Proc. IEEE Int. Conf. on Robotics and Automation, Nice, France, pp. 2719-2714.

[3] Hügli H., Maître G. , Tièche F., Facchinetti C. (1992) "Vision-based behaviours for robot navigation", Proc. 4th Annual SGAICO Meeting, Neuchâtel, Switzerland.

[4] Hügli H., Tièche F., Chantemargue F., Maître G. (1993) "Architecture of an experimental vision-based robot navigation system", Proc. Swiss Vision, Zürich, Switzerland, pp. 53-60.

[5] Slack M. G. (1992) "Autonomous Navigation of Mobile Robots for Real-World Applications", in Interdisciplinarity Computer Vision, SPIE, Vol. 1838, pp. 101-109.

[6] Thorpe C. E. (1992) "Point-CounterPoint: Big Robots vs. Small Robots", in Interdisciplinary Computer Vision, SPIE, Vol. 1838, pp. 78-88.

ALINSPEC PROJECT: AN INTELLIGENT VISION SYSTEM FOR AUTOMATIC INSPECTION OF ALIMENTARY PRODUCTS

A. Barducci
Consorzio CEO - Centro di Eccellenza Optronica
Via Albert Einstein 35, I50013 - Campi Bisenzio (Fi), ITALY

M. Barni
V. Cappellini
Università degli Studi di Firenze
Dipartimento di Ingegneria Elettronica
Via di Santa Marta 3, I50139 - Firenze, ITALY

S. Livi
Officine Galileo
Via Albert Einstein 35, I50013 - Campi Bisenzio (Fi), ITALY

A. Mecocci
Università degli Studi di Pavia
Via Abbiategrasso 209, I27100 - Pavia, ITALY

Abstract. ALINSPEC (Intelligent Vision System for Real Time Automated Inspection of Alimentary Products) is a research project partially supported by EEC, and devoted to the automated inspection of chicken and pig meats for quality control purposes. It comprises several european partners, comprising Universities, Industries and Research Institutions. In this work some preliminary results originated from the research activities related to the project are briefly discussed. The feasibility of achieving an intelligent vision system that performs a fully automated inspection of the considered alimentary products is here proven.

Keywords. Alimentary Inspection, Classification, Computer Vision, Defect Detection.

1 Introduction

Within the Food Industry there is today a growth of interest for higher quality products. Nowadays, quality control operations are mainly performed by means of manual inspection; however, inspector's capabilities can not meet the high-speed production rates achieved by the modern manufacturing facilities. The introduction of machine vision tools can allow a full-rate automated inspection system for alimentary products. As a

E. A. Yfantis (ed.), Intelligent Systems, 705–711.

matter of fact, the use of advanced vision systems in agriculture and alimentary industry is negligible in comparison with other industrial environments: electronic and mechanic industries share over 70% of the vision market, with applications mostly devoted to measurement, gauging and guidance. This means that new methodologies have to be developed and extensively applied, in order to fully exploit image vision capabilities. This paper addresses the problem of the development of a flexible *Intelligent Perception System* (IPS) for real time inspection of alimentary products, devoted to *on line* quality control of defects on chicken and pig meat before packing. For the chicken meat case, the IPS should analyze images representing the chickens after they have been washed and plucked. The IPS is aimed to detect chicken defects (burns, hematoma, fractures ..), together with other relevant features. Use of it should allow the automated monitoring of all the produced samples. Information collected by the IPS should also be used to address the further product processing. As an example, fully damaged chickens have to be discharged, whereas a particular dissection strategy should be suggested when defects only affect a limited part of the chicken. Similar considerations hold for the case of pig legs. The IPS introduction should yield the following benefits: a continuous and exhaustive control of the quality level of the production; an immediate feedback of the IPS towards dissector workers; an automatic management of the current production process, according to production strategy and demand; an immediate analysis of the effects due to a poor production strategy; a deeper knowledge of production line conditions, with consequent planning of machinery maintenance periods; an immediate feedback towards previous stages of the production chain; a growth of the overall quality of the production cycle.

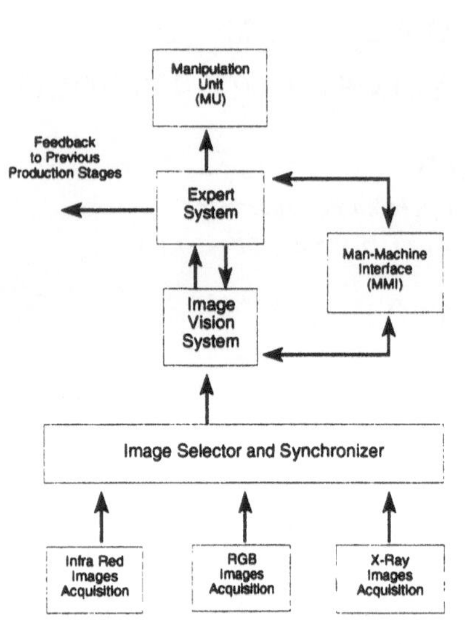

Figure 1: Overall IPS architecture

2 System Layout

In figure 1, the overall architecture of the proposed system is reported. Images are gathered by three different sensors: a classical RGB camera, an infrared sensor and an X-ray acquisition system. Image data are sent to the *Image Vision System* (IVS) via the selector-synchronizer triggers. This block also accounts for the handling of all the acquisition details, image preprocessing inclusive. The main goal of the IVS is to segment the image according to its content (e.g. the body of the chicken must be separated from the background) and to detect any existing defect. Besides, some general features of the product must be measured (e.g. the global color of a chicken skin). The information extracted by the IVS is fed to the *Expert System* (ES) whose aim is to decide the actions to be performed on the chicken (or pig leg). According to the defects revealed by the IVS, the ES can decide to reject the item, or to use it for a suitable application. If the occurrence rate of a particular kind of defect exceeds some predetermined limit, the ES should alert production engineers and suggest a suitable reaction strategy to be undertaken. The IVS and the ES are interfaced to the user via the *Man Machine Interface* (MMI). Through the MMI, personnel controls the inspection system. The control includes the input of parameters such as: system learning sets, system knowledge database, classification procedures, rejection/acceptance parameters and production strategies. Finally, the ES controls the *Manipulation Unit* (MU) which is responsible of materially performing the actions the ES decides to undertake.

3 System Architecture

Acquisition System. Digitized images are provided by three different sensors: a thermal infrared detector, operating at about 3.5 micron, an X-ray sensor and an RGB camera. Data gathered by the camera provide most of the information needed by the IVS in order to reveal a large class of defects. The X-ray sensor has been added to aid the detection of internal defects like bone fractures. The usefulness of the IR detector is still to be verified. The acquisition process is accomplished by means of a proper block which is responsible for synchronizing the image acquisition channel by channel, and for preprocessing the images. To this purpose it is equipped with a specialized hardware carrying out image smoothing and coordinates transformation. The need for input-image smoothing is evident if we consider that within real farmer environments the image acquisition may be affected by noise originated by any kind of mechanical vibration and electromagnetic interference. In our case, simple filters such as medians [1] or average filters [2] represent a satisfactory jointing-point between filtering effectiveness and simplicity. Since IVS processing will be mainly based on the HIS color system [2], the preprocessing accounts also for the transformation from the RGB to the HIS system.

The setting up of the acquisition system must be carefully considered, since it affects the quality of the images to be processed. Various parameters must be kept under finely controlled conditions. Among them it is remarkable the sample position with respect to the viewing camera, which should offer an optimal view for sample segmentation and which is expected to be constant in time. Other relevant parameters are the chromatic response of the RGB camera, the tuning of the lighting source and the rear panel. With regard to the rear panel, an easily distinguishable blue screen has been chosen.

Intelligent Vision System. The IVS constitutes the core of the IPS. Its main tasks are the image segmentation and the defects detection. Image segmentation consists of two stages: the extraction of the chicken or pig leg from the background and its partitioning in anatomic parts.

Defects detection is a very complex task due to the great variety of defects to be revealed and to the lack of a precise definition of them. For example an ammonia burn is described as a "*dark brown-black discoloration with possible damage to the skin epidermis in the breast or hock.*", which is quite a difficult definition to translate in image vision terms. We remark that defect detection requires a number of different processing tools. As an example, burns or hematomas may be recognized by their color features, whereas other syndromes (e.g.: *toad skin*) need texture and structural analysis. In addition to defects detection, the IVS also provides some general information useful for the global characterization of the chicken (pig leg). These information (e.g.: the average skin color) together with the statistics of the production line derived by elementary IVS information, are used by the Expert System to perform a full quality monitoring of the production.

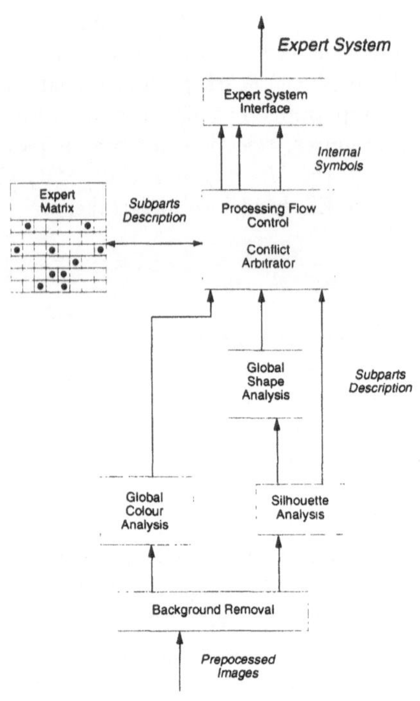

Figure 2: IVS architecture

Expert System. The interpretation of data gathered by the IVS is accomplished by the ES. The ES also decides which actions must be carried out for an optimal management of the production process. The ES consists of four main modules: a *Knowledge Data Base*, describing the static knowledge relative to the specific application environment and to the involved events; a *Flexible Data Base* for the description of the actual interpretation of the environment evolution; a *Hypothesis Initializer/Verifier*, capable of suggesting a set of possible interpretations for each set of events and of updating the interpretation confidence; a *Supervisor*, which coordinates the whole modules set. The Hypothesis Initializer/Verifier is the key unit of the IPS. Starting from data provided by the IVS, it must interpret the occurring events and apply the available knowledge (stored in the ES databases) to update the interpretation confidence. This module tries to recognize predetermined situations and activates a set of actions to guide the interpretation of the environment evolution properly. It focuses the system attention on a specific event or group of events, thus preventing the combinatorial explosion of the number of possible configurations to be considered. At last, the Supervisor module is in charge of activating the processes in the proper order.

Man-Machine Interface. The industry personnel controls the operations performed by the ES and the IVS by means of the MMI. By it, the IPS is provided with many input parameters. Among them: the system learning set; the system knowledge data base; rejection/acceptance parameters; production strategies. As output, the MMI provides system diagnostic.

Manipulation Unit. The task performed by the Manipulation Unit consists in physically accepting or discharging the examined sample according to the decisions taken by the Expert System.

4 IVS Architecture

The Vision System architecture is shown in fig.2. The first action performed by the IVS is the background removal. Its output is sent to the global color and silhouette analysis modules. The former extracts data about global characteristics of the sample. These characteristics will be used by the classification algorithms to tune their detection thresholds and strategies. Sometimes, global features can be regarded themselves as defects. The silhouette analysis module segments the chicken into its major subparts, namely breast, neck, legs and wings, by analyzing the silhouette previously extracted. Then, the global shape module computes some geometrical parameters (e.g.: longitudinal and transverse sizes) of the observed sample. Classifiers are thought as a matrix of chromatic and shape *experts*. For each chicken subpart a suitable set of processing algorithms is developed. Experts activation is controlled by a block devoted to processing flow control and conflict arbitration. Besides, such a block solves possible contentions among experts. Finally, the interface to the ES translates internal symbols into the string representation required by the ES. Along with the defect name a degree of confidence, the defective area and its location on the chicken body are given.

5 Algorithms Development and Testing

The defects detection algorithms have been tested by using the preliminary image database. So far, pig legs, as well as non-RGB data, have not been considered. The images, supplied by the University of Vigo, have been firstly investigated in order to assess their optical quality, and to obtain some pre-classification algorithms for background identification. Then, the chromatic clustering of defective meat regions has been investigated. Another point we have considered is the use of different synthetic channels, derived from the RGB tristimulus. In fact, to meet the speed constraint we have tried to avoid using full 3-D clustering. Instead we investigated the possibility of working either with only one channel or with "uncoupled" channels. Upon statistical analysis we discovered that the sole hue chrominance holds most of the information content originally recorded in the RGB tristimulus.

Background Identification. In order to distinguish the chicken body from the background, two features have been considered: the hue and the intensity. By considering only the hue two well separated clusters appear: a yellowish cluster corresponding to the chicken body and a bluish cluster relative to the background. The area between the clusters is filled by pixels lying on the contour of the chicken. These pixels have intermediate colors between that of the background and those of the chicken skin. It is worthwhile noting that pixels belonging to bluish bruises and to shadowy areas can belong to the sparse area between the clusters or even to the blue cluster, so they cannot be distinguished from the background on the basis of chrominance information only. To get around this problem the information provided by the intensity has been considered. In particular the following considerations has been taken into account:

- pixels relative to healthy skin always belong to the yellowish cluster;
- pixels relative to bluish bruises belong to the sparse area between the clusters or to the blue cluster; they can be distinguished from the background since they are darker;
- with regard to pixels belonging to other defects, three cases are possible: they belong to the yellow cluster; they are darker than the background; they are reddish;
- shadowy pixels are darker than the background.

The background extraction has been accomplished by implementing the above rules by means of a hierarchical decision tree. Finally, some morphological operators [3] have been introduced to regularize the shape of the region defining the chicken shape. The background extraction has been successfully tested on all of the images of the preliminary database.
Chromatic Clustering of Chicken Defects. In order to get a reliable chromatic characterization for the defects clusters a thorough statistical analysis has been carried out on the preliminary database. The following channels have been investigated in order to use the most reliable features: CIE L-a-b [2], H-I-S and three band ratios. Due to the insufficiency of the preliminary database only hematomas, blood accumulations, ammonia burns and tears have been considered. It is noticeable that chromatic clustering does not agree with defects classification. Particularly, hematoma-like syndromes (i.e.: hematomas, blood accumulations and tears) are grouped into two separate clusters: blue and red hematomas (BH and RH). Ammonia burns are split into four different clusters: Dark-Red-Burns (DRB), Light- Red-Burns (LRB), Dark-Blue-Burns (DBB) and Light-Blue-Burns (LBB). Physical causes for this singular defect behavior can be inferred by investigation of the optical properties of the skin [4]. Besides, five other classes have been added for background, healthy skin, shadows, yellow legs and spread blood.
The IVS has been applied to all of the eleven images of the database. Upon inspection of the results some statistical calculations concerning the cluster separability have been carried out. In order to cut the effect of outlier-measures the ratio between the center-cluster-distance and the sum of their standard deviations may be a valuable indicator for cluster separability. Whenever the ratio is large with respect to one an acceptable clusters separation may be achieved, while for ratio values far below one the clusters are undistinguishable. In figure 3 cluster pairs for which the ratio value is less than one are indicated as "weak-pair". It is to notice that only 6 pairs out of 66 are weak. We remark that such weak pairs can be accurately distinguished only by using improved quality images. Based on these results some preliminary expert-classifiers have been coded. The classifiers are hybrid, with embedded knowledge and are specialized for HB and HR detection. They have been tested on the available images; a full description of their functionality may be found in [4]. We remark that classification results fairly agree with the classification of the input image performed by experienced personnel. However, before a definitive judgement can be made the IVS must be tested on a greater number of images.

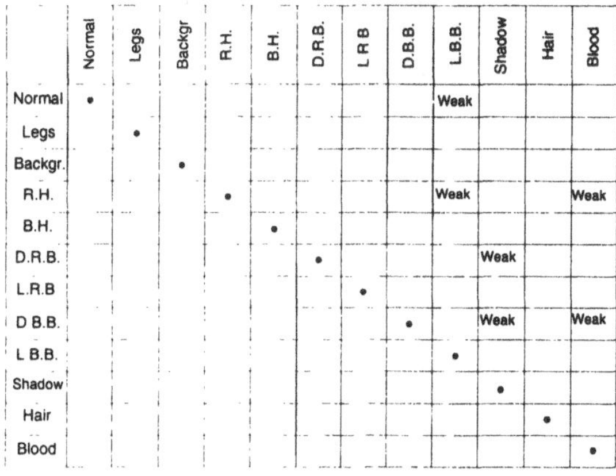

	Normal	Legs	Backgr	R.H.	B.H.	D.R.B.	L R B	D.B.B.	L.B.B.	Shadow	Hair	Blood
Normal	•								Weak			
Legs		•										
Backgr.			•									
R.H.				•					Weak			Weak
B.H.					•							
D.R.B.						•				Weak		
L.R.B							•					
D B.B.								•		Weak		Weak
L B.B.									•			
Shadow										•		
Hair											•	
Blood												•

Figure 3: Clusters separability

6 Conclusions

The problem of automating the inspection of alimentary products such as chicken and pig meats has been examined. The needs of manufacturing industry have been considered, and results related to the ALINSPEC project have been detailed. In addition to the outline of the software and hardware architectures of a possible real time vision system devoted to automatic inspection of alimentary products, some pertinent algorithms of image processing have been shown. These algorithms enable part of the functions which will be characteristic of such a vision system. Particularly, this work shows the feasibility of automating the background extraction, and the recognition of some relevant classes of chicken defects.

7 References

[1] I.Pitas, A.N. Venetsanopoulos, Nonlinear digital filters: principles and applications, Kluwer Academic Publishers, 1990.

[2] W.K.Pratt, Digital image processing, J.Wiley, 1978.

[3] J.Serra, Image analysis and mathematical morphology, Academic Press, 1983.

[4] A.Angotti, A.Barducci, M.Barni, A.Mecocci: "Analysis of Signals Retrievable from Standard RGB Tristimulus for Proper Meat Characterization and Classification"; Internal Report No. 940102, Dipartimento di Ingegneria Elettronica Università degli Studi di Firenze, January 1994.

AN INTELLIGENT SYSTEM FOR GIS INFORMATION MAPPING

G. M. Gallitano, Department of Mathematics and Computer Science
West Chester University, 13-15 University Ave.
West Chester Pennsylvania, 19383

E. A. Yfantis, Computer Science Department
University of Nevada, Las Vegas
Las Vegas, Nevada, 89154

A. Pitchford, Environmental Protection Agency
EAD Division, P.O. Box 93478
Las Vegas, Nevada, 89193-34783

Abstract

The mapping of a geographic point to a local area map is based on triangulation. Due to the error associated with the determination of the exact coordinates of a geographic point it is possible for the same geographic point to be mapped in a local area map by the aid of two different triangles on two different points in the map. A new algorithm for mapping geographic points is introduced in this paper. The algorithm provides consistent mapping of the geographic points to a local area map. Our algorithm is based in the theory of affine transformarmations and the mean square error. This algorithm can be used as a measure of accuracy or measure of quality for local area maps. Thus if our method applies to two different maps represented in the computer the best map of the two is the one with the smaller sum of absolute distances between the points already mapped and the corresponding points mapped by our method.

Keywords: Affine Transformations, Geographic Information Systems, Mean Square Error .

1. Introduction

The usual way of mapping a geographic point to a local area computer map represented in the computer is first to calculate or determine the x, y) coordinates of the point in the actual space and then map the point in the computer map. The usual method employed is the triangulation method, whereby three points of the geographic space and the corresponding points in the map are used to determine the position of the new point in the map. Due to the error introduced in the process of measuring the geographic coordinates of the point different reference points used in the triangulation method could give different mappings of the same geographic point. Thus the position in the screen of a pixel representing a geographic point could depend on the points used in the triangulation method. Our method uses several points, instead of three or four that other methods use. Our method is based in the theory of affine transformation and mean square error. Our method can be used to allocate geographic points on a computer generated map, as well as to evaluate the quality of a map. Very often a number of attributes are associated with geographic data. Such attributes could be concentration of various pollutants, ore concentration, and several other attributes. Based on these measured attributes, surface estimation method

E. A. Yfantis (ed.), Intelligent Systems, 713–717.

can be used (Barnhill, 1985), (Farin, 1993), (Foley, 1987), (Nielson, 1986), (Yfantis et al, 1988), (Yfantis, 1993).

2. Mapping of Geographic Data Using Affine Transformations and Mean Square Error

Let $(x_1, y_1), (x_2, y_2), ..., (x_n, y_n)$, be the (x, y), coordinates of n geographic points and their corresponding points $(x_1', y_1'), (x_2', y_2'), ..., (x_n', y_n')$ on the computer map. These are the points to be used as the basis for mapping any future geographic points on the map. Let $\varepsilon_i = \begin{pmatrix} \varepsilon_{ix} \\ \varepsilon_{iy} \end{pmatrix}$ be the error associated with the allocation of the geographic coordinates of the *ith* point. The affine transformation used to map the point (x_i, y_i) on the screen is of the form:

$$\begin{pmatrix} x_i' \\ y_i' \end{pmatrix} = \begin{pmatrix} a & b \\ c & d \end{pmatrix} \begin{pmatrix} x_i \\ y_i \end{pmatrix} + \begin{pmatrix} e \\ f \end{pmatrix} + \begin{pmatrix} \varepsilon_{ix} \\ \varepsilon_{iy} \end{pmatrix} \tag{1}$$

$$\varepsilon_i = \begin{pmatrix} \varepsilon_{ix} \\ \varepsilon_{iy} \end{pmatrix} \tag{2}$$

The square of the Euclidean norm for the above error is:

$$||\varepsilon_i||^2 = \varepsilon_{ix}^2 + \varepsilon_{iy}^2 \tag{3}$$

The idea here is to determine the coefficients a, b, c, d, e, f, so that we will minimize the mean squre error. The mean square error is:

$$\sum_i ||\varepsilon_i||^2 = \sum_{i=1}^{n} (x_i' - ax_i - by_i - e)^2 + \sum_{i=1}^{n} (y_i' - cx_i - dy_i - f)^2 \tag{4}$$

If we represent the above sum of squared errors as a function of the parameters a, b, c, d, e, f to be estimated then

$$F(a, b, c, d, e, f) = \sum_{i=1}^{n} (x_i' - ax_i - by_i - e)^2 + \sum_{i=1}^{n} (y_i' - cx_i - dy_i - f)^2 \tag{5}$$

The parameters a, b, c, d, e, f which minimize the above function can be found by taking the partial derivatives with respect to each one of these parameters and setting it to zero. The following equations are obtained.

$$\frac{\partial F}{\partial a} = -2 \sum_{i=1}^{n} x_i (x_i' - ax_i - by_i - e) = 0 \tag{6}$$

from this equation we obtain

$$a \sum_{i=1}^{n} x_i^2 + b \sum_{i=1}^{n} x_i y_i + e \sum_{i=1}^{n} x_i = \sum_{i=1}^{n} x_i x_i'. \tag{7}$$

Now

$$\frac{\partial F}{\partial b} = -2\sum_{i=1}^{n} y_i(x_i^{'} - ax_i - by_i - e) = 0 \tag{8}$$

which becomes

$$a\sum_{i=1}^{n} x_i y_i + b\sum_{i=1}^{n} y_i^2 + e\sum_{i=1}^{n} y_i = \sum_{i=1}^{n} y_i x_i^{'}. \tag{9}$$

Also

$$\frac{\partial F}{\partial e} = -2\sum_{i=1}^{n}(x_i^{'} - ax_i - by_i - e) = 0 \tag{10}$$

or

$$a\sum_{i=1}^{n} x_i + b\sum_{i=1}^{n} y_i + ne = \sum_{i=1}^{n} x_i^{'}. \tag{11}$$

The above equation can be written as

$$a\overline{x} + b\overline{y} + e = \overline{x}^{'}. \tag{12}$$

The partial derivative with respect to d is

$$\frac{\partial F}{\partial d} = -2\sum_{i=1}^{n} y_i(y_i^{'} - cx_i - dy_i - f) = 0 \tag{13}$$

or

$$c\sum_{i=1}^{n} x_i y_i + d\sum_{i=1}^{n} y_i^2 + f\sum_{i=1}^{n} y_i = \sum_{i=1}^{n} y_i y_i^{'}. \tag{14}$$

Partial derivative with respect to c is

$$\frac{\partial F}{\partial c} = -2\sum_{i=1}^{n} x_i(y_i^{'} - cx_i - dy_i - f) = 0 \tag{15}$$

or

$$c\sum_{i=1}^{n} x_i^2 + d\sum_{i=1}^{n} x_i y_i + f\sum_{i=1}^{n} x_i = \sum_{i=1}^{n} x_i y_i^{'}. \tag{16}$$

Finally the partial with respect to f is

$$\frac{\partial F}{\partial f} = -2\sum_{i=1}^{n}(y_i^{'} - cx_i - dy_i - f) = 0 \tag{17}$$

or

$$c\overline{x} + d\overline{y} + f = \overline{y}^{'}. \tag{18}$$

The above equations can be written as

$$\begin{pmatrix} \sum x_i^2 & \sum x_i y_i & \sum x_i \\ \sum x_i y_i & \sum y_i^2 & \sum y_i \\ \overline{x} & \overline{y} & 1 \end{pmatrix} \begin{pmatrix} a \\ b \\ e \end{pmatrix} = \begin{pmatrix} \sum x_i x_i' \\ \sum x_i' y_i \\ \overline{x}' \end{pmatrix} \tag{19}$$

$$\begin{pmatrix} \sum x_i^2 & \sum x_i y_i & \sum x_i \\ \sum x_i y_i & \sum y_i^2 & \sum y_i \\ \overline{x} & \overline{y} & 1 \end{pmatrix} \begin{pmatrix} c \\ d \\ f \end{pmatrix} = \begin{pmatrix} \sum x_i y_i' \\ \sum y_i' y_i \\ \overline{y}' \end{pmatrix} \tag{20}$$

From the above equations we obtain

$$e = \overline{x}' - a\overline{x} - b\overline{y} \tag{21}$$

$$aS_x^2 + bS_{xy} = S_{xx'} \tag{22}$$

$$aS_{xy} + bS_y^2 = S_{x'y} \tag{23}$$

$S_x^2, S_y^2, S_{xy}, S_{x'x}, S_{x'y}$ are the biased estimates of the variances of the x coordinates, the y coordinates, the biased estimates of the covariances of the x,y coordinates, the x', x, coordinates, and the x', y, coordinates. From the above equations we obtain

$$a = \frac{S_{xx'}S_{xy} - S_y^2 S_{xy}}{S_x^2 S_y^2 - S_{xy}^2} \tag{24}$$

$$b = \frac{S_x^2 S_{x'y} - S_{xx'}S_{xy}}{S_x^2 S_y^2 - S_{xy}^2} \tag{25}$$

$$f = \overline{y}' - c\overline{x} - d\overline{y} \tag{26}$$

$$c = \frac{S_{xy}S_y^2 - S_{xy}S_{yy}}{S_x^2 S_y^2 - S_{xy}^2} \tag{27}$$

$$d = \frac{S_x^2 S_{y'y} - S_{xy}^2}{S_x^2 S_y^2 - S_{xy}^2} \tag{28}$$

These coefficients can be substituted in the following equation,

$$\begin{pmatrix} x' \\ y' \end{pmatrix} = \begin{pmatrix} a & b \\ c & d \end{pmatrix} \begin{pmatrix} x \\ y \end{pmatrix} + \begin{pmatrix} e \\ f \end{pmatrix} \tag{29}$$

which is the transform of the arbitrary geographic point (x, y) to the pixel with screen coordinates (x', y').

3. Conclusion

An algorithm for mapping geographic points to a local area computerized map was presented here. This algorithm can also be used to evaluate the consistency or quality of already existing maps. Unlike the triangulation method our method provides consistent mapping. Our method is based on the theory of affine transformation and mean square error.

NOTICE

Although the research described in this article has been supported by the United States Environmental Protection Agency, it has not been subjected to Agency review and therefore does not necessarily reflect views of the Agency and no official endorsement should be inferred.

4. References

[1] Barnhill, R. (1985), Surfaces in Computer Aided Geometric Design, A Survey with new results, Computer Aided Geometric Design 2, 1-17.

[2] Farin, G. (1993), Curves and Surfaces for CAGD, Third Edition, Academic Press, Boston, pp. 1-473.

[3] Foley, T. (1987), Weighted Bicubic Spline Interpolation to Rapidly Varying Data, ACM Transactions on Graphics, 6 (1) 1-18.

[4] Nielson, G. (1986), Rectangular nu-Splines, IEEE Computer Graphics and Applications, 6 (2), 35-40.

[5] Yfantis, E. A., Flatman, G. T. and Behar, J. V. (1987), Efficiency of kriging Estimation for Square, Triangular, and Hexagonal Grids, Mathematical Geology, 19 (3), 183-207.

[6] Yfantis, E. A., (1993), A New Quadratic and Biquadratic Algorithm for Curve and Surface Estimation, Computer Aided Geometric Design, vol. 10, no. 6, pp. 509-520.

PITS: An Intelligent Tutoring System Loosely Coupled to External Database Systems

Y. D. Yoo
Department of Information Systems
Dongguk University
Kyungju, Korea

Abstract. This paper describes the details of a pedagogical intelligent tutoring system which is loosely coupled to external programs such as a database program, a worksheet program, and an ASCII file. A modular knowledge base has been developed and implemented in support of this study. The knowledge base consists of two separate components, the database and the rule base. The database exists external to the main system, while the rule base resides in the main system. The PITS system is an experimental prototype designed to demonstrate the functionality of the system. The emphasis has been on the design and implementation of a knowledge base that has a separate database and a rule base, and a progress record file that keeps track of the student's learning progress during a tutoring session. A PRF is a non-volatile file even after a student leaves the system. The PITS system offers several advantages: 1) a convenient way to identify the student's learning progress by using the PRF; 2) easy and speedy maintenance of the rules stored in the modularized rule base; 3) virtually unlimited size of a knowledge base by using the modularization features; and 4)easy and speedy maintenance of the data stored in the external data base.

1 Introduction

Since the concept of Intelligent Tutoring Systems (ITS) was articulated in Carbonell's paper[3], ITS has received considerable attention among Artificial Intelligence (AI) researchers, cognitive scientists, and educators. Three different disciplines (i.e., AI, cognitive science, and education) have evolved from ITS research. Thus, research goals, objectives, theories, terminology, and emphasis are different between the systems. Accordingly, various types of ITS have been developed based on the different view points. It is very difficult for a single discipline to solve ITS problems, which requires a mutual understanding of the three disciplines involved[4]. The problems are scattered on the area of student modeling, tutoring strategy, student interface (communication method), and expert knowledge.

Accordingly, there is no existing system that has all components fully developed, because of the size and com- plexity of ITS. Most researchers tend to focus on a single component of the system. This study also approached from the viewpoint of AI technique.

Designing an ITS from scratch using current programming language (e.g., C, PROLOG, LISP, etc.) is an enormous task and, thus far, there doesn't exist any such system in a complete form. One of the problems addressed by this research is the design of an ITS

E. A. Yfantis (ed.), Intelligent Systems, 719–728.

from existing building blocks (i.e., dBase III Plus, Lotus 1-2-3, VP_Expert), and even more importantly, using an existing interface between these building blocks.

In addition to the above problem, maintenance of the knowledge of ITS (including expert systems) is an issue critical to a successful design. The maintenance of the knowledge depends on how a large volume of data or information can be managed efficiently. The current method of building a production knowledge based system is straight forward. In other words, two components of a knowledge base, facts and rules, are stored together in a single knowledge base.

However, the ITS frequently involves a large amount of data that needs periodic updating. Whenever a small update is necessary, the whole ITS must be reopened, and a small update will be made. When information in a knowledge base is increased, the size of the knowledge base also should be increased in order to manage the infortion. It may not be a significant matter when the size of the knowledge base is small, but, the management of a knowledge base can become a serious issue when their sizes are increased[10].

This paper presents an architecture of ITS that provides solutions to the above problems. The next section reviews the architecture of the traditional ITS and its components.

2 An Architecture of the Traditional ITS

An ITS is a computer aided program that behaves intelligently in teaching. It is that ITS should know what to teach, how to teach and who it teach[6]. These are based on the view point of its contents, and indicate that ITS should include domain knowledge, tutoring strategy, and target student. On the other hand, an ITS is observed based on the view point of its functions. An ITS is a computer system that contains knowledge representation tation formalisms. Such knowledge representation formalisms are derived from the field of AI, and exists in the form of an expert system.

Since the concept of ITS was first introduced in Carbonell's paper in the early 1970, more concepts of ITS have been articulated by AI researchers[6]. Wexler[8] presented "generative CAI" which has an ability to generate problems from a large database representing the subject they taught. A system that interacts with a large database remains still as an critical issue to be solved.

The concept of "reactive learning envrionment" was coined by Brown[1]. Brown used the concept of "tutorial dialogue" to distinguish the difference between ITS and CAI. Koffman and Blount[5] highlighted the importance of recognizing a student's misconcepts. These concept were used in the following ITS applications: the geography tutor SCHOLAR by Carbonell and Collins[3], the electronics trouble shooting tutor SOPHIE by by Brown ad Burton[2], the logic and set theory tutor EXCHECK by suppes and his associates[7].

ITS applications including the above applications show some common elements in their structures, although each application has been focused on the subject differently. The common elements include "tutoring strategy," "learner model," "knowledge base," "learner interface." Therefore, the components of an ideal ITS would include a tutoring module, a learner model module, an expert module, and a learner interface module as shown in Figure 1.

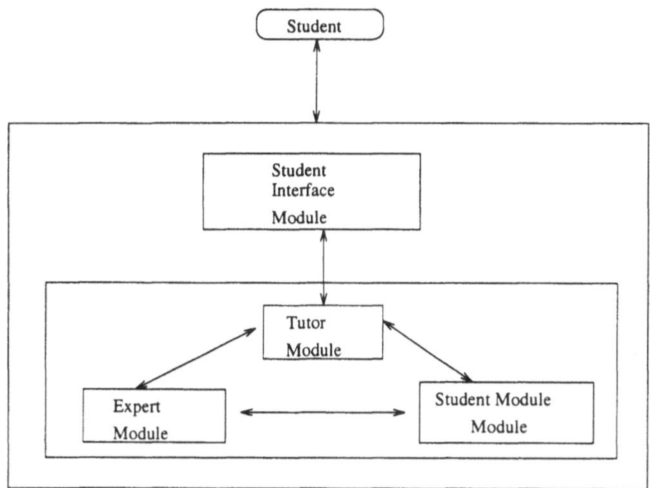

Figure 1: The Structure of ITS

2.1 Analysis on the Traditional ITS

Kearsley[4] argued that the development of ITS applications in the real world is very complex, because of the essential characteristics of ITS. He pointed out that the development of ITS applications should crystallize an appropriate combination of three different disciplines: computer science (AI techniques), psychology (understanding student's behavior), and educations and training (teaching and training methodology or philosophy).

A variety of knowledge representation techniques, inference mechanisms, and teaching strategies have been identified which can be used to support the development of a extended architecture of ITS. However, the study found that only a few applications (e.g., GUIDON, etc.)[9] have concerned the management of the knowledge base. Most reviewed applications in this study focus on tutoring strategy or knowledge representation technique. This is possibly due to the characters of ITS which do not require frequent modifications and a large amount of instructional data. The size of a knowledge base and its maintenance have not been sufficiently treated in most applications, which could be a problem or disadvantage should the size of the knowledge base increases significantly.

A major disadvantage of current ITS applications is that there is no way of identifying a student's learning progress after the learner leaves the system. How can a teacher identify the student's progress unless someone watches the student's working?

The major problems presented in this study thus far have include the followings:

- Maintenance of a knowledge base that contains a large volume of data.
- No way of keeping track of a student's learning progress after his/her leaving the system in current ITS applications.
- No model of ITS interfaces to external database systems and applications.

This paper presents the architecture of a prototype of ITS which might be the solution to the above problems.

3 Design of PITS

To achieve the goal of developing an ITS prototype loosely coupled to an external database and worksheet, three major issues will be considered in this study as follows:

1. The development of an ITS prototype that demonstrates the capability of the proposed model, which could be of potential benefit to the traditional education environment and the industrial training education.
2. The development and implementation of a knowledge base (expert module) that consists of two sub-modules: a rule base and a database. It could be more efficient to store a large amount of instructional data in a separate database, and design a rule base as a set of invariant rules. In this way, the rule base can be kept to a more manageable size.

   A database is a set of lessons of defined domain (e.g., INGRES) that resides outside the main ITS. The benefit of this mechanism is to allow an instructor or a teacher to update the materials such as lessons simply without getting involved in the main ITS.
3. The development and implementation of a student's progress record module that outputs a student's learning process into a progress record file, which resides separately from the main ITS.

The system prototype, named Prototype Intelligent Tutoring System (PITS), has been developed. PITS is a pedagogical ITS that provides a student an opportunity to obtain knowledge of INGRES. The PITS system is capable of the following functions:

1. Diagnose a student's knowledge.
2. Guide a student to an appropriate level of domain knowledge.
3. Instruct a student based on the step 1 and 2.
4. Retrieve a text file for an instruction from an external database.
5. Maintain a record of a student's progress in a non- volatile file.
6. Allow an instructor easy access to a database for update.
7. Offer a friendly user interface using menu options.
8. Store a large amount of facts (sets of text data for training) in external databases, and access the external databases.

A major benefit for an ITS program designer or an instructor in developing his/her own instructional program is that the PITS system requires only the replacement of the set of questions and instructional material stored in external database files (e.g., lessons, slides, etc.) unless he/she wants to change a tutoring strategy. Since the external database files are independent from the main system, the update or replacement of the external data base is simple. For example, a set of questions are stored in the external database (dBase III Plus file). These questions are used for diagnosing a student's knowledge level. To modify or update the questions in the external database, the required process is as follows:

1. Run dBase III Plus program.
2. Open a file (e.g., "question.dbf") that contains a set of questions.
3. Update or modify the questions stored in question.dbf.
4. Close the file questin.dbf and save it.
5. Exit the word processor.

Thus, an ITS program developer can replace the current tutoring instructions (e.g., lessons, questions) by new tutoring instructions in a reasonable time period without any major difficulty. An ITS program developer could use a scanner to read text instructions into the external database. In doing so, he/she could reduce the database development time significantly.

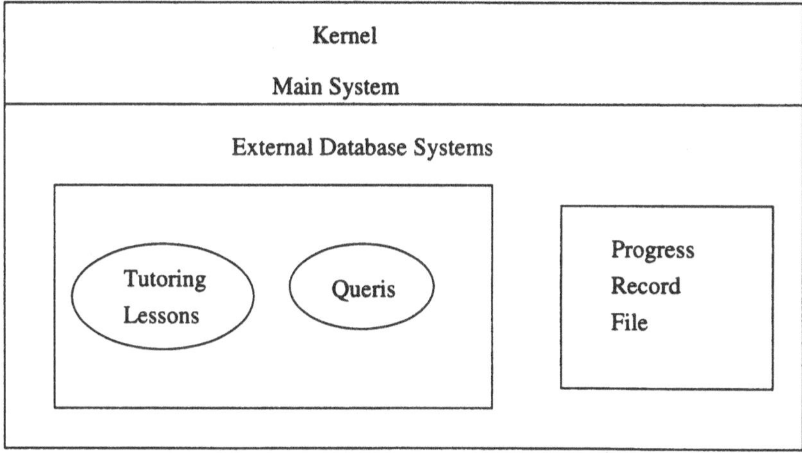

Figure 2: The Global Architecture of the PITS system

In addition, the PITS system does not require any complicated AI knowledge for changing, modifying, or updating instruction materials stored in an external database. This provides users (e.g., ITS program developers) a user-friendly system. However, it may require the ITS program developer to have knowledge of the PITS system in order modify the tutoring strategy which is embedded in the rules. These rules are written in the system shell of VP_Expert. Thus, an ITS program developer may also need to be familiar with the use of VP_Expert to do advanced work such as modifying the source code.

Seen from a high conceptual level, PITS can be regarded as having four major components (Figure 2) that effect interfacing to external systems.

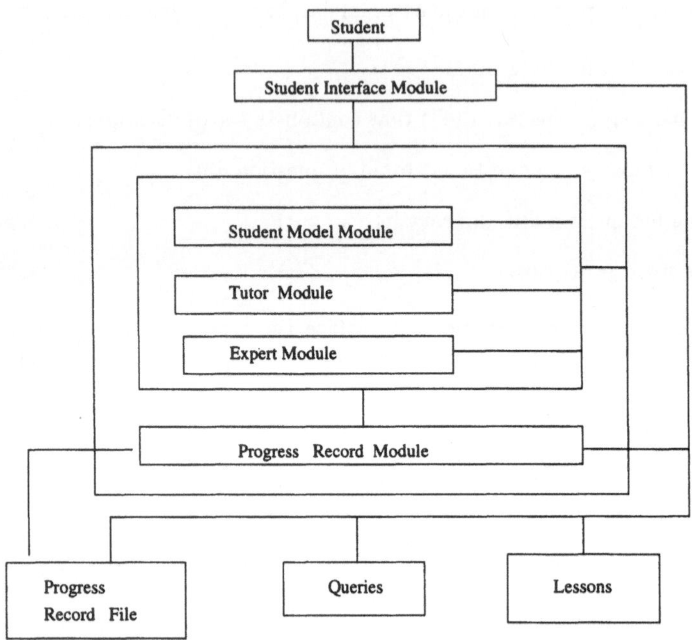

Figure 3: The tructure of an ITS prototype

These components are one main system and three external systems. The main system consists of three modules as follows:

1. The user interface module.

2. The kernel (diagnostic procedures).

3. The progress record module.

One module, the kernel, also contains the modules that implement the intelligent instruction system. They are:

1. The tutor module.

2. The student model module.

3. The expert module.

Each module interacts with one or more other modules at the same level (Figure 3). The large knowledge base of the PITS system is comprised of three external systems that, in turn, comprise the storage mechanism. They are:

1. Lessons
2. Queries
3. Progress record file

In more detail,

1. The queries file contains a set of questions that are used for diagnosing a student's knowledge level during a tutoring session.
2. The progress record file contains a record of a student's progress during a tutoring session. In other words, it contains a record of every interaction that was executed between a student and the PITS system during a tutoring session.
3. The lesson file is a set of text instruction materials. It could be comprised of a series of slides or lessons. These lessons or slides are automatically retrieved and displayed on the screen to instruct a student when necessary.

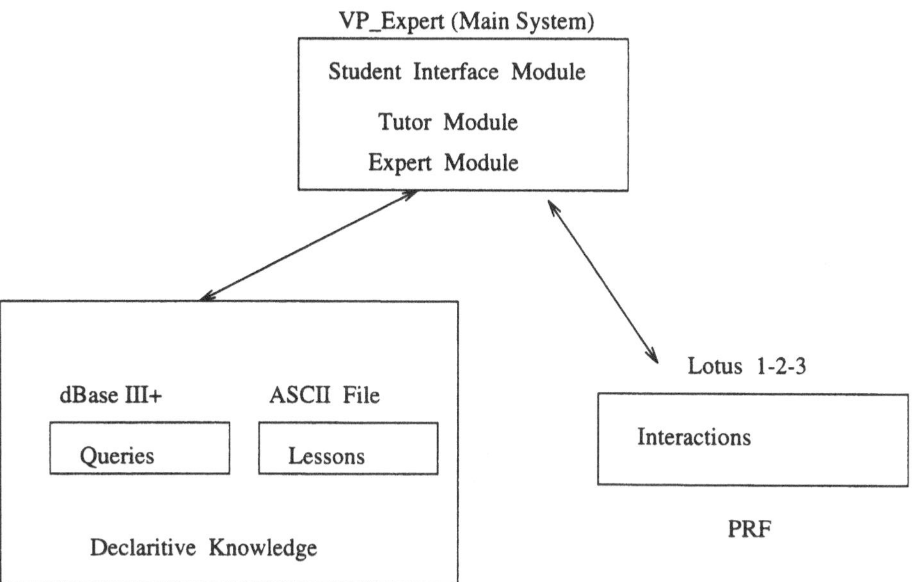

Figure 4: The tructure of the PITS system

4 Implementation and Results

A prototype of the PITS system, was implemented using the VP_Expert, dBase III Plus, and Lotus 1-2-3 on a PC to demonstrate the functionality of the proposed prototype of an ITS. The implementation focused on the integration of the rule base (expert module) in the main system with external programs such as a database file in dBase III Plus, and worksheet in Lotus 1-2-3. The database of the defined domain is open, in the sense that the instructor can always access such database as dBase III Plus, Lotus 1-2- 3, ASCII files to modify and update the contents which constitute the knowledge in the system as shown in Figure 4.

The system begins with identifying the student's name and date for tracking his/her progress record. The system provides a couple of modules so that both the instructor and student can review how the student proceeds during the tutoring. The modules are progress record file (PRF in Figure 5) and statistical report. Optionally, the student can study the necessary part(s) of the subject in the following instances:

1. After reviewing what he/she has done during the tutoring.
2. Without following the system's diagnosis procedure, if he/she knows which lesson(s) need to be learned. This will be the case if the student has used the system before and he/she remembers the point where he/she left.

The PITS system includes a question-answer, mixed- initiative tutoring strategy employed, which is performed via the provided menu options. Even during the procedure of the system's diagnosis, the student can obtain some concepts through the multiple choices by identifying his/her misconceptions about the subject. Some rules from the knowledge base are as follows:

```
RULE 6-1
IF   spacequest=UNKNOWN
THEN WKS from-quest, COLUMN=questions, young
     DISPLAY "{rule 2: from-quest[1]}"
     qdataread=yes;

RULE 6-3
IF   qdataread=yes AND
     from-quest[1] <> 0
THEN DISPLAY"
     No, the pointer indicates that there is a data in
     questions!
     {from-quest[1]}
     press any key to continue~"
     qx=1
     qx=(qx+1)
     DISPLAY "Rule 6-3: {from-quest[qx]}~"
     spacequest=no;
```

No	Menus	Questions	Answers	Valdity
1	4	1	3	-1
2	4	2	4	1
3	0	3		
4	0	4		

Lessons	Queries	Xtrlesson	Name	Date
100	35	101	Young	
101	0	0	Yoo	
				10-1-91

Figure 5: A Progress Record File

5 Conclusions

The goal of this study has been to develop an ITS prototype, to develop and implement its knowledge base that is loosely coupled to the external programs, and to develop and implement a student's progress record module. The results show how the system can be used to share the benefits of other system's mature technology.

The primary contribution of this research is the design of a knowledge base (expert module) that consists of an external database and a rule base, and student progress record module for an ITS prototype. The approach developed here integrates external systems, such as a database system, to an expert system. Therefore, the approach developed in this study focused on 1) the development and implementation of the interactions between the heterogeneous systems and 2) the development and implementation of the system that supports a student's progress record.

The PITS system has been implemented in order to test the proposed ITS prototype discussed in this study. The PITS system is an experimental prototype designed to demonstrate the functionality of the system. The emphases has been on the development and implementation of a knowledge base that has a separate database and a rule base, and a progress record file that keeps track of the student's learning progress during a tutoring session. A PRF is a non-volatile file even after a student leaves the system.

As an implementation techniques, modularization has been employed to implement the PITS system in the restricted environment regarding both hardware and software. The modularization approach has addressed of the problems of search space and reduction of recursion levels.

Consequently, the knowledge base has been modularized into several parts, which has resulted in both advantages and disadvantages. The major advantages include:

- Easy identification of the location of specific rules or facts.
- Reduce the efforts to modify and update of necessary rules or facts, and easy access.
- Rapid implementation by reducing the compilation process of q whole system.
- Virtually, unlimited number of properly modularized sizes of knowledge bases.

In addition, it also provides a rapid and convenient way of maintaining the data stored in the external database. The modularization and separate database have shown a possibility of building a portable knowledge base.

The major disadvantage of the PITS system is that the modularized knowledge base slows down the system performance. However, this disadvantage will be depend on the target system.

6 References

[1] Brown, J. S.(1977) "User of Artificial Intelligent and advanced Computer Technology in education," In R.J. Seidel and M. Rubin (Eds.), Computer and Communications: Implications for education, NY: Academic Press, pp. 253-270.

[2] Brown, J. S. & Burton, R.(1975) Multiple representation of knowledge for tutorial reasoning. In Bobrow and Collins (Eds.), Representation and Understanding, NY: Academic press.

[3] Carbonell, J. R.(1970) "AI in CAI: An artificial intelligence approach to computer aided instruction." IEEE Transactions on Man-Machine Systems MMS, Vol. 11, No. 4, pp. 190-202.

[4] Kearsley, G.(1987) Artificial Intelligent and instruction: Applications and Methods, NY: Addison-wesley.

[5] Koffman, E. B. & Blount, S. E.(1975) "Artificial and automatic programming in CAI," Artificial Intelligence, Vol. 6, pp. 215-234.

[6] Nwana, H. S.(1991) "User modelling and user adapted interaction in an intelligent tutoring," in A. Kobsa (ed.) User Modelling and User-adapted Interaction, Vol. 1, pp. 1-32.

[7] Suppes, P. S.(1981) (ed.), University-level computer assisted instruction at stanford: 1968-1980, Stanford, Calif.: Institute for mathmatical studies in the social sciences.

[8] Wexler, J. D.(1970) "Information networks in generative computer- based instruction," IEEE Transactions on Man-Machine Systems MMS, Vol. 11, pp. 181-190.

[9] Wenger, E.(1987) Artificial Intelligence and Tutoring systems. CA: Morgan Kaufmann, 1987.

[10] Yoo, Y. D.(1992)"A Primitive model of an expert training system," Journal of Information Systems, Youngname MIS, Vol. 1, pp. 149-178.

DATA BASE MANAGEMENT

A Finely-Interleaved Consistency Checking Method for Knowledge-bases

J. V. Harrison
Department of Computer Science
University of Queensland
Brisbane, QLD, 4072 Australia
E-mail: harrison@cs.uq.oz.au

An efficient method to determine consistency in a knowledge-base is described. The standard generation and evaluation steps are interleaved. The method employs an optimized update propagation algorithm to compute the differences between two consecutive knowledge-base states. Updates to both the facts and the rule-base are supported.

Keywords: integrity, deductive database, logic programming

1 Introduction

An integrity constraint (IC) specifies semantic information that is used to restrict a knowledge-base to states that more accurately model the application domain. If an update to the knowledge-base results in a violation of the constraint, the knowledge-base is said to be in an *inconsistent* state. An IC checking method is often employed to prevent the knowledge-base from entering an inconsistent state.

In a practical knowledge-base system, many ICs may be defined. In addition, updates to the knowledge-base may be frequent. A naive IC checking method can enforce consistency by evaluating all ICs after every update. This approach, however, is normally unsatisfactory as the amount of computation required can be extremely large. A more sophisticated IC checking method is required.

Many approaches to this problem have been proposed[NIC78, LLO87, BRY88, SAD88, KUC91, OLI91, VIE91, CEL94a]. As indicated by Celma et al.[CEL94b], all methods are comprised of both a *generation* and an *evaluation* phase. When facts are added to, or removed from, the knowledge-base the generation phase is initiated. During the generation phase, potential changes to the knowledge-base caused by both the fact updates, and also the updates induced using the rules, are identified. During the evaluation phase relevant ICs, which have been constrained or simplified using results from the generation phase, are evaluated. The result of this evaluation determines if the knowledge-base is inconsistent.

E. A. Yfantis (ed.), Intelligent Systems, 731–739.

The difference between the various methods centers on how the phases are implemented. For example, the phases may be performed separately and sequentially, or alternatively, they may be performed stepwise and interleaved. Generation may access facts in the knowledge-base or may only access rules.

These differences can greatly affect the efficiency of the methods. Methods that do not interleave will in many cases perform more computation than necessary to detect inconsistency. The result is significant irrelevant computation during both the generation and evaluation phases.

In some knowledge-base systems the number of facts is an order of magnitude (or more) larger than the number of rules. In this case, one might expect that the IC checking methods that do not need to access the facts to perform the generation phase will be more efficient. However, this is not necessarily true since key information, i.e., variable bindings, can be lost when reasoning solely with the rulebase. This can result in a large overestimate of the potential changes to the knowledge-base coupled with a significant increase in computation being required to perform the generation phase.

An IC checking method that interleaves generation and evaluation is described. The method employs an update propagation algorithm to compute the differences between two consectutive knowledge-base states resulting from updates. The differences are used to both identify and evaluate integrity constraints efficiently. The algorithm accomplishes this by issuing constrained queries on the knowledge-base. The differences identified are strictly in the form of ground facts, which results in minimal cost for performing the subsequent evaluation phase.

In section two, the fundamental concepts used to describe the method are presented. The third section describes the consistency checking method and enhancements that minimize redundant computation. Related work is addressed in section four.

2 Basic Concepts, Definitions and Notation

A knowledge-base comprised of both a relational database, for efficient storage and retrieval of facts, and a set of logical rules for representing both derived and semantic information, i.e., a *deductive database*, is considered. The rules are expressed using recursive Datalog, augmented with *stratified negation*[ULL88]. Derived information is represented using deductive rules and semantic information is represented using integrity constraints.

IC checking methods are invoked after an update to the database is initiated. They reason with two states of the knowledge-base, namely the state in existence before and after the update is initiated. The integrity constraint method described here computes the updates, i.e., Δsets, to the derived information, i.e., IDB relations. This corresponds to the generation phase. During the evaluation phase, the updates are used to identify relevant integrity constraints. They are also used to simplify the ICs before they are evaluated. These phases are interleaved to allow the algorithm to detect a violation before all of the changes to the derived information is computed hence improving efficiency

Each knowledge-base state consists of a set of extensionally defined relations (EDB)

and a set of intensionally defined (IDB or derived) relations. Let relation $P \in$ IDB and be defined using the rules that comprise predicate p. Let $\mathcal{U}_e$ be a set of updates to the EDB. The state of the database before $\mathcal{U}_e$ is performed is referred to as *old*. The state after $\mathcal{U}_e$ is performed is referred to as *new*. Let the function *mat(IDB_Pred,DB_State)* compute an IDB relation defined by the predicate *IDB_Pred* using the EDB indicated by *DB_State* and the IDB. Let the difference between the materialization of an IDB relation in the old state and the materization of the same IDB relation in the new state be termed the "delta set" (abbreviated Δset) for the relation. A delta set can be viewed as the updates that must be made to the old relation to obtain the new relation.

The notation ΔP represents the Δset for IDB relation P. A Δset consists of two distinct (possibly empty) subsets. The first, labeled ΔP_{add}, consists of tuples that must be added to the *old* relation to obtain the *new* relation. The second, labeled ΔP_{rem}, consists of tuples that must be removed from the *old* relation to obtain the *new* relation. These concepts are formalized using the definitions below, which assume that the predicates that define the IDB relations are not updated. Later, we address rule updates.

Definition. Let EDB_{Old} refer to an arbitrary EDB before a set of updates $\mathcal{U}_e$ are performed to the EDB relations. Let EDB_{New} refer to the same EDB after $\mathcal{U}_e$ are performed. Let p denote a predicate representing an arbitrary IDB relation P. Since we defer our discussion of IDB updates until a later section, let $IDB_{New} = IDB_{Old}$.

$$\Delta P_{rem} = mat(p, Old) - mat(p, New) \qquad DB_{Old} = EDB_{Old} \cup IDB_{Old}$$
$$\Delta P_{add} = mat(p, New) - mat(p, Old) \qquad DB_{New} = EDB_{New} \cup IDB_{New}$$
$$\Delta P = \{\Delta P_{rem}, \Delta P_{add}\}$$

□

Note that using the definition above as an algorithm to perform the generation phase would result in a very inefficient implementation since it would require computing all derived facts twice, i.e., once in each database state, after each update to the database. Instead, an *incremental* approach is employed, which is based on the generation and evaluation phases.

The dependency graph[ULL88] $\mathcal{DG}$ for a Datalog program $\mathcal{D}$ can be used to determine the IDB relations defined by $\mathcal{D}$ that may have been updated as a result of the updates to the EDB relations. An IDB relation I may have been updated as a result of updates to the EDB relations if the predicate defining I, namely i, depends, directly or indirectly, on one or more of the set of updated EDB relations $\mathcal{E}_u$.

Definition. Let e_u represent the predicate defining an arbitrary EDB relation E_u where $E_u \in \mathcal{E}_u$. Let $\Rightarrow$ be a path in $\mathcal{DG}$. An IDB relation I, defined by the predicate i, is termed a *candidate* for update if:

$$e_u \Rightarrow i \in \mathcal{DG}$$

An IDB relation is said to be *unaffected* if it is not a candidate. □

A subset (not necessarily proper) of the rules that define a candidate relation can contribute changes after updates to the base relations are introduced.

Definition. A rule defining a candidate predicate that contains one or more literals in the rule body corresponding to either a candidate predicate or an updated EDB relation is termed a *candidate rule.* A rule is termed *unaffected* if it is not a candidate.

The *TPF* algorithm computes the updates for all candidate relations by iterating *propagation* phase followed by a *filtration* phase. During the propagation phase, candidate rules are evaluated when the relations that correspond to subgoals are updated. The evaluation is constrained using bindings taken from these updates. The result of the evaluation is a set of tuples representing possible updates to the candidate relation.

Definition. The set of tuples generated for an IDB relation as a result of a propagation phase is termed an *approximation.* Each tuple in the approximation is termed a *potential IDB update.*

To obtain an approximation from a candidate rule, a query consisting of the literals appearing in the rule body is invoked over either DB_{Old} or DB_{New}. Consider a rule r defining an IDB relation P where both additions and removals have been identified for a relation L corresponding to a literal l appearing in the body of r. To propagate the additions to L to the IDB relation P, the rule body is evaluated using DB_{New}. The evaluation is constrained using bindings from ΔL_{add}. The result of this evaluation is a relation whose schema contains all of the variables that occur in the rule body. The relation is projected onto the set of attributes corresponding to the set of variables that appear in the head literal. A similar procedure is performed to process the removals, however, the rule body is evaluated using DB_{Old} and the evaluation is constrained using bindings from ΔL_{rem}. The result of the projection is the approximation for P.

Note that for both additions and removals the subgoal representing the literal l is removed from the query for efficiency, since the updates to L used to constrain the query bind all variables appearing in subgoal l. This forms a query that tests a tuple that has already been determined to be an actual IDB update and, therefore, represents redundant computation.

If the rule body contains several literals, each representing either a candidate relation or an updated base relation, then a separate query is issued for each literal. In addition, a separate query is issued for the additions to, and the removals from, each relation. In the worst case situation, where a rule defining a predicate p has k subgoals each corresponding to a relation where both addition and removal updates have been identified, $2k$ queries would be issued during the propagation phase to obtain all potential updates for p. Memoing[WAR92] is used in our implementation to minimize redundant computation.

The propagation phase propagates the changes to the extensional relations up through the rules and identifies potential changes to the intensional relations. Potential changes are filtered to identify actual changes. For example, a potential addition represents a derivation of a tuple t. If t is provable in the database state before the updates, then the potential addition is filtered and is *not* reflected as an actual change to the database. Similarly, a potential removal represents the deletion of a derivation for a tuple t and if t is still provable in the database state after the updates, then the filter phase does not identify the potential removal as an actual removal.

Thus, the filtration phase of *PF* refines the approximation of potential updates to IDB relations identified during the propagation phase. Potential IDB updates that cannot be proven are removed from the approximation. Each potential addition is posed as a query to the database using DB_{Old}. Each potential removal is posed as a query to the database using DB_{New}. Tuples returned as a result of the query do not represent a change in the database state so they are deleted from the approximation.

2.1 The TPF Algorithm

The consistency checking method is based on the *TPF* (Tightly-coupled Propagation and Filtration) update propagation algorithm, which is a successor of the *EPF* algorithm[HAR94]. The basic *TPF* algorithm consists of three steps, which are described below. After a candidate rule is identified that includes a reference to the updated relation, the first step is to perform a propagation phase using the rule. The result is an approximation of the changes to the knowledgebase.

It is possible that a potential IDB update p may appear in more than one approximation as a result of multiple recursive invocations of the *TFP* algorithm. Filtration need only be performed once to determine if p is disqualified, or alternatively, if p represents an actual IDB update. To avoid redundant computation, *TFP* performs a check to ensure that p is only filtered once. If a tuple has already been filtered it is removed from the approximation.

The Δsets contain the actual IDB updates that have been identified at any point in time during computation. The Ωsets represent the momo tables that contain tuples that have been rederived but determined not to be actual IDB updates. These checks are implemented in step 2.

Filtration is implemented in step 3. Identification (by unification) and instantiation of relevant constraints are also performed during this step. These constraints are evaluated during this step to implement generation. If a violation is detected, the algorithm terminates reporting a violation. If the algorithm terminates without reporting a violation, then the knowledge-base is consistent. The TPF procedure is given in Figure 1.

2.2 Tighter Coupling of Propagation and Filtration

Although the TPF algorithm eagerly identifies disqualified IDB updates after each rule is evaluated, it is possible to filter earlier. In fact, the filtering can be done after each subgoal is evaluated. This results in minimal irrelevant computation.

This can be accomplished by disqualifying *bindings*, rather than tuples. A binding that is disqualified will not be used when evaluating other subgoals. The algorithm for "inter-rule" filtration is given in figure 2. Note that Q_i represents the ith (out of

Procedure TPF:
Given a set of updates $\mathcal{U}$ to relation R:
For every candidate rule referencing R of the form:
$p(\overline{X}) \leftarrow q_1(\overline{Q_1}), \ldots, r(\overline{Y}), \ldots, q_n(\overline{Q_n})$

Step 1: Compute an Approximation

For each $\mu \in \mathcal{U}$:
Let $\theta = sub(\mu, \overline{Y})$;
Issue query: $\leftarrow (q_1(\overline{Q_1}), \ldots, q_n(\overline{Q_n}))\theta$;

* if *addition*(μ) $\rightarrow$ DB_{New} is queried.
* if *removal*(μ) $\rightarrow$ DB_{Old} is queried.

Let $\mathcal{A} = \mathcal{A} \bigcup \pi_{\overline{X}}$(query_result);

Step 2: Minimize the Approximation

Let $\mathcal{A}' = (\mathcal{A} - \Delta p) - \Omega p$;

Note: Relation Ωp represents disqualified IDB updates (obtained from the memo relations for P).

Step 3: Filter (rederive) Approximation and Evaluate ICs

For each $a \in \mathcal{A}'$:
Let $\alpha = sub(a, \overline{X})$;
Issue query: $\leftarrow p(\overline{X})\alpha$

* if *addition*(μ) $\rightarrow$ DB_{Old} is queried.
* if *removal*(μ) $\rightarrow$ DB_{New} is queried.

if result = *false* then
$\Delta p = \Delta p \bigcup a$;
Instantiate and evaluate relevant ICs using α
Call TPF(a)
else /* *success* */
$\Omega p = \Omega p \bigcup a$;

Figure 1: Basic TPF Algorithm

Step 1 (Revised): Compute an Approximation

$U_0 = u;$

$A_1 = \pi_{\overline{X1}}(U_0);\ D_1 = \pi_{\overline{X1}}(A_1 \bowtie Q_1);\ U_1 = A_1 - D_1;$

$A_2 = \pi_{\overline{X2}}(U_1 \bowtie Q_2);\ D_2 = \pi_{\overline{X2}}(A_2 \bowtie Q_1 \bowtie Q_2);\ U_2 = A_2 - D_2;$

.

.

.

$A_n = \pi_{\overline{Xn}}(U_{n-1} \bowtie Q_n);\ D_n = \pi_{\overline{Xn}}(A_n \bowtie Q_1 \bowtie \ldots \bowtie Q_n);\ U_n = A_n - D_n;$

Figure 2: Enhanced Approximation Algorithm

n) relation in the rule p and u is the update. It presumes that the subgoals of the rule body are ordered starting with the "most bound" where initially the only bound variables are those corresponding to bindings in u.

If after considering each subgoal of the rule, the set of bindings computed, i.e., U_i, is null then the algorithm terminates reporting that the approximation computed using the rule is null. Note that the rule used in this procedure will not have to be used for the subsequent filtration phase. This is because the filtering has already taken place during computation of D_i. Note also that static analysis of the rule, which exploits information regarding known keys, is required to decide if this enhancement is indeed applicable as in certain instances it is not. This analysis is not described here because of space limitations.

3 Related Work

There has been a significant amount of work done in the area of consistency checking in deductive databases. The classic approach developed by Lloyd, Sonenberg and Topor[LLO87] (LST) performs propagation without accessing the stored facts. However, the resulting approximation may not be ground and constraining bindings can be lost. This can cause significant redundant computation during evaluation. This is especially inefficient when the constraints contain constants which would not unify with the approximation.

The approach taken by Celma and Decker[CEL94a] is similar to LST except that the antecedent's of rules used to defined the affected ICs are used to form an expression that is less expensive to evaluate than that of LST. The expression is created using the bodies of rules that were used to propagate towards the IC. This expression will be less expensive to evaluate since at most one rule from each IDB definition will be used to create the expression.

The approach proposed by Olive'[OLI91] is similar to the EPF algorithm but without the memoing optimization. As a result it is less efficient in many cases. However, it offers an optimization that permits the elimination of some rules from consideration during the propagation phase.

The *Delete & Rederive (DRed)* algorithm, which was proposed by Gupta et al.[GUP93], is based on similar concepts to the TPF algorithm described here. *DRed* performs one propagation phase and one filtration phase per stratum. This behavior allows the creation and propagation of disqualified IDB updates between predicates within a stratum. After *DRed* completes this irrelevant propagation, it must then filter, i.e. rederive, each of these erroneously propagated tuples thereby incurring the cost of still more irrelevant computation. The *TPF* algorithm does not propagate disqualified IDB updates like *DRed.* Instead, it detects and removes any irrelevant tuples immediately after they are generated. This pruning eliminates costly erroneous propagation. The net result is a significant increase in efficiency.

4 Summary

An method for constraint checking in a knowledge-base is presented that interleaves generation and evaluation and accesses portions of the fact base to utilize constraining bindings. The method is based on an efficient update propagation algorithm, which offers a tightly-coupled derivation and rederivation procedure that results in minimum redundant computation. For future work we intend to develop additional optimizations and also upgrade our prototype, which is currently implementing the EPF algorithm. We also plan to implement TPF in a parallel environment. Since TPF only queries the knowledge-base, several recursive invocations of the algorithm could execute simultaneously.

Acknowledgements

The author wishes to thank Suzanne W. Dietrich for many informative discussions regarding this and related issues.

References

[BRY88] Bry, F., Decker, H. and Manthey, R., "A Uniform Approach to Constraint Satisfaction and Constraint Satisfiability in Deductive Databases", *Proceedings of the Intl. Conf. on Extending Database Technology,* Venice, 1988, pp. 488–505.

[CEL94a] Celma, M. and Decker, H., "Integrity Checking in Deductive Databases - The Ultimate Method?", *Proceedings of the 5th Australasian Database Conference*, Christchurch, January, 1994, pp. 136–146.

[CEL94b] Celma, M., Garcia, C., Mota, L. and Decker, H., "Comparing and Synthesizing Integrity Constraint Checking Methods for Deductive Databases", IEEE Intl. Conf. on Data Engineering, Houston, USA, February, 1994.

[GUP93] Gupta, A., Mumick, I. S. and Subrahmanian, V. S., "Maintaining Views Incrementally", *Proceedings of the 1993 ACM SIGMOD*, Washington, DC, May 1993, pp. 157–166.

[HAR94] Harrison, J. and Dietrich, S., "Incremental View Maintenance", In *Proceedings of the 5th Australasian Database Conference*, Christchurch, January, 1994, pp. 45-63.

[KUC91] Kuchenhoff, V.,"On the efficient computation of the difference between consecutive database states", In *Proc. of the Second Intl. Conf. on Deductive and Object-Oriented Databases (DOOD)*, Munich, Germany, December 1991.

[LLO87] Lloyd, J. W., Sonenberg, E. A., and Topor, R. W., "Integrity Constraint Checking in Stratified Databases", *Journal of Logic Programming*, 4:331–343, 1987.

[NIC78] Nicolas, J. M. and Yazdanian, K., "Integrity Checking in Deductive Data Bases", In *Logic and Databases*, Gallaire H., Minker, J., (Eds), , Plenum Press, New York, NY, 1978, pp. 325–344.

[OLI91] Olive, A., "Integrity Constraint Checking in Deductive Databases", *Proc. 17th Intl. Conf. Very Large Databases*, Barcelona, September, 1991, pp. 513–523.

[SAD88] Sadri, F. and Kowalski, R., "A Theorem-Proving Approach to Database Integrity", In *Foundations of Deductive Databases and Logic Programming* (ed. Jack Minker), Morgan Kaufmann Pub., Los Altos, CA,1988, pp. 313–362.

[ULL88] Ullman, J., Principles of Database and Knowledge-base Systems, Vol. 1, Computer Science Press, Rockville, MD, 1988.

[VIE91] Vieille, L., Bayer, P. and Kuchenhoff, V., "Integrity Checking and Materialized Views Handling by Update Propagation in the EKS-V1 System", ECRC Technical Report TR-KB-35, ECRC, Munich, Germany, June 1991.

[WAR92] Warren, D. S. , "Memoing for Logic Programs", *Communications of the ACM*, Vol. 35, No. 3, March 1992, pp. 93–111.

Heuristic Query Analysis within a Distributed Deductive Database

Kathleen Neumann
Department of Mathematics / PIC
UCLA
Los Angeles, CA 90024-1555
neumann@math.ucla.edu

Abstract: In this paper an approach is presented for the heuristic ordering of rule execution within a distributed deductive database (DDDB). This paper presents a two phase heuristic analysis. Phase 1 is Argument Instantiation analysis, and Phase 2 is Cost Analysis. The result of this two phase analysis is a linear sequencing that indicates the order in which the "sub-rules" in the DDDB should be executed.

1 Introduction and Background

Because the needs and requirements of database users have continued to grow and change, we are forced to constantly look for new and more efficient ways to handle database access. One way to approach the database problem is through the use of deductive databases and distributed processing. This scheme's view of a distributed deductive database system is one in which data is distributed across many computers, or workstations, which are in a somewhat geographically localized vicinity.

In a Deductive Database, relations represent stored information and rules represent knowledge about how to use this information to deduce new facts. A Distributed Deductive Database (DDDB) is a Deductive Database (DDB) in which facts and rules are distributed among many processors. The problem we are concerned with is how to perform the distribution of rules and relations to minimize cost. In this Distributed Deductive Database model, the rules and relations of the DDB are first compiled into a graph, which is called the Extended Predicate Connection Graph (EPCG). A partitioning process takes place on this graph to divide it into a suitable number of clusters (see Figure 1). Each cluster, or partition, is then assigned to a processing node. By partitioning the EPCG and distributing the partitions, queries are able to be processed in a distributed fashion. Each partition will contain one or more relations as well as the set of rules relevant to those relations. No individual rule node or relation node is partitioned across the network. Also, more than one query can be active within the system at any given point in time. Although there are numerous issues and parts that comprise the whole model, this paper will describe a heuristic for ordering the execution of the rules.

E. A. Yfantis (ed.), Intelligent Systems, 741–748.

Rules:
P0(X,Y):-P1(X,Z) P2(Z,Y)
P2(Z,Y):-P3(Z,W) P4(Z,Y)
P4(Z,Y):-B2(Z,Y)
P6(Y,W):-B3(Y,W)
P1(X,Z):-B0(X,Z)
P3(Z,W):-B1(Z,W)
P5(Z,W):-P2(Z,Y) P6(Y,W)

Base Relations:
B0(X,Z) *B1(Z,W)*
B2(Z,Y) *B3(Y,W)*

*(Note: Nodes P0(X,Y) and P5(Z,W) are the Entry Point Nodes for this DDS *)*

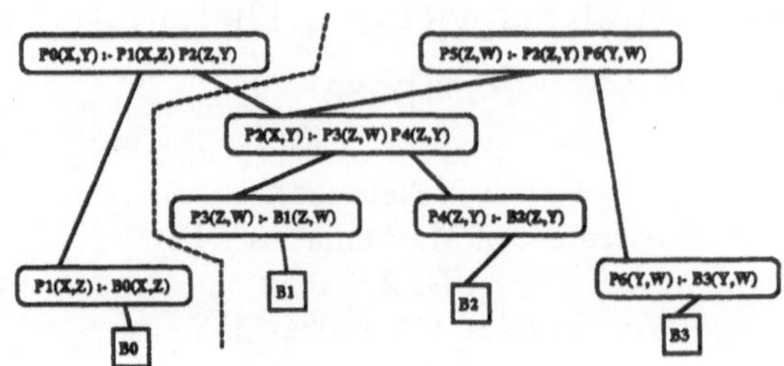

Figure 1: Partitioned EPCG (two partitions)

2 Related Work

This work was inspired by SYGRAF [Kif88]; [Kif86] and by a deductive database processing method presented in [Van86]. However, the present model approaches the problem from a different point of view.

The SYGRAF method is dynamic. So, it is difficult to compare the two in general. However, SYGRAF used a bottom-up processing strategy. Bottom-up processing techniques have well know and well published drawbacks. These drawbacks become even less tolerable in a distributed environment where communication of unnecessary tuples needs to be avoided whenever possible. Also, the use of dynamic filtering can become very complex and hard to understand, and the implications are even more serious in a distributed system.

The proposal of Van Gelder [Van86] is very interesting, yet not general or efficient enough to be implemented effectively in a real world situation. In trying to generalize the model, various problems emerge. First of all, a customized graph is constructed for every query when the query arrives for processing. This results in an extreme amount of overhead in graph construction. I feel that by maintaining a minimal amount of information obtained from the *heuristic query analysis*, the most effective method of rule processing can easily be determined.

Several ideas from Van Gelder have been utilized in the present model. For example, the concept for rule adornments was adapted for use the this model.

3 Heuristic Query Analysis

Since a user's query does not specify access paths to the desired relations, the system must provide a query-processing subsystem which will evaluate a number of strategies for processing the query and select the one that optimizes a given performance measure. The effectiveness of this subsystem affects the performance of the whole system.

A general strategy for distributed query processing is the decomposition of a query into sub-queries to be executed where the data resides. This is in contrast to the collection of all required data at one site, where the whole query is subsequently executed. The presented model employs a method in which sub-queries are executed where the data resides.

The only reason that query optimization is desirable is to make the system operate more efficiently. Efficiency can be expressed in terms of response time, communication cost, cpu time, etc. This optimization strategy strives to meet the efficiency goals in all areas to an acceptable degree.

Regardless if the deductive database system is distributed or centralized, subgoal ordering analysis can be done for all the rules at precompile time if we assume a function free database. The objective is to order subgoals for processing in such a way that solutions are generated in an efficient and the most cost effective way possible. There are several reasons to take advantage of *heuristic subgoal ordering analysis.* In general, a subgoal will have fewer possible solutions, and therefore be more easily solved and require less transmission cost when some of its variables are bound. If all of the subgoals are processed immediately in parallel, the join can be quite difficult and time consuming to perform. By waiting, the search space will be narrower and not as many fruitless possibilities will be tried. In other words, executing all subgoal in parallel may force the computation of the entire relation, when what is desired is a small fraction of that relation. Therefore, we need a method of ordering the subgoals in a linear fashion. The basis for the ordering of the subgoals in the body of a rule is the sharing of variables among the subgoals.

The order in which the subgoals initially appear "written" within the rule body does not have any correlation to the order in which the subgoals are executed. Each rule node in the EPCG where a query may enter the system, an *Entry Point Node*, is decomposed into an appropriate number of subgoals. The optimal subgoal ordering for the evaluation of each of these *Entry Point Nodes* is constructed by integrating the optimal ordering solutions of the subgoals that make up the current *Entry Point Node.* The order in which the subgoals of a rule node are executed is determined by the instantiation pattern of the variables in the head literal and by cost considerations. This subgoal ordering sequence is obtained by applying a set of heuristic rules. The result is a linear sequencing that indicates the order in which the subgoals should be executed or "fired".

To aid in the task of subgoal ordering, the concept of rule adornments is used. The predicate arguments are divided into three classes "b", "d" and "f". Class "b" arguments are bound constants. These are arguments that are known when a query is given to the system. The class "d" arguments are variables whose possible values will be determined, and the class "f" arguments are free variables for which we are looking for possible bindings. These classifications are used to help simplify the task of finding a producer and consumer for each argument.

This query analysis scheme results in a scheduler that relieves the user from the task of query optimization. This heuristic query analysis scheme combines two optimization methods: 1) Argument Instantiation Analysis - from the logic view point, and 2) Cost Analysis - from the database view point.

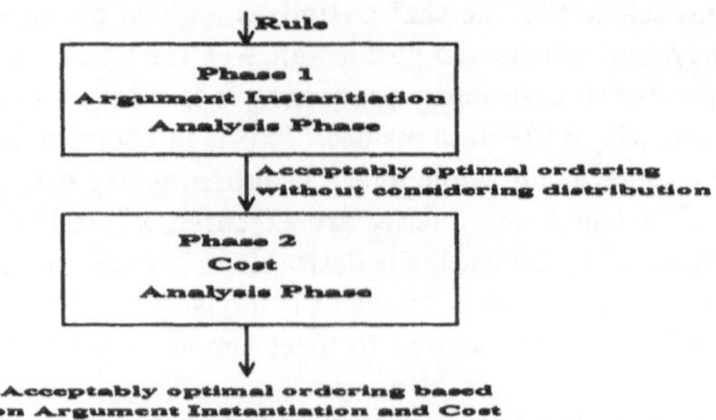

Figure 2: Overall view of *Heuristic Query Analysis*

Figure 2 shows a high level overall view of the ***heuristic query analysis.***

4 Argument Instantiation Analysis Phase

The Argument Instantiation Analysis Phase yields the most efficient ordering when considering instantiated variables, number of variable occurrences, etc. This phase does not take into account any concerns resulting from the rule nodes or base relation nodes distributed on the network. This phase determines the producer for each subgoal and the consumer of each subgoal. In effect, it reduces the search space of a query to one that only derives intermediate solutions that might contribute to the final solution of a query with the given instantiation pattern.

4.1 Argument Instantiation Analysis Phase – two cases

- Case 1: The situation in which all variables within any given rule head are uninstantiated: Determine all subgoal ordering permutations. If this number is unreasonably large, then determine an adequate subset. There are several common methods for finding an acceptable subset.
- Case 2: At least one variable in the head predicate of a rule is instantiated: Use the following guidelines to determine the order.

Argument Instantiation Analysis Guidelines:

1. If a DDDB rule contains only one subgoal, then no matter what the instantiation pattern in the head predicate, the subgoal ordering is always the same: execute Subgoal 1.
2. Any argument that is instantiated in the head, is instantiated with the same value everywhere that it appears in the rule body. No new values are produced for arguments that are initially bound by the user.
3. A producer for every non-bound argument must be determined. The basis for the ordering of subgoals in the rule body is the sharing of arguments. Whenever two or more subgoals have an argument in common, one of the subgoals is the producer for

that argument and is solved before the others, which are now consumers. After the producer subgoal of an argument has been determined, the consumers can be ordered for processing.

4. If an uninstantiated variable appears in the head predicate and only appears once in the body of the rule, then the subgoal in which this argument appears is processed last.

5. Any subgoal which contains existential variables and non-existential arguments is processed before subgoals that also contain the same non-existential arguments. This potentially reduces the search space of the non-existential arguments. Non-existential variables are ones which appear only in body literals not in the head literal.

6. If more than one subgoal could potentially be placed in the ordering list next, then choose the left most one as the next subgoal in the ordering list. Note: left most in this case means the one that appears the most to the left in the original rule body.

4.2 Argument Instantiation Analysis Algorithm

The following algorithm produces "layers" of subgoals. Let N be the number of layers produced by the *Argument Instantiation Analysis Algorithm*, where $N \geq 1$. Subgoals are places in their appropriate layer based on the above guidelines. All of the subgoals in layer 1 must be processed before any of the subgoals in layer 2 can be processed, etc, through layer N. More than one subgoal can potentially be in each layer, but Phase 1 does not specify the processing order within a given layer. The ordering within a layer is done in the *Cost Analysis Phase.*

- Step 1: If the rule contains only one subgoal, then that subgoal is the only one that appears in the ordering list for all instantiation patterns. The algorithm terminates for that rule.

- Step 2: Find all subgoals that share the bound argument(s) with the head literal. If there is more than one subgoal, then apply above rules 4, 5, and then rule 6. Place the subgoals in the ordering list.

- Step 3: From the previous step, mark all other arguments in these subgoals as "free". These subgoals are the producers for those marked arguments.

- Step 4: Find all subgoals that share the "free" arguments from the previous step. These arguments are now marked as "determined". These subgoals can now be placed in the ordering list. If there is more than one subgoal in question at this point, apply above rules 4, 5 and 6 in order to determine the processing order. Place the subgoals in the ordering list.

- Step 5: Go to Step 3 until all subgoals have been ordered.

The result of Phase 1's analysis is an acceptable optimal ordering without considering distribution.

5 Cost Analysis Phase

The Cost Analysis Phase handles concerns resulting from the distribution of EPCG nodes on the network. This phase uses information from the Argument Instantiation Analysis Phase and produces an efficient subgoal ordering in terms of cost. Cost is defined to be: the sum of all costs incurred in processing the operations of the query at various workstations and the cost incurred in inter-workstation communication. Both transmission and local processing costs are linear functions of the amount of data. There are two common ways to view cost. The first way is to view cost in terms of minimizing the amount of data transmitted. The second way is to view cost in terms of minimizing the response time. This scheme is concerned primarily with the amount of data transmitted. The combination of the two phase *heuristic query analysis* is designed to keep response time down to an acceptable level. The cost analysis takes into consideration such things as: the cardinality of relations, transmission cost between nodes, cost of join and union operations, etc.

There are two cases for cost analysis:

- Case 1: Calculate the cost of all permutations of the <all free> variable instantiations from Phase 1: Case 1. Retain only the cheapest ordering scheme.
- Case 2: Calculate the cost of all other possible instantiation pattern ordering schedules that where obtained in Case 2 of Phase 1. For the given instantiation pattern, keep the cheaper of this current ordering or the ordering obtained in Case 1 of Phase 2.

The following variable are used in the cost analysis equations:

Parameters:

X_j Definition: The (cardinality of the relation at node j) / (the sum of the number of distinct values of each bound argument).
i.e.: the value of X_j depends on the number of bound arguments in the current instantiation pattern, of the given predicate, and on the number of distinct values of each bound argument.

C_j Definition: Unit data processing cost at node j. (The average time determined to move a tuple within main memory.)
Typical Value: 20 usec

a_{ij} Definition: Unit data transmission cost between nodes i and j. (The same as the time to perform a random I/O.)
Typical Value: 15 msec

b_{ij} Definition: Transmission startup cost between nodes i and j. Eventually, these values will be kept in a matrix, however, prior to partitioning, all of these values are assumed to be the same.
Typical Value: 15 msec

n_m Definition: Cardinality of intermediate relation at node m.

$g_m(n_m)$ Definition: The cost associated with a join operation.
Typical Value: This is dependent on the semi-join method used.

k Definition: The number of relations within the given rule.

c Definition: Constant used in calculating the cost of materialization. This relates the cost of performing a compare and an insert. [3 usec = time to compare two attribute values in main memory + 20 usec = time to move a tuple in main memory]
Typical Value: 23 usec

IP Definition: The instantiation pattern

$n + n_{11}$ Definition: The cost of joining relation *1* with cardinality n_1 with relation *11*

CP Definition: Constant cost of comparing members of the unioned relation to the deletion of duplicates.
Typical Value: 23 usec

Cost Analysis Algorithm

```
For each Base Relation in the DDDB Do
 {For all possible instantiation patterns Do
  {Calculate the cost for the current instantiation pattern value where:
      Cost(X) = C_j * X_J
  }
 }

For each rule in the DDDB Do
working conceptually bottom up
 {For each possible instantiation pattern (IP) Do
  {Total_Rule_Cost(IP) = 0
   Layer = 1
    For each Layer Do
    {Repeat until all Subgoals in this layer are ordered
     {While more Subgoals Do
      {Calculate the cost of each remaining subgoal where:
         Cost(current subgoal at node i) = a_ij * X + b_ij
           + cost(current subgoal's IP at node j)
      }
     Add to the ordering list the subgoal with the cheapest cost as was
     determined in the previous step.

     Add this cost figure to Total_Rule_Cost(IP)

     Total_Rule_Cost(IP)= Total_Rule_Cost(IP) + Current_cheapest_subgoal's_cost

     Mark any arguments that appear in the most recently ordered subgoal and in
     the remaining unordered subgoal as "bound" in the unordered subgoals

     } end of Repeat Until
    Layer = Layer + 1
    } end for for each layer

   (Add cost of Joins and Unions to Total Cost)

   Total_Rule_Cost(IP)=
```

Total_Rule_Cost(IP) + $\sum_{m=2}^{k}(n_{m-1} * g_m(n_m))$ + $\sum_{m=2}^{k}(c + n_m)$ **(join cost)**

+ $\sum_{u=2}^{unions}(n_l + n_{ll}) + (CP + n_{l+ll}))$ **(union cost)**

} (for each IP)
} (for each rule)

6 Conclusions and Future Work

In this paper, we have focused on the problem of heuristics query analysis. Algorithms for the solution to our particular problem have been developed. Experimental results indicate that the proposed algorithms perform well on a variety of EPCGs. This paper considered only one set of possible heuristics to use. There are numerous other heuristics that could be employed. The worth and potential of these other heuristics are currently being evaluated.

References

[Ber81] Bernstein, Philip A. and Nathan Goodman. 1981 Query Processing in a System of Distributed Databases (SDD-1). *ACM Transactions on Database Systems.* 6(4) (Dec): 602-625.

[Kif86] Kifer, M. and L. Lozinskii. 1986. Filtering Data Flow in Deductive Databases. *Proceedings of the International Conference on Database Theory.* (September): 186-202.

[Kif88] Kifer, M. and L. Lozinskii. 1988. SYGRAF: Implementing Logic Programs in a Database Style. *IEEE Transactions on Software Engineering.* (July):922-934.

[Neu92] Neumann, Kathleen. 1992. Multi-query Processing within a Distributed deductive Database. Ph.D. diss., Northwestern University.

[Qad91] Qadah, G.Z., Jung Kim and Larry Henschen. 1991. Efficient Algorithms for the Instantiated Transitive Closure Queries. *IEEE Transactions of Software Engineering.* (March).

[Sha91] Shao, J., D.A. Bell, M.E.C. Hull. 1991. Combining Rule Decomposition and Data Partitioning In Parallel Datalog Program Processing. *PDIS 91*: 106-115.

[Ull89] Ullman, Jeffery D. 1989. *Principles of Database and Knowledge-Base Systems, Vol I & II.* Rockville, MD: Computer Science Press.

[Van86] Van Gelder, Allan. 1986. A Message Passing Framework for Logical Query Evaluation. *ACM SIGMOD*: 155-165.

[Zha91] Zhang, Weining, Ke Wang, Siu-Cheung Chau. 1991. Data Partition: A Practical Parallel Evaluation of Datalog Program. *PDIS 91*: 98-105.

Methodology for Implementing the Access Control in Behaviorally Object-Oriented Database Systems

Cristian Radu Mark Vandenwauver René Govaerts Joos Vandewalle

Katholieke Universiteit Leuven, Laboratorium ESAT
Kardinaal Mercierlaan 94, B-3001 Heverlee, Belgium

Christian.Radu@esat.kuleuven.ac.be
Mark.Vandenwauver@esat.kuleuven.ac.be

Abstract. Providing security for a general-purpose database system is an expensive operation. An access control mechanism has to offer the level of security that people can afford. The behaviorally object management systems have features which can be used to achieve this goal. We present a methodology of providing security for the behaviorally object management systems. We adopt an efficient security mechanism implemented with access control lists. The granularity of the controlled authorization operations is as fine as the methods in the interfaces of the objects. We implemented this scheme in the framework of the ONTOS [6] object management system.

Keywords. Access control lists, Direct control of the method invocation, Authorization object lattice.

1 Introduction

An object management system (OMS) is a set of programs that access, manipulate, protect and manage a collection of related objects, referred to as the *object database*. The database often needs to be accessed by several applications, which could act on behalf of different users. Therefore the OMS has to guarantee that only the authorized operations are executed and only to those parts of the database the users are entitled to [1]. An *access control mechanism*, also called *security mechanism*, enforces this requirement.

Our paper addresses a methodology of providing security features for general-purpose behaviorally OMS (with respect to Dittrich's taxonomy [2]). We present an efficient security mechanism implemented with access control lists. The database schema is implemented with Type objects, whose instances are persistent objects in the object database. The granularity of the controlled authorization operations is as fine as the methods in the interfaces of these Types. This is achieved by the *direct control of the method invocation.* We stress the modification of the general form of a Type in order to allow the access control at the level of a method. A pseudocode description, in terms related to C++ language and in the framework of ONTOS-OMS, is used to briefly describe the methodology we recommend.

E. A. Yfantis (ed.), Intelligent Systems, 749–755.

2 Security Considerations

From the viewpoint of the security requirements, the behaviorally OMS have suitable features. We mention here the encapsulation and the inheritance.

The encapsulation feature provides an initial layer of security. All the data is stored as values for the attribute variables, which are encapsulated inside objects. They are available *only* through the methods provided in the interface.

Object interfaces are much richer than the simple read/write/execute operations used in relation with files. Therefore there are as many operations as methods in the interface of an object. This implies that the access rights have to be controlled in a correspondingly fine grained way. This can be achieved by *the direct control of the method invocation.* This is the approach we adopt to design the security mechanism in the framework of the behaviorally OMS.

The simple and multiple inheritance feature provides the ability to attach different security mechanisms and policies to individual objects. This is described in [3] and isn't further investigated in this paper.

3 The Access Control Mechanism

Let S be the set of *authorization subjects*, O the set of *authorization objects* - granule of the object database whose privacy and integrity have to be protected - and A the set of possible *authorization operations.* The main function of the access control mechanism, also known as security mechanism, is to decide if the authorization subject $s \in S$ is allowed to execute the authorization operation $a \in A$ with regard to the authorization object $o \in O$. This authorization request is expressed as the triplet (s, o, a). The access is granted or denied depending on the existence of an explicit authorization rule (s', o', a'), on behalf of which the request is considered eligible or not. The details of the validation process of the authorization request against the authorization rule can be found in [4]. In this section we establish the elements of the three authorization sets S, O and A. We show that using *the direct control of the method invocation* the set of authorization objects O can be reduced. For this purpose, we use the concepts of *Authorization Object Schema (AOS)* and *Authorization Object Lattice (AOL)* as they are introduced in the authorization model presented in [4]. We do not consider the problem of the implicit and weak authorization. Therefore, the implication rules along the three domains S, O, A are not part of this paper.

3.1 Authorization Subjects

Usually, in order to reduce the number of authorization subjects that need to be involved in the definition of the explicit authorization rules, the users are grouped according to their roles. Afterwards, hierarchies are established among roles. This allows an efficient way of "ordering" the subjects and consequently of deriving implications among the subject dimension of the authorization. This approach is used in [4, 5], and leads to a restricted set of authorization rules.

We consider in our access control mechanism only three roles for which we associate rights. These are the roles accepted by the UNIX protection mechanism: the owner of the object database (ow), the group to whom the owner belongs (g), the other people (p).

3.2 Authorization Operations

In an OMS, applications are groups of instances that invoke other operations. Our security mechanism is based on the control of the invocation of these operations. It is important to distinguish between two categories:

- **OMS operations** - they are available at the level of the application program which manages the object database.
- **internal operations** - they represent the "atomic operations" on which relies the OMS operations. The internal operations correspond to the *methods* provided in the interfaces of the object classes.

As an example of the OMS operation, *readAttribute* reads the value of an attribute of an object in the database. The *objectName* and the *attributeName* are specified as parameters of the OMS operation. The internal operations involved are the methods in the interface of the object class *className*. This represents the abstract data type from which the *objectName* object was instantiated. The methods in the interfaces of other object classes may be invoked as well. Thus, any OMS operation can be derived from a set of methods which are invoked during the execution of an OMS operation. This implies that *the security mechanism has only to provide means to control the invocation of the methods.* This leads us to consider the method in the interface of the object class as the authorization operation unit. Then, A is the union of all the methods offered in the interface of the object classes, describing the database schema. We do not consider the OMS operation as the authorization operation. Furthermore, we do not analyze and establish hierarchies among authorization operations. An approach based on hierarchies and implications among OMS operations is described in [5].

3.3 Authorization Objects

The set of *authorization objects O* is derived from the set of objects kept with the *OMS database*. This contains *the database schema* and the *object database*. The database schema can be subdivided into the *user's schema* and the *Kernel Schema*. The user's schema is the collection of objects Types and corresponds to a Type hierarchy. The subtype/supertype relationship between Types is maintained as part of the user's schema. There may be additional Kernel Types that are provided by default with the OMS that form the Kernel Schema. Such Types could be *metatypes* - types which are used to define and represent other types and whose instances are object Types - or *aggregate types* (List, Set, Array, Dictionary). The *object database* keeps the instances of the object Types defined in the database schema.

If we consider the database of an university, the user's database schema is formed by the object Types: *Person*, *Student*, and *Professor*. *Person* is a supertype for *Student* and *Professor*. In order to substantiate the discussion about the Kernel Schema we refer to the ONTOS-OMS. Here, the representation of an object Type in the user's schema is a

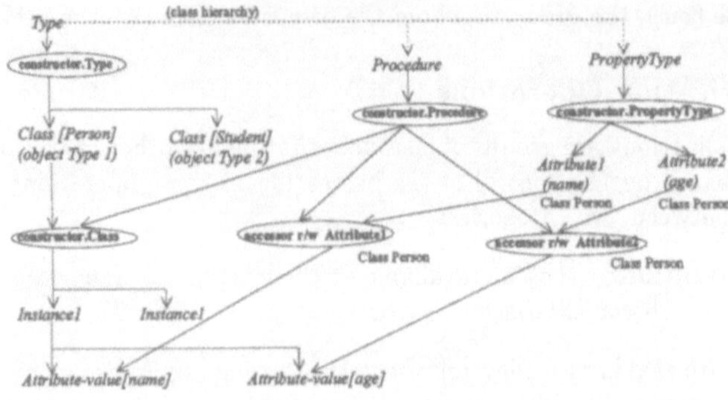

Figure 1: The Authorization Object Lattice (AOL)

C++ class. The AOL of our example is envisaged in Figure 1. This subsequent paragraphs explain the structure of the AOL as well as the authorization objects we consider in O.

We can define a fundamental hierarchy in the OMS database, defining how objects are organized in terms of other objects. This hierarchy is named the *database granularity hierarchy*. Given an object o, this hierarchy defines how o is composed of objects of the lower levels. Because we chose the option of a fine-grained set of authorization operations, to each node of the database granularity hierarchy we associate one and only one *authorization object type*. If we consider that the classes are themselves objects - because they are denotable instances of the Type - we can distinguish between the *instance attributes*, that characterize all instances of the class, and the *class attributes* that characterize the class itself as an object. Hence, we introduce the authorization object type *Attribute*. There is an implication link from the node *Attribute* to the node *Attribute-Value* because any authorization operation defined for an attribute of a class is available on the corresponding attribute values of all instances. Moreover, in behaviorally OMS, the database schema introduces an additional entity namely methods. Correspondingly, an authorization object named *Method* is adopted. We also consider two other distinct authorization object types, namely *PropertyType* and *Procedure*, belonging to the Kernel Schema. Each of the attributes of a class is an instance of the *PropertyType* type. In the same way each method of a class is an instance of the *Procedure* type.

Because of the encapsulation and communication paradigm, the access to the attributes of a class can be controlled through the method invocation. The method invocation can also be used to change the class definition - including the set of attributes and methods of a class - and to generate new instances of the class. Consequently, there are implication links from the node *Method* to the nodes *Attribute* and *Class* in the AOL.

In the AOL, presented in Figure 1, we stress the important role of the method invocation. For example, the invocation of the constructor of the Type creates a new instance object Type. It results that the right of generating new object Types in the user's database schema is equivalent to the right of invoking the constructor of the Type. The right of reading or writing an attribute of an instance is reduced to the right of invoking the data status accessor method of the class from which the instance was created.

4 Implementation

In this section we use the terminology agreed in the framework of the ONTOS object database management system. ONTOS provides a reliable persistent storage facility for C++ objects. This means [6]:

- objects denoted by C++ program variables can have a lifetime that is longer than that of the program that created them;
- C++ programs are allowed to retrieve persistent objects (objects created by other programs) into their program variables.

We consider the OMS database *University*. Moreover, we only mention the object Type denoted *Person* in the user's database schema. The abstract state of the type *Person* is represented by properties as name, age, etc. The set of procedures describing the behavior of the type consists of: constructors/destructors, accessor to read and write the state of the objects. In ONTOS database schema Types correspond to C++ "persistent classes". The persistence of the class is provided by inheritance from a class belonging to the Kernel Schema of ONTOS, denoted by *Object*. In C++ the term *class* describes the state and behavior of objects by defining *data members* and *member functions* used by each instance of the class. The class describing the *Person* type is presented in the left side of Figure 2.

The application program that manages the database *University* defines OMS operations like:

- *createObject(objName)* - creates a new object in the database,
- *deleteObject(objName)* - deletes an object from the database,
- *readAttribute(objName, attrName)* - gets the value of the attribute *attrName* for the object denoted by *objName*,
- *modifyAttribute(objName, attrName, newValue)* - modifies the value of the attribute *attrName* according to the value indicated by *newValue*.

Furthermore, an invocation of an OMS operation, let's say *readAttribute (Person1, age)*, relies on the invocation of the method *intPerson :: get_age()* in the interface of the persistent class *Person* on the instance *Person*1. Therefore we have to control the invocation of the *get_age* accessor.

The main idea in order to achieve the access control mechanism is to transform the structure of the initial *Person* class. Thus the traditional constructors, destructors and accessors - which previously represented the behavior of this class - have to be committed as "protected". A new interface of the type class is defined, the signature of each method embeds a particular identification/authentication information about the subject. The code implementing the method contains a ramification statement which decides if the subject

```
class Person : public Object {
    private :
        int age;
        ...
    public :
        constructor(...);
        destructor(...);
    // classic accessors
        int get_age( );
        // read operation
        ...
};

int Person::get_age( )
{
    return age;
}
```

⇒

```
class ProtectedPerson : public Object {
    private :
        Reference priv_aclDictionary;
        int age;
        ...
        constructor (...);
        destructor (...);
    // classic accessors
        int get_age ( );
        ...
    public :
        prot_constructor (..., ID);
        prot_destructor (..., ID);
    // factual accessors
        int prot_get_age(long ID);
        ...
    // some method accessible only to
    // the owner to deal with the
    // aclDictionary.
};

int ProtectedPerson::prot_get_age(long ID) {
    if ( ID ∈ aclDictionary && access is granted)
        return (get_age( ));
    else
        Exception
}
```

Figure 2: Pseudo code.

is authorized to call the method or not. If the subject is eligible, the method is executed by calling the corresponding protected method. Otherwise, the access to the protected method is denied. At the same time an exception is risen, informing the subject about the refusal. This can also be recorded in the journal of access violations, which can be read by the administrator of the database. The form of the resulting persistent class, that we have denoted by *ProtectedPerson*, is shown in the right side of Figure 2.

The authorization checking relies on a search operation in a dictionary of access rights. For this purpose a reference to a dictionary aggregate object is added to the protected data members of the object class. The dictionary is tagged with respect to the number of the messages which can invoke methods of the interface, and the element corresponding to each tag represents an access control list (acl) dedicated to that method (see Figure 3). Each entry in the acl consists of the identifier of the subject and the permission (+) or denial (-) of the access to the corresponding method. The creation and deletion of this dictionary as well as the insertion and removing of tags and elements in the dictionary is executed only by the owner of the database. He controls the granting of access rights to other subjects. Correspondingly, the OMS operations *SetAclEntry(objectName, SubjectID, AccessValues)* and *GetAclObject(objectName)* are provided at the level of the application program.

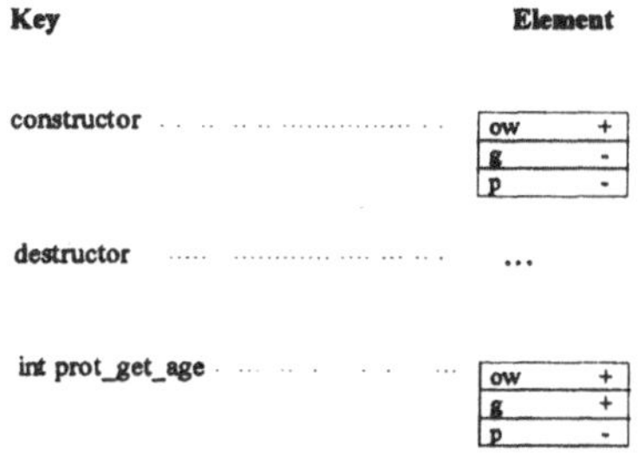

Figure 3: The Access Control List Dictionary

5 Conclusion

The security mechanism we have presented relies mainly on the encapsulation and communication paradigm between objects. These features characterize the behaviorally object-oriented databases. The implementation introduces a modality to access a hidden method through an identification part. The requesting subject is first looked up in the access list kept with the object. This list is realized by means of a dictionary aggregate type. The extension of the access control mechanism with a subject restriction list mechanism could be a theme for future development. We will also extend our work in order to cover both the class composition hierarchy and class hierarchy dimensions of the object data model.

References

[1] E.B. Fernandez, R.C. Summers and C. Wood, "*Database Security and Integrity,*" Addison-Wesley, Reading, Mass., 1984.

[2] K.R. Dittrich, M. Härtrig and H. Pfeferlee, "Discretionary Access Control in Structurally Object-Oriented Database Systems," *Database Security II, Status and Prospects*, C.E. Landwehr (Ed.), pp. 105-122, Elsevier Science Publisher B.V. (North-Holland), 1989.

[3] H. Härtig. O. Kowalski and W. Kühnhauser, "The BIRLIX Security Architecture," *Journal of Computer Security*, No. 2, 1993, pp. 5-21.

[4] F. Rabitti, E. Bertino, W. Kim and D. Woelk, "A Model for Authorization for Next-Generation Database Systems," *ACM Transaction on Database Systems*, Vol. 16, No. 1, March 1991, pp. 88-131.

[5] H.H. Brüggemann, "Rights in an Object-Oriented Environment," *Database Security V: Status and Prospects*, C.E. Landwehr and S. Jajodia (Ed.), pp. 99-115, Elsevier Science Publisher B.V. (North-Holland), 1992.

[6] "ONTOS Developer's Guide", *ONTOS, Inc.*, Release 2.1, 1991.

Conclusion

PROSPECTIVE VIEW ON INTELLIGENT DATABASES

Sang C. Suh
R. Daniel Crieder
Veerasekhar Kandula
Department of Computer Science
East Texas State University
Commerce. Texas 75429 USA

Abstract. Intelligent systems can take many forms such as natural-language translation systems, computer chess games, robotics applications, vision and speech recognition systems, neural networks, and knowledge based systems. Each of these systems offers a varying degree of ability to represent artificially the human knowledge. In attempting to replicate human intelligence, these programs often strive to manipulate vast amounts of information, to offer some degree of problem solving ability, and to augment human's ability to connect both information and concepts in nonlinear ways. We think that the effective use of information today is just a small fraction of what it could be if given the appropriate storage and retrieval mechanisms. In this paper we present our views on developing intelligent databases using different technologies

Keywords. Intelligent DataBases, Expert Systems, Intelligent Systems, Knowledge based systems, Hypermedia, Object Oriented Technology.

1 Introduction

Intelligent systems are present in many forms such as natural-language translation systems, computer chess games, robotics applications, vision and speech recognition systems, neural networks, and knowledge-based or expert systems etc. Each of these systems offers the ability to represent artificially the human mind at work often trying to manipulate large amounts of information. These systems offer some degree of problem solving ability, and also augment a human's ability to connect both information and concepts in nonlinear ways (Larry Bielawski et al., 1991).

Some of the information that is being published electronically is available on on-line databases. Many publishers carry electronic versions of books and journals, even if the final result appears in hard copy or book form. These electronic versions can be indexed and input directly into text databases (Kamran Parsaye et al., 1989). The increasing availability of information in electronic form is leading to the development of full text databases. The growth of information stored in institutions and databases is increasing day by day. And this vast body of information is only useful if there are sufficient methods of information retrieval. Traditional indexing methods may not be strong enough to deal with the organization and description of all this information. Even with necessary indexing, knowing where to find and how to access and retrieve the desired results with the current retrieval systems is difficult especially where the input specification is incomplete.

No one technology will completely solve a given problem, so the need to integrate diverse technologies within a single application is becoming a necessity. Combining hypermedia and knowledge-based systems technologies can supply embedded AI-based functionality within existing applications

E. A. Yfantis (ed.), Intelligent Systems, 757–761.

thereby rendering it more competitive. Conceptually we can see what to do, but the creation of intelligent database functionality turns out to be a complex topic.

2 Models for Intelligent Systems Integration with Databases

When it comes to calculating, sorting, reports generating traditional databases certainly are time saving tools. But they do not replicate human activities such as planning, diagnosis etc. Intelligent systems, on the other hand, behave logically, solve complex problems, provide nonlinear program navigation, make effective use of existing information, user friendly and interactive (Larry Bielawski et al., 1991). Hence, in order for a computer system to be "intelligent", it is desirable to integrate it with intelligent systems, so that the final system will possess at least a subset of the above said characteristics. In this section, we discuss a few basic models for intelligent systems integration with databases.

2.1 Intelligent Front-Ends to Databases and Existing Information

A common way that developers link hypermedia and expert systems to data bases and other information sources is as an intelligent front-end. The intelligent front-ends to existing information have several advantages, whether they help in the control or navigation of the running application or to ensure data/information integrity. The intent of such front-ends is to embed expert system and hypermedia components within the user interface (for example, in rule or hypermedia-based interfaces that include fill-in forms, hot graphics, tables, menus, dialog boxes, and so forth) to perform certain actions or tasks that would normally require human intervention. In such systems, the core structure of the information or data base is left intact, but a new degree of functionality is achieved by virtue of appending the intelligent front-end.

Samples of intelligent front-ends include preprocessing of information, where a knowledge-base component might create an information search-and-retreival strategy, or hypermedia can narrow the overall working domain. At times, these functions are accompanied by the generation of key words that link to data-base-specific function calls or that are used as search strings themselves. The net result is that the user does not have to spend as much time sorting through irrelevant information to find what is needed for the task at hand. Another popular type of intelligent front-end that is particularly applicable to a data-base or spreadsheet program is a data verification utility. With this type of intelligent front-end, the expert system component acts as a filter for incoming information input by the user.

2.2 Integration of Information During Consultation

Once an intelligent system consultation is under way, another benefit of integrating intelligent systems with existing information and data bases is the supplying of answers to questions that are needed during the working session. For example, instead of querying the user directly for information that already exists, the IF part of a rule can initiate a call to an on-line data base or issue a command to read an ASCII file that contains the desired information. This type of integration is useful in industrial and manufacturing batch-processing environments, where several iterations that concern testing or validation processes are required of an intelligent system to reach an aggregate result or conclusion.

2.3 Information Sources for Procedural Knowledge

Often, existing documents within an organization contain much procedural information expressed sequentially. In such cases, developers can directly map both the organization and the presentation of this information into rules within an expert system component that asks questions interactively while the inference mechanism supplies specific advice, recommendations, or conclusions. The benefit of this approach is that the system can behave more dynamically and interactively while efficiently taking into account all of the possible paths toward a single recommendation or conclusion. In addition, by simply adding more rules and factors, developers can modify the intelligent system more easily than the previous manual system that was in place.

2.4 Intelligent Back-ends to Database Programs

Like intelligent front-ends to data-base programs, intelligent back-ends provide additional functionality to traditional data-intensive applications, but in programs like this, developers employ hypermedia and expert system components either to create more effective reports or to help analyze the stored data. In other cases, where the information is textual, developers may use classification algorithms, procedural rules, or statistical procedures to scan the data, looking for possible relationships or patterns that might be initially hidden. To aid in this discovery process, an induction technique is sometimes used to help identify significant patterns or relationships. With this technique, the existing data, what was once just a collection of facts can be transformed into advice, conclusions, and recommendations.

3 Intelligent Databases

Traditional Data Base Management Systems (DBMSs) provide capabilities for data definition and manipulation without any reasoning involved. This makes the traditional DBMSs not suitable for future use where more responsibility is to be placed to the system. In order to satisfy current and future demand on system's capability, more functionality must be incorporated into databases, thus making the databases intelligent.

Intelligent databases are databases that manage information in a natural way, making that information easy to store, access, and use. Three notable features that allow databases to achieve certain level of intelligence when integrated with intelligent systems are that an intelligent database (Parsaye, et al., 1989).

- Provides high-level tools for data analysis, discovery, and integrity control, allowing users to both extract knowledge from, and apply knowledge to data.
- Allows users to interact directly with information as naturally as if they were flipping through the pages of standard text on the topic or talking with a helpful human expert.
- Retrieves knowledge as opposed to data and uses inference to determine what a user needs to know.

To support the intelligent DB model, a three layer architecture is proposed. The three layer architecture of an intelligent database consists of high-level tools, the high-level interface level, and the database support level. The relationship between each level of the architecture is as shown in Figure 1.

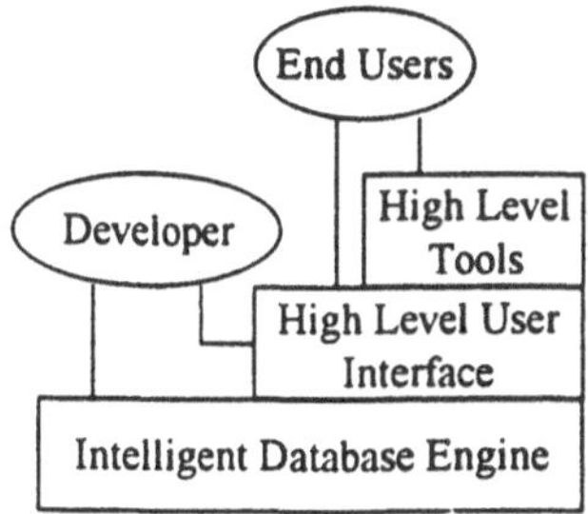

FIGURE 1: Architecture of an Intelligent Database

4 A View on Intelligent Database Models

The motivation behind writing this paper is based on what is required

- to manage large volumes of information available on on-line databases,
- to make information easily and meaningfully available to satisfy users needs, and
- to provide mechanisms that will magnify human capabilities.

In addition to what we thought that an intelligent database should do, the primary characteristics of an intelligent database are that it is easy and natural to use, can handle large amounts of information in a seamless and transparent fashion, and that it allows users to carry out their tasks using an appropriate set of information management tools. Information management tools such as hypermedia and expert systems are essential in a world of information. In our view, the solution to the problem of information overload requires the development of intelligent databases models as shown in Figure 2. The model we propose consists of three major components: front-end interface, back-end interface and an intermediate interface. Each component should be tightly coupled to make the database system more intelligent. It is noteworthy that users interact with the system throughout the operation by being loosely coupled.

One way to build intelligent databases is to integrate existing database applications with intelligent systems. In such cases the data itself becomes the core of the application and retains its information either in primary memory, or on a secondary memory such as on a hard disk, or on a CD-ROM device. The intelligent part of the system can be built around existing information. The information can either be in the form of structured text or data, which can be accessed directly through file input/output operations or through special utility programs. Linking a database and other sources of information to an intelligent system in this way produces a usable effect. The intelligent system can limit the scope of search process and efficiently retrieve information in a way that makes the program more effective than a conventional system. This is an essential function of an intelligent system and is necessary because of the amount of information stored in large databases.

Although intelligent systems can effectively combine hypermedia and expert systems, one technology will often dominate the overall program structure. Furthermore, when intelligent systems are designed around an abundance of information, as in existing text files or databases, the original program or document structure will sometimes dominate, relegating the expert system or hypermedia components to more subservient but, nonetheless, useful roles. As a computer program that relies on knowledge and reasoning to perform a difficult task expert systems prove to be very useful in database applications. Not only do expert systems provide the machine reasoning capability that can make the databases intelligent, but they may be viewed as higher level methods of database programming. With the availability of expert system shells most of which can function as front end tools to databases, we view that the future database management systems will have integrated expert systems for advanced query expressions, customized front ends as well as for dealing expert rules programmed in the same environment.

The associative and nonlinear nature of hypermedia makes it a browsing medium and an interface for knowledge to the databases. Hypermedia is already in effective use in most of the PC based software applications. Object-oriented programming, where the programming techniques allow the development of extensible and reusable modules, is a milestone in the development of post-relational database management systems underway. We foresee that Object-oriented database DBMSs will incorporate all those features that we are looking for in intelligent databases. We foresee an integration of all the afore said technologies in future database management systems which ought to move the traditional database technology into intelligent database technology.

5 Conclusions

In this paper, we have proposed a new model for intelligent database systems. Our views on how the intelligent data base systems should be designed and developed, were illustrated in connection with existing technologies. It is our strong belief that the future directions for data base systems design and development should be focused on intelligent aspects of the technology.

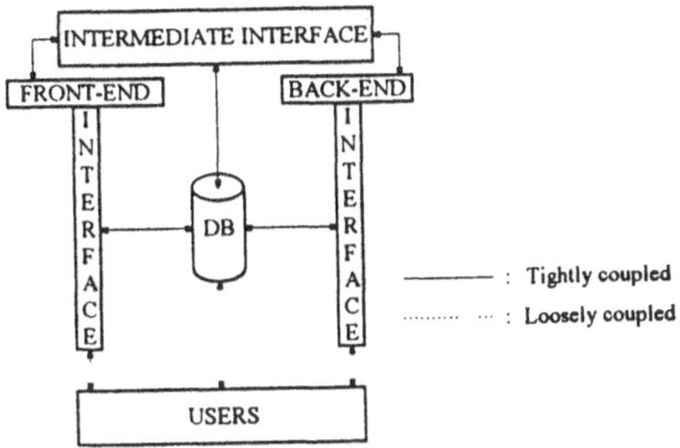

FIGURE 2: Intelligent Database (DB) Model

6 References

[1] Kamran Parsaye, Mark Chignell, Setrag Khoshafian, Harry Wong., Intelligent Databases, John Wiley, 1986, pp 25, pp 368.

[2] Larry Bielawski, Robert Lewand., Intelligent Systems Design, John Wiley, 1991, pp 42-44, 162-165.

[3] Emily Berk, Joseph Delvin, Hvpertext/Hypermedia Handbook, Armadillo Associates, Inc., 1991

[4] Hypertext 1987 Proceedings, pp 177

[5] Jakob Nielsen, Hypertext & Hypermedia, Academic Press, Inc., 1990

[6] Maureen Caudill, "Expert networks", Byte, Oct. 1991, pp 108.

[7] Robert H. Michaelsen, "The Technology of Expert Systems", Byte, may 1990, pp 303.

[8] Jonathan Barker, "Expert Systems Explained", Management Decision, 1990, pp 51.

[9] Frederick Hayes-Roth, Donald A. Waterman, Douglas B. Lenat, Building Expert Systems, Addison-Wesley Publishing Company, Inc., 1983.

[10] Joseph Giarratano, Gary Riley, Expert Systems: Principles and Programming, PWS Publishing Company, 1994.

Adaptive Query Reformulation In Attribute Based Image Retrieval

Gwang S. Jung
Department of Computer Science
Jackson State University
Jackson, MS 39217

Venkat N. Gudivada
Department of Computer Science
Ohio University
Athens, OH 45701

Abstract. In this paper, we describe the design and development of an adaptive Image Retrieval (IR) system. Our approach to image retrieval has a distinctive advantage in that the image and query descriptions need not be highly structured as would be required with the IR based on the traditional database management techniques. This system features an easy to use graphical user interface for both *relevance feedback mechanism* and query (re)formulation. The relevance feedback mechanism enables the system to capture the user's conceptual view of the image data. The query reformulation mechanism is designed for providing facilities by which the user's relevance feedback is effectively utilized to adaptively improve the retrieval effectiveness. The query reformulation algorithm is based on the calculation of functional dependency between each image attribute and the user's relevance feedback using a theoretical framework referred to as Rough Set Theory. Therefore, the manner in which the query is reformulated is systematic and is easily interpreted. Preliminary experimental results show that the reformulated query generated by the proposed method significantly improves the retrieval effectiveness.

Keywords. Image Retrieval, Query Re-formulation, Adaptive User Interface.

1 Introduction

Image Retrieval (IR) problem is concerned with retrieving images that are relevant to users' requests from a large collection of images referred to as the *image database*. Earlier approaches to the IR problem have generally been in one of the two directions. In one direction, IR systems that *emphasize* various kinds of image understanding techniques are developed [2]. A main component is designed for computing indexable features of the images from their low level representations (e.g., raster or vector formats). These features play a prominent role in processing the user queries. In the other direction, IR systems that represent an image by a set of attributes to suit conventional database technology have been developed [3]. Since these systems are developed based on the notion of high level attributes of images, they are referred to as *attribute based* IR systems. Each image is represented by either a set of key words or attribute-value (or property-value) pairs. Attributes represent formatted[1] data that are derived external to the image contents, such as the date of acquisition and the resolution of the image, and/or formatted data that are derived from the image through either image interpretation programs or human involvement. Typically, considerable human effort is involved in deriving semantically meaningful attributes [3]. Most of the currently available IR systems are based on attribute based retrieval techniques.

[1]Formatted data refers to data that are managed by conventional database management systems

E. A. Yfantis (ed.), Intelligent Systems, 763–774.

During the retrieval process, only the keywords or the attribute-value pairs are used, rather than the original image. Further, as a consequence of their design and implementation dependence on conventional database management systems, it is presumed that the users are familiar with the logical structure of the image database and it is possible for the users to specify their queries precisely and completely. However, it is often difficult to construct precise and complete queries by users. Moreover, because of the visual nature of images, images retrieved in response to the same query that is identically formulated by different users can be perceived/interpreted differently by users based on their backgrounds and characteristics.

Therefore, instead of aiming for ideal retrieval, that is, to retrieve all relevant images without retrieving any irrelevant ones, it is desirable to provide facilities by which the users can obtain *optimal retrieval*. Optimal retrieval is defined as the best retrieval that satisfies certain evaluation criteria under some realistic restrictions. One way to achieve optimal retrieval is through *dynamic reformulation* of a query by using the relevance feedback from the user. Relevance feedback is obtained in the form of user's judgments as to whether or not the retrieved images are relevant to his information need. Hence, it is desirable that an IR system be adaptive to the user's retrieval needs and provide facilities for characterizing the user's conceptual interpretation of the image data through interactive retrieval sessions.

In this paper, we describe the design and implementation of an adaptive IR system that features an easy to use graphical interface for query formulation and a *relevance feedback* mechanism by which a user's conceptual view of the image data can be captured and used to modify the query automatically. Toward this goal, an effective method that allows incomplete and imprecise image and query descriptions is developed. This method incorporates techniques similar to those of vector space information retrieval for unstructured natural language texts [6, 7]. The advantages of our description is that the image and query descriptions need not be highly structured as would be required with the most of the current IR systems. An inductive learning module is designed for providing facilities by which the user's relevance feedback is effectively utilized to incrementally/adaptively reformulate the query to improve the retrieval effectiveness. User's relevance feedback is obtained by asking the user to evaluate the relevance of the retrieved images. Our query reformulation algorithm is based on the functional dependency between each image attribute and the user's relevance feedback using a theoretical framework referred to as Rough Set Theory [4]. The importance (or weight) of each image attribute in the reformulated query is modified based on the degree of such functional dependencies. Hence, our query reformulation algorithm is designed systematically, and the query reformulation process is both intuitive and easily understood.

Query reformulation based on user feedback has been investigated in traditional text-based information retrieval [7]. Query reformulation in IR applications is investigated in [1]. However, this method needs statistics which can only be obtained by having global term occurrences in the descriptions of all the images in the database. Whenever the database is updated, such statistics must be recalculated. Furthermore, their query reformulation method is rather ad hoc and is difficult to interpret.

An IR prototype system based on the proposed query reformulation method has been implemented and tested for retrieving images in the hair style domain. Although this method is illustrated in a specific domain, it should be noted that the proposed method is a general one and can be used in several other image retrieval domains. Initial results indicate that the reformulated query generated by our method significantly improves the retrieval effectiveness.

The remainder of the paper is organized as follows. In section 2, Rough Set Theory (RST) is briefly explained and its implication to adaptive query reformulation is described. Representation of image descriptions, user query formulation tools, and the image retrieval algorithm are introduced in section 3. This section also discusses the query reformulation algorithm and the image browsing tool for eliciting user's relevance feedback. Section 4 presents preliminary experimental results and section 5 concludes the paper.

2 Rough Set Theory

RST provides a systematic framework for the study of problems arising from imprecise and insufficient knowledge [4, 5].

2.1 Approximation Space

First, we describe the basic concepts of RST proposed by Pawlak [4]. Let U denote a finite set of objects, and let $R \subseteq U \times U$ be an *equivalence relation* on U. $A = (U, R)$ is called an approximation space. If $(x, y) \in R$, then x is usually related to y in a certain way and we denote this as xRy. If x and y are elements of U and (x, y) is an element of R then we can say that x and y are *indistinguishable* in the approximation space. R is often referred to as an *indiscernibility* relation.

Let $R^* = \{X_1, X_2, \cdots, X_n\}$ denote the partition induced by the equivalence relation R on U, where X_i is an equivalence class (or elementary set) in R^*. For any subset $X \subseteq U$, we can define the lower and upper approximations of X, which are denoted $\underline{A}(X)$ and $\bar{A}(X)$ respectively, in the approximation space $A = (U, R)$, as follows: $\underline{A}(X) = \bigcup_{X_i \subseteq X} X_i$: the union of all elementary sets in A that are contained in X and $\bar{A}(X) = \bigcup_{X_i \cap X \neq 0} X_i$: the union of all elementary sets in A each of which have a non-empty intersection with X. Given a subset $X \subseteq U$ representing a certain concept of interest, we can characterize the concept X in the approximations space $A = (U, R)$ with three regions: Positive Region: $POS_A(X) = \underline{A}(X)$; Boundary Region: $BND_A(X) = \bar{A}(X) - \underline{A}(X)$; Negative Region: $NEG_A(X) = U - \bar{A}(X)$. X is considered *definable* in A if $\underline{A}(X) = \bar{A}(X)$ (or $BND_A(X) = 0$); otherwise, X is said to be *non-definable* or a *rough set*. Intuitively, for example, $POS_A(X)$ consists of several elementary concepts which are closely related to concept X. In other words, concept X can be well explained by the elementary sets in $POS_A(X)$.

2.2 Calculation of Functional Dependency Between Attributes

A simple Knowledge Representation System (KRS), denoted $S = (U, C, D, V, \rho)$, is formally defined as follows: U denotes a set of objects, C is a set of conditional attributes, D is a set of decision (or action) attributes, $\rho : U \times F \rightarrow V$ is an information function, where

$F = C \cup D$, $V = \bigcup_{a \in F} V_a$, and V_a is in the domain of attribute $a \in F$. Note that the restricted function, $\rho_u : F \to V$ defined by $\rho_u(a) = \rho(u, a)$ for every $u \in U$ and $a \in F$, provides the complete information about each object u in S.

An example of a KRS is given in Table 1. Information about images in the hair style domain $U = \{I_1, I_2, I_3, I_4, I_5, I_6\}$ is characterized by means of the conditional attribute set C = {Hair Color, Hair Length, Hair Style} and the decision (or action) attribute set D = {Relevance}. The domains of the attributes are given by: Hair Color = {*blonde, red, brunette*}; Hair Length = {*short, medium, long*}; Hair Style = {*rounded, straight, feathered*}; Relevant = {*yes, no*}.

Given the information represented in a KRS, we want to determine the functional dependency between the conditional and decision attributes. Such a functional dependency can be used for constructing procedural knowledge such as "Under what conditions can a decision take place?" Let us assume that the 6 images in Table 1 are shown to the user, and the relevance feedback/judgment with respect to each of these images given by the user is as shown in the last column of this table. In the context of example presented in Table 1, we are particularly interested in assessing the degree of functional dependency between the conditional attributes and the decision attribute (i.e., the user's relevance judgment) to know why the user makes such a decision. To answer this question, we further need to define the following terminology. For any subset G of conditional attributes C or decision attributes D, the equivalence relation on U can be defined as: $(u_i, u_j) \in \tilde{G}$ if and only if $\rho(u_i, g) = \rho(u_j, g)$ for every $g \in G$. Let $A \subseteq C$, $B \subseteq D$, and let $A^* = \{X_1, X_2, \cdots, X_n\}$ and $B^* = \{Y_1, Y_2, \cdots, Y_n\}$ be the partitions induced by the equivalence relation $\tilde{A}$ and $\tilde{B}$, respectively. The partition B^* as a whole can be approximated or characterized by partition A^*. The quality of such an approximation depends on the relationship of the two subsets of attributes A and B. The measure of dependency of B on A is defined as [4]:

$$0 \leq \gamma_A(B) = |POS_A(B^*)|/|U| \leq 1, \tag{1}$$

where $|\,|$ denotes the cardinality of a set, and $POS_A(B^*) = \bigcup_{Y_i \in B^*} \text{APS}(Y_i)$ in the approximation space $APS = (U, \tilde{A})$. Note that $\gamma_A(B) = 1$ when B is totally dependent of A (i.e., A functionally determines B). If $0 \leq \gamma_A(B) \leq 1$, we say that B *roughly* depends on A. A and B are *totally independent* of each other when $\gamma_A(B) = 0$. In general, the dependency of B on A can be denoted by $A \xrightarrow{\lambda} B$. For instance, from Table 1, the following can be obtained:

{Hair Color, Hair Length, Hair Style } $\xrightarrow{1.0}$ {Relevance},
{Hair Color, Hair Length} $\xrightarrow{1.0}$ {Relevance},
{Hair Color} $\xrightarrow{0.5}$ {Relevance}, {Hair Style} $\xrightarrow{0.33}$ {Relevance}.

This means that the knowledge represented by the values of the condition attributes "Hair Color," "Hair Length," and "Hair Style" are sufficient to determine the relevance of an image to the user. It should also be noted that the condition attribute "Hair Style" is redundant with respect to the decision attribute "Relevant" because the removal of the latter from the KRS would not affect the dependency between the set of condition and decision attributes in this example.

Image	Hair Color	Hair Length	Hair Style	Relevance
I_1	*blonde*	*short*	*rounded*	*yes*
I_2	*brunette*	*short*	*straight*	*no*
I_3	*blonde*	*long*	*straight*	*no*
I_4	*brunette*	*long*	*feathered*	*no*
I_5	*blonde*	*medium*	*feathered*	*yes*
I_6	*red*	*medium*	*rounded*	*no*

Table 1: An Example of a Knowledge Representation System

2.3 The implication of Rough Set Theory to Adaptive Image Retrieval

If images can be represented by a set of objectively measurable attributes, they can be retrieved based on the values of these attributes. However, rarely does a user really knows what the value of these attributes should be in the query in order to fulfill the retrieval need. When there are several attributes, it is often difficult to determine the importance of each attribute to retrieve relevant images.

Initially, the user submits a query which consists of intuitively assigned values to the (decision) attributes. Further, the user is also involved in subjectively specifying which attributes are more important than the others. However, the initial user query may not necessarily be precise and complete to direct the system to retrieve only relevant images. Since the relevance judgment is based on the user's perception of which images are more relevant than others, it is difficult for the system to obtain such a relevance characterization based on the user's initial query. For example, although the user might have indicated high importance to the attribute "Hair Length" in the initial query, the retrieved images that possess a high value for the "Hair Length" attribute may or may not be judged relevant by the user. The importance of each attribute to the user may also change after several retrieved images are reviewed by the user. Hence, it is desirable that the importance of each attribute be interactively calculated by using the user's relevance feedback.

The importance of each attribute can be automatically calculated based on the assumption that the conditional attributes are independent of each other. Given a set of images whose conditional attribute values are known and the user relevance feedback on them is also known, the functional dependency between each conditional attribute and the user's relevance (i.e., the decision attribute) can be automatically calculated by using Equation 1.

3 Adaptive Image Retrieval System

An Adaptive Image Retrieval (AIR) system which incorporates query reformulation based on user relevance feedback has been developed for hair style image domain. In this section, the representation of the images and the user query, and retrieval and query reformulation algorithms of AIR system are discussed.

	Color	Intensity	Curly	Length	Part	Cut
Characteristics	Symbolic	Numeric	Numeric	Numeric	Symbolic	Symbolic
Domain	{blonde, brunette, red}	0-100% darkness	0-100% curly	0-49 inches	{left, right, center, none}	{feathered, straight, round}

Table 2: Attributes and Their Domains in Hair Style Images

3.1 Representation of Images and the User Query

Each database image is logically represented by n objectively measurable attributes. Then, an image can be described by an n dimensional vector as: $I_k = (a_{k1}, a_{k2}, \cdots, a_{kn})$, where a_{kj} represents the value of the attribute j in image k. User query is also represented in a similar manner. However, the importance of each attribute need to be added to the user query representation. A user query q is represented as: $q^T = ((w_1, q_1), (w_2, q_2), \cdots, (w_n, q_n))$, where w_j represents the importance of attribute j and q_j represents the value of attribute j. Each attribute can have either a symbolic value or a real value. For instance, a possible value for the attribute "Hair Color" can be a symbolic value such as *blonde.* A possible value for the attribute "Hair Length" can be a numeric value such as 10.5 inches. In the AIR system, each hair style image is represented by 6 attributes. Table 2 shows these attributes and their domains.

3.2 Retrieval Algorithm

Users queries are formulated using the query formulation tool. This tool is used for eliciting information from the user about the type, value, and importance of the attributes he/she would like to see in an image. The query is then translated into a vector whose components represent the values of the attributes and their importance in the query.

The retrieval function for calculating the retrieval status value (RSV) or similarity of an image I_k with respect to a user query q is defined as follows:

$$RSV_q(I_k) = \sum_{i=1}^{n} w_j \times sim(q_j, a_{kj}),$$

where n represents the number of attributes and w_j represents the importance of the attribute j in the query. $sim(q_j, a_{kj})$ calculates the similarity between the value of attribute j in the user query vector and the value of j in the image representation vector. The retrieval algorithm is shown in Figure 1.

The RSV of each image with respect to the user query is then computed. The images are sorted in descending order of their RSVs. The names of images are retrieved and shown to the user in this order. Image browsing tool is used to view the retrieved images (Figure 2). This tool is also used for eliciting user's relevance feedback.

Algorithm RetrieveImages
$I \leftarrow$ Set of images in the collection
$A \leftarrow$ Set of attributes used to describe the images in the domain
for each $I_k \in I$ **do**
begin
  $RSV_q(I_k) \leftarrow 0.0$
  for each $j \in A$ **do**
  begin
    if j is a symbolic attribute **do**
      if $q_j = a_{kj}$ **then** $sim(q_j, a_{kj}) = 1$ **else** $sim(q_j, a_{kj}) = 0$;
    else if j is a numeric attribute **do**
    begin
      $max\_range \leftarrow$ maximum value of the attribute j
      $sim(q_j, a_{kj}) \leftarrow 1 - \frac{|q_j - a_{kj}|}{\max\_range}$
    end
    $RSV_q(I_k) \leftarrow RSV_q(I_k) + w_j \times sim(q_j, a_{kj})$
  end
end
end RetrieveImages

Figure 1: Retrieval Algorithm

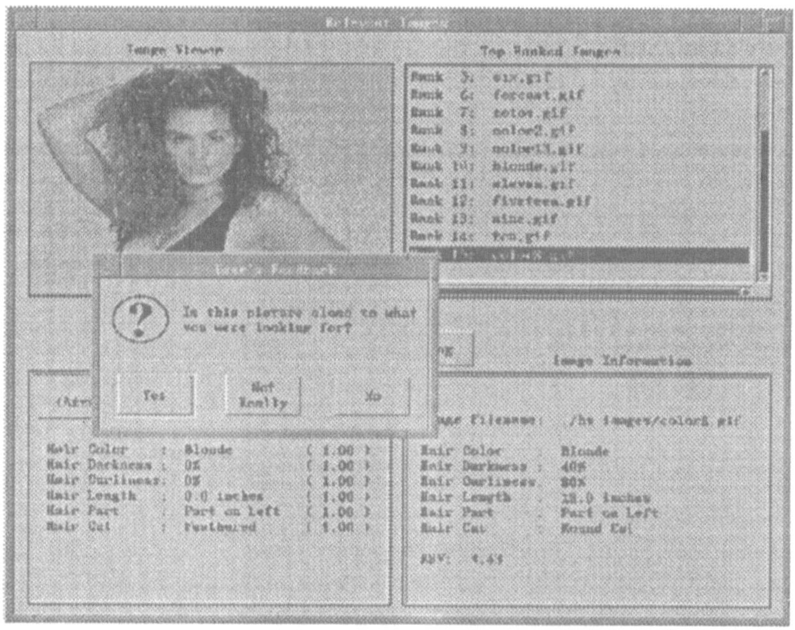

Figure 2: Image Browsing Tool of the AIR System

Image	Color	Intensity	Curly	Length	Part	Cut	Relevance
I_{rank1}	*blonde*	*very-dark*	*none*	*short*	*left*	*round*	*yes*
I_{rank2}	*brunette*	*light*	*light-curly*	*long*	*right*	*round*	*no*
I_{rank3}	*red*	*dark*	*curly*	*long*	*center*	*straight*	*yes*

Table 3: Internal Table Created After the User has Viewed Three Top Ranked Images

Intensity	Curly	Length
very-light ($\leq$ 25%)	*not-curly* ($\leq$ 25%)	*short* ($\leq$ 3 inches)
light ($\leq$ 50%)	*light-curly* ($\leq$ 50%)	*medium* ($\leq$ 6inches)
dark ($\leq$ 75%)	*medium-curly* ($\leq$ 75%)	*long* ($\leq$ 9 inches)
very-dark ($\leq$ 100%)	*heavy-curly* ($\leq$ 100%)	*very-long* ($\leq$ 49 inches)

Table 4: Rule for Converting Values of Real-Valued Attributes into Symbolic Values

3.3 Query Reformulation Algorithm and Learning from User Feedback

As the user views images and provides relevance feedback to the system, the system creates a table internally as depicted in Table 3. Since symbolic (attribute) values are used to categorize images into equivalence classes, we need to define rules for converting real-valued attributes into symbolic values. These conversion rules are shown in Table 4. Attribute weights in the user query are recomputed (hence, query is reformulated) using RST and Table 3. A new set of images is then determined by the system using the reformulated query. The query reformulation algorithm is shown in Figure 3. $FD(j, D)$ represents the functional dependency between the attribute j and decision attribute D (i.e., Relevance). This functional dependency is calculated based on the assumption that the conditional attributes are independent of each other. Since the calculation of $POS_C(D^*)$ involves set manipulation, the computational complexity of this algorithm may be high. However, in reality, since the number of images reviewed by the user is relatively small for the query reformulation, the algorithm runs fast enough for interactive use. The user involvement in query reformulation is at relatively high level of abstraction and the task is quite simple. Rough Set Theory provides the formal foundation for the query reformulation, consequently, the manner in which the query reformulation is performed is systematic and easily justified.

4 Preliminary Experimental Results And Discussion

Two experiments were conducted to test the effectiveness of query reformulation in the AIR system. These trials were set up to show the ability of the system to modify the initial query to the point where attribute weights have reversed themselves due to feedback. The experiment was conducted with an image collection consisting of 80 images in the hair style domain. In both test runs, one specific attribute was chosen and the user assigned weight for that attribute was set to zero. The user was then asked to provide relevance feedback

```
Algorithm QueryReformulaton
    A ← Set of attributes used to describe the images in the domain
    n ← Number of images for which the user has provided relevance feedback
    D ← {Relevance}
    REL ← Set of images considered as relevant by the user
    NREL ← Set of images considered as non-relevant by the user
    for each j ∈ A do
    begin
       C ← {j}
       POS_C(D*) ← ∅
       Obtain elementary sets C*
       for each elementary set e ∈ C* do
          if ((e ⊆ REL) or (e ⊆ NREL)) then
             POS_C(D*) ← POS_C(D*) ∪ e
       FD(j, D) ← |POS_C(D*)|/n
       w_j ← FD(j, D)
    end
end QueryReformulation
```

Figure 3: Query Reformulation Algorithm

based solely on the closeness of the match between the query and the image with regard to the attribute that was chosen. The system should "recognize" the fact that the user is selecting images as relevant based solely upon the attribute whose weight was assigned initially to zero. The system should then "modify" the weight of this attribute to a greater value while decreasing the weights of the other attributes. In both test cases, the system automatically modified the weights of the selected attributes as expected.

In the first experiment (Case 1), the initial user query was formulated as shown in Table 5. The user assigned weights to attributes are shown in the third row of the table. Notice that the user has assigned a weight of 0.0 for the attribute "Length," meaning that the user initially did not care about the hair length in the images to be retrieved. Table 6 shows the top 4 images that were retrieved with respect to the initial query. The relevance feedback was provided by the user based solely on the attribute Length. Table 7 shows that while initially the importance of attribute Length was zero, after user relevance feedback only on four images, this value has been increased to 1.0. Also, notice in Table 6 that the user preferred to have images which contain the value "long" for the attribute Length. Such user preference is well reflected in the weights of the reformulated query as shown in Table 7. Table 8 shows the retrieved images with respect to the reformulated query. As is evident from this table, images which contain the value "long" for the attribute Length are retrieved. This indicates that the system is able to capture user's conceptual preference of certain images over the others.

The second experiment (Case 2) was conducted with the attribute "Curly" as a focus of attention. The weight of the attribute "Curly" has been set to zero to signify that the attribute "Curly" is unimportant to the user. The user preferences are based solely on the

Attribute	Color	Intensity	Curly	Length	Part	Cut
Value	*blonde*	30.0	10.0	16.0	*right*	*feathered*
Weight	1.00	1.00	1.00	0.0	1.00	1.00

Table 5: User's Initial Query for the Experiment: Case 1

	I_{rank1}	I_{rank2}	I_{rank3}	I_{rank4}
Color	*blonde*	*blonde*	*blonde*	*blonde*
Intensity	30.0	40.0	43.0	10.0
Curl	5.0	8.0	2.0	12.0
Length	7.0	14.0	7.0	14.0
Part	*left*	*none*	*right*	*right*
Cut	*feathered*	*feathered*	*round*	*straight*
Relevant	*no*	*yes*	*no*	*yes*

Table 6: User's Relevance Feedback for Case 1

	Color	Intensity	Curly	Length	Part	Cut
Reformulated Query Weight	0.00	0.25	0.00	1.00	0.50	0.50

Table 7: Attributes Weights After Query Reformulation

	I_{rank1}	I_{rank2}	I_{rank3}	I_{rank4}
Color	*brunette*	*red*	*brunette*	*brunette*
Intensity	72.0	95.0	79.0	10.0
Curl	40.0	75.0	72.0	45.0
Length	16.0	16.0	16.0	16.0
Part	*none*	*none*	*none*	*none*
Cut	*feathered*	*feathered*	*round*	*round*

Table 8: Images Retrieved by the Reformulated Query-Case 1

	I_{rank1}	I_{rank2}	I_{rank3}	I_{rank4}
Color	*red*	*blonde*	*brunette*	*red*
Intensity	92.0	86.0	100.0	95.0
Curl	99.5	100.0	98.9	100.0
Length	3.0	13.8	5.0	2.0
Part	*center*	*center*	*none*	*none*
Cut	*straight*	*straight*	*round*	*round*

Table 9: Images Retrieved by the Reformulated Query-Case 2

degree of curl. The images which have heavy curly hair style were selected relevant. The query is reformulated based on these preferences and the top 4 images retrieved are shown in Table 9. Observe that the images which have heavy curly hair style are placed at the top of the rank ordering.

Both the experiments were conducted with only a single attribute as a focus of attention. More experiments need to be conducted with multiple attributes as focus of attention in order to see the retrieval effectiveness of reformulated query in a more general case.

5 Conclusions

In this paper, we have described the design and implementation of an adaptive image retrieval system that incorporates query reformulation mechanism. This mechanism allows imprecise and incomplete user query specifications. Our method incorporates techniques similar to those of vector space information retrieval for an unstructured natural language texts. The advantages of our approach to IR is that the image and query descriptions need not be highly structured as would be required with the IR based on the traditional database management techniques. Furthermore, an inductive learning module is designed for providing facilities by which a user's relevance feedback is effectively utilized to adaptively improve retrieval effectiveness.

The user involvement in query reformulation is at relatively high level of abstraction. Rough Set Theory provides the formal foundation for the query reformulation; consequently, the manner in which the query reformulation is performed is systematic and easily justified.

Although the proposed method is applied in a specific domain, it can be used for retrieving images in other domains. Preliminary experimental results indicate that the reformulated query generated by our method significantly improves the retrieval effectiveness. The initial results warrant further study of the retrieval effectiveness of the proposed approach.

Acknowledgment

This work has been supported partly by ARPA Grant Number: N00174-93-RC-00004 (first author) and partly by Ohio University Research Office under Grant Number: 917 (second author).

References

[1] Al-Hawamdeh, S. et al. (1991),"Nearest Neighbor Searching in a Picture Archive System," *International Conference on Multimedia Information Systems*, McGraw-Hill, pp. 19-34.

[2] Grosky, W. and Mehotra, R. (1989), **IEEE Computer**, Vol. 22, No 12, Guest Editors's Introduction, Special Issue on Image Database Management.

[3] Gudivada, V. and Raghavan, V. (1994), "Picture Retrieval Systems: A Unified Perspective", **TR-19942**, Department of Computer Science, Ohio University, Athens, OH.

[4] Pawlak, Z. (1982), "Rough Sets," *International Journal of Information and Computer Sciences*, Vol. 11, No. 5, pp. 145-172.

[5] Pawlak, Z. and Wong, S.K.M. (1987), "Rough Sets: Probabilistic Versus Deterministic Approach," **Technical Report**, Department of Computer Science, University of Regina, Regina, Saskatchewan, Canada.

[6] Raghavan, V. and Wong, S.K.M. (1986), "A Critical Analysis of Vector Space Model for Information Retrieval," *Journal of the American Society for Information Science*, Vol. 37, No 5, pp. 279-287.

[7] Salton, G. (1989), **Automatic Text Processing**, Addison-Wesley.

Implicit Representation for Extensional Answers in Object-Oriented Systems

Suk-Chung Yoon
Dept. of Computer Science
Widener University
Chester, PA 19013

Il-Yeol Song
College of Information Studies
Drexel University
Philadelphia, PA 19104

Abstract

In the near future, we believe that we will need much more sophisticated answer-finding schemes in an object-oriented database in order to satisfy the needs of truly intelligent information system. In this paper, we introduce a method to apply the intensional query processing techniques of deductive databases to object-oriented databases. So, we can generate intensional answers to represent answer-set abstractly for a given query in object-oriented databases.

Our approach consists of four steps: rule generation, pre-resolution, resolution, and post-resolution. In rule generation, we generate a set of deductive rules based on an object-oriented database schema. In pre-resolution, rule transformation is done to get unique intensional literals and extended term-restricted rules. In resolution, we identify rules that are potentially relevant to a query. In post-resolution, we find relevant resolvents as candidates for intensional answers among potentially relevant resolvents. We also uses the notion of potentially relevant resolvents and relevant resolvents to avoid generating certain meaningless intensional answers.

Keywords: Object-Oriented Database, Deductive Database,
Knowledge-Based Systems, Artificial Intelligence

1. Introduction

Object-oriented databases have received a great deal of attention over the past few years and have been the subject of intense research and development efforts. In the near future, we believe that we will need much more sophisticated answer-finding schemes in an object-oriented database in order to satisfy the needs of truly intelligent information system.

When processing a query in object-oriented databases, a set of database instances (*extensional answers*) are usually returned to users for an answer set. That is, object-oriented databases provide the answer set in the form of an enumeration of objects retrieved from the databases. In certain queries, we may not be interested in the set of objects that satisfy a given query in

E. A. Yfantis (ed.), Intelligent Systems, 775–788.

a particular database state. Instead, we may be interested in the conditions and the characteristics that objects must satisfy, in any state, to belong to the usual extensional answer of a query. Object-oriented databases do not support that capability. However, this situation can be different in a deductive database that supports complex reasoning which is not possible in object-oriented databases. When querying a deductive database, users can get not only the answer of a query as a set of facts but also the answer of a query as set of formulas(*intensional answers*). Deductive databases can generate a set of first order logic formulas as an answer set for a given query. That is, deductive databases deal with two different types of queries which can be distinguished by the kind of answer they are providing.

We can easily find many advantages in intensional query processing. As we get intensional answers as a set of formulas, they are independent of the particular circumstance in the database. Therefore, intensional answers provide a more stable answer than the extensional answers. Giving us exactly what conditions must be fulfilled to get a certain extensional answer, intensional answers provide a more compact and intuitive form than a set of facts could ever do. Intensional answers can be considered as a kind of interpretation or explanation of extensional answers. We can evaluate an intensional answer against the database and get a partial extensional answer. Intensional query processing has an advantage in computation compared with extensional answers because intensional answers can be computed without accessing the database, which greatly reduce the costs of processing a query.

In this paper, we introduce a method to apply the intensional query processing techniques of deductive databases to object-oriented databases. So, we can generate intensional answers to represent answer-set abstractly for a given query in obejct-oriented databases. In a great variety of ways to generate intensional answers, we are interested in an approach to construct resonably simple intensional answers. Each of them states a sufficient condition for a set of objects to belong to the extensional answer of the query. This paper is structured as follows. Section II introduces examples to show the advantages of intensional answers. Section III recalls basic definitions for object-oriented databases and deductive databases. Section IV surveys related works. Section V gives a precise definition for the concept of intensional answer. Section VI presents our approach to convert an object-oriented database schema based on class hierarchy into nonrecursive Horn clause notations and generate candidate intensional answers from logical consequences of nonrecursive Horn clauses and a given query. Section VII discusses possible extensions of our method.

2. Motivating Examples

The following examples illustrate the advantages of intensional answers to a given query in an object-oriented database.
Suppose we have the following object-oriented database schema about a uni-

versity.

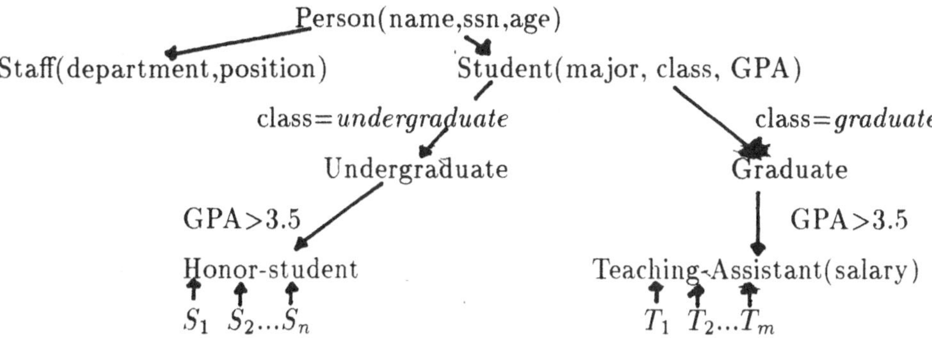

From the class hierarchy, we know that honor students are undergraduate students whose GPA is greater than 3.5 and that teaching assistants are graduate students whose GPA is greater than 3.5.

Now, we suppose we have a query asking "names of all persons whose GPA is greater than 3.5". In our example, the extensional answer might be a long list of all honor students or all teaching assistants $=\{ S_1, S_2, ...S_n, T_1, T_2, ...T_m\}$. All these objects belong to different subclasses of the class Student. However, if a database system can process intensional queries, the answers can be represented as a set of simple classes such as

Ans_1 = { Honor-student: all honor students } and

Ans_2 = { Teaching-Assistant: all teaching assistants }

This example shows that extensional answers would only provide a list of all the objects that satisfy a query, whereas intensional answers distinguish between two cases for extensional answers. Thus, intensional answers are more informative than extensional ones. Intensional answers provide more insight into the nature of the extensional answer.

Suppose that there is an object-oriented database schema about vehicles and their types.

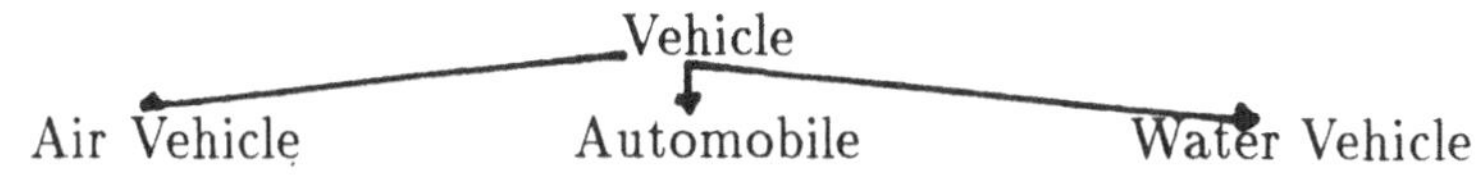

In the database, there is a class vehicle having the various vehicle types as subclasses. A user may query the structure of the database in a very natural way. For example, suppose the user wants to know what are all the vehicle types. In this case, if a database can support intensional answers, the answer can be represented as a set of simple subclasses such as Ans= {Air Vehicle, Automobile, Water Vehicle }. That kind of queries about the structure of the database can be handled easily.

3. Object-Oriented Databases and Deductive Databases

We review the object-oriented data model and the deductive model, primarily derived from (Kim 1990, Gallaire 1984). Object-oriented databases and deductive databases have been areas of active research during the past several years. One common goal of these areas is to extend programming languages with database systems. The goal of objective-oriented databases is to apply object-oriented concepts in object-oriented programmings in modeling data. The goal of deductive databases is to integrate rules and traditional database of facts in logic programming to support complex reasoning.

Object-oriented databases have some features that do not exist in a conventional database. An object-oriented database is a set of object-oriented concepts, including object-identity, encapsulation, inheritance and polymorphism, for modeling data. These concepts are sufficiently powerful to support data-modeling requirements of many types of application. In object-oriented database systems, data are organized in a hierarchy of classes and subclasses. The class hierarchy captures the relationship between a class and its subclass. Subclasses inherit all the attributes of their superclasses and can have some of their own attributes. Object at any level of the hierarchy inherits all the properties of object higher up in the hierarchy. The access scope of a given query on a class, say C, is either the set of instances of C, or the set of instances of the entire class hierarchy rooted at C(that is, all subclasses of C). The result of the query may be the set of objects that belong to different classes within a class hierarchy.

Deductive databases have some features that do not exist in a conventional database. A deductive database is a database in which new facts may be derived from the facts that were explicitly stored by using an inference system. A deductive database comprises an extensional database(EDB) consisting of a set of facts explicitly stored in a physical database, and an intensional database(IDB) consisting of a set of deductive rules. These rules can be used to derive new facts from the facts in the EDB. In a deductive database, there is one advantage: facts, deduction rules and queries can be written in a uniform database language typically based on a first-order logic. A rule has the form(Prolog-like notation) $a : -b_1, b_2, \ldots b_m \; m \geq 0$ where a is an atomic formula and b_i's are literals. All the variables occuring in the rule are assumed to be universally quantified. The literal a is called the *head* of the rule, the b_i's are referred to as the *body* of the rule. Every rule can be represented as a clause that is finite disjunction of one or more literals. . So, our rule can be represented as a clause, and would look like:

$$a \bigvee \neg b_1 \bigvee \neg b_2 \bigvee \ldots \bigvee \neg b_m$$

In this paper, we convert an object-oriented database schema into a set of non-recursive Horn clasuses. Horn clauses are a subset of clauses that have at most one positive literal. A Horn clause is adequate for specifying a large class of database applications. An object-oriented database system can become a deductive object-oriented database system once it can directly support rules and various reasoning concepts. A deductive object-oriented database can get big improvements in productivity and functionality.

4. Research on Intensional Answers

Cholvy and Demolombe(1986) studied the idea of having a set of formulas as an answer set. They provided answers which were independent of these facts and valid in all the states associated with the set of facts. Their answers were a set of formulas defining the conditions that a given tuple must satisfy to be an answer for a given query. In their approach, only non-recursive rules were allowed to make sure that the algorithm for generating answers terminated. Based on the research by Cholvy and Demolombe, Pascual and Chovy(1988) developed an improved algorithm to retrieve an answer set of a given query. The former approach accepted any form of clauses and therefore is inefficient in the case of Horn clauses. The latter approach dealt with only Horn clauses to avoid numerous tests while generating the answer set. Thus, the latter approach accelerates the resolution process. However, the detailed steps to remove redundant resolution steps or meaningless answers were not discussed in Pasual and Cholvy(1988). These are important steps to reduce intensional query processing time. Song and Kim(1991) solved all of these problems left on Pasual and Cholvy's approach. They discussed a three-step intensional query processing scheme based on SLD resolution and discussed an implementation of an intensional query processor in Prolog. They use the notions of extended term-restricted rules, relevant literals and relevant clauses to avoid generating certain meaningless intensional answers. Yoon and Song(1994) generalized Song and Kim's approach and made their ideas efficient by using a graph structure.

Imielinski(1987), Motro(1989), and Pirotte, Roelants and Zimmanyi(1991) took different approaches from those discussed so far. Imielinsky introduced a new concept of an answer for a query. His answer can be composed of atomic facts and general rules built from projection, join and selection, so that they can be incorporated into the answer of a query. Pirotte et al.(1991) used integrity constraints to filter out inadequate answers. Motro(1989) also discussed the method that applies database constraints to generate intensional answers, and Motro and Yuan(1990) informaly discussed a query language incorporating intensional queries.

Our approach is applying Yoon and Song's work to object-oriented databases. Our approach deals with a non-recursive set of Horn clauses in order to get a terminating algorithm for generating the answers. Also, our approach avoids generating meaningless intensional answers.

5. Definition of Intensional and Extensional Answers

We start this section by informally defining extensional answers and intensional answers as follows.

Definition: A *query* Q(X) in a deductive database is a formula where X is a tuple of free variables.

Definition: The *extensional answer* for a query is the set of tuples ā such that Q(ā) can be shown to be true on the extensional database when the deduction rules are taken into account.

Definition: An *intensional answer* to a query Q(X) is a formula A(X), obtained from the intensional database and from the query, that states a condition to be satisfied by tuples of values for x̄ in order to be part of the extensional answer of that query.

A formal definition of the intensional answers is given in Cholvy and Demolombe(1986). We define T as the database theory consisting of a set of facts and rules. Let Q(X) be a query. The intensional answer to a certain query Q(X), ANS(Q) is defined as:

$$\text{ANS(Q)} = \{\ ans_i(X) : T| - \forall X(ans_i(X) \longrightarrow Q(X))\ \}\ (5.1)$$

The literal $ans_i(X)$ is defined so that under the theory T, any element X where X ∈ ans_i(X), satisfy Q(X). That is, for any tuple of values a, if $T| - ans_i(a)$ then $T| - Q(a)$. ans_i(X) now can be any formula in a logical sense. As we are only interested in meaningful answers, which means answers within a defined domain of interest, we try to put some restrictions on the intensional answer set. Above all, this means that our intensional answers consist of only predicates from T. Also, we don't allow contradictory formulas or redundant answers.

So let DP= $\{P_1, \ldots, P_n\ \}$ be a set of predicate symbols either of the IDB or the EDB. Let L(DP) be the first-order language whose predicate symbols are $P_1, \ldots, P_n$. Thus, an intensioanl answer set ANS(Q,DP) to a query Q(X) is defined by:

$$\begin{aligned}\text{ANS(Q,DP)} = \{\ & ans_i(X) : ans_i(X) \in L(DP) \text{ and} \\ & T| - \forall X(ans_i(X) \longrightarrow Q(X)) \text{ and} \\ & (ans_i(\text{X}) \text{ is not the negation of a tautology) and} \\ & (\text{each } ans_i(X) \text{ is not redundant})\}\ (5.2)\end{aligned}$$

A redundant intensional answer is defined as follows: given two intensional answers: $ans_1(X)$ and $ans_2(X)$, $ans_1(X)$ is redundant if there is a following relationship between them $T| - \forall X(ans_1(X) \longrightarrow ans_2(X))$, that is, every tuple that satisfies $ans_1(X)$ also satisfies $ans_2(X)$, but not vice versa. A set of intensional answers is said to be complete if and only if the set of database values obtained from the set of intensional answers is equal to the set of database values in the extensional answer of the query. We need to add an additional condition to (5.2): ANS(Q,DP) must be complete. That is,

$$\forall\, \bar{a} \quad (\bigvee_{\text{v-}ans_i \in \text{ANS(Q,DP)}} \text{v-}ans_i\ (\bar{a})) \equiv \text{v-}ans_E(Q)$$

where v-ans_i(ā) represents the set of database value tuples obtained from the set of an intensional answer and v-$ans_E(Q)$ represents the set of values obtained from the extensional answer.

If a query is formulated as a formula with free variable it is also possible to try to find bindings for these variables during the answering process. An extensional answer set then would be defined as:

$$ans_E(Q) = \{X| - T \bigwedge[\forall X(Q(X) \longrightarrow ans_e(X))]\} \quad (5.3)$$

An extensional answer is defined as a tuple $\bar{a}$ which makes $Q(\bar{a})$ true in the database theroy T.

6. Formalization of Intensional Answers

In this section, we present the method of deriving intensional answers. We divide our approach into four phases: rule generation, pre-resolution, resolution and post-resolution. The four phases are partitioned into two categories: processing that can be done statically once and processing that has to be performed at run time. Rule generation and pre-resolution phases belong to the first category and the remaining two phases belong to the second category. So, the rule generation and the pre-resolution phases are independent of the queries posed to the database and hence are computed once prior to the processing of any query.

6.1 Rule Generation

In this phase, we perform three steps. In the first step, we generate a set of deductive rules based on the hierachy in an object-oriented database. We can classify rules into two different categories: rules to specify the class hierarchy and rules to specify relationship between two classes. The rules in the first category represent the hierarchy between a class and its superclass. For example, there is a class Vehicle having a vehicle type, say Airvehicle, as subclass. Then we have the rule: Airvehicle(x):- Vehicle(x). The rules in the second category represent the semantic knowledge between classes. For example, there are two classes, Automobile and Sportscar. If the number of cylinders in an automobile is greater than 6, then that automobile belongs to a subclass Sportscar. This semantic knowledge involoves two classes of Automobile and Sportscar. In this case, we many need to introduce a new literal called *property* literal. We can represent the above knowledge as follows:

Sportscar(x):- Automobile(x), Num-of-Cylinder(x,y), GT(y,6)

The literal Num-of-Cylinder is a property literal to show that the automobile x has y cylinders and the literal GT is a comparison literal to mean "greater than". After the first step, we get several rules that can be classified in one of two categories.

In the second step, we introduced EDB literals to represent the objects that belong to each class. For example, there is an object, say C1, that belongs to the class Car that has attributes such as id, price, size, num-of-door and gas-mileage. An object in the class Car can be represented in the EDB literal Car such as Car(10,25000,full,4,25).

In the third step, we remove literals that will cause recursion in rules. For example, suppose we have the following two rules:

Rule1: Automobile(x):- Bike(x)

Rule2: Bike(x):- Automobile(x), Num-of-Wheel(x,y), Less(y,3)

During resolution process, we may get another rules that will cause recursion as follows: Bike(x):- Bike(x), Num-of-Wheel(x,y),Less(y,3)
So, we eliminate the literal Automobile in the Rule2. After this step, we will get a non-recursive set of rules. These tranformed rules are the essential components of the IDB.

6.2 Pre-Resolution

In the first step some rule transformations will be done in order to get unique intensional literals and extended term-restricted rules.
Unique intensional literals mean that a literal should either be extensionally or intensionally defined but not both. One can always get rid of this equality of names by renaming the extensional literal to p^* and introducing the rule p:- p^*. The reason to force unique literals is to get a unique resolution tree.
Extended term-restricted rules have the following characteristics:
1. All the variables in the head of the rule appear also in the body
2. The rule does not have any constant in the head.

Converting our rules to extended term-restricted rules provides that all the information in the head of a rule also appears in the body. That keeps us from losing this information while doing resolution because the subsequent resolvent will also contain it. Using non extended term-restricted rules in resolution for the derivation of intensional answers will prevent the derivation of intensional answers for some queries, even though intensional answers exist(Song,1988).

6.3 Resolution

Now we use SLD resolution with the notion of potentially relevant resolvents. One of the difficult problems in providing intensional answers is how to identify rules that are relevant to a query. Identifying relevant resolvents avoids getting meaningless answers. In our approach, we define a resolvent that is not potentially relevant as follows:
Rule 1: A resolvent is not potentially relevant of the resolvent (1) does not contain any intensional literal and (2) contains an extensional literal(s) and a property literal(s) and (3) the extensional literal(s) does not contain attribute about the property literla(s)(that is, the property literal(s) does not specify the attributes of the extensional literal(s)).
Rule 2: A resolvent is not potentially relevant if the resolvent (1) does not contain any intensional literal and (2) contains at least two property literals and/or two comparison literals that have same name and (3) factoring is impossible between those property literals and/or between those comparison literals. If factoring between two property or two comparison literals are possible, then we delete those two literals. After performing factorization, the resolvent may be empty.

Any resolvent which is not potentially relevant is not included in the subsequent resolution process, therefore avoiding the possibility of deriving meaningless intensional answers. Using the notion of relevant resolvents has two advantages over Cholvy and Demolombe(1986) and Pascual and Cholvy(1988).

First, it can eliminate unnecessary rules which are used for the derivation of intensional answers. Second, certain meaningless intensional answers, as in Cholvy and Demolombe, can be avoided. The resolution stage serves to either simplify or discard resolvents. During this stage, users might be able to reduce the number of literals in a resolvent, or users might even be able to discard the resolvent itself totally.

6.4 Post-resolution

In this step, all the potentially relevant resolvents are taken to find a candidate for an intensional answer. Considering only the potentially relevant resolvents insted of all the resolvents ever generated in the whole resolution process is much more efficient and can easily be justified. We use following rules to find relevant resolvents among potentially relevant rsolvents.
Rule 1: If there is a potentially relevant resolvent that is empty, we select the intensional literal as an intensional answer from the parent clauses participating in the resolvent. In this case, we use backtracking technique to find the intensional literal.
Ruel 2: If there is a potentially relevant resolvent that consists only extensional literals, find the intensional literal from the parent clauses participating in the resolvent and generate intensional answer by negating that intensional literal.

An Example

This example shows the aplication of the algorithm introduced in the section above. Suppose we have the following class hierarchy about vehicles.

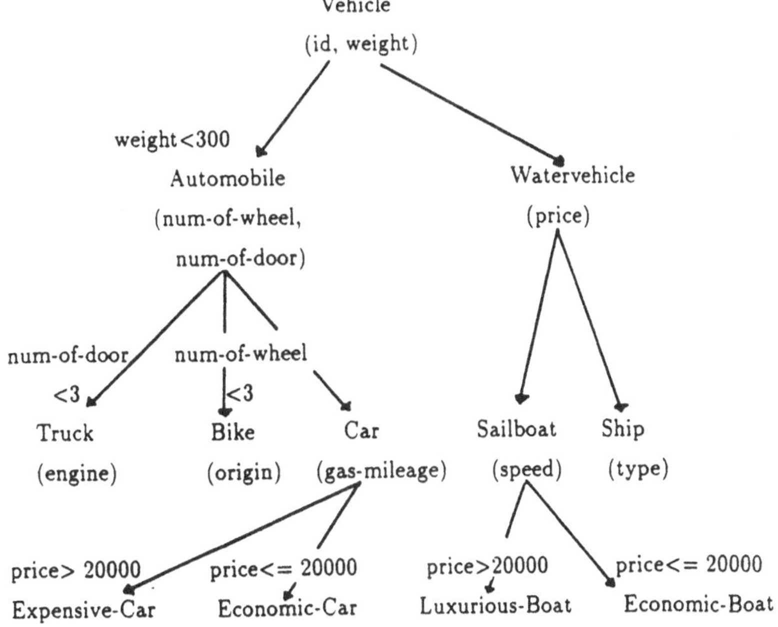

In the rule generation phase, we generate a set of deductive rules based on the above object-oriented database schema. In the first step, we get the following IDB rules that belong to two different categories.

IDB:
1. Rules to specify the class hierarchy:

Rule 1	Vehicle(x):- Airvehicle(x)
Rule 2	Vehicle(x):- Automobile(x)
Rule 3	Vehicle(x):- Watervehicle(x)
Rule 4	Airvehicle(x):- Plane(x)
Rule 5	Airvehicle(x):- Helicopter(x)
Rule 6	Automobile(x):- Truck(x)
Rule 7	Automobile(x):- Bike(x)
Rule 8	Automobile(x):- Car(x)
Rule 9	Watervehicle(x):- Sailboat(x)
Rule 10	Watervehicle(x):- Ship(x)
Rule 11	Car(x):- Expensive-Car(x)
Rule 12	Car(x):- Economic-Car(x)
Rule 13	Sailboat(x):- Luxurious-boat(x)
Rule 14	Sailboat(x):- Economic-boat(x)

2. Rules to specify properties between two classes:

Rule 15	Automobile(x):- Vehicle(x), Weight(x,y), LT(y,300)
Rule 16	Bike(x):- Automobile(x), Num-Of-Wheel(x,y), LT(y,3)
Rule 17	Truck(x):- Automobile(x), Num-Of-Door(x,y), LT(y,3)
Rule 18	Expensive-Car(x):- Car(x), Price(x), GT(y,20000)
Rule 19	Economic-Car(x):- Car(x), Price(x), LE(y,20000)
Rule 20	Luxurious-boat(x):- Watervehicle(x), Price(x,y), GT(y,20000)
Rule 21	Economic-boat(x):- Watervehicle(x), Price(x,y), LE(y,20000)

The comparison literal LT means "less than". The comparison literal LE means "less than and equal". The comparison literal GT means "greater than".

In the second step, we introduce the following EDB literals to represent objects that belong to each class.

EDB:
Vehicle(id, weight)
Airvehicle(id, weight, num-of-seat)
Plane(id, weight, num-of-seat, type)
Helicopter(id, weight, num-of-seat, speed)
Automobile(id, weight, num-of-wheel, num-of-door)

Watervehicle(id, weight, price)
Truck(id, weight, num-of-wheel, num-of-door, engine)
Bike(id, weight, num-of-wheel, origin, num-of-door)
Car(id, weight, price, num-of-wheel, num-of-door, gas-mileage)
Sailboat(id, weight, price, speed)
Ship(id, weight, price, speed)
Expensive-Car(id, weight, price, num-of-wheel, num-of-door, gas-mileage)
Economic-Car(id, weight, price, num-of-wheel, num-of-door, gas-mileage)
Luxurious-Boat(id, weight, price, speed)
Economic-Boat(id, weight, price, speed)

In the third step, we need to eliminate literals that will occur recursion. After that elimination, we will get the following transformed rules.

Rule 15	Automobile(x):- Weight(x,y), LT(y,300)
Rule 16	Bike(x):- Num-Of-Wheel(x,y), LT(y,3)
Rule 17	Truck(x):- Num-Of-Door(x,y), LT(y,3)
Rule 18	Expensive-Car(x):- Price(x,y), GT(y,20000)
Rule 19	Economic-Car(x):- Price(x,y), LE(y,20000)
Rule 20	Luxurious-boat(x):- Price(x,y), GT(y,20000)
Rule 21	Economic-boat(x):- Price(x,y), LE(y,20000)

In the pre-resolution phase, we need to rename the extensional literals to get unique intensional literals. Therefore, we add new rules to the IDB.
New rules:
Vehicle(x) :- *Vehicle**(x)
Airvehicle(x):- *Airvehicle**(x)
Automobile(x):- *Automobile**(x)
Watervehicle(x):- *Watervehicle**(x)
Truck(x):- *Truck**(x)
Car(x):- *Car**(x)
Expensive-Car(x):- Expensive-*Car**(x)
Economic-Car(x):- Economic-*Car**(x)
Luxurious-boat(x):- Luxurious-*boat**(x)
Economic-boat(x):- Economic-*boat**(x)
In this example, every rule is already in extended term-restricted form. We can skip the process to convert rules into extended term-restrcited rules. In the above rules, the predicates with the symbol *, which represents that those predicates are just EDB predicates, are used to simplify our process. For example, we use the predicate *vehicle** to represent the fact that the predicate *vehicle** means the EDB predicate *vehicle**(id,weight). Therefore, whenever we need to search EDB actually, we use the EDB predicate *vehicle**(id,weight).

In the resolution phase, we use the SLD resolution with a query and the IDB and EDB gained from the previous phase. The query is negated and converted into the clause form. Suppose that we have the following query "find all vehicles whose price is greater than 20000"

:- Vehicle(x), Price(x,y), GT(y,20000)
After the first step, we get the following three resolvents:
R1) :- Airvehicle(x), Price(x,y), GT(y,20000)
R2) :- Automobile(x), Price(x,y), GT(y,20000)
R3) :- Watervehicle(x), Price(x,y), GT(y,20000)
For our convenience, we will only consider R2 branch.(We can use same technique for R1 and R3 branches.) After the second step, we get the following resolvents:
R4) :- Truck(x), Price(x,y), GT(y,20000)
R5) :- Bike(x), Price(x,y), GT(y,20000)
R6) :- Car(x), Price(x,y), GT(y,20000)
After the third step, we get the following resolvents:
R7) :- *Truck**(x), Price(x,y), GT(y,20000)
R8) :- *Bike**(x), Price(x,y), GT(y,20000)
R9) :- *Car**(x), Price(x,y), GT(y,20000)
R10):- Expensive-car(x), Price(x,y), GT(x,20000)
R11):- Economic-car(x), Price(x,y), GT(y,20000)
According to rule 1 in section 5.3, the resolvents 7 and 8 are not potentially relevant. We ignore those two resolvents. The resolvent 9 is potentially relevant. After the fourth step, we get the following resolvents:
R12) :- Expensive-*car**(x), Price(x,y), GT(y,20000)
R13) :- Economic-*car**(x), Price(x,y), GT(y,20000)
R14) :- Price(x,y), GT(y,20000), Price(x,y), GT(y,20000)
R15) :- Price(x,y), GT(y,20000), Price(x,y), LE(y,20000)
According to rule 2 in section 5.3, the resolvent 15 is not potentially relevant. After factorization, clause 14 is empty.

After the resolution phase, we have four potentially relevant resolvents R9, R12, R13 and R14. The resolution tree is as follows:

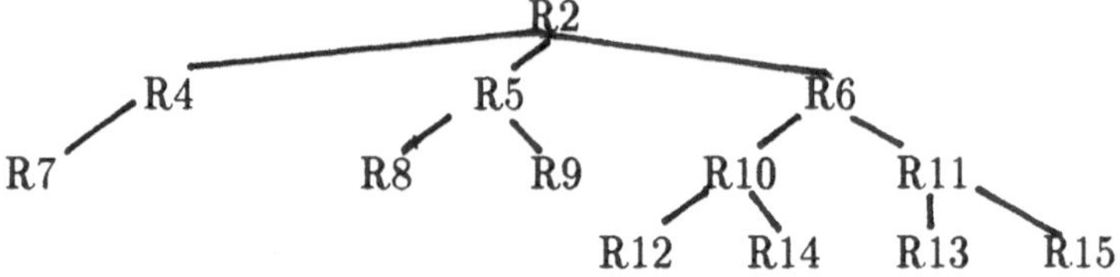

In the post-resolution phase, we apply rules in section 5.4 to those four potentially relevant resolvents to find relevant resolvents for intensional answers. We have an intensional answer from R14, Expensive-car(x). Similarly, we have an intensional answer from the R3 branch, Luxurious-boat(x).

7. Conclusion

Object-oriented database have received a great deal of attention over the past few years and have been the subject of intense research and development efforts.

In this paper, we have presented a method that addresses the important issue of providing more meaningful answers(intensional answers) to queries in an object-oriented database by using SLD-resolution technique. We apply the intensional query processing techniques of deductive databases to object-oriented databases. So, we can generate intensional answers in object-oriented databases. An intensional answer characterizes the extensional answer and provides more insight into the nature of the extensional answer.

Our method is complete because it discovers all of the intensional answers. We avoid intensional answer that have little or no value to the user.

It may be interesting to use different computational methodologies to get intensional answers. Another method besides resolution might help to get intensional answers.

References

Chang, C. and Lee, R. (1973), *Symbolic Logic and Mechanical Theorem Proving*, Academic Press

Cholvy, L. and Demolombe, R. (1986), "Querying a Rule Base", In *Proceedings of the First International Conference on Expert Database Systems*, pp. 365-371

Gallaire, H., Minker, J. and Nicolas, J. (1984), "Logic and Databases: A Deductive Approach", Computing Survey 16(2), pp.153-185

Imielinski, T.(1987), "Intelligent Query Answering in Rule Based Systems", Journal of Logic Programming, Vol. 4, No. 3, pp.229-258

Kim, Won (1990), *Introduction to Object-Oriented Databases*, MIT Press

Motro A.(1989), "Using Integrity Constraints to Provide Intensional Answers to Relational Queries", In *Proceedings of 15th VLDB Conference*, pp. 237-246

Motro A. and Yuan Q. (1990), "Querying Database Knowledge", In *Proceedings of the International Conference on Management of Data*, pp. 173-183

Pascual, E. and Cholvy, L. (1988), "Answering Queries Addressed to The Rule Base of a Deductive Database", In *Procceedings of the Second International Conference on Uncertainty in Knowledge-based Systems*, pp. 138-145

Pirotte, A. and Roelants, D. (1989), "Constraints for improving the generation of intensional answers in a deductive database", In *Proceedings of 5th International Conference on Data Engineering*, pp.652-659

Pirotte, A., Roelants D., and Zimanyi, E (1991), "Controlled Generation of Intensional Answers", In IEEE Transactions on Knowledge and Data Engineering, Vol.3, No. 2, pp. 221-236

Song, I.Y. and Kim, H. J. (1991), "Design and Implementation of a Three-Step Intensional Query Processing SCheeme", Journal of Data Administration, Vol.

2, No, 2, pp. 23-25

Song, I. Y. and Dubin, D. (1991), “Intensional Query Processor in Prolog”, In *Proceedings of ISMM Int'l Symposium on Computer Applications in Design, Simulation and Analysis*, pp. 204-207

Ullman, J. (1988), *Principles of Database and Knowledege-base Systems*, Computer Science Press

Yoon, S.C. and Song, I.Y. (1994), “A General Method for Generating Intensional Answers in am Intelligent Information System”, In *Proceedings of the 1994 ISCA International Conference on Computers and Their Applications*, pp. 94-98

Zdonik, S. and Maier, D. (1990), *Readings in object-oriented database systems*, Morgan Kaufmann Publisher

A General Framework for Building Intelligent Database Applications

Zhan Cui and John Fox
Advanced Computation Laboratory
Imperial Cancer Research Fund
61 Lincoln's Inn Fields
London WC2A 3PX, UK

Abstract. This paper proposes SLOT - a formal specification and design language for building large intelligent database applications. SLOT is an attempt to provide database users with a general framework bringing together recent advances in database technology, techniques in AI and the disciplines of software/knowledge engineering by a combination of logic programming and object-oriented programming. In SLOT, users can divide their applications according to the natural structure of their domains into theories (corresponding to objects in object-oriented programming) explicitly specifying the data structures and their use for each theory. Applications built in SLOT are independent of any real database system, but can be partially automatically mapped onto one. This significantly reduces the time used for remodelling when the actual database evolves or changes.

Keywords. Deductive and Object-Oriented Database, Knowledge Representation, Formal Specification.

1 Introduction

Recent advances in database technology have brought database systems one step closer to knowledge-based systems. Deductive databases (Gallaire, Minker and Nicolas 1984) permit users to specify meta-knowledge about the facts in a database in terms of integrity constraints and deductive laws. Object-oriented databases (Atkinson et al. 1989) permit users to describe a world in terms of high level objects and to associate a set of customised operations with the objects. Deductive and object-oriented databases (DOOD) (Kim, Nicolas and Nishio 1991) combine the advantages of both logic programming and object-oriented programming paradigms.

Database systems are traditionally considered to be data rich and knowledge lean (they can efficiently store large volumes of data but provide few conceptual modelling formalisms for representing complex concepts) while for knowledge-based systems the reverse is true. Research on deductive and object-oriented databases, however, has been bridging the gap between the two technologies. Many data rich and knowledge rich applications (Yoshida et al 1992) are being developed on such hybrid platforms.

The ever increasing complexity of database applications poses new challenges to researchers in both the database and knowledge-based system research communities. More

E. A. Yfantis (ed.), Intelligent Systems, 789–800.

powerful modelling formalisms are required to specify and design such "intelligent" applications. Although DOOD systems provide advanced conceptual modelling techniques, these often deal with only data modelling. However, applications are composed of both data and their uses. The clean separation may be good for the database research, but is undesirable from the user's point of view. Furthermore, database conceptual modelling offers little help for verifying whether an application built in this way meets the requirements, not to mention the correctness of the modelling. This situation can be improved by bringing together recent advances in database technology, techniques from Artificial Intelligence and from software/knowledge engineering (Karbach et al 1991, van Harmelen and Balder 1992 and Barbuceanu 1993). So far relatively little work has been done on using formal methods for the specification of database applications. It would be very desirable to have a general framework to represent both of them (data and their uses) together and leave the system to decide which part should be represented in the database.

To address these problems we have developed a specification and design language called SLOT (Specification Language for Object Theories) for building large database/knowledge base applications in a formal and systematic way. In the next section, we first discuss the general requirements for such a language, and then move onto the various choices made for SLOT. Section 3 presents an overview of SLOT. In sections 4 and 5, we briefly discuss how to map a SLOT specification onto a specific database system and its implementations. Finally, we conclude the paper with a summary and further work in section 6.

2 Development Methodology

Experience in building expert and knowledge based systems has shown that construction of consistent and maintainable knowledge bases on a large scale is difficult. Simple tools for ensuring consistency of terminology, detection of data duplication and inconsistency etc. are useful (Knowledge Engineering Review, 1992) but far from sufficient, for a number of well-recognised reasons, viz: 1). simple syntactic checks do not guarantee that domain data and knowledge are encoded in a meaningful and appropriate structure 2). it is difficult to avoid building assumptions about specific uses of data/knowledge into the encoding, which may impede both reusability and the incorporation of new functions. 3). when data and knowledge base compilation is to be carried out by a number of people it is important that they have an explicit, shared view of the structure

Systematic design methods and implementation tools foster clarity, modularity and a coherent structure for the application knowledge base. Without these a knowledge base can become incoherent with the result that applications become unreliable and progressively difficult to understand and extend.

Experience has shown the value of an explicit, modular knowledge model which defines the functional organisation of a knowledge base. Database designers have shown comparable benefits of explicit, implementation-independent definitions or data models for building large databases. Van Harmelen and Balder (1992) summarise the advantages of using formal languages for explicitly describing the structure and/or behaviour of software systems as: the removal of ambiguity; facilitation of communication and discussion; and the ability to derive properties of the design in the absence of an implementation.

We chose logic as the basic formalism for its well-defined, well-understood semantics and its established position as a tool for formal analysis. Van Harmelen and Balder (1992) have proposed the use of logical formalisms to define data and knowledge modules explicitly as logical "theories". The idea of a theory subsumes simple collections of domain data (datasets), logic programs for deductions on datasets, programs which implement generalised tasks (e.g. experiment planning), strategies for managing procedures involving multiple tasks over time, and so forth. SLOT divides knowledge into modular theories which are arranged in a theory hierarchy through an importing structure. When a theory imports another theory, it inherits information presented in the imported theory. Broadly speaking, each theory is specified by many sorted first-order logic. The interaction between theories is through a message passing mechanism.

Theories in SLOT are modelled in an object-oriented style. We chose object-orientation for its suitability and success for building very large software systems because it supports data-abstraction, encapsulation, information hiding and modularity(Atkinson et al. 1989). The final structure of SLOT is strongly influenced by the $(ML)^2$ language (van Harmelen and Bolder 1992) – a formal modelling language for KADS conceptual models. We have extended the basic structure of $(ML)^2$ with the concepts of object-oriented, deductive and active databases. Other influences on the SLOT structure are from work on combining the logic programming and object-oriented programming paradigms, including Plog (Jenkins and Chester 1993), SILO (Hatzilygeroudis 1993), and L & O (McCabe 1990). Other formal influences on the methodology come from work on knowledge acquisition systems such as MODEL (Barbuceanu 1993), MODEL-K (Karbach et al. 1991), KARL (Fensel et al. 1991) and CODE (Skuce 1993).

Our work also relates to research on sharing knowledge across AI systems, such as Gruber's work on Ontolingua (1993). The major difference is that Gruber's work is geared towards knowledge-based systems while our work also relates to building large database systems. We believe our work also makes a contribution to the problem of database schema evolution (Tresh and Scholl 1992) in that applications developed in SLOT are independent of any real database systems. Thus applications built in SLOT can be partially automatically mapped onto a specific database system.

3 Overview of SLOT

We have extended the basic structure of $(ML)^2$ with the concepts of object-oriented, active and deductive databases, to give the benefits of both logic-programming and object-oriented paradigms. In the following, we overview the general architecture of SLOT and its components briefly and informally. For more formal aspects and details, see Cui (1994).

3.1 The Structure of SLOT

SLOT extends the first-order logic theory by 1). typing variables and functions; 2). organising formulae into a theory hierarchy; 3). handling theory interactions by triggers or production rules (Hajnal, Fox and Krause 1990). The basic unit in SLOT is a *Theory*, which specifies a particular structure, including the sorts of data and operations on data. The extended style of *theory* specification of the SLOT language can be informally represented as follows:

object_theory <theory name >
  import <set of theories>
  signature
    sorts <subsort hierarchy>
    functions <set of sorts → sort mappings>
    predicates<set of predicates and arguments and their sorts>
    triggers<set of trigger events and their sorts>
    constants<set of constants and their sorts>
  logical_dependency
    variables <set of symbols and their sorts>
    axioms<set of facts and rules >
  transaction_language
    variables<set of symbols and their sorts>
    preconditions <set of constraints>
    productions<set of event-condition-action rules>
    postconditions <set of constraints>
end_object_theory

Each theory has four main components: 1). an import statement, which indicates any other object theories that are assumed by the definition. 2). a signature, which defines syntax information, and sort restrictions on the arguments of functions, predicates and triggers making up the theory, 3). a logical sub-theory consisting of facts and/or rules, which define the internal database of the object theory and 4). a transaction language which specifies the ways in which one object theory can interact with others.

3.2 Signatures

In designing SLOT, we have emphasised the importance of the formal specification by using many sorted logic (Cohn 1987). The advantages of many sorted logic are that it improves the readability and conciseness of axiom specifications, and supports automatic procedures to capture syntactical specification errors. The sortal declaration is made mandatory by the signature component of a SLOT theory.

The signature specifies sorts (or types) of functions, constants, predicates and triggers used in a theory. All sorts referred to must be declared explicitly, i.e. the declaration of super sorts does not automatically make it possible to refer to their subsorts. SLOT maintains a system sort hierarchy, and in most cases the declaration of sorts just involves naming sorts from the system sort hierarchy. However, a new sort can be introduced; if so it is the user's responsibility to ensure that a suitable sort-checking predicate is defined. When SLOT encounters a new sort name, it adds the new sort into the system sort hierarchy after making the appropriate checks.

The sort declaration has the format:
$sort_1$ or $(sort_1, sort_2, ..., sort_n)$

where $sort_i$ can be a 'sort name' from the system sort hierarchy or 'sort name'($sort_{i1}$, ..., $sort_{ij}$). The sorts in bracket are subsorts of the sort outside the bracket. When no sort is mentioned before the bracket, all sorts in the bracket are assumed to be direct subsorts of the highest possible sort *top*.

After the sort declaration, all functions, constants, predicates and triggers must be declared in the following formats.

For functions:

<functor name>: $sort_1 \times sort_2 \times ... \times sort_{n-1} \rightarrow sort_n$

where $sort_n$ is the result sort of function <functor name>/$(n-1)$. Constants are zero arity functions, but we declare them separately for its special use in Prolog/SLOT.

For constants:

<constant name>, ..., <constant name>: *sort*

For predicates:

<predicate name>: $sort_1 \times sort_2 \times ... \times sort_n$

Triggers have the same syntactical form as predicates.

When functions, predicates and triggers are declared in a SLOT theory, we say they are introduced. After introduction they can be used in any theory which imports them directly or indirectly. Along any importing path, a function, a predicate or a trigger can be introduced only once, but can be introduced again on different importing paths.

3.3 Axioms

Axioms in SLOT are generally axioms in many sorted, first-order logic. Since full first order logic could not be implemented efficiently, we have chosen to restrict ourselves to Prolog clauses. Therefore by SLOT axioms, we mean facts and rules in Prolog. It should be noted that variables in axioms must be typed. The type declaration of a variable is the same as that of a constant except that constants are prefixed with lower-case letters whereas variables are prefixed with upper-case letters.

Axioms in a theory form a local database (to be referred as dataset in case of confusion) peculiar to that particular theory, that is, they are not visible from outside the defining theory unless they are explicitly imported. Generally speaking, we can write any Prolog axioms. However, we may deliberately restrict ourselves to a subset of Prolog if we want axioms in a theory to be mapped onto a specific database system, because of the limited expressive power of current database systems. This situation should improve as rules in databases become less restricted.

In the above section, we have described the concept of the introduction of predicates and triggers. Another concept in SLOT is the definition of a predicate. Ground facts of a predicate are considered to be definitions of the predicate so are rules with the predicate as their heads. The trigger definition is discussed in the next section. Even though a predicate can be introduced only once along one importing path, it can be defined and redefined on any importing path. The semantics of axioms is somehow complicated by the importing structure. However, their semantics is the same as that of Prolog if only one theory is considered.

In Prolog, facts and rules of dynamic predicates can be freely asserted as well as retracted. This is made difficult in SLOT in favour of a 'pure' declarative style. For example, predicates such as assert and retract in Prolog have been abolished because we want to restrict the use of update predicates, thus reducing the possibility of an inconsistent database state. In their place, we have introduced SLOT update predicates which can be used only in triggers.

3.4 Active behaviours

Triggers are considered to be an especially powerful mechanism for a database to react automatically when certain events occur. These uses are also supported by SLOT. In addition, SLOT triggers are used as a mechanism to pass messages among SLOT theories. Triggers are specified in the transaction language part of a theory, and take the following form:

<Event>: <Conditions> ⇒ <Actions>

This can be read, informally, to mean "if the set of <Conditions> hold at the moment that the event <Event> occurs, then carry out the <Actions> in sequence." A trigger may be a database event, such as a data update event, a "message" (originating from the user or some object theory), or a system event such as a clock-tick or rollback event. <Conditions> are predicates on the object-theory's internal data and <Actions> are operations which may destructively modify the data, send a message to another theory or carry out some other operation with side-effect.

In order to keep the database consistent, constraints can be introduced into a theory. In SLOT, constraints are specified by preconditions and postconditions in the form of Prolog clauses within the transaction language part of a theory. These constraints must hold for any trigger defined in the production to fire. Preconditions are checked before the application of any trigger whereas postconditions are checked after the completion of one cycle of trigger firing.

3.5 Simple Examples

Sor far we have described the syntax of an individual theory, i.e. specification of an object theory. Before we move onto the general organisation of SLOT theories, we will give simple examples to illustrate how theories are specified and used.

Suppose we wish to develop a theory about natural numbers. In the theory, we define the concepts of positive integers and prime numbers. We also want to provide an algorithm to calculate prime numbers. Since the calculation of prime numbers is time-consuming, we may as well store the calculated prime numbers in a dataset. Furthermore, we want to define prime numbers in a subtheory even though we may not need to do so in this particular case. This, however, is very important for the maintainability and readability of a large application.

```
object_theory  nature_number
   import
   signature
     sorts  integer
     predicates
            positive: integer
```

```
   logical_dependency
     variables X, Y: integer
     axioms
            positive(1).
            positive(X) :- Y is X-1, positive(Y).
end_object_theory
```

In this theory, we have declared the sort **integer**. As we do not have functions in this theory, the keyword **functions** is omitted. (We can omit any keyword in a theory specification if there are no contents under that keyword.) The predicate **positive** is introduced in the signature and defined in the axioms as Prolog clauses. Ideally, we should also define integers, but that will take too much space and does not add much to the points we are making.

After defining the theory **natural_number**, we can use all information in it. We, therefore, do not need to declare sort **integer** again nor do we want to introduce and define predicate **positive** in **prime_number**.

```
object_theory  prime_number
   import nature_number
   signature
     predicates
            prime: integer
            prime: integer x integer
     triggers
            add_prime: integer
   logical_dependency
      variables  X, Y, Z: integer
      axioms
            prime(2).
            prime(3).
            prime(X, X).
            prime(X, Y) :-  not mod(X, Y, 0), Z is Y+1, prime(X, Z).
   transaction_language
      variables  X: integer
      productions
            add_prime(X): positive(X), prime(X, 2)
                                 => slot_assert(prime(X)).
end_object_theory
```

prime_number defines a trigger **add_prime**, which first checks that a prime number is a positive integer and verifies it by means of the predicate **prime/2**. If the input is a prime, it will be added to the dataset by **slot_assert**. Note that **slot_assert** does more than simply assert a term into the dataset; it also decides whether an asserted fact should be stored in a specific database. For example, in the case that theory **prime_number** is mapped onto a specific database, the two given prime numbers(2, 3) will be automatically moved into that database so do any subsequent assertions.

3.6 Theory Hierarchy

SLOT applications usually consist of many theories organised in an inheritance hierarchy. The classification into theories facilitates modularity and maintainability of a large application. It also encourages reuse of theory specifications. Indeed, we could develop a library of SLOT theories which can be imported by any other theory. The reuse of a SLOT theory is through the importing structure. An inferior theory (theories at a lower level of the hierarchy) inherits all information in its superiors, i.e. signatures, axioms, triggers, preconditions and postconditions. An inferior theory can override a previous specification by redefinition, such as a specialisation of a predicate, a generalisation or even a completely different definition.

A theory in SLOT consists of all the axioms in its local database, together with all the axioms in any imported theories. It is arguable that a theory should also include all its subtheories. However, this will make the inheritance difficult to understand. Furthermore it sets apart from the standard inheritance theory. We will investigate the possible implications of this in the future. In the current SLOT, the theory **natural_number** does not understand prime numbers, but **prime_number** understands natural numbers. In the case that the prime numbers in **prime_number** are stored as instances of a specific database, the extent of **prime_number** will include all those instances, and the axioms of the theory **natural_number** unless there is a redefinition for the predicate **positive** in the theory **prime_number**.

The inheritance of trigger rules is similar to the inheritance of logical rules and axioms.

3.7 Theory Interaction

The interaction between theories is through message passing in SLOT. There are two basic primitives for this purpose:

tell(<theory name>, <trigger>)

ask(<theory name>, <query>)

tell(<theory name>, <trigger>) activates triggers defined in the object theory <theory name>. When a user sends a message to a designated SLOT theory without expecting a reply, **tell** should be used and the only result will be a side-effect. As **tell** has side-effects, we need to know in which theory the side-effect will take place. When we send a trigger to an object theory, that trigger may be defined in that object theory or its imported theories. When an action such as slot-assert is executed, the effect is normally in the object theory which receives the message, not where the trigger is defined. However, if an action is expected to happen in a specific theory, regardless of where the trigger is called, we have to explicitly mention that specific theory. For example, **slot_assert**(<theory name>, <axiom>) will assert <axiom> in theory <theory name> no matter where **slot_assert** is used. When a side-effect is expected in the object which receives a message, **slot_assert**(self, <axiom>) is used[1].

ask(<theory name>, <query>) is used to query the axioms in a theory and does not have any side-effect. It will activate the SLOT proof procedure to prove <query> in theory

[1]self is a special keyword which will be replaced by the name of the targeted object theory; slot_assert(<axiom>) is a short form for **slot_assert**(self, <axiom>).

<theory name>. Notice that inheritance will be considered in the proving process.

4 Mapping to Database Systems

One of our objectives in designing SLOT is to automatically translate applications developed in SLOT onto specific deductive and object-oriented database systems. This has not been yet fully realised but we have anticipated an incremental approach; as more and more functionalities become available, more and more parts of an application will be mappable onto a database system.

The database system we are currently using is Chimera (Ceri et al. 1993, Sikeler et al. 1993, Jonker et al. 1993, Jonker et al. 1993). Chimera is a novel database language for a deductive and object-oriented database of IDEA project (Intelligent Database Environment for Advanced Applications - Esprit EP6333). The Chimera language integrates an object-oriented data model, a declarative query language based on deductive rules, and an active rule language for reactive processing. Many constructs in Chimera have correspondences in SLOT. In SLOT, we use `compile_to_database`(<theory name>) to translate <theory name> onto Chimera. `compile_to_database` will use the translation rules defined in a special SLOT theory. A different set of translation rules has to be defined for a different database. To date we have successfully translated the data-intensive part of a genetic database onto Chimera (Hearne et al. 1994). The initial results are very encouraging; details can be found in Cui (1994).

When part of the axioms in SLOT are translated onto a database, all the references to those axioms have to be rewritten in the appropriate form, as dictated by the specific target database system. However, this is not necessary. We can design interface predicates in the top level SLOT theory to handle the mapping between references in SLOT and references in databases.

5 Implementation

The current implementation of SLOT is in Eclipse Prolog (Sepia team 1989) and a trigger language, Sceptic (Hajnal et al. 1990), implemented in Eclipse Prolog. We have used two representations: one is the intermediate form for mapping SLOT specification onto a specific database system; the other is in Prolog form, which can be executed in Eclipse Prolog.

The user level support offers a graphical interface in PCE (Wielemaker and Anewierden 1992) with a control panel and an editing area. The control panel consists of various buttons to interact with SLOT. The main use of the editing area is for users to load, write, edit and modify object theories. SLOT will check the syntax and if an error is encountered, the user is informed about the type of the error and its location. SLOT also provides facilities for filling in missing variable declarations, browsing theory hierarchies and the import structure of a particular theory, etc.

To demonstrate SLOT we have developed an 'intelligent' database for molecular biology (Cui, Fox and Hearne 1993, Hearne et al. 1994). Figure 1 shows a snapshot of a typical user session with this application.

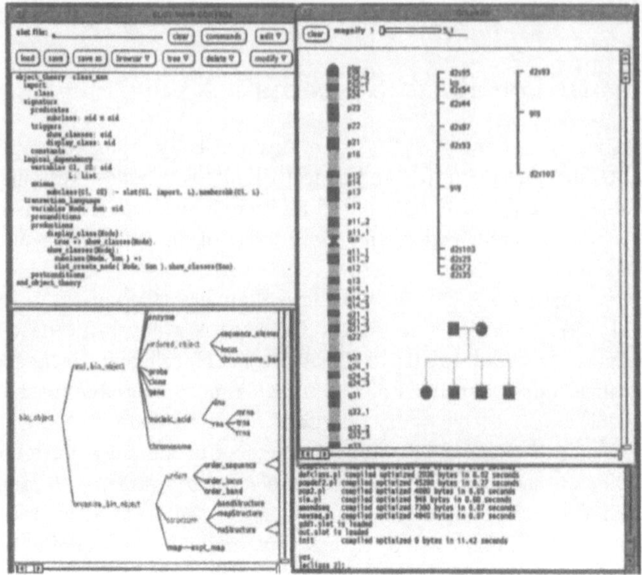

Figure 1: SLOT graphical control window

6 Conclusion and Further Work

We have presented SLOT – a formal specification and design language for building large deductive and object-oriented database applications. In designing SLOT, we have emphasised a common and well understood formalism for both users and database designers to ease the communication between them. We have also emphasised the importance of formal specification and software development methodology by bringing together the recent advances in database technology, techniques in AI and the disciplines of software/knowledge engineerings.

We have implemented a prototype SLOT in a Prolog, and used it to build a prototype genetic database (Cui et al. 1993, Hearne et al. 1994) in the context of IDEA project. This application has been mapped onto an initial release of Chimera testbed – a prototype of deductive and object-oriented database (Sikeler et al. 1993, Jonker et al. 1993). Our initial experience has shown that SLOT is very promising as a general framework for developing knowledge-rich and data-rich applications. In principle one could generate executables in a variety of knowledge representations, raising the possibility that an abstract specification formalism could provide the users with an interchange language for applications written using different tools and under different conventions. We can also build a set of special SLOT theories such as temporal logic, uncertainty reasoning. These standard theories available from SLOT library can be applied to and used by any application developed in SLOT.

Currently, we are working towards improving the automatic translation process and providing a standard SLOT library which defines more common SLOT theories. The mechanism for applying one SLOT theory onto another also needs to be further developed. In the long run, we would like to investigate the possibility of using SLOT to prototype distributed AI systems because an agent in the distributed system can be described by an object theory and the cooperations between agents can be simulated by the SLOT message passing mechanism. It is also possible to implement some of the standard SLOT theories in the spirit of theory resolution (Stickel 1985) or on a parallel platform for efficiency. Some other interesting works include, for instance, the declarative semantics of SLOT triggers, an enhanced graphical interface, and design and development tools.

7 Acknowledgement

The financial support from IDEA project is gratefully acknowledged. We thank Drs Saki Hajnal, Catherine Hearne and Simon Parsons for their comments and suggestions. This paper reflect the opinions of the authors and not necessarily those of the consortium.

References

[1] M. P. Atkinson, F. Bancilhon, D. J. DeWitt, K. Dittrich, D. Maier and S. Zdonik: "The Object-Oriented Database Manifesto", International Conference on Deductive and Object-Oriented Databases, Kyotyo, 1989

[2] M. Barbuceanu: "Models: toward integrated knowledge modeling environments", Knowledge Acquisition, 5, 1993

[3] S. Ceri, R. Manthey, E. Baralis, E. Bertino, C. Draxler, U. Griefahn, D. Montesi, A. Sikeler and L. Tanca: "Consolidated Specification of Chimera, CM and CL", IDEA Deliverable T2.2D2, Politecnico di Milano, November 1993

[4] A. G. Cohn: "A More Expressive Formulation of Many Sorted Logic", Journal of Automated Reasoning 3, 1987

[5] Z. Cui: "SLOT: the program and its use", Technical Report, ICRF, 1994

[6] Z. Cui and J. Fox: "A General Framework for Building Intelligent Database Applications", to be presented in GW International Conference on Intelligent Systems, Las Vegas, 1994

[7] Z. Cui, J. Fox and C. Hearne: "Knowledge-based Systems for Molecular Biology: the role of advanced technology and formal specification", In Proc. of IJCAI Workshop on AI and Genome, Chambery, 1993

[8] D. Fensel, J. Angele and D. Landes: "Knowledge Representation and Acquisition Language (KARL)", In Proc. 11th International Workshop of Expert Systems and their Applications (Volume: Tools and Techniques), Avignon, 1991

[9] H. Gallaire, J. Minker and J-M. Nicolas: "Logic and Databases: a deductive approach", ACM Computing Surveys, 16 (2), 1984

[10] T. R. Gruber: "A Translation Approach to Portable Ontology Specifications", Knowledge Acquisition, 5, 1993

[11] S. Hajnal, J. Fox and P. Krause: "Sceptic User Manual", Technical Report, ACL, ICRF, 1990

[12] F. van Harmelen and J. Balder: "$(ML)^2$: a Formal Language for KADS Conceptual Models", Knowledge Acquisition 4, 1992

[13] I. Hatzilygeroudis: "Knowledge Representation and Reasoning in a System Integrating Logic in Objects", in Proc. of IEEE on Conf. TAI93

[14] C. Hearne, Z. Cui and S. Parsons: "Prototyping a Genetics Deductive Database", In Proc. of ISMB94, Stanford, 1994

[15] M. Jenkins and D. Chester: "A Combined Object-oriented and Logic Programming Tool for AI", in Proc. of IEEE on Conf. TAI93

[16] W. Karbach, A. Voss and U. Drouwen: "Model-K: Prototyping at the Knowledge Level", In Proceedings Expert Systems 91, Avignon, France, 1991

[17] W. Kim, J-M. Nicolas and S. Nishio: "Deductive and Object-Oriented Databases", North Holland, Amsterdam, 1991

[18] The Knowledge Engineering Review (1992), Volume 7, Special Issue, Knowledge Base Verification.

[19] E. G. McCabe: "Logic and Objects", Impeiral College, London, 1990

[20] Sepia Team (ITC): "Sepia 2.13 User Manual", SEP/UM/022, ECRC, April 1989

[21] A. Sikeler, W. Jonker and S. Ceri: "Specification of the Testbed Framework", IDEA Deliverable T3.1D1, ECRC, Munich, November 1993

[22] D. Skuce: "A Multi-functional Knowledge Management System", Knowledge Acquisition, 5, 1993

[23] M. Stickel: "Automated Deduction by Theory Resolution", Journal of Automated Reasoning, 1, 1985

[24] M. Tresch and M. H. Scholl: "Meta Object Management and its Application to Database Evolution", In Proc. 11th Int'l Conf. on the Entity Relationship Approach, Karlsruhe, Germany, 1992

[25] J. Wielemaker and A. Anjewierden: "Programming in PCE/Prolog", SWI, University of Amsterdam, 1992

[26] K. Yoshida, C. Smith, T. Kazic, G. Michaels, R. Taylor, D. Zawada, R. Hagstrom and R. Overbeek: "Toward a Human Genome Encyclopaedia", In Proc. of Int'l conf. of Fifth Gen. Comp. Systems, 307 ICOT, 1992

Retrieval in Image Databases

N. Dimitrova
F. Golshani
Department of Computer Science and Engineering
Arizona State University
Tempe, AZ 85287-5406

Abstract. We describe an integrated retrieval scheme for image databases based on the multi-resolution hierarchical description of image data which accommodates various stages of image interpretation. The multi-resolution image hierarchy consists of the following levels: physical image, set of object descriptors, identified image features, object semantics association level, and semantic description level. The objects at various levels of the image data type hierarchy are obtained by semantic operators: low level operators, object representation operators, and retrieval operators.

Keywords. Image databases, Retrieval Methods, Image Indexing.

1 Introduction

Many diverse applications require the ability to deal with pictorial information. Examples include: geographic information systems, CAD, document storage and retrieval (office automation), medicine, architectural archives, and law enforcement databases. Thus it is no surprise that management of visual information is a crucial task for the success of many traditional and emerging applications.

Image databases, or more generally Visual Information Management Systems (VIMS), sit at the confluence of many technologies, including databases, image processing, computer vision, graphics and visualization, and knowledge representation and modeling. Also significant are findings in the areas of compression and decompression, and memory/storage organization and management. Considering the size of the multimedia objects, particularly images, the two areas above are particularly important for the success of image databases in their intended domains. In addition, although image databases are supposed to have capabilities similar to those of traditional databases, the techniques for indexing, querying, searching and processing are significantly more complex.

While there are numerous issues related to image databases such as acquisition, compression, storage, analysis, and communication issues, in this article we concentrate on the task of speedy retrieval of images. Simply put, we attempt to answer the following question: Given a (large) quantity of images, how can we analyze, classify, index, and process these images so that, when needed, a desired image can be found in the shortest possible

E. A. Yfantis (ed.), Intelligent Systems, 801–812.

time? Obviously, one can browse the collection of images and, eventually, the desired image will be found. However, being fundamentally sequential, browsing is the least efficient method of search. With appropriate pre-processing and indexing, one can formulate a query that specifies the desired image by its properties, and the system will retrieve a set of images that represent the best match. In many cases, we may not be able to obtain the exact result without user intervention. However, the objective is to minimize browsing by formulating queries that very precisely specify the desired image(s). Let us consider an example. Suppose in a criminal investigation, the user is interested in matching the picture of a suspect with the facial images contained in a police database. Without any querying capabilities, one would scan each and every image for possible matching. However, if some pre-processing has been done on the contents of the database, many other attributes and features of the person can be used to narrow the possibilities down to a handful of possibilities as compared to the entire database. The identification of these features and other properties of image contents is of utmost importance.

The paper is organized as follows. Section 1.1 surveys examples of existing models for image databases. Section 2 introduces our representation model for images using a hierarchically structured data type and explains the role of each level of the hierarchy. Sections 2.1 through 2.3 describe the operators that are applicable to the structured image data. A general unifying architecture for using the introduced structured data type and image processing, description and retrieval operators is given in section 3. An example application and the process of image insertion and querying is presented. Section 4 contains some concluding remarks and future directions.

1.1 Existing Models

Query languages and retrieval in image information systems have been influenced by the language support from conventional (hierarchical, network, relational) databases [1]. Pictorial SQL (PSQL) [2], and Query by Pictorial Example (QPE) [3] are examples. The language of PSQL is an extension of SQL which supports user-defined abstract data types that are used for definition of pictorial domains. Spatial comparison operators and functions for computing attributes are defined on each domain. The retrieval is based on the associations between the pictorial and alphanumeric domain. In QPE, the queries are specified using forms (tables) in the style similar to QBE (Query by Example). Another query language that has similar flavor to QBE is PICQUERY [4]. PICQUERY is built on top of a conventional DBMS. It provides support for a comprehensive set of data accessing and manipulation operators for image manipulation and pattern recognition as well as geometric operations.

The query processor of PROBE [5] supports point sets with a geometry filter which produces an approximate answer that is refined further by detailed manipulations on individual spatial objects, such as spatial join and spatial selection.

The Intelligent Image Database System (IIDMS) [6] offers several modes of querying images: query by the name of the image, query by keywords, query by a frame number, query by an iconic example, and query by 2d strings.

The VIMSYS project supports an iterative query model [7, 8]. The user first specifies a domain of interest and then uses an interactive graphical interface to choose and display the objects of interest and their attributes.

The types of information to be managed in an image database are classified into: iconic data (the images themselves), image related data (resolution, format description), information extracted from processing images (numerical, structural features), image world relationship data, and world (application) related data [9]. In order to facilitate representation of these different types of data, most data models in image databases distinguish between logical and physical image representations [10, 11, 12].

The physical representation describes the raw image at a level higher than pixels independent of the semantics of an application. The physical image includes a representation of edges, loops, lines, or connected diagrams. The logical representation has associated semantics that depends on the actual application. The logical representation is normally what the user requires for image retrieval.

In [13], the physical image contains the raw image as well, and the logical image acquires several levels of semantic information as it passes through different stages of processing. The logical and the physical image constitute the so-called *generalized icon.* The generalized icon can contain active index cells as well. An *active index* means that the index can be used to initiate actions, can be dynamically changed, or can be imprecise (approximate). This leads to the notion of *smart images*, which offers potential that has not yet been exploited in image retrieval.

A fast image indexing algorithm is proposed by Gong and others in [14]. Images are indexed by both the numerical index keys generated automatically from the captured primitive features using a set of rules and traditional descriptive keywords entered by users. The image features used for indexing are: color, shape features like circularity and major axis orientation, location of regions, as well as histogram content.

2 Image Data Type

Ideally the image data type should be flexible enough to accommodate various stages of image interpretation. It should support general operations for extracting the semantics of the image as well as the associated domain dependent, and user dependent operations for extracting information. The image data type should also incorporate multiple views for various interpretations of the image, as well as evolving interpretations of the image.

In figure 1 a graphic representation of the image data type is given. We consider the image to be a hierarchically structured data type. The image data type includes the following:

1. Physical image which refers to the digitized image representation. At the physical level this translates into an image stored (or compressed) in an image file format. At the abstract level, image is a function that associates values to pixels on a two dimensional (or multidimensional) grid.

2. Set of extracted boundaries of objects, or regions, or chain codes (basically, object descriptors).

3. Identified image features for object recognition and pattern classification.

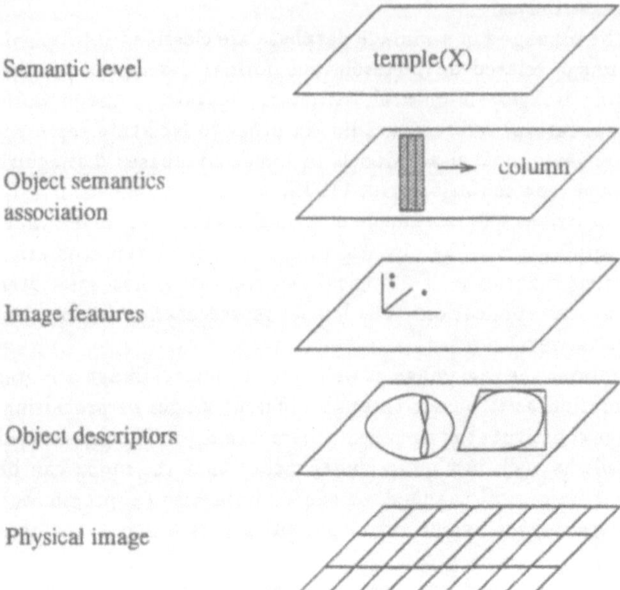

Figure 1: Hierarchical Representation of the Image Data Type

4. Object semantics association: assignment of real world semantics to the objects or features identified in the image.

5. Semantic description level: domain knowledge expressed using predicates, semantic nets, or a scheme that enables further reasoning about the content of the images.

The classification of the operators in our model is done with respect to the functionality of a particular operator and its purpose in the whole system. The operators that work between the levels one and two in the image data type representation are basic enhancement/segmentation image processing operators. Cross level operators at levels 3, 4 and 5 are the semantic level operators. The operators that are applicable to various levels are categorized into:

1. Low level operators (image enhancement and registration operators, edge detection, etc.)

2. Object representation operators

3. Retrieval operators (search operators)

Figure 2 gives an abstract view of the image data type and associated operators. The following sections present more elaborated overview of these operators. Our approach to modeling image data is different from the approach taken in the VIMSYS data model (see [7, 8]) since our base representation of images is just an abstraction of the physical

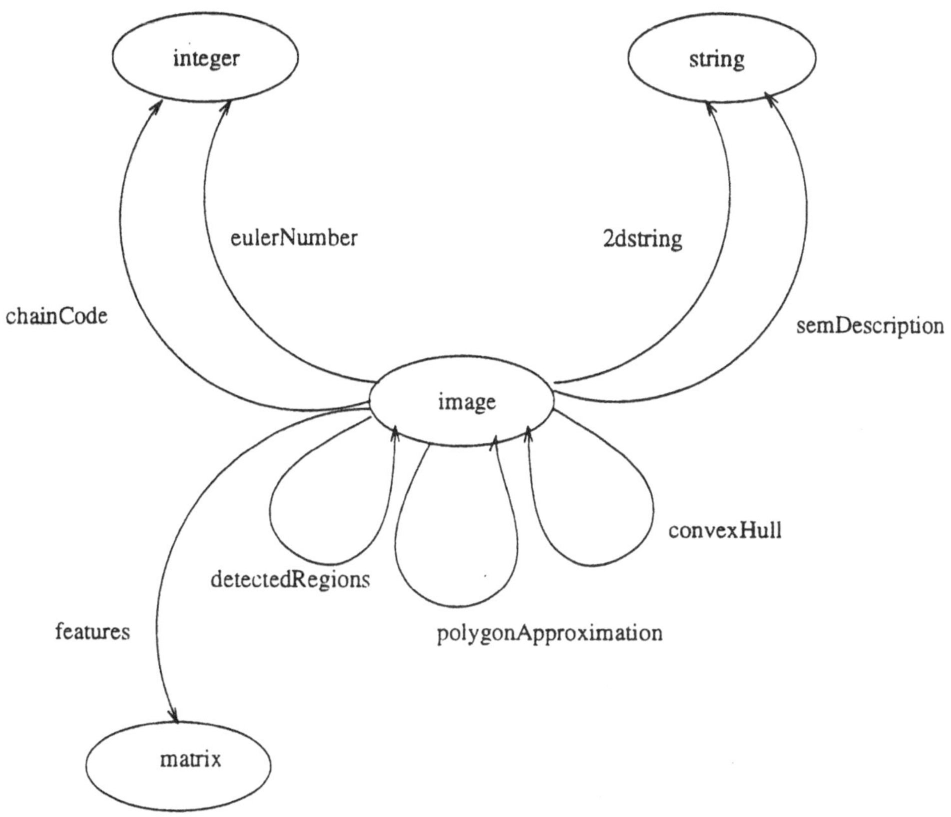

Figure 2: Image Data Type

image and we do not treat the types "integer" and "string" as base representations. In our model we have separated the image descriptors and the features of objects since we assume that features are represented by values of attributes which are derived from the object descriptors.

2.1 Low Level Operators

Image processing operators for image registration, image enhancement, salt and pepper noise removal, contrast operators, spatial filtering, i.e. the preprocessing operators, fall in this category. These operators return the physical image type, that is, the output is of the same type as the input image. The preprocessed image is further used for analysis and region/edge detection.

One important group of operators is based on histogram processing. A histogram is a function that describes the distribution of the colors in an image. Sometimes the distribution of colors has a very small limited range in the beginning of the color pallette, which makes the image look very dark and the details are not discernible. With histogram equalization we can redistribute the colors in a broader dynamic range which reveals more visible details and increases the image contrast.

$$histEqualize(image, transfFunction) \rightarrow image$$

Another set of preprocessing opertors involves transformation of the image from the spatial domain to the frequency domain. One of the most useful operators is the smoothing operator which filters out the high frequencies. This is equivalent to blurring the input image.

Low level operators also include segmentation operators: edge detection and region detection operators. In edge detection the idea is to find boundaries between regions based on intensity discontinuities. Various applications need edge detection operators that work in a particular direction and produce edges of a different width. The output is usually a binary image consisting of the detected edges:

$$edges(image, direction, width) \rightarrow image$$

Segmentation using region detection operators is applied directly towards recognizing regions that consist of pixels which are "similar enough". Both region growing and region splitting approach require an input number of regions expected by the user, or a threshold factor as a similarity measure between regions. The output is a labeled image which contains the number of desired regions in different colors:

$$regions(image, n) \rightarrow image$$

The object recognition and description operators work on images that are obtained using edge detection and region detection operators. Regions, edges, and discerned objects are considered as subsets of rectangular images, i.e. they are objects with random shape.

2.2 Object Representation Operators

The object representation operators serve for populating the higher levels of the image data type. These functions are useful for extracting shape information, boundary, and skeleton of the objects inside the image, as well as for associating real world semantics with the extracted features and objects. Here are some examples of object representation operators:

- Chain code description of an object is a numeric representation of the shape of the object.

  $$chainCode : (image) \rightarrow string$$

  The number of orientations (normally 4 or 8) of the description primitives is implicitly given by the implementation algorithm. The chain code extraction function takes a binary image which consists of the object boundary. The input image is normally produced by an edge extraction operator.

- Polygonal approximation: the objects in the image are represented by the best fit (minimal) polygon.

$$polygonApproximation : image \rightarrow image$$

- Convex hull of a set S, is the smallest convex set that contains the set S.

$$convexHull : image \rightarrow image$$

- Medial axis or skeleton is the reduction of a plane region to a line (graph) representation.

$$medialAxis : image \rightarrow image$$

- Perimeter of a region is the length of its boundary

$$perimeter : image \rightarrow integer$$

- Euler number is a topological descriptor calculated as a difference between the connected components and holes in a figure.

$$eulerNumber : image \rightarrow integer$$

- Signature of the image is a one dimensional function that represents the boundary of an object.

$$signature : image \rightarrow function$$

- An image function can be represented as a series of moment functions.

$$moments : image \rightarrow \mathcal{P}(integers)$$

2.3 Retrieval Operators

Operators in this category can use information stored at any level of the hierarchy. These operators are used for comparison of descriptor and feature values like perimeter comparison, signature comparison, moments comparison, region matching, etc. Their output is either boolean or integer value that shows the similarity level between the input object and the found matching object.

The closest to the top level of the image data type are the predicate level operators. Predicates are of the following type:

$$doricStyle(X)$$

where X is an image object with extracted features like boundary, medial axis, etc.

The predicates are asserted by the user or inferred by the system. We have predicate inference functions and predicate inference manipulation functions:

$$infer : image \rightarrow \mathcal{P}(predicates)$$

$$validate : \mathcal{P}(predicates) \rightarrow \mathcal{P}(Predicates)$$

The function $infer$ might be a complicated composition of lower level type image processing operators as well as semantic association operators. It is an application dependent operator. The *validatc* operator is used for checking the validity of the predicates against the world knowledge that is particular to that implementation. It removes the inconsistencies, and ensures that the inferred predicates make sense with respect to the application at hand.

3 Architecture for Image Retrieval

An image database system that provides the capabilities described above, will have several components as shown in figure 3. Upon acquisition of a new image, the insertion module is invoked. There are several preprocessing operators that the insertion module will apply to the new image. This is a set of operators that we described in the section 2.1. The derived descriptions, and computed features from those descriptions using additional functions and rules are contained in the knowledge base.

The query module accepts the query in any given form: pictorial, numerical and image form. If the input query contains an image, the input image has to go through the same process as the inserted images of enhancement, segmentation, analysis descriptor and feature extraction. The obtained features of the input image are then matched with the corresponding features in the knowledge base and the best matching images are displayed.

Our example database contains images from the ancient Greek architecture. The scanned images are related to the temples of various styles: Doric, Ionian, and Corinthian (for further reference see [15]). We have chosen this application area since the elements of these temples are conceptually consistent in number, kind and relation to one another.

Structurally the temples consist of a stepped platform, the columns and the entablature (which includes everything that is on top of the columns). The basic classification for style recognition is inferred from the shape of the so called *capital* (the top part of the column). The difference between the Doric and Ionian style is that the base and the capital of the Doric columns is plain, while the base and the capital of the Ionian style is more ornate. The capital of the Ionian has a large double scroll, or volute which projects beyond the the width of the shaft. Corinthian capital was invented as an elaborate substitute for the Ionic capital. Its shape is that of an inverted bell covered with leaves of the acanthus plant.

Additional classification feature for determining the period when the temple was built is the height v.s. width ratio. The shaft of the Ionian columns of the later temples is more slender. There are additional ratios that can be extracted and used in the retrieval of these images.

In figure 4 the image hierarchy is shown for an input image during the insertion process. The input image is preprocessed, segmented and prepared for extraction of object descriptors and features. The objects of importance are the capitals of the columns. The regions that are easiest to find in the image are the regions that coincide with the columns. The object descriptors for columns distinguish the shaft from the capital. Once the capital's

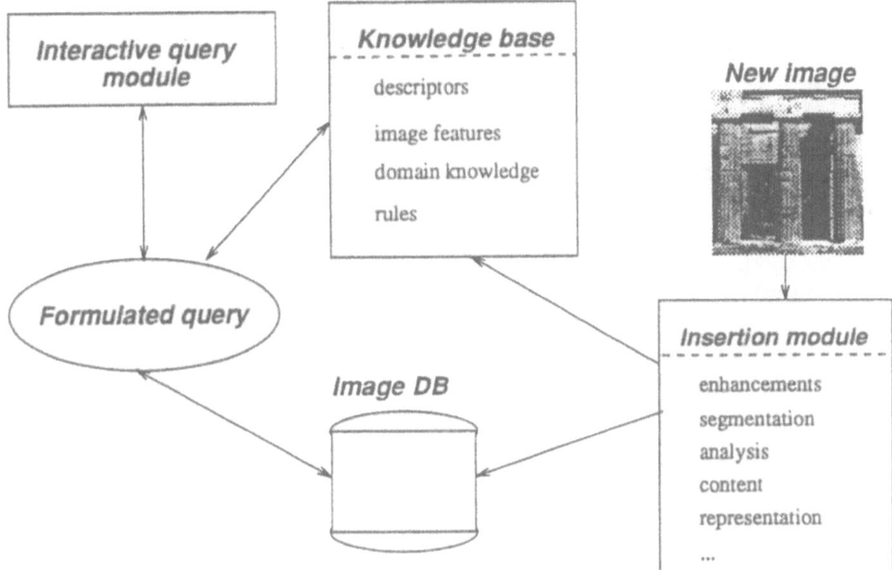

Figure 3: Visual Information System Architecture

descriptors are found we can find the matching shape and also match height vs. width of the shaft ratio.

Using the described retrieval scheme we can answer a query of the following form "retrieve images of any Doric style column that has a ratio 5:1, dates from 500B.C. and its capital resembles the given image". In figure 5 the querying process and the matching images of temples from the Doric style are presented.

4 Conclusions

Image databases are often used only as image repositories. The common methods for image retrieval such as browsing and labeling using keywords are not very efficient. Image retrieval methods based on content representation and semantic analysis are the future steps towards more efficient usage of image databases.

Various image processing techniques have been employed in existing VIMS systems to represent the content of images including color, spatial frequency content, distribution of colors (histogram matching) as well as shape and form of the objects present in the scene. In this paper we have proposed a layered model for extraction of image descriptions and inferring image features for the purpose of image classification. We have categorized the information obtained from the images into five levels and we have also categorized the operations that are performed on the images. Within each category there is a humongous number of choices for the implementation of a particular algorithm for doing a specific task. Our idea is that we can abstract out the main operators within each category and then work

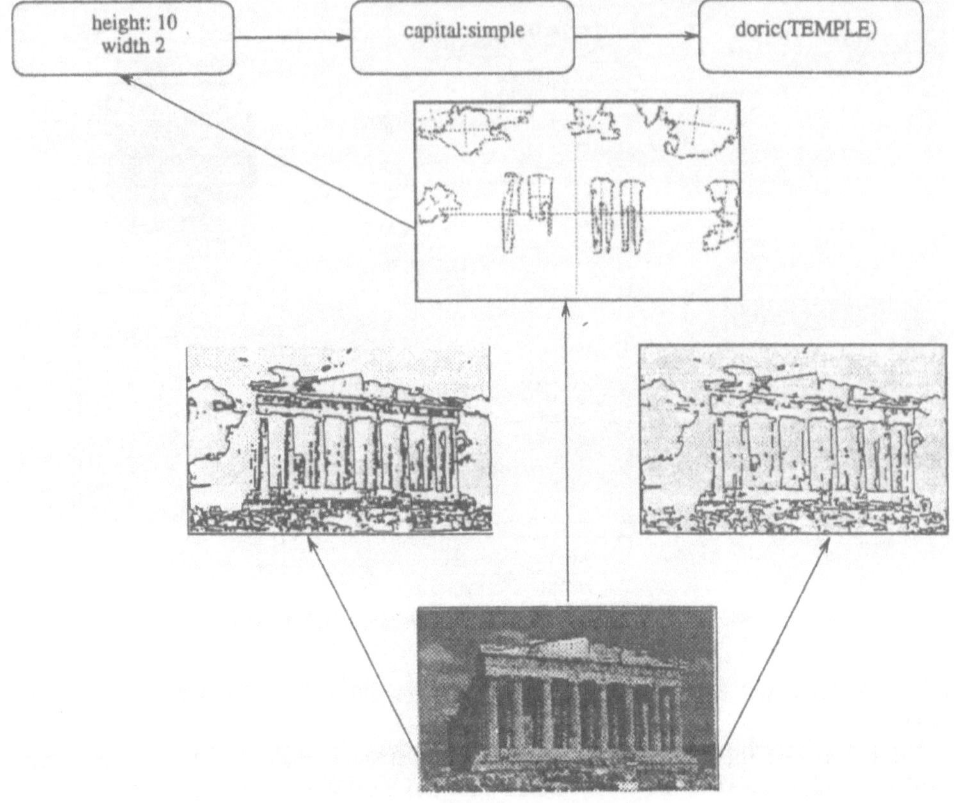

Figure 4: Extraction of features

with a few abstractions. However, it is important to note that there is a necessary sequential process of applying the low level operators and then object representation operators before the retrieval operators can be applied.

Our credo for future advancements in the area of intelligent retrieval of images based on their semantic contents is in the selective use of the available image processing and computer vision technologies. Currently we are working on a prototype system that makes use of the image data hierarchy and the abstracted operators. This work is part of our ongoing SunSet Multimedia Information System project.

Acknowledgements

We would like to thank architect Lira Nikolovska for sharing her expertise in ancient Greek architecture, for the discussions and her insightful comments on the examples in this paper.

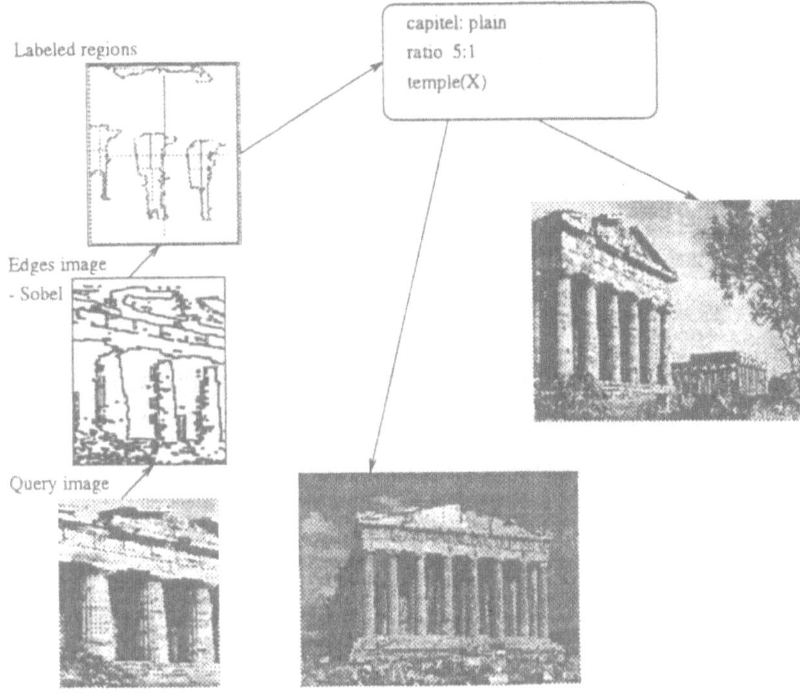

Figure 5: Matching temples from Doric style

References

[1] Rabitti, F. and Stanchev, P., (1989), "GRIM DBMS: a GRaphical IMage DataBase Management System," in Kunii, T. (ed.), *Visual Database Systems*, Elsevier Science Publishers, pp. 415–430.

[2] Roussopoulos, N., Faloutsos, C., and Sellis, T., (1988), "An Efficient Pictorial Database System for PSQL," *IEEE Transaction on Software Engineering*, vol. 14, no. 5, pp. 651–658.

[3] Chang, S.-K., (1982), "A Methodology for Picture Indexing and Encoding," in sun Fu, K. and Kunii, T. (eds.), *Picture Engineering*, Springer-Verlag, pp. 33–53.

[4] Joseph, T. and Cardenas, A., (1988), "PICQUERY: A High Level Query Language for Pictorial Database Management," *IEEE Transaction on Software Engineering*, vol. 14, no. 5, pp. 630–638.

[5] Orenstein, J. and Manola, F., (1988), "PROBE Spatial Data Modeling and Query Processing in an Image Database Application," *IEEE Transaction on Software Engineering*, vol. 14, no. 5, pp. 611–629.

[6] Chang, S., Yan, C., Dimitroff, D. C., and Arndt, T., (1988), "An Intelligent Image Database System," *IEEE Transactions on Software Engineering*, vol. 14, no. 5, pp. 681–688.

[7] Gupta, A., Weymouth, T., and Jain, R., (1991), "Semantic Queries with Pictures: The VIMSYS Model," in *Conference on Very Large Data Bases*, pp. 69–79.

[8] Gupta, A., Weymouth, T., and Jain, R., (1991), "Semantic Queries in Image Databases," in Knuth, E. and Wegner, L. (eds.);*Visual Database Systems II*, Elsevier Science Publishers (North-Holland), pp. 201–215.

[9] Grosky, W. and Mehrotra, R., (1989), "Image Database Management," *IEEE Computer*, vol. 22, no. 12, pp. 7–8.

[10] Jagadish, H. and O'Gorman, L., (1989), "An Object Model for Image Recognition," *IEEE Computer*, vol. 22, no. 12, pp. 33–41.

[11] Mohan, L. and Kashyap, R. L., (1988), "An Object-Oriented Knowledge Representation for Spatial Information," *IEEE Transaction on Software Engineering*, vol. 14, no. 5, pp. 675–681.

[12] Chien, Y., (1980), "Hierarchical Data Structures for Picture Storage, Retrieval and Classification," in Chang, S. K. and Fu, K. (eds.),*Pictorial Information Systems*, Springer-Verlag, pp. 39–74.

[13] Chang, S.-K. and Hsu, A., (1992), "Image Information Systems: Where Do We Go from Here?," *IEEE Transactions on Knowledge and Data Engineering*, vol. 4, no. 5, pp. 431–442.

[14] Gong, Y., Zhang, H., Chuan, H., and Sakauchi, M., (1994), "An Image Database System with Content Capturing and Fast Image Indexing Abilities," in *In Proceedings of the International COnference on Multimedia Computing and Systems*, (Boston, Massachusetts), IEEE, IEEE Computer Society Press, pp. 121–130.

[15] Janson, H. W., (1986), *History of Art*. Harry N. Abrams Incorporated.

Design Issues Of Object Manager For The TOS

T. Al-Sayegh[1], A. Shah[2], I. Faraj[2], F. Fotouhi[3], and W. Grosky[3]

1 National Information Center,
Ministry Of Interior, P.O. Box 69910,
Riyadh 11557, Saudi Arabia
2 Department Of Computer Science,
King Saud University, P.O. Box 51178,
Riyadh 11543, Saudi Arabia
3 Department Of Computer Science,
Wayne State University,
Detroit, Michigan 48202, USA

Abstract. A temporal object system (TOS) was proposed in [3,10], that handles the structural and stature changes of an object in a uniform fashion. The system uses a hybrid approach of class-based and prototype-based approaches, which makes the TOS more flexible than the class-based approach. The TOS provides the aggregation and abstraction to integrate a set of temporal objects to build another high level temporal object, which is referred to as a temporal complex object (TCO). We have applied the TOS to different engineering applications such as vehicle manufacturing environment and construction environment [3,12,13].

Object Manager is a main component of the TOS and its main functions are: (i) to define a new temporal object by creating its birth stage and to validate the temporal condition for the new temporal object, (ii) to update the existing temporal objects by creating a new stage, (iii) to keep record of participant temporal objects for each offstage object and to validate the temporal condition when the offstage is being defined [11]. In this paper, we present the design of the object manager.
Key Words: object-oriented databases, temporal databases, object-manager

1. Introduction

In object-oriented paradigm, an object is defined by two parameters: structure (methods and instance-variables) and state (data-values). In the existing object-oriented database systems, changes to the state of an object are maintained *via* version management [6]. Also, structural changes are supported in most object-oriented database systems. Such changes to a class are referred to as *schema evolution* in the literature [9]. There are three possible scenarios for a class to change its structure. These are:

(type I)	Adding new instance-variables and methods
(type II)	Deleting instance-variables and methods
(type III)	Changes to an instance-variable and method

E. A. Yfantis (ed.), Intelligent Systems, 813–822.

In type I changes, there is no loss of knowledge of a class because the pervious knowledge of the class structure is also retained along with the new one. On the other hand, for type II and type III changes, the history of changes in a class structure is not readily available, as it is overwritten or deleted in the latest version of the class structure. Currently available object-oriented database systems keep only the current version of each class structure. *After* any one of the type II or type III changes, it is necessary to reload a pervious version of the database to retrieve any information from a previous version of a class structure.

In [3, 10], we introduced a temporal object system (TOS) which maintains the history of changes to both the structure and the state of an object in a consolidated and elegant manner. We associate time (time model) to both structure and state of an object. Such an object is referred to as a *temporal object*. A temporal object evolves over time by changing its state and structure. A set of temporal objects that share a common knowledge (i.e., structure and state) is referred to as a *family*. The TOS also facilities the construction of a complex family that is an aggregation of temporal objects from various families. The temporal objects in a complex family are referred to as *temporal complex objects* [4]. A complex family enhances the knowledge sharing of non-homogeneous temporal objects and their transportability. A temporal object system (TOS) is then a collection of families that are defined at different time instances. The object manager is the main module of the TOS that is responsible to do the following functions:

(1) It defines a new temporal object by creating its birth stage and to validate the temporal condition for the new temporal object,
(2) It updates the existing temporal objects by creating a new stage.
(3) It keeps record of participant temporal objects for each offstage object and to validate the temporal condition when the offstage is being defined.

This module provides the basic facilities to create and update temporal objects. In this paper, we report initial work about design of the object manager.

The remainder of this paper is organized as follows: In Section 2, we briefly describe TOS. In Section 3 we report design of Object Manager (OM). Finally, in Section 4 we give our concluding remarks and future research directions.

2. Temporal Object System

In object-oriented paradigm, an object is defined by its structure and state. With the passage of time an object may change its structure and/or its state. By associating time to both the structure and state of an object, we can keep the history of changes to the object. Therefore, we defined a temporal object (TO) to be an ordered set of objects that is constructed at different time instances. A *temporal object* is represented as TO = <(SRt1, STt1), (SRt2, STt2), . . ., (SRtn, STtn)> where $ti \leq ti+1$ for all $1 \leq i < n$, the ordered pair (SRti, STti) is the i-th object of the temporal object which is constructed at the time instance ti with structure SRti and state STti. An i-th object of the temporal object is referred to as its i-th *stage* [6].

A stage is maintained in a prototypical form, i.e., a structure, a state, or a combination of the two [2]. For example, if a temporal object suffers a structural change, then new stage of the temporal object captures only the structural change. We are using time instance as a physical time and time point model. A temporal object may also be referred to as an ordered set of stages. The first stage and last stages of a temporal object are significant because they hold the initial and current knowledge of the temporal object. We refer to these stages as the *birth stage* and the *current stage*, respectively, of the temporal object. A new stage is appended to a temporal object if the structure and/or state associated with its stage changes (see [6] for more details).

A new temporal object can also be created from a temporal object that is referred to as an *offstage object* (see [11] for more details). If an offstage is sharing knowledge of only one temporal object which is referred to a *participant object*, then this is analogous to simple inheritance in the class-based approach, and if an offstage object is sharing knowledge from more than one participant temporal object, then this is similar to the concept of multiple inheritance in the class-based approach (see [16] for more details). An offstage object has some similarities and differences with an object that is defined by using the class-based approach. For example, an offstage object of a family and an object of a class both share same common knowledge *ROF* (root-of-family is defined in the next section) and class structure respectively. In a family state of a temporal object can also be shared by an offstage object, whereas, in the class-based approach it is usually not allowed. The concept of an offstage object enhances the reusability of knowledge within a family. This benefit is not available in the class-based approach.

2.1 FAMILIES OF THE TOS

The concept of a family is used to assemble a group of temporal objects sharing a common context (or common knowledge). All temporal objects within a family can be handled in a similar fashion by responding uniformly to a set of messages. A set of similar structures and/or states defines a *common context* of a family. The common context is referred to as the *Root-Of-Family* (*ROF*) where common knowledge about all its temporal objects is maintained (see [6,7] for more details). Temporal objects of a family can be defined only after the construction of the *ROF* of the family.

In the class-based object-oriented systems, the concept of a class is used to assemble a set of objects that share some common assets as it is with the definition of family in TOS. However, a family encapsulates more features than a class. For example, in a class, the structure of the class is always shared by all its states (or instances), and a change in its structure affects all states, subclass structures and their corresponding states. For example, if one deletes "Color" instance-variable from the class Vehicle of Figure 1, then all its instances are going to delete data value of the instance-variable Color. In a family, however, each temporal object of the family shares the *ROF* only at the time instance of its birth. After that each temporal object is independent and a change in a particular temporal object does not affect the *ROF* or any other object of the family. In other words, the *ROF* of a family is read-only, and it does not change with passage of time. Time associated with a temporal object and *ROF* of family. In TOS, two types of families, *simple families* and *complex families*, can be defined [7,17]. They are described in the next two paragraphs.

A simple family represents an independent object development environment in which temporal objects can be constructed without sharing any knowledge of other families. For example, in Figure 1, the families Engine, Body and Wheel are simple families. Two simple families do not share any knowledge, and in term of knowledge sharing they are mutually disjoint. A simple family is analogous to a class in the class-based approach, which has no super-class (expect *system class* or *root class*).

In existing class-based object-oriented systems, a complex object is defined as an object that has another object as the value of a particular instance-variable [10]. For example, Figure 1 shows the structure of complex object Vehicle. Here, we extend our definition of family to complex family that provides a facility for integration of non-homogeneous temporal objects of different families in order to build another temporal object that is referred to as *temporal complex object* (TCO). The components of a TCO are temporal objects of non-homogeneous families (or independent families), and the temporal objects that take part in the construction of a TCO are called *subobjects* (or *components*) of the TCO.

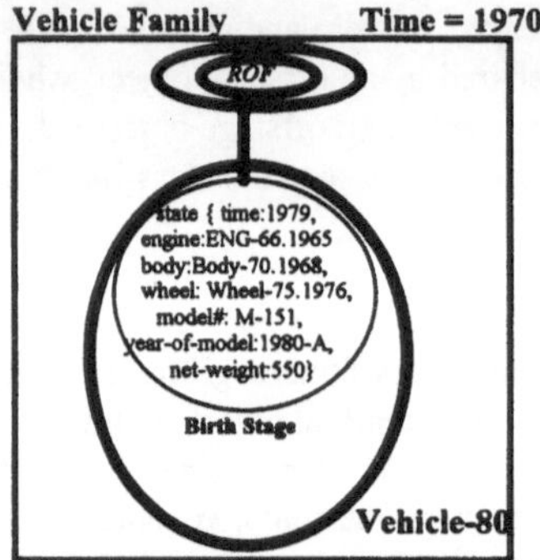

Figure 1. The TCO Vehicle-80 and it's subobjects

A new TCO, TCOc can be defined in a family Fc at a given time instance t1,c with r subobjects from r different simple families. The birth stage of TCO, TCOc may be created at time instance t1,c if the temporal $(t1,k \leq t1,c)$ and $(tFc \leq t1,c)$ is true for all k such that $1 \leq k \leq r$, where tFc is the time instance when the complex family Fc was created, and t1,k is the time instance when birth stage of the k-th subobject was created. This temporal condition ensures that all temporal subobjects and the complex family must exist *before* the existence of the TCO (see details in [7,17]. Figure 1 shows the birth stage of the complex family Vehicle that is an aggregation of three families Engine, Body and Wheel. In Figure 1 and onward figures, double oval, oval, circle, and rectangle represent *ROF*, temporal object, stage, and family, respectively. In Figure 1, the TCO Vehicle-80 is created at time instance 1979 (denoted by Vehicle-80.1979) if the temporal condition, (ENG-66.1965 $\geq$ Vehicle-80.1979) $\wedge$ (Body-70.1968 $\geq$ Vehicle-80.1979) $\wedge$ (Wheel-75.1976 $\geq$ Vehicle-80.1979) $\wedge$ (Vehicle.1970 $\geq$ Vehicle-80.1979) is true, where ENG-66.1965, Body-70.1968 and Wheel-75.1976 are time instances when subobjects ENG-66, Body-70, and Wheel-75 are created, respectively, and Vehicle.1970 is the time instance when family Vehicle is created. *ROF* of the family Vehicle at a time instance 1970 can be defined as follows:

ROF (Vehicle)

Aggregation-of: {Engine, Body, Wheel}

Instance-Variables: {time =1970, model#, year-of-model, net-weight}
Methods: {assemble-it, test-it}

Within the boundary of a simple family, we use the *offspring* technique and among the families we prefer the *copying* technique for knowledge sharing [2]. The aggregation and integration of temporal objects into a TCO can be generate certain conflicts and compatibility problems such as naming and scaling between a TCO and its subobjects. For example, Naming conflicts occur when two or more subobjects of a TCO contain instance variables or methods with the same name such as the instance-variable "weight" that has been defined in subobjects Engine and Body as well as the TCO Vehicle. We are currently investigating these issues.

3. Object Manager Design

The object manager is one of the components of the TOS architecture as shown in Figure 2. The Object Manger (OM) allows to define a new temporal object in a family by creating its birth stage and validates the temporal condition when the new temporal object is being defined. It updates an existing temporal object by creating its new stage (current stage). The OM also keeps a record of participant temporal objects for each offstage object and validates the temporal condition when the offstage object is being defined [11]. The validation of the participant temporal objects of the offstage object is done at the retrieval of the offstage object. The offstage objects are discussed in [10,11].

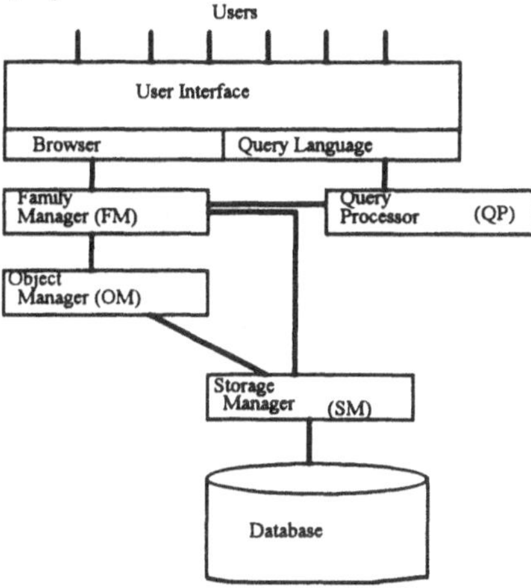

Figure 2: Architecture of the temporal object system (TOS)

The components of the object manager(OM) are referred to as *modules*. These modules deal with different control files that are used to store and retrieve data from. The modules of OM and its other interfacing components of the TOS are shown in Figure 3. The object manager consists of temporal object module, stage module, instance-variable module, and method module. The temporal-object module controls the temporal object, and passes the control to stage module to complete information of temporal objects in storing and retrieving by passing the control to instance-variable and method modules from stage module. As shown in Figure 3, most of the OM's modules are called by the main family module. The family module is also called by the main driver *Root of the Temporal Object System* (RTOS) which is the system object. Before describing details of the OM's modules (Section 3.3), we give a brief description of both the RTOS module and the family module in the next two sections.

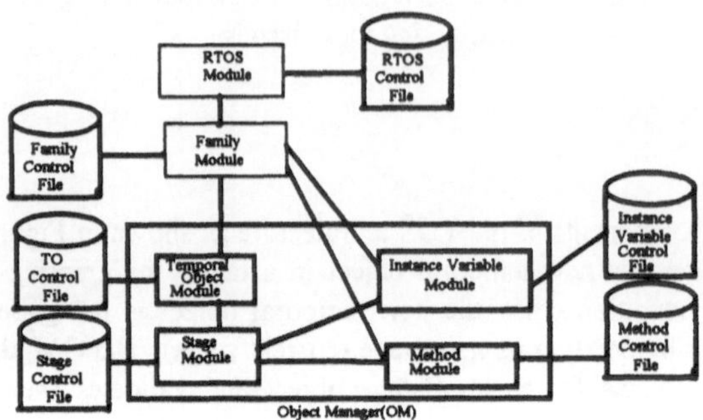

Figure 3. Components of the object manager

3.1. RTOS MODULE

It is a main module of the TOS that controls and takes every command and query to the system. It loads the RTOS-Control-File once the system is being started. The basic function of this module is loading of system primitives and to accept command. This module interfaces with family module for browsing system level information. When a new family is defined in the system, this module checks the temporal condition of the new family before its creation.

3.2. FAMILY MODULE

This module deals with ROF's of the families in the system. It uses Family-Control-File to store and retrieve information. The module has an interface to temporal-object module when a new temporal object is defined in a family after the family has been defined. Also, this module checks the temporal condition before creation of the temporal object. This module interfaces with *instance-variable module* and *method module* to define the instance-variables and the methods of the family, respectively. These two interfaces are combined together to define the root of the family (ROF). The main functions of this module are as follows:

(i) It is responsible to store and retrieve information from Family-Control-File.

(ii) It maintains that part of data dictionary that is related to families. It stores attributes of families such as names and number temporal objects in each family.

(iii) It keeps track of the instance-variables and methods associated with ROF of each family.

(iv) It keeps track of the temporal objects in each family.

3.3 OBJECT MANAGER MODULES

These are four modules of the object manager, they are *Temporal-Object Module, Stage Module, Instance-Variables Module,* and *Method Module*. Details of these four modules are given separately in the following four sections.

3.3.1 Temporal-Object Module

The main task of this module is to access temporal objects. This module interfaces with *stage module* when a new stage (current stage) of a temporal object is defined, or when a new temporal object is defined by creating its birth stage. The module also checks the temporal condition before creation of a new stage. The main functions of this module are as follows:
(i) It stores and retrieves the information of a temporal object in temporal-object-control-file.
(ii) It maintains a part of data dictionary that is related to temporal objects. It stores attributes of temporal object such as names and *OID*, and the module stores or retrieves information of a specific temporal object and/or its stages.
(iii) It keeps track of association of a stages of a temporal object.

3.3.2 Stage-Module

This module is responsible for maintaining stages of the temporal objects by interfacing with *instance-variables module* and *method module* to define the instance-variables and methods, respectively. The main features of this module are as follows:
(i) It stores and retrieves the information of the stages of a specific temporal object in Stage-Control-File.
(ii) It keeps track of the instance-variable and methods associated with each stage.

3.3.3 Instance-Variables Module

This module controls instance-variables, their types and data (if any) for ROF's of families and stages of the temporal object in the system. The main function of this module is to store and retrieve the information of the instance-variables, their types and data (if any) from Instance-Variables-Control-File.

3.3.4 Method Module

Method module is an important module of the OM. It controls the method that is associated with family and/or temporal object. The main features of this module are as follows:
(i) It stores and retrieves the information of the method for family and stages of a temporal object from Method-Control-File.
(ii) It modifies a method.
(iii) It compiles a method and generates its object code.
(iv) It delete a method.
(v) It invokes a method.

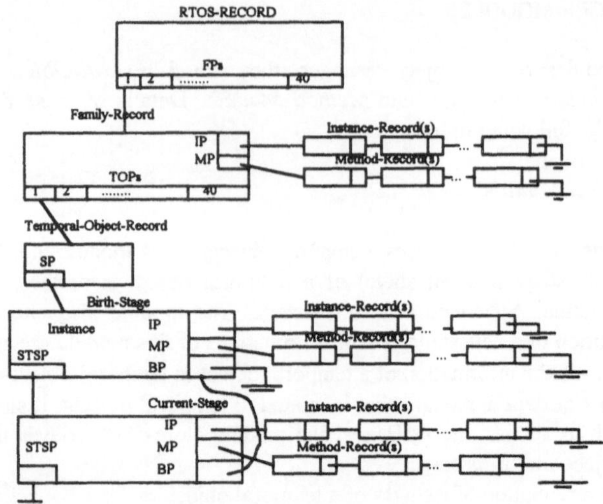

Figure 4: Data structure of the object manager (OM)

3.3.5 Control Files

The modules that are discussed in the previous sections use six different records and fields. They are as follows:
1-RTOS-Control-Record(RCR),
2-Family-Control-Record (FCR),
3-Temporal-object-Control-Record (TOCR),
4-Stage-Control-Record(SCR),
5-Instance-Variables-Control-Record (IVCR), and
6-Method-Control-Record (MCR).
The structure is shown in Figure 4, and fields of each record are given in the next section.

3.3.6 Control File Attributes

Here, we give the fields of each control file. These are the fields that are anticipated at this stage of object manager design.

1. Fields of The RTOS-Control-File
Time of the system creation,
Name of the system,
Last number of family that is created, and
Last temporal object that is created.
2. The Family-Control-File Fields:
Time of family creation,
Family number,
Family name,
Temporal object address at temporal-control-file,

Number of temporal objects,
Instance address at instances-control-file,
Number of instance-variables,
Method address at method-control-file, and
Number of methods.

3. The Temporal-Object-Control-File Fields:
Time of temporal object creation,
Object Identification Number (OID),
Temporal object name,
Stage address at stage-control-file, and
Number of stages.

4. The Stage-Control-File Fields:
Time of stage creation,
Instance address at instances-control-file,
Number of instance-variables,
Method address at method-control-file, and
Number of methods.

5. The Instance-Variables-Control-File Fields:
Time creation of instance-variable,
Instance-variable name,
Instance-variable type, and
Instance-variable data (if any).

6. The Method-Control-File Fields:
Time of method creation,
Method name,
Method internal name.

Note that the first stage (birth stage) of a temporal object is stored to the stage-control-file immediately after successful validation of the temporal condition, the time instance when this stage is created, is also time instance of the temporal object's creation.

4. Conclusion

We have briefly described the main features of the TOS, and design of object manager of the TOS. The object manager is a central and main component of the TOS. Now we are starting its implementation using the programming language C++, while the design work of other components of the TOS is under-progress.

Acknowledgments

We would like to thank Brig. General Saleh AL-Dubayyan, the Director of National Information Center (NIC) of the Ministry Of Interior, Mr. Anwar Bangar, and Mr. AbdulAziz AL-Zaid from NIC for their continuous help and encouragement.

References

[1] Clifford, J. (1982) 'A Model For Historical Databases', *Proc. of Logical Bases For Databases*," France.

[2] Dutta, S. (1989) 'Generalized Events In Temporal Databases', *Proc. of Conf. on Data Eng.*

[3] Fotouhi, F., Shah, A. and Grosky, W. (1992) 'TOS: A Temporal Object System', *Proc. of the 4th Intl. Conf. on Computing & Information (ICCI '92)*, Toronto, Canada.

[4] Fotouhi, F., Shah, A. and Grosky, W. (1992) 'Complex Objects in the Temporal Object System', *IEEE Post-Conference Proc. of the 4th Intl. Conf. on Computing & Information.*

[5] Gadia, S. and Yeung, C. (1991) 'Inadequacy of Interval Time stamps in Temporal Databases', *Information Sciences Journal.*

[6] Katz, R., Chang, E. and Bhatega, R. (1986) 'Version Modeling Concepts For Computer-Aided Design Databases', *Proc. of ACM SIGMOD Conf.*

[7] Ling, D. and Bell, D. (1990) 'Taxonomy of Time Models in Databases', *Information and Software Technology Journal.*

[8] Maier, D. (1986) 'Why Object-Oriented Database Can Succeed Where Others Have Failed', *Proc. of Intl. Workshop on Object-Oriented Database Systems.*

[9] Nguyen, G. and Rieu, D. (1989) 'Schema Evolution in Object-Oriented Database Systems', *Data & Knowledge Engineering Journal*, North-Holland.

[10] Shah, A. (1992) 'TOS: A Temporal Object System', *Ph.D. Dissertation*, Wayne State University, Detroit, Michigan, USA.

[11] Shah, A., Fotouhi, F., Grosky, W., Rana, S. and Vashishta, A. (1993) 'Offstage Objects and their Renovation in the Temporal Object System TOS', *the 3rd Intl. Symposium on Database for Advanced Applications*, Daejeon, South Korea.

[12] Shah, A., Fotouhi, F., Grosky, W., Al-Dhelan, A., and Vashishta, A. (1993) 'A Temporal Object System For a Construction Environment', *Proc. of Conf. the Brazilian Computer Science (SEMISH '93)*, Florianopolois, Brazil.

[13] Shah, A., Fotouhi, F. and Grosky, W. (1994) 'TOS: A Temporal Object System Applied to Vehicle Design & Development', *the International Conference on Data & Knowledge Systems for Manufacturing & Engineering*, Shatin, Hong Kong. (accepted)

Issues in Management of Class-History in Object-Oriented Databases

Abdulrahman Al-Khudair [1], Abad Shah[2], and Hasan Mathkour[2]

[1] Department of Information Systems and Exchanges
Telecom College, P.O. Box 2067, Riyadh 11451, Saudi Arabia

[2] Department of Computer Science
King Saud University, P.O. Box 51178, Riyadh 11543, Saudi Arabia

Abstract. Keeping the history of changes of an object is a desirable feature of current database applications such as CAD/CAM, Computer-Aided Construction, etc. Changes can be managed in object-oriented databases to both parameters (structure and state) of an object. In this paper we address the main issues related to keeping the history of class structure through *class-versioning*. The impact of class-versioning can be in two dimensions, one is to the instances of the class (or the horizontal dimension), and another is to the superclass/subclass of the class (or vertical dimension).
Key Words. object-oriented database, temporal database, class history, schema evolution, object versioning.

1. Introduction

In object-oriented paradigm, all conceptual entities are modeled as objects, and an object is defined by the two parameters *structure* and *state*. The structure of an object provides the structural and behavioral capabilities to that object, that is defined by a set of instance variable (or attributes), methods and/or rules (or integrity constraints). The state of an object assigns data values to the instance variables of the object and methods that operates on them. (In this paper, we will use the terms *instances* and *objects* interchangeably.) If every object carries its own attribute names and methods, then the amount of information that is specified and stored may become unmanageably large. For this reason, as well as for conceptual simplicity, a set of objects sharing the same structure is grouped together into a *class* (or *class definition*). A database schema is a set of class definitions connected by the superclass/subclass relationships that is called *class-hierarchy*. It is represented by a *class lattice* (or a directed acyclic graph). In the existing object-oriented database systems, changes in the state of an object are maintained *via* version management [6]. Also, structural changes are supported in most of object-oriented database systems. Such changes to a class are referred to as *schema evolution* in the literature [2,5]. Although there are several proposals for supporting object history in the relational systems, but there has not been given enough attention toward semantic and object-oriented systems [7]. In this paper we identify the issues that are related to class changes in object-oriented systems that are managed in the these systems non-temporally.

The remainder of this paper is organized as follows. In Section 2 we give possible changes that can happen to a class definition. In Section 3 we identify the aspects of history management in object-

E. A. Yfantis (ed.), Intelligent Systems, 823–831.

oriented databases. In Section 4 we define the horizontal dimension of class-history management and explain the process of creating new class version, propagation of changes on instances (data values), referencing class version, and deleting class version. In Section 5 we define the vertical dimension of class-history management, and explain inheritance problem and the changes to superclass/subclass relationship. Finally, in Section 6 we give our concluded remarks and future work.

2. Basic Non-Historical Class Changes

A class is group of instance-variables and methods that defined together at one place and they are shared by a set of data instance. The possible changes to a class that may occur to the class are as follows [2,3,9, 19,20,22,23]:

- ■ changes to the content of a class
  - ● changes to an attribute of a class
    - ◆ add a new attribute to a class
    - ◆ drop an existing attribute from a class
    - ◆ change the name of an attribute of a class
    - ◆ change the domain of an attribute of a class
    - ◆ change the default value of an attribute
  - ● changes to a method of a class
    - ◆ add a new method to a class
    - ◆ drop an existing method from a class
    - ◆ change the name of a method of a class
    - ◆ change the code of a method in a class

- ■ changes to class-lattice
  - ● add a new class
  - ● drop an existing class
  - ● change the name of a class
  - ● make a class S a superclass (subclass) of a class C
  - ● remove a class S from the superclass (subclass) list of a class C
  - ● change the order of superclasses of a class C
  - ● create new class C as a generalized superclass of n existing classes
  - ● partition a class C into n new classes
  - ● coalesce n classes into one new class

Clearly all above changes to a class or a class-hierarchy are possible in object-oriented databases, but the history of these changes is not retained in currently available object-oriented databases. Current object-oriented database systems keep only the current version of each class definition.

3. Class History Management

Discussion of schema evolution has principally been revolved around the relational and object-oriented paradigms [5]. In a database consisting of data and schema, it is the schema that is usually considered to be stable and the data that changes. However, schemas are not stable as one would expect. Class modification and class versioning, (i.e., class history) and schema versioning have been used to support schema evolution. Class modification is the modification of existing class definitions. Class versioning is the creation of a new version of the class definition. That is, the class definition before the change, is retained, and allowing multiple versions of a class definition to co-exist. Schema versioning is changing the schema as a whole rather than changing individual classes [4]. We are using the terms class-history and class-versioning interchangeably. Issues in the management of schema evolution are the subject of the current research of object-oriented databases [9].

History management in object-oriented databases can be divided into the following aspects:

- **Object Versioning**: The concern here is to maintain the history of a *state* (or instances) of an object, that is, to maintain the history of *data values* of an object.
- **Class Versioning**: The concern here is to maintain the history of *class definition*, that is, class definition before change is retained.
- **Schema Versioning**: The concern here is to maintain the history of whole schema (entire class-lattice) rather than individual classes.

In this paper, we focus our attention on class-history and its relevant issues. In general, the mechanism for class updates in a system with class-history management support can be shown as follows:

$$V_i \xrightarrow{\alpha} V_{i+1} \xrightarrow{\beta} V_{i+2}$$

Where V is class version, i is version number, and α and β are class changes. Note that α and β are changes to current version of a class definition, and they can be any of class changes stated in Section 2. The major motivation for class-history is to keep a record of changes to a class that occur to the class over its lifetime. This record is useful to manipulate different sets of objects under different versions of the class, otherwise instances of the previous class definition cannot communicate consistently and they may generate undesirable errors. For example, if an attribute is dropped from a class definition, then there is no way to see this attribute by the class and hence by the program that is using this class. But in the case of keeping the history of the class, old version can be accessed and its instances will hold the dropped attribute of the new version of the class.

When a change occurs at a time instance to a class its effects are two dimensional, i.e., horizontal and vertical. The change affects all instances (instances that exist at the time instance) of the class. We refer these effects as *horizontal effects*. The change also affects sub-tree under the modified class in the class-hierarchy. We refer this effect as *vertical effect*. In the next two sections we discuss the issues in the management of class-history of horizontal and vertical effects. We refer the management of these two dimensional effects of a change to a class as follows:

- Horizontal Class-History Management.
- Vertical Class-History Management.

4. Horizontal Class-History Management

A new version of a class is created when any change occurs to the class. The new version causes compatibility issues between the new version of the class and its instances, since these instances were compatible with the previous version of the class. We refer to such issues as *horizontal issues* and their management as *horizontal class-history management*. In this section we discuss the issues related to horizontal effects of a change.

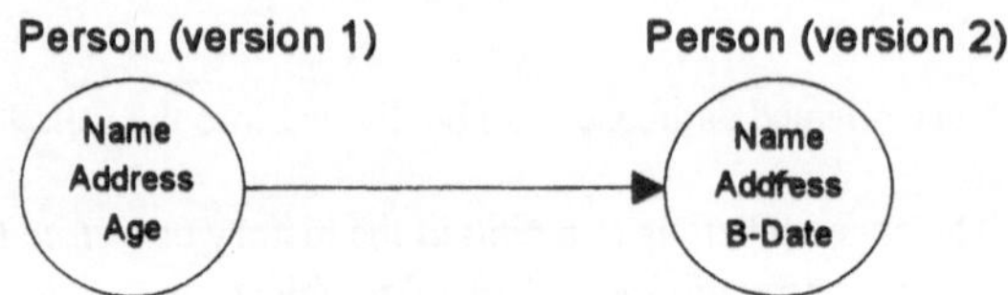

Figure 1. Creation of new class version.

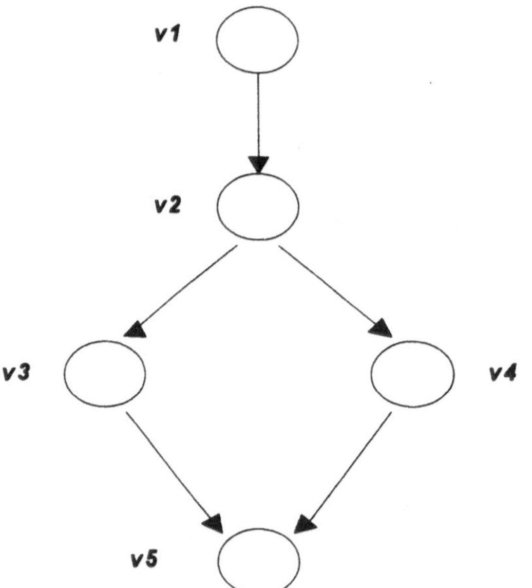

Figure 2. The graph of class versions derivation.

4.1 CREATING NEW CLASS VERSION

We differentiate between the creation of a class itself and the creation of a version of this class. The creation of a class is the process of establishing new knowledge to the database and adding a new class to class-lattice (or schema). The creation of a version of a class is the process of augmenting an existing class (or a class version) and hence creating new version. Figure 1 shows the creation of a new class version of an existing class "Person". In the figure, class Person (version 1) has three attributes: name, address, and age. The new version Person (version 2) of Person (version 1) is created by dropping attribute "age" from Person (version 1) and adding a new attribute "B-Date" in the new version. Person (version 1) continues to exist with the new version, and it can be accessed at any time. Note that Figure 1 shows an example of a change to attributes, other changes such as change to a method of a class and change to class-hierarchy may also be occurred. A versioned class consists of hierarchy of versions called a *version-derivation hierarchy*, if a new version of a class introduces cycles in the version-derivation hierarchy, then the version is rejected. New versions of a class need not be generated in strict linear sequence. It is possible that two versions of a class are derived from the same old version (or a class). It is also possible to have a class version derived from two existing versions of the class. This is shown in Figure 2 where versions v3 and v4 are derived from version v2, and version v5 is derived from versions v3 and v4. Every version of a class has its own unique identifier that distinguishes it from other versions in the scope of the corresponding class.

4.2 PROPAGATION OF CHANGES ON INSTANCES

Perhaps the most important issue in the horizontal class-history management is how to propagate changes to instances of a class. When this goal is achieved the system will be reliable. The creation of new class version is a result of some change to a class and this change is unknown to instances that were creating before incorporating the change to the class. There can be a problem when a program accesses the new class version while the instances are still under the old version definition, this issue is addressed here. No system provides full support for object's evolution resulting from changes in the classes and their relationships [3]. There are many approaches that can be used to manage propagation of changes to instances. Some approaches are described as follows [2,3,4,9,20,23]:

- **Immediate update**: Instances are immediately affected by the change in the class. Disadvantage of this approach is that the performance of the system is affected by frequent and immediate updates. This approach is adopted in GemStone object-oriented database system [3].
- **On-use (or Lazy) update**: Updates to instances are deferred until the instance is used. Disadvantage of this approach is that it requires a permanent propagation mechanism throughout the system's lifetime. This approach is used in Orion object-oriented database system [2].
- **Write-once class**: Disallow class modification when instance of a class is created. Instead, new version of the class must be defined to incorporate change and the old instances are copied to it [23].
- **Schema mapping**: Defer updates to instances indefinitely, or until a reorganization is requested explicitly, maintaining a mapping between the current representation of classes and all previous versions [23].

Including semantic to class changes is another important issue that has received little attention in the current object-oriented systems, except for CLOSQL that can preserve the semantic to class changes [4].

4.3 REFERENCING CLASS VERSION.

Since a new class version introduces some compatibility problem, therefore, references to a class must be updated when a new version of the class is created or deleted. Every version in the version-derivation hierarchy has its own identifier. A change in the class definition may affect programs written to manipulate instances of the class. A program may fail to access an attribute or method that is no longer defined for the new version. In case of addition of attributes and/or methods to a class version, programs written according to the new change may not work on instances created before the change. If an attribute and/or method is deleted in the new version, programs written according to old version may not work on instances created after the change. One method is to include a version number of the required class version. This can be done easily as the version number is always known at the time of writing the code [4].

4.4 DELETING CLASS VERSION

When a class version is required to be deleted from version-derivation hierarchy, we suggest the following steps to delete the class version:

(i) Delete all relationships to superclasses
(ii) Delete all relationships to subclasses
(iii) Delete the class from the version-derivation hierarchy

Some restrictions may be imposed on the deletion of a class version like preventing deletion when the class version have a child class version, i.e., a version that is derived directly from the current version. Another case when there are any existing instances of the class version, this is adopted in GemStone and Encore systems [20].

5. Vertical Class-History Management

Vertical class-history management deals with class versioning and relating these versions to their superclass/subclass relationships. Generating new version of a class implies adding superclass and/or subclass relationships to the version, and all attributes and methods from all superclasses of the new version are inherited, unless there is a conflict. The inheritance mechanism for the versions is an important issue, also there is semantic on changes to superclass/subclass relationships of a class version. These two issues are discussed in the next two sections.

5.1 INHERITANCE PROBLEM

Instances of subclasses under a class version may be affected by class changes. Figure 3 illustrates creation of a new class version and inheritance of the change by its two subclasses (shown as shaded circles). The super-class of the new version is not affected by this change. Issues related to extension of inheritance at version levels is explained in [2]. The idea in this issue is to allow automatic inheritance of access scope (i.e., a set of objects which are accessible to a class version) of a class version into any class version derived from it. This approach avoids copying of instances and change is visible to both new and old versions of a class. However, this approach can be inappropriate if there are a small number of instances of a class where it is desirable visibility of change in new class versions. Another important issue that is also addressed in [2] is the updating problem for the inherited access scope. If a change occur in a class version and by the inheritance that is allowed to all its subclasses, can update support for those inherited instances be achieved ?

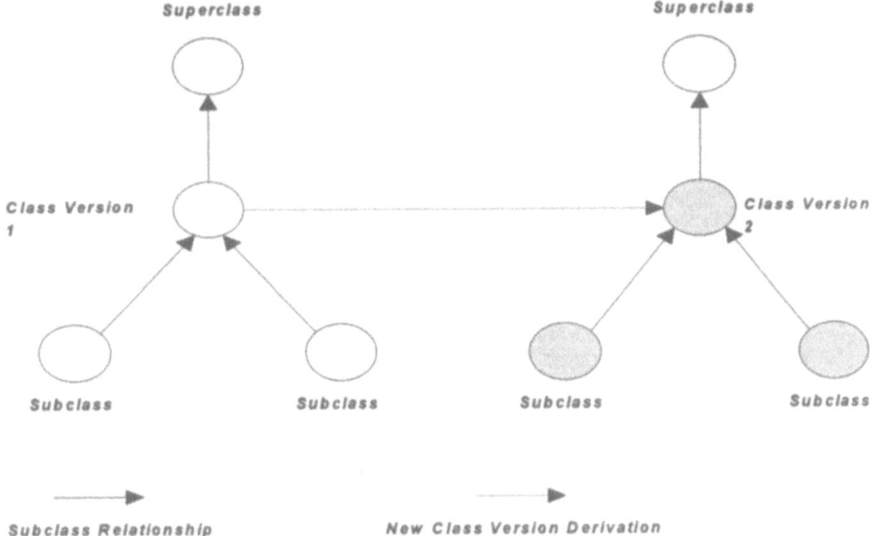

Figure 3. Change in inheritance by new version of class.

5.2 CHANGE TO SUPERCLASS/SUBCLASS RELATIONSHIP

Changing a class relationship and preservation of class-history are also important issues. A change to relationship can be used for the incremental definition of classes or to model a new semantic in a class [3]. Change to superclass of a version of class may result naming conflict in the version. Another effect to the new version or to its subclasses is indirect inheritance from some high level superclass (superclass of a superclass), and a change (for example dropping an attribute) to the immediate superclass, the inheritance from the high level superclass is no longer be valid since the inheritance is done through the immediate super class.

6. Conclusions and Future Work

The management of change in object-oriented databases is an important issue in the current researches. Incorporating the knowledge of time in the database is an essential requirement for the wide range of advanced applications of object-oriented databases. In this paper we identified the major issues that are related to keeping the history of structure of objects through class versioning. The impact of class change is discussed and the relevant consequences to objects and/or programs are addressed.

We are in the process of developing such a management system for object-oriented databases, which will maintain a complete history of changes to a class. This history will be traceable.

References

[1] Thompson, C. (1993) 'A Reference Model for Object Data Management', *Computer Standard & Interfaces*, Vol. 15, No.2-3.

[2] Kim, W. and Chov, H. (1988) 'Versions of Schema For Object-Oriented Data Base', *Proc. of the 14th VLDB Conference*, Los Angeles, California.

[3] Nguyen, G. and Riev, D. (1989) 'Schema Evolution in Object Oriented Data Base Systems', *Data and Knowledge Engineering Journal*, No. 4, North-Holland.

[4] Monk, S. and Sommerville, I. (1993) 'Schema Evolution In OODBs Using Class Versioning', *ACM SIGMOD RECORD*, Vol. 22, No.3.

[5] Roddick, J. (1992) 'SQL/SE - A Query Language Extension for Databases Supporting Schema Evolution', *ACM SIGMOD RECORD*, Vol. 21, No. 3.

[6] Agrawal et al. (1991) 'Object Versioning in Ode', *Proc. of Conf. on Data Eng.*.

[7] Sciore, E. (1991) 'Using Annotations to Support Multiple Kinds of Versioning in an Object-Oriented Database System', *ACM Transactions on Database Systems*, Vol. 16, No. 3.

[8] Taylor, C. (1993) 'Object - Oriented Concepts for Distributed Systems' , *Computer Standards & Interfaces*, Vol. 15, No. 2-3.

[9] Lerner, B. and Habermann, A. (1990) 'Beyond Schema Evolution to Database Reorganization', *Conf. Proc. of ECOOP/OOPSLA*.

[10] Sciore, E. (1989) 'Object Specialization', *ACM Trans. on Info. Sys.*, Vol.7, No.2.

[11] Ling,, D. and Bell, D. (1990) 'Taxonomy of Time Models in Database', *Information and Software Technology*, Vol. 32, No. 3.

[12] Zdonik, S. (1984) 'Object Management System Concepts', the 2nd ACM-SIGOA *Conference on Office Information Systems*, Toronto, Canada.

[13] Maier, D. (1986) 'Why Object Oriented Database Can Succeed Where Others Have Failed', *Proc. of Intl. Workshop on Object-Oriented Database Systems*.

[14] Borning, A. (1986) 'Classes Versus Prototypes in Object-Oriented Languages' *ACM/IEEE Fall Joint Computer Conference*.

[15] Kim, W. (1990) 'Object-Oriented Databases: Definition and Research Directions', *IEEE Transactions on Knowledge and Engineering*, Vol. 2, No. 3.

[16] Martin, N., Navathe, S. and Ahmed, R. (1987) 'Dealing with Temporal Schema Anomalies in History Databases', *Proc. of the 13th VLDB Conference*, Brighton.

[17] Vossen, G. (1991) 'Bibliography on Object-Oriented Database Management', *ACM SIGMOD RECORD*, Vol. 21, No.1.

[18] Khoshaflan, S. (1988/1989) 'A Persistent Complex Object Database Language', *Data and Knowledge Engineering*, Vol. 2.

[19] Zicari, R. (1991) 'A Frame Work For Schema Updates In an Object-Oriented Database Systems', *The O2Book*, Morgan Kaufmann.

[20] Skarra, A. and Zdonik, S. (1986) 'The Management of Changing Types in an Object-Oriented Database', *Proc. of OOPSLA'86.*

[21] Holl, R. and King, R. (1987) 'Semantic Database Modeling: Survey, Applications, and Research Issues', *ACM Computing Survey*, Vol.19, No.3.

[22] Kim, W. (1990) Introduction to Object-Oriented Databases, *MIT Press*.

[23] Cattell, R. (1991) Object Data Management: Object-Oriented and Extended Relational Database Systems, *Addison-Wesley Publishing Company*.

Methodology To Convert A Traditional Design To An Object-Oriented Design

Sami H. Al-Harbi[1], Abad A. Shah[2], Hassan Mathkour[2], and J. C. Agrawal[2]

1. Directorate of Computing, Royal Saudi Air Force,
   P. O. BOX 59787, Riyadh-11535,
   Kingdom of Saudi Arabia

2. Computer Science Department, King Saud University,
   P. O. BOX 51178, Riyadh-11543,
   Kingdom of Saudi Arabia

Abstract. The organizations that have old systems and those systems were developed using conventional methodologies such as structured methodology, have some compiling arguments to convert these systems to object-oriented systems. We are in the process of developing a methodology that converts a conventional design into an equivalent object-oriented design by using analysis, design and documentation of the existing conventional system. This methodology can save both development time and cost of the new object-oriented systems.
Key Words: traditional design, object-oriented design, object-oriented methodology

Introduction

Numerous information systems comprising of very many lines of code were developed during, and before, the 1970s using traditional design methodologies. At some point in the life of these systems there will be a need for a complete redesign into an object-oriented system. For example, the US Department of the Army has many large information systems that were developed using COBOL, with over a million lines of code, using traditional design methodologies. Now, with the mandate from the US Department of Defense to use Ada, and with the arrival of Ada 9X, with it's object-oriented approach, there will be an increasing demand for the use of this design methodology.

In any redesign effort the best source of requirements is the existing system and its design documentation. Their use will reduce the cost of redesign, and produce a more reliable system. Shah and Kaushal [12] initiated such an effort and this paper advances a step further in proposing a methodology for converting traditional design to an object-oriented design.

The remainder of this paper is organized as follows: In Section 2, we give background of the problem. In Section 3 we describe object-oriented approach and give primitives the approach. Section 4 deals with motivation and related work. In Section 5 we give outlines of the proposed methodology. Finally, in Section 6 we give our concluding remarks and future research directions.

E. A. Yfantis (ed.), Intelligent Systems, 833–842.

2. Background

Software development is the key element in the evolution of computer-based information systems and products. During the past four decades, software development evolved from specialized problem solving and ad-hoc programming, through information analysis and systems analysis, to an industry in itself. Many of the software problems encountered during the evolution process were found, and have been created by the culture and characteristics of early ad-hoc programming. Yet, as the software development methodologies became formalized and utilized automation tools, the old systems still were able to provide the requirements. Further development can be achieved as we advance from traditional designs to object-oriented designs [9].

The software development life-cycle integrates methods, tools and procedures for the development of software. A number of different paradigms have been proposed, each exhibiting strengths and weaknesses [4,6]. One of the well-known methodologies used in the development of information analysis is Structured Analysis (SA) [4]. Structured analysis requires hierarchical and functional decomposition of the problem. Data Flow Diagrams (DFD) are used as notational tools in SA. A DFD is an abstract data type (ADT) graph whose nodes represent functions and arcs represent data flow. A DFD specifies data sources, data stores and data transformations. A data store is a conceptual data structure.

3. Object-Oriented Approach

A new approach, object-oriented approach, has been emerging lately for the development of information systems. It is a new way of thinking about problem solving using models organized around the real-world concepts. Many methodologies are proposed based on object-oriented approach to develop information systems [10].

In this approach, an *object* is the fundamental building block that combines both the structural and the behavioral capabilities of the object into a single entity. A *class* (or type) is a template description that specifies common properties and behavior for a group of similar objects and an object is an instance of a class. The classes themselves can be organized into a graph that is referred to a *class-hierarchy.* Such a class-hierarchy allows similar classes to be related together in such a way that commonalties of one class can be inherited (reused) rather than duplicated by classes lower in the hierarchy, thus simplifying the design and implementation of these lower level classes. In the other words, classes are organize a parent-child relationship. The parent class is referred to as *superclass* and child class is referred to as *subclass* [2,14]. A *subclass* is a class that inherits behavior from another class. A subclass usually adds its own behavior to define its own unique kind of object. A *superclass* is a class from which specific behavior is inherited. A class might have only one superclass, or might have several.

The properties and behavior of objects, and hence their commonalties, are described in terms of attributes and operations (methods). An *attribute* is a named property of an object which holds a value and maintains an abstract state for that object. An *operation* identifies an action which may be applied to objects of a class. *Inheritance* is a mechanism which permits classes to share attributes and operations based on relationships of specialization and generalization between them within a class-hierarchy.

A *message* consists of the name of an operation and any required arguments. Limiting object access to strictly defined interface such as the message-send allows another use of abstraction known as *polymorphism*. Polymorphism is the ability of two or more classes to respond to the same message, each in its own way.

Compared to traditional design methodologies, object-oriented design shifts the focus from a procedure oriented approach to an object oriented approach. By focusing on the objects, the object-oriented approach becomes very useful in understanding problems in communicating with application experts in modeling enterprises, in preparing documentation and in designing programs. The object-oriented approach uses concepts of polymorphism, encapsulation and inheritance, that are not generally used in traditional methodologies. This, in turn, boosts the reusability of software components designed with an object-oriented methodology. Software is *reused* when it is used as a part of software other than that for which it was initially designed. Software is *refined* when it is used as the basic for the definition of other software [5].

Naturally, object-oriented methodology improves productivity, provides better control of software complexity and decreases costs in long run. Most developers of information systems are looking towards object-oriented methodology because its benefits. However, such developers face problems in re-training staff, capturing the analysis, and investments made in design decisions for previous designs developed through traditional methodologies.

4. From Traditional Design to Object-Oriented Design

Life-cycle of traditionally and object-oriented software is supposed to start with a requirements specification. Then the software grows into a design, is implemented, and then tested. If it passes its tests, the software graduates to product status. Once attracts the users, and these users find bugs, it must be maintained. When users clamor for more, it must be extended [7]. The main steps of software's life-cycle are as follows:

(1) System requirements,
(2) Analysis,
(3) Design,
(4) Implementation
(5) Testing,
(6) Delivery,
(7) Maintenance, etc.

In traditional software development methodologies, a major portion of the project time is spent on analysis phase, but in object-oriented methodology, a major portion of the project time is spent on design phase.

Shah and Kaushal [12] pointed out two possible approaches for developing a new object-oriented version of an existing information system. The first approach disregards the analysis and design of the existing system and develops an object-oriented design by undertaking a fresh analysis and design steps by following an object-oriented methodology. The second approach uses analysis and design information of the existing system to get an object-oriented design. The first approach is costly in terms of time and money. Rebecca Wirfs-Brock [14], have shown that a large portion of the overall time for an object-oriented software's life-cycle is spent on the design phase and only a small part is spent on implementation and testing. Therefore, it can be seen that the second approach results in a substantial saving time and money.

This paper proposes a new methodology for the second approach, i.e., making use of the analysis and design of the existing system. This methodology is developed by studying and taking into account the strengths and weaknesses of the proposed object-oriented methodologies for an object-oriented design. Booch [1] uses an introductory example that derives objects from a terminator of a context diagram and data stores from data flow diagrams. Gray [7] does the same by using additional rules. Seidewitz and Stark [11] define a process called "abstraction analysis" to begin the transition from data flow diagrams to an object-oriented design. The key is to find a central entity representing the best abstraction for what the system does. To achieve this, one must look for a set of processes and data stores that are most abstract. Krell addresses tasks before objects, advocating strongly that it is important to avoid premature decomposition and consequent isolation of tasks in separate components. The Hierarchical Object-Oriented Design (HOOD) method [13] does significantly differentiate between objects that do, and the objects that do not, represent concurrent execution. One of the HOOD basic design steps as illustrated by Heitz [8] is mapping from data flow diagrams to objects.

Some of the ideas mentioned above do not appear to have reached full maturity, either due to lack of testing, or because of their over-dependence on Ada projects. The approach outlined in this paper is unique, because it starts from the Entity-Relationship Diagram (ERD) [3] which is a powerful tool for designing a system. It has characteristics to relate two entities as a class-hierarchy.

The *ERD* equips database designers with a powerful tool during designing of the conceptual and logical levels of information system [3]. The *ERD* also has a great power of graphical representation, that is helpful in designing high-level schema of a system. For each real-world object in the ERD, three basic concepts define the role and the structure of the object. These concepts are: Entity, Relationship, and Attribute. An entity represents an object in the real world, it can be defined by a set of attributes and a set of data values which are assigned to each attribute from its representative data value domain. An entity is a basic building block of the ERD and entities of a universe of discourse are interrelated to each other through different relationships. Each entity is involved in a relationship with some other entities, or with itself. The cardinality of a relationship can be one-to-one, one-to-many, or many-to-many. For example, in a student registration system the entity "instructor" might have two attributes: the instructor's name and the department that offers the course. The entity "instructor" may have the relationship "teaches" with the entity “course”. The entity "student" may have the relationship "enrolled" with the entity "course". The first relationship may be one-to-many (one instructor may teach many courses). The second relationship is many-to-many, because each student may enroll in several courses and each course may have several students.

5. FTD-OOD Methodology

The aim of the FTD-OOD methodology is to utilize the available structured analysis and design of an existing system to get an object-oriented design with minimum time and cost. The analysis and design details are already available in the forms ERD and DFD of an existing system. We use the structured analysis and design as a front-end to achieve an object-oriented design by using this methodology.

Structured analysis and design of an existing system has two main components: DFD and a conceptual diagram (control flows or context diagram) [3]. We assume that during structured analysis an ERD is formed of the existing system that gives a conceptual representation of the system.

The proposed methodology takes the available ERD and DFD of a system and converts them into a complete class-hierarchy. An ERD outlines classes and their *logical links* in an object-oriented design. The entity and relationship types, along with their logical links, are the basic building blocks for the classes and class-hierarchy. But at this stage, a class contains only instance-variables (or attributes). An ERD does not provide information about functions (or methods) of entities and relationships, or their all type of links among entities and relationships. In other words, the ERD lacks information flow in a system. This information is provided by the system's DFD. Therefore, both an ERD and DFD are necessary to convert a traditional design to an object oriented-design.

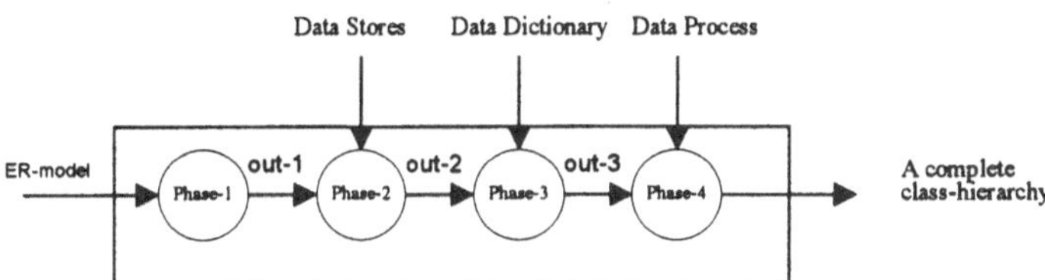

Figure 1. Block diagram of the methodology

Figure 1 outlines the four main phases of our proposed methodology FTD-OOD. The first phase takes an ERD and converts it into a *preliminary-class-hierarchy*. A preliminary-class-hierarchy is a class-hierarchy without methods. It is first skeleton and realization of the final and complete class-hierarchy that is final output the proposed methodology. The second phase takes a data stores from DFD and may add more classes and places them in appropriate places in the preliminary-class-hierarchy. The third phase takes a data dictionary and produces a data structure for instance-variables of the classes. The final phase takes a data process from DFD and converts it into methods, and places them in appropriate classes of the preliminary-class-hierarchy. In Figure 1, out-1 is the output of Phase-1 that is the preliminary-class-hierarchy, out-

2 is the output of Phase-2, out-3 is output of Phase-3, and Phase-4 gives final output the methodology that the complete class-hierarchy of the new system.

Phase-1 takes entities and their relationships from the ER diagram and converts them into a preliminary-class-hierarchy. We propose the following steps to form a preliminary-class-hierarchy.

1. By the definition of entity (see Section 4.0), an entity represents a class of objects.

2. Convert each relationship in a ERD, one-to-one, one-to-many, and many-to-many to an equivalent graph in FTD-OOD methodology. These graphs keep the meaning of the relationship between classes. Figure 2 shows the entities and their relationships after conversion.

3. Some of the relationship has a set of attributes that can created between one or more entities. For example the marriage relationship has a date of marriage attribute that appears once when the relationship was created and disappears from the relationship when it is canceled. In this case, a class will be created that contains those attributes. This class is a subclass and the entities will be the superclasses (see Figure 3).

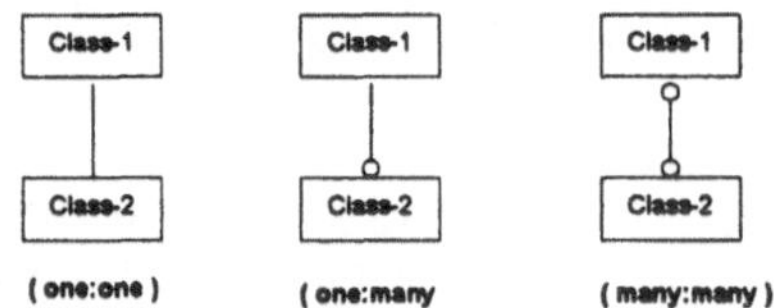

Figure 2. Shapes of the relationship into FTD-OOD methodology

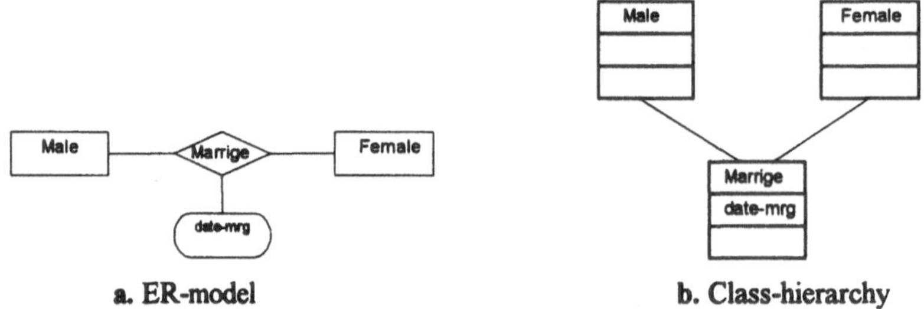

a. ER-model

b. Class-hierarchy

Figure 3. A marriage relationship in ER-diagram and class-hierarchy

4. Every class-hierarchy consists of different levels, the root is at 0-th level, and the leaves are the n-th level. The nature of object-oriented approach usually discourage a cycle in class-hierarchy. To convert a cycle of an ERD we propose the following algorithm:

Function (ERD) : preliminary-class-hierarchy;

begin {main}

1. If ($L(e_i) = L(e_j)$) & ($R(e_i,e_j)$) then

```
      begin
        remove R(e_i,e_j)
2.      If superclass(e_i) <> superclass(e_j) then
          create new R(e_i,superclass(e_j))
      end
3.  If ( L(ei) >= 2 L(ej) ) & ( R(ei,ej) ) then
      remove R(ei,ej)
      call PHASE-4
   end. {main}
```

Where "L" is a number that represents the level of preliminary-class-hierarchy, e_i and e_j entities, and R relationship between the entities e_i and e_j. At Step 1, if there are two entities at the same level and there is a relationship between them, then remove this relationship. After removing the relationship between entities, step 2 checks if the superclass of the first entity is different from the superclass of the second entity, then it creates a new relationship between the first entity and superclass of the second entity. At Step 3 if there is a relationship between tow entities that are at different levels and difference of levels is more than two, then the relationship is removed. Figure 4 represents partial working of the above algorithm, the classes are denoted by letters (A, B, ...), and relationships between the classes are denoted by lines.

The input of Phase-2 is the data stores from the DFD and the preliminary-class-hierarchy. We do not take the data store as a values stored, rather we view them as an abstraction. Figure 4 shows an example of converting a 'Char File' data store to

a class. The values which are stored in 'Char File' data store is converted to the values for Char File class. The process in DFD that stores and retrieves data in/from data store is converted to a method of the class. The designer can add the attributes needed for the class. Figure 5 shows "size" attribute in Char File class.

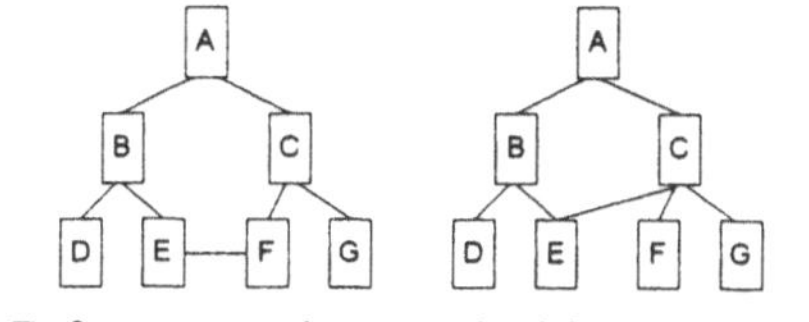

a. Before conversion   b. After conversion

Figure 4. An example of step 2.

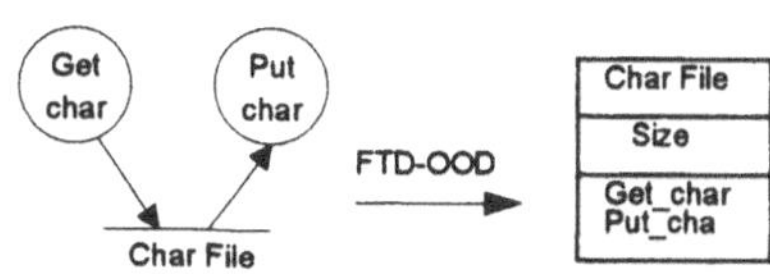

Figure 5. An example of converting data store

After defining a new class, a relationship is created between the new class and one of it's superclasses. Two kinds of classes can be produced from data stores. Passive classes that include only methods, and the second kind is active classes that contain both methods and instance-variables. The basic idea of Phase-2 is to add more information in the classes of the preliminary-class-hierarchy.

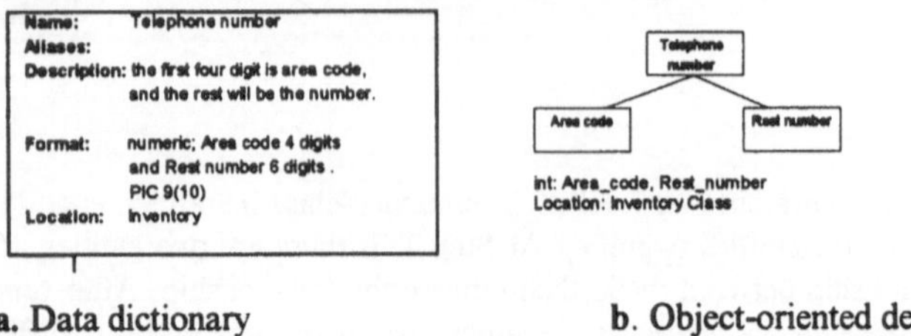

a. Data dictionary b. Object-oriented design

Figure 6. Convert a data dictionary to an object-oriented design

Phase-3 takes data dictionary of the existing system, and produces structure of instance-variables of classes of object-oriented design and the appropriate documentation for these instance-variables. Figure 6 show an example for conversion of a telephone number from a data dictionary to object-oriented design.

In object-oriented approach a method is the implementation of an operation. The designer, therefore, takes the pseudo-code for each process from the lowest-level of data flow diagram and consider it as a class's method in preliminary-class-hierarchy. The inputs for the process is the arguments for the method, and the output from the process will be the output of the method. Initially, the designer identifies methods in DFD and proposes its class. Later, Phase-4 puts each method to the most appropriate class for the method. The following example shows a process specification in pseudo-code form for moving x, y to a new position and shows the equivalent code for the move operation in C++.

process specification	*operation in C++*
name : move	void Shape :: move (int dx, int dy)
input: dx, dy	{
move the points x, y	
by adding new value dx, dy	x = x + dx; y = y + dy; }

The second part of this phase reorganizes the methods in classes of the class-hierarchy. There are four possible cases to reorganize methods based upon the input and output of each method. The cases are as follows:

Case 1: If a method, M_1, receives instance-variables from j number of classes and gives input to a single class, then the method M_1 can be placed either in all input classes or in the output class (see Figure 7(a)). If we place the method in the input classes, it reduces reusability of the method within the system. The message passing overhead to produce the output will be the same if the method is placed either in the input classes, or in the output class. Therefore, we make the input classes as subclasses of the output class and place the method M_1 in the output class C_0 as shown in Figure 7(b).

Case 2: If a method, M_2, exists in the superclass and its subclass (see Figure 8), then placing of the method in the superclass and the subclass inherits the method from its superclass.

Case 3: If the method, M_3, updates one or more its input classes C_1, C_2 ....., C_j (see Figure 9), then we recommend to create a new superclass include the sharing attributes of the input classes and place the method M_3 in the new superclass.

Case 4: If the method, M_4, generates j number of output classes C_1, C_2 ....., C_j without receiving any input (see Figure 10), then we recommend to create new superclass include the sharing attributes of the output classes, and place the method M_4 in the new superclass to improve reusability and reduce the message passing overhead.

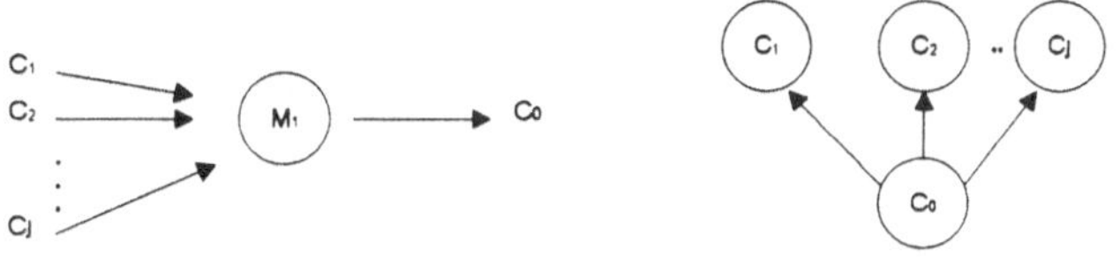

a. J inputs and one output **b**. Subclass-hierarchy for (a)

Figure 7. Conversion of the output class as superclass for input classes

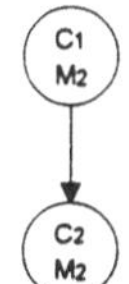

Figure 8. Superclass and it's subclass with the same method

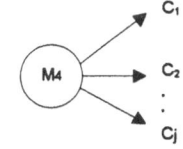

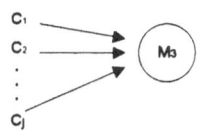

Figure 9. "j" number of inputs and no output **Figure 10**. No input and "j" number of output

6. Conclusions

We have outlined a methodology that transforms an existing traditional information system design into an object-oriented design by using analysis and design of the existing system. The methodology consists of four main phases. The first phase takes an ERD and produces a preliminary-class-hierarchy, the second phase takes the data stores from DFD and produces more classes. The third phase takes a data dictionary and produces a data structure for instance-variables of each class. The final phase takes each data process successively from DFD and places it in an appropriate class. The area is still open and these phases need more improvements and refinements. We are working to refine and improve the methodology, and after that we intend to use the methodology on a case-study.

Acknowledgments

We would also like to thank Lt.Col. Abdulaziz Al-Anazi, Major Musaed Al-Shablan, and Captain Zaid Al-Hazmi of Royal Saudi Air Force for their help. Also, we would like to thank Mr. Trevor Rowe for reviewing this paper and his valuable comments.

References

[1] Booch, G. (1986) 'Object-Oriented Development', *IEEE Transactions on Software Engineering, Volume SE-12*, Number 2.

[2] Capertz, L. and lee, P. (1993) 'Object-Oriented Design: Guide Lines and Techniques', *Information and Software Technology*, Volume 35 No 4.

[3] Chen, P. (1976) 'The Entity-Relationship Model-Towards a Unified View of Data', *ACM Transaction Database Systems*, Volume 1.

[4] DeMarco, T. (1978) Structured Analysis and System Specification, *Yourdon Press*, New York.

[5] Embley, T. (1992) Barry Kurtz, and Scott Woddfield, 'Object-Oriented Systems Analysis: A Model-Driven Approach', *Prentice Hall*, Englewood Cliffs, NJ.

[6] Gomaa, H. (1984) 'A Software Design Method for Real-Time Systems', *Communications of the ACM*, Volume 27, Number 9.

[7] Gray, L. (1988) 'Transitioning from Structured Analysis to Object-Oriented Design', *Proceedings of the Fifth Washington Ada Symposium*.

[8] Heitz, M. (1990) 'Practical Issues of Using HOOD', *Ada UK Tutorials, Ada UK International Conference*, Brighton, England.

[9] Pressman, R. (1987) 'Software Engineering: A Practitioner's Approach', *McGraw Hill*, New York

[10] Rumbaugh, J., Blaha, M., Premerlani, W., Eddy, F. and Lorensen, W. 'Object- Oriented Modeling and Design', *Prentice Hall*, Englewood Cliffs, NJ, 07632.

[11] Seidewitz, E. and Stark, M. (1987) 'Towards a General Object-Oriented Software Development Methodology', *Ada Letters, ACM/SIG Ada*.

[12] Shah, A. and Kaushal, R. (1992) 'A Methodology for Converting an Imperative Design to an Object-Oriented Design', *Proceedings of the 23rd Pittsburgh Conference on Modeling and Simulation*.

[13] European Space Agency (ESA), 'HOOD User Manual', Issue 3.0, WME/89-35/JB, ESA Research and Technology Center, Noordwijk, The Netherlands.

[14] Wirfs-Brock, R., Wilkerson, B. and Wiener, L. (1990) 'Designing Object-Oriented Software', *Prentice Hall*, Englewood Cliffs, NJ.

COMPUTER GRAPHICS AND IMAGE PROCESSING

DEFORMATION OF VOLUMES USING SCATTERED LANDMARK POINTS

T.A. Foley
Department of Computer Science and Engineering
Arizona State University
Tempe, AZ 85287-5406 USA

Abstract: We address the problem of deforming one three dimensional volume to anothe volume. For our applications, landmark points in the domain volume are arbitrarily located and it is desired to form a deformation to another volume so that the scattered landmark points get mapped to another set of corresponding points in the range volume. Applications involve modeling solids, volumetric morphing and mapping one brain volume to a "standard" brain volume. This paper describes some volume deformations that make use of trivariate functions that interpolate real valued data sampled at arbitrary locations in space.

1 Introduction

For free-form deformation of volumetric or solid models, an effective approach is given in [18] which involves the use of tensor product Bezier trivariate mappings of 3D space to 3D space. Their approach requires that the 3D control points form a rectilinear grid in the domain and a logically similar grid in the range space. For some applications, landmark points in the domain volume are arbitrarily located and it is desired to form a deformation to another volume so that the scattered landmark points get mapped to another set of corresponding points in the range volume. One application with scattered landmark points that we have encountered involves mapping one brain volume to a "standard" brain volume. Other applications involve modeling solids and volumetric morphing. This paper describes some volume deformations that make use of trivariate functions that interpolate real valued data sampled at arbitrary locations in space. We first describe some trivariate interpolants and then give extensions for volumetric deformations. Finally, we give examples involving deformations of the brain and observations on how the locations of the landmark points affect the stability of the deformation.

2 Trivariate Interpolation Methods

The problem of fitting a function to data sampled at arbitrarily located positions in a 3D domain arises often in scientific and engineering applications. Data of this type commonly

E. A. Yfantis (ed.), Intelligent Systems, 845–851.

occur in applications where the data site locations are restricted. Some data sets come from experimental tests, the measurement of physical quantities, and from computational values such as those output from a finite element solution of a partial differential equation. Examples occur in meteorology, mining, optics, medicine, oceanography, fluid flow analysis, and numerous other places in the broader fields of physics, chemistry and engineering. For a precise statement of the problem, suppose that we are given N distinct points $p_i = (x_i, y_i, z_i)$ and N real values f_i. We first address the interpolation problem of constructing a smooth function $F(p)$, where $p = (x, y, z)$, that satisfies $F(p_i) = F(x_i, y_i, z_i) = f_i$ for $i = 1, \ldots, N$.

Since the scattered data interpolation problem has so many important applications, a great deal of research has been done in this area for functions of 2 variables and the survey papers [2, 3, 11, 13, 16] contain descriptions of many methods that solve this problem. Some of these methods extend to trivariate functions, and some of these are briefly mentioned in the following paragraphs.

One of the simplest methods to implement and one of the most effective scattered data interpolants is Hardy's multiquadric (MQ) method (see[15]). The multiquadric method is an infinitely differentiable function of the form

$$F(x,y,z) = \sum_{i=1}^{N} a_i B_i(x,y,z) + \sum_{k=1}^{M} c_k Q_k(x,y,z),$$

where $R^2 > 0$, $p = (x,y,z)$, $d_i^2(p) = \|p - p_i\|^2 = (x - x_i)^2 + (y - y_i)^2 + (z - z_i)^2$ and $B_i(p) = \sqrt{d_i^2(p) + R^2}$. The functions $Q_1(x,y,z)$, $Q_2(x,y,z)$, ..., $Q_M(x,y,z)$ form a basis for the space of polynomials of degree $< m$. The coefficients $a_1, ..., a_N$ and $c_1, ..., c_M$ are computed from the system of $N + M$ linear equations

$$\sum_{i=1}^{N} a_i B_i(x_j, y_j, z_j) + \sum_{k=1}^{M} c_k Q_k(x_j, y_j, z_j) = f_j, for j = 1, 2, ..., N, and$$

$$\sum_{i=1}^{N} a_i Q_k(x_i, y_i, z_i) = 0, for k = 1, 2, ..., M.$$

In matrix form, the linear system can be written as $Kx = f$, where $x = (a_1, ..., a_N, c_1, ..., c_M)^t$, $f = (f_1, ..., f_N, 0, ..., 0)^t$ and K is the symmetric matrix described by: the first N rows and columns have $K_{i,j} = B_j(p_i)$, the last M columns of the ith row ($i \leq N$) are $Q_j(p_i)$ for $j = 1, ..., M$, and the lower M by M portion of K are all zero. Plots of this generally effective method and additional discussion can be found in [5, 9].

The accuracy and the visual smoothness of the multiquadric method depends heavily upon the value chosen for R^2 as illustrated in Figure 1. The optimal value of R^2 is problem-dependent, and most early formulas for R^2 involve the number of data points together with the size and shape of the domain containing the data. A critical unsolved problem involving the use of the MQ method is how to compute the optimum value (or even a consistently "good" value) for R^2. In [5], the MQ interpolant with various values of R^2 was applied to data generated by several test functions applied to various xy data sets. In addition to stating several observations on the effects of the parameter, an ad hoc algorithm is also given in [5] that often computes an effective value for the parameter R^2. Although the MQ method performed exceptionally well in [11], it should be noted that the accuracy of this method can be significantly better than previously reported if a near optimal value of R^2 is used.

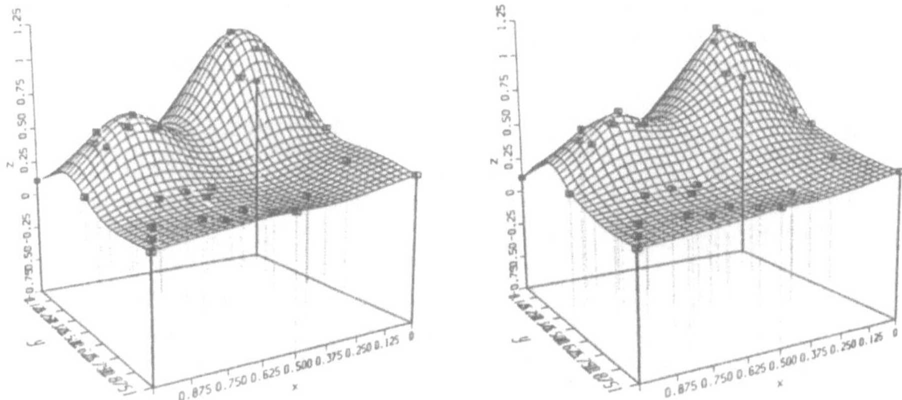

Figure 1. MQ interpolant to 33 points using $R^2 = .1$ in a) and $R^2 = .000001$ in b).

The reciprocal multiquadric (RMQ) interpolant is similar to the MQ interpolant, except that the basis functions are $[B_i(p)]^{-1}$. The bivariate thin-plate spline, see [4, 6, 11], has the same form as MQ, except that $B_i(x,y) = d_i^2(x,y)log(d_i(x,y))$ and linear polynomials $1, x$ and y are the Q_k. This method is known as the thin-plate spline because it is the function in an appropriate Soblev space that minimizes the strain energy in a elastic plate given by

$$\int\int_{R^2} \left(\frac{\partial^2 F}{\partial x^2}\right)^2 + 2\left(\frac{\partial^2 F}{\partial x \partial y}\right)^2 + \left(\frac{\partial^2 F}{\partial y^2}\right)^2 dx\, dy\, ,$$

subject to the interpolation conditions. For a trivariate generalization, simply using d_i as 3D distance and adding the linear function z can be used, although the minimizer of the appropriate spline variation problem yields $B_i(x,y,z) = d_i^3(p)$. Other trivariate interpolation methods are discussed in [1, 7, 9, 17].

In the critical testing of bivariate methods in [11], the multiquadric and thin-plate spline methods consistently produced visually pleasing results with very small observed errors on known test functions. A negative aspect of these radial basis methods is that if there are a large number of data points, the linear system of equations can be ill conditioned and costly to solve. One remedy is to localize the interpolant using techniques similar to those used in [12] for thin plate splines. This technique involves partitioning the data into overlapping rectangular regions, forming local interpolants to the data in each region, and then blending the local interpolants so that the resulting function is a C^1 interpolant. Since the blending functions are locally defined, data points in one region have no effect on the interpolant on non-neighboring regions. Although this localizing approach generally yields effective results, one drawback of this method is the limitation of using a tensor product grid structure to define the local regions. A generalization of this approach is given in [8] which defines local regions that depend on an arbitrary triangulation of an arbitrary collection of points. Weight functions are defined as C^1 composite rational triangular patches and the interpolant is a local rational blending of local radial basis interpolants. For the application of deforming one brain volume to another volume, it should be noted that we generally have less than 200 landmark points, thus localization is not critical.

The bivariate quadratic Shepard's method described in [14] easily be generalized to the form

$$F(x,y,z) = \left(\sum_{i=1}^{n} W_i(x,y,z)L_i(x,y,z)\right) / \sum_{i=1}^{n} W_i(x,y,z)$$

where $L_i(x,y,z)$ is a weighted least squares quadratic fit to the neighboring points of (x_i, y_i, z_i) and

$$W_i(x,y,z) = \left(\frac{(T-d_i)_+}{Td_i}\right)^2,$$

where $d_i = \|(x,y,z) - (x_i,y_i,z_i)\|$ and T is a positive constant dependent on the number of data points and the diameter of the point set. [14] provided default values for the number of neighboring data points used and the value T. These values yielded reasonable results, and [17] provided different default values that improved the results somewhat on the test functions in [11]. An efficient FORTRAN implementation of the quadratic Shepards' method is provided in [17] for functions of two and three variables. Trivariate multistage methods are given in [7] that involve a combination of a local version of the MQ method, followed by a piecewise bicubic function based on a gridded domain, and finally followed by a local version of Shepards method. Generalizing triangle based interpolants to trivariate tetrahedral patches is significantly more complicated and is covered in [1].

3 Volume Deformation Interpolants

Suppose that 3D landmark points q_i are to be the image of the discrete points p_i. If $q_i = (f_{i,1}, f_{i,2}, f_{i,3})$ for $i = 1, \ldots, N$, the problem is to construct a vector valued function $\vec{F}(p)$ that satisfies $\vec{F}(p_i) = q_i$. The trivariate interpolants described earlier can be extended by solving the three scattered data problems that satisfy $F_1(p_i) = f_{i,1}$, $F_2(p_i) = f_{i,2}$ and $F_3(p_i) = f_{i,3}$, for $i = 1, ..., N$. The volumetric deformation can then be denoted by $\vec{F}(p) = (F_1(p), F_2(p), F_3(p))$. It follows that the multiquadric and thin-plate spline methods described earlier can then be written in the form

$$\vec{F}(p) = \sum_{i=1}^{N} \vec{A}_i B_i(p) + \sum_{k=1}^{M} \vec{C}_k Q_k(x,y,z),$$

where the $B_i(p)$ are the same real valued basis functions and the $\vec{A}_i$ and $\vec{C}_k$ are unknown vectors of length 3. To compute the unknown vector coefficients $\vec{A}_i$ and $\vec{C}_k$, we need to solve $N+M$ linear equations with $N+M$ unknowns for each of the 3 vector components so that $\vec{F}(p_i) = q_i$ for $i = 1, \ldots, N$. Fortunately, all of these 3 linear systems of equations have the same $N+M$ by $N+M$ coefficient matrix K noted earlier. If $\vec{A}_i = (a_{i,1}, a_{i,2}, a_{i,3})$

and $\vec{C}_i = (c_{i,1}, c_{i,2}, c_{i,3})$, then they must satisfy

$$K[a_{1,j}, a_{2,j}, \ldots, a_{N,j}, c_{1,j}, \ldots, c_{M,j}]^t = [f_{1,j}, \ldots, f_{N,j}, 0, \ldots, 0]^t$$

for each $j = 1, 2, 3$. Solving three linear systems which involve the same coefficient matrix K can be efficiently computed by finding the LU decomposition of K or by computing K^{-1}.

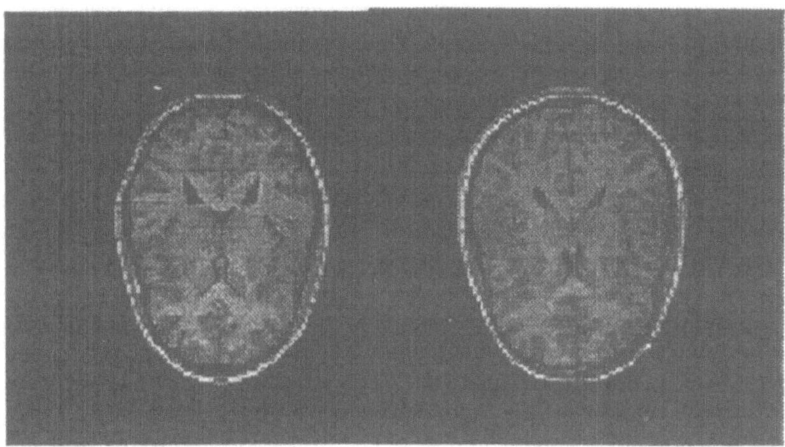

Figure 2. MRI slices of two different brain volumes.

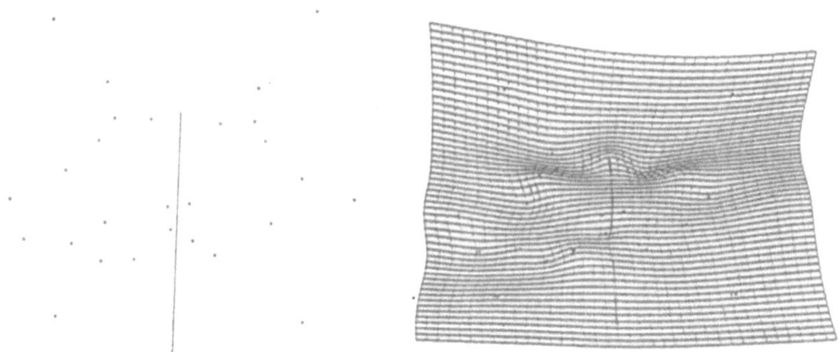

Figure 3. Part a) displays 29 landmark points from Figure 2a), while part b) illustrates a uniform grid in the first region deformed to the second region.

We applied several of these techniques to volumetric data obtained from MRI scans of different brains. For a doctor to make comparisons based on other functional brain data, it was necessary to map each of these brain volumes to a "standard" brain. These mappings use multiquadric and thin-plate spline deformations based upon user selected landmark points in both the patient brain and the "standard" brain. We have generated software for and examples of volume deformations that were shown in the conference presentation. Due to the black and white limitations of this paper, we only give an example involving a 2D deformation based on the two MRI slices shown in Figure 2. A total of $N = 29$ landmark points were interactively located in these slices and a radial basis deformation was computed. Figure 3a represents the 29 landmark points from Figure 2a. Figure 3b illustrates a uniform grid in the first region deformed to the second region by a radial basis method with the corresponding landmark points labeled.

4 Concluding Remarks

In related work in [10], the authors use vector valued interpolation on a sphere to approximate volumetric ray tracing techniques. After a lengthy initialization procedure, approximate ray-traced volumetric images can be computed in less than one second on a general purpose graphics workstation. The initialization involves computing some ray-cast volumetric images from various viewpoints on a sphere surrounding the volumetric data. They give an example involving $N = 38$ ray-cast images consisting of 500 x 500 pixels, which can be considered as a vector of length 250,000. The approximate ray-cast images are computed at an arbitrary viewpoint on the sphere by evaluating a vector valued interpolant of the $N = 38$ ray-cast vectors on the spherical domain.

A related problem involves generating an approximation to the given scattered data, generally in the least squares sense. This is generally more desirable than interpolation when there are a very large number of data points or when the data is noisy. Given the scattered landmark data p_i, q_i, for $i = 1, ..., N$, the multiquadric approximating function is of the form

$$\vec{F}(p) = \sum_{j=1}^{M} \vec{A}_j \sqrt{||p - u_j||^2 + R^2},$$

where $\vec{A}_1, ..., \vec{A}_M$ are chosen to minimize the following least squares error

$$\sum_{i=1}^{N} ||\vec{F}(p_i) - q_j||^2.$$

In general, M is significantly smaller than N and the points $u_1, ..., u_M$ should be somewhat representative of the given data points $p_1, ..., p_N$.

5 Acknowledgments

This work was supported in part by funding provided by the Flinn Foundation and Good Samaritan PET Center to Arizona State University. Thanks are due to Lang Yun and Ping Ning for their assistance.

6 References

[1] P. Alfeld (1989) "Scattered data interpolation in three or more variables," in: Mathematical Methods in CAGD, T. Lyche and L.L. Schumaker, eds., Academic Press, New York, pp. 1–33.

[2] R.E. Barnhill (1977) "Representation and approximation of surfaces," in: Mathematical Software III, J.R. Rice, ed., Academic Press, New York, pp. 69–120.

[3] R.E. Barnhill (1985) "Surfaces in computer aided geometric design: A survey with new results," Computer Aided Geometric Design, Vol 2, pp. 1–17.

[4] F.L. Bookstien (1989) "Principal Warps: Thin-plate splines and the decomposition of deformations," IEEE Transactions on Pattern Analysis and Machine Intelligence, Vol 11, pp. 567–585.

[5] R.E. Carlson and T.A. Foley (1991) "The parameter R^2 in multiquadric interpolation," Computers & Math. Appl., Vol 22, pp. 29–42.

[6] J. Duchon (1975) "Splines minimizing rotation invariant semi-norms in Sobelev spaces," in: Multivariate Approximation Theory, W. Schempp and K. Zeller, eds., Birkhauser Basel, pp. 85–100.

[7] T.A. Foley (1987) "Interpolation and approximation of 3-D and 4-D scattered data, Computers & Math. Appl., Vol 13, pp. 711–740.

[8] T.A. Foley, S. Dayanand and D. Zeckzer (1994) "Localized radial basis methods using rational triangle patches," to appear.

[9] T.A. Foley and D.A. Lane (1990) "Visualization of irregular multivariate data," Proceedings of IEEE Visualization '90, A. Kaufman, ed., IEEE Computer Society, Los Alamitos, CA, pp. 247–254.

[10] T.A. Foley, D. Lane and G.M. Nielson (1990) "Towards animating ray-traced volume visualization," Visualization and Computer Animation J., Vol 1, pp. 2–8.

[11] R. Franke (1982) "Scattered data interpolation: tests of some methods," Math. Comp., Vol 38, pp. 181–200.

[12] R. Franke (1982) "Smooth interpolation of scattered data by local thin plate splines," Computers & Math. Applic., Vol 8, pp. 273–281.

[13] R. Franke (1987) "Recent advances in the approximation of surfaces from scattered data," in: Topics in Multivariate Approximation, L.L. Schumaker, C.C. Chui and F. Utreras, eds., Academic Press, New York, pp. 175–184.

[14] R. Franke and G.M. Nielson (1980) "Smooth interpolation to large sets of scattered data," Intern. J. Numer. Meth. Eng., Vol 15, pp. 1691–1704.

[15] R.L. Hardy (1990) "Theory and applications of the multiquadric-biharmonic method," Computers & Math. Applic., Vol 19, pp. 163–208.

[16] M.J.D. Powell (1991) "The theory of radial basis function approximation in 1990," in: Advances in Numerical Analysis II: Wavelets, subdivision algorithms and radial functions, W. Light, ed., Oxford University Press, Oxford, pp. 105–210.

[17] R.L. Renka (1988) "Algorithm 660: QSHEP3D: Quadratic Shepard method for bivariate interpolation to scattered data," ACM Trans. Math. Soft., Vol 14, pp. 149–150.

[18] T. Sederberg and S. Parry (1986) "Free-form deformation of solid geometric models," Computer Graphics, Vol 20, pp. 151–160.

A HYBRID MOUNTAIN GENERATION ALGORITHM USING SUBDIVISION AND B-SPLINES

B. L. Hagstrom and E. A. Yfantis
Department of Computer Science
University of Nevada, Las Vegas, 89154, USA
e-mail: hagstrom@cs.unlv.edu, yfantis@cs.unlv.edu

Abstract

The use of a bivariate spline interpolation is used as a finishing step in a surface generation algorithm. Fractal surface generation techniques are employed to generate the data set used by the two directional spline interpolation method. The resulting surface estimation is then used as the data set for a graphics rendering of a geographic surface. Related post-processing methods used to produce added realism to the graphic rendering are discussed.

Keywords. Spline Interpolation, Fractal Methods, Surface Estimation, Surface Generation.

1 Introduction

The original work of Hubbard and subsequent improvements by Mandelbrot [3] produced computer generated images of mathematical formulae of stunning beauty and complexity. The use of fractal methods in computer graphics has developed from the realization of the similarity between the images generated and the natural world. The use of fractal methods to produce realistic three dimensional geological landscapes has recieved attention in the past few years due to the technological advances of graphic capable computers and the definition of geometric methods used to model the scenes.

The use of strictly fractal methods for generating a terrain scene produces realistic images, however, the methods produce a data set from which it is impossible to determine certain mathematical properties of the surface. One outstanding difficulty is the computation of surface normals for use in lighting and shading methods. Fractal methods generate a data set that produces a positional sampling of the surface. In order to extend from graphic images to real world terrain, this difficulty must be addressed.

The use of bivariate interpolation methods is a well established practice in differential geometry. Any numerical analysis text will provide the theory behind the methods. Computer aided geometric design (CAGD) has benifitted from the mathematics involved. The text by Farin [6] provides a broad view of the theory behind the methods.

One useful property of interpolation methods is the ability to stipulate the amount of continuity desired in the curve or surface of interest. Another property inherit to the methods is the ability to correctly determine the surface normals. Some of the methods demand information a priori to the actual solution of the systems of equations. Some of the methods are also data dependent in their solution. Methods exist that can circumvent some of these demands [5].

E. A. Yfantis (ed.), Intelligent Systems, 853–861.

We set out to develop a method of combining the randomness and continuity of nature in a surface generation algorithm for computer graphic rendering. In order to facilitate this goal, we have merged the two methods listed into a new surface generation scheme. Our aim is to include proper mathematical methods in the surface generation to facilitate extension from aesthetically pleasing graphic images to correct real world surface modeling.

2 Fractal Surface Models

Many methods exist in fractal geomety for generating surfaces. Our prime interest is in the methods that produce a data set in which we can establish control over the resultant surface. Since we wish a data set that allows ease of manipulation for futher work, our focus fell on methods that produce a grid of data points. This allows the opportunity to include real world data samples into our method. Terrain data may be extracted from topographic mappings and mathematically mapped into the data structures.

The subdivision methods of surface generation include many variations [1,2,4]. Some of the early images suffered from surface defects produced by the methods. Improvements in the algorithm reduced or eliminated some of these defects. The defect of "creasing" can be directly attributed to lack of cross correlation in the data generation. Two algorithms that do not suffer from the lack of cross correlation are outlined in [4]. One is a two dimensional midpoint displacement and the other a spectral synthesis.

For our work we desire a data set $P(x, y, z)$ in a grid pattern. Using the x coordinate as an example, we want arranged data $x_{0,0}, x_{0,1}, x_{0,2}, \cdots, x_{0,n}, x_{1,0}, x_{1,1}, x_{1,2}, \cdots, x_{1,n}, \cdots, x_{n,0}, x_{n,1}, x_{n,2}, \cdots, x_{n,n}$. This desired data set is naturally represented with a two dimensional array. In the subdivision method, $x_{0,0}, x_{0,n}, x_{n,0}, x_{n,n}$ are choosen arbitrary at the start of the algorithm as seed values. In the spectral synthesis method, the data is generated and placed in an array in a manner facilitating inverse fourier transformation. The former method produces a data set directly, the latter produces a data set from the given spectrum.

Since we desire control over the resultant surface, all coordinate values are stored explicitly. This also facilitates rapid rendering after the surface is computed. We control surface properties by "stretching" or "shrinking" the surface by manipulating the coordinate values that control the grid pattern. This allows real terrain sampling mapping into our data structure by allowing a matching of sample intervals.

We include the use of real terrain data for completeness. If the data samples are of fine enough detail, there is no need for modification. However, if the data samples are course grained and finer detail is desired, we incorporate the samples into the data structure and enter the subdivision algorithm at a later state. We justify this method directly from the observation that natural phenomena follows a self similar or self affine property that is inherit in fractal methods. From the given sampling we can determine the general spectum of the surface and produce intermediate data that follows the general scheme. This intermediate data is a fair guess at the actual data and is defended by sampling error inherit in sampling methods.

If we are not able to work from terrain sampling, but are able to determine the spectrum of the surface, we can work from that with the spectral synthesis method. Our data set is generated from the spectrum through the inverse fourier transformation. This method allows us to either choose a specta or to work from a given one.

The spectral synthesis methods used in [4] have an implicit cross correlation from the use of periodic transcendental functions. We developed a method that does not contain a cross correlation for use in generating a surface without a "rolling" quality. This method is used to generate the random appearing effect of waves on a body of water. This method has only aesthetic properties and is used to add realism to the rendering. Correct mathimatical modeling is possible but beyond the scope of this paper.

We used the modified spectral synthesis method we developed to produce realistic appearing waves for our water bodies. The description of the method of data generation is similar to that used in [4] with the modification in the computation of the data. For our water body surface data we generated data from spectra of the form $\frac{1}{f^\alpha + f^\beta}$ and $\frac{1}{f^\alpha . f^\beta}$. The α and β terms are set constant $0 < \alpha, \beta < 2$. This relates to the Brownian fractal motion spectrum which is $\frac{1}{f^2}$. The value of f is position dependent and in our case we used the indices of the data array.

3 B-spline Surface Estimation

B-spline interpolation is a simple method of generating a curved line or surface from some arbitrary data points. B-Spline interpolation accomplishes the construction of a curve passing through the data points by use of knot insertion. The method inserts "work points" to get the resulting curve to pass through the data points. Since we are allowing for surface generation from either data samples that may include sampling error or for purely random surfaces for use in graphic images, we relax from strictly data interpolation.

The derivation of the equations used in producing surface patches between the data points is simple and straight forward. For our purpose we wish the surface to be C^2.

$$P_i(t)|_{t=1} = P_{i+1}(t)|_{t=0}$$

$$\frac{d}{dt}P_i(t)|_{t=1} = \frac{d}{dt}P_{i+1}(t)|_{t=0}$$

$$\frac{d^2}{dt^2}P_i(t)|_{t=1} = \frac{d^2}{dt^2}P_{i+1}(t)|_{t=0}$$

Given four points $P_{i-1}, P_i, P_{i+1}, P_{i+2}$ a curved line results from estimating with these points. The general blending function used in the two dimensional curve estimation from coplanar data is given as $P_i(t)$.

$$P_i(t) = (P_{i-1}, P_i, P_{i+1}, P_{i+2}) \begin{pmatrix} e_3 & e_2 & e_1 & e_0 \\ f_3 & f_2 & f_1 & f_0 \\ g_3 & g_2 & g_1 & g_0 \\ h_3 & h_2 & h_1 & h_0 \end{pmatrix} \begin{pmatrix} t^3 \\ t^2 \\ t^1 \\ t^0 \end{pmatrix}$$

Using the constraints that we have continuity of the first and second derivative, (B-Spline of 3^{erd} degree), along with the above equation we get a series of equations that yield solutions for the variables in the above matrix.

$$\begin{pmatrix} e_3 & e_2 & e_1 & e_0 \\ f_3 & f_2 & f_1 & f_0 \\ g_3 & g_2 & g_1 & g_0 \\ h_3 & h_2 & h_1 & h_0 \end{pmatrix} = \frac{1}{6} \begin{pmatrix} -1 & 3 & -1 & 1 \\ 3 & -6 & 0 & 4 \\ -3 & 3 & 3 & 1 \\ 1 & 0 & 0 & 0 \end{pmatrix}$$

Performing the above matrix multiplication and incrementing over the data points as t increments from 0 to 1 per patch, produces the patch of the curve. Selecting small enough increments for t will produce a smooth rendering of a curve. Very good results were obtained with as few as four increments for t. This method is useful in two or three dimensions where a point P_i is either a two or three dimensional coordinate. The equations produced by the above matrix multiplication are repeated in use for each dimension. Thus $P_i(t_x)$ is produced from x values of the points, $P_i(t_y)$ is produced from y values of the points, and $P_i(t_z)$ similarly.

This method is extended to a bivatiate interpolation to produce a B-Spline surface. For brevity and space, let (M_{Bs}) be the above solution matix. The bivatiate interpolation is conducted by the equations produced by completing the following matrix multiplication.

$$P_{ij}(t,u) = (u^3, u^2, u^1, u^0)(M_{Bs})^T \begin{pmatrix} P_{i-1,j-1} & P_{i-1,j} & P_{i-1,j+1} & P_{i-1,j+1} \\ P_{i,j-1} & P_{i,j} & P_{i,j+1} & P_{i,j+1} \\ P_{i+1,j-1} & P_{i+1,j} & P_{i+1,j+1} & P_{i+1,j+1} \\ P_{i+2,j-1} & P_{i+2,j} & P_{i+2,j+1} & P_{i+2,j+1} \end{pmatrix} (M_{Bs}) \begin{pmatrix} t^3 \\ t^2 \\ t^1 \\ t^0 \end{pmatrix}$$

The surface points are estimated from the data by traversing over the data points and incrementing u and t from 0 to 1 for each patch.

To allow for stricter modeling of real terrain surfaces, the outlined method is useful since we can modify the resultant surface properties by using other basis matrices in place of M_{Bs}. For completeness we include some of the more common matrices. Each carries with it slightly different mathematical properties and the desired surface characteristics may be included by simply replacing the B-Spline basis matrix M with the basis matrix from one of the other methods. Some small amount of overhead is necessary in some of the other methods.

$$\begin{pmatrix} e_3 & e_2 & e_1 & e_0 \\ f_3 & f_2 & f_1 & f_0 \\ g_3 & g_2 & g_1 & g_0 \\ h_3 & h_2 & h_1 & h_0 \end{pmatrix} = \begin{pmatrix} 2 & -2 & 1 & 1 \\ -3 & 3 & -2 & -1 \\ 0 & 0 & 1 & 0 \\ 1 & 0 & 0 & 0 \end{pmatrix} = (M_{Hermite})$$

$$\begin{pmatrix} e_3 & e_2 & e_1 & e_0 \\ f_3 & f_2 & f_1 & f_0 \\ g_3 & g_2 & g_1 & g_0 \\ h_3 & h_2 & h_1 & h_0 \end{pmatrix} = \begin{pmatrix} -1 & 3 & -3 & 1 \\ 3 & -6 & 3 & 0 \\ -3 & 3 & 0 & 0 \\ 1 & 0 & 0 & 0 \end{pmatrix} = (M_{Bezier})$$

$$\begin{pmatrix} e_3 & e_2 & e_1 & e_0 \\ f_3 & f_2 & f_1 & f_0 \\ g_3 & g_2 & g_1 & g_0 \\ h_3 & h_2 & h_1 & h_0 \end{pmatrix} = (\frac{1}{\delta}) \begin{pmatrix} -2\beta_1^3 & 2(\beta_2+\beta_i^3+\beta_1^2+\beta_1) & -2(\beta_2+\beta_1^2+\beta_1+1) & 2 \\ 6\beta_1^3 & -3(\beta_2+2\beta_1^3+2\beta_1^2) & 3(\beta_2+2\beta_1^2) & 0 \\ -6\beta_1^3 & 6(\beta_1^3-\beta_1) & 6\beta_1 & 0 \\ 2\beta_1^3 & \beta_2+4(\beta_1^2+\beta_1) & 2 & 0 \end{pmatrix}$$

$$= (M_\delta)$$

In the above matrices $\beta_1, \beta_2, a_i, b_i, c_i, a_{i+1}, b_{i+1}, c_{i+1}$ are provided parameters to the computation. The variation of these parameters controls the properties of the resultant surface. The properties are in relation to the data points that the resultant surface is being computed from. Tighter or looser estimation to the data points is allowed with these methods.

The surface normal computation is correctly performed by taking the cross product of the the tangent vectors of the surface. We get the tangent vectors of the surface from the following parial derivatives.

$$\frac{\partial}{\partial u}P(u,t) = \frac{\partial}{\partial u}(U \cdot M^T \cdot G \cdot M \cdot T) = \frac{\partial}{\partial u}(U) \cdot M^T \cdot G \cdot M \cdot T$$

$$\frac{\partial}{\partial t}P(u,t) = \frac{\partial}{\partial t}(U \cdot M \cdot G \cdot M^T \cdot T) = U \cdot M \cdot G \cdot M^T \cdot \frac{\partial}{\partial t}T$$

4 B-spline Fractal Surface

We have successfully integrated the two surface estimation methods listed in the previous two sections. The B-Spline method of surface estimation relies on an existing collection of data points. It is quite natural to use the fractal sub-division method of surface generation the create the data set for the B-Spline method to work from.

A B-Spline implementation of four increments for the creation of a patch produced satisfactory results. From these observations, we implemented our method by creating a data set one forth the size of the desired surface size. We then used these data points as input to a B-Spline surface estimation implementation to construct the final data set. From this final data set, surface normals are calculated, lighting calculations preformed and the surface rendered.

We maintain our grid by setting the x and z components of the surface grid at static intervals. We then run a bivariate B-Spline estimation using the data computed from the fractal mid-point subdivision method. We only run the method once to produce the y values in the final matrix. In this manner we retain control over the smoothness of the resulting surface. We can change the overall surface property with a simple change in grid size. This is an indirect method of scaling the data. Instead of sizing the data to a desired range, we change the ratio of height to length and width. Thus we may model a surface of desired visible characteristic and then simply map into a range of units if so desired.

We have also allowed further manipulation of the surface by using the differential geometry methods. We implemented the Beta-Spline methods and have done some preliminary exploration of the images through various parameter changes. Surface properties are easily modified in this manner. Thus we now have more control over the resultant surface.

We have observed the following properties from the merging of the methods. The surfaces generated from the strict fractal methods and from our hybrid method are similar in terrain characters. The mountain peaks, vally floors, and water bodies are generally shaped and located the same in both methods. On a large data value surface, an uninformed observer may believe that only cosmetic methods were changed in the rendering. On a small data volume surface the differences become obvious. The surface discontinuities of the fractal method, along with the lack of smooth surface normal continuity, creates discontinuities in lighting and shading of the model. The triangular mesh becomes distractful. Our hybrid algorithm does not suffer this problem.

The surface itself reaches a characteristic shape during the fractal subdivision process. This shape may be provided by real world data samples. The fractal method continues to fill in the missing areas in the characteristic shape. One the characteristic shape point is reached and enough intermediate data is incorporated, further finishing of the surface with the fractal method continues to enhance any anomolies. These anomolies are difficult or impossible to conpensate for with cosmetic methods. Thus a switch to another method of finishing is indicated.

The spline method used as a finishing process serves to avoid some problems. The spline surface estimation methods act as a low pass filter. Datum that would cause anomolies in the finished surface are smoothed out. Since the surface is not finished at the time of the spline implementation, this smoothing effect is deeper than the cosmetic smoothing methods. Thus we prevent surfae anomolies from forming and becoming imbedded in the resultant surface. Our Resultant surface has asthetically pleasent qualities from any viewing angle. The spline methods provide mathematical properties that allow surface manipulation in the characteristic surface.

The locality of control in the spline methods allows manipulation in the surface without affecting the entire surface. Surface characters such as erosion patterns, riverbeds, fault lines, cliffs, sink holes, and environmental impact on the surface from large scale geologic projects. Man made projects such as mining, road construction, river damming, and irrigation channel constrution may be modeled into the characteristic surface. Our method is conducive to these changes since they may be incorporated in the characteristic surface. Since these characters are imbedded into the surface as the model evolves, we circumvent the difficulty of incoporating these artifacts at the cosmetic level. Our hybrid algorithm acts as the link between asthetic computer graphics and real world terrain models.

Figure 1 shows a prepared surface data set. This set is used as a common starting point for both the fractal finishing and the spline finishing methods. The prepared surface was constucted with a fractal sub-division method. This method is allowed to finish the surface in the fractal finished figure. Figure 2 shows the resultant surface from fractal finishing. Figure 3 shows the resultant surface from spline finishing.

5 Rendering Methods

The rendering technique of describing triangles was used. In this manner the grid pattern of a two dimensional matrix was further subdivided during the rendering of the surface. Two coordinates from one column are used with a coordinate from the next column to produce one triangle. The two coordinates from the next column are used with a coordinate from the present column . The column is incremented and the procedure repeated. When the

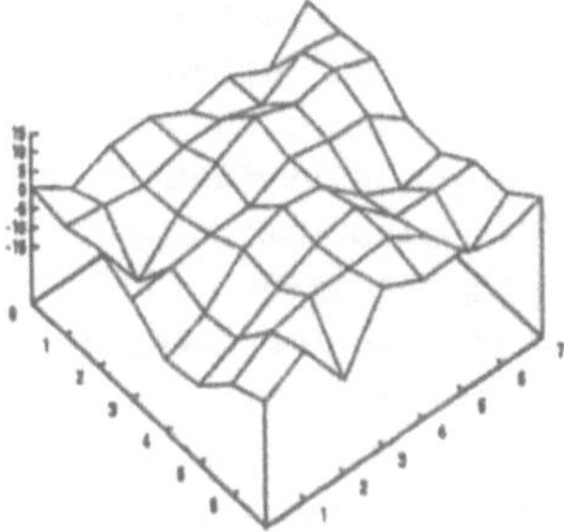

Figure 1: Common starting point for finishing methods

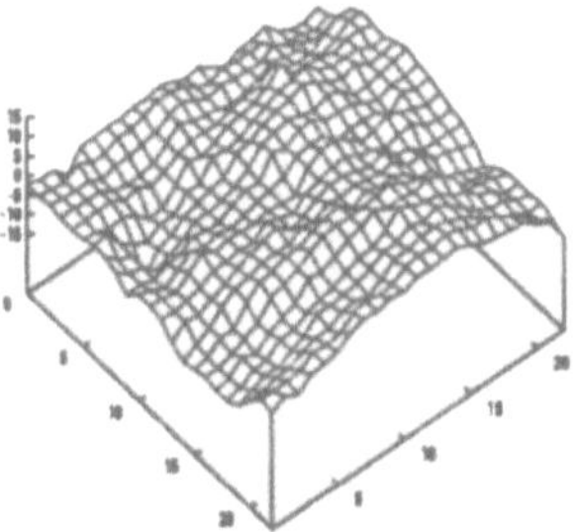

Figure 2: Fractal finished surface

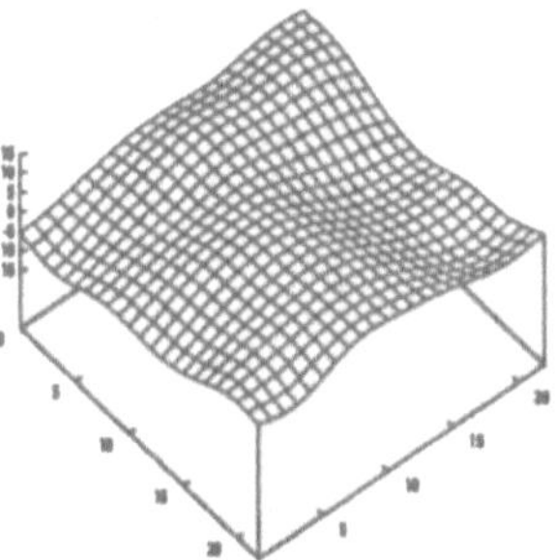

Figure 3: B-spline finished surface

end of the column is reached, the next column is rendred. In order to avoid the rendering from appearing like a collection of colored connected triangles, gouruad shading is used. In order to use gouruad shading, surface normals must be used.

From the use of the differential geometry methods, we can now correctly compute the surface normals for use by the lighting computations. To obtain the surface normals, we must take the cross product of the surface tangents. This method is mathematically correct but has been correctly observed to be computationally expensive. Advances in hardware now allows the use of the correct methods. This is the desired method if mathematical accuracy is demanded. If mathematical accuracy is not an issue in the rendered surface, there are some methods we found to be satisfactory at high levels of data amounts.

The computation of a surface normal for a flat plane segment is simply the cross product of two vectors lying in the plane. Since we are rendering small triangles to build our surface, it is simple to compute the normal for each triangle. However, using a single normal for a surface in the lighting calculations produces a flat shaded object. In order to "curve" our triangles into a smoother surface, we used various methods to average the normal of a given point by all the normals from the surfaces tangent to that point.

The triangle method discussed above produces at most six triangles meeting at an interior point. By averaging the six normals to those triangles we "estimate" the normal at the point. We have acheived very good results by avoiding this averaging and simply using the normal for a connected triangle at two of the points, along with the actual normal for the triangle being rendered at the third point. This method relies on the gouruad shading itself to cause a smoothing effect on the rendered surface. Care must be taken to render the boundry triangles.

6 Conclusions

We have merged two methods of surface estimation into a single algorithm. The fractal sub-division method provides raw data points for a B-Spline method of surface estimation. Since the fractal methods can be controlled with the parameter used to perturb the surface we can produce raw data of desired characteristics. Our method also contains two more levels of control by the "stretching" or "shrinking" methods possible with grid positions. The use of B-Splines as a finishing step eliminates any undesirable artifacts that may be produced by the various methods of fractal surface generation.

Our methods also allow easy modification to include desired properties that may not be included in the B-Spline approach. Our methods also allow for the use of real sampling data to be an integral part of the graphic image generation. We have opened some paths for extending graphic methods to correctly model real world terrain and natural phenomena.

Thus we have developed a method to produce surfaces that will contain the desired features and will do so from any viewing angle. The use of fractal methods assures us of having the random quality of nature and the use of B-Spline methods assures us of having the continuity of nature. Having merged the best of both methods we are able to produce geological images with natural appearance and underlying mathematical soundness. Since our target surface was mountainous terrain with water surface bodies, this was the coloring scheme used. Other target surfaces can also be modeled with this method and colored accordingly.

7 Further Research

We have simply merged two forms of mathmatical methods used in surface generation and surface estimation. Advancement of the method includes abstraction to allow for specific surface properties to be modeled. New methods of data set generation are possible in the front end of the method. Since the surface is finished using interpolation techniques, it is possible to use methods of data generation that in themselves would not produce satisfactory surfaces.

Since the fractal methods may be viewed as a data set producer and the differential geometry methods may be viewed as a data set consumer, new surface generation methods are possible for natural phenomena of varying characteristics. Modifications in the data set production, consumption, or both allow for better methods of modeling graphic objects. The generation of these graphic objects is better facilitated with this method. The image space can now be further explored.

8 Acknowledgments

This research was supported by the DOE, Las Vegas lab. The authors would like to thank Drs. P. Gottlieb, J. Gauthier, and the reviewers for their valuable suggestions.

9 References

[1] Miller, Gavin S. P. (1986) "The Definition and Rendering of Terrain Maps" Computer Graphics, Vol. 20 No. 4 August. SIGGRAPH '86 Conference Proceedings.

[2] Foley, VanDam, Feiner, Hughes. Computer Graphics, Principles and Practice. Addison-Westly Systems Programming Series.

[3] Sørensen P. (1990) "The Fruits of Fractals" Computer Graphics World May, pp 54-63.

[4] Voss R. F. "The Science of Fractal Images" Peitgen, Saupe Ed. Springer-Verlag New York Inc.

[5] Yfantis E. A. "A New Quadratic and Biquadratic Algorithm for Curve and Surface Estimation" Computer Aided Geometric Disign V 10 pp 509-520.

[6] Farin G. Curves and Surfaces for Computer Aided Geometric Design, A Practical Guide. Academic Press Inc. Harcourt Brace Jovanovich, Pub.

[7] Bartels R. H., Beatty J. C., Barsky B. A. (1987) "An Introduction To Splines For Use In Computer Graphics & Geometric Modeling" Morgan Kaufmann Pub.

VIEW VARIATIONS IN ANGLES

RAASHID MALIK
TAEGKEUN WHANGBO

Department of Electrical Engineering and Computer Science
Stevens Institute of Technology
Hoboken, New Jersey 07030

Abstract: The appearance and measure of an angle between the edges of an object in a scene varies with view orientation. Some measures of a known angle however appear "more likely" than others. In "most cases" we would expect that the measures of a viewed angle to be close to the actual angle. In this paper we quantify terms such as "most cases" and "more likely" by expressing them using probability concepts. The orientation or view angles are modeled as random variables as is the measure of the view of the angle in the scene. We derive the probability density function of this angle when measured in images that are orthographical projections. The utility of this density function is demonstrated in a sequence of experiments to identify specific triangles in images. The results of these experiments establish the potential of using view density functions for object recognition.

Key words : view variations, 3D object recognition, model-base object recognition, computer vision, image understanding, scene analysis, angle pdf, statistical decision

I. Introduction

Most schemes for recognizing objects in two dimensional images are model-based systems in which recognition requires matching features in the input image with sets of model features. Geometric features of a polyhedral object's image such as area, line length, perimeter, angle etc., are defined by projections of 3-D points and line segments that define the silhouette of the object. A problem of significance in three dimensional object recognition is the fact that there are no general case view-invariant features for any number of 3-D points, for perspective, weak perspective, or orthographic projections [1]. Therefore as the view orientation relative to an object becomes less constrained, recognition of 3-D objects using model-based methods becomes more difficult. This difficulty is caused by the fact that features of an object's image vary with the changes in view orientation. This paper describes the quantification of perceived variation in certain geometric features of objects in images so as to rationalize the matching procedure in recognition.

The spatial arrangement of landmarks (or easily recognized icons) on an object may constitute a unique characteristic of that object. For example the angle between the wing tips and the nose cone of an aircraft may comprise a sufficient property in distinguishing amongst a given class of aircraft. In a class of polyhedral objects the angles at a given vertex may form a distinct and characteristic alignment of faces. For many classes of objects it is possible to identify spatial arrangement of icons that may be used to uniquely identify elements within

E. A. Yfantis (ed.), Intelligent Systems, 863–872.

the class. In the aircraft example the icons (or landmarks) were the wings and cone tip and in the polyhedral object case the icons were the object's vertices which would appear as junctions in an edge enhanced image of the object. The relative spatial placement of these icons may be used in identifying an object provided there is knowledge of how these spatial placement reveal themselves in an image. We show how knowledge about spatial placement variation in images can be captured and utilized for object recognition.

This paper describes the variation in the angle formed between rays connecting three icons. For a polyhedral object this would be the angle between two edges on the surface of the object. The measure of this angle in an image of course varies with view orientation. We analytically derive the probability density function of this measured angle in orthographically projected images. The orthographic projection is quite accurate when the distance from the object to the camera is much larger than the object dimensions. The orthographic projection can be simulated by using the observation sphere [6]. The results allow us to predict the likelihood of the appearance of a known physical angle in an image and to gauge (and assign a numeric value to) the possibility of an arbitrary angle in an image being the impression of a spatial arrangement in the scene. The significance of these densities arises from the possible utilization of these densities in object recognition and location. In situations of ambiguity, the likelihood of an angle measurement (which is derived from the probability density of its projected angle) may be used to resolve identification.

In a survey of the literature, only a limited number of references to the variation of features with the view orientation were discovered. Zhang *et al* [2] proposed the view independent model construction using the statistical inference between model features (component parts). Malik [3] derives probabilistic density functions of a measured length of adjacent edges in an image and uses it to develop a decision scheme for identifying polyhedral objects. Burns *et al* [1] describe the variation of orientation, size, and position (two components) of one projected line segment with respect to another. They also prove the non-existence of general-case view invariants for any number of 3-D points. Ben-Arie [4] proposes two novel probabilistic models of imaged angles and distances, and applies them to 3-D object recognition. The authors in reference [5],[6] derive the variation in the angles of a cube. They also derive the probability of imaging cube in non-general views.

II. Isotropic Viewpoint Density:

An orthographic projection model places the object in the scene at the center of an extended observation sphere as shown in fig. 1. The camera, or viewpoint, is positioned isotopically somewhere on the surface of this sphere. Isotropic positioning implies that surface patches of equal area anywhere on the sphere have equal chances of containing this viewpoint.

A viewpoint (v) is defined in terms of θ, ϕ; the polar and azimuth angles respectively. The isotropic *viewpoint density* is [3]:

$$f_{\theta,\phi}(\theta,\phi)=\begin{cases}\dfrac{1}{4\pi}\sin\theta & ;0\le\theta\le\pi,\ \ 0\le\phi\le 2\pi \\ 0 & ;\ \textit{elsewhere}\end{cases} \tag{2.1}$$

An isotropic orientation should be assumed if no other a priori information about the camera position relative to the target is available. Other a priori densities are discussed in [3].

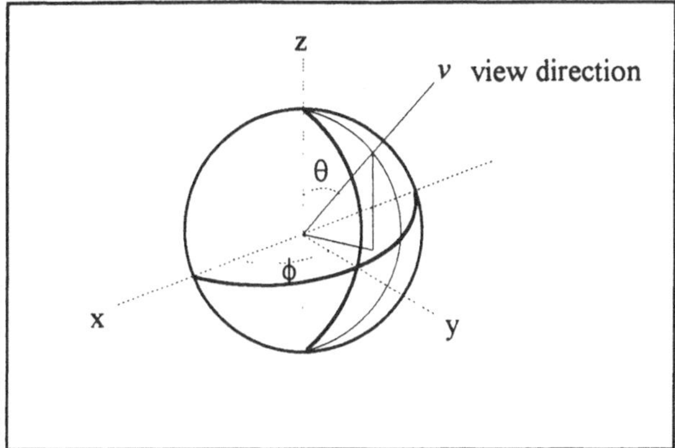

Fig. 1. Observation sphere

III. The Density Function

A triangle is placed at the center of the observation sphere such that the surface is in the y-z plane as shown in fig. 2. The angle **a** is the angle of interest and is assumed to be less than 90° for the mathematical convenience, although all the following mathematical derivations are symmetric in the case of **a**>90°. If the triangle is the face of an object, then the angle **a** can be seen only from the hemisphere above the y-z plane. This placement is selected to simplify the algebraic manipulations that follow. It will not affect the result or the aim of our analysis which is the derivation of the probability density of the projection of the wedge angle **a** in an image. The density function of θ, ϕ is (from equation 2.1)

$$f_{\theta,\phi}(\theta,\phi)=\begin{cases}\dfrac{1}{2\pi}\sin\theta & ;0\le\theta\le\pi,0\le\phi\le\pi\\ 0 & ;elsewhere\end{cases}$$

Since θ and ϕ are independent variables, the marginal densities of θ and ϕ are

$$f_{\theta}(\theta)=\frac{\sin\theta}{2} \quad ;0\le\theta\le\pi, \qquad f_{\phi}(\phi)=\frac{1}{\pi} \quad ;0\le\phi\le\pi$$

Let A be the corresponding projected angle of **a** in an image. A spherical triangle is constructed at a given view orientation by the three great circles 1,2, and 3 as shown in fig. 2. From the spherical triangle, using the sine and cosine laws, the projected angle A is derived as

$$A=\tan^{-1}\left(\frac{\sin\phi}{\dfrac{\sin\theta}{\tan a}-\cos\theta\cdot\cos\phi}\right) \qquad ;0\le a\le\frac{\pi}{2},0\le\phi\le\pi,0\le\theta\le\pi$$

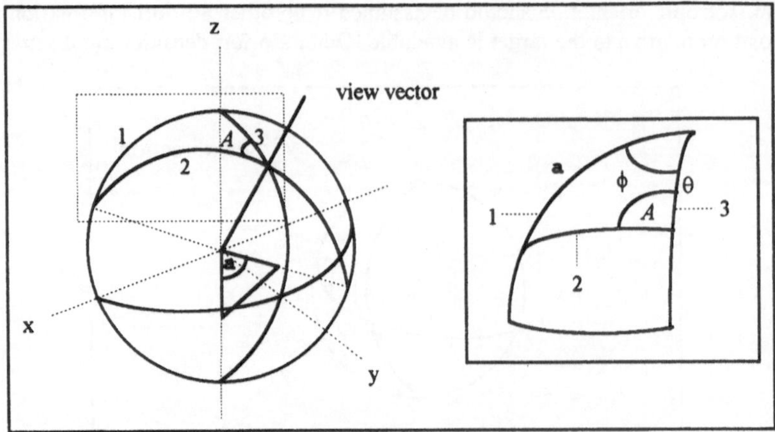

Fig. 2. Great circles (1-3) and Spherical triangle

Now the joint density function, $f_{A,\theta}(A,\theta)$, is obtained as (see reference [6] for detail derivations)

1) when $0 \le \theta \le \mathrm{a}$

$$f_{A,\theta}(A,\theta) = \frac{1}{2\pi}\left(\frac{\cot \mathrm{a}\cdot\cot A\cdot\csc^2 A\cdot\sin^2\theta}{\left(\csc^2 A-\sin^2\theta\right)\cdot\sqrt{\csc^2 A-\csc^2 \mathrm{a}\cdot\sin^2\theta}}+\frac{\csc^2 A\cdot\sin\theta\cdot\cos\theta}{\csc^2 A-\sin^2\theta}\right)$$

$$;(0\le A\le\pi)$$

2) when $\mathrm{a} \le \theta \le \frac{\pi}{2}$

$$f_{A,\theta}(A,\theta) = \frac{1}{\pi}\left(\frac{\cot \mathrm{a}\cdot\cot A\cdot\csc^2 A\cdot\sin^2\theta}{\left(\csc^2 A-\sin^2\theta\right)\cdot\sqrt{\csc^2 A-\csc^2 \mathrm{a}\cdot\sin^2\theta}}\right) \quad ;\ 0\le A\le\left(\frac{\tan \mathrm{a}}{\sqrt{\sin^2\theta-\cos^2\theta\cdot\tan^2 \mathrm{a}}}\right)$$

From the joint density function, the probability density function, $f_A(A|\mathrm{a})$, is approximated as

$$\bar{f}_A(A|\mathrm{a}) = \begin{cases} -k_1\dfrac{\csc^2 A}{2\pi}\log\left(\dfrac{\sin^2\mathrm{a}-\sin^2 A}{\sin^2\mathrm{a}}\right) & ;0\le A<\mathrm{a} \\ -k_2\dfrac{\csc^2 A}{2\pi}\log\left(\dfrac{\cos\mathrm{a}-\cos A}{1+\cos\mathrm{a}}\right)^2 & ;\mathrm{a}<A\le\pi \end{cases} \tag{3.1}$$

$$where \quad k_1 = \frac{\mathrm{a}+2\cot\mathrm{a}\cdot\log(1+\sin\mathrm{a})}{\mathrm{a}+\cot\mathrm{a}\cdot\log(2\cot\mathrm{a})},\quad k_2 = \frac{\pi-\mathrm{a}-2\cot\mathrm{a}\cdot\log(1+\sin\mathrm{a})}{\pi-\mathrm{a}-\cot\mathrm{a}\cdot\log\left(2\sin\mathrm{a}+\tan\left(\frac{\mathrm{a}}{2}\right)\right)}$$

The comparison between the numerical integration of $f_A(A|\mathrm{a})$ and the approximated function for all range of the measured angle A at various fixed angle **a** is shown in fig. 3. It shows that they are very close at all tested angles.

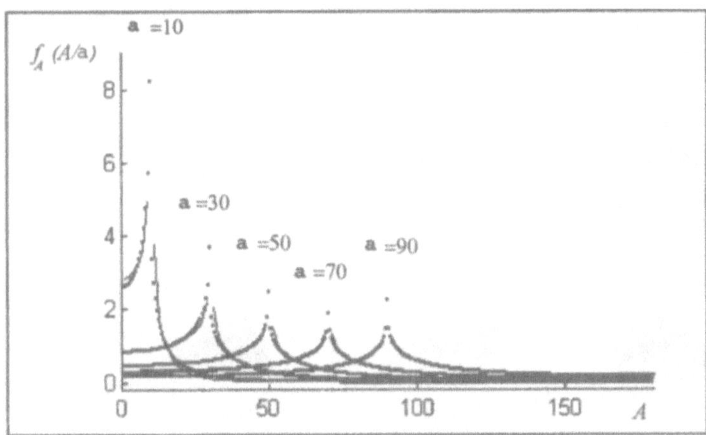

Fig. 3. Plot of $f_A(A|a)$ varying **a** (solid line : numerical integration, dotted line: approximation)

IV. Experiments

The usefulness of the derived probability density function was tested in a series of experiments. The angle in an image may be between any three distinguishable landmarks. In order to avoid the errors associated with landmark detection and location within an image a simple and easily recognizable target was selected for the experiments so that we could focus on analyzing the utility of the density function. The model targets selected for the experiments were the group of triangles shown in figure 4. A Panasonic WV1600 TV camera with a 50mm (f/3.5) lens was used for imaging and a Data Translation DT2803 frame grabber was used to capture and digitize the image (6 bits/pixel, 240 x 256 resolution). The data extracted from the raw image was rescaled to accommodate the 4/3 pixel aspect ratio of the frame grabber. The simplicity of the target made it easy to recognize and locate the vertices of the triangle in the image. The image was binarized and the vertices detected using horizontal and vertical projections.

4.1. *Decision rule*

Once the location (x-y coordinates) of the vertices of the triangle in the image are known the angles of the triangle may be computed using trigonometry. Suppose the three angles in the image are A, B and C. The ordering of the angles is selected to be clockwise and consistent with the model data. (Of course a counter-clockwise ordering could also have been selected without effecting the results. If the model data is counter-clockwise ordered then angles extracted from the image should be clockwise ordered). The third angle is linearly dependent on the other two angles and need not be used in an optimum decision rule (see reference [7] [pp. 220] on the Theorem of Irrelevance). An optimum decision rule however requires knowledge of the joint density which has not yet been derived. Therefore all three angle measures may be significant. Now there are 12 distinct angles and 19 distinct angle_pairs in model triangles. Notice the angle_pairs (a_i b_i), and angle_triple (a_i b_i c_i) are indexed, i=1 to 19.

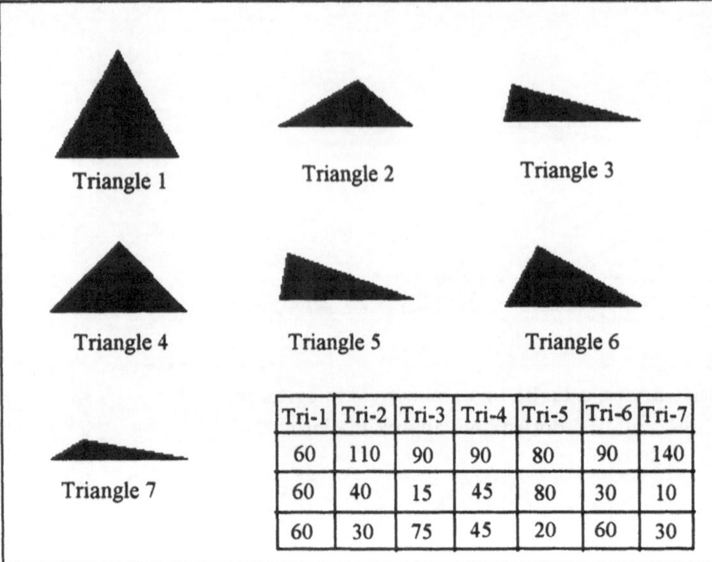

Tri-1	Tri-2	Tri-3	Tri-4	Tri-5	Tri-6	Tri-7
60	110	90	90	80	90	140
60	40	15	45	80	30	10
60	30	75	45	20	60	30

fig 4. 7 Model triangles and their angles

Now a decision rule is required to determine the best matching between (A B) and (a_i b_i), or between (A B C) and (a_i b_i c_i), where i=1 to 19.

The closest analog to an optimum decision rule is to compute:

$$\overline{f}_A(A|a_i),\ \overline{f}_B(B|b_i) \text{ and } \overline{f}_C(C|c_i), \text{ where i=1 to 19}$$

using Eqn. 3.1 and to use the product of the densities as a likelihood function.
We have a choice of selecting this function as

$$l2_i(A,B) = \overline{f}_A(A|a_i)\cdot\overline{f}_B(B|b_i) \tag{4.1}$$

or using the triple product:

$$l3_i(A,B,C) = \overline{f}_A(A|a_i)\cdot\overline{f}_B(B|b_i)\cdot\overline{f}_C(C|c_i) \tag{4.2}$$

One decision rule is therefore:

> Given an angle_pair measurement, A and B, from a triangle in an image, select T(i) as the most appropriate model triangle if,
> $l2_i(A,B) \geq l2_j(A,B)$ for all $j \neq i$

An alternative decision rule is:

> Given an angle_pair measurement, A, B and C, from a triangle in an image, select T(i) as the most appropriate model triangle if,
> $l3_i(A,B) \geq l3_j(A,B)$ for all $j \neq i$

4.2 *Samples*

Various samples were selected and imaged on the target structure sketched in figure 5. The samples used in the experiments are labeled and described in table 1. The images captured using these samples are shown in figure 6.

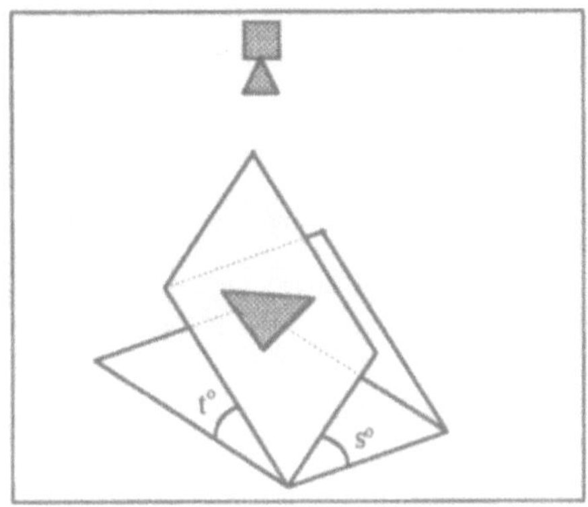

Fig. 5. A sketch of the target area showing the target orientation angle t & s.

Sample	Triangle	t°	s°	Measured angles		
a	2	16	40	31.6770	101.8908	46.4322
b	1	38	44	73.0989	55.3940	51.5071
c	3	35	40	63.0794	99.9854	16.9352
d	4	38	45	45.6658	49.0547	85.2795
e	7	35	43	24.1325	145.3833	10.4842
f	1	40	55	54.0058	82.2607	43.7335
g	5	21	32	73.0507	84.2267	22.7225
h	1	21	40	69.1531	55.2777	55.5693

table. 1. Sample triangles with their measured angles after approximate orientations.

4.3 *Results*

The vertices in the images in figure 6 were extracted and the angles computed (see table 1). For all model angle_pairs the likelihood function (eqn 4.1) was evaluated. A tabulation of all these values is shown in table 2.

Consider the Sample-a values shown in the second column of the table. The likelihood function values evaluated using the nineteen different angle_pairs are shown. The largest value of the likelihood function in this case is 0.829 and generated by angle_pair i=4 which corresponds triangle T(4)=2. The next most likely in this case is triangle 4 with a likelihood value of 0.3983 The decision, of course, is triangle 2. We can see from table 1 that this is indeed the correct decision. In fact the scheme gave incorrect results only in the case of Sample-c, Sample-f and Sample-g. It should be remembered that the method is statistical and will not always give the correct result. The method becomes less reliable, as is to be expected, the sharper the target orientation angles (t and s).

A substantial improvement occurs when we use the alternative decision rule which uses the triple product (eqn 4.2) of the densities. Table 3 lists the likelihood values when $l3_i(A,B,C)$ is used. This scheme only fails with sample-f which has the largest orientations.

A more powerful scheme based on the joint density of an angle_pair has been developed. This scheme will be outlined in later papers.

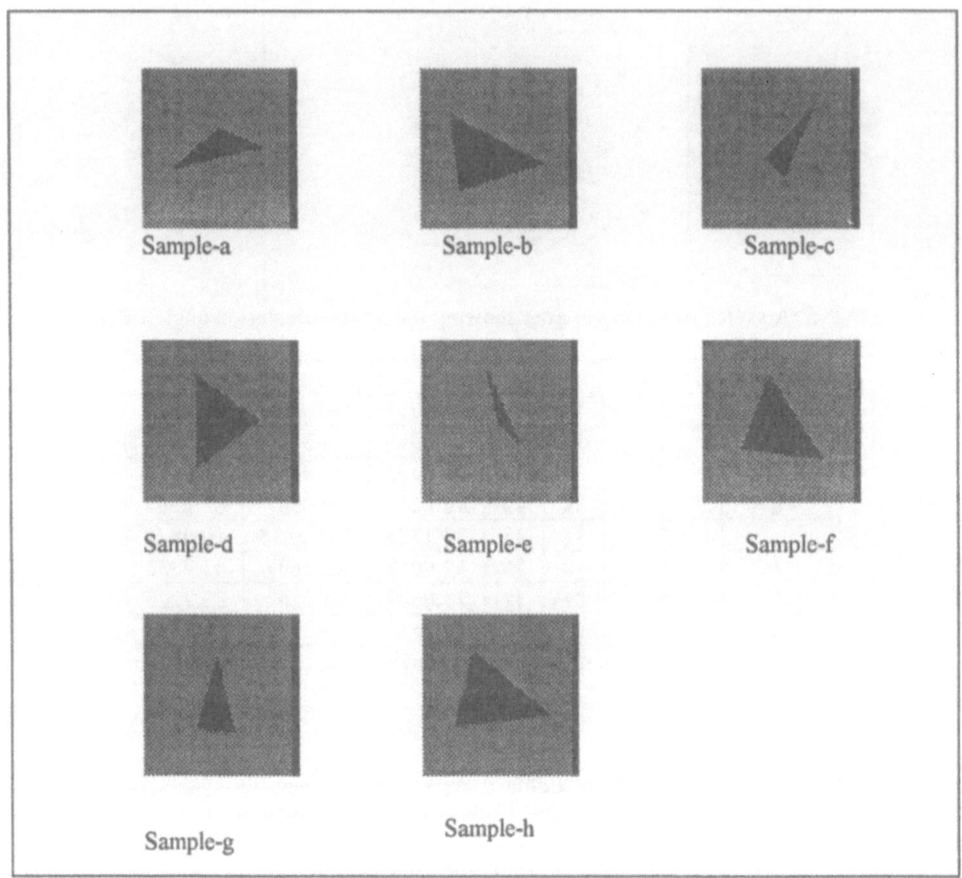

Fig. 6. Images of sample triangles after certain orientations.

Angle_pair	Samp-a	Samp-b	Samp-c	Samp-d	Samp-e	Samp-f	Samp-g	Samp-h
1 (1)	0.0734	0.4161	0.1566	0.4052	0.0329	0.2542	0.1257	0.5145
2 (2)	0.0080	0.0799	0.0129	0.0726	0.0040	0.0177	0.0244	0.0733
3 (2)	0.0452	0.0367	0.0124	0.2307	0.0187	0.0316	0.0122	0.0430
4 (2)	0.8290	0.0146	0.0804	0.0486	0.4850	0.0607	0.0286	0.0169
5 (3)	0.0029	0.0290	0.0051	0.0207	0.0015	0.0065	0.0102	0.0256
6 (3)	0.0832	0.0150	0.0145	0.0435	0.0787	0.0461	0.0190	0.0172
7 (3)	0.1567	0.2710	0.3246	0.0860	0.0533	0.2820	0.7484	0.1961
8 (4)	0.0172	0.2252	0.0307	0.1935	0.0084	0.0407	0.0635	0.1989
9 (4)	0.0702	0.1144	0.0328	1.2811	0.0305	0.0907	0.0324	0.1355
10 (4)	0.3983	0.0580	0.1951	0.3480	0.1220	0.3742	0.1615	0.0680
11 (5)	0.0890	0.2587	0.1685	0.1058	0.0357	0.3706	0.5482	0.2166
12 (5)	0.0063	0.0720	0.0116	0.0462	0.0031	0.0144	0.0249	0.0607
13 (5)	0.1657	0.0198	0.0274	0.0595	0.1627	0.1136	0.0422	0.0228
14 (6)	0.0085	0.0917	0.0151	0.0672	0.0042	0.0194	0.0304	0.0809
15 (6)	0.2266	0.0939	0.0258	0.2503	0.1117	0.0688	0.0283	0.1075
16 (6)	0.2345	0.1194	0.5223	0.1419	0.0779	0.5500	0.3326	0.1485
17 (7)	0.0003	0.0025	0.0005	0.0022	0.0002	0.0007	0.0009	0.0023
18 (7)	0.0072	0.0039	0.0013	0.0184	0.0078	0.0030	0.0013	0.0045
19 (7)	0.1982	0.0050	0.0202	0.0169	1.6057	0.0188	0.0087	0.0058
Decision	2	1	6	4	7	6	3	1

Table. 2. Product of densities in all angle pair.

Angle_pair	Samp-a	Samp-b	Samp-c	Samp-d	Samp-e	Samp-f	Samp-g	Samp-h
1 (1)	0.0452	0.3097	0.0595	0.1089	0.0121	0.1449	0.0502	0.4827
2 (2)	0.0028	0.0209	0.0131	0.0050	0.0035	0.0075	0.0309	0.0155
3 (2)	0.5931	0.0072	0.0540	0.0056	0.3036	0.0564	0.0213	0.0065
4 (2)	0.0057	0.0050	0.0012	0.0673	0.0018	0.0038	0.0012	0.0062
5 (3)	0.0011	0.0119	0.0013	0.0111	0.0004	0.0023	0.0028	0.0115
6 (3)	0.0168	0.0221	0.5871	0.0020	0.1272	0.0353	0.5477	0.0131
7 (3)	0.0188	0.0037	0.0024	0.0348	0.0127	0.0100	0.0033	0.0046
8 (4)	0.0212	0.1564	0.0175	0.0277	0.0045	0.0694	0.0393	0.1031
9 (4)	0.4898	0.0403	0.1115	0.0499	0.0659	0.6369	0.1001	0.0352
10 (4)	0.0158	0.0282	0.0055	1.0254	0.0049	0.0196	0.0056	0.0361
11 (5)	0.0151	0.0333	0.3855	0.0038	0.0539	0.0739	0.7545	0.0228
12 (5)	0.0530	0.0070	0.0064	0.0442	0.0368	0.0348	0.0102	0.0087
13 (5)	0.0020	0.0253	0.0027	0.0343	0.0007	0.0044	0.0060	0.0233
14 (6)	0.0052	0.0682	0.0057	0.0181	0.0016	0.0111	0.0122	0.0759
15 (6)	0.0830	0.0313	0.5310	0.0098	0.0695	0.2328	0.4213	0.0314
16 (6)	0.0511	0.0231	0.0043	0.2004	0.0181	0.0149	0.0049	0.0287
17 (7)	0.0001	0.0007	0.0005	0.0002	0.0002	0.0003	0.0011	0.0005
18 (7)	0.0116	0.0002	0.0163	0.0002	5.6041	0.0013	0.0033	0.0002
19 (7)	0.0003	0.0002	0.0000	0.0016	0.0003	0.0001	0.0000	0.0002
Decision	2	1	3	4	7	4	5	1

Table 3. Product of densities in all angle_triple

V. Conclusions

We have derived the probability density of the measure of an angle in an orthographic projection. We have demonstrated the use of this density to distinguish amongst the triangles in a group. The method is not limited to two dimensional objects because a triangle captures the spatial arrangement of identifiable landmarks in any three dimensional scene. We have not come across any equivalent scheme for distinguishing amongst samples that are so similar while simultaneously imposing no constraints on the view of the sample. The importance of being able to distinguish similar shapes arises from the fact that surface segments (such as planar faces) of distinct objects are often similar.

This method, however, needs to be extended to incorporate higher dimensional knowledge (i.e. two or more angles) and to model the errors that arise from incorrect positioning of landmarks in an image.

VI. References

[1] J.B. Burns, R.S. Weiss, E.M. Riseman, "View Variation of Point-Set and Line-Segment Features," IEEE Tans. Pattern analysis and Machine Intelligence, Vol. 15, No. 1, 1993, pp. 51-68.

[2] S. Zhang, G.D. Sullivan, K.D. Baker, "The Automatic Construction of a View-Independent Relational Model for 3-D Object Recognition," IEEE Tran. on Pattern Analysis and Machine Intelligence, Vol. 15, No. 6, 1993, pp. 531-543.

[3] R. Malik, "Polyhedral Object Recognition using View Density," I4-OC. Proc. of the 1991 IEEE International Conference on System, Man and Cybernetics, vol. 1, Oct. 1991, pp. 111-116

[4] J. Ben-Arie, "The Probabilistic Peaking Effect of Viewed Angles and Distances with Application to 3-D Object Recognition," IEEE Tran. on Pattern Analysis and Machine Intelligence, Vol. 12, No. 8, 1990, pp. 760-774.

[5] R. Malik, T. Whangbo, "Junction View Densities in Images," Proc. of IEEE Southwest Symposium on Image Analysis and Interpretation, April 1994, pp. 142-147

[6] R. Malik, T. Whangbo, "Angle View Densities," Dept of EECS, Stevens Inst. of Tech., Technical Report # 9400

[7] J.M. Wozencraft, I.M. Jacobs, *Principles of Communication Engineerinng*, JohnWiley & Sons, 1965.

A SIMPLE AND EFFICIENT THINNING METHOD

Chung-E Wang

Department of Computer Science
California State University, Sacramento
Sacramento, CA 95819-6021
E-mail: wang@csus.edu

Abstract. In this paper, we consider a simple but efficient method for thinning digitalized images. We implement the method into sequential and parallel algorithms for the thinning problem.

1 Introduction

Consider a digitalized image, i.e., a two-dimensional array of pixels. The object, which forms the foreground of the image is represented by a set of 1-valued pixels while the background corresponds to a set of 0-valued pixels. Thinning is to remove pixels from an image until all the lines or curves are of unit width. The resulting set of lines and curves is called the skeleton of the object. Thinning is an important preprocessing step for many image processing and pattern recognition algorithms, since thinned images are easier to process and thus produce savings in both time and space complexity.

Most thinning algorithms in the literature [1-13] are based on the contour method which iteratively deletes boundary pixels from objects until the remaining pixels form thin lines along the medial axes of these objects. All of these algorithms use a similar approach for determining the deletability of all pixels. They all use some kind of 3 × 3 local operations which only examine eight neighbors of a pixel to determine whether the pixel is deletable. It's difficult to understand how those 3 × 3 local operations work and it's complex to implement the contour tracing and the contour generation.

Instead of identifying deletabe pixels, we find pixels of the skeleton of an image. To do so, we first compute distances of pixels from edges of the image. Then we identify those pixels having an equal distance from two opposite edges of the image as pixels of the skeleton. Note that our idea is similar to the idea of Blum's analog technique [4]. The obvious advantage of our method is that the basic idea behind is very natural. Despite the need of looking beyond the 3 × 3 neighborhood, our method can be easily and efficiently implemented into sequential and parallel algorithms.

E. A. Yfantis (ed.), Intelligent Systems, 873–876.

2 The Sequential Thinning Algorithm

According the definition of the skeleton, for each pixel, we first compute the left, right, up, and down depths, i.e., distances to the left, right, top, and bottom edges, of the pixel. For objects of simple shapes such as retangle and square, these four depths are sufficient to identify pixels of the skeleton. However, for objects of complex shapes such as S curve and L shape, we need to compute four generalized depths, one for each direction, for each pixel.

The generalize depth of a direction of a pixel is defined as the pixel's shortest distance from an edge within the 90° span of that direction. With this new definition, skeleton pixels can be identified as pixels with an equal generalized depth for two opposite directions and not too close to an edge. The last condition is important because all pixels on an edge have a generalized depth of 0 for more than two directions.

Since a line with an even number of pixels doesn't have a midpoint, we relax the condition of "eaual generalized depth" to a conditon of "the diffenence of the generalized depths of two opposite dircetions is less than or equal to 1". However, after we relax the condition, some line might have two midpoints and thus the resulting skeleton won't have the unit width. To solve the problem, we use a loop to remove extra midpoints. The condition we use to remove extra midpoint is based on our experiments. Even though it's not perfect, it's enough to remove 95% of the extra midpoints. Figure 1 shows a detailed algorithm not including the final step of removing extra midpoints.

3 The Parallel Thinning Algorithm

The main issue of implementing our method into a parallel algorithm is to compute those depths parallelly. In [2], Hillis and Steele showed how to compute all partial sums, i.e., prefix sums, of an array. The way we compute left, right, up, and down depths is exactly the same as that of Hilli and Steele's algorithm for partial sums.

To compute generalized depths, we expand the idea of Hillis and Steele's algorithm for partial sums. A detailed algorithm for computing generalized left depths is shown in Figure 2. Note that the algorithm uses a SIMD parallel computer with a shared memory. Moreover, each pixel (r,c) has an assigned processor (r,c) working on the generalized left depth of the pixel.

4 Conclusion

We have presented a simple method for the thinning problem. We also showed how to implement the method into efficient sequential and parallel algorithms. The result of our algorithms is as good as the result of any thinning algorithm in the literatur. The only dsiadvantage of our method is that our method is sensitive to the quality of the image to be thinned. Noise and distortion might affect the results of our algorithms.

```
Initialize arrays L_Dpth, R_Dpth, U_Dpth, D_dpth, G_Ldpth, G_R_Dpth, G_U_Dpth, and G_D_Dpth to zeros.

for (c=2; c<=n; ++c)
   for (r=1; r<=n; ++r)
   if (Img[r,c]==1)
   {  L_Dpth[r,c] = L_Dpth[r,c-1]+1;
      G_L_Dpth[r,c] = G_L_Dpth[r,c-1];
      if (G_L_Dpth[r-1,c-1]<G_L_Dpth[r,c])G_L_Dpth[r,c]=G_L_Dpth[r-1,c-1];
      if (G_L_Dpth[r+1,c-1]<G_L_Dpth[r,c])G_L_Dpth[r,c]=G_L_Dpth[r+1,c-1];
   }

for (c=n-1; c>0; --c)
   for (r=1; r<=n; ++r)
   if (Img[r,c]==1)
   {  R_Dpth[r,c] = R_Dpth[r,c+1]+1;
      G_R_Dpth[r,c] = G_R_Dpth[r,c+1];
      if (G_R_Dpth[r-1,c+1]<G_R_Dpth[r,c])G_R_Dpth[r,c]=G_R_Dpth[r-1,c+1];
      if (G_R_Dpth[r+1,c+1]<G_R_Dpth[r,c])G_R_Dpth[r,c]=G_R_Dpth[r+1,c+1];
   }

for (r=2; r<=n; ++r)
   for (c=1; c<=n; ++c)
   if (Img[r,c]==1)
   {  U_Dpth[r,c] = U_Dpth[r-1,c]+1;
      G_U_Dpth[r,c] = G_U_Dpth[r-1,c];
      if (G_U_Dpth[r-1,c-1]<G_U_Dpth[r,c])G_U_Dpth[r,c]=G_U_Dpth[r-1,c-1];
      if (G_U_Dpth[r-1,c+1]<G_U_Dpth[r,c])G_U_Dpth[r,c]=G_L_Dpth[r-1,c+1];
   }

for (r=n-1; r>0; --r)
   for (c=1; c<=n; ++c)
   if (Img[r,c]==1)
   {  D_Dpth[r,c] = D_Dpth[r+1,c]+1;
      G_D_Dpth[r,c] = G_D_Dpth[r+1,c];
      if (G_D_Dpth[r+1,c-1]<G_D_Dpth[r,c])G_D_Dpth[r,c]=G_D_Dpth[r+1,c-1];
      if (G_D_Dpth[r+1,c+1]<G_D_Dpth[r,c])G_D_Dpth[r,c]=G_D_Dpth[r+1,c+1];
   }

for (c=1; c<=n; ++c)
   for (r=1; r<=n; ++r)
   {  UD_ave = (G_D_Dpth[r,c]+G_U_Dpth[r,c]+1)/2;
      LR_ave = (G_L_Dpth[r,c]+G_R_Dpth[r,c]+1)/2;
      if (Img[r,c]==1)
         if (!((abs(G_D_Dpth[r,c]-G_U_Dpth[r,c])<2)&&(R_Dpth[r,c]>UD_ave+1)&&(L_Dpth[r,c]>UD_ave+1) ||
             (abs(G_L_Dpth[r,c]-G_R_Dpth[r,c])<2)&&(U_Dpth[r,c]>LR_ave+1)&&(D_Dpth[r,c]>LR_ave+1)) )
            Img[r,c] =0;
   }
```

Figure 1. Algorithm for Finding Skeleton Pixels.

```
for all processor (r,c) in parallel do
   G_L_Dpth[r,c] = 0;

for (i = 0; i<log2n; ++i)
{  for all processor (r,c) in parallel do
      if ((Img[r,c]==1)&&(c-2i>0))
      {  G_L_Dpth[r,c] = G_L_Dpth[r,c-2i]+2i;
         if ((r-2i>0)&&(G_L_Dpth[r-2i,c-2i]>G_L_Dpth[r,c]+2i))
            G_L_Dpth[r,c] = G_L_Dpth[r-2i,c-2i]+2i;
         if ((r+2i<=n)&&(G_L_Dpth[r+2i,c-2i]>G_L_Dpth[r,c]+2i))
            G_L_Dpth[r,c] = G_L_Dpth[r+2i,c-2i]+2i;
      }
}
```

Figure 2. Compute Generalized Left Depth in Parallel

5 References

[1] Arcelli, C. A condition for digital points removal, *Signal Processing 1*, 4 (1979), 283-285.

[2] Arcelli, C. and Di Baja, G. A width-independent fast thinning Algorithm, *IEEE Trans. Patt. Anal. and Match. Intell. PAMI-7*, 4 (July 1985), 463-474.

[3] Bel-Lan, A. and Montoto, L. A thinning transform for digital images, *Signal Processing 3*, (1981), 37-47.

[4] Blum, H. A transformation for extracting new descriptors of shape. *Symposium on Models for the Perception of Speech and Visual Form*, MIT Press, Cambridge, Mass., 1964.

[5] Chin R., Wan H., Stover, D. and Iverson, R. A one-pass thinning algorithm and its parallel implementation. *Comp. Vis. Graphics Image Processing 40 (1987)*, 30-40.

[6] Golay, M. Hexagonal parallel pattern transformations, *IEEE Trans. Comput. C-18*, 8 (Aug. 1969) 733-740.

[7] Guo, Z. and Hall, R. Parallel thinning with two-subiteration algorithms, *CACM 32*, 3 (1989), 359-373.

[8] Hall, R. Fast Parallel thinning algorithms: Parallel speed and connectivity preservation, *CACM 32*, 1 (1989), 124-131.

[9] Hilditch, C. Linear skeletons from square cupboards, In *Machine Intelligence IV*, B. Mertzer and D. Michie, Eds. University Press, Edinburgh, 1969, 403-420.

[10] Hillis, D., and Steele, G. Data parallel Algorithms, *CACM 29*, 12 (1986), 1170-1183.

[11] Holt, C., Stewart, A., Clint, M., and Perrott, R. An improved parallel thinning algorithm, *CACM 30*, 2 (1987), 156-160.

[12] Kwok, P. A thinning algorithm by contour generation, *CACM 31*, 11 (1988), 1314-1324.

[13] Lu, H. and Wang, P. A comment on "A fast parallel algorithm for thinning digital patterns," *CACM 29*, 3 (1986), 239-242.

[14] Pavlidis, T. *Algorithms for graphics and processing*, Computer Science Press, Rockville, Md., 1982.

[15] Rosenfeld, A. A characterization of parallel thinning algorithms, *Inform. Contr. 29*, 3 (1975), 286-291.

[16] Stefanelli and Rosenfeld Some parallel thinning algorithms for digital pictures, *J. ACM 18*, 2 (1971), 255-264.

[17] Tamura, H. A comparison of line thinning algorithms from digital geometry viewpoint, In *Proceedings of 4th International Conference on Pattern Recognition*, Kyoto, Japan, 1978, 715-719.

[18] Xu, W. and Wang, C. CGT: A fast thinning algorithm implemented on a sequential computer, *IEEE Trans. System, Man and Cybernetics 17*, 5 (1987), 847-851.

[19] Zhang, T.Y. and Suen, C.Y. A fast parallel algorithm for thinning digital patterns, *CACM 27*, 3 (1984), 236-239.

Using Proper Object-Oriented Design To Enhance Portability Of Graphical Applications

Jason Brown
Dorit Eisenberger
Center For Communications and Information Technology
Duquesne University
Pittsburgh, PA 15282 USA

Abstract. This paper describes how applications can be developed in a portable manner between two incompatible graphical user interfaces: NeXTstep[tm] and Microsoft Windows[tm]. Through the judicious use of objects and a good division of labor within the application, developers can increase the portability of their applications even when they are heavily dependent on a particular graphical user interface. As the design of intelligent systems rise in computational complexity, the ability of the developer to port these systems to state of the art hardware and operating systems will become increasingly more important.

Keywords. Graphics, Object Oriented, Graphical User Interface

1 Introduction

NeXTstep[tm] has proven to be a very powerful and user friendly environment, in use by many individuals at Duquesne. And with its wealth of object-oriented development tools, it allows for the rapid development and deployment of custom applications within the university. Development under NeXTstep[tm] occurs primarily in the form of courseware created jointly by the faculty and the academic computing group of the university's computer center.

Recently, the faculty have been expressing an interest in making it possible for their students to work with the custom software on their own personal computers while not giving up the benefits of development and use under NeXTstep[tm]. Keeping all of the primary development on one platform increases the return on investment of the development knowledge as well as reducing future maintenance issues. This presents the challenge of supporting both environments as efficiently as possible.

The objective then becomes the identification of a method of development that reduces the amount of duplicated effort required to support projects in more than one graphical user interface. Building on an object-oriented framework beginning with NeXTstep[tm], allows the final product to be more easily ported to Microsoft Windows[tm] once the application has matured.

E. A. Yfantis (ed.), Intelligent Systems, 877–883.

2 Designing Objects For Portability

Object-oriented design and development enhance portability across platforms with differing graphical user interfaces. However, object-oriented methodologies are not enough by themselves, they must be used properly. In order to achieve the greatest degree of portability in a graphical application, distinct separation of the computational engine from the user interface is very important. Previously developed applications which did not strictly adhere to this model, have now become mostly un-portable even though they were developed using object-oriented tools.

Object-oriented design teaches us to dissect our world into discrete classes of objects, each containing some state and the knowledge to manipulate that state. Objects then become sentient entities able to store information and respond to requests for that information. This encapsulation is one of the features that gives objects the potential to be portable. An object that implements a well defined method with a well defined response is free to implement that method however it makes sense on a given platform. The rest of the object-world need not, and should not care.

However, a collection of loose objects is no more a system than a loose pile of stones is a house. Some object must be present to impose some type of order to the system. Traditionally we refer to this object as the controller, and it is responsible for managing the interactions between the other objects in a system.

Figure 1 shows part of a traditional system which draws some type of graph in a view on a window. The Controller object acts as the nervous system for the application, intercepting events from the user and dispatching commands to the GraphView object. In this example, the GraphView handles two tasks: The computation of the values in the graph, and the display of this information. This structure evolves from traditional object-oriented design whose goal is to encapsulate all of the behavior and knowledge of an object into a single entity.

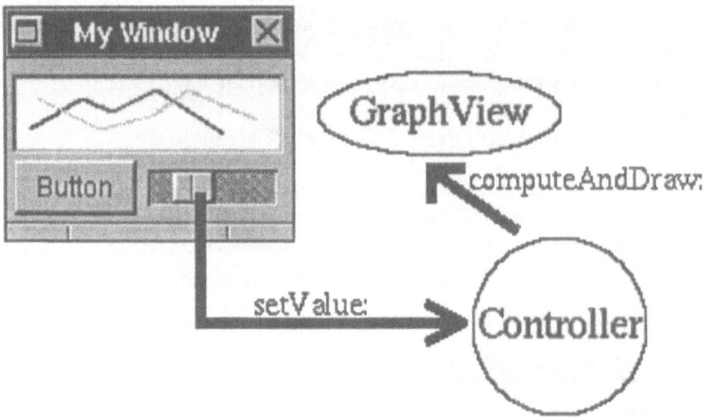

Figure 1

However, the benefits of this encapsulation come at the cost of portability. In Figure 2, we further atomize the system, breaking down the GraphView object into a computational engine and a view object. When porting to another graphical user interface, rather than having to recreate the complete GraphView object, only the View object must be implemented. Separating the graph's computational logic into it's own distinct class results in a totally portable computational object that can be transparently moved from one platform to another.

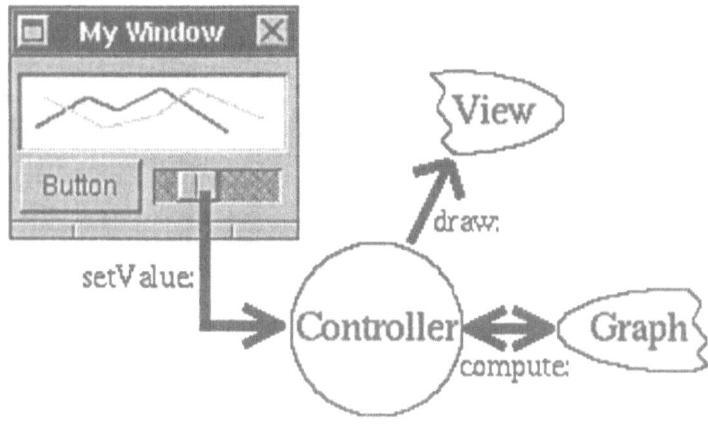

Figure 2

As in the first example, the Controller still maintains the chain of communication among the objects in the system. The portability of the Controller is somewhat limited in that it is often tightly coupled to the underlying event mechanism and is therefore portable only at the logic level. Fortunately, traditional logic is highly portable and easily recreated.

3 The Application

Our methodology will be demonstrated on an application developed for the Chemistry Department at Duquesne University. The application, titled Equilibrium, allows the user to adjust two parameters relevant to reaction kinetics and then compute the equilibrium state of the reaction. The distribution of chemical species present as reactant and product are then shown graphically. This information can then be saved to a text pane for later evaluation.

4 Messaging Architecture

In the Equilibrium application, the user alters the values of the input parameters via the two horizontal sliders and the two white text fields. As this happens events are sent to a controller object which we now refer to as the ViewManager. When the user requests a solution to the reaction, by pressing the Simulate button, the ViewManager messages an object we call the Engine. Once the ViewManager has passed the appropriate values to the Engine it computes the solution and makes the values available to the ViewManager. The ViewManager then notifies the necessary interface objects so that the display can be updated to reflect the current values.

5 Categorizing the Object Types

As in most graphical applications, there are three levels of abstraction in our Equilibrium system: computational, event handling, and visual. Computational objects tend to be the most portable as they rely the least on the interface aspects of the environment. Event handling objects pose difficulties in portability as they imply some prior knowledge of the event handling mechanism in use. Visual objects are by far the least portable as they rely on the graphics drawing primitives available. Being able to define clear lines of separation between these types of objects helps to map out those structures that must have their own implementations on each platform.

The computational object in our system is represented by the Engine. The Engine stores all of the knowledge necessary for computing the solutions to the reaction. A well defined set of methods allows the ViewManager to request the services of the Engine without regard for the implementation.

The ViewMananager assumes the role of event handler. Fielding actions from the user interface, it communicates with the Engine and the other objects coordinating their operations.

The visual objects in the system are composed of the standard user interface controls such as buttons and sliders, as well as the custom balance object we'll refer to as the BalanceView. User interface development tools provide ready access to the majority of the standard controls. The BalanceView is the only visual object which must have its own implementation on each platform.

6 Development Process

The migration from NeXTstep™ to Microsoft Windows™ occurs through the use of a set of commercial tools built to reuse as much of the existing application as possible. The tools used include Stepstone's Objective-C Compiler[1], version 4.3.2, Digitalk's Smalltalk/V for Windows[2], version 2.0, Object share Systems' WindowBuilder[3], Borland's C/C++ Compiler for Windows[4], version 4.0, and the Berkely Productivity Group's BPG Smalltalk Interface to Objective-C[5], release 1.1.

The Objective-C code from any computational objects is first translated into ANSI C using Stepstone's Objective-C Compiler and then recompiled using Borland's C/C++ compiler. This step creates a windows DLL (Dynamic Link Library) which will be loaded by the Smalltalk runtime in order to make the Objective-C classes available.

The user interface portion of the application is then rebuilt using WindowBuilder and Smalltalk/V. WindowBuilder allows you to graphically layout the user interface and define the messages each control sends to the ViewManager in response to user actions. From this information WindowBuilder automatically generates a template ViewManager class. The logic contained in the NeXtstep™ version of the ViewManager is then re-coded in these new action methods in the Smalltalk/V ViewManager. See Listings 1 and 2 for an example of a ViewManager method implemented in Objective-C and Smalltalk/V. These listings demonstrate the portability of traditional logic from one language to another.

```
- adjustDeltaG:sender
 {
 char buf[100];

 if (sender==deltaGField) {
          if ([sender doubleValue]>[deltaGSlider maxValue]) {
                  sprintf(buf,"%.1f",[deltaGSlider maxValue]);
                  [sender setStringValue:buf];
                  [deltaGSlider takeDoubleValueFrom:sender];
                  [moveResultsButton setEnabled:NO];
                  return self;
          } else {
                  if ([sender doubleValue] < [deltaGSlider minValue]) {
                          sprintf(buf,"%.1f",[deltaGSlider minValue]);
                          [sender setStringValue:buf];
                          [deltaGSlider takeDoubleValueFrom:sender];
                          [moveResultsButton setEnabled:NO];
                          return self;
                  }
                  [deltaGSlider takeDoubleValueFrom:sender];
          }
          sprintf(buf,"%.1f",[deltaGSlider doubleValue]);
          [sender setStringValue:buf];
 } else {
          sprintf(buf,"%.1f",[deltaGSlider doubleValue]);
          [deltaGField setStringValue:buf];
 }
 [moveResultsButton setEnabled:NO];

   return self;
}
```

Listing 1: Objective-C Version of a Method that alters the Delta G Parameter

```
adjustDeltaG: aPane

|slider field|
slider := (self paneNamed:#deltaGSlider).
field := (self paneNamed:#deltaGField).

(aPane = field)
ifTrue:[
(((aPane contents) asFloat)>(((slider max)/10) asFloat))
ifTrue:[
          aPane contents:((((slider max)/10) asFloat) asString).
          slider contents:(((aPane contents) asFloat)*10).
          (self paneNamed:#moveResultsButton) disable.
          ^self
]
```

```
ifFalse:[
        (((aPane contents) asFloat)<(((slider min)/10) asFloat))
        ifTrue:[
                aPane contents:((((slider min/10) asFloat) asString).
                slider contents:(((aPane contents) asFloat)*10).
                (self paneNamed:#moveResultsButton) disable.
                ^self
        ].
        slider contents:(((aPane contents) asFloat)*10).
].
]
ifFalse:[
field contents:((((slider contents)/10) asFloat) asString)
].
(self paneNamed:#moveResultsButton) disable.
^self
```

Listing 2: Smalltalk/V Version of a Method that alters the Delta G Parameter

In order for the Smalltalk/V ViewManager to communicate with the Objective-C classes contained in the Windows DLL one additional Smalltalk/V class, a class representing the DLL, must be created. BPG provides a class, ObjectiveCDLL, that must be subclassed and customized based on the Objective-C classes that are present in the Windows DLL.

In the subclass of ObjectiveCDLL four class methods must be implemented: *fileName*, *open*, *initializeRemoteSelectors*, and *initializeProxyObjects*. *fileName* returns the DOS filename of the Windows DLL and is used by the *open* method to locate the DLL. *open* first initializes the superclass, ObjectiveCDLL, and then calls the methods *initializeRemoteSelectors* and *initializeProxyObjects*. *initializeRemoteSelectors* sets up the Smalltalk/V environment with the information necessary for Smalltalk/V to send messages to the classes contained in the Windows DLL. This information includes the argument types and return types for the public methods in each Objective-C class in the DLL. *initializeProxyObjects* creates instances of the SmallTalk/V class, ClassProxy, which act as stand-ins for the Objective-C classes contained in the Windows DLL. These instances of ClassProxy make up the glue that allows Smalltalk/V to communicate with the Objective-C classes as though they were native Smalltalk/V classes.

7 Summary

Object-oriented design is not a single set of rules but rather a conglomeration of methodologies assembled to assist the developer in meeting the goals of the system. If portability is one of the goals of a system, this architecture should be part of the system designer's toolkit. Although some platform specific flexibility is sacrificed in order to use this design architecture, the benefits to be gained rapidly begin to out-pace any losses as the complexity of the computational objects in the system rise.

As graphical user interfaces become more and more commonplace it will be imperative that applications built in these environments adhere to a strategy that will allow them to achieve maximum portability. Object-oriented tools coupled with proper design, represent the vehicle to reach this goal.

8 References

[1] Stepstone Objective-C Compiler, The Stepstone Corporation, 75 Glen Road, Sandy Hook, CT 06482

[2] Digitalk Smalltalk/V, Digitalk Corporation, 5 Hutton Drive, Santa Ana, CA 92707

[3] Window Builder, Object Share Systems, 5 Town & Country Village, Suite 735, San Jose, CA 95128

[4] Borland C++, Borland, 100 Borland Way, Scotts Valley, CA 95066

[5] Small Talk Interface to Objective-C, Berkely Productivity Group, 35032 Maidstone Court, Newark, CA 94560

A Fast Fourier Method for Mountain Generation

Sai Prasad V. Pallati
E. A. Yfantis
Department of Computer Science
University of Nevada
Las Vegas, NV 89154 USA

Abstract. Fractals give an excellent description of the various complex natural forms and help in modelling natural phenomena in computer graphics. This paper presents one of the methods used to create fractals - fast Fourier filtering. This is used to generate mountains in particular; also some extensions have been suggested, whereby different sets of mountains can be obtained by modifying some parameters.

Keywords and Phrases : Fractals, Modelling of Natural Phenomena, Spectral Synthesis, Random Walks, fBm.

1 Introduction

Computer graphics has played a pivotal role in the rapid acceptance of fractal geometry as a new discipline. In fact, fractal geometry now makes a prominent contribution to the realistic rendering and modelling of natural phenomena in computer graphics. Mandelbrot's fractal geometry not only provides a description but also a mathematical model for many of the complex forms in nature such as mountains, coastlines and clouds. The most essential quality of these fractals, however, is their simplifying invariance under magnification.

Fractional Brownian motion(fBm) is the model most used for characterizing shapes and processes with a fractal appearance. Indeed, fBm seems to be the most wide-spread method of any type for modelling many natural processes(e.g. modelling of mountains, clouds etc.,). Fast Fourier method for mountain generation, is the purest interpretation of the concept of fBm. By changing the spectrum, the appearance of the mountain varies with respect to its roughness.

Though the main emphasis in this paper had been laid on the fast Fourier method, a brief description of the Mid-point displacement also has been given.

E. A. Yfantis (ed.), Intelligent Systems, 885–895.

2 Midpoint displacement method

Approximation of a random fractal with some resolution is used as input whereby the algorithm produces an improved approximation with the resolution increased by a certain factor. This process is repeated with the outputs used as new inputs until the desired resolution is achieved. This can be easily understood from the following example.

We start with a triangle, then go to the midpoint of each side of the triangle and displace it along the line at right angles to the line. The amount of each displacement is determined by applying a Gaussian random multiplier to a proportion of the line length. Next we connect each displaced midpoint to the two nearest apexes of the triangle. We then connect together each pair of displaced midpoints. Finally, we throw away the original sides of the triangle.

The result of this process is that we have replaced the original triangle with four new triangles. We then apply the same process to each of the four new triangles, generating four more triangles from each, so that we then have sixteen triangles. All these steps are shown in Fig. 1.

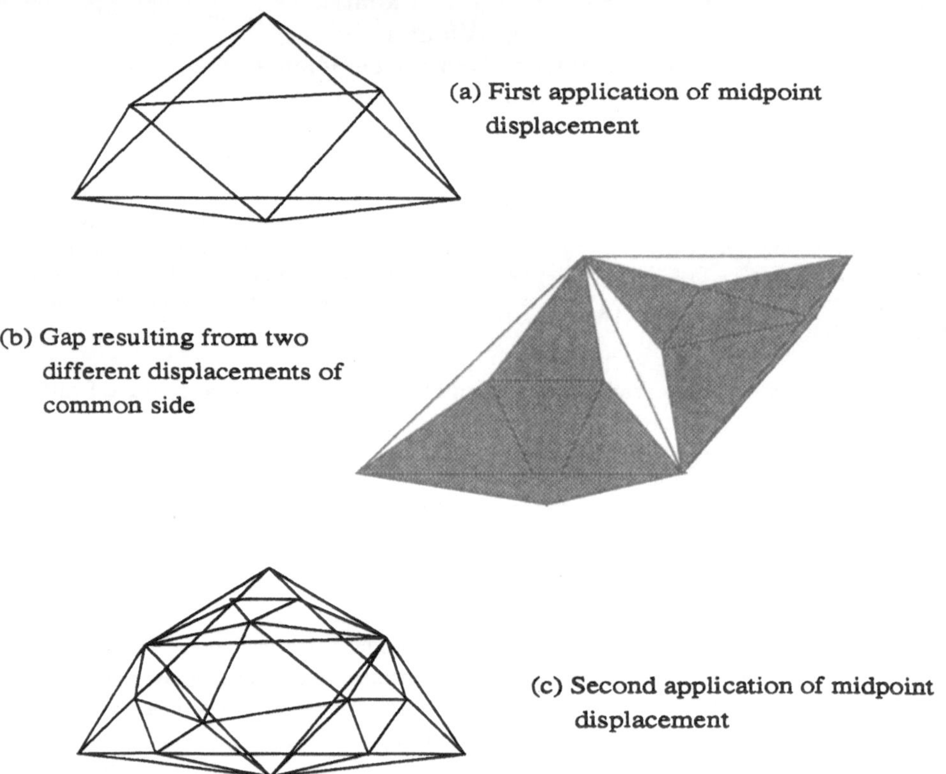

(a) First application of midpoint displacement

(b) Gap resulting from two different displacements of common side

(c) Second application of midpoint displacement

Figure 1: Midpoint Displacement of Triangle Sides

Figure 1(b) depicts the *creasing* problem common to this method. When the midpoint of a side common to two triangles is displaced by two different displacements, *creasing* occurs. One of the methods to avoid this problem is, use the coordinates of the undisplaced midpoint of the line that we are working on to generate a unique number. This number is used as a seed for the random number generator, so that when the same line occurs, this number is used to get the same displacement.

3 Fast Fourier method

Here the fractal scenes were generated by filtering white noise with a $\frac{1}{f}$ noise (power law) filter. Before proceeding further into this method, let us recollect some of the aspects related to random processes.

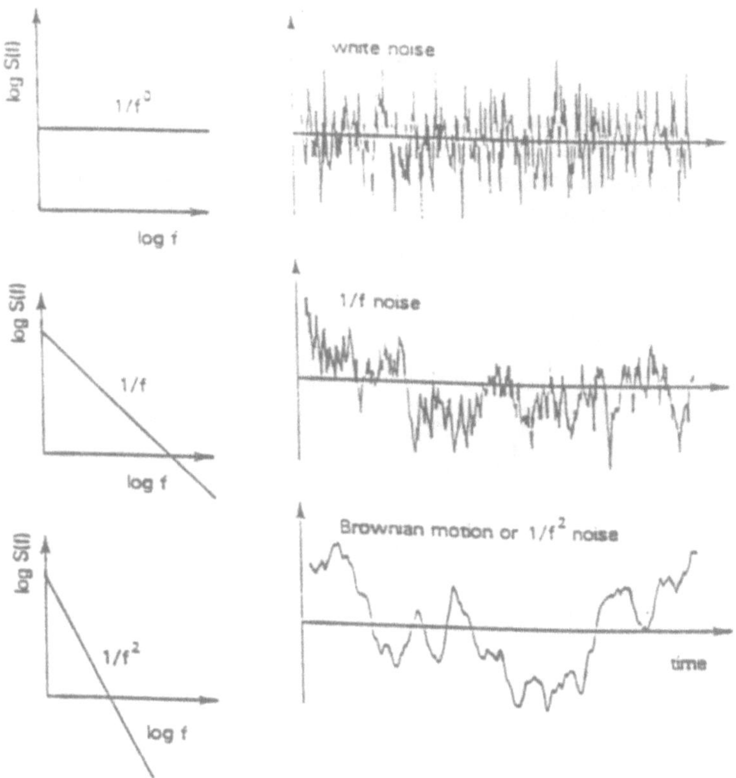

Figure 2: Typical noises and their spectral densities, $S(f)$

Fig.2 displays samples of typical random functions $X(t)$ and their spectral densities $S(f)$. The spectral density $S(f)$ provides information about the time correlations of $X(t)$.

A fractional Brownian motion, $X_H(t)$, is a single valued function of one variable, t (usually time). Its increments $X_H(t_2) - X_H(t_1)$ have a Gaussian distribution with variance

$$\left\langle |X_H(t_2) - X_H(t_1)|^2 \right\rangle \propto |t_2 - t_1|^{2H} ,$$

where the brackets $<$ and $>$ denote ensemble averages over many samples of $X_H(t)$ and the parameter H has a value $0 < H < 1$. As shown in the figure the traces of the fBm reminds us of a mountanous horizon. Formally, it is the increments of fBm (the differences between successive times) that gives the noise of Fig. 2. The scaling behavior of the different traces shown in Fig. 3 is characterized by a parameter H.

For lower values of H the traces are rough and for higher values they are relatively smooth. H relates the typical change in X, $\Delta X = X(t_2) - X(t_1)$, to the time difference $\Delta t = t_2 - t_1$, by the simple scaling law

$$\Delta X \propto \Delta t^H$$

In the usual Brownian motion or random walk, the sum of independent increments or steps leads to a variation that scales as the square root of the number of steps. Thus, $H = 1/2$ corresponds to a trace of Brownian motion. The fBm traces shown in the Fig. 3 repeat statistically only when the t and X direction are magnified by different amounts. If t is magnified by a factor r (t becomes rt), then X must be magnified by a factor r^H (X becomes $r^H X$).

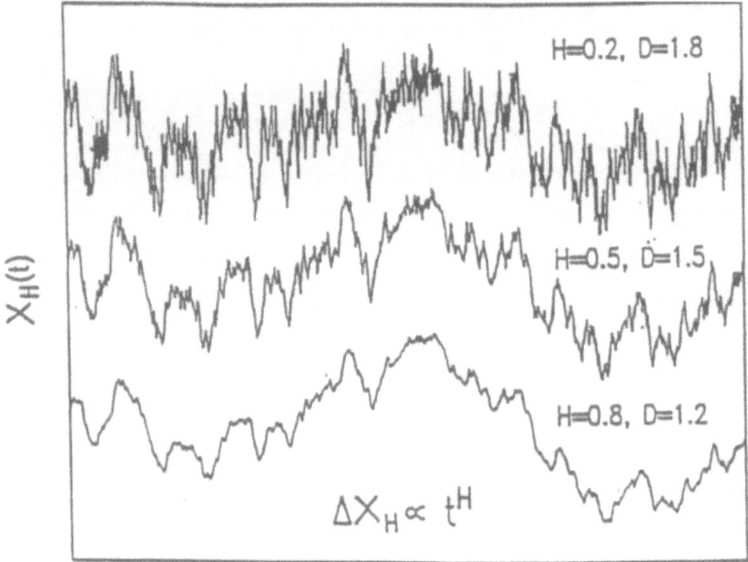

Figure 3: Fractional Brownian motion traces

If $X(f)$ is the Fourier coefficient of $X(t)$ at frequency f ,

$$X(f) \propto \int X(t) \exp^{-2\pi i f t} dt$$

then

$$S(f) = \frac{|X(f)|^2}{\Delta f}$$

where Δf is the effective bandwidth of the Fourier integral.

An alternative characterization of the time correlations of $X(t)$ is given by the auto-correlation or pair-correlation function

$$G(\tau) = \langle X(t)X(t+\tau)\rangle - \langle X(t)^2\rangle$$

$G(\tau)$ provides a measure of how the fluctuations in a quantity $X(t)$ are correlated between times t and $t+\tau$. $S(f)$ and $G(\tau)$ are not independent. In most cases they are related by the Weiner-Khintchine relations:

$$S(f) \propto \int G(\tau) \cos(2\pi\tau) dt$$

and

$$G(\tau) \propto \int S(f) \cos(2\pi f) df$$

The white noise $w(t)$ in Fig.2(a) is the most random. The future is completely independent of the past and its $G(\tau)$ has the form $G(\tau) \propto \delta(\tau)$. Consequently, its spectral density $S(f) =$ *a constant* with equal power at all frequencies f, like white light.

The integral of white noise $w(t)$ (or the summation of random increments) produces a Brownian motion $X(t) = \int w(t)dt$ or random walk as shown in Fig. 2(c). Thus, $X(t)$ corresponds to the random diffusion of the particle and the average distance travelled in a time T obeys the usual diffusion law,

$$\Delta X(T) = \left\langle |X(t+T) - X(t)|^2 \right\rangle^2 \propto T^{1/2}.$$

Although the appearance is much more correlated and $S(f) \propto \frac{1}{f^2}$, whether $X(t)$ increases or decreases in the future is independent of its entire past.

3.1 Fractional Brownian Motion(fBm) and $\frac{1}{f}$ noise

Although the white noise in Fig. 2(a) and Brownian motion in Fig. 2(c) are well understood mathematically and physically, they are characteristic of relatively few naturally occuring fluctuation phenomena. Many natural time series look much more like $\frac{1}{f}$ noise in Fig. 2(b). In most cases the physical reason for this behavior remains a mystery.

The most effective mathematical model of such behavior is fractional Brownian motion of fBm as developed by Mandelbrot and Wallis. fBm is an extension of the central concept of Brownian motion. fBm is a good starting point for understanding random walks on fractals.

Fractional Brownian motion can be extended to any dimensions. To generate a surface, the single variable t is replaced by coordinates x and y in the plane to give $X_H(x, y)$ as the surface altitude at position x, y. In this case the altitude variations of a hiker following any straight line path at constant speed in the xy-plane is fBm. If the hiker travels a distance Δr in the xy-plane ($\Delta r^2 = \Delta x^2 + \Delta y^2$), the typical altitude variation, ΔX, is given by

$$\Delta X \propto \Delta r^H .$$

The fractal dimension D will be greater than the topological dimension . Here the fractal dimension D of a surface is

$$D = 3 - H \text{ for a fractal landscape } X_H(x, y).$$

This generalization of fBm can continue to still higher dimensions to produce, for example, a self affine fractal temparature or density distribution $X_H(x, y, z)$. Here, the variations of an observer moving at constant speed along any straight line path in xyz-space generate a fBm with scaling given by $\Delta X \propto \Delta r^H$, where now $\Delta r^2 = \Delta x^2 + \Delta y^2 + \Delta z^2$. Here the fractal dimension is

$$D = 4 - H \text{ for a fractal cloud } X_H(x, y, z).$$

To summarize, a statistically self affine fBm, X_H, provides a good model for many natural scaling processes and shapes. So, as a function of two variables ($E = 2$) fBm is a good mathematical model for fractal landscapes and random surfaces. Here, the scaling property may be characterized equivalently by either H or the fractal dimension D.

As discussed above, random functions in time $X(t)$ are often characterized by their spectral densities $S_X(f)$. As shown in Fig. 2, the spectral exponent β changes with the appearance of the noise trace similar to the variation with H of the traces of fBm from Fig. 3. So another characterization is provided by the spectral exponent β. The connection between the 3 equivalent charaterizations, D, H and β of a fBm function of E variables is given by

$$D = E + 1 - H = E + (3 - \beta)/2.$$

For H in the range $0 < H < 1$, D spans the range $E < D < E + 1$, and $1 < \beta < 3$.

The following subsection deals with the method for producing a finite sample of fBm as a noise(E = 1), a landscape(E = 2), or a cloud(E = 3).

3.2 Algorithm

The main idea behind this method is to transform a Gaussian noise to the frequency domain by passing it through a $1/f$ filter, and then transforming the result back to the time domain. For this purpose, a pseudo-random number generator is used to produce a "white noise" $W(t)$. This $W(t)$ is filtered with a transfer function $T(f)$; this results in $X(t)$, whose spectral density is

$$S_X(f) \propto |T(f)|^2 \, S_W(f) \propto |T(f)|^2.$$

Thus, to generate a $\frac{1}{f^\beta}$ noise from a $W(t)$ requires

$$T(f) \propto \frac{1}{f^{\beta/2}}.$$

A continous function of time, $X(t)$ may be approximated by a finite sequence of N values, X_n, defined at discrete times $t_n = n\Delta t$, where n runs from 0 to $N-1$ and Δt is the time between successive values. Complex Fourier coefficients, x_m, define the discrete Fourier tranform (DFT), X_n, as :

$$X_n = \sum_{m=0}^{N-1} X_m e^{\frac{2\pi i m n}{N}}$$

where n ranges from 0 to $N-1$.

The complex Fourier coefficients $X_m(U_m - iV_m)$ are calculated as follows : for all U_m and V_m generate two independent normals with mean 0 and variance 1 (say r_m and s_m). If the index $m < N/2$, then multiply each of these random numbers by $\sqrt{TS_m/2}$, where S_m is the spectrum and is given as $1/f^\beta$ and let the output be U_m and V_m. If the index m is $N/2$, then multiply the random number by $\sqrt{TS_{n/2}}$ to get $U_{N/2}$ and $V_{N/2}$ is 0.

Here we have some similarities, so the complex Fourier coefficients are calculated upto $N/2$ (Nyquist frequency), then the symmetries $U_{N-M} = U_m$ and $V_{N-M} = -V_m$ are applied to generate the complex sequence. Now the sequence X_n is calculated with a fast Fourier transform (FFT) algorithm. This algorithm requires an order of $N \log N$ operations to produce a series of N points.

This procedure has been extended to functions of 2 coordinates to generate a fractal surface or a fractal mountain $X_H(x,y)$. The corresponding fractal or mountain is approximated on a finite $N * N$ grid to give $X_H(x_{n1}, y_{n2})$. The 2-dimensional complex FFT can be used to evaluate the series :

$$X_{n_1 n_2} = \sum_{m_2=0}^{N_2-1} \sum_{m_1=0}^{N_1-1} X_{m_1 m_2} e^{2\pi i(\frac{m_1 n_1}{N_1} + \frac{m_2 n_2}{N_2})}$$

where n_1 ranges from 0 to $(N_1 - 1)$ and n_2 ranges from 0 to $(N_2 - 1)$.

Here the complex Fourier coefficients $X_{m_1 m_2}$ are calculated in almost the same way as explained above. The spectrum usually depends on two frequency variables m_1 and m_2 corresponding to the x and y directions. But since all the directions in the xy-plane are equivalent with respect to the statistical properties, the spectrum depends only on $\sqrt{m_1^2 + m_2^2}$. So the 2-dimensional spectrum is given as :

$$S_{m_1 m_2} \propto \frac{1}{(m_1^2 + m_2^2)^{2H+2}}$$

where H ranges between 0 and 1.

The random fractals or mountains generated by this method are illustrated in figures 4, 5 and 6.

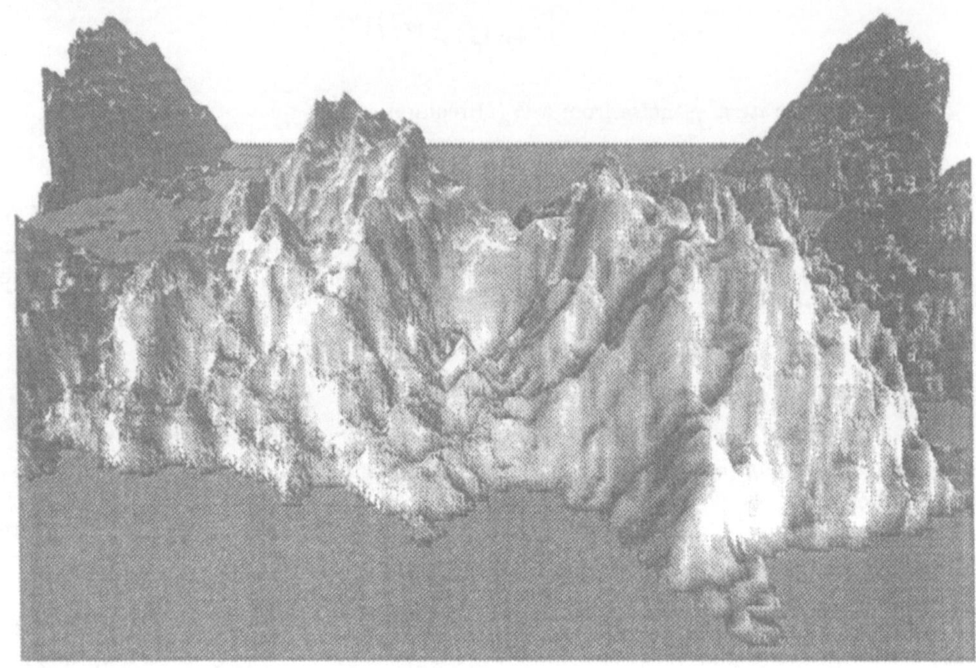

Figure 4: Mountain generated by Fast Fourier method ($H = 0.8$)

The figures shown in Fig.4, Fig.5 and Fig.6 are generated by using different spectr a. It clearly shows how some of the features of the mountain, like its roughness, v ary by changing different parameters. Also, different seed values for the random nu mber generator produce different mountains. The mountains shown in these figures ar e got by changing the parameter H (where H is the fractal codimension). As the valu e of H decreases, the mountain becomes more spikier.

For a 3-dimensional fBm, $X_H(x, y, z)$, the spectrum is $S_{m_1 m_2 m_3} = 1/(m_1^2 + m_2^2 + m_3^2)^{2H+3}$ where H ranges from 0 to 1. This can be used to produce fractal flakes and clouds.

Due to the nature of Fourier transform, the generated samples are periodic. In this case one can compute twice or four times as many points as actually needed and the n discard a part of the sequence.

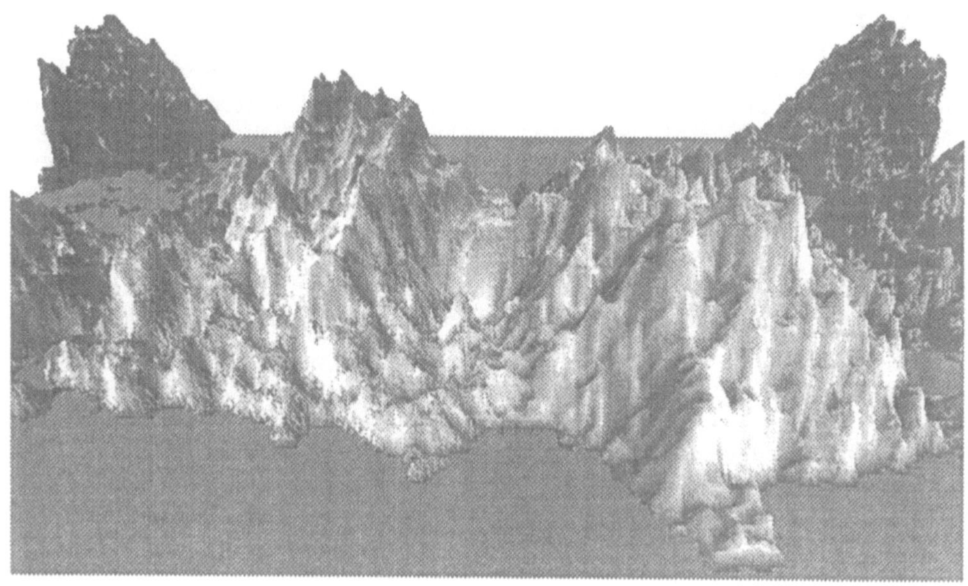

Figure 5: Mountain generated by Fast Fourier method ($H = 0.7$)

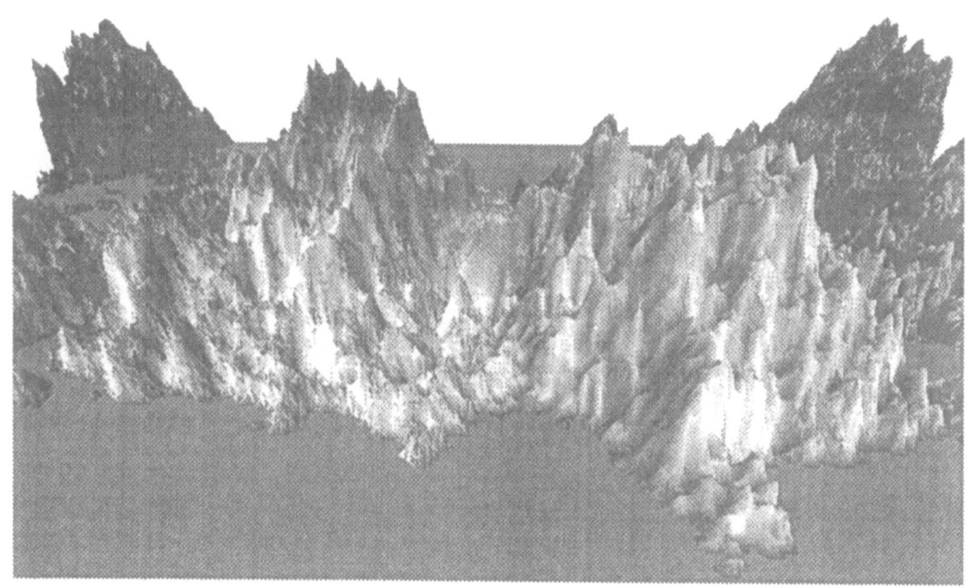

Figure 6: Mountain generated by Fast Fourier method ($H = 0.6$)

4 Conclusions and Further Study

The Fast Fourier Transformation (FFT) technique has been one of the most influential tools for the generation of high quality random fractals. But as it first needs a good estimation of the spectral density of the fractal, this method can be difficult to apply for synthesis purposes. So an investigation for transformation methods other than FFT is needed. Most of the fractal methods that generate a fractal by using only its global properties do not allow local changes. This is a drawback with the FFT too. One of the methods that holds a better potential is the wavelet transforms, as it provides an increased number of parameters and there is an increased possibility that the spectral density might be locally varied. There is also scope for further study of alternative fractal models beyond fBm.

References

[1] Bracewell, R. (1965) The Fourier Transform and Its Applications, McGraw-Hill Book Co., New York.

[2] Carpenter, L. C. (1986), "Computer Rendering of Fractal Curves and Surfaces", ACM Siggraph, Dallas, August 18-22, 9-15.

[3] Cochran, W.T. et al. (1967), "What is the Fast Fourier Transform ?", Proc. IEEE 55, 1664-1677.

[4] Encarnacao, J.L, Peitgen, H.O, Sakas, G. and Englert, G. (1992), Fractal Geometry and Computer Graphics, Springer-Verlag, New York.

[5] Gonzalez, R.C. and Wintz, P. (1987), Digital Image Processing, Addison-Wesley, Reading, Mass.

[6] Jens Feder (1988), Fractals, Physics Dept., Univ. of Oslo, Oslo, Norway, Plenum Press, New York.

[7] Mandelbrot, B.B. (1982), The Fractal Geometry of Nature, W.H.Freeman and Co., New York.

[8] Miller, Gavin S. P. (1986) "The definition and Rendering of Terrain Maps", Computer Graphics, Vol. 20 No. 4 August. SIGGRAPH '86 Conference Proceedings.

[9] Peitgen, H.O, Saupe, D. (1988), The Science of Fractal Images, Springer-Verlag, New York.

[10] Voss, R. F. (1985), "Random fractal forgeries", Fundamental Algorithms for Computer Graphics, R. A. Earnshaw (ed.), Springer-Verlag, Berlin, 805-835.

[11] Yfantis, E. A., Flatman, G. T. and Englund, E. J. (1988), "Simulation of Geological Surfaces Using Fractals", 20 (6), 667-673.

[12] Yfantis, E. A., Gallitano, G. M. and Flatman, G. T. (1992), "A Chaotic Algorithm for Surface Estimation", Journal of Landscape and Urban Planning, vol. 21, pp. 309-311.

[13] Yfantis, E. A. and Frazer, D. W. (1992), "An Algorithm for Mountain Generation ", Proc. 5th International Conference on Engineering Computer Graphics and Descriptive Geometry, August 17-21, 1992, Melbourne, Australia, vol. 2, pp.534-537.

A NEW DIFFERENTIAL PULSE CODE MODULATION COMPRESSION ALGORITHM

T. Pike, E. A. Yfantis, Computer Science Department
University of Nevada, Las Vegas, 89154
Z. Psyllakis, University of Patras, Greece

Abstract

A new lossless algorithm which is a variation of the Differential Pulse Code Modulation is presented in this research report. Unlike the Differential Pulse Code Modulation where we use constant coefficients to predict pixels, the calculation of the coefficients used in our algorithm are based on the minimization of the mean square error, the intrinsic hypothesis, and the use of the semivariogram. Lossless algorithms can be used alone, but they could also be used as a part of a losses algorithm.

Keywords: Intrinsic Hypothesis, Kriging, Semivariogram.

1. Introduction

This algorithm is an improvement of the differential pulse modulation algorithm. The coefficients used in this algorithm are based on the minimization of the mean squared error,the intrinsic hypothesis, and the use of the semivariogram (Yfantis, et al, 1993a,1993b). (Yfantis, et al. 1993c,1993d). The algorithm is a lossless compression algorithm but can also be used as the coding part of a lossy algorithm. A typical lossy compression algorithm consists of three parts. The first one is the transformation part, the second is the quantization, and the third is the coding part. (Yfantis et al, 1992a, 1992b). The interest on efficient image compression algorithms has increased recently, due to the fact that the need for image transfer, image storage, image recording and animation, has increased. Digital images occupy relatively large computer space, thus storage, retrieval time, and transfer time via the network, can all be reduced by image compression. The new multimedia technology, digital photography, the plethora of video equipment, the ability to convert analog images to digital, computerized medicine, the plethora of NASA images, and various images obtained via satellites, the need for storing movie and other video tapes, they all call for image processing, image storage, and image compression. Like any other differential pulse code modulation k pixels within a causal neighborhood of the current pixel are used to make a linear prediction of the pixel's value. In our theory and application we use three neighboring pixels, namely the one north, the one west, and the one northwest, to estimate the current pixel. The probability distribution of the original pixels has a large variance, where the probability distribution of the difference of the true pixels from the estimated pixels has a very small variance and therefore is easy to code at a high compression ratio.

2. Our Differential Pulse Code Modulation Algorithm

In our algorithm we use kriging and the three pixels of the upper left corner of the image to estimate the rest of the pixels of the image. For each pixel we take the difference between the true value and the estimated value and we obtain the differential value for

E. A. Yfantis (ed.), Intelligent Systems, 897–901.

the pixel. The dependence between differential values of neighboring pixels is not as high as that between the corresponding pixels. The reduction of pixel dependence and also the fact that the probability density function of the differential values has much smaller variance than the variance of the original pixels, enables us to compress the differential values efficiently. A pixel value is equal to the estimated value plus the differential value.

Let x_{ij}, $i = 1, ..., m$, $j = 1, ..., n$, be the pixel value for the i, j pixel, where the screen resolution is m, by n, pixels. If $i = 1$, and $j > 2$, then an estimate x^*_{1j}, of x^*_{1j}, is obtained as a linear combination of the previous two pixels in line one. Thus:

$$x^*_{1j} = a_1 x_{2j-2} + a_2 x_{1j-2} + a_3 x_{1j-1}, 2 \leq j \leq m \tag{1}$$

The parameters are constrained so that their sum is equal to 1. Hence

$$a_1 + a_2 + a_3 = 1 \tag{2}$$

The idea here is to estimate the parameters, subject to the above constraints, so that we will minimize the mean square error. The mean square error is:

$$\begin{aligned} E(x^*_{1j} - a_1 x_{2j-2} - a_2 x_{1j-2} - a_3 x_{1j-1})^2 = \\ a_1^2 E(x_{1j} - x_{2j-2})^2 + a_2^2 E(x_{1j} - x_{1j-2})^2 + \\ a_3^2 E(x_{1j} - x_{1j-1})^2 + 2a_1 a_2 E(x_{1j} - x_{2j-2})(x_{1j} - x_{1j-2}) \\ +2a_1 a_3 E(x_{1j} - x_{2j-2})(x_{1j} - x_{1j-1}) + 2a_2 a_3 E(x_{1j} - x_{1j-2})(x_{1j} - x_{1j-1}) \end{aligned} \tag{3}$$

The above equation can be written as:

$$\begin{aligned} E(x^*_{1j} - a_1 x_{2j-2} - a_2 x_{1j-2} - a_3 x_{1j-1})^2 = \\ 2a_1^2 \gamma(h\sqrt{5}) + 2a_2^2 \gamma(2h) + 2a_3^2 \gamma(h) \\ +2a_1 a_2 (\gamma(2h) + \gamma(h\sqrt{5}) - \gamma(h)) + 2a_2 a_3 \gamma(2h) \\ +2a_1 a_3 (\gamma(h) - \gamma(h\sqrt{5}) + \gamma(h\sqrt{2})) \end{aligned} \tag{4}$$

Therefore we have to find the parameters a_1, a_2 that minimize the function:

$$\begin{aligned} f(a_1, a_2) = 2a_1^2 \gamma(h\sqrt{5}) + 2a_2^2 \gamma(2h) + 2(1 - a_1 - a_2)^2 \gamma(h) \\ +2a_1 a_2 (\gamma(2h) + \gamma(h\sqrt{5}) - \gamma(h)) + 2a_2 (1 - a_1 - a_2) \gamma(2h) \\ +2a_1 (1 - a_1 - a_2)(\gamma(h) - \gamma(h\sqrt{5}) + \gamma(h\sqrt{2})) \end{aligned} \tag{5}$$

The values of a_1, a_2 minimizing the above equation are solutions to the equations $\frac{\partial f(a_1,a_2)}{\partial a_1} = 0$ and $\frac{\partial f(a_1,a_2)}{\partial a_2} = 0$. The $\frac{\partial f(a_1,a_2)}{\partial a_1} = 0$ after some calculations implies:

$$\begin{aligned} a_1(4\gamma(h\sqrt{5})) - 2\gamma(h\sqrt{2})) + a_2(2\gamma(h\sqrt{5}) - \gamma(h\sqrt{2})) = \\ \gamma(h\sqrt{5}) - \gamma(h\sqrt{2}) + \gamma(h) \end{aligned} \tag{6}$$

Also the $\frac{\partial f(a_1,a_2)}{\partial a_2} = 0$ after some calculations derives:

$$a_1(2\gamma(h\sqrt{5}) - \gamma(h\sqrt{2})) + 2a_2 \gamma(h) = 2\gamma(h) - \gamma(2h) \tag{7}$$

The calculation of the parameters along the first column is similar to the calculation of the parameters along the first row.

In order to estimate a pixel we use three neighboring pixels, one directly above it, one to the left of it, and one in the upper left diagonal, as shown in figure 1. The estimate x^*_{ij}, of x_{ij}, is:

$$x^*_{ij} = a_1 x_{ij-1} + a_2 x_{i-1j-1} + a_3 x_{i-1j} \tag{8}$$

As before, the idea is to find the parameters that minimize the mean square error:

$$E(x_{ij} - a_1 x_{ij-1} + a_2 x_{i-1j-1} + a_3 x_{i-1j})^2 \tag{9}$$

subject to the condition,

$$a_1 + a_2 + a_3 = 1 \tag{10}$$

and using the semivariogram approach with a spherical semivariogram with zone of influence greater than or equal to 20.

Figure 1: The three points denoted by a solid circle are used to estimate the point denoted by x

Figure 2: The Image our Compression Algorithm was Applied to.

From the mean square error expression after some calculations we obtain

$$\begin{aligned} E(x_{ij} - a_1 x_{ij-1} + a_2 x_{i-1j-1} + a_3 x_{i-1j})^2 = \\ a_1^2 E(x_{ij} - x_{ij-1})^2 + a_2^2 E(x_{ij} - x_{i-1j-1})^2 + \\ a_3^2 E(x_{ij} - x_{i-1j})^2 + 2a_1 a_2 E(x_{ij} - x_{ij-1})(x_{ij} - x_{i-1j-1}) + \\ 2a_1 a_3 E(x_{ij} - x_{ij-1})(x_{ij} - x_{i-1j}) + 2a_2 a_3 \gamma(h\sqrt{(2)}) \end{aligned} \tag{11}$$

If we denote the mean square error as a function f of its parameters then after some calculations the mean square error can be expressed as:

$$f(a_1,a_2) = 2[a_1^2\gamma(h) + a_2^2\gamma(h\sqrt{2}) + (1-a_1-a_2)^2\gamma(h) + a_1a_2\gamma(h\sqrt{2}) + a_1(1-a_1-a_2)(2\gamma(h)-\gamma(h\sqrt{2})) + a_2(1-a_1-a_2)\gamma(h\sqrt{2})] \qquad (12)$$

The values for which the above function is minimized are solutions to the equations $\frac{\partial f(a_1,a_2)}{\partial a_1} = 0$ and $\frac{\partial f(a_1,a_2)}{\partial a_2} = 0$. The $\frac{\partial f(a_1,a_2)}{\partial a_1} = 0$ after some calculations derives

$$2a_1 + a_2 = 1 \qquad (13)$$

Now the $\frac{\partial f(a_1,a_2)}{\partial a_2} = 0$ after some calculations gives us

$$a_1\gamma(h\sqrt{2}) + 2a_2\gamma(h) = 2\gamma(h) - \gamma(h\sqrt{2}) \qquad (14)$$

The solution of the above two linear equations with two unknowns is:

$$a_1 = \frac{\gamma(h\sqrt{2})}{4\gamma(h) - \gamma(h\sqrt{2})}$$
$$a_2 = \frac{4\gamma(h) - 3\gamma(h\sqrt{2})}{4\gamma(h) - \gamma(h\sqrt{2})} \qquad (15)$$

In all the above equations h is the distance between two consecutive pixels belonging in the same scan-line or the same column, and therefore h can be considered to be equal to one. The spherical semivariogram is of the form

$$\gamma(s) = 1.5\frac{s}{a} - 0.5\frac{s^3}{a^3}, 0 \leq s \leq a \qquad (16)$$
$$(17)$$

a is the zone of influence Using the above semivariogram model with zone of influence equal to 20 or more pixels, which is a conservative number for the zone of influence of an image we obtain

$a_1 = 0.55, a_2 = -0.10$, therefore the $a_3 = 0.55$. When the pixels to be estimated are along the first scan-line of the image or along the first column of the image then the parameters found are $a_1 = 0.15, a_2 = 0.24, a_3 = 0.61$.

The algorithm was applied to an image provided by CompressTech Inc., we will refer to it as chick and it is shown in figure 2. ratio was found to be 2.3433 to 1.

3. Conclusion

A lossless algorithm based on kriging was introduced. Our algorithm belongs in the Differential Pulse Code Modulation family of algorithms. Our contribution is based on the fact that we are only using the three left corner pixels of the image in order to estimate all the image pixels. Also the parameters used in the pixel estimation are based on kriging with a spherical semivariogram with a conservative zone of influence of 20 pixels. The compression ratio obtained by applying our algorithm to the image shown in figure 2, is 2.3433 to 1. The image in figure 2 was provided to us by CompressTech Inc.

4. Acknowledgments

This research was supported by CompressTech Inc. The authors would like to thank Leo Roussell, Daniel Cordova both of CompressTech Inc., and the reviewers for their valuable suggestions.

5. References

[1] E. A. Yfantis, M. Au, and G. Miel, (1992a), "An efficient Image Compression Algorithm for Computer Animated Images", Journal of Electronic Imaging, 1(4), pp. 381-387.

[2] E. A. Yfantis, S. Baker, and G. M. Gallitano, (1992b), "An Image Compression Algorithm and its Implementation," Finite Fields, Coding Theory, and Advances in Communications and Computing, Lecture Notes in Pure and Applied Mathematics vol. 141, pp. 417-427, Marcel Dekker, Inc., N.Y.

[3] E. A. Yfantis, M. Y. Au, (1993a), "An Image Compression Algoridon Kriging", Image and Video Processing, Vol. 1903, pp. 215-227.

[4] E. A. Yfantis, and F. S. Makri, (1993b), "An Index Dependence Algorithm for Increasing Parallelization", Proceedings of Ninth confereneon Systems Engineering, July 14-16, 1993, pp. 305-309.

[5] E. A. Yfantis, M. Au, and F. S. Makri, (1993c), "Image Compression and Kriging", Geostatistics for the Next Century, Kluwer Academic Publishers,pp. 156-161.

[6] E. A. Yfantis, (1993d), "Reply to Comments on Image Compression", Geostatistics for the Next Century, Kluwer Academic Publishers, pp. 168-170.

ALGORITHMS

The Problem of Partitioning with Duplications and its Applications

J. Haralambides
Dept. of Math and CS
Barry University
Miami Shores, FL 33161 USA

S. Tragoudas
Dept. of Computer Science
Southern Illinois University
Carbondale, IL 62901 USA

Abstract. We study the problem of partitioning a graph $G = (V, E)$ into two sets V_1 and V_2 such that $0 \leq |V_1 \cap V_2| \leq d$, for an integer d, and such that $\sum_{u \in V_1 - V_2, v \in V_2 - V_1}(u, v)$ is minimized. We show that this problem is NP-hard in general, and remains NP-hard if $|V_1| = |V_2|$, or if we insist that, for any bipartition, $v_1 \in V_1$ and $v_2 \in V_2$ for two specific nodes $v_1, v_2 \in V$. We construct polynomial time algorithms for the special cases of (a) edge-weighted series-parallel graphs and (b) edge-unweighted solid grids. The algorithms for the above special cases offer attractive solutions to problems in the areas of Hypermedia Organization and VLSI Layout. Furthermore, duplication is characterized as a means of improving network integrity by providing alternative interconnections.

Keywords. Algorithms, VLSI Layout, Graph Partitioning, Solid Grids, Series-parallel Graphs, Integrity.

1 Introduction

In this paper, we first consider the problem of partitioning a graph $G = (V, E)$ of n nodes and m edges into two sets V_1 and V_2 such that $|V_1 \cap V_2| \leq d$, for some $d < V$, and the sum $\sum_{u \in V_1 - V2, v \in V_2 - V_1}(u, v)$ is minimized. We call this problem the d *min-cut* problem. The d min-cut problem is equivalent to either one of the following two problems:

(a) Remove up to d nodes v_i, $1 \leq i \leq d$, and partition the nodes in $V - \bigcup_i\{v_i\}$ into disjoint sets $V_1' \subset V$ and $V_2' \subset V$. The final sets of the partition are going to be $V_1 = V_1' \bigcup(\bigcup_i\{v_i\})$ and $V_2 = V_2' \bigcup(\bigcup_i\{v_i\})$. Nodes v_i are selected to be common in both sides of the partition. Therefore, edges (v_i, v_j), $v_j \in V$, do not count in the partition cost and can be removed.

(b) Allow up to d nodes v_i, $1 \leq i \leq d$, to be duplicated, and partition the nodes in $V - \bigcup_i\{v_i\}$ into disjoint sets $V_1' \subset V$ and $V_2' \subset V$. The final sets of the partition are going to be $V_1 = V_1' \bigcup(\bigcup_i\{v_i\})$ and $V_2 = V_2' \bigcup(\bigcup_i\{v_i'\})$, where v_i' is the copy of v_i. Let $a \in \{v_i, v_i'\}$. Edges (a, v_j), , $v_j \in V$, do not count in the partition cost and can, therefore, be removed.

Throughout this paper, we are going to use the second equivalent formulation. The special case when $d = 0$ has been extensively studied in the literature and is the well known *min-cut* problem. Consider the variation of the *min-cut* problem where we insist that node v_1 be in V_1 and node v_2 be in V_2. This is the v_1/v_2 *min-cut* problem which is solvable in $\tilde{O}(n \cdot m)$ time using maximum flow techniques [6]. Clearly, the algorithm for the v_1/v_2

E. A. Yfantis (ed.), Intelligent Systems, 905–912.

min-cut problem can be used as a subroutine to solve the min-cut problem in polynomial time. The $(d, v_1/v_2)$ problem is defined in a way analogous to the v_1/v_2 min-cut problem by allowing up to d node duplications. Here we assume that v_1 and v_2 cannot be duplicated.

In contrast to the above min-cut problems, we show in Section 2 that both the d min-cut and the $(d, v_1/v_2)$ min-cut problems are NP-hard for any integer d and polynomially solvable for any fixed values of d or $n - d$. If either $|V_1|$ or $|V_2|$ is fixed, both problems are also polynomially solvable.

Subsequently, we concentrate on a variation of the partitioning problem where we insist that sets V_1 and V_2 have equal sizes, i.e., $|V_1| = |V_2|$, or balanced sizes, i.e., the ratio $|V_1|/|V_2|$ does not depend on the input. If $|V_1| = |V_2|$, we call this partitioning problem the d-bisection problem. In addition, consider the variation where $|V_1| = k$. We call the latter problem the *(k,d) min-cut* problem on G. Throughout this paper, we call an optimal cut c for this problem the optimal *(k,d)-cut*. In Section 2, we show that the (k, d) min-cut problem can be solved polynomially if either k is fixed or $n - d$ is fixed. It is NP-hard for integer values of k or integer values of d. We also show that the d-bisection problem is polynomially solvable if $n - d$ is fixed. It is NP-hard if d is either an integer or fixed.

The d-bisection and the (k, d) min-cut problems find important applications in hypertext partitioning and VLSI Layout. In the area of VLSI Layout, and in particular Placement and Routing, bipartitioning with overlapping has the meaning of duplicating those components of the circuit that will minimize the interconnections between the two sets of the partition. This allows for a well defined placement of nodes in cohesive modules and, furthermore, it simplifies the wiring phase.

The problem finds another application in the context of hypertext environments [7]. Here the nodes are entries (possibly text) which must reside on menus of certain content capacity. We do not insist on maintaining an entry only in one menu. (This is natural for menu entries that provide links to overlapping definitions i.e., biochemistry may appear as an entry in both the areas of biology and chemistry.) Thus, we can move from a menu $\mathcal{M}_1$ where entry A resides to another menu $\mathcal{M}_2$ containing A at practically no cost. However, if we are searching for entry A in menu $\mathcal{M}_2$ while being in menu $\mathcal{M}_1$ which does not contain A, we move by paying a predefined navigation cost. It is, therefore, clearly desirable for entries to be allocated in menus so that the total navigation cost is minimized. Clearly, a solution to the d-bisection problem is directly applicable here. Overlapping definitions are accessed through link duplication. Furthermore, the integrity of the hypermedia system is increased since the extra links will reduce the possibility of fragmentation or node inaccessibility for the system [2].

The above applications inspired us to examine the d-bisection and the (k, d) min-cut problems on special cases of graphs that have also been examined for the 0 bisection problem [1, 8, 9]. In Section 3, we present an $O(d^2n^2)$ time algorithm for the d-bisection and the (k, d) min-cut problems for edge-weighted series-parallel graphs, and in Section 4, we give a summary of results for the same problems on edge-unweighted solid grid graphs. More detailed analysis for the algorithm on solid grids can be found in [5]. We define series-parallel graphs and solid grids below.

The grid graph (or grid) is defined as the finite induced subgraph of the two-dimensional infinite grid. Nodes and edges of the outer face are called boundary nodes and edges, respectively. A grid graph G is called *solid*, if it contains no *holes*. A hole is a finite connected component of the complement of G, with respect to the infinite grid.

Series-parallel graphs can be defined in terms of series or parallel edges. Two edges $e_1 = (i_1, j_1)$ and $e_2 = (i_2, j_2)$, where $e_1 \neq e_2$, are in series if $j_1 = i_2$ and $i_1 \neq j_2$. They are in *parallel* if $i_1 = i_2$ and $j_1 = j_2$. Then, a series-parallel graph can be defined recursively as follows: A single edge is a parallel graph. Substituting an edge of a series-parallel graph with two edges in series or parallel results in a series-parallel graph. Outerplanar graphs and trees are not series-parallel graphs under this definition but they can be easily transformed with the insertion of zero-weight edges.

Both series-parallel graphs and solid grids are often encountered in VLSI. Furthermore, hypermedia can be organized as a series-parallel graph, in which case the application is straightforward. The algorithms are generalizations of the techniques in [1] and [8], respectively, for the 0-bisection problem. Both algorithms are based on dynamic programming and use the equivalent problem formulation, described at the beginning of this section, where we allow up to d node duplications.

2 Complexity analysis for variations of the problem

We first examine the d-bisection problem. We have shown (the proof is omitted here) that:

Theorem 2.1 *The d-bisection problem on a graph $G = (V, E)$ is NP-hard for any fixed or integer value of d.*

However, the d-bisection problem is polynomially solvable by exhaustive enumeration, if $n - d$ is fixed. Next, we briefly consider complexity issues for the (k, d) min-cut problem. The problem behaves similarly to the d-bisection problem if k is part of the input. (The proof is omitted here.) Furthermore, if k is fixed, the (k, d) min-cut problem is solvable in polynomial time by exhaustive enumeration. (Observe that $d \leq k$ and, thus, d is also fixed.)

In the remainder of the section, we concentrate on problems $(d, v_1/v_2)$ and d min-cut that do not impose restrictions on the sizes $|V_1|$ and $|V_2|$ of sets V_1 and V_2. Both problems can be solved in $O(n^d)$ time, by exhaustive enumeration, if either d or $n - d$ is fixed. We show, however, that the $(d, v_1/v_2)$ min-cut problem is NP-hard for integer values of d.

Specifically, we prove that the *decision* $(d, v_1/v_2)$ min-cut problem is NP-complete. The problem asks whether there exist at most d node duplications so that the number of the interconnections between V_1 and V_2 is no more than an integer K. We reduce from the Node Cover problem [4].

Theorem 2.2 *The decision $(d, v_1/v_2)$ problem is NP-complete, for any integer value of $d < V$.*

Proof: Clearly, $(d, v_1/v_2)$ is in NP. We reduce polynomially as follows: Let n and m be the number of nodes and edges of graph G, respectively. We construct a graph $G' = (V', E')$ consisting of $2 + 10 \cdot m + 4 \cdot (m+1) \cdot n$ nodes and $12 \cdot m + 8 \cdot (m+1) + 4 \cdot m = 24 \cdot m + 8$ edges in the following manner.

Every node $u \in V$ has a corresponding node $u \in V'$ connected to $m+1$ nodes $u_i^{v_1}$, $1 \leq i \leq m+1$ with edges $(u, u_i^{v_1})$, and to nodes $u_j^{v_2}$, $1 \leq j \leq m+1$ with edges $(u, u_j^{v_2})$. We also have the connections $(u_i^{v_1}, v_1)$, $(u_j^{v_2}, v_2)$, for all possible values of u, i, and j.

For each edge $e \in E$ we introduce nodes e_1, e_2, e^{v_1}, e^{v_2}, and nodes e^1, e^2, $e_1^{v_1}$, $e_2^{v_1}$, $e_1^{v_2}$, and $e_2^{v_2}$. We connect these nodes with the edges (e_1, e^1), (e_1, e^2), (e_2, e^1), (e_2, e^2), (e_1, e^{v_1}), (e_2, e^{v_2}), $(e^{v_1}, e_1^{v_1})$, $(e^{v_1}, e_2^{v_1})$, $(e_1^{v_1}, v_1)$, $(e_2^{v_1}, v_1)$, $(e^{v_2}, e_1^{v_2})$, $(e^{v_2}, e_2^{v_2})$, $(e_1^{v_2}, v_2)$, $(e_2^{v_2}, v_2)$, for all values of e. We call nodes $u_i^{v_1}$, $u_j^{v_2}$, e^1, e^2, $e_1^{v_1}$, $e_2^{v_1}$, $e_1^{v_2}$, and $e_2^{v_2}$, for all values of u and e, *auxiliary* nodes.

Finally, for every edge $e = (u, v) \in E$, we have edges (u, e_1), (u, e^{v_1}), (v, e_2), (v, e^{v_2}). (The decision whether node u is connected to nodes e_1 and e^{v_1} or nodes e_2 and e^{v_2} is arbitrary. A symmetric connection gives the same results.)

We set d, the number of duplications to be k, and K, the upper bound on the minimum edge cut separating v_1 and v_2, to be $2\cdot(m+1)\cdot(n-k)+m$. This completes the construction.

We now claim that the Node Cover problem on graph $G = (V, E)$ is satisfiable if and only if the $(d, v_1/v_2)$ decision problem on $G' = (V', E')$ is satisfiable for a value of $K = 2\cdot(m+1)\cdot(n-k)+m$. Clearly, if the Node Cover is feasible, then one can duplicate the corresponding nodes in V' and the $(d, v_1/v_2)$ problem is satisfiable. (Observe that since the k nodes form a node cover on G, our construction guarantees that after the duplication of the k nodes the minimum cut separating v_1 and v_2 cuts at most m edges of the form (e_1, e^{v_1}) or (e_2, e^{v_2}).)

Asume now that the $(d, v_1/v_2)$ decision problem is satisfiable. Observe that the value of K guarantees that the duplicated nodes must be nodes that correspond to nodes in V since every duplication of an auxiliary node reduces the cut by one, and every duplication of an e_1 or e_2 node reduces the cut by two. Furthermore, the value of K implies that we have no more than m edges of the form (e_1, e^{v_1}) or (e_2, e^{v_2}) in the cut. This implies that duplicated nodes in V' correspond to nodes in V that form a node cover on graph G. □

We can also appropriately modify the proof of Theorem 2.2 to show that the decision version of the d min-cut problem is NP-complete. We have shown that (the proof is omitted here due to space limitations):

Theorem 2.3 *The d min-cut decision problem is NP-complete.*

3 The d-bisection problem on series-parallel graphs

We are going to utilize two structures, namely the two-terminal series-parallel graphs (TTSP) and the binary structure tree (BST) associated with them (see also [3, 10]). The BST is used to decompose the graph into smaller subgraphs of constant size whose optimal partition cost can be computed using a partition table. Then, we merge the partition tables for smaller graphs to obtain the partition table for increasingly larger graphs. Observe that the merging of smaller graphs is done in a restrictive form, specifically, either as a series or a parallel connection. This allows the algorithm to produce optimal results in polynomial time.

We now give some definitions and basic results for the $d = 0$ case. Given a series-parallel graph G with n vertices, a *partition vector* is an (n+1)-vector P, where $P[k]$, $0 \leq k \leq n$, represents the size of an optimal $(k, 0)$-cut for G. As mentioned above, we compute partition vectors starting at the leaves of the BST (a single edge) and moving toward its root. The partition vector at the root of the BST gives the optimal $(k,0)$-cut of G. Subgraphs which are siblings in the BST share only one node (series connection) or two nodes (parallel connection). Shared vertices must be handled carefully. For that, *constraint sets* are introduced. Any node of a subgraph H_1 shared by any other subgraph H_2 must be in the constraint set for H_1. Nodes in constraint sets are restricted to a particular side of the partition. To obtain an optimal $(k,0)$-cut all possible combinations for these nodes must be examined. Since for a constraint set C, nodes can be in any one of two sides, $2^{|C|}$ partition vectors must be constructed to cover all possible cases. It is, therefore, essential, to maintain a constant number of nodes in the constraint sets.

The set of all possible partition vectors for a subgraph is called a *partition table*. Given a graph $G = (V, E)$, a *partial assignment* on a vertex set A, $A \subseteq V$, is a function f_A such that for each $v \in A$, $f_A(v) = v^R$ (v^L), if v is placed in the right (left) side of the partition. Two partial assignments f_A and g_B are *consistent* if $\forall v \in A \cap B$, $f_A(v) = g_B(v)$. Partition vectors define partial assignments since they restrict nodes in the constraint set to a particular side. We can merge partition vectors if the partial assignments of their respective contraint sets are consistent.

Assume that we create partition tables for the subgraphs of G produced by the BST. Since merging two subgraphs G_1 and G_2 can be done only as a series or a parallel connection, the cardinality of the constraint set C for the resulting subgraph G is at most two. Let G_1 and G_2 be the left and right child of an interior node in the BST, respectively. Let also C_1 and C_2 be their respective constraint sets. We distinguish two cases:

a) The interior node in the BST represents a series connection. Then, the new constraint set is $C = (C_1 \cup C_2) - (C_1 \cap C_2)$, with $|C| = 2$. The shared node is not part of the constraint set of G (it cannot be shared with any other subgraph).

b) The interior node in the BST represents a parallel connection. Then, C is the same with C_1 and C_2, thus, $|C| = 2$.

The above, shows that $2^{|C|} = 4$ partition vectors are sufficient for any partition table constructed, hence, it guarantees a solution in polynomial time. The optimality and correctness of the algorithm are shown in [1].

In our case, we allow a maximum of d nodes to be duplicated. The partition matrices now become two-dimensional to accommodate the optimal (k,d)-cuts for different values of d. Given a subgraph $H = (V_h, E_h)$, the first dimension of the matrix is $|V_h| + 1$ as in the $d = 0$ case. The second dimension is equal to the maximum number of duplications allowed in a subgraph of $|V_h|$ nodes, which is $\min\{d, |V_h|\} + 1$. The (d,k) entry in the matrix shows the value of the optimal k-cut with the k nodes fixed in the left side of the partition, and exactly d duplications allowed. We assign ∞ to entries for which partitions under the constraints are impossible.

We construct partition tables starting at the leaf level of the BST. In our approach, duplication is allowed but multiplication is not, since it is meaningless in bipartitions into overlapping sets. Thus, while merging two subgraphs we must be careful not to allow node multiplication. This rule could be violated in case we merge two subgraphs H_1 and H_2, each one containing a copy of the same node. Observe, however, that we face this problem only with nodes that H_1 and H_2 share. We solve this problem by maintaining for each entry in the matrices for H_1 and H_2 a list of all the shared nodes that are possible candidates for duplication. Fortunately, since series-parallel graphs are TTSP graphs, these nodes are at most two, the terminals of the subgraphs. More importantly, this upper bound can be maintained throughout the merging process for each entry of the resulting partition matrices. We have two cases:

a) A series connection between H_1 and H_2 with respective terminals (a, b) and (b, c). If b must be duplicated in both subgraphs, we discard the combination of the corresponding matrix entries as impossible. In all other cases, and given the choice, we duplicate b, since it can no longer be a shared node. Then, in the worst case the new entries contain at most two nodes, a and c.

b) A parallel connection between H_1 and H_2 with the same pair of terminals (a, b) for both subgraphs. If either a or b must be duplicated in both subgraphs, we discard the combination of the corresponding entries as impossible. Otherwise, in the worst case the new entries contain at most two nodes, a and b.

Lemma 3.1 *Merging two subgraphs in series or in parallel can be done in polynomial time with no node multiplications allowed.*

We have shown that our algorithm does not allow node multiplications. Furthermore, the constant size of the list of shared nodes for each entry ensures polynomial time complexity for the computation of new matrix entries. We can also maintain, for each entry, a record of all the node duplications performed so far to achieve the optimal value of the entry. This information is essential, if we want to specify which d' nodes, $d' \leq d$, were duplicated to obtain the optimal (k, d)-cut.

Theorem 3.1 *The algorithm for partitioning TTSP graphs with at most d node duplications is correct.*

Proof: By induction and the fact that multiplication is forbidden. □

Theorem 3.2 *The time complexity of our algorithm is* $O(d^2n^2)$.

Proof: Constructing the BST, and (b) removing the constraints each take $O(n)$ time, where n is the number of nodes of our series-parallel graph. We compute the optimal (k, d)-cut, $d > 0$, by using all the combinations i, j, such that $i + j = d$. That means that $O(dn^2)$ time is needed for each one of the d rows of the two-dimensional partition matrix for a total of $O(d^2n^2)$ time. □

4 The d-bisection problem on solid grids

It has been shown [8] that the bisection problem for solid grid graphs with no weight on the edges can be solved in polynomial time using a dynamic programming technique. We extend this technique for the d-bisection problem on grids. We work on the dual of a solid grid graph where the cut can be considered as a set of edge-disjoint cycles. Cycles containing the infinite face, can be characterized for convenience as paths between boundary edges.

The paths that participate in an optimal (k,d)-cut, are all paths between boundary edges. We call an optimal (k,d)-cut an *optimal simple* (k,d)*-cut*, if it is a single path. The idea of the dynamic programming approach is to find all optimal simple cuts for values of $k \in \{1..n-1\}$ and $d' \in \{0..d\}$. These cuts are then combined to form the optimal (k,d)-cut. Optimal cuts produced by single paths that do not lie solely within the graph are not considered. Such cuts are called *impertinent*. If an optimal simple (k,d)-cut does not intersect the boundary, it is called *pertinent*. We have shown that impertinent cuts can never participate in the optimal solution. Thus, impertinence needs be checked of all optimal cuts. We have shown:

Lemma 4.1 *For given values of k and d, we can check impertinence of a path between a pair of boundary edges* e_1, e_2 *in* $O(dn^3)$ *time.*

We define $L_0(k,d',e_1,e_2)$ to be the length of an optimal simple (k,d')-cut if the path between boundary edges e_1 and e_2 that defines the cut is pertinent, and ∞ if it is impertinent (it cannot participate in the optimal cut). Then, we show that:

Lemma 4.2 *We can compute all the values for* L_0 *in* $O(dn^6)$ *time.*

We define $L(k,d,e_1,e_2)$ to be the length of the optimal (k,d)-cut consisting of only simple cuts, with enpoints on the boundary segment between e_1 and e_2, discarding nodes between the simple cuts and the rest of the boundary. L represents the actual cost of the cut. If e_1 and e_2 are adjacent clockwise, $L((n+d)/2,d,e_1,e_2)$ is the bisection width of the grid. To complete the recurrence formulae, we define $L'(k,d,e_1,e_2)$ to be the length of the optimal (k,d)-cut, assuming there is no path between e_1 and e_2 in the cut. We have:

$$L'(k,d,e_1,e_2) = min_{0\le i\le k, 0\le d'\le d, e \text{ on } e_1-e_2}\,[L(i,d',e_1,e) + L(k-i,d-d',e,e_2)], \text{ and}$$

$$L(k,d,e_1,e_2) = min\{min_{k\le j\le n, 0\le d'\le d} L'(j-k,d-d',e_1,e_2)], L'(k,d,e_1,e_2).$$

We can now use the analysis on the time complexities of L_0 costs, plus the computational overhead due to the recurrence formulae to state the following theorem:

Theorem 4.1 *The cost of an optimal* (k,d)*-cut in a solid grid can be computed in* $O(d^2n^7)$ *time.*

5 Conclusions

We examined the graph bipartitioning problem into sets V_1 and V_2 where we allow the cardinality of the intersection $V_1 \cap V_2$ to be at most d and so that we minimize the number of the edges connecting nodes in $V_1 - V_2$ to nodes in $V_2 - V_1$. We have shown that the problem is in some cases more difficult than the classic graph bipartitioning problem where $d = 0$, i.e., V_1 and V_2 are disjoint. For example, our problem is NP-hard if we do not impose restrictions on the size of sets V_1 and V_2, and when d is part of the input. Clearly, if $d = 0$, the latter problem is solvable using flow techniques.

We then concentrated on the case in which either $V_1 = V_2$ or $|V_1|$ is an integer k. This is an NP-hard problem even if d is fixed. The problem finds applications in Hypermedia and VLSI. We constructed polynomial time algorithms for important special cases of graphs, namely, edge-weighted series-parallel graphs, and edge-unweighted solid grids. These special cases of graphs have also been studied in [1, 8] for the graph bipartitioning problem, for $d = 0$. Our algorithms are generalizations of the algorithms in [1, 8].

References

[1] Bui, T. N., Jones, C. (1989) "Sequential and parallel algorithms for partitioning simple classes of graphs", Technical Report CS-89-45, Department of Computer Science, The Pennsylvania State University, University Park, PA.

[2] Clark, L. H., Entringer, R. C., and Fellows, M. R. "Computational Complexity of Integrity", to appear in J. Combin. Math. and Combin. Computing.

[3] Duffin, R. J. (1965) "Topology of Series-Parallel Networks", Journal of Mathematical Analysis and Applications 10, pp. 303-318.

[4] Garey, M. R., Johnson, D. S. (1979) "Computers and Intractability: A Guide to the Theory of NP-completeness", W. H. Freeman and Co., New York.

[5] Haralambides, J. and Tragoudas, S. (1994) "Bipartitioning into Overlapping Sets", to appear in the Proc. of the 6th International Conference of Computing and Information, ICCI'94, Peterborough, Canada.

[6] Lawler, E. L. (1976) "Combinatorial Optimization: Networks and Matroids", Holt, Reinehart, and Winston.

[7] Maurer, H. (1991), personal communication.

[8] Papadimitriou, C., Sideri, M. (1991) "The Bisection Width of Grid Graphs", Proc. of the 1st ACM-SIAM Symposium on Discrete Algorithms, pp. 405-410.

[9] Sherwani, Naveed (1993) "Algorithms for VLSI Physical Design Automation", Kluwer Academic Publishers, Boston.

[10] Valdes, Tarjan, J. R., and Lawler, E. L. (1982) "The Recognition of Series-Parallel Digraphs", SIAM Journal on Computing 11, No. 4, pp. 298-312.

Pr/t Net Method for Robot Planning[1)]

*Jaegeol Yim and **Tadao Murata

*Department of Computer Science
Dong-kook University at KyungJu
KyungJu City, Kyung-Book, Korea, 780-714

**Department of Electrical Engineering and Computer Science
University of Illinois at Chicago
Chicago, Illinois 60680

Abstract. We introduce an algorithm to construct a predicate-transition(pr/t) net model for a robot plannig problem. Once modeled as a pr/t net, the robot planning problem becomes the pr/t net reachability problem. Given two markings M_0 and M_g on a pr/t net, finding a firing sequence which transforms M_0 into M_g is called the reachability problem. We have found a few general purpose admissible and monotonic heuristic functions which can be used to solve the reachability problem. We also have found a couple of heuristic functions for the Blocks World problem. One of them is accurate function. We have implemented an iterative deepening A^* algorithm using our heuristic functions, and present the experimental results for our heuristics.

Keywords. Pr/t net, Reachability, A.I. Search, Heuristic Function, Robot Planning.

1 Introduction

The predicate-transition(pr/t) net was first defined by Genrich [1] and can be conveniently used in modeling first-order predicate logic. A sentence written in first-order predicate logic can always be translated into a set of clauses. Most of the automated reasoning algorithms [2] work on a database consisting of clauses. Therefore, Murata et al [3] defined a (pr/t) net which is a subset of Genrich's for modeling a logic program written in clausal form. The process of constructing a pr/t net model for a robot planning problem is very similar to the process of constructing a pr/t net model for a logic program shown in [3]. In a pr/t net model, applying an action to search for a plan in a robot planning system can be simulated by firing transitions, and deriving a plan becomes answering whether a certain marking, the goal marking, is reachable from the initial marking in a pr/t net model. This problem, the reachability problem of a pr/t net, is of exponential complexity, and there are no known

1 This paper was supported by NON DIRECTED RESEARCH FUND, Korea Research Foundation, 1993

E. A. Yfantis (ed.), Intelligent Systems, 913–921.

efficient reachability test algorithms available.

The reachability test is in fact a graph search. Unfortunately the cost of solving this problem is exponentially expensive. AI search techniques[4, 5] employ the use of heuristics in order to improve the efficiency of a search process. Heuristics are used to guide the search process through the state space graph in a best-first manner by exploring the most promising states first. A heuristic function based on domain knowledge is used to estimate the distance from unexplored states to the goal state. The state which appears to be closest to the goal is the next state that will be explored. We have found several admissible and monotonic heuristic functions which can be used to solve the reachability test problem. We have implemented an iterative deepening A*[6] algorithm using our heuristic functions, and present the experimental results for our heuristics.

2. Algorithm to construct Pr/t net model for Robot planning

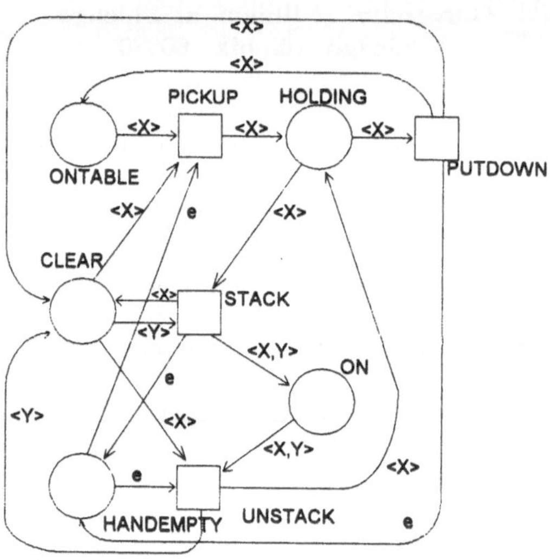

Figure 1. Pr/t net model for Block World planning problem.

We can construct a pr/t net model for a robot planning problem. Predicates used in the action definitions are represented as places. Each action is represented as a transition. Preconditions and delete conditions are represented as the input places of the transition. Add conditions and any predicate, which is an element of the precondition but not an element of the delete condition, are represented as output places of the transition. An algorithm to construct a pr/t net model for a planning problem is shown in Table 1. The pr/t net model for the Blocks World problem[7] is shown in Figure 1. The current state of the world is represented as a distribution of tokens. For example, if the following is true in the current state: ONTABLE(B), ONTABLE(C), CLEAR(A), CLEAR(C), HANDEMPTY and ON(A, B), then the distribution of tokens can be represented as a marking M_0: M_0 = ({<B>, <C>}, {<A>, <C>}, {e}, { }, {<A, B>}). For convenience, we designate a token, $Token_i$, in place p_i by writing $p_i(Token_i)$. For example, ONTABLE(<C>) and CLEAR(<C>) are two different tokens.

The robot's actions are simulated by firing transitions. Therefore, a plan is the firing sequence vector σ such that M_g is reachable from M_0 by σ, denoted by $M_0[\sigma>M_g$, where M_g is the marking for the goal state. A marking M_n is said to be reachable from a marking M_0 if there exists a sequence of firings that transforms M_0 to M_n.

Table 1. Pr/t net construction algorithm

Input : A set of actions
Output : A pr/t net G=<U, V, P, T, F, G>

Step 1 : Define U, V, P, T, G as sets, F as a multiset. All of these sets are any mutually disjoint sets and initially empty. Define i as an integer variable and initialize i := 1.

Step 2 : Scan i-th action and perform the following:

T := T ∪ $\{t_i\}$
P := P ∪ $\{p_{i1}, ..., p_{ij}\}$; (* p_{ij}'s are predicates in action i *)
V := V ∪ $\{x_{i1}, ..., x_{ik}\}$; (* x_{in}'s are variables in action i *)
U := U ∪ $\{k_{i1}, ..., k_{ig}\}$; (* k_{in}'s are constants in action i *)
If a predicate $p_{ij}(w_{i1},..., w_{im})$ is a precondition, then
F := F ∪ $\{(p_{ij}, t_i)\}$;
G := G ∪ $\{L(p_{ij}, t_i) = <w_{i1}, ..., w_{im}>\}$;
(* $L(p_{ij}, t_i)$ = e, if the arity of p_{ij} is zero. *)
If $p_{ij}(w_{i1},..., w_{im})$ is an add condition or precondition but not a delete condition, then
F := F ∪ $\{(t_i, p_{ij})\}$;
G := G ∪ $\{L(t_i, p_{ij}) = <w_{i1}, ..., w_{im}>\}$;
(* $L(t_i, p_{ij})$ = e, if the arity of p_{ij} is zero. *)

Step 3 : i := i+1; If no actions are left, then goto step 4, else goto Step 2.

Step 4 : If there are two or more identical elements in F (which corresponds to the situation of parallel arcs in the graphical representation), then modify the sets F and G in the following way:

(1) Convert F from a multiset into a set.
(2) Construct a new element from all the elements which have the same $L(p_{ij}, t_i)$ (or $L(t_i, p_{ij})$) in G by retaining the same $L(p_{ij}, t_i)$ (or $L(t_i, p_{ij})$) and forming the formal sum of all the $<w_{i1}, ..., w_{im}>$'s. Put the new element into G and delete all the old ones.

Step 5 : Delete any place with arc and arc label if it is corresponding to a predicate of arithmetic or a relational operation, e.g. plus(x, y, z), times(x, y, z). greater(x, y), etc. Then put an arithmetic expression, like z=x+y, z=x*y, or x>y, into the corresponding transition as the transition inscription.

---

3. Heuristic functions

The efficiency of the IDA* algorithm depends on its heuristic function. We have found five general purpose heuristic functions which can be used in solving the reachability test, and two heuristic functions which can be used in solving Blocks World problem.

3.1. Heuristic h_0
As a heuristic function with zero knowledge, we define h_0 which always returns 0. Heuristic h_0 is obviously admissible and monotonic. Since h_0 always return 0, IDA* using h_0 becomes a DFID search.

3.2. Heuristic h_1
Computing h_0 is very simple but unfortunately it does not have any heuristic power. As an improved heuristic function, we define h_1. Given the goal marking, M_g and the current marking M_i, h_1 first computes the difference between M_g and M_i. The difference between M_g and M_i is denoted by $M_d = M_g - M_i$, and is defined by the sets of tokens which appear in M_g but not in M_i. The heuristic value, $h_1(M_i, M_g)$ is defined to be 0 if there exists a transition which has output places to all the places corresponding to non-empty elements of M_d, otherwise it is 1.

3.3. Heuristic h_2
Given markings M_i and M_g, consisting of tokens g_1, g_2, ..., let $d^*(M_i, g_i)$ be the approximate distance from M_i, to g_i, where g_i is a token which appears in M_g. Ideally, $d^*(M_i, g_i)$ should be the minimum number of transitions to be fired from M_i to produce g_i. However, computing the exact number of transitions in an optimal path is very time consuming. Therefore we define:

$d^*(M_i, g_i)$ = the number of transitions to be fired from M_i to produce g_i, ignoring conflicts between transitions.

Before defining h_2, we need to define h_2^* as:

$$h_2^*(M_i, M_g) = \mathrm{Max}(d^*(M_i, g_1), d^*(M_i, g_2), \ldots, d^*(M_i, g_n)),$$

where n is the number of tokens that appear in M_g. Implementing this function for a pr/t net is much more complicated than the impression readers might have from this example. Evaluation of h_2^* involves computing subgoals for every goal token. Computing a subgoal requires considering a unifier. Thus, h_2^* is a very time consuming function.

Theorem 1. Heuristic h_2^* is admissible. Proof is omitted.

Theorem 2. Heuristic h_2^* is monotonic, i.e., $c(M_i, M_j) + h_2^*(M_j, M_g) \geq h_2^*(M_i, M_g)$, where $c(M_i, M_j)$ is the real cost to reach M_j from M_i. Proof is omitted.

By combining h_1 with h_2^*, we define h_2:

$$h_2(M_i, M_g) = h_1(M_i, M_g) + h_2^*(M_i, M_g).$$

Recall that if h_1 is 1 then the goal marking cannot be reached by one transition firing, and d^* is an approximation of the cost to produce one goal token. Therefore, h_2 is admissible and monotonic.

3.4. Heuristic h_m

Suppose the task assigned to a robot is to fetch box A and box B. Let a1 denote the cost of bringing box A, and b1 denote the cost of bringing box B. The cost to fulfill the task of fetching both boxes A and B is obviously not less than a1 or b1. Thus, Maximum(a1, b1) can be used as an underestimate of the cost to fulfill the whole task. Similarly, given a pr/t net with a goal marking M_g consisting of goal tokens g_1, g_2, ..., in certain places, the cost to achieve M_g will not be less than the cost of achieving the most expensive goal token. Let the optimal cost of achieving g_i be denoted by $|s_i|$, which is the minimum number of transitions that must be fired to produce g_i. Then, the maximum value among $|s_i|$ is less than or equal to the cost of achieving M_g, and it can be used as an underestimate of the cost to achieve the goal M_g. Heuristic function h_m is to calculate the maximum value among $|s_i|$:

$$h_m(M_i, M_g) = \mathrm{Max}(|s_1|, |s_2|, ..., |s_n|),$$

where M_i is the current marking, and n is the number of tokens in M_g.

Theorem 3. Heuristic function h_m is admissible, i.e., $h_m(M_i, M_g) \le c(M_i, M_g)$ where $c(M_i, M_g)$ is the cost of the optimal firing sequence σ by which M_g is reachable from M_i.

Proof: This follows from the fact that $|\sigma| \ge |s_1|$, $|\sigma| \ge |s_2|$, ..., and $|\sigma| \ge |s_n|$.

Theorem 4. Heuristic function h_m is monotonic, i.e., $c(M_i, M_j) + h_m(M_j) \ge h_m(M_i)$.
Proof is omitted.

Computing h_m is as hard as solving the planning problem itself, and it cannot be used in practice.

3.5. Heuristic h_3

In this section, we are introducing our last general purpose heuristic function. This function is quite accurate and fast. The idea of this function is based on the fact that the maximum number of tokens a certain place can obtain by one firing can be found by examining its input arc's labels.

Let us define a few terms which we need in describing h_3. Given M_i and M_g, let $M_d = M_g - M_i$ be a marking of tokens which appear in M_g but not in M_i. For every place, let the number of corresponding tokens appearing in M_d be denoted by c_1, c_2, c_3, ..., and so on. Let each one of $a_1(p_i)$, $a_2(p_i)$, $a_3(p_i)$ ... denote p_i's input arcs such that $a_j(p_i)$ is different from $a_k(p_i)$ if j is not the same as k. Recall that L is defined in Section 2 as a function which returns the label of a given arc. The cardinality, or the size of a label l_i is denoted by $||l_i||$, and designates the number of constants and variables constituting the label, l_i. For example, if l_i is <X>, then $||l_i||$ is 1, whereas if l_i is <X>+<Y>, then $||l_i||$ is 2. Now it is time to define a function, ub_i, which returns the maximum number of tokens a given place, p_i, can possibly obtain by one firing:

$ub_i = \mathrm{Max}(||L(a_1(p_i))||, ||L(a_2(p_i))||, ... ||L(a_n(p_i))||)$, where n is the number of input arcs of p_i.

The number of tokens in M_d, denoted by c_1, c_2, ..., and so on are the number of tokens we have to put at p_1, p_2, ..., and so on. Thus c_i/ub_i is a least bound of the number of firings we need to put c_i tokens at the place p_i. We define h_3^* as the maximum value among the divisions:

$h_3^*(M_i, M_g) = Max(c_1/ub_1, c_2/ub_2, ..., c_n/ub_n)$, where n is the number of places in the net.

Theorem 5: Heuristic function h_3^* is admissible.
<Proof> omitted.

We define h3 recursively using h_3^* as follows:

$$h_3(M_i, M_g) + g(M_i) = h_3^*(M_i, M_g) \quad \text{if } M_i = M_0,$$
$$= Max[\ h_3^*(M_i, M_g) + g(M_i),\ h_3(parent(M_i), M_g) + g(parent(M_i))\ ], \text{ O.W.}$$

where, $g(M_i)$ is the real cost to reach M_i from M_0.

Theorem 6: h3 is monotonic and admissible.

3.6. Heuristic h_4

Our heuristic function h_4 is for the blocks world problem and defined as follows: Given M_i and M_g, compute $M_d = M_g - M_i$. For every place, let the number of corresponding tokens appearing in M_d be denoted by c_1, c_2, c_3, ... Let Max(c_1, c_2, ...) be denoted by c. Then, heuristic function h_4 is defined as:

$$h_4(M_i, M_g) = c * 2 - 1.$$

3.7. Heuristic h_5

When the goal is to pile up blocks, many ONs, such as ON(A, B) and ON(B, C), appear in the goal state description. A predicate appearing in the goal state description is called a subgoal. If two different subgoals share one box(like B in this example: ON(A, B) and ON(B, C)) as a constant term, then one subgoal of them must always be achieved first. Fact 1 states which one should be done first.

FACT 1: Let A, B and C be arbitrary boxes. If both ON(A, B) and ON(B, C) appear in a goal state description, then ON(B, C) must be achieved earlier than ON(A, B) in an optimal plan. Proof is omitted.

We write Done-earlier(ON(B, C), ON(A, B)) to denote the relationship between ON(A, B) and ON(B, C) mentioned in FACT 1. Done-earlier is a transitive relation. The fact that the relation Done-earlier is transitive is stated in FACT 2.

FACT 2: ON(A, B), ON(B, C) and ON(C, D) are subgoals, then ON(C, D) must be done earlier than ON(A, B). Proof is omitted.

Let us define Done-later as the reverse relation of Done-earlier, i.e. Done-later(ON(A, B), ON(B, C)) if and only if Done-earlier(ON(B, C), ON(A, B)).

Further, let us define reflexive and transitive relations, Above and Below. Every box is above and below itself, i.e. Above(A, A) and Below(A, A). If ON(A, B), then Above(A, B) and Below(B, A). If Above(A, B) and Above(B, C) are true, then Above(A, C) is true. Since every two boxes in a pile has the relation of Above, <Boxes in a pile, Above> is a linearly ordered set. Let us call the maximum element of this set Top, and the minimum one Bottom.

Now, let us consider Done-earlier and Above relations together. Suppose Done-earlier(ON(C, D), ON(A, B)) is true for two arbitrary subgoals ON(C, D) and ON(A, B). If Below(A, D) or Below(B, D) while ON(C, D) has been achieved, then in order to achieve ON(B, C) and ON(A, B) later, we have to delete ON(C, D) somewhere. Therefore, an optimal plan never performs an action to achieve ON(C, D) when Below(A, D) or Below(B, D) is true. That is we have to clear B and clear A before achieving ON(C, D) in this case. This is FACT 3.

FACT 3: If Done-earlier(ON(C, D), ON(A, B)) is true for two arbitrary subgoals ON(C, D) and ON(A, B), and Below(A, D) or Below(B, D) is true in the current state, then an optimal plan achieves CLEAR(A or B) before achieving ON(C, D).

So far we have not considered a situation where a box is being held. Now let us consider what to do with the box being held. Acording to the definition of our Blocks World, only one box can be being held. We want to stack this box somewhere only if this contributes to achieving the goal, otherwise we want to put down the box on thc table because a box on the table never hinders the robot from doing anything, and a box on the table with clear top is ready to be manipulated by the robot. So, our mission is just to identify the case where stacking the box, x, on somewhere contributes to achieving the goal. The following is the necessary and sufficient condition for STACK(x, y) to contribute to achieving the goal when HOLDING(x) is true.

1. ON(x, y) is a subgoal. and
2. Every subgoal, a, that satisfies Done-earlier(a, ON(x, y)) has been achieved. and
3. CLEAR(y) is true. and
4. If Above(k, x) in the goal, then Below(k, y) is not true in the current state.

These are necessary and sufficient conditions because of:
1: obvious, 2: FACT 2, 3: obvious, 4: FACT 3.

An algorithm to calculate the cost to deal with the box, x, in the hand is shown in Table 2.

<Table 2> An algorithm to deal with the box, x, in robot's hand.
Unholding-cost(M_i, M_g)

If the above 1, 2, 3 and 4 are true, then do STACK(x, y) and increase cost by 1. O.W. PUTDOWN(x) and increase cost by 1.

Considering FACT 2, FACT 3 and the fact that PUTDOWN(x) is the best choice whenever stacking x on something does not contribute achieving the goal, we can exactly estimate the cost to achieve the goal from the current state. (Restriction: goal must be making one pile, current state can be arbitrary) Heuristic function h_5 is an algorithm to exactly estimate the cost and it is shown in Table 3.

<Table 3> Heuristic function h_5: Arguments are M_i and M_g.

```
 1. cost = 0;   // initialize //
 2. Construct GoalArray  // GoalArray is an array representation of M_g  //
 3. Construct CurrentArray  // Two dimensional Array representation of M_i  //
 4. If GoalArray is a part of CurrentArray, then return cost.
 5. If HOLDING(x), then call Unholding-cost(M_i, M_g).
 6. Find i, j, such that (bottom = GoalArray[0]) == CurrentArray[i, j].
 7. For(k=0;  k<j; k++)  {
      if(CurrentArray[i, k] is an element of GoalArray),
          then {  Add the cost to clear CurrentArray[i, j] and
                  the cost to putdown(CurrentArray[i, j]);
                  break;     }   // end of then //
    } // end of for  //
 8. Find again i, j, such that bottom == CurrentArray[i, j].
                                         // Skip achieved subgoals //
 9. for(k=0; GoalArray[k] == CurrentArray[i, j]; k++, j++);
10. For every box from CurrentArray[i, j] to the top,
         putdown the box and increase cost by 2.
11. for(k=j;  Untill GoalArray[k] reaches to Top; k++)  {
         locate GoalArray[k] at CurrentArray
         if it is not clear
                 putdown the boxes above it and increase cost by 2 for every box;
         cost=cost+2;
         Move this box to CurrentArray[i, j], j=j+1;
    }  // end of for //
12. return(cost);
```

4. Test results

In order to verify the heuristics power, we have implemented IDA* using the heuristic functions introduced in Section 3. Our experiments were done on the pr/t net shown in Figure 1. Even though the number of nodes explored by h_2 is much less than the number of nodes explored by h_0 and h_1, the amount of CPU time spent by h_2 is much greater than the amount of time spent by h_1 and h_3. In fact, h_2 spends more than h_0 which is our zero knowledge heuristic. This result confirms that our heuristic function must be simple to calculate.

The IDA* using h_3 can solve 10 blocks problem. Since h_5 calculates the exact cost of an optimal solution, we can just cut off all the nodes whose h_5 value is greater than the h_5 value of the initial marking, and we call this strategy Straight Forward. CPU times spent by Straight Forward is shown in figure 2.

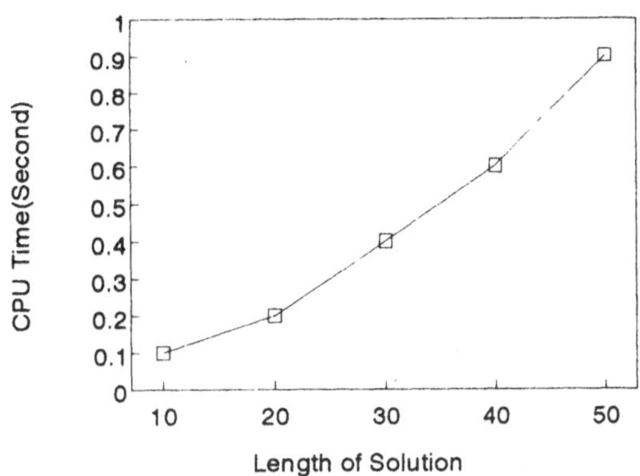

Figure 2. CPU time vs. length of solution for Straight Forward.

5. Conclusion

We have discussed a few general purpose and a couple of special purpose admissible and monotonic heuristic functions. These functions have been tested using the IDA* algorithm. The test results show that we can reduce both CPU time and the number of nodes explored by using a heuristic function. Our heuristic function h_3 is general purpose and as powerful as that IDA* equipped with it can solve 10 blocks world problem. Our h_5 is special purpose and can be used for only piling up blocks problem. It is an accurate function, and using it we can solve any piling up blocks problem.

6. References

[1] H.J. Genrich and K. Lautenbach, "System modeling with high-level Petri nets," *Theoretical Computer Science*, 13, pp. 109-136, 1981.

[2] J. A. Robinson, "A machine-oriented logic based on the resolution principle," *JACM*, 12(1), pp. 23-41, 1965.

[3] T. Murata and D. Zhang, "A predicate-transition net model for parallel interpretation of logic programs," *IEEE Trans. on Software Engrg*, Vol. 14, No. 4, pp. 481-497, April 1988.

[4] J. Pearl, *Heuristics*, Reading, MA: Addison-Wesley, 1984.

[5] P. Hart, R. Duda and B. Raphael, "A formal basis for the heuristic determination of minimum cost paths," *IEEE Trans. on SSC 4* (1968), pp. 100-107.

[6] R.E. Korf, "Depth-first iterative-deepening: An optimal admissible tree search," *Artificial Intelligence* 27, pp. 97-109, 1985.

[7] M. R. Genesereth and N. J. Nilsson, *Logical Foundations of Artificial Intelligence,* Morgan Kaufmann, 9187.

RECOGNITION OF HAND PRINTED DIGITS USING MULTIPLE PARALLEL METHODS

J. R. PARKER
Laboratory for Computer Vision
Department of Computer Science
University of Calgary
Calgary, Alberta, Canada

Abstract
Given the number and variety of methods offered for hand printed character recognition, it may very well be that there is no single method that can be called the 'best'. Many diverse algorithms each have strengths and weaknesses, good ideas and bad. One way to take advantage of this variety is to apply many methods to the same recognition task, and have a voting scheme to merge the results. The weaknesses should more or less cancel out, giving high recognition rates under many sets of conditions. Here, a voting scheme is proposed and used to join the results from three hand printed digit recognition algorithms.

I. Introduction

The recognition of hand printed characters has been the subject of a great deal of research due to the interesting nature of the problem, and to the enormous utility of a solution[1,2,5,7,10, 20]. One reason for this is that a great deal of information is in hand printed form, and this will continue to be true. The problem is difficult because each individual writer produces a unique set of characters *each time they write.* The characters are usually similar enough to others, and to the standard alphabet, that recognition by humans is possible - we have a great capacity to ignore the variations and concentrate on the common elements. To a computer the differences are large enough to pose a difficult problem.
A wide variety of methods have been devised in a effort to solve the problem, and success rates vary greatly. Syntactic, structural, and statistical methods, which are quite different in approach, have all had some degree of success. Given the current availability of CPU power and memory, it may be interesting to ask: 'why not use them all'? Or at least, why not combine a set of diverse methods, running in parallel, to gain the advantages of each method[3,8,17]? The result should be successful over a wider range of inputs than any individual method would be.

As an example of such a system, a multiple/parallel scheme is presented here. Three different systems for recognizing hand printed digits are combined with a voting scheme to give a system with a higher recognition rate than any of the components. Digits were chosen as a useful subset of the general problem - there are only ten possible results, and a successful system could be applied to zip codes, serial numbers, identification numbers of many kinds, and commercial markings and codes.

2. Method 1 - Use of Vector Templates

Template matching techniques in many forms have been applied to the problem of recognizing hand printed digits using a computer[5,11]. The basic idea is that each digit has a particular

E. A. Yfantis (ed.), Intelligent Systems, 923–931.

shape that can be captured in a small set of models, usually stored as raster images. An incoming (unknown) digit, also in raster form, is compared against each template, and the one that matches most closely is selected as belonging to the same digit class as the unknown. This can work very well for machine printed characters, which have the property of being very uniform, but does not generally work for hand printed characters, which show a very high degree of variability. One possibly good idea is to represent the template digits as vectors[23]. This is done in some computer typography systems, where the fonts are stored in vector form[9]. For this application vectors that form the skeleton of the characters will be used rather than the outline, which would be used in a typography system.

The templates are stored as sets of four integers: the starting and ending row and column on a standard grid. All templates have the same size: 10 by 10; this means that all coordinates in any template have an integer value between 0 and 9 inclusive. Given a scale and rotation, then, all templates in the collection would be modified in a consistent way. An of a vector template appears in Figure 1, which shows a template for the digit '2'. Figure 1a shows the vector coordinates, which were obtained manually from a line drawing of a '2' on a 10x10 grid. This is drawn as lines (Figure 1b) using the original scale, and also using a new scale: 20x10 (Figure 1c). Templates can be created either with a pencil and graph paper, or by generating templates from extracted digit images. In either case, a library of templates is needed before recognition can proceed. This corresponds to a 'learning' phase for the algorithm.

The procedure for matching a vector template is a little more involved than for the usual raster variety. An incoming image is first pre-processed in any desired fashion[4,6,15,21] and then is thresholded[12, 14]. The width of the lines in the image is then estimated using horizontal and vertical scans. A histogram containing the widths of the black portions of the image on all slices is produced, and the mode of this histogram has been found to be a close enough approximation to the actual line width. While the line width is being computed, the actual extent of the digit image is also found so that the templates can be scaled. This is saved as the coordinates of the upper left and the lower right pixels of the bounding box.

Now a raster template is created from a vector one. First, the template is scaled in the X and the Y directions independently (the same scale factors can be applied to all templates). The factors include an adjustment that results in a correct scaling accounting for the thickness of the line. Now the template vectors are *drawn* into an otherwise clear image the same size as the input

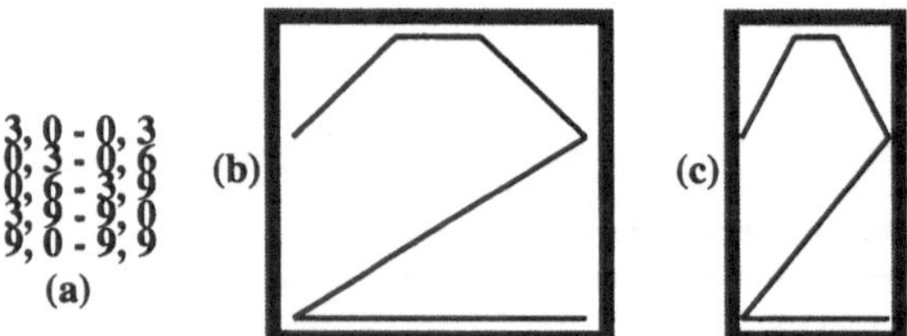

Figure 1 - An example vector template. (a) The coordinates of the vector endpoints for a '2' template. (b) Vectors drawn on a 10x10 grid. (c) Drawn on a 20x10 grid.

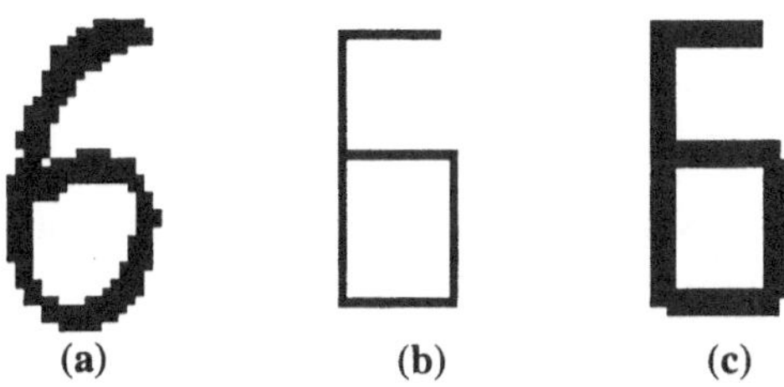

Figure 2 - Matching using a vector template. (a) Input image (to be matched) (b) Scaled vector template (c) Thickened vector template (Not a good match, here)

image, producing an initial raster template that represents the scaled skeleton. Finally, each pixel is 'grown' equally on all sides to give a line width comparable to that found in the input image. The result is a raster template with some similar properties to those found in the input image. Figure 2 illustrates the process of generating a raster template from a vector one.

The first step in matching the input to the template is to locate those pixels that are black in both images. These have a distance between them of zero, and are ignored in future processing. Next, each black pixel in the image has its nearest corresponding pixel in the template located and marked. The 8-distance between these pixels is noted, and a sum of these distances is computed. After all image pixels have been assigned corresponding pixels in the template the total distance is an initial measure of similarity. This is repeated in the reverse direction, computing the sum of the distances between pixels in the template and the corresponding pixel in the input, and the measure of similarity is the sum of the two total distances. This can be normalized to a per-pixel distance and stored as the measure for the numeral having the same class as the template. The

Figure 3 - The final stages of the template match. (a) Pixels that overlap between the template and the input. (b) Distance map between template and image (darker pixels are farther apart)

class having the smallest such measure over all templates is chosen as the class of the input digit image. Figure 3 shows the overlapping pixels and a distance map for the example begun in Figure 2.

The results are, at this time, only at an acceptable level. Template scaling and drawing takes 0.019 seconds per template, which amounts to 0.55 seconds for all of the 28 different templates. This could easily be done in parallel which would reduce the real time needed. Pixel pairing and matching takes and average of 0.015 seconds per template, or 0.84 seconds input image. The average time to recognize a digit without using parallel processing is 4.4 seconds on a Sun workstation.Recognition rates are relatively good, although at the time of publication we have had only a little data available to us for testing; Data was obtained from various sources, including ICDAR CD rom [22] and scanned locally obtained documents. The results are:

Table 1: Vector Template Digit Recognition Rates

0	1	2	3	4	5	6	7	8	9
99%	94%	98%	96%	94%	92%	90%	93%	95%	92%

3. Method 2 - Use Of Convex Deficiencies

While most character recognition systems are concerned with the pixels belonging to the characters themselves, there are good arguments to be made for analyzing the size, shape, and position of the background regions surrounding the character image. Certainly the number and position of the *holes* has been used - an '8', for example, has two holes, one in the upper part and one in the lower part of the character image, and a '0' has one in the middle. However, there are other such features that might be used to classify images - for example, a numeral '2' has a left-facing concave region in the top half of the image and a right-facing one in the bottom half. A more complete analysis of these *convex deficiencies*[13] may permit the development of a classification scheme based on the background regions.

For use with digital character images a relatively crude scheme has been developed. From each background pixel in an input image we attempt to draw a line in each of the four major directions (up, down, left, right). If at least three of these lines encounter an object pixel, then the original pixel is labelled with the direction in which an object pixel was *not* encountered - this is called the *open* direction. If none of the four directions are open then the pixel concerned is part of a hole, and is labelled with a zero. Figure 4 shows the direction labels, and illustrates the process of locating and labelling the convex deficiencies. Some of the holes identified in this way are not holes at all, but are part of a convoluted region in the image. A real hole will have boundary pixels that all belong to the object - if this is not true then the hole is false, and it is converted into a region labelled by its most common non-object neighbor.

After all of the regions are labelled, they are counted and measured. Very small regions, relative to the number of pixels in the object, are ignored, and the largest four regions are used to classify the image. Sometimes a relatively simple relationship exists. For example, 99% of all zeros can be identified by the large central hole and lack of other convex deficiencies. The next more com-

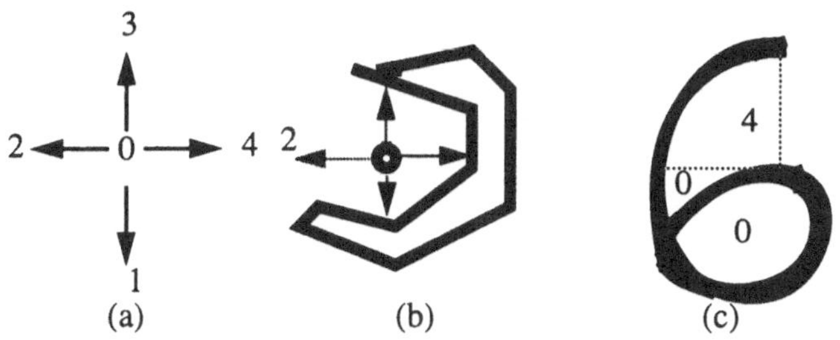

Figure 4 - Locating convex deficiencies. (a) Codes for directions. (b) Background pixels are tested to find open directions (this is a 2) (c) Regions found for a '6'. One of the holes is not real, and will be connected to the 4 region.

plex scheme uses relative positions of the regions; for example, an '8' has two holes, one above the other, a left-facing region on the left of the two holes, and a right facing region on the right of the holes. The most complex schemes require shape information in addition to size and position. As an example, some '7' digits and some '3' digits both have a large left-facing region on the left side of the image. However, this region is convex for the '7' but non-convex for the '3', as seen in Figure 6. This fact can be used to discriminate between some '7's and '3's.

Here is a sample set of digit descriptions in terms of convex deficiencies. It is not complete; sometimes a dozen different descriptions are needed for a single digit. Still, it should serve to give the flavor of the kind of description that will work.

Digit	Open Direction	Size	Location	Shape
0	0	middle	center	convex
1	None	---------	-------------	----------
2	2	middle	upper left	----------
	4	middle	lower right	----------
3	2	large	left-of	non-convex
	4	small	right	----------
4	3	middle	upper	----------
5	4	middle	upper right	----------
	2	middle	lower left	----------
6	4	middle	above	----------
	0	middle	lower	----------
7	2	middle	left	----------
8	0	middle	upper	----------
	0	middle	lower	----------
	2	small	left	----------
	4	small	right	----------
9	0	middle	not-centre, above	----------
	4	middle	lower	----------

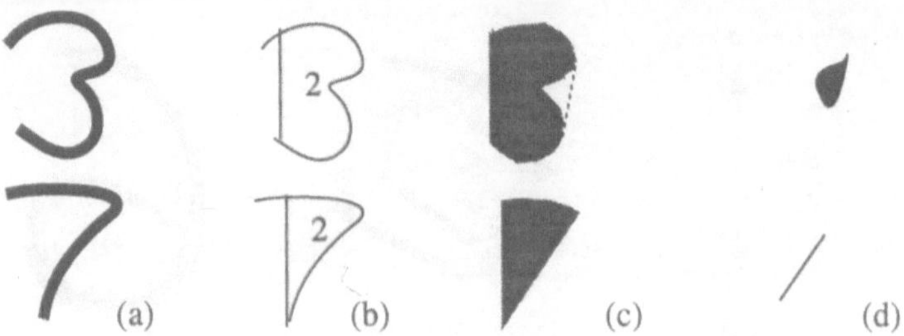

Figure 5 - The use of shape. (a) Two digits. (b) Both have large 2-regions (open left) (c) Find the convex hull of these regions. (d) Convex deficiencies = hull - object. Top region is not convex (large deficiency) while the lower one is (small deficiency)

The recognition rate achieved using this method is acceptable, averaging about 94%. In detail:

Table 2: Recognition Rate - Convex Deficiencies

0	1	2	3	4	5	6	7	8	9
99%	94%	98%	96%	94%	90%	90%	93%	95%	92%

This approach has the advantage of not requiring a thinning step, although smoothing the outline does help, and a closing step may improve the rates for '8', '6', and '9'.

4. Method 3 - Properties of the Object Outline

In a couple of interesting articles, Shridhar et al [8,18,19] describe a collection of topological features that can be used to classify hand printed numerals. Most of these features are properties of the outline, or *profile*, of the numeral. For instance, a digit '8' might be described as having a smooth profile on both the left and the right sides, and as having the width a minimum in the center region. Not all hand printed '8' digits would be recognized by this description, and certainly some other digits might also have this description; the idea is to provide a sufficient number of descriptions for each digit that a high recognition rate can be achieved.

Figure X shows the left and right profile of a sample digit '8'. The profile is found by first scaling the input image to a standard size, then recording for each row the location of the first and last object (black) pixel. The distance of the object from the left hand side of the image defines the left profile, and the right profile is defined similarly for the right hand side. Quite a large number of useful measures can be based on the profiles, including:

- W_{max}, the maximum width of the digit.
- R, the ratio of height to maximum width.

- The locations of the maximum and minimum in both the left and right profiles (L_{max}, L_{min}, R_{max}, R_{min}).
- The maximum and minimum in the first difference of the left and right profiles over a given set of rows (these locate discontinuities).

This set of features is not comprehensive. In all, 48 features are used and others could be defined. (See [8] for a complete list and description)

In the training phase all 48 features are computed for each sample numeral and a feature vector is created in each case. The features are binary, being either TRUE or FALSE; for example, feature number 43 is TRUE if the width of they character at row 20 is greater than or equal to the width at row 40 (W(20) >= W(40)). Then all of the resulting bit strings for each digit are searched for common elements, and the features in common for each digit class are stored in a library. Matching is performed by extracting the profiles of the input image and measuring and saving the bit string (feature vector) that results. This string is matched against the common elements of the templates - this is obviously very fast, since only bit operations are involved. A perfect match of a library bit string against an input string results in the corresponding digit class being assigned to the input digit. The results from this method are, again, only acceptable at this stage. The recognition rates for our samples are:

Table 3: Object Outline Method - Recognition Rates

	0	1	2	3	4	5	6	7	8	9
% Right	94	95	96	100	95	100	84	94	90	94
% Error	1	4	1	0	5	0	10	5	4	6
% Reject	5	1	3	0	0	0	6	1	6	0

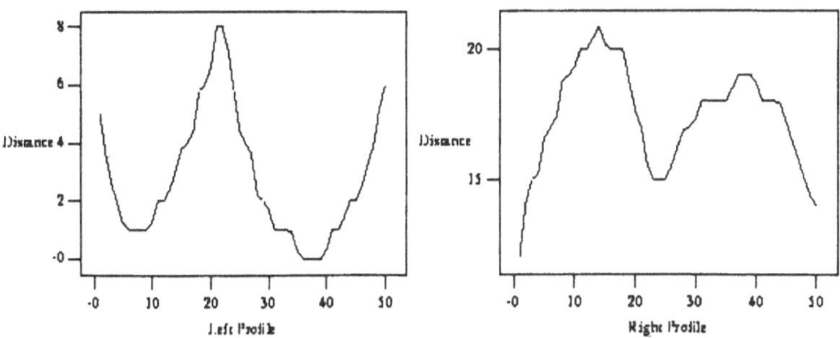

Figure 6 - The left (a) and right (b) profile of a numeral '8'

5. Merging The Multiple Methods

The three numeral recognizers can run either sequentially or, if the hardware permits, in parallel. The latter enjoys the advantage of a quicker real execution time, although more computation effort is expended. There is an effort involved in communication of image data and results along communication links that is often ignored. An object-oriented distributed processing kernel has been devised by our lab, and this permits the programs to be sent to remote workstations to be executed; all of the methods can easily be run in parallel. It may be desirable to have more than three methods, perhaps as many as ten. Still, the basic idea can be demonstrated using only the algorithms mentioned so far. If each of the methods gets a single vote, rejections count as abstentions, and an image is classified by a clear majority of votes, the results are:

Table 4: Multiple/Parallel - Recognition Rates

	0	1	2	3	4	5	6	7	8	9
% Right	99	98	99	98	97	100	98	97	96	97
% Error	0	0	0	0	0	0	1	2	1	1
% Reject	1	2	1	2	3	0	1	1	3	2

This amounts to an overall error rate of 0.5%. One would think that this would only improve as more methods were added to the system. We have also considered a weighted voting system, where the vote coming from each method was weighted with the probability (based on previous runs on test images) that the digit in question was correctly classified. To experiment with these ideas, a simple simulation was devised that would recognize digits with pre-assigned probabilities in the range 0.9-1.0; this is the same range as for the methods discussed previously. The simulation tested both simple and weighted voting for up to 20 different parallel methods. The somewhat surprising result was that, for both weighted and non-weighted votes, there was no improvement after using five different methods. The simulations did not take into account the observation that an image that fails using one method is more likely to fail using others. Future efforts will add new methods to the system, and test the hypothesis that robustness is improved as a result.

This research has been supported by the Natural Sciences and Engineering Research Council of Canada.

6. References

[1] Blackwell, K.T. et al, **A New Approach To Hand-Written Character Recognition**, *Pattern Recognition*, Vol 25 No. 6, 1992.

[2] Brown, R.M., Fay, T.H., and Walker, C.L., **Handprinted Symbol Recognition System**, *Pattern Recognition*, Vol. 21 No. 2, 1988. pp 91-118.

[3] Cai, Z., **A Handwritten Numeral Recognition System Using a Multi-microprocessor**, *Pattern Recognition Letters*, 12, 1991. pp 503-509.

[4] Casey, R.G., **Moment Normalization of Handprinted Characters**, *IBM Journal of Research and Develpment*, Sept. 1970. pp 548-557.

[5] Gader, P. et al, **Recognition of Handwritten Digits Using Template and Model Matching**, *Pattern Recognition*, Vol. 24 No. 5, 1991. pp 421-431.

[6] Holt, C.M. et al, **An Improved Parallel Thinning Algorithm**, *CACM* Vol 30 No. 2, 1987. pp 156-160.

[7] Huang, J.S. and Chuang, K, **Heuristic Approach To Handwritten Numeral Recognition**, *Pattern Recognition*, Vol. 19 No. 1, 1986.

[8] Kimura, F. and Shridhar, M., **Handwritten Numeral Recognition Based On Multiple Algorithms**, *Pattern Recognition*, Vol. 24 No 10, 1991.

[9] Knuth, D.E., **Computer Modern Typefaces", volume E of Computers & Typesetting**, *Addison-Wesley, Reading, MA*. 1986.

[10] Lam, L. and Suen, C.Y., **Structural Classification and Relaxation Matching of Totally Unconstrained Handwritten Zip-Code Numbers**, *Pattern Recognition*, Vol. 21 No. 1. pp. 19-31, 1988.

[11] Mori, S., Yamamoto, K., and Yasuda, M., **Research on Machine Recognition of Handprinted Characters**, *IEEE Transactions on Pattern Analysis and Machine Intelligence*, Vol. PAMI-6 No. 4, July 1984. pp 386-405.

[12] Otsu, N., **A Threshold Selection Method From Grey-level Histograms**, *IEEE Transactions on Systems, Man, and Cybernetics*, Vol. 9 No. 1, 1979. Pp 377-393

[13] Parker, J.R., **Practical Computer Vision Using C**, *John Wiley & Sons*, N.Y., 1994.

[14] Parker, J.R., **Grey Level Thresholding In Badly Illuminated Images**, *IEEE- PAMI*, Vol 13, No. 8, 1991.

[15] Parker, J.R. and Jennings, C., **Defining the Digital Skeleton**, *SPIE Vision Geometry I*, Boston, MA., 1992.

[16] Salkauskas, K. and Lancaster, P., **Curve and Surface Fitting, An Introduction**, A*cademic Press*, 1981.

[17] Suen, C.Y., et. al, **Computer Recognition of Unconstrained Handwritten Numerals**, *Proc. IEEE*, Vol. 80 No. 7, July 1992.

[18] Shridhar, M. and Badrelin, A, **Context-directed Segmenttation Algorithm for Handwritten Numeral Strings**, *Image and Vision Computing*, Vol. 5 No. 1, Feb, 1987.

[19] Shridhar, M. and Badrelin, A, **Recognition of Isolated and Simply Connected Handwritten Numerals**, *Pattern Recognition*, Vol. 19 No. 1, 1986.

[20] Srihari, S.N., **Recognition of Handwritten and Machine-printed Text for Postal Address Interpretation**, *Pattern Recognition Letters*, 14 (1993). pp 291-302.

[21] Zhang, Y.Y. and Suen, C.Y., **A Fast Parallel Algorithm for Thinning Digital Patterns**, *CACM* Vol. 27 No. 3, 1984. pp 236-239.

[22] Japanese Technical Committee for Optical Character Recognition, **ETL Character Database**, *ICDAR '93*, Tsukuba, Japan, October, 1993.

[23] Parker, J.R., **Vector Templates and Handprinted Character Recognition**, *University of Calgary Department of Computer Science Research Report* #94/532/01, January 1984.

DESIGNING FAIL–SOFT SYSTEMS FOR DISTRIBUTED COMPUTING

CONSTANTINE STIVAROS

Computer Science, Fairleigh Dickinson University, Madison, NJ 07940

Abstract

This paper is concerned with the design of fail-soft systems and a measure a reliability that characterizes gracefully degradable systems. In an effort to maximize reliability, thus guarantee graceful degradation in a probabilistic manner, we try to optimize the interconnections that underlie a system along with the locations of its components that are subject to failure. A state (subsystem) of the system is operating if the currently operating nodes comprise a connected subsystem. The probability of the system being in a operating subsystem is the reliability measure of the system. We develop the criteria for optimally reliable systems, that maximize reliable operation for any given set of operating probabilities. These criteria for optimal configurations are applied in certain classes of systems such as centralized, a restricted class of multi-layered and in dense systems.

Keywords: systems reliability, fail-soft, combinatorial optimization, optimal graphs.

1 Introduction

The ability of a system to continue operating despite component failures constitutes its *graceful degradation*. One measure of such behavior is the enumeration of its *reliability* parameter that is directly dependent on the interconnections of the system's components. Graceful degradation characterizes *fail-soft systems* and plays a vital role in the development of highly dependable distributed systems.

Fail-soft systems are desirable in distributed processing of large scale systems; failures of processors may interrupt communications between the operating ones which would otherwise re–allocate the unfinished jobs among themselves. If functions can be distributed among several processors with a proper protocol, then the failure of some processors will not halt the system, but will only slow it down. For example, if we have ten processors and one fails, then each of the operating nine processors must pick up a share of the work of the failed processor. Thus, the entire system runs only 10 percent slower (90% reliable), rather than shutting down, providing service proportional to the level of non–failed hardware.

A similar scenario crops up in communication networks that deploy packet-switching techniques; failures of certain nodes may disconnect the system, resulting in loss of communication paths. There is also a number of design applications in the defense industry such as in space-borne systems and in communication of missile sites.

E. A. Yfantis (ed.), Intelligent Systems, 933–944.

In this paper we define a system to be "reliable" if it has the ability to degrade gracefully. We attempt to show how the modeling of the underlying topology of the system can influence positively such behavior. We study a very early stage of the design, trying to embed reliability in the system's interconnections and analyze the requirements for this. There are numerous reliability models, evolved from probabilistic graphs which study related reliability issues; this model is concerned with the state of the system when failures have rendered some nodes inoperable while links are considered perfect. In trying to keep the operating parts as connected as possible we select topologies/interconnections that can best tolerate failures and guarantee continuing performance.

The design aspect of this reliability model is concerned with the following problem: Suppose that out of n locations, it is only feasible to establish perfectly reliable communication links between only e pairs of locations. Select these pairs, or links, so that the reliability of the system is maximized when nodes fail. Given their operating probabilities, the selection of these edges may be done in the following manner: first a topology for G is chosen that is optimal for "uniform" failures; then the locations of the components at the nodes are determined based on a "best" allocation function.

We examine three types of systems common in the modeling of operating systems and communications networks with varying bulk of interconnections: Centralized, Multi-layered and Dense. These types of systems although have been traditionally associated with distributed communication, have not been examined under this measure of reliability. In the next two sections, we present the defining aspects of the measure, and we obtain conditions for *maximally reliable systems* or *max-R* systems. In section 4 we prove the requirements for local optimality of the max-R property ("*locally max-R*") of the three systems. In section 5, their *max-allocation* function is derived and in section 6 we characterize their max-R behavior.

We expect the reader to be familiar with graph-theoretic terms, such as class of graphs, induced subgraph, connectivity, trees, matching.

2 Modeling the System

A system consisting of nodes that are interconnected with links is identified by an undirected graph G. Let $Sys(n, e)$ be the <u>class</u> of systems whose underlying graphs have n nodes and e edges. The edges operate perfectly but the nodes fail with independent and known probabilities. We assume that when a node has failed all of its connecting links become inactive and removed from the graph.

Consider a system of a graph $G \in Sys(n, e)$ with a node set $V(G) = \{v_1, v_2, \dots, v_n\}$ and a vector of n non-decreasing probability values $\vec{P} = (p_1, p_2, \dots, p_n)$. These are the operating probabilities for the system's components that will be placed at the nodes of the graph. This placement is done according to an **<u>allocation function</u>** π, a mapping from V to $\{1, \dots, n\}$ that allocates to node v the probability $p_{\pi(v)}$.

As was introduced, the reliability of the system characterizes its ability to degrade gracefully, measured by the connectedness of the operating nodes. Therefore, define the **<u>reliability</u>** of a system G given a vector $\vec{P}$ and allocation π, denoted as $R(G; \vec{P}; \pi)$, to be the probability that despite random node failures the operating nodes of G are connected.

We use the term operating subsystem of G, to denote any non-empty subset S of nodes that induce a connected subgraph of G and Θ to denote all such subsystems. The probability of an operating subsystem $S \in \Theta$ contributes $\prod_{v \in S} p_{\pi(v)} \prod_{v \notin S}(1 - p_{\pi(v)})$ to the reliability of G and thus, we can form an algebraic formulation for the reliability measure by enumerating all operating subsystems of G:

$$R(G;\vec{P};\pi) = \sum_{S \in \Theta} \prod_{v \in S} p_{\pi(v)} \prod_{v \notin S} (1 - p_{\pi(v)})$$

An explicit expression for a given large system is rather hard to derive; we often make use of the factoring formula which expresses reliability of a system in terms of the reliabilities of systems embedded in it. The factoring formula was first introduced in [8] for a similar study of networks. In this expression a node v of the system serves as the *factor* by considering its operating status. For a graph G and a set S of vertices, $G - S$ is the subgraph of G induced on the vertices of $G - S$. Also, the graph $G - v$ is obtained by removing v from G, and G/v is the graph obtained if we remove v from G and replace the subgraph induced by its neighborhood $N(v)$ in G by a clique. Assuming an allocation π that maps p_v to v and a vector $\vec{P}$ we refer to $R(G;\vec{P};\pi)$ as $R(G)$ for ease of notation:

Theorem 2.1 (Factoring formula) *Let p_v be the operating probability of node v of a system $G \in Sys(n,e)$. Then*

$$R(G) = (1 - p_v)R(G - v) + p_v[R(G/v) - \prod_{u \in N(v)} (1 - p_u)R(G - v - N(v))] + p_v \prod_{u \neq v}(1 - p_u)$$

The recursive structure of the factoring formula suggests an immediate exponential algorithm for the computation of $R(G)$. We use here the factoring formula to establish the optimality of certain systems.

There is a simple case where nodes operate with the same probability p, called the uniform case. Since any allocation yields the same result, we ignore π and denote the reliability as $R(G;p)$. The probability of an operating subsystem S is now $p^i(1-p)^{n-i}$, if $|S| = i$ and G has n nodes. Therefore, the expression for $R(G;p)$ is a polynomial in p:

$$R(G;p) = \sum_{i=1}^{n} s_i(G)p^i(1-p)^{n-i} \tag{1}$$

where the coefficient $s_i(G)$ is the number of connected induced subgraphs of G (operating subsystems) having exactly i nodes.

3 Conditions for optimal degradation

A system G is termed maximally reliable or max-R in $Sys(n,e)$ if there exists an allocation π such that for any given vector $\vec{P}$ of probabilities:

$$R(G;\vec{P};\pi) \geq R(G';\vec{P};\pi'), \text{ for all systems } G' \in Sys(n,e) \text{ and all allocations } \pi' \text{ under } \vec{P}$$

It is worth noting a few points here:

- a max-R system may not exist in all classes
- the allocation of $\vec{P}$ to the nodes of a max-R system has to be the best such.

Certainly, a max-R system has to remain optimal in the uniform case. In this case we call a system $G \in Sys(n, e)$ **locally max-R** if it maximizes the reliability in its class over any uniform vector or, if

$$R(G;p) \geq R(G';p), \text{ for all } G' \in Sys(n,e) \text{ and all } 0 < p < 1$$

It is clear that such a property of a system should be studied first since global optimality implies local. Two necessary conditions are known (see [8, 1, 4, 5]) for a locally max-R system; these can be derived directly from the coefficients of polynomial (1) depending on the range of p:

(a) when p is sufficiently close to 1, $R(G;p)$ is maximized only if G has the maximum possible connectivity κ in its class and the maximum value of $s_{n-\kappa}$ among all systems in its class with maximum κ. Such a system is called **super–κ**.

(b) when p is sufficiently small, $R(G;p)$ is maximized only when s_3 is maximized over $Sys(n, e)$. We call a system **super–3** if, $s_3(G) \geq s_3(G')$ for all $G' \in Sys(n, e)$.

Thus, two necessary conditions for a system to be locally max-R are that it be super–κ and super–3 in its class. For the characterization of locally max-R systems one has to first address the issue of systems that are both super–κ and super–3. It is the case though, that such systems do not always exist:

Theorem 3.1 *Locally max-R systems do not exist for all n and e.*

Proof: Consider the class $e = n$. The unique super–κ system is the cycle C_n on n nodes having $\kappa = 2$. However, the authors of [2, 3] show that the unique super–3 is the star system plus an edge, having $\kappa = 1$. Therefore, in this class there is no system that is both super–κ and super–3. □

4 Locally Max-R Systems

4.1 Centralized systems

The most economical system in terms of interconnections is the centralized or **star** system. It has a central node (*center*) and all other nodes are connected via direct links to it. We establish next the local optimality of a star system whose underlying graph is $K_{1,n-1}$.

Theorem 4.1 *The centralized system S, is the unique locally max-R system in the class $e = n-1$ (includes tree–systems).*

Proof: If $t_i(G)$ is the number of subgraphs of G that are trees with i nodes, then any induced connected subgraph with i nodes has at least one tree that spans these nodes:

$$s_i(G) \leq t_i(G) \leq \binom{e}{i-1} \text{ for all } i = 3, \ldots, n$$

Now consider the star S with n nodes and e edges, where $e = n - 1$. Since all edges of S are incident upon a node, every coefficient s_i in the polynomial of S realizes the above upper-bound. Therefore since each $s_i(S)$ is maximized, from polynomial (1) we conclude that S is locally max-R in its class. For uniqueness, it is enough to observe that any other system in the same class must have at least two independent edges, which implies that t_3 and s_3 of its polynomial are not maximum. □

4.2 Restricted multi-layered systems

Multi-layered topologies are common for systems that have several layers of nodes (processors, nets) and only the nodes in different layers are connected with links. Versions of this interconnection can also be found in VLSI circuits, usually known as *Channel* or *Match-box* [6].

A multi-layered system consists of partitions of nodes in a way that only nodes in different partitions are fully connected. Its natural model is a **complete multipartite** graph $K_{k_1,k_2,\ldots,k_r}$, a graph whose complement consists of a union of cliques $K_{k_1} \cup K_{k_1} \cup \ldots \cup K_{k_r}$.

A restricted multi-layered system has equal size partitions, or partitions with size–difference at most one between any two. The underlying restricted complete multipartite structures are then the **regular** (with $k_i = k_j$) and **almost-regular** graphs (with $|k_i - k_j| \leq 1$).

We prove next that these restricted multi-layered systems are locally max-R. For this, let G_1 and G_2 be the regular and almost–regular complete multipartite graphs:

$$G_1 = K_{k,\ldots,k},\ G_2 = K_{\underbrace{k,\ldots,k}_{r},\underbrace{k+1,\ldots,k+1}_{s}}$$

To prove this claim, we make use of the next lemma:

Lemma 4.1 *For any graph G with n nodes, let $\mu_i(G)$ be the number of disconnected induced subgraphs of G with i nodes. Then,* $\mu_i(G) \geq \frac{1}{i} \sum_{v \in V} \binom{n-1-d_v}{i-1}$

The proof of the lemma rests in the following observation: if v is a node of G then there are $n - 1 - d_v$ nodes that v has no edge to, with d_v being the degree of v. Then we can always form a disconnected induced subgraph in G with i nodes by selecting $i - 1$ non-neighbors of v along with v itself. Since there are $\binom{n-1-d_v}{i-1}$ such subgraphs from v only, for the whole graph we obtain the sum of the right-hand side. The lemma follows if one considers that the sum is minimum when the selected i nodes form an independent set, i.e. each subgraph is counted i times, once from each of its nodes. □

The local optimality of the restricted multi-layered systems follows from the next theorem:

Theorem 4.2 *The regular and almost-regular complete multipartite graphs are the unique locally max-R in their respective classes.*

Proof: Since $s_i(G) + \mu_i(G) = \binom{n}{i}$, for any graph G, we can rewrite the polynomial of (1) as:

$$R(G,p) = 1 - \sum_{i=1}^{n} \mu_i G)p^i(1-p)^{n-i}$$

Hence, maximizing each s_i is equivalent to minimize the μ_i's in the class of G. Consider first $G_1 = K_{k,\ldots,k}$. The only disconnected induced subgraphs of G_1 are of size $i \leq k$ and are formed by selecting i nodes from the same partition (independent set). Having r partitions with k nodes each, we have $\mu_i(G_1) = r\binom{k}{i}, i = 2,3,\ldots,k$ and $\mu_i(G_1) = 0$, for $i = k+1,\ldots,n$. From lemma (4.1), using the lowest bound for the μ_i's with $n = rk$ and $d_v = (r-1)k$, the above becomes:

$$\mu_i(G_1) \geq \frac{1}{i} rk \binom{k-1}{i-1} = r\binom{k}{i}$$

Therefore, the graph G_1 achieves the lowest bound for each μ_i in its class, for its degree sequence. In a similar argument we can show the same for the graph G_2.

In order to show that the degree sequences of G_1, G_2 result to the minimum-possible lowest bound for μ_i's in their classes, observe that if $d_i, d_j \geq 0$ with $d_i - d_j \leq 1$, then $\binom{d_i}{k} + \binom{d_j}{k} < \binom{d_i+1}{k} + \binom{d_j-1}{k}$. From this, we conclude that the bound of lemma (4.1) is minimized uniquely when all the degrees are equal or any two differ by no more than one (i.e. almost-equal); these are exactly the degree sequences of the graphs G_1 and G_2. Thus, the graphs G_1, G_2 minimize uniquely each μ_i coefficient, resulting to the unique locally max-R systems in their respective classes. □

4.3 Dense systems

We turn next to the classes of very "dense" systems, i.e. systems whose underlying graphs have number of edges close to the complete graph. The next theorem presents and proves the locally max-R systems having up to $\lfloor \frac{n}{2} \rfloor$ missing edges from the complete graph (K_n) or, the class $e \geq \binom{n}{2} - \lfloor \frac{n}{2} \rfloor$.

We define D_n to be the system in the above class with graph $K_n - M$, where M denotes a matching (independent set of edges).

Theorem 4.3 *The system D_n is the locally max-R system in its class.*

Proof. Since $s_1 = n$ and $s_2 = e$, it suffices to show that D_n maximizes the s_i coefficients of the polynomial for each $i = 3,4,\ldots,n$. An obvious upper bound for each s_i is $\binom{n}{i}$.

Consider the system D_n with its graph $K_n - M$, where $|M| \leq \lfloor \frac{n}{2} \rfloor$. All the nodes of $K_n - M$ have degree $n-1$ or $n-2$. Then each induced subgraph on 3 or more nodes is always connected, so that each $s_i(K_n - M)$ realizes the upper bound. Thus, D_n is locally max-R in its class. For uniqueness, suppose that $G \in (n,e)$ and $G \not\simeq K_n - M$. Then G has a node of degree $n-3$ or less, so that there exists a set of three or more nodes which induces a disconnected graph. Hence, at least one s_i of G is not maximized. □

5 Optimal Allocations

Searching for max-R systems the allocation issue has to be resolved first. Having the candidate topology G for max-R (the locally max-R system) we have to determine the best placement of the given component/probabilities to the nodes of G in a way that it maximizes its reliability. For a given system G and vector $\vec{P}$, a **max-allocation** π is an allocation function with the property

$$R(G;\vec{P};\pi) \geq R(G;\vec{P};\pi'), \text{ for any other } \pi' \neq \pi$$

Since there is a finite number of possible permutations such a max-allocation for G always exists. However, there may not be a max-allocation for vectors of varying probabilities, given the multinomial nature of the reliability measure in this case.

The max-allocation function has an interesting behavior for a given system. A more theoretical study is the subject of [9] for node-allocations and [10] for edge-allocations.

5.1 Centralized systems

Let a star graph $K_{1,n-1}$ on n points, and a fixed vector $\vec{P}$ with ordered non-decreasing probabilities. We claim that a max-allocation π allocates probability p_n to the center node. For suppose allocation π allocates probability p_v to its center node v, and probability p_u to a "leaf" node u. Applying the factoring formula (2.1) on v and u we can write $R(S;\pi)$ in terms of p_v and p_u:

$$R(S;\pi) = (1-p_v)p_u \prod_{S-v-u} (1-p_i) + (1-p_v)(1-p_u)R(S-v-u) + p_v$$

Obtain an allocation π' from π by swapping the probabilities p_v and p_u and compute the difference:

$$R(S;\pi) - R(S;\pi') = (p_v - p_u)(1 - \prod_{S-v-u} (1-p_i))$$

then, it is clear that since $R(S;\pi) > R(S;\pi') \iff p_v > p_u$, the center node has to be more reliable. This fact can accommodate as well any arbitrary probability vector.

5.2 Restricted multi-layered systems

In order to compute the max-allocation for the restricted multi-layered graphs we choose to examine first the regular complete bipartite graph $K_{r,r}$ which is an instance of the graph G_1.

Let G be the regular complete bipartite graph $G = G_A + G_B$ where each of the bipartitions G_A, G_B is the graph $K_{r,r}$. Label $1, 2, ..., r$ the nodes in G_A and $r+1, r+2, ..., 2r$ in G_B. It is easier for this graph to express its reliability via its complement "unreliability" measure $\overline{R}(G) = 1 - R(G)$ (the probability that the operating nodes induce a disconnected induced subgraph).

Assuming an allocation π that - for ease of notation - maps p_i to node i of G, for $i = 1, 2, ..., 2r$, observe that an unoperating subsystem of G can only be formed with nodes from the same bipartition. Therefore:

$$\begin{aligned}\overline{R}(G) &= \overline{R}(G_B)\prod_{i\leq r}(1-p_i)+\overline{R}(G_A)\prod_{r<i\leq 2r}(1-p_i)-\prod_{i\leq 2r}(1-p_i)\\ &= \prod_{i\leq r}(1-p_i)+\prod_{r<i\leq 2r}(1-p_i)-\prod_{i\leq 2r}(1-p_i)-\\ &\quad \prod_{i\leq r}(1-p_i)R(G_B)-\prod_{r<i\leq 2r}(1-p_i)R(G_A) \qquad (2)\end{aligned}$$

Since G_A and G_B consist of r isolated nodes, their reliability is simply that each of their nodes operates alone with the rest being in a failed state, of the form $\sum_{i\in A,B} p_i \prod_{j\neq i}(1-p_j)$. Then, by converting back to reliability, equation (2) becomes:

$$R(G) = 1-(\prod_{i\leq r} q_i + \prod_{r<i\leq 2r} q_i) + \prod_{i\leq 2r} q_i - \sum_{i\leq 2r} p_i \prod_{\substack{j\leq 2r\\ j\neq i}} q_j$$

In the above we have let $q_i = 1 - p_i$. For a given set of n probabilities, the max-allocation function must minimize the term $\prod_{i\leq r} q_i + \prod_{r<i\leq 2r} q_i$ in the above expression of $R(G)$, since all the other terms are independent of allocation. The authors of [9] show that optimizing this term for an arbitrary vector of probabilities is NP-complete; therefore we try next to demonstrate the max-allocation for this graph G under selected vectors of probabilities.

Consider vectors $\vec{P}$ of $2r$ probabilities, where the p_i's (or equivalently the q_i's) appear in pairs: $p_1, p_1, p_2, p_2, \ldots, p_r, p_r$.

Lemma 5.1 *Given a set of probabilities $q_1, q_1, q_2, q_2, \ldots, q_r, q_r$, the expression $\Sigma = \prod_{[r\ q_i s]} q_i + \prod_{[r\ q_j s]} q_j$ is minimized when the two products are equal, i.e. when both products contain the same q_i's.*

The proof of the lemma rests in straight algebra by considering that the value of Σ can only increase for any switching of its terms (see [7]). According to this lemma the allocation that maximizes the reliability of $K_{r,r}$ under vectors $\vec{P}$ that contain "pairs of probabilities" allocates the same probabilities to each of the bipartitions of the nodes. It is this fact that we use for arguing about max-R systems in section 6.2.

5.3 Dense systems

We attempt to determine the max-allocation π for the graph $K_n - M$ that underlies D_n, over a vector $\vec{P}$. The lemma that follows shows that full degree nodes have to be made more reliable than the rest. It can be proved easily by applying the factoring formula on the nodes u and v:

Lemma 5.2 *Let a graph G with a full-degree node v and a node u not of full-degree. If π is an allocation of G under $\vec{P}$, and π' is defined by switching the probabilities on the nodes v and u i.e. $\pi'(v) = \pi(u)$, $\pi'(u) = \pi(v)$ and $\pi'(w) = \pi(w)$ whenever $w \neq u, v$, then*

$$R(G;\pi) > R(G;\pi') \Leftrightarrow p_{\pi(v)} > p_{\pi(u)}$$

Consider now the graph $K_n - M$. If the matching M is not complete, then there is at least one full degree node. By lemma (5.2), given a vector $\vec{P}$, a max-allocation π for $K_n - M$ allocates the highest probabilities to the full degree nodes. Having fixed the, say $n-k$, full degree nodes of $K_n - M$, we next determine the max-allocation of the rest of the probabilities $\{p_1, p_2, \ldots, p_k\}$ to the remaining k nodes, each having degree $n-2$.

Lemma 5.3 *A max-allocation π of the k probabilities $p_1 \leq p_2 \leq \ldots \leq p_k$ to the remaining k nodes of $K_n - M$, allocates probabilities to the end points of a matching edge whose difference is as large as possible, that is, there is an ordering of the matching edges, say $\{u_1, v_1\}, \ldots, \{u_{\frac{k}{2}}, v_{\frac{k}{2}}\}$ such that $\pi(u_i) = i$ and $\pi(v_i) = k - i + 1$, for each $i = 1, 2, \ldots, \frac{k}{2}$.*

Proof. Consider any allocation ψ that violates the condition satisfied by π, i.e. there is a matching edge $\{u, v\}$ and an index i such that $\psi(u) = i$ and $\psi(v) = r \neq k - i + 1$. Pick the smallest such i so that there is another matching edge $\{u', v'\}$ with $\psi(u') = k - i + 1$ and $\psi(v') = s \neq i$. By minimality of i it follows that:

$$p_i \leq p_s \text{ and } p_r \leq p_{k-i+1} \tag{3}$$

Obtain a new allocation ψ' by switching the probabilities p_r and p_{k-i+1}. We prove that $R(D_n; \psi') \geq R(D_n; \psi)$: Observe that the graph $K_n - M$ fails to be in a operating subsystem when only the end-points of a matching edge operate or when all nodes fail. Thus, we are led to:

$$R(D_n; \psi') - R(D_n; \psi) = (p_{k-i+1} - p_r)(p_s - p_i) \prod_{j \neq i, k-i+1, r, s} (1 - p_j)$$

Because of the conditions (3) we conclude from the above expression that $R(D_n; \psi') \geq R(D_n; \psi)$. Continuing in this fashion, we ultimately arrive at an allocation having the property given in the lemma. Thus π is a max-allocation. To see that π is also unique up to a permutation of matching edges it is enough to retrace the argument in the case when $0 < p_1 < p_2 < \ldots < p_n < 1$ for then the inequalities obtained in the proof are strict. □

6 On Max-R systems

6.1 Centralized systems

The only candidate in this class (trees) is the star, the locally max-R graph. Given a vector $\vec{P}$ of n non-decreasing probabilities, let π the max-allocation (that maps p_n to the center node).

Theorem 6.1 *The star with its max-allocation is the unique max-R system in the class of tree-systems.*

Proof. We prove the above by induction on n: for $n = 1$ or 2, $K_{1,n-1}$ is the unique tree and the allocation π is the only one possible. Assume that the theorem is true for trees with $n-1$ nodes, and consider an arbitrary tree with n nodes T_n, under the same allocation function π

over the vector $\vec{P}$ according to the labelling of its nodes. Select a leaf-node x of T_n, and let r be the node adjacent to x.

Let now the star $K_{1,n-1}$ with its max-allocation π that allocates p_n to the center node. Select a leaf-node with probability $p_{\pi(x)}$-for ease of notation call it x again-and let c be the center node. Applying the factoring formula on x for both $K_{1,n-1}$ and T_n:

$$\begin{aligned} R(K_{1,n-1}) - R(T_n) &= R(K_{1,n-1} - x) - R(T_n - x) \\ &+ p_{\pi(x)}[(1 - p_{\pi(r)})R(T_n - x - r) - (1 - p_n)R(K_{1,n-1} - x - c)] \end{aligned} \quad (4)$$

Observing that $K_{1,n-1} - x - c$ consists of $n-2$ "trivial" operating subsystems while $T_n - x - r$ must have at least one edge (if not, then $T_n \equiv$star), we can show by enumerating the operating subsystems that the expression in the square brackets is positive. Using the hypothesis it follows that $R(K_{1,n-1}) > R(T_n)$, which completes the induction. □

6.2 Restricted multi-layered systems

For the restricted multi-layered systems we show that the regular complete bipartite graph $K_{r,r}$ is not the most reliable graph in its class, even under its max-allocation. Since the regular complete multipartite graphs are the unique locally max-R graphs that implies immediately that there are no max-R systems in this class.

Theorem 6.2 *There are no max-R systems in the class of the restricted multi-layered.*

Proof: Consider the graph $G = K_{3,3}$ with nodes labeled $1, 2, 3$ in one bipartition and nodes $4, 5, 6$ in the other. Also, a vector $\vec{P}$ of three pairs of probabilities $\vec{P} = (p_1, p_1, p_2, p_2, p_3, p_3)$. The max-allocation π of this graph allocates the probabilities p_1, p_2, p_3 to the nodes of each bipartition, by lemma (5.1). Let H be the graph obtained from G if we remove the edge $\{1, 4\}$ and add the edge $\{1, 2\}$, keeping the same allocation. Computing the difference their reliabilities we get:

$$R(H; \vec{P}; \pi) - R(G; \vec{P}; \pi) = p_1(1 - p_2)(1 - p_3)[(1 - p_3)(p_2 - p_1) - 2p_1p_3(1 - p_2) - p_1p_2]$$

This expression becomes positive for vectors having p_2 much larger than p_1 and relatively small p_3. For example, if $p_1 = 0.05, p_2 = 0.65, p_3 = 0.05$ then $\Delta = 0.0089 > 0$. Therefore, there are vectors of probabilities for which $K_{3,3}$ - under its max-allocation - results to lower reliability value than that of a certain other graph. □

6.3 Dense systems

The same conclusion holds for this class, as in the restricted multi-layered:

Theorem 6.3 *There is no max-R system in the class of systems D_n with $e \geq \binom{n}{2} - \lfloor \frac{n}{2} \rfloor$.*

Proof. Consider the graph $K_n - M$ of the system D_n, where M is a matching of edges; we know it to be locally max-R. Let π be the max-allocation (section 5.3) to the nodes of $K_n - M$.

Suppose that $\{v,u\},\{w,z\} \notin E(K_n - M)$ where u,v,w,z are nodes of $K_n - M$. Obtain the graph G from $K_n - M$ by deleting the edge $\{v,z\}$ and adding the edge $\{v,u\}$, and give G the same allocation π. Then we get

$$R(G) - R(D_n) = p_{\pi(v)}[(p_{\pi(u)} - p_{\pi(z)}) - p_{\pi(w)}p_{\pi(u)}(1 - p_{\pi(z)})] \cdot \prod_{i \neq v,u,w,z} (1 - p_{\pi(i)})$$

The above expression is seen to be positive for vectors $\vec{P}$ having $p_{\pi(i)} = \frac{1}{2}$ for $i \neq u$ and $p_{\pi(u)} > \frac{5}{8}$
□

7 Conclusions

We have presented a model for the design and analysis of the reliability of gracefully degradable systems. It finds its application in distributed fail-soft systems and in parallel ones. Under this measure we examined three major classes, the centralized systems, the restricted multi-layered and systems with dense interconnections. We proved their local optimality under uniform failures of the nodes. Of these types of systems, only the star interconnection in the centralized systems achieves "global" optimality with arbitrary failures. Max-R optimality is a difficult property for a system to achieve. It is left as an open question whether other classes contain max-R systems.
□

References

[1] D. Bauer, F. Boesch, C. Suffel, R. Tindell, *Combinatorial optimization problems in the analysis and design of probabilistic networks*, NETWORKS 15, 1985.

[2] F. Boesch, X. Li, C. Stivaros, C. Suffel, *On graphs having the maximum number of three-point induced connected subgraphs*, Stevens Institute Research Report in Computer Science # 8817, 1988.

[3] F. Boesch, X. Li, C. Stivaros, C. Suffel, *Reliable graphs with Unreliable nodes*, Proceedings of Sixth International Conference on the Theory and Application of Graphs, Vol. 1, pp.159-177, John Wiley & Son, 1991.

[4] H. Frank, *Maximally reliable node-weighted graphs*, Proceedings of 3rd Annual Conference on Information Sciences and Systems, Princeton University, 1969.

[5] H. Frank, *Some new results in the design of survivable systems*, Proceedings of 12th Annual Midwest Circuit Theory Symposium, University of Texas, Austin, 1696.

[6] C. Stivaros, *A Model for Testing Reliable VLSI Routing Architectures*, Proceedings of the 11th IEEE VLSI Test Symposium, pp.337-339, Atlantic City, April 1993, IEEE Computer Society Press.

[7] C. Stivaros, *The Reliability of the Complete Multipartite Systems*, CONGRESSUS NUMERANTIUM JOURNAL, vol. 88, pp. 193-206, Utilitas Mathematica 1992.

[8] C. Stivaros, C. Suffel, *Uniformly Optimal Networks in the Residual Node Connectedness Reliability Model*, CONGRESSUS NUMERANTIUM JOURNAL, vol. 81, pp. 51-64, Utilitas Mathematica 1991.

[9] C. Stivaros, K. Sutner *Computing Optimal Assignments for Residual network Reliability*, DISCRETE APPLIED MATHEMATICS JOURNAL, under review.

[10] C. Stivaros, K. Sutner *Optimal Link Assignments for All–Terminal Network Reliability*, CONGRESSUS NUMERANTIUM JOURNAL, under review.

Distributed Programming Using Objects - a Case Study

D. Molaro
J. R. Parker
Laboratory for Computer Vision
Department of Computer Science
University of Calgary
Calgary, Alberta, Canada

Abstract. Recent developments have made it possible to use the networks of moderately powerful computers that now exist in many offices as distributed processors with supercomputing power. This paper presents a simple toolkit for distributing any C++ program in which inheritance is used to distribute a C++ object around the network, thus freeing the programmer to focus on the actual application. It is portable to any system supporting the socket abstraction. It was tested through the implementation of several algorithms on a network of UNIX workstations, and the results are described. In addition, a discussion of the techniques available to the systems programmer in coping with the object-oriented model of software construction is provided. Finally, the limitations of distributed computer systems are examined.

Keywords. Distrubuted Processing, Object Oriented Programming, High performance computing, Image Processing.

1 Introduction

Many computationally intensive problems now solved on supercomputers could be solved less expensively if they could somehow be distributed across a network[2,3,5,7]. This paper presents a toolkit for distributing parallel processes across a network of moderately powerful non-homogeneous UNIX workstations. It has two objectives:

- Decrease program run time by using more than one processor.
- Decrease development time for these programs by eliminating the time that application programmers might otherwise spend trying to figure out the low level tools UNIX provides for writing parallel programs.

This paper describes a methodology for breaking problems into a distributed design and a C++ object that distributes data and code throughout a workstation network. We call the model of distributed processing supported by our system the *connected object paradigm*[6]. Three sample programs that use distributed objects will be used to demonstrate the system: a thresholding problem, an FFT, and a thinning algorithm. The intent is to permit software developers to concentrate on the system under development while taking advantage of the performance enhancements offered by distributed processing.

2 The Connected Object Paradigm

In the connected object paradigm, distributed objects are conceptually the same as other objects in that both have methods and data. However, a distributed object also has the ability to transmit and receive itself to and from a remote host. The connection model of distributed processing relies on the notion of the topo-

E. A. Yfantis (ed.), Intelligent Systems, 945–951.

logical computer, in which the processing nodes are linked so as to exploit parallelism to solve a particular problem. Any connection scheme can be imposed, and the topology of the computer can change during program execution.

The idea behind our connection model is simple; some of the objects in the system will be designated as distributed objects. When the application to be distributed is started on several machines, any distributed objects on one machine will link to those on other machines, and data in the object on the first machine, by calling the transmit()/receive() methods, is kept consistent with data in the corresponding object in the remote machine. In our distributed objects system, the user specifies the topology and resource scheduling when an object is created. The user also decides what other objects each object will connect to, thus specifying the topology. Finally, the user determines on which computer each object will execute, thereby determining the resource allocation; This is all done at run time making dynamic configurations possible. Currently remote processes are started by hand, in the future this would be handled by the internet damon[9].

An application programmer would use the system by stating what objects should be distributed. The remote and local programs would link via a communications channel whenever a distributed object was created. Concurrency scheduling results from the objects calling *transmit* or *receive* methods. Not all the data in the object needs to be distributed, of course, so the object can specify what to send in an *encode* and *decode* method. This also allows the application program to transmit compressed data and to send data in an external representation format so that a program can distribute objects across big-endian and little-endian byte ordering processors.

The connection model is implemented in a system that aims to enable a programmer to connect the nodes of the system in an arbitrary fashion. As well, the facilities for data access across the network should be as flexible as with any other data structure. The communications channel should be used efficiently, especially since it tends to be the bottleneck in distributed systems. The system should furthermore be portable to heterogeneous systems, and the design should be elegant and simple to use and maintain. This first version of the distributed objects system was implemented using TCP/IP STREAM sockets in the UNIX environment in C++[10]. UNIX was chosen due to its availability, network interface, and standardization among several vendors; TCP/IP was chosen due to its availability and wide acceptance; C++ supports the object oriented paradigm and provides a good interface to system level routines.

The connection model is implemented as a a C++ class that is designed to be extended into a functional computation object by using inheritance. The class that inherits the connection model has methods for transmitting and receiving objects of this class around the network. The class definition is:

```
class Connection
{
public:
enum ConnectionType {Client, Server};
Connection(char *server, char *service);
Connection(char *service);
Boolean pending();
Boolean receive();
Boolean transmit();
}
```

Here is a general design methodology, intended as a guide for distributing computationally taxing algorithms over a computer network:

1. Define the problem:
2. Develop an object oriented implementation:
3. Identify the objects that contain the computation:
4. Determine which of the objects in the program can execute in parallel.
5. Pick a target topology.
6. Isolate the heavy computation:

7. Split the objects into client and server halves:
8. Inherit the communications facilities:

Example - Thresholding an Image

The first example of a distributed image processing computation is doing a simple threshold of an input image[1]. The program to achieve this is to use remote processors to do the computation. This example is such a simple computation it will not benefit from being run in parallel it is however a good demonstration of the technique. Using the outline presented above the process of creating a distributed program looks like:

1. Define the problem:

   The program is to take a greyscale image and threshold value as input and produce a binary image such that values below the threshold are set to OFF and values above the threshold are set to ON.

2. Develop an object oriented implementation:

   The program is to treat each pixel of the image as an object. This object contains a single pixel and responds to the method threshold(int value) which returns ON or OFF according to the value of the pixel.

3. Identify the objects that contain the computation:

   The only computation involved in this program is the calculation of the threshold.

4. Determine which of the objects in the program can execute in parallel.

   The results of each threshold computation does not depend on any other computation in the system. Therefore it is possible to execute all of the Pixel::threshold() computations in parallel.

5. Pick a target topology.

   The topology of the system is to be a star configuration of processors. The central unit dispatches jobs out and collects their results. The outlying units do the computation and then return their results.

6. Isolate the heavy computation:

   The pixel object currently takes the value and threshold of the pixel it represents, the location should be included so that the central processor does not have to keep that information and can inquire it from the object.

7. Split the objects into client and server halves:

   The pixel object is spilt into two halves: the server half has a constructor, destructor and threshold() method, client object has the same methods. The threshold() method in the server half packages up the object and sends it to the client object where the computation is done and the results returned.

8. Inherit the communications facilities:

   The distributed object used in this program consists of:

```
class PixelProcessor : public Connection
{
public:
Processor(char *server, char *service);
Processor(char *service)
Boolean Busy;
void setValue(int x, int y, int value,int threshold);
void computeThreshold();
State getValue();
};
```

This type of example could be expanded on and achieve a valuable speedup; If for example an algorithm that computed some value about every pixel in the image and was an N^2 operation, It would be reasonable to expect that with the minor extension of transmitting the entire image to each oracle ahead of time this type of model could be very successful in speeding up computation[4].

3. Example - The Fast Fourier Transform

The second sample use of our system is a two-dimensional Fast Fourier Transform in parallel. This is the type of "raw" computation for which supercomputers are often used. The same process that was used to distribute the thresholding algorithm was applied.

Briefly then the method used for distributing the 2D FFT is:

The object to be distributed is a vector of data that the FFT is to be run on. The processors are configured in a star Each the vectors can be executed in parallel.

The vector object is split into a server which takes a vector of information and transmits it to a client; the client accepts the vector does the required computation and returns the result back to the server.

In this example, the one-dimensional Fast Fourier Transform (FFT) we chose to base our system is presented in "Numerical Recipes In C"[8].

Limitations of Computation

If the computation is assumed to take no time and network overhead is assumed to be zero the computation time is only limited by the speed of the communications channel. In all cases Ethernet has been used; Ethernet can theoretically transmit data at 10 Mega-Bits per second. Each individual vector consists of 8192 bytes. There are 1024 entries each containing a complex number, with the complex and real part represented as floats. The image consists of 1024 such vectors. Therefore the computation will require transmitting the data both to and from the processing node resulting in 16777216 bytes to be transmitted/received. The minimum time required to compute the Fourier transforms is therefore:

$$\frac{(((1024 \times 4) + (1024 \times 4)) \times 1024\,(Bytes)) \times 2}{1250000\,(BytesPerSecond)} = 13.4 Seconds$$

Experimental Results

Theoretical FFT computation Rate Vs. Actual FFT computation Rate

Number of Processors	FFT(1024) Time(sec)(Maximum)	FFT(1024) Time(sec)(30%)	FFT (1024) Time(sec)(Actual)
1 (No-Communication)	81	81	81
1	94	125.7	120
2	54	85.2	89
3	40	71.7	84
4	34	65.0	68
5	30	60.9	64
6	27	58.2	66
7	25	56.3	58
8	24	54.8	58

Results and Analysis

The FFT experiments were run on a synthetic image using various numbers of processors, All of the processing nodes participating were SUN SparcStataion 2's. Ethernet was used as the transport medium between the processing units. The difference between the theoretical maximum and the recorded values is presented in table 1.

The first result is from executing the non-distributed implementation on a single CPU. The second and subsequent results are from using distributed processors communicating with a server node. Significant gains in processing time can be achieved can be achieved from using a small number of processors, but the gains quickly peter out as the communications channel becomes saturated. This example while achieving significant gains quickly saturates the network and would be unusable in anything but a dedicated environment.

4. Example - Distributed Zhang Suen Thinning.

Zhang Suen thinning[11] is a common image processing technique; It is also computationally expensive. This example has to do with a real world problem where computation speed of the algorithm is critical to the success of the overall system. The first step in digitizing well-logs is to thin the images down. Well-logs are long strip charts containing data recorded from instruments lowered down oil-wells; there is considerable commercial interest in constructing automatic systems to digitize them. A function of the type of document here is that they may be 9.0 inches wide and sometimes as much as 150 feet long; about half a billion pixels when scanned at 200 dpi. While it would be possible to process these types of images on a single processor even on the fastest available workstation the computation can take considerable time.

Zhang suen thinning does also require some local knowledge of information. The method chosen to distribute the computation is to tile the image and on each pass of the algorithm exchange boundary information to keep the images in sync with one another. This also has the side effect of keeping the processors synchronized.

Implementation

The Zhang-Suen thinning algorithm remains unchanged, but on the completion of each pass of the algorithm the neighboring strips are queried for updated edge information. The distributed object in this example is a line of pixels:

```
class Neighbor : public Connection
{
public:
Neighbour(char*server,char*service):
Connection(server,service) {  };
Neighbour(char*service):Connection(service)     {  };
void sendData(unsigned char *q,int length);
unsigned char *getData(int *length);
};
```

Theoretical Limits:

The time required to communicate between each strip is negligible. Performance of this system by computing speed of the participating nodes rather than network bandwidth. Consider an image 2048 wide 1024 pixels high that must be thinned. Each pass of the thinning algorithm would be expected to take approximately 20 seconds on a SUN Sparc Station 2. The Top and bottom rows of the image would consume 256 bytes. The required bandwidth consumed by each of the processors:

$$\frac{2048 Bits \times 2 Transmissions}{20 Seconds} = 205 BitsPerSecond$$

In the previous example Ethernet was seen to run at approximately 30% efficiency, in this calculation we shall assume the same:

$$\frac{205 BitsPerSecondPerNode}{300000 BitsPerSecond} = 15000 Nodes\,(Approx)$$

It would seem that approximately 15000 workstations would be required to saturate an average Ethernet. The computation speed therefore -- not communications bandwidth -- should determine performance.

Results and analysis

The program was run on a sample image of 1600 pixels wide by 9152 pixels long. This would represent approximately 45 inches of paper well-log. Typical well logs are in the range of 100 - 600 inches, with many examples well over 1000 inches.

Distributed Zhang Suen Thinning -- Experimental Results

Number of Processors	Time to complete	Aggregate Time	Overall Speedup
1	3676	3676	1.00x
2	1849	3698	1.99x
3	1251	3753	2.94x
4	932	3728	3.94x
5	755	3775	4.88x
6	637	3822	5.77x
7	547	3829	6.72x
8	480	3840	7.66x

The computation speeds up very close to linearly with the number of processors. This example seems ideally suited to working in a distributed environment.

6 Conclusions and Future Work

The connection based model is an effective tool for the distribution of complex problems in an arbitrarily structured topological computer. The system is easy to use, and little knowledge of systems programming is required. The system effectively distributes computing around a network and does not restrict topology. It should be apparent from the analysis that network computers are not limited by the types of processors involved but rather by the speed at which data can be transmitted from node to node. The overall result is that an easy to use package has been produced that allows the distribution of hard problems across a network of computers.

This system is being used at the University of Calgary for vision and image processing problems with great success, and it is anticipated that, with an increase in the network bandwidth the computational power of our net would compare favorably with that of a small supercomputer.

In the future we are looking into using the inetd autostart facilites of the UNIX operating system to simplify the start-up process, also we are looking for host environments such as ATM to test the system.

Copies of this system are being made freely available. (contact molaro@cpsc.ucalgary.ca). This work was sponsored by a grant from the Natural Sciences and Engineering Research Council of Canada.

7. References

[1] Deriche, R., *Fast Algorithms for Low-level Vision*, IEEE PAMI, Vol. 12, Jan 1990. Pp. 78-87

[2] Goscinski, A.,*Distributed Operating Systems The Logical Desgin*, Addison-Wesley Publishing Company, 1991

[3] Jennings, C., Parker, J.R., Molaro, D., *Comparative Performaces of HPC Systems for Seismic and Image Processing*, SS'93 High Performance Computing: New Horizons, Calgary, Alberta, June 6-9, 1993. Pp 357-364.

[4] Jennings, C., D. Molaro. and Parker, J.R., *Distributed Force-based Thinning and a General Distribution Method*, Dept. of Computer Science Research Report No. 93/505/10, February, 1993.

[5] Karpoff, W. and Lake B. *PARDO - a deterministic, scalable programming paradigm for distributed memory parrellel computer systems and workstation clusters*. SS'93 High Performance Computing: New Horizons, Calgary, Alberta, June 6-9, 1993. Pp 145-152.

[6] Molaro, D., Jennings, C., *A Simple Interface to Distributed Programming*, SS'93 High Performance Computing: New Horizons, Calgary, Alberta, June 6-9, 1993. Pp279-286.

[7] Parker, J.R. and Ingoldsby, T.R., *Design and Analysis of a Multiprocessor for Image Processing*, Journal of Parallel and Distributed Computing, Vol. 9 pp 297-303.

[8] Press, W.H. et al, *Numerical recipes in C*, Cambridge University Press, Cambridge, 1988.

[9] Rieken B. and Weiman L. *Adventures in UNIX network applications programming*, John Wiley & Sons Inc. 1992.

[10] Stevens R. *Advanced Programming in the UNIX environment*, Addison Wesley Pulishing Company, 1992.

[11] Zhang, T.Y., and Suen, C.Y., *A Fast Parallel Algorithm for Thinning Digital Patterns*, CACM Vol 27 No 3, March 1984. Pp 236-239.

A DISTRIBUTED DEADLOCK DETECTION ALGORITHM

BRIAN M. JOHNSTON
SYSTEM COMPUTING SERVICES
UNIVERSITY AND COMMUNITY COLLEGE SYSTEM OF NEVADA
LAS VEGAS, NV 89154

AJOY KUMAR DATTA
DEPARTMENT OF COMPUTER SCIENCE
UNIVERSITY OF NEVADA, LAS VEGAS
LAS VEGAS, NV 89154

Abstract

The problem of distributed deadlock detection has undergone extensive study. The contributions of this research are the development of a distributed deadlock detection algorithm and the formal verification of this algorithm. Formal verification of deadlock detection algorithms in distributed systems is an area of research that has largely been ignored. Instead, most proposed distributed deadlock detection algorithms have used informal or intuitive arguments, simulation or just neglect the entire aspect of verification of correctness. As a consequence, many of these algorithms have been shown incorrect. This research will abstract the notion of deadlock in terms of a temporal logic of actions and discuss the invariant and eventuality properties.

1 Introduction

A collection of autonomous processes spatially separated, that communicate through some communications network are commonly referred to as a distributed computer system. Processes that operate in this environment share no common memory nor do they have global clock. These processes share information regarding the operation of the distributed system by passing messages. Distributed computer systems make no assumptions about the particular hardware contained in the system. Therefore, a wide variety of architectures can be used.

Processes or transactions can move through five legitimate states: running, pending, waiting, halted and deadlocked. When a transaction has all needed data objects or needs no access to any data objects, it will be actively processing. After processing completely, a transaction will move to its final state which is halted. However, if this transaction needs part of the database for completion, it will make a request to its data manager and move into the pending state. Pending implies that a request has been made and no response has com back. Once a decision has been made, the transaction will either be granted the lock request and return to the running state or blocked by another transaction and forced into the wait state. If the system is abstracted in terms of a transaction wait-for graph, a deadlock occurs when a cycle forms among a subset of the waiting transactions.

1.2 Related Work

Current work in the area of distributed deadlock detection has focused primarily on the development of algorithms. Most of the algorithms proposed fail to address the area of formal verification. Most algorithms present informal arguments of correctness of the algorithm. Consequently, they are prone to errors. [9]

E. A. Yfantis (ed.), Intelligent Systems, 953–963.

For example, Sinha and Natarajan have developed an edge-chasing algorithm for detecting deadlocks. They offered an informal description of the correctness of this algorithm. The authors concluded that the algorithm met the criteria for correctness. In 1989, this algorithm was modified to correct for the errors in the original paper. Again, this modification was informally shown to be correct. However, in 1990 the modified algorithm was shown to have errors. Kshemkalyani and Singhal demonstrated the flaws of the modified Sinha algorithm and proposed a solution. Not only was a solution offered, but a rigorous proof using temporal logic was produced. [4, 11, 12, 14]

The next section will detail the proposed algorithm. In section 3 temporal logic will be discussed and a formal proof of correctness will be presented. An example will be given in section 4.

2 Proposed Algorithm

The proposed algorithm of this research falls within the category of edge-chasing algorithms. Each site in the network carries a unique site identifier called *Site_ID*. Within the network a site maintains a certain portion of the database. Each site owns some data objects and maintains a few transactions. Each data object is identified by a unique identifier denoted by *Data_obj*. Every data object controlled by a site

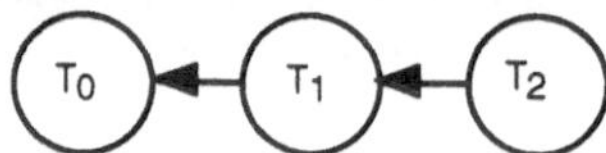

Transaction	Wait_for	Held_by	Request_Q
T_0	*nil*	*nil*	T_1
T_1	T_0	T_0	T_2
T_2	T_0	T_1	*nil*

Figure 2.1 Difference between Wait_for(Ti) and Held_by(Ti)

has a variable associated with it called *Locked_by*. The variable *Locked_by* determines the current state of the data object. If the data object is not locked by any transaction, *Locked_by* will store nil, otherwise it stores the identification of the locking transaction.

Each transaction has a unique site identifier denoted by *T_ID*. A transaction can use data objects within its own site or make explicit requests for a data object in another site. As each site has a unique *Site_ID*, and every transaction within a site has a unique *T_ID*, the *T_ID* can be considered to be unique throughout the network (see Figure 2.1).

Assumption 1 : A transaction can have at most one outstanding lock request.

In case a transaction needs more than one data object, the second data object can be requested only after the first data object has been granted.

Each transaction T_i at site S_i has the following data structure: a variable called *Wait_for(T_i)*, a variable called Held_by(T_i), and a queue of requesting transactions Request_Q(T_i). If the current transaction is not waiting for any other transaction then *Wait_for(T_i)* is set to nil, else, it denotes which transaction is at the head of the locked data object. Held_by(T_i) is set to nil if the current transaction is executing, otherwise, it stores the transaction that is holding the data object required by the current transaction. Request_Q(T_i) is a tuple (T_j,D_i), where T_j is the requesting transaction and D_i is the particular data object held by T_i.

The difference between *Wait_for(T_i)* and held_by(T_i) can be well understood with the help of an example.

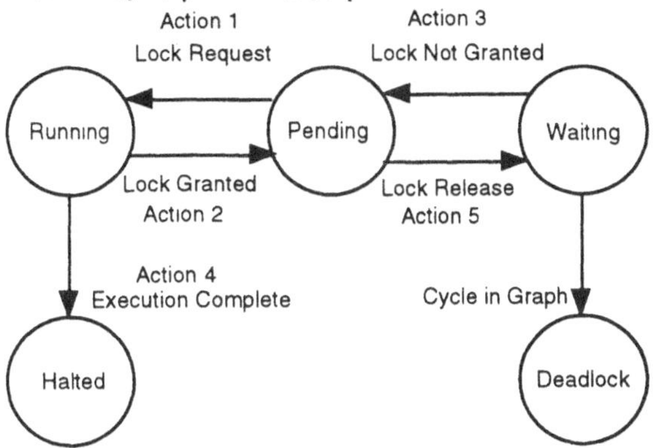

Figure 2.2 Actions of a transaction

As shown in Figure 2.2, Transaction T2 is waiting for a data object held by transaction T1, which is further waiting on transaction T0. Thus Held_by(T2) is T1, while Held_by(T1) is T0. As described above, Wait_for(T1) and Wait_for(T2) are equal to T0.

Suppose a transaction T_i makes a lock request for a data object Dj. If Dj is free then Dj is granted to T_i and Locked_by(Dj) is set to T_i. If Dj is not free then Dj sends a not granted message to T_i along with the transaction identifier locking Dj. T_i joins the Request_Q(T_j) and sets its Wait_for equal to Wait_for(T_j). Now T_i initiates an update message to modify all the Wait_for variables which are affected by the changes in Locked_by variable of the data objects. Update message is a recursive function call that will continue updating all elements of every Request_Q in the chain.

When a transaction T_j receives the update message it checks if its Wait_for value is the same as the new Wait_for value. If it is not the same then the value is modified. Now, a check for deadlock is performed. If a deadlock is not detected then the update message is forwarded, else deadlock is declared and deadlock resolution is initiated.

The transaction detecting the deadlock is chosen as the transaction to be aborted. This transaction sends a clear message to the transaction holding its requested data object. It also allocates every data object it held to the first requester in its Request_Q and enqueues remaining requesters to the new transaction.

The transaction receiving the clear message purges the tuple in its Request_Q having the aborting transaction as an element.

```
{Transaction Ti makes a lock request for data object Dj}
Begin
        send lock_request(Ti) to Dj;
        wait for granted / not granted;
        if granted then
        begin
                Locked_by(Dj) := Ti;
                Held_by(Ti) := {empty set};
        end
        else {Suppose Dj is being used by transaction Tj}
        begin
                Held_by(Ti) := Tj;
                Enqueue(Ti, Request_Q(Tj));
                if Wait_for(Tj) = nil then
                        Wait_for(Ti) := Tj
                else
                        Wait_for(Ti) := Wait_for(Tj);
                update(Wait_for(Ti), Request_Q(Ti));
        end;
End;

{Data object Dj receiving a lock_request(Ti)}

Begin
        if Locked_by(Dj)=nil then
                send granted
        else
        begin
                send not granted to Ti;
                send Locked_by(Dj) to Ti;
        end;
End;


{Transaction Tj receiving an update message}

Begin
        if Wait_for(Tj) ≠ Wait_for(Ti) then
                Wait_for(Tj) := Wait_for(Ti);
        if Request_Q(Tj) - Wait_for(Tj) = nil then
                update(Wait_for(Ti), Request_Q(Tj))
        else
                DECLARE DEADLOCK;
End;
```

3 Correctness

Temporal logic provides operators that allow one to reason how the truth of a given proposition can vary throughout time. Pnueli proposed the use of temporal logic in the verification of programs in 1977. In this chapter, temporal logic will be developed, explained and the edge-chasing deadlock detection algorithm from the previous chapter will be verified correct. This will be done by detailing precisely how temporal logic works within the framework of first-order logic. Once the axioms and operators of temporal logic have been devised, properties of this logic will be detailed. [1,3, 6, 8 10, 13]

3.1 Properties of Temporal Logic

This distributed system is abstracted by defining actions. An action is any possible change in the system state. Actions represent the relation between the old states and new states of the system. Actions are always considered to be atomic. Any change of state in this system require message passing or processing. So, the actions of this system are simply the events which cause a transition from one state to another.

Let G=(V,E) be the transaction wait-for graph of this system, where V={T_ID | T_ID not halted} and E={(T_ID$_i$,T_ID$_j$) | Transaction i is waiting on Transaction j}. Let G_i be a connected component of the graph. All G_i are trees until deadlock situation occurs in which case a cycle will exist. A cycle is formed when a transaction, T_i, must wait for a transaction, T_j, that is already waiting direct or indirectly on T_i. Let such a graph be denoted $C(G_i)$. Further define $|C(G_i)|$ to be the length of the cycle.

Let state(T_ID) be a predicate defined on any given transaction in the system in the following manner: state(T_ID)= {state of the given transaction at a reference point in time}.

Now, six temporal logic of actions can be defined.

Action 1: Lock request is made by T_i
state(T_i)=running O(state(T_i))=pending

Action 2: Receive message lock request granted
state(T_i)=pending O(state(T_i))=running

Action 3: Receive message lock request not granted
state(T_i)=pending O(state(T_i))=waiting

Action 4: Transaction enter Phase 2 of locking protocol
state(T_i)=running ◊(state(T_i))=halted

Action 5: Transaction T_j releases data object needed by T_i
state(T_i)=waiting ◊(state(T_i))=running

Action 6: Transaction T_i receives an update message
state(T_i)=pending wait until state change
state(T_i)=waiting Propagate update message or Declare deadlock

Where O is the next time operator and ◊ is the eventually operator

In addition to these six actions, the criteria for correctness will be established through the invariant and eventuality properties.

Invariant or safety properties of a distributed system is defined in the following manner: A distributed program does nothing wrong. Invariant properties are initially true and stable. A stable property of system is one that once it holds, it will hold forever. The first criterion for correctness of distributed deadlock detection algorithms is: No phantom deadlocks are detected.

Eventuality or liveness properties ensure that a distributed program does what it is suppose to do. In deadlock detection, the liveness property dictates that all deadlocks are detected in a finite amount of time.

3.2 Verification of Proposed Algorithm

The goal of the verification will be to show that the proposed algorithm satisfies both criteria for correctness. The first aspect to be considered is the safety property. This is the property that guarantees that the program does nothing wrong.

In distributed deadlock detection, no phantom deadlocks represents a safety property. To determine that the proposed algorithm provides this safety property, no phantom deadlocks will be shown to be invariant.

Suppose a given property P must be shown to be invariant, the following conditions must be met: P must hold initially and P holds regardless of the actions of the system during execution. The second condition is known as stability.

3.2.1 Proof of Safety Property

At time=0 all transactions in the system will have their variables initialized to nil. At this point in time no transaction will receive any messages. So no transaction will be able to declare deadlock. Therefore, initially no phantom deadlocks will be found.

Suppose this distributed system has only one transaction. So, in this case no true deadlock can exist. For deadlock to be declared by the algorithm in this system, the system must successfully execute actions 1 and 3 to be in a position to declare a deadlock, phantom or not. If the transaction needs a particular data object, then a lock request will be made. This is the first action of the system. When the transaction moves to the pending state, the transaction is in a busy wait state pending the notification of the result of its request. At some point in time the result will be returned. This message generates one of two possible actions in the system. A granted message is action 2. This will move the transaction back to the running state, and no deadlock can be declared. Action 3 indicates a not granted message was returned to the requesting transaction. A not granted message implies that another transaction has locked that particular data object. However this contradicts the assumption that only one transaction is in the system. Similarly, action 5 can not be applied to a one transaction system. Action 6 assumes the receipt of an update message from another transaction, however, it was assumed that only one transaction was in the system. When action 4 occurs in this system a transaction is entering phase two of the locking protocol; so no phantom deadlocks will be detected. Now, the base step is valid and no phantom deadlocks will be declared.

Suppose this is a stable property for up to an n transaction system. Since the property is assumed to be stable for up to n transactions, there will be at most n-1 edges in any waiting chain of transactions. If they were n edges then a cycle would exist.

When action 1 is executed in the system, a transaction moves to the pending state and will wait for a message to be received. At this point the other n transactions will be running, pending or waiting. Since, this transaction is in the pending state, it must wait before forwarding any update messages it receives. Consequently, no deadlock will be declared. The transaction which has made the request for the resource will at some point receive notification regarding its request. If the transaction receives a granted message it will transition back to the running state and there will be less than n edges in the wait_for graph. So, by assumption no deadlock will be declared. Suppose that the transaction receives a not granted message. So, it will move to the waiting state as the result of action 3. The wait for graph will have precisely n edges in this instance (if there were more a cycle would exist). This transaction, as the result of this action, will begin propagating an update message. Suppose that every transaction in this system has a value for wait_for. That would imply a deadlock exists. So in this system, there exists a transaction whose wait_for variable must be nil. Similarly, there exists a transaction in the system whose Request_Q is nil. Therefore, the update message will reach a transaction that has a nil Request_Q and the message will stop. So, no deadlock will be declared as a consequence of action 3. Suppose n+1 transactions are waiting and n edges exist in the graph, when action 5 occurs a transaction will move back to the running state. With only n-1 edges now in the graph the inductive assumption holds. Action 6 considers what will happen upon the receipt of an update message. A transaction which receives an update message will propagate that message to its Request_Q. Action 6 has a similar argument as Action 3. So, no phantom deadlocks are declared.

3.2.2 Proof of Liveness Property

Suppose in this system a deadlock will occur in a finite amount of time. Consider a distributed system with two transactions. So, |C(G)| is precisely 2 when the system is deadlocked. There are two possible cases for a deadlock to develop in this system. The first instance can occur when both transaction are in the running state and simultaneously request resources that the other transaction has a locked.

Let the system be at time T_i when such a request occurs. So, action 1 will occur in the system for both transactions. So at T_{i+1}, both transactions will have sent their requests. Without loss of generality, assume that at T_j, where j>i+1, that both transactions receive a message indicating that the resource was not granted. So, each transaction will execute Action 3 at T_{j+1}. Both will be in the waiting state and a cycle will exist in this transaction wait-for graph. Each transaction must distributed the information regarding their state change to waiting. This is accomplished by propagating the update message at T_{j+2}. In a finite amount of time, one or both transactions will receive the message from the other regarding this state change (denoted by T_k, where $k>j+2$). So, the difference between the Request_Q and Wait_for variables will not be nil (if they were nil, one transaction would be in the running state). Therefore, a deadlock would be declared at time T_{k+1}.

The other case can occur when a transaction is in the waiting state and the other transaction needs a resource from it to complete. At T_j, the running transaction makes a request for a resource and moves to the pending state. In a finite time this transaction will receive a not granted message from the other transaction. At that point, the transaction will move to the waiting state, forming the deadlock in the system. This transaction must then propagate the information regarding the state change to the other transaction. Once the other transaction receives the update message, it will declare deadlock because its Request_Q and Wait_for difference is not nil. Therefore, in the base step deadlock is declared.

Now, assume a system with n transactions will declare deadlock for all $|C(G)|\leq n$. Consider a system with n+1 transactions. Again, there are two possible cases, simultaneous requests that result in deadlock or two requests that are separated in time that generate a deadlock.

Let $|C(G)|=n-1$ and T_i be the time in the system. When simultaneous requests are made the n and n+1 edges are added to the transaction wait-for graph. Thus, forming the cycle. When this occurs two transactions begin propagating update messages in the system. At T_{i+n+1} the two transactions making the request to form the cycle will receive an update message and will execute Action 6. So, both will discover the deadlock.

Suppose n transactions are waiting for the n+1 transaction to complete. However, that transaction makes a request that generates a cycle when $|C(G)|=n+1$. The system has deadlock. When that transaction makes that particular request and eventually receives the not granted message it will be propagating the update message through the system. At T_{i+n+1} the transaction which formed the cycle will receive an update message. This transaction will discover that its Request_Q and Wait_for difference are not nil and declare a deadlock.

Therefore, the proposed algorithm detects all deadlocks in a finite time. So, both criteria for correctness have been verified and the proposed algorithm is correct.

4 Example

Transaction	Wait_for	Held_by	Request_Q
T_0	nil	nil	T_1
T_1	T_0	T_0	T_2
T_2	T_0	T_1	T_3
T_3	T_0	T_2	T_4,T_6
T_4	T_0	T_3	T_5
T_5	T_0	T_4	nil
T_6	T_0	T_3	nil

Figure 4.1 Example before Deadlock

Consider a distributed database with seven transactions as shown in Figure 4.1. The state of each transaction is also shown in the figure. However, it does not necessarily imply that each transaction resides in the same site.

Transaction	Wait_for	Held_by	Request_Q
T_0	T_0	T_3	T_2
T_1	T_0	T_0	T_2
T_2	T_0	T_1	T_3
T_3	T_0	T_2	T_4,T_6,T_0
T_4	T_0	T_3	T_5
T_5	T_0	T_4	nil
T_6	T_0	T_3	nil

Figure 4.2 Example after Deadlock

shows the state of the system before the occurrence of deadlock.

When transaction T_0 makes a request to transaction T_3, a cycle is created and the state of the above system changes, as shown if Figure 4.2 T_0 joins the Request_Q of T_3. To will update its Wait_for to reflect the current state and will propagate the update message to all elements in its own Request_Q. This continues until T_3 discovers that Wait_for(T_3) intersected with Request_Q(T_3) is not nil. Now, T_3 declares deadlock.

5 Conclusion

This paper addresses one of the most important topic in distributed deadlock detection and distributed algorithms in general, formal verification. The techniques of formal verification through temporal logic can be applied to a wide variety of distributed algorithms.

The emphasis of this paper was to describe temporal logic and use that structure to formally verify the proposed algorithm. By first defining the six legitimate actions of the system and the three properties which exist in temporal logic, first order logic was extended to include temporal properties. Working with safety and liveness properties within the system and temporal logic, it was possible to demonstrate the correctness of the proposed algorithm.

Most distributed deadlock detection algorithms developed, do not address the formal verification of correctness. Consequently, the major advantage to formal verification, is the ability to develop and verify correct distributed deadlock detection algorithms.

BIBLIOGRAPHY

[1] B. Banieqbal, H. Barringer & A. Pnueli, *Temporal Logic in Specification.* Lecture Notes in Computer Science, Springer-Verlag 1987.

[2] P. Bernstein, V. Hadzilacos, & N. Goodman, *Concurrency Control and Recovery in Database Systems.* Addison-Wesley, 1987.

[3] K. M. Chandy & J. A. Misra, Deadlock absence proofs for networks of communicating processes, *Information Processing Letters*, 9,4, November 1979, 185-189.

[4] A. N. Chounhary, et. al. A modified priority based probe algorithm for distrbuted deadlock detection and resolution, *IEEE Trans. Software Engineering*, 15,1, January 1989, 10-17.

[5] A. K. Elmagarmid, *Database Transaction Models For Advanced Applications.* Morgan Kaufann 1992.

[6] E. A. Emerson, *Chapter 16:Handbook of Theoretical Computer Science Volume B, Formal Models and Semantics.* editted by J. van Leeuwen. MIT Press, 1990.

[7] J. N. Gray, *Note on database operating systems, in Operating Systems: An Advanced Course.* Lecture Notes in Computer Science. Spring-Verlag, 1978.

[8] B. T. Hailpern, *Verifying Concurrent Processes Using Temporal Logic.* Lecture Notes in Computer Science, Springer-Verlag, 1982.

[9] R. C. Holt, Some deadlock properties of computer systems. *ACM Computing Surveys 4*, 3 (September 1972), 179-196.

[10] F. Kroger, *Temporal Logic of Programs* EATCS Monographs on Theoretical Computer Science, Springer-Verlag, Berlin, Germany, 1987.

[11] A. Kshemkalyani, Characterization and Correctness of Distributed Deadlock Detection and Resolution, PhD. Dissertation, Ohio State University 1991.

[12] A. Kshemkalyani and M. Singhal, Invariant-based verification of distributed deadlock detection Algorithm

[13] Z. Manna, & A. Pnueli, *The Temporal Logic of Reactive and Concurrent Systems*. Springer-Verlag, 1992.

[14] M. K. Sinha and Natarajan N., A priority based distributed deadlock detection algorithm, *IEEE Trans. Software Engineering*, SE-11,1, January 1985, 67-80.

INDEX

Abductive Logic Programming 27
Access Control Lists 749
Acting 307
Adaptive Abstraction 681
Adaptive Interfaces 691
Affine Transformations 713
Alimentary Inspection 705
Artificial Neural Networks 638
Authorization Object Lattice 749
Autonomous Agents 610
Autonomous Mobile Robot 699
Back-Propagation 531,555,13,77
BBI Architectures 405
BDI Architectures 307
Cascade-Correlation 497
Category Theory 194
Cellular Automata 537,4
Chaos 457
Chemical Mass Balance 627
Classification 4,706
Combinatorial Optimization 933
Computer Vision 705,863
Concurrent Logic Programming 27
Conversion 333
Deadlock Detection 953
Deductive Database 731,775,789
Default Logic 20
Digital Signal Processing 3
Degree of Generalization 531
Direct Control of the Invocation 749
Disjunctive Logic Programs 291
Distributed Deductive Database 741
Distributed Processing 945
Domain Volume 845
Dynamic Topologies 617
Event Inhibition 691
Evolutianary Computation 393
Evolutianary Search 497
Extended Logic Programs 437
External Data Base 719
Facet List 175
Fail-Soft 933
Failure Modes 665
Feedforward Neural Networks 446
Formal Specification 789
Fractal Methods 854
Fractals 885
Fuzzy Databases 239
Fuzzy-Neural-Networks 13
Fuzzy Systems 321

Fuzzy Process Controllers 319
Genetic Algorithms 451,467,567,583,609,628
Geographic Information System 713
Graph Partitioning 905
Graphical User Interface 877
Graphics 877
Heterogeneous Inductive Learning 341
Heuristic Analysis 741
Heuristic Classification 39
Heuristic Function 913
Heuristic Search 167
High Performance Computing 945
Hopfield Network 573
Hybrid Learning Systems 657
Hybrid Architecture 699
Hypermedia 351
Image Compression 567
Image Databases 801
Image Indexing 801
Image Processing 517,945
Image Retrieval 763,801
Information Filtering 507
Information Overload 507
Integrity 731
Intelligent Decision Support .69
Intelligent Databases 757
Intelligent Kernel 175
Intelligent Systems 757
Intelligent Tutoring Systems 109,719
Interactive Discourse 61
Intrinsic Hypothesis 897
Inversion 583
Knowledge Representation. 437
Knowledge Retrieval 85
Kriging 897
Language Translation 421
Linguistic Geometry 45
Logic Programming 20,294,731
Mean Squre Error 713
MATUM 69
Meta-Architectures 363
Metalogic Programming 261
Metric 5
MIMD 567
Modelling of Natural Phenomena 885
Morphing 845
Morphological Analysis 221
Multimodal Functions 78
Multiversion Information Retrieval 537
Natural Languages 327
Nearest Neighbour 283
Networks of Paths 45
Network Planning 483
No-Causality-In-Function Principle 665

Object Oriented Database 775,809
Object Oriented Design 475
Object-Manager 809
Object Oriented Technology 757
Object Versioning 823
Occam 249,261
Optimal Graphs 933
Paraconsistency 143,437
Pattern Recognition 231
Phoneme Recognition 525
Portfolio Management 195
Quantum Event 119
Query Re-formulation 763
Rainfall Forecasting 574
Random Walk 885
Reachability 913
Receptor Model 627
Recurrent Neural Networks 581
Robot Planning 913
Scene Analysis 863
Schema Evolution 823
Seismic Refraction 467
Seismic Waveform 585
Semantic Networks 375
Semivariogram 897
Singleversion Information Retrieval 537
Software Architecture 681
Spectral Synthesis 885
Speech Recognition 525
Spline Interpolation 853
Static Analysis 95
Surface Generation 855
System Reliability 933
Taxonomy 135
Telecmmunicatio Services 691
Temporal DataBases 809
Text Planning 61
Thining Problem 873
Transcript Evaluation 589
Transputer 567
Training 447
Type Free Logic 159
Uncertainty Management 69
Unification 209
Vector Quantization 555
Vector Templates 923
Vibration Waveforms 600
View Variations 863
VLSI Layout 905
Waste Water Treatment 55

THEORY AND DECISION LIBRARY

SERIES D: SYSTEM THEORY, KNOWLEDGE ENGINEERING AND PROBLEM SOLVING

1. E.R. Caianiello and M.A. Aizerman (eds.): *Topics in the General Theory of Structures.* 1987 ISBN 90-277-2451-2
2. M.E. Carvallo (ed.): *Nature, Cognition and System I.* Current Systems-Scientific Research on Natural and Cognitive Systems. With a Foreword by G.J. Klir. 1988 ISBN 90-277-2740-6
3. A. Di Nola, S. Sessa, W. Pedrycz and E. Sanchez: *Fuzzy Relation Equations and Their Applications to Knowledge Engineering.* With a Foreword by L.A. Zadeh. 1989 ISBN 0-7923-0307-5
4. S. Miyamoto: *Fuzzy Sets in Information Retrieval and Cluster Analysis.* 1990 ISBN 0-7923-0721-6
5. W.H. Janko, M. Roubens and H.-J. Zimmermann (eds.): *Progress in Fuzzy Sets and Systems.* 1990 ISBN 0-7923-0730-5
6. R. Slowinski and J. Teghem (eds.): *Stochastic versus Fuzzy Approaches to Multiobjective Mathematical Programming under Uncertainty.* 1990 ISBN 0-7923-0887-5
7. P.L. Dann, S.H. Irvine and J.M. Collis (eds.): *Advances in Computer-Based Human Assessment.* 1991 ISBN 0-7923-1071-3
8. V. Novák, J. Ramík, M. Mareš, M. Černý and J. Nekola (eds.): *Fuzzy Approach to Reasoning and Decision-Making.* 1992 ISBN 0-7923-1358-5
9. Z. Pawlak: *Rough Sets.* Theoretical Aspects of Reasoning about Data. 1991 ISBN 0-7923-1472-7
10. M.E. Carvallo (ed.): *Nature, Cognition and System II.* Current Systems-Scientific Research on Natural and Cognitive Systems. Vol. 2: On Complementarity and Beyond. 1992 ISBN 0-7923-1788-2
11. R. Slowiński (ed.): *Intelligent Decision Support.* Handbook of Applications and Advances of the Rough Sets Theory. 1992 ISBN 0-7923-1923-0
12. R. Lowen and M. Roubens (eds.): *Fuzzy Logic.* State of the Art. 1993 ISBN 0-7923-2324-6
13. L. Kitaimik: *Fuzzy Decision Procedures with Binary Relations.* Toward a Unified Theory. 1993 ISBN 0-7923-2367-X
14. J. Fodor and M. Roubens: *Fuzzy Preference Modelling and Multicriteria Decision Support.* 1994 ISBN 0-7923-3116-8
15. E.A. Yfantis (ed.): *Intelligent Systems.* Third Golden West International Conference. Edited and Selected Papers. 1995 ISBN 0-7923-3422-1 (Set of 2 volumes)

KLUWER ACADEMIC PUBLISHERS – DORDRECHT / BOSTON / LONDON